AF664400

FOOD MICROBIOLOGY

THE AUTHORS

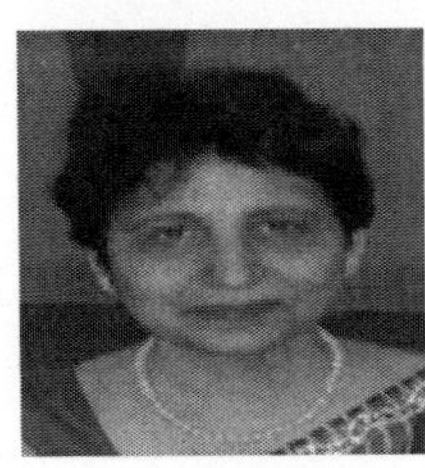

Dr. (Mrs.) Neelam Khetarpaul presently working as Emeritus Professor (ICAR) and former Dean and HOD (Foods and Nutrition), College of Home Science at CCS Haryana Agricultural University, Hisar is the recipient of many awards *viz.*, Distinguished Teacher Award-2000, Young Investigator Award, Ms. Manju Utreja Gold Medal and cash award for doing the best research work in the University and Best Research Paper Award by AFST (I) Hisar Chapter. She is the recipient of various Visiting Fellowships abroad funded by different national and international agencies and visited USA, Australia, UK, Hungary and Netherlands for academic pursuits. She was the country Coordinator of Indo-Netherlands Tailoring Biotechnologies project programme being funded by the Department of Biotechnology, Ministry of Science and Technology, Govt. of India and the Wageningen University, The Netherlands. She has published more than 260 research papers in various journals of national and international repute and 23 books in the discipline of Food Science and Human Nutrition.

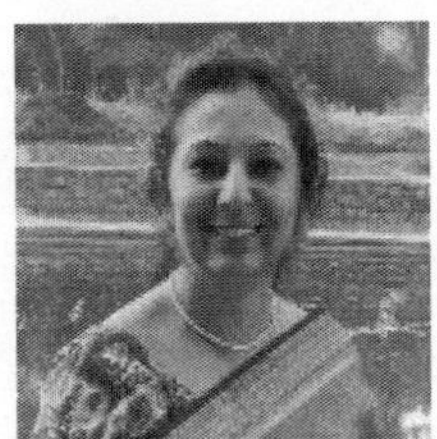

Dr. Anshu Sharma is currently teaching as Assistant Professor in the Food & Nutrition and Food Technology Department at Institute of Home Economics, University of Delhi. Prior to this, she has worked with Nutrition Foundation of India, for almost two decades, on various projects which were funded by many governmental and non-governmental organizations. She has also been a visiting faculty in various institutes. She was the recipient of the Young Scientists' Senior Award at the XXXVIII Annual conference of Nutrition Society of India in 2006. She had been the editor of NFI Bulletin for two decades and is life-member of National Organizations like NSI and IDA. She has quite a few publications in peer reviewed journals and is a co-author of the Food-Exchange list brought out by the Institute of Home Economics.

FOOD MICROBIOLOGY

2nd *Revised & Enlarged Edition*

— *Authors*—

Prof. Neelam Khetarpaul

Dr. Anshu Sharma

2025

Daya Publishing House®

A Division of

Astral International Pvt. Ltd.

New Delhi – 110 002

First Edition: 2005

Second Revised and Enlarged Edition, 2025

ISBN: 978-93-5919-537-7

Published by : **Daya Publishing House®**
A Division of
Astral International Pvt. Ltd.
– ISO 9001:2015 Certified Company –
4736/23, Ansari Road, Darya Ganj
New Delhi-110 002
Ph. 011-43549197, 23278134
E-mail: info@astralint.com
Website: www.astralint.com

Digitally Printed at : **Replika Press Pvt. Ltd.**

Preface

The rationale for revising this book has been the many changes in 'Food Microbiology' forced upon us by the revolution in microbial pathogenesis, food fermentations, changing life styles of consumers, the need for risk assessment, risk evaluation and risk management, food safety, food hygiene and sanitation, incorporation of HACCP programs and finally societal changes in health consciousness. Although some of these areas have been addressed by review papers or specialized texts, but these are often hard to find or are too detailed for the non-specialist. Thus, purpose of writing this text on 'Food Microbiology' was to compile and cover all the recent advances in condense, concise, scientific but simple and readable form.

The book is divided into 20 chapters which mainly cover history and development of food microbiology, role of microorganisms in food, effect of food processing and preservation on microorganisms, bacterial food poisoning, non-bacterial illness, enumeration of microorganisms in food, methods of detecting microorganisms, microflora of different foods, food spoilage, food borne illnesses, fermented foods, role of probiotics in health and disease, food hygiene and sanitation, cleaning and sanitizing food premises and food equipment, food safety and HACCP, food borne illness: implications for the future *etc.* Substantial changes have been made in most of the chapters and in their logical arrangement in this new edition. In addition, new chapters have also been added. Tables, figures, and data have been used only when they are necessary for better understanding. The authors chose the most current works on these aspects that could provide a greater overview and where necessary to the primary literature. The authors have collected latest proved and updated detailed authentic information from journals, books, Web Med *etc.* Efforts have been made to include the recent topics in this book.

We believe that this is a valuable text designed to examine the many current issues related to food microbiology in context of the changing global scenario and would serve as a reference for graduate and postgraduates of Foods and Nutrition, Food Science and Microbiology, food scientists and technologists involved in the area of teaching and research. We hope to get great response from our readers as in our previous publications. I shall regard my efforts amply rewarded if the book is received with the enthusiasm by the readers.

We wish to express our sincere thanks to Mr. Anil Mittal and Mr. Prateek Mittal of Daya and Astral Publishing House for their continued interest in bringing out this revised edition.

Neelam Khetarpaul

Anshu Sharma

Contents

Chapter 1

Introduction

Today food microbiology has become the science, which mainly deals with occurrence and significance of bacteria, algae, protozoa and fungi. Food microbiology comprehends the study of microorganisms that colonise, modify, and process, or contaminate and spoil food. It is one of the most diverse research areas within microbiology. It comprises a wide variety of microorganisms including spoilage, probiotic, fermentative, and pathogenic bacteria, molds, yeasts, viruses, prions, and parasites. It deals with foods and beverages of diverse composition, combining a broad spectrum of environmental factors, which may influence microbial survival and growth. Food microbiology includes microorganisms that have beneficial or deleterious effects on food quality and safety and may therefore be of concern to public health. Microbiology is important to food safety, production, processing, preservation, and storage. Food microbiology papers published during the past decade have been characterized by multidisciplinary interests that have confirmed the increasing amount of evidence that has implicated microorganisms in different areas, including food technology, food safety and hygiene, food poisoning, food genomics, and, more generally, food omics, functional foods, and probiotics, besides emerging methodologies that have been applied to food analyses.

Micro-organisms are present everywhere on earth, which includes humans, animals, plants and other living creatures, soil, water, and atmosphere, and they can multiply everywhere except in the atmosphere. Together, their numbers far exceed all other living cells on this planet. They were the first living cells to inhabit the earth over 3 billion years ago; and since then, they have played important roles, many of which are beneficial to the other living systems. Among the micro-organisms, some molds, yeasts, bacteria, and viruses have both desirable and undesirable roles in our food. Their role in food can be either desirable (food bioprocessing) or undesirable (food borne diseases and food spoilage). Many pathogenic micro-organisms (bacteria, molds and viruses) can contaminate foods during various stages of their handling, between production and consumption. Consumption of these foods can cause food borne diseases. Food borne diseases can be fatal and may also cause large economic losses. Foods of animal origin are associated, more with food borne diseases than foods of plant origin. Effective intervention technologies are being developed and implemented to ensure the safety of consumers against food borne diseases. New methods are also being developed to effectively and rapidly identify the pathogens in contaminated foods.

Many food-grade beneficial micro-organisms are used to produce different types of fermented foods using raw materials from animal and plant sources. Consumption of these foods has increased greatly over the last 15 to 20 years and is expected to increase further in the future. There have been great changes in the production and availability of these micro-organisms (starter cultures) to meet the large demand. In addition, novel and better strains are being developed by using genetic engineering techniques.

Microbial enzymes are also being used to produce food and food additives. By employing genetic recombination techniques, and using diverse microbial sources enzymes of higher purity and activity are obtained. Many types of additives from microbial sources are being developed and used in food. Some of these include single-cell proteins, essential amino acids, colour compounds, flavour compounds, stabilizers and organic acids.

Antimicrobial metabolites (*e.g.*, bacteriocins and organic acids like acetic, propionic and lactic acids) of desirable micro-organisms are being developed and used in foods in place of preservatives of non-food (chemical) origin to control pathogenic and spoilage micro-organisms in food. Economic production of these antimicrobial compounds and their effectiveness in food systems have generated wide interest.

Consumption of foods containing live cells of bacteria (probiotics) and that have apparent health benefits has generated interest among consumers. The role of these bacteria for health and bacterial efficacy benefits is being critically investigated. Probiotic foods are becoming very popular.

Keeping in view the importance of microorganisms in food safety, food processing and preservation, fermented foods *etc.*, today food microbiology, the science, which mainly deals with occurrence and significance of bacteria, algae, protozoa and fungi has become very popular among researchers.

All healthy animals carry a complex microbial flora, part of which may be very specialized and adapted to growth and survival on its host and part of which may be transient, reflecting the immediate interactions of the animal with its environment.

The increased productivity of the land and the improved nutrition that resulted led to population growth and a probable increase in the average lifespan. The availability of food surpluses also liberated some from daily toil in the field stimulated the development of specialized crafts, urban centres and trade–in short, civilization.

In our everyday lives we are perhaps less aware of the presence of microorganisms in the atmosphere than anywhere else, unless we happen to suffer from an allergy to the spores of molds or actinomycetes, but, although they cannot grow in it, the atmosphere forms an important vehicle for the spread of many microorganisms and the subsequent contamination of foods.

Within and on the bodies of all living creatures, as well as in soil and water, microorganisms build up and change molecules, extracting energy and growth substances. They also help to control population levels of higher animals and plants by parasitism and pathogenicity.

When plants and animals die, their protective antimicrobial systems cease to function so that, sooner or later, decay beings liberating the smaller molecules for re-use by plants. Without human intervention, growth, death, decay and regrowth would form an intricate web of plants, animals and microorganisms, varying with changes in climate and often showing apparently chaotic fluctuations in populations of individual species, but inherently balanced in numbers between producing, consuming and recycling groups.

The evolutionary pressure to produce macroscopic fruiting bodies, which is seen in the mushrooms and toadstools, has produced a structure which provides its own microclimate of high relative humidity so that these fungi can go on firing their spores into the air even in the middle of a dry day.

The foods that we eat are rarely if ever sterile, they carry microbial associations whose composition depends upon which organisms gain access and how they grow, survive and interact in the food over time. The microorganisms present will originate from the natural micro-flora of the raw material and those organisms introduced in the course of harvesting/slaughter, processing, storage and distribution. The numerical balance between the various types will be determined by the properties of the food, its storage environment, properties of the organisms themselves and the effects of processing.

In most cases this microflora has no discernible effect and the food is consumed without objection and with no adverse consequences.

In any event, the world's food supply will need to increase to keep pace with population growth and this has its own environmental and social costs in terms of the more intensive exploitation of land and sea resources. One way of mitigating this is to reduce the substantial pre- and post-harvest losses which occur, particularly in developing countries where the problems of food supply are often most acute. It has been estimated that the average losses in cereals and legumes exceed 10 per cent whereas with more perishable products such as starchy staples and vegetables the figure is more than 20 per cent–increasing to an estimated 25 per cent for highly perishable products such as fish. In absolute terms, the US National Academy of Sciences has estimated the losses in cereals and legumes in developing countries as 100 million tonnes, enough to feed 300 million people.

To this day, we are not free from concerns about over-population. Globally there is sufficient food to feed the world's current population. World grain production has more than managed to keep pace with the increasing population in recent years and the World Health Organization's Food and Agriculture Panel considers that current and emerging capabilities for the production and preservation of food should ensure an adequate supply of safe and nutritious food up to and beyond the year 2010 when the world's population is projected to rise to more than 7 billion.

There is however little room for complacency. Despite overall sufficiency, it is recognized that a large proportion of the population is malnourished. The principal cause of this is not insufficiency however, but poverty which leaves an estimated one-fifth of the world's population without the means to meet their daily needs. Any long-terms solution to this must lie in improving the economic status of those in the poorest countries and this, in its train, is likely to bring a decrease in population growth rate similar to that seen in recent years in more affluent countries.

Foods, by their very nature, need to be nutritious and metabolizable and it should be expected that they will offer suitable substrates for the growth and metabolism of microorganisms.

Perhaps one of the most hostile environments for many microorganisms is the atmosphere. Suspended in the air, the tiny microbial propagule may be subjected to desiccation, to the damaging effects of radiant energy from the sun and the chemical activity of elemental gaseous oxygen (O_2) to which it will be ultimately exposed. Many microorganisms, especially Gram-negative bacteria, do indeed die very rapidly when suspended in air and yet, although none is able to grow and multiply in the atmosphere, a significant number of microbes are able to survive and use the turbulence of the air as a means of dispersal.

Viable microorganisms may be found in a very wide range of habitats, from the coldest of brine ponds in the frozen wastes of polar regions, to the almost boiling water of hot springs. Indeed, it is now realized that actively growing bacteria may occur at temperatures in excess of 100°C in the thermal volcanic vents, at the bottom of the deeper parts of the oceans, where boiling is prevented by the very high hydrostatic pressure. Microorganisms may occur in the acidic wastes draining away from mine workings or the alkaline waters of soda lakes. They can be isolated from the black anaerobic salts of estuarine muds or the, purest waters of biologically unproductive, or oligotrophic, lakes. In all these and many other, habitats microbes play an important, part in the recycling of organic and inorganic materials through their roles in the carbon, nitrogen and sulfur cycles. They thus play an important part in the maintenance of the stability of the biosphere.

The surfaces of plant structures such as leaves, flowers, fruits and especially the roots, as well as the surfaces and the guts of animals all have a rich microflora of bacteria, yeasts and filamentous fungi. This natural, or normal flora may affect the original quality of the raw ingredients used in the manufacture of foods, the kinds of contamination which may occur during processing and the possibility of food spoilage or food associated illness. Thus, in considering the possible sources of microorganisms as agents of food spoilage or food poisoning. It will be necessary to examine the natural flora of the food materials themselves, the flora introduced by processing and handling and the possibility of chance contamination from the atmosphere, soil or water.

Food has a long association with the transmission of disease. Regulations governing food hygiene can be found in numerous early sources such as the Old Testament and the writings of Confucius, Hinduism and Islam. Such early writers had at best only a vague conception of the true causes of foodborne illness and many of their prescriptions probably had only a slight effect on its incidence. Even today, despite our increased knowledge, 'Foodborne disease is perhaps the most widespread health problem in the contemporary world and an important cause of reduced economic productivity,' (WHO 1992). The available evidence clearly indicates that biological contaminants are the major cause.

Microbes can however play a positive role in food. They can be consumed as foods in themselves as in the edible fungi, mycoprotein and algae. They can also effect desirable transformations in a food, changing its properties in a way that is beneficial.

Food microbiology is unashamdely an applied science and the food microbiologist's principal function is to help assure a supply of whole some and safe food to the consumer. To do this requires the synthesis and systematic application of our knowledge of the microbial ecology of foods and the effects of processing to the practical problem of producing, economically and consistently, foods which have good keeping qualities and are safe to eat.

Chapter 2

History and Development of Food Microbiology

As a discipline, Food Microbiology does not have a precise beginning. It is very difficult to pinpoint the precise beginning of human awareness of the presence and role of microorganisms in foods but the available evidence indicates that this knowledge preceded the establishment of bacteriology or microbiology as a science. The era prior to the establishment of bacteriology as a science may be designated the prescientific era. This era may be further divided into what has been called the food-gathering period and the food-producing period. The former covers the time from human origin over 1 million years ago up to 8,000 years ago. During this period, humans were presumably carnivorous, with plant foods coming into their diet later in this period. The food-producing period dates from about 8,000 to 10,000 years ago and, of course, includes the present time. Agriculture and animal husbandry were adopted by the early civilizations around 8000 B.C. It is also during this period that foods were first cooked. It is presumed that the problems of spoilage and food poisoning were encountered early in this period. Events which stretched over several centuries ultimately led to the recognition of the significance and role of microorganisms in foods. Food borne disease and food spoilage have been part of the human experience since the dawn of our race. Although the actual cause of these problems remained a mystery for thousands of years, but many early civilizations discovered and applied effective methods to preserve and protect their food.

Important Development of Early Food Microbiology

Some of the more significant dates and events in the history of food microbiology including food preservation, food spoilage, food poisoning *etc.* are as under which explored the way for the establishment of food microbiology in the 20th century:

- ☆ Between 8000 and 1000 B.C. drying, cooking, baking, smoking, salting, sugaring with honey, ice, storage in pits, fermentation with fruits, grains and milk *etc.* were done to preserve the food.
- ☆ The first boiler pots are thought to have originated in the Near East about 8,000 years ago. The arts of cereal cookery, brewing, and food storage, were either started at about this time or stimulated by this new development.
- ☆ 7000 BC: The first evidence of beer manufacture has been traced to ancient Babylonia as far back as 7000 BC. Evidence showed that the Babylonians manufactured beer (fermentation). Wine appeared in about 3500 BC. In early civilizations (and even today in underdeveloped countries where modern sanitation is lacking), alcoholic beverages like beer and wine were much safer to consume than the local water supply, because the water was often contaminated with intestinal microorganisms that caused cholera, dysentery and other serious diseases.
- ☆ 6000 BC: The first apparent reference to food spoilage was recorded in history during this period. With the advent of prepared foods, the problems of disease transmission by foods and of faster spoilage caused by improper storage made their appearance.

- 5000BC: The practice of making pottery was brought to Western Europe about 5000 BC from the Near East.
- 3500 BC: Wines are known to have been prepared by the Assyrians by 3500 BC.
- 3000 BC: Egyptians manufactured cheese (fermentation) and butter (fermentation, low a_w). Again, fermented foods such as cheese and sour milk (yogurt) were safer to eat and resisted spoilage better than their raw agricultural counterparts. Several cultures also learned to use salt (low a_w) to preserve meat and other foods around this time.
- 3000 BC: The Sumerians of about 3000 BC are believed to have been the first great livestock breeders and dairymen and were among the first to make butter. Salted meats, fish, fat, dried skins, wheat, and barley are also known to have been associated with this culture.
- 3000 BC: Milk, butter, and cheese were used by the Egyptians as early as 3000 BC. Between 3000 BC and 1200 BC, the Jews used salt from the Dead Sea in the preservation of various foods. The Chinese and Greeks used salted fish in their diet, and the Greeks are credited with passing this practice on to the Romans, whose diet included pickled meats. Mummification and preservation on of foods were related technologies that seem to have influenced each other's development.
- 1500 BC: Fermented sausages were prepared and consumed by the ancient Babylonians and the people of ancient China as far back as 1500 BC.
- Another method of food preservation that apparently arose during this time was the use of oils such as olive and sesame. It has also been pointed out that the use of oils leads to high incidences of staphylococcal food poisoning.
- 1000 BC: The Romans excelled in the preservation of meats other than beef by around 1000 BC and are known to have used snow to pack prawns and other perishables, according to Seneca. The practice of smoking meats as a form of preservation is presumed to have emerged sometime during this period, as did the making of cheese and wines. It is doubtful whether people at this time understood the nature of these newly found preservation techniques. It is also doubtful whether the role of foods in the transmission of disease or the danger of eating meat from infected animals was recognized.
- 1100 AD: Few advances were apparently made toward understanding the nature of food poisoning and food spoilage between the time of the birth of Christ and 1100 AD.
- 943 AD: Ergot poisoning (caused by *Claviceps purpurea*, a fungus that grows on rye and other grains) caused many deaths during the Middle Ages. Over 40,000 deaths due to ergot poisoning were recorded in France alone in 943 AD, but it was not known that the toxin of this disease was produced by a fungus.
- 1156-1248: Meat butchers are mentioned for the first time in 1156, and by 1248 the Swiss were concerned with marketable and nonmarketable meats. In 1276, a compulsory slaughter and inspection order was issued for public abattoirs in Augsburg. Although people were aware of quality attributes in meats by the thirteenth century, it is doubtful whether there was any knowledge of the causal relationship between meat quality and microorganisms.
- 1626: The Theory of Spontaneous Generation (Abiogenesis): After the discovery of microorganisms by Leeuwenhoek, scientists began investigations about the origin of microbes. Since organic matter decomposes quickly outside the living body, it was assumed that microorganisms were arising by spontaneous generation. Francesco Redi (1626), supported spontaneous generation theory. He boiled the meat and covered the mouth of the flask with wire gauze. The flies attracted due to the odour of meat, laid eggs on the wire gauze, that later developed into maggots. Thus, he established that origin of maggots was from meat and not from fly. Additionally, John Needham (1749), an Irish priest, observed the appearance of microorganisms in putrefying meat and interpreted this as spontaneous generation.
- 1658: Perhaps the first person to suggest the role of microorganisms in spoiling foods was A. Kircher, a monk, who as early as 1658 examined decaying bodies, meat, milk, and other substances and saw what he referred to as "worms" invisible to the naked eye. Kircher's descriptions lacked precision, however, and his observations did not receive wide acceptance.
- 1665: Even though early human cultures discovered effective ways to preserve food (fermentation, salt, ice, drying and smoking), they did not understand how these practices inhibited food spoilage or food borne disease. Their ignorance was compounded by a belief that living things formed spontaneously from non-living matter (Theory of Spontaneous Generation).

An Italian physician by the name of Francesco Redi demonstrated that maggots on putrefying meat did not arise spontaneously but were instead the larval stages of flies (put meat in container capped with fine gauze so that flies couldn't get access to deposit eggs). This was the first step away from the doctrine of spontaneous generation.

- ☆ 1676: Food microbiology evolved as tools and techniques revealed problems and solutions. When Antonie van Leeuwenhoek first observed 'animalcules' in 1676 with his microscope it was a tool that revealed there was microscopic life around us.
- ☆ 1683: Anton van Leeuwenhoek from the Netherlands examined and described bacteria through a microscope. Leeuwenhoek was the first to observe yeast cells. At about the same time, the Royal Society was established in England to communicate and publish scientific work, and they invited Leeuwenhoek to communicate his observations. He did so for nearly 50 years until his death in 1723. As a result, Leeuwenhoek's reports were widely disseminated and he is justifiably regarded as the person who discovered the microbial world.
- ☆ 1700: In the early 1700s Leeuwenhoek, the inventor of the microscope was first to observe these microscopic living forms. He referred to them as "wee beasties".
- ☆ 1765: Theory of Biogenesis: Italian named L. Spallanzani priest tried to disapprove the theory of spontaneous generation of life. He showed that beef broth that had been boiled for an hour and sealed remained sterile and did not spoil. Spallanzani performed this experiment to disprove the doctrine of the spontaneous generation of life. He proposed the theory of biogenesis. He said that every form of life takes its origin from their parents, germ cells or seeds. This theory of biogenesis was later proved and supported by Louis Pasteur. However, he did not convince the proponents of the theory because they believed that his treatment excluded oxygen, which they felt was vital to spontaneous generation.
- ☆ 1778: The history of thermal canning necessitates a brief biography of Nicolas Appert (1749–1841). This Frenchman worked in his father's wine cellar early on, and he and two brothers established a brewery in 1778.
- ☆ 1784: Nicolas Appert opened a confectioner's store in Paris that was later transformed into a wholesale business. His discovery of a food preservation process occurred between 1789 and 1793. He established a cannery in 1802 and exported his products to other countries. The French navy began testing his preservation method in 1802, and in 1809 a French ministry official encouraged him to promote his invention. In 1810, he published his method and was awarded the sum of 12,000 francs. This, of course, was the beginning of canning as it is known and practiced today. This event occurred some 50 years before L. Pasteur demonstrated the role of microorganisms in the spoilage of French wines, a development that gave rise to the rediscovery of bacteria.

  In 1810, Nicholas Appert placed food in glass jars, sealed them and heated them in boiling water. He did not know at the time that what was killing spoilage bacteria and invented what is known today as "canning".

  The effect of heat on spoilage was known years before Nicholas Apert won his 12 000 Franc prize, but he developed a reliable method for food preservation. Centuries prior, humans had discovered that salting or drying would prevent food spoilage; however, food microbiologists began understanding the mechanisms of how these methods worked.
- ☆ 1780: Scheele identified lactic acid as the principal acid in sour milk.
- ☆ 1782: Canning of vinegar was introduced by a Swedish chemist.
- ☆ 1795: The French government offered 12,000 francs to anyone who could develop a practical way to preserve food. A French confectioner named Nicholas Appert was issued the patent after showing that meat could be preserved when it was placed in glass bottles and boiled. This was the beginning of food preservation by canning.
- ☆ 1800: Although processes of food spoilage and methods of food preservation and food fermentation have been recognized since ancient times, it was not until the 1800s that the relationship between foods and micro-organisms was established.
- ☆ 1813: Use of SO_2 as a meat preservative is thought to have originated around this time.
- ☆ 1837: Schwann proposed that the yeast which appeared during alcoholic fermentation was a microscopic plant.
- ☆ 1837: Schwann showed that heated infusions remained sterile in the presence of air, which he supplied by passing it through heated coils into the infusion. It is interesting to note that although Spallanzani and Schwann each used

heat to preserve food, neither man apparently realized the value of turning these observations into a commercial method for food preservation. Although both of these men demonstrated the idea of the heat preservation of foods, neither took advantage of his findings with respect to application. The same may be said of D. Papin and G. Leibniz, who hinted at the heat preservation of foods at the turn of the eighteenth century.

- 1839: Tin cans came into wide use in the United States.
- 1840: Fish and fruit were first canned.
- 1853: R. Chevallier-Appert obtained a patent for sterilization of food by autoclaving.
- 1854: Pasteur began wine investigations. Heating to remove undesirable organisms was introduced commercially in 1867- 1868.
- 1855: Grim wade in England was the first to produce powdered milk.
- 1857: When Louis Pasteur, a Professor of Chemistry at the University of Lille, France, demonstrated that fermentation was caused by yeasts, not spontaneous germination, and later that heat could inactivate microbes, he was building on the work of earlier scientists, such as Girolamo Fracastoro, Agostino Bassi, Friedrich Henle, and others. PBut Pasteur understood the mechanism that heat killed bacteria and publicized it.

  He also worked on souring of wine and beer and found that this alcohol spoilage is due to the growth of undesirable organisms, while the desirable microorganisms produce alcohol by a chemical process called "Fermentation". He showed that wine did not spoil, if it is heated to 50-60°C for a few minutes. This method is called "Pasteurization", now widely used in dairy units, to kill pathogenic microorganisms in milk. He is a founder of "Germ theory of disease" as he visualized those diseases are caused by microorganisms. In course of his research, he discovered the importance of sterilization and discovered steam sterilizer, autoclave and hot air oven. He also established the importance of cotton wool plugs for protection of culture media from aerial contamination. He differentiated between aerobic and anaerobic bacteria and coined the term "anaerobic" to refer to the organisms that do not require oxygen for growth

  The first person to appreciate and understand the presence and role of microorganisms in food was Pasteur. Pasteur showed that the souring of milk was caused by the growth of organisms in it in 1837. In the 1860s, Louis Pasteur showed that certain microorganisms were responsible for fermentation and decay in wine and beer. He developed the application of heat to reduce the number of bacteria which now bears his name (*e.g.*, Pasteurization). This heat process is now used for a variety of foods and is called pasteurization. Because of the importance of his work, Pasteur is known as the founder of food microbiology and microbiological science as his contribution led to the development of Microbiology as a separate scientific discipline. He proved the theory of "Biogenesis" and disproved the "Theory of spontaneous generation" (Abiogenesis), experimentally by using swan-necked flasks. Pasteur passed the untreated and unfiltered air in to boiled nutrient broth, germs settled in the goose neck and no microbes appeared in the solution. Thus, he disproved that living organisms appear from non-living matter. He demonstrated that air doesn't have to be heated to remain sterile using his famous swan-necked flasks.

  Similarly, Robert Koch built on the work of others, developed a cadre of microbiologists, and provided tools for future food microbiologists. Robert Koch was a German country Doctor who later became the Professor of hygiene and Director of institute of infective diseases at Berlin. He perfected many bacteriological techniques and known as "Father of Practical Bacteriology". Before Pasteur and Koch, people had practical knowledge about food spoilage and fermentations from experience. Brewers and bread makers knew that successful ferments could be transferred to new batches.

- From the time of Pasteur, microbiological discoveries and developments began to proceed more rapidly. Microbes were implicated in several diseases, heat-resistant spores were discovered, toxins were identified, and by the late 1800s, governments began to enact legislation to protect the quality of food. At about the same time, the dairy industry was driven to implement microbiological control over milk because of several notorious outbreaks of milk-borne typhoid fever, diphtheria, tuberculosis and brucellosis. Public health authorities established requirements that addressed animal health, sanitation, pasteurization (which had an immediate and very effective impact on the problems), and refrigeration, with all of these steps reinforced by bacterial standards. As a result, pasteurized milk was one of our safest foods by the mid-1900s.

- 1857: Milk was incriminated as a transmitter of typhoid fever by W. Taylor of Penrith, England.
- 1857-1876: Pasteur showed that micro-organisms were responsible for the chemical changes that take place in foods and beverages. Their observations laid the foundation for the development of food microbiology as we know it today. Soon after these early discoveries were made, knowledge about the role that micro-organisms play in food preservation, food spoilage and food preservation, food spoilage and food poisoning accelerated rapidly until food microbiology gradually emerged as a discipline in its own right.
- 1865: The artificial freezing of fish on a commercial scale was begun in the United States in 1865. Eggs followed in 1889.
- 1873: The first reported study on the microbial deterioration of eggs was carried out by Gayon.
- Lister was first to isolate *Lactococcus lactis* in pure culture.
- 1880: The pasteurization of milk was begun in Germany.
- 1882: Krukowitsch was the first to note the destructive effects of ozone on spoilage bacteria.
- 1888: Miquel was the first to study thermophilic bacteria.
- 1888: Gaertner first isolated Salmonella enteritidis from meat that had caused 57 cases of food poisoning.
- 1888: Pasteur developed a vaccine against rabies (Hydrophobia), which made a greatest impact in medicine. He obtained the causative agent of rabies by serial intracerebral passage in rabbits and the vaccine was prepared by drying pieces of spinal cord. He tested with a boy named Joseph Meister and he saved his life. In 1888, Pasteur institute was established for mass antirabic treatment. Pasteur gave the general term "Vaccine" (Vacca=cow) in honour of Jenner's cow pox vaccine, to various materials used to induce active immunity.
- 1890: Mechanical refrigeration for fruit storage was begun in Chicago.
- 1895: S.C. Prescott and W. Underwood traced the spoilage of canned corn to improper heat processing for the first time.
- 1895: In 1895, the Massachusetts Institute of Technology determined that mysterious spoilage of some canned foods resulted from the failure to apply sufficient heat to destroy microorganisms. In the U.S. many food industries hesitated to adopt industry wide microbiological standards until they were economically threatened by the publicity which surrounded outbreaks of food borne disease.
- 1896: Van Ermengem first discovered *Clostridium botulinum*.
- 1906: 1906 Bacillus cereus food poisoning was recognized
- 1907: E. Metchnikoff and co-workers isolated and named one of the yogurt bacteria,.
- 1908: Sodium benzoate was given official sanction by the United States as a preservative in certain foods.
- 1915: *Bacillus coagulans* was first isolated from coagulated milk by B. W. Hammer.
- 1920: Several nasty outbreaks of botulism in the early 1920s finally prompted the U.S. canning industry to adopt a very conservative heat treatment, known as the 12D process, that reduces the probability of survival of the most heat resistant *C. botulinum* spores to one in a billion (10^{-12}). This practice continues today, and since 1925, the canning industry has produced more than a trillion containers with only 5-6 known incidents of botulism. Most of these incidents involved faulty containers, not under processing.
- 1928: The first commercial use of controlled atmosphere storage of apples was made in Europe (first used in New York in 1940).
- 1937: Paralytic shellfish poisoning was recognized.
- 1941: George W. Beadle and Edward L. Tatum studied the relationship between genes and enzymes using the bread mold, Neurospora. Precursors were ornithine, citrulline and arginine. They proposed one gene, one polypeptide hypothesis.
- 1943: Salvadore Luria and Max Delbruck demonstrated spontaneous gene mutations in bacteria (not directed by the environment).
- 1950: The D value concept came into general use.
- 1955: Similarities between cholera and *Escherichia coli* gastroenteritis in infants were noted by S. Thompson.
- 1960: The production of aflatoxins by *Aspergillus flavus* was first reported.
- 1960-1969: Type E botulism caused death in the early 1960s. Salmonella continued to be an important pathogen in food.
- 1970-1979: A new and unfamiliar agent, *Escherichia coli* 027:H_2O, which caused

outbreak of gastroenteritis in 1971. Disease outbreaks caused by *Yersinia enterocolitica* and *Campylobacter jejuni* were heard of near the end of the decade, but received little attention except from specialists in foodborne disease control.

- 1980-1989: *C. jejuni* emerged as the leading cause of gastroenteritis in U.S. *Y. enterocolitica* was also identified in several outbreaks of gastroenteritis, most of them from dairy products. *E. coli* O157:H7 first appeared in 1982 and caused serious outbreaks. *Aeromonas hydrophila* also was recognized in the early part of the decade as a possible cause of foodborne disease. Beside the conventional *V. cholerae* and *V. parahaemolyticus*, other Vibrio species also attracted attention. *L. monocytogenes* also re-emerged as an important foodborne disease.

Space has restricted the inclusion of the history of individual pathogens, control programs, stabilizing foods, and controversies over issues, such as irradiation and nitrite. A half century ago Salmonella, *S. aureus*, and *C. botulinum* were the foodborne pathogens. Prions, Campylobacter spp., and Shiga toxin-producing *Escherichia coli* (STECs) were not on the list. Since that time, we have learned that benign organisms such as *E. coli* and *Citrobacter* can acquire virulence factors and become pathogenic. Perhaps, in the future, microbiologists will assay for virulence factors, not genera and species. A century from now, microbiologists will smile on our ignorance.

This history of food microbiology is incomplete because food microbiology is evolving rapidly with new tools and techniques for discovering, inactivating, and preventing foodborne diseases.

On-Going New Developments

The twentieth century has seen the maturing of food microbiology as a science. With increasing population growth and the increased demands for health and quality from our foods, the food microbiologist is an important contributor to our quality of life.

Basic information on the physiological, biochemical, and biological characteristics of microorganisms in foods such as microbial interactions in food environments and microbial physiology has helped open new frontiers in food microbiology. Presently, topics of major interest to the food microbiologist include: Food fermentations/probiotic fermentation, Laboratory management, Water quality, Food preservation, Food hygiene, Food-borne disease, Food spoilage, Quality control *etc.* Not all groups of micro-organisms are of equal interest to the food microbiologist. Bacteria come very much on top of the list with molds and yeasts also of considerable importance and viruses less so. Protozoa and algae have minimum direct impact on the production, processing and consumption of food. Food-borne disease can be caused by some protozoa and others belonging to this group are important in the treatment of wastes. Algae are used to produce alginates; some have the potential for use in the production of single-cell protein and some marine species produce toxins that might enter our food along with sea foods.

Food microbiology papers published during the past decade have been characterized by multidisciplinary interests that have confirmed the increasing amount of evidence that has implicated microorganisms in different areas, including food technology, food spoilage, foodborne diseases, food poisoning, food safety and hygiene, food genomics, and, more generally, food omics, functional foods, and food fermentations/probiotics, besides emerging methodologies that have been applied to food analyses.

Several articles have focused on fermented foods, such as bread, cheese, wine, and others. Even if these foods have already been studied extensively in the past, the use of new technologies and omics approaches to implement the knowledge of how the microbiota affects quality and safety attributes of these foods and beverages has been encouraged, and this trend will be confirmed also in the future. These fermented foods represent a particular niche that could be rich with new positive and beneficial microbial strains influencing food quality and safety and that can also improve human health among other aspects.

Several studies dealing with health-promoting benefits associated with the consumption of fermented foods and beverages are being conducted. Global fermented foods, classified into nine major groups on the basis of raw materials, can be represented by more than 5,000 varieties being consumed around the world by billions of people. In the last 20 years, culture-independent methods have emerged as a convenient complement for analyzing the microbiota of fermented foods.

Probiotics research and innovation in functional food production deserves particular attention. Many articles have focused on the survival of potential probiotic bacteria in the gastro-intestinal tract (GIT), the microbial adhesive capacity and colonization of the gut, the safety status of probiotic strains, as well as gut microbiome homeostasis maintenance by competitively inhibiting the growth of pathogens or producing antimicrobial compounds. New probiotic strains are being screened for natural bioactive substances, immunomodulation capacity, as well as anticancer and other health benefits. A new era within probiotics research has started with an increasing interest in the use of gut commensal bacteria as potential probiotics, such as strains belonging to the genera *Bacteroides, Clostridium, Bifidobacterium*, and *Faecalibacterium*, which predominate in the human gut microbiome (Langella *et al.*, 2019).

In general, the main relevant topics in the next years, related to fermented foods, can be summarized: microbiota involved in product fermentation; a selection of technological, protective, and probiotic starters as well as safety concerns related with their use; the genomic and metabolomics characterization of microorganisms with a technological impact on fermented products; the control and inhibition of pathogens and spoilage organisms; the relationship between technological procedures and microbiota of fermented foods; and traditional and ethnic fermented foods and beverages.

Lot of work is being done in the area of Microbiological Risk Analysis (MRA) on foods which comprises three interrelated activities: risk assessment, risk communication, and risk management. New technologies used in the food production, processing, distribution, storage, and consumption introduce new problems. Food microbiology deals with these problems and tries to effectively solve them.

As we eliminate microbes from food, we create an environment free of competition which may allow opportunities for other microorganisms to grow and cause disease. For this reason, there is considerable interest in identifying safe bacteria (*e.g.*, lactic acid bacteria) which, when deliberately added to food, would inhibit growth of pathogens but would not rapidly spoil the product themselves (though some lost shelf life seems inevitable).

The stress response in food microbes has been the focus of more than 400 articles published in the last 10 years, and more than 20 RTs have targeted microbial resistance. A wide range of food bacteria, pathogens or not, have been described to possess many adaptive mechanisms and specific stress responses that are useful to guarantee and improve fitness under specific environments. An important bacterial stress response is related to cross-protection, which plays a significant role in minimally processed foods. Future research should focus on the implementation of new methodologies for analytical methods able to detect and enumerate viable—but not culturable—cells as well as their stress responses and adaptation (Ruiz *et al.*, 2017).

Bacterial pathogens associated with foodborne disease worldwide include *Salmonella enterica, Campylobacter jejuni, Escherichia coli, Listeria monocytogenes, Cronobacter sakazakii, Vibrio cholerae,* and *Vibrio parahaemolyticus.* The subjects have ranged from foodborne pathogenic bacteria resurgences to pathogenesis and control strategies, antibiotic resistance, enteric virus, bacterial toxin, stress responses, applications of protective cultures, and bacteriocins for foods preservation, and new methods for the study foodborne pathogens.

Application of protective cultures and bacteriocins for food preservation" (Hammami *et al.,* 2019) focused on antimicrobial substances produced by LAB inhibiting foodborne pathogens and spoilage microorganisms. These studies are in progress and will increase further in the near future, offering a promising alternative to chemical preservatives to ensure the quality and safety of ready-to-eat, extended-shelf-life, fresh-tasting, and minimally processed foods. Similar remarks can be made regarding *Saccharomyces* and non-*Saccharomyces* yeasts for alcoholic fermentation and *Bacillus* spp. for alkaline fermented foods. However, new starter cultures should be identified that involve natural adaptation and evolution, such as a direct selection of mutants with the desired properties, adaptative laboratory evolution, and genetic methods besides genome sequencing of wild type strains, for guiding safety assessments and strain-improvement activities (Johansen, 2018).

New biological and non-biological innovation technologies, new compounds and treatment strategies, and advances in DNA sequencing technologies, with the characterization of bacterial genomes, have emerged for the control of foodborne pathogens; this must also be pursued further in the near future (Chen and Alali, 2018).

Our technical struggle has been to satisfy the demands of society for good quality foods without causing poisoning. This is an area where we have been peculiarly successful, and should be proud of the achievements of workers in food microbiology. Unfortunately, during the twentieth century, so far new microbial threats have always been a surprise. Operating globally leads to a set of conditions which often conspire against the best intent of the food microbiologist. These conditions can be categorised as the increased vulnerability of the population (including ageing, immuno-suppression, environmental exposure to pollutants, chemicals, poverty and crowding); technological advances (including antimicrobials and pesticides, land usage and agricultural change and urbanisation); the human impact of increased travel and contact of different cultures and customs, genetic make-up and 'cultural vectors'. As we enter the next millennium, it is a good time to ask what the future holds. None of us know, but there are a few issues that can be addressed which might signpost the way forward.

Looking forward, conventional food sources may be complemented by edible microbial biomass derived from bacteria, yeasts, filamentous fungi, or microalgae. Nowadays, these groups of microorganisms are evaluated as an important and good source of proteins, vitamins, and beneficial bioactive compounds. It appears that the human population will increase up to about 9–12 billion people by the year 2100, and microorganisms could be an integral part of the sustainable production system.

Science never stands still and today's certainties always become tomorrow's myths. Food microbiology is an important discipline. Everyone must eat, everyone wishes to enjoy eating without being poisoned, yet the need is to feed more and more people with finite resources. Demands upon us will increase in the next few years and it would be nice if we stopped becoming the 'victim' of the next organism which we did not see coming and were able to pro-actively take steps to deal with threats in the food chain. The "Food Microbiology" section of "Frontiers in Microbiology" could provide good guidance through its RTs on new advances in microbiology for the improved utilization, production, and supply of food in the food industry and related fields, which can help to ensure global food safety and security.

Chapter 3

The Role of Microorganisms in Foods

Introduction

Many times we feel that microorganisms are trying to ruin our food sources, by destroying plants and animals, including man; this is by no means their primary role in nature.

During this period, humans were presumably carnivorous, with plant foods coming into their diet later in this period. It is also during this period that foods were first cooked.

Since human food sources are of plant and animal origin, it is important to understand the biological principles of the microbial flora associated with plants and animals in their natural habitats and respective roles.

The microbial spoilage of foods may be viewed simply as an attempt by the food flora to carry out what appears to be their primary role in nature. This should not be taken in the teleological sense. In spite of their simplicity when compared to higher forms, microorganisms are capable of carrying out many complex chemical reactions essential to their perpetuation. To do this, they must obtain nutrients from organic matter, some of which constitutes our food supply.

The first man to suggest the role of microorganisms in spoiling foods was A. Kircher, a monk, who as early as 1658 examined decaying bodies, meat, milk and other substances and saw what he referred to as "worms" invisible to the naked eye. Kircher's descriptions lacked precision, however and his observations did not receive wide acceptance. In 1765, L. Spallanzani showed that beef broth that had been boiled for an hour and sealed remained sterile and did not spoil. Spallanzani performed this experiment to disprove the doctrine of the spontaneous generation of life. However, he did not convince the proponents of the theory since they believed that his treatment excluded oxygen, which they felt was vital to spontaneous generation. In 1837 Schwann showed that heated infusions remained sterile in the presence of air, which he supplied by passing it through heated coils into the infusion. While both of these men demonstrated the idea of the heat preservation of foods, neither took advantage of his findings with respect to application. The same may be said of D. Papin and G. Leibniz, who hinted at the heat preservation of foods at the turn of the eighteenth century.

The role of microorganisms can be understood easily by studying Nitrogen Cycle (Figure 3.1).

Results from many laboratories show that untreated foods may be expected to contain varying numbers of bacteria, molds, or yeasts and the question often arises as to the safety of a given food product based upon total microbial numbers. The question should be twofold: What is the total number of microorganisms present per g or ml? and what *types* of organisms are represented in this number? It is necessary to know which organisms are associated with a particular food in its natural state and which of the organisms present are not normal for that particular food. It is, therefore, of value to know the general distribution of bacteria in nature and the general types of organisms normally present under given conditions where foods are grown and handled.

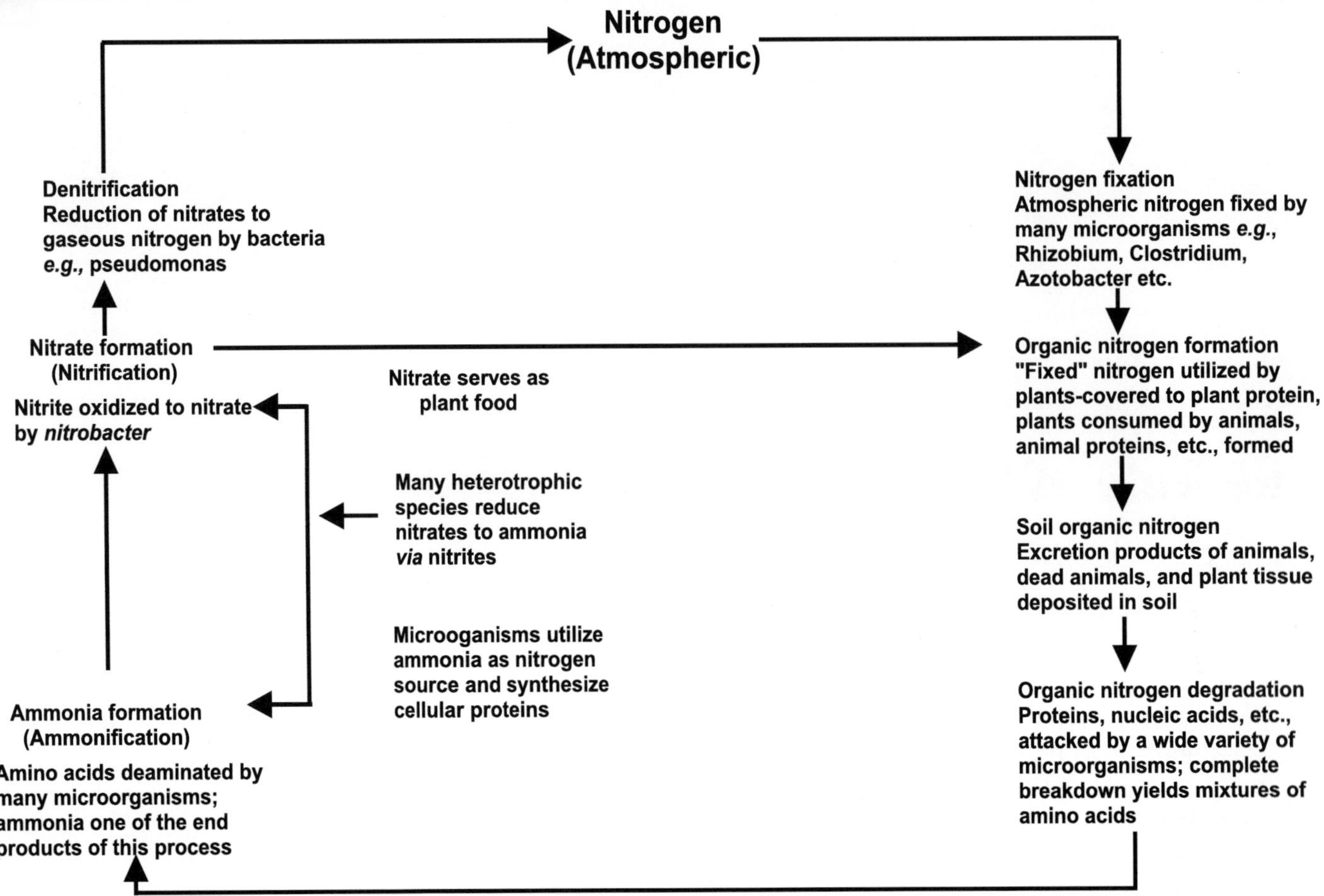

Figure 3.1: Nitrogen Cycle in Nature.

Some of the most common genera of molds associated with foods are:

Alternaria *Aspergillus*
Botrytis *Byssochlamys*
Cephalosporium *Cladosporium*
Colletotrichum *Fusarium*
Geotrichum *Gloeosporium*
Helminthosporium *Monilia (Neurospora)*
Mucor *Penicillium*
Rhizopus *Sporotrichum*
Thamnidium *Trichothecium*

Many genera of bacteria are known to occur in foods. They are as follows:

Acetobacter *Acinetobacter*
Aeromonas *Alcaligenes*
Alteromonas *Bacillus*
Brochothrix *Campylobacter*
Citrobacter *Clostridium*
Corynebacterium *Enterobacter*
Erwinia *Escherichia*
Flavobacterium *Lactobacillus*
Leuconostoc *Micrococcus*
Moraxella *Pediococcus*
Proteus *Pseudomonas*
Salmonella *Serratia*
Shigella *Staphylococcus*
Streptococcus *Vibrio*
Yersinia

The other most common organisms that we find in foods are yeasts. They are as follows:

Brettanomyces *Candida*
Debaryomyces *Endomycopsis*
Hansenula *Kloeckera*
Kluyveromyces *Mycoderma*
Rhodotorula *Saccharomyces*
Sacchacomycopsis *Schizosaccharomyces*
Torulopsis (Torula) *Trichosporon*

Food Environment

The genera of microorganisms to be found on food utensils depend upon the types of foods handled, the care of these utensils, their storage and other factors. If vegetables are handled in a given set of utensils, one would expect to find some or all of the organisms associated with vegetables. When utensils are cleaned with hot or boiling water, the remaining flora would normally be those best able to withstand the effects of this treatment.

It may be assumed that at one time all microorganisms existed in water. The drying of surface soils gives rise to dust that, when disseminated by winds, carries adhering microorganisms to many places including other areas of the soil, rivers and oceans. The formation of clouds over large bodies of water and the subsequent rainfall over land as well as other waters has the same effect. It is not surprising, then, that soil and water microorganisms are often one and the same.

The following genera of food-borne bacteria generally found in soils and water may be expected in foods: *Acinetobacter, Alcaligenes, Bacillus, Citrobacter, Clostridium, Corynebacterium, Enterobacter, Micrococcus, Proteus, Pseudomonas, Serratia* and *Streptomyces,* among others.

Many of the molds listed above occur in soils and also in waters. Molds are in general very widespread in nature, where they participate in the degradation of both plant and animal matter as well as cause many diseases of plants and animals. Among those that are nearly always present in soils are: *Aspergillus, Rhizopus, Penicillium, Trichothecium, Botrytis, Fusarium* and others.

Any one or all of the genera of bacteria, yeasts and molds cited earlier in this chapter may be found in animal feeds. The types of organisms to be found would depend, of course, on the source of the feeds the treatment given to them to destroy microorganisms, the containers in which they are stored and the like. Animal feeds are of great importance in the spread of food-poisoning *Salmonella.* Organisms from this source have been shown to be rapidly disseminated throughout processing plants where feeds are handled.

We can find several genera of bacteria in the intestinal tract of man and animals. The most notable of these is the genus *Escherichia*, which has as its natural habitat in the intestinal tract of man and other mammals. Species of other genera common to the intestinal tract include *Clostridium, Citrobacter*, *Enterobacter* and *Pseudomonas*. From the intestinal tracts of animals, intestinal microorganisms find their way directly to the soil and water. And from the soil they may find their way onto plants, dust, utensils and so on. Molds are not thought to be transmitted by faecal sources, though the yeast genus *Candida* is very often found in the intestinal tract of man.

The microflora on the hands and outer garments of food handlers generally reflects the environment and habits of the individuals. This flora would normally consist of organisms found on any object handled by the individual as well as some of those picked up from dust, water, soil and the like. In addition, there are several genera of bacteria that are specifically associated with the hands, nasal cavities and mouth. Among these are the genera *Micrococcus* and *Staphylococcus*, the most notable of which are the staphylococci, which are found on hands, arms, in nasal cavities, the mouth and other parts of the body. While the genera *Salmonella* and *Shigella* are basically intestinal forms, they may be deposited onto foods and utensils by food handlers if sanitary practices are not followed by each individual. Any number of molds and yeasts may be found in the hands and garments of food handlers depending upon the immediate history of each individual.

The most important food-poisoning and gastroenteritis-causing bacteria belong to the genera *Staphylococcus, Salmonella, Clostridium* and *Campylobacter*. The staphylococci are associated with the nasal cavities of man and animals as well as with other parts of the body. Salmonellae are indigenous to the intestinal tract of man and animals but may enter foods from other sources contaminated, from fecal matter. The *Clostridia* are basically soil forms while *Campylobacter* spp. are animal associated.

The types of organisms to be found in air and dust, with the exception of some of the pathogens, include the genera of bacteria, the eighteen genera of molds and many of the yeasts. Although *Staphylococcus* and *Salmonella* spp. may at times be found in air and dust yet these are not the major sources of these organisms. Notable among the bacterial genera in air and dust are *Bacillus* and *Micrococcus* spp. all of which are able to endure dryness to varying degrees. Notable among the yeasts is the genus *Torulopsis* and many mold genera may be found from time to time.

Unlike the true bacteria and most yeasts, mold grow in the form of a tangled mass which spreads rapidly and may cover several inches of area in two to three days. The total of the mass or any large single portion of it is referred to as *mycelium*. The mycelium is composed of branches or filaments referred to as *hyphae*. At the time of asexual reproduction, *sporangiophores* or *conidiophores* are sent up which bear, usually at their tips, sporangia or conidia. In those molds that produce sporangia, the spores are borne by these structures and are responsible for the various colors displayed by molds. Conidia represent unprotected spores. In addition to these characteristic and common asexual spores, some molds produce other asexual spores. *Chlamydospores* result when a thick wall develops around any cell of the mycelium. These structures are somewhat resistant to adverse environmental conditions.

Arthrospores or *oidia* result from fragmentation by some molds that produce a septate mycelium. Chlamydospores and arthrospores are a bit more difficult to destroy than other parts of the mold mycelium and they sometimes cause concern in the food industry. Molds also reproduce by sexual means where they form either ascospores. oospores, or zygospores.

Most molds of importance in foods are now placed in the *Fungi imperfecti* group. These are molds whose sex cycle are not known. Most or all are thought to be related to the *Ascomycetes* group.

Common Food-borne Molds and Yeasts

Molds grow in the form of a tangled mass, which spreads rapidly. Figure 3.2 indicates the common food-borne molds.

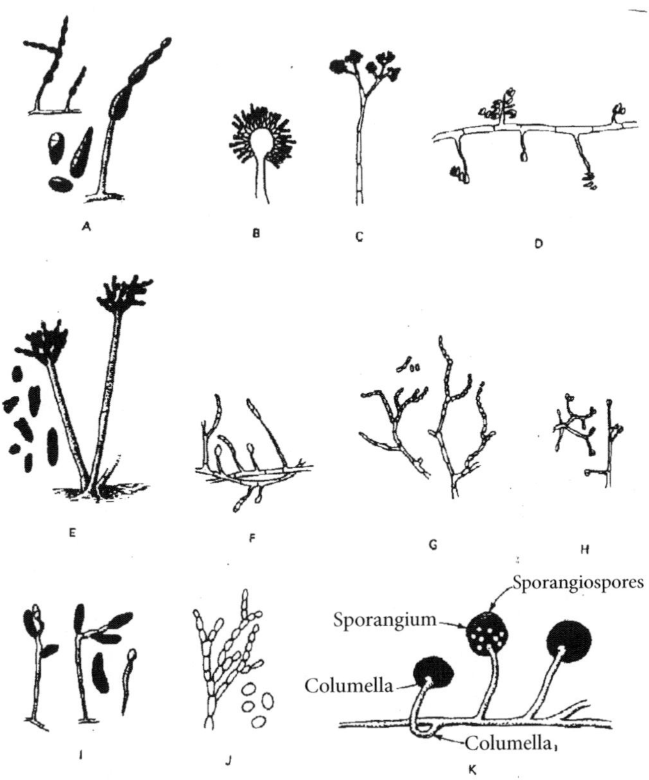

Figure 3.2: Common Food-borne Molds.

Now let us study about some of the food-borne molds.

Sclerotinia

These organisms produce long, slender and often pigmented conidiophores. The mycelium is septate and the conidia are borne on apical cells. The conidia are gray in mass and one-celled. Black irregular sclerotia are frequently produced. They cause a "gray mold" condition on many plants and plant foods. They are important causes of market diseases of fruits and vegetables. The most important crop plants affected by *Sclelrotinia sclerotiorum* are sunflower, soybean, oilseed rape, lettuce, cabbage, carrots *etc.*

Some species cause rots of vegetables and fruits, where they are present in the conidial stage. The lemon-shaped conidia are in chains, with a "plug" separating conidia. In the preceding discussion of molds important in foods, genera that only occasionally grow on specific foods have been omitted. They will be mentioned with those foods. For the descriptions of these genera, reference should be made to text and reference books on fungi listed at the end of this chapter

Gibberella

Gibberella species are destructive plant pathogens, although many are more familiar under their Fusarium anamorph names. Some Gibberella species have broad host ranges and worldwide distributions, whereas other species appear to have more restricted host ranges and geographic distribution. These molds produce an extensive mycelium that is cottony in culture with tinges of pink, purple, or yellow. The conidia are canoe shaped and borne singly or in chains. They are important in the spoilage of many vegetables and fruits including the "neck rot" of bananas. Some produce T–2 toxin, vomitoxin and zearalenone. Many Gibberella species produce an impressive array of toxins and other biologically active metabolites.

Oidium

Oidium is also known as 'powdery mildew' and mildew as 'downy mildew'. These are yeast-like fungi that are usually but not always white. The mycelium is separate and reproduction occurs by fragmentation of mycelium into arthrospores. These organisms are sometimes, referred to as the "dairy mold" since they impart flavour and aroma to many types of cheese. They are also referred to as "machinery molds," since they build up on food-contact equipment in food-processing plants, especially tomato canning plants. They display a dimorphic form and a feathery appearance and are capable of infecting ripe fruits.

Cephalothecium

These forms produce separate mycelia that bear long, slender and simple conidiophores. The conidia appear singly, apically and sometimes in groups of chains. Some are pink on foods such as fruits and vegetables.

Byssochlamys

Members of this genus of *Ascomycetes* produce clusters of asci each of which contains eight ascospores. The asci appear to be without a covering wall or ascocarp. The ascospores of these organisms are heat resistant, resulting in spoilage of some high-acid canned foods. They can grow under conditions of low Eh. They exist in soils and can be recovered from ripening fruits, especially grapes.

They are the most heat resistant of all spoilage molds.

Colletotrichum

This genus belongs to the order *Melanconiales*. They produce simple but elongate comdiophores and hyaline conidia that are one-celled, ovoid, or oblong. The acervuli produced are disc or cushion shaped, waxy and generally dark in color. They are common contaminants of fruits and vegetables.

Helminthosporium

These molds produce a light to dark mycelium in culture. Their conidiophores are short or long, septate, simple or branched. The conidia are borne successively on new growing tips; they are dark and typically contain more than three cells. This genus contains both plant pathogens and saprophytes.

Species of this genus are for the most part plant pathogens but may grow saprophytically on vegetable materials.

Monilia

These molds produce white or gray mycelia that bear branched conidiophores. The conidia are pink or tan in mass. Some species are the imperfect stages of *Neurospora*. Some of those whose perfect states are *Monilinia* or *Sclerotinia* spp. cause brown rot of fruits. *N. sitophila* is often referred to as the "red bread mold."

Mucor

These molds produce non-septate mycelia that give rise to conidiophores that bear columella and a sporangium at the apex. The spores are smooth, regular and borne within the sporangium. They may be found, growing on a large number of foods.

Mucor are involved in the spoilage of some foods and the manufacture of others. A widely distributed species is M. racemosus. M. rouxii is used in the "Airiyta" process for the saccharification of starch, and mucors help ripen some chesses (*e.g.*, Gammelost) and are used in making certain Oriental foods.

Penicillium

These organisms produce septate mycelia that bear conidiophores arising singularly or sometimes in synnemata with branches near the apex to form a brushlike, conidia-bearing apparatus. The conidia form by pinching off from phialides. Typical colours on foods are blue to blue-green. These molds are important in the making of some cheeses. Some are important in the production of antibiotics (penicillin, *e.g.*). They are widespread in soil, air, dust and many other places and may be found on foods such as breads, cakes, fruits and preserves.

Penicillium is a genus of ascomycetous fungi of major importance in the natural environment as well as food and drug production. It produces penicillin, a molecule that is used as an antibiotic, which kills or stops the growth of certain kinds of bacteria inside the body. The thallus (mycelium) typically consists of a highly branched network of multinucleate, septate, usually colourless hyphae. Many branched conidiophores sprout on the mycelia, bearing individually constricted conidiospores. The conidiospores are the main dispersal route of the fungi, and often green. Sexual reproduction involves the production of ascospores, commencing with the fusion of an archegonium and an antheridium, with sharing of nuclei. The irregularly distributed asci contain eight unicellular ascospores each. Species of Penicillium are ubiquitous soil fungi preferring cool and moderate climates, commonly present wherever organic material is available. Saprophytic species of Penicillium and Aspergillus are among the best-known representatives of the Eurotiales and live mainly on organic biodegradable substances. They are commonly known as molds and are among the main causes of food spoilage. Many species produce highly toxic mycotoxins. Some species have a blue colour, commonly growing on old bread and giving it a blue fuzzy texture.

Aspergillus

Aspergillus is a genus consisting of several hundred mold species found in various climates worldwide. Aspergillus was first catalogued in 1729 by the Italian priest and biologist Pier Antonio Micheli. Viewing the fungi under a microscope, Micheli was reminded of the shape of an aspergillum (holy water sprinkler), from Latin spargere (to sprinkle), and named the genus accordingly. Today 'aspergillum' is also the name of an asexual spore-forming structure common to all Aspergilli; around one-third of species are also known to have a sexual stage. Aspergillus species are common contaminants of starchy foods (such as bread and potatoes), and grow in or on many plants and trees.

These molds produce upright conidiophores that are simple and terminate in a globose or clavate swelling. The conidia are one-celled, globose and variously colored in mass. These molds appear yellow to green to black on a large number of foods. They produce septate mycelia. Some species produce aflatoxins and ochratoxins while others are employed as commercial sources of proteases and citric acid. Some species are field and, some are storage fungi. Aspergilli may be found on cakes, fruits and vegetables.

After studying about molds, let us study about some types of yeasts.

Yeasts are microscopic organisms that may be differentiated from the common bacteria by their larger cell size, their oval, elongate, elliptical, or spherical cell shapes and by their production of buds during the process

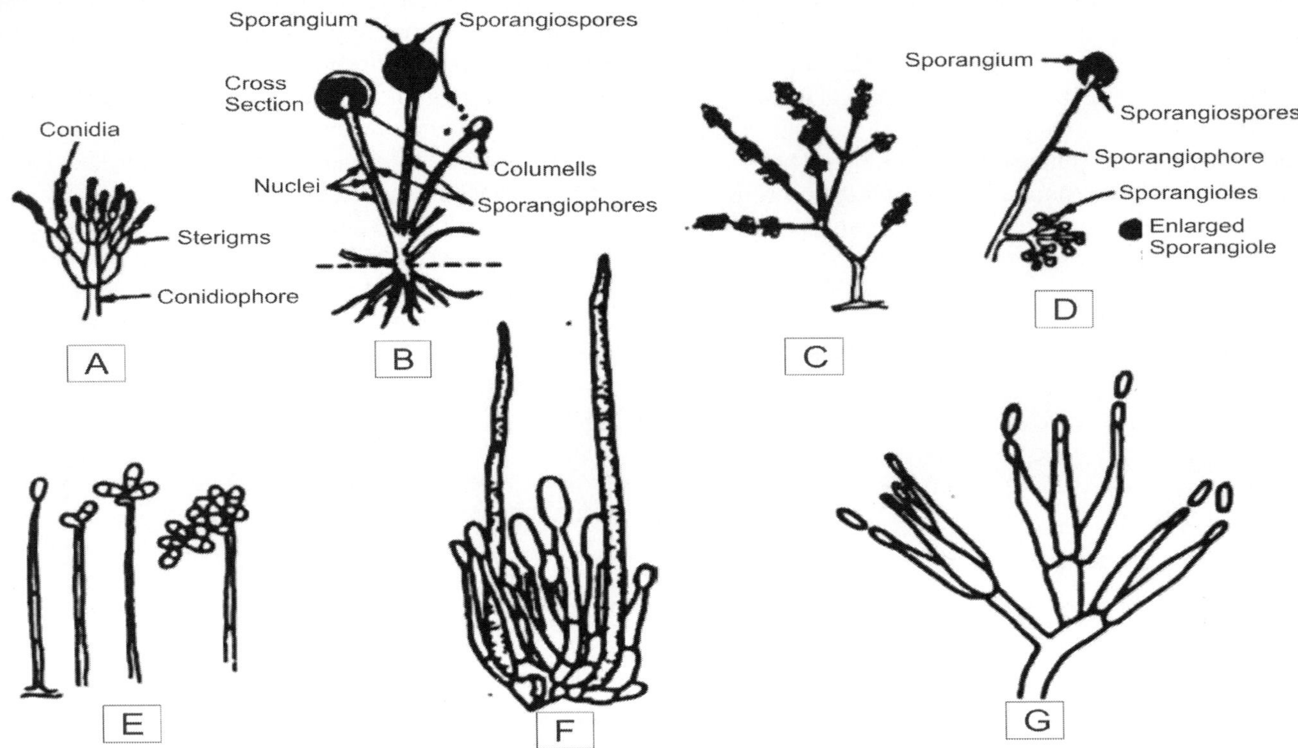

Figure 3.3: Common Food-borne Molds.

of division. The sizes of yeast cells vary, some ranging from 5 to 8 µm in diameter, while others may be as large as 100 µm in length. In general, older yeast cells tend to be smaller in size than young growing cells. Yeasts can grow over rather wide ranges of pH, alcohol and sugar concentrations. Some have been reported to grow at a pH as low as 1.5 and in up to 18 per cent ethanol. Many grow in the presence of 55 per cent or more sucrose. These organisms produce pigments of many colors, with red and black pigment producers being common. A microscopic slide of growing yeast cells generally reveals cells in varying stages of budding with some showing several buds. The true yeasts or *ascosporogenous* yeasts reproduce by sexual reproduction involving copulation to varying degrees. True yeasts also produce asexual spores and *chiamydospores.* The latter are durable and tend to be produced when a culture encounters unfavorable conditions of growth. The ascosporogenous yeasts presented below all belong to the family Endomycetaceae.

Yeasts are often placed into grouping based upon some particular function or activity. For example, *film yeasts* are those that grow at the surface of certain acid products such as sauerkraut and pickles. Some film yeasts are species and strains of the genera *Candida* and *Hansenula*. These organisms are capable of oxidizing acids and alcohols as source of energy. *Top yeasts* are those that carry out the conversion of sugars to alcohol at the top of a vessel, while *bottom yeasts* are those that can do the same but from the bottom of the vessel. *Apiculate* or lemon-shaped yeasts are undesirable in wine fermentations, where they produce off-flavors.

The *asporogenous* yeasts (false yeast, wild yeasts) do not display sexual reproduction and are all placed in the family *Cyptococcaceae*. Some of these are yeast-like organisms which are sometimes placed in the *Fungi imperfecti* group along with some of the more common molds. The wild yeasts are difficult to distinguish morphologically and culturally but may be identified serologically. While some are strains of *Saccharomyces cerevisiae* or *S. carlsbergensis,* some are representatives of many other genera including *Candida, Debaryomyces* and *Pichia.*

Debaryomyces

These are ascosporogenous yeasts that sometimes produce a pseudomycelium. They generally reproduce by multipolar budding and also by sexual means. This genus is often found on the surface of spoiling foods such as fresh and cured meats, sausage, pickle brine and wines.

Hansenula

These are ascosporogenous yeasts that produce spherical, elongate, or oval cells and often a pseudomycelium. They reproduce by multipolar budding and by sexual means. When the latter occurs, hat-shaped spores are produced inside of the asci. They are common on citrus fruits, grapes, grape products and in olive brine and fruit juice concentrates.

This is one of the most prevalent genera in dairy products. *D. hansenii* is an important food spoilage species, it can grow in 24 per cent NaCl and at a^w as low as 0.65.

Kloeckera

These are non-sporeforming yeasts common on fruits, where they are disseminated and also consumed by fruit flies. They possess both fermentative and oxidative abilities. Some cause off-flavor and turbidity in wines.

These are imperfect apiculate or lemon-shaped yeasts. *K. apiculata* is common on fruits and flowers and in the soil.

Kluyveromyces (Fabospora)

These are ascosporeforming yeasts. The ascospores are reniform to oblong with asci containing one to many spores, which are released at maturity. Vegetative reproduction is by multilateral budding. They are vigorous fermenters of sugars including lactose. *K. marxianus* produces ethanol directly from D-xylose and *K. fragilis* and *K. lactis* produce β-galactosidase (lactase).

Kluyveromyces is a genus of ascomycetous yeasts in the family *Saccharomycetaceae*. Some of the species, such as *K. marxianus*, are the teleomorphs of Candida species.

Brettanomyces

Brettanomyces is a non-spore forming genus of yeast in the family *Saccharomycetaceae*, and is often colloquially referred to as 'Brett'. The genus name Dekkera is used interchangeably with Brettanomyces, as it describes the teleomorph or spore forming form of the yeast. These are acid-producing asporogenous yeasts that produce oval, elongate, spherical, or olive-shaped cells. The yeast is acidogenic and when grown on glucose rich media produce large amounts of acetic acid. Reproduction is by multipolar budding that often leads to the formation of chains of cells. Brettanomyces is important to both the brewing and wine industries due to the sensory compounds it produces. In the wild, Brettanomyces lives on the skins of fruit. Some species carry out an after-fermentation in certain European beers and ales while others have been isolated from spoiled pickles.

Chapter 4

Effect of Food Preservation on Microorganisms

Heat Process Employed in Food Preservation

Foods are subject to thermal processes in a number of different contexts. Often, their main objective is not destruction of microorganisms in the product, although this is an inevitable and frequently useful side effect.

Credit for discovering the value of heat as a preservative agent goes to the French chef, distiller and confectioner, Nicolas Appert. In 1795 the French Directory offered a prize of 12000 francs to anyone who could develop a new method of preserving food. Appert won this prize in 1811 in a book called the 'Art of Conserving all kinds of Animal and Vegetable Matter for several Years'. Appert developed a technique based on packing foods in glass bottles, sealing them, and then heating them in boiling water. A similar technique was used by the Englishman Saddington in 1807 to preserve fruit and for which he too received a prize, this time of five guineas, from the Royal Society of Arts. British patents describing the use of iron or metal containers were issued to Durand and de Heine in 1810 and the firm of Donkin and Hall established a factory for the production of canned foods in Bermondsey, London around 1812.

Appert demonstrated that this process would keep food from spoiling for extended periods of time, provided the seals were not broken. It was used to preserve soups, meats, vegetables, juices, various dairy products, jams, jellies, and syrups. Appert held the view that the cause of food spoilage was contact with air and that the success of his technique was due to the exclusion of air from the product. This view persisted with sometimes disastrous consequences for another 50 years until Pasteur's work established the relationship between microbial activity and putrefaction. Today, the two types of heat processes employed to destroy microorganisms in food *i.e.*, pasteurization and appertization, bear the names of these eminent figures.

Table 4.1: Heat Processes Applied to Foods

Heat Processes	*Temperature*	*Objective*
Cooking baking boiling frying grilling	≤ 100°	Improvement of digestibility, *e.g.* starch gelatinization, collagen breakdown during cooking of meat Improvement of flavour Destruction of pathogenic microorganisms
Blanching	< 100°C	Expulsion of oxygen from tissues inactivation of enzymes
Drying/ Concentration	< 100°C	Removal of water to enhance keeping quality
Pasteurization	60–80°C	Elimination of key pathogens and spoilage organisms
Appertization	> 100°C	Elimination of microorganisms to achieve 'commercial sterility'

Now let us try to understand about pasteurization.

Pasteurization, the term given to heat processes typically in the range 60–80°C and applied for up to a few minutes, is used for two purposes. First is the elimination of a specific pathogen or pathogens associated with a product. This type of pasteurization is often a legal requirement introduced as a public health measure when a product has been frequently implicated as a vehicle of illness. Notable examples are milk, bulk liquid egg and ice cream mix, all of which have a much improved safety record 'as a result of pasteurization'. The second reason for pasteurizing a product is to eliminate a large proportion of potential spoilage organisms, thus extending its shelf-life. This is normally the objective when acidic products such as beers, fruit juices, pickles and sauces are pasteurized.

Where pasteurization is introduced to improve safety, its effect can be doubly beneficial. The process cannot discriminate between the target pathogen(s) and other organisms with similar heat sensitivity so a pasteurization which destroys say *Salmonella* will also improve shelf-life. The converse does not normally apply since products pasteurized to improve keeping quality are often considered as being intrinsically safe due to other factors such as low pH. This may be less true than was previously thought following recent outbreaks associated with unpasteurized fruit juices.

On its own, the contribution of pasteurization to extension of shelf-life can be quite small, particularly if the pasteurized food lacks other contributing preservative factors such as low pH or a_w. Thermoduric organisms such as spore formers and some Gram-positive vegetative species in the genera *Enterococcus, Microbacterium* and *Arthrobacter* can survive pasteurization temperatures. They can also grow and spoil a product quite rapidly at ambient temperatures, so refrigerated storage is often an additional requirement for an acceptable shelf-life.

Appertization is simply defined as the heat-treatment of food at certain temperature levels that inhibit or kill pathogenic microorganisms present in the food. Unlike pasteurization which uses temperatures below 100°C to kill microbes in food, appertization (which was discovered by Nicolas Appert in the 18th century) is generally a food preservation technique that is used to heat food at temperatures of 40°C which is lower than that used in pasteurization. In appertization, foods or food products are exposed to varying low temperatures and time which makes the food microbiologically fit for consumption. Food products are heated at this temperature in containers or cans in order to bring the microbial load in the food to levels that are generally safe for human consumption. Hence, *appertization* refers to processes where the only organisms that survive processing are non-pathogenic and incapable of developing within the product under normal conditions of storage. As a result, appertized products have a long shelf-life even when stored at ambient temperatures. The term was coined as an alternative to the still widely used description *commercially sterile* which was objected to on the grounds that sterility is not a relative concept; a material is either sterile or it is not. An appertized or commercially sterile food is not necessarily sterile–completely free from viable organisms. It is however free from organisms capable of growing in the product under normal storage conditions as foods have been so treated at temperature levels that significantly reduce or limit the possibility of microbial growth in the food product during storage. This process of appertization is used in food industries such as in canning to produce canned food products. The rationale behind appertization is based on the fact that viable organisms that survive during processing will not be able to grow under normal storage conditions; and these organisms will be below certain levels after a particular temperature (*e.g.* 40°C) is applied to the food and at a specific time period during its processing. Thus, for a canned food in temperate climates, it is not a matter of concern if viable spores of a thermophile are present as the organism will not grow at the prevailing ambient temperature. However, appertization can still be carried out at high temperatures (*e.g.* above 100°C) for some food products.

High-temperature short time (HTST) is applied in both pasteurization and appertization techniques to destroy pathogenic and spoilage microorganisms present in foods including the spores of some food-borne pathogens such as *Clostridium* species. **HTST** is defined as the heating of milk and other liquid food products at high temperatures and at varying time intervals that can range from seconds to minutes and hours in order to inhibit or kill pathogenic microorganisms and other spoilage microbes present in them.

Many types of methods are used in food preservation. Let us try to understand some of these methods.

Table 4.2 indicates mechanisms of principal food preservation procedures.

Table 4.2: Mechanisms of Principal Food Preservation Procedures

Procedure	*Factor Influencing Growth or Survival*
Cooling, chill distribution and storage	Low temperature to retard growth
Freezing, frozen distribution and storage	Low temperature and reduction of water activity to prevent growth
Drying, curing and conserving	Reduction in water activity sufficient to delay or prevent growth
Vacuum and oxygen-free 'modified-atmosphere' packaging	Low oxygen tension to inhibit strict aerobes and delay growth of facultative anaerobes
Carbon dioxide-enriched 'modified-organisms by carbon dioxide	Specific inhibition of some micro-atmosphere' packaging
Addition of acids	Reduction of pH value and sometimes additional inhibition by the particular acid
Lactic fermentation	Reduction of pH value *in situ* by microbial action and sometimes additional inhibition by the lactic and acetic acids formed and by other microbial products, *e.g.* ethanol, bacteriocins
Emulsification	Compartmentalization and nutrient limitation within the aqueous droplets in water-in-oil emulsion foods

Addition of preservatives	Inhibition of specific groups of microorganisms
Pasteurization and appertization	Delivery of heat sufficient to inactivate target microorganisms to the desired extent
Radurization, radicidation and Radappertization	Delivery of ionizing radiation at a dose sufficient to inactivate target microorganisms to the desired extent
Application of high hydrostatic pressure Pascalization	Pressure-inactivation of vegetativebacteria, yeasts and moulds

Heat Resistant Spores

Various microorganisms exhibit heat sensitive reactions. Now let us study about psychrotrophs and mesophiles.

Generally, psychrotrophs are less heat resistant than mesophiles, which are less heat resistant than thermophiles; and Gram-positives are more heat resistant than Gram-negatives. Most vegetative cells are killed almost in instantaneously at 100°C and their D values are measured and expressed at temperatures appropriate to pasteurization.

Bacterial spores are usually far more heat resistant than vegetative cells; thermophiles produce the most heat resistant spores while those of psychrotrophs and psychrophiles are most heat sensitive. Since spore inactivation is the principal concern in producing appertized foods, much higher temperatures are used in appertization processes and in the measurement of spore D values.

Yeast ascospores and the asexual spores of moulds are only slightly more heat resistant than the vegetative cells and will normally be killed by temperatures at or below 100°C *e.g.* in the baking of bread. Ascospores of the mould *Byssochlamys fulva* and a few other ascomycetes do show a more marked heat resistance and can be an occasional cause of problems in canned fruits which receive a relatively mild heat process.

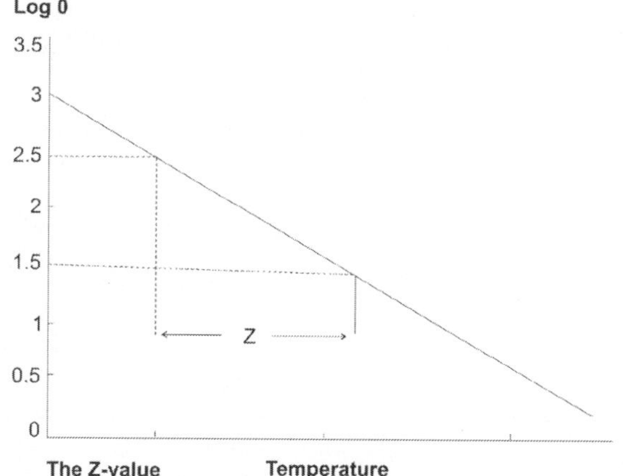

Figure 4.1: The Z-Value.

The heat resistance exhibited by the bacterial endospore is due mainly to its ability to maintain a very low water content in the central DNA-containing protoplast; spores with a higher water content have a lower heat resistance. The relative dehydration of the protoplast is maintained by the spore cortex, a surrounding layer of electronegative peptidoglycan which is also responsible for the spore's refractile nature. The exact mechanism by which it does this is not known, although it may be some combination of physical compression of the protoplast by the cortex and osmotic extraction of the water. As the cortex is dissolved during germination and the protoplast rehydrates, so the spore's heat resistance declines. Suspension of a germinated spore population in a strong solution of a non-permeant solute such as sucrose will reverse this process of rehydration and restore the spore's heat resistance. The total picture is probably more complex than this however, since other features of the spore such as its high content of divalent cations, particularly calcium, are thought to make some contribution to heat resistance.

Thermal sensitivity as measured by the D value can vary with factors other than the intrinsic heat sensitivity of the organism concerned. This is most pronounced with vegetative cells where the growth conditions and the stage of growth of the cells can have an important influence. For example, stationary phase cells are generally more heat resistant than log phase cells. Heat sensitivity is also dependent on the composition of the heating menstruum; cells tend to show greater heat sensitivity as the pH is increased above 8 or decreased below 6. Fat enhances heat resistance as does decreasing a_w through drying or the addition of solutes such as sucrose. The practical implications of this can be seen in the more severe pasteurization conditions used for high sugar or high fat products such as ice cream mix and cream compared, with that used for milk. This effect is quite dramatic in the instance of milk chocolate where the D_{70} value of *Salmonella* Seftenberg 775W has been measured as between 6 and 8 hours compared with only a few seconds in milk. A more specialized example of medium effects on heat sensitivity occurs in brewing where the ethanol content of beer has been shown to have a profound effect on the heat sensitivity of a spoilage *Lactobacillus*; an observation that has implications for the pasteurization of low-alcohol beers.

Table 4.3 indicates the microbial heat resistance.

Table 4.3: Microbial Heat Resistance

Vegetative Organisms (z ~ 5°C)	*D (mins)*	
Salmonella sp.	D_{65}	0.02–0.25
Salmonella Seftenberg	D_{65}	0.8–1.0
Stophylococcus aureus	D_{65}	0.2–2.0
Escherichia coli	D_{65}	0.1

Vegetative Organisms (z ~ 5°C)	D (mins)	
Yeasts and moulds	D_{65}	0.5–3.0
Listeria monocytogenes	D_{60}	5.0–8.3
Campylobacter jejuni	D_{55}	1.1
Bacterial Endospores (z ~ 10°C)	D_{121}	
B. stearothermophilus		4–5
C. thermosaccharolyticum		3–4
Desulfotomaculum nigrificans		2–3
B. coagulans		0.1
C. botulinum types A and B		0.1–0.2
C. sporogenes		0.1–1.5
C. botulinum type E	D_{80}	0.1–3.0
	D_{110}	< 1 second

The primary assumption which gives rise to log-linear kinetics is that at a constant temperature each cell has an equal chance of inactivation at any instant. This can be explained on theoretical grounds if there is a single target molecule in each cell whose inactivation causes death. If this is not the case and there is, say, a large number of the same target molecules present or a large number of different targets, then several inactivation events would be required for cell death and a lag would be observed in the thermal death curve.

Damage to DNA has been identified as the probable key lethal event in both spores and vegetative cells. In spores, however, inactivation of germination mechanisms is also important. If this inactivation can be bypassed in some way, then apparently dead spores may be cultured. This has been demonstrated by the inclusion of lysozyme in recovery, media where the enzyme hydrolyses the spore cortex, replacing the spore's own inactivated germination system.

Deviations from log-linear kinetics in the thermal death of vegetative cells probably reflect a greater multiplicity of target sites for thermal inactivation such as the cytoplasmic membrane, key enzymes, RNA and the ribosomes. This type of damage can be cumulative rather than instantly lethal.

Individual inactivation events may not kill the cell but will inflict sub-lethal injury making it more vulnerable to other stresses. If however injured cells are allowed time in a non-inhibitory medium, they can repair and recover their full vigour. Examples of sub-lethal damage can be seen when cells do not grow aerobically but can be cultured anaerobically or in the presence of catalase, or when selective agents such as bile salts or antibiotics, which are normally tolerated by the organism, prove inhibitory.

Two other factors also contribute to deviations from log-linear behaviour in vegetative organisms. Individual cells within a population may exhibit a broader range of heat resistance than is seen with spores and, since vegetative cells are not metabolically inert, they may also respond and adapt to a heating regime modifying their sensitivity.

Lethal Effect of Different Processes

Heating processes are neither uniform nor instantaneous. To be able to compare the lethal effect of different processes it is necessary for us to have some common currency to describe them. For appertization processes this is known as the F value; a parameter which expresses the integrated lethal effect of a heat process in terms of minutes at a given temperature indicated by a subscript. A process may have an F_{121} value of say 4, which means that its particular combination of times and temperatures is *equivalent* to instantaneous heating to 121°C, holding at that temperature for four minutes and then cooling instantly, it does not even necessarily imply that the product ever reaches 121°C. The F value will depend on the *z* value of the organism of concern; if $z = 10$°C then 1 minute at 111°C has an $F_{121} = 0.1$, if $z = 5$°C then the F_{121} value of the same conditions will be 0.01. It is therefore necessary to specify both the *z* value and the temperature when stating F. For spores *z* is commonly about 10°C and the F_{121} determined using this value is designated F_0,

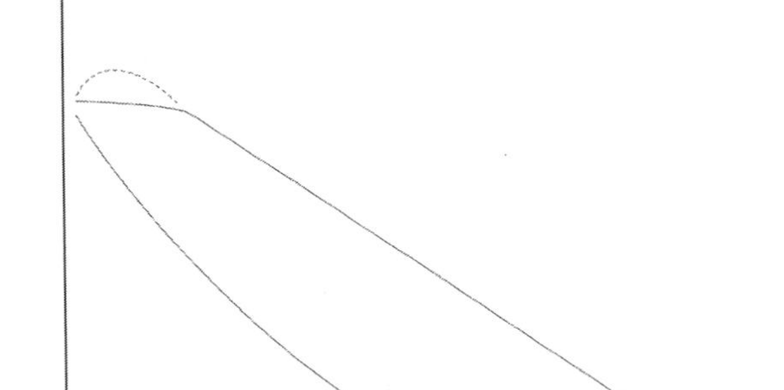

Figure 4.2: Deviation from loge-linear Death Kinetics.

To determine the F_0 value required in a particular process one needs to know the D_{121} of the target organism and the number of decimal reductions considered necessary.

$$F = D_{121}(\log N_0 - \log N)$$

In this exercise, the canner will have two objectives, a safe product and a stable product. From the point of view of safety in low acid canned foods (defined as those with a pH > 4.5) *Clostridium botulinum* is the principle concern. The widely accepted minimum lethality for a heat process applied to low-acid canned foods is that it should produce 12 decimal reductions in the number of surviving *C. botulinum* spores ($\log N_0 - \log N = 12$). This is known as the 12D or botulinum cook. If D_{121} of *C. botulinum* is 0.21 minutes then a botulinum cook will have an F_0 of 12 × 0.21 = 2.52 min. The effect of applying a process with this F_0 to a product in which every can contains one spore of *C. botulinum* ($N_0 = 1$) will be that a spore will survive in one can out of every 10_{12}.

The canner also has the objective of producing a product which will not spoil at an unacceptably high rate. Since spoilage is a more acceptable form of process failure than survival of *C. botulinum*, the process lethality requirements with respect to spoilage organisms do not need to be so severe. In deciding the heat process to be applied, a number of factors have to be weighed up.

Most canners would regard an acceptable spoilage rate due to under-processing as something around 1 in 10_5–10_6 cans and this is normally achievable through 5–6 decimal reductions in the number of spores with spoilage potential (the USFDA use 6D as their yardstick). PA3679, *Clostridium sporogenes* is frequently used as an indicator for process spoilage and typically has a D_{121} of about 1 min. This will translate into a process with an F_0 value of 5–6; sufficient to produce about 24–30 decimal reductions in viable *C. botulinum* spores–well in excess of the minimal requirements of the botulinum cook.

Having decided the F value required, it is necessary to ensure that the F_0 value actually delivered by a particular heating regime achieves this target value. To do this, the thermal history of the product during processing is determined using special cans fitted with thermocouples to monitor the product temperature. These must be situated at the slowest heating point in the pack where the F_0 value will be at a minimum. The precise location of the slowest heating point and the rate at which its temperature increases depend on the physical characteristics of the can contents. Heat transfer in solid foods such as meats is largely by conduction which is a slow process and the slowest heating point is the geometric centre of the can. When fluid movement is possible in the can, heating is more rapid because convection currents are set up which transfer heat more effectively. In this case the slowest heating point lies on the can's central axis but nearer the base.

The slowest heating point is not always easy to predict. It may change during processing as in products which undergo a sol-gel transition during heating, producing a broken heating curve which shows a phase of convection heating followed by one of conduction heating. In most cases heating is by conduction but some can contents show neither pure convection nor pure conduction heating and the slowest heating point must be determined experimentally.

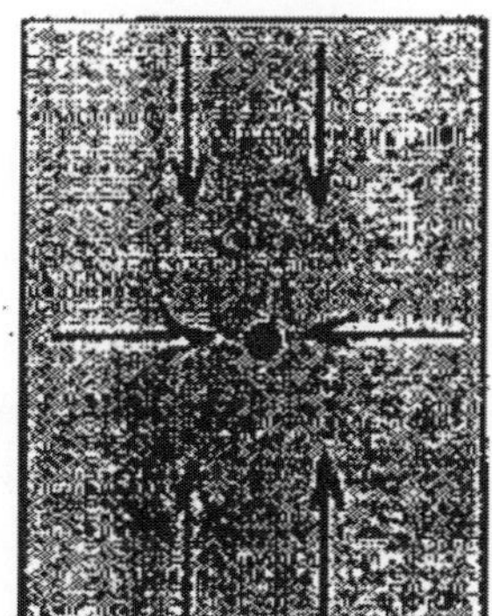

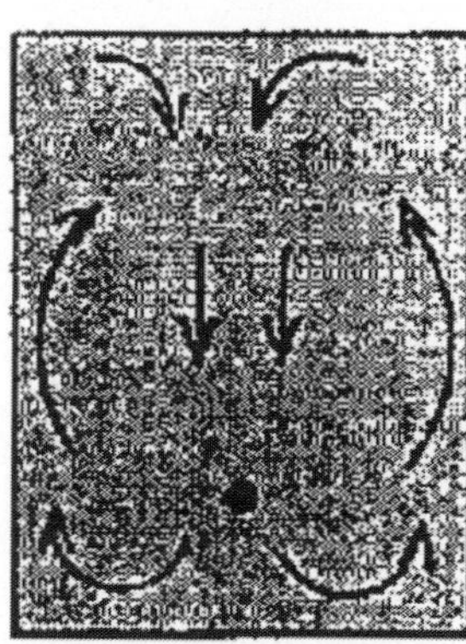

Figure 4.3: Conduction and Convection Heating in Cans. λ Denotes slowest heating point.

Movement of material within the can improves heat transfer and will reduce the process time. This is exploited in some types of canning retort which agitate the cans during processing to promote turbulence in the product.

Process confirmation can also be achieved by microbiological testing in which inoculated packs are put through the heat process and the spoilage/survival rate determined. Heat penetration studies though give much more precise and useable information since inoculated packs are subject to culture variations which can affect resistance and also recovery patterns.

A change in any aspect of the product or its preparation will require the heat process to be re-validated and failure to do this could have serious consequences. An early example of this was the scandal in the mid-nineteenth century when huge quantities of canned meat supplied to the Royal Navy putrefied leading to the accusation that the meat had been bad before canning. It transpired that the problem arose because cans with a capacity of 9–14 lb were being used instead of the original 2–6 lb cans. In these larger cans the centre of the pack took longer to heat and did not reach a temperature sufficient to kill all the bacteria. More recently, replacement of sugar with an artificial sweetener in hazelnut puree meant that spores of *C. botulinum* surviving the mild heat process given to the product were no longer prevented from growing by the reduced a_w.

Microbial Activity in Canned Foods

If a canned food contains viable microorganisms capable of growing in the product at ambient temperatures, then it will spoil. The incidence of spoilage in canned foods is low, but when it occurs it must be investigated properly. The spoilage of can could be due to biological or chemical reasons or combination of both. The biological spoilage is primarily due to microbial growth while chemical spoilage is due to hydrogen produced due to reaction of acid in food and iron on can.

Swollen cans often indicate a spoiled product. During spoilage, cans may progress from normal to flipper, to springer, to soft swell, to hard swell. However, spoilage is not the only cause of abnormal cans. Other causes such as overfilling, buckling, denting, or closing while cool may also be responsible. Microbial spoilage and hydrogen, produced by the interaction of acids in the food product with the metals of the can, are the principal causes of swelling. High summer temperatures and high altitudes may also increase the degree of swelling. Organisms may be present as a result of an inadequate heat process, underprocessing, or of post process contamination through container leakage. Spoilage by a single spore former is often diagnostic of underprocessing since rarely would such a failure be so severe that vegetative organisms would survive. Some microorganisms that grow in canned foods, however, do not produce gas and therefore, cause no abnormal appearance of the can; nevertheless, they cause spoilage of the product.

Microbial spoilage may occur due to leakage of cans. It occurs because of manufacturing defects, punctures or rough handling. Bacteria are introduced into can by either in holes or improper seams. In this type, the microorganisms are not usually heat resistant and wide array of organisms had been found to cause spoilage as it is post processing contamination. Microbes may also get entry into can due to cold water, used to cool cans after heat treatment. Leakage may also be responsible for release of vacuum, which can favour the growth of microorganisms. Presence of low heat resistance organisms usually indicates leakage of can. Underprocessing of cans includes sub-optimal heat treatment, faulty retort operations, excessive microbial load and contamination in product, change in consistency of the product.

The microbial spoilage of canned food is classified as caused by thermophilic bacteria and mesophilic organisms. Spoilage by these types of bacteria is most prevalent in under processed heat treated canned foods. Their spores survive the heat treatment and undergo vegetative cell formation and subsequent growth in canned conditions. Most common spoilages of microbial origin are known as flat sour spoilage, Thermophilic anaerobic spoilage and putrefaction.

A normal sound can will either be under vacuum with slightly concave ends or have flat ends in those cases where the container is brimful. Spoilage often manifests itself through microbial gas production which causes the ends to distend and a number of different terms are used to describe the extent to which this has occurred. This type of spoilage (TA) is caused by thermophilic anaerobe not producing hydrogen sulfide. *Clostridium thermosaccharolyticum* is the main organism involved. It produces acid and gas in foods. The spore-forming anaerobes *Clostridium* can be either predominantly proteolytic or saccharolytic but both activities are normally accompanied by gas production causing the can to swell. Spoiled food produces sour or cheesy smell.

Cans may sometimes swell as a result of chemical action. Defects in the protective lacquer on the inside of the can may allow the contents to attack the metal releasing hydrogen. These hydrogen swollen cans often be distinguished from microbiological spoilage since the appearance of swelling occurs after long periods of storage and the rate at which the can swells is usually very slow.

In cases where microbial growth occurs without gas production, spoilage will only be apparent once the pack has been opened. This type of spoilage is known as flat sour spoilage. One characteristic of this spoilage is that ends of can remain flat during souring. Because of this condition, the detection of spoilage from outside is not possible thereby culturing of contents become necessary to detect the type of organisms. Flat sour spoilage is caused by souring bacteria. Main organisms involved are *Bacillus*, while it occurs more frequently in low acid foods. *Bacillus* spp. has ability to produce acid without gas formation. *Bacillus* species, with the exceptions of *B. macerans* and *B. polymyxa*, usually break down carbohydrates to produce acid but no gas giving a type of spoilage known as a 'flat sour', which describes the characteristics of both the can and the food.

The heat process a product receives is determined largely by its acidity: the more acidic a product is, the milder the heat process applied. Although more complex schemes have been described, the essential classification of canned foods is into low acid (pH > 4.5, or 4.6 in the United States) and acid foods (pH < 4.5 or 4.6).

For years, laboratories connected with the canning industry and others have *studied Clostridium botulinum*, its heat-resistance and processing recommendations for low-acid foods. While studying the growth requirements for *Clostridium botulinum*, it was found that the dividing line of acidity between products in which the organism would grow, and those in which it would not grow, was about pH 4.6; below this level, growth of the organism in a favourable medium is inhibited. Under other conditions, such as one in which nutrient value is low, growth may be inhibited, regardless of pH. As a practical matter, this means that products of pH levels higher than pH 4.6 must be processed under steam pressure, at temperatures considerably higher than 212°F (100°C), usually higher than 240°F (116"C), in order to ensure destruction of spores, while products at pH 4.6 or lower may be safely processed in an open bath at 212°F (100°C)

A product's acidity also determines the type of spoilage that may result from underprocessing since it can prevent the growth of some spoilage organisms. At normal ambient temperatures (< 38°C) only mesophilic species will grow. Typical examples would be *C. botulinum, C. sporogenes* and *B. subtilis* in low acid products and *C. butyricum* and *C. pasteurianum* in products with a pH below 4.5.

Cans are cooled rapidly after processing to prevent spoilage by thermophiles. Thermophilic spores are more likely to survive the normal heat process but would not normally pose a problem. If however a large assemblage of cans is allowed to cool down naturally after retorting, the process will be slow and the cans will spend some time passing through the thermophilic growth range. Under these conditions surviving thermophilic spores may be able to germinate and grow, spoiling the product before it cools. This may also occur if cans are stored at abnormally high ambient temperatures (> 40°C) and canned foods destined for very hot climates may receive a more stringent process to reduce thermophilic spoilage.

Thermophilic organisms commonly associated with spoilage of low acid canned foods are the saccharolytic organism *C. thermosaccharolyticum, B. stearothermophilus* and *Desulfotomaculum nigrificans*. The spores of these organisms are destroyed at optimal heat treatment; thus presence of this organism usually indicates under processing in terms of heat treatment. The last of these causes a type of spoilage known as 'sulfur stinker'. It produces hydrogen sulfide which does not usually distend the can but does give, the product an objectionable smell and reacts with iron from the can to cause blackening. Cans usually remain flat due to solubility of H_2S in water.

Table 4.4: Spoilage Manifestations in Low-Acid Products

Type of Organism	*Classification*	*Manifestations*
Flat sour bacteria	Can flat Product	Possible loss of vacuum on storage. Appearance not usually altered. pH markedly lowered, sour. May have slightly abnormal odor. Sometimes cloudy liquor.
Thermophilic anaerobe bacteria	Can swells Product	May burst. Fermented, sour, cheesy or butyric
Sulfide spoilage bacteria	Can flat Product	H_2S gas absorbed by product odor. Usually blackened. "Rotten egg" odor.
Putrefactive anaerobe bacteria	Can swells Product	May burst. May be partially digested. pH slightly above normal. Typical putrid odor.
Aerobic sporeformers (odd types)	Can flat	Usually no swelling, except in cured meats when nitrate and sugar present. Coagulated evaporated milk, black beets.
Bacillus themoacidurans (flat sour, tomato juice)	Can flat Product slight pH change.	Little change in vacuum Off-odor.
Butyric anaerobes(tomatoes, tomato juice)	Can swells Product	May burst. Fermented, butyric odour.
Non-sporeformer bacteria (mostly lactic types)	Can swells Product	Usually burst, but swelling may be mostly arrested. Acid odour.

Presence of non-spore formers in cans indicates post processing contamination. The organisms which are heat resistant are generally found. These organisms are more prominent namely, *Micrococcus, Lactobacillus, Enterococcus, Streptococcus, thermophilus, Leuconostoc, Microbacterium etc.* Presence of these organisms indicates leakage of container. Cooling water is one of the important sources of contamination and coliforms may also gain entry into the can through this leakage and spoil the canned product.

Spoilage by yeasts, molds and non-spore forming bacteria is not common in low acid canned foods. Yeasts and their spores are not thermo tolerant, so they are not found in properly heat treated cans. Their presence indicates faulty processing treatment or post processing contamination through leakage. Fermentative yeasts are more prominent and they often produce carbon dioxide and cause swelling of cans. Occasionally, film yeasts also showed growth on the surface of the canned food products.

Molds *viz., Aspergilus* and *Penicillium* are common molds found responsible in spoiling canned food. They can grow even at high sugar concentration level and acidification during processing or storage prevents the growth of these molds. Some molds are resistant to normal heat treatment applied during processing and it is generally found in homemade canned foods where heating and sealing is not done under total aseptic conditions.

The presence of yeasts or molds in canned food would indicate: (a) gross under sterilization; or (b) contamination due to defective seam. These organisms are readily controlled by relatively short processes at temperatures below 212°F (100°C).

Leakage is the most common cause of microbiological spoilage during post processing of canned foods. During processing, cans are subjected to extreme stress, particularly when the hot can is cooled down rapidly from processing temperatures. If the can is leaking, water gets sucked into the can while can cooling, microorganisms present in water get into it and spoil the food. Sometimes, leaked cans close by itself and due to gases produced inside, the cans bulge. Such a spoilage is called as 'leaker spoilage'. In such cans, a variety of microorganisms (mixture of non-spore formers or spore formers) are present.

Cans are the most common containers used for retorted products, although glass jars, rigid plastic containers and soft pouches are also sometimes used. Cans are usually made of two or three parts: the three-part can consists of a base, body and lid while in two part cans the body and base are made from a single piece of metal. In a three-part can the body seam is electrically welded but the lid on all cans is held in place by a double seam. The correct formation and integrity of this seam are crucial to preventing leakage and monitoring seam integrity is an important aspect of quality control procedures in canning.

Table 4.5: Laboratory Diagnosis of Bacterial Spoilage

	Under Processed	*Leakage*
Can	Flat or swelled. Seams generally normal	Swelled; may show defects*
Product Appearance Odor	Sloppy or fermented Normal, sour or putrid, but generally consistent from can to can	Frothy fermentation; Viscous Sour, faecal; generally varying from can to can.
pH	Usually fairly constant	Wide variation.
Microscopic and Cultural	Cultures show spore-forming rods only, Growth at 98°F and or 131 °F. May be characteristic on special media, *e.g.*, acid agar for tomato juice. If product misses retort completely, rods, cocci, yeast, or molds or any combination of these may be present.	Mixed cultures, generally rods and cocci. Growth only at usual temperatures.
History	Spoilage usually confined to certain portions of pack. In acid products, diagnosis may be less clearly defined; similar organisms may be involved in understerilization and leakage.	Spoilage scattered.

*Leakage may not be due to can defects but to other factors such as contamination of cooling water or rough handling, *e.g.*, can unscramblers, rough conveyor system.

Figure 4.4 indicates the can double seam. Let us try to understand more details about double seam dimensional terminology.

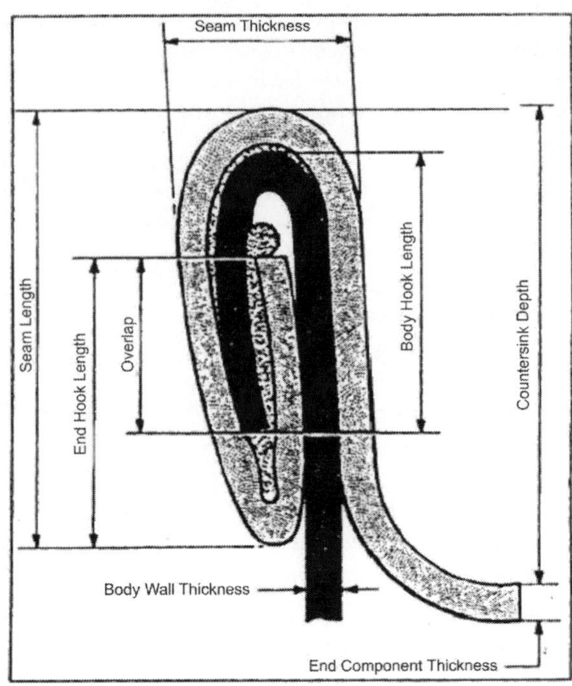

Figure 4.4: Double Seam Dimensional Terminology.

The negative pressure created in the can under these conditions could lead to microorganisms on the container's surface or in the cooling water being sucked inside through a small defect in the seam. The defect in the hot can that allowed leakage to occur may seal up and be undetectable when the can is cool so leaker spoilage can cause cans to blow. Since the microorganisms enter the can after processing, there is no restriction on the type of organism capable of causing leaker spoilage, therefore the presence of a mixed culture or non-sporing organisms is almost certainly a result of can leakage.

There have been occasional reports or pre-process spoilage in canned foods where there was an unacceptable delay between preparing the product and heat processing. During this time spoilage may occur although the organisms responsible will have been killed by the heat process. The UHT processing is very important in avoiding microorganisms.

In UHT processing the food is heat processed before it is packed and then sealed into sterilized containers in a sterile environment. This approach allows more rapid heating of the product, the use of higher temperatures than those employed in canning, typically 130–140°C and processing times of seconds rather than minutes. The advantage of using higher temperatures is that the *z* value for chemical reactions such as vitamin loss, browning reactions and enzyme inactivation is typically 25–40°C compared with 10°C for spore inactivation. This means that they are less temperature sensitive so that higher temperatures will increase the microbial death rate more than they increase the loss of food quality associated with thermal reactions.

Resistance of Microorganisms by UV Radiation and Microwaves

Unlike the other forms of radiation, we will discuss, microwaves act indirectly on microorganisms through the generation of heat. Microwaves are electromagnetic (EM) waves, which are synchronized perpendicularly oscillations of electric and magnetic fields that propagate at the speed of light in a free space. The microwave region of the e.m. spectrum occupies frequencies between 10^9 Hz up to 10^{12} Hz and so has a relatively low quantum energy. For the two frequencies used in food processing, 2 450 MHz and 915 MHz, this is around 10^{-18} ergs or 10^{-6}eV. When a food containing water is placed in a microwave field, the dipolar water molecules align themselves with the field. As the field reverses its polarity 2 or 5 × 10_9 times each second, depending on the frequency used, the water molecules are continually oscillating. This kinetic energy is transmitted to neighbouring molecules leading to a rapid rise in temperature throughout the product. In foods with a high salt content, surface heating due to ions acquiring kinetic energy from the microwave field can also contribute, but this is generally of minor importance.

Microwaves are generated using a magnetron, a device first developed in the UK during research into radar during the Second World War. Magnetrons are used both commercially and domestically, but their biggest impact has been in the domestic microwave oven and in catering where their speed and convenience have enormous advantages. The efficiency of microwaves to interact with water and food is determined by several factors namely dielectric properties, electromagnetic penetration, food shape, volume, surface area and composition (Fu, 2006; Chaplin, 2015). To understand the properties of the interaction of microwaves with food, it is essential to know the dielectric properties of food (Sosa-Morales *et al.*, 2010). Dielectric properties of a material are described as the ability of the material to absorb, transmit and reflect electromagnetic effect (Puligundla *et al.*, 2013). A material that shows these abilities is categorized as 'lossy dielectric material'. Food materials generally consist of a mixture of organic materials, water, and salt (Chandrasekaran *et al.*, 2013), are neither poor electrical conductors nor good electrical conductors, but they have the ability to store and dissipate energy, putting them under the group of lossy dielectric material. Dielectric properties of food vary with the composition of the food, particularly moisture and salt percentages whether of free or bound manner state in the food. Besides, the changing composition of carbohydrate, fat, and protein also contributes considerably. The distribution of these ingredients in the food exerts a strong influence on the mode of heating (Fu, 2006).

The interaction of microwaves with the food molecules creates different volumetric heating effect. Different food composition types have been observed to have influence in microwave heating in relation to the thermal inactivation of microorganism. Fat food materials have been observed to have the highest influence due to its low specific heat compared to carbohydrate and protein food materials as all the pathogenic bacteria was thermally inactivated below 50s. In a research study, all pathogens were fully thermally inactivated at 60s which indicated 60s microwave heating time should be ascribed as the microbiologically safe microwave heating time qualitatively. No pathogenic bacteria were observed to gain thermal resistance towards microwave heating. Microwave heating is able to reduce the food processing time, but at the same time, microwave heating has some major drawbacks especially non-uniform temperature distribution which could not fully thermally inactivate foodborne pathogens if present in the food materials through contamination, thus making the food unacceptable for microbiological safety concerns (New *et al.*, 2019).

The principal problem associated with the domestic use of microwaves is non-uniform heating of foods, due to the presence of cold spots in the oven and the non-uniform dielectric properties of the food. This is due to the fact that wave is reflected against the lower edge of the product leading to interference phenomena. This results in cold spots (transmitted and reflected waves cancel each other) and/or hot spots (transmitted and reflected waves add up). These cold spots in some microwaved foods are of great concern over the risks associated with consumption of inadequately heated meals which has led to more explicit instructions on microwaveable foods. These often specify a tempering period after heating to allow the temperature to equilibrate. For most cases, product size is very large compared to the penetration depth; thus, the energy of the EM wave will be completely absorbed before being able to reach the core or the bottom of the product. All the parts of the product not crossed by the EM wave will not undergo heating. Smaller products will heat faster than large ones, because MWs are able to penetrate the entire product. However, small products are also sensitive to variations in EM fields that exist in the oven, particularly in a multimode cavity. Another concern is if the product is made of several components, the component with the highest dielectric constant tends to concentrate the energy, and its temperature will increase strongly compared to the other components which leads to selective overheating. The dependence of dielectric properties with the temperature often leads to increase the local temperature of the already hot points. As foods are generally low thermal conductivities, this phenomenon leads to overheating (thermal runaway). The shape of the product also plays a major role in the heterogeneity of MW heating. All these factors should be kept in mind before microwave heating of food and accordingly frequency and time of heating should be decided.

Microbial destruction by microwave radiation has great potential in the pasteurization of foods. Its short heating and exposure time is less destructive to food than longer conventional heating. Microwave has been extensively applied to inactivate microorganisms in liquids, food, and surfaces. Microwave heating is known to inactivate many microorganisms, such as *Escherichia coli*, *Streptococcus faecalis*, *Clostridium perfringens*, *Staphylococcus aureus*, *Salmonella*, and *Listeria* spp. Bacterial and mold spores, as well as the bacteriophage PL-1, which is specific to *Lactobacillus casei*, have also been reported to be sensitive to microwave radiation.

The study of microbial destruction during an MW heating has attracted a lot of interest. Two mechanisms have been proposed for inactivation of microorganisms by microwaves. The first states that microwaves inactivate microorganisms entirely by heat through mechanisms comparable to other biophysical processes induced by heat, such as denaturation of enzymes, proteins, nucleic acids, or other vital components, as well as disruption of membranes (Heddleson and Doores, 1994). There is no question as to the validity of this mechanism. A second proposed mechanism for inactivation by microwaves involves non-thermal effects. Several theories have been proposed to explain how electromagnetic fields can inactivate m.os. at sublethal temperature conditions. This effect would be due to the interaction between microwaves and certain cellular constituents. Four predominant theories have been used to explain non-thermal inactivation by microwaves or "cold pasteurization": selective heating, electroporation, cell membrane rupture, and magnetic field coupling (Kozempel *et al.*, 1998). The selective heating theory states that solid microorganisms are heated more effectively by microwaves than the surrounding medium and are thus killed more readily. Electroporation is caused when pores form in the membrane of the microorganisms due to electrical potential across the membrane, resulting in leakage. Cell membrane rupture is related in that the voltage drop across the membrane causes it to rupture. In the fourth theory, cell lysis occurs due to coupling of electromagnetic energy with critical molecules within the cells, disrupting internal components of the cell.

These mechanisms have been studied extensively since the 1970s by a number of researchers. Culkin and Fung (1975) reported earlier studies that suggested microwave heating at 2450 MHz caused greater destruction of *Aspergillus*, *Penicillium*, *Rhizopus*, aerobic microorganisms, *Salmonella* and *Proteus* in foods than heating alone. Culkin and Fung (1975) exposed *E. coli* and *Salmonella typhimurium* in soups to 915 MHz microwaves and then determined survivors in the top, middle and bottom regions of the product. The temperatures were measured using temperature-sensitive strips. They found that the greatest survival in the soups was in the top layer, which was also shown to have the lowest temperature. A series of studies by Khalil and Villota (1988; 1989a, b) suggested non-thermal effects of microwave heating. They first determined that *Bacillus stearothermophilus* spores in various media (water, NaCl, sucrose, phthalate, or phosphate buffers) had lower D100 °C values when 2450 MHz microwaves were used compared to using a heated water bath. The experiment appears to have involved 6 tubes at a single temperature with no replication. In addition, the come-up times, although a small part of the overall heating times (microwaves: 58–83 s out of 90–190 min, conventional heat: 100–135 s out of 113–240 min), were not considered. Heddleson and Doores (1994) disputed the above conclusions due to inaccuracies in temperature measurement. Khalil and Villota (1988) further studied the effect of microwaves (2450 MHz assumed) on injury of *S. aureus* FRI-100. They heated cells at a sublethal temperature of 50°C and maintained microwave temperature using recirculated cooled kerosene.

Apart from the described studies, most research has concluded that there is little or no non-thermal effect of microwaves on microorganisms and that inactivation of microorganisms is due only to heat. Goldblith and Wang (1967) heated suspensions of *E. coli* and *Bacillus subtilis* under conventional heating and with microwaves at 2450 MHz. The degree of inactivation of both microorganisms was identical with conventional and microwave heating. Vela and Wu (1979) heated various bacteria, fungi and bacteriophages in 2450 MHz microwaves in water and as lyophilized cultures. There was no inactivation of dry cultures even after extended exposure. Similarly, Jeng and others (1987) found no difference in inactivation of *B. subtilis* spores under conventional or microwave (2450 MHz) heating in automated computer-controlled temperature monitoring systems. Kazbekov and Vyacheslavov (1978) found that thymidine and thymine uptake, leakage of potassium and hydrogen ions, and uptake of DNA by cells of *E. coli* or *B. subtilis* under low power microwaves were typical of that shown for heating. Fujikawa and others (1992) found no major differences in inactivation kinetics of *E. coli* in phosphate buffer between microwaves and conventional heating. Welt and others (1993a) demonstrated no difference between conventional and microwave inactivation of *Clostridium sporogenes* PA3679 at 90, 100 and 110°C. A suspension of spores that was exposed to microwaves, but continuously cooled in silicone tubing demonstrated no detectable inactivation.

To summarize, despite many studies on microbial destruction by microwave radiation, the mechanism of destruction is not fully understood. It is generally thought that the destruction of microorganisms is mainly due to a thermal effect of microwave exposure. The destruction of microorganisms by microwave at temperatures lower than the thermal destruction point has been observed but it is still a subject of debate (Woo *et al.*, 2000; Laguerre and Hamoud-Agha, 2019).

Microwaves have been slow to find industrial applications in food processing, although they are used in a number of areas. Microwaves have been used to defrost frozen blocks of meat prior to their processing into products such as burgers and pies thus reducing wear and tear on machinery. There has also been a limited application of microwaves in the blanching of fruits and vegetables and in the pasteurization of soft bakery goods and moist (30 per cent H_2O) pasta to destroy yeasts and moulds. In Japan, microwaves have been used to pasteurize high-acid foods, such as fruits in syrup, intended for distribution at ambient temperature. These are packed before processing and have an indefinite microbiological shelf-life because of the heat process and their low pH. However, the modest oxygen barrier properties of the pack have meant that their biochemical shelf-life is limited to a few months.

Identification of Pathogens Resistant to Microwaves

Numerous studies address the effect of microwave heating on pathogenic microorganisms in foods. Bacteria reported to be inactivated by microwave heating include *Bacillus cereus*, *Campylobacter jejuni*, *Clostridium perfringens*, pathogenic *Escherichia coli*, *Enterococcus*, *Listeria monocytogenes*, *Staphylococcus aureus*, and *Salmonella* (Gedikli *et al.*, 2010; Shin *et al.*, 2010). The nematode *Trichinella spiralis*, the organism that causes trichinosis, may also be inactivated (Zimmerman 1983). Foodborne pathogens have been shown to be inactivated by microwave heating in various poultry, beef, fish and pork products, milk and eggs; however, Heddleson and Doores (1994), reported that "in-home pasteurization" of milk was "problematic" and "potentially dangerous" due to non-uniform heating and lack of standardization of home microwave ovens.

It is very difficult to precisely compare the effectiveness of microwave heating to conventional heating based on the literature, because of the different techniques employed or the lack of detail in the methods or materials used, especially in relation to temperature monitoring (Heddleson and Doores 1994). A recurring conclusion in the literature is that non-uniform heating by microwaves may lead to survival of foodborne pathogens, including *Salmonella* and *L. monocytogenes*, in certain locations of foods heated at selected internal locations to endpoint temperatures that would normally be lethal (Schnepf and Barbeau 1989; Harrison and Carpenter 1989). For example, several studies have demonstrated that the measured internal temperature of poultry does not indicate the extent of inactivation of surface-inoculated *Salmonella* on poultry due to lower temperatures at the product surface (Schnepf and Barbeau 1989).

There do not appear to be any obvious "microwave-resistant" foodborne pathogens. Various studies have shown increased resistance of *S. aureus*, *C. perfringens*, or *Enterococcus faecalis* but not necessarily to the point that these could be labeled as resistant. As with conventional heating, bacteria are more resistant to thermal inactivation by microwave heating than yeasts and molds and bacterial spores are more resistant than vegetative cells.

Effects of Critical Process Factors on Inactivation

As with other thermal processes, the primary factors that determine safety are temperature and time (that is, integrated time-temperature history). A number of critical process factors affect time-temperature history. Some of these critical process factors are moisture, ionic content, microwave frequency, product parameters (including mass, density, geometry), specific heat, and the temperature achieved. It is important to note that in the context of microwave processing, these critical process factors do not change the rate of inactivation per se. Rather, these factors change the spatial distribution of microwave absorption and, therefore, the spatially distinct heating rate and time-temperature history. The spatial distribution of time-temperature history, in turn, changes the distribution of the extent of inactivation within the food, thus generally changing the total inactivated population within a given food sample. For example, the population of cells heated for 47 s at 700 W in a microwave oven in phosphate buffer were reduced by 99.8 per cent, while those in 1 per cent sodium chloride were reduced only by 62.4 per cent (Heddleson *et al.*, 1994; New *et al.*, 2019). Such a difference is attributed to the effect of salt in decreasing the penetration of microwaves. Less microwave penetration leads to a lower internal temperature and a lesser destruction in the interior regions, resulting in an overall lower destruction.

Table 4.6: Critical Process Factors in Microwave Heating

Food	Shape, size, composition (moisture, salt, and so on), multiple components (as in a frozen dinner), liquid against solid
Package	Presence of metallic elements such as aluminum foil, susceptor
Process	Power level, cycling, presence of hot water or air around the food, equilibration time
Equipment	Dimensions, shape and other electromagnetic characteristics of the oven, frequency, agitation of the food, presence of mode stirrers and turntables

Now let us try to understand more things about the effect of UV radiations on the microorganisms and their resistance.

Ultraviolet processing involves the use of radiation from the ultraviolet region of the electromagnetic spectrum for purposes of disinfection. UV radiation has wavelengths below 450 nm (v $\simeq 10^{15}$ Hz) and a quantum energy of 3-5 eV (10^{-12} ergs). Typically, the wavelength for UV processing ranges from 100 to 400 nm. This range may be further subdivided (Bolton 1999) into UVA (315 to 400 nm) normally responsible for changes in human skin that lead to tanning; UVB (280 to 315 nm) that can cause skin burning and eventually lead to skin cancer; UVC, (200 to 280 nm) called the germicidal range since it effectively inactivates bacteria and viruses, and the vacuum UV range (100 to 200 nm) that can be absorbed by almost all substances and thus can be transmitted only in a vacuum. The quanta contain energy sufficient to excite electrons in molecules from their ground state into higher energy orbitals making the molecules more reactive. Chemical reactions thus induced in micro-organisms can cause the failure of critical metabolic processes leading to injury or death. Only quanta providing energy sufficient to induce these photochemical reactions will inhibit micro-organisms, so those wavelengths that are most effective give us an indication of the sensitive chemical targets within the cell. The germicidal properties of UV irradiation are mainly due to DNA mutations induced through absorption of UV light by DNA molecules. This mechanism of inactivation results in a sigmoidal curve of microbial population reduction. The greatest lethality is shown by wavelengths around 260 nm which correspond to a strong absorption by nucleic acid bases.

To achieve microbial inactivation, the UV radiant exposure must be at least 400 J/m_2 in all parts of the product. Critical factors include the transmissivity of the product, the geometric configuration of the reactor, the power, wavelength and physical arrangement of the UV source(s), the product flow profile, and the radiation path length. UV may be used in combination with other alternative processing technologies, including various powerful oxidizing agents such as ozone and hydrogen peroxide, among others.

Applications include disinfection of water supplies and food contact surfaces. Recently, interest has increased in using UV to reduce microbial counts in juices.

The resistance of microorganisms to UV is largely determined by their ability to repair such damage, although some organisms such as micrococci also synthesize protective pigments. Generally, the resistance to UV irradiation follows the pattern.

Gram-negatives < Gram-positives Yeasts < Bacterial spores < Mould spores << Viruses.

Death of a population of UV-irradiated cells demonstrates log-linear kinetics similar to thermal death and, in an analogous way, D values can be determined. These give the dose required to produce a tenfold reduction in surviving numbers where the dose, expressed in ergs or μWs, is the product of the intensity of the radiation and the time for which it is applied.

Table 4.7: UV Resistance of some Food-borne Microorganisms

Species	*D (ergs × 10^2)*	*Species*	*D (ergs × 10^2)*
E. coli	3–4	*Proteus vulgaris*	3–4
Serratia marcescens	3–4	*Shigella flexneri*	3–4
Pseudomonas fluorescens	3–4	*Bacillus subtilis* (vegetative cells)	6–8
Bacillus subtilis (spores)	8–10	*Micrococcus luteus*	10–20
Staphylococcus aureus	3–4	*Aspergillus flavus*	50–100
Mucor racemosus	20–50	*Penicillium roquefortii*	20–50
Rhizopus nigrificans	> 200	*Saccharomyces cerevisiae*	3–10

Determination of UV D values is not usually a straightforward affair since the incident radiation can be absorbed by other medium components and has very low penetration. Passage through 5 cm of clear water will reduce the intensity of UV radiation by two-thirds. This effect increases with the concentration of solutes and suspended material so that in milk 90 per cent of the incident energy will be absorbed by a layer only 0.1 mm thick. This low penetrability limits application of UV radiation in the food industry to disinfection of air and surfaces.

Wavelengths below 200 nm are screened out by surrounding the lamp with an absorbent glass since these wavelengths are absorbed by oxygen in the air producing ozone which is harmful. The output of these lamps falls off over time and they need to be monitored regularly.

Air disinfection is only useful when the organisms suspended in air can make a significant contribution to the product's microflora and are likely to harm the product; for example, in the control of mould spores in bakeries. UV lamps have also been mounted in the head space of tanks storing concentrates, the stability of which depends on their low a_w. Fluctuations in temperature can cause condensation to form inside the tank. If this contacts the product, then areas of locally high a_w can form where previously dormant organisms can grow, spoiling the product. Process water can be disinfected by UV; this avoids the risk of tainting sometimes associated with chlorination, although the treated water will not have the residual antimicrobial properties of chlorinated water. UV radiation is commonly used in the depuration of shellfish to disinfect the water recirculated through the depuration tanks. Chlorination would not be suitable in this situation since residual chlorine would cause the shellfish to stop feeding thus stopping the depuration process.

Surfaces can be disinfected by UV, although protection of microorganisms by organic material such as fat can reduce its efficacy. Food containers are sometimes treated in this way and some meat chill store rooms have UV lamps to retard surface growth. UV can however induce spoilage of products containing unsaturated fatty acids where it accelerates the development of rancidity. Process workers must also be protected from UV since the wavelengths used can cause burning of the skin and eye disorders.

Radiation Resistance

Food irradiation has been identified as a safe technology to reduce the risk of food borne illness as part of high-quality food production, processing, handling, and preparation. FDA approved food irradiation technologies include gamma radiation from cobalt-60 and accelerator produced electrons and x-rays. Food irradiation's history of scientific research, evaluation, and testing spans more than 60 years. The process has been approved by many countries around the world and it has been endorsed or supported by numerous national and international food and health organizations and professional groups. Food irradiation utilizes a source of ionizing energy that passes through food to destroy harmful bacteria and other organisms. It is a safe and effective approved process that can be used to disinfest or delay the maturation of fruits and vegetables, improve the microbiological safety of shellfish, eggs, raw meat and poultry, spices, and seeds used for sprouting. Foods that undergo irradiation include Beef, pork, poultry, eggs in shells, shellfish, such as shrimp, lobster, crab, oysters, clams, mussels, scallops; Fresh fruits and vegetables, including seeds for sprouting (such as alfalfa sprouts); Spices and seasonings *etc.* Often referred to as "cold pasteurization," food irradiation offers negligible loss of nutrients or sensory qualities in food as it does not substantially raise the temperature of the food during processing. Food irradiation does not replace proper food production, processing, handling, or preparation, nor can it enhance the quality of or prevent contact with food borne bacteria after irradiation. The manufacturers are required to identify irradiated food sold to consumers with an international symbol (Radura) and terminology describing the process on product labels. In addition, food irradiation facilities are thoroughly regulated and monitored for worker and environmental safety.

Ionization radiations are used to ensure food safety and quality. The ionizing irradiation is the process of applying ionizing energy to a material, such as food, to sterilize or preserve them by destroying microorganisms, parasites, insects, and other pests. Ionizing radiations can remove bound electrons from an atomic or molecular structure, leaving the structure to become electrically charged or ionized. The major ionizing radiations used in food industries are electromagnetic gamma (γ) rays obtained from radioactive forms of the elements cobalt (^{60}Co) or cesium (^{137}Cs), and the corpuscular beta rays (β) that are the electrically formed electrons in a cathodic tube, which are taken by an electron accelerator to form a stream of high-speed electrons propelled to foods.

Although patents describing the use of ionizing radiation in the treatment of food appeared soon after the discovery of radioactivity at the turn of the century, it was not until after the Second World War that food irradiation assumed commercial potential. This was largely due to technological advances during the development of nuclear weapons, but also to a strong desire *to* demonstrate that nuclear technology could offer the human race something other than mass destruction.

The use of ionizing radiation technology, also named ionization or irradiation, to sterilize the single-use medical equipment started in the late 1960s. Since then, it has become a major application for laboratory material sterilization and improvement of material quality at very high doses ranging from 40 to 100 kGy. At the same time, the Agro-Food Industries (AFI) became interested in this process, which seemed to have certain advantages, such as preserving food quality and higher preservation potential, as compared to conventional heat treatments. The ionizing radiation started being used on various foodstuffs to improve microbiological safety and/or extend the shelf life. The FDA and the World Health Organizations reviewed numerous short-term and long-term feeding studies in animals and *in vitro* studies prior to approving the use of ionizing radiation to be used as a food safety technology. Irradiation of ground beef, 1.5 kGy for refrigerated and 3.0 kGy for frozen product, resulted in a 6 log inactivation of *E.coli* O157:H7, and a 3 log reduction of *Salmonella* with little or no effect on ground beef quality. An ionizing radiation dose of 1 kGy can be used to disinfest and delay the maturation of fruits and vegetables, and would result in a 4 log reduction of *E. coli* O157:H7 and a 2 log reduction of *Salmonella*. Hence, ionizing radiations can be used effectively to avoid microorganisms.

Ionizing treatments of food now has been approved in more than 60 countries. The international and national bodies define a maximum dose limit for each approved food product.

Ionizing radiation can affect microorganisms directly by interacting with key molecules within the microbial cell, or indirectly through the inhibitory effects of free radicals produced by the radiolysis of water. In living cells, *e.g.*, yeasts, molds and bacteria, the irradiation affects through the induction of genomic, biochemical, physiological and morphogenetic changes. Consequently, it has long been accepted that the mode of action of ionizing radiation consists of a more or less pronounced denaturation of the cellular material. One of the main effects is the direct energy dissipation and damage to macromolecules such as nucleotides by breaking single and double bonds and inducing the cell apoptosis.

Table 4.8: Use of different Levels of Irradiation Dose for Food Preservation

Dose (kGy)	*Usage for Food Products*
0.05–0.15	Anti-germination for potatoes, onions, garlic, shallots, *etc.*
0.15–3	Pesticide and insecticide properties (by the destruction of embryos and larvae, and sterilization of adults) for cereals, fruits, and pulses. In this dose range, some meat parasitic worms, *e.g.*, Trichinella in pork, can be eliminated, and also the ripening of fresh fruits and vegetables is delayed.
2–10	Treatment of meat and meat products, ready meals, fish products (fish, shellfish, frog's legs), fresh fruit and vegetables, spices and herbs, and various ingredients (gums, additives) to improve and guarantee a hygienic quality by eliminating pathogenic microorganisms and/or extend their shelf life by reducing the microbial population of spoilage.
10–60	Food is sterilized and then can be stored for up to 2 years at room temperature in sealed plastic packaging

Source: Munir, M.T. and Federighi, M. 2020. Control of Foodborne Biological Hazards by Ionizing Radiations. Foods, 9, 878; doi:10.3390/foods9070878

The direct effects of ionizing radiation on microbial cells arise from the deposition of energy directly into biomolecules. This may result in the radiolysis of bonds and radical production within molecules, due to the ejection of electrons. The biological implications of direct damage are generally considered to be relatively minor compared to indirect effects. For instance, only ~20 per cent of damage in DNA is incurred from direct energy transfer.

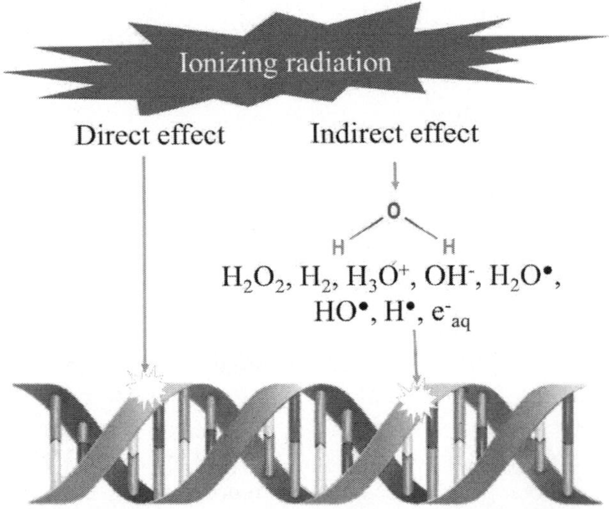

Figure 4.5: Schematic Presentation of the Effect of Ionizing Radiations on the Nucleic Acid.

The indirect effect considers the impact of secondary reactive species produced by ionizing radiation. Hence, in biological systems, as the cytosol has a high water content, the indirect effect primarily concerns the aqueous solution chemistry of water radiolysis products. Such products contribute to ~80 per cent of the radiation damage observed in DNA. Removal of oxygen also increases microbial resistance 2–4 fold and it is thought that this may be due to the ability of oxygen to participate in free radical reactions and prevent the repair of radiation induced lesions. Furthermore, indirect radiochemical effects play role in cell inactivation, for example, the formed free radicals and reactive oxygen species can damage nucleic acids and other cellular material or compartments *e.g.*, membranes and key enzymes. As with UV irradiation, the main site of damage in cells is the chromosome. The hydroxyl OH• radical is the most common disruptor due to its strong oxidizing power. Hydroxyl radicals cause single and double-strand breaks in the DNA molecule as a result of hydrogen abstraction from deoxyribose followed by β-elimination of phosphate which cleaves the molecule. They can also hydroxylate purine and pyrimidine bases. This hydroxyl OH• radical is responsible for irreversible damage to the biological molecules near its formation. Other highly reactive entities such as H3O+, H• and e-hydrated (the hydrated electron) are also formed during the radiolysis of water and play an important role in cellular damage.

When the generation of reactive species exceeds the capacity of endogenous scavengers (*e.g.*, glutathione, superoxide dismutase (SOD), catalase and peroxidase) to neutralize them, oxidative stress is incurred. Hence, the mechanisms to reduce and reverse oxidative stress and the direct and indirect damage incurred are vital components of radiation resistance.

Resistance to ionizing radiation depends on the ability of the organism to repair the damage caused. Inactivation kinetics are generally logarithmic, although survival curves often appear sigmoidal exhibiting a shoulder and a tail to the phase of log-linear death. The shoulder is usually very slight but is more pronounced with bacteria. Table 4.9 indicates the radiation resistance of some food-borne microorganisms.

Table 4.9: Radiation Resistance of some Food-borne Microorganisms

Species	*6D Dose (KGy)*	*Species*	*6D Dose (KGy)*
E. coli	1.5–3.0	*Salmonella enteritidis*	3–5
S. typhimurium	3–5	*Vibrio parahaemolyticus*	< 0.5–1
Pseudomonas fluorescens	0.5–1	*Bacillus cereus*	20–30

Species	6D Dose (KGy)	Species	6D Dose (KGy)
B. stearothermophilus	10–20	*C. botulinum* type A	20–30
Lactobacillus spp.	2–7.5	*Micrococcus* spp.	3–5
Demococcus radiodurans	> 30	*Aspergillus flavus*	2–3
Penicillium notatum	1.5–2	*S. cerevisiae*	7.5–1
Viruses	> 30		

Irradiation is a physical treatment used to control postharvest diseases and appears to offer one of the most viable alternatives to the use of chemicals. In particular, food irradiation has the advantage of being a much more precisely controlled process than heating, since penetration is deep, instantaneous and uniform. The penetrating power and uniformity of irradiation and its derived effectiveness on the inactivation of pathogens is better than that of fungicides. Irradiation is essentially a nonthermal process, with minimal to no heating of fresh produce, thus eliciting negligible sensory changes in the produce. Hence, it retains the fresh character of the product as low level irradiation produces no detectable sensory change in most products. Ionizing radiation does not render food radioactive because it is not strong enough to affect the neutrons in the nucleus. Therefore, ionizing radiation is considered a safe, energy efficient and environmentally safe process. While irradiated products are still not widely popular in developed countries, ionizing irradiation of fresh produce is used to control quarantine pests rather than postharvest diseases. The maximum US FDA permitted dose of 1 kGy to treat fruits and vegetables is too low for effective fungal pathogen control, whereas higher effective doses are often phytotoxic to treated fresh produce.

In recent years, there has been considerable development of the commercial use of irradiation of fresh produce. Many countries have taken advantage of this technology as a replacement for methyl bromide for plant quarantine purposes. Irradiation not only eliminates postharvest diseases, but also kills quarantine pests in produce aimed for export (Smith and Pillai, 2004). Although the use of irradiation is a well-organized and promising technology, several obstacles should be considered before commercial use.

Food irradiation is not without its disadvantages, but a lot of the concerns originally voiced have proved to be unfounded. In 1981 an expert international committee of the FAO/WHO and the International Atomic Energy Authority recommended general acceptance of food irradiation up to a level of 10 kGy. They held the view that it constitutes no toxicological risk. Further toxicological examinations of such treated foods are therefore not required.' In addition to safety concerns, irradiated products have other issues such as resistant strains, cost and nutrient content of treated produce. One common concern about irradiating fresh produce is that long-term use of irradiation promotes development of radiation-resistant strains of fungi and bacteria. It had been thought that irradiation could lead to pathogens becoming more virulent but, apart from one or two exceptions, it has been found that where virulence is affected it is diminished. In the exceptions noted, the effect was slight and not sufficient to compensate for the overall reduction in viable numbers. No example has been found where a non-pathogenic organism has been converted to a pathogen as a result of irradiation. In addition, some studies have shown that irradiation increases mycotoxin levels because of the absence of competitive inhibition by mould growth, leading to toxin-producing strains eventually dominating the mycoflora (Ribeiro *et al.,* 2011). Another disadvantage is that irradiated fresh produce can be more expensive owing to the upfront costs of an irradiation facility. Irradiation of fresh produce generally costs more than any other conventional technology due to the initial costs of the treatment facility. However, its increased use will lower the per-unit cost of the treatment. Irradiation can also have a negative effect on the physical and/or physiological conditions of the fruit, for example, causing softening, discolouration, and the development of 'off-flavours'. A single irradiation dose required to kill postharvest pathogens can negatively affect the physical properties, such as skin colour, texture and firmness, of the stored fruits and vegetables. A combined treatment with other agents can decrease the required dose of radiation and thereby decrease its effect on physical properties of the commodity. In addition, irradiation causes a temporary response that significantly modifies the metabolism of fresh produce, increasing the enzyme activities and gene expression. Moreover, irradiation can deplete fruits and vegetables of nutrition such as vitamins, proteins, carbohydrates, lipids and undesirably changes the taste at high dosages (Mostafavi *et al.,* 2012).

The failure of low doses of radiation to produce appreciable chemical change in the product has been an obstacle to the development of simple tests to determine whether a food has been irradiated. Although availability of such a test is not essential for the control of irradiation, it is generally accepted that it would facilitate international trade in irradiated food, enhance consumer confidence and help enforce labelling regulations. A number of promising methods have been identified including thermoluminescence, electron spin resonance, high-field nuclear magnetic resonance and chemical analysis for specific markers such as hydroxylated tyrosine in high protein foodstuffs. One proposed test is based on differences between irradiated and non-irradiated foods in the ratio between the viable plate count and the quantity of Gram-negative endotoxin detected using the Limulus amoebocyte lysate test.

Morphological, biochemical and other changes which may impede isolation and identification and increased radiation resistance have been noted as a result of repeated cyclic irradiation. However, these experiments were performed under the most favourable conditions and for this to occur in practice would require extensive microbial regrowth after each irradiation: a condition that is readily preventable by good hygienic practices and is most unlikely to be met.

The levels of radiation proposed for foods are not sufficient to induce radioactivity in the product and there is no evidence that consumption of irradiated foods is harmful. Food irradiation facilities do require stringent safety standards to protect workers but that is already in place for the irradiation of other materials such as the sterilization of medical supplies and disposables.

By far the greatest obstacle to the more widespread use of food irradiation is not technical but sociological in the form of extensive consumer resistance and distrust. Much of this is based on inadequate information and raise propaganda and parallels very closely earlier arguments over the merits or milk pasteurization. Among the same objections raised then were that pasteurization would be used to mask poor quality milk and would promote poor practices in food preparation. While it has to be agreed that those who take the most cynical view of human nature are often proved correct, this did not prove to be the case with milk pasteurization where the production standards and microbiological quality of raw milk are now higher than they have ever been.

Depending on the lethality required, food irradiation can be applied at two different levels. At high levels it can be used to produce a safe shelf stable product in a treatment known as *radappertization*. Though this has been investigated in the context of military rations, it is unlikely to be a commercial reality in the foreseeable future. *C. botulinum* spores are the most radiation resistant known, so very high doses are required to achieve the minimum standard of a 12D reduction (45 kGy) for low-acid foods. In the event of a process failure, the growth of more resistant, non-pathogenic clostridia would not act as a warning as it can in thermal processing. High radiation doses are also more likely to produce unacceptable sensory changes and the product has to be irradiated in the frozen state to minimize migration of the radiolytic species that cause such changes. These considerations would not apply when the food was inhibitory to the growth of *C. botulinum* as a result of low pH or the presence of agents such as curing salts.

Two terms are used to distinguish different types of radiation pasteurization. *Radicidation* is used to describe processes where the objective is the elimination of a pathogen, as, for example, in the removal of *Salmonella* from meat and poultry. *Radurization* applies to processes aiming to prolong shelf-life. This distinction may be thought a little over elaborate since, as with thermal pasteurization, irradiation treatments are relatively non-discriminating and will invariably improve both safety and shelf-life. Applications of food irradiation can be understood in Table 4.9.

We can also control microorganisms by applying different pressures.

Interest in the application of high pressures in food processing, sometimes called *pascalizaiton,* lapsed until the 1980s when progress in industrial ceramic processing led to the development of pressure equipment capable of processing food on a commercial scale and a resurgence of interest, particularly in Japan.

High hydrostatic pressure acts primarily on non-covalent linkages, such as ionic bonds, hydrogen bonds and hydrophobic interactions and it promotes reactions in which there is an overall decrease in volume. It can have profound effects on proteins, where such interactions are critical to structure and function, although the effect is variable and depends on individual protein structure. Some proteins such as those of egg, meat and soya form gels and this has been employed to good effect in Japan where high pressure has been used to induce the gelation of fish proteins in the product surimi. Other proteins are relatively unaffected and this can cause problems when they have enzymic activity which limits product shelf-life. Pectin esterase in orange juice, for instance, must be inactivated to stabilize the desired product cloudiness but is very stable to pressures up to 1000 MPa. Non-protein macromolecules can also be affected by high pressures so that pascalized starch products often taste sweeter due to conformational changes in the starch which allow salivary amylase greater access.

Table 4.9: Application of Food Irradiation

Application	*Commodity*	*Dose (kGy)*
Inhibition of sprouting	Potatoes	0.1–3
	Onions	
	Garlics	
	Mushrooms	
Decontamination of food ingredients	Spices	3–10
	Onion powder	
Insect disinfestation	Grains	0.2–7
Destruction of parasites	Meats	0.3–0.5
Inactivation of *Salmonella*	Poultry	3–10
	Eggs	
	Shrimps and frog's legs	
Delay in fruit maturation	Strawberries	2–5
	Mangoes	
	Papayas	
Mould and yeast reduction		1–3

Adverse effects on protein structure and activity obviously contribute to the antimicrobial effect of high pressures, although the cell membrane also appears to be an important target. Membrane lipid bilayers have been shown to compress under pressure and this alters their permeability. As a general rule vegetative bacteria and fungi can be reduced by at least one log cycle by 400 MPa applied for 5 min.

Bacterial endospores are more resistant to hydrostatic pressure, tolerating pressures as high as 1200 MPa. Their susceptibility can be increased considerably by modest increases in temperature, when quite low pressures (100 MPa) can produce spore germination, a process in which the spores lose their resistance to heat and to elevated pressure.

Hydrostatic processing has a number of appealing features for the food technologist. It acts instantly and uniformly throughout a food so that the processing time is not related to container size and there are none of the penetration problems associated with heat processing. With the exceptions noted above, adverse effects on the product are slight; nutritional quality, flavour, appearance and texture resemble the fresh material very closely. To the consumer, it is a 'natural process with none of the negative associations of processes such as irradiation or chemical preservatives.

Safety Issues of Irradiated Fresh Produce

Food irradiation remains unpopular compared to conventional technologies. Nuclear disasters and their derived radioactive contamination have given consumers a negative perception of irradiated fresh produce. For this reason, more than 50 years of extensive research has been devoted to the safety and effectiveness of irradiation as a food sterilization technique to kill pathogens in food and extend its shelf life. There are three primary categories of safety that need to be considered namely, radiological safety (existence of radioactivity in the irradiated food), microbiological safety (occurrence of radiation-resistant microbial populations), and toxicological safety (effect of irradiated food on humans). The safety and efficacy of irradiation has been proved and accepted by the World Health Organization (WHO), the FAO, the International Atomic Energy Agency (IAEA), Codex Alimentarius, and the governments of different countries commending the process (Smith and Pillai, 2004). It was concluded that there are no microbiological risks for consumers, and that chemical substances generated during irradiation in fresh produce are not significantly higher than those induced by heat treatment. These documents have been widely adopted internationally, and specific applications are being applied in over 55 countries worldwide (US Food and Drug Administration, 2012; European Union, 2007).

Despite the fact that irradiated fresh produce has been carefully evaluated, there is a lack of acceptance of food irradiation by the public, although health organizations worldwide have concluded that irradiated fruits or vegetables pose no health hazard. In addition, people fail to differentiate between irradiated fresh produce and radioactive fresh produce. Education and outreach to consumers may be an initial requirement for commercial application of irradiation for commodities. The determination of appropriate irradiation conditions (irradiation dose, other agents) will enable the application of effective treatments against postharvest diseases at the commercial scale. However, there are some barriers to the commercial application of irradiation. It is necessary to prepare strategies that will help the consumers to overcome their fear of irradiated products. There is no doubt that if the challenges such as the ineffectiveness of permitted doses for treatment of fruits and vegetables and the phytotoxicity associated with effective doses to irradiation treatment are handled well, irradiation will be an ideal technology for the control of postharvest diseases in fruits, and vegetables, and will lead to its use for export markets in the future.

Acids and Esters as Food Preservatives

To preserve pickles and sauces, organic acids can be used. They differ from the other acids and esters described here in that they are usually present in amounts sufficient to exert an effect on flavour and on product pH, thus potentiating their own action by increasing the proportion of undissociated acid present.

Benzoic acid is an organic acid first used in foods almost 100 years ago. Benzoic acid (E210 or INS210), a colorless aromatic carboxylic acid, is naturally present in plant (fruits, nuts, spices and vegetable), fungal and animal tissues, but it can also be produced by microorganisms during food processing and/or be added as a food additive. Benzoic acid is the compound with the antimicrobial properties, and is found naturally in cherry bark, cranberries, greengage plums, prunes, cinnamon, cloves, tea and anise but is prepared synthetically for food use. Rich natural sources of benzoic acid are strawberries (up to 29 mg/kg), cayenne pepper and mustard seeds (up to 10 mg/kg), cloves, salvia, thyme and nutmeg (up to 50 mg/kg) and cinnamon (up to 335 mg/kg) (del Olmo *et al.*, 2017).

Benzoic acid has a low taste threshold and low volatility and wide antimicrobial spectrum. Inhibition by benzoic acid appears multifactorial. The free acid form of benzoic acid is poorly soluble in water and the sodium salt (sodium benzoate) is often used because of its greater solubility. Sodium benzoate is the sodium salt of benzoic acid and is preferred over benzoic acid in many food applications because it is 180 times more soluble in water. There is a marked pH effect for this preservative: the lower the pH, the more effective it is. Sodium benzoate will only work if the food product has a pH below 4.5; that is, if the food is naturally acidic or has been acidified. For example, at pH 3.0 one requires approximately 0.05 per cent of the compound to achieve the same antimicrobial effect as pH 4.0 and 0.1 per cent benzoate. Optimum functionality occurs when the pH is between 2.5 and 4.0. About 0.1 per cent of sodium benzoate is usually sufficient to preserve a product that has been properly prepared and adjusted to pH 4.5 or below. Sodium benzoate is a white granular or crystalline powder, odorless, inexpensive (at the usage level) and should be stored in a cool, dry place in watertight containers, if possible. It should be used at low levels to avoid possible off-flavors in some products. The maximum level allowable by PFA act is 0.075 per cent. Sodium benzoate is used in fruit products, jams, relishes, beverages, dressings, salads, pie and pastry fillings, icings, olives and sauerkraut, and is against yeasts, some bacteria (foodborne pathogens but not spoilage bacteria) and some molds.

Sodium benzoate and *benzoic acid* are related because sodium benzoate produces benzoic acid once it is dissolved in water. Its antimicrobial activity is principally in the undissociated form. The ability of the undissociated molecule to interfere with membrane energetics and function appears to be of prime importance since growth inhibition has been shown to parallel closely the inhibition of amino acid uptake in whole cells and membrane vesicles. Some inhibition may also result from benzoic acid once it is inside the cell as a number of key enzyme activities have also been shown to be adversely affected. Since it is a relatively strong acid (p*K*a 4.19), it is effective only in acid food. It is a commonly used antimicrobial preservative in food and beverages, especially in carbonated beverages, pickles, sauces, and jelly as it presents its strongest antibacterial activity at pH 2.5–4.0. The effect results from greater permeability of the unionized form into microorganisms. It has inhibitory effects on the proliferation of bacteria and yeasts, a major cause of food spoilage. As a consequence, its practical use is to inhibit the growth of spoilage yeasts and mould. Activity against bacteria has been reported but they show greater variability in their sensitivity. Benzoic acid and sodium benzoate are used as antimicrobial agents in edible coatings.

Although the addition of benzoic acid can extend the shelf life of drinks and prevent nutritional losses, yet its excessive intake may cause diarrhea, abdominal pain, and other symptoms and even interfere with the intermediate metabolic processes of the body. Thus, the maximum allowed concentrations of benzoic acid in every kind of food are restricted by legislation. According to the US Food and Drug Administration, the restriction of BA addition in general types of food is kept as 1000 mg kg^{-1}. In contrast, the Chinese laws keep BA addition at 200 mg kg^{-1} as far as carbonated drinks are concerned. The maximum amount used in foods ranges from 0.2 to 2.0 g/kg. According to Wood *et al.* (2014), the major food groups contributing to dietary intake of benzoic acid are a wide variety of foods permitted at the following levels; various foods 200–1000 mg/kg (prepared salads, confectionery, *etc.* 1500 mg/kg; food supplements, preserved vegetables 2000 mg/kg; liquid egg 5000 mg/kg; cooked seafood 2000–6000 mg/kg) and soft drinks 150 mg/kg, alcohol-free beer 200 mg/kg (Sacramental grape juice 2000 mg/kg, liquid tea concentrates 600 mg/kg). The acceptable daily intake (ADI) for benzoic acid is 5 mg/kg body weight.

Benzoic acid itself has low toxicity, but there has been concern because of a potential reaction that converts it to benzene. Although benzene is a toxic and carcinogenic compound, the reaction causing this change has a very low chance of occurring in food. Typically, benzene is rapidly converted to hippuric acid in the body and excreted in the urine.

Parabens

Parabens (additives E 214-219) are man-made chemicals often used in small amounts as preservatives in cosmetics, pharmaceuticals, foods, and beverages. They are antimicrobial compounds chemically derived from benzoic acid. Chemically, parabens are alkyl esters of *para-hydroxybenzoic acid, which are used as antimicrobial compounds. The esters include methyl, ethyl, propyl, butyl and benzyl forms.* Paraben esters have antimicrobial activity against molds and yeasts. Typical food products which contain parabens for preservation, include beer, sauces, desserts, soft drinks, jams, pickles, frozen dairy products, processed vegetables and flavouring syrups. In recent years, antibacterial and/or antioxidative effect in vegetable or plant oil (*e.g.*, carrot, tomato, rapeseed oil, olive oil) receives great attention, and these products are known to include paraben and p-hydroxybenzoic acid. Some fruits, such as blueberries, cloudberry, yellow passion fruit *etc.* contain parabens as a naturally occurring preservative but at very low concentrations. For example, the concentration of methylparaben in Andrographis paniculata is much lower is only 0.0008 per cent of its weight. Thus, paraben intake from plant sources is negligible. Cosmetics and pharmaceuticals also represent the largest use of parabens. The concentration of parabens in cosmetic formulations can reach up to 0.8 per cent that is nearly 1000 times more than natural sources. Due to the low level of accumulation in plants, all industrially used parabens are produced synthetically.

The two most common esters are methyl and propyl parabens, which are approved for food use under the GRAS classification. The maximum concentration allowed is 0.1 per cent. They are used in processed vegetables, baked goods, fats and oils, seasonings, sugar substitutes, jams, jellies, beverages and frozen dairy products in concentrations up to 0.1 per cent; these parabens, as well as ethyl paraben, can be used in food packaging (Soni *et al.,* 2005). They are most active against yeasts and molds.

Under EU Directive 95/2/EC, they are allowed in a limited number of foods in combination with either sorbates or sorbates and benzoates. Methyl-, ethyl-, and propyl parabens are permitted in the European Union as food additives in four categories of processed foods by Directive 95/2/EC. They are authorised for use quantum satis (*i.e.* according to *good manufacturing practice* based on the level required to achieve the desired technological effect) for the surface treatment of dried meat products; with a Maximum Permitted Level of 1g/kg in jelly coatings of meat products such as paté; in confectionery, excluding chocolate, at levels of 0.3g/kg; and in liquid dietary food supplements (2 g/kg).

In 2005, in the European Union, the former Scientific Committee on Consumer Products (SCCP, now called the Scientific Committee for Consumer Safety or SCCS) affirmed the safety of methylparaben and ethylparaben, concluding the two parabens could be safely used at levels up to 0.4 per cent. In 2010, the SCCS concluded the sum of the individual concentrations of propyl- and butylparaben should not exceed 0.19 per cent based on a "conservative choice for the calculation of the Margin-of-Safety (MoS) of butyl- and propylparaben." An expert science panel at the European Food Safety Authority (EFSA) concluded that a group ADI (Acceptable Daily Intake) of 0-10 mg/kg body weight per day could now be established for methyl and ethyl parabens and their sodium salts, but not propyl paraben due to research study that questioned its impact on sperm production and decreased testosterone levels in juvenile male rats. However, these results were not replicated in larger studies conducted under Good Laboratory Practice (GLP) conditions and evaluating the same endpoints. The SCCS also concluded the available data was not adequate to evaluate the human risk of isopropyl- or isobutylparaben.

Parabens are white powders with faint odour and fair solubility in water at room temperature. The solubility is greatly increased by heating the water to 160°F-180°F. Methyl paraben is more soluble in water but less effective against molds than propyl paraben. To balance these differences, mixtures of 2 to 3 parts of methyl paraben with 1 part propyl paraben are normally used.

Parabens (*para*-hydroxybenzoic acid esters) differ from the other, organic acids described here in the respect that they are phenols rather than carboxylic acids. They are much weaker acids with pK_a values of 8.5 and so are predominantly uncharged even at neutral pH. This, means that they can be used effectively in non-acidic foods. Their antimicrobial activity increases with the length of the ester group carbon chain, although this also decreases their water solubility and may lead to poor performance in some foods where partition into the fatty phase may occur. Some gram-negatives are resistant to the higher homologues and this has been ascribed to the cell's outer membrane acting as a barrier.

Parabens appear to act mainly at the cell membrane eliminating the DpH component of the protonmotive force and affecting energy transduction and substrate transport. In contrast to other weak acid preservatives, there is little evidence suggesting that parabens interfere directly with specific enzymic activities. Propylparaben had been considered more active against most bacteria than methylparaben. The stronger antibacterial action of propylparaben may be due to its higher solubility and high permeability bacterial membrane, which may allow it to reach cytoplasmic targets in greater concentrations. However, since a majority of the studies on the mechanism of action of Parabens propose that their antibacterial action is linked to the membrane, it is likely that its greater lipid solubility disrupts the lipid bilayer, thus interfering with bacterial membrane transport processes and perhaps triggering the leakage of intracellular constituents (Tade *et al.,* 2018).

In animal studies, PHBA is a major metabolite of parabens, but it is also normally found in human urine as a decomposition product of the endogenous amino acid tyrosine and in a wide variety of plant foods. Parabens consumed with food are fully metabolized in human beings. Upon entering the body, parabens are supposed to first be absorbed in the intestines, followed by the hydrolysis into Para-hydroxybenzoic acid (PHBA) in the liver, which is then excreted in urine. They have been generally recognized as safe (GRAS) because PHBA is less toxic than the parent compounds and the excretion process usually takes place within 24 hours. *In vitro* studies suggest that parabens may be metabolized to PHBA in the intestine and subsequently re-esterified to another paraben or to the parent paraben (Lakeram *et al.,* 2006). Parabens do not accumulate in the skin or other body organs.

Important advantages of parabens are their effectiveness at higher pH values, from 3 up to 8, and stability to high and low temperatures, even to steam sterilization. Applications include bakery products (formulated without yeast), beverages, flavour extracts, food colours, fruit products, jams, jellies, preserves (artificially sweetened), gelatin, marinated and smoked fish, pickles, salad dressings, syrups, wine and olives. Principal applications for parabens include the jelly coatings of meat products, surface treatment of dried meat products, cereal- or potato-based snacks and coated nuts, confectionery (excluding chocolate), and liquid dietary food supplements. Despite these properties, parabens, E214 to 219, are not as widely used by food manufacturers as other antimicrobial agents, probably due to higher cost and flavour objections.

Sorbic Acid and its Salts

Sorbic acid is a natural compound that is also made synthetically. Sorbic acid derives its name from *Sorbus aucuparia*, because it was from berries of this tree that it was first isolated. W. Van Hoffman was the first to isolate sorbic acid from the berries of the mountain ash tree in the year 1859. The antimicrobial (preservative) properties of sorbic acid were recognized in the 1940's. In the late 1940's and 1950's it became commercially available. Since then, sorbic acid has been extensively tested and due to its antimicrobial properties, it is added to a variety of foods to preserve freshness. In wine, its use was legalized in France in 1959 and in Germany in 1971. Sorbic acid and its salts (generally called sorbate) are now used as preservatives in a variety of foods in many countries.

Sorbic acid is an unsaturated six- carbon fatty acid, 2,4-hexadienoic acid. It has a p*Ka* of 4.8 and shows the same pH dependency of activity as other organic acids. It is commercially produced as a powder or granules. It has a characteristic acrid odour and acid taste. The carboxyl (COOH) group in sorbic acid is very reactive and can form salts with calcium, sodium, and potassium. The potassium salt of sorbic acid is commercially available as a powder or granules. Its molecular weight is 150.22 and it is very soluble in water. Sorbic acid is active against yeasts, moulds and catalase-positive bacteria but, interestingly, is less active against catalase-negative bacteria. This has led to its use as a selective agent in media for *Clostridia* and lactic acid bacteria and as a fungal inhibitor in lactic fermentations. This preservative can be used in low water content food such as baked goods, cheese, dried fruits, meat and fatty media. Sorbic acid and particularly calcium sorbate can be used as active substances in fungistatic wrappers.

Sorbic acid is the compound with the antimicrobial properties but its salts (sorbates) are used in many cases due to differences in solubility and are available as potassium sorbate, sodium sorbate or calcium sorbate. Sorbic acid is sparingly soluble in water, sodium sorbate has better solubility, and potassium sorbate is very freely soluble and can be used to produce 50 per cent stock solutions. The soluble sorbates are preferred when it is desired to use the preservative in liquid form, or when aqueous systems are to be preserved. Because of their physiological inertness, their effectiveness even in the weakly acid pH range and their neutral taste, sorbic acid and its salts have become the leading preservatives in the food sector throughout the world over the past 40 years. Sorbic acid, potassium sorbate, and calcium sorbate approved worldwide are often now successfully used as standard products in many branches of the food industry. They are the substitute for the benzoic acid as a traditional preservative. As they are acidic preservatives, it is better to use them at pH 5–6. They are used as preservatives in a wide range of food, including dairy, meat, fish, vegetables, fruit, bakery, emulsions, beverages, and so on. Their advantages include broad antimycotic and antibacterial spectrum of activity; lack of effect on organoleptic properties of the foods; good activity in less acidic conditions (compared with propionate or benzoate) and safety. Sorbate has been granted generally recommended as safe status and has an acceptable daily intake of 25 mg kg^{-1} body weight that is higher than most other preservatives.

As with the other weak acids, the membrane is an important target for sorbic acid, although inhibition of a number of key enzymes of intermediary metabolism, such as enolase, lactate dehydrogenase and several Krebs cycle enzymes, has been shown. In contrast to its use as a selective agent for *Clostridia*, some studies have shown that sorbic acid inhibits the germination and outgrowth of *C. botulinum* spores. At one time this attracted some interest in the possibility that sorbic acid could be used as an alternative or adjunct to nitrite in cured meats. Sorbic acid inhibits the growth of *C. botulinum* and further reduces nitrosamine formation. To counter the use of nitrite at high concentrations in meat curing, fractional replacement of nitrite by sorbic acid has been put forth.

Table 4.10 indicates the information about organic acid food preservatives.

Table 4.10: Organic Acid Food Preservatives

Preservative	*ADI (mg kg^{-1} Body Wt)*	*Typical Usage and Levels*	*(mg kg^{-1})*
E200 Sorbic acid E201 Sodium salt E202 potassium salt E203 Calcium salt	25	Salad dressing bakery products fruit desserts	< 2000
E210 Benzoic acid E211 Sodium salt E212 Potassium salt E213 Calcium salt	5	Cider, soft drinks, fruit products, bottled sauces	< 3000
E260 Acetic acid	No limit	Pickles, sauces (Chutneys)	Up to per cent levels (1 per cent = 10000 mg kg^{-1})

E270 Lactic acid	No limit	Fermented meats dairy and vegetable products-Sauces and dressings drinks	up to per cent levels (1 per cent =10000 mg kg^{-1})
E280 Propionic acid E281 Sodium salt E282 Calcium salt E283 Potassium salt	10	Bakery goods Cheese spread	1000–5000
Parabens	10		
E214 *p*-Hydroxy-benzoic acid ethyl ester E215 Sodium salt		Bakery goods, pickles, fruit products, sauces	< 2000
E216 *p*-Hydroxybenzoic acid n-propyl ester E217 Sodium salt		Bakery goods, pickles, fruit products, sauces	< 2000
E218 *p*-Hydroxybenzoic acid methyl ester E219 Sodium salt		Bakery goods, pickles, fruit products, sauces	< 2000

Potassium sorbate is much more soluble in water than the sorbic acid. Potassium sorbate will produce sorbic acid once it is dissolved in water and is the most widely used food preservative in the world. It is effective up to pH 6.5 but effectiveness increases as the pH decreases.

Potassium sorbate is a white crystalline powder, inexpensive (at the usage level), with basically no noticeable flavour at normal usage concentrations. It has about 74 per cent of the antimicrobial activity of the sorbic acid, thus requiring higher concentrations to obtain the same results that pure sorbic acid provides. Potassium sorbate is effective against yeasts, moulds, and select bacteria, and is widely used at 0.025 to 0.10 per cent levels in cheeses, dips, yogurt, sour cream, bread, cakes, pies and fillings, baking mixes, doughs, icings, fudges, toppings, beverages, margarine, salads, fermented and acidified vegetables, olives, fruit products, dressings, smoked and salted fish, confections and mayonnaise.

Sorbic acid and its potassium salt are now used in many countries in the production of sweet white wines. In wine processing, sorbates are used to prevent refermentation. Maximum level allowable by law is 0.1 per cent.

It is important to know that the addition of sodium benzoate and/or potassium sorbate to a food product will raise the pH by approximately 0.1 to 0.5 pH units depending on the amount, pH, and type of product. Additional adjustment of the pH might be needed to keep the pH at a safe level.

In many food products, sorbate and benzoate are used together to provide greater protection against a wider variety of microorganisms. This only makes sense if the pH of the product is below 4.5.

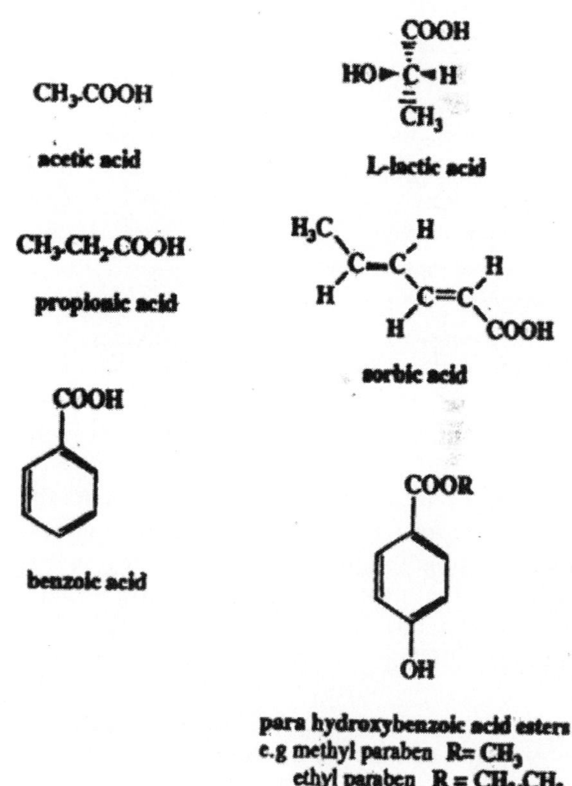

Figure 4.6: Structures of Organic Acid Food Preservatives.

Propionic Acid and Propionates

Propionic acid, also known as propanoic acid, is a naturally occurring three-carbon carboxylic acid that can be used as a preservative in food. It can be found naturally in dairy products such as butter and cheese. It occurs naturally in Swiss cheese at 1 per cent. It is used to inhibit the growth of mold in baked goods and cheeses. As early as 1906, Watkins recognized that it was effective against ropy bacteria in bread. Since the 1930s, propionates have been widely used to preserve bread in the United States. Propionic acid (PA) is used as a preservative in foods such as cheeses, baked goods, or additive for artificial fruit flavours. The U.S. Environmental Protection Agency considers it safe and therefore, has no limitation on its use.

Propionic acid and its salts, sodium and calcium propionates, collectively known as 'propionates'. They are produced by the reaction of propionic acid with the respective hydroxides or carbonates:

- Sodium propionate is produced from propionic acid and sodium hydroxide in hydrogen peroxide and water.
- Calcium propionate is produced by the reaction of propionic acid and calcium oxide in water in the presence of a flocculent.

They are approved by USFDA as GRAS (Generally Recognized As Safe) substances for food use also by PFA Act in India. Their antimicrobial action is directed to molds and rope bacteria, with almost no effect on yeast, thus making them an ideal choice for products that use commercial yeast as an ingredient. It is primarily used as a mold inhibitor which is fungistatic rather than fungicidal. It is effective against molds with concentration ranging from 0.05 to 0.25 per cent.

Propionic acid is a volatile acid. Unlike with its salts, potassium, sodium and calcium propionate in powder or granular form, it is an oily liquid, soluble in water, with a slight pungent, disagreeable, rancid odour. It is miscible with water and soluble in most organic solvents. It is a weak acid and its PKa value is 4.88, less acidity than benzoic acid and sorbic acid. It is also corrosive and flammable, thus requiring special handling. It occurs in a number of plants and is also produced by the activity of propionibacteria in certain cheeses. It is produced naturally during the fermentation of some cheeses such as Swiss cheese, in concentrations as high as 1 per cent, thus inhibiting the growth of molds. The additive is produced by chemical synthesis following two possible synthetic routes, the one starting with ethylene, carbon monoxide and water, the other starting with propionaldehyde, oxidised to propionic acid in air in the presence of a catalyst, both followed by distillation.

The sodium and calcium salts are transparent and white crystals with a mild cheeselike flavour. The sodium form is more soluble in water than the calcium salt. Sodium propionate is recommended in baked products that use baking powder and baking soda instead of yeast as the leavening agent, because the presence of calcium ions (if you were to use calcium propionate) disrupts the leavening process. Calcium propionate is preferred in baked foods that use yeast, such as breads and rolls, because it does not interfere with its fermentation and the nutritional value is increased by the added calcium. In some applications such as cakes, however, it may not be a good preservative option, as the high use level and its available calcium interfere with the chemical leavening. In contrast, sodium propionate will delay fermentation of yeast and is not recommended for use in breads or rolls, but it is preferred for the preservation of cakes.

Like other preservatives, propionates effectiveness is affected by the pH of the food, with 5.5 pH being the upper effective limit. It is important to know that the addition of sodium and calcium propionate to a food product will raise the pH by approximately 0.1 to 0.5 pH units depending on the amount, pH and type of product. Additional adjustment of the pH might be needed to keep the pH at a safe level. For example, calcium propionate is most active in the pH range below 5.5. Therefore, it is common to use acids to adjust the pH to optimize the activity. Moreover, salts of benzoic or sorbic acid are recommended for use in products with higher pH levels, such as in many chemically leavened sweet baked goods. In tortillas, calcium propionate and potassium sorbate are commonly used together, to achieve a broad spectrum of mold inhibition while maintaining product quality.

Propionates are used mainly as mold and rope inhibitors in bread; although they are also useful in cheese, non-alcoholic beverages, confections, fillings, frostings, fresh dough, pizza crust, puddings, gelatins, jams, jellies and some meat products. Propionic acid has found extensive use in the bakery field where it inhibits mould and the bacterium *Bacillus mesentericu*, the spores of which can survive the baking process. Propionates are also particularly effective in inhibiting *L. monocytogenes* at the pH level associated with comminuted meat and poultry that receive no further heat treatment prior to consumption by the consumer. Propionates can be used at low concentrations to prevent growth of *L. monocytogenes* and have the advantage of not imparting any flavour on the product. When used in combination with nitrite/nitrate they can provide control against a broader range of microorganisms and potentially result in reduced nitrite/nitrate use levels.

Propionic acid's mechanism of preservation is similar with that of benzoic acid and sorbic acid. The optimal pH range for its antimicrobial effectiveness is from 2.5 to 5.5. It is the undissociated form of propionic acid that has antimicrobial activity. The undissociated propionic acid penetrates into the microbial cell membrane and enters into cytoplasm. And then, propionic acid accumulates and dissociates prone which decreases the internal pH of the microbial and finally prevents the growth of the microbial or even kills them.

Propionates naturally occur in foods such as butter, cheese and other dairy foods and have been used as food preservatives widely for more than 60 years. The safety of propionates has been reviewed by the Joint FAO/WHO Expert Committee on Food Additives (JECFA), the European Food Safety Authority (EFSA) and the US Food and Drug Administration (US FDA) and all of these agency reviews have confirmed or reconfirmed that there are no safety concerns in respect to the use of propionates as preservatives in the currently permitted uses and usage levels. Typical usage level of propionic acid and propionates is 0.1 to 0.4 per cent. Federal regulations limit the maximum level for flour, white bread and rolls at 0.32 per cent based on the weight of the flour; for whole wheat products at 0.38 per cent based on the weight of the flour; and for cheese products at 0.3 per cent. However, calcium propionate is effective at inhibiting growth of mold and ropy bacteria when its dose relative to the number of microbial cells present is adequate to block cell metabolism. If the baked good is produced in an environment without effective current good manufacturing practices (cGMP), the dose may not be effective in inhibiting microbial growth.

Antibacterial Action of Nitrite

The antibacterial action of nitrite was first described in the 1920s though it had long been employed unknowingly in the production of cured meats where it is also responsible for their characteristic colour and flavour. It is used along with a mixture of salt, sugar, spices, and ascorbate for curing meats. In early curing processes, nitrite was produced by the bacterial reduction of nitrate present as an impurity in the crude salt used, but now nitrate, or more commonly nitrite itself, is added as the sodium or potassium salt. Nitrite contributes to the development of the characteristic colour, flavour and texture improvement in addition to preservative effects. Sodium nitrite is quite soluble in water and is more effective below neutral pH (below 7.0). Along with salt, nitrite exhibits stronger antimicrobial action.

Nitrite is consumed in the diet, through vegetables and drinking water. It is also added to meat products as a preservative. The potential risks of this practice are balanced against the unique protective effect against toxin-forming bacteria such as *Clostridium botulinum.* Today, nitrite is also being used to meet consumer requirements with respect to product safety (protection against multiplication of *Clostridium botulinum*) and sensory characteristics connected with cured meats. Correspondingly, the process of meat curing has been traditionally coupled with processed meats in order to reach proper colour, texture, flavour, safety, and shelf-life characteristics that make the products diverse. It is known that the bactericidal species is not nitrite itself, but a compound or compounds derived from it during food preparation. Of a range of nitrosyl compounds tested, the anion of Roussin's black salt $[Fe_4S_3(NO)_7]^-$ was the most inhibitory to *C. sporogenes*. This compound is active against both anaerobic and aerobic food-spoilage bacteria, while some other compounds are selective, indicating multiple sites of action. There are numerous possible targets for inhibition in the bacterial cells, including respiratory chains, iron–sulfur proteins and other metalloproteins, membranes and the genetic.

Sodium nitrite has long been known to have antimicrobial properties as a food preservative. Nitrite may also contribute to host defense against *Helicobacter pylori* and *Clostridium difficile*. The antimicrobial action of the nitrite moiety is due in part through generation of NO, inactivation of Fe-S containing proteins, and inhibition of bacterial respiration.

Nitrite can also inhibit the growth of several aerobic and anaerobic microorganisms such as not only pathogen *C. botulinum* but contribute to the management of other microorganisms too including *L. monocytogenes*, *Salmonella enterica serovar* Typhimurium and *Listeria spp*. Nitrite inhibits the growth of such pathogenic bacteria (*C. botulinum*), which under anaerobic condition produces the most lethal neurotoxin. Salt (NaCl), besides controlling *C. botulinum*, co-operates with nitrite and other factors such as acidity, meat type, heat treatment. It enables effectively control of the outgrowth of spores and thus widely helps in controlling *C. botulinum*. Salt at 5 per cent (wt/vol) was indicated to completely inhibit *C. botulinum* growth in its optimal growth conditions (Wójciak *et al.*, 2019). Nitrite is able to inhibit bacteria more effectively at low pH or higher acidity. The characteristics of curing with nitrite that make it an effective antibotulinal compound depend on the interactions of nitrite with several other variables. The variables that nitrite interacts with include salt, pH, heat treatment, spore level, nitrite input level during manufacture, and residual nitrite amount in the meat (Archer, 2002). The characteristics of the competing flora, accessibility of iron in the product, and other supplements present in the meat including ascorbate, erythorbate, phosphate, *etc.* are other additional factors. Furthermore, cooking, curing, and storage temperatures are the other important factors. In the literature, it has been found that the impact of nitrate on controlling *C. botulinum* toxin production was restricted; nitrite at the amount of 50 ppm resulted in two toxic samples out of 110 tested, and nitrite levels above the amount of 50 ppm resulted in zero toxic samples (Pierson and Smoot, 1982). This particular observation is in accordance with Lövenklev *et al.* (2004), who discovered that the amount of 45 ppm sodium nitrite effectively suppressed *C. botulinum* gene representation. Cui *et al.* (2010) used a combination of sodium nitrite and spice extracts and found that sage extract inhibits the growth of *C. botulinum* and clove and nutmeg can inactivate bacteria when combined with 10 ppm of sodium nitrite. Christieansa *et al.* (2018) showed that a 47 per cent reduction in the nitrite concentration provided the same sanitary effect against *Salmonella* and *Listeria* as the regulatory inclusion level in French dry-fermented sausage. Nitrite in meat (50–150 mg/kg) can increase the pace of the development of various aerobic and anaerobic microorganisms. In January 2016, the Food Chain Evaluation Consortium discussed the requests of European Commission and inferred in the report that a fair inclusion level of 100 ppm of added nitrite would be adequate for the majority of products, without having a significant effect on colour, flavour, and microbiological safety (Wójciak *et al.,* 2019).

Another property of nitrite is its ability to retard the oxidation process during storage period and the subsequent warmed-over and rancid flavours developed during thermal processing of meat and meat products (Sindelar *et al.,* 2007). The antioxidant activity of nitrite is attributed to the potential of nitric oxide to bind to and stabilize heme iron of meat pigments during the meat curing process. Nitric oxide, being a free radical, can also trigger lipid autoxidation by chelate free radicals including peroxyl radicals. The nitrite binds free irons and stabilizes the heme iron, which can reduce lipid oxidation by limiting prooxidant activity of iron. Nitrite can form another antioxidant compound, for example, S-nitrosocysteine. Dethmers and Rock (1975) stated that the addition of nitrite above the amount of 50 ppm to Thuringer sausage lowered the off-flavour development and bettered the flavour quality, whereas treatments with no nitrite added were believed to be the most rancid ones due to the poor flavour quality. Doolaege *et al.* (2012) investigated the effects of different inclusion levels of sodium nitrite (40, 80, and 120 ppm) combined with different inclusion levels of rosemary extract (0, 250, 500, and 750 ppm) and reported that the concentration of sodium nitrite added to liver pate could be reduced to 80 ppm when rosemary extract is added at 250, 500, and 750 ppm concentrations without any negative effect on lipid oxidation and colour parameters. Therefore, nitrite is a hurdle technology which effectiveness depends on several other hurdle technologies including sodium chloride (accelerating the autoxidation of oxymyoglobin and promote peroxynitrite formation), ascorbate (increasing ONOO- synthesis), and Aw. In this environment, certain species are more resistant than others to acidic, oxidative, and nitrative stresses. The most resistant are gram-negative aerobic/facultative anaerobic bacteria (*Escherichia coli, Salmonella*), and the most fragile are gram-positive anaerobic bacteria (*Clostridium botulinum*).

Early workers showed that a level of 200 mg kg^{-1} at pH 6.0 was sufficient to inhibit strains of *Escherichia, Flavobacterium, Micrococcus, Pseudomonas* and others, although *Salmonella* and *Lactobacillus* species were more resistant. Of most practical importance though is the ability of nitrite to inhibit spore-forming bacteria such as *Clostridium botulinum* which will survive the heat process applied to many cured meats. To achieve this commercially, initial levels of nitrite greater than 100 mg kg^{-1} are used. The mechanism of its action is poorly understood partly due to the complexity of the interaction of several factors such as pH, salt content, presence of nitrate or nitrite and the heat process applied to the cured meat. Descriptive mathematical models of these interactions have however been produced which quantify the precise contribution of nitrite to safety (Adams *et al.,* 2015).

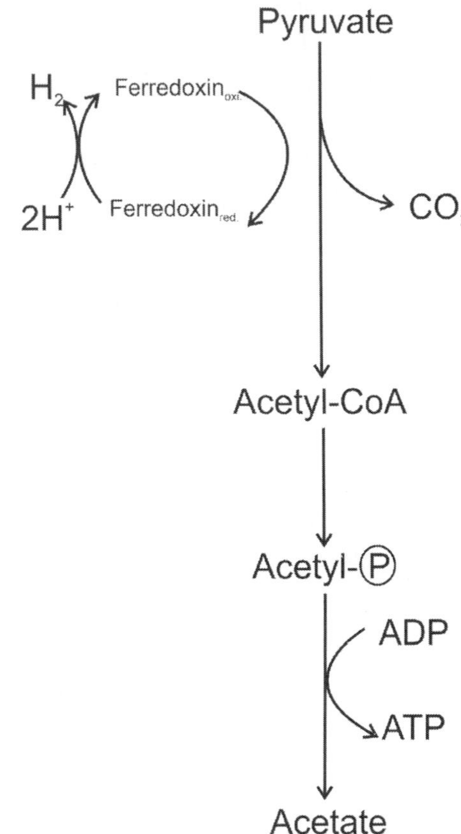

Figure 4.7: The Phosphoroclastic System.

Bacterial inhibition by nitrite increases with decreasing pH, suggesting that nitrous acid (HNO_2, pK_a 3.4) is the active agent. In the case or spores, it appears that nitrite acts by inhibiting the germination and outgrowth of heated spores and by reacting with components in the product to form other inhibitory compounds. The latter effect was first noted in the 1960s by Perigo who observed that when nitrite was heated in certain bacteriological media, the resulting medium proved more inhibitory to *Clostridia* than when filter-sterilized nitrite was added after heating. *Clostridia* are very sensitive to these 'Perigo factors' which, differ from nitrite in displaying activity that is independent of pH. However, they do not seem to be formed in meat and their effect in bacteriological media could be removed if meat was added. The presence of 'Perigo-type factors' has been reported in heated cured meats but these are only produced by severe heating and have minor antibacterial activity (Adams *et al.,* 2015).

Studies into the nature of Perigo and Perigo-type factors have looked particularly at the production of Roussin's salts; complex salts of iron, nitrosyl and sulfydryl groups. Although these compounds have not been shown to be present in cured meats in sufficient quantity to cause the inhibition observed, their formation may give an indication of the way nitrite itself interferes with bacterial metabolism. It has been proposed that the biochemical mechanism of inhibition involves nitrite reacting with iron and sulfhydryl groups of key cell constituents. Iron-containing proteins such as ferredoxins are very important in electron transport and energy production in *Clostridia*. For example, the phosphoroclastic system is used by *Clostridia* to generate additional ATP by substrate-level phosphorylation. Pyruvate, produced by glycolysis, is oxidized to acetate via acetyl-CoA and acetyl phosphate which phosphorylates ADP to produce ATP. Ferredoxin acts as a carrier for the electrons removed in the oxidation step and which are ultimately used to reduce hydrogen ions to hydrogen gas. Support for this hypothesis has come from the observation that nitrite addition leads to an accumulation of pyruvate in *C. sporogenes* and *C. botulinum* (Adams *et al.,* 2015).

Awareness of the problem has led to changes in production practices for cured meats such as the use of low levels of nitrite in preference to nitrate and the increased use of ascorbic acid which inhibits the nitrosation reaction. These measures have produced significant reductions in nitrosamine levels.

Mention should also be made here of the other contributions made by nitrite to the quality of cured meats. Reduction of nitrite to nitric oxide produces the characteristic red colour of cured meats. The nitric oxide co-ordinates to the haem ferrous ion in the muscle pigment myoglobin converting it to nitrosomyoglobin. When raw cured meats such as bacon are cooked this pigment decomposes to produce nitrosylhaemochrome which has the pink colour also seen in cooked cured hams. Only small quantities of nitrite are required to produce the cured meat colour: theoretically 3 mg kg^{-1} is sufficient to convert half the myoglobin present in fresh meat, but because of competing reactions, 25 mg kg^{-1} is required to give a stable colour Adam *et al.* (2015).

Nitrite also contributes to the typical cured meat flavour. Taste panels can distinguish cured meats where nitrite has not been used but the precise reasons for this are not known. It is thought that nitrite acts as an antioxidant to inhibit lipid degradation in the meat although this may only be part of the story.

Figure 4.8 explains the reactions of nitrite with meat myoglobin.

Current European Union (EU) regulations concerning the use of nitrite and nitrate differ with respect to both the method of curing used and the product that is cured. According to Directive 2006/52/EC of the European Parliament and of the Council and Commission Regulation (EU) No 1129/2011, all in all, 150 mg nitrite per kg is permitted to be inserted into all meat products plus 150 mg nitrate per kg for non-cooked meat products. Finally, maximum levels of 300 mg nitrite and nitrate per kg can be used for ripening ham production. In cooked meat products, no more than 150 mg nitrite per kg is allowed. However, the maximum allowed nitrite level is 100 mg/kg in the sterilized meat products.

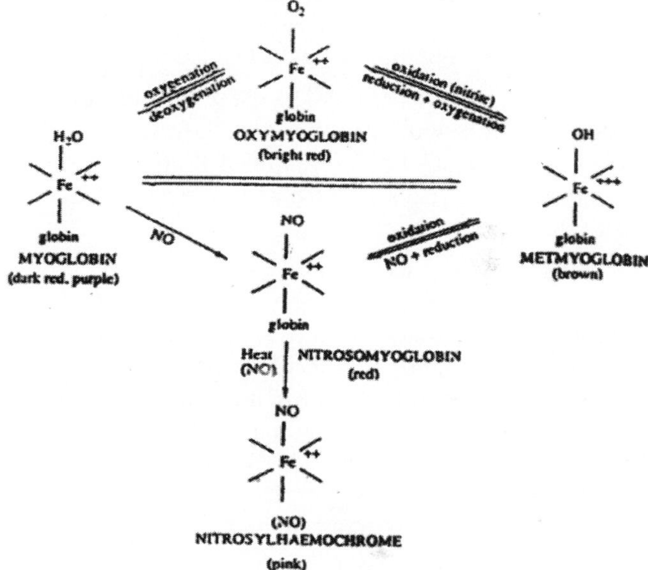

Figure 4.8: Reactions of Nitrite with Meat Myoglobin

Nitrite is difficult to replace as a preservative, because it can simultaneously perform many functions. Further studies are needed to investigate the complex effect of various reduced levels of nitrite and potential alternative compounds and/or technologies that can substitute nitrite.

Sulfur Dioxide and Sulphites

Sulphur dioxide and its derivatives have been widely used in foods as a food preservative. It serves both as an antioxidant and reducing agent and prevents enzymatic and non-enzymatic reactions, leading to microbial stability. The commonly used forms are sulphur dioxide gas and sodium, potassium and calcium salts of sulphite, bisulphite or metabisulphite, which are powders.

In food system and industry, they are well-known additives and rarely contaminants (in fermented products, SO_2 is formed by microbial reduction of sulfate), used and scheduled as preservatives. Among their manifold functions, they act as inhibitors of microbial growth, reducers of food deterioration, enhancers of colour, bleaching agents, enzyme inhibitors, antioxidants, and oxygen scavengers (Carocho *et al.,* 2014. D'Amore *et al.,* 2020). They are included in the FAO/WHO Codex Alimentarius document, "Codex General Standard for Food Additives" (GSFA, Codex STAN 192- 1995) that lists all authorized food additives, and establishes their maximum concentrations relating to each foodstuff. The Codex Alimentarius Committee assigned a number code to sulfites, following the International Numbering System (INS). These codes match with E-numbers assigned by European additives classification (FAO/WHO Codex Alimentarius, 2019). It is used as preservative and antioxidant, due to its capacity to release sulphur dioxide in acid conditions (FAO/WHO Expert Committee on Food Additives, 2009; FSSAI, 2011; NHFPC, 2015).

Sulphur dioxide gas (SO_2) is one of the oldest known fumigants and a wine preservative. Sulphur dioxide is one of the gas which has long enjoyed a reputation for its disinfecting properties. Let us try to understand about its disinfecting action.

Sulphur dioxide, the gaseous form is produced either by burning sulphur or by its release from the compressed liquefied form. It is a colourless, suffocating, pungent-smelling, non-flammable gas and highly soluble in cold water. It readily dissolves in water to establish a pH-dependent equilibrium similar to CO_2. In solution, it hydrates to sulfurous acid (H_2SO_3), which dissociates in turn to form bisulfite (HSO_3^-) and sulfite (SO_3^{2-}) ions. The bisulfite ion is quite reactive by means of ionic and free radical mechanisms.

$$SO_{2(gas)} = SO_2 + H_2O == H_2SO_3 = H_+ + HSO_3- = 2H_+ + SO_{32-}$$

Sulfurous acid (H_2SO_3) is a dibasic acid with pK_a values of 1.86 and 6.91.

The antimicrobial activity of sulfur dioxide is believed to be due to the molecular sulphur dioxide moiety of that part which remains unbound (the so-called 'free' sulphur dioxide). Hence, the unionized forms of SO_2 which can readily penetrate the cell have the greatest antimicrobial activity. It has been reported that they are between 100 and 1000 times more active than the bisulfite anion. Since the unionized forms predominate at low pH values, it follows that SO_2 is used to best effect in acidic foods. Less sulphur dioxide is needed in juices of high acidity, for instance 15 mg l^{-1} of free sulphur dioxide at pH 3.0 has the same antimicrobial effect as 150 mg l^{-1} at pH 4.0. At neutral pH, SO_2 is present as a mixture of the relatively inactive bisulfite (HSO_{3-}) and sulfite (SO_{32-}) ions, although salts of these anions prove the most convenient way of handling the preservative in the food industry.

SO_2 is a reactive molecule and can disrupt microbial metabolism in a number of ways. As a reducing agent, it can break disulfide linkages in proteins and interfere with redox processes. It can also form addition compounds with pyrimidine bases in nucleic acids, sugars and a host key metabolic intermediate. One disadvantageous consequence of this reactivity is its ability to destroy the vitamin thiamine in foods and the once widespread practice of using it in meat and meat products has now been prohibited, with the exception of British fresh sausage.

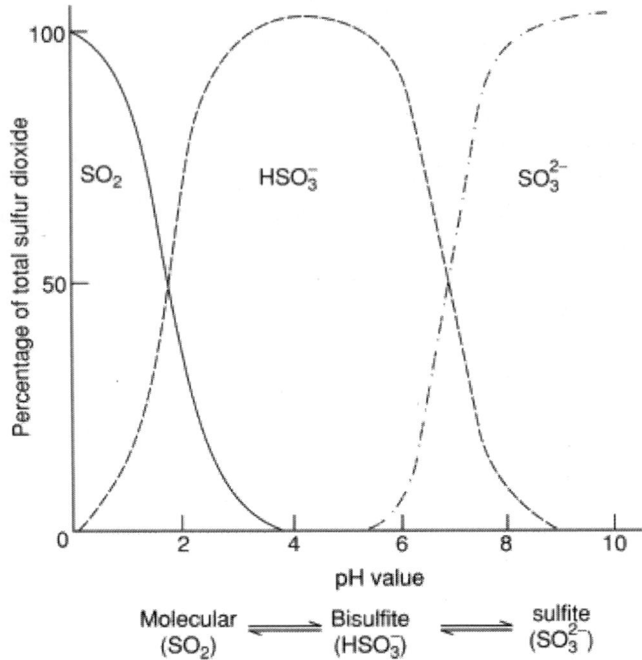

Figure 4.9: Percentage of Sulfite, Bisulfite and Molecular Sulfur Dioxide as a Function of pH in Aqueous Solution.

(*Source*: Hammond, S. M. and Carr, J. G. (1976) The antimicrobial activity of SO_2 – with particular reference to fermented and non-fermented fruit juices. In: Skinner, F. A. and Hugo, W. B. (eds) *Inhibition and Inactivation of Vegetative Microbes (S.A.B. Symposium Series No. 5)*, pp. 89–110. London: Academic Press.)

Sulfur dioxide is active against bacteria, yeasts and moulds, although some yeasts and moulds are more resistant. Gram-negative bacteria are most susceptible and in British fresh sausage where sulfite is permitted up to a level of 450 mg kg^{-1}, the Gram-negative spoilage flora normally associated with chilled meats is replaced by one dominated by Gram-positive bacteria and yeasts.

Sulfur dioxide can be used in the form of a gas or in aqueous solution as sodium sulfite or disulfite. As a disulfite, it acts less rapidly, but is more easily controlled and results in a few flavour changes. Of deep-frozen products, apples and apricots are most suited to sulphur dioxide treatment. Fruit slices or halved fruits are soaked for 3–4 min in solutions containing 0.4–0.5 per cent sulphur dioxide.

Sulfur dioxide has been used in the manufacture of wine for thousands of years to destroy bacteria, molds, and wild yeasts without harming the yeasts used in fermentation. In winemaking, the tolerance of the wine yeast *Saccharomyces cerevisae* to SO_2 levels around 100 mg l^{-1} is exploited to control the growth of wild yeasts and acetifying bacteria. Seasonal surpluses of soft fruits are also preserved by the addition of high levels of SO_2 to permit jam production throughout the year. Grapes and cut fruits are exposed to fumes of burning sulfur before dehydration or transportation.

Sodium, potassium and calcium sulfites, hydrogen sulfites (bisulfites), and metabisulfites exhibit almost identical preserving and antioxidant properties to sulphur dioxide. Thus, anhydrous sodium sulfite is relatively stable but reacts rapidly with oxygen in the presence of moisture; all dissolved oxygen is removed in a few seconds. Commercial sodium bisulfite consists mainly of sodium metabisulfite. Sodium and potassium metabisulfites are irritants and possible sensitizers.

Sulfur dioxide, sulfites, bisulfites, and metabisulfites are used as antiseptics and bleaching agents in breweries and other fermentation industries, and are used to retard spoilage of meat, egg yolks, grains, dried and fresh fruits, and other products. Sodium sulfite is allowed as a bleaching agent for gelatin, beet sugars, and food starch, and sodium and potassium sulfites can be used to assist caramelization

Dried fruits and vegetables routinely report the highest levels of sulphites, due to their preservative abilities, while maintaining colour and texture. This makes them especially popular in lighter coloured fruits and vegetables that risk turning brown. Another key application area is wine, cider and beer. While these beverages have levels of naturally occurring sulphites, more are added to products to help preserve them and kill bacteria. Sulphites are also routinely found in soft drinks and fruit juices. Other uses include, but are not limited to condiments, sausages and burger meat, jam, prawns and dried soup mixes.

The researchers observed that sulphur dioxide is the most active form. Although characterized by lower antimicrobial power, also bisulfite has good activity. Moreover, also the bound forms of SO_2, sulfonates, have antimicrobial activity (especially on yeasts), but it is about 1/30 of the free form. The effect of sulfites is dose-dependent and may be microbiostatic (extension of microbial lag phase) or microbicidal (leading to microbial death). It is carried out by the reaction with disulfide bonds (S–S) and thiol (S–H) groups of microbial proteins/enzymes, but also by interference with the synthesis of microorganism DNA. First, sulfites pass through microorganism cell membrane mainly by passive transport. The reaction with disulfide bonds leads to the formation of thiol and S-sulfonates causing several proteins conformational changes. Consequently, both enzymes and proteins located on the cell surface and involved in key processes of microorganism life cycle (transport, glycolysis, cell respiration, general metabolism) are inactivated, with partial or complete loss of activity (Simon and Kroneck, 2013). Moreover, several microbial cofactors, coenzymes and prosthetic groups, such as thiamine, NAD+, flavin, heme, folic acid, and pyridoxal, necessary for many enzymatic reactions, can be degraded by the action of sulfites. Another mechanism particularly effective in yeast cells is the cleavage of adenosine triphosphate (ATP). In fact, a rapid fall in ATP levels was observed in *Saccharomyces cerevisiae*, just prior to microorganism death mediated by sulfites (Gould, 2000). The bacteria growth is inhibited most effectively than yeasts and molds. In fact, sulfites have a broad spectrum of activity, against most of the Gram-negative and several Gram-positive bacteria.

Although sulfite is safe, sulphur dioxide is harmful in large quantities and even small amounts can trigger an attack in a certain class of asthmatics. Low pH foods are a particularly high risk for producing gaseous SO_2. Sulfites ingestion has been correlated with several adverse and toxic reactions, such as hypersensitivity, allergic diseases, vitamin deficiency, and may lead to dysbiotic events of gut and oral microbiota. In many countries, these additives are closely regulated and in meat products the legislation restricts their usage.

Due to the possible harmful consequences deriving from ingestion of sulfites, monitoring and surveillance of these additives are extensively recommended by WHO, FAO, EFSA, FSANZ, FDA, and other governmental authorities, worldwide, which turned their attention especially on meat products, because several cases of abuses and unlawful usage were reported (Carrabs *et al.*, 2017; Timbo *et al.*, 2010; Tomasevic *et al.*, 2018)

Natural Food Components as Preservatives

The uncertainty voiced by consumer organisations and pressure groups over the use of food additives including preservatives has already been referred to. One approach to reassuring the consumer has been recourse to methods of preservation that can be described as 'natural'. The whole area though is riddled with inconsistency and contradiction; it can be argued that any form of preservation which prevents or delays the recycling of the elements in plant and animal materials is unnatural.

On the other hand, there is nothing more natural than strychnine or botulinum toxin. Smoking of foods might be viewed as a natural method of preservation. Its antimicrobial effect is a result of drying and the activity of woodsmoke components such as phenols and formaldehyde which would probably not be allowed were they to be proposed as chemical preservatives in their own right.

The use of natural food components possessing antimicrobial activity such as essential oils and the lactoperoxidase system in milk have attracted some attention in this respect. Attention has also been paid to the bacteriocins produced by food-grade microorganisms such as the lactic acid bacteria. Nisin is an already well-established example and its use can be extended by expedients such as inclusion of whey fermented by a nisin-producing strain of *Lactococcus lactis* as an ingredient in formulated products like prepared sauces.

Intermediate Moisture Foods

Foods that can be preserved in a simple way at ambient temperatures (by restricting water mobility) and that are moist enough to be consumed without rehydration traditionally have come to be known as intermediate moisture (IM) foods. Their shelf stability at ambient temperatures is attributed mainly to adjustment and control of water activity. Moisture content of IM foods varies between dried foods with levels of less than 7 per cent, which can be stored at room temperatures, and fresh foods with levels of 60–80 per cent, and above which need to be preserved by an additional method and still may have a limited shelf life, albeit higher than that of the fresh counterpart. Usual ranges of water levels are between 10 and 50 per cent and water activities (a_w) vary from 0.65 to 0.90. Therefore, IM foods can be classified as partially dehydrated foods with suitable concentrations of dissolved solids to inhibit the growth of bacteria, molds, and yeasts and to control undesirable enzymatic activity. Fruits, vegetables, fish, and meats are processed successfully processed into IM-range products.

National Centre for Coordination of Research on Food and Nutrition, France during 1980's proposed a comprehensive definition for Intermediate Moisture Foods as, 'Food products of soft texture, subjected to one or more technological treatments, consumable without further preparation and with shelf stability of several months, assured without thermal sterilization, nor freezing or refrigeration, but an adequate adjustment of their formulation; composition, pH, additives *etc.* and mainly water activity (a_w) which must be between 0.6 and 0.84 (measured at 25°C)'. Water activity is a thermodynamic property which is defined as the ratio of vapor pressure of water in a food system and vapor pressure of pure water at the same temperature.

IMF typically contain 20 to 50 weight- per cent [w/w] moisture content. The a_w of IMF is between 0.70 and 0.85, and the products are generally considered as microbiologically stable at room temperature (Vermeulen *et al.*, 2012; Vermeulen *et al.*, 2015). In general, IMF maintain some certain initial characteristics (colour, texture and flavour) of fresh food products. The unique features that make IMF appeal to consumers include conceived microbial safety, desirable odours, high nutritional values, ready to eat (RTE). Production of IMF with properties close to fresh foods yet having extended the shelf life to satisfy the demand of the consumer is vitally important for the food industry (Qiu *et al.*, 2019).

Traditional IMF are processed by the withdrawal of water, by adsorption or desorption, and/or by the addition of conventional humectants, such as salt (sodium chloride) and sugar (sucrose). Although some traditional IMF, such as dried fruits, are simply dried to the desired a_w, most IMF are prepared by a combination of drying with added humectants. The application of IMF technology has been very much successful in preserving fruits and vegetables without refrigeration. The addition of high amount of sugar in fruits during processing will create a protective layer against microbial contamination after the heat processing. The sugar acts as a_w depresser limiting bacterial growth. These foods are more stable at room temperature. Bakery products are prepared by adding water and other ingredients to low-a_w cereals, followed by various processing and heating procedures to yield a_w in the range of IMF. Examples of IMF include jams, jellies, candies, baked goods, honey, dried fruits and vegetables.

Novel IMF are often processed by using humectants, such as glycerol, propylene glycol, and sorbitol, for a_w adjustment. Combinations of several humectants, each at lower concentrations, are often used to reduce off-flavour problems. In the production of novel IMF, pasteurization and blanching are commonly used to reduce initial levels of microorganisms and to inactivate enzymes. Antimycotics, such as sorbic and benzoic acid and their salts; methyl and propyl paraben; or calcium propionate, assist in the control of mold and yeast growth. Other chemical additives may also be incorporated to slow down physical and chemical deterioration during storage.

Different preservation methods *viz.*, dehydration, osmotic dehydration and dry infusion may be used to adjust water activity during the preparation of IMFs. Water content of such products may vary between 25-50 per cent to get an acceptable intermediate moisture level in products which are softer, moist and ready to eat. Different preservation methods that can be used to reduce the water activity in food products ae categorized as partial drying, osmotic drying, dry infusion, moist infusion and blending of formulations.

Drying is an important step to reduce water in IMF. Hot air drying is one of the most commonly used drying methods, because of low cost and simple operation. Partial drying is the simplest and cheapest method of food preservation. In most parts of the country, open sun drying is still being practised for drying fruits, vegetables, raisins, fish and similar types of food products. However, on commercial level, large size dehydrators, ovens and dryers can be used to reduce water content and control the water activity (a_w).

Water-activity-lowering agents are widely used to prepare fruits, vegetables or meat products of immediate moisture contents for extended shelf life (Finn *et al.,* 2015). Conventional water-activity-lowering agents include salt and phosphates. However, it was also demonstrated that some of those effective conventional water-activity-lowering agents were expensive and usually influenced the flavour of IMF while providing no nutritional benefits. In addition, they need to be labelled, overuse may be harmful to human health (Torti *et al.,* 2016). It is important to select suitable water-activity-lowering agents within an acceptable concentration to avoid their influence on the flavour of food products. Recently many research efforts have been concentrated on the water-activity-lowering agents to solve these problems, among which application of the combination of water-activity lowering-agents is the most promising.

Thermal processing is the most effective means to in activate pathogenic and spoilage micro-organisms. Thermal processing can be divided into pasteurization (≤100 °C) and commercial sterilization (> 100 °C). Thermal processing has been extensively used in the food industry for production of ready-to-eat high moisture products, but commonly used for IMF. With increasing food safety concerns associated with low moisture food and IMF, thermal pasteurization can be an effective option to ensure food safety of pre-packaged IMF. Nonthermal methods for microbial reduction of food products have recently received considerable interest. These technologies are carried out at around the ambient temperature and thus, eliminate the adverse effects on food quality associated with the traditional thermal methods. Among these nonthermal methods, CP, HPP and coating technologies are considered to be promising and novel technologies.

Plasma treatment (PT) is a relatively new decontamination technology compared to traditional thermal decontamination technologies in the food industry. It can efficiently inactivate microorganisms, including biofilms, bacterial spores, fungi and bacteria (Segat *et al.,* 2016). Plasma is an ionized gas that consists of charged particles, electric fields, UV photons, and reactive species, all of which are considered to be effective to inactive microorganisms and ensure the safety of food products. Low-temperature plasma (LTP) has also been termed such as atmospheric cold plasma (ACP), non-thermal plasma (NTP), atmospheric pressure plasma (APP) or simply cold plasma (CP) by various researchers. Formerly, the PT was carried out in vacuum environment, but now an atmospheric plasma system has been designed to reduce the cost, increase the treatment speed and apply in industry. LTP has been applied to various food products to inactive microorganism and achieve longer shelf life.

High pressure processing (HPP) is an innovative non thermal method for the treatment of food products, especially juices and muscle-containing products such as fish and meats. The mechanism is based on the sublethal damage of microbial cells at high pressure combined with low temperature and hold time (Sevenich *et al.,* 2016). The thermal load of HPP is lower than other thermal treatments, HPP has several benefits over other methods. The quality and nutritional quality of food products are not greatly influenced and the shelf life of food products can be prolonged due to the reduction of microorganism. HPP technology is expensive, it has been used in developed countries for production of meats and juices that require cold storage. The effect of HPP on microbial inactivation in IMP is not well documented and requires further research. In addition, more research is needed on integrity of food packages and sensory properties of specific products.

Recent studies have revealed that thermal resistance of bacterial pathogens, *e.g., Salmonella* and *Listeria monocytogenes*, increases sharply with reduction in water activity of food matrices measured at treatment temperatures (Taylor *et al.,* 2018). Thus, much higher processing temperature or longer time are needed for thermal pasteurization of IMF as compared to the high moisture counterparts, even for the same target bacterial pathogens. Such processes may cause thermal degradation of heat sensitive quality attributes. Thermal technologies for preserving IMF should be carefully selected to achieve a balance between shelf life and the food quality. After meeting microbial safety requirements serious consideration should be given to the quality changes of food products during processing and storage. Optimization of pasteurization condition should be carried out by considering the thermal kinetics of the target pathogens, thermal inactivation of quality-related enzymes, and related quality parameters (Peng *et al.,* 2017; Liu *et al.,* 2018)). Storage tests should also be carried out at normal distribution and storage conditions to ensure microbial safety and acceptance of food products within their maximum shelf life.

Osmotic dehydration (OD) is another effective traditional method to reduce the amount of water in fresh foods in preparation of IMF. This method of drying is based on the principle of 'osmosis' in which different humectants or osmotic agents such glucose, salt, sucrose, glycerol, sorbitol, glucose syrup, organic acids and proteins, corn syrup, and fructo-oligosaccharides are commonly used. This method is used in the preparation of candid foods and preserves where sugar is used as humectant. Osmotic agents should be carefully selected for different food materials. For example, salt is used for vegetable processing. But because of its salty taste, it is not widely applied to the dehydration of fruit. Sucrose has commonly been used as osmotic agent for fruit. Osmotic agents such as honey, ethanol, high fructose corn syrup and maple syrup are receiving considerable attention due to their high dehydration rates, good rehydration properties, and little damage to texture of food products.

Osmotic dehydration partially removes water from food products, leading to minimally processed food of high quality with a relative long shelf life. The difference in the osmotic pressure of the food and the solution used acts as the driving force between the two components. Thus, water in the food material diffuses into solution and humectants diffuses into food material. However, a major disadvantage is that OD takes a long time which may cause partial quality degradation in some products. Novel technologies have been explored to overcome these disadvantages. Among these are microwave assisted osmotic dehydration (MAOD), vacuum impregnation (VI), ultrasound assisted osmotic dehydration (UAOD) and osmodehydro-freezing.

Osmodehydro-freezing has been found to cause less unwanted changes and improve the quality of food products at low temperatures. In addition, osmodehydro-freezing is less cost and energy consumption than freezing alone. Similarly, dehydro-freezing is another approach to producing frozen fruits with high nutrition value and desirable sensory qualities. It has been reported that dehydro-freezing strawberries showed more colour and vitamin C retention when compared to conventional strawberries. A good organoleptic quality and the strawberry tissue integrity were also obtained. Partial removal of water prior to freezing has been reported to enhance the final quality of cucumbers. Zhao *et al.* (2014) reported that OD of mangoes prior to freezing reduced quality losses of the frozen foods. It particularly helps the retention of product color, texture, Vitamin C content, and reduces dip loss upon thawing, as compared to conventional frozen methods.

Microwave assisted osmotic dehydration (MAOD) is another novel combination treatment that can enhance moisture transfer in food products and reduce processing time. Researchers also combined microwave with OD to achieve superior quality foods. The application of microwave causes a rise of temperature in products and osmotic solution, thus increasing the mass transfer rate.

Dry infusion (adsorption) method is recommended for the preparation of high quality of dehydrated preserve products. This is also an energy intensive method of high-quality dehydration where fruits are first dehydrated and then soaked in required humectant until the desired water activity is achieved. In Moist infusion (desorption) method, solid food pieces are equilibrated in a solution of lower a_w, often with heating. Drying may be used to reduce further a_w (*e.g.*, intermediate-moisture fruits). IMF processed by adsorption normally have slower microbial growth and lipid oxidation due to a lower moisture content than IMF prepared by desorption with identical a_w. This is because of sorption hysteresis *i.e.* the phenomenon in foods whereby the moisture content of a food is lower during equilibration by adsorption than by desorption. In spite of these hysteresis effects and the waste of residual infusion solutions, the desorption method is generally used more, due to its advantages of lower cost, enzyme inactivation, and better retention of flavour and structural integrity. With this method, it is possible to reduce food weight by 50 per cent by the simultaneous counter current diffusion of water out of the food and solute into the food.

Blending of formulation method is the most recent method of dehydration and its ease of use makes it more popular amidst food producers. Here, the food is mixed with required ingredients including humectants and the food mixture is applied to required cooking, baking or extrusion techniques to achieve the required water activity in the final product. It is time saving, energy efficient and more flexible method to get a desired type of product including preserves, confectionery items, snacks and pet foods.

In order to protect consumers from foodborne diseases, microbial inhibition has been an important consideration in new package developments. Long shelf life can be achieved by modified atmosphere packaging (MAP) and active packaging (AP) through changing the ratio of gases composition in the package made from semipermeable materials or by incorporating chemical substances. Edible coating can extend the shelf life of foods by using edible antibacterial material directly coated on the surface of IMF. The coating also serves as a barrier to oxygen (O_2) and water.

Intermediate moisture foods (IMFs) having *a*w between 0.85 and 0.6, the range which corresponds roughly to a moisture content of 15–50 per cent, prohibit the growth of Gram-negative bacteria as well as a large number of Gram-positives, yeasts and moulds, giving the products an extended shelf-life at ambient temperature. When spoilage does occur, it is often a result of incorrect storage in a high relative humidity environment. In correctly stored products growth of xerophilic moulds, osmophilic yeasts or halophilic bacteria may occur, depending on the product and in many IMFs, the shelf-life is further protected by the inclusion of antifungal agents such as sulphur dioxide or sorbic acid. Recent outbreaks caused by salmonella in low moisture environment have caused food safety concerns for both low moisture foods and IMF (Syamaladevi *et al.,* 2016).

At the a_w of IMFs, pathogens are also prevented from growing. Although *Staph. aureus* is capable of growing down to an a_w of 0.83, it cannot produce toxin and is often effectively inhibited by the combination of a_w with other antimicrobial hurdles.

There are a number of traditional IMFs such as dried fruits, cakes, jams, fish sauce and some fermented meats. Sweetened condensed milk is one interesting example. Milk is homogenized, heated to 80°C and sugar added before it is concentrated in a multi-effect vacuum evaporator at 50–60°C. When the product emerges from the concentration stage it is cooled and seeded with lactose crystals to induce crystallization of the lactose. This gives sweetened condensed milk its characteristic gritty texture. Although the product is packed into cans and has an almost indefinite shelf-life, it is not an appertized food. Its stability is a result of its high sugar content (62.5 per cent in the aqueous phase) and low a_w (< 0.86). Spoilage may sometimes occur due to growth of osmophilic yeasts or, if the can is under-filled leaving a headspace, species of *Aspergillus* or *Penicillium* may develop on the surface.

Some years ago, our developing understanding of the stability of IMFs led to considerable interest in applying the same principles to the development of new shelf stable foods. Novel humectants such as glycerol, sorbitol and propylene glycol were often used to adjust a_w in these products in addition to the solutes salt and sugar. They were not however well received in the market for human food because of acceptability problems, although a number of successful pet food products were developed. One interesting observation made during this work is that products with the same water activity differ in their keeping quality depending on how they are made. Traditional IMFs are generally made by a process of desorption whereby water is lost from the product during processing but a number of the new IMFs used an adsorption process in which the product is first dried and its moisture content readjusted to give the desired a_w.

The water activity of a product can be reduced by physical removal of liquid water either as vapour in drying, or as a solid during freezing. It is also lowered by the addition of solutes such as salt and sugar. Freezing has already been discussed in this chapter and so here we will confine ourselves to drying and solute addition. The primal role of these techniques in food preservation has been alluded to in a number of places. It was the earliest food preservation technique and, until the 19_{th} century, water activity reduction played some part in almost all the known procedures for food preservation.

Nature provided early humans with an object lesson in the preservative value of high solute concentrations in the form of honey produced by bees from the nectar of plants. The role of salt in decreasing a_w accounts for its extreme importance in the ancient economy as evidenced today in the etymology of the word, salary and of place names such as Salzburg, Nantwich, Moselle and Malaga. It can also be seen in the extraordinary hardship people were prepared to endure (or inflict on others) to ensure its availability; to this day the salt mine remains a by-word for arduous and uncomfortable labour.

Solar drying, while perhaps easy and cheap, is subject to the vagaries of climate. Drying indoors over a fire was one way to avoid this problem and one which had the incidental effect of imparting a smoked flavour to the food as well as the preservative effect of chemical components of the smoke.

Salting and drying in combination have played a central role in the human diet until very recently. One instance of this is the access it gave the population of Europe to the huge catches of cod available off Newfoundland. From the end of the 15_{th} century, salted dried cod was an important item in trans-Atlantic trade and up until the 18_{th} century accounted for 60 per cent of all the fish eaten in Europe. It remains popular today in Portugal and in the Caribbean islands where it was originally imported to feed the slave population. Other traditional dried and salted products persist in the modern diet such as dried hams and hard dry cheeses but the more recent development and application of techniques such as refrigeration, MAP and heat processing and the preference for 'fresh' foods has meant that their popularity has declined. Nevertheless, this should not obscure the important role that low a_w foods still play in our diet in the form of grains, pulses, jams, bakery products, dried pasta, dried milk, instant snacks, desserts, soups, *etc.*

Microorganisms that were in the product before drying or were introduced during processing can survive for extended periods. This is most important with respect to pathogens if they were present in hazardous numbers before drying or if time and temperature allow them to resume growth in a product that is rehydrated before consumption. There have been a number of instances where the survival of pathogens or their toxins has caused problems in products such as chocolate, pasta, dried milk and eggs. Generally *Salmonella* and *Staphylococcus aureus* have been the principal pathogens involved-there have been 17 major outbreaks associated with these organisms and dried milk in the last 40 years, but spore formers are particularly associated with some other dried products such as herbs or rice.

The hysteresis effect in water sorption isotherms means that although products made using the two techniques will have the same initial a_w they will have different moisture contents and so will eventually equilibrate to different a_w values. It was found that products made by desorption and having the higher water content were also more susceptible to microbial spoilage.

Solar drying is still widely practised in hot climates for products such as fruits, fish, coffee and grain. The traditional technique of spreading the product out in the sun with occasional turning often gives only rudimentary or, sometimes, no protection from contamination by birds, rodents, insects and dust. Rapid drying is essential to halt incipient spoilage; this is usually achievable in hot dry climates, though in tropical countries with high humidity drying is usually slower so that products such as fish are often pre-salted to inhibit microbial growth during drying,

There are a number of procedures for mechanical drying which are quicker, more reliable, albeit more expensive than solar drying. The drying regime must be as rapid as possible commensurate with a high-quality product so factors such as reconstitution quality must also be, taken into account. With the exception of freeze-drying where the product is frozen and moisture sublimed from the product under vacuum, these techniques employ high temperatures. During drying a proportion of the microbial population will be killed and sub-lethally injured to an extent which depends on the drying technique and the temperature regime used. It is however no substitute for bactericidal treatments such as pasteurization. Although the air temperature employed in a drier may be very high, the temperature experienced by the organisms in the wet product is reduced due to evaporative cooling.

As drying proceeds and the production temperature increases, so too does the heat resistance of the organisms due to the low water content. This can be seen for example in the differences between spray dried milk and drum dried milk. In spray drying, the milk is pre-concentrated to about 40–45 per cent solids before being sprayed into a stream of air heated to temperatures up to 260°C at the top of a tower. The droplets dry very rapidly and fall to the base of the tower where they are collected. In drum drying, the milk is spread on the surface of slowly rotating metal drums which are heated inside by steam to a temperature of about 150°C. The film dries as the drum rotates and is scraped off as a continuous sheet by a fixed blade close to the surface of the drum. Although it uses a lower temperature, drum drying gives greater lethality since the milk is not subject to the same degree of pre-concentration used with spray-dried milk and the product spends longer at high temperatures in a wet state. Spray drying is however now widely used for milk drying because it produces a whiter product which is easier to reconstitute and has less of a cooked flavour. Milk is pasteurized before drying although there are opportunities for contamination during intervening stages. Most of the organisms which survive drying are thermoduric but Gram-negatives may survive and have on occasion been the cause of food poisoning outbreaks.

The limited lethality of drying processes and the long storage life of dried products means that manufacturers are not exempted from the stringent hygiene requirements of other aspects of food processing. Good quality raw materials and hygienic handling prior to drying are essential. Outbreaks of *Staph aureus* food poisoning have been caused by dried foods which were stored at growth temperatures for too long prior to drying allowing the production of heat resistant toxin which persisted through to the final product. The dried product must also be protected from moisture by correct packaging and storage in a suitable environment otherwise pockets of relatively high a_w may be created where microbial growth can occur.

Now let us study about the microbial activity in butter and milk.

Butter is an emulsion of water droplets in a continuous fat phase in contrast to milk which is an emulsion of fat globules in a continuous water phase. It has a higher fat content than milk (80 per cent) and uses pasteurized cream as its starting point. Typically, the cream is pasteurized using an HTST process of 85°C for 15 sec. and held at 4–5°C for a period to allow the fat globules to harden and cluster together. In making a conventional ripened cream butter, the starter culture is added at this stage and the cream incubated at around 20°C to allow flavour production to take place, A more recent method developed at NIZO, the Dutch Dairy Research Institute, employs a concentrated starter added to sweet-cream butter after manufacture. Phase inversion, the conversion from a fat-in-water emulsion to a water-in-fat emulsion, is achieved by the process of churning. During this process fat globules coalesce, granules of butter separate out and considerable amounts of water are lost from the product in the form of buttermilk. The buttermilk phase retains most of the microorganisms from the cream and numbers may show an apparent increase due to the breaking of bacterial clumps.

Traditional farmhouse butter making used wooden butter churns and these were originally scaled up for the earliest commercial butter making. However, the impossibility of effectively cleaning and sanitizing wood has led to its replacement by churns made of stainless steel or aluminium-magnesium alloys. After the butter has formed, the buttermilk is drained off, the butter grains washed with water and, in the case of sweet-cream butter, salt is added usually at a level of 1–2 per cent. The butter is then 'worked' to ensure further removal of moisture and an even distribution of water and salt throughout the fat phase. In properly produced butter the water is distributed as numerous droplets (> 10_{10} g^{-1}) mostly less than 10 μm in diameter. Since the butter should contain at most around 10_3 cfu g^{-1}, most of these droplets will be sterile. In those that do contain microorganisms, the nutrient supply will be severely limited by the size of the droplet. If the butter is salted, the salt will concentrate in the aqueous phase along with the bacteria which will therefore experience a higher, more inhibitory salt level. For example, bacteria in a butter containing 1 per cent salt and with a moisture content of 16 per cent would experience an effective salt concentration of 6.25 per cent.

Figure 4.10 indicates the processed followed in butter making.

Few microorganisms survive pasteurization so the microbiological quality of butter depends primarily on the hygienic conditions during subsequent processing, particularly the quality of the water used to wash the butter. Good microbiological quality starting materials are essential though, as preformed lipases can survive pasteurization and rapidly spoil the product during storage. Butter spoilage is most often due to the development of chemical rancidity but microbiological problems do also occur in the form of cheesy, putrid or fruity odours or the rancid flavour of butyric acid produced by butterfat hydrolysis. *Pseudomonas* are the most frequently implicated cause and are thought to be introduced mainly in the wash water. Psychrotrophic yeasts and moulds can also cause lipolytic spoilage and these are best controlled by maintaining low humidity and good air quality in the production environment and by ensuring the good hygienic quality of packaging materials. In this respect aluminium foil wrappers are preferred to oxygen-permeable parchment wrappers as they will help discourage surface mould growth.

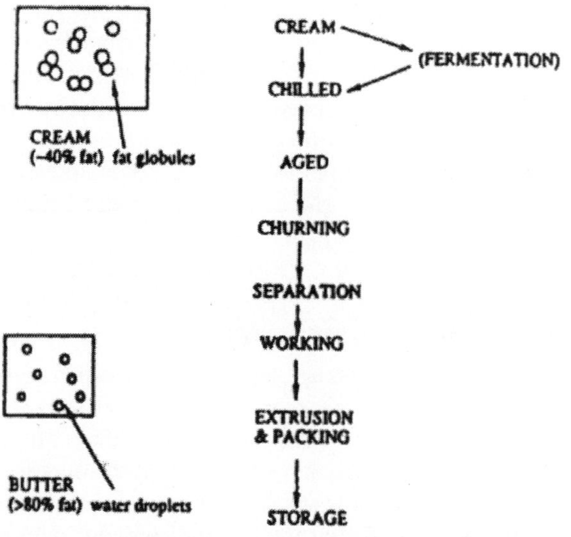

Figure 4.10: Butter Making.

Margarine relies on a similar compartmentalization for its microbiological stability, but uses vegetable fat as its continuous phase. Although skim milk is often included in the formulation, it is possible to make the aqueous phase in margarine even more deficient nutritionally than in butter, thus increasing the microbiological stability further. With the move towards low fat spreads containing 40 per cent fat, the efficacy of this system is more likely to breakdown. A higher moisture content means that the preservative effect of salt or lactic acid, which is often included, is diluted and that microorganisms can grow to a greater extent in the larger aqueous droplets.

From various research studies conducted recently, it can be concluded that water content and a_w have a close relationship with the quality (flavour, colour, taste and so on) of IMF. Traditional methods have some limitations such as time consuming and lower efficiency. The quality of the final products is poor. In order to produce high quality IMF as well as achieve high energy and drying efficiency, more attention must be paid to innovative technologies and novel combined dewatering methods. Combined technologies play important roles in reducing water of IMF as they can make the best use of each drying method and overcome the disadvantages of drying methods alone. Nowadays, the consumer's demand for IMFs is expanding, to which combined drying methods for energy saving and products with high quality have been applied. As for the preservation of IMF, novel microbial reduction, packing methods and the combination of multiple hurdles should be considered. Non-thermal methods have attracted considerable attention recently for maintaining a balance between quality and shelf life of IMF. But some technologies are uncommon and the cost are high, and the inactivation mechanisms for pathogens are unclear. More systematic research is required to fully utilize the potential of the latest novel technologies for preserving IMF.

Chilled Foods

The foods stored at temperatures near, but above their freezing point typically 0–5°C, are called chilled foods.

This commodity area has shown a massive increase in recent years as traditional chilled products such as fresh meat and fish and dairy products have been joined by a huge variety of new products including complete meals, prepared and delicatessen salads, dairy desserts and many others.

Low temperatures can also cause physiological changes in microorganisms that modify or exacerbate spoilage characteristics. Two such examples are the increased production of phenazine and carotenoid pigments in some organisms at low temperatures and the stimulation of extracellular polysaccharide production in *Leuconostoc* spp. and some other lactic acid bacteria. In most cases, such changes probably represent a disturbance of metabolism due to the differing thermal coefficients and activation energies of the numerous chemical reactions that comprise microbial metabolism.

The ability of organisms to grow at low temperatures appears to be particularly associated with the composition and architecture of the plasma membrane. As the temperature is lowered, the plasma membrane undergoes a phase transition from a liquid crystalline state to a rigid gel in which solute transport is severely limited. The temperature of this transition is lower in psychrotrophs and psychrophiles largely as a result of higher levels of unsaturated and short chain fatty acids in their membrane lipids. If some organisms are allowed to adapt to growth at lower temperatures, they increase the proportion of these components in their membranes.

Though mesophiles cannot grow at chill temperatures, they are not necessarily killed. Chilling will produce a phenomenon known as cold shock which causes death and injury in a proportion of the population but its effects are not predictable in the same way as heat processing. The extent of cold shock depends on a number of factors such as the organism (Gram-negatives appear more susceptible than Gram-positives), its phase of growth (exponential-phase cells are more susceptible than stationary phase cells), the temperature differential and the rate of cooling (in both cases the larger it is, the greater the damage) and the growth medium (cells grown in complex media are more resistant).

The principal mechanism of cold shock appears to be damage to membranes caused by phase changes in the membrane lipids which create hydrophilic pores through which cytoplasmic contents can leak out. An increase in single-strand breaks in DNA has also been noted as well as the synthesis of specific cold-shock proteins.

Since chilling is not a bacteriocidal process, the use of good microbiological quality raw materials and hygienic handling are key requirements for the production of safe chill foods. Mesophiles that survive cooling, albeit in an injured state, can persist in the food for extended periods and may recover and resume growth should conditions later become favourable. Thus, chilling will prevent an increase in the risk from mesophilic pathogens, but will not assure its elimination. There are however pathogens that will continue to grow at some chill temperatures and the key role of chilling in the modern food industry has focused particular attention on these. Risks posed by these organisms may increase with duration of storage but this process is likely to be slow and dependent on the precise storage temperature and composition of the food.

Some foods are not amenable to chill storage as they suffer from cold injury where the low temperature results in tissue breakdown which leads to visual defects and accelerated microbiological deterioration. Tropical fruits are particularly susceptible to this form of damage.

We have studied about chilling of foods. Now we can also deal with freezing process.

As with chilling, freezing will not render an unsafe product safe–its microbial lethality is limited and preformed toxins will persist. Frozen chickens are, after all, an important source of *Salmonella*.

Survival rates after freezing will depend on the precise conditions of freezing, the nature of the food material and the composition of its microflora, but have been variously recorded as between 5 and 70 per cent. Bacterial spores are virtually unaffected by freezing, most vegetative Gram-positive bacteria are relatively resistant and Gram-negatives show the greatest sensitivity. While frozen storage does reliably inactivate higher organisms such as pathogenic protozoa and parasitic worms, food materials often act as cryoprotectants for bacteria so that bacterial pathogens may survive for long periods in the frozen state. In one extreme example *Salmonella* has been successfully isolated from ice cream stored at -23°C for 7 years.

The extent of microbial death is also determined by the rate of cooling. Maximum lethality is seen with slow freezing where, although there is little or no cold shock experienced by the organisms, exposure to high solute concentrations is prolonged. Survival is greater with rapid freezing where exposure to these conditions is minimized. Food freezing processes are not designed however to maximize microbial lethality but to minimize loss of product quality. Formation of large ice crystals and prolonged exposure to high osmotic pressure solutions during slow cooling also damage cells of the food material itself causing greater drip loss and textural deterioration on thawing, so fast freezing in which the product is at storage temperature within half an hour is the method of choice commercially. The rate of freezing in domestic freezers is much slower so, although microbial lethality may be greater, so too is product quality loss.

Foods begin to freeze somewhere in the range –0.5 to –3°C, the freezing point being lower than that of pure water due to the solutes present. As water is converted to ice during freezing, the concentration of solutes in the unfrozen water increases, decreasing its freezing point still further so that even at very low temperatures, *e.g.* –60°C, some water will remain unfrozen. The temperatures used in frozen storage are generally less than –18°C.

Slow thawing at lower temperature is generally preferred. It does have some lethal effect as microbial cells experience adverse conditions in the 0 to –10°C range for longer, but it will also allow psychrotrophs to grow. Provided the product is not subject to contamination after thawing, the microflora that develops will differ from that on the fresh material due to the selective lethal effect of freezing. Lactic acid bacteria are often responsible for the spoilage of defrosted vegetables whereas they generally comprise only about 1 per cent of the microflora on fresh chilled produce which is predominantly Gram-negative.

Freezing and defrosting may make some foods more susceptible to microbiological attack due to destruction of antimicrobial barriers in the product and condensation, but defrosted foods do not spoil more rapidly than those that have not been frozen. Injunctions against refreezing defrosted products are motivated by the loss of textural and other qualities rather than any microbiological risk that is posed.

Freezer burn is another non-microbiological quality defect that may arise in frozen foods, where surface discolouration occurs due to sublimation of water from the product and its transfer to colder surfaces in the freezer.

This can be prevented by wrapping products in a water-impermeable material or by glazing with a layer of ice.

Low temperature is not the only inhibitory factor operating in frozen foods; they also have a low water activity produced by removal of water in the form of ice. As far as microbiological quality is concerned, this effect is only significant when frozen foods are stored at temperatures where microbial growth is possible (above –10°C). In this situation, the organisms that grow on a product are not those normally associated with its spoilage at chill temperatures but yeasts and moulds that are both psychrotrophic and tolerant of reduced water activity. Thus, meat and poultry stored at –5 to –10°C may slowly develop surface defects such as black spots due to the growth of the mould *Cladosporiwn herbarum*, white spots caused by *Sporotrichum carnis* or the feathery growth of *Thamnidium elegans.*

Microorganisms are affected by each phase of the freezing process.

Modification of Atmosphere Surrounding

Modified atmospheres exert their effect principally through the inhibition of fast-growing aerobes that would otherwise quickly spoil perishable products. Obligate and facultative anaerobes such as *Clostridia* and the *Enterobacteriaceae* are less affected. Thus, keeping quality is improved but there is generally little effect on pathogens, if present and the technique is invariably applied in conjunction with refrigerated storage.

Packaging is needed to maintain the integrity and quality of IMF during distribution and in storage. Additional benefits can be added by carefully selecting appropriate packaging technologies. In practice three different procedures are used to modify the atmosphere surrounding a product: vacuum packing, modified-atmosphere packing or gas flushing and controlled atmospheres. Here we will, discuss some of their important characteristics although other aspects will be dealt with under specific commodities.

An essential feature of all three techniques is that the product is packed in a material which helps exclude atmospheric oxygen and retain moisture. This requires that it should have good barrier properties towards oxygen and water and be easily sealed. The packaging materials used are usually plastic laminates in which the innermost layer is a plastic such as polyethylene which has good heat sealing properties. Mechanical closures on packs are far less effective as they often leave channels through which high rates or gas exchange can occur. Overlying the layer or polythene is usually another layer with much better gas barrier properties. No plastics are completely impermeable to gases, although the extent of gas transmission across a plastic film will depend on the type or plastic, its temperature, the film thickness and the partial pressure difference across the film. In some cases, it can also be affected by factors such as humidity and the presence of fat. Polyvinylidene chloride, PVDC, is a material commonly used as a gas barrier; the oxygen permeability of a 25µm thick film is 10 $cm_3\, m-_2 (24h)^{-1}$ atm^{-1} compared with values of 8500 and 1840 for low density and high-density polythene, respectively. Higher rates of transfer occur with CO_2 for which the permeability values are about five times those for oxygen. If a film is required to exclude oxygen transfer completely, then a non-plastic material such as aluminium foil must be included. This is seen for example in the bags used to pack wines. In addition to the sealing- and gas barrier-layers, laminates may also contain an outer layer such as nylon which gives the pack greater resistance to damage.

In *vacuum packing* the product is placed in a bag from which the air is evacuated, causing the bag to collapse around the product before it is sealed. Residual oxygen in the pack is absorbed through chemical reactions with components in the product and any residual respiratory activity in the product and its microflora. To achieve the best results, it is important that the material to be packed has a shape that allows the packaging film to collapse on to the product surface entirely -without -pockets and without the product puncturing the film.

Vacuum packing has been used for some years for primal cuts of red meats. At chill temperatures, good quality meat in a vacuum pack will keep up to five times longer than aerobically stored meats. The aerobic microflora normally associated with the spoilage of conventionally stored meats is prevented from growing by the high levels of CO_2 which develop in the pack after sealing and the low oxygen tension. The microflora that develops is dominated by lactic acid bacteria which are metabolically less versatile than the Gram-negative aerobes, grow more slowly and reach a lower ultimate population.

In recent years vacuum packing has been increasingly used for retail packs of products such as cooked meats, fish and prepared salads. It has been used less often for retail packs of red meats since the meat acquires the purple colour of myoglobin in its unoxygenated form. This does not appeal to consumers even though oxygenation occurs very rapidly on opening a vacuum pack and the meat assumes the more familiar bright red, fresh meat appearance of oxymyoglobin. Cured meats, on the other hand, are often vacuum packed for display since the cured meat pigment nitrosomyoglobin is protected from oxidation by vacuum packing.

The expanding range of chilled foods stored under vacuum and the availability of vacuum packing equipment for small-scale catering and domestic use has prompted concern about increasing the risk from psychrotrophic *Clostridium botulinum*. A number of surveys have been conducted to determine the natural incidence of *C.*

botulinum in these products and the consensus is that it is very low. In one recent example, workers failed to isolate *C. botulinum* or detect toxin in more than 500 samples analysed. When they deliberately inoculated these products with *C. botulinum* spores and incubated at the abuse temperature of 10°C, only in the case of vacuum packed whole trout was toxin produced within the declared shelf-life of the product. Nevertheless, misuse of the technique does have the potential for increasing risk and a Government Committee has recommended that all manufacturers of vacuum packing machinery should include instructions alerting the user to the risks from organisms such as *C. botulinum*.

In a variant of vacuum packing, known as *cuisine sous-vide* processing, food is vacuum packed before being given a pasteurization treatment which gives it a longer shelf-life under chill storage. The technique was developed in the 1970s in France and is said to give an improved flavour, aroma and appearance. It is used for the manufacture of chilled ready meals for various branches of the catering industry and *sous-vide* meals are also available in the retail market in some European countries. They have been slow to appear on the UK market due to the lack of appropriate UK regulations and concern over their microbiological safety with respect to psychrotrophic *C. botulinum*. It has been recommended that *sous-vide* products with an intended shelf-life of longer than 10 days at < 3°C should receive a minimum heat process equivalent to 90°C for 10 minutes; 70°C for 100 minutes should be sufficient for products with shorter shelf-lives.

Table 4.11 indicates about the MAP gas mixtures which are used with foods.

Modified atmosphere packing (MAP) has attracted much attention recent years as a valid method to inhibit microbial growth and enzymatic spoilage. It has been widely used as an effective method to extend shelf-life of fresh produce, by limiting the exchange of respiratory gases in packages made of semi permeable plastic films. Oxygen (O_2), carbon dioxide (CO_2) and nitrogen (N_2) are most commonly used gases in MAP. CO_2 is an important component in the gas mixture because of its antibacterial activity. N_2 is an inert gas which does not have any antibacterial ability and is usually served as a filler gas in the gas mixture. MAP relies on the modification of the atmosphere inside the package to obtain a balance between gas transmission through package and gas consumption via respiratory process. In *modified atmosphere packing* (MAP) a bulk or retail pack is flushed through with a gas mixture usually containing some combination of carbon dioxide, oxygen and nitrogen. The composition of the gas atmosphere changes during storage as a result of product and microbial respiration, dissolution of CO_2 into the aqueous phase and the different rates of gas exchange across the packing membrane. These changes can be reduced by increasing the ratio of pack volume to product mass although this is not often practicable for other reasons.

Table 4.11: MAP Gas Mixtures Used with Foods

Product	*Per cent CO_2*	*Per cent O_2*	*Per cent N_2*
Fresh meat	30	30	40
	15–40	60–85	–
Cured meat	20–50	0	50–80
Sliced cooked roast beef	75	10	15
Eggs	20	0	80
	0	0	100
Poultry	25–30	0	70–75
	60–75	5–10	> 20
	100	0	0
	20–40	60–80	0
Pork	20	80	0
Processed meat	0	0	100
Fish (white)	40	30	30
Fish (oily)	40	0	60
	60	0	40
Cheese (hard)	0–70		30–100
Cheese	0	0	100
Cheese; grated/sliced	30	0	70
Sandwiches	20–100	0–10	0–100
Pasta	0	0	100
	70–80	0	20–30
Bakery	0	0	100
	100	0	0

Active Packaging (AP) is another effective packaging method for the preservation of IMF by adding a chemical compound to the packaging materials. The most commonly used chemical is O_2 absorber which can change the headspace gas composition and absorb the O_2 to < 0.01 per cent within 1–4 days at room temperature.

In *controlled-atmosphere storage* (CAP), the product environment is maintained constant throughout storage. It is used mainly for bulk storage and transport of foods, particularly fruits and vegetables such as the hard cabbages used for coleslaw manufacture. CAP is used for shipment of chilled lamb carcasses and primal cuts which are packed in an aluminium foil laminate bag under an atmosphere or 100 per cent CO_2. It is more commonly encountered though with fruits such as apples and pears which are often stored at sub-ambient temperatures in atmospheres containing around 10 per cent CO_2.

Chapter 5

Microorganisms in Meat and Meat Products

Introduction

In general, the numbers and types of microorganisms present in a finished food product are influenced by the following:

1. The general environment from which the food was originally obtained,
2. The microbiological quality of the food in its raw or unprocessed state,
3. The sanitary conditions under which the product is handled and processed and
4. The adequacy of subsequent packaging, handling and storage conditions in maintaining the flora at a low level.

In producing good-quality market foods, it is important to keep microorganisms at a low level for reasons of aesthetics, public health and product shelf-life. Other than those foods that have been made sterile, all foods should be expected to contain a certain number of microorganisms of one type or another. Ideally, the numbers of organisms should be as low as is possible under good conditions of production. Excessively high numbers of microorganisms in fresh foods present cause for alarm. It should be kept in mind that the inner parts of healthy plant and animal tissues are generally sterile and that it is theoretically possible to produce many foods free of microorganisms. This objective becomes impractical, however, when mass production and other economic considerations are realized. The number of microorganisms in a fresh food product, then, may be taken to reflect the overall conditions of raw product quality, processing, handling, storage and so forth. This is one of the most important uses made of the standard plate count in food microbiology. The question immediately arises as to the attainability of low numbers under the best and most economic production conditions known. With few exceptions, it is difficult to know what is the lowest number of microorganisms attainable under good production conditions because of the many variables that must be considered. With advances in modern technology, it has been possible to reduce the microbial load in a large number of foods over what was possible even 10 years ago.

The microbiology of various groups of foods is treated below. With each group, the sources of organisms are listed and discussed, along with their incidence as reported by various workers. It should be noted that the numbers reported for a particular product by different investigators do not always reflect the microbiology of that product under ideal conditions of production, handling and storage. The chapters on food spoilage and food poisoning should be consulted for additional information on the types of organisms associated with the foods discussed.

The microbial types and numbers on the meat can be affected by the pre slaughter conditions of the animals. Pathogenic microbial species contaminating meat from the gastrointestinal tract are *Salmonella* enteric strains, *Yersinia enterocolitica*, *Campylobacter jejuni*, *Aeromonas hydrophila*, *Listeria monocytogenes*, and *Escherichia coli*. Fresh meat storage in refrigerator is spoiled by psychrotrophic aerobes and facultative anaerobes. Meat spoilage results in the development of off-odour known as putrefaction. The vacuum- and gas-packaged meat and its products can be spoiled by psychrotrophic *Lactobacillus*. Spoilage of meat products occurs generally in three types: sliminess, souring, and greening. Biogenic amines are formed by decarboxylation of amino acids during manufacturing and storage periods of sausage through the action of substrate-specific enzymes produced by microorganisms: *Bacillaceae*, *Enterobacteriaceae*, *Lactobacillus*, *Pediococcus*, and *Streptococcus*. Poultry is very high on the list of food poisoning outbreaks. Poultry is ideally suited as a culture medium for bacteria.

Microorganisms in Meat

Since antiquity, meat has been used as a source of food due to its nutritive value. A large number of the products, prepared from the meat of food animals, are consumed worldwide. Meat and meat products play a vital role in the maintenance of human health by providing all essential nutrients such as protein, vitamins, and minerals. It is pertinent to mention that about 30 per cent of zinc in our diet comes from meat and meat products. If meat and meat products are not processed appropriately, they will be spoiled quickly by a variety of microbes. In this context, it has been reported that protein and vitamins (especially A and B_{12}) in meat cannot be substituted by plant sources. In India, approximately, 12,000 tons of frozen products are produced and consumed annually. Today, large population of the world depends on meat and meat products as a source of food (Pal *et al.,* 2018). On the contrary, meat and meat products can act as an important vehicle for many microbes, which can either cause spoilage or food poisoning. Microbial spoilage of food can result from the failure or inability to control organisms at one or more stages of food production. The microbes are ubiquitous in distribution and hence, contamination can occur at any stage of food chain originating from farm to fork.

Three major types of microorganisms are generally present:

1. Molds
2. Yeast
3. Bacteria: They are the major source of contamination

Type and number of microbes contribute to the rate of meat spoilage. Organisms which eventually grow and cause spoilage are those which find existing conditions favourable.

Many different organisms may be present, usually one or maybe up to three multiply rapidly enough to cause spoilage. Bacteria grow faster than yeast and yeast outgrows mold.

Important genera found on meat during refrigerated storage are:

1. *Pseudomonas*
2. *Achromobacter*
3. *Micrococcus*
4. *Lactobacillus*
5. *Streptococcus*
6. *Leuconostoc*
7. *Pediococcus*
8. *Flavobacterium*
9. *Proteus*
10. *Salmonella*
11. *Listeria*
12. *E. Coli*
13. *Clostrium perfringens*
14. *Staphlococcus*
15. *Coliforms*

Deterioration of Meat

Meat spoilage is a metabolic process where the change in sensory characteristics deems it to be unacceptable for human consumption. The process of meat spoilage is often rapid, especially when driven by bacteria, which can multiply to high densities within a short period of time; for example, the generation time for *Escherichia coli* is every 20 min. The growth of microbes in meat is governed by a number of intrinsic and extrinsic factors. Meat undergoes constant chemical and/or physical changes in all types of storage conditions. These changes are a combined effect of deterioration, which occurs naturally in meat, and degradation caused by the storage environment. Both factors can lead to the development of oxidative rancidity and microbial deterioration, which are detrimental to the storage life of any meat product.

The process of meat spoilage is highly dependent on a number of environmental factors and conditions, as each microorganism may require niche survival conditions to optimize its growth, source its nutrition and reproduce. Altered conditions of nutrient availability, oxygen demand, temperature, and other competing organisms all affect microbial growth. Meat spoilage typically occurs under conditions of optimal water availability, low oxygen, and low temperature conditions, where growth conditions are ideal for microbial growth. Microbial spoilage can cause chemical and physical changes which alter the sensory characteristics of meat and meat product.

Chemical Changes

Accelerate lipid oxidation through *lipases*–lipid hydrolyzing enzymes secreted by microorganisms which hydrolyze triglycerides and phospholipids.

Chemical reactions that cause offensive sensory changes in foods are mediated by a variety of microbes that use food as a carbon and energy source. Microbial growth also can lead to rancidity. These organisms include prokaryotes (bacteria), single-celled organisms lacking defined nuclei and other organelles, and eukaryotes, single-celled (yeasts) and multicellular (molds) organisms with nuclei and other organelles. Lipolytic bacteria may cause some lipolysis and also may accelerate the oxidation of the fats. Rancidity of fats may be caused by lipolytic species of *Pseudomonas* and *Achromobacter* or by yeasts.

Oxidative rancidity is the result of a series of natural chemical reactions occurring in meat fat after exposure to air. The oxidation of unsaturated fats in meat takes place chemically in air and may be catalyzed by light and copper. The process of fat deterioration in the presence of oxygen eventually results in the formation of objectionable odours and flavours, which constitute spoilage of the product. Fat from certain animal species is more susceptible to oxidative rancidity. This is due primarily to differences in the physical composition and types of fats present in these species. For example, pork fat contains a higher proportion of unsaturated fatty acids that are more susceptible to oxidation as compared to beef or lamb fat. Consequently, pork has a shorter storage life. Cured meats have a greater potential for developing oxidative rancidity than fresh meats because of the presence of such pro-oxidants as nitrite, sodium chloride, and copper, iron, manganese and cobalt impurities often found in the curing ingredients. The addition of antioxidants (such as BHA, BHT, propyl gallate or citric acid) during the curing process, however, tends to prolong the storage life of these products.

Physical Changes

1. Slime formation.
2. Colour changes.
3. Gas production–sulfur production–blown bags.
4. Souring.

Microbial deterioration results from growth of microorganisms introduced through the meat handling chain. If not controlled, microbial growth can severely deteriorate the nutritional quality, appearance, odour and flavour of meat. The development of a sticky meat surface, slime formation, discoloration and sour odours and flavours often are indications of microbial spoilage; these changes occur due to the breakdown of fat, protein and carbohydrates in meat by the microorganisms. *Lactobacillus* spp. and *Leuconostoc* spp. are almost the largest group which causes sensory changes, such as souring, the production of H_2S, gas and slime.

1. Changes Caused by Bacteria under Aerobic Condition

- **Surface slime:** Spoilage of meat due surface slime is produced by species of *Pseudomonas, Acinetobacter,* Moraxella, *Alcaligens, Micrococcus, Streptococcus, Leuconostoc,* and *Bacillus.* Some species of *lactobacillus* also showed to produce slime. The thin film formation on meat provided microbes' nutrients from the substrate.
- **Changes in colour of meat pigments:** The red colour of meat, called its 'bloom', may be changed to green, brown, or grey, as a result of the production of oxidizing compounds, *e.g.,* peroxides, or of hydrogen sulfide, by bacteria. Species of *Lactobacillus* (mostly heterofermentative) and *Leuconostoc* are reported to cause the greening of sausage.
- **Surface colour discolouration due to pigment forming bacteria**: Red spot may be caused by *Serratia marcescens* or other bacteria with red pigment. *Pseudomonas syncyanea* can impart blue colour to the surface; *Micrococcus* or *Flavobacterium* imparts yellow colour
- **Changes in fats:** Lipolytic bacteria may cause some lipolysis and also may accelerate the oxidation of the fats. Rancidity of fats may be caused by lipolytic species of *Pseudomonas* and *Achromobacter* or by yeasts.
- **Off odours and off tastes:** "Cold-storage flavour" or taint is an indefinite term for a stale flavour. Taints, or undesirable odours and tastes, which appear in meat as a result of the growth of bacteria on the surface often, are evident before other signs of spoilage. Souring of meat may be caused due to volatile acids, *e.g.,* formic, acetic, butyric, and propionic, or even to growth of yeasts. Actinomycetes may be responsible for a musty or earthy flavour.
- **Phosphorescence:** Phosphorescence, this uncommon defect is caused by phosphorescent or luminous bacteria, *e.g., Photobacterium* spp., growing on the surface of the meat. The red spot may be caused by *Serratia marcescens* or other bacteria with red pigments. *Pseudomonas syncyanea* can impart a blue colour to the surface. Yellow discolorations are caused by bacteria with yellow pigments, usually species of *Micrococcus* or *Flavobacterium. Chromobacterium lividum* and other bacteria give greenish-blue to brownish-black spots on stored beef. The purple 'stamping ink' discoloration of surface fat is caused by yellow-pigmented cocci and rods. When the fat becomes rancid and peroxides appear, the yellow colour changes to a greenish shade and later become purplish to blue.

Table 5.1: Changes in Fresh Meat Caused by Microorganisms

Condition	Kind of Defects	Microorganisms
Aerobic condition	Surface slime	*Pseudomonas, Moraxella, Streptococcus, Bacillus, micrococcus*
	The red color of meat called "bloom" caused by the production of an oxidizing compound	*Lactobacillus, Leuconostoc*
	Oxidative rancidity	*Pseudomanas* spp, *Archromobacter*
	Red spot	*Serratia marcescens,*
	Blue color	*Pseudomonas syncyanea*
	Greenish blue or brownish black spot	*Chromobacterium lividum*
	Stickiness, whiskers, Green patches	Mold
Anaerobic condition	Putrefaction	*Clostridium* spp, *Alcaligennes, Proteus*
	Souring	Lactic acid bacteria

2. Changes Caused by Molds under Aerobic Conditions

- **Stickiness:** Budding growth of molds causes the surface of the meat sticky to the touch known as stickiness. Such types of spoilage result the unusual appearance of meat.
- **Black spot:** This type of spot is commonly caused by *Cladosporium herbarum*, but other molds with dark pigments may also be responsible.
- **White spot:** *Sporotrichum carnis* is the most common cause of white spot, although any mold with wet, yeast like colonies, *e.g.*, *Geotrichum*, can cause white spot.
- **Green patches:** *Penicillium* such as *P. expansum*, *P. asperulum*, and *P. oxalicum* are responsible for the green patches on meat.
- **Decomposition of fats:** Hydrolysis of fats due to lipase produced by molds causes the oxidation of fats.
- **Off odours and off tastes:** Musty flavour to meat is caused by mold due to their growth. Sometimes the defect is given a name indicating the cause, *e.g.*, 'thamnidium taint'.
- **Whiskers:** When meat is stored in frozen condition, mycelial growth may take place without sporulation.
- **Spots of surface:** Under aerobic conditions, yeasts may grow on the surface of meats, causing sliminess, lipolysis, off odours and tastes, and discolorations-white, cream, pink, or brown-due to pigments in the yeasts. Spoilage by yeasts and molds usually are localized to a great extent and can be trimmed off without harm to the rest of the meat.

3. Changes Caused by Bacteria under Anaerobic Conditions

Facultative and anaerobic bacteria are able to grow within the meat under anaerobic conditions and cause spoilage. Few are listed below:

- **Souring:** Souring of meat caused by formic, acetic, butyric, propionic, and higher fatty acids or other organic acids such as lactic or succinic is produced by microbes under anaerobic conditions. Souring can result from either from action of the meat's own enzymes during aging or ripening, anaerobic production of fatty acids or lactic acid by bacterial action, or proteolysis without putrefaction, caused by facultative or anaerobic bacteria and sometimes called 'stinking sour fermentation'. Acid and gas formation caused by *Clostridium* species and the *Coliform* bacteria on carbohydrates. Vacuum packed meats, especially those in gastight wrappers, commonly support the growth of lactic acid bacteria.
- **Putrefaction:** Decomposition of protein is known as putrefaction; it is produced by anaerobic microbes with foul-smelling compounds such as hydrogen sulfide, mercaptans, indole, skatole, ammonia, and amines. Generally, *Clostridium* spp. is responsible microbe, but facultative bacteria may also cause putrefaction or assist in its production. *Pseudomonas*, *Proteus*, *Clostridium* and *Alcaligenes* cause putrefaction in meat and meat products. The putrefaction caused by *Clostridia* is usually accompanied by gas formation.
- **Taint:** "Taints" or undesirable odours and tastes, that appear in meat.

Microbial Spoilage of Meat

Common type of microbial spoilage of meat can be classified on the basis of whether they occur under aerobic or anaerobic condition and whether they are caused by bacteria, yeast, or mold.

1. Spoilage of Meat under Aerobic Conditions

i) Bacterial Spoilage of Meat

Surface spoilage

- It is caused by *Pseudomonas, Acenatobacter, Streptococcus, Leuconostoc, Bacillus* and *Micrococcus.*

- ☆ Temperature and available moisture influence type of microorganisms causing slime.

Change in Colour of Meat

- ☆ Red color of meat may be changed into green brown or grey due to production of oxidising agent, H_2S, *etc.* by microorganisms. For example, *Lactobacillus* and *Leuconostoc* cause greening of sausage.

Change in Fat

- ☆ Fat of meat may become rancid due to lipase producing microorganisms such as *Pseudomonas* and *Achromobacter.*

Surface Colour Due to Pigmented Bacteria

- ☆ *Serratia marcescens* give red spots.
- ☆ *Pseudomonas syncyanea* give blue colour, *Chromobacterium lividum* gives greenish blue to brownish black color, *Flavobacterium* give yellow colour.

Phosphorescence

- ☆ It is caused by luminous bacteria *e.g. Photobacterium* growing on surface of meat.

Off Odours and Off Taste

- ☆ Undesirable odour and taste called taint are caused by many bacteria due to production of volatile acids such as formic acid, acetic acid, butyric acid *etc.*
- ☆ Actinomycetes give musty or earthy flavour.

ii) Fungal Spoilage of Meat

Stickiness

- ☆ Many molds grow on surface of meat and make it sticky to touch.
- ☆ Whiskers: when meat is kept at temperature near freezing, molds grow slowly without sporulation on surface producing while cottony growth.
- ☆ It may be caused by *Thamnidium, Mucor mucedo, Mucor racemosus etc.*

Black Spot

- ☆ It is caused by *Cladosporium herbarum.*

White Spot

- ☆ It is caused by *Sporotrichum carnis.*

Green Spot

- ☆ It is caused by *Penicillium* species.

Change in Fat

- ☆ Many molds produce lipase and cause hydrolytic rancidity of fat.

Off Odour and Off Taste

- ☆ Many molds give musty flavour to meat in the vicinity of their growth.

By Yeast

- ☆ Under aerobic condition, yeast grow on surface of meat causing sliminess, rancidity of fat, off odour and taste and discolorations like white, pink, brown spots.

2. Spoilage of Meat under Anaerobic Conditions

In anaerobic condition, anaerobic or facultative anaerobic bacteria spoil the meat.

Souring

- ☆ It is caused by formic acid, acetic acid, butyric acid, propionic acid, higher fatty acids and other organic acids. *e.g.*, lactic acid produced by bacteria. Souring may also be caused by foods own enzyme.

Putrefaction

- ☆ It refers to the anaerobic decomposition of protein with production of offensive smelling compounds such as H_2S, mercapatans, indole, skatole *etc.* It is usually caused by *Clostridium* species but species of *Pseudomonas proteus* and Alkaligens may cause putrefaction.

Taint

- ☆ It refers to any undesirable odour or taste.

Sources of Microbial Contamination in Meat and Meat Products

Microbial contaminants rather common than any other form of contaminants as food animals itself harbour them. Microbial status of fresh meat depends on animal rearing, transportation, slaughtering and cutting and packaging, besides hygiene and processing conditions of the slaughter plant. The natural surface flora of meat animals usually is not important as the contaminating microorganisms from their intestinal or respiratory tracts. However, hides, hooves and hair contain not only large numbers of microorganisms from soil, manure, feed and water but also important kinds of spoilage organisms. The skin of many meat animals may contain *Micrococci*, *Staphylococci* and *Streptococci*. *Staphylococci* on the skin or from the respiratory tract may find their way onto the carcass and then to the final raw product. The faeces and faecal contaminated products of animals can contain many enteric organisms including *Salmonella*. People working in meat processing plants also can act as vector of many foodborne **pathogenic bacteria**, but this represents only little importance.

Body controls microbial numbers through defensive mechanisms of white blood cells and antibody defenses. Upon exanguination, these mechanisms are lost. Hence, major points of meat contamination are:

1. Non-sterile knives used for exsanguination.
2. Head, leg, hide and offal removal during slaughter.
3. Water sources for cleaning.
4. Surface contact during storage.
5. Contact during fabrication.
6. Contact during handling and processing.
7. Clothing and hands of personnel.
8. Air
9. Water
10. Walls and doors

The transformation of animals into meat involves several operations such as (a) handling and loading of animals on the farm, (b) transporting animals to slaughterhouses, (c) off-loading and holding of animals and (d) slaughtering of animals. Some of the pre slaughter handling that influence the spoilage of meat includes nutrition, transportation, marketing, lair aging and stunning. Poor operational techniques and facilities in any of these operations will result in unnecessary suffering and injuries to animals which can lead to loss of meat, reduced meat quality and spoilage of meat. Therefore, prevention of contamination after slaughtering during meat cutting and processing is essential. Storage time can be extended through hygienic slaughtering and clean handling of the carcass.

The main causes of meat and meat products spoilage after slaughtering and during processing and storage are microorganisms, lipid oxidation and autolytic enzymatic spoilage. Different technical operations are involved in slaughtering *viz.*, stunning, bleeding, skinning, evisceration and carcass splitting. Inadequacy at one stage will result in a rigorous negative impact on the product and/or process in the following stage. In addition to the hygiene and storage temperature, the acidity of the meat and the structure of the muscular tissue also affect the rate of meat spoilage. For example, liver will spoil faster than the firm muscular tissue of beef. After few hours of slaughtering of animals, muscles become firm and rigid, a condition known as rigor mortis occurs. The process of rigor mortis depends on the stress induced on the animals during the slaughtering process. Raw meat quality is reported to be severely affected by the stress conditions during slaughtering process and the slaughtering methods.

Contamination of Meat

- The healthy inner flesh of the meat contains few or no micro-organisms although they have been found in lymph node and bone marrow.
- Upon, death of animal, invasion of tissue by contaminating microorganisms takes place.
- In meat, microorganisms come from external source or from meat animal itself.
- Factors that influence invasion of meat by contaminating microorganisms are:
    - The load of organism in intestine. The greater the load, the greater will be the invasion of meat tissue. Therefore, starvation of animal for 24hrs before slaughter has been recommended.
    - The physiological condition of the animal immediately before slaughter:
        - The method of killing or bleeding: The better and more sanitary bleeding method contributes less contamination.
        - The rate of cooling: The rapid cooling will reduce the rate of invasion of meat by contaminants.
- Microorganisms are spread in the meat through the blood and lymph vessel, connective tissue spaces and by grinding in ground meat.

Preslaughter handling of livestock and post slaughter handling of meat play an important part in deterioration of meat quality. The glycogen content of animal muscles is reduced when the animal is exposed to pre-slaughter stress which changes the pH of the meat, to higher or lower levels, depending on the production level of lactic acid. Lactic acid is produced due to the breakdown of glycogen content of animal muscles via an anaerobic glycolytic pathway as shown in Fig.5.1. Higher levels of pH (6.4-6.8) result in Dark, Firm and Dry (DFD) meat. Long term stress causes DFD meat which has a shorter shelf life. Severe short term stress results in a Pale, Soft and Exudative (PSE) meat. PSE meat has a pH lower than normal ultimate value of 6.2 which is responsible for the breakdown of proteins, providing a favourable medium for the growth of bacteria (Dave and Ghaly, 2011).

The intrinsic and extrinsic factors affecting the shelf life of meat and meat products are summarized in Table 5.2.

Table 5.2: Factors Affecting the Shelf-Life of Meat

Intrinsic Factors	*Extrinsic Factors*
Species, breed, age and feeding of the animal of origin	Quality management system
Initial microbiota	Packaging system
Chemical properties (pH, a_w, redox potential, peroxide value)	Temperature control
Product composition	Processing conditions and hygiene
Antimicrobial components	Storage types

Intrinsic Factors	Extrinsic Factors
Biopreservation systems (bacteriocinogenic LAB cultures and/or their bacteriocin)	Relative humidity Atmospheric gas composition and ratio

Based on (McDonald and Sun, 1999; Dave and Ghaly, 2011; Cenci-Goga, 2012; Kalschne *et al.,* 2014).

There are three main mechanisms for meat and meat products spoilage after slaughtering and during processing and storage:

(a) Microbial spoilage,

(b) Lipid oxidation and

(c) Autolytic enzymatic spoilage.

Meat is a nutritious, protein-rich food which is highly perishable and has a short shelf-life unless preservation methods are used. Fat, protein, minerals, carbohydrate and water are the constituents of meat; the quality of meat and meat products degrade as a result of digestive enzymes, microbial spoilage and fat oxidation. Lipid oxidation, protein degradation and the loss of other valuable molecules are the consequence of meat spoilage process. Proteins and lipids can break down resulting in the production of new compounds causing changes in meat flavour, tenderness, juiciness, odour and texture. It is therefore, important to understand the causes of spoilage of meat and meat product in order to develop optimum preservation techniques to maintain the freshness of these food products.

Microbial Spoilage

Meat and meat products provide excellent growth media for a variety of microflora (bacteria, yeasts and molds) some of which are pathogens. The intestinal tract and the skin of the animal are the main sources of these microorganisms; more the load of mos, greater the invasion of tissues. As already explained above, the composition of microflora in meat depends on various factors: (a) preslaughter husbandry practices (free range Vs intensive rearing), (b) age of the animal at the time of slaughtering, (c) handling during slaughtering, evisceration and processing, (d) temperature controls during slaughtering, processing and distribution (e) preservation methods, (f) type of packaging and (g) handling and storage by consumer (Cerveny *et al.,* 2009). All these sources affect the initial microbial load which is a major factor in determining shelf-life.

Table 5.3 and 5.4 present the major genera of bacteria, yeasts and molds found in meat and poultry products before spoilage.

Table 5.3: Genera of Bacteria most Frequently found on Meats and Poultry

Genus	Gram Reaction	Fresh Meats	Fresh Livers	Poultry
Acinetobacter	–	xx	x	xx
Aeromonas	–	xx		x
Alcaligenes	-	x	x	x
Arcobacter	-	x		
Bacillus	+	x		x
Brochothrix	+	x	x	x
Campylobacter	–			xx
Carnobacterium	+	x		
Caseobacter	+	x		
Citrobacter	–	x		x
Clostridium	+	x		x
Corynebacterium	+	x	x	xx
Enterobacter	–	x		x
Enterococcus	+	xx	x	x
Erysipelothrix	+	x		x
Escherichia	-	x	x	
Flavobacterium	-	x	x	x
Hafnia	-	x		
Kocuria	+	x	x	x
Kurthia	+	x		
Lactobacillus	+	x		
Lactococcus	+	x		
Leuconostoc	+	x	x	
Listeria	+	x		xx
Microbacterium	+	x		x
Micrococcus	+	xx	xx	xx
Moraxella	-	xx	x	xx
Paenibacillus	+	x		x
Pantoea	-	x		x
Pediococcus	+	x		
Proteus	-	x		x
Pseudomonas	-	xx		xx
Psychrobacter	-	xx		x
Salmonella	-	x		x
Serratia	-	x		x
Shewanella	-	x		
Staphylococcus	+	x	x	x
Vagococcus	+			xx
Weissella	+	x	x	
Yersinia	-	x		

x: Known to occur; xx: Most frequently reported.

Source: Jay *et al.,* 2005.

Table 5.4: Genera of Mold and Yeast most Often found on Meats and Poultry

Genus	Fresh and Refrigerated Meats	Poultry	Genus	Fresh and Refrigerated Meats	Poultry
Molds					
Alternaria	x	x	*Aspergillus*	x	x
Aureobasidium	x		*Cladosporium*	xx	x
Eurotium	x		*Fusarium*	x	
Geotrichum	xx	x	*Monascus*	x	
Monilia	x		*Mucor*	xx	x
Neurospora	x		*Penicillium*	x	x
Rhizopus	xx	x	*Sporotrichum*	xx	
Thamnidium	xx				
Yeasts					
Candida	xx	xx	*Cryptococcus*	x	x
Debaryomyces	x	xx	*Hansenula*	x	
Pichia	x	x	*Rhodotorula*	x	xx
Saccharomyces		x	*Torulopsis*	xx	x
Trichosporon	x	x	*Yarrowia*		xx

Source: Jay *et al.*, 2005.

Mold species include *Cladosporium*, *Sporotrichum*, *Geotrichum*, *Penicillium* and *Mucor* while yeasts species include *Candida* spp., *Cryptococcus* spp. and *Rhodotorula* spp., While studying the microbial quality of some meat products obtained from local markets in Egypt, researchers reported many types of fungi belonging to several genera such as *Aspergillus*, *Candida*, *Cladosporium*, *Eupenicillium*, *Eurotium*, *Geotrichum*, *Mucor*, *Penicillium*, *Rhototorula* besides aflatoxin B1. These researchers also isolated *Clostridium perfringens* and *Staphylococcus aureus*.

Pathogenic bacteria including *Escherichia coli*, *Listeria monocytogenes*, and *Staphylococcus aureus* may be present in meat products including poultry, pork, lamb, mutton, and beef (Table 5.3). Bacteria species include *Pseudomonas*, *Micrococcus*, *Streptococcus*, *Sarcina*, *Lactobacillus*, *Salmonella*, *Escherichia*, *Clostridium* and *Bacillus*. *Compylobacter* spp., *Escherichia coli* and *Salmonella* spp. which are often present in fresh meat and poultry. According to Remenant and Zagorec (2018**)**, the dominant spoilage microorganisms on meat and meat products are *Firmicutes*, *Actinobacteria*, and *Proteobacteria*. Among these, *Brochothrix thermosphacta* and *Carnobacterium* spp. and other lactic acid bacteria including *Lactobacillus* and *Leuconostoc* are the most prevalent on both raw meat and packaged meat products, and form a large part of the core meat community on meats.

Comminuted meats such as ground beef invariably contain more numbers of microorganisms than non-comminuted meats such as steaks. This has been reported rather consistently for over 70 years and there are several reasons for the generally higher counts in these types of products than one finds on whole meats. First, commercial ground meats generally consist of trimmings from various cuts. These pieces have been handled excessively and consequently normally contain more microorganisms than meat cuts such as steaks. Second, ground meat provides a greater surface area, which itself accounts in part for the increased flora. It should be recalled that as particle size is reduced, the total surface area increases with a consequent increase in surface energy. Third, this greater surface area of ground meat favours the growth of aerobic bacteria, the usual low-temperature spoilage flora. Fourth, in some commercial establishments, the meat grinders, cutting knives and storage utensils are rarely cleaned as often and as thoroughly as is necessary to prevent the successive build-up of microbial numbers. This may be illustrated by data obtained by the author from a study of the bacteriology of several areas in the meat department of a large grocery store. The blade of the meat saw and the cutting block were swabbed immediately after they were cleaned on three different occasions with the following mean results. The saw blade had a total log/in_2 count of 5.28, with 2.3 coliforms, 3.64 enterococci, 1.60 staphylococci and 3.69 micrococci. The cutting block had a mean log/in_2 count of 5.69, with 2.04 coliforms, 3.77 enterococci, < 1.00 staphylococci and 3.79 micrococci. These are among the sources of the high total bacterial count to comminuted meats. Fifth, one heavily contaminated piece of meat is sufficient to contaminate others as well as the entire lot as they pass through the grinder. This heavily contaminated portion is often in the form of lymph nodes which are generally embedded in fat. These organs have been shown to contain high numbers of microorganisms and account in part for hamburger meat having a generally higher total count than ground beef.

Various foodborne pathogens such as *Escherichia coli* O157:H7, *Listeria* spp., *Salmonella enteritidis* and *Shigella* species have been reported from meat samples in retail meat shops, while microbiological examination of meat handling equipments in retail shops revealed *Staphylococcus* and *Shigella* spp.

L. monocytogenes occurs in fresh meat and meat products but human listeriosis is predominantly associated with ready-to-eat meat products that have a long refrigerated shelf life (Sofos, 2008). Presence of *L. monocytogenes* in food processing environments is specifically noted previously, with deli meats alone (particularly sliced and packaged) responsible for almost 83 per cent of all human listeriosis cases in the United States (Buchanan *et al.*, 2017).

S. aureus was detected in about 35.0 per cent of retail meat samples such as raw fresh meat, quick frozen meat, and ready-to-eat meat, which were collected from 39 cities all over China during 2011–2016 (Wu *et al.*, 2018); however, meat and meat products do not significantly contribute to foodborne intoxication with *Staphylococcal* enterotoxins. Soyiri and workers (2008) also recovered *Staphylococcus aureus*, *Bacillus cereus*, *Clostridium perfringens* and *Escherichia coli* from beef samples.

Lactic acid bacteria (LAB) also play an important role in the spoilage of refrigerated meats and compete with other spoilage organisms such as *Enterobacteriaceae* and *Pseudomonas* under appropriate conditions. The growth of lactic acid bacteria on bologna-type sausage was retarded 2 and 4-fold with decreases in temperature from 7-2°C and from 7-0.6°C, respectively (Borch *et al.,* 1996).

B. thermosphacta has been shown to be the most ubiquitous and abundant species in meats, and can grow on meats during both aerobiosis and anaerobiosis, making it an important member of the spoilage-related flora due to off-odour production (Ercolini *et al.,* 2006; Chaillou *et al.,* 2015).

The most frequently identified bacterial pathogens associated with consumption of beef products are *Salmonella* spp., *Bacillus cereus, Campylobacter* spp., *Clostridium perfringens, Staphylococcus aureus, Escherichia coli*, Listeria *monocytogenes*, *Yersinia enterocolitica* and *Vibrio parahaemolyticus* (Biswas *et al.,* 2011).

Carnobacterium has been found to predominate among the bacterial community on chilled vacuum-packaged beef during the whole storage life, and the species *Carnobacterium divergens* with more tolerant to acidic environment had a higher prevalence in vacuum-packaged beef compared to *Carnobacterium maltaromaticum* (Kaur *et al.*,2017; Zhang *et al.,* 2018). The diversity of the prevalence of spoilage microbiota in meats is ascribed to the different migration routes of samples or contamination ways during the collection from each area. The psychrotrophic spoilage microbiota of fresh meat and meat products was found to be predominantly derived from environmental, mainly water-borne organisms rather than the animal-derived microbiota from the skin and the gut of the animals (Chaillou *et al.,* 2015).

Escherichia coli, since its discovery by Theobald Escherich in 1885 has been receiving much greater importance due to its pathogenicity by certain strains both in man and animals. Worldwide contamination of this group of bacteria occurred in meat through soiling of the carcass and plant environment with faecal materials during slaughter process. The incidence of *E. coli* not very variable in domestic or export beef meat regardless the fat content in trimmings. The average *E. coli* Counts (ECC) in Indian buffalo meat were 1.1log cfu g^{-1}. In another study,153 buffalo meat samples were screened, among which 24.78 per cent samples were positive for *E. coli*. However, enumeration data regarding various groups of *E. coli* is sparse.

Prevalence rate of verotoxic *E. coli* in meat and meat products has been recorded at an alarming rate in India. In an Indian study, it has been reported that 89.19 per cent of 37 *E. coli* isolates were found verotoxic by vero-cell cytotoxicity assay. Similarly, *E. coli* strains isolated from different meat and meat products revealed 15.90 per cent isolates to be verotoxigenic. Hazarika *et al.* (2005) reported that 27 per cent isolates of *E. coli* were verotoxigenic (VTEC) in vero-cell cytotoxicity assay. They further concluded that majority of VTEC isolates from meat and meat products of buffaloes were found positive for vt_2 gene (77.42 per cent) followed by vt_1 (16.13 per cent), while both vt_1 and vt_2 were detected only in 6.45 per cent of the VTEC isolates.

Shiga-toxigenic *E. coli* (STEC) are usually associated with beef and cause severe disease including blooding diarrhoea and hemolytic-uremic syndrome. According to a report of European Food Safety Authority, Shiga-toxigenic *E. coli* (STEC) mainly occurs on beef, which are associated with 30 per cent of illnesses in the Europe (Koutsoumanis *et al.,* 2020).

Some members of the family Enterobacteriaceae have been found to be common in fresh and frozen beef, pork and related meats. Of 442 meat samples examined by Stiles and Ng, 86 per cent yielded enteric bacteria with all 127 ground beef samples being positive. The most frequently found were *Escherichia coli* biotype I (29 per cent), *Serratia liquefaciens* (17 per cent) and *Enterobacter agglomerans* (12 per cent). A total of 721 isolates (32 per cent) were represented by *Citrobacter freundii, Klebsiella pneumoniae, Enterobacter cloacae,* and *E. hafniae*. In an examination of 702 foods for faecal coliforms by MPN representing ten food categories, the highest number was found in the 119 ground beef samples, with geometric mean by AOAC procedure being 59/g. Mean number for 94 pork sausage samples was 7.9/g. From 32 samples of minced goat meat, mean coliform count was log 2.88, mean Enterobacteriaceae was 3.07, while log APC mean was 6.57. Similarly, Hayes *et al.* (2003) also found *Enterococcus* spp. to be the most dominant bacteria on 971 of the 981 samples (99 per cent) of all meat (chicken, turkey, pork and beef) in the state of Iowa. About 97 per cent of pork samples contained *Enterococci* with 54 per cent of isolates identified as *Enterococcus faecalis* and 38 per cent as *Enterococcus faecium*, 3.4 per cent as *Enterococcus hirae*, 2.4 per cent as *Enterococcus durans*, 0.8 per cent as *Enterococcus casseliflavus*, 0.4 per cent *Enterococcus gallinarum* and 1 per cent as unidentified. All of beef samples contained *enterococci* with 65 per cent of isolates identified as *Enterococcus faecium*, 17 per cent as *Enterococcus faecalis*, 14 per cent as *Enterococcus hirae*, 2 per cent as *Enterococcus durans* 0.7 per cent, as *Enterococcus casseliflavus*, 0.4 per cent *Enterococcus gallinarum* and 0.9 per cent as unindentified.

Streptococcal species are of faecal origin and are better indicators of food sanitary quality, especially for frozen foods. It has been reported that potential meat contamination of *Streptococcus faecalis* occurred during slaughtering and butchering of food animals. It has been further revealed that knife trimming do not contribute to enhancing **microbiological quality** of dressed carcasses, except aesthetic values.

Storage conditions affect the type of microbes found in meat and meat products. It has been reported that *Pseudomonas* spp., *Moraxella* spp., *Psychrobacter* spp., *Acinetobacter* spp. and Gram-negative psychrotrophic members of the family Enterobacteriaceae are frequently present on refrigerated meat product. Generally, *Brochothrix thermosphacta* and lactic acid bacteria are the bacteria that cause spoilage of refrigerated meat. Pathogenic microorganisms found in refrigerated meats include *C. botulinum type E, Yersinia enterocolitica, enteropathogenic Escherichia coli, Listeria monocytogenes, and Aeromonas hydrophila* as they are capable of growing at temperatures below 5°C. The mold genera associated with spoilage of refrigerated meat are *Alternaria, Cladosporium, Geotrichum, Mucor, Monilia, Penicillium, Sporotrichum, and Thamnidium*; and yeast genera associated with spoilage of refrigerated meat are *Candida, Torulopsis, Debaryomyces,* and *Rhodotorula.*

The cured meat has a long shelf-life compared to fresh and raw meat however they are not immune to spoilage.

Cured meat is the meats in which they are preserved by aging, drying, canning, brining, or smoking for enhancement of flavour and to extending its shelf life. Some examples of cured meats are sausage, bacon, salami, ham, canned meat, dry spicy meat, meat pickles, kebab, meatballs, meat patty, *etc.* The bacterial spoilage causing organisms in processed and cured meats are lactic acid bacteria such as *Lactobacillus sake, Lactobacillus curvatus, Leuconostoc gelidium, Leuconostoc carnosum, Leuconostoc mesenteroides*), *Acinetobacter, Bacillus, Micrococcus, Serratia,* and *Staphylococcus.* The spoilage causing mold found in cured meat includes *Aspergillus, Penicillium, Rhizopus,* and *Thamnidium.* The spoilage causing yeast found in cured meat includes *Candida, Debaryomyces, Torula, Torulopsis,* and *Trichosporon.* Other spoilage causing microorganism found in cured meats include *Clostridium* spp, *Hafnia* spp, *Weisella* spp, *Shewanella* spp, *Pseudomonas* spp, *Enterococcus* spp, *etc.* The pathogenic microorganisms found in cured meat include *Escherichia coli, Salmonella, Staphylococcus aureus, Listeria monocytogenes, Clostridium botulinum,* and *Toxoplasma gondii.* Microbial growth in cured meat can result in slime formation, structural components degradation, decrease in water holding capacity, off odours, and texture and changes in appearances.

Table 5.5: Some Defects Caused by Microorganism in different Kinds of Meat

Products	*Defects*	*Microorganisms*
Vacuum packed meat	Sulfide odour	*Clostridium* spp., *Hafnia* spp.
	H_2S greening	*Shewanella* spp.
	Blown Pack	*Clostridium* spp., Lactic acid bacteria
Cured meat	Moldy odour	*Aspergillus, Penicillium, Rhizopus*
	Greening	*Pediococcus, Streptococcus*
	Souring	*Micrococcus, Pseudomonas*
	Slimy	*Leuconostoc*
Modified atmosphere packaging meats	Souring, off-odour	*Leuconostoc, Lactobacillus*
	Souring	*B. thermosphacta*
Refrigerated packaged meat	Off-flavours, slime, putrefaction	*Pseudomonas, Acinetobacter, Moraxella*
	Sour, slime, and flavour change	Lactic acid bacteria

Source: Magar, S. T. 2021. Microbial spoilage of meat and meat products. Microbe notes. https://microbenotes.com/preservation-of-meat-and-meat-products/+

It was indicated that psychrotrophic lactic acid bacteria, *Enterococci*, *Micrococci* and yeasts were predominately found in raw, salted-cured products such as corned beef, uncooked hams and bacon due to their resistance to curing salts. In another study, the growth of *Enterobacteriaceae* and *Pseudomonas* were more prevalent on modified atmosphere packed meat (especially on pork) than on vacuum packed meat, their growth being favoured by storage at 5°C. *Pseudomonas* spp. growth rate was considerably slow at 0°C, but increased at 2°C and affected the shelf life of meat. It was also noticed that *Salmonella* growth below 7°C was slow, which increased above 7°C and affected the shelf life of meat (Cerveny *et al.*, 2009; Sentence1991; Garcia-Lopez *et al.*,1998; Dave and Ghaly, 2011).

The earliest, most detailed study is that of Craven and Mercuri, who found that when ground beef or chicken was extended with 10 or 30 per cent soy, APCs of these products increased over unextended controls when both were stored at 4°C for up to 8-10 days. While *coliforms* were also higher in beef-soy mixtures than in controls, this was not true for the chicken-soy blends. In general, APCs were higher at the 30 per cent level of soy than at 10 per cent. These findings have been supported by several investigators. In one study in which 25 per cent soy was used with ground beef, the mean time to spoilage at 4°C for the beef-soy blend was 5.3 days compared to 7.5 days for the unextended ground beef. In another study using 10, 20 and 30 per cent soy, APC increased significantly both with time and concentration of soy in the blend.

Just why bacteria grow faster in the meat-soy blends than in non-soy controls is not clear. The soy itself does not alter the initial flora and the general spoilage pattern of meat-soy blends is not unlike that of all-meat controls. One notable difference is a slightly higher pH (0.3–0.4 unit) in soy-extended products and this alone could account for the faster growth rate. This was assessed by Harrison *et al.,* by using organic acids to lower the pH of soy blends to that of beef. By adding small amounts of a 5 per cent solution of acetic acid to 20 per cent blends, spoilage was delayed about 2 days over controls, but not all of the inhibitory activity was due to pH depression alone. With 25 per cent fat in the ground meat, bacterial counts did not increase proportionally to those of soy-extended beef. It is possible that soy protein increases the surface area of soy-meat mixtures so that aerobic bacteria of the type that predominate on meats at refrigerator temperatures are favoured, but data along these lines are wanting.

The incidence of *C. perfringens* in a variety of American foods was studied by Strong *et al.,* They recovered the organism from 16.4 per cent of raw meats, poultry and fish tested; from 5 per cent of spices; from 3.8 per cent of fruits and vegetables; from 2.7 per cent of commercially prepared frozen foods; and from 1.8 per cent of home-prepared foods. Others have found low numbers of this organism in both fresh and processed meats. In ground beef, *C. perfringens* at 100 or less/g was found in 87 per cent of ninety-five samples, while forty-five of the ninety-five (47 per cent) samples contained this organism at levels < 1.000/g. One group was unable to recover *C. perfringens* from pork carcasses, hearts and spleens, but 21.4 per cent of livers were positive. Commercial pork sausage was found to have a prevalence of 38.9 per cent.

Table 5.6: Examples of Foodborne Pathogens and Spoilage Bacteria Relevant in Meat and Meat Products

Product Category	*Target Pathogen*	*Other Pathogens that may be Relevant*	*References*
Beef	STEC, *Salmonella*	*Listeria, Yersinia, C. perfringens, Staphylococcus aureus*	Chaillou *et al.,* 2015; Garcia-Hernandez *et al.,* 2015; Jackson *et al.,* 2013; Liu *et al.,* 2015; Mayrhofer, Paulsen, Smulders, and Hilbert, 2004; Midelet and Carpentier, 2002; Waters *et al.,* 2011
Pork	*Salmonella, Campylobcter jejuni* and *C. coli*	*Listeria, Yersinia, C. perfringens, S. aureus*	Antunes, Reu, Sousa, Peixe, and Pestana, 2003; Jackson *et al.,* 2013; Mayrhofer *et al.,* 2004; Waters *et al.,* 2011
Poultry	*Salmonella, Campylobacter jejuni*	*Listeria, Yersinia, C. perfringens, S. aureus*	Akbar and Anal, 2013; Alonso-Hernando, Prieto, García-Fernández, Alonso-Calleja, and Capita, 2012; Antunes *et al.,* 2003; Capita *et al.,* 2001; Chai *et al.,* 2017; Corry *et al.,* 2001; Gudbjörnsdóttir *et al.,* 2004; Jackson *et al.,* 2013; Keener *et al.,*
			2004; Liu *et al.,* 2012; Mayrhofer *et al.,* 2004; Rouger *et al.,* 2017; Waters *et al.,* 2011; Waters *et al.,* 2011
Other meats	STEC (ruminants), *Salmonella*	*Listeria, C. perfringens, Bacillus*	Chaillou *et al.,* 2015; Jackson *et al.,* 2013; Liu *et al.,* 2015; Mor-Mur and Yuste, 2010; Singh, Bhilegaonkar, and Agarwal, 2005; Tewari, Singh, and Singh, 2015b; Waters *et al.,* 2011
Ready-to-eat meats, cured or uncured	*Listeria monocytogenes*	*S. aureus, C. botulinum* group II (uncured RTE meats), *Bacillus*	Gómez *et al.,* 2015; Matarante, Baruzzi, Cocconcelli, and Morea, 2004; Tewari and Abdullah, 2015a; Tewari *et al.,* 2015b; Wu *et al.,* 2018
Fermented and dried meats	STEC, *Salmonella*	*Listeria monocytogenes, Staphylococcus aureus*	Chaillou *et al.,* 2015; Mor-Mur and Yuste, 2010; Wang, Ren, and Zhan, 2017
Product category	Spoilage organisms		
Fresh meat (aerobic or modified atmosphere)	*Aeromonas Brochothrix; Pseudomonas; Buttiauxella, Hafnia, Serratia, Shewanella*	*Carno-bacterium*	Arnaut-Rollier *et al.,* 1999; Arnaut-Rollier, Zutter, and Hoof, 1999; Hilton *et al.,* 2004; Praveen *et al.,* 2016; Rouger *et al.,* 2018; Säde, Murros, and Bjorkroth, 2013
Vacuum-packaged meat	*Carnobacterium; Leuconostoc, Latilactobacillus*	*Brochothrix*	Rouger *et al.,* 2018
Ready-to-eat meats	*Carnobacterium; Leuconostoc, Latilactobacillus*	*Brochothrix*	Miller, Liu, and McMullen, 2015
Fermented and dried meats	*Staphylococcus* spp.; *Compani lactobacillus*; osmotolerant yeasts and fungi	–	Wang, *et al.,* 2017; Zheng *et al.,* 2020

Note: Means species was not determined.

Source: Hui *et al.,* 2020. Control of pathogenic and spoilage bacteria in meat and meat products by high pressure: Challenges and future perspectives. Comprehensive reviews in food science and food safety. 19(20): https://doi.org/10.1111/1541-4337.12617

Both *bacilli* and *clostridia* may be found in meats of all types. In a study of the incidence of putrefactive anaerobe (PA) spores in fresh and cured pork trimmings and canned pork luncheon meat, Steinkraus and Ayres found these organisms to occur at very low levels, generally less than 1/g. In a study of the incidence of clostridial spores in meats. Greenberg *et al.,* found a mean PA spore count/g of 2.8 from 2,358 meat samples. Of the 19,727 PA spores isolated, only 1 was a *C. botulinum* spore and it was recovered from chicken. The large number of meat samples studied by these investigators consisted of beef, pork and chicken, obtained from all parts of the United States and Canada. The significance of PA spores in meats is due to the problems encountered in the heat destruction of these forms in the canning industry. With regard to the microbiological quality of soy products, the geometric mean APC of 1,226 sample units of seasoned product was found to be 1,500jg, with fungi, coliforms, *E. coli and S. aureus* counts of 25, 3, 3 and 10/g, respectively.

When meat animals are slaughtered for human consumption, meat from the carcasses is removed by meat cutters. However, the most economical way to salvage the small bits and pieces of lean meat left on carcass bones is by mechanical means (mechanical deboning). Mechanically deboned meat (MDM) is removed from bones by machines. The production of MDM began in the 1970s, preceded by chicken meat in the late 1950s and fish in the late 1940s. During the deboning process, small quantities of bone powder become part of the finished product and the 1978 USDA regulation limits the amount of bone (based on calcium content) to no more than 0.75 per cent (the calcium content of meat is 0.01 per cent). MDM must contain a minimum of 14 per cent protein and no more than 30 per cent fat. The most significant parametrical difference between MDM and conventionally processed meat relative to microbial growth is the higher pH of the former, typically 6.0–7.0. The Increased pH is due to the incorporation of marrow in MDM.

While most studies on the microbiology of MDM have shown these products to be not unlike those produced by conventional methods, some have found higher counts. The microbiological quality of deboned poultry was compared to other raw poultry products and while the counts were comparable, MPN coliform counts of the commercial MDM products ranged from 460 to > 1,100/g. Six of fifty-four samples contained *Salmonellae*, four contained *C. perfringens*, but none contained *S. aureus*. The APC of hand-boned lamb breasts was found to be 680,000 while for mechanically de boned lamb allowed to age for 1 week, APC was 650,000/g. Commercial samples of mechanically deboned fish were found to contain tenfold higher numbers of organisms than conventionally processed fish but different methods were used to perform the counts on fish frames and the mechanically deboned flesh. These investigators did not find, *S. aureus* and concluded that the spoilage of MDF was similar to that for the traditionally processed products. In a more recent study, MDM was found to support the more rapid growth of psychrotrophic bacteria than lean ground beef.

Several studies have revealed the absence of *S. aureus* in MDM, reflecting perhaps the fact that these products are less handled by meat cutters. In general, the mesophilic flora count is a bit higher than that for psychotrophs and fewer gram negatives tend to be found. Field concluded that with good manufacturing practices, MDM should present no microbiological problems and a similar conclusion was reached by Froning relative to deboned poultry and fish.

It has been reported that a favourable pH for the growth of spoilage bacteria for meat is in the range of 5.5-7.0. Slime formation, structural components degradation, off odours and appearance change were found in meat as a result of microbial growth within this pH range. The methylamine, dimethylamine and trimethylamine have been commonly detected during bacterial spoilage by researchers. It has been observed that microbial metabolism produces fatty acids, ketones and alcohols, which exhibit a variety of fruity and sweety odours. Generation of hydrogen sulphide, methylsulphide and dimethylsulphide exhibit putrid and sulphury odours. The diamines, cadaverine and putrescine (the metabolic by-products of meat spoilage) have been studied as indicators of meat spoilage (Jay *et al.,* 2005).

Mechanisms for Meat and Meat Products Spoilage

Meat is a susceptible commodity bound to decompose in time. Compared to a multitude of foodstuffs, meat represents one of the most perishable: first, for the presence of chemical and enzymatic activities, and second, because it constitutes a perfect pabulum for the growth of a wide variety of microorganisms, especially as a result of its nutrient composition, high water content and moderate pH.

Microbial growth, oxidation and enzymatic autolysis are the three basic mechanisms responsible for the spoilage of meat. In addition to lipid oxidation and enzyme reactions, meat spoilage is almost always caused by microbial growth. The breakdown of fat, protein and carbohydrates in meat results in the development of off-odours, off flavours and slime formation, which determine disagreeable meat for human consumption. Chemical food spoilage occurs when different components in the meat react with each other or with some added component which alter the meat's sensory characteristics. Examples of this include: oxidation; enzymatic browning; and nonenzymatic browning.

Autoxidation of lipids and the production of free radicals are natural processes which affect fatty acids and lead to oxidative deterioration of meat and off-flavour development. Lipid hydrolysis can take place enzymatically or non-enzymatically in meat. After slaughtering of animals, the fatty acids in tissues undergo oxidation when the blood circulation stops and metabolic processes are blocked. Lipid oxidation is the reaction of oxygen with double bonds of fatty acids. It involves three stage free radical mechanisms: initiation, propagation and termination. Oxidation of lipids in meat depends on several factors including fatty acid composition, the level of the antioxidant vitamin E (α tocopherol) and prooxidants such as the free iron presence in muscles. Polysaturated fatty acids are more susceptible to lipid oxidation. Hydroperoxides are produced due to the lipid oxidation of highly unsaturated fatty acid fractions of membrane phospholipids, which are susceptible to further oxidation/decomposition. Their breakage causes secondary reaction products such as pentanal, hexanal, 4- hydroxynonenal and malondialdehyde (MDA) as well as other oxygenated compounds such as aldehydes, acids and ketones. These secondary products can cause loss of colour and nutritive value due to severe effects on lipids, pigments, proteins, carbohydrates and vitamins and are directly related to carcinogenic and mutagenic processes.

The enzymatic hydrolysis of fats is termed lipolysis or fat deterioration and is governed by specific enzymes such as lipases, esterases and phospholipase. Lipases are present in the skin, blood and tissue of animals. During lipolysis, lipases split the glycerides forming free fatty acids which are responsible for common off-flavour, frequently referred to as rancidity. The main enzymes involved in meat lipid hydrolysis are phospholipase A1 and phospholipase A2. In muscle cells of slaughtered animals, enzymatic actions take place naturally and they act as catalysts for chemical reactions that finally end up in meat self-deterioration. Softening and greenish discoloration of the meat results due to tissues degradation of the complex compounds (carbohydrates, fats and protein) in the autolysis process. The non-enzymatic hydrolysis is caused by heme proteins such as hemoglobin, myoglobin and cytochrome which are susceptible to oxidation and produce hydroperoxides. During heme catalysis, a Fe2+ protoporphyrin complex (P–Fe2+), like myoglobin, will be oxidized to P–Fe3+. The formed superoxide radical anion O_2 - reacts with H+ and will yield H_2O_2. Hydrogen peroxide will then oxidize P–Fe3+to the oxene species P–Fe = O. The free iron redox cycle contributed by ascorbic acid is the main initiator of lipid peroxidation in fresh muscle foods and it significantly affects the oxidation of oxymyoglobin.

In the autolysis process, the complex compounds (carbohydrates, fats and protein) of the tissues are broken down into simpler ones resulting in softening and greenish discoloration of the meat. These autolysis changes include proteolysis and fat hydrolysis which are prerequisite for microbial decomposition. Excessive autolysis is termed "souring". Post-mortem autolysis takes place in all animal tissues but at different rates in different organs, quicker in glandular tissue such as the liver and slower in striated muscle. The enzymes calpains, cathepsins and aminopeptidases are found to be responsible for the post mortem autolysis of meat through digestion of the z- line proteins of the myofibril. Among these enzymes, calpains has been described as a preliminary contributor to the proteolytic tenderization process of meat. Cathepsins were, also, found to contribute to tenderization at low pH. Proteolytic enzymes are active at low temperatures (5°C) which lead to deterioration of meat quality due to growth of microbes and biogenic amines production.

Besides lipid oxidation and autolytic enzymatic reactions, spoilage of meat can be considered the result of microbial activity of a wide variety of microorganisms. The high-water content ($a_w > 0.99$), the pH that corresponds to the optimal range for microbial growth (5.5–6.5), the availability of energy-yielding nutrients (*e.g.*, glucose, ribose, amino acids, and nucleosides) as well as vitamins and minerals, account for meat being a foodstuff with a short shelf-life. Microbial quality of raw meat results from the physiological status of the animal at slaughter, processing, transportation, preservation and storage conditions. On the basis of knowledge of a few chemical and physical parameters it is possible to predict which microorganisms will grow and dominate during the storage of the meat. Temperatures and packaging atmospheres are the most important factors affecting microbial growth and selection during storage of fresh meat; therefore, the spoilage is caused only by a fraction of species and strains of the initial microbial association. The refrigeration selects psychrotrophic species while further selection is introduced by the type of packaging. In particular, the availability of oxygen affects microbial growth and metabolism. Therefore, depending on the affinity of each species to oxygen, bacteria differ in their competitive growth potential under aerobic or anaerobic conditions. Their spoilage potential depends on which groups or microorganisms will dominate the meat matrix, and on their ability to produce spoilage-associated compounds such as esters, ketones, aldehydes, sulfur compounds, amines and volatile fatty acids. When microbial spoilage occurs as slime and/or off-odours and flavours, the so called specific spoilage organisms (SSO) include the microorganisms that have contributed to the spoilage and others that have grown but not necessarily caused unpleasant changes.

The spoilage potential of a microorganism is determined by its ability to produce the metabolites that are associated with the spoilage. However, it is also important to consider the interaction between microbial growth and enzyme activities. The microbial ecology associated to the spoilage of meat in different storage conditions has been reviewed. Microbial development and consumption of meat nutrients by bacteria result in a consequent release of undesired metabolites. The volatile organic compounds (VOCs) that are generated during meat storage can have an olfactory impact and can lead to rejection of the product when their concentration increase significantly as a result of microbial development. The microbial populations for causing spoilage are known as belonging to the groups of *Enterobacteriaceae*, lactic acid bacteria (LAB), *Brochothrix thermosphacta*, *pseudomonads* and some *clostridia*. In addition, several pathogens can develop in meat, potentially affecting meat safety. The rate and the type of spoilage seem to depend on the concentration of glucose, lactic acid, nitrogenous compounds and free amino acids present in meat, resulting the principal precursors of those microbial metabolites responsible for spoilage. These compounds are attacked in various order to produce different catabolic by-products depending on the microbial species and their oxygen affinity. The volatile fraction of the microbial catabolites (VOCs) most commonly identified in meat during storage include organic acids, volatile fatty acids, ethyl esters, sulfur compounds, ketones, aldehydes, alcohols, ammonia and other molecules. All these molecules will affect the sensory quality of both fresh and cooked meat. Depending on their olfactory thresholds and on the masking and synergic effects within volatiles or between volatiles and non-volatile compounds, they can cause off-odours that will render meat spoiled.

The glucose in raw meat is the first substrate preferably utilized by most bacteria developing in meat during chill storage in any packaging conditions. When the concentration of glucose is depleted other substrates such as lactate, gluconate, glucose-6-phospate, pyruvate, propionate, formate, ethanol, acetate, amino acids, nucleotides, urea and water-soluble proteins are metabolized by almost all the bacteria of the meat microbiota. Glucose has been found to be the precursor of many off-odours during meat storage such as acetate, acetoin, diacetyl, acetic acid, iso-butyric acid, iso-valeric acid, 2-methylbutyric acid, 3- methylbutanol, 2 methylpropanol and ethanol. The changes of glucose and lactate and their products could be used for describing or predicting the degree of spoilage. In fact, it is demonstrated that their concentration can affect the type and the rate of spoilage and they seem to be the principal precursors of the microbial metabolites that are responsible for the olfactory detection of spoilage. This is particularly evident in the case of meat stored under aerobic conditions where *pseudomonads* are the major spoilage microorganisms. *Pseudomonads* catabolize sequentially D-glucose and L and D-lactic acid, with the oxidation of glucose and glucose-6-P via the extracellular pathway leading to a transient accumulation of D-gluconate and an increase in gluconate-6-P concentration. Subsequently, *Pseudomonas* spp. utilize lactate, pyruvate and gluconate in aerobic conditions and pyruvate, gluconate, acetate and amino acids in anaerobic. When glucose is depleted the *Pseudomonas* spp. can attack amino acids with consequent production of volatiles malodorous sulphides, esters and amines. Other Gram-negative bacteria belonging to the family of *Enterobacteriaceae* preferentially utilize glucose prior to degrading amino acids.

Sugar uptake is of extreme importance for LAB proliferation in meat and for the associated type of spoilage. LAB occurring in meat is either obligatory heterofermentative producing lactic acid, acetic acid, CO_2 and ethanol or facultatively heterofermentative producing two moles of lactate from glucose (Figure 5.1). Facultatively heterofermentative LAB possess also an inducible phosphoketolase active when pentoses are present and catalyzing their conversion into lactate and acetate without gas formation. These ribose fermenting *Lactobacillus* under glucose limitation switch their sugar metabolism from homofermentative to heterofermentative determining a significant production of acetic acid as it is the case in meat. During aerobic storage of meat, in presence of low concentration of glucose, most spoilage LAB may oxidize lactate and its catabolic precursor pyruvate into acetate. The increase in acetate may have a negative impact on sensory quality of spoiled meat because of its sharper vinegar-like flavour. Moreover, some LAB may utilize other carbohydrates and amino acids present in the meat to support their growth. In particular, studies in *vitro* showed that *L. sakei* is able to hydrolyze arginine to support growth in the presence of low content of glucose forming ammonia and biogenic amines such as putrescein and spermine.

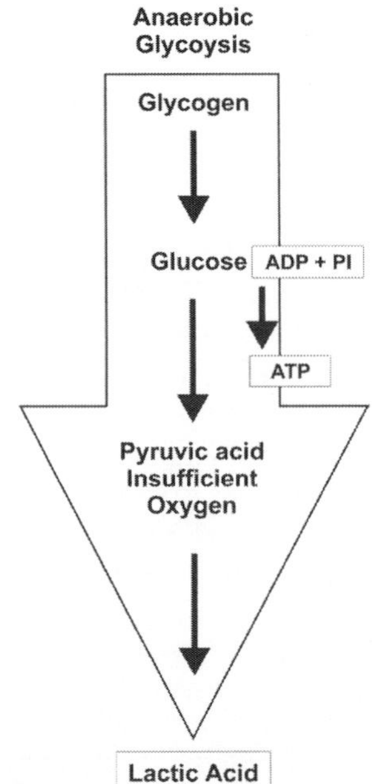

Fig. 5.1 Anaerobic glycolytic pathway

Br. thermosphacta in anaerobic conditions use preferentially glucose as substrate while in air it is able to use also ribose, glycerol and amino acids producing many VOCs. Under low oxygen and glucose availability *Br. thermosphacta* is a homofermentative bacterium, while it shifts to heterofermentative under high oxygen and/or glucose limiting conditions. Finally, some psychrotrophic *Clostridia* associated with spoilage of meat stored in vacuum-packages use preferentially glucose as substrate with a butyric fermentation producing butyric acid, butanol, carbon dioxide, hydrogen and causing a discoloration of meat.

Pseudomonas spp. and *Carnobacterium* spp. appear as the bacteria most involved in the production of alcohols. Many metabolic pathways are involved in the biosynthesis of the alcohols that are found in meat; proteolytic activity and amino acid metabolism, methyl ketone reduction, as well as reduction of aldehydes coming from lipid oxidation. The most common alcohols found in naturally spoiled meat and in model meat systems inoculated with different spoilage microorganisms are 3 methyl-1-butanol, 1-octen-3-ol, 2-ethyl-1- hexanol, 2,3-butanediol, butanol, 1-heptanol, 1-hexanol and 3- phenoxy-1-propanol. These molecules are reported as produced by a wide range of bacteria. These molecules have been found mainly in meat stored in VP and air, whereas the 1-octen-3-ol was also found in meat stored in MAP. Some of these compounds are indicated to cause possible off-odour in meat. In particular, 2-ethyl-1-hexanol is associated with resin, flower and green; 1-heptanol with fragrant, woody, oily, green, fatty, winey, sap and herb; 1-octen-3-ol with mushrooms; 1-hexanol with chemical wine, fatty, fruity and weak metallic while 3 methyl-1-butanol can be associated with a whiskey like odour.

Br. thermosphacta and *Carnobacterium* spp. are the major producers of volatile fatty acids during storage of fresh meat. Acetic and hexanoic acids appear frequently in meat stored in different conditions and have been associated with vinegar and pungent odour, and sour, sweaty, cheesy and goaty odour in cheese, respectively. Volatile fatty acids occurring in meat may originate from hydrolysis of triglycerides and phospholipids and can also be an initial substrate for oxidation with the formation of final odorous products. Fatty acids having generally between 2 and 6 carbon atoms come from the degradation of amino acids and can also arise from ketones, esters and aldehydes by oxidation. Volatile fatty acids such as butanoic, hexanic and nonanoic are mostly associated with *Clostridia*, *Serratia* spp., *Pseudomonans* spp., *Carnobacterium* spp. and other LAB.

Volatile sulfur compounds play an important role in the off flavour of meat, in fact sensory data suggest that the formation of these compounds is responsible, at least in part, for the putrid, sulfury odour which becomes superimposed at advanced stage of storage in air on the earlier developing fruity odours. Sulfur compounds essentially originate from degradation of sulfur-containing amino acids (methionine and cysteine) and result from cleavage of a bond between carbon and sulfur by amino acids-demethiolase. These components are described as having strong garlic and very ripe cheese odours with low perception thresholds in cheese. The most common S-compound in meat is dimethylsulphide, followed by dimethyldisulphide, dimethyltrisulphide and methyl thioacetate, mostly associated with the presence of *Pseudomonas* spp. Dimethylsulphide, dimethyldisulphide and dimethyltrisulphide are S-compounds obtained from methionine metabolism, in particular, from catabolism of methionine to methanthiol, which is then further chemically or enzymatically oxidated to form the polymeric organic sulfides. Dimethylsulphide has been associated with sulfurous, onion, sweet, corn vegetable, cabbage, tomato, green and radish-like odour, dimethyldisulphide has been related to sulfurous, vegetable, cabbage, onion aroma and dimethyltrisulphide to sulfurous, cooked onion and savory meaty odour. Sulfur compounds can be at least partly responsible for the odours detected in the vacuum packed (VP) and the recognized causative organisms are *S. proteamaculans*, *S. liquefaciens*, *P. fragi*, *C. maltaromaticum*, *H. alvei*, *L. curvatus*, *L. sakei*, *Cl. algidicarnis*, Cl. putrefaciens and *Cl. Perfringens*.

Pseudomonas spp., *Carnobacterium* spp. and *Enterobacteriaceae* are the microorganisms able to produce the highest number of aldehydes. Aldehydes are reported as particularly important components in imparting fatty flavour; the short-chain aldehydes tend to be rather sharp or acid with the fattiness increasing with chain length and degree of unsaturation. They are derived from hydrolysis of triglycerides and fatty acid metabolism such as β-oxidation of unsaturated fatty acids or lipid auto-oxidation. Aldehydes can also arise from amino acids by transamination, leading to an intermediate imide. Aldehydes associated to spoilage microorganisms during the storage of fresh meat in air, VP and MAP have been reported in research studies. The most common aldehydes found in naturally spoiled meat and in inoculated model meat systems are: hexanal, nonanal, heptanal, benzaldehyde and 3-methylbutanal. These aldehydes are important contributors to meat flavour profiles because they have low odour thresholds. Straight-chain aldehydes, such as hexanal, nonanal and heptanal are characterized by green-grass and soppy odour and can be produced by both Gram-positive and Gram-negative bacteria. When their concentrations exceed certain thresholds, these molecules are very unpleasant in fact, high concentration of hexanal causes a rancid aroma in stored meat.

Ketones have been reported to be produced during chill storage of fresh meat in aerobic conditions, VP and MAP. Some methyl ketones can derive from a lipolytic activity but also from several other possible pathways, such as alkane degradation by bacteria through a unique α-oxidation, with no change in the carbon skeleton alkane oxidation sequence. The presence of ketones in stored fresh meat is mainly associated with the presence of *Pseudomonas* spp., *Carnobacterium* spp. and *Enterobacteriaceae*. Acetoin is the ketone mostly found in fresh meat stored in different conditions. This compound can originate from the glucose catabolism of *Br. thermosphacta*, *Carnobacterium* spp. and *Lactobacillus* spp. and from the microbial degradation of aspartate. It is an important component in terms of flavour and its detection odour threshold is 800 mg/L. The accumulation of acetoin and diacetyl was related to detection of creamy, dairy and cheesy odours, which were not sufficiently intense or unpleasant for the meat to be regarded as spoiled, but rather as not fresh. Other ketones frequently found in inoculated fresh meat and naturally spoiled meat stored in air and VP are diacetyl, 2-heptanone, 3- octanone and 2-butanone. Diacetyl produces butter, sweet, creamy and pungent caramel flavour; 2-heptanone a cheesy, creamy, fruity, spicy, sweet herbal coconut wood flavour; 3-octanone can be responsible for musty, mushroom, ketonic, moldy and cheesy fermented with a green, vegetative nuance and 2-butanone has been related to acetone-like ethereal fruity camphor.

Esters are formed through esterification of alcohols and carboxylic acids found in meat following microbial esterase activity. Forty-five esters were detected in studies during the storage of fresh meat in air, VP and MAP and among these twenty-seven were produced only by *P. fragi*. Major esters found in naturally spoiled meat and inoculated model meat systems are: ethyl acetate, ethyl butanoate, ethyl-3-methylbutanoate, ethyloctanoate, ethyl hexenoate and ethyl decanoate. The esters with greater odour impact are ethyl butanoate with fruity juicy, fruit pineapple and cognac odour, ethyl hexenoate with sweet, fruity, pineapple, waxy, green and banana odour, ethyl3-methylbutanoate with fruity, sweet, apple and pineapple odour and ethyl octanoate with fruity, wine, waxy, sweet apricot, banana, brandy and pear.

Microbial, chemical and enzymatic activities can be controlled by low temperature storage and chemical techniques in the industry. Proper handling, pre-treatment and preservation techniques can improve the quality of meat and meat products and increase their shelf life. Combination of chemical additives (TBHQ and ascorbic acid) and low temperature storage (5°C) in darkness are well recognized techniques for controlling the spoilage (microbial, enzymatic and oxidative) of meat and meat products.

Different Meats and their Usage

Livers, kidneys, hearts and tongues of bovine, porcine and bovine origins, differ from the skeletal muscle parts of the respective animals in having both higher pH and glycogen levels, especially in the case of liver. The pH of fresh beef and pork liver ranges from 6.1 to 6.5 and that of kidneys from 6.5 to 7.0. Most investigators have found generally low numbers of microorganisms on these products, with surface numbers ranging from $\log_{10}$ 1.69 to 4.20/cm_2 for fresh livers, kidneys, hearts and tongues. The initial flora has been reported to consist largely of gram-positive cocci, coryneforms, aerobic sporeformers, *Moraxella-Acinetobacter* and *Pseudomonas* spp. In a detailed study by Hanna *et al.*, micrococci, streptococci and coryneforms were clearly the three most dominant groups on fresh livers, kidneys and hearts. In one study, coagulase-positive staphylococci, coliforms and *C. perfingens* counts ranged from $\log_{10}$ 0.19–1.37/cm_2 but no *Salmonellae* were found.

Vacuum-packaged beef and pork livers and beef kidneys had lower counts after 7- and 14-day storage at 2°C, or 7 days at 5°C, than those stored in air or PVC film. As is the case for vacuum-packaged non-organ meats, pH values of livers decreased during storage for up to 28 days concomitant with increases in homo- and heterofermentative lactic acid bacteria in livers and kidneys.

In the conventional processing of meats (cold boning), carcasses are chilled after slaughter for 24 + h and processed in the chilled state (post rigor). Hot boning (hot processing) involves the processing of meats generally within 1–2 h after slaughter (pre-rigor) while the carcass is still "hot."

In general, the microbiology of hot-boned meats is comparable to that of cold-boned meats, but some differences have been reported. One of the earliest studies on hot-boned hams evaluated the microbiological quality of cured hams made from hot-boned meat (hot-processed hams). These hams were found to contain a significantly higher APC (at 37°C) than cold-boned hams and 67 per cent of the former yielded *staphylococci* to 47 per cent of the latter (115). Mesophiles counted at 35°C were significantly higher on hot-boned prime cuts than comparable cold-boned cuts, both before and after vacuum-packaged storage at 2°C for 20 days. *Coliforms*, however, were apparently not affected by hot boning.

In a study of hot-boned carcasses held at 16°C and cold-boned bovine carcasses held at 2°C for up to 16 h post-mortem, no significant differences in mesophilic and psychrotrophlc counts were found. Both hot-boned and cold-boned beef initially contained low bacterial counts, but after a 14-days storage period the hot-boned meats contained higher numbers than the cold-boned. These investigators found that the temperature control of hot-boned meat during the early hours of chilling is critical and in a later study found that chilling to 21°C within 3–9 h was satisfactory.

In a study of sausage made from hot-boned pork, significantly higher counts of mesophiles and lipolytics were found in the product made from hot-boned pork than in the cold-boned product, but no significant differences in psychrotrophs were found.

The effect that delayed chilling might have on the flora of hot-boned beef taken about 1 h after slaughter was examined by McMillin *et al.,* Portions were chilled for 1, 2, 4 and 8 h after slaughter and subsequently ground, formed into patties, frozen and examined. No significant differences were found between this product and a cold-boned product relative to *coliforms, staphylococci,* psychrotrophs and mesophiles. A numerical taxonomy study of the flora from hot-boned and cold-boned beef both at the time of processing and after 14 days of vacuum storage at 2°C revealed no statistically significant differences in the flora. The predominant organisms, after storage, for both products were *streptococci* and *lactobacilli,* while in the freshly prepared hot-boned product (before storage) more *staphylococci* and *bacilli* were found. Overall, though, the two products were comparable.

Restructured lamb roast made from 10 per cent and 30 per cent MDM and hot-boned meat was examined for microorganisms and overall the two uncooked products were of good quality. The uncooked products had counts < 3.0 × 10^4/g, with generally higher *numbers in products* containing the higher amounts of MDM. *Coliforms* and fecal coliforms especially were higher in products with 30 per cent MDM and this was thought to be caused by contamination of shanks and pelvic regions during slaughtering and evisceration. Not detected in either uncooked product (in 0.1 g) were *Staphylococcus aureus* and *Clostridium perfringens,* while no *salmonellae, Yersinia enterocolitica,* or *Campylobacter jejuni* were found in 25 g samples. Cooking reduced cell counts in all products to < 30/g.

A summary of the work of ten groups of investigators made by Kotula on the effect of hot-boning on the microbiology of meats revealed that six found no effect, three found only limited effects, while only one found higher counts. Kotula concluded that hot-boning per se has no effect on microbial counts. Hot-boning is often accompanied by prerigor pressurization consisting of the application of around 15,000 psi for 2 min. This process improves muscle color, overall shelf appearance and increases tenderization. It appears not to have any effect on the microbial flora.

It is estimated that at least 80 per cent beef in the United States in the mid-1980s leaves the packing plant in vacuum packages and because of the longer shelf-life of this product, its microbiological quality is of great interest. Vacucm packaging is achieved by placing meats into plastic bags or pouches followed by the removal of air using a vacuum packaging machine and the closure of bag or pouch, often by a heat sealer. From the microbiological standpoint, the most dramatic effect of this process is the change in the gaseous environment of the product. While not all O_2 is removed before sealing, some of what remains is consumed by the aerobic flora and the meat itself, resulting in an increased level of CO_2, which is inhibitory to the flora. The relative quantities of O_2 and CO_2 in the headspace of stored vacuum packages is controlled largely by the degree to which the plastic package impedes the flow of these gases.

A large number of investigators have shown that when O_2-permeable packaging is used, refrigerated fresh meats undergo gram-negative spoilage accompanied by increased pH and foul odors, with *Pseudomonas* spp. being the predominant organisms.

It has been shown that the storage life of vacuum-packaged beef is inversely related to film permeability, with the longest shelf life (> 15 weeks) obtained with a "zero" O_2 film and the shortest (2–4 weeks) with a highly permeable film (920 cc $O_2/m_2/24$ h/atm at 25°C, R.H. 100 per cent). Growth of *Pseudomonas* spp. increases with increasing film permeability. On the other hand, if O_2-impermeable packaging is used, the growth of lactic acid bacteria and sometimes that of *Brochothrix thermosphacta* is favored because of increased levels of CO_2 and a lowered Eh. These organisms typically effect a decrease in pH and create an unfavorable environment for most food-borne pathogens and gram-negative bacteria. With an initial pH of 5.45–5.65, vacuum-packaged beef inoculated with lactic acid bacteria showed pH of 5.30–5.45 after 24–35 days at 5°C. With vacuum-packaged cooked luncheon meats held for 16 days at 5°C, plate counts on LBS agar essentially equalled the APC on APT agar, indicating that the flora was composed essentially of lactics. After 15 weeks, 0.6-0.8 per cent lactic acid was produced and the pH of some products decreased to < 5.0.

The findings by Henry *et al.,* for vacuum-packaged lamb chops and steaks are typical of those with other vacuum-packaged fresh meats.

The relative dominance of the flora of vacuum-packaged meats by lactobacilli versus *B. thermosphacta* is an interesting one. As noted earlier, the latter organism is more sensitive to nitrites than the former so that its relative absence in nitrite-treated products would be expected. Lamb chops with a storage life of 2 weeks at –1°C in air showed a storage life increase of 8 weeks in O_2-free atmospheres and *B. thermosphacta* was the major organism in each of a combination of gas atmospheres including nitrogen, O_2 and CO_2. It has been reported that lactobacilli markedly inhibit the growth of *B. thermosphacta* in vacuum-packaged beef when the two organisms are present in approximately equal numbers but not on beef incubated in air. In the latter study, up to 75 per cent CO_2 did not inhibit the ultimate growth of *B. thermosphacta*. However, when the initial flora of vacuum-packaged meats contains significant numbers of *B. thermosphacta*, this organism grows faster than the homo- or heterofermentative lactics and becomes dominant.

Reports vary on the relative dominance by *B. thermosphacta* and the lactic acid bacteria in vacuum-packaged meats. In one study, vacuum-packaged luncheon meats with lactobacilli as the dominant flora were considered acceptable up to 21 days with counts of 10^8/g. These investigators found that more homofermentative lactics were required to produce product offness than heterolactics because of the end products produced by th latter. While about 9 days were required for *B. thermosphacta* and heterofermentative lactics to reach a count of 10^8/g, homofermentative lactics required 12–20 days. With N_2– and CO_2-packaged frankfurters stored at –4° to 7°C for 49 days, *B. thermosphacta* decreased from about 48 per cent to about 5 per cent, while lactobacilli increased from around 6 per cent to 94–96 per cent under CO_2. With raw pork stored at 2°C for 14 days, about 30 per cent of the flora consisted of lactobacilli, about 18 per cent of *B. thermosphacta* and the remainder of *pseudomonas* types. The initial counts were approximately 10^3/g, land the CO_2 content in packages was around 12.3 per cent.

In a study of the populations of lactic acid bacteria on vacuum-packaged beef, 18 of 177 psychrotrophic lactic acid bacteria were *Leuconostoc mesenteroides* and the remainder were lactobacilli, atypical streptobacteria, or atypical betabacteria. From vacuum-packaged beef loins stored at 4°C, fewer pseudomonads and more lactobacilli were found after 3 weeks than in controls. While some of the reported variations may result from product differences, it is not easy to classify correctly the lactic rods to genus and species; and until a common taxonomic scheme is followed by all investigators, variations in the lactic flora of vacuum-packaged meats as reported by different investigators may be expected to continue.

Microorganisms in Seafood

Seafood consumption has increased worldwide in recent decades, reaching a high during the past decade: the average American now eats approximately 16.5 pounds of seafood each year, compared with 10 to 12 pounds during the 1980s. However, along with the nutrients and benefits derived from seafood consumption come the potential risks of eating contaminated seafood. Seafood is one of the most highly perishable food products because of the chemical effects of atmospheric oxygen and the growth of spoilage microorganisms. Chemicals, metals, marine toxins, and infectious agents have been found in seafood. The major cause of seafood spoilage is microbial growth and metabolic activity which result in the formation of amines, sulphides, alcohols, aldehydes, ketones, and organic acids with unpleasant and unacceptable off-flavours. Sulphurous, ammoniacal, or fishy odours are some of the main organoleptic changes taking place during spoilage development. Infectious agents associated with food-borne illness include bacteria, viruses, and parasites, and the illnesses caused by these agents range from mild gastroenteritis to life-threatening syndromes.

Seafood includes mollusks (*e.g.*, oysters, clams, and mussels), finfish (*e.g.*, salmon and tuna), marine mammals (*e.g.*, seal and whale), fish eggs (roe), and crustaceans (*e.g.*, shrimp, crab, and lobster). Some seafood commodities are inherently more risky than others owing to many factors, including the nature of the environment from which they come, their mode of feeding, the season during which they are harvested, and how they are prepared and served. Contamination of seafood by pathogens with a human reservoir can occur when growing areas are contaminated with human sewage. Outbreaks of seafood-associated illness linked to polluted waters have been caused by calicivirus, hepatitis A virus, and *Salmonella enterica* serotype Typhi. Additionally, seafood may become contaminated during handling, processing, or preparation. Contributing factors may include storage and transportation at inappropriate temperatures, contamination by an infected food handler, or cross-contamination through contact with contaminated seafood or seawater. Adequate cooking kills most pathogens; however, unlike other foods, such as meat and poultry, that are usually fully cooked, seafood is often consumed raw or prepared in ways that do not kill organisms.

The recent establishment of the SSO (specific spoilage organism) concept has contributed significantly to our understanding of seafood spoilage. The growth of different SSOs depends on several parameters such as food product, type of preservation, temperature, atmosphere, and salt content, among others. During storage, the microflora changes owing to different abilities of the microorganisms to tolerate the preservation conditions. Only a fraction of the initial microbiota of seafood known as specific spoilage organisms (SSOs), which is favoured by storage conditions (*e.g.*, atmosphere, temperature), prevails over the rest of the microbiota, reaching high populations and producing corresponding metabolites. The seafoods like oysters, shrimp, *etc.* contain many types of microorganisms. These organisms depend greatly upon the quality of water from which, they are harvested.

Assuming good-quality waters, most of the organisms are picked up during the various stages of processing. In the case of breaded raw shrimp, the breading process would be expected to add organisms if not properly done or if the ingredients are of poor microbiological quality. In their study of 91 samples of shrimp of various types, Silverman *et al.*, found that all precooked samples except one had total counts of $<\log 4.00$/g. Of the raw samples, 59 per cent had total counts below log 5.88 while 31 per cent were below log 5.69/g. In a study of 204 samples of frozen, cooked and peeled shrimp, 52 per cent had total counts < log 4.70/g and 71 per cent had counts of log 5.30 or less/g.

In newly caught marine seafood from temperate waters, microflora is formed mainly by aerobic rods-shapes, anaerobic facultative and psychrotrophic Gram-negative bacteria, whose growth is possible at 0 and optimal at around 25. The majority belongs to the Gammaproteobacteria *viz., Pseudomonas, Shewanella, Acinetobacter, Aeromonas, Vibrio, Moraxella, Psychrobacter, Photobacterium, etc.* The same bacterial genus can be found in tropical marine seafood, but Cram-positive bacteria, *Enterobacteriaceae* and *Vibrionaceae* are often dominant. Generally, *Pseudomonas* spp., *S. putrefaciens, S. baltica or Aeromonas* spp. were common dominant spoilage bacteria in iced sea salmon; gutted sea bass; chilled fresh Mediterranean swordfish; tropical prawns; large yellow croaker. *Pseudoalteromonas* and *Vibrio* were dominant microorganisms in shucked oysters during iced storage, as spoilage proceeded, enterococci, lactobacilli, and yeasts dominated at the later stages, the spoilage patterns of Mollusca shellfish differ in most species of seafood as they contain high levels of carbohydrate in the form of glycogen (Table 5.7).

Table 5.7: Main Spoilage Microorganisms in different Seafood and Seafood Products

Spoilage Bacteria	*Seafoods and Seafood Products*	*References*
Shewanella spp.	Gutted sea bass Iced sea salmon Air stored swordfish Refrigerated shrimp Refrigerated large yellow croaker	Parlapani *et al.*, 2015 Hozbor *et al.*, 2006 Pantazi *et al.*, 2008 Zhu *et al.*, 2019 Zhu *et al.*, 2016
Pseudomonas spp.	Gutted sea bass Air stored swordfish	Parlapani *et al.*, 2015 Pantazi *et al.*, 2008
Aeromonas spp.	Iced sea salmon	Hozbor *et al.*, 2006
Photobacterim phosphoreum	MAP/VP stored raw salmon VP packaged squid mantle	Mace *et al.*, 2012 Paarup *et al.*, 2002
Enterobacteriaceae	MAP/VP packaged swordfish VP packaged pressurised squid mantle	Pantazi *et al.*, 2008 Paarup *et al.*, 2002
LAB	MAP/VP stored raw salmon MAP/VP packaged swordfish VP packaged pressurised squid mantle	Mace *et al.*, 2012 Pantazi *et al.*, 2008 Paarup *et al.*, 2002

Cases of *Vibrio* infections have a marked seasonal distribution; most occur during summer and early fall, corresponding to the period of warmer temperatures. Almost all cases of food-borne *Vibrio* infection are associated with a recent history of seafood consumption, primarily raw oyster consumption. Like other organisms found in water, vibrios can be concentrated in the tissues of filter-feeding bivalve mollusks. Measures to prevent food-borne *Vibrio* infections include consumer education regarding the dangers of eating raw or undercooked shellfish, particularly among persons with medical conditions, such as liver disease, that predispose them to severe illness. Thoroughly cooking shellfish and preventing raw seafood from cross-contaminating other foods are effective measures for consumers to reduce risk.

Consumption of raw bivalves or cooked Crustacea subjected to cross-contamination or time/temperature abuses has been associated with cases of infection due to *Vibrio* spp., which are ubiquitous in aquatic environments. Most of the species are mesophilic, and their numbers tend to increase during warm seasons (>15 °C). Seafood-borne diseases are primarily caused by *Vibrio parahaemolyticus, V. vulnificus* and *V. cholerae, V. parahaemolyticus* and *V. cholerae* both cause gastrointestinal disease, while *V. vulnificus* causes septicemia. These pathogenic *Vibrio* spp. are abundant in warm seawater environments. Marine *Vibrio* species are known to attach themselves to the exoskeletons of Crustacea and metabolize them as carbon/energy sources. *Vibrio* organisms can be found at levels of 102–103 cells/g in shellfish and 104–108 cells/g in the intestines of fish; their presence and numbers are influenced by factors such as temperature, salinity and algal density.

Routes of *Salmonella* transmission include food-borne and waterborne routes, person-to-person contact, and contact with animals, particularly reptiles. Most cases of nontyphoidal salmonellosis are caused by ingestion of contaminated seafood, and outbreaks have been associated with a wide range of food vehicles in the United States. Seafood-associated outbreaks have been caused by fish, shrimp, oysters, and clams. Studies to determine the prevalence of *Salmonella* spp. in oysters from domestic bays and testing of sampled oysters from domestic seafood samples by the U.S. Food and Drug Administration (FDA) have demonstrated the presence of salmonellae in a variety of fish and shellfish, including seafood intended for consumption without further preparation upon distribution, requiring minimal cooking, and shellfish eaten raw (Iwamoto *et al.,* 2010). *Pathogens including Shigella, Staphylococcus aureus, Clostridium perfringens, Clostridium botulinum* and *Bacillus cereus have also been reported in infected seafoods.*

Fish and shellfish can acquire *Salmonella* from polluted waters. Historically, sewage contamination of shellfish harvest beds led to large shellfish-associated outbreaks of *Salmonella* serotype Typhi infections. Additionally, seafood can become contaminated with *Salmonella* during storage and processing. *Salmonella* infection can be prevented by adequate cooking, proper storage and processing after harvest, and avoidance of cross-contamination during seafood handling.

Psychrobacter spp. and *Pseudoalteromonas* spp. have been reported to be the dominant microbiota of cooked brown shrimp and enhanced spoilage by breaking down lipids and hydrolysing amino acids and proteins. The major spoilage bacterial isolates from spoiled cooked and whole tropical shrimp stored under MAP were *C. maltaromaticum* and *S. baltica*. LAB and *Brochothrix* spp. were dominant bacteria in the latter storage period of the VP-packed cold-smoked salmon, whereas *Brochothrix* spp. rather than LAB were responsible for spoilage. Differently, in another study, reserachers identified *L. sakei* and *S. liquefaciens*-like as the most spoiling bacteria. Besides, psychrotrophic marine vibrio and Photobacterium spp. were reported to be dominant microflora.

In a study of haddock fillets, most microbial contamination was found to occur during filleting and subsequent handling prior to packaging. These investigators showed that total count increased from log 5.61 in the morning to log 5.65 at noon and to log 5.94/g in the evening for one particular processor. According to their study, results obtained in other companies were generally similar if the night time clean-up was good. In the case of shucked, soft-shell clams, the same general pattern of build-up was demonstrated from morning to evening. The mean clostridial count for both haddock fillets and soft-shell clams was less than 2/g, with clams being slightly higher than haddock fillets for these organisms, although both were low. Total counts on fresh perch fillets produced under commercial conditions were found to average log 5.541g with yeast and mold counts of about log 2.69/g.

Clams may be expected to contain the organisms that inhabit the waters from which they are obtained. Of sixty clam samples from the coast of Florida, 43 per cent contained salmonellae, which were also found in oysters at a level of 2.2/100 g oyster meats. Hard-shell clams have been shown to retain *S. typhimurium* more efficiently than *E. coli*.

In a study of the flora of raw Pacific shrimp taken from docks, *Moraxella* spp. constituted 30–60 per cent followed by types I and II *pseudomonas* (8–22 per cent), *Acinetobacter* (4–24 per cent), *Flavobacterium-Cytophaga* (7–16 per cent) and *Pseudomonas* types III and IV (8–22 per cent); but following blanching and machine peeling, *Acinetobacter* spp. represented 16–35 per cent, *Pseudomonas* types III and IV 2–76 per cent and *Flavobacterium-Cytophaga* 3–37 per cent. The initial flora of herring fillets has been found to be dominated by *Alteromonas putrefaciens* and *Pseudomonas* spp., with the latter dominating at 2°C and *A. putrefaciens* more predominant at 2–15°C. In a *C. perfringens* survey of 287 retail samples of fresh fish and shellfish, 10 per cent were positive for this organism.

In general, frozen seafood as well as other frozen products have lower microbial counts than the comparable fresh products. In a study of 597 fresh and frozen seafood from retail stores, APC geometric means for the 240 frozen products ranged from log 3.54–4.97/g and from 4.89–8.43/g for the 357 fresh products. For coliforms, geometric mean MPN counts ranged from 1 to 7.7 cells/g for frozen and from 7.8 to 4,800/g for fresh. By MPN, only 4.7 per cent of the 597 were positive for *E. coli*, 7.9 per cent were positive for *S. aureus* and 2 per cent were positive for *C. perfringens*. All were negative for *salmonellae* and *Vibrio parahaemolyticus*.

Lightly preserved RTE seafood products, such as cold smoked fish or bivalves, can allow the survival and growth of *Listeria monocytogenes*, as no bactericidal step is included in the process. During cold smoking, seafood is subjected to smoke at a temperature where the product undergoes only incomplete heat coagulation of protein, so the processing parameters are considered insufficient to provide a listericidal effect in the product. Listeriosis is a serious foodborne illness that affects susceptible populations, such as the elderly, pregnant women and immunocompromised individuals, and can lead to death. Due to its long incubation period, it is frequently difficult to identify the food associated with outbreaks.

Plate counts are generally higher on seafood when incubated at 30°C than at 35°C and this is reflected in results from fresh crabmeat, clams and oysters evaluated by Wentz *et al.*, APC geometric means for 896 crabmeat samples at 35°C was log 5,15 and 5.72 at 30°C; for 1,337 shucked oysters 5.59 at 35° and 5.95 at 30°; and for 358 soft-shell clams, log APC was 2.83 at 35° and 4.43 at 30°. This was also seen in raw in-shell shrimp and frozen raw lobster tails, where geometric mean APC for shrimp at 35°C was log 5.48 and 5.90/g at 30°C, while for lobster tail, 4.62 at 35° and 5.15/g at 30°C.

Many investigators have suggested that these products should be produced with total counts not to exceed log 5.00/g. In a study of 48 meat pies, 84 per cent had APC < log 5 while in another study of 188 meat pies, 93 per cent had counts less than log 5.00. Accordingly, a microbiological criterion of log 5.00 seems attainable for such products. Table 5.8 indicates the quality of seafood products and about microbial group.

Table 5.8: Quality of Seafood Products, and Microbial Group

Products	Number of Samples	Microbial Group/Target
Frozen catfish fillet	41	APC 32°C: 105/g or less
	41	MPN coliforms: < 3/g
	41	MPN *S. aureus*: < 3/g
Frozen salmon steaks	43	APC 32°C: 105/g or less
	43	MPN coliforms: < 3/g
	43	MPN *S. aureus*: < 3/g
Fresh clams	53	APC 32°C: 105/g or less
	53	MPN coliforms: < 3/g
	53	MPN *S. aureus*: < 3/g
Fresh oysters	59	APC 32°C: 107/g or less
	59	MPN coliforms: 1,100 or less/g
	59	MPN *S. aureus*: < 3/g
Shucked oysters (retail)	1,337	APC 30°C: 106/g or less
	1,337	MPN coliforms: 460 or less/g
	1,337	MPN fecal coliforms: 460 or less/g
Blue crabmeat (retail)	896	APC 30°C: 106/g or less,
	896	MPN coliforms: 1,100/g or less
	896	MPN *E. coli*: < 3/g
	896	MPN *S. aureus*: 1,100/g or less
Hard-shell clams (wholesale)	1,124	APC 30°C: 106/g or less
	1,130	MPN coliforms: 460/g or less
	161	MPN fecal coliforms: < 3/g
Soft-shell clams (wholesale)	351	APC 30°C: 106/g or less
	363	MPN coliforms: 460/g or less
	75	MPN fecal coliforms: < 3/g
Retail frozen breaded raw shrimp	27	APC: 6.00 or less/g
	27	Coliforms: 3.00 or less/g
	27	Presence of *E. coli*
	27	Presence of *S. aureus*
Fresh channel catfish	335	APC: ≤ 7.00/g
	335	Fecal coliforms: 2.60/g
	335	Presence of salmonellae
Frozen channel catfish	342	APC: ≤ 7.00/g
	342	Fecal coliforms: 2.60/g
	342	Presence of salmonellae
Frozen, cooked, peeled shrimp	204	APC: < 4.70/g
	204	APC: 5.30 or less/g
	204	Coliforms: none or < 0.3/g
	204	Coliforms: < 3/g
Peeled shrimp (raw)	1,468	APC 30°C: 10/g or less
	1,468	MPN coliforms: 64/g or less
	1,468	MPN *E. coli*: < 3/g
Peeled shrimp (cooked)	1,464	APC 30°C: 105/g or less
	1,464	MPN coliforms: < 3/g
	1,464	MPN *E. coli*: < 3/g
Lobster tail (frozen, raw)	1,315	APC 30°C: 106/g or less
	1,315	MPN coliforms: 64 or less/g
	1,315	MPN *E. coli*: < 3/g

In a recent study of 1,290 frozen tuna pot pies, geometric mean APC at 35°C was log 3.20 while at 30°C it was log 3.38/g. *Coliforms* averaged 5/g, *E. coli* < 3/g and *S. aureus* < 10/g.

ICMSF Recommended Microbiological Limits for Seafoods have been given in Tables 5.9 and 5.10.

Table 5.9: Recommended Microbiological Limits for Seafoods (ICMSF, 1986)

Product	Test	n	c	Limit per gram or per cm^2	
				m	M
Fresh and frozen fish and cold-smoked	APC	5	3	5×10^5	10^7
	E. coli	5	3	11	500
Precooked breaded fish	APC	5	2	5×10^5	10^7
	E. coli	5	2	11	500
Frozen raw crustaceans	APC	5	3	10^6	10^7
	E. coli	5	3	11	500
Frozen cooked crustaceans	APC	5	2	5×10^5	10^7
	E. coli	5	2	11	500
	Staph. aureus	5	0	10^3	-
Cooked, chilled, and frozen crabmeat	APC	5	2	10^5	10^5
	E. coli	5	1	11	500
	Staph. aureus	5	0	10^3	-
Fresh and frozen bivalve molluscs (These criteria are to be used only for molluscs from approved harvesting areas where waters are free from enteric bacteria or virus contamination and there is no significant contamination by toxic metals or toxic and carcinogenic chemicals which may be accumulated by animals. Molluscs from non-approved areas should enter trade only after processing by a treatment to destroy enteric bacteria and viruses and may be tested for dangerous chemicals before distribution)	APC	5	0	5×10^5	-
	E.Coli	5	0	16	-

Table 5.10: (Supplement) Additional Tests to be Carried out when Appropriate (ICMSF, 1986)

Product	Test	n	c	Limit per gram or per cm²	
				m	M
Fresh and frozen fish and cold-smoked (For fish known to derive from inshore or inland waters of doubtful bacteriological quality, particularly in warm-water areas and where fish are to be eaten raw, it may be desirable to test for *Salmonella* and *V. parahaemo-lyticus*. Smoked fish may be tested for *Staph. aureus*)	*Salmonella*	5	0	0	–
	V. parahaemo -lyticus	5	2	10^2	10^3
	Staph. aureus.	5	2	10^3	10^4
Precooked breaded fish (*Staph. aureus* may also be tested for cooked fish products likely to be mishandled.)	*Staph. aureus*	5	1	10^3	10^4
Frozen raw crustaceans (*Staph. aureus* may be tested in breaded products. *Salmonella* and *V. parahaemolyticus* tests may be applied to products from warm-water areas and *V. parahaemolyticus* tests in products from temperate regions in summer, if likely to be eaten raw.)	*Salmonella*	5	0	0	–
	V. parahaemo-lyticus	5	1	10^2	10^3
	Staph. aureus	5	2	10^3	10^4
Frozen cooked crustaceans (*Salmonella* and *V. parahaemolyticus* may be tested for in cooked crustaceans originally harvested from waters and processed in regions of known high environmental hazard.)	*Salmonella*	10	0	0	–
	V. parahaemo-lyticus	5	1	10^2	10^3
Cooked, chilled, and frozen crabmeat (Crabmeat from animals harvested from waters above 15°C may be tested for *V. parahae-molyticus*)	*V. parahaemo-lyticus*	10	1	102	10^3
Fresh and frozen bivalve molluscs (*Salmonella* should be tested for when there is concern for bacteriological safety and *V. parahaemolyticus* should be tested for in molluscs from endemic areas and harvested from warm waters)	*Salmonella*	20	0	0	–
	V. parahaemo-lyticus	10	1	10^2	10^3

n: Number of representative sample units; c: Maximum number of acceptable sample units with bacterial counts between m and M; m: Maximum recommended bacterial counts for good quality products; M: Maximum recommended bacterial counts for marginally acceptable quality products. Plate counts below "m" are considered good quality. Plate counts between "m" and "M" are considered marginally acceptable quality, but can be accepted if the number of samples does not exceed "c." Plate counts at or above "M" are considered unacceptable quality (ICMSF, 1986).

Spoilage microorganisms are responsible for various sensory deterioration in seafoods, but these sensory descriptors are not easily associated with enzymatic functions or metabolic pathways. Several studies have reported the detection or measurement of molecules (biogenic amines and volatile compounds) in seafood spoiled by known microorganisms. However, it is difficult to correlate the production of spoilage related metabolites to the functions of spoilers, and the studies investigating the metabolism and physiology of bacteria responsible for seafood spoilage are much less abundant, not to mention the studies on these spoilage phenomena caused by spoilers employing QS system.

Chapter 6

Factors Influencing Growth of Microorganisms in Food Products

In most cases, micro-organism utilizes our food supply as a source of nutrient for their growth. When microorganisms grow in the food, they cause varying degrees of change in the food's attributes as a result of metabolic activity. Some of these changes, like those taking place during fermentation, are desirable, while others, like those resulting in food spoilage and food poisoning, are undesirable. The microbial flora of food consists of microorganisms associated with the raw material, processing and those surviving pasteurization and preservation treatments. The ability of microorganisms to grow or multiply in a food is determined by the food environment as well as the environment where the food is stored.

Biodeterioration takes place in close relation to the chemical and physical factors affecting the microorganisms. The growth and development of microorganisms is stimulated by external stimuli, *i.e.*, environmental factors. Microorganisms display a relatively wide range of tolerance to changes in environmental conditions. Under the right conditions, microorganisms develop very well. The main growth factors are temperature, humidity, concentration of hydrogen ions in the environment, oxidoreduction potential, water activity in the environment, and hydrostatic pressure. These factors influencing the biodeterioration may be mutually dependent.

The most important factors that influence microbial growth in foods can be summarized in the following categories as given in Table 6.1.

Table 6.1 Factors influencing the development of microbes in food

Intrinsic Factors	*Extrinsic Factors*	*Implicit Factors*	*Process Factors*
Nutrient content	Relative humidity	Metabiotic	Physical treatments
pH	Storage temperature	Symbiotic	Use of chemicals
Oxidation–reduction potential	Gaseous atmosphere	Commensalism	Commensalism
Water activity	Presence of other microorganisms	Antagonism	
Biological structure		Predation	
Antimicrobial content		Competition	

Source: Osman Erkmen and T. Faruk Bozoglu, 2016, Food Microbiology: Principles into Practice, First Edition

(i) Intrinsic factors: These parameters are natural or inherent properties of food. They greatly affect the number and types of microorganism that will colonize the food and food product. Intrinsic parameters affect only microorganisms, not to the food itself. Factors related to the food itself, the "intrinsic factors," include nutrient content, water activity, pH value, redox potential, presence of antimicrobial constituents and mechanical barriers to microbial invasion

(ii) **Extrinsic factors**: Factors related to the environment in which the food is stored, the "extrinsic factors," including the storage temperature, composition of gases, activities of microorganisms and relative humidity in the atmosphere surrounding the food

(iii) **Implicit factors**: Factors related to the microorganisms themselves, the "implicit factors," including interactions between the microorganisms contaminating the food and between these microorganisms and the food, *e.g.*, their abilities to utilize different nutrient sources, tolerate stresses, and produce promoters or inhibitors of growth of other microorganisms, *etc.*

(iv) **Processing factors**: These factors include treatments such as heating, cooling, and drying that affect the composition of the food and also affect the types and numbers of microorganisms that remain in the food after treatment.

(v) interaction between the above-described factors can also affect the growth of microorganisms in foods in a complicated way; the combined effects may be additive or synergistic.

Intrinsic Factors

pH and Acidity

The pH is a function of the hydrogen ion concentration in the food:

$$pH = -\log10\ [H+]$$

pH is a term used to describe the acidity or alkalinity of a solution. At pH 7, there is an equal amount of acid (hydrogen ion: H +) and alkali (hydroxyl ion: OH-), so the solution is "neutral". pH values below 7 are acidic, while those above 7 are alkaline. pH expresses the H + concentration logarithmically, that is, in multiples of 10. For example, at pH 5 there are 10 times as many H + as at pH 6; at pH 3 there are 100 times as many H + as at pH 5, and so on.

Another useful term relevant to the pH of foods is the pKa. The term pKa describes the state of dissociation of an acid. At equilibrium, pKa is the pH at which the concentrations of dissociated and undissociated acid are equal. Strong acids have a very low pKa, meaning that they are almost entirely dissociated in solution. For example, the pH (at 25 °C (77 °F]) of a 0.1 M solution of HCl is 1.08 compared to the pH of 0.1 M solution of acetic acid, which is 2.6. This characteristic is extremely important when using acidity as a preservation method for foods. Increasing the acidity of foods, either through fermentation or the addition of weak acids, has been used as a preservation method since ancient times. Organic acids are more effective as preservatives in the undissociated state. Lowering the pH of a food increases the effectiveness of an organic acid as a preservative. The type of organic acid employed can dramatically influence the microbiological keeping quality and safety of the food.

Another important characteristic of a food to consider when using acidity as a control mechanism is its buffering capacity. The buffering capacity of a food is its ability to resist changes in pH. Foods with a low buffering capacity will change pH quickly in response to acidic or alkaline compounds produced by microorganisms as they grow. Meats, in general, are more buffered than vegetables by virtue of their various proteins. Titratable acidity (TA) is a better indicator of the microbiological stability of certain foods, such as salad dressings, than is pH. Titratable acidity is a measure of the quantity of standard alkali (usually 0.1 M NaOH) required to neutralize an acid solution. It measures the amount of hydrogen ions released from undissociated acid during titration. Titratable acidity is a particularly useful measure for highly buffered or highly acidic foods. Weak acids (such as organic acids) are usually undissociated and, therefore, do not directly contribute to pH. Titratable acidity yields a measure of the total acid concentration, while pH does not, for these types of foods.

In their natural state, most foods such as meat, fish, and vegetables are slightly acidic while most fruits are moderately acidic. A few foods such as egg white are alkaline. Table 6.2 lists the pH ranges of some common foods.

Table 6.2: pH Ranges of some Common Foods

	Food	*pH Range*
Dairy Products	Butter	6.1 - 6.4
	Buttermilk	4.5
	Milk	6.3 - 6.5
	Cream	6.5
	Cheese (American mild and cheddar)	4.9; 5.9
	Yogurt	3.8 - 4.2
Meat and Poultry	Beef (ground)	5.1 - 6.2
	Ham	5.9 - 6.1
	Veal	6.0
	Chicken	6.2 - 6.4
Fish and Shellfish	Fish (most species)	6.6 - 6.8
	Clams	6.5
	Crabs	7.0
	Oysters	4.8 - 6.3
	Tuna Fish	5.2 - 6.1
	Shrimp	6.8 - 7.0
	Salmon	6.1 - 6.3
	White Fish	5.5

	Food	*pH Range*
Fruits and Vegetables	Apples	2.9 - 3.3
	Apple Cider	3.6 - 3.8
	Bananas	4.5 - 4.7
	Figs	4.6
	Grapefruit (juice)	3.0
	Limes	1.8 - 2.0
	Honeydew melons	6.3 - 6.7
	Oranges (juice)	3.6 - 4.3
	Plums 2.8 - 4.6	2.8-4.6
	Watermelons	5.2 - 5.6
	Grapes	3.4 - 4.5
	Asparagus (buds and stalks)	5.7 - 6.1
	Beans (string and lima)	4.6 - 6.5
	Beets (sugar) 4.2 - 4.4	
	Broccoli 6.5	6.5
	Brussels Sprouts	6.3
	Cabbage (green)	5.4 - 6.0
	Carrots	4.9 - 5.2; 6.0
	Cauliflower	5.6
	Celery	5.7 - 6.0
	Corn (sweet)	7.3
	Cucumbers	3.8
	Eggplant	4.5
	Lettuce	6.0
	Olives (green)	3.6 - 3.8
	Onions (red)	5.3 - 5.8
	Parsley	5.7 - 6.0
	Parsnip	5.3
	Potatoes (tubers and sweet)	5.3 - 5.6
	Pumpkin	4.8 - 5.2
	Rhubarb	3.1 - 3.4
	Spinach	**5.5 - 6.0**
	Squash	5.0 - 5.4
	Tomatoes (whole)	4.2 - 4.3
	Turnips	5.2 - 5.5

Source: ICMSF 1980, p 109-110; Jay 2000, p 39.

From Table 6.2, it can be observed that generally, fruits have lower pH values than vegetables, meat, poultry, fish, and dairy products. Hence, fruits are more susceptible to spoilage by yeasts and molds, while bacteria are the most important spoilage agents of the other groups of food. According to the pH values of the foods, they are generally divided into four groups:

Group 1: Low-acid foods with a pH of more than 5.3 include corn, meat, fish, and milk. All types of microorganisms can grow on them, especially bacteria.

Group 2: Medium-acid foods with a pH range of 5.3–4.5 include bananas, yogurt, and pumpkin. All types of microorganisms can grow on them.

Group 3: Acid foods with a pH range of 4.5–3.7 include tomatoes, orange juice, sugar beet, and grapes; they are susceptible to the growth of yeasts and molds, and a few types of bacteria. Most pathogenic bacteria do not grow on these foods.

Group 4: High-acid foods with a pH below 3.7 include apples, grapefruit juice, and limes; they are susceptible to the growth of yeasts and molds. Most bacteria, particularly the pathogenic ones, do not grow on such foods.

The pH of foods can be natural, due to microbial fermentation, or result from acids added to the food. Microbial growth leads in most cases to a change in the pH of the food. When lactic acid bacteria, for example, grow in the milk, they consume its lactose and produce lactic acid, which lowers the pH of the milk. Also, the changes in pH can transform food into one that can support the growth of pathogens. Different foods tend to spoil in different ways. For instance, carbohydrate-rich foods often undergo acid hydrolysis when they spoil, which usually reduces the pH and reduces the risk of pathogen growth. This principle is used in the dairy and lactic meat fermentations. On the contrary, protein-rich foods tend to increase in pH when they spoil, making them possibly less safe as the pH rise to the zone where more pathogens can grow. While the pH of many vegetables is in a range suitable for the growth of pathogenic bacteria, some, *e.g.*, fully ripe tomatoes, are in a pH range (3.9–4.4) that prevents or retards growth. Yeasts and molds, on the other hand, have a competitive advantage over bacteria that may access bruised tissues of acidic vegetables and many fruits because they can to grow at a lower pH range (2.2–5.0) characteristic of much of this produce. Spoilage of fruits is often caused by specific molds or groups of molds and yeasts.

pH has a profound effect on the growth of microorganisms. Every organism has a minimal, maximal and optimal pH for growth. Some organism can grow better at low pH or acidic pH, some can grow in alkaline pH and while other grow at somewhat neutral pH. pH influence both the growth rate and types of organism that will predominant the food. In general yeast and mould are more acid tolerant than bacteria. Table 6.3 lists the approximate pH ranges for growth in laboratory media for selected organisms relevant to food. Most bacteria grow best at about pH 7 and grow poorly or not at all below pH 4. Yeasts and molds, therefore, predominate in low pH foods where bacteria cannot compete. The lactic acid bacteria are exceptions; they can grow in high acid foods and actually produce acid to give us sour milk, pickles, fermented meats, and similar products. Some strains, called Leuconostoc contribute off-flavours to orange juice.

Table 6.3: Approximate pH Values for the Growth of Selected Microbes in Food

Microorganism	*Minimum*	*Optimum*	*Maximum*
Clostridium perfringens	5.5 to 5.8	7.2	8.0 to 9.0
Vibrio vulnificus	5.0	7.8	10.2
Bacillus cereus	4.9	6.0 to 7.0	8.8
Campylobacter spp.	4.9	6.5 to 7.5	9.0
Shigella spp.	4.9	9.3	
Staphylococcus aureus			
Growth	4.0	6.0 to 7.0	10.0
Toxin	4.5	7.0 to 8.0	9.6
Salmonella (most).	4.5 to 5.0	6.0 to 7.5	8.0 to 9.6
S. typhi	4.0 to 4.5	6.5 to 7.2	8.0 to 9.0
Escherichia coli	4.3-4.4	6.0-8.0	9.0-10.
L. acidophilus	4.0 to 4.6	5.5 to 6.0	7.0
L. plantarum	3.5	5.5 to 6.5	8.0
C. perfringens	5.0 to 5.5	6.0 to 7.6	8.5 to 9.0
Clostridium botulinum	4.8 to 5.0	0 6.0 to 8.0	8.5 to 8.8
Propionibacterium	4.7	6.2 to 7.0	7.5
Pseudomonas (most)	5.6	6.6 to 7.0	8.0
Leuconostoc cremoris	5.0	5.0	5.0
	5.5 to 6.0	5.5 to 6.0	5.5 to 6.0
	6.5	6.5	6.5
Saccharomyces cerevisiae	2.0 to 2.4	4.0 to 5.0	-
S. rouxii	1.5	3.5 to 5.5	8.5 to 10.5
Aspergillus niger	1.2	3.0 to 6.0	-
A. oryzae	1.6 to 1.8	-	9.0 to 9.3
Penicillium	1.9	4.5 to 6.7	9.3

Source: Banwart, G.J. 1989. Basic Food Microbiology, Comprehensive Reviews in Food Science and Food Safety, Springer.

In general, pathogens do not grow, or grow very slowly, at pH levels below 4.6; but there are exceptions. Many pathogens can survive in foods at pH levels below their growth minima. It has been reported that *C. botulinum* was able to produce toxin as low as pH 4.2, but these experiments were conducted with high inoculum levels (10^3 -10^4 CFU/g up to 10^6 CFU/g), in soy peptone, and with the presence of *Bacillus* spp. The panel did not consider these results to be relevant to the foods under consideration in this report. It should also be noted that changes in pH can transform a food into one which can support growth of pathogens. For example, several botulism outbreaks have been traced to foods in which the pH increased due to mold growth. These are important considerations when determining the shelf life of a food formulation. Based on a comprehensive review of the literature, the panel concluded that a pH of 4.6 is appropriate to control spore-forming pathogens.

Ferreira and Lund (1987) reported that six out of 13 strains of *Salmonella* spp. representing 12 serovars could grow at pH 3.8 at 30 °C (86 °F) within 1-3 d, and at 20 °C (68 °F) in 3-5 d, when using HCl as an acidulant. Other reports note that certain acids at pH 4.5 inactivate salmonellae. The panel therefore concluded that using a pH minimum of 4.0 for *Salmonella* spp. would not be scientifically substantiated for foods subject to Food Code requirements. Based on a comprehensive review of the literature data, the panel also concluded that it would be scientifically valid to use a pH minimum of 4.2 to control for *Salmonella* spp. and other vegetative pathogens.

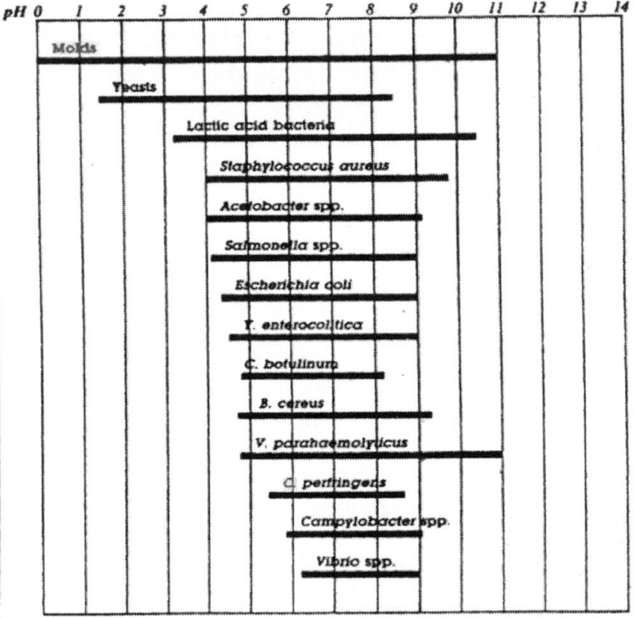

Figure 6.1: Approximate pH Growth Ranges for some Food-borne Organisms.

As with other intrinsic properties, when analyzing multicomponent foods, the pH should be measured not only for each component of the food but also for the interface areas among components and for any potential microenvironment.

Adverse pH affects at least two aspects of a respiring microbial, cell: the functioning of its enzymes and the transport of nutrients into the cell. The cytoplasmic membrane of microorganisms is relatively impermeable to H+ and –OH ions. Their concentration in the cytoplasm therefore probably remains reasonably constant despite wide variations that may occur in the pH of the surrounding medium. The intracellular pH of resting baker's yeast cells was found by Conway and Downey to be 5.8. While the outer region of the cells during glucose fermentation was found to be more acidic, the inner cell remained more alkaline. On the other hand, Pena *et al.,* did not support the notion that the pH of yeast cells remains constant with variations in pH of medium. It appears that the internal pH of most all cells is near neutrality. Bacteria such as *Sulfolobus* and *Methanococcus* may be exceptions, however. When microorganisms are placed in environments below or above neutrality, their ability to proliferate depends upon their ability to bring the environmental pH to a more optimum value or range. When placed in acid environments, the cells must either keep H+ from entering or expel H+ ions as rapidly as they enter. Such key cellular compounds as DNA and ATP require neutrality. When most microorganisms grow in acid media, their metabolic activity results in the medium or substrate becoming less acidic, while those that grow in high pH environments tend to affect a lowering of pH. The amino acid decarboxylases that have optimum activity at around pH 4.0 and almost no activity at pH 5.5 cause a spontaneous adjustment of pH toward neutrality when cells are grown in the acid range. Bacteria such as *Clostridium acetobutylicum* raise substrate pH by reducing butyric acid to butanol, while *Enterobacter aerogenes* produces acetoin from pyruvic acid to raise the pH of its growth environment. When amino acids are decarboxylated, the increase in pH occurs from the resulting amines. When grown in the alkaline range, a group of amino acid deaminases that have optimum activity at about pH 8.0 cause the spontaneous adjustment of pH toward neutrality as a result of the organic acids that accumulate.

With respect to the transport of nutrients the bacterial cell tends to have a residue negative charge. Nonionized compounds can, therefore, enter cells while ionized cannot. At neutral or alkaline pH, organic acids do not enter, while at acid pH values these compounds are non-ionized and can enter the negatively charged cells. Also, the ionic character of side chain ionizable groups is affected on either side of neutrality, resulting in increasing denaturation of membrane and transport enzymes.

Among the other effects that are exerted on microorganisms by adverse pH is that of the interaction between H_+ and the enzymes in the cytoplasmic membrane. The morphology of some microorganisms can be affected by pH. The length of the hyphae of *Penicillium chrysogenum* has been reported to decrease when grown in continuous culture where pH values increased above 6.0. Pellets of mycelium rather than free hyphae were formed at about pH 6.7. Extracellular H_+ and K– may be in competition where the latter stimulates fermentation, for example, while the former represses it. The metabolism of glucose by yeast cells in an acid medium was markedly stimulated by K_+. Glucose was consumed 83 per cent more rapidly in the presence of K_+ under anaerobic conditions and 69 per cent more under aerobic conditions.

As with other factors, pH usually interacts with other parameters in the food to inhibit growth. The pH can interact with factors such as a_w, salt, temperature, redox potential, and preservatives to inhibit growth of pathogens and other organisms. The pH of the food also significantly impacts the lethality of heat treatment of the food. With respect to temperature, pH of substrate becomes more acid as temperature increases. Less heat is needed to inactivate microbes as the pH is reduced. Concentration of salt has definite effect on pH-growth rate. The addition of 0.2 M NaCl broadened the pH growth range of *Alcaligenes faecalis*. A similar result was noted for *Escherichia coli* by these investigators. When salt content exceeds this optimal level, the pH growth range is narrowed. An adverse pH makes cells much more sensitive to toxic agents of a wide variety and young cells are more susceptible to pH changes than older or resting cells.

When microorganisms are grown on either side of their optimum pH range, an increased lag phase results. The increased lag would be expected to be of longer duration if the substrate is a highly buffered one in contrast to one that has poor buffering capacity. In other words, the length of the lag phase may be expected to reflect the time necessary for the organisms to bring the external environment within their optimum pH growth range. Analysis of the substances that are responsible for the adverse pH is of value in determining not only speed of subsequent growth but also minimum/maximum pH at which an organism may grow. In a study of the minimum pH at which salmonellae would initiate growth, Chung and Goepfert found the minimum pH to be 4.05 when hydrochloric and citric acids were used but 5.4 and 5.5 when acetic and propionic acids were used, respectively. This is undoubtedly a reflection of the ability of the organisms to alter their external environment to a more favourable range in the case of hydrochloric and citric acids than with the other acids tested. It is also possible that factors other than pH come into play in the varying abilities of organic acids as growth inhibitors.

Moisture and Water Activity (a_w)

Microorganisms need water in an available form to grow in food products. The control of the moisture content in foods is one of the oldest exploited preservation strategies. Micro-organism has an absolute demand for water, however, the exact amount of water needed for growth of microorganisms varies. This parameter helps us to understand the movement of water from the environment to the cytoplasm or from cytoplasm to the environment. The water requirement of microorganisms is expressed in physical form, called water activity (a_w). Food microbiologists generally describe the water requirements of microorganisms in terms of the water activity (a_w) of the food or environment.

Water activity is defined as the ratio of the vapour pressure of water present in food substrate(solution) to the vapour pressure of pure water at the same temperature.

$$a_w = P/Po$$

where,

P = vapour pressure of water present in food and Po = Vapour pressure of pure water at some temperature

The a_w of pure water is 1.00 and the a_w of a completely dehydrated food is 0.00. The a_w of a food on this scale from 0.00 - 1.00 is related to the equilibrium relative humidity above the food on a scale of 0 - 100 per cent. Thus, per cent Equilibrium Relative Humidity (ERH) = a_w x 100. The a_w of a food describes the degree to which water is "bound" in the food, its availability to participate in chemical/biochemical reactions, and its availability to facilitate growth of microorganisms. Most fresh foods, such as fresh meat, vegetables, and fruits, have a_w values that are close to the optimum growth level of most microorganisms (0.97 - 0.99). It has been reported that reducing water activity below 0.7 prevents microbial spoilage. Even though the food would not spoil from microorganisms, other deteriorative reactions can still occur. To efficiently preserve a food product, water activity would have to be lowered to a range where the rate of deteriorative reactions is minimized.

Microorganisms commonly have optimum and minimum levels of a_w for growth depending on other growth factors in their environments. The food is classified into three categories based on its a_w range

- High-moisture foods (a_w above 0.85)
- Intermediate-moisture foods (a_w 0.60–0.85)
- Low-moisture foods (a_w below 0.60)

High-moisture foods with water activity above 0.85 are highly perishable foods as they are susceptible to the growth of spoilage and pathogenic microorganisms, especially the fast-growing bacteria. Microbial spoilage of intermediate moisture foods is a relatively slow process, caused mainly by yeasts and molds. Low moisture foods are not spoiled by microorganisms unless their moisture content is raised by any means. Microbial growth stops at water activity below the minimal value, but the microbes don't die immediately and may remain dormant in the food for prolonged periods. Almost all fresh foods, such as fresh meat, vegetables, and fruits, have a_w values that are close to the optimum growth level of most microorganisms (0.97-0.99). The addition of solutes such as salt or sugar, physical removal of water through drying or baking, or binding of water to various macromolecular components in the food is some of the means to manipulate a_w in foods (Urmila *et al.,* 2017). Weight for weight, these food components will decrease a_w in the following order: ionic compounds > sugars, polyhydric alcohols, amino acids and other low-molecular-weight compounds > high molecular-weight compounds such as cellulose, protein or starch.

Table 6.4 depicts a_w (app.) values for selected foods at their normal moisture content and the phenomenon that takes place at that particular a_w range. Water activity is inversely related to osmotic pressure; if a solution has high osmotic pressure, its a_w is low. Most spoilage bacteria do not grow below a_w = 0.91, whereas spoilage molds can grow as low as 0.80.

Table 6.4: Approximate a_w Values for Selected Food at their Normal Moisture Content

a_w	*Food*	*Phenomenon*
0.95 - 1.00	Meat (fresh and cooked), fresh poultry, fresh fish, milk, eggs, fruit (fresh and canned in syrup), vegetables (fresh and canned in brine), bread, liver, sausage, pudding, fresh fruit and vegetable juices, some types of cheese	Inhibition of *Pseudomonas, Bacillus, C. perfringens,* and some yeasts
0.90 - 0.95	Cured meat products, some hard cheese, mayonnaise, low-calorie jams, baked cake, refrigerated biscuit dough	Lower limit for bacterial growth; *Salmonella, V. parahaemolyticus, C. botulinum,* and *Lactobacillus.* Some yeast can grow. Inhibition of spoilage bacteria.
0.85 - 0.90	Aged hard cheeses, dry ham, fruit juice concentrates, maple syrup (saturated sucrose solution; a_w 0.86)	Inhibition of many types of yeast and molds.Inhibition of *S. aureus*
0.80 - 0.85	Sweetened condensed milk, parmesan cheese, fruit cake, high-moisture prunes, rice, beans, peas	Lower limit for most enzyme activity and growth of most fungi
0.75 - 0.80	Heavily salted fish, molasses, jams, marmalades, conserves, marzipan, glace fruits, prunes, fondants (saturated NaCl solution; a_w 0.75)	Lower limit for halophilic bacteria
0.65 - 0.75	Some dried fruit (dates, figs, sultanas) and nuts, rolled oats, malt extract, chocolate candies	Lower limit for most xerotolerant molds
0.60 - 0.65	Some dried fruits, honey, caramels, toffee. Maillard reactions can occur.	Lower limit for growth of Osmophilic or xerophilic yeasts and molds
Below 0.60	Dry pasta, spices, biscuits, crackers, dried milk, dried whole egg, dried vegetables, flour, instant coffee, cereals, sugar, noodles	

Source: Christian, 2000; Banwart, 2004; Montville and Matthews, 2005 and Food Microbiology: Principles into Practice, First Edition, Osman Erkmen and, T.Faruk Bozoglu, 2016.

The a_w of most fresh foods is above 0.99. In general, bacteria require higher values of a_w for growth than fungi, with gram-negative bacteria having higher requirements than gram positives. Table 6.5 indicates the needed a_w values for growth of microorganisms.

Table 6.5: Needed a_w Values for Growth of Microorganisms

Organisms	a_w	*Organisms*	a_w
Groups		**Groups**	
Most spoilage bacteria	0.90	Halophilic bacteria	0.75
Most spoilage yeasts	0.88	Xerophilic molds	0.61
Most spoilage molds	0.80	Osmophilic yeasts	0.60
Specific Organisms		**Specific Organisms**	
Clostridium botulinum, Type E	0.97	*Candida scottii*	0.92
Pseudomonas spp.	0.97	*Trichosporon pullulans*	0.91
Acinetobacter spp.	0.96	*Candida zeylanoides*	0.90
Escherichia coli	0.96	*Endomyces vernalis*	0.89
Enterobacter aerogenes	0.95	*Staphylococcus aureus*	0.86
Bacillus subtilis	0.95	*Alternaria citri*	0.84
Clostridium botulinum		*Penicillium patulum*	0.81
Types A and B	0.94	*Aspergillus glaucusa*	0.70
Candida utilis	0.94	*Aspergillus conicus*	0.70
Vibrio parahaemolyticus	0.94	*Aspergillus echinulatus*	0.64
Botrytis cinerea	0.93	*Saccharomyces rouxii*	0.62
Rhizopus stolonifer	0.93	*Monascus bisporus*	
Mucor spinosus	0.93	(*Xeromyces bisporus*)	0.61

Microorganisms respond differently to a_w depending on a number of factors. Microbial growth, and, in some cases, the production of microbial metabolites, may be particularly sensitive to alterations in a_w. Microorganisms generally have optimum and minimum levels of a_w for growth depending on other growth factors in their environments. One indicator of microbial response is their taxonomic classification. For example, Gram (-) bacteria are generally more sensitive to low a_w than Gram (+) bacteria.

Certain relationships have been shown to exist between a_w, temperature and nutrition. First, at any temperature, the ability of microorganisms to grow is reduced as the a_w is lowered. Second, the range of a_w over which growth occurs is greatest at the optimum temperature for growth; and third, the presence of nutrients increases the range of a_w over which the organisms can survive. The specific values given above, then, should be taken only as reference points, since a change in temperature or nutrient content might permit growth at lower values of a_w.

Table 6.5 lists the approximate minimum a_w values for the growth of selected microorganisms relevant to food. It should be noted that many bacterial pathogens are controlled at water activities well above 0.86 and only *S. aureus* can grow and produce toxin below a_w 0.90. It must be emphasized that these are approximate values because solutes can vary in their ability to inhibit microorganisms at the same a_w value. To illustrate, the lower a_w limit for the growth of *Clostridium botulinum* a_w A has been found to be 0.94 with NaCl as the solute versus 0.92 with glycerol as the solute. Just as yeasts and molds grow over a wider pH range than bacteria, the same is true for a_w. The lowest reported value for foodborne bacteria of any type is 0.75 for halophilic (literally, "salt-loving") bacteria, while xerophilic ("dry-loving") molds and osmophilic (preferring high osmotic pressures) yeasts have been reported to grow at a_w values of 0.65 and 0.60, respectively.

Table 6.6: Relationship between Water Activity and Concentration of Salt Solutions

Water Activity	*Sodium Chloride Concentration*	
	Molal	*Per cent, w/v*
0.995	0.15	0.9
0.99	0.30	1.7
0.98	0.61	3.5
0.96	1.20	7
0.94	1.77	10
0.92	2.31	13
0.90	2.83	16
0.88	3.33	19
0.86	3.81	22

With regard to specific compounds used to lower water activity, results akin to those seen with adsorption and desorption systems have been reported. In a study on the minimum a_w for the growth and germination of *Clostridium perfringens*, Kang *et al.*, found the value to be between 0.97 and 0.95 in complex media when sucrose or NaCl were used to adjust a_w but 0.93 or below when glycerol was used. In another study, glycerol was found to be more inhibitory than NaCl to relatively salt-tolerant bacteria but less inhibitory than NaCl to salt-sensitive species when compared at similar levels of a_w in complex media. In their studies on the germination of *Bacillus* and *Clostridium* spores, Jakobsen and Murrell observed strong inhibition of spore germination when a_w was controlled by NaCl and $CaCl_2$, but less inhibition when glucose and sorbitol were used and very little inhibition when glycerol, ethylene glycol, acetamide, or urea was used. The germination of clostridial spores was completely inhibited at a_w 0.95 with NaCl, but no inhibition occurred at the same a_w when urea, glycerol, or glucose was employed. In another study, the limiting a_w for the formation of mature spores by *B. cereus* strain T was shown to be about 0.95 for glucose, sorbitol and NaCl but about 0.91 for glycerol. Both yeasts and molds have been found to be more tolerant to glycerol than to sucrose. Using a glucose minimal medium and *Pseudomonas fluorescens*, Prior found that glycerol permitted growth at lower a_w values than either sucrose or NaCl. It was further shown by this worker that the catabolism of glucose, sodium lactate and DL-arginine was completely inhibited by a_w values greater than the minimum for growth when a_w was controlled with NaCl. The control of a_w with glycerol allowed catabolism to continue at a_w values below that for growth on glucose. In all cases where NaCl was used by this investigator to adjust a_w, substrate catabolism ceased at an a_w greater than the minimum for growth, while glycerol permitted catabolism at lower a_w values than the minimum for growth.

It must be kept in mind, however, that a_w is influenced by other environmental parameters such as pH, temperature of growth and Eh. In their study of the effect of a_w on the growth of *Enterobacter aerogenes* in culture media, Wodzinski and Frazier found that the lag phase and generation time were progressively lengthened until no growth occurred with a lowering of a_w. The minimum a_w was raised, however, when incubation temperature was decreased. When both the pH and temperature of incubation were made unfavorable, the minimum a_w for growth was higher. The interaction between a_w, pH and temperature on the growth of molds on jam was shown by Homer and Anagnostopoulos. The interaction between a_w and temperature was the most significant.

There are some definite effects of lowered a_w on microorganisms. Some bacteria accumulate proline as a response to low a_w and increases in some "pool" amino acids have been reported to occur in some salt-tolerant *Staphylococcus aureus* strains. Using a defined medium, Christian found that *Salmonella oranienburg* at a_w of < 0.97 had the added requirement for the amino acid proline and Christian and Waltho showed that proline stimulated respiration at reduced values of a_w. With *S. aureus* MF-31 in 10 per cent NaCl, proline was shown to accumulate by transport while glutamine accumulated via synthesis. As to the trigger for proline synthesis, K_+ accumulates inside of cells as a_w is lowered and catalyzes the formation of proline precursors. However, proline synthesis is not stimulated by osmotic stress among enteric bacteria but instead by transport from the culture medium. Osmophilic yeasts accumulate polyhydric alcohols to a concentration commensurate with their extracellular a_w. According to Pitt, the xerophilic fungi accumulate compatible solutes or osmoregulators as a consequence of the need for high internal solutes if growth at low a_w is to be possible. In a comparative study of xero-tolerant and non-xerotolerant yeasts to water stress. Edgley and Brown found that *Sacchromyces rouxii* responded to low a_w controlled by polyethylene glycol by retaining within the cells increasing proportions of glycerol. However, the amount did not change greatly, nor did the level of arabitol change appreciably by a_w. On the other hand, a nontolerant *S. cerevisiae* responded to a lowering of a_w by synthesizing more glycerol but retaining less. The *S. rouxii* response to low a_w was at the level of glycerol permeation/transport, while that for *S. cerevisiae* was metabolic. It appears from this study that a low a_w forces *S. cerevisiae* to divert a greater proportion of its metabolic activity to glycerol production accompanied by an increase in the amount of glucose consumed during growth.

While yeasts concentrate polyols as "osmoregulators" and enzyme protectors, halophilic bacteria operate under low a_w conditions by virtue of their ability to accumulate KCl in the same general manner. In the case of halophilic bacteria, KCl is a requirement while osmophilic yeasts have a high tolerance for high solute concentrations.

It is known that the growth of at least some cells may occur in high numbers at reduced a_w values, while at the same time certain extracellular products are not produced. For example, reduced a_w results in the cessation of enterotoxin B production by *S. aureus* even though high numbers of cells are produced at the same time. In the case of *Neurospora crassa*, low a_w resulted in nonlethal alterations of permeability of other cell membrane leading to loss of several essential molecules. Similar results were observed with electrolytes or non-electrolytes.

When formulating foods using a_w as the primary control mechanism for pathogens, it is useful to employ microbiological challenge testing to verify the effectiveness of the reduced a_w when target a_w is near the growth limit for the organism of concern; because a_w limits vary with different solutes or humectants, other measures may provide more precise moisture monitoring for certain products. For example, factors other than a_w are known to control the antibotulinal properties of pasteurized processed cheese spreads (Tanaka and others 1986). Also, a_w may be used in combination with other factors to control pathogens in certain food product. Care should be taken when analyzing multi components foods, because effective measurements of a_w may not reflect the actual value in a microenvironment or in the interface among the different components. In these cases, the a_w should be measured at the interface areas of the food, as well as in any potential microenvironment.

Nutrient Content

Microorganisms require certain basic nutrients for growth and maintenance of metabolic functions. The kinds and proportional of nutrient in food are all important in determining which microorganism (microorganisms) is most likely to grow. These nutrients include water, a source of energy, nitrogen, vitamins, and minerals. Varying amounts of these nutrients are present in foods. Meats have abundant protein, lipids, minerals, and vitamins. Most muscle foods have low levels of carbohydrates. Plant foods have high concentrations of different types of carbohydrates and varying levels of proteins, minerals, and vitamins. Foods such as milk and milk products and eggs are rich in nutrients. The role of water has been discussed above.

Foodborne microorganisms can derive energy from carbohydrates, alcohols, and amino acids. In general, the simple compound is utilized first by the measuring microorganisms. The carbohydrate (simple sugar) is most commonly utilized as an energy source. Most microorganisms will metabolize simple sugars such as glucose. Others can metabolize more complex carbohydrates, such as starch or cellulose found in plant foods, or glycogen found in muscle foods. In general, mould can grow in the higher concentration of sugar, yeast in fairly high concentration but most bacteria grow best in the low concentration of sugar. Some microorganisms can use fats as an energy source; but fats are attacked, by a relatively small number of microbes in foods.

Protein-rich food like meat, egg, fish *etc.* are always spoiled by protolytic organism because they can utilize protein as a source of energy if sugar is not available. The primary nitrogen sources utilized by heterotrophic microorganisms are amino acids. Amino acids serve as a source of nitrogen and energy and are utilized by most microorganisms. Some microorganisms are able to metabolize peptides and more complex proteins. Other sources of nitrogen include, for example, urea, ammonia, creatinine, and methylamines. In fact, protein-rich food promotes more growth of bacteria then yeast and mould.

Examples of minerals required for microbial growth include phosphorus, iron, magnesium, sulfur, manganese, calcium, and potassium. In general, small amounts of these minerals are required; thus, a wide range of foods can serve as good sources of minerals. In general, the Gram (+) bacteria are more fastidious in their nutritional requirements and thus are not able to synthesize certain nutrients required for growth. For example, the Gram (+) foodborne pathogen *S. aureus* requires amino acids, thiamine, and nicotinic acid for growth. Fruits and vegetables that are deficient in B vitamins do not effectively support the growth of these microorganisms.

Microorganisms may require B vitamins in low quantities and most all natural foods tend to have an abundant quantity for those organisms that are unable to synthesize their essential requirements. In general, gram-positive bacteria are the least synthetic and must, therefore, be supplied with one or more of these compounds before they will grow. The gram-negative bacteria and molds are able to synthesize most or all of their requirements. Consequently, these two groups of organisms may be found growing on foods low in B vitamins. Fruits tend to be lower in B vitamins, than meats and this fact along with the usual low pH and positive Eh of fruits all help to explain the usual spoilage of these products by molds rather than bacteria.

The Gram (-) bacteria are generally able to derive their basic nutritional requirements from the existing carbohydrates, proteins, lipids, minerals, and vitamins that are found in a wide range of food (Jay, 2000). An example of a pathogen with specific nutrient requirements is *Salmonella enteritidis*. Growth of *Salmonella enteritidis* may be limited by the availability of iron. For example, the albumen portion of the egg, as opposed to the yolk, includes antimicrobial agents and limited free iron that prevent the growth of *Salmonella enteritidis* to high levels. Clay and Board demonstrated that the addition of iron to an inoculum of *Salmonella enteritidis* in egg albumen resulted in growth of the pathogen to higher levels compared to levels reached when a control inoculum (without iron) was used.

The microorganisms that usually predominate in foods are those that can most easily utilize the nutrients present. Generally, the simple carbohydrates and amino acids are utilized first, followed by the more complex forms of these nutrients. The complexity of foods in general is such that several microorganisms can be growing in a food at the same time. The rate of growth is limited by the availability of essential nutrients. The abundance of nutrients in most foods is sufficient to support the growth of a wide range of foodborne pathogens. Thus, it is very difficult and impractical to predict the pathogen growth or toxin production based on the nutrient composition of the food.

Biological Structure

The natural covering of some foods provides excellent protection against the energy of microorganism and spoilage of food by such microorganisms. Plant and animal derived foods, especially in the raw state, have biological structures that may prevent the entry and growth of pathogenic microorganisms. Examples of such physical barriers include testa of seeds, skin of fruits and vegetables, shell of nuts, animal hide, egg cuticle, shell, and membranes. Plant and animal foods may have pathogenic microorganisms attached to the surface or trapped within surface folds or crevices. Intact biological structures thus, can be important in preventing entry and subsequent growth of microorganisms. Several factors may influence penetration of these barriers. The maturity of plant foods will influence the effectiveness of the protective barriers. Physical damage due to handling during harvest, transport, or storage, as well as invasion of insects can allow the penetration of microorganisms (Mossel and others, 1995; Jay, 2000). During the preparation of foods, processes such as slicing, chopping, grinding, and shucking will destroy the physical barriers. Thus, the interior of the food can become contaminated and growth can occur depending on the intrinsic properties of the food. For example, *Salmonella* spp. have been shown to grow on the interior of portions of cut cantaloupe, watermelon, honeydew melons, and tomatoes, given sufficient time and temperature (Urmila *et al.,* 2017).

Fruits are an example of the potential of pathogenic microorganisms to penetrate intact barriers. After harvest, pathogens will survive but usually not grow on the outer surface of fresh fruits and vegetables. Growth on intact surfaces is not common because foodborne pathogens do not produce the enzymes necessary to break down the protective outer barriers on most produce. This outer barrier restricts the availability of nutrients and moisture. One exception is the reported growth of *E. coli* O157:H7 on the surface of watermelon and cantaloupe rinds. Survival of foodborne pathogens on produce is significantly enhanced once the protective epidermal barrier has been broken either by physical damage, such as punctures or bruising, or by degradation by plant pathogens (bacteria or fungi). These conditions can also promote the multiplication of pathogens, especially at higher temperatures. Infiltration of fruit was predicted and described by Bartz and Showalter based on the general gas law, which states that any change in pressure of an ideal gas in a closed container of constant volume is directly proportional to a change in temperature of the gas. In their work, Bartz and Showalter describe a tomato; however, any fruit, such as an apple, can be considered a container that is not completely closed. As the container or fruit cools, the decrease in internal gas pressure results in a partial vacuum inside the fruit, which then results in an influx from the external environment. For example, an influx of pathogens from the fruit surface or cooling water could occur as a result of an increase in external pressure due to immersing warm fruit in cool water (Urmila *et al.*, 2017). Internalization of bacteria into fruits and vegetables could also occur due to breaks in the tissues or through morphological structures in the fruit itself, such as the calyx or stem scar. Although infiltration was considered a possible scenario, the panel concluded that there is insufficient epidemiological evidence to require refrigeration of intact fruit.

The egg is another good example of an effective biological structure that, when intact, will prevent external microbial contamination of the perishable yolk; contamination is possible, however, through transovarian infection. For the interior of an egg to become contaminated by microorganisms on the surface, there must be penetration of the shell and its membranes. In addition, the egg white contains antimicrobial factors. When there are cracks through the inner membrane of the egg, microorganisms penetrate into the egg. Factors such as temperature of storage, relative humidity, age of eggs, and level of surface contamination will influence internalization. For example, conditions such as high humidity and wet and dirty shells, along with a drop in the storage temperature will increase the likelihood for entry of bacteria. If eggs are washed, the wash water should be 12 °C (22 °F) higher than the temperature of the eggs to prevent microbial penetration. After washing, the eggs should be dried and then cooled. The Food and Drug Administration (FDA) published a final rule that applies to shell eggs that have not be processed to destroy all live *Salmonella* before distribution to the consumer. The rule mandates that eggs should be kept dry and chilled below 7.2 °C (45 °F) to prevent growth of *Salmonella enteritidis.* Heating of food as well as other types of processing will break down protective biological structures and alter such factors as pH and a_w. These changes could potentially allow the growth of microbial pathogens.

Redox Potential

The oxidation-reduction or redox potential of a substance is defined in terms of the ratio of the total oxidizing (electron accepting) power to the total reducing (electron donating) power of the substance. In effect, redox potential is a measurement of the ease by which a substance gains or loses electrons. The redox potential (Eh) is measured in terms of millivolts. A fully oxidized standard oxygen electrode will have an Eh of +810 mV at pH 7.0, 30 °C (86 °F), and under the same conditions, a completely reduced standard hydrogen electrode will have an Eh of -420 mV. The Eh is dependent on the pH of the substrate; normally the Eh is taken at pH 7.0 (Jay 2000, p 45-7).

The major groups of microorganisms based on their relationship to Eh for growth are aerobes, anaerobes, facultative aerobes, and microaerophiles. Examples of foodborne pathogens for each of these classifications include *Aeromonas hydrophila*, *Clostridium botulinum*, *Escherichia coli* O157:H7, and *Campylobacter jejuni*, respectively. Generally, the range at which different microorganisms can grow are as follows: aerobes +500 to +300 mV; facultative anaerobes +300 to -100 mV; and anaerobes +100 to less than -250 mV (Ray 1996, p 69-70). For example, *C. botulinum* is a strict anaerobe that requires an Eh of less than +60 mV for growth; however, slower growth can occur at higher Eh values.

The relationship of Eh to growth can be significantly affected by the presence of salt and other food constituents. For example, in one study with smoked herring, toxin was produced in inoculated product stored at 15 °C (59 °F) within three days at an Eh of +200 to +250 mV. In this case, the major oxidant would be trimethylamine oxide, which becomes the electron acceptor for *C. botulinum*. The anaerobe *Clostridium perfringens* can initiate growth at an Eh close to +200 mV; however, in the presence of increasing concentrations of certain substances, such as salt, the limiting Eh increases.

The Eh values for various foods can be highly variable depending on changes in the pH of the food, microbial growth, packaging, the partial pressure of oxygen in the storage environment, and ingredients and composition (protein, ascorbic acid, reducing sugars, oxidation level of cations, and so on). Another important factor is the poising capacity of the food. Poising capacity, which is analogous to buffering capacity, relates to the extent to which a food resists external affected changes in Eh. The poising capacity of the food will be affected by oxidizing and reducing constituents in the food as well as by the presence of active respiratory enzyme systems. Fresh fruits and vegetables and muscle foods will continue to respire; thus, low Eh values can result (Morris, 2000).

The measurement of redox potential of food is done rather easily, either for single or multicomponent foods. For multicomponent foods, in addition to measurement of each component, the redox potential of the interface areas and microenvironments should be considered. However, difficulties arise in taking accurate measurements and in accounting for the differences throughout the food and the equilibrium at the point of measurement. According to Morris (2000): "This imposes the further requirements 1) that the measuring electrode be so prepared and calibrated that it gives stable and reproducible readings, and 2) that a foodstuff is tested in a manner that does not cause any change in the potential that is to be measured. it would be unwise to use redox potential information in isolation to predict food safety, or to rely exclusively on control of redox potential as the means of preventing growth of specific microorganisms." Redox measurements could possibly be used in combination with other factors to evaluate the potential for pathogen growth. However, the limitations discussed above make it a rather difficult and variable factor that could result in erroneous conclusions in the absence of other comprehensive information.

Effect of Aerobes on Redox Potential

Microorganisms affect the Eh of their environments during growth just as they do pH. This is true especially of aerobes, which can lower the Eh of their environment. As aerobes grow, O_2 in the medium is depleted, resulting in the lowering of Eh. Growth is not slowed, however, as much as might be expected due to the ability of cells to make use of O_2-donating or hydrogen-accepting substances in the medium. The result of this is that the medium becomes poorer in oxidizing and richer in reducing substances. The Eh of a medium can be reduced by microorganisms by their production of certain metabolic by-products such as H_2S, which has the capacity to lower Eh to –300 mv. Since H_2S reacts readily with O_2, it will accumulate only in anaerobic environments.

As noted above, Eh is dependent upon the pH of the substrate and the direct relationship between these two factors is the rH value defined in the following way:

$$Eh = 2.303 \frac{RT}{F} (rH - 2pH)$$

where,

R = 8.315 joules,

F = 96,500 coulombs, and

T = Absolute temperature.

Therefore, the pH of a substrate should be stated when Eh is given. Normally Eh is taken at pH 7.0 (expressed Eh'). When taken at pH 7.0, 25°C and with all concentrations at 1.0 M. Eh = Eh'_0. In nature, Eh tends to be more negative under progressively alkaline conditions.

Among naturally occurring nutrient, ascorbic acid and reducing sugars in plants and fruits and –SH groups in meats are of primary importance. The presence or absence of appropriate quantities of oxidizing/reducing agents in a medium is, therefore, or obvious value to the growth and activity of all microorganisms.

While the growth of anaerobes is normally believed to occur at reduced values of Eh, the exclusion of O_2 may be necessary for some anaerobes. When *Clostridium perfringens, Bacteroides fragilis* and *Peptococcus magnus* were cultured in the presence of O_2, inhibition of growth occurred even when the medium was at a negative Eh of –50 mv. These investigators found that growth occurred in media with an Eh as high as + 325 mv when no O_2 was present.

With regard to the effect of Eh on lipid production by *Saccharomyces cerevisiae*, it has been shown that anaerobically grown cells produce a lower total level, a highly variable glyceride fraction and decreased phospholipid and sterol components than aerobically grown cells. The lipid produced by anaerobically grown cells was characterized by a high content (up to 50 per cent of total acid) of 8: 0 to 14: 0 acids and a low level of unsaturated fatty acid in the phospholipid fraction. In aerobically grown cells, 80–90 per cent of the fatty acid component was associated with glyceride and the phospholipid was found to be 16: 1 and 18: 1 acids. Unlike aerobically grown cells, anaerobically grown *S. cerevisiae* cells were found to have a lipid and sterol requirement.

Naturally Occurring and Added Antimicrobials

Some foods intrinsically contain naturally-occurring antimicrobial compounds that convey some level of microbiological stability to them. There are a number of plant-based antimicrobial constituents, including many essential oils, tannins, glycosides, and resins, that can be found in certain foods. Specific examples include eugenol in cloves, allicin in garlic, cinnamic aldehyde and eugenol in cinnamon, allyl isothiocyanate in mustard, eugenol and thymol in sage, and carvacrol (isothymol) and thymol in oregano. The hydroxycinnamic acid derivatives (*p*-coumaric, ferulic, caffeic and chlorogenic acids) found in fruits, vegetables, tea, molasses and other plant sources all show antibacterial and some antifungal activity. Other plant-derived antimicrobial constituents include the phytoalexins and the lectins. Lectins are proteins that can specifically bind to a variety of polysaccharides, including the glycoproteins of cell surfaces. Through this binding, lectins can exert a slight antimicrobial effect. The usual concentration of these compounds in formulated foods is relatively low, so that the antimicrobial effect alone is slight. However, these compounds may produce greater stability in combination with other factors in the formulation.

Some animal-based foods also contain antimicrobial constituents. Examples include lactoferrin, conglutinin and the lactoperoxidase system in cow's milk, lysozyme in eggs and milk, and other factors in fresh meat, poultry and seafood. Lysozyme is a small protein that can hydrolyze the cell wall of bacteria. The lactoperoxidase system in bovine milk consists of three distinct components that are required for its antimicrobial action: lactoperoxidase, thiocyanate, and hydrogen peroxide. Gram (-) psychotrophs such as the pseudomonads have been shown to be very sensitive to the lactoperoxidase system. Consequently, this system, in an enhanced form, has been suggested to improve the keeping quality of raw milk in developing countries where adequate refrigeration is scarce. Similar to the plant-derived antimicrobial compounds, the animal-derived compounds have a limited effect on ambient shelf life of foods.

It is also known that some types of food processing result in the formation of antimicrobial compounds in the food. The smoking of fish and meat can result in the deposition of antimicrobial substances onto the product surface. Maillard compounds resulting from condensation reactions between sugars and amino acids or peptides upon heating of certain foods can impart some antimicrobial activity. Smoke condensate includes phenol, which is not only an antimicrobial, but also lowers the surface pH. Some processors also lower the surface pH with liquid smoke to achieve an unsliced shelf-stable product. Some types of fermentations can result in the natural production of antimicrobial substances, including bacteriocins, antibiotics, and other related inhibitors. Bacteriocins are proteins or peptides that are produced by certain strains of bacteria that inactivate other, usually closely related, bacteria. The most commonly characterized bacteriocins are those produced by the lactic acid bacteria. The antibiotic nisin produced by certain strains of *Lactococcus lactis* is one of the best characterized of the bacteriocins. Nisin is approved for food applications in over 50 countries around the world. Nisin's first food application was to prevent late-blowing in Swiss cheese by *Clostridium butyricum*. Nisin is a polypeptide that is effective against most Gram (+) bacteria but is ineffective against Gram (-) organisms and fungi. Nisin can be produced in the food by starter cultures or, more commonly, it can be used as an additive in the form of a standardized preparation. Nisin has been used to effectively control spore-forming organisms in processed cheese formulations, and has been shown to have an interactive effect with heat. For example, an Fo process for conventional low acid canned foods may be in the 6 - 8 range, but with the addition of nisin, can be reduced to a Fo of 3 for inactivating thermophilic spores. There are a number of other bacteriocins and natural antimicrobials that have been described, however, these have found very limited application in commercial use as food preservatives because of their restricted range of activity, limited compatibility with the food formulation or their regulatory status.

In addition to naturally-occurring antimicrobial compounds in foods, a variety of chemical preservatives and additives can extend the shelf life of food and/or inhibit pathogens, either singly or in combination. The most frequently used preservatives in various foods include nitrate, nitrite, sulfur dioxide, acetic acid, propionic acid, sorbic acid, benzoic acid, BHA and BHT, smoke, nisin, parabens *etc.* The selection and use of these preservatives is typically governed by food law regulation of a country or region of the world. A number of criteria should be followed when selecting a preservative for a specific food application. Ideally, the preservative should have a wide spectrum of activity against the target spoilage organisms and pathogens expected to be encountered in the food. The preservative must be active for the desired shelf life of the food and under the expected formulation conditions in the food. It should cause minimal organoleptic impact on the food and should not interfere with desirable microbiological processes expected to occur in the food, such as the ripening of cheese or leavening of baked goods.

Added antimicrobial compounds can have an interactive or synergistic effect with other parameters of the formulation. One example is the interaction with pH. Many preservatives have an optimum pH range for effectiveness. Other factors include a_w, presence of other preservatives, types of food constituents, presence of certain enzymes, processing temperature, storage atmosphere, and partition coefficients. The effective use of combinations of preservatives with other physico-chemical parameters of a food formulation can stabilize that food against spoilage organisms or pathogens. Leistner systematically developed the "hurdle concept" to describe these effects (Leistner, 1995). The hurdle concept states that several inhibitory factors (hurdles), while individually unable to inhibit microorganisms, will, nevertheless, be effective in combination. A classic example of applying the hurdle concept is the anti-botulinal stability of certain shelf stable processed cheese formulations. Combinations of moisture, total salt, and pH have been shown to allow for the safe storage of these products at room temperature for extended time even though the individual factors, taken singly, would not support that practice. In combination products, the effectiveness of an antimicrobial may be altered by other factors including the potential for migration of the antimicrobial to other components of the food and the different food parameters at the interface areas.

There are a number of food formulations that, either by addition of preservatives or through the application of the hurdle concept do not require refrigeration for microbiological stability or safety. However, in the absence of a well-defined and validated microbiological model, it is usually difficult to evaluate the microbiological safety of these products. In the majority of these cases, the application of appropriate microbiological challenge testing is the most effective tool for judging the suitability of these formulations for non-refrigerated storage

Competitive Microflora

The potential for microbial growth of pathogens in temperature-sensitive foods depends on the combination of the intrinsic and extrinsic factors, and the processing technologies that have been applied. Within the microbial flora in a food, there are many important biological attributes of individual organisms that influence the species that predominates. These include the individual growth rates of the microbial strains and the mutual interactions or influences among species in mixed populations.

Growth

In a food environment, an organism grows in a characteristic manner and at a characteristic rate. The length of the lag phase, generation time, and total cell yield are determined by genetic factors. Accumulation of metabolic products may limit the growth of particular species. If the limiting metabolic product can be used as a substrate by other species, these may take over (partly or wholly), creating an association or succession. Due to the complex of continuing interactions between environmental factors and microorganisms, a food at any one point in time has a characteristic flora, known as its association. The microbial profile changes continuously and one association succeeds another in what is called succession. Many examples of this phenomenon have been observed in the microbial deterioration and spoilage of foods. As long as metabolically active organisms remain, they continue to interact, so that dominance in the flora occurs as a dynamic process. Based on their growth-enhancing or inhibiting nature, these interactions are either antagonistic or synergistic.

Competition

In food systems, antagonistic processes usually include competition for nutrients, competition for attachment/adhesion sites (space), unfavourable alterations of the environment, and a combination of these factors. Early studies demonstrated that the natural biota of frozen pot pies inhibited inoculated cells of *S. aureus*, *E. coli* and *Salmonella typhimurium*. Another example of this phenomenon is raw ground beef. Even though *S. aureus* is often found in low numbers in this product, staphylococcal enterotoxin is not produced. The reason is that the *Pseudomonas-Acinetobacter-Moraxella* association that is always present in this food grows at a higher rate, outgrowing the staphylococci. Organisms of high metabolic activity may consume required nutrients, selectively reducing these substances, and inhibiting the growth of other organisms. Depletion of oxygen or accumulation of carbon dioxide favours facultative obligate anaerobes which occur in vacuum-packaged fresh meats held under refrigeration. *Staphylococci* are particularly sensitive to nutrient depletion. *Coliforms* and *Pseudomonas* spp. may utilize amino acids necessary for staphylococcal growth and make them unavailable. Other genera of *Micrococcaceae* can utilize nutrients more rapidly than *staphylococc*i. *Streptococci* inhibit *staphylocci* by exhausting the supply of nicotinamide or niacin and biotin. *Staphylococcus aureus* is a poor competitor in both fresh and frozen foods. At temperatures that favour staphylococcal growth, the normal food saprophytic biota offers protection against staphylococcal growth through antagonism, competition for nutrients, and modification of the environment to conditions less favourable to *S. aureus*. Changes in the composition of the food, as well as changes in intrinsic or extrinsic factors may either stimulate or decrease competitive effects.

Effects on Growth Inhibition

Changes in growth stimulation have been reported among several foodborne organisms, including yeasts, micrococci, streptococci, lactobacilli and Enterobacteriaceae. Growth stimulating mechanisms can have a significant influence on the build-up of a typical flora. There are several of these mechanisms, a few of which are listed below:

- Metabolic products from one organism can be absorbed and utilized by other organisms.
- Changes in pH may promote the growth of certain microorganisms. An example is natural fermentations, in which acid production establishes the dominance of acid tolerant organisms such as the lactic acid bacteria. Growth of molds on high acid foods has been found to raise the pH, thus stimulating the growth of *C. botulinum*.
- Changes in Eh or a_w in the food can influence symbiosis. At warm temperatures, *C. perfringens* can lower the redox potential in the tissues of freshly slaughtered animals so that even more obligately anaerobic organisms can grow.

- ✰ There are some associations where maximum growth and normal metabolic activity are not developed unless both organisms are present. This information can be used in the hurdle concept to control microorganisms in temperature sensitive foods.

Extrinsic Factors

Extrinsic parameters are environmental factors, in which food and food products are kept. Extrinsic parameters are substrate independent and affect both micro-organism (mos) as well as food. Unlike intrinsic parameters, extrinsic parameters can be maintained and regulated well. The extrinsic parameters include:

i) Temperature of storage

ii) Relative humidity of the environment

iii) Presence and concentration of gases

iv) Presence and activities of micro-organism

Storage Temperature and Time

Microorganisms grow over a very wide range of temperatures (between -34 °C to 100 °C), and each microorganism has a defined temperature range in which they grow, with a minimum, maximum, and optimum. It is highly important parameter that affects the spoilage of highly perishable food. Therefore, it would be well to consider at this point the temperature growth ranges for organisms of importance in foods as an aid in selecting the proper temperature for the storage of different types of foods.

Temperature of storage is the single most important parameter that affects the spoilage of highly perishable foods and this fact has been emphasized by the work of Olley and Ratkowsky and their co-workers. According to these investigators, spoilage can be predicted by a spoilage rate curve. The general spoilage curve has been incorporated into the circuitry of a temperature function integrator which reads out the equivalent days of storage at 0°C and thus, makes it possible to predict the remaining shelf life at 0°C. It has been shown that the rate of spoilage of fresh, poultry at 10°C is about twice that at 5°C and that at 15°C is about three times that at 5°C.

Temperature has dramatic impact on both the generation time of an organism and its lag period. Over a defined temperature range, the growth rate of an organism is classically defined as an Arrhenius relationship. The log growth rate constant is found to be proportional to the reciprocal of the absolute temperature:

$$G = -\mu/2.303\ RT$$

where,

G = Log growth rate constant

μ = Remperature characteristic (constant for a particular microbe)

R = Gas constant

T = Temperature (°K)

The above relationship holds over the linear portion of the Arrhenius plot. However, when temperatures approach the maxima for a specific microorganism, the growth rate declines more rapidly than when temperatures approach the minima for that same microorganism. Instead of using the above Arrhenius law equation, the following was developed to describe the relationship between temperature and growth rate of microorganisms between the minimum and optimum temperatures:

$$r = b(T - T0)$$

where,

r = The growth rate;

b = The slope of the regression line; and

$T0$ = A conceptual temperature of no metabolic significance.

This relationship predicts more accurately the growth rates of microorganisms at low temperatures. At low temperatures, two factors govern the point at which growth stops:

1. Reaction rates for the individual enzymes in the organism become much slower, and
2. Low temperatures reduce the fluidity of the cytoplasmic membrane, thus interfering with transport mechanisms

At high temperatures, structural cell components become denatured and inactivation of heat-sensitive enzymes occurs. While the growth rate increases with increasing temperature, the rate tends to decline rapidly thereafter, until the temperature maximum is reached.

The relationship between temperature and growth rate constant varies significantly across groups of microorganisms. Four major groups of microorganisms have been described based on their temperature ranges for growth: thermophiles, mesophiles, psychrophiles, and psychrotrophs. Tables 6.7 and 6.8 list the temperature ranges for these four groups and for pathogens of concern. The optimum temperature for growth of thermophiles is between 55 to 65 °C (131 to 149 °F) with the maximum as high as 90 °C (194 °F) and a minimum of around 40 °C (104 °F). Mesophiles, which include virtually all human pathogens, have an optimum growth range of between 30 °C (86 °F) and 45 °C (113 °F), and a minimum growth temperature ranging from 5 to 10 °C (41 to 50 °F). Psychrophilic organisms have an optimum growth range of 12 °C (54 °F) to 15 °C (59 °F) with a maximum range of 15 °C (59 °F) to 20 °C (68 °F). There are very few true psychrophilic organisms of consequence to foods.

Psychrotrophs such as *L. monocytogenes* and *C. botulinum type E* are capable of growing at low temperatures (minimum of - 0.4 °C [31 °F] and 3.3 °C [38 °F], respectively, to 5 °C [41 °F]), but have a higher growth optimum range (37 °C [99 °F] and 30 °C [86 °F], respectively) than true psychrophiles. Psychrotrophic organisms are much more relevant to food and include spoilage bacteria, spoilage yeast and molds, as well as certain foodborne pathogens.

Growth temperature is known to regulate the expression of virulence genes in certain foodborne pathogens. For example, the expression of proteins governed by the Yersinia enterocolitica virulence plasmid is high at 37 °C (99 °F), low at 22 °C (72 °F), and not detectable at 4 °C (39 °F). Growth temperature also impacts an organism's thermal sensitivity. *Listeria monocytogenes*, when held at 48 °C (118 °F) in inoculated sausages, has an increase of 2.4-fold in its D value at 64 °C (147 °F). It must be emphasized that the lag period and growth rate of a microorganism are influenced not only by temperature but by other intrinsic and extrinsic factors as well. *Salmonellae* do not grow at temperatures below 5.2 °C (41 °F). The intrinsic factors of the food product, however, have been shown to impact the ability of salmonellae to grow at low temperatures. *Staphylococcus* aureus has been shown to grow at temperatures as low as 7 °C (45 °F), but the lower limit for enterotoxin production has been shown to be 10 °C (50 °F). In general, toxin production below about 20 °C (68 °F) is slow. For example, in laboratory media at pH 7, the time to produce detectable levels of enterotoxin ranged from 78 - 98 h at 19 °C (66 °F) to 14-16 h at 26 °C (79 °F) (ICMSF 1980, p 10). Less favourable conditions, such as reduced pH, slowed enterotoxin production even further.

Table 6.7: Temperature Ranges for Prokaryotic Microorganisms

Group	*Temperature °C (°F)*		
	Minimum	*Optimum*	*Maximum T*
Thermophiles	40 - 45 (104 - 113)	55 - 75 (131 - 167)	60 - 90 (140 - 194)
Mesophiles	5 - 15 (41 - 59)	30 - 45 (86 - 113)	35 - 47 (95 - 117)
Psychrophiles	-5 - +5 (23 - 41)	12 - 15 (54 - 59)	15 - 20 (59 - 68)
Psychrotrophs	-5 - +5 (23 - 41)	25 - 30 (77 - 86)	30 - 35 (86 - 95)

Source: Table 1.1 in ICMSF 1980, p 4.

Table 6.8: Approximate Minimum, Maximum and Optimum Temperature Values in °C (°F) Permitting Growth of Selected Pathogens Relevant to Food

Organism	*Minimum*	*Optimum*	*Maximum*
Bacillus cereus	5 (41)	28 - 40 (82 - 104)	55 (131)
Campylobacter spp.	32 (90)	42 - 45 (108 - 113	45 (113)
Clostridium botulinum types A and B*	10 - 12 (50 - 54)	30 - 40 (86 - 104)	50 (122)
Clostridium botulinum type E**	3 - 3.3 (37 - 38)	25 - 37 (77 - 99)	45 (113)
Clostridium perfringens	12 (54)	43 - 47 (109 - 117)	50 (122)
Enterotoxigenic Escherichia coli	7 (45)	35 - 40 (95 - 104)	46 (115)
Listeria monocytogenes	0 (32)	30 - 37 (86 - 99)	45 (113)
Salmonella spp.	5 (41)	35 - 37 (95 - 99)	45 - 47 (113 - 117)
Staphylococcus aureus growth	7 (45)	35 - 40 (95 - 104)	48 (118)
Staphylococcus aureus toxin	10 (50)	40 - 45 (104 – 113)	46 (115)
Shigella spp.	7 (45)	37 (99)	45 - 47 (113 - 117)
Vibrio cholerae	10 (50)	37 (99)	43 (109)
Vibrio parahaemolyticus	5 (41)	37 (99)	43 (109)
Vibrio vulnificus	8 (46)	37 (99)	43 (109)
Yersinia enterocolitica	-1 (30)	28 - 30 (82 - 86)	42 (108)

* Proteolytic; ** Non-proteolytic.

Source: ICMSF 1996; Lund and others 2000; Doyle and others 2001.

Table 6.9: The Relationship of pH and Temperature to Growth Rate of *Clostridium perfringens* (Welchii) F2985/50

Incubation Temperature	*Hours to Visible Turbidity in RCM Broth at pH*	
	5.8	*7.2*
15 °C (59 °F)	>700	>700
20 °C (68 °F)	74	48
25 °C (77 °F)	30	24
30 °C (86 °F)	24	8
37 °C (99 °F)	5	5

Source: Table 1.3 in ICMSF 1980, p 10.

As the temperature increases above the optimum, the growth rate declines much more sharply as a result of the irreversible denaturation of enzymes and proteins, and breakdown of the cytoplasmic membrane. At the temperatures above the maximum, these changes are sufficient to kill the microorganism.

As the temperature decreases from the optimum, growth rates lowers due to the slowing enzyme reactions within the cell. Many microorganisms grow at lower temperatures by increasing the quantity of unsaturated fatty acids in their membrane lipids, and psychrophiles usually have higher levels of unsaturated fatty acids than mesophiles. Increasing the degree of unsaturation in fatty acids decreases its melting point. Membranes containing higher levels of unsaturated fatty acid will remain fluid and functional at lower temperatures. However, the cytoplasmic membrane fluidity is reduced at low temperature in mesophile, and thermoduric microorganisms and nutrient transport through membrane are reduced.

With regard to bacteria, psychrotrophic species and strains are found among the following genera: *Alcaligenes, Alteromonas, Brochothrix, Corynebacterium, Flavobacterium, Lactobacillus, Micrococcus, Pseudomonas, Streptococcus, Streptomyces* and others. The psychrotrophs found most commonly on foods are those that belong to the genera *Alcaligenes, Pseudomonas* and *Streptococcus.*

These organisms grow well at refrigerator temperatures and cause spoilage of meats, fish, poultry, eggs and other foods normally held at this temperature. Standard plate counts of viable organisms on such foods are generally higher when the plates are incubated at about 7°C for at least 7 days than when incubated at 30°C and above. Mesophilic species and strains are known among all genera and may be found on foods held at refrigerator temperature. They apparently do not grow at this temperature but do grow at temperatures within the mesophilic range if other conditions are suitable. It should be pointed out that some organisms can grow over a range from 0° and 30°C or above. One such organism is *Streptococcus faecalis.*

Most thermophilic bacteria of importance in foods belong to the genera *Bacillus* and *Clostridium*. While only a few species of these genera are thermophilic, they are of great interest to the food microbiologist and food technologist in the canning industry.

Just as molds are able to grow over wider ranges of pH, osmotic pressure and nutrient content, they are also able to grow over wider ranges of temperature than bacteria. Many molds are able to grow at refrigerator temperatures, notably some strains of *Aspergillus, Cladosporium* and *Thamnidium*, which may be found growing on eggs, sides of beef and fruits. Yeasts grow over the psychrophilic and mesophilic temperature ranges but generally not within the thermophilic range.

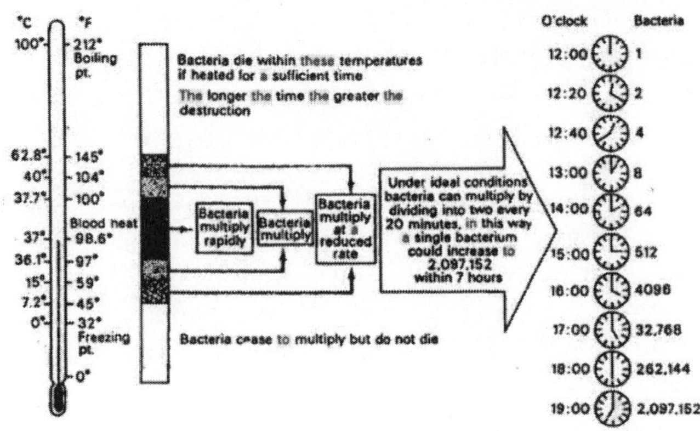

Figure 6.2: Effect of Temperature and Time on the Growth of Bacteria.

Thus, all microorganisms have a defined temperature range in which they grow, with a minimum, maximum, and optimum. The quality of the food product must also be taken into account in selecting a storage temperature. In every case, the success of storage temperature depends to a great extent upon the RH of the storage environment and the presence or absence of gases such as CO_2 and O_3. Hence, an understanding of the interplay between time, temperature, and other intrinsic and extrinsic factors is crucial to selecting the proper storage conditions for a food product.

Relative Humidity

Relative humidity is the ratio expressed as the percentage of moisture in the air to the moisture present in food under the saturation condition at temperature and pressure. The RH of the storage environment is important for the growth of microorganisms on the surface area of food. The relative humidity in which a food is held will influence on the water activity of that product and influence on the growth of microorganisms on the surface of a product. When foods with low a_w values are placed in an environment of high RH, foods pickup moisture until equilibrium has been established. Likewise, food with a high a_w loses moisture when placed in an environment of low RH. If food has low a_w, it will need a storage condition of low RH to maintain that low a_w at the surface of the product.

There is a relationship between RH and temperature that ought to be borne in mind in choosing proper storage environments for foods. In general, the higher the temperature, the lower the RH, and vice versa. At high humidity, the growth of bacteria is expected to be higher than at lower humidity. Also, the temperature can affect humidity by altering the amount of liquid water present. This is expected since humid conditions can provide ideal environments for microbial growth. Foods that bear surface spoilage from molds, yeasts, and certain bacteria should be stored under conditions of low RH. Improperly wrapped meats like whole chickens and beef cuts suffer a lot of surface spoilage in the refrigerator before deep spoilage happens, due to the generally high RH of the refrigerator and also the undeniable fact that the meat-spoilage biota is essentially aerobic. Though it is possible to lessen the chances of surface spoilage in certain foods by storing under low conditions of RH, it should be remembered that the food itself will lose moisture to the atmosphere underneath such conditions and thereby become undesirable. In choosing the right environmental conditions of RH, consideration must be given to both the possibility of surface growth and the desirable quality to be maintained in the foods in question. By altering the gaseous atmosphere, it is possible to retard surface spoilage without lowering the RH.

Presence and Concentration of Gases in the Storage Environment

Many scientific studies have demonstrated the antimicrobial activity of gases at ambient and sub-ambient pressures on microorganisms important in foods. Presence of different gases and its varying concentration may significantly affect the colonizing microorganisms on the food *i.e.* surface spoilage is prevented by altering the gaseous composition. Carbon dioxide (CO_2), ozone (O_3), and oxygen (O_2) are gases that are directly toxic to certain microorganisms.

Oxygen is one of the most important gases which affects both food products as well as microorganisms. Oxygen gas when comes in contact with food, influence redox potential of food and finally the microbial growth. All microorganisms are sensitive to molecular oxygen at some partial pressure. The low solubility of oxygen in aqueous media ensures that many environments are inhabited by predominantly anaerobic organisms. It is possible that oxygen itself is a toxic agent, perhaps by acting as an alternative electron acceptor depriving the cell of reductant for biosynthetic reactions. It may oxidize labile thiol groups, for example, those present in low-potential carriers such as ferredoxin, or inactivate specific enzymes, such as nitrogenase in nitrogen-fixing bacteria. Oxygen may be toxic as one of its metabolites; these include singlet oxygen, the superoxide anion, hydrogen, and the hydroxyl free radical. The absence of the enzyme superoxide dismutase accounts for the sensitivity of anaerobes to oxygen. Generally, increased oxygen concentrations promote higher leakage of reactive oxygen species (superoxide and $H_2 O_{2)}$ from the respiratory chain affecting metalloenzymes and DNA that, in turn, cause impaired growth and elevated mutagenesis.

High-pressure CO_2 has a severe lethal effect on microorganisms. It has been reported that living cell numbers reduced remarkably within 24 h at CO_2 partial pressures ≥1000 kPa. It has also been demonstrated in another study that the cell counts of *E. coli* were decreased by N3 log (10) (CFU/ml) after 15 min high-pressure CO_2 treatment. Using *Pseudomonas aeruginosa* and *Bacillus subtilis* as model species, it was found that high pressure CO_2 reduced the survival ratio of Gram-negative and Gram-positive bacteria by about seven orders of magnitude. The destruction of cell membrane integrity by CO_2 results in the loss of cell protection function, thereby damaging microbial cells and even causing cell lysis and death.

Ozone (O_3) is the other atmospheric gas that has antimicrobial properties, and it has been tried over several decades as an agent to extend the shelf life of certain foods. It is effective against a variety of microorganisms, but because it is a strong oxidizing agent, it should not be used on high-lipid-content foods since it would cause an increase in rancidity. The cell target for O_3 is the membrane, where it disrupts permeability functions. Ozone is effective in reducing pathogens on several food products. A typical concentration used is 0.1–0.5 ppm, which is effective against Gram-positive and Gram-negative bacteria as well as viruses and protozoa.

Gases inhibit microorganisms by two mechanisms. The first mechanism involves a direct toxic effect that can inhibit growth and proliferation. The inhibitory mechanism of carbon dioxide (CO_2), ozone (O_3), and oxygen (O_2) depends upon the chemical and physical properties of the gas and its interaction with the aqueous and lipid phases of the food. O_3 and O_2 generate oxidizing radicals that are highly toxic to anaerobic bacteria and can have an inhibitory effect on aerobes depending on their concentration. Carbon dioxide is very effective against obligate aerobes and, at high levels, will deter other microorganisms. A second inhibitory mechanism is accomplished by modifying the gas composition, which has indirect inhibitory effects by altering the ecology of the microbial environment. The various technologies used to modify the gas composition in the storage environment of food are Modified Atmospheric Packaging (MAP), Controlled Atmospheric Packaging (CAP), Controlled Atmospheric Storage (CAS and vacuum packaging. Altering the atmosphere also alters the competitive environment, and the atmospheres that exert a negative effect on the growth of one particular microorganism may promote the growth of another. This result could have positive or negative consequences relying upon the native pathogenic microflora and their substrate. Nitrogen replacement of oxygen gas is an example of this indirect antimicrobial activity.

The preservation principle of antimicrobial atmospheres has been applied to fruits and vegetables, raw beef, chicken and fish, dairy foods including milk and cottage cheese, eggs, and a variety of prepared, ready-to-eat foods.

The Concentration of CO_2 in the Atmosphere

In general, the inhibitory effects of CO_2 increase with decreasing temperature due primarily to the increased solubility of CO_2 at lower temperatures and the pH of meats stored in high CO_2 environments tends to be slightly lower than that of air-stored controls due to carbonic acid formation. Gram-negative bacteria are more sensitive to CO_2 than gram positives with pseudomonas being among the most sensitive and the lactic acid bacteria and anaerobes being among the most resistant.

With steaks stored in 100 per cent CO_2, significantly lower counts were seen 16–27 days after slaughter compared to steaks stored in 100 per cent N_2 or 100 per cent O_2 or in air. Steaks stored at 1°C in 15 per cent CO_2 + 85 per cent air, or in 15 per cent CO_2 + 85 per cent O_2 for up to 20 days had consistently lower counts (about one-third) than those stored in air. The CO_2 + O_2 mixture was slightly more effective than the CO_2 + air mixture. When lamb chops were stored at –1°C in 80 per cent O_2 + 20 per cent N_2, or 80 per cent air or O_2 + 20 per cent CO_2, or 80 per cent N_2 or H_2 + 20 per cent CO_2, psychrotrophic organisms decreased successively when compared to those stored in air. Using slices of lean beef inoculated with pseudomonas and *Moraxella-Acinetobacter* spp. and stored at 5°C in 70–85 per cent O_2 + 15 per cent CO_2, shelf life was increased by 9 days compared to storage in air. In a study of pork stored at 4° or 14°C, the time needed for the APC at 4°C to reach 5 × 10_6 organisms/cm_2 was about three times longer when stored with 5 atm CO_2 and about fifteen times longer than in air. With poultry stored in air, APC of drip after 16 days at 10°C was log 9.40, while in 20 per cent CO_2 log APC was 6.14. Carbon dioxide was less effective against coliforms in the latter study with log numbers of 3.28 and 3.21 for air and 20 per cent CO_2 storage, respectively.

Carbon dioxide atmosphere storage is effective with fish. Using 80 per cent CO_2 + air, log numbers after 14 days at 35°C were approximately 6.00/cm_2 compared to air controls with numbers > 10.5 cm_2. The pH of CO_2-stored products after 14 days decreased from around 6.75 to around 6.30, while controls increased to around 7.45. The shelf life of rockfish and salmon at 4.5°C was extended by 20–80 per cent CO_2. At least 1 log difference in bacterial counts over controls was obtained when trout and croaker were stored in CO_2 environments at 4°C. When fresh shrimp or prawns were packed in ice in an atmosphere of 100 per cent CO_2, they were edible for up to 12 weeks and bacterial counts after 14 days were lower than air-packed controls after 7 days. When cod fillets were stored at 2°C, air-stored samples spoiled in 6 days with aerobic plate counts (APC) of log_{10} 7.7, while samples stored in 50 per cent CO_2 + 50 per cent O_2 or 50 per cent CO_2 + 50 per cent N_2 or 100 per cent CO_2 did not show bacterial spoilage until, respectively, 26, 34 and 34 days, with respective APCs of 7.2, 6.6 and 5.5/g. It was suggested that the use of 50 per cent CO_2 + 50 per cent O_2 is technically more feasible than the use of 100 per cent CO_2. While the practical upper limit of CO_2 for red meats is around 20 per cent, higher concentrations can be used with fish since they contain lower levels of myoglobin.

Now let us try to understand some more information about the concentrations of CO_2 in meat packs.

The overall effect of high concentrations of CO_2 in meat packs is to shift the flora from a heterogeneous one consisting of gram-negative bacteria to one consisting primarily of lactobacilli and other lactic acid bacteria. In a study by Blickstad *et al.*, > 90 per cent of the flora of pork consisted of *Pseudomonas* spp. after aerobic storage for 8 days at 4°C and 3 days at 14°C, but after 5 atm CO_2 storage, the lactobacilli dominated at both temperatures. This effect can be seen, where smoked pork loins and frankfurter sausage were held from 48 to 140 days at 4°C in the presence of 100 per cent CO_2, 100 per cent N_2 and in vacuum. While pH did not decrease with the pork loins, decrease did occur with frankfurters in CO_2 and in vacuum, reflective of the domination by the lactobacilli. The shelf life of both products increased in the order, vacuum < N_2 < CO_2. Several investigators have found the shelf life of raw meats to increase in the order N_2 < vacuum < CO_2. In another study, the initial flora of normal, low-pH beef consisted of 13 per cent *Pseudomonas fluorescens* and 87 per cent non-fluorescent *Pseudomonas* spp. After vacuum storage for 21 days at 4°C, 96 per cent of the flora consisted of homofermentative lactic acid bacteria, while after 51 days in 100 per cent CO_2, 100 per cent of the flora consisted of these organisms. In a study employing inoculated, high-pH slices of meat stored in a mixture of 20 per cent CO_2 + 80 per cent air, *Enterobacter* and *Brochothrix thermosphacta* were unaffected at 3°C compared to air-stored controls.

With regard to the relative sensitivity of microorganisms to CO_2 inhibition, the fungus *Sclerotium rolfsii* was inhibited by 0.12 atm CO_2 while *Streptococcus cremoris* required ~ 11 atm. *Proteus* and *Micrococcus* spp. are less sensitive than the pseudomonas but not as resistant as *S. cremoris*. From their study of CO_2 atmospheres on the flora of normal, low-pH beef, Erichsen and Molin found resistance to CO_2 as follows: *Pseudomonas* spp. < *B. thermosphacta* < lactic acid bacteria. Storage of meat in 100 per cent CO_2 extended shelf life the greatest, followed by vacuum packaging. To effect a 50 per cent reduction in growth rate compared to controls, 0.5 atm of CO_2 was required for *Pseudomonas fragi*, 1.3 atm for *Bacillus cereus* and 0.6 atm for *S. cremoris.*

Since extremely high pressures of CO_2 are required to prevent the germination of *C. sporogenes* and *C. perfringens* spores, the storage of meats at temperatures of growth of *C. botulinum* in 1 atm CO_2 may present concerns relative to the possible germination of botulinal spores. In a culture medium flushed with CO_2, germination of types A and B spores of C.

With regard to mechanism of CO_2 inhibition of microorganisms, two explanations have been offered. King and Nagel found that CO_2 blocked the metabolism of *Pseudomonas aeruginosa* and appeared to affect a mass action on enzymatic decarboxylations. Sears and Eisenberg found that CO_2 affected the permeability of cell membranes and Enfors and Molin found support for the latter hypothesis in their studies on the germination of *Clostridium sporogenes* and *C. perfringens* endospores. At 1 atm CO2, spore germination of these two species was stimulated while *B. cereus* spore germination was inhibited. As was shown by others, CO2 is more stimulatory at low pH than high. With 55 atm CO_2, only 4 per cent germination of *C. sporogenes* spores occurred, while with *C. perfringens* 50 atm reduced germination to 4 per cent. These authors suggested that CO_2 inhibition was due to its accumulation in the membrane lipid bilayer such that increased fluidity results. In an earlier study, the generation time for eight strains of *C. perfringens* in 100 per cent CO_2 and 100 per cent N_2 grown in thioglycollate medium at 1 atm was about the same 12.9–17.2 and 12.9–16.9, respectively. With *P. fluorescens* in a minimal medium, CO_2 pressure of 100 mm Hg stimulated growth but at higher concentrations, the growth rate declined. In a complex medium at 30°C, maximum inhibition of this organism was achieved at 250 mm Hg. These investigators along with Enfors and Molin found that CO_2 inhibition increases as temperature is lowered.

Presence and Activities of Other Micro-Organisms

Inhibition or destruction of one population of micro-organism by the presence of other population of microorganisms present in the same habitat is the microbial interference. Some micro-organisms produce substances/metabolites (like secondary metabolites), that are either lethal or inhibitory to others and these embrace antibiotics, bacteriocins, hydrogen peroxide, and organic acids. Antibiotics are secondary metabolites produced by microorganisms that inhibit or kill a wide spectrum of other microorganisms. Most of the useful ones are produced by molds and bacteria of the genus *Streptomyces*, and a few by *Bacillus* and *Paenibacillus* spp. Many of the clinically useful agents now in use are synthetic products. Some of the examples are given as under:

Monensin inhibits gram-positive bacteria, and thus its long-term use has the potential of shifting the gastrointestinal tract bacterial biota from one that is normally gram-positive to one that is more gram- negative. Like nisin, monensin is also an ionophore that destroys the selective permeability of cell membranes.

Subtilin is produced by some strains of *Bacillus subtilis*. Like nisin, it is effective against gram-positive bacteria, stable to acid and possesses enough heat resistance to withstand destruction at 121°C for 30–60 minutes. Subtilin is effective in canned foods at levels of 5–20 ppm in preventing the outgrowth of germinating endospores, and its site of action is the same as for nisin.

Antibiotic natamycin (also known as pimaricin, tennecetin, and myprozine) is a polyene that is quite effective against yeasts and molds but not bacteria. Natamycin is the international nonproprietary name, as it was isolated from *Streptomyces natalensis*. Natamycin appears to act in the same manner as other polyene antibiotics *i.e.*, by binding to membrane sterols and inducing distortion of selective membrane permeability. Because bacteria do not possess membrane sterols, their lack of sensitivity to this agent is thus explained.

Tylosin antibiotic is a nonpolyene macrolide, more inhibitory than nisin or subtilin. As a macrolide, it is most effective against gram-positive bacteria. It inhibits protein synthesis by associating with the 50S ribosomal subunit and shows at least partial cross-resistance with erythromycin.

The first use of nisin as the food was shown by Hurst 88 to prevent the spoilage of Swiss cheese by *Clostridium butyricum*. Nisin is produced by some strains of *Lactococcus lactis*, and it is a lantibiotic (contains the rare amino acids, meso-lanthionine, and 3-methyl-lanthione). It is the prototype of foodborne bacteriocins. The compound is effective against gram-positive bacterium, primarily spore formers, and is ineffective against fungi and gram-negative bacterium. *Enterococcus faecalis* is one of the most resistant gram-positives.

Implicit Factors

Antagonisms

In antagonism, the growth of one organism inhibits or suppresses the growth of the second organism, and most of the associations between microorganisms are antagonistic. Many microorganisms generate organic acids and alcohols that are inhibitory to some of their competitors. Some produce antibiotics or bacteriocins, which possess highly specific antimicrobial activity, often against closely related species. Some microbes can gain a competitive advantage by exploitation or hoarding an essential mineral or vitamin that is needed by its competitors. Some pseudomonad produces siderophores, an iron-chelating compound, thereby preventing the growth of competitors that require iron.

Synergisms

A synergistic association exists when two or additional microorganisms grow together, producing an impact that none of the individual microbes may produce alone. Few genuine synergisms have been documented. It has been known for a long time that *Pseudomonas syncyanea* and *Lactococcus lactis* will produce a blue color in milk only when growing together. In a mixed started (containing *Streptococcus lactis* and *Leuconostoc spp.* like *Leuco.dextranicum*, *L.citrovorum*), the production of characteristic flavour due to the conversion of citrate to volatile compounds by *Leuconostocs* is possible only at low pH. The lowering of pH due to lactic acid production is brought about by *S.lactis*.

Metabiosis

Metabiotic associations are essentially "sequential synergisms," in which the growth of one microorganism produces environmental conditions favourable for the growth of a second microorganism, which in turn can create favourable conditions for a third microorganism, and so on. Raw milk provides a superb example of extended metabiosis. *Lactococcus lactis* and a few coliforms are the first microorganisms to grow in raw milk. They turn out lactic acid, which creates a favourable environment for aciduric lactobacilli. When the accumulated acidity of the milk stops the proliferation of lactobacilli, oxidative yeasts and molds begin to grow and oxidize the lactic acid, thereby raising the pH of the milk and allowing the proteolytic bacteria growth. The production of sauerkraut is additionally a splendid example of metabiosis.

The growth of aerobic, oxidative microorganisms can remove oxygen and reduce the O/R potential of food, thereby creating anaerobic conditions that favour the growth of vastly different microbes. Even a straightforward spoilage pathway will exemplify a metabiotic association. The amino acid arginine can be metabolized by lactic acid bacteria to ornithine, which in turn is metabolized by enteric bacteria to the foul-smelling amine putrescine. In Swiss cheese-making, the lactic acid bacteria convert lactose to lactic acid, which in turn is utilized by propionibacteria to produce propionic acid (the compound responsible for characteristic flavour production in Swiss cheese).

Commensalism

Commensalism is that the interaction between populations during which one gains from the interaction, and the other is unaffected. Commensalism benefits the symbiont but has neutral effects for the host. In the microbial world, commensalism is mostly related to nutrition that is when metabolic products of one microbial population can be used by other microbial populations with no specific gain to the first population. Examples include the production of reduced redox-sensitive species, which, when oxidized, can fuel the growth of other microbial populations. Thus, methanogens produce methane, which can be oxidized by methanotrophs, and sulfate reducers produce sulfide, which can be oxidized by a variety of sulfide-oxidizing organisms.

Amensalism

It defines a relationship in which the activity of one population is harmful to another. For microbes, this typically results when the products of one type of metabolism are detrimental to another. Examples include the production of oxygen by cyanobacteria, inhibiting anaerobic organisms, or the production of inhibitory organic compounds as metabolic by products. For example, ethanol, a fermentation product, is inhibitory to many microorganisms, particularly at higher concentrations. The production of acid during sulfide oxidation, particularly in surface sediments where the pH can be driven very low, creates an extreme environment inhibitory to a wide variety of microorganisms.

Amensalism is the opposite of commensalism, where one organism is harmed while the other is unaffected. A good example of this interaction is when one organism produces an associate antibiotic against another organism. Such an interaction is usually the premise of biological control. For example, some isolates of the microorganism *Pseudomonas* fluorescens will suppress the fungal pathogen *Gaeumannomyces graminis*, chargeable for 'take-all' in wheat (Triticum aestivum).

Predation

A predator is an organism that feeds on different organisms, and plenty of varieties of protozoans are the principal predators of prokaryotes in nature. Protists actively engulf prokaryotes (or other food particles) in a process known as phagocytosis, in which a food particle (*e.g.*, prokaryote) is ''consumed'' in special feeding organelles located at the cell surface. Prokaryotes are delivered to the protist by filter-feeding, direct interception, or passive diffusion. Filter feeding is accomplished by the active transport of water through a filter of cilia, or rigid tentacles on the surface of the protist, which strain small cells (and other food particles) from the environment. In direct interception, fluid flow within the medium carries particles to the surface of the feeding protist. When feeding by passive diffusion, food particles migrate to the protist, either by Brownian motion or through the prokaryote's motility. Once ingested, food particles form a vacuole, which fuses to membrane-bound enzyme sacs called lysosomes, accomplishing the digestion of the particle. Numerous different prokaryotic parasites are disease-causing agents in plants and animals, and some prokaryotes can parasitize other prokaryotic organisms. For example, the Gram-negative bacteria *Bdellovibrio* actively hunt and kill its prey to accomplish its parasitic lifestyle. In the attack phase, it is a flagellated non-reproductive cell that enters the periplasmic space of other gram-negative host cells. There it loses its flagellum and grows, feeding off the host, into a reproductive septate filament. After growth ceases, the filament separates, lysing the host and releasing many individual attack cells, ready to repeat the cycle.

Processing Steps

The current definition of "potentially hazardous foods" considers the effect of processing in much the same way that it considers pH and a_w: it divides foods into two categories. Low-acid canned foods in a hermetically-sealed container do not require temperature control for safety. This rigid definition fails to address less processed foods, in less robust packaging, which still would not require temperature control for safety. Consider a baked product, such as a pie, with a pH of 5.5 and a_w of 0.96. Since this product is baked to an internal temperature >180 °F (82 °C) to set the product structure of the pie, it will not contain any viable vegetative pathogens. Any pathogenic spores that survive the baking process will be inhibited by the pH and a_w values. If the product is cooled and packaged under conditions that do not allow recontamination with vegetative pathogens, the product is safe and stable at room temperature until consumed, or until quality considerations (that is, staling) make it unpalatable. Scientifically sound criteria for determining whether foods require time/temperature control for safety should consider:

1. Processes that destroy vegetative cells but not spores (when product formulation is capable of inhibiting spore germination);
2. Post-process handling and packaging conditions that prevent reintroduction of vegetative pathogens onto or into the product before packaging; and
3. The use of packaging materials that while they do not provide a hermetic seal, do prevent reintroduction of vegetative pathogens into the product.

Processing Factors

Various processing factors that influence microbial growth include treatments such as heating, cooling, and drying that affect the composition of the food and also affect the types and numbers of microorganisms that stay within the food after treatment.

Microbial cell death has been associated with either structural damage or physiological dysfunctions. Among structural damage, disruption of the envelopes, DNA conformational changes, ribosome alterations, or protein aggregation are the most frequently described. Also, physiological disorders, such as membrane selective permeability alterations or loss of function of key enzymes, have been proposed as events leading to cell death. Perhaps the most important difficulties that researchers encounter in this area are that several of these lesions may occur simultaneously when the cells are subjected to an agent, and it is, difficult to attribute loss of viability of the cell to a single event.

Heat causes membrane damage, loss of nutrients and ions, ribosome aggregation, DNA strand breaks, inactivation of essential enzymes, protein coagulation, *etc.* In other words, almost every cellular structure is somehow affected by elevated temperatures, and it is very difficult to discern which events are leading to cell death. It is well established that the critical target for irradiation is the chromosome. The effects of ionizing irradiation on bacterial cells are classified as direct and indirect. Direct actions comprise the events caused by the absorption of radiation energy by the target molecules, whereas indirect actions are those derived from the interaction between the reactive species formed by the radiolysis of water, such as the hydroxyl radical, and the target molecules. The hydroxyl radical OH• reacts with the sugar-phosphate backbone of the DNA chain giving rise to the elimination of hydrogen atoms from the sugar. This causes the cutting of the phosphate ester bonds and the subsequent appearance of singlestrand breaks. Double strand breaks occur once two single-strand breaks take place in every chain of the double helix at an adjacent distance.

The mechanism of inactivation of bacterial cells by ultrasound under pressure has also been described. Most authors agree that the cavitation phenomenon is responsible for the lethal effects of ultrasound. When bubbles implode under an intense ultrasonic field, very high pressures and temperatures are generated, and consequently, strong mechanical forces and free radicals are formed. Free radicals could therefore inactivate bacterial cells in a similar mode as that described for IR. It has been instructed that ultrasonic waves may injure the external layers of the spore, facilitating its rehydration and consequently reducing its extreme heat resistance.

The high pressure causes tighter packing of the acyl chains within the phospholipid bilayer of membranes and promotes membrane transition from liquid crystalline to gel section, in a very similar approach as a temperature downshift. Although phase transition of membrane lipids is not necessarily lethal to bacteria, it has been demonstrated that the composition and state of the bacterial cell membrane before pressure treatment affect bacterial resistance to HHP. The pressure at which phase transition occurs would be higher in cells with a more fluid membrane, but it is not known in which circumstances, if any, cell damage is linked to phase transition.

Damage to the cytoplasmic membrane after pressurization has also been repeatedly reported, through the loss of osmotic responsiveness, loss of intracellular material, and formations of buds and vesicles of lipidic origin. In conclusion, HHP inactivation seems to be multitarget in nature. The membrane is a key target, but in some cases, additional damaging events such as extensive solute loss during pressurization, protein coagulation, key enzyme inactivation, and ribosome conformational changes, together with impaired recovery mechanisms, seem also needed to kill bacteria. Bacterial spores are extremely resistant to HHP, being able to withstand up to 1000 MPa for long treatment times unless they are in the germinated state.

Membrane structural or functional alteration is usually accepted as the reason behind cell death by PEF. Free charges tend to accumulate within the inner and outer surface of the membrane generating a transmembrane potential of about ten mV. When an external electric field is applied, as in PEF treatment, a higher amount of free charges of opposite charge accumulate at both membrane surfaces, resulting in compression of the membrane. When the external electric field exceeds a critical value or threshold, the membrane is unable to withstand the electrocompression, and pores are formed. The size and amount of pores depend on the electric field strength and the duration of the treatment.

Alternative theories propose that permeabilization is the consequence of the dipolar reorientation of the membrane phospholipids under an electric field. It has also suggested that the formation of hydrophilic pores would lead to a localized Joule heating phenomenon that could be responsible for the denaturation of proteins and phase changes in the membrane. Sublethal or repairable injury has been detected for irradiated, pressurized, and PEF-treated cells. For each agent, the mechanism of inactivation is different, and so is the nature and the magnitude of the sublethal injury.

Other Factors

Intended End-Use of Product

In addition to carefully assessing how the product is produced and distributed, it is important to consider how the food will ultimately be prepared, handled, and/or stored by the end user. A food product that does not require time/temperature control for safety at one point in the food production or distribution chain may require time/temperature control at another point, depending on its intended use. For example, a thermally processed food that is hot-filled into its final packaging may not require refrigeration if spore-forming pathogens are not capable of outgrowth. However, once the food item is taken out of its original packaging, it may require time/temperature control for safety if the product is likely to be recontaminated during its intended use.

Product History and Traditional Use

The panel struggled with the concept of product history and traditional use as a means to determine the need for time/temperature control for safety. For example, there are foods which have a long history of safe storage use at ambient temperatures, yet have formulations, pH, and a_w that would designate them as "temperature controlled for safety" (TCS) foods. Paramount among them is white bread, but products such as intact fruits and vegetables, other breads, bottled waters, and some processed cheeses have a history of being stored and used at ambient temperatures with no public health impact. In addition, moisture protein ratios (MPR) for shelf- stable fermented sausages were developed to ensure process control values for these sausages that also have a traditional history of safety as a non-TCS food. Moreover, an evaluation of the food characteristics provides a scientific explanation for the products to be safely stored at ambient temperatures. For example, baking of bread controls the growth of pathogens in the interior, and the low a_w precludes the growth of pathogens on the outer surface, so that it can be stored safely at ambient temperatures. Clearly these products' traditional uses and histories provide a valid justification for a decision to be made based on history. Care must be observed, however, as this traditional history can be influenced by the intrinsic and extrinsic factors and any changes in product end- use, processes, formulation, physical structure, processing, distribution, and/or storage. Changes in any of these parameters may invalidate the sole use of history as a basis for decisions on whether a food needs temperature control for safety. The panel recognizes that the use of history as a factor to decide whether a product needs time/temperature control for safety can be subjective. As a guidance, one should determine whether the food in question or any of its ingredients have been previously implicated as a common vehicle of foodborne disease as a result of abuse or storage at ambient temperature. Of particular importance are the microbiological agents that may be of concern based on food formulation, or that may be responsible for illnesses associated with the food and the reported contributing factors that have led to documented illnesses. Has adequate temperature control been clearly documented as a factor that can prevent or reduce the risk of illness associated with the food? As intrinsic or extrinsic factors change (for example, MAP or greatly extended shelf life), historical evidence alone may not be appropriate in determining potential risk. Therefore, for a product to be identified as non-TCS based on history and traditional use, the intrinsic and extrinsic factors affecting microbial growth need to have remained constant. Lastly, product history alone should not be used as the sole factor in determining whether a food needs time/temperature control for safety. This decision requires a valid scientific rationale such as that provided above for white bread.

Thus, all the above-discussed factors (intrinsic, extrinsic, implicit, processing and other) influence the growth of the microorganism to a different extent in some way or the other, and hence control of these factors is very much essential in the growth of microorganism.

Interactions of Factors

Traditional food preservation techniques have used combinations of pH, a_w, atmosphere, numerous preservatives, and other inhibitory factors. Microbiologists have often referred to this phenomenon as the "hurdle effect". For example, certain processed meat products and pickles may use the salt-to-moisture ratio (brine ratio) to control pathogens. USDA recognizes this strategy in designating as shelf-stable semi-dry sausages with a moisture-protein ratio of less than or equal to 3.1:1 and pH less than or equal to 5.0.

In salad dressings and mayonnaise-type products, the acid-to-moisture ratio along with pH is the governing factor for pathogen control. An acid: moisture ratio > 0.70 in combination with a pH < 4.1 is often used as the pathogen-control target level for these products. Usually, these ratios are combined with other factors such as pH or added antimicrobials to effect pathogen control. It is the interaction of these factors that controls the ability of pathogens to proliferate in foods.

Despite this long-standing recognition of the concept of hurdle technology (the possible synergistic effect of combining different inhibitory factors), the current definition of potentially hazardous foods only considers pH and a_w independently, and does not address their interaction. The panel believes that these interactions have to be taken into consideration.

Scientific advances in predictive food microbiology over the last two decades have repeatedly shown that different inhibitory factors that might not prevent pathogen growth when considered singly will prevent pathogen growth when used in concert. Models that address the interaction of other factors (for example, atmosphere, preservatives) have been published, but are not nearly as numerous as models using pH and a_w. Individual companies have shown, however, that in-house models incorporating preservative effects can be useful tools in reducing the need for extensive challenge testing and assessing risk. However, a general model for foods to cover all interactions of atmospheric gases and/or preservative combinations with pH and a_w does not currently exist. In order to design effective combinations of factors, an understanding of the pathogen (vegetative or spore-forming) and of the mechanisms by which individual factors exert their impact are necessary.

Chapter 7

Bacterial Food Poisoning

The association between the consumption of food and human diseases was recognized very early and it was Hippocrates (460 B.C.) who reported that there is a strong connection between food consumed and human illness. Food-borne diseases are major health problems both in developed and developing countries, but developing countries tend to suffer from the largest share of the burden of food-borne diseases. According to the WHO, 30 per cent of the population suffer from food-borne diseases each year in developed countries, and up to 2 million deaths are estimated per year in developing countries.

According to World Health Organization (WHO), food-borne diseases are defined as diseases of infectious or toxic nature which are caused by the consumption of food or water. These are acute illnesses associated with the recent consumption of food. The food involved is usually contaminated with a disease pathogen or when a toxigenic pathogen establishes itself in a food product and produces a toxin, which is then ingested by the human host. Such food contains enough pathogens or toxicant necessary to make a person sick.

Food borne diseases are classified into:

- ☆ Food borne infections (ingestion of food containing pathogens)
- ☆ Food borne intoxications (toxin produced by the pathogens causes food poisoning)
- ☆ Toxico-infections (producing toxins while growing in the human intestines)

Food borne infections are caused by the entrance of pathogenic microorganisms contaminating food into the body, and the reaction of the body tissues to their presence. These can either be fungal, bacterial, viral or parasitic. In foodborne infections, since an incubation period is usually involved, the time from ingestion until symptoms occur is much longer than that of foodborne intoxications. Food borne infections tend to have long incubation periods and are usually characterized by fever. More than 200 different food-borne diseases have been identified. The most severe cases tend to occur in the very old, in the very young, in those who have compromised immune system function, and in healthy people exposed to a very high dose of an organism.

Bacterial foodborne infections occur when food, that is contaminated with bacteria, is eaten and the bacteria continues to grow in the intestines, setting up an infection which causes illness. Bacterial food borne infections include cholera, salmonellosis, typhoid fever, shigellosis, Yersiniosis, *Escherichia coli* infection, Campylobacteriosis, Vibrio parahemolyticus and Listeriosis. Mycotic food borne infections include Candida spp., Sporothrix spp., Wangiella spp. *etc.* Viral food borne infections include hepatitis A, Norwak virus and poliomyelitis virus.

Food intoxication results from consumption of toxins (or poisons) produced in food by bacterial growth. Toxins, not bacteria, cause illness. Toxins may not alter the appearance, odour or flavour of food. Common kinds of bacteria that produce toxins include *Staphylococcus aureus* and *Clostridium botulinum*. In the case of *Clostridium perfringens*, illness is caused by toxins released in the gut when large numbers of vegetative cells are eaten.

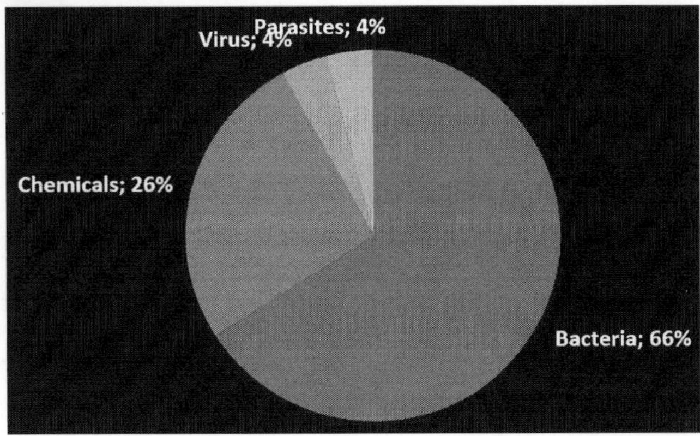

Figure 7.1: Foodborne Causative Agents.

Bacterial Foodborne Illnesses

Bacteria are the most common cause of foodborne diseases and exist in a variety of shapes, types and properties. Bacterial food-borne illnesses are among the most widely spread global public health problems in recent times. They are the causative agents of two-thirds of human food-borne diseases worldwide with high burden in developing countries. Bacteria are the causative agents of foodborne illness in 60 per cent of cases requiring hospitalization. The international impact of foodborne illness is difficult to estimate. However, about 2.1 million children in developing countries die due to diarrheoal-related illnesses annually. It is suspected that food or water is the vehicle for many of these illnesses. Because food is biological in nature and is capable of supplying consumers with nutrients, it is equally capable of supporting the growth of contaminating microorganisms.

Food borne infections vs intoxication

Infections	Intoxications
Bacterial / Viral / parasite /protozoa	toxins (natural / preformed bacterial / chemical)
Invade and /or multiply in lining of intestine	No invasion or multiplication
Incubation period- hours to days	Incubation period- minutes to hours
S/s – Diarrhoea, nausea, vomitting, abdominal cramps, fever	S/s – Vomitting, nausea, diarrhea, weakness, resp. failure, numbness, sensory/motor dysfunction
Communicable-spreads from person to person	Not communicable
Factors-inadequate cooking, cross contamination, poor personal hygiene, bare hand contact	Factors-inadequate cooking, improper handling temperaturescan

Food Borne Pathogenic Bacteria

The most common food borne pathogenic bacteria have been listed as under:

1. *Salmonella* spp.
2. *Clostridium botulinum*
3. *Staphylococcus aureus*
4. *Campylobacter jejuni*
5. *Yersinia enterocolitica* and *Yersinia pseudotuberculosis*
6. *Listeria monocytogenes*
7. *Vibrio cholerae* O1
8. *Vibrio cholerae* non-O1
9. *Vibrio parahaemolyticus* and other *vibrios*
10. *Vibrio vulnificus*
11. *Clostridium perfringens*
12. *Bacillus cereus*
13. *Aeromonas hydrophila* and other spp.
14. *Plesiomonas shigelloides*
15. *Shigella* spp.
16. Miscellaneous enterics
17. *Streptococcus*

Enterovirulent *Escherichia coli* Group (EEC Group)

1. *Escherichia coli*–enterotoxigenic (ETEC)
2. *Escherichia coli*–enteropathogenic (EPEC)
3. *Escherichia coli* O157: H7 enterohemorrhagic (EHEC)
4. *Escherichia coli*–enteroinvasive (EIEC)

These are discussed in detail as under.

Botulism

The popularity of botulinum neurotoxin (BoNT), also known as "Botox," to reduce facial wrinkles and the potential to use *C. botulinum* and its neurotoxin as agents of intentional contamination have sparked renewed interest in this pathogen, historically associated with home-canned or prepared products.

Seven types (A, B, C, D, E, F and G) of botulism caused by *Clostridium botulinum,* an anaerobic, gram-positive, spore-forming rod are recognized, based on the antigenic specificity of the toxin produced by each strain. The organism and its spores are widely distributed in nature. They occur in both cultivated and forest soils, bottom sediments of streams, lakes, and coastal waters, and in the intestinal tracts of fish and mammals, and in the gills and viscera of crabs and other shellfish.

Food Borne Botulism

Foodborne botulism is the name of the disease (actually a foodborne intoxication) caused by the consumption of foods containing the neurotoxin produced by *C. botulinum.* Foodborne botulism (as distinct from wound botulism and infant botulism) is a severe type of food poisoning caused by the ingestion of foods containing the potent neurotoxin formed during growth of the organism. The toxin is heat labile and can be destroyed if heated at 80°C for 10 minutes or longer. The incidence of the disease is low, but the disease is of considerable concern because of its high mortality rate if not treated immediately and properly. Most of these reported food borne outbreaks are associated with inadequately processed, home-canned foods, but occasionally commercially produced foods have also been involved in outbreaks. Sausages, meat products, canned vegetables and seafood products have been the most frequent vehicles for human botulism.

The first outbreaks of botulism were described in 1735. After this, another one occurred in 1793 in Wildbad, Wurttemburg, 13 people fell ill and 6 later died after eating Blunzen, a type of sausage made by packing blood and other ingredients into a pig's stomach. The sausage had been boiled and then smoked, after which it was considered stable at room temperature for several weeks and suitable for consumption without reheating.

Several further incidents of *Wurstvergiftung*, or sausage poisoning, were recorded in the years that followed, usually associated with sausages that contained animal components other than muscle tissue. This prompted a local district medical officer, Justinius Kerner, to undertake a study of the disease which became known as botulism (Latin: *botulus* = sausage), Kerner noted several important features including the facts that heating was an essential precondition for the development of toxicity in sausages and that small sausages or those containing air pockets were less likely to become toxic. The name "botulism" was derived in 1870, by the German physician Muller from the Latin word *botulus*, which means "sausage".

It was not until 1896 that the microorganism responsible was isolated and described by van Ermengem, Professor of Bacteriology at the University of Ghent and former pupil of Robert Koch. This was a result of his investigation into an outbreak of botulism where 34 members of a music club in Belgium ate raw, unsmoked ham. Several noted that the ham had a slightly 'off' flavour akin to rancid butter but was otherwise unremarkable. About a day later, 23 of the group fell ill and 3 died within a week.

Van Ermengem established that botulism resulted from the consumption of food containing a heat-labile toxin produced by an obligately anaerobic, spore-forming bacillus which he called *Bacillus botulinus*. He further demonstrated that toxin would not be produced in the presence of sufficient salt, that it was resistant to mild chemical agents and was not uniformly active against all animal species. To express better both its spindle shape (*kloster* is Greek for spindle) and its anaerobic metabolism the pathogen name was later changed in *Clostridium botulinum*.

C. botulinum will not grow in acidic conditions (pH less than 4.6), and therefore the toxin will not be formed in acidic foods (however, a low pH will not degrade any pre-formed toxin). Combinations of low storage temperature and salt contents and/or pH are also used to prevent the growth of the bacteria or the formation of the toxin. Though spores of *C. botulinum* are heat-resistant, the toxin produced by bacteria growing out of the spores under anaerobic conditions is destroyed by boiling (for example, at internal temperature greater than 85 °C for 5 minutes or longer). Therefore, ready-to-eat foods in low oxygen-packaging are more frequently involved in cases of foodborne botulism.

Foodborne botulism occurs when *C. botulinum* grows and produces toxins in food prior to consumption. *C. botulinum* produces spores and they exist widely in the environment including soil, river and sea water. The botulinum toxin has been found in a variety of foods, including low-acid preserved vegetables, such as green beans, spinach, mushrooms, and beets; fish, including canned tuna, fermented, salted and smoked fish; and meat products, such as ham and sausage. The food implicated differs between countries and reflects local eating habits and food preservation procedures. Occasionally, commercially prepared foods are involved.

Traditionally, foodborne botulism has been associated with under processed and abused sausages or home canned foods; however, in recent years botulism has been acquired through the consumption of contaminated foods such as meat, potato salad, sauteed onions, garlic sauce, cheese, yogurt, bean paste, and olives. Botulism is found to occur wherever foods and their processing offer conditions suitable for survival and growth of the causative organism. It was identified with icthyism, a paralytic illness associated with the consumption of raw, salted fish, known in Russia since and in 1904 an outbreak of botulism in Darmstadt Germany was caused by canned white beans.

Processors and consumers must take preventive measures to eliminate *C. botulinum* or to inhibit growth and toxin production by this microorganism to prevent outbreaks of botulism from occurring. Instances of botulism resulting from improper canning are primarily associated with home canning, not industrial processes. This results in a limited number of cases annually, but does not mean that attention should be diverted from this pathogen

Physiological diversity within the species *C. botulinum* is recognized by its division into four groups and molecular studies based on DNA homology and ribosomal RNA sequences have confirmed this grouping. Group I strains are culturally indistinguishable from the non-toxigenic species *Clostridium sporogenes* which can sometimes serve as a useful and safe model in laboratory studies. They are strongly proteolytic and will often betray their presence in food by partial disintegration of the product and a slight rancid or cheesy odour. Unfortunately, despite these warning signs the potency of the toxin is such that the amount ingested on sampling the food has often proved sufficient to cause illness. Group I strains are not psychrotrophic and are therefore of little concern in adequately refrigerated products. They do, however, produce the most heat-resistant spores and can pose a problem when foods that depend upon a heating step for their stability and safety are under processed.

In contrast, Group II strains represent a greater potential hazard in chilled foods. They are non-proteolytic with native protein, can grow and produce toxin down to about 3°C and produce spores with a low resistance to heat. They also tend to be more susceptible to inhibition by salt. The rate of growth and toxin production at the lower temperature limit is slow and will be reduced still further by any other factors adverse to growth. Experimental studies have indicated that storage periods of 1–3 months are necessary for toxin production at 3.3°C, although this period can be markedly reduced at higher temperatures still within the chill range. Vacuum-packed herrings inoculated with 100 spores per pack became toxic after 15 days storage 5°C.

Most cases of botulism in humans are due to toxin types A, B or E and the incidence of other types in human illness is extremely rare. Group III strains producing toxin types C and D are usually associated with illness in animals and birds. Type G toxin has been incriminated in human illness largely as a result of its isolation at autopsy from people who had died suddenly and unexpectedly. Since botulism was not necessarily the cause of death in these cases and there have been no reports of the presence of type G in foods, its role in food-borne illness is questionable. It has been proposed that the Group IV strains which produce type G toxin and some non-toxigenic clostridia be re-designated as a new species, *Clostridium argentinense*.

Although it is found occasionally, growing in the alimentary tract of birds and mammals, *C. botulinum* is essentially a soil saprophyte. It occurs widely, although the geographical distribution is not uniform. Surveys conducted in the United States found type A to be the most common in the Western States, rare in the Mississippi Valley but less so along the Eastern Seaboard where type B was predominant. This distribution was reflected in outbreaks of botulism in the United States in the period 1950–1979; when 85 per cent of those west of the Mississippi were due to type A toxin and 63 per cent of those to the east were due to type B. In European soils type B tends to be more common than type A.

The minimum pH at which *C. botulinum* will be grown depends very much on factors such as temperature, water activity and the acid used to adjust the pH. The consensus has long been that a pH around 4.7 represents an absolute minimum and this fact has had important practical implications for the canning industry. Non-proteolytic strains have a lower acid tolerance and are generally inhibited at pH 5.0–5.2. Reports have appeared of growth and toxin production at pH values as low as 4.0 in protective, high-protein containing media but this does not reflect the situation in acid canned foods which are generally low in protein. In cases where botulism has occurred in foods where acidity is an important protective hurdle, such as canned fruits, it has been as a result of other organisms, yeasts or moulds, growing in the product and increasing the pH.

Figure 7.2 indicates the production and activation of botulinum toxin.

Botulism is an example of bacterial food poisoning in its strictest sense: it results from the ingestion of an exotoxin produced by *Clostridium botulinum* growing in the food. The botulinum toxins are neurotoxins; unlike enterotoxins, which act locally in the gut, they affect primarily the cholinergic nerves of the peripheral nervous system.

Experiments in animals have shown that toxin ingested with food and surviving inactivation is absorbed in the upper part of the small intestine and reaches the bloodstream via the lymphatics. It binds to the nerve ending at the nerve-muscle junction, blocking release of the acetylcholine responsible for transmission of stimuli, thus producing a flaccid paralysis.

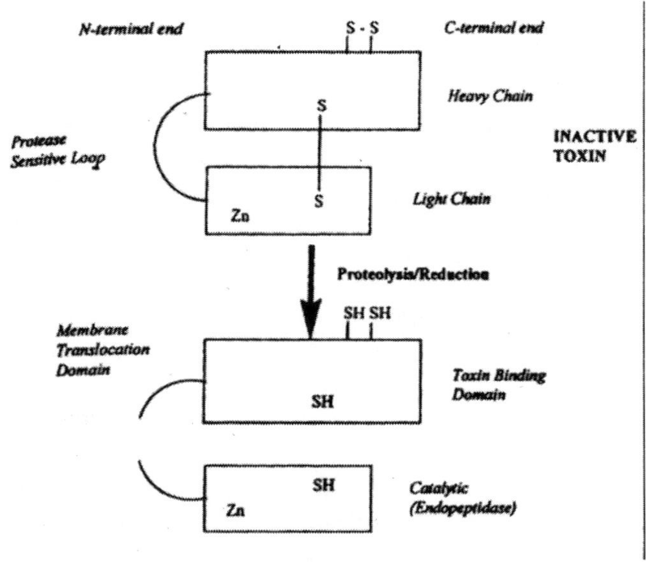

Figure 7.2: Production and Activation of Botulinum Toxin

Botulinum neurotoxin is synthesized during cellular growth and is subsequently released during cell lysis, where proteolytic cleavage activates the molecule.

There are four categories of botulism, which include the classic foodborne botulism derived from the ingestion of preformed toxin in foods, wound botulism resulting from toxin production after organism growth in an infected wound, infant botulism from toxin elaboration in the intestinal tract of infants, and botulism due to intestinal colonization in older children and adults with intestinal disorders or complications resulting in a lack of microbial competition. Botulinum neurotoxin introduced in any of these categories is transported via the bloodstream to neuromuscular junctions, where the toxin irreversibly binds to receptors on peripheral nerve endings and subsequently is internalized into the nerve cell.

Initial symptoms of botulism occur anything from 8 h to 8 days, most commonly 12-48 h, after consumption of the toxin-containing food. Symptoms include vomiting, constipation, urine retention, double vision, difficulty in swallowing (dysphagia), dry mouth and difficulty in speaking (dysphonia). The patient remains conscious until, in fatal cases, shortly before the end when the progressive weakness results in respiratory or heart failure. This usually occurs 1–7 days after the onset of symptoms. Surviving patients may take as long as 8 months to recover fully.

The clinician can do little to mitigate the effect of toxin already adsorbed at the neuromuscular junction, although neuromuscular blockade antagonists such as 4-aminopyridine have produced transient improvements. Survival is therefore critically dependent on early diagnosis and treatment, principally by alkaline stomach washing to remove any remaining toxic food, intravenous administration of specific.

The mortality rate is usually high (20–50 per cent), but will depend on a variety of factors such as the type of toxin (type A usually produces a higher mortality than B or E), the amount ingested, the type of food and the speed of treatment.

The botulinum toxins are the most toxic substances known with a lethal dose for an adult human in the order of 10_{-8} g. They are high molecular mass (150 kDa) proteins and can be inactivated by heating at 80°C for 10 min. In culture, they are produced during logarithmic growth as complexes and released into the surrounding medium on cell lysis. In the smallest of these complexes, the M complex, neurotoxin is accompanied by a similar-sized protein with no apparent biological activity, while in the larger L complex, an additional haemagglutinin component is also present. It appears that the neurotoxin is synthesized as a single chain protoxin which is activated by proteolytic cleavage to produce a molecule consisting of light (M_r 50 kDa) and heavy (M_r 100 kDa) chains linked by a disulfide bridge. Where the organism does not itself produce appropriate proteolytic enzymes, protoxin can be activated by the gut enzyme trypsin.

Botulism is currently classified into four categories: (1) classical foodborne botulism and intoxication caused by the ingestion of preformed BoNT in contaminated food; (2) wound botulism, the rarest form of botulism, which results from the elaboration of BoNT in vivo after growth of C. botulinum in an infected wound; (3) infant botulism, in which botulinal toxin is elaborated in vivo in the intestinal tract of an infant who has been colonized with C. botulinum; and (4) an "undetermined" classification of botulism for those cases involving individuals older than 12 months in which no food or wound source is implicated.

Infant botulism differs from the classical syndrome in that it results from colonization of the infant's gut with *C. botulinum* and production of toxin *in situ*. It was first described in 1976 and is most frequently reported in the United States, although cases have occurred in Australia, Canada, Europe and South America. Until 1987 only two cases had been reported in the UK, one type A and the other involving the rare type F toxin.

It occurs mostly in infants aged 2 weeks to 6 months, particularly around the time that non-milk feeds are introduced. At this stage the infant's gut microflora is not fully developed and is less able to out-compete and exclude *C. botulinum*. Since it only requires the ingestion of viable spores, environmental sources other than food may be involved and those foods that do act as vehicles need not be capable of supporting growth of the organism. Of the various potential environmental sources such as soil, cistern water, dust and foods, honey is the one dietary reservoir of *C. botulinum* spores thus far definitively linked to infant botulism by both laboratory and epidemiologic studies. Honey has been associated with several cases of infant botulism in the USA and some surveys have found viable spores of *C. botulinum* in 10 per cent of the samples examined. Consequently, it is thought inadvisable to feed honey to children less than a year old. The number of confirmed infant botulism cases has increased significantly as a result of greater awareness by health officials since its recognition in 1976. It is now internationally recognized, with cases being reported in more countries.

Clinical symptoms of infant botulism consist of constipation that occurs after a period of normal development. This is followed by poor feeding, lethargy, weakness, pooled oral secretions, and wail or altered cry. Loss of head control is striking. The illness is characterized by neuromuscular symptoms related to those of classical botulism. Recommended treatment is primarily supportive care. Antimicrobial therapy is not recommended. Infant botulism is diagnosed by demonstrating botulinal toxins and the organism in the infants' stools. Diagnosis can be confirmed by the isolation of the organism and its toxin from the faeces. Although implicated in a small proportion (4 per cent) of cases of sudden infant death syndrome in the United States, the mortality rate is low in treated cases.

C. botulinum will often constitute only a small proportion of the total microflora so enrichment or pre-incubation is necessary to improve the chances of isolation. Sometimes enrichment cultures are heated prior to incubation to eliminate non-spore forming anaerobes. However, depending on the heating regime used 80°C for 10 min is commonly cited, this may also eliminate the less heat resistant strains of *C. botulinum* and is therefore, often omitted.

Figure 7.3 indicates the mouse neutralization assay for botulinum toxin.

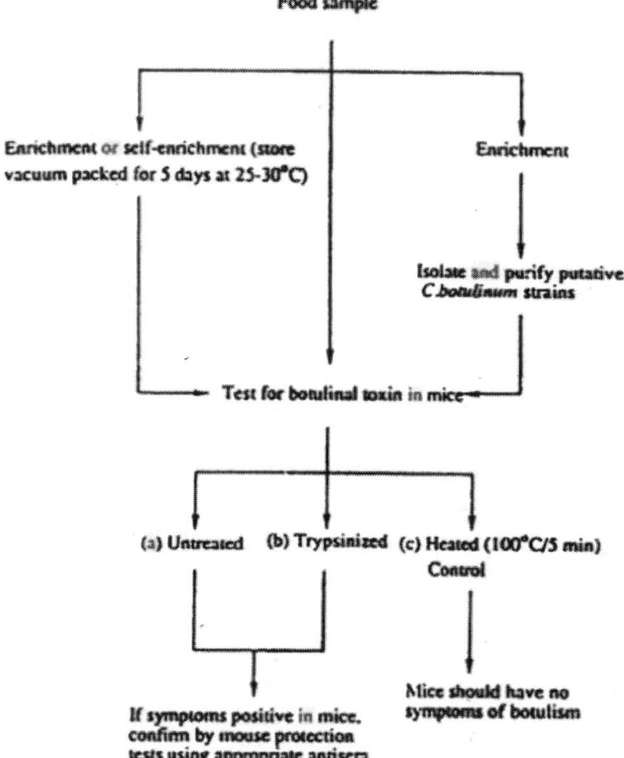

Figure 7.3: Mouse Neutralization Assay for Botulinum Toxin.

Despite the development of a range of *in vitro* immunoassay procedures for toxin, the mouse neutralization test remains the most sensitive (a typical lethal dose of toxin for a mouse is a few picograms). However, the distressing nature of the test guarantees its eventual replacement as soon as immunoassay amplification systems have been sufficiently improved.

Wound botulism is the rarest form of botulism. The illness results when *C. botulinum* by itself or with other microorganisms infects a wound and produces toxins which reach other parts of the body via the blood stream. Foods are not involved in this type of botulism.

Undetermined category of botulism involves adult cases in which a specific food or wound source cannot be identified. It has been suggested that some cases of botulism assigned to this category might result from intestinal colonization in adults, with in vivo production of toxin. Reports in the medical literature suggest the existence of a form of botulism similar to infant botulism, but occurring in adults. In these cases, the patients had surgical alterations of the gastrointestinal tract and/or antibiotic therapy. It is proposed that these procedures may have altered the normal gut flora and allowed *C. botulinum* to colonize the intestinal tract.

Now let us learn about the common features which are discernible in outbreaks of botulism.

The food has been contaminated at source or during processing with spores or vegetative cells of *C. botulinum*. The food receives some treatment that restricts the competitive micro flora and, in normal circumstances, should also control *C. botulinum*. Conditions in the food (temperature, pH, E_h, a_w) are suitable for the growth of *C. botulinum*.

The food is consumed cold or after a mild heat insufficient to inactivate toxin. Since low-acid canned foods can fulfil all the above criteria, it has been necessary for the canning industry to introduce stringent process control measures to ensure safety. When canned foods are produced as a small-scale, domestic activity however, greater variability and less rigorous control are clearly potential sources of problems. In the United States, where home-canning is more widely practised than elsewhere, inadequately processed products, particularly vegetables, are the most common cause of botulism. Between 1899 and 1981 there were 522 outbreaks associated with home-canned products, including 432 involving vegetables. This compares with 55 outbreaks over the same period caused by commercially canned products; the majority occurring before 1925.

A variety of foods have been associated with botulism in the UK and they are frequently home-produced rather than commercial products. Fortunately, the incidence is extremely low, though slightly higher rates have been reported in some other European countries.

Fish can be contaminated with *C. botulinum*, particularly type E, from the aquatic environment and uncooked fish products have been responsible for several outbreaks of type E botulism.

Smoked fish consumed without reheating has generally been hot-smoked so control of *C. botulinum* depends on microbial inactivation by heat plus the inhibitory effects of salt, smoke constituents and surface drying. With the advent of refrigeration, the severity of the salting and smoking stages has been reduced in line with the perceived consumer preference for a less strongly flavoured product. In the early 1960s, two outbreaks of type E botulism in North America associated with vacuum-packed, hot-smoked fish caused considerable alarm and led to Canada banning the importation of all types of packaged fish. A similar outbreak in Germany in 1970 was caused by smoked trout from a fish farm. At first it was feared that vacuum packing, an emerging technology at that time, was responsible by providing an anaerobic environment in which *C. botulinum* could flourish.

The long association of botulism with meat products in Europe has already been noted and inadequate curing of meats still gives rise to occasional problems in some European countries. Outbreaks of botulism in the UK are relatively infrequent. The largest outbreak occurred in 1989 when 27 people fell ill and one died. In this outbreak the vehicle was hazelnut yoghurt. The pH of yoghurt is too low for toxin production *in situ*, but the toxin (type B) had been produced in the hazelnut puree which was inadequately heat processed.

Soil contamination is a major source of *C. botulinum* in foods and one to which vegetables, particularly root crops, are inevitably prone. Three, outbreaks of type A botulism in the United States have been attributed to potato salad where cooked or partly cooked potatoes had been stored for several days at ambient temperatures and under anaerobic conditions before further processing. In 1987 an airline passenger in Europe contracted type A botulism from a prepacked vegetable salad. Important features in these outbreaks were temperature abuse and anaerobiosis, created by vacuum packing or wrapping in aluminium foil. They further indicate the importance of ensuring that vegetables, particularly those such as mushrooms that are frequently eaten raw, are not stored at ambient temperatures in hermetically sealed packs.

Fish production that are consumed raw after a fermentation process have also caused occasional problems, for example I-sushi. In 1986 In the Canadian Northwest Territories, an outbreak of type E botulism was recorded after consumption of a meal comprising raw fish, seal meat and fermented seal flipper. The latter had been prepared by packing the product in a plastic bucket, covering with seal fat and leaving it outside the house to ferment. The process differed from normal in that the product was stored for 7 days instead of the usual three and the weather had been unseasonably warm. It was claimed that the seal flipper had an unusual taste and subsequent investigation established the presence of *C. botulinum* type E in the product. In Europe, the Norwegian fermented trout rak-orret has also been responsible for outbreaks of botulism.

Hence, the types of foods involved in botulism vary according to food preservation and eating habits in different regions. Any food that is conducive to outgrowth and toxin production, that when processed allows spore survival, and is not subsequently heated before consumption can be associated with botulism. Almost any type of food that is not very acidic (pH above 4.6) can support growth and toxin production by *C. botulinum*. Botulinal toxin has been demonstrated and reported in a considerable variety of foods, such as canned corn, peppers, green beans, soups, beets, asparagus, mushrooms, ripe olives, spinach, tuna fish, chicken and chicken livers and liver pate, and luncheon meats, ham, sausage, stuffed eggplant, lobster, and smoked and salted fish.

A very small amount (a few nanograms) of toxin can cause illness. Onset of symptoms in foodborne botulism is usually 18 to 36 hours after ingestion of the food containing the toxin, although cases have varied from 4 hours to 8 days. Early signs of intoxication consist of marked lassitude, weakness and vertigo, usually followed by double vision and progressive difficulty in speaking and swallowing. Difficulty in breathing, weakness of other muscles, abdominal distention, and constipation may also be common symptoms.

Botulinum toxin causes flaccid paralysis by blocking motor nerve terminals at the myoneural junction. The flaccid paralysis progresses symmetrically downward, usually starting with the eyes and face, to the throat, chest and extremities. When the diaphragm and chest muscles become fully involved, respiration is inhibited and death from asphyxia results. Recommended treatment for foodborne botulism includes early administration of botulinal antitoxin and intensive supportive care (including mechanical breathing assistance).

Although botulism can be diagnosed by clinical symptoms alone, differentiation from other diseases may be difficult. The most direct and effective way to confirm the clinical diagnosis of botulism in the laboratory is to demonstrate the presence of toxin in the serum or faeces of the patient or in the food which the patient consumed.

Since botulism is foodborne and results from ingestion of the toxin of *C. botulinum*, determination of the source of an outbreak is based on detection and identification of toxin in the food involved. Currently, the most sensitive and widely used method for detecting toxin is the mouse neutralization test. This test takes 48 hours. Culturing of specimens takes 5-7 days.

All people are believed to be susceptible to this foodborne intoxication. The incidence of the disease is low, but the mortality rate is high if not treated immediately and properly. There are generally between 10 to 30 outbreaks a year in the United States. Some cases of botulism may go undiagnosed because symptoms are transient or mild, or misdiagnosed as Guillain-Barre syndrome.

Typical methods used to prevent botulism include reduction of vegetative cell and/or spore contamination in the food by heat treatment (*e.g.*, pressure cooking for canning) to obtain "commercial sterility" and use of lower heat treatments (pasteurization) in combination with other control measures. Nitrite and salt (in a brine solution) are control measures used in addition to heat treatment for low-acid canned meats (*e.g.*, canned ham). In addition to these controls, refrigeration is an important control for perishable vacuumpackaged meats. Acidification is used for other food products such as pickles, mayonnaise, and canned fruit products. Reduction of moisture level (drying) below an a_w of 0.93 is employed in certain foods. Frozen storage is yet another significant factor in controlling pathogen growth and subsequent toxin formation.

The excellent safety record of cured meats relative to botulism outbreaks has been largely attributed to the use of nitrite as a curing ingredient. Many studies have been published on the efficacy of sodium nitrite in inhibiting *C. botulinum* growth and toxin production in perishable cured meats such as wieners, bacon, canned ham, luncheon meat, and canned comminuted meat. In perishable cured meats, safety cannot be totally attributed to nitrite alone, but a variety of factors and their interaction with nitrite provide protection against botulinal growth and toxin production. Factors such as heat treatment, acidity (pH), salt, and bacterial spore level influence the effect of nitrite on *C. botulinum* growth and toxin production. Other curing adjuncts, such as ascorbic acid and sodium erythorbate, may also influence the efficacy of nitrite. The greater incidence of botulism due to improper home processing and storage of foods, particularly home-canned foods, compared to commercially processed foods reflects the food industry's greater awareness and control of the key factors in inhibition of *C. botulinum* growth and toxin production. Consumers must recognize that certain control measures, such as proper refrigeration and frozen storage, must be maintained from purchase until consumption. If a commercial product is mishandled, control measures are compromised and then botulism may result.

Mycobacterium Species

Mycobacteria are classified according to the host species, *i.e.* human, bovine, murine, avian and piscine, and are strict aerobes with nutritionally-exacting growth requirements. They are gram-positive (though difficult to stain), pleomorphic 1-4μm in length, acid-fast rods that do not produce flagella, capsules or spores and do not grow on ordinary microbiological media. Mycobacterium species are generally non-fastidious.

A special feature of mycobacteria is the chemical composition of their cell walls. These have a high lipid content made up of esterified mycolic acids, complex branched-chain, hydroxy lipids with the general formula $R_1CHOH.CHR_2.COOH$ (where R_1 and R_2 are very long aliphatic chains), and, as a result, the wall is very hydrophobic and waxy. This confers a number of important properties on the organisms. For example, uptake of nutrients from aqueous solution is impeded making them very slow growing so that it often takes more than a week for growth to be apparent on solid media. They are also very resistant to drying and therefore, can persist and remain infectious in the environment for long periods. The cell wall is more resistant to degradation by lysosomal enzymes in phagocytes enabling the pathogenic mycobacteria to survive and grow in macrophages. Its hydrophobic nature also makes the cells rather difficult to stain. However, once stained they are very resistant to decolourization and have the characteristic diagnostic property of 'acid fastness'. This was first noted by Ehrlich in 1882 and is detected using the Ziehl-Neelsen staining procedure in which cells are stained with hot carbol fuchsin and mycobacteria, if present, will resist subsequent decolourization with acid alcohol.

Most forms of tuberculosis are chronic taking months or even years before recovery or death. The commonest clinical signs include fever chills and weight loss, but other symptoms present depending on the organs involved.

Mycobacterium paratuberculosis causes paratuberculosis, otherwise known as Johne's disease, in cattle and it has been suggested that it may be implicated in the etiology of Crohn's disease in humans. This remains to be established, but if so, consumption of infected milk may be a possible route of transmission.

The genus *Mycobacterium* consists largely of harmless environmental organisms but is best known as the cause of two of the most feared and ancient of human diseases, tuberculosis (TB) and leprosy. TB, described by John Bunyan as 'Captain of these men of death', can sometimes be food-borne and is therefore of more concern to us here.

There is archaeological evidence to suggest that TB was endemic in much of the world from ancient times but with the rise in urbanisation between the 18_{th} and 20_{th} centuries it became epidemic in many areas, killing millions. Death rates in Europe and the United States peaked in the 19_{th} century when it has been estimated that 30 per cent of all deaths under the age of 50 in Europe were due to TB. By the late 20_{th} century, a combination of improved social conditions, childhood immunization, screening and effective chemotherapy had reduced the incidence of TB in the developed world to the point where public health officials talked confidently of eliminating the disease altogether. This optimism proved unfounded as we have seen increasing numbers of cases since the late 1980s in groups such as AIDS patients and the socially disadvantaged, as well as the emergence of drug resistant strains. In the world's poorer countries, tuberculosis has always remained an important cause of morbidity and mortality. In 1990 the WHO and International Union against Tuberculosis and Lung Disease estimated that one-third of the world's population was infected with the tubercle bacillus and there were 7–8 million new cases each year.

Human illness is primarily associated with *Mycobacterium tuberculosis* which is thought to account for 98 per cent of cases of pulmonary TB and 70 per cent of non-pulmonary forms. It is spread person to person by aerial transmission of droplets produced by an infected person, coughing, sneezing or spitting. *Mycobacterium bovis* is very closely related to *Myco. tuberculosis* but causes tuberculosis in cattle and other animals as well as in humans. It too is spread by respiratory aerosols between animals and from animals to humans but can also be transmitted to humans by milk and, to a lesser extent, by meat from tuberculous animals.

In food borne tuberculosis, *M. bovis* enters the body through the intestinal tract and the primary infection usually occurs at the mesenteric lymph nodes. The bacteria are engulfed by macrophages and are then isolated in nodules called tubercles or granulomas which are mainly composed of a dense accumulation of activated macrophages and lymphocytes. For many people this is as far as the infection proceeds, the development of the tubercle is checked by surrounding it with a fibrous wall and it then calcifies to a yellow gritty mass. In others, however, illness ensues when the tubercle liquefies causing local tissue necrosis and releasing the bacteria to spread infection around the body.

Tuberculosis lesions can be identified in meat animals by post-mortem inspection of carcasses but disease can also be identified in the live animal (and humans) using the tuberculin test. In this, the animal exhibits delayed hypersensitivity to injection of tuberculin, a protein preparation from *Myco. bovis.* In clinical specimens, mycobacteria can be identified directly on the basis of their acid-fast reaction in the Ziehl-Neelson stain when the organisms appear red and the surrounding tissue blue. The organisms can be cultured on simple media but are very slow growing.

In North America, compulsory pasteurization regulations were introduced in a number of the large cities from about 1910 and had a marked effect reducing the incidence of bovine TB in children. In the UK, there was considerable resistance to the introduction of milk pasteurization, but the available evidence suggests that its later and more gradual introduction had a similar effect. By 1944, all London's milk was pasteurized and the death rate from abdominal tuberculosis in children was 4 per cent of what it had been in 1921 when there was no pasteurization. In contrast, the death rate in 1944 in rural areas, where pasteurization was less extensively practised, was 10 times the London rate.

Though there has been considerable recent concern about the worldwide resurgence of TB, the contribution of food-borne transmission is probably insignificant in the developed world where it is effectively controlled by the testing and elimination of infected cattle, rigorous meat inspection and milk pasteurization. This may not be true in many developing countries such as those of Africa where the extent of human tuberculosis caused by *Myc. bovis* is not known and there is widespread consumption of unpasteurized milk and, in some areas, raw meat products.

Mycobacterium paratuberculosis and Milk

Mycobacterium paratuberculosis known as "Johne's bacillus" was first isolated and described by Johne and Frothingham in 1895 during the investigation of a massive infiltration of the intestinal tract of cattle that had resulted in chronic diarrhoea. Now known as *Mycobacterium paratuberculosis* the organism was originally believed to be an isolate of *Mycobacterium avium* but has also been known as *Mycobacterium johnei* or *Mycobacterium enteritidis.*

More recent studies have confirmed that *M. paratuberculosis* is related to the *M. avium* "complex" with which it shares a number of characteristics. In particular, *M. paratuberculosis* shares more antigens with *M. avium* than with *M. tuberculosis* or with *M. bovis*, there is a similarity in the composition of cell wall mycolic acids and the overall structure of cyclopropane rings is similar in *M. paratuberculosis* and *M. avium*. Due to the hydrophobic nature of their cell walls, *M. paratuberculosis* cells are prone to form clumps. *M. paratuberculosis* does not grow in the presence of 5 per cent sodium chloride but is able to grow in microbiological media at pH 5.5.

Mycobacteria comprise slow-growing and fast-growing species; fast-growing organisms (forming colonies in <7 days) are generally regarded as non-pathogenic whilst the slow-growing organisms are often regarded as the causal agents of some chronic infections. *M. paratuberculosis* falls within the slow-growing species.

The disease caused by *M. paratuberculosis*, known as Johne's disease or paratuberculosis, is an incurable, chronic, infectious, granulomatous enteritis of ruminants, characterised clinically by diarrhoea, weight loss, debilitation and, ultimately, death. Johne's disease is one of the most widespread bacterial diseases of domestic animals throughout the world. It is most common in cattle and to a lesser extent in sheep and goats. Asymptomatic carriers also exist. Currently, Johne's disease is not classed as a zoonosis (*i.e.*, it is not a disease of animals that can be naturally transmitted to humans).

As the causative organism of Johne's disease, *M. paratuberculosis* may be found in dairy animals. Infected cattle may shed *M. paratuberculosis* in their faeces at levels up to 108 cfu/g. *M. paratuberculosis* has been cultured from the milk of 35 per cent of clinically infected cattle and 11.6 per cent of asymptomatic carriers, the latter having been found to contain 2-8 cfu/50ml of milk (Sweeney *et al.*, 1992). *M. paratuberculosis* is resistant to drying and may survive in faeces on pasture land for one year or so.

M. paratuberculosis contamination of commercial raw milk supplies must therefore be assumed to occur.

However, standards of animal husbandry and veterinary practice vary widely in different countries, as a result of which there may be wide variations in the levels of *M. paratuberculosis* present in commercial milk supplies. Indeed, there is considerable variation in the overall hygienic quality of raw milk in different countries; it is widely acknowledged that the most hygienic raw milk in the world is produced by the UK and some Scandinavian countries and this has been the case for many years.

Relatively little is known about the incidence and survival characteristics of *M. paratuberculosis* during the processing of milk and manufacture of dairy products. No surveys of dairy products have been reported although the potential for its presence and survival in unpasteurised dairy products exists. Other potential sources of human infection might also be expected, including water supplies, raw vegetables and undercooked meat, but these sources do not appear to have been investigated.

In 1996, an International Dairy Federation symposium on the bacteriological quality of raw milk acknowledged the potential transmission of *M. tuberculosis* and *M. bovis* to humans via raw milk although these organisms were considered to be primarily animal pathogens and "may not be seen as classical foodborne bacteria" (Hahn, 1996). Nevertheless, milk-borne transmission of *M. tuberculosis* and *M. bovis* to humans had been a significant problem before the days of widespread pasteurisation and tuberculin testing of dairy herds. Goats' milk, which is often drunk unpasteurised, may also contain *M. paratuberculosis* and may therefore pose a potential source of human infection.

Whilst the potential for survival of *M. paratuberculosis* in processed dairy products is unknown, some indications might be drawn from the known survival of the related organism *M. tuberculosis*, the tubercle bacillus, which has been reported to survive in soft cheese for 3 months and in hard cheese for up to 10 months (Anon., 1980). It might therefore not be unreasonable to expect *M. paratuberculosis* to have similar survival characteristics, although there is currently no direct evidence for this. Clearly, the possibility must be considered that *M. paratuberculosis* might exist in cheeses and other dairy products, especially those made from unpasteurised milk.

Association of *M. paratuberculosis* with Crohn's Disease

Crohn's disease was first described by Crohn and colleagues in 1932. Over the years, many theories regarding its aetiology have been suggested including: allergic reaction, auto-immune disease, defective mucosal barrier, dietary factors, measles virus and, of course, *M. paratuberculosis*. However, it was Dalziel in 1913 who first suggested a possible association between *M. paratuberculosis* and chronic intestinal enteritis in humans. He described several patients with chronic intestinal enteritis, and drew parallels with the recently described disease in cattle called paratuberculosis (Johne's disease). He suggested that "the histological characteristics of the two diseases were so similar as to justify a proposition that they may be the same".

To date, the aetiology of Crohn's disease still eludes the scientific community. However, investigators keep coming back to the question of whether or not Crohn's disease could be a mycobacterial disease. Dr Rod Chiodini (Rhode Island Hospital and Brown University, Providence, RI, USA) and Prof. John Hermon-Taylor (St. George's Hospital Medical School, London, UK) are two of the most prominent proponents of an association between *M. paratuberculosis* and Crohn's disease. Indeed, it was Chiodini and his co-workers who, in 1984, successfully cultured the first two strains of *M. paratuberculosis* in the USA from patients with Crohn's disease. Since then, this organism has been sporadically cultured from humans with Crohn's disease and a total of 10 isolates of *M. paratuberculosis* had been cultured from patients with Crohn's disease in the USA, Australia, the Netherlands and France (Chiodini, 1989).

In some cases, cell wall-deficient forms of *M. paratuberculosis*, called sphaeroplasts, have been found. Some investigators speculate that formation of sphaeroplasts is the reason why *M.paratuberculosis* cannot be seen in acid-fast stained tissue sections from patients with Crohn's disease, in contrast to Johne's disease tissues where abundant acid-fast bacilli are observed, and therefore why it has proved so difficult to culture the organism from clinical specimens from patients with Crohn's disease.

When subjected to restriction fragment length polymorphism (RFLP) molecular typing, isolates of *M. paratuberculosis* from patients with Crohn's disease have been shown to be of the bovine type rather than of ovine-caprine origin, as their RFLP patterns were indistinguishable from those of strains isolated from cattle (François *et al.*, Krishnamoorthy, 1997). However, it must be stressed that relatively few strains were examined; only 4 human isolates were compared with 8 bovine strains and 1 from a caprine source.

The discovery of the insertion element IS900 in *M. paratuberculosis* in the late 1980s was a major breakthrough in *M. paratuberculosis* research. The validity of PCR techniques for the detection of clinical infections has been questioned (McFadden and Fidler, 1996). In view of these difficulties, there would appear to be uncertainty as to the statistical significance of the reported rates of detection of *M.paratuberculosis* in healthy and diseased human tissues and it might therefore be unreasonable to conclude, on the basis of current evidence, that *M.*

paratuberculosis is more likely to be associated with diseased than with healthy tissues.

Heat Resistance

Because *M. paratuberculosis* can occur in the milk of cattle and goats with Johne's disease as well as in the milk of asymptomatic carriers, and since a possible association exists between *M. paratuberculosis* and Crohn's disease, interest in the thermal inactivation of this organism in milk at pasteurisation temperatures has arisen. Milk pasteurisation was originally established to inactivate *Coxiella burnetti* (the organism responsible for Q-fever) and Mycobacterium tuberculosis (the organism responsible for tuberculosis, TB), these being the most heat-resistant vegetative pathogens likely to be present in raw milk. Results of two studies (Chiodini and Hermon-Taylor, 1993; Grant *et al.,* 1996) suggested that *M. paratuberculosis* may survive heat treatment at 63°C for 30 min (equivalent to holder pasteurisation) and 71.7°C for 15s (equivalent to high-temperature, short-time pasteurisation) if present in high numbers (10^4 to 10^7 cfu/ml) prior to heat treatment. A more recent study (Grant *et al.,* 1998) investigated the efficacy of HTST pasteurisation when low numbers ($<10^3$ cfu/ml) of *M. paratuberculosis* were present in milk. The results showed that *M. paratuberculosis* could not be detected following pasteurisation if <10 cfu/ml were present in the raw milk.

The concept that the number of microbes surviving any given heat process is proportional to the number of organisms heated applies to all food-borne organisms and it has long been known that commercial processes are designed to inactivate the levels of organisms normally encountered in foods. Indeed, renewed debate on the efficacy of commercial milk pasteurisation took place in the 1980s when Listeria monocytogenes became linked with food-borne illness. It was concluded that the commercial process would only fail to inactivate that organism if very high levels ($>10^6$ cfu/ml) were present in the raw milk, which was considered unlikely.

A number of additional studies on the pasteurisation of cows' milk containing *M. paratuberculosis* were presented at the fifth meeting of the International Association for Paratuberculosis in Wisconsin, USA. The results of these further studies support the findings of Chiodini and Hermon-Taylor (1993) and Grant *et al.* (1996) that *M. paratuberculosis* may survive current HTST pasteurisation time-temperature conditions if present in high numbers prior to heat treatment. In addition, researchers at the University of Wisconsin, Madison reported D values (thermal destruction values) for *M. paratuberculosis* in milk at pasteurisation temperatures (Sung and Collins, 1998) indicating that a heat treatment of 72°C for 15s would not be sufficient to kill 10^6 cfu/ml of *M. paratuberculosis* in milk. A thermal death time plot was also constructed. This tends to suggest that *M. paratuberculosis* is more heat resistant than either *Cox.burnetti* or *M. tuberculosis*. However, some of the evidence regarding its ability to survive pasteurisation is controversial and there is a need for further research.

Some attempt has been made to determine why *M. paratuberculosis* may be able to survive pasteurisation. When the thermal inactivation kinetics of the organism were studied, non-linear thermal inactivation curves were obtained for *M. paratuberculosis* which exhibited significant 'tailing' (Grant *et al.,* 1996). A vital staining procedure has shown that, at heating times corresponding to the 'tail' region, any viable cells present are always located within clumps of cells (Grant *et al.,* 1997). Indeed, researchers at the University of Wisconsin have found that the D values of clumped *M. paratuberculosis* cells are higher than those of single cells. These findings lend some support to the theory that it is the organism's tendency to clump which may assist its survival during pasteurisation. However, further investigation is necessary before definitive conclusions can be drawn.

In most of the thermal inactivation studies to date, laboratory-scale pasteurising equipment has been used to simulate the commercial pasteurisation of milk, artificially inoculated with M. *paratuberculosis.* Care must be taken in extrapolating the results of such studies to commercial milk processing. Two studies employing pilot-scale 'commercial' pasteurisers have yielded conflicting data regarding the ability of *M. paratuberculosis* to survive HTST pasteurisation (Hope *et al.,* 1997; Stabel *et al.,* 1997). Stabel *et al.* (1997), who used an Armfield laboratory-scale pasteuriser, reported that *M.paratuberculosis* (strains ATCC 19698, Ben and Kay) did not survive heating at 72°C for 15s, even when present at levels of 10_8 cfu/ml. In contrast, Hope *et al.* (1997), who employed a similar laboratory-scale pasteuriser and studied strain ATCC 19698 and a variety of bovine field isolates of *M.paratuberculosis,* were able to culture viable *M. paratuberculosis* from 4 of 8 batches of milk heated at 72-73°C for 15s and 1 of 4 batches of milk heated at 82-90°C for 15s when $>10_4$ cfu/ml were present prior to heat treatment. However, neither group reported survival of viable *M. paratuberculosis* when milk samples, initially spiked with $<10_4$cfu/ml were subjected to heating at 72°C for 25s. Thus, the heat process applied to milk in laboratory systems may not be comparable with that used commercially, even though the time-temperature profiles appeared similar.

No attempt appears to have yet been made to pasteurise naturally infected milk obtained from cattle or goats with Johne's disease, in order to determine whether *M. paratuberculosis* growing *in vivo* responds to heating in the same way as strains cultured *in vitro*. It is possible that, in practice, raw milk is naturally-contaminated only with low levels of *M. paratuberculosis* that would not survive a commercial heat process.

The dairy industry, worldwide, is naturally concerned about the possibility that *M. paratuberculosis* might survive pasteurisation. However, in view of the currently tenuous link between Crohn's disease and Johne's disease, it believes that the apparently low risk of foodborne transmission of *M.paratuberculosis* is acceptable until further information becomes available (Hammer *et al.,* 1997). Similarly, the United Kingdom Ministry of Agriculture, Fisheries and Food considers that, on the basis of what is currently known about *M. paratuberculosis* there is no need for anyone to change their dietary habits (Anon., 1998). Clearly, many questions remain unanswered about the incidence, survival and heat-resistance of *M. paratuberculosis* in milk and dairy products. Considerable further research is needed before definitive conclusions can be drawn.

According to IFST, it concludes that:

1. Until the results of the proposed survey are known, the dairy industry has no need to regard its current pasteurisation processes as inadequate;
2. There is a need for further quantitative investigations to determine clearly the heat resistance of *M. paratuberculosis* and to ascertain the incidence of Johne's disease in dairy cattle, sheep and goats, worldwide, as well as the incidence of asymptomatic carriers;
3. Other potential sources of human infection such as water supplies, undercooked meat and raw vegetables should also be investigated.

Human Pathogens

Salmonella

Most *Salmonellas* are regarded as human pathogens, though they differ in the characteristics and the severity of the illness they cause. Typhoid fever is the most severe and consequently was the earliest *Salmonella* infection to be reliably described. This is credited to Bretonneau, the French physician who is also regarded as the founder of the doctrine of the aetiological specificity of disease. During his life, he published only one paper on typhoid, or 'dothinenterie' as he called it, in 1829 and his treatise on the subject was only published in 1922 by one of his descendants.

In 1856, the English physician William Budd concluded that each case of typhoid is epidemiologically linked to an earlier case and that a specific toxin is disseminated with the patients faeces. To support his proposition he demonstrated that treating the excreta of victims with chlorinated lime (bleaching powder) reduced the incidence of typhoid. The typhoid bacillus was first observed by the German bacteriologists Eberth and Koch in 1880 and four years later Gaffky succeeded in its cultivation. The paratyphoid bacilli, responsible for the clinically similar condition, paratyphoid fever, were first isolated by Achard and Bensaud (1896) and by Gwyn (1898) and confirmed as culturally and serologically distinct from the typhoid bacillus by Schottmilller in 1901. Other *Salmonellas* were isolated during the same period; Salmon and Smith (1885) isolated *Bacillus cholerae-suis* from pigs with hog cholera, a disease now known to be viral in origin and similar bacteria were isolated from cases of food-borne infection and animal disease. The genus *Salmonella* was finally created in 1900 by Lignieres and named in honour of D.E. Salmon, the American veterinary pathologist who first described *Salmonella cholerae-suis.*

In Europe in 1989 the annual incidence of salmonellosis was around 50 per 100000 inhabitants in most countries, though actual figures varied from below 10 in the case of Luxembourg to more than 120 in Hungary and Finland. Data collected by the European surveillance system Enter-net reported an increase in the annual total for salmonellosis in 12 European countries from 41870 in 1995 to 55278 in 1997. Since the contribution of "England and Wales to these figures was 29314 in 1995 and 32596 in 1997 it suggests that the effectiveness of data collection and/or food hygiene are far from uniform across these countries.

The disease is grossly underreported because it is generally a self-limiting gastroenteritis which may be misdiagnosed as intestinal influenza by the patient or the physician. As a consequence, estimates of the true incidence of disease are based on assumptions derived from epidemiological evidence. Clearly, salmonellosis continues to be an important cause of foodborne disease worldwide.

Salmonellas are now established as one of the most important causes of food-borne illness worldwide. Some of the salmonella species involved in food poisoning include; *Salmonella typhimurium*, *Salmonella enteritidis*, *Salmonella dublin*, *Salmonella softenburg*, Salmonella virchow, *Salmonella montevideo*, *Salmonella infantis*, and *Salmonella newport.* These species are also involved in causing diarrhoea in animals. Some salmonella serotypes are associated with human disease and are not known to affect animals *e.g., S. typhi* and *Salmonella paratyphi.*

The salmonella are small, gram negative, non-spore forming rods that are indistinguishable from the *Escherichia coli* under the microscope or ordinary nutrient media. They are widely distributed in the nature with humans and animals being their primary reservoirs. Salmonella food poisonings results from ingestion of food containing appropriate strains of this genus in significant numbers. Some significant changes have occurred in the taxonomy of salmonella. The genus salmonella is considered to have a two species named salmonella enteric and salmonella bongori. One of these subspecies, which is relatively unimportant as a cause of human infection and accounts for less than 1 per cent of *Salmonella* serovars, has been proposed for elevation to species status as *S. bongori.* Serotyping differentiates the strains and they are referred as to by, for example *S. enterica* serotype *typhimurium* or as *S. typhimurium.*

The Kauffman-White serotyping scheme has proved the most useful technique for differentiating within the genus. This describes organisms on the basis of their somatic (O) and flagellar (H) antigens and by capsullar (Vi) antigens (possessed by *S. typhi, S. dublin* and occasional strains of *S. paratyphi* C). In 1941 the scheme contained 100 serovars and the number has since risen to the current level of around 2400.

The taxonomic nomenclature of the genus is rather different from that of other genera. Many of the different serovars were named as if they were distinct species. The earliest to be described were given species epithets derived from the disease they caused, either in humans (*S. typhi, S. paratyphi* A and B), or in animals (*S. typhimurium, S. choleraesuis,* or *S. abortusovis*). Limitations in this approach led to the use of serovar names based on the geographical location of the first isolation, for example *S.dublin, S. montevideo, S. minneapolis* and even *S. guildford.* This has some advantage over the use of long serological formulae but since 1966 has only been applied to serovars of subspecies (*S. enterica* subsp. *enterica*) which accounts for more than 59 per cent of the 2400 serovars known and the vast majority (> 99 per cent) of human isolates.

The most recent proposal to introduce some taxonomic rectitude is to use the non-italicized serovar name after the species name so that *S. typhimurium* becomes *S. enterica* subsp. *enterica* ser. Typhirnurium or more concisely, *Salmonella typhirnurium*. By retaining the old serovar name much of the potential for confusion inherent in other schemes is reduced.

Biotyping according to biochemical characteristics has sometimes proved useful in epidemiological investigations where it can supplement phage typing or sub-divide a large group of otherwise untypable strains. This has, proved most useful for *S. typhimurium* where Duguid's scheme based on 15 biochemical tests has identified 184 full biotypes.

There are three main ways as to how *Salmonella* can enter the food supply to cause illness. Animals harbour Salmonella, making meats, poultry, eggs, and milk often implicated vehicles. Salmonella, which are introduced into the environment, possibly through manure and litter, may persist and contaminate fruits and vegetables on the farm. Cross-contamination in the food service environment or the home, often between raw poultry and ready-to-eat (RTE) products, such as raw vegetables, can also cause salmonellosis.

Environmental sources of the organism include water, soil, insects, factory surfaces, kitchen surfaces, animal faeces, raw meats, raw poultry, and raw seafoods, to name only a few. Raw meats, poultry, eggs, milk and dairy products, fish, shrimp, frog legs, yeast, coconut, sauces and salad dressing, cake mixes, cream-filled desserts and toppings, dried gelatin, peanut butter, cocoa, and chocolate are the foods reported to be associated with this food borne illness. They can be disseminated via faeces to soil, water, foods and feeds and hence, to other animals (including humans).

Some Salmonella are host-adapted (*e.g., Salmonella pullorum* in poultry, *Salmonella dublin* in cattle); however, most Salmonella are not. Although any Salmonella is a potential pathogen for humans, most foodborne salmonellosis is caused by non-host-adapted serotypes. Much human salmonellosis is directly related to human association with animals, both wild and domestic. Foods of animal origin are vehicles for salmonellosis. Salmonella was isolated in 19–54 per cent of cattle carcasses, 1.9 per cent of beef samples at retail and 4.2 per cent of retail chicken samples.

The animal-to-man link is only one factor in the epidemiology of human salmonellosis. Contaminated red meat and poultry provide a nidus (source of infection) for Salmonella, which man nurtures through mishandling. Furthermore, inedible parts of the animal are processed to yield important components of livestock feeds. As a result of poor manufacturing practices (postprocessing contamination), these render animal by-products to become recontaminated with Salmonella, which, in turn, are carried into the feeds. The consumption of these feeds by livestock, followed by animal-to-animal transmission, completes the Salmonella cycle. This is not to suggest, however, that contaminated feeds are the only source of Salmonella in livestock. Epidemiological evidence indicates that there is a direct link between the presence of Salmonella in meat and poultry and human salmonellosis. Man induces salmonellosis through improper food handling practices and perpetuates salmonellosis through recontamination of rendered animal by-products, which are incorporated into livestock feeds.

Historically, egg products (dried, frozen or liquid eggs, with or without added ingredients, as defined in the Egg Products Inspection Act; U.S. Congress, 1970) were a significant source of human salmonellosis in the United States. In 1990, the U.S. Dept. of Agriculture (USDA) required that eggs produced by flocks previously implicated in human disease be tested for Salmonella (FDA/CFSAN, 2003). Mandatory pasteurization of egg products was responsible for greatly reducing eggs as a major cause of salmonellosis, and consumer awareness of the need to refrigerate eggs also contributed in the decreased incidence of disease.

Consumption of raw milk may also cause human salmonellosis. In one study, Salmonella was isolated in 6.1 per cent of bulk raw milk samples. Between 1972–2000, 16 outbreaks of salmonellosis resulted from raw milk consumption (CDC, 2003). Milk-borne salmonellosis was particularly prevalent in Scotland, where, prior to 1983, the sale and consumption of raw milk was common. Milk outbreaks in England and Wales were primarily associated with raw milk.

Animal products are not the only sources of human salmonellosis. Produce can serve as a vehicle of Salmonella as well, becoming contaminated either on the farm or through cross-contamination with contaminated products. Consumption of produce contaminated with Salmonella has recently caused several outbreaks. In 2000, 361 individuals in England and Wales acquired salmonellosis after eating lettuce, and although the lettuce appeared to come from one of three farms, investigators could not conclusively find the source of contamination (Horby *et al.,* 2003). The strain, *S. enterica* subsp. enterica serotype Typhimurium DT 104, is resistant to several antibiotics, and the increase in its prevalence poses challenges in treatment of the infection. Raw sprouts, particularly alfalfa and mung bean, have been involved in 15 salmonellosis outbreaks since 1995 (Thomas *et al.,* 2003). Seeds may be contaminated, and the warm, moist conditions for sprouting allow growth of the pathogen. On-farm contamination of cantaloupes resulted in 24 cases of salmonellosis in 1997 (Mohle-Boetani *et al.,* 1999). Contamination of cantaloupes on one farm caused multistate outbreaks of salmonellosis each spring between 2000 and 2002 (CDC, 2002). Since produce may be eaten raw, different control measures are necessary to prevent illness when the pathogen is introduced on the farm.

While Salmonella may survive in contaminated foods as a result of improper cooking, it is more common that cross-contamination of foods after cooking is the source of Salmonella. Foodservice workers or in-home food preparers may transfer salmonellae from raw products to cooked or other uncontaminated foods as a result of unsanitary preparation practices (*e.g.,* failure to wash hands) between handling of these foods. Salmonella can also be transferred from contaminated raw foods to equipment surfaces, such as knives, cutting boards, and counter tops, and then from equipment to previously uncontaminated foods. Once contamination occurs, the situation may be further complicated by improper storage of the product before serving (*e.g.,* kept at room temperature, improperly refrigerated, or held in warmers within the growth range for Salmonella). Although responsible for fewer outbreaks, contamination of foods by infected workers cannot be ignored as a cause of foodborne salmonellosis. Some infected individuals may excrete Salmonella for weeks, months, and, occasionally, years with little or no evidence of disease. Improper hygiene practices by these individuals may lead to either contamination of foods or direct person-to person contamination.

The primary habitat of salmonella species is the gastrointestinal tract of food animals (especially poultry), wild animals, rodents, pets, birds, reptiles and insects without the display of any apparent illness. Although their primary habitat is intestinal tract, they may be found in other parts from time to time. *Salmonella* organisms may penetrate and pass from gut lumen into epithelium of small intestine where inflammation occurs; there is evidence that an enterotoxin may be produced, perhaps within the enterocyte.

As an intestinal form, the organisms are excreted in faeces from which they may be transmitted by insects and other living creatures to large number of places. For epidemiological purposes, the salmonella can be placed into three groups; the first are those that infect humans only. This includes, *S. typhi*, *S. paratyphi* A, *S. paratyphi* C. this group includes the agents of typhoid and paratyphoid fevers, which are the most severe of disease caused by salmonella. The second was the host-adapted serovars (some of which are human pathogens and may be contracted from food), included are *S. gallirinum* (poultry), *S. dublin* (cattle), *S. abortus* equi (equine), *S. abortusovis* (sheep) and *S. choleraesuis* (swine). The third is unadapted serovars (no host preference). There are pathogenic for humans and other animals. The epidemiology of the salmonella is complex, which often make animals control of the disease is difficult. Animals are the reservoir of food born disease of salmonella.

Most *Salmonellas* infect a range of animal species but some serotypes are host adapted such as *S. enteritidis* PT4, *S. pullorum* and *S. gallinarum* in poultry and *S. choleraesuis* in pigs. In these cases, direct animal-to-animal transmission can be more important and vertical transmission may occur–parents infecting offspring. For example, *S. enteritidis* PT4 can pass from breeding flocks to newly hatched broiler and egg-laying chicks via transovarian infection of the egg or its shell.

Various *Salmonella* species have long been isolated from the outside of egg shells. The present situation with *S. enteritidis* is complicated by the presence of the organism inside the egg, in the yolk. This and other information strongly suggest vertical transmission, *i.e.,* deposition of the organism in the yolk by an infected layer hen prior to shell deposition. Foods other than eggs have also caused outbreaks of *S. enteritidis* disease.

Salmonellas are responsible for a number of different clinical syndromes, grouped here as enteritis and systemic disease.

Gastrointestinal infections are predominantly associated with those serotypes which occur widely in animals and humans. They can range in severity from asymptomatic carriage to severe diarrhoea and are the most common type of salmonellosis.

S. typhi and *S. paratyphi* A, B, and C produce typhoid and typhoid-like fever in humans. Various organs may be infected, leading to lesions. The fatality rate of typhoid fever is 10 per cent compared to less than 1 per cent for most forms of salmonellosis. *S. dublin* has a 15 per cent mortality rate when septicaemic in the elderly, and *S. enteritidis* is demonstrating approximately a 3.6 per cent mortality rate in hospital/nursing home outbreaks, with the elderly being particularly affected.

Salmonella often enters the host by ingestion, and even with several system to mediate acid resistance, few survive the stomach and move into the small intestine. Normal flora protects against colonization of administration of oral antibiotics facilitates establishment of infection. Entry of salmonella usually occurs without mucosal damage in systemic infections, but enteric infection is characterized by local damage without septicemia-salmonella infection with M cells in payer's patches is facilitated by fimbrial adhesions. This is followed by ruffling of the target cell membrane which result in internalization of the bacteria in membrane bound vacuoles. The ruffles facilitate uptake of the bacteria in membrane bound vacuoles or vesicles which often coalesce. The organisms replicate in these vesicles and are eventually released from the cells, which sustains only mild or transient damage. The complex invasion process is mediated by the product of a number of chromosomal genes, whereas growth within host a cell depends on the presence of virulence plasmids.

The incubation period of the salmonella is 12-36 hours. The clinical sign includes diarrhoea, which may be watery, greenish and foul smelling. This may be preceded by headache and chills. Other findings include prostration, muscle weakness and moderate fever. In most cases the symptoms resolve in 2-3 days without any complication. The bacterium induces responses in the animal that is infecting, this is what typically causes symptoms, rather than any direct toxin product. Symptoms are usually gastrointestinal, including nausea, vomiting, abdominal cramps and bloody diarrhoea with mucous, headache, fatigue and rose spots are also possible. These symptoms can be severe, especially in young children and elderly. Symptoms last generally up to a week, and can appear 12-72 hours after ingesting the bacterium. Reactive arthritis, sickle-cell anaemia and osteomyelitis due to salmonella infection is much more common that in the general population. For diagnosis of human illness, serological identification of culture isolated from stool is done.

Other forms of salmonellosis generally produce milder symptoms. Infective dose includes as few as 15-20 cells; depends upon age and health of host, and strain differences among the members of the genus.

Salmonella septicemia has been associated with subsequent infection of virtually every organ system. Under chronic consequences, post-enteritis reactive arthritis and Reiter's syndrome have also been reported to occur generally after 3 weeks. Reactive arthritis may occur with a frequency of about 2 per cent of culture-proven cases. Septic arthritis, subsequent or coincident with septicemia, which occurs can be difficult to treat.

All age groups are susceptible, but symptoms are most severe in the elderly, infants, and the infirm. AIDS patients suffer salmonellosis frequently (estimated 20-fold more than general population) and suffer from recurrent episodes.

Salmonellosis may be confused clinically with staphylococcal intoxication, but there are important distinctions. Salmonella has a longer incubation period than staphylococci (usually 12–36 hr vs. 2–4 hr) and is usually accompanied by fever, which is absent in staphylococcal intoxication. And unlike Salmonella food poisoning, the acute symptoms of staphylococcal food poisoning normally disappear within 24 hours.

Detection of salmonella can be provided only by isolation of the agent from stool or vomit in human, feed samples in cases of animals, and samples concerned food items like milk and milk-products samples. In general, for culture and isolation, the use of selective enrichment media such as salmonella-shigella, agar, hektoen enteric agar or deoxycholate agar and broth for enrichment before sub-culture to these sugar agars after 24 hours is usual procedure. Selenite enrichment broth or tetrathionate broth can be used to isolate highly selective for salmonella, especially S. enterica serovar Typhi. Agar and plates are incubated at 37° C overnight and growth identified by biochemical tests and slide agglutination tests

At any one-time human illness is usually associated with a limited number of serotypes; in the UK only about 200 serotypes may be reported in anyone year. Currently *S. enteritidis, S. typhimurium* and *S. virchow* are the most common, the first two accounting for about three-quarters of laboratory reports.

More probably the cause of *Salmonella* diarrhoea is multifactorial. Evidence has been obtained that diarrhoeagenic *Salmonellas* also produce enterotoxins which stimulate fluid secretion in assays such as the ligated ileal loop test. In particular, a heat-labile toxin has been purified from *S. typhimurium* which bears some structural and serological similarity to *E. coli* LT and cholera toxin. However, some of the evidence in this area is contradictory and resolution of the problem will probably require optimization of the conditions for toxin production and more sophisticated tests for enterotoxin activity.

As a general rule, the infectious dose of *Salmonella* is high, of the order of 10_6 cells, but this will vary with a number of factors such as the virulence of the serotype, the susceptibility of the individual and the food vehicle involved. A number of outbreaks have occurred where epidemiological evidence points to an infective dose as low as 10–100 cells. This appears to be particularly associated with more susceptible individuals such as children and the elderly and with fatty foods such as cheese, salami and chocolate. In an outbreak in Canada where the vehicle was cheddar cheese it was found to contain 1.5–9.1 per 100 g. It seems likely that the high fat content in some foods affords the bacteria some protection from stomach acidity. A low infective dose (< 200) was also indicated in a water-borne outbreak in early 1970s. In this case fat was clearly not a factor, but the more rapid transit of water through the stomach may have served a similar purpose.

Different control measures exist depending on the mode of contamination of the food. Reduction of the incidence of Salmonella contamination of foods requires a number of approaches to the problem, beginning at the farm and going right through to the kitchen.

Preventive and control measures should include the following:

- Efficient refrigeration and hygienic handling of food.
- Consumption of properly cooked meat.
- Complete thawing of frozen meats and adequate cooking.
- Heat processing of meat, milk, fish and poultry to destroy salmonella organisms in food.

There is a critical need to develop method to control the spoilage or poisoning of food by salmonella ordinary farms by instituting bio-security and bio-containment practices in addition to enhanced food processing method, preparation and storage practices. The most common method of eliminating Salmonella from food products is heat processing. Salmonella is heat sensitive, and ordinary pasteurization or cooking conditions are generally sufficient to kill it in high-moisture foods. As with other microorganisms, the heat resistance of Salmonella is markedly increased as water activity (a_w) decreases. Effective heat processing of food of animal origin, which includes pasteurization of milk and eggs, irradiation of meat and poultry thermal processing; good hygiene practices during production of food; vaccination of egg-producing flocks and food producing animals. At food service establishment, safe food preparation practices should include thorough cooking and reheating of food and boiling of milk, adequate refrigeration, prevention of cross contamination; cleaning and disinfection of food preparation surfaces; exclusion of pets and other animals from food handling areas *etc.*

Several approaches have been taken to reduce the carriage of Salmonella by animals. Vaccination of laying chickens significantly reduced the per cent of eggs positive for *Salmonella enteritidis*. Other research has examined the effect of probiotics on the intestinal microflora of chickens. In this approach, the goal is to colonize the chicken with known microorganisms to "competitively exclude" Salmonella. Successful reduction in the per cent of the flock positive for Salmonella was achieved when the feed was supplemented with yeast or when chickens were both sprayed with a patented mucosal starter culture and fed the culture through water.

Typhoid Fever

Enteric fevers include typhoid and paratyphoid fevers caused by *Salmonella typhi* and *Salmonella paratyphi* A, B and C, respectively. The serotypes are similar to other salmonella bacteria, but unlike them, they are essentially parasites of man. *S. typhi* possesses capsular (vi). antigen in addition to the usual O and H antigens found in other serotypes.

The typhoid and paratyphoid bacilli are essentially human parasites and are acquired mostly from human sources, namely, patients and carriers. The bacteria can be transmitted by the contamination of water, milk or food by flies. Only a few organisms are needed to cause disease.

Generally, the incubation period is usually 2 weeks, but might vary between 3 and 28 days for typhoid fever and between 1 and 15 days for the paratyphoid fevers. The enteric fevers are generalized septicaemic infections with a frequent, if not constant bacteraemia during the first two weeks of the disease. Invasive *Salmonellas* penetrate the intestinal epithelium and are then carried by the lymphatics to the mesenteric lymph nodes. After multiplication in the macrophages, they are released to drain into the blood stream and are then disseminated around the body. They are removed from the blood by macrophages but continue to multiply within them. This eventually kills the macrophages which then release large numbers of bacteria into the blood stream causing a septicaemia. In this, the first phase of the illness the organism may be cultured from the blood. There is a slow onset of symptoms including fever, headache, abdominal tenderness and constipation and the appearance on the body of rose red spots which fade on pressure.

During the second stage of the illness, the organism reaches the gall bladder where it multiplies in the bile. The flow of infected bile reinfects the small intestine causing inflammation and ulceration. The fever persists but with the onset of a diarrhoea in which large numbers of the bacteria are excreted with the characteristic 'pea soup' stools and, to a lesser extent, with the urine. In more serious cases, haemorrhage of the ulcers may occur and perforation of the intestine leading to peritonitis; In milder cases, the ulcers heal and fever falls with recovery after 4–5 weeks.

Unlike the more localized enteric infections, typhoid is usefully treated with antibiotics such a chloramphenicol, ampicillin and amoxycillin. After remission of symptoms, a carrier state can persist for several months and occasionally years as parts of the gall bladder are colonized and bacteria are discharged intermittently with the bile into faeces. This occurs more commonly in women and the elderly and the have been a number of typhoid carriers who have achieved some notoriety as a result of their condition and its consequences. These include the 'Strasbourg Master Baker's Wife', the 'Folkestone Milker' and, probably best known of all, 'Typhoid Mary'. Mary Mallon worked as a cook in a number of households and institutions in the New York area at the beginning of the century. She first attracted the attention of the authorities when she disappeared after an outbreak of typhoid fever in a family for whom she had been working. When she was eventually tracked down by following a trail of outbreaks in places she worked, she was forcibly detained in the New York City Hospital for three years. Despite an undertaking not to work as a cook or handle food on her release, she disappeared again, assumed a false name and started work as a cook. In 1915 she was working at another New York hospital when a typhoid outbreak occurred in which eight people died. Methods for the isolation and identification of *Salmonellas* in foods have arguably received more attention than those for any other food-borne pathogen. Using traditional cultural techniques, a five-stage procedure has emerged as the widely accepted norm.

Pre-enrichment in a non-selective medium increases the recovery rate of *Salmonellas* by allowing the repair of cells which have been sub lethally damaged. Such damage can result from any exposure to adverse conditions that might occur during food processing, such as chilling, freezing, or drying and increases the cell's sensitivity to selective agents used in media in subsequent stages of the isolation procedure. Failure to include a resuscitation step could therefore result in the non-detection of cells that might recover and cause infection if the food is mishandled.

The selective enrichment stage is intended to increase the proportion, *Salmonella* cells in the total microflora by allowing them to proliferate while restricting growth of other microorganisms present. To this end a number of different media have been proposed employing selective agents such as bile, brilliant green, malachite green, tetrathionate and selenite. The most widely used are selenite-cystine broth, which contains cystine to stimulate growth of *Salmonellas*; Muller-Kauffman tetrathionate broth, containing tetrathionate, brilliant green, and bile; and Rappaport-Vassiliadis (RV) broth, which contains malachite green, magnesium chloride and a slightly reduced pH as selective factors. Since they differ in their selectivity, two broths are usually used in parallel; commonly a combination of the less selective selenite-cystine broth and one of the others.

From the selective enrichment of broths, cultures are streaked on to selective and differential solid media. Once again it is usual to use two different media in parallel. The selective agents used are bile salts or deoxycholate and/or brilliant green and the diagnostic reaction is usually provided by the inability to most *Salmonellas* to ferment lactose and/or the production of hydrogen sulfide. In choosing the media to use, it is advisable to select two based on different diagnostic reactions to ensure that atypical strains, for instance lactose-positive ones, will not be missed.

Figure 7.4 indicates the traditional cultural protocol for isolation of *Salmonella* from food.

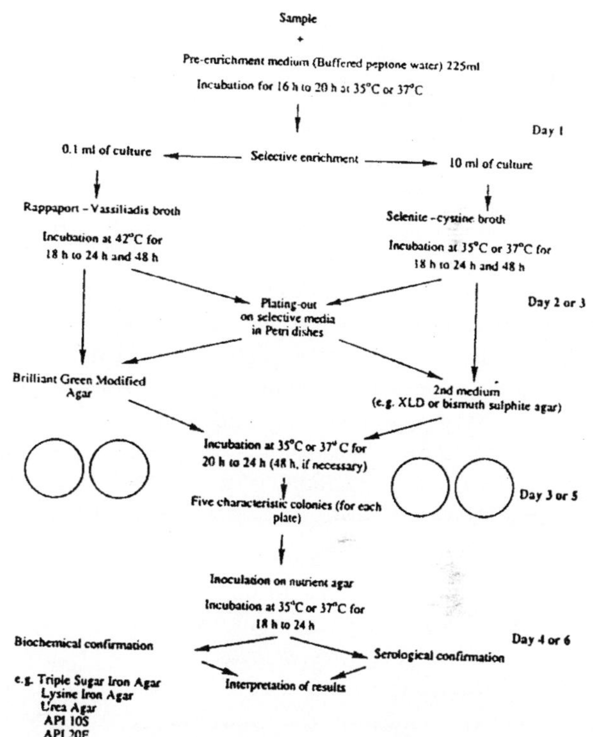

Figure 7.4: Traditional Cultural Protocol for Isolation of Salmonella from Food.

The whole protocol is rather complex and lengthy, requiring at least four days for a negative result. In view of this, a number of procedures have been described which attempt to simplify the procedure and reduce the elapsed time involved. Two of these employ the motility of *Salmonellas* which means that they would fail to detect non-motile *Salmonellas* (incidence < 0.1 per cent).

In one, a conventional pre-enrichment culture is inoculated into an elective medium, *Salmonellas* swim into a compartment containing a selective medium and from there into one containing a diagnostic medium. A diagnostic medium giving the appropriate colour change is then tested for ability to agglutinate latex particles coated with *Salmonella* antibodies. A positive result indicates a presumptive *Salmonella*, which must then be confirmed by conventional serological and biochemical testing using a sub-culture from the diagnostic medium. With this technique, presumptive identification of a *Salmonella* is obtained within 42 h compared with 3–4 days by the traditional cultural method.

Control measures for typhoid fever should include:

- Hygienic control of food and water supplies
- Detection and treatment of chronic carriers
- Vaccination using TAB-vaccine. The vaccine contains a mixed culture of *S. typhi*, and *S. paratyphi*. The vaccine protects for 5-7 yrs.

Zoonotic Infection

The word 'Zoonosis' (Pleural: Zoonoses) was introduced by Rudolf Virchow in 1880 to include collectively the diseases shared in nature by man and animals. Later WHO in 1959 defined that Zoonoses are "those diseases and infections which are naturally transmitted between vertebrate animals and man". Zoonoses include only those infections where there is either a proof or strong circumstantial evidence for transmission between animals and man. Historically, zoonotic diseases had a tremendous impact on the evolution of man, especially those cultures and societies that domesticated and bred animals for food and clothing. Zoonoses are among the most frequent and dreaded risk to which mankind is exposed. Zoonoses occur throughout the world transcending the natural boundaries.

Among those zoonoses recognized today as particularly important are anthrax, plague, brucellosis, Bovine tuberculosis, leptospirosis, salmonellosis, spotted fever caused by Rickettsiae, rabies, several common athropod borne viral infections (arboviral infection), certain parasitic diseases, especially cysticercosis, hydatid disease, trypanosomiasis and toxoplasmosis.

Salmonellosis is described as a zoonotic infection since the major source of human illness is infected animals. Transmission is by the faecal-oral route whereby intestinal contents from an infected animal are ingested with food or water. A period of temperature abuse which allows the salmonellae to grow in the food and an inadequate or absent final heat treatment are common factors contributing to outbreaks.

Meat, milk, poultry and eggs are primary vehicles; they may be undercooked, allowing the *Salmonellas* to survive, or they may cross-contaminate other foods that are consumed without further cooking. Cross-contamination can occur through direct contact or indirectly via contaminated kitchen equipment and utensils.

Human carriers are generally less important than animals in the transmission of salmonellosis. Human transmission can occur if the faecally contaminated hands of an infected food handler touch a food which is then consumed without adequate cooking, often after an intervening period in which microbial growth occurs. This was the cause of a major outbreak affecting an international airline in 1984. The outbreak involved 631 passengers and 135 crew and was due to contamination of an aspic glaze by a member of catering staff who returned to work after illness but was still excreting *Salmonella* Enteritidis PT4.

Direct person-to-person spread by the faecal-oral route is also possible but is usually restricted to institutional outbreaks such as occur in hospitals, old people's homes and nurseries.

Food animals may acquire *Salmonella* infection on the farm from wild birds and rodents, but the principal sources are other animals, which may be symptomless excreters and contaminated feeding stuffs. Measures that can be taken to minimize transmission between animals on the farm include good animal husbandry, protection of feeds and water from contamination, hygienic disposal of wastes and maintenance of a generally clean environment. Transfer of *Salmonella* between animals is particularly associated with situations where animals may be stressed and crowded such as during transport at markets and when in lairage at the slaughterhouse. It is best minimized by avoidance of overcrowded conditions, ensuring a clean environment and otherwise limiting the stress to animals on such occasions.

An important factor in maintaining the cycle of *Salmonella* infection in food animals has been the practice of using animal by-products as animals' feeds such as meat and bone meal. The heat process that these materials undergo in their conversion to feeds should destroy any *Salmonellas* present. Nevertheless, they are subject to post-process contamination either in the plant or on the farm by contact with unprocessed material or with bird and rodent faeces. The importance of animal feeds as a source of *Salmonella* should decline following the 1996 ban on feeding mammalian-derived protein to all farm animals.

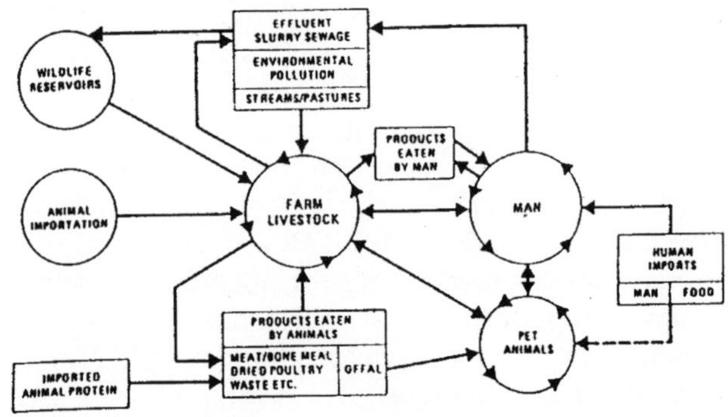

Figure 7.5: The Salmonella Cycle of Infection.

Contamination of eggs with *Salmonellas* is a long-recognized problem but in most cases, this was due to contamination of the eggshell exterior with faecal material in the hen's cloaca or after laying in the nest or battery. The shell could then contaminate the contents when the egg was broken. This is a particular problem when breaking large quantities of eggs where it is difficult to avoid some contamination with shell fragments. During the Second World War, there were a number of outbreaks attributed to dried whole egg powder and a survey conducted in 1961/2 found 16 per cent of frozen whole egg samples to contain *Salmonellas*. This led to the introduction in 1963 of regulations requiring that liquid whole egg be pasteurized at 64.4°C for 2.5 min. The prescribed heat treatment also inactivates the yolk enzyme α-amylase and provides the basis of a simple test to ensure that the regulations have been complied with.

In the more recent cases however, contamination of the yolk of intact hen's eggs has also been indicated. In the UK and in other European countries, particularly Spain, this problem has been associated with *S. enteritidis* PT4 but other phage types have been reported to cause similar problems in the United States. It is thought that these organisms infect the bird's ovaries and oviduct and thereby contaminate the egg contents. The temperatures reached in the yolk during mild cooking procedures such as 'soft boiling' or light frying are probably insufficient to kill the organisms and the fat content of the yolk may protect the organism from gastric acidity.

Salmonella typhimurium DT 104

The most common *Salmonella* involved in food borne illnesses is *Salmonella enteritidis*, followed by *Salmonella typhimurium*. However, whilst the overall incidence of human salmonellosis has apparently become constant, illness caused by *Salmonella typhimurium* DT 104 is gradually increasing. Susceptible DT104 was first reported in the 1960s in humans and subsequently as multidrug-resistant DT104 (MDR DT104) in the early 1980s in humans and birds from the United Kingdom. Another instance of human MDR DT104 was reported in Hong Kong in the late 1970s. Isolates from agricultural animals were first reported in the United Kingdom in 1988 and in the United States in 1990. MDR DT104 rapidly emerged globally in the 1990s and became the most prevalent phage type isolated from humans and animals in many countries. Previous epidemics with MDR phage types of *S. typhimurium*, such as DT29, DT204, DT193, and DT204c, were mostly restricted to cattle, whereas MDR DT104 spread among all domestic animals, including cattle, poultry, pigs, and sheep. A decline in MDR DT104 has been reported in the last decade.

Multidrug resistant *Salmonella typhimurium* DT104 was first identified in the 1980s; following its identification it became one of the most commonly isolated serovars of Salmonella from cases of human and animal clinical infection over the following two decades in Europe and North America.

Salmonella typhimurium DT 104 is attracting attention from the public health authorities and the media because it is causing increasing numbers of cases of salmonellosis in man. *Salmonella typhimurium* DT 104 is now the most commonly reported phage type of *Salmonella typhimurium*. Its unique feature is that it is resistant to many of the commonly used antibiotics including ampicillin, chloramphenicol, streptomycin, sulphonamides, and tetracyclines. Strains resistant to these substances (designated as R-type ACSSuT) accounts for 58 per cent of all isolates of *Salmonella typhimurium* DT 104 whilst only 5 per cent of *Salmonella typhimurium* DT 104 isolates are sensitive to all of the commonly used antibiotics. Indeed, some organisms are now appearing that are also resistant to trimethoprim and to quinolone antibiotics such as ciprofloxacin (known as R-types ACSSuTTm and ACSSuTCp).

Apart from its multiple resistance to antibiotics, *Salmonella typhimurium* DT 104 has similar characteristics to other *Salmonellae* and is no more resistant to food processing conditions such as acids, drying, preservatives or disinfectants than other serotypes. There is some evidence to suggest that it might be slightly more heat-resistant than many strains of *Salmonellae*.

Sources in the Food Chain and Foods Involved

Salmonella typhimurium DT 104 is primarily associated with cattle but it has spread to a range of food animals, especially pigs and chickens. As with many other *Salmonellae*, *Salmonella typhimurium* DT 104 may infect animals without showing any signs of illness although cases of clinical disease are also reported. *Salmonella typhimurium* DT 104 appears to be spread from farm to farm by raw water and once a farm becomes contaminated with *Salmonella typhimurium* DT 104 it is very difficult to eradicate, primarily because it survives in dry as well as in wet environments. Dogs and cats are known to have acquired the infection (Wall *et al.*, 1996) and we must expect that it will continue to spread throughout the environment.

The organism has been reported in beef, pork, chicken and cereals. Epidemiological studies have shown that burgers, sausages and sausage rolls have been associated with illness.

Until recently, there has been little information as to whether foods sourced from other EU countries or elsewhere are also becoming increasingly contaminated with antibiotic-resistant *Salmonella typhimurium* DT 104.

However, US statistics had shown a dramatic increase in the proportion of multi-resistant isolates of all *Salmonella typhimurium* from 7 per cent in 1990 to 28 per cent in 1995, 83 per cent of the latter being *Salmonella typhimurium* DT104 (Anon, 1997; Benson and Munro, 1997).

The worst scenario imaginable would be that the organism will become very widespread throughout the food chain and result in a huge increase in human salmonellosis, similar to the increase observed in the early 1990s when *Salmonella enteritidis* became endemic in poultry flocks. In the case of *Salmonella typhimurium* DT 104, such a situation would be very much more difficult to control because, whereas *Salmonella enteritidis* was confined to poultry and eggs almost exclusively, *Salmonella typhimurium* DT 104 already occurs more widely.

From the present knowledge, it seems that the methods used by food factories making comminuted meat products to control *Salmonella*e and other infectious pathogens must be vigorously applied if we are to control this 'emerging' pathogen. This means:

1. Ensuring that suppliers of all raw meats operate effective GMP and HACCP programmes in order to control growth of the organism and prevent build-up of contamination within the factory
2. Safe cooking and handling instructions on retail packs of raw products
3. Good food hygiene and safe cooking procedures in all catering services.

Bacillus cereus and Other *Bacillus* spp.

Bacillus cereus is a toxin-producing gram-positive, facultatively aerobic spore former whose cells are large rods and whose spores do not swell the sporangium. These and other characteristics, including biochemical features, are used to differentiate and confirm the presence *B. cereus*, although these characteristics are shared with *B. cereus* var. mycoides, *B. thuringiensis* and *B. anthracis*. Differentiation of these organisms depends upon determination of motility (most *B. cereus* are motile), presence of toxin crystals (*B. thuringiensis*), hemolytic activity (*B. cereus* and others are beta hemolytic whereas *B. anthracis* is usually nonhemolytic), and rhizoid growth which is characteristic of *B. cereus* var. mycoides.

The bacteria are commonly found in the environment, are often found in soil and vegetation, and can be present in foods. They can quickly multiply at room temperature. All people are believed to be susceptible to *B. cereus* food poisoning. *B. cereus* food poisoning is the general description, although two recognized types of illness are caused by two distinct metabolites. The diarrheal type of illness is caused by a large molecular weight protein, while the vomiting (emetic) type of illness is believed to be caused by a low molecular weight, heat-stable cereulide, a small peptide stable for 20 min at 121°C. *B. cereus* has also been implicated in infections of the eye, respiratory tract, and in wounds. The pathogenicity of *B. cereus*, whether intestinal or non-intestinal, is intimately associated with the production of tissue-destructive exoenzymes. Among these secreted toxins are four hemolysins, three distinct phospholipases, an emesis-inducing toxin, and proteases.

B. cereus illness is related to many foods *i.e.*, beef, turkey, rice, beans, vegetables. Specifically, the diarrheal illness is often related to a wide variety of foods including meats, milk, vegetables, cereal dishes (corn, corn starch) and fish. The vomiting-type outbreaks have generally been associated with rice products; however, other starchy foods such as fried or boiled rice dishes, potato, pasta and cheese products have also been implicated. Food mixtures such as sauces, puddings, soups, casseroles, pastries, and salads have frequently been incriminated in food poisoning outbreaks.

The spores of *B.cereus* normally occur on many foods from harvest through primary processing and generally survive cooking. The microorganism is not considered a hazard at the very low levels typically present in food. However, problems develop when food (especially cooked foods which are devoid of most heat-sensitive microbial competitors) is held at 10–55°C (50–131°F) for a long period of time. Under these conditions, the microorganism grows to large numbers, releasing toxin during growth in the food and in the intestinal tract following ingestion. Although ingestion of more than 10^5 cells/g is required to produce illness, ingesting large numbers of *B. cereus* does not always produce illness. The presence of large numbers of *B. cereus* (greater than 10^6 organisms/g) in a food is indicative of active growth and proliferation of the organism and is consistent with a potential hazard to health. To date, all outbreaks of both types of *B. cereus* illness have resulted because of improper holding temperatures or slow cooling of large amounts of food. Because these illnesses are intoxications, they are not infections and are not transmitted. It does not appear that processing of raw agricultural commodities contributes to the incidence of *B. cereus* illness.

Bacillus cereus illness is caused by the ingestion of food contaminated with either the enterotoxigenic *B. cereus* or with the emetic toxin. In non-gastrointestinal illness, reports of respiratory infections similar to respiratory anthrax have been attributed to *B. cereus* strains harbouring *B. anthracis* toxin genes. The symptoms of *B. cereus* diarrheal type food poisoning mimic those of *Clostridium perfringens* food poisoning. The onset of watery diarrhoea, abdominal cramps, and pain occurs 6-15 hours after consumption of contaminated food. Nausea may accompany diarrhoea, but vomiting (emesis) rarely occurs. Symptoms persist for 24 hours in most instances. The diarrheal syndrome results following ingestion of large numbers of the microorganism.

The emetic syndrome is likely to result from ingestion of preformed toxin in the food. The emetic type of food poisoning is characterized by nausea and vomiting within 0.5 to 6 h after consumption of contaminated foods. Occasionally, abdominal cramps and/or diarrhoea may also occur. Duration of symptoms is generally less than 24 h. The symptoms of this type of food poisoning parallel those caused by *Staphylococcus aureus* foodborne intoxication. Some strains of *B. subtilis* and *B. licheniformis* have been isolated from lamb and chicken incriminated in food poisoning episodes. These organisms demonstrate the production of a highly heat-stable toxin which may be similar to the vomiting type toxin produced by *B. cereus*.

Although no specific complications have been associated with the diarrheal and vomiting toxins produced by *B. cereus*, other clinical manifestations of *B. cereus* invasion or contamination have been observed. They include bovine mastitis, severe systemic and pyogenic infections, gangrene, septic meningitis, cellulitis, panophthalmitis, lung abscesses, infant death, and endocarditis.

All people are considered to be susceptible to *B. cereus* infection. Mortality is rare. The emetic enterotoxin has been found to be associated with a few cases of liver failure and death in otherwise healthy people. There is a newly found cytotoxin that has been associated with a *B. cereus* strain that caused one confirmed outbreak and three deaths. The infective dose or the number of organisms most commonly associated with human illness is 10^5 to 10^8, but pathogenicity arises from the preformed toxin, not from the bacteria themselves.

Confirmation of *B. cereus* as the etiologic agent in a foodborne outbreak requires either (1) isolation of strains of the same serotype from the suspect food and faeces or vomitus of the patient, (2) isolation of large numbers of a *B. cereus* serotype known to cause foodborne illness from the suspect food or from the faeces or vomitus of the patient, or (3) isolation of *B. cereus* from suspect foods and determining their enterotoxigenicity by serological (diarrheal toxin) or biological (diarrheal and emetic) tests. The rapid onset time to symptoms in the emetic form of disease, coupled with some food evidence, is often sufficient to diagnose this type of food poisoning. A variety of methods have been recommended for the recovery, enumeration and confirmation of *B. cereus* in foods. More recently, a serological method has been developed for detecting the putative enterotoxin of *B. cereus* (diarrheal type) isolates from suspect foods. Recent investigations suggest that the vomiting type toxin can be detected by animal models (cats, monkeys) or possibly by cell culture.

B. cereus related food poisoning is not a notifiable disease in most countries; therefore, incidence data is limited. It is recognized that there may be significant under reporting of *B. cereus* illness due to the generally mild, short duration and self-limiting symptoms, however, fatal incidences have been reported. The Centers for Disease Control (CDC) website states that there were 619 confirmed outbreaks of *Bacillus*-related poisoning from 1998 through 2015, involving 7385 illnesses. In this timeframe, there were 75 illnesses and three deaths due to confirmed *Bacillus*-related illness. These statistics refer to all *Bacillus*-related illness, and not just *B. cereus* related illness. In 1980, 9 outbreaks related to *B.cerus* food illnesses were reported to the Centers for Disease Control and included such foods as beef, turkey, and Mexican foods. In 1981, 8 outbreaks were reported which primarily involved rice and shellfish. Other outbreaks go unreported or are misdiagnosed because of symptomatic similarities to *Staphylococcus aureus* intoxication (*B. cereus* vomiting-type) or *C. perfringens* food poisoning (*B. cereus* diarrheal type).

It was estimated that *B. cereus* caused 0.7 per cent of foodborne illness among the 31 major pathogens in the US. In a study for *B. cereus* outbreaks were most often attributed to rice dishes (50 per cent); fried rice was the most common type of rice dish (68 per cent). Rice dishes were most commonly cooked and served immediately (42 per cent) or were part of large, solid masses of food (33 per cent). Twenty-four per cent of *B. cereus* outbreaks were associated with meat or poultry dishes. Meat or poultry dishes were cooked and served immediately (50 per cent), roasted (33 per cent), or part of liquid or semisolid mixtures (17 per cent).

For preventive measures, good hygiene should be observed in food establishments and by food handlers; proper cooking of foods should be done to destroy spores; keep food at low temperature; and fast cooling of food should be done.

Gram-positive Rod-shaped Anaerobe

Clostridium perfringens previously known as *Clostridium welchii*, belongs to the family Bacillaceae and is an important cause of foodborne disease. It is probably the most extensively studied anaerobic bacterial human pathogen. It is a gram-positive, rod-shaped anaerobe which forms oval subterminal spores. It differs from most other clostridia in that the relatively large rods (1 × 3–9 μm) are encapsulated and non-motile. Though a catalase-negative anaerobe, *C. perfringens* will survive and occasionally grow in the presence of oxygen. They produce protein toxins and form spores resistant to various environmental stresses such as radiation, desiccation, and heat. It is widely distributed abundantly in the environment, and frequently occurs in the intestines of humans and many domestic and feral animals, vegetation and sewage. Spores of the organism persist in soil, sediments, and areas subject to human or animal faecal pollution.

Food poisoning strains have a variety of origins including human and animal faeces, abattoirs, sewage and flies. Spores produced by these organisms can resist boiling for 4 or more hours. If the spores are present as contaminants on raw meat, they may resist boiling or steaming, and on slow cooling the spores will germinate into rapidly multiplying bacterial cells, which produce large amounts of toxin. The enterotoxigenic strains of *C. perfringens* cause food borne intoxication due to *Clostridium perfringens* enterotoxin (CPE) produced in the gastrointestinal tract by the organism.

Growth occurs over the temperature range 12 to 50°C although it is very slow below about 20°C. At its temperature optimum, 43–47°C, growth is extremely rapid with a generation time of only 7.1 min at 41°C. Vegetative cells show no marked tolerance to acid (minimum pH 5, optimum 6.0–7.5), have a minimum a_w for growth of 0.95–0.97, depending on the humectant and will not grow in the presence of 6 per cent salt.

The heat resistance of vegetative cells is comparable to that of non-spore forming bacteria with D values at 60°C in beef of a few minutes, D values of spores at 100°C show a wide inter-strain variation with recorded values from 0.31 min to more than 38 min. This may, in part be due to differences in the culture methods used since some workers included lysozyme in their media to improve the recovery of heat-damaged spores.

Five toxin-producing types of *C. perfringens* have been identified (A through E based on the production of four lethal extracellular toxins *viz.*, alpha, beta, epsilon, and iota), and all produce an alpha-toxin (phospholipase) that plays a role in myonecrosis. Type B strains produce beta- and epsilon-toxins, type D strains also produce epsilon-toxin, and type E strains produce an iota-toxin Perfringens food poisoning is most commonly caused by organisms producing type A enterotoxin. Other types of enterotoxins (B to G) do not normally cause food born disease. As virtually all food poisoning outbreaks are caused by type A strains, it is not necessary to determine the toxin type in such outbreaks.

C. perfringens are estimated to be the second most common bacterial causes of foodborne illness in the US, causing one million illnesses each year. Local, state, and territorial health departments voluntarily report *C. perfringens* outbreaks to the US CDC through the Foodborne Disease Outbreak Surveillance System. From 1998 to 2010, 289 confirmed outbreaks of *C. perfringens* illness were reported with 15,208 illnesses, 83 hospitalizations, and eight deaths. The number of outbreaks reported each year ranged from 16 to 31 with no apparent trend over time.

Almost all reported cases of foodborne gastroenteritis in the US that involve *C. perfringens* are a result of type A infection after the ingestion of highly contaminated foods with greater than 10^6 –10^7 viable vegetative cells, which undergo sporulation in the small intestine and produce enterotoxin. The enterotoxin produced during sporulation is released with the spores during cell lysis. After release, the enterotoxin binds to epithelial cells, causing cytotoxic cell membrane damage and subsequent alteration of permeability, leading to diarrhoea and abdominal cramping.

A more serious but rare illness is also caused by ingesting food contaminated with Type C strains. The latter illness is known as enteritis necroticans or pig-bel disease. Necrotic enteritis (pig-bel) caused by *C. perfringens* is often fatal. This disease also begins as a result of ingesting large numbers of the causative bacteria in contaminated foods. Deaths from necrotic enteritis (pig-bel syndrome) are caused by infection and necrosis of the intestines and from resulting septicemia but this disease is very rare.

Now let us study some details about food poisoning by *C. perfringens.*

Perfringens food poisoning is the term used to describe the common foodborne illness caused by *C. perfringens*. In most instances, the actual cause of poisoning by *C. perfringens* is temperature abuse of prepared foods. *C. perfringens* food poisoning is more likely to occur where food is being prepared some time in advance of consumption and that adequate refrigeration is the key to its control. Small numbers of the organisms are often present after cooking and multiply to food poisoning levels during cool down and storage of prepared foods. Meats, meat products, fish, pork, fruits, vegetables, spices and gravy are the foods most frequently implicated. The heating of such foods is inadequate to destroy heat resistant endospores. Upon cooling and warming the endospores germinate and grow.

Most cases (> 70 per cent in the United States and > 87 per cent in England and Wales) are associated with meat products such as stews, meat gravies, roast joints and pies. This is partly due to the frequent association of the organism with meats, but major contributory factors are the low redox potential, mode of preparation and consumption which can give *C. perfringens* the opportunity to multiply to dangerous levels. It is a common infectious cause of outbreaks of foodborne illness in the western countries, especially outbreaks in which cooked beef is the implicated source. Cured meats are rarely involved in *C. perfringens* food poisoning. This is a fine example of the hurdle concept in action; individual preservative factors such as salt content, nitrite level and heat processing are insufficient on their own to assure safety but effectively control growth of *C. perfringens* in combination.

Institutional feeding (such as school cafeterias, hospitals, nursing homes, prisons, *etc.*) where large quantities of food are prepared several hours before serving is the most common circumstance in which perfringens poisoning occurs. The young and elderly are the most frequent victims of perfringens poisoning. Except in the case of pig-bel syndrome, complications are few in persons under 30 years of age. Elderly persons are more likely to experience prolonged or severe symptoms.

C. perfringens food poisoning is generally a self-limiting, non-febrile illness characterized by nausea, abdominal pain, diarrhoea and less commonly vomiting. Onset is usually 8 to 24 h after consumption of food containing large numbers of the vegetative organism; the median count of *C. perfringens* in foods implicated in UK outbreaks is $7 \times 10_5$ g^{-1} and the required ingested dose has been variously estimated at 10_6–10_8cfu. In otherwise healthy individuals, medical treatment is not usually required and recovery is complete within 1–2 days, but may last up to two weeks, although occasional fatalities occur in the very old or debilitated.

The common form of perfringens poisoning is characterized by intense abdominal cramps and diarrhea which begin 8-22 hours after consumption of foods containing large numbers of those *C. perfringens* bacteria capable of producing the food poisoning toxin. The illness is usually over within 24 hours but less severe symptoms may persist in some individuals for 1 or 2 weeks. A few deaths have been reported as a result of dehydration and other complications.

Ingested vegetative cells that survive the stomach's acidity pass to the small intestine where they grow, sporulate and release an enterotoxin. The enterotoxin is synthesized by the sporulating cells, although low levels of production have been observed in vegetative cultures. The toxin is closely associated with the spore coat, but is not thought to be an important structural component and is released into the intestinal lumen on lysis of the sporangium.

Toxin production can also occur *in vitro.* Low levels have been detected in foods, including a sample involved in an outbreak and a few reported cases with incubation periods less than 2 h suggest that, on occasion, ingestion of pre-formed toxin may cause illness. This is generally held to be rare however.

The enterotoxin is a 35 kDa protein with an isoelectric point of 4.3. It is inactivated by heating in saline at 60°C for 10 min and is sensitive to some proteolytic enzymes. It acts like cholera toxin by reversing the flow of Na_+, Cl_- and water across the gut epithelium from absorption to secretion, though it does so by a different mechanism. Rather than increase the level of intracellular cyclic nucleotides, it acts at the cell membrane. It first binds to specific protein receptors on the epithelial cell and is then inserted into the cell membrane producing morphological changes within a few minutes. It changes cell permeability, inhibits synthesis of cell macromolecules and produces pores in the membrane of the cell which eventually dies as a result of membrane damage. Illness tends to be relatively mild and short-lived since the toxin affects primarily the cells of the villus tip, which are the oldest intestinal cells and therefore replaced much sooner than younger cells.

In the investigation of outbreaks, enrichment culture is rarely necessary since *C. perfringens* will invariably be present in high numbers in implicated foods or clinical samples. Similarly, in routine quality assurance of foods, there is generally little value in being able to detect very low numbers in view of its ubiquity in the environment. In the examination of foods, the total count (vegetative cells plus spores) is determined but with faecal specimens a spore count, obtained after heating a suspension at 80°C for 10 min, is also performed.

The most commonly employed selective plating media used to enumerate *C. perfringens* employ antibiotic(s) as the selective agent and sulfite reduction to produce black colonies as the differential reaction. The most popular combinations are tryptose/sulfite/cycloserine (TSC) medium and oleandomycin/polymyxin/sulfadiazine/perfringens (OPSP), incubated anaerobically for 24 h at 37°C. A better diagnostic reaction is obtained if pour plates are used since colonies on the agar surface of spread plates can appear white.

A number of methods are available for the detection of enterotoxin. Traditional biological tests such as the ligated ileal loop and mouse challenge have been superseded by more sensitive, rapid, convenient and humane serological techniques. A commercially available kit employs reverse passive latex agglutination. The name derives from the fact that in a standard agglutination assay, soluble antibody reacts with a particulate antigen such as bacterial cells. In a reversed passive latex agglutination assay, a soluble antigen reacts with antibody attached to latex particles. These play no part in the reaction and are therefore passive, but they do provide a visual signal when they cross-link as a result of the antigen-antibody reaction.

Preventive measures should include proper cooking of food and eating freshly prepared foods; thorough washing and sanitation of containers; hygienic handling of cooked food; fast cooling of cooked food; storing food in small quantities will enhance cooling; proper reheating of cold cooked food before consumption; and storage of leftovers or unused foods in freezer.

Non-sporeforming Gram-negative Rods

Shigellas are non-motile non-sporeforming, gram-negative rods. Shigella is a species of enteric bacteria that causes disease in humans and other primates. The bacteria primarily cause a human disease, but has been found in some primates. Shigellas are facultative anaerobes, similar to enterics such as *E. coli.*

The genus *Shigella* was discovered as the cause of bacillary dysentery by the Japanese microbiologist Kiyoshi

Shiga in 1898. The normal habitat for Shigella is the intestinal tract of humans and other primates. The genus Shigella is a member of the family Enterobacteriaceae and possesses four serogroups that have been traditionally treated as species: serogroup A as *Shigella dysenteriae*, serogroup B as *Shigella flexneri*, serogroup C as *Shigella boydii*, and, serogroup D as *Shigella sonnei*. Whereas serogroups A, B, and C consist of 38 serotypes, serogroup D possesses only one. All the four species are regarded as human pathogens though they differ in the severity of the illness they cause. *Sh. dysenteriae* has been responsible for epidemics of severe bacillary dysentery in tropical countries but is now rarely encountered in Europe and North America where *Sh. sonnei* is more common. *Sh. sonnei* causes the mildest illness, while that caused by *Sh. boydii* and *Sh. flexneri* is of intermediate severity. All strains of shigella possess potent exotoxins which are carbohydrate-lipid protein complexes. The infective dose is smaller than that of salmonellae, except *S. typhi*.

Shigella spp. are closely related to *E. coli* in their DNA homology and share some biochemical characteristics as well as reactivity to some of the same antibodies, but despite these similarities, their differentiation should be considered clinically significant based, at least in part, on differences in symptoms expressed by infected individuals.

Data suggest that Shigella is responsible for less than 10 per cent of reported foodborne illnesses per year, infecting approximately 300,000–450,000 people annually (FDA/CFSAN 2003). Laboratory reports of *Sh. sonnei* infections in England and Wales rose to 9830 in 1991 compared to 2319 and 2228 in the previous two years. In the United States annual reports over recent years have ranged between 300 000 and 450 000. *Shigellas* are spread primarily person-to-person by the faecal-oral route although food-borne outbreaks have, been recorded. Some experts consider that the problem of food-borne shigellosis is greatly underestimated. Approximately 20 per cent of all shigellosis cases in the US are related to international travel (*i.e.*, travellers' diarrhoea), with *S. sonnei* being the most prevalent and *S. flexneri* being the second most common in developed countries. However, in developing countries, *S. flexneri* and *S. dysenteriae* type 1 are the most common serogroups, with *S. dysenteviae* type 1 having been involved in a lengthy epidemic in southern Africa and major epidemics in other parts of Africa, in Asia and in Central America. These epidemics have resulted in high morbidity and mortality rates, especially in malnourished children, immuno-compromised individuals, and the elderly.

Human cases and carriers are the only important sources of infection. Shigella transmission can occur through direct person-to-person spread by the faecal-oral route or from contaminated food and water. There have been some documented cases of foodborne shigellosis also. The bacteria leave the body in stool of an infected person and infects another person through contaminated hands, food, water, or objects (toys, pens *etc.*). The main source of Shigella involved in outbreaks is asymptomatic carriers or persons recovering from disease. Shigella may persist in the intestinal tract for months. The minimal infectious dose can be transmitted directly from contaminated fingers, since intermediate bacterial replication is not required to achieve the low infectious dose. Any type of food can transmit the shigella pathogens to cause disease in man. Flies can spread shigella germs when they get into contact with infected stool and then contaminate drinking water or food. Shigella organisms may remain viable in tap water for as long as 6 months, and in sea water for 2 to 5 months.

In developed countries, most cases are transmitted by faecal-oral spread from people with symptomatic infection. In developing countries, both faecal-oral spread and contamination of common food and water supplies are important mechanisms of transmission symptom

Shigellas as are generally regarded as rather fragile organisms which do not survive well outside their natural habitat which is the gut of humans and other primates. They have not attracted the attention that other food-borne enteric pathogens have, but such evidence as is available suggests that their survival characteristics are in fact similar to other members of the Enterobacteriaceae. They are typical mesophiles with a growth temperature range between 10–45°C and a heat sensitivity comparable to other members of the family. They grow best in the pH range 6–8 and do not survive well below pH 4.5. A number of studies have reported extended survival times in foods such as flour, pasteurized milk, eggs and shell-fish.

Shigella rarely occurs in animals; principally a disease of humans except other primates such as monkeys and chimpanzees. The organism is frequently found in water polluted with human faeces.

Shigella spp. are found most frequently in environments of compromised sanitation and poor hygiene, and although the primary route of transmission is by person-to-person contact, shigellosis can occur after the ingestion of focally contaminated water or food. Shigella spp. have not been associated with one specific type of food; foods associated with outbreaks of shigellosis have included milk, salads, chicken, shellfish, and other fresh produce served at a wide range of establishments including restaurants, homes, schools, sorority houses, commercial airlines, cruise ships and military mess halls. In various research studies, salads (potato, tuna, shrimp, macaroni, and chicken), raw vegetables, milk and dairy products, and poultry have been reported to be the associated foods with this poisoning. Contamination of these foods is usually through the faecal-oral route. Faecally contaminated water and unsanitary handling by food handlers are the most common causes of contamination.

The disease is caused when virulent *Shigella* organisms attach to, and penetrate, epithelial cells of the intestinal mucosa. After invasion, they multiply intracellularly, and spread to contiguous epitheleal cells resulting in tissue destruction. Some strains produce enterotoxin and Shiga toxin (very much like the verotoxin of *E. coli* 0157: H7). Shiga toxins block mRNA transcription and induce apoptosis in endothelial cells

The species are distinguished on the basic of biochemical tests and both serotyping and phage typing schemes are available for further sub-division of species.

Shigellas cause bacillary dysentery in humans and other higher primates. Studies with human volunteers have indicated that the infectious dose is low; of the order of 10–100 organisms. The incubation period can vary between 7 h and 7 days although food-borne outbreaks are commonly characterized by shorter incubation periods of up to 36 h. Symptoms are of abdominal pain, vomiting and fever accompanying diarrhoea which can range from a classic dysenteric syndrome of bloody stools containing mucus and pus, in the cases of *Sh. dysenteriae, Sh. flexner;* and *Sh. boydii,* to a watery diarrhoea with *Sh. sonnei.* The severity of the disease may vary from very mild to severe diarrhoea with bloody stools, mucus secretia, and dehydration. Reactive arthritis, hemolytic uremic syndrome and Reiters disease are sequelae associated with shigellosis. Illness lasts from 3 days up to 14 days in some cases. Frequently, an asymptomatic carrier state may develop during convalescence, lasting from a few days to several months. Milder form of the illness are self-limiting and require no treatment but *Sh. dysenteriae* infections often require fluid and electrolyte replacement and antibiotic therapy. Infection by some strains results in a 10– 15 per cent mortality rate. Human-volunteer studies indicate that ingestion of as few as 10–100 microorganisms can induce illness.

Shigella infection is diagnosed through testing of a stool sample. First a stool sample must be obtained from the potentially infected person, and then the sample is placed on a medium to encourage the growth of bacteria. If and when there is growth, the bacteria are identified, usually by looking at the growth under a microscope.

Lack of interest in *Shigella* as a food-borne pathogen has meant that laboratory protocols for its isolation and identification from foods are relatively underdeveloped. A pre-enrichment procedure has been described based on resuscitation on a non-selective agar before over-laying with selective media. Selective enrichment in both Gram-negative broth and selenite broth has been recommended. Selective plating media used are generally those employed for enumerating the Enterobacteriaceae or *Salmonella* although neither are entirely satisfactory.

Rapid techniques for identification based on immunoassays which detect the virulence marker antigen and on the polymerase chain reaction to detect the virulence plasmid by DNA/DNA hybridization have also been applied.

The principles necessary for control of Shigella at the preparation and service level are the same as those necessary for control of other more widely recognized pathogens such as *Salmonella* and *Campylobacter*. The increasing awareness of consumers and employees of foodservice establishments of the potential for foodborne diseases should do much to reduce the incidence of shigellosis in the developed countries. In the developing countries, control of shigellosis, as always, will depend on improved hygiene and waste-handling practices. Proper cooking and hygienic food handling thus can prevent Shigella infections to a large extent. There is currently no vaccine for Shigellosis prevention, but there is current research that appears promising. The most effective method for prevention is frequent and vigorous hand washing with warm, soapy water and ensuring clean drinking water sources and proper sewage disposal in developing nations.

The *Staphylococci*

The staphylococci were first described by the Scottish surgeon, Sir Alexander Ogston as the cause of a number of pyogenic (pus forming) infections in humans. In 1882, he gave them the name *Staphylococcus* (Greek: *staphyle,* bunch of grapes; *coccus,* a grain or berry), after their appearance under the microscope.

The first description of food poisoning caused by *Staphylococci* is thought to be that of Vaughan and Sternberg who investigated a large outbreak of illness in Michigan believed to have been caused by cheese contaminated with *Staphylococci*. Clear association of the organisms, with food-borne illness had to wait until Barber (1914) demonstrated that *Staphylococci* were able to cause poisoning by consuming milk from a cow with staphylococcal mastitis. In 1930, Dack showed that staphylococcal food poisoning was caused by a filterable enterotoxin.

There are currently 27 species and 7 subspecies of the genus *Staphylococcus;* enterotoxin production is principally associated with the species *Staph. aureus,* although it has also been reported in others including *Staph. intermedius* and *Staph. hyicus.*

As a relatively mild, short-lived type of illness, staphylococcal food poisoning is perhaps more likely to be under-reported than others. Most reported cases are associated with outbreaks and only a few sporadic cases are detected. In the United States between 1983 and 1987, *Staphylococci* accounted for 7.8 per cent of the 600 bacterial food poisoning outbreaks that were recorded. Equivalent figures for England and Wales over the same period were 1.9 per cent out of a total of 2815 outbreaks.

Outbreaks of staphylococcal food poisoning in the UK peaked during the 1950s at 150 outbreaks per year but have since declined to an annual level of 5-10 outbreaks in the period 1990 to 1996. Studies in the United States and the UK have found that poultry products and cold, cooked meats are the most common vehicles. Salted meats such as ham and corned beef are particularly vulnerable since the *Staph. aureus* is unaffected by levels of salt that will inhibit a large proportion of the competitive flora. Buffet meals where such meats are served are a common scenario for outbreaks as the food is necessarily prepared some time in advance and too often stored at ambient temperature or inadequately chilled.

In 2000, between 14,000 and 15,000 people in Japan suffered from *Staphylococcus aureus* food poisoning after consuming milk or yoghurt products from the country's largest dairy producer. While much research attention is now focused on the so-called emerging pathogens, this outbreak appears to have been caused by one or perhaps two of the 'traditional' culprits. Tests on samples of the implicated low-fat milk from the firm's Osaka factory showed evidence of contamination by *Staphylococcus aureus* and the organism was subsequently found in a valve in one of the company's milk storage tanks. Further investigation revealed *Bacillus cereus* in valves of tanks used to store low same fat milk and yoghurt products (*Food Safety and Security*, August 2000). The symptoms reported by sufferers of food poisoning are consistent with ingestion of toxin from *Staphylococcus aureus*.

Canned foods also offer *Staph. aureus* a congenial, competitor free environment and post-process leakage contamination of cans has been an occasional cause of outbreaks.

Other outbreaks have been caused by hard cheeses, cold sweets, custards and cream-filled bakery products. In Japan, rice balls that are moulded by hand are the commonest vehicle while in Hungary, it is ice cream.

Investigation of the plant which produced the offending product revealed a number of deficiencies. More recent but as yet unconfirmed reports suggest that contaminated skim milk powder from the company's plant in Hokkaido may have been the original source of the problem. A power failure at the Hokkaido plant resulted in long delays in processing skim milk powder which could have been reworked into other powders shipped to the plant at Osaka. This pattern of events would be consistent with *Staphylococcus aureus*, if present in the original milk, producing large quantities of enterotoxin which would not be destroyed by subsequent processing. The powder was apparently not sold separately but used only as a manufacturing ingredient.

Staphylococcus aureus are nonmotile, gram-positive cocci that appear singly or in pairs, tetrads, short chains, or characteristic "grapelike" clusters. Staphylococci are catalase-positive, oxidase-negative, facultative anaerobes that, with the exception of *Staphylococcus saccharolyticus* and *Staph. aureus* subsp. anaerobius, grow more rapidly under aerobic conditions. Their ability to ferment glucose can be used to distinguish them from the strictly respiratory genus *Micrococcus,* although there are species in both genera where this distinction is not clear cut due to low acid production by some *Staphylococci* and production of small amounts of acid under anaerobic conditions by some micrococci. Enterotoxin production is adversely affected by anaerobic conditions far more than growth. Only the strains that produce enterotoxin can cause food poisoning. The food handler with an active lesion or carriage later initiates infection

Staphylococcus aureus is a typical mesophile with a growth temperature range between 7 and 48°C and an optimum at 37°C under otherwise optimal conditions. The range of temperature over which enterotoxin is produced is narrower by a few degrees and has an optimum at 35–40°C. The organism has unexceptional heat resistance with a D_{62} of 20–65 s and a D_{72} of 4.1 s when measured in milk using log-phase cultures. Heat resistance has been shown to vary considerably though and D values were found to increase three-fold when stationary-phase cultures were tested.

Growth occurs optimally at pH values of 6–7 with minimum and maximum limits of 4.0 and 9.8–10.0, respectively. The pH range over which enterotoxin production occurs is narrower with little toxin production below pH 6.0 but, as with growth, precise values will vary with the exact nature of the medium.

A characteristic of *Staphylococci aureus* which is a particularly important consideration in some foods is its tolerance of salt and reduced a_w. It grows readily in media containing 5–7 per cent NaCl and some strains are capable of growth in up to 20 per cent NaCl. It will grow down to an a_w of 0.83 where it has a generation time of 300 min. Once again, the range over which enterotoxin production occurs is more limited with a minimum a_w recorded of 0.86.

The Staphylococcus species are host adapted with about one-half of the known species inhabiting humans solely and other animals. Humans and animals are the primary reservoirs. The principal habitat of the *Staphylococci* is the skin, skin glands, nose, throat, hair and the mucous membranes of warm-blooded animals. Several species are associated with particular hosts, for example *Staph. hyicus* with pigs and *Staph. gallinarum* with chickens. *Staph. aureus* is more widespread but occurs most frequently on the skin of higher primates. In humans, it is particularly associated with the nasal tract where it is found in 20–50 per cent of healthy individuals. The two most important sources to foods are nasal carriers and individuals whose hands and arms are inflicted with boils and carbuncles, who are permitted to handle foods. Outside the body, *Staph. aureus* can survive for long periods of time in a dry state, and have been isolated from air, dust, sewage, milk, food or on food equipment, environmental surfaces and water, making it one of the most resistant non-spore-forming pathogens. It can be isolated from faeces and sporadically from a wide range of other environmental sites such as soil, marine and fresh water and plant surfaces. In addition to environmental sources of infection, some reported *Staph. aureus* containing foods include ground beef, pork sausage, ground turkey, salmon steaks, oysters, shrimp, cream pies, milk, and delicatessen salads.

Any food which requires handling in preparation may therefore easily become contaminated. Infected wounds, lesions, and boils of food handlers may also be sources of contamination, as well as coughing and sneezing by individuals with respiratory infections. *S. aureus* also commonly occurs on the skin and hides of animals, and may thus contaminate foods from these animals as a result of cross-contamination during slaughter. The presence of small numbers of *Staph. aureus* on foods is not uncommon. A variety of foods can support the growth of *S. aureus*. Foods which support growth best include proteinaceous foods, such as meat and meat products, poultry, fish and fish products, milk and dairy products, cream sauces, salads (ham, chicken, potato, *etc.*), puddings, custards, and cream-filled bakery products. It can be isolated from raw milk where levels may sometimes be elevated as a result of *Staph. aureus* mastitis in the producing herd. As a poor competitor, it normally poses no problem in these situations since it does not grow and is eliminated by cooking or pasteurization. There have however been outbreaks caused by milk products such as dried milk and chocolate milk where growth and enterotoxin production occurred in the raw milk and the enterotoxin, but not the organism survived subsequent pasteurization.

Staphylococci are usually outnumbered by competitive harmless microorganisms in raw foods because they do not compete well with other bacteria, and it is probably for this reason that raw foods are less frequently implicated as the cause of staphylococcal food poisoning. Cooking, however, eliminates the normal competitive bacteria of raw foods; therefore, it is in prepared foods such as meat, potato, and macaroni salads, custards, and cream-filled bakery products that growth of contaminating staphylococci may be permitted.

Though normally a harmless parasite of human body surfaces where it plays a useful role metabolizing skin products and possibly preventing skin colonization by pathogens, *Staph. aureus* can cause minor skin abscesses such as boils and more seriously, as an opportunistic pathogen when the skin barrier is breached or host resistance is low. It can also be associated with more serious infections like bacteraemia, pneumonia, osteomyelitis, cerebritis, meningitis, and abscesses of muscle, urogenital tract, central nervous system, and various abdominal organs. Toxic shock syndrome, a condition resembling septic shock and resulting from the production of toxic shock syndrome toxin 1, has been attributed to *Staph. aureus* infection.

Humans are the major reservoir for *Staph. aureus*, and contamination of food can occur through direct contact, indirectly by skin fragments, or through respiratory tract droplets, with most staphylococcal food poisoning cases being traced to food contamination during preparation because of inadequate refrigeration, inadequate cooking or heating, or poor personal hygiene. If food is stored for some times at room temperature, the organism may in the food can produce toxin. Since large numbers, typically > 10_6 g^{-1}, are required for the production of enough toxin to cause illness, contamination is necessary but is not alone sufficient for an outbreak to occur. In particular, temperature and time conditions must also be provided that allow the organism to grow. Toxin production occurs at growth temperature 12-44 °C, pH 4.2 and salt concentration of ≤10 per cent. No toxin production occurs at temperatures below 12°C, pH < 4.2 and > 10 per cent salt.

Staph. aureus toxins are strictly neurotoxins. The bacteria produce enterotoxin while multiplying in food. Several antigenically different protein enterotoxins exist. To date, SE A, B, C1, C2, C3, D, E, G, H, I, and J have been identified. Enterotoxin A is the most toxic and the one most commonly involved in staphylococcal food poisoning outbreaks followed by D. Data from outbreaks involving enterotoxin A indicate that less than 1µg of toxin can result in illness. These enterotoxins are heat stable, with type B being most heat resistant. Staphylococcal enterotoxins are included in a larger family of toxins, known as pyrogenic toxins, that have the unique ability to act as superantigens, thereby stimulating an extraordinarily high percentage of T cells. They are difficult to inactivate with heat, because temperatures required to inactivate them are higher than those needed to kill the organism. Staphylococcal enterotoxin A is more heat sensitive than enterotoxins B or C and requires heating at 80 or 100 °C for 180 or 60 s, respectively, to cause a loss in serological reactivity. They are insensitive to pH changes (pH stable) and resistant to most proteolysis enzymes (trypsin, chymotrypsin, renin, and pepsin). The enterotoxins are also not affected by irradiation.

Staphylococcal food poisoning (staphyloentero-toxicosis; staphyloentero-toxemia) is the name of the condition caused by the enterotoxins which some strains of *S. aureus* produce. The onset of symptoms in staphylococcal food poisoning is usually rapid and, in many cases, acute, depending on individual susceptibility to the toxin, the amount of contaminated food eaten, the amount of toxin in the food ingested, and the general health of the victim.

Enterotoxin stimulates Central Nervous Systems (CNS) vomiting center in the brain via the vagus and sympathetic nerves and inhibit water and sodium absorption in the small intestine. If these nerves are severed then vomiting does not occur. It is not known how the toxin induces diarrhoea but it has been shown not to stimulate adenylate cyclase diarrhoea activity. Staphylococcal enterotoxins, along with the toxic syndrome toxin and others, are classed as bacterial super antigens relative to *in vivo* antigen recognition in contrast to conventional antigens.

Table 7.1: Factors Permitting Growth and Enterotoxin Production by *Staphylococcus aureus*

Factor	*Growth*		*Enterotoxin Production*	
	Optimum	*Range*	*Optimum*	*Range*
Temperature °C	35–37	7–48	35–40	10–45
pH	6.0–7.0	4.0–9.8	Ent. A 5.3–6.8 others 6–7	4.8–9.0
NaCl	0.5–4.0 per cent	0–20 per cent	0.5 per cent	0–20 per cent
Water activity	0.98–>0.99	0.83–>0.99	>0.99	0.86–>0.99
Atmosphere	Aerobic	Aerobic-Anaerobic	5–20 per cent DO_2	Aerobic-Anaerobic
E_h	> +200mV	< -200 to > +200mV	> +200mV	?

Unlike other foodborne illnesses, which usually have longer incubation periods, onset of staphylococcal foodborne illness may occur between 30 min and 8 hr following consumption of the toxin-containing food. Most illnesses, however, occur within 2– 4 hr. Staphylococcal enterotoxins (SE) cause severe gastroenteritis (inflammation of the intestinal tract lining). Common symptoms of staphylococcal intoxication include nausea, vomiting, retching, abdominal cramping, sweating, chills, muscular cramping, and/or prostration, weak pulse, shock, shallow respiration, and subnormal body temperature. Recovery from his intoxication (which is rarely fatal) usually occurs uneventfully within 24–48 hr. In severe cases dehydration, masked pallor and collapse may require treatment (intravenously) infusion. Death from staphylococcal food poisoning is very rare, although such cases have occurred among the elderly, infants, and severely debilitated persons. All people are believed to be susceptible to this type of bacterial intoxication; however, intensity of symptoms may vary.

S. aureus intoxications are often associated with institutions such as schools and prisons where food is often prepared in mass quantities and held until consumption. Lack of sanitation by workers and improper time-temperature combinations can lead to contamination of the product and growth of the microorganism to levels at which toxin is produced. Accordingly, outbreaks can affect a large number of people.

In the diagnosis of staphylococcal foodborne illness, proper interviews with the victims and gathering and analyzing epidemiologic data are essential. Incriminated foods should be collected and examined for staphylococci. The presence of relatively large numbers of enterotoxigenic staphylococci is good circumstantial evidence that the food contains toxin. The most conclusive test is the linking of an illness with a specific food or in cases where multiple vehicles exist, the detection of the toxin in the food sample(s). In cases where the food may have been treated to kill the staphylococci, as in pasteurization or heating, direct microscopic observation of the food may be an aid in the diagnosis. A number of serological methods for determining the enterotoxigenicity of *S. aureus* isolated from foods as well as methods for the separation and detection of toxins in foods have been developed and used successfully to aid in the diagnosis of the illness. Phage typing may also be useful when viable staphylococci can be isolated from the incriminated food, from victims, and from suspected carrier such as food handlers.

The most successful and widely used selective plating medium for *Staph. aureus* is the one devised by Baird-Parker in the early 1960s. It combines the virtues of a high degree of selectivity, a characteristic diagnostic reaction and the ability to recover stressed cells. Lithium chloride and tellurite act as selective agents while egg yolk and pyruvate assist in the recovery of damaged cells. Reduction of the tellurite by *Staph. aureus* gives characteristic shiny, jet-black colonies which are surrounded by a zone of clearing, resulting from hydrolysis of the egg-yolk protein lipovitellenin. Colonies also often have an inner white margin caused by precipitation of fatty acid.

Colonial appearance on Baird-Parker (B-P) agar gives presumptive identification of *Staph. aureus* which is often confirmed by tests for the production of coagulase and thermostable nuclease.

Coagulase is an extracellular substance which coagulates human or animal blood plasma in the absence of calcium. It is not specific to *Staph. aureus* but is also produced by *Staph. intermedius* and *Staph. hyicus*. *Staph. intermedius* is unable to reduce tellurite and therefore produces white colonies on B-P agar, but *Staph. hyicus*, which is found on the skin of pigs and poultry, requires a series of further biochemical tests to distinguish it reliably from *Staph. aureus*.

The presence of coagulase can be demonstrated using EDTA-treated rabbit plasma in the tube coagulase test. More rapid test kits are available, based on the detection of bound coagulase (also known as clumping factor) and/or protein A, which reacts with the Fc part of IgG molecules. Detection is by agglutination of erythrocytes or latex particles coated with fibrinogen or plasma and colonies from selective media can be tested directly, without any intermediate sub-culturing. Coagulase production can also be detected directly in an egg yolk-free modification of B-P agar containing pig or rabbit plasma.

Table 7.2: Characteristics of Staphylococcal Enterotoxins

Enterotoxin Seotype	*Molecular Mass (D)*	*pl Value of Major Forms*	*pH Range of Isoelectric Forms*	*Toxin Gene Location*
A	27 100	8.1; 7.3	6.5–8.6	Chromosomal
B	28 366	8.3; 8.6	7.8–9.6	Chromosomal/ plasmid
C1	27 496	8.8; 9.2	7.9–9.2	Chromosomal/ plasmid
C2	27 531	5.6; 6.9	5.5–9.2	NK
7.5; 7.6				
C3	27 438	8.2	6.6–8.3	NK
D	26 360	8.4	7.9–9.3	Plasmid
E	26 425	8.0	7.0–8.5	Chromosomal

Four biotypes of *Staph. aureus* are recognized but the use of biotyping is limited since nearly all of the strains isolated from human sources belong to biotype A. Phage typing schemes are used with *Staphylococci aureus;* most food poisoning strains belonging to serogroup III.

Since the enterotoxins will survive heat processes that eliminate the producing organism toxin detection in a food is a more reliable indication of hazard than viable counting procedures. A number of immunoassay techniques for staphylococcal enterotoxins are available. Early immunoprecipitation techniques such as the microslide gel diffusion test are less sensitive and require lengthy extraction and concentration procedures to isolate sufficient enterotoxin for detection. ELISA techniques which will detect 0.1–1.0 ng toxin g^{-1} food and reverse passive latex agglutination tests with a sensitivity of 0.5 ng ml^{-1} are now available and more widely used.

Food born bacterial illness by bacteria are most commonly prevented and controlled by proper cooking and preparing of food as well as storing. For examples adequate refrigeration of food, improve personal hygiene, adequate cooking and heating processing. The control method or measures also includes; a) education of those who prepare the food at home and other food handlers, so that they have to take proper personal measures; b) prohibiting individuals with absences or other skin lesions from handling food; c) placing of food in cold place at 4 degrees centigrade or lower of all food in order to prevent bacterial multiplication and the formation of toxin. Foods must be kept at room temperature for as little time as possible.

Escherichia coli

Escherichia coli are bacterium which belongs to family enterobactereriae and are gram negative non-spore forming rod up to 3 μm in length. It may or may not be mobile; some rods are flagellated and some are not. The organism is a facultative anaerobe and ferments simple sugars such as glucose to form lactic, acetic, and formic acids; the optimum pH for growth is 6.0 to 8.0; however, growth can occur as low as pH 4.3 and as high as 9 to 10 pH.

E. coli is an almost universal inhabitant of the gut of humans and other warm-blooded animals where it is the predominant facultative anaerobe though only a minor component of the total microflora. Generally, a harmless commensal, it can be an opportunistic pathogen causing a number of infections such as Gram-negative sepsis, urinary tract infections, pneumonia in immunosuppressed patients and meningitis in neonates. Its common occurrence in faeces, ready culturability, generally non-pathogenic character and survival characteristics in water led to the adoption of *E. coli* as an indicator of faecal contamination and the possible presence of enteric pathogens such as *S. typhi* in water. This usage has been transferred to foods where greater circumspection is required in interpreting the significance of positive results.

Strains of *E. coli* were first recognized as a cause of gastroenteritis by workers in England investigating summer diarrhoea in infants in the early 1940s. Until 1982, strains producing diarrhoea were classified into three types based on their virulence properties: enteropathogenic *E. coli* (EPEC), enteroinvasive *E. coli* (EIEC) and enterotoxigenic *E. coli* (ETEC). They are not very common causes of food-borne illness in developed countries, but an important cause of childhood diarrhoea in less developed countries. ETEC is also frequently associated with so-called traveller's diarrhoea. However, since 1982, enterohaemorrhagic *E. coli* (EHEC) particularly associated with serotype 0157: H7 has been recognized as the cause of a number of outbreaks of haemorrhagic colitis and haemolytic uraemic syndrome, particularly in North America, where foods such as undercooked ground meat and raw milk have been implicated. An exponential rise in isolations of 0157: H7 was reported in Canada between 1982 and 1986 and a study in the UK between 1985 and 1988 suggested that the increased reporting of isolations there represented a real increase. The number of cases in the UK has continued to increase reaching more than 1100 per year in 1995 and 1996 and other European countries have also reported increased isolation rates.

Two further types of *E. coli* are recognized as causes of diarrhoea. Termed enteroaggregative *E. coli* (EaggEC) and diffusely adherent.

E. coli (DAEC), they have characteristic patterns of adherence to Hep-2 cells in culture. Epidemiological evidence has associated EaggEC with protracted diarrhoea in children, though challenge trials with adult volunteers suggest that only certain strains displaying enteroaggregative behaviour can cause diarrhoea. DAEC has been associated with diarrhoea in children in Mexico.

Enterotoxigenic *Escherichia coli*

Enterotoxigenic *Escherichia coli* (ETEC) has four recognized classes (collectively referred to as the EEC group) that cause gastroenteritis in humans. Among these are the enterotoxigenic (ETEC) strains. They comprise a relatively small proportion of the species and have been etiologically associated with diarrheal illness of all age groups from diverse global locations. Infants and travellers to underdeveloped countries are most at-risk of infection. The organism frequently causes diarrhoea in infants in less developed countries and in visitors there from industrialized countries. The etiology of this cholera-like illness has been recognized for about 20 years.

Gastroenteritis is the common name of the illness caused by ETEC, although travellers' diarrhoea is a frequent sobriquet. Illness caused by ETEC usually occurs between 12 and 36h after ingestion of the organism. The most frequent clinical syndrome of infection includes watery diarrhoea, abdominal cramps, low-grade fever, nausea and malaise. The disease is usually self-limiting, persisting for 2-3 days, although in developing countries it is a common cause of infantile diarrhoea where it can cause serious dehydration. In infants or debilitated elderly persons, appropriate electrolyte replacement therapy may be necessary.

ETEC is considered a serious foodborne disease hazard in countries not having high sanitary standards and practices. Contamination of water with human sewage may lead to contamination of foods. Infected food handlers may also contaminate foods. These organisms are infrequently isolated from dairy products such as semi-soft cheeses.

A relatively large dose (100 million to 10 billion bacteria) of enterotoxigenic *E. coli* is probably necessary to establish colonization of the small intestine, where these organisms proliferate and produce toxins which induce fluid secretion. With high infective dose, diarrhoea can be induced within 24 hours. Infants may require fewer organisms for infection to be established.

During the acute phase of infection, large numbers of enterotoxigenic cells are excreted in faeces. These strains are differentiated from nontoxigenic *E. coli* present in the bowel by a variety of *in vitro* immunochemical, tissue culture, or gene probe tests designed to detect either the toxins or genes that encode for these toxins. The diagnosis can be completed in about 3 days. With the availability of a gene probe method, foods can be analyzed directly for the presence of enterotoxigenic *E. coli*, and the analysis can be completed in about 3 days. Alternative methods which involve enrichment and plating of samples for isolation of *E. coli* and their subsequent confirmation as toxigenic strains by conventional toxin assays may take at least 7 days.

The ingested organism resists expulsion from the small intestine with the rapidly flowing chyme by adhering to the epithelium through attachment or colonization factors in the form of fimbriae on the bacterial cell surface. These can have different morphology and be either rigid (6–7 nm diameter) or flexible structures composed of 14–22 kDa protein subunits. They are mannose resistant, *i.e.,* they mediate haemagglutination in the presence of mannose and particular colonization fimbriae are restricted to certain O: H serotypes. They are encoded on plasmids which frequently also encode for the diarrhoeagenic toxins.

Two toxin types are produced: the heat-stable toxins (ST), which can withstand heating at 100°C for 15 min and are acid resistant and the heat-labile toxins (LT) which are inactivated at 60°C after 30 min and at low pH. LTI bears a strong similarity to cholera toxin; it consists of five B subunits (Mr 11.5 kDa) which are responsible for binding of the toxin to the epithelial cells and an A subunit (Mr 25 kDa) which is translocated into the epithelial cell where it activates adenylate cyclase. The subsequent increase in cAMP levels then inhibits Na_+, Cl_- and water absorption by the villus cells and stimulates their loss from intestinal crypt cells thus leading to profuse watery diarrhoea. LTI toxin produced by certain ETEC strains has similar biological activity to LTI but does not cross react with antiserum to LTI or cholera toxin.

Two types of ST have been recognized; the most common, ST_A, is a low molecular weight, poorly antigenic polypeptide produced from a 72 amino acid precursor. The reported size of the extracellular molecule has ranged from 18–42 amino acids and its resistance to heat, low pH and proteolytic digestion probably derive from its compact three-dimensional structure. It acts by stimulating the production of cGMP by guanylate cyclase in epithelial cells. The mechanism of action of ST_B, which can be distinguished from ST_A by its inability to produce fluid secretion in the intestines of suckling mice, is not known but does not appear to operate through the stimulation of cyclic nucleotide production.

In the last decade, four major common-source outbreaks of ETEC gastroenteritis occurred in the U.S. In late 1975 one-third of the passengers on two successive cruises of a Miami-based ship experienced diarrheal illness. A CDC investigation found ETEC to be the cause, presumably linked to consumption of crabmeat cocktail. In early 1980, 415 persons eating at a Mexican restaurant experienced diarrhea. The source of the causative organism was an ill food handler. In 1981, 282 of 3,000 personnel at a Texas hospital acquired ETEC gastroenteritis after eating in the hospital cafeteria. No single food was identified by CDC.

Escherichia is the type genus of the Enterobacteriaceae family and *E. coli* is the type species of the genus. It is a catalase-positive, oxidase-negative, fermentative, short,

Gram-negative and non-sporing rod. Genetically, *E. coli* is closely related to the genus *Shigella,* although characteristically it ferments the sugar lactose and is otherwise far more active biochemically than *Shigella* spp. Late lactose fermenting, non-motile, biochemically inert strains of *E. coli* can however be difficult to distinguish from *Shigella.*

E. coli can be differentiated from other members of the Enterobacteriaceae on the basis of a number of sugar-fermentation and other biochemical tests. Classically an important group of tests used for this' purpose are known by the acronym IMViC.

Since production of mixed acids and acetoin are alternative pathways for the metabolism of pyruvate, most species of the Enterobacteriaceae are either VP positive or methyl red positive. In the IMViC tests, most strains of *E. coli* are indole and methyl red positive and VP and citrate negative. The tests are still used for identification purposes but now-a-days usually as part of the larger range of tests available in modem miniaturized test systems.

E. coli is a typical mesophile growing from 7–10°C up to 50°C with an optimum around 37°C, although there have been reports of some ETEC strains growing at temperatures as low as 4°C. It shows no marked heat resistance, with a D value at 60°C of the order of 0.1 min and can survive refrigerated or frozen storage for extended periods. A near-neutral pH is optimal for growth but growth is possible down to pH 4.4 under otherwise optimal conditions. The minimum a_w for growth is 0.95.

A serotyping scheme for *E. coli* based on lipopolysaccharide somatic O, flagellar H and polysaccharide, capsular K antigens was proposed by Kauffman in the 1940s. As currently applied in the O: H system, principal serogroups are defined by O antigens and then subdivided into serotype on the basis of H antigens. Strains of each category of pathogenic *E. coli* tend to fall within certain O: H serotypes, so the scheme plays an important role in detecting pathogens as well as in epidemiological investigations. Selective techniques for *E. coli* mostly exploit the organism's tolerance of bile and other surfactive compounds, a consequence of its natural habitat, the gut. Aniline dyes and the ability of many strains to grow at temperatures around 44°C are also used as selective agents.

The first selective and differential medium was that originally devised by MacConkey in 1905. It has been variously modified since but its essential characteristics have remained unchanged. Bile salts (and sometimes the aniline dye, crystal violet) act as inhibitors of Gram-positive and some fastidious Gram-negative bacteria. Lactose is included as a fermentable carbohydrate with a pH indicator, usually neutral red. Strong acid producers like *Escherichia, Klebsiella* and *Enterobacter* produce red colonies, non-lactose fermenters such as *Salmonella, Proteus* and *Edwardsiella,* with rare exceptions produce colourless colonies. MacConkey agar is not however strongly selective and will support the growth of a number of non-Enterobacteriaceae including Gram-positives such as *Enterococci* and *Staphylococci.*

Eosin/methylene blue agar is a popular selective and differential medium in North America. The aniline dyes eosin and methylene blue are the selective agents but also serve as an indicator for lactose fermentation by forming a precipitate at low pH. Strong lactose fermenters produce green-black colonies with a metallic sheen.

A biochemical feature of *E. coli* increasingly being used in diagnostic media is β-glucuronidase activity, which is possessed by around 95 per cent of *E. coli* strains but' by only a limited number of other bacteria. A fluorogenic or chromogenic glucuronide is incorporated into a conventional medium and enzyme activity detected by the production of colour or fluorescence. Most widely used is the fiuorogen 4-methylum- belliferyl-β-D-glucuronide (MUG) which is hydrolysed to produce fluorescent 4-methylumbelliferone.

Suspect colonies from selective and differential media can be confirmed by further biochemical testing.

Detection of *E. coli* O157: H7 is based on phenotypic differences from most other serotypes: its inability to ferment sorbitol on MacConkey-sorbitol agar and absence of β-glucuronidase activity in most strains. Presumptive *E. coli* O157: H7 from these tests must then be confirmed serologically for which a latex agglutination kit is commercially available.

Enteropathogenic *E. coli* (EPEC)

When the properties of ETEC and EIEC were established, it was noted that these strains were rarely of the same serotypes first associated with *E. coli* diarrhoea in the 1950s. Subsequent investigation of some of these earlier strains in most cases failed to demonstrate the property of enteroinvasiveness or the ability to produce ST or LT and yet they retained the ability to cause diarrhoea in volunteers.

Symptoms of EPEC infection, malaise, vomiting and diarrhoea with stools containing mucus but rarely blood, appear 12–36 h after ingestion of the organism. In infants, the illness is more severe than many other diarrhoeal infections and can persist for longer than two weeks in some cases. Pathogenesis appears to be related to the ability of EPEC strains to adhere closely to the enterocyte membrane. Binding to the enterocytes occurs in three stages: non-intimate association mediated by pili, attachment or signal transduction and then intimate contact. During this process the bacteria facilitate their own binding by producing a series of changes in the underlying enterocytes.

The attachment stage is accompanied by increased levels of intracellular Ca_{2+}, release of inositol phosphates and activation of tyrosine kinase, an enzyme which phosphorylates tyrosine residues on intracellular proteins. Following this the enterocytes accumulate filamentous actin as they form pedestal-like surface structures on which the bacteria rest. This results in deformation and loss of some microvilli; events which are thought to cause diarrhoea by disrupting the balance between absorption and secretion in the small intestine.

Enterohaemorrhagic *E. coli* (EHEC)

EHEC, sometimes also known as Verotoxin-producing *E. coli* (VTEC), was first described in Canada where in some areas it rivals *Campylobacter* and *Salmonella* as the most frequent cause of diarrhoea. *E. coli* 0157: H7 is the most common EHEC serotype report, although others do occur. Non-motile (H negative) 0111 and 0157 are more common in Australia for example. EHEC has attracted attention not only because food-borne transmission is more common than with other diarrhoeagenic *E. coli,* but because the illness it causes can range from a non-bloody diarrhoea, through haemorrhagic colitis, to the life-threatening conditions haemolytic uraemic syndrome (HUS) and thrombotic thrombocytopaenic purpura (TTP).

Enteroinvasive *E. coli* (EIEC)

Infection by EIEC results in the classical symptoms of an invasive bacillary dysentery normally associated with *Shigella.* Like *Shigella,* EIEC invades and multiplies within the epithelial cells of the colon causing ulceration and inflammation, though EIEC strains do not produce Shiga toxin. Clinical features are fever, severe abdominal pains, malaise and often a watery diarrhoea which precedes the passage of stools containing blood, mucus and faecal leukocytes.

Now let us study about outbreak of EHEC. Outbreaks caused by EHEC serotype 0157: H7 have mostly involved undercooked ground meat products and occasionally raw milk. Cattle, seem to be an important reservoir of infection and 0157: H7 has been isolated from 0.9–8.2 per cent of healthy cattle in the UK. Other surveys have isolated the serotype from 3.7 per cent (6/164) samples of retailed fresh beef and a significant percentage (1–2 per cent) of other fresh meat products such as pork, poultry and lamb.

There have been a number of very large outbreaks around the world and their public impact has often been dramatic. Six hundred people became ill and four children died in a major US outbreak in 1993 caused by undercooked beef hamburgers. This caused a major public outcry over meat hygiene and resulted in, amongst other things, the introduction of new meat labelling regulations.

In August 1997, a cluster of cases in Colorado prompted the largest food recall in US history when more than 12000 tons of ground beef were recalled.

A large outbreak in Scotland in 1996 had a similar impact in the UK. Nearly 500 were affected and 20 elderly patients died. The cause was thought to be cross contamination of cooked meats from raw meat in a butcher's shop and the resultant enquiry produced a tightening of regulation.

The failures that led to these outbreaks were generally simple breakdowns of basic food hygiene. With both raw milk and ground beef products, the primary cause has been a failure to heat process/cook the products adequately. While it is true that intact cuts of meat such as steaks can often be consumed safely when the interior is undercooked, this is because microbial contamination is usually a surface phenomenon. Communication of the meat, however, will mix surface contaminants into the middle of products and they will therefore need cooking throughout to ensure microbial safety. The USDA has produced regulations specifying that the centre of beef hamburgers should reach on cooking: 71.1°C (160°F) instantaneously for consumers and 68.3°C (155°F) for 16 seconds in food service operations.

Outbreaks of EHEC have been reported with other foods. Lettuce has been associated on several occasions and unpasteurised apple juice was the vehicle in a large outbreak in the US. In the summer of 1996, an epidemic in Japan involved over 9000 cases and 12 deaths in children. The largest outbreak during the epidemic, in Sakai City, involved 5700 people and was associated with contaminated radish sprouts and the same vehicle was implicated in a further outbreak the following year. Alfalfa sprouts were also implicated in an outbreak in the US.

Outbreaks caused by acidic foods such as apple juice and fermented sausages and laboratory studies with mayonnaise, remind us of the potential for bacteria to survive for prolonged periods at pH values that do not permit growth, particularly when the product is refrigerated. EHEC does appear to have a more marked ability to survive at low pH values than some other bacteria and this may also account for the relatively low infectious dose, 2–2000 cells, recorded in outbreaks.

Faecal contamination of water supplies and contaminated food handlers have been most frequently implicated in outbreaks caused by EPEC, EIEC and ETEC. A number of foods have been involved, including a coffee substitute in Romania in 1961, vegetables, potato salad and sushi. In the United States, mould-ripened soft cheeses have been responsible for outbreaks in 1971, associated with EIEC in which more than 387 people were affected and in 1983, caused by ETEC (ST). *E. coli* would not be expected to survive well in a fermented dairy product with a pH below 5 but, where contamination is associated with mould-ripening, the local increase in pH as a result of lactate utilization and amine production by the mould would allow the organism to grow.

The prevention/avoidance of food borne illness caused by *E. coli* can be prevented by the same method as prevention of other food borne illness caused by bacteria. However, because of the consequences to young children, special precaution needed to observe. The heat sensitivity of this organism is such that cases should not occur when food properly cooked. In the cases of ground beef, the recommendation is that it cooked to 160° F or that the core temperature be brought to a minimum of 155° C for at least 15 second and that the juices are clear. Because of unevenness of food cooking at 155-160° F provides a measure of safety. Once cooked, the food meats should not be held between 40° F and 140° F for more than 3-4 hours.

Bacterium Monocytogenes

Listeria monocytogenes was described by Murray, which caused infection to laboratory rabbits, while he was carrying research. The organism is widespread in nature and is a transient constituent of the intestinal flora excreted by 1-10 per cent of healthy humans. It is extremely hardy and can survive for many years in the cold in naturally infected sources.

Prior to the 1980s, listeriosis, the disease caused predominantly by *Listeria monocytogenes*, was primarily of veterinary concern, where it was associated with abortions and encephalitis in sheep and cattle. Interestingly, evidence indicates that veterinary listeriosis was frequently foodborne. As a result of its wide distribution in the environment, its ability to survive for long periods of time under adverse conditions, and its ability to grow at temperatures as low as –0.4°C, Listeria has since become recognized as an important foodborne pathogen.

As an important veterinary problem, it causes two main forms of disease: a meningoencephalitis most common in adult ruminants such as sheep and cattle and a visceral form more common in monogastrics and young ruminants which attacks organs other than the brain, causing stillbirth, abortion and septicaemia. Listeriosis in sheep increased in Britain from 86 recorded incidents in 1979 to 423 in 1988. This was partly due to the increased size of the national flock over that period but has also been attributed to changes in silage-making techniques towards greater use of big-bale ensilage. In big-bale ensilage, the silage is made in large plastic bags rather than in a single large clamp. Rupture of the bag or inadequate sealing at its neck can allow mould growth to occur on the lactic acid present increasing the silage pH to a value at which *L. monocytogenes* can flourish.

The widespread concern it has caused is largely attributable to the realization that food is a major source of the infection (a possibility first suggested as long ago as 1927), the psychrotrophic character of the organism and the high mortality rate of the illness. Reported incidence of human listeriosis increased in several countries during the 1980s, but remains generally low when compared to other food borne infections such as salmonellosis. For example, in England and Wales reported cases of listeriosis peaked in the late 1980s at around 300 per year while reports of *Salmonella* and *Campylobacter* infections numbered nearly 27500 and 29000, respectively.

The transmission of listeriosis by food was first convincingly demonstrated in an outbreak that occurred in the Maritime Provinces of Canada in 1981. The outbreak involved 41 cases in all. Of the 34 perinatal cases, there were 9 stillbirths, 23 neonatal cases with a mortality rate of 27 per cent and 2 live births of healthy infants. The mortality rate in adult cases was 28.6 per cent. Coleslaw was implicated as the result of a case control study and *L. monocytogenes* serotype 4b (the outbreak strain) was isolated from a sample of coleslaw in a patient's refrigerator. It was not possible to isolate the organism at the manufacturer's plant but it transpired that a farmer who supplied cabbages to the manufacturer also kept sheep, two of whom had died of listeriosis. The cabbage had been grown in fields fertilized by fresh and composted manure from the sheep and the harvested cabbages had been stored in a large shed through the winter–factors thought to account for the introduction of the organism and its multiplication to dangerous levels.

Raw vegetables, in the form of garnish containing celery, tomatoes and lettuce, were also implicated on epidemiological grounds in an outbreak that occurred in eight Boston hospitals in 1979. Surveys in the UK, the United States, Australia and elsewhere have reported a high frequency of isolation of *L moncytogenes* from meats and meat products, where serotype I generally predominates. A number of sporadic cases of listeriosis have been associated with products such as pork sausage, turkey frankfurters, cook-chill chicken and chicken nuggets. *L monocytogenes* is relatively resistant to curing ingredients and has been found in a range of delicatessen meats such as salami, ham, corned beef, brawn and pate. In an Australian survey 13.2 per cent of samples were, found to be positive, largely as a result of cross-contamination in the shop. In Britain in 1989/90, high levels on vacuum-packed ham and on pate, from which serotype 4b was isolated, prompted the recall of both products from the market. Pork tongues in aspic were identified as the original source of a large outbreak in France caused by serotype 4b. Between March and December 1992, 279 cases were reported with 63 deaths and 22 abortions.

Dairy products such as raw and pasteurized milk and soft cheeses have been associated with a number of major outbreaks of listeriosis. The overall incidence of *L. monocytogenes* in raw milk derived from surveys in Australasia, Europe and the United States averages at around 2.2 per cent, although one Spanish study reported an incidence in excess of 45 per cent. Pasteurized milk was responsible for an outbreak in Massachusetts in 1983 involving 42 adult and 7 perinatal cases with an overall mortality rate of 29 per cent. The milk had come from farms where bovine listeriosis is known to have occurred at the time of the outbreak. f. It was the absence of evidence of improper pasteurization at the dairy that gave rise to the concern that *L monocytogenes* might display marked heat resistance in some instances.

Soft cheeses are also frequently contaminated with *L. monocytogenes.* In 1985 there was an outbreak in California in which a Mexican-style soft cheese which had been contaminated with raw milk was the vehicle. Reported cases of listeriosis dropped in 1990 and 1991 to 118 and 131, respectively; a decrease attributed to the effect of Department of Health advice to the immunocompromised and pregnant to avoid soft cheeses and to reheat certain chilled foods adequately and to withdrawal of contaminated pate from a single manufacturer. Provisional figures for human listeriosis in England and Wales in 1996 and 1997 were 115 and 118, respectively. In the United States, the Center for Disease Control (CDC) has estimated an annual incidence of around 1700 cases resulting in 450 adult death and 100 foetal and post-natal deaths.

One of the largest and deadliest multi-state outbreaks of listeriosis in the US occurred in late summer of 2011. The incident marked the first time that Listeria spp. contamination had been linked to whole cantaloupe and one of the few times it had been linked to fresh produce. 146 Individuals had become ill after being infected with the outbreak strain of listeria; 29 deaths and 1 miscarriage had also been attributed to the infection. FDA inspectors cited unsanitary conditions such as old, corroded, and difficult-to-clean equipment and standing pools of water and the absence of processing steps to cool the melons before cold storage as likely contributors to contamination.

L. monocytogenes is a gram-positive, facultatively anaerobic, catalase-positive, oxidase-negative and non-sporeformer. The coccoid to rod shaped cells (0.4–0.5 µm × 0.5–2.0 µm) cultured at 20–25°C possess peritrichous flagella and exhibit a characteristic tumbling motility. Colonies on tryptose agar viewed under oblique illumination-have a characteristic blue-green sheen.

L. monocytogenes elaborates a 58 kDa β-haemolysin, listerolysin O, which acts synergistically with the haemolysin produced by *Staphylococcus aureus* to give enhanced haemolysis on blood agar. This reaction forms the basis of a useful diagnostic test to distinguish *L. monocytogenes* from *L. innocua* and is known as the CAMP test after Christie, Atkins and Munch-Peterson who first described the phenomenon with group B *Streptococci*.

L. monocytogenes will grow over a wide range of temperature from 0–42°C with an optimum between 30 and 35°C. Below about 5°C, growth is extremely slow with lag times of 1 to 33 days and generation times from 13 to more than 130 h being recorded. The thermal survival characteristics of *L. monocytogenes* have received considerable attention following an outbreak in the United States associated with pasteurized milk and the suggestion that the organism could survive commercial pasteurization conditions. Despite some conflicting data in the literature, it appears that the heat resistance of *L. monocytogenes* is similar to that of other non-sporeforming gram-positives with atypical D_{60} of a few minutes and a D_{70} of a few seconds. It was proposed that *L. monocytogenes* cells in contaminated milk are protected from heat by their intracellular location within milk leucocytes but subsequent studies have failed to demonstrate any significant effect.

Models of the thermal inactivation of *L. monocytogenes* in milk have indicated that conventional HTST pasteurization achieves a reduction of 5.2 log cycles in the number of survivors; an acceptable safety margin assuming low numbers of the organism present on the incoming milk.

L. monocytogenes can grow in the pH range of 4.39–9.4 in a good growth medium. Environments with pH values outside that range are unfavourable for survival. Growth of all strains is inhibited at pH values below 5.5 but the minimum growth pH is dependent on both strain and acidulant and has been variously reported as between 5.6 and 4.4. *L. monocytogenes* is also quite salt tolerant being able to grow in 10 per cent sodium chloride and survive for a year in 16 per cent NaCI at pH 6.0. However, 12.5 per cent sodium chloride inhibited the growth of *L. monocytogenes* at pH 6.0 between 8 and 30°C.

The organism is ubiquitous in the environment. It has been isolated from fresh and salt water, soil, sewage sludge, plants, decaying vegetation and silage (pH 5.5) in which bacteria can multiply. Its prolonged survival in the environment has been demonstrated in one study where the level of *L. monocytogenes* in sewage sludge sprayed on to agricultural land remained unchanged for more than 8 weeks. Listeria can survive for long periods in both soil and plant materials. Ingestion of contaminated silage by ruminants has been linked to the occurrence of Listeria in milk. Asymptomatic faecal carrier occurs in man and animal species. They can grow at temperature range of 3-4.

Biological oxidation during wastewater treatment favours the growth of Listeria in sewage. Sludge provides a favorable environment for survival and growth of *L. monocytogenes* for many months. *L. monocytogenes* in concentrations of 103 – 104 cells/mL have been found associated with sewage plant effluents, while sewage sludges have shown higher concentrations. This makes the use of sewage plant effluent or waters receiving sewage plant effluent for irrigation of edible crops dangerous. The use of sludge for fertilization of edible crops is equally hazardous.

Its ubiquity in the environment suggests that human exposure to *L. monocytogenes* must be frequent. Incidence of infection is however low since invasive infection will result only if a susceptible individual is exposed to a sufficiently high dose of a virulent strain. The bacterium has been found in a variety of raw foods, such as uncooked meats and vegetables, as well as in foods that become contaminated after cooking or processing. Foods associated with this food borne illness include cheese, poultry, sauerkraut, salads, sea foods, meat and meat products contaminated by Listeria. Delicatessens and other ready-to-eat foods are important in causing Listeria food poisoning.

Unlike many other foodborne pathogens, Listeria multiplies in cold environments such as refrigerators. It can quickly spread in damp buildings, dripping off pipes or ceilings onto food. Once Listeria bacteria get into a food-processing factory, they can live there for years, sometimes contaminating food products. It can spread from the site of infection in the intestines to the central nervous system and the foetal-placental unit. It can cause meningitis (inflammation of the membrane surrounding spinal cord and brain), gastroenteritis (inflammation of mucous membranes of stomach and intestine), and septicemia (systemic spread of bacteria and toxins in the blood) can result from infection. It has 13 serotypes, including 1/2a, 1/2b, 1/2c, 3a, 3b, 3c, 4a, 4ab, 4b, 4c, 4d, 4e, and 7; among them, serotypes 1/2a, 1/2b, and 4b have been associated with the vast majority of foodborne infections. *Listeria monocytogenes* is one of the leading causes of death from food-borne pathogens especially in pregnant women, newborns, the elderly, and immuno-compromised individuals.

Estimates of the minimum infective dose are always fraught with difficulty and this is certainly the case with *L. monocytogenes.* It is thought to be relatively high since foods implicated in outbreaks have. been found to contain numbers in excess of 10_2 cfu g^{-1}.

Incubation periods for the disease have varied from 1 day to as long as 90 days with a typical incubation period of a few weeks; a situation which makes the identification of food vehicles difficult if not often impossible.

In pregnant women, it most commonly features as an influenza-like illness with fever, headache and occasional gastrointestinal symptoms, but there may be an associated transplacental foetal infection which can result in abortion, stillbirth, or premature labour. Infections in pregnant women can be devastating to the foetus, resulting in miscarriages, stillbirths, and birth defects

Listeriosis in the newborn can be an early-onset syndrome, which occurs at birth or shortly afterwards, or a late-onset disease appearing several days to weeks after birth. Early-onset illness results from in *utero* infection, possibly through the aspiration of infected amniotic fluid and is characterized by pneumonia, septicaemia and widely disseminated granulomas (abscesses). Meningitis is rare.

The symptoms of listeriosis usually last 7-10 days, with the most common symptoms being fever, muscle aches, and vomiting. Diarrhoea is another, but less common symptom. If the infection spreads to the nervous system it can cause meningitis, an infection of the covering of the brain and spinal cord

In the late-onset syndrome, meningitis is more common, 93 per cent (39 of 42) of late-onset cases in Britain between 1967 and 1985 had evidence of infection of the central nervous system. Infection may occur from the mother during passage through the birth canal, but some may also be acquired after delivery. A study in the UK found a lower mortality rate, for late-onset disease (26 per cent) than for early-onset listeriosis (38 per cent).

Adult listeriosis has a high mortality rate, figures calculated using data for 1989 gave values around the world of between 13 and 34 per cent. Early treatment with antibiotics, normally ampicillin, with or without an aminoglycoside, or chloramphenicol, is essential but in the most severe forms, the prognosis remains poor.

L. monocytogenes is a facultative intracellular pathogen which like *Mycobacterium, Brucella* and others can survive and multiply in cells of the monocyte-macrophage system. The organism penetrates the gut either by crossing the Peyer's patches or by invading enterocytes. This process of endocytosis is promoted by a number of virulence factors including internalin, an 800 amino acid bacterial surface protein encoded by a chromosomal gene *inlA,* other products of the so-called *inl* family of genes and p60, a 60 kDa extracellular protein. Internalization results in the bacteria being enclosed in a phagosome. In order to multiply intracellularly the organism must survive in the phagosome and escape rapidly before it fuses with a lysosome, a process which would kill the bacteria. Pathogenic *L monocytogenes* produces listeriolysin O (LLO), a 58 kDa haemolysin which breaks down the lipid bilayer of the phagosomal membrane allowing the bacteria to escape from the phagosome. Most *L. monocytogenes* infecting epithelial-type cells do this and are released into the cytoplasm. Only 10 per cent of those invading monocytes are successful, the remaining 90 per cent being destroyed.

During intracellullar replication, actin polymerizes around the bacterial surface, a process which propels the bacterial cell around the host cell and into adjacent cells, thereby spreading the organism while avoiding the host's immune system. The bacterial cells reach the mesenteric lymph nodes and are disseminated around the body *via* the blood. The liver plays an important role in eliminating the organism and controlling the infection. Infection of hepatocytes causes an intense inflammatory reaction. Polymorphonuclear cells (neutrophils) destroy infected hepatocytes forcing them to release the bacteria which are in their turn destroyed. Where infection is not controlled in the liver, it can further disseminate by the blood to the central nervous system or placenta causing more severe illness.

L. monocytogenes may be serotyped according to a scheme based on somatic and flagellar antigens. This is of limited epidemiological value since the majority of human cases of listeriosis are caused by just three of the thirteen serotypes identified. Phage typing and molecular typing techniques can however be used to assist in epidemiological investigations.

Its widespread distribution in the environment and its ability to grow on most non-acid foods offer *L. monocylogenes* plenty of opportunity to enter the food chain and multiply.

Low-temperature enrichment at 4°C is the traditional technique for isolating *L. monocytogenes* from environmental samples, but the increased interest in routine isolation of the organism from foods has led to its replacement by more rapid, selective enrichment procedures based on antibiotic cocktails as selective agents and incubation at near-optimal growth temperatures.

Selective agars have like-wise relied on a combination of selective agents such as lithium chloride, phenylethanol and glycine anhydride and antibiotics. Identification of presumptive *Listeria* colonies was based on microscopic examination of plates illuminated from below at an incident angle of 45° (Henry illumination), when they appear blue-grey to blue-green. Some media avoid the use of this technique by incorporating aesculin and ferric ammonium citrate so that *Listeria* colonies appear dark brown or black as a result of their ability to hydrolyse aesculin.

Confirmation of *L. monocytogenes* requires further biochemical testing including sugar-fermentation tests to distinguish it from other *Listeria* species and, in particular, the CAMP test to differentiate *L. monocytogenes* from *L. innocua*. Specific miniaturized test kits have been produced to simplify this procedure including one which replaces the CAMP test, which is not always easy for the inexperienced to interpret, with one for acrylamidase activity (*L. monocytogenes,* negative; *L. innocua,* positive). Enzyme-linked immunosorbent assay (ELISA) and gene probe kits are also available.

The main means of prevention is through the promotion of safe handling, cooking and consumption of food. This includes washing raw vegetables and cooking raw food thoroughly, as well as reheating leftover or ready-to-eat foods like hot dogs until steaming hot. Preventing listeriosis as a food illness requires effective sanitation of food contact surfaces. Alcohol and Quaternary ammonium is an effective topical sanitizer against Listeria. Refrigerated foods in the home should be kept below 4°C (39.2°F) to discourage bacterial growth. Preventing listeriosis also can be done by carrying out an effective sanitation of food contact surfaces.

US Risk Ranking of Ready-to-eat Foods with Regard to *Listeria monocytogenes*

The USDA and FDA have jointly released a draft risk ranking for ready-to-eat foods that might be contaminated with *Listeria monocytogenes*. The US risk assessment takes into account three age-based groups:

Perinatal

These are pregnancy associated cases where the mother experiences a foodborne *L. monocytogenes* infection during pregnancy exposing her foetus to the pathogen.

Elderly

This group includes people 60 or more years of age. People in this age bracket are considered to have increased susceptibility to listeriosis.

Intermediate Age

This group includes the entire remaining population. Healthy people appear not to be very susceptible to severe illness from *L. monocytogenes*. There are certain subpopulations within this group that are more susceptible to listeriosis. However, there is insufficient data to separate this group further.

The aim of the risk ranking which took two years to complete was to identify which ready-to-eat foods pose the greatest risk to consumers particularly those susceptible to the organism. Twenty products were included in the ranking and those included foods in the categories of seafood, fruit and vegetables, dairy, meats and combination foods.

When foods were considered by the ranking body on a per-serving basis, the foods ranked most likely to transmit listeriosis were pâté, fresh soft cheeses, smoked seafood and deli meats. This ranking accords closely with warnings to susceptible groups issued by local health authorities.

When the amount of each food consumed each year by Americans *i.e.,* the consumption pattern, was included in the ranking, deli meats rose from fourth to first. They were followed by deli salads (sixth on a per-serving basis) and pasteurised milk–presumably as a result of post-pasteurisation contamination. Pâtés, soft cheeses and smoked seafood moved down the ranking list moderately.

The report concludes that while the exposure assessment suggests that US consumers are exposed to low levels of *L. monocytogenes* on a regular basis, the likelihood of acquiring listeriosis is extremely small. Risk factors to be taken into account include:

1. How much of the food is consumed
2. How many organisms are on the food
3. Whether the organism can grow
4. The temperature at which the food is stored
5. How long the food is stored before it is eaten.

Vibrois spp.

The genus Vibrio, belonging to the family Vibrionaceae, contains more than 35 species, of which nearly half have been described in the last 20 years and more than one-third are pathogenic to humans. The three Vibrio species of primary significance in foodborne illnesses are *V. cholerae* (serogroups 01, non-01, and recently, 0139), *V. parahaemolyticus*, and *V. vulnificus*. Another species, *V. mimicus* sp. novo, formerly included in the species *V. cholerae*, also is a recognized pathogen.

V. parahaemolyticus, which was first shown to be an enteropathogen in 1951, is responsible for 50–70 per cent of outbreaks of food-borne gastroenteritis in Japan. *V. fluvialis* has been isolated from sporadic cases of diarrhoea in some countries, particularly those with warm climates, although its exact role is uncertain since other enteropathogens were often present in the stool samples. *V. mimicus,* which produces diarrhoea, is distinguishable from *V. cho1erae* only by its ability to produce acid from sucrose and acetoin from glucose, *V. vulnificus* does not usually cause diarrhoea but severe extra-intestinal infections such as a life-threatening septicaemia. Patients normally have some underlying disease and have eaten seafood, particularly oysters about a week before the onset of illness.

Vibrios are Gram-negative pleomorphic (curved or straight), short rods which are motile with (normally) sheathed, polar flagella. Catalase and oxidase-positive cells are facultatively anaerobic and capable of both fermentative and respiratory metabolism. Sodium chloride stimulates the growth of all species and is an obligate requirement for some. The optimum level for the growth of clinically important species is 1–3 per cent. *V. parahoemolyticus* grows optimally at 3 per cent NaCl but will grow at levels between 0.5 per cent and 8 per cent. The minimum a_w for growth of *V. parahaemolyticus* varies between 0.937 and 0.986 depending on the solute used.

Growth of enteropathogenic *Vibrios* occurs optimally at around 37°C and has been demonstrated over the range 5–43°C, although <"10°C is regarded as a more usual minimum in natural environments. When conditions are favourable, *Vibrios* can grow extremely rapidly; generation times of as little as 11 min and 9 min have been recorded for *V. parahoemolyticus* and the non-pathogenic marine vibrio, *V. natrigens*, respectively.

V. parahaemolyticus is generally less robust at extremes of temperatures than *V. cholerae.* Numbers decline slowly at chill temperatures below its growth minimum and under frozen conditions a 2-log reduction has been observed after 8 days at –18°C. The D_{49} for *V. parahaemolyticus* in clam slurry is 0.7 min compared with a D_{49} for *V. chlorae* of 8.15 min measured in crab slurry. Other studies have recorded higher D values for *V. parahaemolyticus,* for instance 5 min at 60°C produced only 4–5 log reductions in peptone/3 per cent. NaCl. Pre-growth of the organism in the presence of salt is known to increase heat resistance.

V. parahaemolyticus and other *Vibrios* will grow best at pH values slightly above neutrality (7.5–8.5) and this ability of *Vibrios* to grow in alkaline conditions up to a pH of 11.0 is exploited in procedures for their isolation. *Vibrios* are generally viewed as acid sensitive although growth of *V. parahaemolyticus* has been demonstrated down to pH 4.5–5.0.

The natural habitat of *Vibrios* is the marine and estuarine environment. *V. cholerae* can be isolated from temperate, sub-tropical and tropical waters throughout the world, but seem to disappear from temperate waters during the colder months. Long-term survival may be enhanced by attachment to the surfaces of plants and marine animals and a viable but non-culturable form has also been described where the organism cannot be isolated from the environment using cultural techniques even though it is still present in an infective form.

Table 7.3 indicates about the *Vibrio* species associated with human diseases.

Table 7.3: *Vibrio* Species Associated with Human Diseases

Species	Disease
V. cholerae, 01	Cholera, wound infection
V. cholerae, non-01	Diarrhoea, gastroenteritis, wound infection, secondary septicaemia
V. mimicus	Diarrhoea, gastroenteritis, wound infection
V purahuemolylicus	Gastroenteritis, wound infection, otitis media.
V. fluviulis	Diarrhoea
V. furnissii	Diarrhoea
V. hollisae	Diarrhoea
V. vulnificus	Wound infection, primary septicaemia, secondary septicaemia
V. ulginolyticus	Wound infection, otitis media
V. damsela	Wound infection

V. parahaemolyticus is primarily associated with coastal inshore waters than the open sea. It cannot be isolated when the water temperature is below 15°C and cannot survive pressures encountered in deeper waters. The survival of the organism through winter months when water temperatures drop below 15°C has been attributed to its persistence in sediments from where it may be recovered even when water temperatures are below 10°C.

V. parahaemolyticus grow at temperatures between 5 and 44 °C, with an optimum temperature and pH for growth between 30 and 37 °C and 7.6 and 8.6, respectively; the organism will grow in an environment at pH 4.8–11.0, in sodium chloride concentrations of 0.5–10.0 per cent, and in environments with a minimum aw of 0.94; however, it prefers a concentration of sodium chloride in the range of 2 to 4 per cent and a α_w of 0.981.

V. parahaemolyticus food poisoning is invariably associated with fish and shellfish. Occasional outbreaks have been reported in the United States and Europe, but in Japan it is the commonest cause of food poisoning. This has been linked with the national culinary habit of consuming raw or partially cooked fish, although illness can also result from cross-contamination of cooked products in the kitchen. Though the organism is only likely to be part of the natural flora of fish caught in coastal waters during the warmer months, it can readily spread to deep-water species through contact in the fish market and it will multiply rapidly if the product is inadequately chilled.

The risk of *V. vulnificus* infection associated with raw oysters is most effectively controlled by cooking the product before consumption. Susceptible individuals, such as those described previously and the immunosuppressed should avoid consumption of uncooked shellfish.

The reported incubation period for *V. parahaemolyticus* food poisoning varies from 2 h to 4 days though it is usually 9–25 h. Gastroenteritis is typically associated with consumption of raw, inadequately cooked, or cooked but recontaminated seafood. After a 4 to 96 h incubation period, symptoms of *V. parahaemolyticus* induced gastroenteritis include nausea, vomiting, headache, abdominal cramps, slight fever, chills, and watery diarrhoea that is occasionally bloody. Additional symptoms, after exposure to contaminated water, may include infected wounds, eyes, and ears. Illness persists for up to 8 days. *V. parahaemolyticus* is more enteroinvasive than *V. cholerae* and penetrates the intestinal epithelium to reach the *lamina propria.* A dysenteric syndrome has also been reported from a number of countries including Japan.

Although symptoms are usually self-limiting, lasting only 2–3 days, severe cases may result in dysentery, primary septicemia, or cholera-like illness with the possibility of death. The presence of a pre-existing condition (*e.g.*, liver disease, alcoholism, diabetes mellitus, antacid medication, peptic ulcer disease, immune disorder, *etc.*) greatly enhances the likelihood of developing a clinical syndrome such as gastroenteritis, wound infection, or septicemia.

Pathogenicity of *V. parahaemolyticus* strains is strongly linked to their ability to produce a 22 kDa, thermostable, extracellular haemolysin. When tested on a medium known as Wagatsuma's agar, the haemolysin can lyse fresh human or rabbit blood cells but not those of horse blood, a phenomenon known as the Kangawa reaction. The haemolysin has also been shown to have enterotoxic, cytotoxic and cardiotoxic activity.

Most (96.5 per cent) strains from patients with *V. parahaemolyticus* food poisoning produce the haemolysin and are designated Kanagawa-positive (Ka_+) while 99 per cent of environmental isolates are Ka–. Volunteer feeding studies have found that ingestion of 10_7–10_{10} Ka– cells have no effect whereas 10_5–10_7 Ka_+ cells produce illness. A number of other virulence factors have been described but have been less intensively studied.

Most environmental isolates of both *V. cholerae* and *V. parahaemolyticus* are non-pathogenic. The majority of the *V. cholerae* are non-01 serotypes and even those that are 01 tend to be non-toxigenic. Similarly, 99 per cent of environmental strains of *V. parahaemolyticus* are non-pathogenic.

V. vulnificus is a highly invasive organism that causes a primary septicaemia with a high fatality rate (50 per cent). Most of the cases of food-borne transmission identified occurred in people with pre-existing liver disease, diabetes or alcoholism. Otherwise, healthy individuals are rarely affected and, when they are, illness is usually confined to a gastroenteritis. In food-borne cases, the symptoms of malaise followed by fever, chills and prostration appear 16–48 h after consumption of the contaminated food, usually seafoods, particularly oysters. Unlike other *Vibrio* infections, *V. vulnificus* infections require treatment with antibiotics such as tetracycline.

Cholera Caused by *V. cholerae*

Historically, cholera has been one of the diseases most feared by mankind. It is endemic to the Indian subcontinent where it is estimated to have killed more than 20 million people this century. During the 19_{th} century there were a number of epidemics of Asiatic cholera, spread from the Indian subcontinent throughout Europe and the Americas. It spreads inexorably across Europe at a rate of about eight kilometres a, day reaching England in 1831, where it thrived in the appalling over-crowded, insanitary conditions of the burgeoning towns and cities. The approach of a second outbreak in 1848 prompted Parliament to establish the Central Board of Health which began the long task of improving sewerage and water supply systems. Similar apprehension of an approaching cholera outbreak in 1866 inspired the foundation of a similar Board in New York in the United States.

Pacini (1854) is credited with the first description of the etiological agent of cholera when he observed large numbers of curved bacilli in clinical specimens from cholera patients in Florence. His findings were not however generally accepted because of the widespread occurrence of similar but harmless *Vibrios* in the environment. It was Robert Koch who firmly established the causal link between *Vibrio cholerae* and cholera when working in Egypt in 1886.

Koch isolated what is now known as the classical *V. cholerae* biotype which was responsible for most outbreaks of cholera until 1961. The *El Tor* biotype, first isolated in 1906 by Gotschlich from pilgrims bound or Mecca at the El Tor quarantine station in Sinai, Egypt, is responsible for the current (7_{th}) pandemic. This started in Celebes in Indonesia in 1961, reached Africa in 1970 and the Americas in 1991. Of the 594694 cases reported to the WHO in 1991, 391220 were in South and Central America.

It was recognized in the 1930s that both biotypes are agglutinated by a single antiserum designated O1. Other strains of *V. cholerae* do not react with this antiserum and are termed non-agglutinable, or more correctly non-O1 strains, though some do produce cholera toxin. In 1993 a new serotype, O139, was associated with epidemic cholera in India and Bangladesh and has also been isolated from cholera patients in Thailand.

V. cholerae divided into the serogroups O1 and non-O1. O1 strains, can be further classified into the classical (non-haemolytic) or El Tor (haemolytic) biotypes; each of which can be subdivided by serotyping into one of three groups: Ogawa, Inaba or Hikojima. Clinical strains of *V. parahaemolyticus* can be serotyped for epidemiological purposes using a scheme based on 11 thermostable O antigens and 65 thermolabile K (capsular) antigens.

Previous illness with *V. cholerae* O1 does not confer immunity and the disease is now endemic. The organism, *V. cholerae* serogroup O139 (Bengal), causes characteristic severe cholera symptoms. In the U.S., *V. cholerae* O139 has been implicated in one case, a traveller returning from India. The strain has not been reported in U.S. waters or shellfish.

V. cholerae serogroup O1 and O139 may be recovered from foods by methods similar to those used for recovering the organism from the faeces of infected individuals. Pathogenic and non-pathogenic forms of the organism exist, so all food isolates must be tested for the production of cholera enterotoxin.

Vibrio cholerae Serogroup O1 bacterium is responsible for Asiatic or epidemic cholera. No major outbreaks of this disease have occurred in the United States since 1911. However, sporadic cases occurred between 1973 and 1991, suggesting the possible reintroduction of the organism into the U.S. marine and estuarine environment. The cases between 1973 and 1991 were associated with the consumption of raw shellfish or of shellfish either improperly cooked or re-contaminated after proper cooking. Environmental studies have demonstrated that strains of this organism may be found in the temperate estuarine and marine coastal areas surrounding the United States.

In 1991 cholera was reported for the first time in this century in South America, starting in Peru. The outbreaks quickly grew to epidemic proportions and spread to other South American and Central American countries, and into Mexico.

The South American strain of V. cholerae O1 has been isolated from Gulf Coast waters, presumably transmitted by ships off-loading contaminated ballast water. Hence, cholera may also be transmitted by shellfish harvested from nonpolluted waters since *V. cholerae* O1 is part of the autochthonous microbiota of these waters. All people are believed to be susceptible to this infection, but individuals with damaged or undeveloped immunity, reduced gastric acidity, or malnutrition may suffer more severe forms of the illness.

The optimum temperature for growth of *V. cholerae* is between 30 and 37 °C, although growth can occur between 10 and 43 °C. *Vibrio cholerae* grow at pH 5.0–9.6 but prefer a pH of 7.6. They grow at a a_w of at least 0.97 but prefer 0.984. Optimum growth occurs in an environment with a sodium chloride concentration of 0.5 per cent, although *V. cholerae* growth can occur at concentrations of 0.1–4.0 per cent

Cholera is regarded primarily as a water-borne infection as it is generally a disease spread by poor sanitation, resulting in contaminated water supplies. Food which has been in contact with contaminated water can often serve as the vehicle. Consequently, a large number of different foods have been implicated in outbreaks, particularly products such as washed fruits and vegetables which are consumed without cooking. Foods coming from a contaminated environment may also carry the organism, for example seafoods and frog's legs. This is clearly the main mechanism for the spread of cholera in poor communities in South America. The excellent sanitation facilities in the U.S. are responsible for the near eradication of epidemic cholera. Sporadic cases occur when shellfish harvested from faecally polluted coastal waters are consumed raw. In the cholera pandemic in South and Central America, an uncooked fish marinade, in lime or lemon juice, *ceviche* has been associated with some cases.

V. cholerae typically gain entrance into the human body through ingestion of a contaminated food, such as mollusks (raw oysters) or crustaceans eaten raw, undercooked, or even contaminated after cooking, or exposure of an open wound to a contaminated water source. Conditions resulting from *V. cholerae* O1 infection range from asymptomatic to the most severe form known as "cholera gravis" and in part depend on which biogroup is involved, because 75 per cent of the El Tor biogroup and 60 per cent of the classic biogroup lead to asymptomatic infections. Additionally, the El Tor biogroup results in severe disease in 2 per cent of the infected individuals and mild or moderate disease in 23 per cent whereas the classic biogroup produces severe disease in 11 per cent, of individuals and mild or moderate disease in 30 per cent.

Cholera usually has an incubation period of several hours to 5 days, depending on inoculum size and the amount of food ingested; can vary from mild, self-limiting diarrhoea to a severe, life-threatening disorder. The infectious dose in normal healthy individuals is large when the organism is ingested without food or buffer, of the order of 10_{10} cells, but is considerably reduced if consumed with food which protects the bacteria from stomach acidity. Human volunteer feeding studies utilizing healthy individuals have demonstrated that approximately one million organisms must be ingested to cause illness. Antacid consumption markedly lowers the infective dose. Studies conducted in Bangladesh indicate that 10_3–10_4 cells may be a more typical infectious dose. Individuals with low stomach acidity (hypochlorohydric) are more liable to catch cholera.

Symptoms include muscle cramping caused as a result of severe dehydration (fluid loss up to 500–1000 ml/h) resulting from vomiting, increased peristalsis followed by loose stools progressing to watery stools, and mucus-flecked diarrhea that is characteristic of cholera. Patients may produce up to 20-30 stools per day, losing many litres of water and electrolytes. Patients therefore exhibits extreme dehydration, urine is suppressed, the skin becomes wrinkled, the eyeballs are sunken and the voice becomes weak and husky. Blood pressure falls, the heart sounds become barely audible and the pulse become rapid and weak. In addition to dehydration, other complications may include hypovolemic shock, hypoglycemia, and metabolic acidosis. Death may occur in 12 to 24 hrs due to rapid dehydration.

Cholera is a non-invasive infection where the organism colonizes the intestinal lumen and produces a potent enterotoxin. In severe cases, the hypersecretion of sodium, potassium, chloride and bicarbonate induced by the enterotoxin results in a profuse, pale, watery diarrhoea containing flakes of mucus, described as rice water stools. The diarrhoea, which can be up to 201 day^{-1} and contains up to 10_8 *Vibrios* ml^{-1}, is accompanied by vomiting, but without any nausea or fever.

Unless the massive losses of fluid and electrolyte are replaced, there is a fall in blood volume and pressure, an increase in blood viscosity, renal failure and circulatory collapse. In fatal cases, death occurs within a few days. In untreated outbreaks the death rate is about 30–50 per cent but can be reduced to less than 1 per cent with prompt treatment by intravenous or oral rehydration using an electrolyte/glucose solution.

For transmission, man is the only natural host of the cholera *Vibrios*. Spread of infection is from person-to-person, through contaminated water or foods. Shrimps and vegetables are the most frequent carriers. Cholera is an infection of crowded poor class communities and it tends to persist in such areas. Cholera outbreaks occur either as explosive epidemics usually in non-endemic areas or as protracted epidemic waves in endemic areas

The enrichment media used for *Vibrios* exploit their greater tolerance for alkaline conditions. In alkaline peptone water (pH 8.6–9.0) the incubation period must be limited to 8 h to prevent overgrowth of the *Vibrios* by other organisms. Tellurite/bile salt broth (pH 9.0–9.2) is a more selective enrichment medium and can be incubated overnight.

For diagnosis, a vibrio immobilization test with dark field microscopy can be done. In the acute stage, *vibrios* are abundantly present in watery stool (upto 10^7 to 10^9 organisms per ml). Bacteriological examination can be done in stool, suspect water and food. The most commonly used selective and differential agar used for *Vibrios* is thiosulfate/citrate/bile salt/sucrose agar (TCBS). The medium was originally designed for the isolation of *V. parahaemolyticus* but other enteropathogenic *Vibrios* grow well on it, with the exception of *V hollisae. V. parahaemolyticus, V. mimicus* and *V. vulniftcus* can be distinguished from *V. cholerae* on TCBS by their inability to ferment sucrose which results in the production of green colonies. *V. cholerae* produces yellow colonies. Individual species can then be differentiated on the basis of further biochemical tests. Precise identification of biotype and serotype can be done using serological (*e.g.* CFT) and phage sensitivity tests.

Following control measures should be followed:

1. Provision of potable water
2. Proper sewage disposal
3. Proper cooking and hygienic handling of food
4. Observation of personal hygiene
5. Vaccination -The heat killed, phenol preserved vaccine has protection that lasts for 3 to 6 months.

Synopsis of Most Common Bacterial Genera

Some of the microorganisms cause foodborne diseases, either from poisoning, intoxication or toxicoinfection. Let us study about some food-borne bacteria and their role in the alteration of food.

Table 7.4: Pathogens that Cause Infection

Bacteria	*Disease/Medical Complications*	*Food Products Involved*
Salmonella enterica serovar Typhi and *Salmonella enterica serovar Paratyphi*	Typhoid and paratyphoid fever.	Undercooked pork, beef and poultry, contaminated eggs, and milk.
Salmonella spp.	Salmonellosis (*Salmonella* Typhimurium, *Salmonella* Enteritidis).	Undercooked poultry, cauliflowers, and tomatoes.
Vibrio vulnificus	Septicemia in people with underlying diseases or people who are taking immunosuppressive drugs or steroids.	Seafood, usually oysters.
Mycobacterium bovis	Cervical lymphadenopathy, intestinal lesions, chronic cutaneous tuberculosis.	Contaminated milk.
Mycobacterium avium, subspecies *paratuberculosis*	Crohn's disease.	Pasteurized milk.
Listeria monocytogenes	Meningitis, encephalitis, sepsis in pregnant women, intrauterine or cervical infection that can lead to miscarriage or birth of a dead child.	Raw beef, pork, poultry, vegetables and milk, cheese, ice cream, smoked fish, and raw fish.
Clostridium botulinum	Paralysis of arms, legs, trunk, and respiratory muscles.	Mixture of oil and nonacid garlic, potatoes cooked at high temperatures, and stews.
Bacillus cereus	Fried rice syndrome.	Rice cooked at high temperatures, sauces, soups, and puddings.
Staphylococcus aureus	Toxic shock syndrome.	Meat and meat products cooked at high temperatures, poultry, and salads with mayonnaise.

Table 7.5: Pathogens that Cause Intoxication

Bacteria	*Disease/Medical Complications*	*Food Products Involved*
Escherichia coli O157:H7	Hemorrhagic colitis, Hemolytic *uremic syndrome in children.*	Hamburgers, nonpasteurized milk, contaminated water, spinach, and lettuce.
Shigella spp.	Hemolytic Uremic Syndrome.	Salads, lettuce, raw vegetables, and milk.
Aeromonas spp.	Meningitis, peritonitis, myocarditis, hemolytic uremic syndrome, necrotizing fasciitis in wounds.	Meat (beef, sheep, pork and chicken), vegetables, eggs, fish, seafood, and prepared food.
Cronobacter sakazakii	Permanent neurological or developmental deficits; death.	Powdered infant formula.
Vibrio parahaemolyticus	Gastroenteritis, septicemia and wound infection. Severe infections in immunocompromised people.	Raw or undercooked seafood, usually oysters.
Clostridium perfringens	Clostridial necrotizing enteritis.	Meat juice, stews, cooked beans, meat cooked at high temperatures.
Campylobacter spp.	Campylobacteriosis, arthritis, meningitis, *Campylobacter jejuni* can cause Guillain-Barré Syndrome.	Cheese made with raw milk and chicken meat.
Yersinia enterocolitica	Yersiniosis, enterocolitis, pseudoappendicitis, mesenteric lymphadenitis, infections in wounds, joints and the urinary tract, and Reiter's syndrome.	Nonpasteurized milk, tofu, nonchlorinated water, undercooked meat, oysters and fish.
Vibrio cholerae serogroup O1 or serogroup O139	Cholera.	Contaminated water and raw seafood.
Vibrio cholerae serogroup no-O1	Less severe than Cholera; gastrointestinal infections, sepsis.	Raw, semicooked or recontaminated fish and shellfish after cooking.

Table 7.6: Pathogens that Cause Toxico-infection

Name	*Biological Effect*
Cholera toxin (Ctx) (A-5B)	It activates the adenylyl cyclase; increases the levels of intracellular cAMP promoting fluid and electrolytes secretion in the intestinal epithelium, causing diarrhea. It is a potent exotoxin.
Thermolabile toxin (LT) (A-5B)	Similar effect as the Cholera toxin.

Name	Biological Effect
Thermostable toxin (ST)	The binding of ST to the guanylyl cyclase receptor results in an increase of cyclic GMP, affecting the flow of electrolytes. It promotes water and electrolytes secretion from the intestinal epithelium by causing diarrhoea.
Shiga toxin (A-5B)	Inactivates the ribosomal subunit 60S and inhibits protein synthesis causing the death of susceptible cells.
Botulinum toxin (A/B)	It is a neurotoxin consisting of a heavy and a light chain linked by a disulfide bond. It is a Zn^{++}-dependent protease. It inhibits the presynaptic release of acetylcholine from peripheral cholinergic neurons, resulting in flaccid paralysis. The neurotoxin exists in seven different serotypes (A-G).
CPE enterotoxin	Lethal, cytotoxic and enterotoxic activity. Stimulates the adenylyl cyclase allowing the increase of cAMP in epithelial cells, which causes diarrhoea.
Alpha-toxin	It produces gas gangrene. It has phospholipase (PLC), sphingomyelinase, hemolytic, and dermonecrotic activities. The mature protein is organized into two domains; the amino-terminal, which contains the PLC activity, and the carboxyl-terminal binding that depends on calcium. Depending on the lipid composition of the cell membrane, the Alpha-toxin may be hemolytic in the presence of calcium.
Beta-toxin	It forms selective pores for monovalent cations in lipid bilayers and sensitive cells membranes, so it functions as a neurotoxin capable of producing arterial constriction.
Epsilon-toxin	Produced and secreted by a prototoxin that, when it suffers a specific proteolytic cleavage, it acquires its maximum biological activity. Activation can be catalyzed by proteases such as trypsin, chymotrypsin, and a zinc-dependent metalloproteinase.
Iota-toxin	It has dermonecrotic, cytotoxic, enterotoxic activities, and it causes intestinal histopathological damage. This toxin is binary and consists of a binding peptide (Ib) and an enzymatic peptide (ADP-ribosyltransferase) (Ia). The first one is necessary to internalize the second one. The Iota-toxin requires proteolytic removal of a propeptide fragment, which allows the Ib unit to be inserted into the membrane and to interact with the Ia portion to form a heptameric pore that allows the K^+ and Na^+ ions to escape; in addition to the Ia portion entrance into the cell where it ribosylates the G-actin to depolymerize the actin filaments, with the consequent destruction of the cytoskeleton. The Iota-toxin is generally activated by the effect of the proteases present in the intestinal tract.
Toxin A/Toxin B	It modifies the Rho, a subfamily of GTP-binding proteins that regulate cytoskeletal actin. The deamination of the glutamine residue at position 63 of Rho to a glutamic acid produces a dominant-active Rho protein incapable of hydrolyzing the GTP, resulting in cellular necrosis and bloody diarrhea associated with colitis.
Enterotoxins (A, B, C1, C2, D and E, G, H, I, J)	Enterotoxins are thermostable; they differ in toxicity.Staphylococcal enterotoxins are superantigens that cause massive activation of the immune system, including lymphocytes and macrophages; the exact role in emesis is not known.
Toxic Shock Syndrome Toxin (TSST-1)	Superantigen that acts on the vascular system causing inflammation, fever, and shock.
Cereulide	Thermostable peptide, toxic for the mitochondria when acting as a potassium ionophore.
HBL, NHE, Citotoxin K or CytK	HBL is a three-component hemolysin; two protein subunits, L2 and L1 (cytolytic components), and a B protein (favors binding to the host cell), apart from the hemolytic effect, it is cytotoxic, dermonecrotic and causes vascular permeability. NHE also consists of three components (NheA is a cytolytic component and NhB and NheC favor binding to cells of small intestine). Both toxins are organized into operons (*hbl* and *nhe*), where the genes encoded the NHE components are transcribed together. CytK forms pores in the epithelial cells membrane, and it has necrotizing and cytotoxic activity.

Proteus

These organisms belong to the family Enterobacteriaceae and consist of six species. Like the other members of this family, they are gram-negative rods that are aerobic and often show pleomorphism. All are motile while all but one species hydrolyze urea. They are found in the intestinal tract of man and animals and on decaying materials in general. They may be isolated from spoiled eggs and meats, especially those allowed to spoil above refrigerator temperatures. For most species, G + C content is 38–42 moles per cent.

Pseudomonas

These important bacteria belong to the family Pseudomonadaceae and while eighty-seven species are recognized, over two hundred others have been reported. They are short, gram-negative and aerobic rods that usually produce a single polar flagellum. Many psychrotrophic species and strains as well as mesophiles exist in this genus. They are widely distributed in nature in soils and water, on plants and in the intestinal canal of man and other animals. These are by far the most important bacteria in the low-temperature spoilage of foods such as meats, poultry, eggs and seafoods. Some strains produce fluorescent pigments (pyocyanin and/or/fluorescin). Most are capable of oxidizing glucose to gluconic acid, 2-ketogluconic acid, or other intermediates. Many are plant pathogens, causing leaf spot, leaf stripe and related diseases. Many of those that cause food spoilage do not produce the water-soluble pigment but may fluoresce under ultraviolet light. The G + C content of DNA ranges from 58–70 moles per cent.

Salmonella

This important genus of bacteria belongs to the family Enterobacteriaceae and while only five species are recognized, over two thousand serovars (serotypes) are known. with new ones added yearly. They are gram-negative, short rods that are aerobic and do not produce pigments on culture media. Most ferment glucose and other simple sugars with the production of acid and gas. They generally do not ferment lactose although some do. These are basically intestinal forms, as are most of the Enterobacteriaceae genera. They may be found widely distributed in nature. The etiologic agents of typhoid and paratyphoid fevers belong to this group as well as those that cause food-borne salmonellosis in man. All species and strains of this genus are undesirable in foods. The G + C content of DNA = 50–53 moles per cent.

Citrobacter

This genus belongs to the family Enterobacteriaceae and three species are recognized. The slow, lactose-fermenting and gram-negative rods previously designated paracolons and the Bethesda-Ballerup group have been placed in this genus. All members can use citrate as their sole carbon source. They are common in fecal matter, water and some foods.

Clostridium

This is an important group of anaerobic bacteria that belong to the family Bacillaceae. Some seventy-six species are now recognized. They are gram-positive and sporeforming rods. The genus contains some thermophilic species that are of great importance in the canning industry due to the extreme heat resistance of their endospores. The etiologic agents of tetanus, gas gangrene, perfringens food poisoning and botulism are members of this genus. Some species are of commercial importance in the production of certain solvents. They are very widely distributed in nature, soils, water, the intestinal tract of man and animals and other places. They may be found in many foods where they may or may not grow. Many species are strongly proteolytic as are many of the *Bacillus* spp. in the same family. The G + C content of DNA = 23–43 moles per cent.

Table 7.7: Summary of some Common Foodborne Illnesses Caused by Bacteria

Organism	*Common Name of Illness*	*Onset Time after Ingesting*	*Signs and Symptoms*	*Duration*	*Food Sources*
Bacillus cereus	*B. cereus* food poisoning	10-16 hrs	Abdominal cramps, watery diarrhoea, nausea	24-48 hours	Meats, stews, gravies, vanilla sauce
Campylobacter jejuni	Campylobacteriosis	2-5 days	Diarrhoea, cramps, fever, and vomiting; diarrhoea may be bloody	2-10 days	Raw and undercooked poultry, unpasteurized milk, contaminated water
Clostridium botulinum	Botulism	12-72 hours	Vomiting, diarrhoea, blurred vision, double vision, difficulty in swallowing, muscle weakness. Can result in respiratory failure and death	Variable	Improperly canned foods, especially home-canned vegetables, fermented fish, baked potatoes in aluminum foil
Clostridium perfringens	Perfringens food poisoning	8–16 hours	Intense abdominal cramps, watery diarrhoea	Usually 24 hours	Meats, poultry, gravy, dried or precooked foods, time and/or temperature-abused foods
Cryptosporidium	Intestinal cryptosporidiosis	2-10 days	Diarrhoea (usually watery), stomach cramps, upset stomach, slight fever	May be remitting and relapsing over weeks to months	Uncooked food or food contaminated by an ill food handler after cooking, contaminated drinking water
Cyclospora cayetanensis	Cyclosporiasis	1-14 days, usually at least 1 week	Diarrhoea (usually watery), loss of appetite, substantial loss of weight, stomach cramps, nausea, vomiting, fatigue	May be remitting and relapsing over weeks to months	Various types of fresh produce (imported berries, lettuce, basil)
E. coli (Escherichia coli) producing toxin	E. coli infection (common cause of "travelers' diarrhea")	1-3 days	Watery diarrhoea, abdominal cramps, some vomiting	3-7 or more days	Water or food contaminated with human feces
E. coli O157:H7	Hemorrhagic colitis or E. coli O157:H7 infection	1-8 days	Severe (often bloody) diarrhoea, abdominal pain and vomiting. Usually, little or no fever is present. More common in children 4 years or younger. Can lead to kidney failure	5-10 days	Undercooked beef (especially hamburger), unpasteurized milk and juice, raw fruits and vegetables (*e.g.* sprouts), and contaminated water

Organism	Common Name of Illness	Onset Time after Ingesting	Signs and Symptoms	Duration	Food Sources
Listeria monocytogenes	Listeriosis	9-48 hrs for gastrointestinal symptoms, 2-6 weeks for invasive disease	Fever, muscle aches, and nausea or diarrhoea. Pregnant women may have mild flu-like illness, and infection can lead to premature delivery or stillbirth. The elderly or immunocompromised patients may develop bacteremia or meningitis	Variable	Unpasteurized milk, soft cheeses made with unpasteurized milk, ready-to-eat deli meats
Salmonella	Salmonellosis	6-48 hours	Diarrhoea, fever, abdominal cramps, vomiting	4-7 days	Eggs, poultry, meat, unpasteurized milk or juice, cheese, contaminated raw fruits and vegetables
Shigella	Shigellosis or Bacillary dysentery	24-48 hrs	Abdominal cramps, fever, and diarrhoea. Stools may contain blood and mucus	4-7 days	Raw produce, contaminated drinking water, uncooked foods and cooked foods that are not reheated after contact with an infected food handler
Staphylococcus aureus	Staphylococcal food poisoning	1-6 hours	Sudden onset of severe nausea and vomiting. Abdominal cramps. Diarrhoea and fever may be present	24-48 hours	Unrefrigerated or improperly refrigerated meats, potato and egg salads, cream pastries
Vibrio parahaemolyticus	V. parahaemolyticus infection	4-96 hours	Watery (occasionally bloody) diarrhoea, abdominal cramps, nausea, vomiting, fever	2-5 days	Undercooked or raw seafood, such as shellfish
Vibrio vulnificus	V. vulnificus infection	1-7 days	Vomiting, diarrhoea, abdominal pain, bloodborne infection. Fever, bleeding within the skin, ulcers requiring surgical removal. Can be fatal to persons with liver disease or weakened immune systems	2-8 days	Undercooked or raw seafood, such as shellfish (especially oysters)

Adapted from U.S. Food and Drug Administration. Foodborne illnesses: what you need to know. http://www.fda.gov/food/resourcesforyou/consumers/ucm103263.htm.

Corynebacterium

This genus belongs to the coryneform group and contains some thirty-two species. They are aerobic, gram-positive rods that often show granules and club-shaped swellings. They are non-sporeformers. This group contains the etiologic agent of diphtheria and other species common on the body. Most are mesophiles with some being psychrotrophs. These organisms are widely distributed among certain plants such as wheat, beans and tomatoes.

Escherichia

As stated above, this group belongs to the family Enterobacteriaceae and contains three species. They are short, gram-negative rods that are indistinguishable from *Enterobacter* on culture media and under the microscope. They differ from *Enterobacter* spp. in being MR + and VP–. Their main habitat appears to be the intestinal tract of man and other animals from which they may be found in soils, waters and many other places in nature. *E. coli* is the most important member of the coliform group along with *E. aerogenes.* Their presence in large numbers in foods is generally taken to indicate fecal contamination. The G + C content of DNA = 50–51 moles per cent.

Flavobacterium

This genus contains eighteen species. They are gram-negative rods that generally produce yellow to red pigments on agar. They are widely distributed in soils and water and on fish and plants. They may be isolated from decaying plant materials. Many are psychrophilic while most are mesophiles. For motile species, the G + C content of DNA = 63–70 moles per cent, while for some non-motile species 30–42 moles per cent.

Aeromonas

This genus is represented by three species and eight subspecies and belongs to the Vibrionaceae. They are gram-negative rods that are facultatively anaerobic and display respiratory and fermentative metabolism. They are proteolytic and produce copious quantities of gases from those sugars fermented. Growth is over the pH range 5.5–9.0. They are found in the aquatic environment, where they are normal inhabitants of the intestines of fish. Some are fish pathogens and some *A. hydrophila* strains are human pathogens. The G + C content of their DNA is 57–65 moles per cent.

Alcaligenes

This genus consists of nine species. They are gram-negative rods that sometimes stain variable or gram-positive. They do not ferment sugars, but produce alkaline reactions that are also produced in litmus milk. They do not produce pigments. They are widely distributed in nature in decomposing matter of all types, in raw milk, poultry products, the intestinal tract and so on. The G + C content of DNA = 58–70 moles per cent.

Alteromonas

These organisms are common in marine fish and are closely related to *Alcaligenes* and *Pseudomonas.* While the G + C content of DNA of *Pseudomonas* spp. is 57–70 per cent, the range for *Alteromonas* is 43.2–48 per cent. Most require Na– for growth and they are non-fermentative. They are gram- negative, straight or curved rods, motile by means of a single polar flagellum, strict aerobes, lack arginine dihydrolase and do not accumulate polyhydroxybutyrate. Some, such as *A. putrefaciens* and *A. nigrifaciens,* were formerly classified as pseudomonas.

Bacillus

This genus belongs to the family Bacillaceae. Some thirty-one species are recognized. Most are aerobic, gram-positive rods that produce endospores. They often exist in long chains on cultural media. Most are mesophiles with some being psychrotropic and some thermophilic in nature. The thermophilic members are of great importance in the canning industry due to the extreme heat resistance of their spores. This genus contains one pathogen for man and other vertebrates–*B. anthracis,* which causes anthrax and one food-poisoning species, *B. cereus.* Some species are insect pathogens. The non-pathogens are widely distributed and may be found in air, dust, soil, water and on utensils and in various foods. They are important in the spoilage of many foods held above refrigerator temperatures. The G + C content of DNA = 32–62 moles per cent.

Acetobacter

This genus contains three species and nine subspecies. They are gram-negative, rod-shaped cells that are strict aerobes. They are commonly found in fermented grain mash, vinegar, beer, wines and souring fruits and vegetables. Some species such as *A. aceti* subsp. *aceti* oxidize ethanol to acetic acid and thereby give rise to vinegar, perhaps their greatest industrial use. The G + C content of DNA = 55–64 moles per cent.

Leuconostoc

This genus belongs to the family Streptococcaceae. It contains at least six species. They are gram-positive, spherical to oval, catalase-negative, heterofermentative organisms. They are widely distributed among plants, from which they find their way into milk and dairy products. Some cause problems in sugar refineries, where they form slime in sugar lines. Some species are employed in dairy starter cultures while others are often found in cured meat products. Some synthesize the medically important polymer dextran. The G + C content of DNA of most = 38–42 moles per cent.

Micrococcus

This group belongs to the family Micrococcaceae and contains at least nine species. They are gram-positive cocci that are catalase positive in contrast to the *Streptococci.* Some produce non-pigmented colonies while others produce pink to orange-red to red pigments on culture media. Most or all can tolerate high levels of salt. They are widely distributed in nature, on the skin of man and the hides of animals, as well as in dust, soil, water and many foods. Most are mesophiles while some are capable of growth in the psychrophilic range. Several species are associated with dairy products, through which they enter processed meats such as frankfurters. The formerly recognized genera *Gafjkya* and *Sarcina* have been reduced to species of this genus. The G + C content of DNA = 66–75 moles per cent.

Moraxella

This genus is a member of the family Neisseriaceae as is the genus *Acinetobacter.* They are very short and plump gram-negative rods that differ from *Acinetobacter* in being sensitive to penicillin, oxidase positive and having a G + C content of DNA of 40–46 moles per cent. Their metabolism is oxidative and they do not form acid from glucose. Because of the similarity of the two groups, some investigators use the designation "*Acinetobacter-Moraxella* group."

Pediococcus

These homofermentative cocci belong to the family Streptococcaceae and differ microscopically from *Streptococci* and *Leuconostocs* by their arrangement in pairs and tetrads resulting from cell division in two planes. Like the other lactic acid genera, they are widespread in nature, especially on plants.

Brochothrix

Only one species (*B. thermosphacta*) is recognized in this genus and it was placed originally in the genus *Microbacterium.* The genus is placed tentatively in the family Lactobacteriaceae but differs from typical lactobacilli in being catalase positive. Exponential-phase cells are rods while stationary-phase cells are coccoids, typical of coryneform bacteria. Glucose metabolism is fermentative and the G + C content of DNA is 35 moles per cent. *B. thermosphacta* is found most on processed meats stored in gas-impermeable packages, where it is sometimes the dominant organism.

Campylobacter

Four species and four subspecies are represented by this genus of slender, spirally curved rods which are 1.5–3.5 μm in length and 0.2–0.4 μm in width. They are motile by a corkscrewlike motion and possess polar flagellae at one or both ends. They are inactive on carbohydrates, are microaerophilic to anaerobic and have a G + C content of DNA of 30–35 moles per cent. Once classified as *Vibrio* spp., they can be found in the oral cavity and intestinal canal and some are human and animal pathogens.

Streptococcus

This genus belongs to the family Streptococcaceae and consists of twenty-seven species. They are gram-positive, catalase-negative cocci that often appear as spherical to ovoid forms. All produce small colonies when growing on culture media, as do the lactobacilli. They are non-pigmented and microaerophilic in nature. Some are associated with the upper respiratory tract of man and other animals, where they may cause diseases such as scarlet fever and septic sore throat. Others are found in the intestinal tract of man and animals and tend to be rather widespread on plants and plant parts and in dairy products. While most are mesophilic, some grow within the psychrotrophic range. Some cause mastitis in cattle while others are important in dairy starter cultures. One species, *S. lactis,* is the most common cause of sour milk and is important in the manufacture of cheese. The presence of some species in foods in large numbers may indicate faecal contamination. The G + C content of DNA = 33–42 moles per cent.

The genus Streptococcus is defined by a combination of antigenic, hemolytic, and physiological characteristics into Groups A, B, C, D, F, and G. Groups A and D can be transmitted to humans via food.

Group A

One species with 40 antigenic types (*S. pyogenes*).

Group D

Five species (*S.faecalis, S.faecium, S.durans, S.avium* and *S. bovis*).

Group A

They cause septic sore throat and scarlet fever as well as other pyogenic and septicemic infections. Sore and red throat, pain on swallowing, tonsilitis, high fever, headache, nausea, vomiting, malaise, rhinorrhea; occasionally a rash occurs, onset 1-3 days; the infectious dose is probably quite low (less than 1,000 organisms). For diagnosing the human illness, culturing of nasal and throat swabs, pus, sputum, blood, suspect food, environmental samples are done. Streptococcal sore throat is very common, especially in children. Usually, it is successfully treated with antibiotics. Complications are rare and the fatality rate is low.

Food sources include milk, ice cream, eggs, steamed lobster, ground ham, potato salad, egg salad, custard, rice pudding and shrimp salad. In almost all cases, the foodstuffs were allowed to stand at room temperature for several hours between preparation and consumption. Entrance into the food is the result of poor hygiene, ill food handlers, or the use of unpasteurized milk.

Group D

May produce a clinical syndrome similar to staphylococcal intoxication. Diarrhoea, abdominal cramps, nausea, vomiting, fever, chills, dizziness in 2-36 hours. Following ingestion of suspect food, the infectious dose is probably high (greater than 107 organisms. Diarrheal illness is poorly characterized, but is acute and self-limiting. Associated food sources include sausage, evaporated milk, cheese, meat croquettes, meat pie, pudding, raw milk and pasteurized milk. Entrance into the food chain is due to under processing and/or poor and unsanitary food preparation.

Group A infections are low and may occur in any season, whereas Group D infections are variable.

All individuals are susceptible to Streptococcal infections. No age or race susceptibilities have been found. Suspected food is examined microbiologically by selective enumeration techniques which can take up to 7 days. Group specificities are determined by Lancefield group-specific antisera.

Outbreaks of septic sore throat and scarlet fever were numerous before the advent of milk pasteurization. Salad bars have been suggested as possible sources of infection. Most current outbreaks have involved complex foods (*i.e.*, salads) which were infected by a food handler with septic sore throat. One ill food handler may subsequently infect hundreds of individuals. Outbreaks of Group D are not common and are usually the result of preparing, storing, or handling food in an unsanitary manner.

It has been proposed that *S. faecalis* and *S. faecium* be placed in a new genus, *Enterococcus.* If this proposal is adopted, the new genus would be characterized by a G + C content of DNA of 31–45 moles per cent, with *E. faecalis* being the type species. DNA hybridization studies reveal a close relationship between these two species but not to other members of the genus *Streptococcus.*

Enterobacter

These organisms belong to the family Enterobacteriaceae and are represented by nine species They are short gram-negative, non-pigment-forming rods that grow well on many cultural media and ferment glucose and lactose with the production of acid and gas. In their sugar fermentations these organisms produce two or more times as much CO_2 as hydrogen gas. The most distinctive taxonomic features are: MR– and VP+ (MR = methyl red; VP = Voges-Proskauer reaction). They are widely distributed in nature, being especially present on plants and grain and in water and the intestinal tract. This genus is one of the coliform genera along with *Escherichia.* The G + C content of DNA = 52–59 moles per cent.

Erwinia

This genus belongs to the family Enterobacteriaceae. Some twenty-one species are now recognized. They are gram-negative rods that are motile. Many species grow within the psychrotrophic range and many produce pigments of various shades of red when growing on culture media and certain foods. They are characteristically associated with plants where they cause dry necroses, galls, wilts and soft rots. These are the most important bacteria in the cause of market diseases of fruits and vegetables. The G+C content of DNA=53.6–54.1 moles per cent.

The Genus *Brucella*

The genus *Brucella* is named after Sir David Bruce who in 1887 recognized it as the causative organism of undulant fever (brucellosis, Malta fever, Mediterranean fever). Each of the four species that are human pathogens is associated with a particular animal host, *B. abortus* (cattle), *B. melitensis* (sheep and goats), *B. suis* (pigs) and *B. canis* (dogs). Brucellosis is principally contracted from close association with infected animals and is an occupational disease of farmers, herdsman, veterinarians and slaughterhouse workers. It can also be contracted by consumption of milk or milk products from an infected animal, although the risk is lower. The illness has been effectively eliminated from the United States, Scandinavia, the UK and are other countries by campaigns of eradicate the organism in the national dairy herds through a programme of testing, immunization of young calves and compulsory slaughter of infected cattle.

Brucella are Gram-negative, catalase-positive, oxidase-positive, short rods (0.3 µm × 0.4 µm) which are non-motile and usually occur singly, in pairs, or, rarely, in short chains. It grows optimally around and is killed by heating at 63°C for 7–10 min. When shed in the milk of an infected animal it can survive for many days provided the acidity remains low (< 0.5 per cent as lactate).

Brucellosis is a protracted and debilitating illness characterized by an incubation period of from one to six weeks followed by a chronic, relapsing fever with accompanying lassitude, sweats, headache, constipation, anorexia, pains in the limbs and back and weight loss. After the temperature has returned to normal for a few days, another bout of fever may ensue and such episodes occur a number of times over several months. Treatment is commonly with a mixture of tetracycline and streptomycin.

It is a facultative parasite and can live intracellularly or in extra-cellular body fluids. During the febrile stage, caused by circulating endotoxin, the organism may be isolated from the bloodstream but in the majority of laboratory-confirmed cases diagnosis is based on serological tests rather than cultural techniques.

Table 7.8: Differential Characteristics of *Brucella* Species

	B. arborus	*B. canis*	*B. melitensis*	*B. suis*
5 per cent CO_2 required	+	–	–	–
H_2S produced	+	–	–	+
Urease	Weak	Strong	Weak	Strong
Growth on dye medium:				
Basic fuchsin (1 : 105)	+	–	+	–
Thionin (1 : 105)	–	+	+	+

The organism grows slowly and cultures are normally incubated for three weeks before they are considered negative. In view of this, testing foods for the organism is not practically feasible or useful. Cattle are tested for the presence of antibodies to the organism in the 'Ring Test'. Stained antigen is mixed with the test milk, if antibodies to *Brucells* are present (indicative of infection) then they will cause the antigen to clump and rise with the milkfat on standing to form an intense blue-violet ring at the top of the milk.

Although brucellosis has sometimes been associated with the consumption of inadequately cooked meat from an infected animal, raw milk or cream are the principal food vehicles. *Brucella* is readily killed by normal milk pasteurization conditions so there is no risk from pasteurized milk or products made from it. Cheeses made from the unpasteurized milk can sometimes pose a problem since the organism can survive the cheese-making processes and subsequent storage in the product.

Campylobacter Infection and Foodborne Campylo-bacteriosis

Campylobacters occur widely as part of the normal intestinal flora of many warm-blooded animals including chickens and turkeys. In addition, the organism occurs in raw water and raw milk. Campylobacters have been known as the cause of diseases in animals since 1909, but they have been generally recognised only recently as a cause of the human disease, campylobacteriosis. Campylobacters enter the human food chain at slaughter of the animals and this is believed to be the main source of infection for man. The illness in man involves diarrhoea, sometimes with a bloody stool, which may last from 2-10 days. It is usually self-limiting.

In the countries where records are kept, campylobacters are now reported to be the leading cause of bacterial diarrhoea in man. The increase in reports in the Western world must in part be attributed to increased awareness, better laboratory techniques and improved surveillance. Campylobacteriosis is a very widespread diarrhoeal disease in the developing world, commonly affecting babies and young children. Campylobacters occur widely in the intestinal tract of many animals especially chickens and turkeys.

During slaughtering and preparation of raw birds a high percentage of poultry carcasses may become contaminated with the campylobacters in the intestinal track especially *Campylobacter jejuni*. Consequently, undercooked poultry, meat and offals are a source of campylobacter infections. Raw milk and poorly treated water supplies are also important sources of campylobacter infections. Factory manufactured foods have not been shown to be a cause of campylobacter food poisoning.

The main known facts about campylobacters, the disease and the preventive measures are given below.

Campylobacteriosis in Man

The illness is a gastrointestinal infection caused by the bacterium *Campylobacter jejuni* or sometimes *C. coli*. The infectious dose required to cause foodborne disease may be very low, *i.e.* only a few hundred cells. Babies, young children and debilitated people are the most susceptible. Symptoms vary from mild (with very little signs of illness even though campylobacter may be present in the stool) to severe (with bloody diarrhoea as the most characteristic symptom). Other symptoms are: fever, nausea, abdominal cramps and (seldom) vomiting. The duration of the illness is usually 2-10 days, but symptoms. particularly abdominal cramps, may recur for up to 3 months after the infection. Complications such as septicaemia, may sometimes arise.

In man, transmission is via contaminated food or water. The main route is thought to be either under-cooking of raw poultry or cross-contamination from raw to ready-to-eat food. Person-to-person transmission is known but thought to be rare in the Western world. Household pets with diarrhoea have often been shown to be the source of infection for man.

The disease is usually self-limiting, so antibiotic treatment is only indicated in severe cases. Erythromycin is the drug of choice, with ciprofloxacin as a suitable alternative in adults. Cases of septicaemia are best treated with gentamicin, but erythromycin, chloramphenicol and tetracycline may also be used. The faeces often remain positive for 2-7 weeks, but long-term carriage is rare. Treatment with erythromycin may significantly shorten the duration of excretion.

Campylobacter (and others) in the Kitchen

Campylobacter spp. are major causes of concern with regard to gastrointestinal disease including that which is foodborne.

It is noted that there was only the one outbreak of campylobacteriosis included amongst the number of incidents meaning that most of those reported were sporadic and probably involved only one person or a few people. Since number of cases reported is likely to be a small percentage total cases, it is not surprising that estimates of total numbers of foodborne illness annually in vary widely from country to country. Infection with enteric campylobacters usually leads to diarrhoea and severe cramping. It occurs most frequently in young children and young adults. Initial symptoms are usually followed by abdominal pain and general malaise. The disease is usually limited to 5–8 days but an association exists with certain long-term neurological symptoms and Guillain Barré Syndrome (*Journal of Applied Microbiology* 90 2001 45S).

Campylobacters are considered to be part of the normal intestinal flora of a wide range of domestic and wild animals. Faecal contamination of meat occurs during slaughter and subsequent operations and human infection is usually through consumption of undercooked meat or other cross-contaminated food products.

Much effort has gone into food safety enhancement measures in the food manufacturing and retail/service industries in recent years. While the benefits of this effort have yet to be translated into an identifiable reduction in levels of foodborne illness, the combination of regulatory and industry actions has resulted in much greater awareness in the industry of the role it must play in reducing foodborne illness.

Consumers too have an important role to play in reducing the incidence of foodborne disease. While this role may be less well defined, minimisation of cross-contamination in the domestic kitchen is one area where education of the public should prove beneficial.

Raw foods, whether they be vegetable or animal products, can be contaminated with potential human pathogens including *Campylobacter* spp. In developed countries, one food with potential to cross-contaminate in the kitchen is raw poultry meat. There is little doubt that *Campylobacter* spp. will be spread around the environment when foods such as chicken are handled in either commercial or domestic kitchens (*Journal of Applied Microbiology* 90 2002 115S). While insufficient work has been published to confirm this situation as a major source of *Campylobacter* infection in humans, it is a situation over which the householder (as well as the commercial kitchen manager) can exercise some control. A recent study (*International Journal of Food Microbiology* 85 2003 213) investigates factors affecting the survival of *Campylobacter*, *Salmonella* and *Escherichia coli* during what is described as a typical washing-up process and the subsequent potential for transfer of bacteria to sites in the kitchen. Air-drying in organic matter, as would occur if dirty dishes were left for some time prior to washing-up, showed that *Campylobacter* survived very poorly in comparison with *Salmonella* and *E. coli*. This confirms earlier studies which showed that *Campylobacter* is very sensitive to desiccation whereas *Salmonella* and *E. coli* can survive for much longer periods.

The survival of these bacteria during a simulated washing-up was also investigated. Higher levels of bacteria than could be expected in practice were used but this was necessary to permit an accurate description of the inactivation that occurs. The main result to come from this part of the study was that a small elevation in washing-up water temperature could bring about a large increase in bacterial death rate.

The researchers recommend that washing-up water is used at the maximum possible temperature (using gloves to achieve this) and, if possible, a dishwashing machine which can reach much higher temperatures should be used.

A proportion of dishes that had been inoculated with *E. coli*, *Salmonella* or *Campylobacter* remained contaminated after a 'typical' washing-up process. Not all bacteria were physically removed or inactivated. The potential risk associated with the survival of *Campylobacter* during washing-up and cross contamination in the kitchen environment was considered to be small due to its sensitivity to detergent (0.1 per cent) and desiccation. Subsequent studies focused on *Salmonella* and *E. coli*.

When bacterial cultures were used to contaminate organic matter smeared onto dishes, *E. coli* transferred onto washing-up sponges and onto surfaces wiped with a contaminated sponge but *Salmonella* was not recovered from surfaces. Fresh sponges were used for these experiments so the possible build-up of pathogens in washing-up cloths that may occur in kitchens (*Journal of Hygiene* 93 1984 365) was not considered.

With higher initial levels of bacteria, both *E. coli* and *Salmonella* were recovered from kitchen surfaces and a higher proportion of swabs was contaminated. Bacteria transferred from dishes onto food more frequently when the dish had been towel dried rather than air dried but the overall frequency of transfer was low.

Frequent replacement or decontamination of tea towels and washing-up sponges is recommended.

The Disease in Animals-*Campylobacter* in Cattle and Sheep

Campylobacter is the leading bacterial cause of acute gastrointestinal disease in many countries. The disease is characterised by the usual symptoms of foodborne illness including diarrhoea, vomiting and cramping. Recovery is usually complete in 4–5 days but a small proportion of subjects may suffer longer-term consequences.

Campylobacter species including *C. jejuni*, the most common species associated with gastrointestinal illness, are part of the natural intestinal flora of a wide range of birds and animals. Transmission to humans is usually by contamination of food or water although in some countries unpasteurised milk is also implicated. Chicken, poultry and other foods are thought to be the most likely sources of infection in developed countries (Food and Agriculture Organization/World Health Organization, 2003). During 2002, Communicable Diseases Intelligence in Australia commenced a multistate study into risk factors for *Campylobacter* infection. Preliminary results indicated that one of the major risk factors for illness was consumption of chicken.

While *Campylobacter* spp. are relatively easily killed by heat and proper cooking of flesh foods will always destroy them, much of the foodborne illness associated with *Campylobacter* is thought to stem from cross-contamination from uncooked to cooked food. Nineteen commercial cattle and sheep properties in New South Wales and Queensland were selected to cover all production systems for red meat: six dairy cattle properties, four feedlot beef cattle properties, four pasture beef cattle properties, two prime lamb properties and three mutton-sheep properties. *Campylobacter* spp. were found in all production systems and 14 of the 19 herds or flocks tested. Within individual properties there was an apparent higher prevalence in cattle than in sheep with *Campylobacter* being most frequently isolated from feedlot cattle. The median prevalences were: for dairy cattle 6 per cent, feedlot beef cattle 58 per cent, pasture beef cattle 2 per cent, mutton sheep 0 per cent and prime lambs 8 per cent.

For the four feedlots sampled, stocking density and weather conditions were identified from questionnaires as possible contributors to the number of animals shedding *Campylobacter*. The feedlot which had the lowest prevalence of *Campylobacter* (12 per cent) also had the lowest cattle density and was the only feedlot where dry weather conditions prevailed. Of the pastured animals, dairy cattle had the highest stocking rates and also the highest prevalence of *Campylobacter*.

L. monocytogenes or *Yersinia enterocolitica*, both important agents of human foodborne disease, were isolated from any of the properties.

C. jejuni causes abortion in sheep. Dogs and cats can also suffer from campylobacter diarrhoea and they may transmit it to humans if hygiene is poor.

Reservoirs of *Campylobacter* Infection

The main sources of campylobacters are:

1. Chickens
2. Turkeys
3. Raw milk

Healthy pigs, cattle, dogs, cats and wild birds are also sources. Surface waters such as rivers, lakes, *etc.*, can be contaminated, and inadequately chlorinated water supplies have been known to cause widespread outbreaks of campylobacteriosis.

Raw Food Materials Likely to be Contaminated

These include:

1. Chicken
2. Turkey and other poultry
3. Less commonly pork, beef
4. Raw milk.

Survival of *Campylobacter*

Campylobacters do not survive well in food processing environments but they do survive for several months in frozen minced meat and poultry. *Campylobacters* are very heat sensitive: for example, a ten-fold reduction in count takes approximately 6 seconds at 60 degrees Celsius. *Campylobacters* are also very sensitive to drying.

There is some evidence that *Campylobacters* may survive in viable non-culturable forms but the role of this stage in transmission is not known. As with all other infectious pathogens, campylobacters survive better under chilled conditions than at ambient. It is therefore vital to keep *Campylobacters* out of all chilled foods.

Irradiation readily destroys campylobacters and it can be used to greatly reduce the incidence on poultry; this may provide a control measure for the future.

Growth Characteristics of *Campylobacters*

The *Campylobacters* that cause enteritis in man will not grow below 30°C, and so they will not multiply on chilled food or on ambient stable foods stored below 30 degrees Celsius. The optimum growth temperature of these strains is 42°C and the maximum 47°C.

Campylobacters are acid sensitive and will not grow below pH 4.9. *Campylobacter jejuni* grows best in the pH range 6.5-7.5 and has a pH range of 4.9 to 9 0.

Campylobacters are very sensitive to salt; 2.0 per cent salt is sufficient to inhibit them, even under otherwise optimum growth conditions.

Management of Food Handlers

If a food handler is shown to have campylobacteriosis, then he/she must be excluded from work whilst symptoms persist. Such patients may return to work as soon as all symptoms have ceased, even if they are still carrying *Campylobacters*, provided that their stools are normally formed and their personal hygiene is good.

Implications for the Food Industry

Campylobacter jejuni and *Campylobacter coli* must be anticipated as transient contaminants on all kitchen surfaces and food processing equipment used for raw food materials, especially raw poultry. Although, at the moment, knowledge is incomplete as to how *Campylobacters* are transmitted, on present evidence the risk of transmission by factory heat-processed foods is very low. Certainly, it is very much lower than that from salmonellosis for the following reasons:

1. *Campylobacters* do not easily multiply in food and certainly do not grow below 30°C; thus, the organisms are unlikely to 'build-up' in a factory processing area.
2. *Campylobacters* appear to die rapidly on dry foods; this is in contrast to the prolonged survival of *Salmonellae* in dried materials.
3. The frequency of human carriers is low, and long-term carriers are uncommon, < 1 per cent. Thus, the threat from infected food handlers is even lower than that from *Salmonellae*.

With this scenario, it is vital to seek effective ways to control *Campylobacters*; it is not cost effective for food manufacturers to look for them in raw materials, factory environments or finished products.

Foodborne Campylobacteriosis–and How to Safeguard Against it

Campylobacters will not readily grow in food, so the control measures should focus on the prevention of contamination and cross-contamination. Consumers should be aware of the risk associated with consuming raw or undercooked food of animal origin. *Campylobacter jejuni* is readily destroyed by pasteurisation so the main risk is from contaminated 'high risk' products. In this respect, the application of HACCP to production, processing and distribution of poultry products is important. Consumers should be aware of the risk of eating raw or under-cooked food of animal origin; they should be advised to consume pasteurised milk, to cook meat/poultry thoroughly and to obtain water only from approved sources. The risks associated with cross-contamination from surfaces and utensils used in the preparation of raw poultry in the kitchen must also be highlighted through better education in safe food handling. Another risk factor is pets in and around the home, so they should be kept out of the kitchen.

The Genus *Yersinia*

The genus *Yersinia* is named after the French bacteriologist Alexandre Yersin who in 1894, first described the organism responsible for the bubonic plague. It was created to accommodate former members of the genus *Pasteurella* that were clearly members of the Enterobacteriaceae. In 1964, comparison of *B. enterocoliticum* with a number of closely related isolates, identified by other workers as *Pasteurella* spp., led Frederiksen to propose the creation of the new species *Yersinia enterocolitica*. Further definition within the genus has occurred with the creation of seven new species from non-pathogenic strains previously described as '*Yersinia enterocolitica*-like'. Those most commonly isolated from foods, *Y. frederiksenii, Y. intermedia, Y. kristensenii, Y. mollaretii* and *Y. bercovierii* can be readily distinguished from *Y. enterocolitica* on the basis of a few biochemical tests.

The importance of *Y. enterocolitica* as a cause of food-borne illness varies between countries. In England and Wales, laboratory reports of *Y. enterocolitica* infections, mostly sporadic cases, increased from 45 in 1980 to more than 590 in 1989 when it outnumbered cases of both *Staph. aureus* and *Bacillus* food poisoning. Although the increase may simply reflect improved detection and reporting, a much smaller increase was observed over the same period in reports of *Y. pseudotuberculosis*. Since this is detected by the same methods, it suggests that the figures indicate a genuine increase in incidence.

Yersiniosis is most common in the cooler climates of northern Europe, particularly in Belgium and in North America where a number of large outbreaks have been reported. It also displays a different seasonal variation from most other food-borne pathogens with a peak in reported cases occurring in the autumn and winter.

Yersinia enterocolitica is a member of the Enterobacteriaceae; an asporogenous, short (0.5–1.0 by 1–2 μm) Gram-negative rod which is facultatively anaerobic, catalase-positive and oxidase-negative. It can grow over a wide range of temperature, from –1°C to +40°C, with an optimum around 29°C and has a number of temperature-dependent phenotypic characteristics. For example, it is non-motile at 37°C, but motile with peritrichous flagella below 30°C. Like other psychrotrophs, though able to grow at chill temperatures, it does so slowly and at 3°C has been found to take 4 days to increase by 2-1og cycles in broth media.

Illness caused by *Y. enterocolitica* occurs most commonly in children under seven years old. It is a self-limiting enterocolitis with an incubation period of 1–11 days and lasting for between 5 and 14 days, although in some cases it may persist for considerably longer. Symptoms are predominantly abdominal pain and diarrhoea accompanied by a mild fever; vomiting is rare. Sometimes the pain resulting from acute terminal ileitis and mesenteric lymphadenitis (inflammation of the mesenteric lymph nodes) is confined to the lower right hand side of the body and prompts a mistaken diagnosis of appendicitis and subsequent surgery. A problem of post-infection complications such as arthritis and erythema nodosum (a raised, red skin lesion) can occur in adults, the latter particularly in women. This appears to be mainly associated with serotypes O_3 and O_9 and is therefore, more common in Europe.

Ingested cells of pathogenic *Y. enterocolitica* which survive passage through the stomach acid adhere to the mucosal cells of the Peyer's patches (gut-associated lymphoid tissue). Adhesion is mediated through bacterial outer membrane proteins that are encoded for on a 40–48 MDa plasmid possessed by all pathogenic *Y. enterocolitica*. The plasmid is essential but not the sole prerequisite for virulence since cell invasion is controlled by chromosomal genes. The adhered cell is taken up by the epithelial cell by endocytosis where it survives without significant multiplication and can exert cytotoxic activity. Released into the *lamina propria*, it invades phagocytic cells and multiplies extracellularly producing a local inflammatory response. Damage to the absorptive epithelial surface results in malabsorption and a consequent osmotic fluid loss characterized by diarrhoea. Other plasmid-encoded characteristics are thought to contribute to this process, such as the production of outer membrane proteins that confer, resistance to phagocytosis, autoagglutination at 37°C and resistance to serum.

A heat-stable enterotoxin (9000–9700 Da) is produced by *Y. enterocolitica* but its role in pathogenesis, if any, is unclear. It bears some similarity to *E. coli* ST immunologically and in its ability to induce fluid accumulation in ligated ileal loops and to stimulate guanylate cyclase activity. Elaboration of enterotoxin in the gut is unlikely since production usually ceases at temperatures above 30°C. Production in inoculated foods has been shown, but the observed incubation period is inconsistent with a food-borne intoxication.

A number of techniques other than biotyping and serotyping have been described which claim to distinguish pathogenic from environmental strains of *Y. enterocolitica* relatively simply and are therefore, more within the capacities of routine laboratories. These include the ability of pathogenic strains to autoagglutinate at 37°C, their dependency on calcium for growth at 37°C and their uptake of congo red dye and they are usually associated with the presence of the 40–48 MDa virulence plasmid. These tests are not completely reliable due to a number of problems such as the expression of plasmid-encoded phenotype in culture, occurrence of atypical strains and the possibility of plasmid loss during isolation. Pigs are recognized as chronic carriers of those *Y. enterocolitica* serotypes most commonly involved in human infections. The organism can be isolated most frequently from the tongue, tonsils and, in the gut, the caecum of otherwise apparently healthy animals. Despite this, pork has only occasionally been shown to be the vehicle for yersiniosis, although a case *control* study in Belgium, which has the highest incidence of yersiniosis, implicated a national prediction for eating raw pork. In 1988/9 an outbreak of yersiniosis in Atlanta involving 15 victims (14 children) was strongly associated with the household production of pork chitterlings.

A number of outbreaks of yersiniosis have been caused by contaminated milk including the largest hitherto recorded which occurred in 1982 in Tennessee, Arkansas and Mississippi in the United States. In this instance, pigs were implicated as the original source of contamination, but not demonstrated to be carriers of the same 013 serotypes causing the infection. It was presumed that the organism was transferred from pigs, via mud, onto crates used to transport waste milk from the dairy to the pig farm. The crates were returned to the dairy and inadequately washed and sanitized before being used again to transport retail milk. As a consequence the outside of packs was contaminated with *Y. enterocolitica* which was transferred to the milk on opening and pouring. It was subsequently demonstrated that the organism involved could survive for at least 21 days on the outside of milk cartons held at 4°C.

Contaminated water used in the production of beansprouts and in the packaging of tofu (soya bean curd) was responsible for two outbreaks in the United States in 1982.

A large number of procedures or the isolation and detection of *Y. enterocolitica* have been developed. Enrichment procedures usually exploit the psychrotrophic character of the organism by incubating at low temperature, but this has the disadvantage of being slow with the attendant possibility of overgrowth by other psychrotrophs present. Some workers have included selective agents in their enrichment media but some strains, such as serotype 08, are reported to be sensitive to selective agents. The most commonly used enrichment media are phosphate buffered saline (PBS) or tryptone soya broth (TSB) most usually incubated at 4°C for 21 days. *Y. enterocolitica* and related species are more alkali resistant than many other bacteria so the pH of, enrichment media is sometimes adjusted to 8.0–8.3 or cultures subjected to a short post-enrichment alkali treatment.

Table 7.9 indicates the relationship between bio-sero-phage type of *Yersinia enterocolitica* host and geographical distribution.

Table 7.9: Relationship between Bio-sero-phage Type of *Yersinia enterocolitica* Host and Geographical Distribution

Biotype	*Serotype*	*Phage Type*	*Host*	*Syndrome*	*Country*
1	08	X	Man	Gastroenteritis	U S A , Canada
2	09	X3	Man	Gastroenteritis	Europe, Japan
			Pigs	Healthy	Europe, Japan
3	01	II	Chinchillas	Systemic infection	Europe
4	0_3	VIII	Man	Gastroenteritis	Europe, Japan
			Pigs	Healthy	Europe, Japan
		IXA	Man	Gastroenteritis	South Africa
			Pigs	Healthy	South Africa
		IXB	Man	Gastroenteritis	Canada
			Pigs	Healthy	Canada
5	0_2	XI or II	Hares	Death	Europe
			Goats		

Now-a-days scombrotoxic fish poisoning is found here and there. Fish is almost always the food vehicle, particularly the so-called scombroid fish such as tuna, bonito and mackerel, but non-scombroid fish such as sardines, pilchards and herrings have also been implicated. In some cases, canned fish has been responsible indicating that the toxic factor(s) is heat stable.

It is a chemical intoxication with a characteristically short incubation period of between 10 min and 2 h. Symptoms include a sharp, peppery taste in the mouth, itching, dizziness, flushing of the face and neck, often followed by a severe headache, feverishness, diarrhoea, nausea and vomiting. A rash may develop on the face and neck and cardiac palpitations may occur.

The symptoms are those of histamine toxicity and can be alleviated with antihistamines. Histamine is produced by bacterial amino acid, decarboxylases acting on histidine which occurs in high concentrations in the tissues of dark-fleshed fish. The bacteria themselves increase in numbers as a result of long storage at inappropriate temperatures and, freshly caught fish have not been implicated in this type of poisoning.

When making judgements on the risk of scombrotoxic fish poisoning posed by particular products, regulatory authorities usually rely on a measure of the histamine content of the fish. In the United States, the level of histamine deemed hazardous in tuna is 50 mg per cent.

Table 7.10: Guideline for Histamine Levels in Fish

Histamine Level (mg per cent)	*Status*
1	Freshly caught fish
< 5	Normal and safe for consumption
5–20	Mishandled and possibly toxic
0–100	Unsatisfactory and probably toxic
> 100	Toxic and unsafe for consumption

The problem is not as clear cut as it may at first seem, however, it has, for instance, not proved possible to reproduce the symptoms in volunteers fed histamine and cases have also been reported where the fish contained low levels of histamine. Although histamine poisoning can occur when clearance of dietary histamine from the body is slowed by monoamine oxidase inhibitors, histamine is generally metabolized efficiently in the human gut and not adsorbed *per se.* Among the explanations offered are that other biologically active amines are present in the fish which potentiate the toxicity of histamine or that algal toxins may be involved causing the release of endogenous histamine in the body.

The *Aeromonas*

Currently, *Aeromonas* (principally *A. hydrophila,* but also *A. caiae* and *A. sobria*) has the status of a food borne pathogen of emerging importance. Like *Listeria monocytogenes, Plesiomonas* and *Yersinia enterocolitica,* it has attracted attention primarily because of its ability to grow at chill temperatures, prompting the concern that any threat it might pose will increase with the increasing use of chilled foods. Present uncertainty over its significance however, is reflected in much of the information available which does not, as yet, present a coherent picture. It was first isolated from drinking water by Zimmerman in 1890 and the following year from frog's blood by Sanarelli. They called their isolates *Bacillus punctata* and *Bacillus hydrophilus,* respectively and it was not until the 1930s that the genus *Aeromonas* was first described. Although the taxonomy is still not settled, more recent studies have led to the recognition of two major groups within the genus: the Salmonicida group, which contains the non-motile *Aeromonas salmonicida* and several sub-species and the Hydrophila-Punctata group containing a number of motile species, including *A. hydrophila, A. sobria* and *A. caviae.*

Epidemiological investigations in several countries have reported higher rates of isolation of aeromonads from patients with diarrhoea than from control groups, although this does not necessarily indicate a causal relationship. In one study, the incidence of *A. hydrophila* in Amercian travellers to Thailand with diarrhoea was significantly higher than in unaffected individuals. Interestingly, isolation rates in the Thai popualtion were similar for both groups of patients suggesting that *Aeromonas* may be a cause of 'travellers' diarrhoea' in these regions.

Good supporting evidence from sources other than epidemiological studies has proved difficult to obtain. In a feeding trial involving 50 volunteers with doses as high as 5 × 10_{10}, only two cases of diarrhoea resulted, although a laboratory accident has been reported where approximately 10_9 cells were ingested by a worker while mouth pipetting who later suffered acude diarrhoea. Aeromonads are Gram-negative, catalase-positive, oxidase-positive rods which ferment glucose. They are generally motile by a single polar flagellum.

Table 7.11: Reported Lag Times and Growth Rates of Psychrotrophic Pathogens at Chill Temperatures

	Lag Time (Days)			*Generation Time (h^{-1})*	
Temperature (°C)	0–1	2–3	5	0–1	4–5
Aeromonas hydrophila	> 22	6–10	3–4	> 49	9–14
Listeria monocytogenes	3–33	2–8	1–3	62–131	13–25
Yersinia enterocolitica	3	2.4	–	25	25

A. hydrophila is neither salt (< 5 per cent) nor acid (min. pH 6.0) tolerant and grows optimally at around 28°C. Its most significant feature with regard to any threat it may pose in foods is its ability to grow down to chill temperatures, reportedly as low as –0.1°C in some strains. Its principal reservoir is the aquatic environment such as freshwater lakes and streams and wastewater systems. The numbers present will depend on factors such as the nutrient level and temperature but can be as high as 10_8 cfu ml^{-1} in a relatively nutrient rich environment such as sewage. Although it is not resistant to chlorine, it is found in potable water, where it can multiply on the low level of nutrients available in piped water systems. It has also been isolated from a wide range of fresh foods and is a transient component of the gut flora of humans and other animals.

Apart from their possible role in gastroenteritis, food and water are also probably the source of the severe extra-intestinal *Aeromonas* infections associated with immunocompromised individuals.

Aeromonads of the Hydrophila group have been isolated from a wide range of fresh foods including fish, meat, poultry, raw milk and salad vegetables as well as water. The ability of some strains to grow at very low temperatures can lead to the development of high numbers under chill conditions and they can be an important part of the spoilage flora of chilled meats.

They are unlikely to survive even mild cooking procedures but may be introduced as post-process contaminants from uncooked produce or contaminated water.

Chapter 8

Enumeration of Micro-organisms in Food

Increasing attention is being paid to microbiological quality control in food processing establishments throughout the world. Hence, it is frequently necessary to conduct a microbiological examination of food to determine its quality. This may be necessary to estimate its shelf life, its suitability for human consumption or to confirm that it meets some established microbiological criterion.

Now a days, there is an increasing trend to consume ready-to-eat foods. The microbiological testing of ready-to-eat foods should be appropriate to the type of food sample being examined and to the processing it has received. In other foods, knowledge of the total count may be useful but it is often of more value to obtain an estimate of the numbers of a particular component of the total flora. Examples of this could be moulds in a cereal, psychrotrophic bacteria in a chilled food, anaerobes in a vacuum-packed food, or yeasts in a fruit beverage.

A quite different reason for a microbiological examination of a food may be to identify the cause of spoilage or the presence of a pathogen where a food has been implicated in foodborne illness. There are various methods for microbiological examination of foods. To estimate the numbers of viable bacteria in large numbers of food samples, an efficient, simple, rapid and relatively inexpensive method is desirable.

Conventional testing methods make use of growth media and cultivation to enumerate and isolate cells of microorganisms from food samples brought into suspension with diluents. Isolated colonies of microbes are then subjected to a series of biochemical, physiological and serological tests in order to find species identity and subtypes of the microorganisms in question. These methods can provide both quantitative and qualitative information on the importance of microorganisms present in foods, and have been verified over the years. However, testing methods relying on cultivation are generally slow and give results after a period of several days; moreover, they use a lot of material and labour. Modern quality management and control systems, such as good manufacturing practice (GMP) and hazard analysis and critical control point (HACCP) systems require methods and techniques that can be used on-line and give results in real-time. Hence, food microbiologists seek more rapid, sensitive and specific methods to get adequate information in due time to monitor the safety and quality of products. In recent decades a number of improved conventional and alternative non-traditional methods and techniques have been developed that appear suitable for early detection and characterization of microorganisms significant in foods. This chapter will briefly summarize the very broad and growing field of microbiological testing methods.

The Counting of Microcolonies

Observing microbial colonies is the standard method for determining the microbe titer and investigating the behaviours of microbes. Microbiological research techniques often rely on the accurate determination of the number of colony forming units (CFUs). Bacterial growth is an essential indicator for the selection of antibiotics, toxicology tests, and the evaluation of food and drug safety. Counting visible microbial colonies (which can be sampled from various sources, such as water, air, and soil) grown on semi-solid agar-based growth media is the conventional method for quantitative microbiological analysis of a broad spectrum of prokaryotic and eukaryotic microbes.

Microscope colony count methods involve the counting of microcolonies that develop in agar layered over microscope slides. The first was that of Frost, which consisted of spreading 0.1 ml of a milk-agar mixture over a 4 cm_2 area on a glass slide. Following incubation, drying and staining, microcolonies are counted with the aid of a microscope.

In another method, 2 ml of melted agar are mixed with 2 ml of warmed milk and after mixing, 0.1 ml of the inoculated agar is spread over a 4 cm_2 area. Following staining with thionin blue, the slide is viewed with the 16 mm objective of a wide-field microscope.

Let us study about cultural and microscopic methods which have been given as under:

1. Cultural and Microscopic Methods

The four basic type of methods included in this category are:

Standard Plate Count (SPC)

SPC is useful in determining the viable number of microorganisms in foods. The procedure consists of diluting the organisms in a series of sterile dilution blanks. From blanks, measured amounts of diluted organism are transferred by spread plating/pour plating over the appropriate agar plates and incubated for 24-48 h. Thereafter plates are examined and number of colonies is counted to calculate the number of viable cells or colony forming units per unit volume of material under test. The counts obtained in this method depend on some factors: nature of food flora and material, plating medium used, pH, a^w and Eh of the medium, type of diluents and temperature of incubation. The disadvantages of the method are that the technique is cumbersome, material intensive and time consuming.

Most Probable Number (MPN) Method

In this method, dilutions of food samples are prepared as for the SPC. Three serial aliquots or dilutions are then planted into nine or fifteen tubes of appropriate medium for the three- or five-tube method, respectively. Numbers of organisms in the original sample are determined by use of standard MPN tables. The method is statistical in nature and MPN results are generally higher than SPC.

This method was introduced by McCrady in 1915. This is a qualitative method for determination of the presence of coliforms in potable water. The selective medium used is MacConkey's medium which contains a bile salt inhibitory for growth of non-intestinal lactose fermenting bacteria. Since the method is statistical in nature, MPN results are generally higher than SPC. The test can be performed in two formats- three tube and five tube method. It is not a precise method of analysis–the 95 per cent confidence intervals for a three-tube test range from 21 to 395. When the three-tube test is used, 20 of the 62 possible test combinations account for 99 per cent of all results, while with the five-tube test, 49 of the possible 214 combinations account for 99 per cent of all results. In a collaborative study on coliform densities in foods, a three-tube MPN value of 10 was found to be as high as 34, while in another phase of the study, the upper limit could be as high as 60. Although Woodward concluded that many MPN values are improbable, this method of analysis has gained popularity over the past decade. Among the advantages, it offers are the following:

1. It is relatively simple.
2. Results from one laboratory are more likely than SPC results to agree with those from another laboratory.
3. Specific groups of organisms can be determined by use of appropriate selective and differential media, and
4. It is the method of choice for determining faecal coliform densities.

Among the drawbacks to its use is the large volume of glassware required (especially for the five-tube method), the lack of opportunity to observe the colonial morphology of the organisms, and its lack of precision.

Direct Microscopic count (DMC)

Another quick and precise method for determination of microorganisms in liquid foods is the use of haemocytometer or Neubauer counting chamber. The method is quick and requires expertisation. In the method appropriately diluted samples are transferred on the haemocytometer using a Pasteur pipette and counted directly under the microscope. The haemocytometer is itched with a grid system into number of squares (25 large squares). The number of microorganisms present in the large square are counted and multiplied with the factor for large square and 32 dilution factor to determine the total number of microorganisms in the sample. Among the advantages of DMC are: it is rapid and simple and cell morphology can be assessed. Among its disadvantages are: it is microscopic method and therefore fatiguing to the analyst; both viable and nonviable cells are enumerated; food particles interfere with the microbial counts; and DMC counts are invariably higher than counts by SPC.

Agar Droplet Method

In the agar droplet method of Sharpe and Kilsby, the food homogenate is diluted in tubes of melted agar (at 45°C). For each food sample, three tubes of agar are used, the first tube being inoculated with 1 ml of food homogenate. After mixing, a sterile capillary pipette (ideally delivering 0.033 ml/drop) is used to transfer a line of 5 × 0.1 ml droplets to the bottom of an empty petri dish. With the same capillary pipette, three drops (0.1

ml) from the first 9 ml tube are transferred to the second tube and after mixing, another line of 5 × 0.1 ml droplets is placed next to the first. This step is repeated for the third tube of agar. Petri plates containing the agar droplets are incubated for 42-h and colonies are enumerated with the aid of a 10 × viewer. Results using this method from pure cultures, meats and vegetables compared favourably to those obtained by conventional plate counts; droplet counts from ground meat were slightly higher than plate counts. The method was about three times faster and 24-h incubations gave counts equal to those obtained after 48 h by the conventional plate count. Dilution blanks are not required and only one petri dish/sample is needed.

Dry Film Method

A dry film method consisting of two plastic films attached together on one side and coated with plate count agar ingredients and a cold-water-soluble jelling agent has been developed by the 3M Company (Petrifilm). If found successful on further testing, this method will simplify greatly the SPC. For use, 1 ml of diluent is placed between the two films and spread over the nutrient area by pressing. Following incubation, microcolonies appear red because of the presence of a tetrazolium dye in the nutrient phase. The method was compared to conventional SPC on 196 samples of raw milk and the correlation coefficient was 0.96. A Petrifilm-VRB has been compared to violet red bile agar (VRBA) and MPN for coliform enumeration on 120 samples of raw milk. Plate counts by Petrifilm-VRB compared favorably to VRBA counts and both were comparable to MPN results. A plot of coliform counts by Petrifilm-VRB against VRBA yielded a regression line with slope of 0.959. In another study employing 108 milk samples, the dry film method was compared with SPC and a standard plate loop method. The correlation coefficient between dry film and SPC was 0.946 and the mean log difference was – 0.177. These investigators found the film method to be a suitable alternative to SPC.

2. Alternative Methods

Cultural methods are relatively labour intensive and require time for adequate growth to occur. Many food microbiologists also consider that the traditional enumeration methods are not only too slow but lead to an overdependence on the significance of numbers of colony-forming units.

Food manufacturers require information about the microbiological quality of commodities and raw materials rapidly and it could be argued that an assessment of microbial activity is as important as a knowledge of numbers. A number of methods have been developed which aim to give answers more quickly and hence are often referred to as 'Rapid Methods'. These are discussed as under:

I. Chemical Methods

i) Thermostable Nuclease

This method can be used to detect the presence of *S. aureus* in foods. This is possible because of the high correlation (95 per cent) between the production of coagulase and thermostable nuclease by enterotoxin producing *S. aureus* strains. The method is a spectrophotometric method and the amount of extractable thermostable nuclease increases with the numbers of cells in foods. The advantages of testing for heat stable nuclease as an indicator of *S. aureus* growth and activity includes:

1. Because of its heat stable nature, the enzyme will persist even if the bacterial cells are destroyed by heat, chemicals, or bacteriophage, or if they are induced to L-forms;
2. The heat-stable nuclease can be detected faster than enterotoxin (about 3 h versus several days),
3. The nuclease appears to be produced by enterotoxigenic cells before enterotoxins appear;
4. The nuclease is detectable in unconcentrated cultures of food specimen while enterotoxin detection requires concentrated samples; and
5. The nuclease of concern is stable to heat as are the enterotoxins.

There is high correlation between coagulase positive enterotoxigenic strain and thermostable nuclease producing strain.

ii) Limulus Lysate for Endotoxins

Gram negative bacteria are characterized by their production of endotoxins which consists of LPS layer of the cell envelope. The Limulus amoebocyte lysate (LAL) test employs a lysate protein obtained from the blood cells of the horseshoe crab. The lysate protein is the most sensitive substance known for endotoxins (3-300 times more sensitive to endotoxins than the U.S. Pharmacopeia rabbit pyrogen test). The LAL test is performed by adding aliquots of food suspensions or other test materials to small quantities of a lysate preparation followed by incubation at 37°C for 1 h. The presence of endotoxins caused gel formation of the lysate material. The LAL test has been found to be suitable for the rapid evaluation of the hygienic quality of milk relative to the detection of coliforms before and after pasteurization. Since both viable and nonviable gram-negative bacteria are detected by LAL, simultaneous plating is necessary to determine the numbers of cfu's. The method has been applied successfully to detect the 34 microbial spoilage of ground beef; to monitor milk and milk products; microbial quality of raw fish; and cooked turkey rolls.

iii) Dye Reduction Method

A group of tests which depend on the response of a number of redox dyes to the presence of metabolically active micro-organisms. The redox dyes are able to take up electrons from an active biological system and this results in a change of colour.

In dye-reduction method, three redox dyes *i.e.*, methylene blue, resazurin and triphenyltetrazolium chloride are commonly used. To conduct dye-reduction test, properly prepared supernatants of foods are added to standard solutions of either dye for reduction from blue to white for methylene blue and from slate blue to pink or white for resazurin. The time for dye reduction to occur is inversely proportional to the number of organisms in the sample.

From 1937, and until relatively recently, the methylene blue test was a statutory test for grading the quality of milk in England and Wales. Changes in the technology of handling bulk milk, especially refrigeration, have made this test less reliable and it is no longer a statutory test because results show little correlation with the numbers of psychrotrophic bacteria.

Because the reduction of resazurin takes place in two stages, from blue to pink to colourless, there is a wider range of colour that can be scored using a comparator disc and the ten-minute resazurin test is still useful for assessing the quality of raw milk at the farm or dairy before it is bulked with other milk.

Triphenyltetrazolium salts and their derivatives are initially colourless and become intensely coloured, and usually insoluble, after reduction to formazans. Triphenyltetrazolium chloride itself is most widely used as a component of diagnostic and selective agar media on which some bacterial colonies will become dark red to maroon as formazan becomes precipitated within the colony.

The crystals of the formazan produced from 2-(*p*-iodo-phenyl)-3-(*p*– nitro-phenyl)-5-phenyl-tetrazolium chloride (INT) are so intensely coloured that they are readily seen in individual microbial cells under the microscope and their presence may be used to assess the viability of cells.

Methylene blue and resazurin reduction by 100 cultures was studied in milk; with two exceptions, a good agreement was found between numbers of bacteria and time needed for reduction of the two dyes. In a study of resazurin reduction as a rapid method for assessing ground beef spoilage, reduction to the colorless state, odor scores and SPC correlated significantly. One of the problems of using dye reduction for some foods is the existence of inherent reductive substances. This is true of raw meats and Austin and Thomas reported that resazurin reduction was less useful than with cooked meats. For the latter, approximately 600 samples were successfully evaluated by resazurin reduction by adding 20 ml of a 0.0001 per cent resazurin solution to 100 g of sliced meat in a plastic pouch. Another way of getting around the reductive compounds in fresh meats is to homogenize samples by Stomacher rather than by Waring blender. Using Stomacher homogenates, raw meat was successfully evaluated by resazurin reduction when Stomacher homogenates were added to a solution of resazurin in 10 per cent skim milk. Stomacher homogenates contained less disrupted tissue and consequently lower concentrations of reductive compounds. The method of Holley *et al.*, was evaluated further by Dodsworth and Kempton who found that raw meat with an SPC > 10_7 bacteria/g could be detected within 2 h.

Dye-reduction tests have a long history of use in the dairy industry for assessing the overall microbial quality of raw milk. Among their advantages are: they are simple, rapid and inexpensive; and only viable cells actively reduce the dyes. Disadvantages are: not all organisms reduce the dyes equally; and they are not applicable to food specimens that contain reductive enzymes unless special steps are employed as noted above.

iv) ATP Determination/Bioluminescence Method

The body is a complex organism, and as such, it takes energy to maintain proper functioning. In addition to providing energy, the breakdown of ATP through hydrolysis serves a broad range of cell functions, including signalling and DNA/RNA synthesis. ATP synthesis utilizes energy obtained from multiple catabolic mechanisms, including cellular respiration, beta-oxidation, and ketosis.

Adenosine-triphosphate (ATP) is the primary source of energy in all living organisms. Adenosine triphosphate is found in all living cells and is the universal agent for the transfer of free energy from catabolic processes to anabolic processes. The structure of ATP is a nucleoside triphosphate, consisting of a nitrogenous base (adenine), a ribose sugar, and three serially bonded phosphate groups. ATP is an excellent energy storage molecule to use as "currency" due to the phosphate groups that link through phosphodiester bonds. These bonds are high energy because of the associated electronegative charges exerting a repelling force between the phosphate groups. A significant quantity of energy remains stored within the phosphate-phosphate bonds. It disappears within 2 h after cell death and the amount per cell is generally constant, with values of 10–18 to 10–17 mole/bacterial cell which corresponds to around 4 × 104 M ATP/105 cfu of bacteria. Through metabolic processes, ATP becomes hydrolyzed into ADP, or further to AMP, and free inorganic phosphate groups. The process of ATP hydrolysis to ADP is energetically favourable, yielding Gibbs-free energy of -7.3 cal/mol. ATP must continuously undergo replenishment to fuel the ever-working cell. Many feedback mechanisms are in place to ensure the maintenance of a consistent ATP level in the cell. The enhancement or inhibition of ATP synthase is a common regulatory mechanism.

Among procaryotes, ATP in exponentially growing cells is regularly around 2–6 n mole ATP/mg dry weight regardless of mode of nutrition. In the case of rumen bacteria, the average cellular content was found to be 0.3 fg/cell, with higher levels found in rumen protozoal cells. The complete extraction and accurate measurement of cellular ATP can be equated to individual groups of microorganisms in the same general way as endotoxins for gram-negative bacteria. Almost constant level of ATP in the living cell is the basis of bioluminescence rapid method for detecting bacteria.

The microbial ATP (mATP) bioluminescence assay is a technique for measuring levels of total generic bacteria. This technique has a history of use in a variety of industrial applications, including foods and sanitation or hygiene monitoring. Assaying mATP has been shown to be efficacious for detecting generic microbial loads directly on beef carcass tissue inoculated with faeces. An advantage in using mATP as a microbial monitor of carcass processing is rapidity. To be an effective (HACCP) monitoring tool in a processing plant, a carcass assay should require substantially less than I h to complete. Previous work validating an mATP test for carcass tissue required approximately 55 min to complete the assay.

Bioluminescence is a very rapid and sensitive method for bacterial detection. This assay is based upon the assumption that living cells of a given type contain a reasonably constant amount of adenosine 5'triphosphate (ATP), which is lost rapidly upon cell death. One of the simplest ways to measure ATP is by use of the firefly luciferin-luciferase system. A number of quite different living organisms have evolved a mechanism for producing light by the activity of enzymes known as luciferases on substrates known as luciferins.

These reactions require the presence of ATP and magnesium ions and produce one photon of light at the expense of the hydrolysis of one molecule of ATP through a series of intermediates. An ATP molecule facilitates the formation of an enzyme- substrate complex which is oxidized by molecular oxygen to an electronically excited state.

The excited state of the molecule returns to the lower energy ground state with the release of a photon of light and dissociates to release the enzyme luciferase again.

Hence, the principle of the ATP measurement is its reaction with luciferin catalysed by luciferase enzyme (Miller and Gladston, 1989): luciferase luciferin. Use of firefly luciferase to assay adenosine triphosphate (ATP) extracted from microorganisms provides an easy means to enumerate microbes within minutes. In the presence of ATP, luciferase emits light, which is measured with a liquid scintillation spectrometer/luminometer. The amount of light produced by firefly luciferase is directly proportional to the amount of ATP added and thus microbial number. The average bacterium contains around 10^{-15} g ATP per cell. Present reagents permit detection of 10^3 cells per tube. Luminometers currently on the market detect about 10^{-12} g ATP. This is a rapid method for assessing biomass in activated sludge and estimation of microorganisms in foods.

These reactions are summarized below:

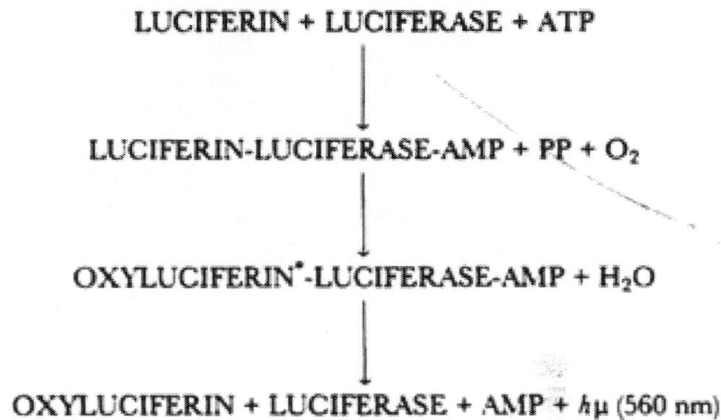

Luciferin (reduced) + ATP + O_2 $\xrightarrow{\text{Luciferase}}$ luciferin (oxidized) + AMP + CO_2 + PPi + photon

One photon of light is produced per molecule of ATP hydrolysed which can be measured using a photometer (Karl, 1980), giving a sensitivity of about 10^{-4} mol of ATP. The light emitted is proportional to the amount of ATP present when all the other reactants are in excess. Knowing the concentration of ATP in a sample, an estimate of the microbial content can be made. This is one of the simplest ways to measure ATP *i.e.* by use of the firefly luciferin-luciferase system. In the presence of ATP, luciferase emits light, which is measured with a liquid scintillation spectrometer or a luminometer. The amount of light produced by firefly luciferase is directly proportional to the amount of ATP added. Cellular ATP is extracted generally by boiling and other details of its assay have been presented by Kimmich *et al.,* and Karl.

Proper extraction of ATP from the microbes is an essential part of any protocol, as is the removal of non-microbial ATP from, for example, somatic cells also present in samples. The technique may be applied to a wide range of samples, for example food and beverages and clinical samples such as urine. The ATP assay gives a global measure of microbial numbers, *i.e.* it is not species specific unless a species separation step is included in the protocol (Stanley, 1989).

A new microbial ATP bioluminescence assay was shown to be an accurate and rapid method to determine the levels of generic bacterial contamination on beef (*n* =

400 and pork ($n = 320$) carcasses sampled in commercial processing plants. Based on *in vitro* faecal dilution studies, the rapid microbial ATP (R-mATP) assay is as accurate as the standard plate count method for estimating bacteria in bovine or porcine faecal samples.

Because instruments are now available which can accurately measure low levels of light emission and pure luciferin and luciferase from, for example, fireflies can be manufactured, the reaction can be used as a very sensitive assay for ATP. Sensitive instruments using photomultiplier tubes can detect as little as 10^2-10^3 fg (fg = femtogram = 10^{-15} g) which corresponds to as few as 10^2— 10^3 bacterial cells. The Lumac Biocounter (Lumac B.V., Schesberg, The Netherlands) and the 'Unilite' (Biotrace Ltd., Bridgend, UK) are two instruments developed for the estimation of microbial biomass based upon the bioluminescence principle. The 'Unilite' is the first detection system of its kind which is specifically designed to measure an entire swab directly, with no dilution or transfer losses. Due to the sensitivity of the technique, both the Biocounter and the 'Unilite' can detect low levels of microorganisms (in the region of 10^3 cells ml-1 assuming an ATP content of 4 nmol mg-1 dry weight and an average dry weight of bacteria 6-4 × 10^{11} cells g-1 (Thierry and Chicheportiche, 1988). With some samples, results can be obtained in 10 minutes, but in the case of milk or urine samples, the assay procedure takes longer as it is necessary to pre-treat the samples to remove non-bacterial ATP present in somatic cells.

The application of ATP measurement as a rapid method for estimating microbial numbers has been developed and used more in clinical microbiology than other areas. In the clinical laboratory, it has been employed to screen urine specimens. The numbers of bacteria in urine are considered to be of etiologic importance if they occur at a level of 10_5 cfu/ml urine. In one study of 348 urine specimens, only 89.4 per cent agreement with culture methods was found. Using 10_5 organisms/ml or above to indicate bacteriuria and < 10_5 as negative, the ATP assay yielded 7 per cent false positive and 27 per cent false negative results. Thore *et al.,* assayed 2,018 urine specimens for ATP by the luciferase method setup to distinguish between samples containing > 10_5 cfu/ml and those containing < 10_5 cfu/ml. Ninety-two per cent of those with > 10_5 cfu and 88 per cent of those with < 10_5 were correctly classified by the ATP assay. The assay was a 15 min test and samples were set at 13.5 nM ATP to define the limit between negative and positive results. These investigators and others cited by them found ATP analysis for bacteriuria to agree well with plate count results. In a comparative study of plate counts, the LAL test and ATP measurement to rapidly assess contamination of intravenous fluids, the LAL test was found to be more sensitive than ATP, but the authors suggested that ATP measurements could be used.

The successful use of the method for bacteriuria and for assessing biomass in activated sludge suggests that it should be of value for foods. It lends itself to automation and represents an excellent potential method for the rapid estimation of microorganisms in foods. Although the method gives very good results with pure cultures, when applied to foods it is essential to ensure that non-microbial ATP, which will be present in foods in considerably larger quantities than the microbial ATP, has been destroyed or that the micro-organisms have been separated from interfering food components. Here comes the major problem that has to be overcome for food use which is the removal of non-microbial ATP. The method was suggested to be of value for food use by Sharpe *et al.,* provided that microbial ATP levels are as great or greater than the intrinsic ATP in the food itself. They found that intrinsic levels decreased with food storage. Non-microbial ATP can be selectively removed by treating with an ATPase after disrupting the somatic cells of animal and plant origin with a mild surfactant. The next stage is to destroy the ATPase activity and then extract the microbial ATP using a more powerful surfactant. The alternative, of removing microbial cells from the food before the ATP assay, can be achieved by centrifugation or filtration of liquid foods but is very much more difficult from suspensions of solid foods. Even when they are successfully separated there are problems arising from the different amounts of ATP in different microbial cells.

A method for removing specific groups of bacteria, which is being actively researched, is to use antibodies or lectins attached to magnetic beads. Lectins are a class of plant proteins which recognize and bind to specific carbohydrate residues which may be exposed on the outer surface of a micro-organism. The bacteria adhere to the beads via the antibody-antigen, or lectin-carbohydrate reaction and can be removed from the food suspension by a powerful magnet acting through the walls of the container. After the food materials have been poured away and the cells washed, they can be released into suspension for assay by removing the magnet.

This need for often complex sample preparation has meant that, rapid and sensitive though it is, ATP measurement is not widely used for routine monitoring of microbial contamination of foods.

It is however being increasingly used to monitor hygiene in food-processing operation plant. Instruments are now available where a swab taken from equipment can be assayed directly for ATP giving a virtually immediate measure of surface contamination.

In these cases, it is not necessary to distinguish between microbial and non-microbial ATP since the presence of either at high levels would indicate poor hygiene. This speed and simplicity make ATP determination the most overtly microbiological test that can be applied for the routine-monitoring of critical control points as part of the HACCP technique of quality assurance.

The utility of employing ATP analysis for foods was questioned by Williams too who called attention to the following:

1. One yeast strain was found to contain 300 times more ATP than the average for bacterial cells (some strains may contain as much as 1,000 times more). Thus, yeast cells may contain 100 times more ATP than bacterial cells and sub-lethally stressed micro-organisms may contain very low levels of ATP and yet be capable of recovering and growing on a food during long-term storage.
2. Sterile 0.1 per cent peptone diluent gave readout values equivalent to about 200,000; organisms/ml;
3. Heat-killed bacteria contain ATP; and
4. ATP is found in food itself and there are interfering substances in foods such as potato starch and milk.

The utility of the method for measuring rumen biomass has been questioned because of the discrepancy between the bacterial and protozoal content of ATP.

The problems noted above have been overcome to a large extent. Thore *et al.,* used Triton X–100 and apyrase to selectively destroy non-bacterial ATP in urine specimens and found that the resultant ATP levels were close to values observed in laboratory cultures with detection at 10_5 bacteria/ml.

In meats, the problem of non-microbial ATP was addressed by Stannard and Wood by use of a three-stage process consisting of centrifugation, use of cation exchange resin and filtration to get rid of food particles and collect bacteria on 0.22 μm filters. ATP analyses were carried out on bacteria eluted from the filter membranes and 70–80 per cent of most microorganisms were recovered on the filters. Linear relationship was shown between microbial ATP and bacterial numbers over the range 10_6–10_9 cfu/g. By the methods employed, results on ground beef were obtained in 20–25 min. In another study, seventy-five samples of ground beef were evaluated and a high correlation was found between $\log_{10}$ APC and $\log_{10}$ ATP when samples were incubated at 20°C. In this study, the amount of ATP/cfu ranged from 0.6 to 17.1 fg with fifty-one of the seventy-five samples containing ≤ 5.0 fg of ATP.

ATP analysis was combined with benzalkon-crystal violet (BC-ATP) to predict keeping quality of pasteurized milk and results were compared to those from two commonly used shelf-life tests. The BC-ATP results obtained within 24 h were as good as those obtained from the shelf-life tests requiring 5 and 10 days for results. These investigators found that at an ATP content of < 1,000 relative light units, it was always accompanied by bacterial counts of < 10_6/ml.

At the present time, ground beef can be extracted and read in 20 min and as few as 10_4 organisms/g can be detected by currently available instrumentation. However, these methods may not be useful when applied to seafoods that may contain luminescent bacteria such as *Photobacterium fisheri* and *Vibrio harveye*, for these organisms produce luciferase.

Hence, the bioluminescent assay of ATP is rapid and simple and may be used as an estimate of microbial numbers. It therefore shows great potential as a technique to provide information on the microbiological quality of a food within a few minutes, in comparison with conventional techniques, which provide results retrospectively. However, despite the advantages of speed and sensitivity, no food microbiologists are using the technique for routine quality control and hygiene monitoring. There is a need for future research to increase the acceptance of ATP assays within the food industry.

v) Radiometry

The automation in the clinical microbiology has eased the diagnosis and treatment of the patients by providing the information with rapidity, accuracy, and uniformity. The radiometric methodology is now utilized in variety of microbiological techniques such as to detect bacteria in blood, rapid antibiotic susceptibility test, detection of bacteria in food and cosmetics, detection of white cell metabolism and sterility testing of radiopharmaceuticals. This technique was developed and used by NASA to search life on planets (Mars). The results with the radiometric method of detection of bacteria has been reported by number of workers.

The radiometric detection of microorganisms is based upon the incorporation of $_{14}$C-labelled metabolite in a growth medium so that when the organisms utilize this metabolite, $_{14}CO_2$ is released and measured by use of a radioactivity counter. For organisms that utilize glucose, $_{14}$C-glucose is usually employed. For those that cannot utilize this compound, others such as $_{14}$C-formate or $_{14}$C-glutamate are used. The overall procedure consists of using capped 50- ml serum vials to which are added anywhere from 12 to 36 ml of medium containing the labelled metabolite. The vials are made either aerobic or anaerobic by sparging with appropriate gases and are then inoculated. Following incubation, the headspace is tested periodically for the presence of $_{14}CO_2$. The time required to detect the labelled CO_2 is inversely related to the number of organisms in a product.

The use of radiometry to detect the presence of microorganisms was first suggested by Levin *et al.,* It is confined largely to clinical microbiology but some applications have been made to foods and water. One of the earliest non-clinical uses was the detection of coliforms in water and sewage. These investigators showed that a direct relationship existed between coliform numbers and the amount of $_{14}CO_2$ produced by the organisms. The experimental detection of *S. aureus, S. typhimurium* and spores of putrefactive anaerobe 3679 and *Clostridium botulinum* in beef loaf was studied by Previte. The inocula employed ranged from about 10_4–10_6/ml of medium and the detection time ranged from 2 hr for *S. typhimurium* to 5–6 hr for *C. botulinum* spores. For these studies, 0.0139 μCi of $_{14}$C-glucose/ml of tryptic soy broth was employed. In another study, Lampi *et al.,* found that 1 cell/ml of *S. typhimurium* or *S. aureus* could be detected by a radiometric method in 9 h. For 10^4 cells, 3 to 4 h were required. With respect to spores, a level of 90 of P.A. 3679 was detected in 11 h while 10^4 were detectable within 7 h. These and other investigators have shown that spores required 3–4 h longer for detection than vegetative cells. From the findings of Lampi *et al.,* the radiometric detection procedure could be employed as a screening procedure for foods containing high numbers of organisms, for such foods produced results by this method within 5–6 h, while those with lower numbers required longer times.

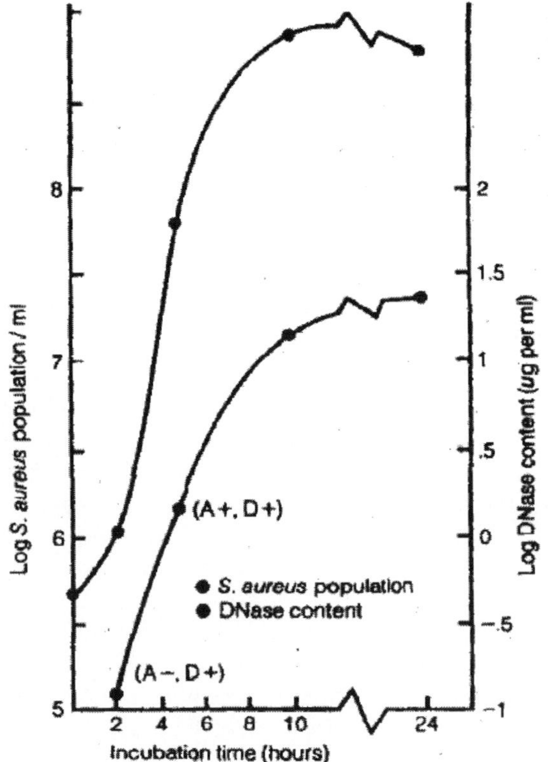

Figure 8.1: Growth of S. aureus (196E) and Production of DNAse and enterotoxin in Brain Heart Infusion Broth at 37ºC. DNAse and enterotoxin D were detectable within 2 h at a population of 2 × 106, whereas enterotoxin. A was detected after 4 h at higher cell populations. DNAse was detectable in unconcentrated cultures, enterotoxins in fifty-fold concentrates.

The detection of non-fermenters of glucose by this method is possible when metabolites such as labelled formate and/or glutamate are used. These investigators further showed that a large number of food-borne organisms can be detected by this method in 1 to 6 h. The radiometric detection of 1 to 10 coliforms in water within 6 h was achieved by Bachrach and Bachrach by employing $_{14}$C-lactose with incubation at 37°C in a liquid medium. It is conceivable that a differentiation can be made between faecal *E. coli* and total coliforms by employing 45.5°C incubation along with 37°C incubation.

Radiometry has been used to detect organisms in frozen orange juice concentrate. The investigators used $_{14}$C-glucose, 4 yeasts and 4 lactic acid bacteria and at an organism concentration of 10_4 cells, detection was achieved in 6–10 h. Of 600 juice samples examined, 44 with counts of 10_4/ml were detected in 12 h and 41 of these in 8 h. No false negatives occurred and only two false positives were noted. The method was used for cooked foods to determine if counts were < 10^5 cfu/ml and the results were compared to APC. Of 404 samples consisting of seven types of foods, around 75 per cent were correctly classified as acceptable or unacceptable within 6 h. No more than five were incorrectly classified. The study employed $_{14}$C-glucose, glutamic acid and sodium formate.

A method for the detection of salmonellae in foods based on salmonellae poly H antibodies preventing salmonellae from producing $_{14}CO_2$ from ($_{14}$C) dulcitol has been developed. It was shown capable of detecting the organisms in 27 h compared to 4–5 days by culture methods. The agreement with culture methods on fifty-eight samples of food was 91 per cent. The radiometric assay flask contained poly H serum and a control flask without poly H. Positive tests were determined by the difference in radioactivity of the two flasks following incubation at 37°C for 7 h. Results were achieved with an inoculum of 10_4–10_5 cells. The potential use of radiometry for field and surface samples is suggested by a procedure employed by Schrot *et al.,* By this method, membrane filter samples are collected and moistened with a small quantity of labelled medium in a closed container. The evolved $_{14}CO_2$ is trapped by Ba $(OH)_2$-moistened filter pads which are assayed later by a radioactivity counter.

vi) Glucuronidase Assay for *E. coli*

The enzyme β-glucuronidase (GUD) is produced by 97 per cent of *E.coli* strains and upto 50 per cent of salmonellae and shigellae. In the presence of the substrate 4-methylumberlliferone-β-Dglucuronide (MUG), GUD produces a fluorogenic end product which is visible in UV light. Thus, it is a rapid and more efficient way to detect *E.coli* in foods; one *E.coli* cell could be detected in 20 h. While most positive reactions occurred in 4 h, some weak GUD-positive strains required up to 16 h for reaction. The main advantage of this method is that fluorescence appears before gas production from lactose. Also, some salmonellae and shigellae are GUD positive does not invalidate the method since these organisms are of greater significance in foods than coliforms.

II. Electrical Methods

When micro-organisms grow, their activity changes the chemical composition of the growth medium and this may also lead to changes in its electrical properties. Measuring this effect has become the basis of one of the most widely used alternative techniques of microbiological analysis.

The electrical properties most frequently monitored are conductance (G), capacitance (C) and impedance (Z), the latter being influenced by both capacitance and resistance (R) as well as the frequency of the alternating current applied (f). The conductance is simply the reciprocal of resistance, *i.e.*

$$G = 1/R \qquad \text{(i)}$$

The relationship between impedance, resistance and capacitance is given by:

$$Z^2 = X^2 + (1/2\pi fC)^2 \qquad \text{(ii)}$$

It is possible to take frequent measurements of the electrical properties of a growth medium by growing organisms in cells supplied with two metal electrodes. By saving the data obtained for subsequent analysis on a computer, large numbers of samples can be monitored at the same time.

Central to the successful application of the method is the choice or design of a medium which will both support rapid growth of the micro-organisms to be monitored, and will change its electrical properties as a result of their growth.

With a suitable medium, the traces obtained resemble the bacterial growth curve, although this analogy can be misleading as the curves are not superimposable. In practice it requires quite a large number of bacteria to initiate a signal, usually about 10^6-10^7 cfu ml^{-1} in the cell.

The time it takes to reach this number and produce a signal (referred to as the detection time) will therefore include any lag phase plus a period of exponential growth and will depend on both the initial number and the growth rate.

Thus, for a particular organism/medium/temperature combination, the detection time will be inversely related to the logarithm of the number and activity of the organisms in the original sample. By obtaining detection times using samples where the microbial population is known, calibration curves relating detection time and microbial numbers can be drawn so that count data can be derived from detection times. One such example for Salmonella enteritidis is presented in Figure 8.2.

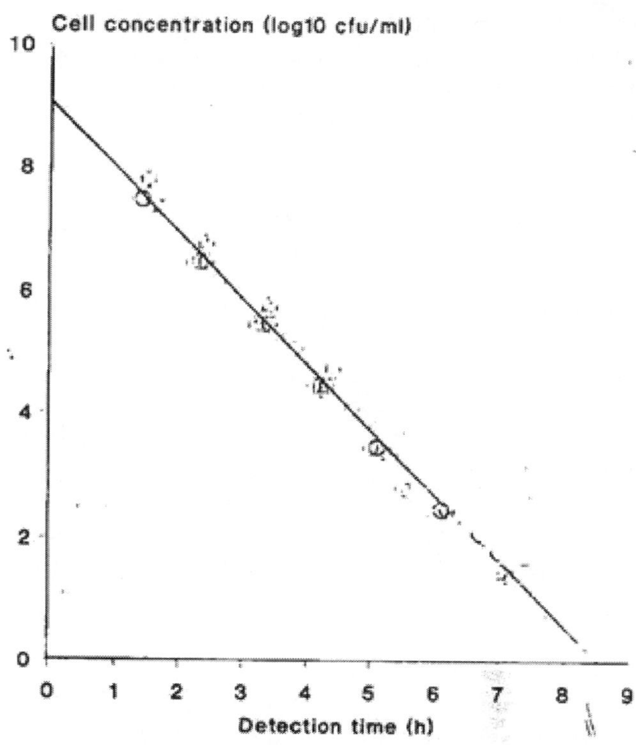

Figure 8.2: Calibration Curve for *Salmonella enteritidis* using a Bactometer.

Some claim that the only value in converting detection times to counts is that the food microbiologist derives a sense of security from having data in a familiar form. Since electrical methods measure microbial activity directly, detection time may be a more appropriate measure of the potential to cause spoilage than a viable plate count.

In the food industry the potential for simultaneously testing many samples makes electrical methods a useful means for assessing the quality of raw materials and products. They have the additional advantage that the worse the microbiological quality, the shorter is the detection time, and the sooner the manufacturer knows that there may be a problem.

In modern instruments, which can accommodate more than 500 samples, the results can be displayed using an unambiguous quality colour code of acceptable (green), marginally acceptable (orange), and unacceptable (red).

By carefully designing the medium to contain selective agents, and diagnostic compounds which will give a strong signal when they are metabolized, it is possible to use electrical methods to estimate the activity (and hence, by calibration, numbers) of specific groups of organisms.

Rapid Methods for the Detection of Specific Organisms and Toxins

The high rate of foodborne diseases in many emerging countries presents major food safety concerns and difficulties; therefore, it is important to detect the responsible pathogens to reduce foodborne diseases. There are two groups of bacteria, named Gram-positive and Gram-negative bacteria, and the difference between them is in the cell structure of their wall. Nearly 95 per cent of pathogenic bacteria are Gram-negative bacteria, and the rest are Gram-positive bacteria. The toxins that are produced by the pathogens are required to be monitored for food safety. There are some traditional methods available for pathogen and endotoxin detection which depend on culturing the microorganisms on agar plates. All the conventional methods are laborious and take 2–3 days to get initial results. It takes up to one week to determine the specific pathogen microorganisms. Therefore, different methods of pathogen detections are in high demand in the food industry to avoid the spread of any bacterial diseases from food poisoning. The detection methods can be categorized into the following groups: conventional methods and various sensor-based methods.

In conventional methods, the detection of pathogens mostly depends on the identification of precise microbiological and biochemical constituents. The culturing technique gives accurate results due to its high selectivity and sensitivity. Various microorganisms grow on the food sample and, depending on the growth of the microorganisms, the specific pathogens are identified. The process requires selective plating, pre-enrichment, selective enrichment, and identifications, which take a few days to produce the results. This detection method is monotonous and lengthy. The conventional methods used to detect foodborne pathogen are time consuming and laborious. Hence, a variety of methods have been developed for rapid detection of foodborne pathogens as it is required in many food analyses.

Recently, different rapid methods with high sensitivity and specificity have been developed to overcome the limitations of conventional methods for the detection and identification of foodborne pathogens. Furthermore, researchers are still developing novel methods with improvements in terms of rapidity, sensitivity, specificity and suitability for *in situ* analysis and distinction of the viable cell. Rapid detection methods are important, particularly in food industry, as they are able to detect the presence of pathogens in raw and processed foods immediately. Rapid methods are also sensitive enough to detect pathogens that present in low numbers in the food. Sensitivity is important because a single pathogen present in food has the risk to cause infection. Rapid methods are more time-efficient, labour-saving and able to reduce human errors. Nevertheless, each of the rapid method has its own advantages and limitations (Law *et al.*, 2015).

Rapid detection methods can be categorized into nucleic acid-based, biosensor-based and immunological-based methods.

Nucleic Acid-Based Methods

Nucleic acid-based methods operate by detecting specific DNA or RNA sequences in the target pathogen. This is done by hybridizing the target nucleic acid sequence to a synthetic oligonucleotide (probes or primers) which is complementary to the target sequence. There are many bacterial pathogens such as *Clostridium botulinum, Vibrio cholerae, Staphylococcus aureus*, and *Escherichia coli* O157 which produce toxins that cause foodborne diseases. The toxin-related genes in these pathogens can be detected by nucleic acid-based methods. Nucleic-acid based methods detect the specific genes in the target pathogens, therefore preventing ambiguous or wrongly interpreted results. The recent nucleic acid-based methods described are simple polymerase chain reaction (PCR), multiplex polymerase chain reaction (mPCR), real-time/quantitative polymerase chain reaction (qPCR), nucleic acid sequence-based amplification (NASBA), loop-mediated isothermal amplification (LAMP) and microarray technology.

Nucleic acid-based methods have high sensitivity and they are widely used for the detection of foodborne pathogens, but these methods require trained personnel and specialized instruments. Alternative nucleic acid-based methods such as NASBA and LAMP are available for the detection of foodborne pathogens and their toxins. NASBA and LAMP are relatively sensitive, specific and cost efficient. They do not require thermocycling system therefore they are useful especially in low resource settings.

Polymerase Chain Reaction (PCR)

The polymerase chain reaction (PCR) is very popular detection method for detection of pathogens. It is now being increasingly used for monitoring of foods for potential food pathogens, so that the safety of these foods could be ensured from public health point of view. Using PCR, the target DNA sequence can be selectively amplified, thus increasing the amount of target available for detection. PCR was invented about 30 years ago and it allows the detection of a single bacterial pathogen that is present in food by detecting a specific target DNA sequence.

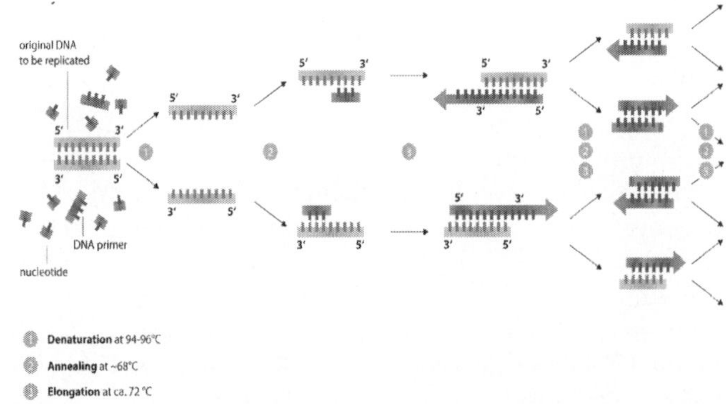

Figure 8.3: Polymerase Chain Reaction.

The PCR technique is based on the reiteration of three-step process: denaturation of dsDNA into single strands, annealing extension primers homologous to the target DNA sequence to be amplified to the single stranded DNA, and enzymatic extension of primers using a thermostable enzyme such as Taq DNA polymerase. A typical amplification is 20-40 cycles and results in million-fold amplification of the original target DNA. The DNA from the single cell would be detectable following PCR amplification. Application of such methods in Food Quality Assurance and Public Health Labs can go a long way in protecting the health of the public. The crucial feature of sample preparation in the direct analysis of food without cultural enrichment is the isolation of target DNA with high reproducibility and efficiency. Construction of the actual PCR assay requires nucleotide sequence information and knowledge about the chosen target gene to guarantee maximum specificity and sensitivity.

By using different formats of PCR, it is now possible to simultaneously detect more than one food pathogen in the food. PCR-based detection of food pathogens has completely revolutionized the area of diagnostics as the results are available in less than 24 h to enable the food processing units to take appropriate follow up action if a pathogen is detected in a food to avoid any risk to the consumers.

Specific bacteria based on their nucleic acid sequence, protozoa, and viruses are targeted when PCR is used for pathogen detection. PCR have been used in the detection of numerous foodborne pathogens like *Listeria monocytogenes*, *Escherichia coli* O157:H7, *Staphylococcus aureus, Campylobacter jejuni, Salmonella* spp. and *Shigella* spp.

Different PCR methods are available for pathogen detection, called, reverse transcript PCR (RT-PCR), real-time PCR, and multiplex PCR. However, the method requires costly instruments, and amplification, isolation, and quantification of DNA technology make it a complex method to perform. It also requires trained personnel to operate the whole procedure.

Multiplex PCR (mPCR)

Multiplex PCR offers a more rapid detection as compared to simple PCR through the simultaneous amplification of multiple gene targets. The basic principle of mPCR is similar to conventional PCR. However, several sets of specific primers are used in mPCR assay whereas only one set of specific primers are used in conventional PCR assay. Primer design is very important for the development of mPCR, as the primer sets should have similar annealing temperature in order to produce a successful mPCR assay (Zhao *et al.,* 2014). Besides, the concentration of primers is also important in mPCR. This is because interaction may occur between the multiple primer sets in mPCR that results in primer dimers, thus, the concentration of primers may need to be adjusted to ensure the production of reliable PCR products. Other important factors for a successful mPCR assay include the PCR buffer concentrations, the balance between magnesium chloride and deoxynucleotide concentrations, the quantities of DNA template, cycling temperatures and Taq DNA polymerase. Previously, mPCR was used to detect around two to three pathogens only. Now, mPCR is more advanced and it can detect up to five or more pathogens simultaneously.

Real-Time or Quantitative PCR (qPCR)

Real-time PCR or quantitative PCR is different from simple PCR whereby it does not require agarose gel electrophoresis for the detection of PCR products. This method is able to monitor the PCR products formation continuously in the entire reaction by measuring the fluorescent signal produced by specific dual-labelled probes or intercalating dyes. The fluorescence intensity is proportional to the amount of PCR amplicons.

Real-Time PCR can be applied to traditional PCR applications as well as new applications that would have been less effective with traditional PCR. With the ability to collect data in the exponential growth phase, the power of PCR has been expanded into applications such as: viral quantitation, quantitation of gene expression, array verification, drug therapy efficacy, DNA Damage measurement, quality control and assay validation, pathogen detection and genotyping. The most popular real-time PCR based system used in food authentication is the Qualicons BAX® detection system, which is a fast and accurate test for screening food and environmental samples for pathogens e.g, *E. coli* O157:H7, Listeria monocytogenes, Enterobacter sakazakii, Salmonella, Campylobacter coli *etc.*

Advantages of using Real-Time PCR a) include the following:

a) Traditional PCR is measured at End-Point (plateau), while Real-Time PCR collects data in the exponential growth phase.
b) An increase in reporter fluorescent signal is directly proportional to the number of amplicons generated.
c) The cleaved probe provides a permanent record amplification of an Amplicon.
d) Increase dynamic range of detection.
e) No-post PCR processing.
f) Detection is capable down to a 2-fold change.

Nucleic Acid Sequence-Based Amplification (NASBA)

ASBA is developed by Compton (1991) in the early 90s and it operates by amplifying nucleic acids under isothermal conditions, unlike PCR which requires thermocycling system. NASBA is normally used for the amplification of RNA whereby the single-stranded RNA template is converted into complementary DNA (cDNA) by the reverse transcriptase during the reaction. NASBA reaction occurs at around 41°C, involving two specific primers and three enzymes: avian myeloblastosis virus (AMV) reverse transcriptase, T7 RNA polymerase and RNase H. The NASBA amplicons can be detected by agarose gel electrophoresis. Real-time NASBA has been used for the detection of various foodborne pathogens such as *Salmonella enterica, Vibrio cholerae, Staphylococcus aureus, Campylobacter jejuni*, and *Campylobacter coli.* NASBA offers high-throughput analysis and it has been commercialized as kits.

Loop-Mediated Isothermal Amplification (LAMP)

LAMP is a novel nucleic acid amplification method developed by Notomi *et al.* (2000) which provides a rapid, sensitivity and specific detection of foodborne pathogens. LAMP is based on auto-cycling strand displacement DNA synthesis carried out by *Bst* DNA polymerase large fragment under isothermal conditions between 59°C and 65°C for 60 min. LAMP has been used for the detection of various foodborne pathogens due to its rapidity and sensitivity. LAMP is proven to be more specific and sensitive as compared to PCR assays for the detection of foodborne pathogens. This is because LAMP uses four primers targeting six specific regions and it provides rapid amplification, greater yield of amplification products and lower detection limits than PCR assays. Till date, commercial LAMP kits are available for the detection of *Listeria, Salmonella, Campylobacter, Legionella*, and verotoxin-producing *Escherichia coli.*

Oligonucleotide DNA Microarray

The recent progress in multi-gene detection technology includes the microarray technology. Microarrays were originally used for the study of gene expression, but oligonucleotide DNA microarray has been widely used in the field of foodborne pathogen detection. Microarrays are made up of glass slides or chips coated with up to hundreds of specific oligonucleotide probes and these probes are chemically synthesized short sequences range from 25 to 80 bp. Each oligonucleotide probe is able to target a specific part of a gene sequence. In this method, the sample nucleic acid fragments (DNA, mRNA or cDNA) are labelled with fluorescent dye, and then they are denaturated to generate single-stranded fragments. These fragments will hybridize to the array through binding to their corresponding oligonucleotide probes. The results are obtained through the visualization of the fluorescence signal produced from the probe-sample complex. The fluorescence intensity is proportional to the concentration of each labelled nucleic acid fragment. In general, DNA oligonucleotide microarray allows simultaneous identification of multiple foodborne bacterial pathogens. Therefore, it is capable of high-throughput analysis and it also has the potential to be automated.

Biosensor-Based Methods

Biosensor is an analytical device that consists of two main elements: a bioreceptor and a transducer. The bioreceptor responsible for recognizing the target analyte can either be a:

1. Biological material: enzymes, antibodies, nucleic acids and cell receptors, or
2. Biologically derived material: aptamers and recombinant antibodies, or
3. Biomimic: imprinted polymers and synthetic catalysts.

The transducer that converts the biological interactions into a measurable electrical signal can be optical, electrochemical, mass-based, thermometric, micromechanical or magnetic (Zhao *et al.*, 2014).

Biosensors are easy to operate and they do not require sample pre-enrichment, unlike nucleic-acid based methods and immunological methods which require sample pre-enrichment for concentrating the pathogens before detection. The recent biosensors that commonly used for the detection of foodborne pathogens are optical, electrochemical and mass-based biosensors (Zhang, 2013; Zhao *et al.*, 2014).

Biosensors-based methods are easy to operate and they do not require trained personnel, furthermore the techniques can be used for the detection of foodborne pathogens without sample pre-enrichment. However, improvement in food matrixes detection is still needed for these methods for on-site detection.

The commercially available biosensors for the detection of foodborne pathogens are mostly optical-based biosensors. The commercial biosensors offer varying degrees of automation. The commercialization of biosensors is slower than other rapid methods due to several factors such as cost consideration, quality assurance, stability issues, sensitivity issues and instrumentation design. There are difficulties in the methods of producing inexpensive and reliable sensors, the storage of biosensors, the stabilization of biosensors, methods of sensor calibration and total integration of the sensor system.

Immunological-Based Methods

The detection of foodborne pathogens by immunological-based methods is based on antibody-antigen interactions, whereby a particular antibody will bind to its specific antigen. The binding strength of a particular antibody to its antigen determines the sensitivity and specificity of immunological-based methods. Immunological-based methods involve the use of polyclonal and monoclonal antibodies. Enzyme-linked immunosorbent assay (ELISA) and lateral flow immunoassay is among the immunological-based methods which has recently been used for the detection of foodborne pathogens. Because of the potential specificity of immunoassays using polyclonal or monoclonal antibodies, there has been considerable effort devoted to developing their application in food microbiology. Commercial immunoassay kits are now available for detecting a variety of foodborne micro-organisms and their toxins, including mycotoxins.

Although a number of different formats are used in immunoassays, their essential feature is the binding of antibody to antigen. A commonly used protocol is that of the sandwich ELISA (enzyme-linked immunosorbent assay) which is the most effective form of ELISA; it involves two antibodies. The primary antibody is usually immobilized onto the walls of the microtiter plate wells. The target antigen like bacterial cells or bacterial toxins from the food sample binds to the immobilized primary antibody and the remaining unbound antigens are removed. After that, an enzyme-conjugated secondary antibody is added and it will bind to the antigen and the remaining unbound antibodies are removed. The complex consisting antigen sandwiched between two antibodies is formed and it can be detected by adding a colourless substrate which will be converted into a coloured form in the presence of the enzyme (Zhang, 2013). There are different types of enzymes which can be used in ELISA, some of the most commonly used enzymes include horseradish peroxidase (HRP), alkaline phosphatase and beta-galactosidase (Figure 8.4). ELISA is also commonly used for the detection of toxins present in foods such as *Clostridium perfringens* α, β, and ε toxin, staphylococcal enteroxins A, B, C, and E, botulinum toxins and *Escherichia coli* enterotoxins.

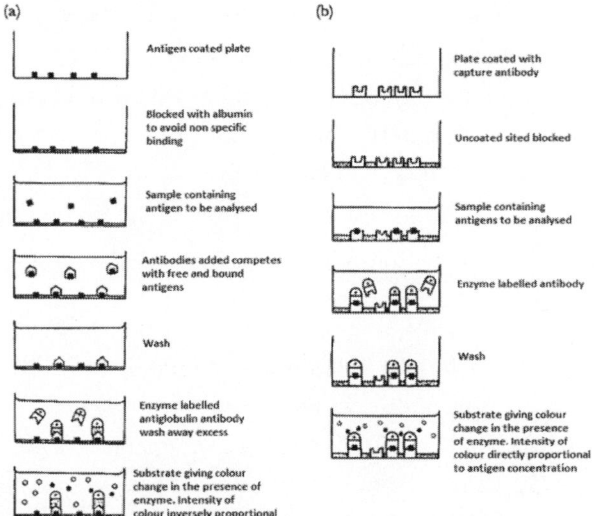

Figure 8.4: Sandwich ELISAs (a) Competitive Sandwich Elisa; (b) Direct Sandwich Elisa.

Commercial ELISAs are available for such organisms as *Salmonella* and *Listeria monocytogenes* but they still require the presence of at least 10^5— 10^6 organisms. Detection of smaller numbers therefore depends on some form of enrichment or concentration by one of the separation methods briefly mentioned above, so that although the immunoassay itself may be rapid the whole analytical protocol may take almost as long as conventional procedures. Some advantage can be gained from the automation of the assay and a number of instruments are commercially available. There may also be some concern over the specificity of immunoassays.

While striving for antibodies that are sufficiently broad in their specificity to recognize all strains of the desired target organism, it is difficult to avoid the problem of cross reactivity with organisms other than those under investigation.

Lateral Flow Immunoassays

ELISA offers a sensitive and accurate detection of foodborne pathogens. However, the operation of ELISA requires specialized equipment and trained personnel (Zhao *et al.*, 2014). Hence, other immunological detection methods which are rapid, cheap, simple and reliable are required. Lateral flow immunoassays such as dipstick and immunochromatographic strips have been developed for rapid on-site detection of foodborne pathogens. Lateral flow immunoassay device is made up of four sections which are arranged orderly on a plastic backing, with sample pad starting at the bottom, followed by conjugate pad, nitrocellulose membrane and then absorbent pad. The sample fluid will migrate along the four sections of lateral flow immunoassay via capillary action. The sample fluid encounters and mixes with the conjugate, which can be antibody or antigen labelled by a colour particle, at the conjugate pad and then pass through the lines in the nitrocellulose membrane that immobilized with antibody or antigen. The colour particle can bind to the antibody or antigen immobilized at test line depending on the analytes present in the sample. The colour can be visualized approximately two to 10 min after the addition of sample.

The detection of foodborne pathogens by lateral flow immunoassay employs labels such as monodisperse latex, colloidal gold, carbon and fluorescent tags. Commercial immunochromatographic test strips are also available for the detection of foodborne bacterial pathogens. Lateral flow immunoassays are simple and fast, but they are designed for individual tests rather than high-throughput screening.

Immunological methods work best in the absence of interfering molecules in the samples such as non-targeted cells, DNA or proteins. Combination of several rapid methods for the detection of a particular foodborne pathogen is also possible as the use of only one detection method may not be sufficient to confirm the detected pathogen. Further studies on the effect of different combinations of rapid methods for foodborne pathogen detection are required in order to develop the most effective and accurate detection method.

This section has been discussed further in detail in next chapter.

Food Analysis

Prior to the mid to late 1970s, microorganisms were extracted from food specimens for plating almost universally by use of mechanical blenders (Waring type). Around 1971, the Colwell Stomacher was developed in England by Sharpe and Jackson and this device is now the method of choice in many laboratories for homogenizing foods for counts. The Stomacher is a relatively simple device which homogenizes specimens in a special plastic bag by the vigorous pounding of two paddles. The pounding effects the shearing of food specimens and microorganisms are released into the diluent. Three models of the machine are available, but the model 400 is most widely used in food microbiology laboratories. It can handle samples (diluent and specimen) of 40 to 400 ml.

The Stomacher has been compared to high-speed blenders for food analysis by a large number of investigators. Plate counts from Stomacher-treated samples are similar to those treated by blender. The instrument is generally preferred over blending for the following reasons:

1. The need to clean and store blender containers is obviated.
2. Heat buildup does not occur during normal operational times (usually 2 min).
3. The homogenates can be stored in the Stomacher bags in a freezer for further use.
4. The noise level is not as unpleasant as that of mechanical blenders.

In a study by Sharpe and Harshman the Stomacher was shown to be less lethal than a blender to *Staphylococcus aureus, Streptococcus faecalis* and *Escherichia coli.* One investigator reported that counts by using a Stomacher were significantly higher than when a blender was used, while other investigators obtained higher overall counts by blender than by Stomacher. The latter investigators showed that the Stomacher is food-specific, being better than high-speed blending for some types of foods but not for others. In another study, SPC determinations made by Stomacher, blender and shaking were not significantly different, although significantly higher counts of gram-negative bacteria were obtained by Stomacher than by either of the other two methods.

Another advantage of the Stomacher over blending is the homogenization of meats for dye-reduction tests. Holley *et al.*, showed that the extraction of bacteria from meat by using a Stomacher does not cause extensive disruption of meat tissue and consequently fewer reductive compounds were present to interfere with resazurin reduction, while with blending the level of reductive compounds released made resazurin reduction results meaningless.

The examination of foods for the presence, types and numbers of microorganisms and/or their products is basic to food microbiology. In spite of the importance of this, none of the methods in common use permits the determination of exact numbers of microorganisms in a food product. Although some methods of analysis are better than others, every method has certain inherent limitations associated with its use.

The four basic methods employed for "total" numbers are:

1. Standard plate counts (SPC) for viable cells
2. The most-probable numbers (MPN) method as a statistical determination of viable cells
3. Dye-reduction techniques to estimate numbers of viable cells that possess reducing capacities, and
4. Direct microscopic counts (DMC) for both viable and non-viable cells.

All of these are discussed in this chapter along with their uses in determining microorganisms from various sources. In addition, variations of these basic methods for examining the microbiology of surfaces and of air are presented along with a summary of methods and attempts to improve their overall efficiency.

The SPC is by far the most widely used method for determining the numbers of viable cells or colony-forming units (cfu) in a food product. When total viable counts are reported for a product, the counts should be viewed as a function of at least some of the following factors:

1. Sampling methods employed.
2. Distribution of the organisms in the food sample.
3. Nature of the food flora.
4. Nature of the food material.
5. The pre-examination history of the food product.
6. Nutritional adequacy of the plating medium employed.
7. Incubation temperature and time used
8. pH, a_w and E_h of the plating medium.
9. Type of diluent used.
10. Relative number of organisms in food sample.
11. Existence of other competing or antagonistic organisms.

In addition to the limitations noted, plating procedures for selected groups are further limited by the degree of inhibition and effectiveness of the selective and/or differential agents employed.

While the SPC is more often determined by pour plating, essentially comparable results can be obtained by surface plating. By the latter method, prepoured and hardened agar plates with dry surfaces are employed. The diluted specimens are planted onto the surface of replicate plates and with the aid of bent glass rods ("hockey sticks") the 0.l ml inoculum/plate is carefully and evenly distributed over the entire surface. Surface plating offers advantages in determining the numbers of heat-sensitive psychrotrophs in a food product because the organisms do not come in contact with melted agar. It is the method of choice when the colonial features of a colony are important to its presumptive identification and for most selective media.

Counting Grid for Spiral Plater

The spiral plater is a mechanical device that distributes the liquid inoculum on the surface of a rotating plate containing a suitable poured and hardened agar medium. The dispensing arm moves from the near center of the plate toward the outside depositing the sample in an Archimedes spiral. The attached special syringe dispenses a continuously decreasing volume of sample so that a concentration range of up to 10,000: 1 is affected on a single plate. Following incubation at an appropriate temperature, colony development reveals a higher density of deposited cells near the center of the plate with progressively fewer toward the edge.

The enumeration of colonies on plates prepared with a spiral plater is achieved by use of a special counting grid. Depending upon the relative density of colonies, colonies that appear in one or more specific area(s) of the superimposed grid are counted. In this particular example, a total sample volume of 0.0018 ml was deposited and the two grid areas counted contained forty- four and sixty-one colonies, respectively, resulting in a total count of $6.1 \times 10_4$ bacteria/ml.

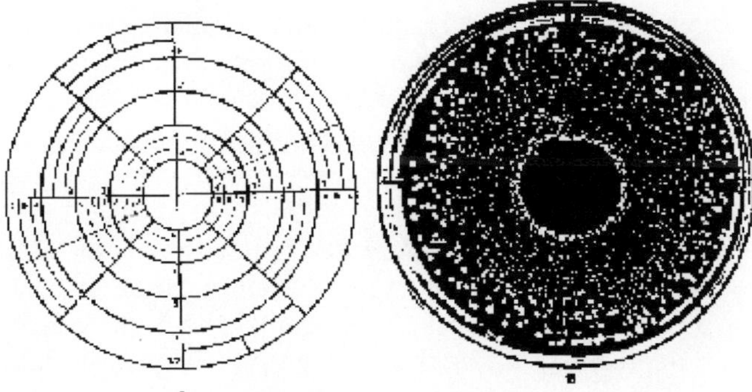

Figure 8.5: (A) Special Counting Grid for Spiral Plater; (B) Growth or Organisms on an Inoculated Spiral Plate; and Areas of Plate Enumerated.

The spiral plating device herein described was devised by Gilchrist *et al.,* although some of its principles were presented by earlier investigators among whom were Reyniers and Trotman. The method has been studied by a rather large number of investigators and compared to other methods of enumerating viable organisms, It was compared to the SPC method by using 201 samples of raw and pasteurized milk; overall good agreement was obtained. A collaborative study from six analysts on milk samples showed that the spiral plater compared favorably, with the SPC. A standard deviation of 0.109 was obtained by using the spiral plater, compared to 0.110 for the SPC. In another study, the spiral plater was compared, with three other methods (pour, surface plating and drop count) and no difference was found between the methods at the 5 per cent level of significance. In yet another study, the spiral plate maker yielded counts as good as those by the droplette method. Spiral plating is an official AOAC method.

Among the advantages of the spiral plater over standard- plating are: less agar is used; fewer plates, dilution blanks and pipettes are required; and three to four times more samples/h can be examined. Also, 50–60 plates/h can be prepared and little training is required for its operation. Among the disadvantages is the problem that food particles may cause blocking the dispensing stylus. It is more suited for use with liquid foods such as milk. A laser-beam counter has been developed for use with the plater. Because of the expense of the device, it is not likely to be available in laboratories that do not analyze large numbers of plates. The method is further described by AOAC and in the *Bacteriological Analytical Manual.*

Importance of Food Contact Surfaces

The need to maintain food contact surfaces in a hygienic state is of obvious importance. The primary problem that has to be overcome when examining surfaces or utensils for microorganisms is the removal of a significant percentage of the resident flora. Although a given method may not recover all organisms, its consistent use in specified areas of a food processing plant can still provide valuable information as long as it is realized that not all organisms are being recovered. The most commonly used methods for surface assessment in food operations are presented below.

Swabbing is the oldest and most widely used method for the microbiological examination of surfaces not only in the food and dairy industries but also in hospitals and restaurants. The swab-rinse method was developed in 1917 by W.A. Manheimer and T. Ybanez. Either cotton or calcium alginate swabs are used. If one wishes to examine given areas of a surface, templates may be prepared with openings corresponding to the size of the area to be swabbed, for example, 1 in_2 or cm_2. The sterile template is placed over the surface and the exposed area is rubbed thoroughly with a moistened swab. The exposed swab is returned to its holder (test tube) containing a suitable diluent and stored at refrigerator temperatures until plated. The diluent should contain a neutralizer, if necessary. When cotton swabs are used, the organisms must be dislodged from the fibers. When calcium alginate swabs are used, the organisms are released into the diluent upon dissolution of the alginate by sodium hexametaphosphate. The organisms in the diluent are enumerated by a suitable method such as SPC, but any of the culture methods in this chapter may be used. Also, a battery of selective and differential culture media may be used to test specifically for given groups of organisms. In a recent innovation in the swab rinse method presented by Koller, 1.5 ml or fluid is added to a flat surface, swabbed for 15 sec over a 3 cm area and volumes of 0.1 and 0.5 ml collected in microliter pipettes. The fluid may be surface or pour plated using plate count agar or selective media.

Concerning the relative efficacy of cotton and calcium alginate swabs, most agree that higher numbers of organisms are obtained by use of the latter. Using swabs, some researchers recovered as little as 10 per cent of organisms from bovine carcasses; 47 per cent of *Bacillus subtilis* spores from stainless steel surfaces; and up to 79 per cent from meat surfaces. Swab results from bovine carcasses were on the average 100 times higher than by contact plate method and the deviation was considerably lower. The latter investigators found the swab method to be best suited for flexible, uneven and heavily contaminated surfaces. The case of removal of organisms depends upon the texture of the surface and the nature and types of flora. Even with its limitations, the swab-rinse method remains a rapid, simple and inexpensive way to assess the microbiological flora of food surfaces and utensils.

The replicate organism direct agar contact (rodac) method employs special petri plates which are poured with 15.5–16.5 ml of an appropriate plating medium resulting in a raised agar surface. When the plate is inverted, the hardened agar makes direct contact with the surface. Originated by Gunderson and Gunderson in 1945, it was further developed in 1964 by Hall and Hartnett. When surfaces are examined that have been cleaned with certain detergents, it is necessary to include a neutralizer (lecithin, Tween 80 and so on) in the medium. Once exposed, plates are covered, incubated and the colonies enumerated.

Perhaps the most serious drawbacks to this method are the covering of the agar surface by spreading colonies and its ineffectiveness for heavily contaminated surfaces. These can be minimized by using plates with dried agar surfaces and by using selective media. The rodac plate has been shown to be the method of choice when the surfaces to be examined are smooth, firm and non-porous. While it is not suitable for heavily contaminated surfaces, it has been estimated that a solution which contaminates a surface needs to contain at least 10 cells/ml before results can be achieved either by contact or by swabs. The latter investigators found that the contact plate removed only about 0.1 per cent of surface flora. This suggests that 10 cfu/cm_2 detected by this method are referable to a surface that actually contains about 10_4 cfu/cm_2. When stainless steel surfaces were contaminated by *B. subtilis* endospores, 41 per cent were recovered by the rodac plate compared to 47 per cent by the swab method. In another study, swabs were better than contact plates when the contamination level was 100.

Now let us study about some surface methods.

Ultrasonic Devices

Ultrasonic devices have been used to assess the microbiological contamination of surfaces, but the surfaces to be examined must be small in size and removable so that they can be placed inside a container immersed in diluent. Once the container is placed in an ultrasonic apparatus, the energy generated affects the release of microorganisms into the diluent. A more practical use of ultrasonic energy may be the removal of bacteria from cotton swabs in the swab-rinse method.

Spray Gun

A spray gun method was devised by Clark based on the impingement of a spray of washing solution against a circumscribed area of surface and the subsequent plating of the washing solution. Although the device is portable, a source of air pressure is necessary. It was shown to be much more effective than the swab method in removing bacteria from meat surfaces.

Direct Surface

A number of workers have employed direct surface agar plating methods, in which melted agar is poured onto the surface or utensil to be assessed. Upon hardening, the agar mold is placed in a petri dish and incubated. Angelotti and Foter proposed this as a reference method for assessing surface contamination and it is excellent for enumerating particulates containing viable microorganisms. It was used successfully to determine the survival of *Clostridium sporogenes* endospores on stainless steel surfaces. Although effective as a research tool, the method does not lend itself to routine use for food plant surfaces.

Sticky Film

The sticky film method of Thomas has been used with some success by Mossel *et al.,* The method consists of pressing sticky film or tape against the surface to be examined and pressing the exposed side on an agar plate. It was shown to be less effective than swabs in recovering bacteria from wooden surfaces. An adhesive tape method has been employed successfully to assess microorganisms on meat surfaces. In a recent study, the swab, rodac and adhesive tape (Mylar) methods were compared for the examination of pork carcasses and the correlation between adhesive tape and rodac was better than that between adhesive tape and swab or between rodac and swab. Plastic strips attached to pads containing culture media have been used to monitor microorganisms on bottles.

Injured Cells in Foods

When microorganisms are subjected to environmental stresses such as sub-lethal heat and freezing, many of the individual cells undergo metabolic injury resulting in their inability to form colonies on selective media that uninjured cells can tolerate. Whether or not a culture has suffered metabolic injury can be determined by plating aliquots separately on a non-selective and a selective medium and enumerating the colonies that develop after suitable incubation. The colonies that develop on the non-selective medium represent both injured and uninjured cells, while only the uninjured cells develop on the selective medium. The difference between the number of colonies on the two media is a measure of the number of injured cells in the original culture of population. These investigators subjected the organism to 52°C for 15 min in a phosphate buffer at pH 7.2 to inflict cell injury. The plating of cells at zero time and up to 15 min of heating on non-selective trypticase soy agar (TSA) and selective TSA + 7.0 per cent NaCl (stress medium; TSAS) reveals only a slight reduction in numbers on TSA, while the numbers on TSAS were reduced considerably, indicating a high degree of injury relative to a level of salt that uninjured *S. aureus* can withstand. To allow the heat-injured cells to repair, the cells were placed in nutrient broth (recovery medium) followed by incubation at 37°C for 4 h. With hourly plating of aliquots from the recovery medium onto TSAS, it can be seen that the injured cells regained their capacity to withstand the 7.0 per cent NaCl in TSAS after the 4 h incubation.

The existence of metabolically injured cells in foods and their recovery during culturing procedures is obviously of great importance not only from the standpoint of pathogenic organisms but for spoilage organisms as well. The data cited above suggest that if a high-salt medium had been employed to examine a heat-pasteurized product for *S. aureus,* the number of viable cells found would have been lower than the actual number by a factor of 3 log cycles. Injury of food-borne microorganisms has been shown by a large number of investigators to be induced not only by sublethal heat and freezing but also by freeze drying, drying, irradiation, aerosolization, dyes, sodium azide, salts, heavy metals, antibiotics, essential oils and other chemicals such as EDTA and sanitizing compounds.

The recognition of sublethal stresses on food-borne microorganisms and their effect upon growth under varying conditions dates back to the turn of the century. However, a full appreciation of this phenomenon did not come until the late 1960s. During the early part of the 1960s, it was observed that an initial rapid decrease in numbers of a metabolically injured organism was followed by only a limited recovery during the resuscitation process ("Phoenix phenomenon"). The increased nutritional requirement of bacteria that had undergone heat treatment was noted by Nelson in 1943 (Nelson also reviewed the work of others up to that time). Gunderson and Rose noted the progressive decrease in numbers of coliforms from frozen chicken products that grew on violet red bile agar (VRBA) with increasing storage time of products. Hartsell inoculated foods with *Salmonellae*, froze the inoculated foods and then studied the fate of the organisms during freezer storage. More organisms could be recovered on highly nutritive non-selective media than on selective media such as MacConkey, deoxycholate, or VRB agars. The importance of the isolation medium in recovering stressed cells was noted also by Postgate and Hunter and by Harris. In addition to the more exacting nutritional requirements of food- borne organisms that undergo environmental stresses, these organisms may be expected to manifest their injury via increased lag phases of growth, increased sensitivity to a variety of selective media agents, damage to cell membranes and TCA-cycle enzymes, breakdown of ribosomes and DNA damage. While damage to ribosomes and cell membranes appear to be common consequences of sublethal heat injury, not all harmful agents produce identifiable injuries.

Metabolically injured cells can recover, at least in *S. aureus,* in no-growth media and at a temperature of 15°C but not 10°C. In some instances at least, the recovery process is not instantaneous for it has been shown that stressed coliforms do not all recover to the same degree but that the process takes place in a step-wise manner. Not all cells in a population suffer the same degree of injury. Hurst *et al.,* found dry-injured *S. aureus* cells that failed to develop on the non-selective recovery medium (TSA) but did recover when pyruvate was added to this medium. These cells were said to be severely injured in contrast to injured and uninjured cells. It has been found that sublethally heated *S. aureus* cells may recover their NaCl tolerance before certain membrane functions are restored. It is well established that injury repair occurs in the general absence of cell.

The protection of cells from heat and freeze injury is favoured by complex media and menstra or certain specific components thereof. Milk provides more protection than saline or mixtures of amino acids and the milk components that are most influential appear to be phosphate, lactose and casein. Sucrose appears to be protective against heat injury, while glucose has been reported to decrease heat protection for *S. aureus.* Non-metabolizable sugars and polyols such as arabinose, xylose and sorbitol have been found to protect *S. aureus* against sublethal heat injury but the mechanism of this action is unclear.

Table 8.1: Summary of Some Methods Employed to Recover Food-borne Microorganisms Subjected to Various Stresses

Organisms	*Method of Injury*	*Stress Medium*	*Recovery Medium and Method*
Clostridium perfringens	0.1 *N* NaOH	TSN agar	TSN agar + lysozyme
C. perfringens spores	Ultra-high temp.	Polymyxin and neomycin	TSN agar + lysozyme; no antibiotics
C. botulinum Type E	Sublethal heat	PYE + 0.07 per cent bile salts	PY agar
C. botulinum Type E	Irradiation	Polymyxin and neomycin	TP-EY medium + polymyxin and neomycin
C. botulinum spores	Irradiation	TYT + 5 per cent NaCl	TYT thioglycolate medium
Coliforms	Freezing	VRBA	TSA surface plate; VRBA overlay
Entero-bacteriaceae	Dried foods	E.E. medium	TSPB, 1–6 hr at 19–25°C
Escherichia coli	Freeze drying	Antibiotics and Nadeoxycholate	TSYA and minimal salts
E. coli	Organic and in-organic acids	VRBA	Nutrient agar at 37°C
E. coli	Frozen foods	VRBA and DCA	TSB for 1 hr at 25°C
E. coli	Freezing and subletha heat	VRBA	3,3'-thiodipropionic acid supple-mented media
Pseudomonas spp.	Freezing	Minimal agar	Trypticase soy agar
Salmonella typhimurium	Sublethal heat	EMBA + 2 per cent NaCl	TSA at 37°C for 24 hr
S. anatum	Freeze drying	XLPA	XLPA + 0.25 per cent Nadeoxycholate
S. anatum	Freezing	XLPA	XLPA + 0.2 per cent Nadeoxycholate
S. typhimurium	Sublethal heat	TSA + yeast extract	M–9 agar at 37°C
S. senftenberg	Sublethal heat	TSYA	M–9 medium at 37°C for 2 hr
Salmonella spp.	Irradiation	SS and DCA	BGA; NB + TT + BG (MPN)
Shigella sonnei	Freezing	Synthetic medium	Blood heart inf., NA plates
Staphylococcus aureus	Sublethal heat	SM #110	Plate count agar
S. aureus	Sublethal heat	TSA + 7.5 per cent NaCl	5 per cent glucose or galactose
S. aureus	Sublethal heat	SM #110	Sodium pyruvate medium
S. aureus	Sublethal heat	TSA + 7.5 per cent NaCl	Trypticase soy broth, 4 hr
S. aureus	Freeze drying	PCA +7.5 per cent NaCl	Plate count agar at 20–50°C

A summary of some of the many recovery or resuscitation methods employed to induce recovery of injured cells is presented in Table 8.1. The consequences of not employing a recovery step have been reviewed by Busta. The use of TSB with incubations ranging from 1 to 24 h at temperatures from 20° to 37°C is widely used for various organisms. The enumeration of sub- lethally heated *S. aureus* strains on various media has been studied. In one of these studies, seven staphylococcal media were compared on their capacity to recover nineteen strains of sub-lethally heated *S. aureus* and the Baird-Parker medium was found to be clearly the best of those studied including non-selective TSA. Similar findings by others led to the adoption of this medium in the official methods of AOAC for the direct determination of *S. aureus* in foods that contain > 10 cells/g. The greater efficacy of the Baird-Parker medium has been shown to result from its content of pyruvate. The use of this medium following recovery in an antibiotic-containing, non-selective medium has been suggested. While this approach may be suitable for *S. aureus* recovery, some problems may be expected to occur with the widespread use of antibiotics in recovery media to prevent cell growth. It has been shown that heat-injured spores of *C. perfringens* are actually sensitized to polymyxin and neomycin and it is well established that the antibiotics that affect cell wall synthesis are known to induce L-phase variations in many bacteria.

Pyruvate is well established as an injury repair agent not only for injured *S. aureus* cells but for other organisms such as *E. coli*. Higher counts are obtained on media containing this compound when injured by a variety of agents. When added to TSB containing 10 per cent NaCl, higher numbers of both stressed and non-stressed *S. aureus* were achieved and the repair-detection of freeze- or heat-injured *E. coli* was significantly improved by pyruvate.

Catalase is another agent that increases recovery of injured aerobic organisms. First reported by Martin *et al.*, it has been found effective by many other investigators. It is effective for sublethally heated *S. aureus, Pseudomonas fluorescens, Salmonella typhimurium* and *E. coli*. It is effective also for *S. aureus* in the presence of 10 per cent NaCl and for water-stressed *S. aureus*. Another compound, shown to be as effective as pyruvate for heat-injured *E. coli*, is 3,3'-thiodipropionic acid.

Radiation injury of *Clostridium botulinum* type E spores by 4 kGy resulted in the inability to grow at 10°C in the presence of polymyxin and neomycin. The injured cells had a damaged post-germination system and formed aseptate filaments during outgrowth, but the germination lytic system was not damaged. The radiation injury was repaired at 30°C in about 15 h on TPEY agar without antibiotics. When *C. botulinum* spores are injured with hypochlorite, the L-alanine germination sites are modified resulting in the need for higher concentrations of alanine for repair. The L-alanine germination sites could be activated by lactate and hypochlorite-treated spores could be germinated by lysozyme, indicating that the chloride removed spore coat proteins. More detailed information on spore injury has been provided by Foegeding and Busta.

Sublethally heat-stressed yeasts are inhibited by some essential oils (spices) at concentrations as low as 25 ppm (M). The spice oils affected colony size and pigment production.

Special plating procedures have been found by Speck *et al.,* and Hartman *et al.,* to allow for recovery from injury and subsequent enumeration in essentially one step. The procedures consist of using the agar overlay plating technique with one layer consisting of TSA, onto which are plated the stressed organisms. Following a 1- to 2-h incubation at 25°C for recovery, the TSA layer is overlaid with VRBA and incubated at 35°C for 24 h. The overlay method of Hartman *et al.,* involved the use of a modified VRBA. The principle involved in the overlay technique could be extended to other selective media, of course. An overlay technique has been recommended for the recovery of coliforms. By this method, coliforms are plated with TSA and incubated at 35°C for 2 h followed by an overlay of VRBA.

In his comparison of eighteen plating media and seven enrichment broths to recover heat-stressed *Vibrio parahaemolyticus,* Beuchat found that the two most efficient plating media were water blue-alizarin yellow agar and arabinose-ammonium-sulphate-cholate agar; arabinose-ethyl violet broth was the most suitable enrichment broth.

Laboratory Accreditation

From what has already been said it should be clear that there can be a number of different ways of detecting the same organism in a food matrix. The choice of method used can be governed by several factors and the relative merits of different methods is a topic of constant investigation and debate.

This can however lead to the situation where differences in a result reported by two laboratories simply reflect the different method used.

In addition to problems arising from intrinsic differences in the performance of different methods, the same method in different laboratories can be subject to variation introduced by factors such as differences in procedures, equipment and its calibration.

Some possible examples would include autoclave temperature profile when sterilizing media, time and temperature of incubation, sources of medium components and, of course, competence and experience of laboratory personnel.

A number of approaches are adopted to avoid such potential problems. Several national and international bodies approve standard methods for conducting certain analyses and one of these should be adopted for routine work and strictly adhered to wherever possible.

Testing laboratories also often participate in quality assurance schemes where a central body distributes standard samples for analysis, often specifying the precise time this should be conducted and the method to be used.

Results are reported back, collated and a report circulated to participating laboratories which can then judge their performance against that of others. Finally, laboratories can seek some form of independent recognition.

There are quality systems such as the Good Laboratory Practice scheme and standards such as BS 5750 (ISO 9000) which are concerned with the quality of management within the organization but which do not set a particular level of quality or competence to be achieved.

There are also schemes of laboratory accreditation more concerned with the quality of performance in specific tests. In the UK this accreditation is usually sought through the National Measurement Accreditation Service (NAMAS) which accredits laboratories over a whole range of activities not just microbiological testing.

Most countries have their own equivalent organization such as NATA (Australia), DANAK (Denmark), ILAB (Ireland), and STERLAB (The Netherlands). The accrediting body inspects the laboratory and its procedures to ensure that tests are carried out consistently and correctly using approved methods with suitable quality control measures in place.

Among the features investigated are the training and qualifications of staff, the suitability of equipment and procedures for its calibration and maintenance, participation in a proficiency testing scheme and the presence of full documentation prescribing the laboratory's operating procedures.

Obtaining laboratory accreditation can be a costly exercise but, if achieved, provides independent testimony to a laboratory's proficiency and will give increased confidence to potential customers.

Chapter 9

Modern Methods of Detecting Microorganisms

Bacterial concentration detection is a very important issue in different fields, such as food quality assurance, clinical analysis and environmental monitoring.

In food industries, there are very strict limitations with respect to the bacterial presence in consumable food products. Procedures such as pasteurization, thermal and chemical procedures are commonly used in food industries to eliminate bacteria from their products. However, there is a possibility that some bacteria will survive these initial processes or may be introduced into the food during additional sample processing or handling. These surviving bacteria can then multiply in the food product over the next several days, leading to food spoilage and/or food borne illnesses depending on the type of bacteria. To prevent this, food products are regularly tested to detect the presence of any live bacteria. Here, the detection of live or viable bacteria is of importance, as these food products carry dead bacteria which have been killed during the treatment processes and do not pose any threat to food spoilage or consumers' health.

Microbial monitoring is important both to assure that the total bacterial concentration does not exceed thresholds set by national and international regulations, and to guarantee the absence of particular pathogens, such as *Salmonella typhimurium* and *Escerichia coli* 0157:H7, representing a threat to human health even at very low concentrations. Hence, microbial monitoring is the first control point in the prevention of diseases caused by foodborne pathogens. Effective detection and inspection methods are necessary to control pathogens in food products.

The current conventional microbiological methods have been a standard practice for the detection and identification of pathogens in food for nearly one century and continue to be a reliable standard for ensuring food safety. The reference technique to measure bacterial concentration is Standard Plate Count (SPC), a reliable and accurate method that, however, features long response times (usually from 24 to 72 hours). Such methods of detection used by the industries are slow, labour intensive, typically taking about ~2-3 days to get the results, by which time many of these products are already shipped to the market and consumed by its consumers. This not only affects the health of its consumers but also leads to huge economic losses to the industries due to product recalls. An interesting fact to note here is that the presence of even as few as 1 – 10 viable pathogenic bacteria in the food products is sufficient to cause significant deterioration of health in the consumers. In fact, many regulatory authorities have strict regulations on the allowable number of bacteria in a given product. These conventional methods are therefore not suitable for modern food quality assurance in the food industries to make a timely response to possible risks.

Alternative techniques for bacterial concentration measurements have been proposed in the literature in recent years. As a result, over the past 25 years, numerous rapid methods have been developed to reduce the assay time. Approaches that have been studied or are currently being studied include miniaturized biochemical tests, physicochemical methods that measure bacterial metabolites, highly specific nucleic acid-based tests, antibody-based methods, and some fully automated instrumental diagnostic systems. Many of these rapid methods developed in earlier years have been extensively reviewed by many researchers (Swaminathan and Feng, 1994; Van Der Zee and Huis in't Veld, 1997; Vasavada, 1997;

Ivnitsk *et al.,* 1999; and Hall, 2002). To date, well studied rapid methods such as enzyme linked immunosorbent assay (ELISA) and polymerase chain reaction (PCR) have reduced the assay time to 10–24 h and 4–6 h, respectively, and have achieved detection limits varying from 101 to 106 cfu/ml (cfu = colony forming units). Recently, various biosensors have been developed for pathogenic bacteria detection because of their improved speed, sensitivity and reliability. Sensitivity of a conventional biosensor is between 103 and 104 cfu/ml with an assay time around 2 h under ideal conditions. Despite the significant reduction in assay time and improvement in detection limit of these rapid methods, there are still key issues and concerns that need to be considered in the development of rapid methods for detection of foodborne pathogens. Amongst these issues are: the number of samples per operation (high throughput format), differentiation of live and dead cells (since dead cells are usually not pathogenic), automation, cost, simplicity, training, and accuracy. Ideally, a useful rapid method is expected to be able to detect low cell numbers since the presence of even a single pathogenic organism in the body or food may be an infectious dose. For example, the infectious dosage of pathogens such as *E. coli* O157:H7 and Salmonella is as low as 10 cells; and the existing coliform standard for *E. coli* in water is 4 cells/100 ml. For many food samples, the demand for detection limit is less than 1 cell per 25 g of food. To date, an ideal rapid and automated system that meets all the needs mentioned above does not exist. The current strategy for developing a rapid method is to choose and combine some functions that can meet specific applications. Generally, low cell numbers take a long time to detect, therefore, sometimes, a compromise or trade-off between detection limit and detection time has to be considered.

Physical Methods

i) Impedance Microbiology and its Basis

An interesting alternative to the previous methods is Impedance Microbiology (IM), a relatively old technique that estimates bacterial concentration by measuring electrical changes in the growth medium induced by bacterial metabolism. Although the concept of electrical impedance measurement of microbial growth was advanced by G. N. Stewart in 1899, it was not until the 1970s that the method was employed for this purpose. Thereafter, an increasing number of research studies were conducted ang many research papers have been published to promote impedance microbiology as a rapid method for detection of bacteria within 24 h. In 1992, the impedance method was approved by the Association of Official Analytical Chemists International (AOAC) as a first action method for screening Salmonella in food samples and was approved as a final action method for the detection of Salmonella in food by AOAC in 1996 (AOAC, 1996).

Over time, much work has been done in the field of media engineering, since direct impedance microbiology is based on the monitoring of impedance change in the medium. The principles of medium design, fundamental to traditional microbiology, are equally important to impedance microbiology. First, a medium must support the selective growth of the target bacteria, which provides the selectivity to the impedance microbiological methods. Second, the medium must be formulated to provide optimal impedance signals. For instance, one can predict that weakly buffered media would allow a greater conductance change than more strongly buffered media.

There are several fundamental differences between the impedance technique and standard methods for the enumeration of microorganisms. The impedance technique relies on the measurement of metabolic changes whereas traditional plate count procedures depend on the production of a visible biomass. Since the impedance technique depends on the rapid detection of metabolic change, factors such as time and temperature become critical parameters in the assay.

In the past few years, some new aspects have been added to this subject which have made the impedance technique a more valuable technique for studies of bacterial growth detection. These new aspects include the use of different electrode systems and the analysis of impedance components using equivalent circuits for better improvement the detection systems. The advances in microfabrication technologies have launched the use of microfabricated microarray electrodes in impedance detection and the miniaturization of impedance microbiology into a chip format, which has shown great promising in rapid detection of bacterial. The integration of impedance technique with biosensor technology has led to the recent development of impedance biosensors that is expending rapidly for bacteria detection. Impedance biosensors for bacteria detection are based on impedance analysis of the electrical properties of bacterial cells when they are attached to or associated with the electrodes. The impedance biosensor methods have substantially reduced the assay time down to between 30 min and 2 h compared with growth-based impedance methods. Particularly, significant progress has been made in developing detection platforms which exploit recent advances in microfabrication and electro mechanical nanotechnology. The dimensional compatibility of these microfabricated biosensors with the target bacterial cells has enabled them to detect the binding of bacterial cells on their surfaces without any amplification step.

To date, impedance biosensors for bacterial detection are comparable with current well studied rapid methods such as enzyme linked immunosorbent assay (ELISA) and polymerase chain reaction which have achieved detection limits varying from 10^1 to 10^6 cfu/ml and other various biosensors that have achieved detection limit of 10^3 to 10^4 cfu/ml with the assay time around 2 h under ideal conditions.

The impedance microbiological methods are perhaps the most successful of all the recently introduced rapid methods in automation. Several commercial analytical instruments are based on the principles of classic impedance microbiology. These systems include Bactometer (Bio Merieux, Nuertingen, Germany), the Malthus systems (Malthus Instruments Ltd., Crawley, UK), rapid automated bacterial impedance technique (RABIT) (Don Whitley Scientific Ltd., Shipley, UK), and Bac-Trac (Sy-Lab, Purkersdorf, Austria).

Principle and Methodology of Classical Impedance Microbiology

As a principle of transduction, one common impedance method for detection and/or quantification of bacterial growth is impedance microbiology. It is based on the measurement of changes in electrical impedance of a culture medium or a reaction solution resulting from the bacterial growth. This growth-based impedance technique allows us to distinguish between viable and dead cells. Such a method has been developed as a rapid method that can detect bacteria within 24 h.

Impedance is the opposition to flow of an alternating electrical current in a conducting material. In other words, impedance is the apparent resistance in an electric circuit to the flow of alternating current corresponding to the actual electrical resistance to a direct current. When microorganisms grow in culture media, they metabolize substrates of low conductivity into products of higher conductivity and thereby decrease the impedance of the media. When the impedance of broth cultures is measured, the curves are reproducible for species and strains and mixed cultures can be identified by use of specific growth inhibitors.

In impedance microbiology, the impedance change is typically measured using a pair of electrodes submerged in the growth medium or the reactant solution. The measurement can be performed in two ways, direct or indirect measurement. In the direct technique, a pair of metal electrodes is immersed in the medium that is inoculated with the testing bacteria. The impedance change caused by bacterial metabolism in the medium is monitored with time. The impedance change in the medium is mainly produced by the release of ionic metabolites from live cells. There are two main origins of ion release by bacteria into their growth environment. One is energy metabolism (catabolism) in which bacteria consumes oxygen and sugars and produces carbon dioxide and organic acids. For example, the conversion of a non-ionized glucose to two molecules of lactic acid will increase the conductivity of the culture medium. Further metabolism will take the lactic acid and three oxygen molecules to carbonic acid. The smaller and more mobile bicarbonate ion is a more effective ionic conductor than the lactate ion (Don Whitley Scientific, 1999). Hydrogen ions are nearly seven times more effective as an ionic conductor than sodium ions (Eden and Eden, 1984). The other origin is ion exchange through the cell membrane. Ions (such as K^+ and Na^+) are actively transported across ion channels embedded in the cell membrane, which serves to regulate the membrane potential and the osmotic difference between the interior and exterior of the cells. While the energy metabolism is the major origin of the ion release from cells to the environment, the ion exchange process is a small contribution to the ion release from cells. What is clear, is that these ion release processes cause changes in the ionic composition of the medium and consequent changes in the conductivity of the medium, which is the basis for the measurement of relative impedance changes.

Unlike the direct technique, the indirect technique does not directly measure the impedance change in the bacterial growth medium. Instead, the electrodes are immersed in a separate solution (usually a potassium hydroxide solution) instead of the inoculated growth medium. The gases (mainly CO_2) produced from bacterial metabolism are absorbed by the potassium hydroxide solution, which leads to a decrease in the conductance of the alkaline solution.

To detect bacteria, the impedance systems measure the relative or absolute changes in conductance, capacitance, or impedance, at regular time intervals during the growth of bacteria at a given temperature. The measured electrical signals are then graphically plotted on the ordinate against the incubation times on the abscissa, producing impedance growth curves. A typical impedance growth curve is illustrated in above Figure 9.1A. The impedance value is quite stable in the first region of the impedance growth curve, and then it starts decreasing. The time corresponding to a point at which the decrease in impedance value exceeds a threshold value is defined as the detection time, t_d (Figure 9.1A). Generally, the detection time does not appear until the bacteria number reaches approximately 10^6–10^7 cfu/ml (as determined by the plating method). The impedance value finally reaches a plateau, where the bacteria has grown to a high concentration of 10^8 cfu/ml or greater, and all the resources in the medium have been metabolized to end products. The shape of the impedance growth curve matches very well with the typical bacterial growth phases, which includes the lag phase where bacteria metabolize but do not multiply, the log or exponential growth phase where the bacteria multiply exponentially, and the stationary phase where the bacterial cell number remains relatively constant (Talaro, 2005).

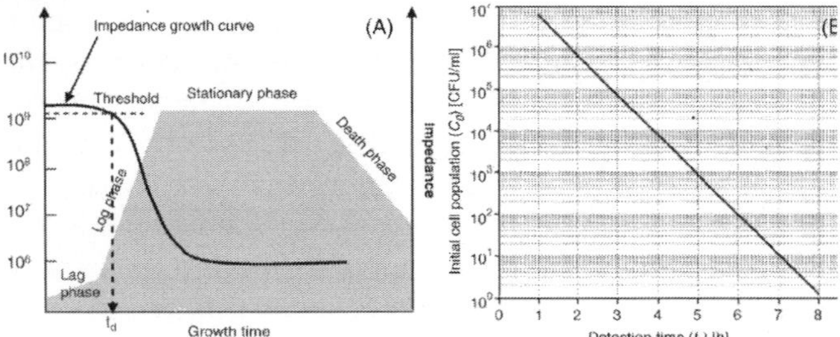

Figure 9.1: A: A generalized impedance growth curve with the threshold and the detection time, along with the typical bacterial growth phases of a bacterial growth curve. B: The relationship between the detection time and the initial cell concentration (*Source*: Gomez *et al.,* 2002).

When monitoring the growth of microorganisms, the conducting material is a microbiological medium. Impedance changes occur in the medium as its chemical composition changes due to growth and metabolic activity of microorganisms. In impedance microbiology (IM), the Sample Under Test (SUT) (possibly diluted in an enriching medium) is placed in direct contact with a couple of electrodes at a temperature favouring bacterial growth and the SUT electrical parameters are measured at regular time intervals. Since bacterial metabolism transforms uncharged or weakly charged compounds present in the growth medium in highly charged ones, the SUT electrical parameters change with the bacterial growth. During growth, microorganisms convert larger molecules to smaller, more mobile metabolites, thus changing the impedance. In practice, they remain almost constant (at a baseline value) until the bacterial concentration reaches a critical threshold (in the order of 10^7 colony forming units (CFU)/ml), when they start to significantly deviate from the baseline.

When the microbial population reaches a threshold of 10^6-10^7 cells/ml, an exponential change in impedance signal is observed. The time required for this exponential change is inversely proportional to the initial cell concentration and is defined as impedance detection time (IDT). In other words, we can say that detection time (IDT) is a linear function of the logarithm of the SUT bacterial concentration. Thus, the IDT can be used to estimate the initial SUT bacterial concentration once the system has been properly calibrated for the particular sample type and bacterial strain. IM can be used as a quick alternative to SPC to measure bacterial concentration with time response in the range 2 to 12 hours (depending on the SUT level of contamination). Furthermore, the impedance microbiology technique can be implemented in the form of devices embedded in an industrial machine or portable instruments for in-situ measurements. The impedance method is highly versatile. It could be used to perform almost all microbiological tasks needed by the food industry, if proper protocols were to be developed.

Although a successful impedance assay does not depend on the presence of identical groups of microorganisms in successive samples, the generation times for all isolates should be similar. Thus, in order to achieve the best possible correlation between standard methods and impedance detection time (IDT), the generation times for all organisms should be brought as close together as possible under the experimental conditions employed.

Certain media are more useful in impedance testing than others. This may be due to the fact that organisms utilize different metabolic pathways in different media. Since some end products yield a stronger impedance signal than others, the selection of an appropriate medium is crucial.

The method achieves faster identification of microbial problems (*e.g.,* several hours vs./ml. several days), is less labour intensive than traditional techniques, and provides automatic recording and presentation of data. The technique has been shown capable of detecting as few as ten to one hundred cells. Cell populations of 105–106/ml can be detected in 3–5 h; and 104–105/ml in 5–7 h. The times noted are required for the organisms in question to attain a threshold of 106–107 cells.

Although classic impedance microbiology was established a long time ago and has been developed as a rapid automated method for bacteria detection in 24 h, impedance technique is now entering into a new stage as a chip-based method. Advances in microfabrication have paved the way for miniaturization of impedance microbiology into microdevices and biochips, which have been proved to be very successful in maximizing the impedance signal, minimizing the volume of testing sample, increasing sensitivity, and saving assay time for the detection of foodborne pathogenic bacteria. Attractively, impedance technique is one of the few techniques that allows growth-based detection of bacteria, which can differentiate dead cell vs. live cells. Differentiation and detection of live cells is more useful than detection of both live and dead cells, since dead cells are usually not pathogenic. The ability to differentiate live and dead cells has the great promises to significantly improve food safety and security and to benefit health care.

Impedance-splitting Methods

While most impedance microbiological methods measure only the conductance of the medium using a pair of electrodes submerged in the inoculated medium at a fixed frequency, several studies have found that the total impedance during bacterial growth actually consists of two components which can be measured at different frequency ranges. One component contributed by the medium, is known as medium or electrolyte impedance, and the other component contributed by the electrode and the electrode/electrolyte interface, is known as electrode or interface impedance. These two contributing components can be distinguished by changing the frequency. Hause *et al.* (1981) observed that the electrode impedance dominated at low frequency (b100 Hz), while impedance measured at 10,000 Hz was dominated by the media effect. Felice *et al.* (1992) investigated medium and interface contributions to the impedance during bacterial growth by using the frequency dependent properties of Warburg's model. They used stainless wire electrodes for both tetrapolar and bipolar impedance measurement at several frequencies between 18 Hz and 18 kHz, by which medium impedance was obtained at high frequencies (N5 kHz) and interface impedance was negligible at such high frequencies.

The identification of electrode and medium impedance components has led to the development of impedance-splitting methods for the detection of bacteria. SY-lab (Neupurkersdorf, Austria) has developed a commercial instrument, BacTrac™, by which impedance measurements are made at 1 kHz for both the electrode impedance (E-value) and medium impedance (M-value). Rapid detection of *Bacillus stearothermophilus* has been achieved using the impedance-splitting method (Flint and Brooks, 2001).

Applications of Impedance Microbiology

Most applications of impedance microbiology in earlier years have been extensively reviewed by previous researcher (Silley and Forsythe,1996; Valentinuzzi *et al.*, 1996; Wawerla *et al.*, 1999; and Yang and Bashir, 2008. The technique has been widely used for analysis of biological material; from differentiation of red blood cells extracted from fish and human leukocytes to detection of DNA in droplets.

Impedance techniques have the great potential to solve the two key issues/concerns in detection of foodborne pathogens: rapidity and detection limit. Rapidity includes assay time and number of samples per operation (Fung, 1992). While microchip-based impedance microbiology can effectively reduce the assay time, high throughput microchips would greatly increase the number of samples per operation. It is possible to design the impedance measurements in a 96 to 1536 microplate format, which would allow continuous monitoring and high throughput screening. Miniaturized devices reduce the handling effort and cost, and provide a faster, space-saving and more economical accomplishment.

Impedance microbiological techniques have been used to detect many bacteria species, such as *Enterobacteriaceae, coliforms, Listeria spp.*, and *L. monocytogenes* in various samples. Impedance techniques can also be used for monitoring the growth behaviour of bacteria. The impedance growth curves under different conditions have been found to be characteristics of bacteria species.

Much work has been done to apply the impedance technique to the detection of different bacteria in different samples using the existing commercial instruments. For example, the Bactometer system was used to detect total microbial load in meat and fish as reported by Russell *et al.* (1994) and Van Spreeken and Stekelenburg (1986), respectively. Detection of *E. coli* in potable water and shellfish was reported by Colquhoun *et al.* (1995) and Dupont *et al.* (1996), respectively. Edmiston and Russell (1998) conducted a study to evaluate a conductance method for rapidly enumerating *E. coli* from broiler carcasses using the Bactometer system. Selective detection of *Staphylococcus aureus* with impedance measurement was carried out in a nutrient broth containing 10 ppm of nalidixic acid and 10 ppm of acriflavine (Glassmoyer and Russell, 2001).

Impedance microbiology has been successfully tested for bacterial concentration measurements in various types of samples (ice-cream, meat, vegetables, raw-milk, fresh water beer, *Escherichia coli* in human urine samples and to test the efficiency of various antibiotics and chemical preservatives (Yang and Bashir, 2008).

Impedance has been evaluated by a large number of investigators as a means of monitoring the overall microbial quality of various foods. Two hundred samples of pureed vegetables were assessed and a 90–95 per cent agreement was found between impedance measurements and plate count results relative to unacceptable levels of bacteria. Impedance analysis required 5 h and the method was reported to be applicable to cream pies, ground meat and other foods. The microbiological quality of pasteurized milk was assessed by using the impedance detection time (IDT) of 7 h or less, which was equivalent to an aerobic plate count (APC) of 10,000/ml for more bacteria. Of 380 samples evaluated, 323 (85 per cent) were correctly assessed by impedance. Using the same criterion for 27 samples of raw milk, 10 h were required for assessment. From a collaborative study of raw milk involving six laboratories, impedance results varied less than SPC results between laboratories. Impedance results have been compared to plate counts and Moseley test results to predict shelf life of pasteurized whole milk. IDTs at 18° and 21°C correlated better with Moseley results than plate counts as shelf-life predictors and results were obtained by impedimetry in 25–38 h, compared to 7–9 days for Moseley test results. The findings of these investigators suggested that if the IDT was ≤ 6.1 h, the potential shelf life would be ≤ 9 days while an IDT of ≥ 12.4 h indicated a potential shelf life of ≥ 9 days.

From another study with raw milk, impedance was found useful when a 7 h cut-off time (10^5 cfu/ml) was used to screen samples. Findings from this study were compared to SPC data. As a monitor of the post-pasteurization contamination of cream, an IDT of > 20 h was found to denote cream of acceptable microbiological quality.

By use of impedance, the brewing industry test for detecting spoilage organisms in beer was shortened from 3 weeks or more to only 2–4 days. Yeasts growing in wort caused an increase in impedance while bacteria caused a decrease. For raw beef, IDTs for forty-eight samples were plotted against log bacterial counts and a regression coefficient of 0.97 was found. The IDT for meats was found to be about 9 h.

Impedance has been used to Impedance has been used to classify frozen orange juice concentrate as acceptable (< 10^4 cfu/ml) or unacceptable (> 10^4 cfu/ml). By using cut-off times of 10.2 h for bacteria, 15.8 h for yeasts, 96 per cent of 468 retail samples could be correctly classified. The method has been employed to detect starter culture failure within 2 h. In this application, complete starter failure occurred with approximately 10^5 phages/ml, resulting in the inhibition of normal impedance decrease caused by growth of normal lactic starter organisms. The relative level of contamination of meat surfaces by impedance has been assessed. With 10_7 cells and above/cm_2, detection could be made accurately within 2 h.

A relatively large number of studies have been conducted with impedance as a rapid method to detect coliforms. Fecal coliforms could be detected in 6.5–7.7 h when 100 log- or exponential-phase cells were inoculated into a selective medium and incubated at 44.5°C. Employing a new selective medium for coliforms, impedance was evaluated on seventy samples of ground beef and 79 per cent of the impedimetric results fell within the 95 per cent confidence limits of the three-tube MPN procedure for coliforms and fewer than 100 to 21,000 cells/g could be detected with results obtained within 24 h. In another study of coliforms in meats, the investigators found necessary to develop a new selective medium that yielded impedance signals more consistent with cfu results. From an inoculum of ten coliforms into the new medium, the average IDT was 3.8 h. In ninety-six meat samples, a correlation coefficient of 0.90 was found between impedance and corrected coliforms counts on violet red bile agar. An IDT of 6.5 h was required for meat samples with 10^3 coliforms and the authors suggested that an impedance signal in 5.5 h or less denoted meat samples with coliforms > 10^3/g while the inability to detect by impedance in 7.6 h denoted coliform levels < 10^3/g. For coliforms in raw and pasteurized milk and two other dairy products, an IDT of < 9 h indicated coliforms were > 10/ml while an IDT of > 12 h indicated < 10 cells/ml.

Table 9.1 indicates the levels of toxins or organisms by physical, chemical and immunologic methods of analysis.

Table 9.1: Levels of Toxins or Organisms by Physical, Chemical and Immunologic Methods of Analysis

Methods	*Toxin or Organism*	*Sensitivity*
Impedance	Coliforms in meats	103/g in 6.5 h
	Coliforms in culture media	10 in 3.8 h
	Fecal coliforms at 44.5°C	100 in 6.5–7.7 h
Microcalorimetry	*S. aureus* cells	2 cells in 12–13 h
	S. aureus	Minimum HPR a~ 104 cells/ml
ATP measurement	Total flora of meats	104 cells/g in 20–25 min.
Radiometry	Frozen orange juice flora	104 cells/g in 6–10 h
	Salmonellae in foods; at 37° C with poly H antiserum	104–105 cells/g in 7 h
	Coliforms in water	1–10 cells in 6 h
Glutamate decarboxylase	*E. coli*	50,000 cells/ml in < 10 h
ONPG	*E. coli*	1 cell in < 20 h
Felix-01 phages	Salmonellae	< 5 cells/ml in 24 h
Fluorescent antibody	Salmonellae	106 cells/ml
	Staph. enterotoxin B	ca. 50 ng/ml
Thermostable nuclease	From *S. aureus*	10 ng/g
	From *S. aureus*	2.5–5 ng
Limulus lysate test	Gram-negative endotoxins	2–6 pg of *E. coli* LPS
Radioimmunoassay	Staph. enterotoxins A, B, C, D and E in foods	0.5–1.6 ng/g
	Staph. enterotoxin B in non-fat dry milk	2.2 ng/ml
Radioimmunoassay	Staph. enterotoxins A and B	0.1 ng/ml for A; 0.5 ng/ml for B
	Staph. enterotoxin C_2	100 pg
	E. coli ST_a enterotoxin	50–500 pg/tube
	Aflatoxin M_1 in milk	5–50 ng/assay
	Aflatoxin B_1	0.5–5.0 ng
	Ochratoxin A	20 ppb
	Bacterial cells	500–1,000 cells in 8–10 min.
Electroimmunodiffusion	*C. perfringens* enterotoxin	10 ng
	Botulinal toxins	3.7–5.6 mouse LD50/0.1 ml
Micro-Ouchterlony	*S. aureus* enterotoxins A and B	10–100 ng/ml
	C. perfringens type A toxin	500 ng/ml

Methods	*Toxin or Organism*	*Sensitivity*
Passive immune hemolysis	*E. coli* LT enterotoxin	< 100 ng
Aggregate-hemaggluti-nation	*B. cereus* enterotoxin	4 ng/ml
Latex agglutination	*E. coli* LT enterotoxin	32 ng/ml
Single radial immuno-diffusion	*S. aureus* enterotoxins	0.3 µg/ml
Hemagglutination-inhibi-tion	Staph. enterotoxin B	1.3 ng/ml
Reverse passive hemagg-lutination	Staph. enterotoxin B	1.5 ng/ml
	C. perfringens type A toxin	1 ng/ml
ELISA	Staph. enterotoxin A in wieners	0.4 ng
	Staph. enterotoxins A, B, and C in foods	0.1 ng/ml
	Staph. enterotoxins A, B, C, D and E in foods	≥ 1 ng/g
	Botulinal toxin type A	About 9 mouse LD50/ml with monoclonal antibody
	Botulinal toxin type A	50–100 mouse i.p. LD50
	Botulinal toxin type E	100 mouse LD50
	Aflatoxin B1	25 pg/assay
	Aflatoxin B1	10 pg/ml
	Aflatoxin M1 in milk	0.25 ng/ml in 3 h
	Aflatoxin B1	< 1 pg/assay
	Ochratoxin A	25 pg/assay
	Salmonellae	105 cells/ml
	Salmonellae	104–105 cells/ml

In general, impedance methods measure changes with the conductance of the growth medium. The effect of microbial growth on changes in polarization capacitance and medium conductivity has been evaluated by use of an electrobacteriological impedance model. This model consists of a pair of electrodes inserted in a growth medium containing microorganisms. Upon passage through the medium of a small-amplitude AC current, total conductance (G_T) is measured. Conductance generally increases from 10 to 100 per cent during growth, apparently from the increase in ion pairs as a result of metabolic activity. The parameter G_{pol} refers to capacitance polarization effect (electrode capacitance) and G_{sol} to medium conductance. The former is more useful in high-conductive media while the latter is useful in low-conductive media where changes are correlated with bacterial growth. Employing two media, yeast growth resulted in 20–100 per cent changes in G_{pol} while G_{sol} changes were only 1–4 per cent. Changes in pH were shown to affect G_{pol} more than G_{sol} is believed by these investigators to be quite suitable for monitoring yeast growth and microbial growth in general.

Impedance has been compared to radiometry for the recovery of heat-stressed fecal coliforms from meat loaf. Radiometry detected in about 17 h and impedance in about 18 h. Either method was found for suitable for determining in cooked food meats, the criterion of 0 fecal coliforms/g.

Further Studies

Single cell level detection needs to be explored using impedance biosensors. Despite the significant improvement in reducing the assay time and lowering detection limit, to date, detection at a single cell level within a short assay time has not been achieved yet. As demonstrated in micromachined impe-dance spectroscopy flow cytometer (Gawad *et al.*, 2001), a micro-device can be designed to measure the spectral impedance of individual cells, which have shown the possibility for individual cell detection in a high throughput way by impedance technique.

Impedance biosensors also offer the promises in developing biosensor array for multiplex analysis of different bacteria simultaneously, providing label-free, on-line and high through-out devices for bacteria detection.

By successfully integrating micro and nano-fluidics with biosensors, many of the unit operations associated with sample preparation, such as separation (cell or biomolecule), mixing, incubation, concentration, *etc.*, may be performed directly in a lab-on-a-chip format. Also, the use of microliter and nanoliter volumes implies the use of smaller sample volumes, smaller quantities of usually costly reagents, and greater analytical sensitivity.

ii) Microcalorimetry

Microcalorimetry is the study of small heat changes; the measurement of the enthalpy change was involved in the breakdown of growth substrates. The heat production that is measured is closely related to the cell's catabolic activities. There are two types of calorimeters *i.e.*, batch and flow. Most of the early work was done with batch-type instruments. The thermal events measured by microcalorimetry are those from catabolic activities as noted above. One of the most widely used microcalorimeters for microbiological work is the Calvet instrument, which is sensitive to a heat flow of 0.01 cal/h from a 10 ml sample. It is a rapid method for identification and characterization of food-borne organisms, but results vary according to the history of the organism, inoculum size and fermentable substrates.

Microcalorimetric results vary according to the history of the organism, inoculum size, fermentable substrates and the like. One group of investigators found the variations such that the identification of microorganisms by this method was questioned, but in a later study in which a synthetic medium was used. Perry *et al.,* successfully characterized commercial yeast strains. The utility of the method to identify yeasts has been questioned but by use of flow microcalorimetry yeasts could be characterized. The latter method is one in which a microcalorimeter is fitted with a flow-through calorimetric vessel. By use of a chemically defined medium containing seven sugars, thermograms were produced by nine lactic acid bacteria (belonging to the genera *Streptococcus, Leuconostoc* and *Lactobacillus*) distinctive enough to recommend the method for their identification. All cultures were run at 37°C except *S. cremoris,* which was run at 30°C and results were obtained within 24 h.

Microcalorimetry has been applied to the study of spoilage in canned foods to differentiate between the Enterobacteriaceae, to detect the presence of *Staphylococcus aureus* and to estimate bacteria in ground meat. Boling *et al.,* were able to differentiate between seventeen species of ten genera of *Enterobacteriaceae* by inoculating approximately 500 cells into BHI broth and recording the temperature changes over an 8–14 h incubation period. The thermograms produced were distinctive for each organism. In detecting *S. aureus,* Lampi *et al.,* achieved results in 2 h using an initial number of 10_7–10_8 cells/ml and in 12–13 h when only 2 cells/ml were used. Russell *et al.,* examined over 250 cultures representing twenty-four genera and found that most of the organisms typically produced maximum outputs of 40–60 µcal/(sec) (ml) and returned to baseline in 5–7 h thereafter. Some profiles developed within 3 h while others required up to 14 h. A monitoring use of microcalorimetry is suggested by the work of Beezer *et al.,* who used flow microcalorimetry to determine the viability of recovered frozen cells of *Saccharomyces cerevisiae* within 3 h after thawing. When applied to comminuted meat, the peak exothermic heat production rate (HPR) could be recorded within 24 h for meats containing 10_5–10_8 cfu/g and results by microcalorimetry correlated well with plate count results. With 10_2 cfu/ml, a measurable HPR was produced after 6 h, with peak HPR at 10 h.

Isothermal microcalorimetry (IMC) is a highly sensitive calorimetry that is used to study thermal processes at a set temperature in the range of 5–95°C. Isothermal microcalorimetry (IMC) is a laboratory method for real-time, continuous measurement of the heat flow rate (µJ/sec = µW) and cumulative amount of heat (J) consumed or produced at essentially constant temperature by a specimen placed in an IMC instrument.

IMC instrumentation can detect signals in the order of 10^{-6}°C in ΔT, far exceeding the sensitivity in conventional DSC (on the order of 10^4 times increased sensitivity). Thermal reactions that lead to physical and/or chemical stability may be studied at higher temperatures where the processes proceed at significant kinetic rates. The introduction of IMC in the 1980s (*e.g.,* thermal activity monitor, TAM) has made it possible to monitor heat processes and deduce stability information of pharmaceuticals at temperatures of interest directly.

The heat flow from an IMC experiment can be represented as:

$$\frac{dQ}{dt} = \Delta H \cdot C_0 \cdot \frac{d\alpha t}{dt}$$

where ΔH is the enthalpy of the process under investigation, C_0 is the initial reactant concentration, α is the extent of conversion, ranging from 0 to 1, and $d\alpha/dt$ is the rate of the reaction.

IMC has been used to study most typical physical chemical processes of pharmaceuticals, including crystallization, polymorphic transformation, dehydration/hydration, and chemical reaction such as drug decomposition and excipient compatibility in formulation design. The volume of the IMC cells allows the humidity of the headspace to be controlled through a humidity-stat such as a salt solution. Therefore, it can be useful to study the hygroscopicity, dehydration/hydration of drugs, and other pharmaceutical components.

One drawback of IMC is that it is incapable of temperature scanning. To that end, micro DSC is capable of performing both isothermal and nonisothermal calorimetric studies. Because of the increased sample size and the sensitivity requirements, the scanning rate is generally small (up to about 1°C/minutes), and the temperature ranges are generally limited (from about –40°C to anywhere between 100°C and 200°C). However, to the plus, the instrument allows for a very slow scanning rate such as 0.001°C/minutes to be controlled. Micro DSC has been used to study the phase transition of biological samples, hydration/dehydration of hydrates, and crystallization of amorphous materials under controlled RH, as well as drug degradations.

iii) Flow Cytometry

Flow cytometry is the science of measuring components (cells) and the properties of individual cells in liquid suspension. In essence, suspended cells one by one are brought to a detector by means of a flow channel. Fluidic devices under laminar flow define the trajectories and velocities that cells traverse the detector and among the cell properties that can be detected are fluorescence, absorbance and light scatter. By use of flow sorting, individual cells are selected on the basis of their measured properties and from 1–3 or more global properties of the cell can be measured. Flow cytometers and cell sorters make use of one or more excitation sources such as argon, krypton, or helium-neon ion lasers and one or two fluorescent dyes to measure and characterize several thousand cells/sec. When a dye is used, its excitation spectrum must match the light wavelengths of the excitation source. Two dyes may be used in combination to measure, for example, total protein and DNA content. In these instances, both dyes must excite at the same wavelength and emit at different wavelengths so that the light emitted by each dye is measured separately. The early history of flow cytometry has been reviewed by Horan and Wheeless.

While most studies have been conducted on mammalian cells, both DNA and protein have been measured in yeast cells. Typically, yeast cells are grown, fixed and incubated in an RNAse solution for 1 h. Cell protein may be stained with fluorescein isothiocyanate and DNA with propidium iodide. Following necessary washing, the stained cells are suspended in a suitable buffer and are now ready for application to a flow cytometer. The one used by Hutter *et al.,* was equipped with a 50 mW argon laser. Yeast cells were excited at different wavelengths with the aid of special optical filters. By this method, baker's yeast was found to contain 4.6×10^{-14} g of DNA/cell and the protein content/cell was found to be 1.1×10^{-11} g.

The possible application of flow cytometry to the analysis of foods for their content of microorganisms is suggested by the study of Van Dilla *et al., 1983* in which bacteria were stained with fluorescent dyes, one of which bound preferentially to DNA rich in guanine-cytosine (G-C) while the other bound to adenine-thymine (A-T) rich DNA. Thus, the rapid identification of bacteria in food based upon their specific A-T/G-C ratios is possible.

In addition to the simultaneous measurement of protein and DNA in cells and the enumeration of cells based upon G-C and A-T content, flow cytometry has been used to distinguish between living and dead cells by dual staining; to determine the ploidy of yeast cells; to differentiate between spores and vegetative cells in *Bacillus* spp.; and to separate pathogenic and non-pathogenic amoebae.

II. Enrichment Serology and Radioimmunoassay (RIA)

The use of enrichment serology (ES) is a more rapid method for recovering salmonellae from foods than the conventional culture method (CCM). Originally developed by Sperber and Deibel in 1969, it is carried out in four steps: pre-enrichment in a non-selective medium for 18 h; selective enrichment in selenite-cystine and/or tetrathionate broth for 24 h; elective enrichment in M broth for either 6–8 h or 24 h; and agglutination with polyvalent H antisera at 50°C for 1 h (Figure 9.2). Results can be obtained in 50 h (depending upon elective enrichment time used) compared to 96–120 h by CCM. A modified ES method has been proposed involving a 6 h pre-enrichment, thus making it possible to obtain results in 32 h.

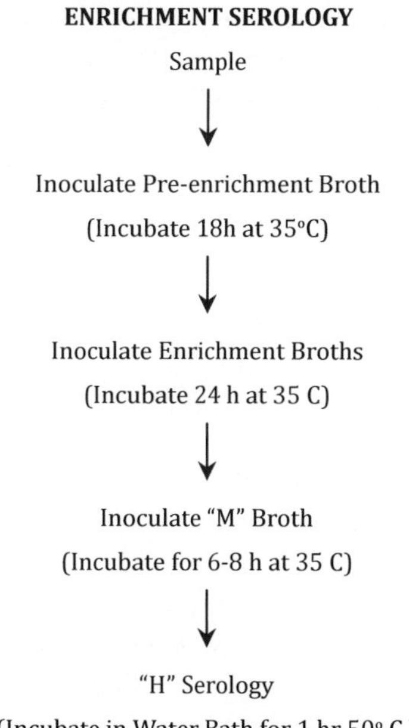

Figure 9.2: Enrichment Serology Methodology for Salmonella Detection.

Two studies comparing the cultural procedure and the ES procedure for detection of salmonellae have been published in 1969 by Sperber and Deibel, and Fantasia *et al.* (1969). An excellent correlation was obtained between the cultural and ES procedures. Sperber and Deibel's study involved 105 samples representing 44 different materials. Samples included animal feeds, condiments, environmental samples from a food plant, and food products. Of the 105 samples analysed for the presence of salmonellae, 37 positives and 68 negatives were obtained by each procedure. Results obtained by the cultural method and the ES procedure were the same for each individual sample. Employing the ES procedure on 105 samples of dried foods and feeds, all positive samples by CCM were positive by ES. Similar findings were achieved when 689 samples were examined by ES and CCM.

The second accelerated method considered for the detection of salmonellae is the Fluorescent Antibody Procedure (FA) which is shown in Fig. 9.3. The FA technique has been proposed by many investigators as a rapid, economical screening method to detect salmonellae. The application of this technique for examination of foods was first suggested by Arkhangel'ski and Kartoshova in 1962. Two years later, Georgala and Boothroyd employed the FA technique to examine raw meats for the presence of Salmonella. Their analysis indicated the potential of the FA method. Soon after the initial suggestion, the FA technique was found suitable for detecting Salmonella in egg products, by Silliker *et al.* (1966); animal feeds and feed ingredients by Laramore and Moritz, 1969; and other foods by Insalata *et al.,* in 1967.

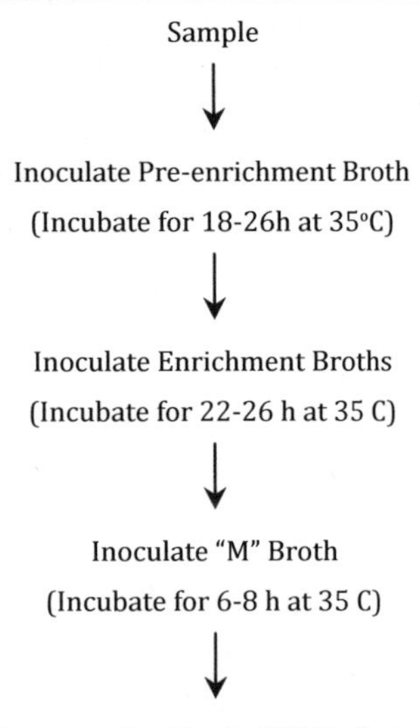

Figure 9.3: Fluorescent Antibody (FA) Methodology for Salmonella Detection.

The conventional FA methodology for detecting salmonellae is also similar to the cultural method. The same pre-enrichment broth and selective enrichment broths are employed. This technique branches from the cultural procedure after the selective enrichment step in a fashion similar to the ES procedure. For the FA technique, 0.001 ml of each 24-h old culture in selenite cystine and tetrathionate broths is placed on a clean microscope slide and allowed to air dry. Dried specimens are then fixed by flooding the slide with a solution of ethyl alcohol-chloroform-formalin at concentrations of 110:60:30, respectively. After being fixed for 3 min, smears are rinsed with 95 o/o ethyl alcohol. The slide is allowed to air dry and then is stained. The FA staining procedure is a matter of personal choice; a direct or indirect staining procedure can be employed.

Smears which have been stained by either the direct or indirect method can be examined with a microscope equipped with an oil-immersion lens, dark-field condenser, and a mercury arc lamp as the light source. To attain the desired wave length of light at 435 nm, a BG 12 exciter filter is used along with heat-absorbing filters and barrier filters. An immuno-fluorescent microscope can also be used. The immuno-fluorescent microscope employs a 100-watt halogen light source in conjunction with a special fluorescein isothiocyanate interference filter.

The ES method was compared with three FA procedures for salmonellae detection in 347 food samples and while 52 were positive by CCM, 51 were positive by ES. In the latter study, only one false negative was found by ES and the authors preferred this method with a 6-h elective enrichment over CCM. In yet another comparison of ES with CCM using 2,208 samples, 95 per cent of positive samples by CCM were positive by ES when elective enrichment was carried out for 24 h. ES was found to be as good as FA. The 6 h elective enrichment was unsuitable for comminuted meat products. When the ES and FA methods were combined, a greater accuracy was achieved than by use of either alone. Employing 126 samples, 66 were salmonellae-positive by CCM and FA and 64 by ES. The latter exhibited a relatively low incidence of false positives while FA showed a higher incidence. The ES method was later shown capable of detecting 97–99 per cent of salmonellae-positive samples. Using a modified ES as noted above on 3,486 samples, 96.7 per cent were correctly analyzed. Overall, the ES method provides results in 32–50 h compared to 92–120 for CCM; results are comparable to both CCM and FA; and no specialized equipment or training is needed. Possible disadvantages to its use is the need for a minimum of about 10_7 cells/ml and its lack of response to non-motile salmonellae. The latter can be overcome by use of a slide agglutination test from the elective enrichment broth employing polyvalent O antiserum.

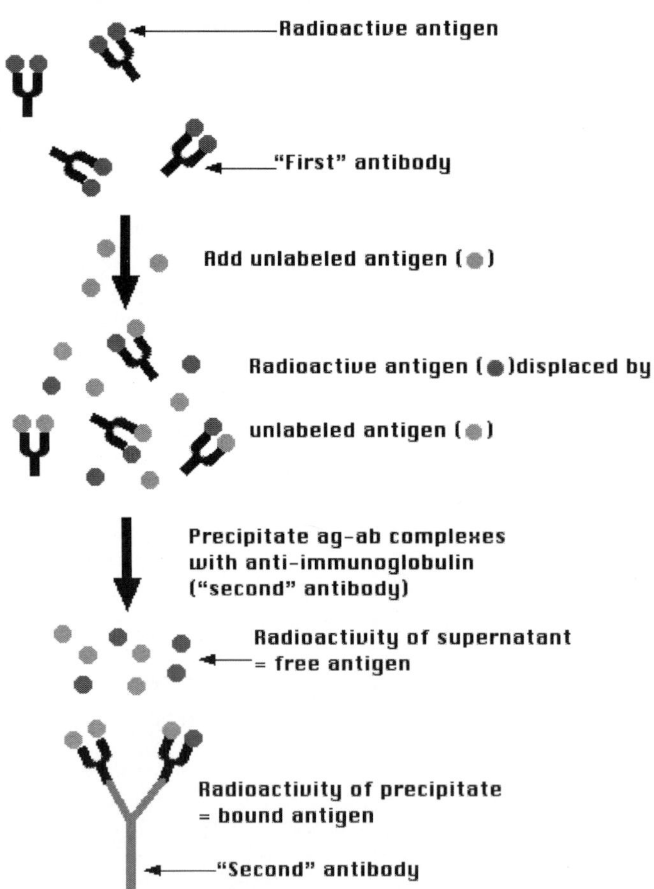

Figure 9.4: Schematic Diagram of Radioimmuno Assay Technique.

Radioimmunoassay (RIA) is an *in vitro* assay that measures the presence of an antigen with very high sensitivity. Basically, any biological substance for which a specific antibody exists can be measured, even in minute concentrations. This method was developed by Rosalyn

Sussman Yalow, Roger Guillemin, and Andrew Schally at the Veterans Administration Hospital in the Bronx, New York. The technique was introduced in 1960 by Berson and Yalow as an assay for the concentration of insulin in plasma. This revolutionary development earned Dr. Yalow the Nobel Prize for Medicine in 1977. Although the RIA technique is extremely sensitive and extremely specific, requiring specialized equipment, it remains among the least expensive methods to perform such measurements. It requires special precautions and licensing, since radioactive substances are used.

Its principle is that it involves competition between radiolabelled and unlabelled antigen for binding with specific antibodies. So, antigen-antibody complexes are formed and radioactivity is measured. This technique consists of adding a radioactive label to an antigen, allowing the labelled antigen to react with its specific antibody and measuring the amount of antigen that combined with the antibody by use of a counter to measure radioactivity (Figure 9.4). Solid-phase RIA refers to methods that employ solid materials or surfaces onto which a monolayer of antibody molecules binds electrostatically. The solid materials used include polypropylene, polystyrene, bromacetylcellulose and so on. The ability of antibody-coated polymers to bind specifically with radioactive tracer antigens is essential to the basic principle of solid-phase RIA. When the free-labelled antigen is washed out, the radioactivity measurements are quantitative. The label used by many workers is ^{125}I.

Johnson *et al.,* developed a solid-phase RIA procedure for the determination of *S. aureus* enterotoxin. Band found the procedure to be five to twenty times more sensitive than the immunodiffusion technique. These investigators found the sensitivity of the test to be in the 1–5 ng range employing polystyrene and counting radioactivity with an integral counter. Collins *et al.,* employed RIA for enterotoxin B with the concentrated antibody coupled to bromacetylcellulose. Their findings indicated the procedure to be 100-fold more sensitive than immunodiffusion and to be reliable at an enterotoxin level of 0.01 μg/ml. Staphylococcal enterotoxin A was extracted from a variety of foods including ham, milk products, crab meat and so on by Collins *et al.,* and measured by RIA all within 3–4 h. They agreed with earlier workers that the method was highly sensitive and useful to 0.001 μg/ml and quantitatively reliable to 0.01 μg/ml of enterotoxin A.

By iodination of enterotoxins, solid-phase RIA can be used to detect as little as 1 ng toxin/g. When protein A was used as immunoabsorbent to separate antigen-antibody complex from unreacted toxin, a sensitivity of < 1.0 ng/g for *Staphylococcal* enterotoxin A (SEA), SEB, SEC, SED and SEE was achieved within one working day. In another study, 0.1 ng/ml of SEA and 0.5 ng/ml of SEB could be detected when protein A was used. A sensitivity of 100 pg for SEC_2 was achieved by use of a double-antibody RIA.

An RIA method has been developed for *E. coli* ST_a that detected 50 to 500 pg toxin/tube. The method measured ST both from human and porcine strains and no cross-reactivity with ST_b was noted. Using solid- phase sandwich RIA, colonies of *E. coli* that produced LT and colonies of *V. cholerae* that produced cholera toxin (CT) could be differentiated from non-toxigenic colonies. The method detected 5 to 25 pg of pure toxin and was more sensitive than ELISA. A microtiter solid-phase RIA for LT was developed and found to be comparable in sensitivity to Y-1 adrenal cells.

For mycotoxins, an RIA method has been developed for ochratoxin A with a sensitivity of 20 ppb. A method has been developed for aflatoxin M_1 (AFM_1 in milk with a sensitivity of 5–50 ng/assay, but it was subject to interference by whole milk. An ELISA method detected as little as 0.25 ng AFM_1 in 3 h. The action level of the US Food and Drug Administration for AFM_1 in milk is 0.5 ng/ml.

The RIA technique lends itself to the examination of foods for other biological hazards such as endotoxins, paralytic shellfish toxins and the like. The detection and identification of bacterial cells within 8–10 min has been achieved by use of $_{125}$I-labeled homologous antibody filtered and washed on a Millipore membrane. Multibacterial species have been detected in one operation when mixtures of homologous antibodies were used. Indirect and direct methods may be used with the former requiring only one labeled globulin, while with the latter a labeled antibody to each organism in a mixed population is needed.

The main drawbacks to radioimmunoassay are:

- The expense and hazards of preparing and handling the radioactive antigen.
- The cost of equipment and reagents.
- Short shelf-life of radiolabelled compounds.
- The problems associated with the disposal of radioactive waste
- Both ^{125}I or ^{131}I emit gamma radiation that requires special counting equipment;
- The body concentrates iodine atoms — radioactive or not — in the thyroid gland where they are incorporated in thyroxine (T4).

Despite these drawbacks, RIA has become a major tool in the clinical laboratory where it is used to assay

- Plasma levels of:
  - most of our hormones;
  - digitoxin or digoxin in patients receiving these drugs;
  - certain abused drugs

- For the presence of hepatitis B surface antigen (HBsAg) in donated blood;
- Anti-DNA antibodies in systemic lupus erythematosus (SLE).

Enzyme-Linked Immunosorbent Assay

Since the development of radioimmunoassay (RIA) in 1960, there has been a rapid increase in immunoassay techniques using radioactive labels. However, radioactive labels have been gradually replaced with enzyme labels because of safety concerns associated with radioactivity since the study by Avrameas in 1969, who coupled antigens or antibodies and enzymes using glutaraldehyde. Currently, ELISA has a higher number of immunoassays compared to RIA. Essentially synonymous with ELISA are enzyme-multiplied immunoassay technique (EMIT) and indirect enzyme-linked antibody technique (ELAT).

There are four main general steps to completing an ELISA immunoassay. These steps are:

1. Coating (with either antigen or antibody)
2. Blocking (typically with the addition of bovine serum albumin [BSA])
3. Detection
4. Final read

A typical ELISA is performed with a solid-phase (polystyrene) coated with antigen and incubated with antiserum. Following incubation and washing, an enzyme-labelled preparation of anti-immunoglobulin is added. After gentle washing, the enzyme remaining in the tube or microtiter well is assayed to determine the amount of specific antibodies in the initial serum. Detection is carried out by the addition of a substrate that can generate a colour. There are many substrates available for use in ELISA detection. However, the most commonly used enzyme is horseradish peroxidase (HRP) and alkaline phosphatase (ALP). The amount of enzyme present is ascertained by the colorimetric determination of enzyme substrate. The substrate for HRP is hydrogen peroxide and results in a blue colour change. ALP measures the yellow color of nitrophenol after room temperature incubation periods of 15 to 30 minutes and usually uses P-Nitrophenyl-phosphate (pNPP) as its substrate. Between each of the above four steps is a "wash" of the plate using a buffer, such as phosphate-buffered saline (PBS) and a non-ionic detergent, to remove unbound material. The wells are washed two or more times during each wash step, depending on the specific protocol being followed. In the ELISA protocol, usually, a serial dilution of concentrations is placed in the wells of the plate. After the results are measured, a standard curve from the serial dilutions data is plotted with a concentration on the x-axis using a log scale and absorbance on the y-axis using a linear scale. Table 9.2 depicts the advantages and disadvantages of ELISA.

What Enzyme-Linked Immunosorbent Assay (ELISA) does

- ELISA is a biochemical assay that uses antibodies and an enzyme-mediated colour change to detect the presence of either antigen (proteins, peptides, hormones, *etc.*) or antibody in a given sample.
- Both "indirect" and "sandwich" ELISAs allow detection of antigen or antibody at very low concentrations.
- The competitive method detects compositional differences in complex antigen mixtures with high sensitivity, even when the specific detecting antibody is present in relatively small amounts.
- Multiple and portable ELISA is a ready-to-use, low-cost lab kit that is ideal for large population screening in low-resource settings.

LIMITATIONS

- The enzyme-mediated colour change will react indefinitely. Over a sufficiently long period of time, the colour strength will inaccurately reflect the amount of primary antibody present, yielding false-positive results.
- To detect a given antibody or antigen, a known reciprocal antigen or antibody must be generated.
- Nonspecific binding of the antibody or antigen to the plate will lead to a falsely high-positive result.

CONCLUSION

- EIA/ELISA is a powerful method not only for general biomedical research but also as a diagnostic tool.
- It allows detection of all types of biological molecules at very low concentrations and quantities.
- Although it has its limitations, EIA/ELISA remains an important tool in both clinical and basic research, as well as in clinical diagnostics.

Among the applications of the ELISA technique in food microbiology is its use to detect *Staphylococcal* enterotoxins, botulinal toxins, salmonellae, shigellae, mycotoxins, viruses, toxoplasma antibodies, and antibodies to gram negative endotoxins.

Table 9.2: Advantages and Disadvantages of ELISA

Advantages	*Disadvantages*
Simple procedure	Labour-intensive and expensive to prepare antibody
Easy to perform with simple procedure	Sophisticated techniques and expensive culture media are required
High specificity and sensitivity	High possibility of false positive/ negative

Advantages	*Disadvantages*
ELISA is based on antigen–antibody reaction	Insufficient blocking of immobilized antigen results in false results
High efficiency	Antibody instability
Simultaneous analysis can be performed without complicated sample pre-treatment	Refrigerated transport and storage are required as an antibody is a protein
Generally safe and eco-friendly	
Radioactive substances and large amounts of organic solvent are not required	
Cost-effective assay	
Reagents are relatively low cost	

Source: Sakamoto, S. *et al.*, 2018. Enzyme-linked immunosorbent assay for the quantitative/qualitative analysis of plant secondary metabolites. J Nat Med. 72(1): 32–42.

There are four major types of ELISA:

- Direct ELISA (antigen-coated plate; screening antibody)
- Indirect ELISA (antigen-coated plate; screening antigen/antibody)
- Sandwich ELISA (antibody-coated plate; screening antigen)
- Competitive ELISA (screening antibody)

Direct ELISA

In 1971, Engvall and Perlmann and Van Weemen and Schuurs were the first to develop direct ELISA (Fig. 9.5), which was the base style for other types of ELISA. Both direct and indirect ELISAs begin with the coating of antigen to the ELISA plates. The first binding step involves adding antigen to the plates, which is incubated for one hour at 37°C or can be incubated at 4°C overnight. Once the incubation step is completed, the next step is to wash the plates of any potential unbound antibody and block any unbound sites on the ELISA plate using agents like BSA, ovalbumin, gelatin, casein, aprotinin, or other animal proteins to avoid the non-specific adsorption of other proteins. This second step is important because it prevents the binding of any non-specific antibodies to the plate and minimizes false-positive results. After adding the buffer, the plate is rewashed, and a selected enzyme-conjugated primary detection antibody is added. The plate is further incubated for one hour.

In a direct ELISA, the primary detection antibody binds directly to the protein of interest. Next, the plate is rewashed to remove any unbound antibody and followed by the addition of a substrate/chromophore, such as alkaline phosphatase (AP) or Horseradish Peroxidase (HRP) to the plate, which results in a colour change. The colour change of the sample occurs by either the hydrolysis of phosphate groups from the substrate by AP or by the oxidation of substrates by HRP. The advantages of using direct ELISA include eliminating secondary antibody cross-reactivity, and due to fewer steps, it is rapid compared to indirect ELISA. Direct ELISA is suitable for the qualitative analysis of macromolecules. Its disadvantages include its low sensitivity compared to the other types of ELISA and its high cost of reaction.

Indirect ELISA

Indirect ELISA systems have been developed on the basis of direct ELISA to evaluate the presence of antibody in antisera (Fig. 9.6). The steps of the indirect ELISA are identical to the direct ELISA, except for an additional wash step and the types of antibody added after the buffer is removed. Indirect ELISA requires two antibodies, a primary detection antibody that sticks to the protein of interest and a secondary enzyme-linked antibody complementary to the primary antibody. The primary antibody is added first, followed by a wash step, and then the enzyme-conjugated secondary antibody is added and incubated. After this, the steps are the same as the direct ELISA, which includes a wash step, the addition of substrate, and detection of a colour change. The signal increases with an increasing amount of the immobilized target antigen.

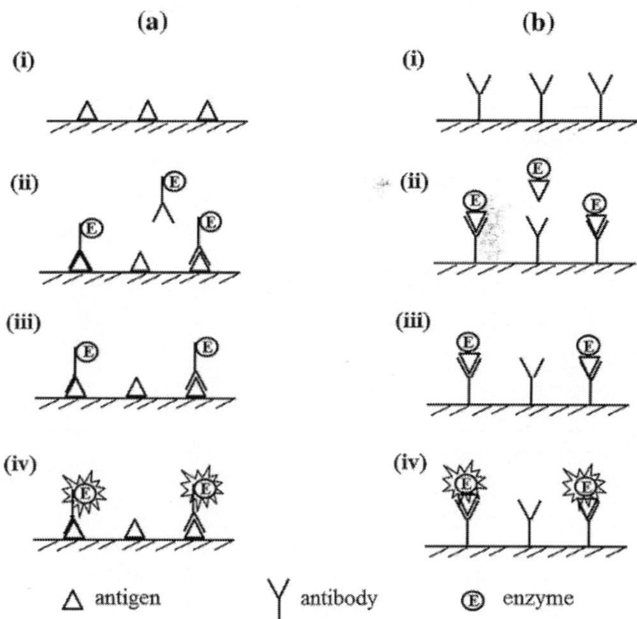

Figure 9.5: Direct ELISA to Detect Antigen (a) and Antibody (b).

(i) Attach antigen/antibody to solid phase. (ii) Incubate with enzyme-labelled antibody/antigen. (iii) Wash unbound enzyme-labelled antibody/antigen out. (iv) Develop colour with substrate

Indirect ELISA is suitable for measuring macromolecules. With the use of antisera as the primary antibody, the presence of a disease-associated antibody in the antisera can be evaluated; thus, indirect ELISA is effectively used to diagnose endocrine diseases.

The indirect ELISA has a higher sensitivity when compared to the direct ELISA. It is also less expensive and more flexible due to the many possible primary antibodies that can be used. The only major disadvantage with this type of ELISA is the risk of cross-reactivity between the secondary detection antibodies.

Sandwich ELISA

Unlike direct and indirect ELISA, the sandwich ELISA begins with a capture antibody coated onto the wells of the plate. It is termed a "sandwich" because the antigens are sandwiched between two layers of antibodies (capture and detection antibodies).

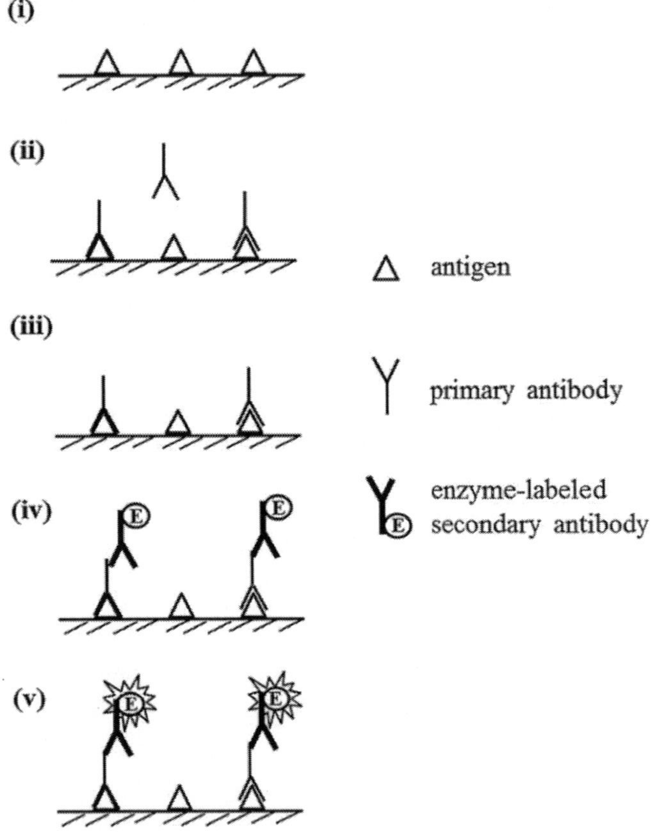

Figure 9.6: Indirect ELISA to Analyze Antibody.

(i) Attach antigen to solid phase. (ii) Incubate with primary antibody. (iii) Wash unbound primary antibody out. (iv) Incubate with enzyme-labeled secondary antibody. (v) Develop color with substrate

In this system, the target antigen is detected via anchoring between two antibodies, which recognize different epitopes, or the so-called sandwich system (Figure 9.7). Sandwich ELISA starts from the immobilization of an antibody, called a capture antibody, on the microtiter plate. After blocking the plate surface to avoid non-specific adsorption of other proteins, the antigen in the sample is allowed to react with the immobilized capture antibody, and the antigen bound to the capture antibody is then sandwiched with an enzyme-labelled antibody for colour development.

After adding the capture antibody to the plates, the plates are then covered and incubated overnight at 4°C. Once the coating step is complete, the plates are washed with PBS, then buffered/blocked with BSA. The buffer washes are carried out for at least 1-2 hours at room temperature. Finally, the plate is washed with PBS once again before the addition of the antigen. The antigen of interest is then added to the plates to bind to the capture antibody and incubated for 90 min at 37°C. The plate is rewashed, and the primary detection antibody is then added to the plate and incubated for another 1 to 2 hours at room temperature, followed by a buffer wash. Then the secondary enzyme-conjugated antibody is added and incubated for another 1 to 2 hours. The plate is rewashed, and the substrate is added to produce a colour change.

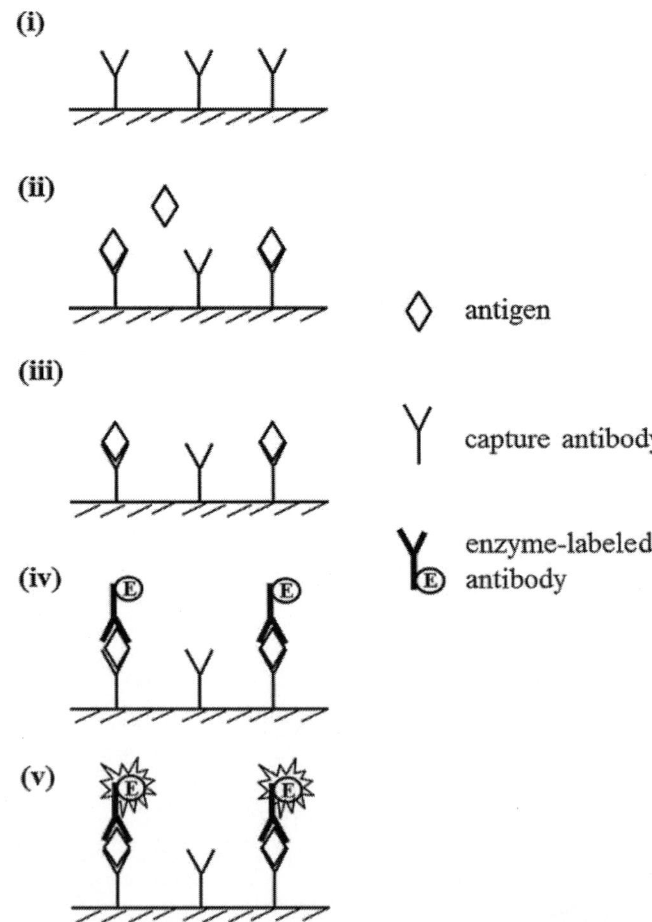

Figure 9.7: Sandwich ELISA for Specific Detection of Antigen.

(i) Attach capture antibody to solid phase. (ii) Incubate with target antigen. (iii) Wash unbound target out. (iv) Incubate with enzyme-labelled antibody. (v) Develop colour with substrate

The sandwich ELISA has the highest sensitivity among all the ELISA types. As two antibodies containing different epitopes are required against the target antigen, sandwich ELISA is generally suitable for measuring macromolecules with some exceptions. More recently, Boscolo *et al.*, reported a sandwich ELISA method for marine biotoxins, *e.g.*, palytoxins with a molecular weight of 2680 Da. The sandwich was formed by using two antibodies obtained from the same antigen with different antibodies, *i.e.*, MAb and PAb, which were used as the capture and primary antibodies, respectively.

A highly specific assay can be obtained via a sandwich system because of the use of two antibodies. However, it is an expensive and labor-intensive process to prepare two antibodies. In addition, one more step is required in the sandwich system because immobilization is necessary for capture antibody, which increases the assay time (Table 9.3).

Open Sandwich ELISA (OS-ELISA)

Advances in DNA technology have enabled the development of unique and interesting immunoassays based on the interaction of variable regions of heavy (V_H) and light (V_L) chains, which are binding regions for antigens. In the presence of an antigen, the interaction between V_H and V_L regions is enhanced to form a ternary complex. In the aforementioned report, OS-ELISA started from coating of a solid-phase microtiter plate with streptavidin. After blocking, the V_L region conjugated with biotin was allowed to react with streptavidin to immobilize the V_L region. In the next process, the phage-displayed V_H region was incubated with hen egg lysozyme, which was used as an antigen. Finally, the phage-displayed V_H regions forming a ternary complex were detected by the HRP-labelled antibody to develop colour. The obtained signal increases with increasing amount of antigen. Currently, this OS-ELISA has been modified to be easier and more effective, and several studies on OS-ELISA for measuring both the macromolecule and hapten have been reported.

Competitive ELISA

In 1973, Belanger developed competitive ELISA (Figure 9.8) to detect rat α-fetoprotein, which involved the development of indirect ELISA and sandwich ELISA. The key event of competitive ELISA is the competitive reaction between targets (antigen or antibody) in the sample and enzyme-labelled targets (antigen or antibody) against corresponding immobilized antibody or antigen. To detect the antigen in competitive ELISA, an enzyme-labelled antigen is used to compete with the target antigens against the immobilized antibody (Figure 9.8b). Hence, the higher the amount of antigen in the sample, the lower the amount of enzyme-labelled antigen that binds to the antibody. That is, with an increasing amount of target antigen, the signal decreases. In this case, competitive ELISA is suitable for measuring macromolecules only because a labelling enzyme is required to measure the antigen. If the antigen is a low molecular weight compound (*e.g.*, hapten), resultant hapten–enzyme conjugates are not recognized by the immobilized antibody, leading to failure of the analysis. To detect the antibody, the antigen is immobilized, and the competition between the antibody in the sample and enzyme-labelled antibody is observed (Figure 9.8a). In this case, both macromolecules and hapten can be detected when hapten is exposed on the surface of the microtiter plate.

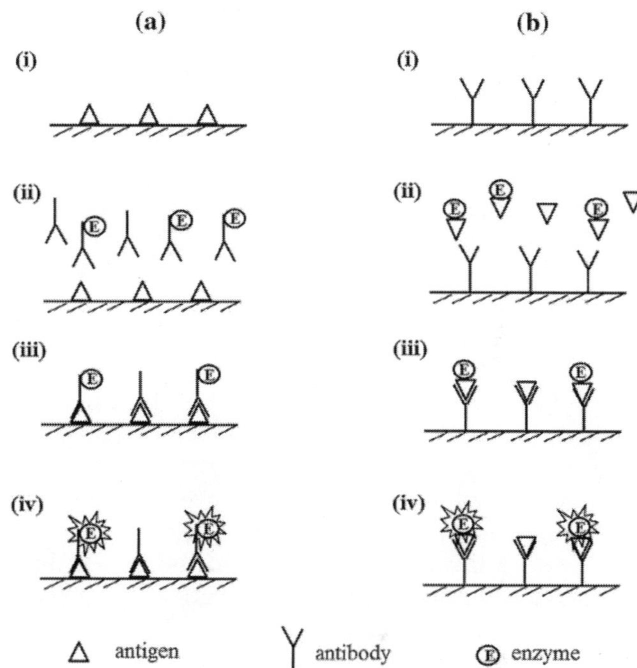

Figure 9.8: Competitive ELISA to Detect Antigen (a) and Antibody (b).

(i) Attach antigen/antibody to solid phase. (ii) Incubate antibody/antigen with enzyme-labelled antibody/antigen. (iii) Wash unbound enzyme-labelled antibody/antigen out. (iv) Develop colour with substrate

Direct and competitive ELISA methods are simple because only one antibody is required. However, the labelling step is required for each of the ELISA methods, possibly leading to inactivation of the antibody.

The competitive ELISA tests for the presence of an antibody specific for antigens in the test serum. This type of ELISA utilizes two specific antibodies, an enzyme-conjugated antibody and another antibody present in the test serum (if the serum is positive). Combining the two antibodies into the wells will allow for a competition for binding to antigen. The presence of a color change means that the test is negative because the enzyme-conjugated antibody bound the antigens (not the antibodies of the test serum). The absence of colour indicates a positive test and the presence of antibodies in the test serum. The

Table 9.3: Characteristics of Various Types of ELISA

	Direct ELISA	*Competitive ELISA*	*Indirect ELISA*	*Indirect competitive ELISA*	*Sandwich ELISA*
Advantage	Simple because only one antibody is used	Higher sensitivity and versatility than direct methods owing to usage of PAb that recognizes different epitopes of primary antibody			High specificity as two antibodies possessing different epitopes are used
Dis-advantage	Labeling antibody is necessary for each ELISA, which may result in inactivation of antibody		Nonspecific signal is induced through cross-reactivity of secondary antibody		To prepare two different antibodies is labour-intensive and expensive
Target	Macro-molecules	Macro-molecules (Hapten)	Macro-molecules	Macro-molecules (Hapten)	Generally, Macro-molecules
Signal (as target antigen increase)	Increase	Decrease	Increase	Decrease	Increase

competitive ELISA has a low specificity and cannot be used in dilute samples. However, the benefits are that there is less sample purification needed, it can measure a large range of antigens in a given sample, can be used for small antigens, and has low variability.

Indirect Competitive ELISA

Indirect competitive ELISA (icELISA) involves the combination of indirect ELISA and competitive ELISA (Figure 9.9). The target antigen is immobilized on a solid phase of the microtiter plate and is blocked. Subsequently, free target antigen and antibody are allowed to incubate and there is a competition between the immobilized antigen and free antigen against antibodies. The primary antibody that binds to the immobilized antigen is detected by the enzyme-labelled secondary antibody. Similar to the case in competitive ELISA, in icELISA, the signal decreases with increasing amount of the free antigen. icELISA can be applied for measuring both the macromolecules and hapten when hapten is exposed on the surface of the microtiter plate.

The use of enzyme-labelled secondary antibodies in indirect methods (*e.g.*, indirect ELISA and icELISA) exhibit advantages over direct methods (direct and competitive ELISA) with respect to sensitivity and versatility. Polyclonal antibody is a type of enzyme-labelled secondary antibody that recognizes different epitopes of the primary antibody, leading to increased sensitivity as compared to direct methods. In addition, a universal secondary antibody can be used if the original animal species of the primary antibody are unified. Thus, the secondary antibody is commercially available, leading to high versatility. Indirect ELISA clearly exhibits disadvantages with respect to the secondary antibody, *i.e.*, the cross-reaction of the secondary antibody should be considered (Table 9.3).

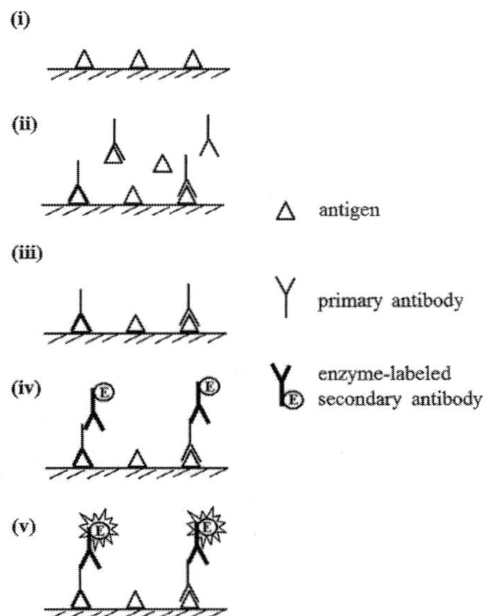

Figure 9.9: Indirect Competitive ELISA to Detect Antigen

(i) Attach antigen to solid phase. (ii) Incubate free target antigen with primary antibody. (iii) Wash unbound free target antigen and primary antibody out. (iv) Incubate with enzyme-labeled secondary antibody. (v) Develop color with substrate

Multiple and Portable ELISA

Multiple and portable ELISA is a new technique that uses a multicatcher device with 8 or 12 immunosorbent protruding pins on a central stick that can be immersed in a collected sample. The washings and incubation with enzyme-conjugated antigens and chromogens are performed by dipping the pins in prefilled microwells with reagents. The main advantage of these ready-to-use lab kits is that they are relatively inexpensive, can be used for large population screening, and do not require skilled personnel or laboratory equipment, making them an ideal tool for low-resource settings (Balsam *et al.*, 2013). Clinical applications include point-of-care detection of infectious diseases, bacterial toxins, oncologic markers, and drug screening.

Tests for the Gram-negative Bacteria Endotoxins

Gram-negative bacteria are characterized by their production of endotoxins, which consist of a lipopolysaccharide (LPS) layer (outer membrane) of the cell envelope. The LPS is pyrogenic and responsible for some of the symptoms that accompany infections caused by gram-negative bacteria.

The *Limulus* amoebocyte lysate (LAL) test employs a lysate protein obtained from the blood (actually haemolymph) cells (amoebocytes) of the horseshoe crab (*Limulus polyphemous*). The lysate protein is the most sensitive substance known for endotoxins. Of six different LAL preparations tested from five commercial companies, they were found to be from 3–300 times more sensitive to endotoxins than the US Pharmacopoeia rabbit pyrogen test. The LAL test is performed by adding aliquots of food suspensions or other test materials to small quantities of a lysate preparation followed by incubation at 37°C for 1 h. The presence of endotoxins causes gel formation of the lysate material. While most applications of this method have been made in clinical microbiology and in the pharmaceutical industry, food applications have been made.

The first food application was the use of LAL to detect the microbial spoilage of ground beef. Endotoxin titers increase in proportion to viable counts of gram-negative bacteria. Since the normal spoilage of refrigerated fresh meats is caused by gram-negative bacteria, the LAL test is a good rapid indicator of the total numbers of gram-negative bacteria. The method has been found to be suitable for the rapid evaluation of the hygienic quality of milk relative to the detection of coliforms before and after pasteurization. For raw and pasteurized milk, it represents a method that can be used to determine the history of a milk product relative to its content of gram-negative bacteria. Since both viable and nonviable gram-negative bacteria are detected by LAL, a simultaneous plating is necessary to determine the numbers of cfu's. The method has been applied successfully to monitor milk and milk products; microbial quality of raw fish and cooked turkey rolls. In the latter, LAL titers and numbers of Enterobacteriaceae in vacuum-packaged rolls were found to have a statistically significant linear relationship. These investigators suggested that the test could be used to predict spoilage in such products.

LAL titers for foods can be determined either by direct serial dilutions or by MPN with results by the two methods being essentially similar. To extract endotoxins from foods, the Stomacher has been found to be generally better than the use of Waring blenders or the shaking of dilution bottles.

In this test, the proclotting enzyme of the *Limulus* reagent has been purified. It is a serine protease with a molecular weight of about 150,000 daltons. When activated with Ca_{2+} and endotoxin, gelation of the natural clot table protein occurs. The *Limulus* coagulogen has a molecular weight of 24,500. When it is acted upon by the *Limulus* clotting enzyme, the coagulogen releases a soluble peptide of about 45 amino acids residues and an insoluble coagulin of about 170 amino acids. The latter interacts with itself to form the clot which involves the cleavage of -arg-lys- or -arg-gly- linkages.

Commercial substrates are available that contain amino acid sequences similar to coagulogen. The chromogenic substrates used for endotoxin consist of these linked to *p*-nitroaniline. When the endotoxin-activated enzyme attacks the chromogenic substrate, free *p*-nitroaniline results and can be read at 405 nm. The amount of the chromogenic compound liberated is proportional to the quantity of endotoxin in the sample. Employing a chromogenic substrate, Tsuji *et al.,* have devised an automated method for endotoxin assay, and the method was shown sensitive to as little as 30 pg of endotoxin/ml.

Assuming that the quantity of endotoxin/gram-negative bacterial cells is fairly constant and assuming further that cells of all genera contain the same given quantity, it is possible to calculate the number of cells (viable and non-viable) from which the experimentally determined endotoxin was derived. With a further assumption that the ratio of gram-negative to gram-positive bacteria is more or less constant for given products, one can make a 1–h estimate of the total numbers of bacteria in food products such as fresh ground beef. Low values by this procedure are more meaningful than high values and as noted above, the latter need to be confirmed by other methods.

Overall, the value of the LAL test lies in the speed at which results can be obtained. Foods that have high LAL titers may be candidates for further testing by other methods, while those that have low titers may be placed immediately into categories of lower risk relative to numbers of gram-negative bacteria.

Now let us learn more things about *Salmonellae* and other gram-negative bacteria.

An EIA method for *Salmonellae* was developed by Swaminathan and Ayres employing a conjugate of IgG fraction of salmonellae polyvalent flagellar antibodies and horseradish peroxidase. When reacted with *Salmonellae* and then treated, with enzyme substrate, the cell wall and flagella were stained. The stained *Salmonellae* could be differentiated from non-salmonellae by light microscopy. A comparison of the above method with a culture method on 142 food samples showed overall agreement of 92.2 per cent. False positives by the EIA, technique amounted to 6.4 per cent. When compared to FA, 95.8 per cent agreement was found between the two methods. Using a similar method in which samples from enrichment cultures were first reacted with H-specific IgG followed by the addition of goat anti-rabbit antibody conjugated to alkaline phosphatase, Minnich *et al.,* found the method sensitive to about 10_5 *Salmonellae*/ml and advantages over the immunofluorescence and tube agglutination methods were noted. A direct EIA for *Salmonellae* using polyclonal antibodies was developed in which the bacterial cells are attached to the walls of polystyrene microtitration plates employing a capture-antibody techniques. The substrate used was 4-methylumbelliferyl-β-D-galactoside and positive results are indicated by fluorescence under long-wave UV light. Results were obtained in 3 working days and the sensitivity threshold was found to be 10_7 cells/ml.

One of the newest variations in the ELISA technique is the introduction of monoclonal antibodies. Employing immunoglobulin A monoclonal antibodies from a myeloma in which the myeloma protein binds to a flagellar determinant of *Salmonellae*, Robison *et al.* studied 100 *Salmonellae* and found that 95 per cent were detectable with the monoclonal antibody method. The method could detect as few as 10_6 cells/ml and the investigators suggested that results could be achieved in 36 h after culture initiation. It was successful in detecting *Salmonellae* added to infant formula. A method highly specific for *Salmonellae* was further developed employing monoclonal IgA antibodies attached to polycarbonate-coated metal beads. Selective enrichment is not required and < 10_6 cells/ml can be detected with results obtained in 2 days. By contrast, the *Salmonellae* agglutination test requires about 10_7 cells/ml for results. A commercially produced kit encompassing the Mattingly-Gehle procedure was evaluated with a variety of meat and poultry products by comparing it with two conventional culture methods. All samples positive by the culture methods were positive by the ELISA kit method and results were obtained in 2–3 days.

The ELAT technique used by Krysinski and Heimsch to detect *S. typhimurium* cells employed the use of cellulose acetate filter membranes that were spotted with broth-enriched samples. After the membranes were fixed, they were treated with an appropriate enzyme-conjugated antibody followed by incubation and then the enzyme substrate. While about $5 \times 10_7$ cells/ml are needed for detection by enrichment serology or FA, these authors found that only 10_4–10_5 cells were needed for results by ELAT.

With regard to the relative sensitivity of ELISA to *Salmonellae* antibody, Carlsson *et al.,* found the procedure capable of measuring at least ten times lower concentrations than passive hemagglutination or quantitative precipitation methods. The method employed by these investigators required 26 to 28 h for results. Vos *et al.,* compared ELISA with passive HA for quantitative detection of antibodies to endotoxin and found ELISA to be two times more sensitive for LPS antibody than HA. In a similar manner, Keren found that ELISA produced an antibody titer of 1 : 100,000 to LPS antigens of *Shigella flexneri* while passive hemagglutination gave a titer of only 1 : 640. These findings suggest that ELISA methods are capable of measuring equally lower levels of antigens. For more information on ELISA methods for *Salmonellae* in foods, refer to studies of Mattingly *et al.,* and Swaminathan *et al.*

Many methods are used for the detection and quantitation of bacterial toxins and enterotoxins, let us study some of them.

Gel diffusion methods have been widely used for the detection and quantitation of bacterial toxins and enterotoxins. The four most used are single-diffusion tube (Oudin), micro-slide double diffusion, micro-Ouchterlony slide and electroimmunodiffusion. They have been employed to measure enterotoxins of *Staphylococci* and *C. perfringens*; and the toxins of *C. botulinum*. Although they should be usable for any soluble protein to which an antibody can be made, they require that the antigen be in precipitable form. Perhaps the most widely used is the Crowle modification of the Ouchterlony slide test as modified by Casman and Bennett and Bennett and McClure. The micro-Ouchterlony method can detect 0.1–0.01 μg of staphylococcal enterotoxin, which is the same limit for the Oudin test. The double-diffusion tube test can detect levels as low as 0.1 μg/ml, but the incubation period required for such low levels is 3–6 days. This immunodiffusion method requires that extracts from a 100 g sample be concentrated to 0.2 ml. While other methods such as RIA and RPH are more sensitive and rapid than the gel diffusion methods, the latter continue to be widely used. Their reliability within their range of sensitivity is unquestioned.

While gel diffusion methods generally require at least 24 h for results, two comparable serologic methods yield results in 2-4 h-hemagglutination- inhibition (HI) and reverse passive hemagglutination (RPH). Unlike the gel diffusion methods, antigens are not required to be in precipitable form for these two tests.

In the HI test, specific antibody is kept constant and enterotoxin (antigen) is diluted out. Following incubation for about 20 min, treated sheep red blood cells (SRBCs) are added. Hemagglutination (HA) occurs only where antibody is not bound by antigen. HA is prevented (inhibited) where toxin is present in optimal proportions with antibody.

In contrast to HI, antitoxin globulin in RPH is attached directly to SRBCs and used to detect toxin. When diluted toxin preparations are added, the test is read for HA after incubation for 2 h. HA occurs only where optimal antigen-antibody levels occur. No HA occurs if no toxin or enterotoxin is present.

Lysis Inhibition Test (LIT)

This test has been employed to detect *E. coli* strains that produce heat-labile enterotoxin (LT). The test employs sheep erythrocytes, complement, anti-LT antibodies and LT. Enterotoxin-sensitized SRBCs are hemolyzed in the presence of anti-LT antisera and complement. Immune hemolysis is inhibited if the anti-LT antisera is pre-exposed to soluble LT before the addition of LT-sensitized SRBCs. In testing seventy-five *E. coli* isolates that were positive by the Y-1 adrenal assay, all were positive by LIT. The test is run in microtiter plates and purified LT is required.

Passive Immune Hemolysis (PIH)

PIH is similar to LIT and has been employed also to detect *E. coli* strains that produce LT. This technique employs complement-mediated lysis of LT-sensitized SRBCs by antitoxin. PIH is more sensitive than LIT and does not require purified LT. Like LIT, the test is run in microtiter plates and has been shown to be sensitive to < 100 ng While the PIH test correlated well with the Y-1 adrenal assay, for human strains of LT producing *E. coli,* correlation was poor with strains of non-human origin.

Chapter 10

Non-Bacterial Illness

Introduction

Not only bacteria but also many non-bacterial agents cause illness by different ways.

Foods may act as vehicles for other disease causing agents such as helminths, nematodes protozoa and viruses as well as toxic metabolites of fungi and algae (Table 10.1). Each of these is a specialist area and cannot be dealt with in the same detail as bacteria but a food microbiologist should be aware of the occurrence and significance of these non-bacterial agents of food-borne illness.

Unlike for major bacterial and viral pathogens, established and readily available vaccines do not exist to prevent parasitic protozoan infections. A better understanding of the factors that influence immunity to these diseases may provide a foundation to design novel public health interventions.

Table 10.1a: Non-bacterial Disease-Causing Agents

Parasitic Protozoa and Worms			
1.	*Giardia lamblia*	2.	*Entamoeba histolytica*
3.	*Cryptosporidium parvum*	4.	*Cyclospora cayetanensis*
5.	*Anisakis* sp. and related worms	6.	*Diphyllobothrium* spp.
7.	*Nanophyetus* spp.	8.	*Eustrongylides* sp.
9.	*Acanthamoeba* and other free-living amoebae	10.	*Ascaris lumbricoides*
11.	*Trichuris trichiura*		
Viruses			
1.	Hepatitis a virus	2.	Hepatitis E virus
3.	Rotavirus	4.	Norwalk virus group
5.	Other viral agents		
Natural Toxins			
1.	Ciguatera poisoning	2.	Shellfish toxins (PSP, DSP, NSP, ASP)
3.	Scombroid poisoning	4.	Tetrodotoxin (Pufferfish)
5.	Mushroom toxins	6.	Aflatoxins
7.	Pyrrolizidine alkaloids	8.	Phytohaemagglutinin (Red kidney bean poisoning)
9.	Grayanotoxin (Honey intoxication)		
Other Pathogenic Agents			
1.	Prions		

1. Parasitic Protozoan and Worms

A parasite is an organism that lives in or on an organism of another species. Many types of organisms have adopted a parasitic mode of existence; that is, they require a host for their own survival. Three major groups of parasites which can cause disease in humans have been recognized. They are:

- Protozoa (belonging to the kingdom Protista)
- Helminths
- Ectoparasites

Protozoa and helminths largely affect the gut, while ectoparasites include lice and mites that can attach to or burrow into the skin, staying there for long periods of time.

A. Protozoan Parasites

Protozoa are tiny single-celled organisms that multiply inside the human body. Parasitic protozoan infections represent a major health burden in the developing world and contribute significantly to morbidity and mortality.

The name 'proto-zoa' literally means 'first animals' and early classification systems grouped the protozoa as basal members of the animal kingdom. However, they were recognized as a discrete assemblage on the basis of their unicellularlity and were assigned to the taxon Protozoa (but still invariably figured as the trunk of the animal tree of life). More recently, the protozoa have been classified together with several algal and fungal groups in the kingdom Protista (protozoa representing the motile protists).

Protozoa are eukaryotic organisms (with a membrane-bound nucleus) which exist as structurally and functionally independent individual cells (including those species which are gregarious or form colonies). Most protozoa are microscopic organisms, only a few grow to a size large enough to be visible to the naked eye. As unicellular eukaryotes, protozoa display all the same essential life activities as higher metazoan eukaryotes: they move about to survive, feed and breed.

The majority of protozoa and helminths tend to be non-pathogenic or result in very mild illness. Some, however, can cause severe disease in humans.

Table 10.1b: Protozoan Parasites

Entamoeba histolytica	Protozoan	Amoeba	Lobose Amoeba
Giardia duodenalis	Protozoan	Flagellate	Diplomonad
Trypanosoma brucei	Protozoan	Flagellate	Kinetoplastid
Trypanosoma cruzi	Protozoan	Flagellate	Kinetoplastid
Leishmania spp.	Protozoan	Flagellate	Kinetoplastid
Eimeria spp.	Protozoan	Apicomplexan	Coccidian
Isospora spp.	Protozoan	Apicomplexan	Coccidian
Cryptosporidium spp.	Protozoan	Apicomplexan	Coccidian
Toxoplasma gondii	Protozoan	Apicomplexan	Cyst-Forming Coccidian
Plasmodium spp.	Protozoan	Apicomplexan	Haemosporidian
Babesia spp.	Protozoan	Apicomplexan	Piroplasm
Balantidium coli	Protozoan	Ciliate	Trichostome
Ichthyophthirius multifiliis	Protozoan	Ciliate	White Spot Ciliate
Thelohania spp.	Protozoan	Microsporan	Pansporoblasto

Transmission of the enteric protozoa typically occurs through the faecal-oral route. The intestine is densely populated by commensal bacteria well situated to influence the behaviour of the protozoan parasites with which they directly interact. The potential influence of the microbiota on parasites is not, however, limited to the intestinal protozoa. Protozoa that live in the blood or tissue of humans may also be affected by the interplay between the gut microflora and the host metabolism and immune system (Burgess *et al.*, 2017). Human microbiota has an impact on the parasitic protozoa that infect the intestine (*Entamoeba histolytica, Giardia, Cryptosporidium, Blastocystis hominis*) or vagina (*Trichomonas vaginalis*) or cause systemic infections (*Plasmodium falciparum*). Changes in the composition of the intestinal microbiota may increase resistance to infection at mucosal sites, as well as alter systemic immunity to these parasites (Figure 10.1).

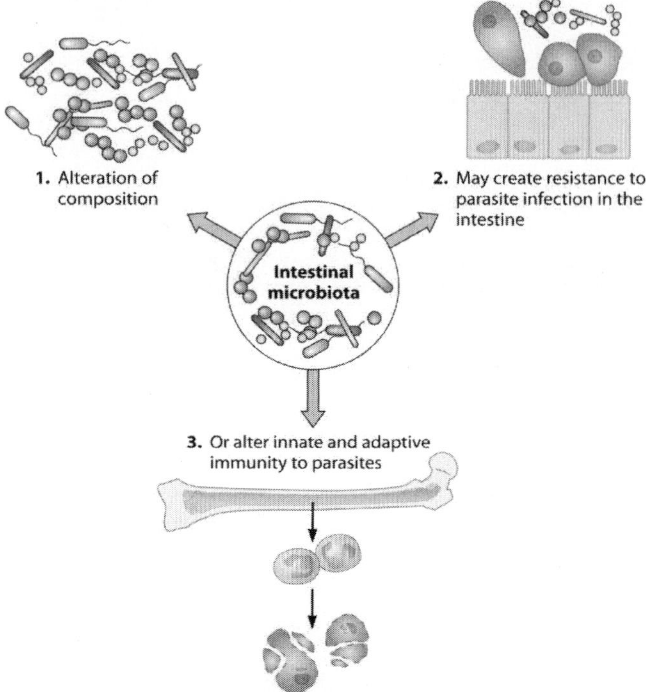

Figure 10.1: Host Intestinal Microibiota and Interactions with Host and Parasite.

Changes in the composition of the intestinal microbiota (image 1) may increase resistance to parasite infection at mucosal sites, such as the intestine, by mechanisms such as decreased virulence or parasite adherence (image 2). Changes in the microbiota may also alter systemic immunity to parasites by alterations of granulopoiesis or adaptive immunity (image 3). A better understanding of the mechanisms underlying microbiota-mediated protection may help explain clinical variability and help treat parasitic protozoan infections.

Source: Burgess, S.L., Gilchrist, C.A., Lynn, T.C. and Petri, Jr, W.A. 2017. Parasitic Protozoa and Interactions with the Host Intestinal Microbiota. Infect Immun.85(8): e00101-17.

Worldwide, diarrhoea is currently the second leading cause of death in children younger than 5 years of age and is associated with around 500,000 deaths per year. Although diarrhoea can be caused by many pathogens, in a large proportion of cases, the causal organism is a parasitic protozoan. The most common intestinal protozoan parasites associated with diarrhoea are *Giardia intestinalis, Entamoeba histolytica, Cyclospora cayetanenensis*, and *Cryptosporidium* spp. The diseases caused by these intestinal protozoan parasites are known as giardiasis, amoebiasis, cyclosporiasis, and cryptosporidiosis, respectively, and they are associated with diarrhoea. In 2010, an estimated 357 million cases of illness with at least one of three enteric protozoa, Entamoeba, Cryptosporidium, and Giardia, resulted in 33,900 deaths and the loss of 2.94 million disability-adjusted life years (Torgerson *et al.*, 2015). In a recent study of moderate-to-severe diarrhoea in African and Asian children, Cryptosporidium spp. were some of the top diarrhoea-associated pathogens.

Spread of these protozoan parasites in developing countries mostly occurs through faecal contamination as a result of poor sewage and poor quality of water. Food and water-borne outbreaks of these protozoan parasites have occurred, and the infectious cyst form of the parasites is relatively resistant to chlorine. Other species of protozoan parasites can also be found in the human gut, but they are not pathogenic, except Microsporidia sp.

i) *Giardia intestinalis*

Giardiasis is an infection of the small intestine that is caused by the parasite, *Giardia duodenalis*, also known as *Giardia lamblia* and *Giardia intestinalis*. It is the most prevalent parasitic cause of diarrhoea in the developed world, and this infection is also very common in developing countries. it is estimated that prevalence of giardiasis among worldwide population is 20 per cent to 40 per cent.

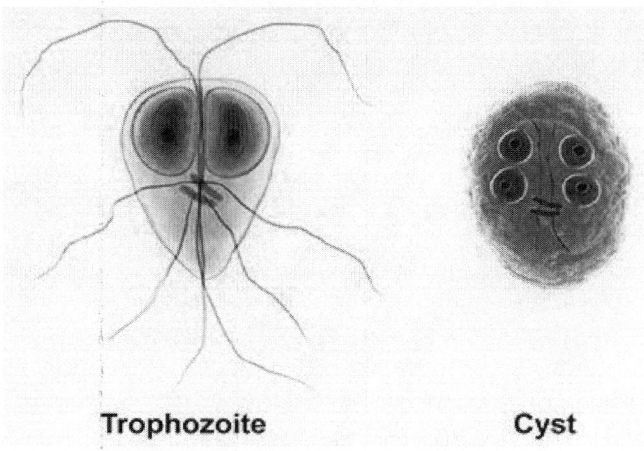

Figure 10.2a: *Giardia intestinalis.*

The protozoa giardia, for example, has a classic two-stage life cycle. *Giardia lamblia* exists in two forms, an active form called a trophozoite, and an inactive form called a cyst. In the first active stage, called trophozoite, the parasite swims around and consumes nutrients from the small bowel. The active trophozoite attaches to the lining of the small intestine with a "sucker" and is responsible for causing the signs and symptoms of giardiasis. The trophozoite cannot live long outside of the body, therefore it cannot spread the infection to others. The inactive cyst, on the other hand, can exist for prolonged periods outside the body.

In the second stage, it develops into a non-moving inactive cyst. Cysts excreted in faeces can contaminate the water supply, and ingesting contaminated food or water results in transmission. Close human to human contact and unsanitary living conditions can promote transmission. The organism survives in food and water as cysts but, although it can be cultured in the laboratory, it does not normally grow outside its host.

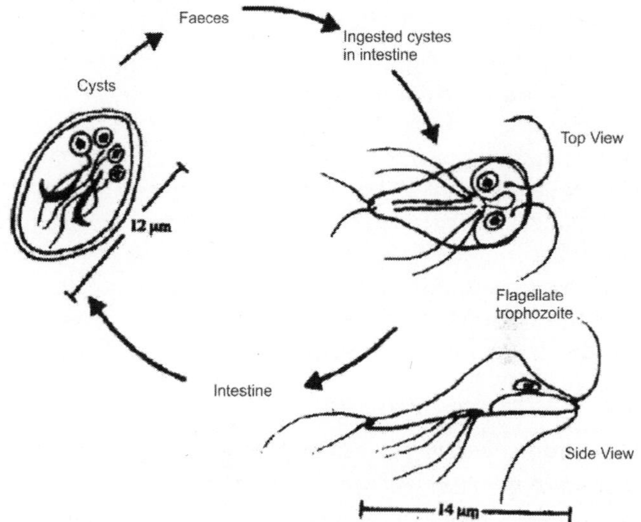

Figure 10.2b: *Giardia lambilia* Trophozoite.

The infective dose may be very low as it takes ingestion of only ten cysts to cause infection. Once cysts are ingested, the gastric juices activate the cyst, and the cyst develops into the disease-causing trophozoite which is characterized by the possession of eight flagella and two nuclei. Trophozoites are important not only because they cause the symptoms of giardiasis, but also because they produce the cysts that exit the body in the faeces and spread the infection to others.

Cysts of *Giardia* are present in the faeces of infected people. Thus, the infection is spread from person to person by contamination of food with faeces, or by direct faecal-oral contamination. Cysts also survive in water, for example in freshwater lakes and streams. Domestic mammals (for example, dogs, cats, calves) and wild mammals (for example, beavers) can become infected with *Giardia*; however, it is not clear how often domestic or wild mammals transmit giardiasis to humans. Giardiasis also has occurred as outbreaks from recreational water sources such as swimming pools, water parks, and hot tubs, most likely because of an infected user rather than a source of water that was contaminated.

Giardia cysts have been found on salad vegetables such as lettuce and fruits such as strawberries and could occur on any food which is washed with contaminated water or handled by infected persons not observing good hygienic practice. Although the cysts are resistant to chlorination processes used in most water treatment systems, they are killed by the normal cooking procedures used in food preparation.

Symptoms of giardiasis can include severe or chronic diarrhoea, abdominal cramps, fatigue, weakness and weight loss. The organism is not particularly invasive and it is not clear how the symptoms of diarrhoea, abdominal cramps and nausea are caused but it is possible that a protein toxin is involved.

Diagnosis of giardiasis is best accomplished by detection of Giardia antigen in stool, since the classic microscopic examination is less sensitive and specific. A recent comparison of nine different antigen-detection tests demonstrated that all had high sensitivity and specificity, except one. Giardia-specific antigen-detection tests are now also commercially available from several diagnostic companies, and their performance is quite good, except a few. In addition to antigen-detection tests, PCR-based test for the detection of G. intestinalis has also been reported.

ii) *Entamoeba histolytica*

Amoebiasis is the second leading cause of death from parasitic disease worldwide. The causative protozoan parasite, *Entamoeba histolytica*, is a potent pathogen. It can live in the large intestine (colon) without causing damage to the intestine. In some cases, it invades the colon wall, causing colitis, acute dysentery, or chronic diarrhoea. The infection can also spread through the bloodstream to the liver. In rare cases, it can spread to the lungs, brain, or other organs.

This condition occurs worldwide and can affect anyone, although it is most common in tropical areas that have crowded living conditions and poor sanitation. It is a major health problem in China, South East and West Asia and Latin America, especially Mexico. Amoebiasis affects about 15 per cent of Indian population. It has been reported throughout India.

This single-celled protozoan usually enters the human body when a person ingests cysts through contaminated food or water. It can also enter the body through direct contact with faecal matter. The cysts are a relatively inactive form of the parasite that can live for several months in the soil or environment where they were deposited in faeces. The microscopic cysts are present in soil, fertilizer, or water that has been contaminated with infected faeces. Food handlers may transmit the cysts while preparing or handling food. Transmission is also possible during anal sex, oral-anal sex, and colonic irrigation. The infected person can then spread the disease by releasing new cysts into the environment through infected faeces.

When cysts enter the body, they lodge in the digestive tract. They then release an invasive, active form of the parasite called a trophozite. The parasites reproduce in the digestive tract and migrate to the large intestine. There, they can burrow into the intestinal wall or the colon. Secreting proteinases that dissolve host tissues, killing host cells on contact, and engulfing red blood cells, *E histolytica* trophozoites invade the intestinal mucosa, causing amoebic colitis, bloody diarrhoea and tissue destruction. In some cases, amoebas breach the mucosal barrier and travel through the portal circulation to the liver, where they cause abscesses consisting of a few *E histolytica* trophozoites surrounding dead and dying hepatocytes and liquefied cellular debris. Amoebic liver abscesses grow inexorably and, at one time, were almost always fatal, but now even large abscesses can be cured by one dose of antibiotic. New models of disease have linked *E histolytica* induction of intestinal inflammation and hepatocyte programmed cell death to the pathogenesis of amoebic colitis and amoebic liver abscess.

Table 10.2: Specific Components of the Microbiota during Human Protozoan Infection

Protozoan	*Microbiota Component*	*Influence*
E. histolytica	Prevotellaceae	Predicted infection
E. histolytica	Prevotella copri	Predicted diarrhea
Cryptosporidium	Proteobacteria, Firmicutes, *Escherichia coli* CFT073, *Bacillus* spp., *Clostridium* spp.	Increased relative abundance in Cryptosporidium-negative subjects
G. duodenalis	Bifidobacterium	Increased relative abundance in Giardia-positive subjects
Blastocystis	*Clostridia, Enterobacteriaceae*	Increased Clostridia levels but lower Enterobacteriaceae levels in Blastocystis-positive subjects
T. vaginalis	*Lactobacilli, Mycoplasma, Parvimonas, Sneathia*	Decreased lactobacilli and increased Mycoplasma, Parvimonas, and Sneathia abundances in *T. vaginalis*-positive subjects
Plasmodium falciparum	*Bifidobacterium, Streptococcus*	Higher proportion of Bifidobacterium and Streptococcus organisms in a low-infection-risk group

Source: Burgess, S.L., Gilchrist, C.A., Lynn, T.C. and Petri, Jr, W.A. Infect Immun. 2017. Parasitic Protozoa and Interactions with the Host Intestinal Microbiota. 85(8): e00101-17.

In addition to being a potentially lethal disease, invasive amoebiasis has important social and economic consequences. Temporarily incapacitating infections, which are frequent in adult males in the wage-earning age group, may require several weeks of hospitalization and 2-3 months for full recovery. Amoebiasis may cause clinical problems in persons with immunodeficiency, homosexuals and immigrants from certain tropical countries, and travellers.

iii) *Cryptosporidium parvum*

In 1907, Tyzzer was the first to describe the intestinal protozoan parasite Cryptosporidium in the gastric mucosa of mice. The importance of Cryptosporidium infection was appreciated only in 1955 when there was an outbreak in a turkey flock. The minimum dose of causing Cryptosporidium infection is as low as ten oocysts, and hence, it is considered as highly infectious. Until 1980, it had not gained importance in causing human infection. In 1982, the first case of cryptosporidiosis in acquired immune deficiency syndrome (AIDS) patient was reported. Since then, many cases of cryptosporidiosis have been reported in immunocompromised patients. Now, it is one of the emerging protozoan parasites according to the Centre of Disease Control and Prevention. It has emerged as an important cause of diarrhoeal illness worldwide, particularly in young children those under age 5 years but the disease can also affect healthy people of any age. However, most clinical problems are encountered in patients who are profoundly immunocompromised. The most common species that causes human infection are the *Cryptosporidium parvum* and *Cryptosporidium hominis*. The special structure which is present in them such as rhoptries and micronemes are responsible for their virulence and pathogenicity.

Two species, *Cryptosporidium hominis* and *Cryptosporidium parvum*, account for most of these laboratory-confirmed cases. Species distinction between *C. hominis* and *C. parvum* is quite recent and for several years both parasites were referred to as *C. parvum* (sometimes genotypes 1 and 2). Of the two species accounting for most disease in humans, *C. hominis* seems to be largely host adapted to humans while *C. parvum* can result in infection in both humans and animals. Less commonly, other species such as *C. meleagridis*, *C. canis*, and *C. felis*, and unusual genotypes have also been reported in patients with diarrhoea but their acquisition is not fully understood.

In the UK, Cryptosporidium is the commonest protozoal cause of acute gastroenteritis, with 3000-6000 laboratory confirmed cases annually, although this is almost certainly an underestimation of the disease burden. Cryptosporidiosis may show a seasonal distribution and at peaks it may be the commonest enteric pathogen being isolated from children. In countries where water supplies are often contaminated, malnutrition is evident, sanitation non-existent and close contact with animals is normal, infection rates can be high.

Cryptosporidiosis can be transmitted from animals, human to human, water, food, and tends to cause waterborne outbreaks. Large waterborne outbreaks highlight the parasite's clinical and economic importance. The clinical problems associated with Cryptosporidium are increasingly becoming recognised internationally, and the parasite was included in the World Health Organization's Neglected Diseases Initiative 2004. These neglected diseases are defined as those that "exhibit a considerable and increasing global burden, and impair the ability of those infected to achieve their full potential, both developmentally and socioeconomically".

C. cayetanensis. Some other species of the genus Cyclospora have been described in non-human primates and most probably are host-specific. It is assumed that all human infections are caused by *C. cayetanensis*.

Cyclospora cayetanensisis, with a direct faecal–oral transmission cycle is globally distributed and an important cause of foodborne outbreaks of enteric disease in many developed countries, mostly associated with the consumption of contaminated fresh produce. Because oocysts (8–10 μm in diameter, with little or no size variation) are excreted unsporulated and need to sporulate in the environment, direct person-to-person transmission is unlikely. Infection by *C. cayetanensis* is remarkably seasonal worldwide, although it varies by geographical regions.

High percentages of asymptomatic carriers have been noted in endemic areas, suggesting that persistent contact with oocysts during the first years of age induces protective immunity against disease that will last into adulthood in endemic areas. However, some outbreaks have been reported among local populations, such as naval recruits in Peru, which challenge this theory. An explanation to these results would be that the acquired immunity is not long lasting and either diminishes over time or that the geographic distribution of *C. cayetanensis* is unequal, leaving some populations unprotected. It seems clear that the persistent contact with oocyst confers some resistance against the disease.

Gender does not have a significant effect on the Cyclospora infection rate in different geographical areas. The epidemiology of *C. cayetanensis* infections is affected by socio-economic status. Poverty and low socio-economic status are considered as risk factors for infection. Cyclosporiasis is common in impoverished endemic areas where water and food sanitation are poor or non-existent, particularly affecting children. A correlation between socio-economic status and infection was observed in Venezuela. In this country, most cases of cyclosporiasis were clustered in areas of extreme poverty where residences were without a latrine or toilet; contact with faecal-contaminated soil was also strongly associated with infection. Higher rates of infection associated with deficient sanitary facilities were also observed in other endemic regions. Even in developed countries like Germany, workers without proper toilet facilities were believed to be the source of an outbreak. Studies in low socio-economic areas have shown high prevalence, such as in Chennai, India (22.2 per cent in 256 samples collected).

Socio-economic status also influences the age pattern of infection by *C. cayetanensis*. In very poor areas of Peru, *C. cayetanensis* prevalence was highest in young children (aged 2–4 years); the infection was almost never detected after 11 years of age and adults usually did not have any clinical infection. In contrast, in middle to upper class families who lived in dwellings with suitable sanitation, children rarely appeared to become infected and infections occurred mainly in adults.

Cyclospora cayetanensis affects both immunocompetent and immunocompromised persons. However, it is more severe in immunocompromised hosts, in particular in HIV-infected patients for whom it is an important cause of diarrhoea. Clinical illness due to *C. cayetanensis* in immunocompromised patients is prolonged, severe and is associated with a high rate of recurrence that can be attenuated with long-term suppressive therapy. The average duration of diarrhoea for HIV-infected patients is longer than that for HIV-negative patients (199 days vs. 57.2 days) and other manifestations, such as acalculous cholecystitis has been reported in HIV-positive patients. Since the 1990s, there have been many reports on *C. cayetanensis* infection in immunocompromised patients worldwide—mainly in AIDS/HIV-infected people, but also in immunosuppressed persons, including those undergoing cancer treatments.

Resident Foreigners, Expatriates and Traveler's Diarrhea Resident foreigners and travellers to endemic areas are at a high risk of infection with clinical disease and are more susceptible to the disease than the indigenous population, including children. Between 1997 and 2008, 33.5 per cent of laboratory confirmed cases of infection in the USA were related to travel. In Canada, 71 per cent of reported cyclosporiasis cases between 2000–2010 were among travellers. The remainder were domestically acquired, presumably foodborne. Most cases of cyclosporiasis reported in Europe and Australia have been acquired by visitors in regions where the parasite is endemic. Countries visited were: Bali, Bangladesh, Bolivia, Bulgaria, Cambodia, China, Colombia, Costa Rica, Cuba, the Dominican Republic, Gabon, the Greek Islands, Guatemala, Hong-Kong, India, Indonesia, Java, Lebanon, Madagascar, Malaysia, Mexico, Morocco, Nepal, Nigeria, Pakistan, Peru, Puerto Rico, Singapore, the Solomon Islands, South-East Asia, Sri Lanka, Thailand, Turkey and Vietnam

A foodborne infection of *C. cayetanensis* is the main route of transmission of the parasite. Contaminated fresh produce such as berries and leafy greens, which are difficult to clean thoroughly and are consumed without prior processing to inactivate or remove the oocysts are the major source of infection. Pasteurized foods or foods thoroughly heated before consumption have not been associated with illness. The cases identified in the USA prior to 1995 were all thought to be linked to imported food or were associated with people who had travelled to areas where the disease was endemic. During the period of 2004–2009, 37.8 per cent (70/185) of the cases were classified as domestically acquired. In 2018, for the first time, *C. cayetanensis* was identified in produce grown in the USA. Consumption of pre-cut salads, leafy greens, Arugula/baby arugula, baby spinach, spring mix, lettuce heads, parsley, leek, green onion, fresh strawberry juice, raw fruits (avocado, lettuce, cabbage, carrot, tomato, banana and mango), ready to eat-packaged salads, winter grown cabbage, sprouts, blueberries, and cherry tomatoes, mustard leaves, basil leaves, mint and herbs (basil, coriander sativum and coriander, Vietnamese mint, marjoram, persicaria) have been reported to be the source of contamination with *C. cayetanensis* in various countries. The contamination of ready to eat and pre-packaged/bulked vegetables is an indication that the current sanitation processes do not guarantee food safety when dealing with certain parasites of faecal origin.

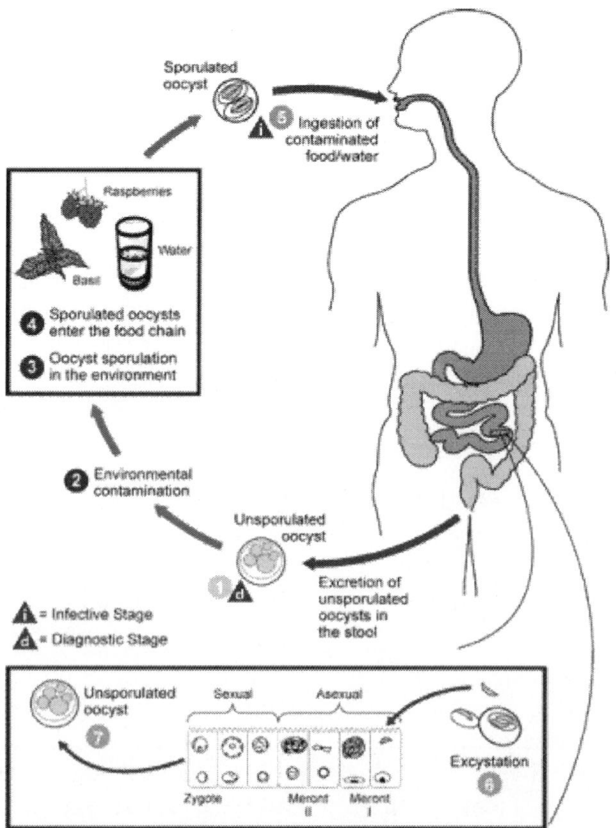

Figure 10.3 Lifecycle of *C. cayetanensis*.

Human transmission of *C. cayetanensis* is by the faecal–oral route. Water contaminated with faecal matter may act as a vehicle of transmission for *C. cayetanensis* infection. The source of drinking water has been determined as a risk factor for cyclosporiasis in endemic areas. More importantly, *C. cayetanensis* oocysts have been detected in several types of water including chlorinated water, and wastewater in endemic areas and in non-endemic areas which suggests the potential spread of the parasite via drinking and recreational water.

The life cycle of *C. cayetanensis* does not require dual or multiple hosts for completion. Because immature, unsporulated oocysts are shed in faeces and need to sporulate before becoming infectious, a vehicle is involved rather than direct person-to-person spread. Unsporulated oocysts are spheroidal, 8–10 µm in diameter, with little or no size variation. The oocyst wall is colorless, thin (11 years) and among adults, clinical symptoms are not always present. The causes for this age distribution pattern are not clear, but may be related to predominant modes of exposure, shared foods and water from which very young children are relatively protected. Sporulation occurs outside the host. The oocysts are robust and can survive for long periods in the environment; transmission can occur through food and water, contaminated by human faeces or sewage.

Infections have an incubation time of 1–2 weeks. Symptoms are primarily gastrointestinal, including diarrhoea and general malaise. The disease is self-limiting in most immunocompetent patients, but it may present as a severe, protracted or chronic diarrhoea in some cases, and may colonize extra-intestinal organs in immunocompromised patients. One important biological feature of C. cayetanensis is that, unlike *Cryptosporidium* and *Giardia*, freshly excreted oocysts are non-infectious and would require several days or weeks under favourable environmental conditions to sporulate and become infectious. After an initial episode of cyclosporiasis, the likelihood of diarrhoea and the duration of symptoms decreases significantly with each subsequent infection

Treatment is available and consists of antibiotic therapy. Most outbreaks are associated with the consumption of fresh produce. Preventative measures are based on safety practices for fresh produce, as well as improved agricultural practices. Further research is needed to understand many unknown epidemiological aspects of this parasitic disease.

iv) *Plasmodium* spp.

Other important protozoa are the plasmodium species; plasmodium parasites belong to the Apicomplexa group of protozoa. Plasmodium parasites cause malaria which is an acute febrile illness which is spread to people through the bites of infected female Anopheles mosquitoes. There are 5 parasite species (*Plasmodium falciparum, Plasmodium malariae, Plasmodium vivax, Plasmodium ovale* and *Plasmodium knowlesi*) that cause malaria in humans, and 2 of these species *i.e.*, P. falciparum and P. vivax that pose the greatest threat. *P. falciparum* is the deadliest malaria parasite and the most prevalent on the African continent. *P. vivax* is the dominant malaria parasite in most countries outside of sub-Saharan Africa.

Plasmodium parasites are transmitted to humans by female mosquitoes of the Anopheles genus. Worldwide approximately 40 anopheline species have been documented to transmit parasites to humans. These species vary remarkably in their capacity to transmit Plasmodium to humans. This variability relates to basic habits, including tendency to seek out humans for a blood meal, types of water sources preferred for laying eggs, and predilection for entering human dwellings, as well as basic biologic capacity to support the development of human Plasmodia to development of mature infection that can be transmitted during the taking of a blood meal.

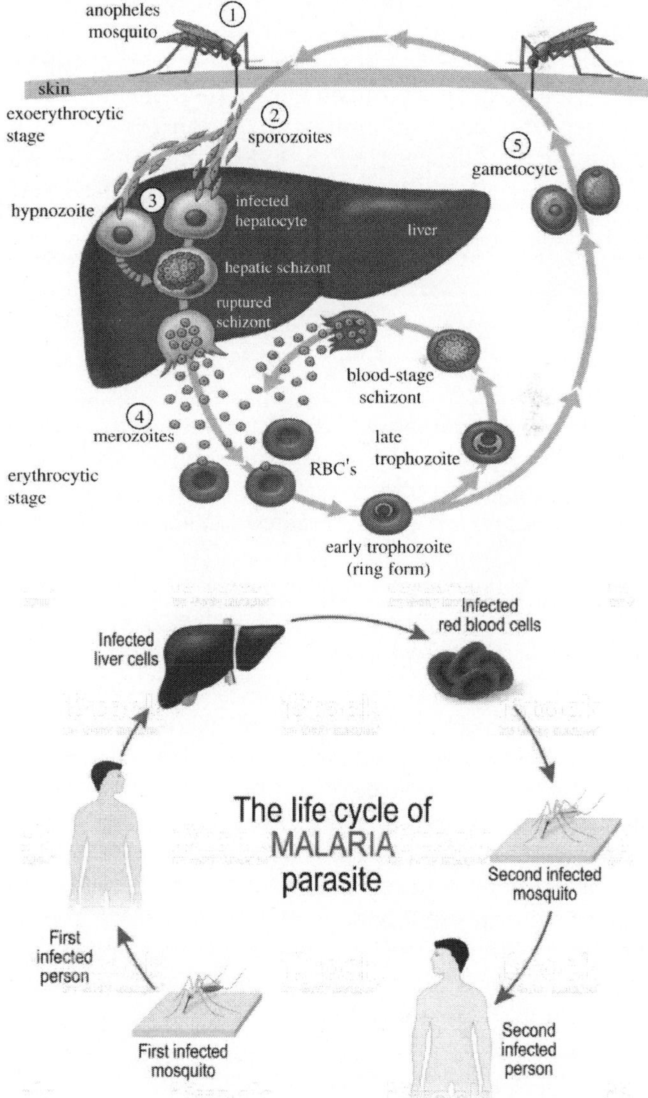

Figure 10.4: Lifecycle of Plasmodium.

Several identifiable factors characterize the capacity of Anopheles mosquitoes to transmit Plasmodium. The adult female mosquito can survive under ideal conditions up to 50 to 60 days. Practically, however, female anophelines survive on the average of 20 to 25 days, and consequently must be infected by malaria parasites in their early blood meals, taken every 3 to 4 days. Maturation of the parasites ingested with the blood meal requires 8 to 12 days before the mosquito has infective sporozoites in her salivary glands.

Plasmodium needs two hosts to complete its cycle. During the hematophagy process, the female Anopheles mosquito inoculates the sporozoites into the human. Sporozoites infect liver cells and multiply by schizogony, leading to the formation of merozoites. After multiplication in the liver (exoerythrocytic schizogony), the parasite invades the red blood cells and reproduces by schizogony (erythrocytic schizogony). Merozoites invade red blood cells. Annular trophozoites become schizonts that will give rise to new merozoites. Some parasites are sexually differentiated (gametocytes). When the parasite is in the blood, the clinical manifestations of the disease appear. Male (microgametocyte) and female (macrogametocyte) gametocytes are ingested by an Anopheles mosquito during hematophagy. The multiplication of the parasite in the mosquito is known as the sporogonic cycle. Microgametes fertilize macrogametes and develop zygotes. The zygote becomes a mobile oochinate, invades the wall of the intestine, and becomes an oocyst. The oocyst develops sporozoites inside, which once released ascend to the salivary glands. Inoculation of sporozoites into a new host perpetuates the malaria cycle.

The first symptoms of malaria include fever, headache and chills which usually appear 10–15 days after the infective mosquito bite and may be mild and difficult to recognize as malaria. Left untreated, P. falciparum malaria can progress to severe illness and death within a period of 24 hours. Malaria causes the most deaths of all parasitic diseases. In 2017 it was estimated malaria resulted in 435,000 deaths globally, most of them young children in sub-Saharan Africa.

Children are more likely to present with non-specific and gastrointestinal symptoms such as fever, lethargy, malaise, nausea, vomiting, abdominal cramps, and somnolence. They are more likely to develop hepatomegaly, splenomegaly, and severe anemia without major organ dysfunction than adults. In a case of severe malaria, they present with more frequent seizures (in 60 per cent to 80 per cent), hypoglycemia, and concomitant sepsis but are less likely to develop pulmonary edema and renal failure than adults.

The clinical features of infection in pregnancy vary from asymptomatic to severe, depending on the degree of (incomplete) immunity that a woman had acquired by the time she got pregnant. In semi-immune pregnant women, only a few infections result in fever or other symptoms. Malaria in pregnancy has a devastating effect not only on maternal health but has been associated with increased infant mortality due to low birth weight caused by either intrauterine growth restriction or preterm labor or both. *P. falciparum* infections are proven to be associated with complications such as maternal anemia, low birth weight, miscarriage, stillbirths, and congenital malaria. It is more likely for a pregnant woman in the second or third trimester to develop severe malaria with complications such as hypoglycemia and pulmonary edema, compared to non-pregnant adults.

Medical treatment should be taken once the malaria has been diagnosed to prevent its complications. Preventive and control measures should also be taken to avoid its spread.

v) *Entamoeba histolytica*

Several protozoan species in the genus Entamoeba colonize humans, but not all of them are associated with disease. *Entamoeba histolytica* is well recognized as a pathogenic amoeba, associated with intestinal and extraintestinal infections.

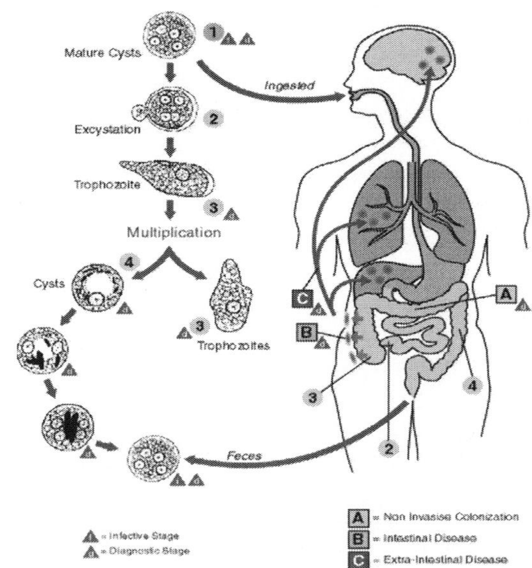

Figure 10.5: Life Cycle of *Entamoeba histolytica*.

E. histolytica is the pathogenic form and can cause amoebic colitis and extraintestinal amoebiasis. *E. dispar* is considered to be non-pathogenic and causes no signs of disease. Although 90 per cent of *E. histolytica* infections are asymptomatic, nearly 50 million people become symptomatic, with about 100,000 deaths yearly. Infection occurs worldwide, with a higher prevalence in countries of low socioeconomic status and poor public health. Countries with a high rate of infections include India, Africa, Mexico, and Central and South America. For example, a three-year study conducted in Bangladesh showed that 2.2 per cent of dysentery cases were caused by *E. histolytica* in preschool children.

Risk factors for infection are mostly related to faecal-oral transmission and have been due to poor hand hygiene, drinking of contaminated water, intake of contaminated food, defecation into water sources such as rivers, and being in close proximity with animals. In developed countries such as the United States, amebiasis infections are rare, accounting for at least 5 deaths per year, and are commonly seen in individuals that have had exposure to endemic areas such as immigrants or recent travellers.

There are two forms *E. histolytica* can take: the cyst form, which can survive in the environment for a prolonged period, and the trophozoite stage, which is the active and invasive form. After the ingestion of the cyst form, trophozoites can be formed, which can invade and penetrate intestinal mucosa destroying epithelial cells and inflammatory cells.

An appropriately robust layer of colonic mucin may be protective against symptomatic infection, whereas trophozoite attachment to the intestinal epithelium results in penetration of the organism into the submucosal layer, where extensive tissue destruction can take place in the form of apoptosis and lysis of cells, hence the name "histolytica."

Symptoms of classic amoebic dysentery usually begin insidiously over 1 to 3 weeks after infection. Diarrhea is almost universal and typically consists of numerous small-volume stools that can contain mucus or frank blood, or both. Stools are almost always heme positive if not grossly bloody. Abdominal pain and tenesmus are common; fever is present in approximately 30 per cent of cases. Some patients have a chronic course characterized by weight loss, intermittent loose stools, and abdominal pain. Rare presentations of amoebic dysentery include amebomas, which can mimic malignancy, and perianal ulcerations or fistulae. Severe disease can occur in the form of fulminant colitis or toxic megacolon; the latter almost universally requires colectomy.

Amoebic colitis generally affects males and females of all ages equally. There are reports of increased risk of infection from gay or bisexual males due to the risk of faecal-oral contamination through oral and anal sex. Factors that are associated with increased risk for complicated infection and mortality are associated with the pregnancy, corticosteroid treatment, malignancy, malnutrition, and alcoholism. Amoebic liver abscess infections are at least 3 times more likely to affect middle-aged men between the ages of 18 and 50.

Amoebic liver abscess formation is the most common extraintestinal complication. It can be seen in months to years after an individual has had exposure to an endemic area. Symptoms include fever and right upper quadrant pain. Examination findings may include hepatomegaly with hepatic tenderness. Less than 10 per cent of patients have jaundice. Common laboratory findings include leukocytosis without eosinophilia, elevated alkaline phosphatase, transaminitis, and elevated erythrocyte sedimentation rate.

To reduce the chances of this parasitic infection, one should follow the undermentioned practices especially when travelling to underdeveloped countries:

- ✰ Avoid drinking tap water. Drink bottled beverages and brush teeth with bottled water. Avoid ice in drinks as they are commonly made from tap water.
- ✰ Avoid eating raw fruits and vegetables cleaned with tap water.
- ✰ Eat well-cooked food when traveling, avoiding undercooked fish or meats.
- ✰ Avoid eating street vendors in carts or stands.
- ✰ Frequently wash hands after using the restroom, touching animals, before eating, or handling trash.

B. Helminths

Helminths are the most common parasites infecting humans. The world's population numbers approximately 7 billion, with probably a similar number of human helminth infections. Helminths are transmitted to humans through food, water and soil, arthropod and molluscan vectors.

The word 'helminth' is a general term meaning 'worm', but there are many different types of worms. Helminths are worms with many cells. Prefixes are therefore used to designate types: platy-helminths for flat-worms and nemat-helminths for round-worms. As a general rule, helminths cannot multiply inside the human body.

Three major assemblages of parasitic helminths are recognized as the Nemathelminthes (nematodes) and the Platyhelminthes (flatworms), the latter being subdivided into the Cestoda (tapeworms) and the Trematoda (flukes). Nematodes (roundworms), cestodes (tapeworms), and trematodes (flatworms) are among the most common helminths that inhabit the human gut. All helminths are multicellular eukaryotic invertebrates with tube-like or flattened bodies exhibiting bilateral symmetry. They are triploblastic (with endo-, meso- and ecto-dermal tissues) but the flatworms are acoelomate (do not have body cavities) while the roundworms are pseudocoelomate (with body cavities not enclosed by mesoderm). In contrast, segmented annelids (such as earthworms) are coelomate (with body cavities enclosed by mesoderm).

Table 10.3: Helminth Parasites

Ascaris lumbricoides	Helminth	Nematode	Roundworm
Trichuris trichiura	Helminth	Nematode	Whipworm
Enterobius vermicularis	Helminth	Nematode	Pinworm
Ancylostoma duodenale	Helminth	Nematode	Old world hookworm
Necator americanus	Helminth	Nematode	New world hookworm
Strongyloides stercoralis	Helminth	Nematode	Threadworm
Dracunculus medinensis	Helminth	Nematode	Guinea worm
Onchocerca volvulus	Helminth	Nematode	Filarial worm
Echinococcus granulosus	Helminth	Cestode	Dog tapeworm
Taenia saginata	Helminth	Cestode	Beef tapeworm
Taenia solium	Helminth	Cestode	Pork tapeworm
Fasciola hepatica	Helminth	Trematode	Liver fluke
Schistosoma spp.	Helminth	Trematode	Blood fluke

Nematodes (Roundworms)

They have long thin unsegmented tube-like bodies with anterior mouths and longitudinal digestive tracts. They have a fluid-filled internal body cavity (pseudocoelum) which acts as a hydrostatic skeleton providing rigidity (so-called 'tubes under pressure'). Worms use longitudinal muscles to produce a sideways thrashing motion. Adult worms form separate sexes with well-developed reproductive systems. The eggs or larvae of these nematodes usually develop in soil before being transmitted to the human host. For this reason, these nematodes are often called soil-transmitted helminths. A good example are hookworms which infest humans by penetrating the skin from contaminated soil. So, wearing appropriate footwear is an important way to prevent hookworm transmission.

Cestodes (Tapeworms)

They have long flat ribbon-like bodies with a single anterior holdfast organ (scolex) and numerous segments. They do not have a gut and all nutrients are taken up through the tegument. They do not have a body cavity (acoelomate) and are flattened to facilitate perfusion to all tissues. Segments exhibit slow body flexion produced by longitudinal and transverse muscles. All tapeworms are hermaphroditic and each segment contains both male and female organs.

Trematodes (Flukes)

They have small flat leaf-like bodies with oral and ventral suckers and a blind sac-like gut. They do not have a body cavity (acoelomate) and are dorsoventrally flattened with bilateral symmetry. They exhibit elaborate gliding or creeping motion over substrates using compact 3-D arrays of muscles. Most species are hermaphroditic (individuals with male and female reproductive systems) although some blood flukes form separate male and female adults.

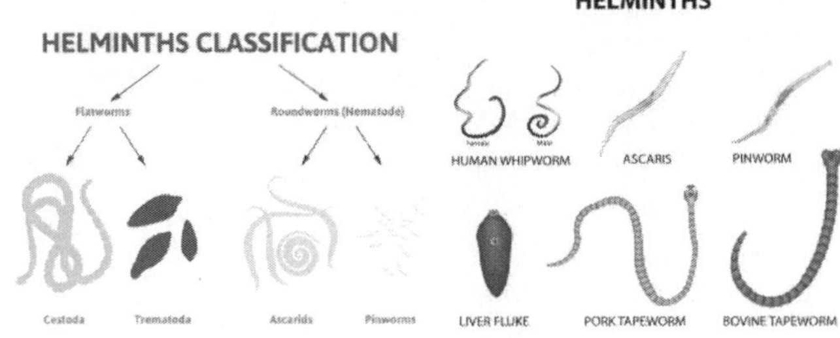

Figure 10.6: Helminths.

Unlike other pathogens (viruses, bacteria, protozoa and fungi), helminths do not proliferate within their hosts. Worms grow, moult, mature and then produce offspring which are voided from the host to infect new hosts. Worm burdens in individual hosts (and often the severity of infection) are therefore dependent on intake (number of infective stages taken up). Worms develop slowly compared to other infectious pathogens so any resultant diseases are slow in onset and chronic in nature. Although most helminth infections are well tolerated by their hosts and are often asymptomatic, subclinical infections have been associated with significant loss of condition in infected hosts. Other helminths cause serious clinical diseases characterized by high morbidity and mortality. Clinical signs of infection vary considerably depending on the site and duration of infection. Larval and adult nematodes lodge, migrate or encyst within tissues resulting in obstruction, inflammation, oedema, anaemia, lesions and granuloma formation. Infections by adult cestodes are generally benign as they are not invasive, but the larval stages penetrate and encyst within tissues leading to inflammation, space-occupying lesions and organ malfunction. Adult flukes usually cause obstruction, inflammation and fibrosis in tubular organs, but the eggs of blood flukes can lodge in tissues causing extensive granulomatous reactions and hypertension.

As far as their life cycle is concerned, helminths form three main life-cycle stages *viz.*, eggs, larvae and adults. Adult worms infect definitive hosts (those in which sexual development occurs) whereas larval stages may be free-living or parasitize invertebrate vectors, intermediate or paratenic hosts. Nematodes produce eggs that embryonate in utero or outside the host. The emergent larvae undergo 4 metamorphoses (moults) before they mature as adult male or female worms. Cestode eggs released from gravid segments embryonate to produce 6-hooked embryos (hexacanth oncospheres) which are ingested by intermediate hosts. The oncospheres penetrate host tissues and become metacestodes (encysted larvae). When eaten by definitive hosts, they excyst and form adult tapeworms.

Trematodes have more complex life-cycles where 'larval' stages undergo asexual amplification in snail intermediate hosts. Eggs hatch to release free-swimming miracidia which actively infect snails and multiply in sac-like sporocysts to produce numerous rediae. These stages mature to cercariae which are released from the snails and either actively infect new definitive hosts or form encysted metacercariae on aquatic vegetation which is eaten by definitive hosts.

Helminth eggs have tough resistant walls to protect the embryo while it develops. Mature eggs hatch to release larvae either within a host or into the external environment. The four main modes of transmission by which the larvae infect new hosts are faecal-oral, transdermal, vector-borne and predator-prey transmission.

Faecal-oral Transmission

Faecal -oral transmission of eggs or larvae passed in the faeces of one host and ingested with food/water by another (*e.g.*, ingestion of Trichuris eggs leads directly to gut infections in humans, while the ingestion of Ascaris eggs and Strongyloides larvae leads to a pulmonary migration phase before gut infection in humans).

Transdermal Transmission

This type of transmission of infective larvae in the soil (geo-helminths) actively penetrate the skin and migrate through the tissues to the gut where adults develop and produce eggs that are voided in host faeces (*e.g.*, larval hookworms penetrating the skin, undergoing pulmonary migration and infecting the gut where they feed on blood causing iron-deficient anaemia in humans).

Vector-borne Transmission

Transmission of larval stages are taken up by blood-sucking arthropods or undergo amplification in aquatic molluscs (*e.g.*, Onchocerca microfilariae ingested by blackflies and injected into new human hosts, Schistosoma eggs release miracidia to infect snails where they multiply and form cercariae which are released to infect new hosts).

Predator-prey Transmission

This type of transmission of encysted larvae within prey animals (vertebrate or invertebrate) are eaten by predators where adult worms develop and produce eggs (*e.g.*, Dracunculus larvae in copepods ingested by humans leading to guinea worm infection, Taenia cysticerci in beef and pork being eaten by humans, Echinococcus hydatid cysts in offal being eaten by dogs).

Helminths can infect every organ and organ system. Prevalent in the intestines, they are found in the liver, lungs, blood and occasionally the brain and other organs. As with other parasites, the major risk factors are poor sanitation and shared living quarters. General symptoms include diarrhoea, abdominal pain, weight loss and weakness.

These infections are most prevalent in tropical and subtropical regions of the developing world where adequate water and sanitation facilities are lacking. Recent estimates suggest that *A.* lumbricoides can infect over a billion, *T. trichiura* 795 million, and hookworms 740 million people. Other species of intestinal helminths are not widely prevalent. Intestinal helminths rarely cause death. Instead, the burden of disease is related to less mortality than to the chronic and insidious effects on health and nutritional status of the host. In addition to their health effects, intestinal helminth infections also impair physical and mental growth of children, thwart educational achievement, and hinder economic development.

Trichinella Species

The nematode parasite *Trichinella*, which people most commonly associate with pork, has historically been a serious concern as a foodborne pathogen. It is distributed worldwide and has been found in virtually all carnivorous and omnivorous animals as well as less commonly in certain herbivores. *Trichinella spiralis* was first observed in 1835 by Paget, a British medical student, in the muscles of a man during post-mortem dissection. The newly discovered worm was further described and named *Trichina spiralis* by Owen (1835). The name change to *T. spiralis* was made in 1896 by Railliet because the genus Trichina had previously been attributed to a group of small flies.

Globally, the most important species of the meatborne parasite *Trichinella* are *Tr. spiralis*, *Tr. nativa*, and *Tr. britovi*, although occasionally other species have been implicated in human infections. Trichinella spiralis is the classical agent of human trichinellosis and is transmitted almost exclusively through pork. With the recognition of trichinellosis as an important foodborne disease, reports of its occurrence increased dramatically, with thousands of cases reported each year in Europe. During the period 1986–2009, more than 65 000 human cases were reported worldwide, over half of which occurred in eastern Europe.

The life cycle of Trichinella is relatively simple but unusual for a helminth parasite in that all stages of development occur within a single host. Epidemiologically, the most important feature of its life cycle is its obligatory transmission through ingestion of meat containing intracellular larvae (trichinae); there is no true free-living stage. After ingestion of infected meat, larvae are digested free from the meat in the stomach, pass into the small intestine, and invade the epithelial cells lining the upper small intestine. Here, within 4–6 days, they develop into sexually mature males and females and mate. Their offspring are released from the female worm live (newborn larvae) and migrate via the circulation system throughout the host's body; they invade successfully only striated muscle. Over the next 10–14 days, they develop into fully intracellular larvae, capable of infecting another host.

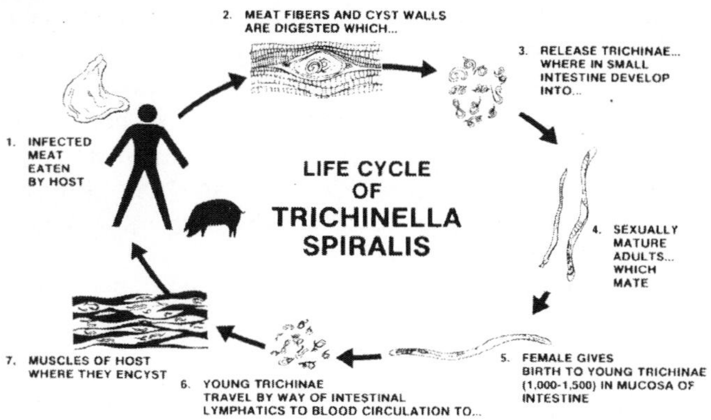

Figure 10.7: Life Cycle of *Trichinella spiralis.*

Trichinellosis is transmitted between animals by predation and scavenging, to pigs by feeding uncooked meat scraps or meat products or by eating rats, and humans by consuming insufficiently cooked meat or meat products, such as raw or rare pork prepared as sausages, smoked ham from an infected animal

The most common symptoms of Trichinella in humans include abdominal discomfort with nausea, diarrhea, vomiting, fatigue, and fever. The invasive phase of the disease is characterized by high fever, facial edema, and muscular pain. Cutaneous rash and subconjunctival hemorrhages may also occur. Myocarditis develops in 5 per cent –20 per cent of all infected persons. The risks of disease are greater in the elderly where they are likely to develop severe complications such as myocarditis or encephalitis which can lead to death.

Prevention has to be by breaking the cycle within the pig population and by adequate cooking of pork products. The United States Department of Agriculture recommends that all parts of cooked pork products should reach at least 76.7°C. Freezing will destroy encysted larvae but in deep tissue it may take as long as 30 days at –15°C. Curing, smoking and the fermentations leading to such products as salami do all eventually lead to the death of encysted *Trichinella* larvae.

The control of these parasites in the human food chain is affected by careful meat inspection and the role of the professional Meat Inspector, supported by legislation is very important. Badly infected animals may be recognized by ante-mortem inspection and removed at that stage. The presence of these parasites in animals usually gives rise to macroscopic changes in tissues and organs which can be recognized by meat inspection after slaughter.

C. Ectoparasites

The term ectoparasites generally refers to organisms such as ticks, fleas, lice and mites that can attach or burrow into the skin and remain there for long periods of time. Scabies, for example, a contagious skin disease marked by itching and small raised red spots, is caused by the human itch mite. Scabies usually is spread by direct, prolonged, skin-to-skin contact. Head lice are small, wingless insects that live and breed in human hair and feed by sucking blood from the scalp.

Table 10.4: Arthropod Parasites

Ctenocephalides canis	Insect	Flea	Dog flea
Lucilia cuprina	Insect	Fly	Sheep blowfly
Pediculus humanus	Insect	Louse	Human body louse
Sarcoptes scabei	Arachnid	Mite	Human scabies mite
Ixodes holocyclus	Arachnid	Tick	Paralysis tick

Ectoparasitoses (infestations with parasites that live on or in the skin) are usually considered as vexing disorders and do not attract much clinical attention. But depending on the socio-economic setting, these infections can carry substantial morbidity and affect much of a population. Ectoparasitic infestations can be sporadic, endemic, or epidemic. The prevalence of ectoparasitoses in the general population is low, but can become high in vulnerable groups. For example, tungiasis usually affects usually less than 1 per cent of the population in an endemic area, but in economically disadvantaged communities, prevalence in children can reach 76 per cent. In some indigenous populations in the Amazon rain forest nearly all individuals have head lice.

The occurrence of parasitic skin diseases is influenced by meteorological variables. In the semi-arid climate of Northeast Brazil, for instance, the incidence of the four major ectoparasitoses varies by season. In temperate climates, the incidence of scabies increases in the winter.

In developing countries, particularly in poor people, polyparasitism is frequent. Even when common, parasitic skin diseases are easily overlooked. In a primary health-care centre serving a poor neighbourhood with an overall prevalence of ectoparasitoses of 63 per cent, the physicians did not recognise scabies in 52 per cent, tungiasis in 94 per cent, and pediculosis in 100 per cent of those infested by the corresponding ectoparasite. Carriers of ectoparasites are likely to be stigmatised, depending on the sociocultural setting. Although ectoparasites elicit various types of immune responses, there is no evidence for protective immune mechanisms in the human host.

The arthropod ectoparasites can threaten human health directly by burrowing into, feeding, dwelling, and reproducing in human skin and orifices (mites, fleas, and flies) or by blood or tissue juice sucking (fleas, lice, mites, and ticks). The arthropod ectoparasites can also threaten human health indirectly by infectious disease transmission (fleas, mites, and ticks). Ticks are the most versatile ectoparasitic arthropods and can transmit a variety of infectious diseases (viral, bacterial, and protozoan) and even inject paralytic toxins (tick paralysis) during their prolonged blood meals.

2. Virus

Viruses are infectious units with diameters of about 16 nm (circoviruses) to over 300 nm (poxviruses). Their small size makes them ultrafilterable, *i.e.*, they are not retained by bacteria-proof filters. Viruses have evolved over long time period, and have adapted to specific organisms or their cells. The infectious virus particles, or virions, are composed of proteins and are surrounded in some species of viruses by a lipid membrane, which is referred to as an envelope; the particles contain only one kind of nucleic acid, either DNA or RNA.

VIRAL STRUCTURE

- ✰ Virus particle = virion
- ✰ Protein which coats the genome = capsid
- ✰ Capsid usually symmetrical
- ✰ Capsid + genome = nucleocapsid
- ✰ May have an envelope
- ✰ Capsid protects NA. Its antigenic.
- ✰ In non-enveloped virus capsid initiates attachment to host cells for entry.

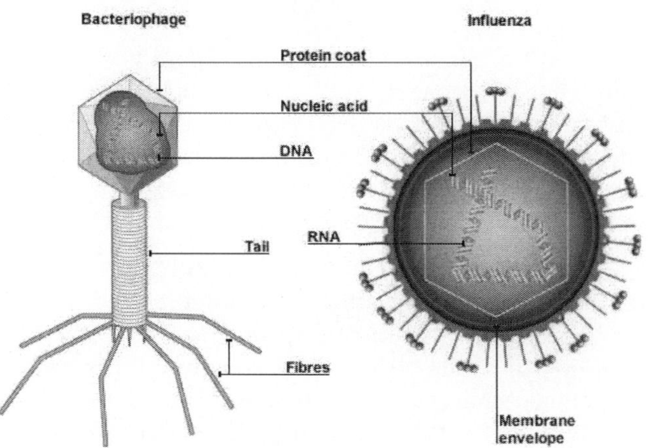

Figure 10.8: Viral Structure.

Viruses do not reproduce by division, such as bacteria, yeasts or other cells, but they replicate in the living cells that they infect. In them, they develop their genomic activity and produce the components from which they are made. They encode neither their own protein synthesis machinery (ribosomes) nor energy-generating metabolic pathways. Therefore, viruses are intracellular parasites. They are able to re-route and modify the course of cellular processes for the optimal execution of their own reproduction. Besides the genetic information encoding their structural components, they additionally possess genes that code for several regulatory active proteins (such as transactivators) and enzymes (*e.g.*, proteases and polymerases).

General Properties of Virus

- ✰ Viruses do not have cellular organization.
- ✰ One type of nucleic acid, either DNA or RNA, but never both.
- ✰ Obligate intracellular parasites.
- ✰ Lack the enzymes necessary for protein and nucleic acid synthesis and are dependent for replication on the synthetic machinery of host cells.
- ✰ Multiply by a complex process and not by binary fission
- ✰ Range in sizes from 16– 300 nanometers
- ✰ Unaffected by antibiotics.

Viral infections are the leading cause of gastroenteritis globally and may also cause enterically transmitted hepatitis and illness after migrating from the human intestine to other organs. Various viruses have been implicated in foodborne illness, with two types of virus, Norovirus and Hepatitis A, causing the most significant burden of foodborne illness and outbreaks, as they are highly contagious. Rotavirus is one of the major causes of diarrhoea in children and Hepatitis E, while primarily associated with waterborne infections, has been associated with foodborne outbreaks. Food borne transmission is important in the epidemiology of these four viruses, in addition to person-to-person contact and environmental transmission. Adenovirus, Astrovirus, Sapovirus, tick borne encephalitis and Avian Influenza H5N1 may also cause viral infections where food is a vector.

Foodborne viruses originate from the human intestine and are excreted in high numbers in the faeces or through emesis. The low infectious dose and robust survival in food and various environments facilitates the spread of viral infections. Food may be intrinsically contaminated during primary production or contaminated when prepared or handled by an infected food handler. Symptoms are typically nausea, vomiting, diarrhoea and abdominal pain.

Viruses behave differently to bacteria in infectivity, persistence and epidemiology and this a concern because current food safety guidelines usually either have not been validated for foodborne viruses, therefore there is no understanding of the efficacy of the controls on viruses, or the control measures are not effective in controlling these viruses and there are currently no reliable indicator organisms for viruses. Good agricultural and manufacturing practices are necessary to avoid introducing viruses into raw materials and food handling environments, as well as the consideration of viruses during the development of HACCP. Enteric viruses often produce mild symptoms, are difficult to culture and there are limited global pathogenic viral surveillance systems globally, so it is expected that as diagnostic methods and surveillance systems improve more viral diseases are likely to emerge as illnesses with foodborne transmission.

We shall now discuss the following important foodborne viruses:

- Hepatitis A virus
- Hepatitis E virus
- Rotavirus
- Norwalk virus group

i) Hepatitis A

First discovered in 1973 by Feinstone, a spherical 27 nanometer particle was seen on immune electron microscopy in the faecal sample of hepatitis A patients. A member of the picornavirus family, the hepatitis A virus (HAV) is an RNA virus responsible for 1.4 million cases per year globally, with an estimated 7134 deaths in 2016; almost half of these cases were reported in Asia. In the United States, the annual incidence rate was reported to be 2 cases per 100000 people, in 2006. Recent outbreaks of the disease have shown a 294 per cent increase in infections between 2016-2018 compared to 2013-2015.

Hepatitis A is a liver infection which is typically acquired through faecal-oral transmission, either from direct person-to-person contact or consumption of contaminated food or water. Foods can be contaminated with the virus through contact with raw sewage, such as is the case with shellfish, or through contact with contaminated water. HAV replicates in the liver, is excreted in bile, and is shed in the stool of infected people in high concentrations 2–3 weeks before and 1 week after onset of illness. Peak infectivity occurs during the 2 weeks prior to onset of clinical signs and symptoms (jaundice or elevated liver enzymes). Most persons cease to be infectious 1 week after jaundice appears. Although virus is present in serum of an infected person, its concentration is several orders of magnitude less than in faeces. Infected children and infants may excrete virus longer than adults.

The minimum infectious dose required for HAV infection in humans is unknown, but presumably is 10–100 virus particles with the incubation period being dependent upon the number of particles ingested (fewer particles means a linger incubation period). The mean incubation period of hepatitis A varies from 10 to 50 days (average 28 days) with a long communicability period from early in the incubation period to about a week after the development of jaundice. The middle of the incubation period (2 weeks before symptoms develop) presents the greatest period of communicability, well before the first presentation of symptoms during peak shedding of the virus. Symptomatic hepatitis A infection is clinically indistinguishable from other types of acute viral hepatitis, but is usually mild and self-limited. Fulminant hepatic failure occurs in less than 1 per cent of cases. The likelihood of symptomatic illness from HAV infection is directly related to age. In children younger than 6 years of age, most (70 per cent) infections are asymptomatic. In older children and adults, infection is usually symptomatic, with jaundice occurring in more than 70 per cent of patients.

Clinical manifestations vary, but may include the abrupt onset of fever, malaise, anorexia, nausea, and abdominal discomfort, followed within a few days to a week by dark urine, pale stools, and jaundice. Clinical illness usually resolves within 2–3 months (85 per cent of cases), and complete recovery is seen within 6 months for nearly all cases. About 10 per cent -15 per cent of patients present with a relapsing course within a 6-mo period of the initial infection. The symptoms during the relapse are usually less severe than the initial infection. Notably, on extremely rare occasions a type 1 autoimmune hepatitis has been observed in genetically predisposed patients. The spectrum of infections can range from asymptomatic patients without jaundice, symptomatic patients with jaundice, cholestasis with prolonged jaundice, to relapsing infections or acute liver failure.

Virus may be excreted in stool during a relapse. Consequently, persons experiencing relapsing hepatitis A should be considered infectious. There is no specific treatment for hepatitis A. Disease is usually self-limiting and treatment and management of HAV infection are supportive. HAV infection does not result in chronic infection or chronic liver disease. However, HAV infection can complicate chronic liver disease.

No specific treatment is available for HAV and the management is mainly symptomatic. The primary focus remains on improving sanitary conditions to minimize the transmission in the community. Historically, immunoglobulins have been used in the prevention of HAV infections. With the availability of an effective vaccine, the use of immunoglobulins has been largely abandoned except in infants below the age of 12 mo. Post exposure prophylaxis with hepatitis A vaccine has been approved since 2007 for immunocompetent patients without chronic liver disease, who are between the ages of 12 mo and 40 years of age. Current guidelines recommend HAV vaccination for individuals with increased risks of exposure to HAV (such as from injecting drug use, oral-anal sex, travel to or residence in endemic areas, frequent clotting factor or blood transfusions) or with increased risks of fulminant disease (such as those with chronic hepatitis).

Anyone who has not been vaccinated or previously infected can get infected with the hepatitis A virus. In areas where the virus is widespread (high endemicity), most hepatitis A infections occur during early childhood. Risk factors include:

- Poor sanitation;
- Lack of safe water;
- Living in a household with an infected person;
- Being a sexual partner of someone with acute hepatitis A infection;
- Use of recreational drugs;
- Sex between men; and
- Travelling to areas of high endemicity without being immunized.

The spread of hepatitis A can be reduced by adequate supplies of safe drinking water, proper disposal of sewage within communities and personal hygiene practices such as regular handwashing before meals and after going to the bathroom.

Examples of Hepatits A Outbreak

The largest viral foodborne disease outbreak ever documented occurred in Shanghai, People's Republic of China, in 1988, with almost 300,000 cases associated with consumption of clams harvested from newly discovered clam beds in sewage-polluted waters. A risk assessment applied to two HAV outbreaks in Valencia, Spain, associated with frozen coquina clams imported from Peru in 1999 (184 cases) and 2008 (100 cases) showed the likelihood of infection. Both outbreaks lasted several months, with clams from different shipments being consumed. Based on national epidemic data, HAV cases were also likely associated with these clams in Peru.

The most widespread HAV outbreak in the USA affected at least 640 people with four deaths in north-eastern Ohio and south-western Pennsylvania in late 2003 arising from green onions served in salsa at a restaurant in Monaca, Pennsylvania; the onions were likely contaminated at source in Mexico.

In 2010, 59 cases of non-travel-associated illness with genotype I, subtype B (IB) strains of HAV in France were linked with most cases consuming semidried tomatoes from one of three different sandwich shop chains, traced back to a supplier in France that imported frozen semidried tomatoes from Turkey. Three other HAV outbreaks caused by highly similar IB strains were associated with eating semidried tomatoes in Australia in 2009 and in the Netherlands in 2010.

Foodborne Outbreaks due to Hepatis A Globally

- Clam associated outbreak in Shanghai in 1988 with 280,000 people affected due to contamination and inadequate cooking.
- More than 50 residents in South Cambridgeshire, UK in 1991 where the vehicle was believed to be bread.
- 213 cases in Maine, United States in 1997 attributed to frozen strawberries
- November 2005 in France 111 cases were reported due to consumption of Oysters
- 2011, 7 people believed to have contracted HAV from sun dried tomatoes in UK.
- In 2013, 165 people became ill in a multistate US outbreak attributed to pomegranate arils.
- In 2016 an outbreak in the US was linked to frozen strawberries.
- In 2018 six EU countries reported confirmed that may have been associated with a single food product.

An outbreak of HAV genotype I subtype B infection in 165 individuals occurred in ten US states in 2013 associated with pomegranate arils imported from Turkey where this genotype is common. Sixty-nine individuals were hospitalized, two had fulminant hepatitis, and one required a liver transplant.

In June 2013, frozen berry mix containing fruit from the USA, Argentina, Chile, and Turkey purchased by approximately 240,000 people in the USA was subject to recall after at least 158 people were infected with HAV, 69 of whom were hospitalized; the isolated HAV strain is known to circulate in North Africa and the Middle East but rarely in the USA.

ii) Hepatitis E

Hepatitis E virus (previously labeled epidemic or enterically transmitted non-A, non-B hepatitis), is an enterically transmitted virus that occurs primarily in India, Asia, Africa, and Central America, where it is most common cause of acute hepatitis. The hepatitis E virus (HEV) is a small 32 to 34 nm, nonenveloped, HAV-like virus with a 7600-nucleotide, single-strand, positive-sense RNA genome. The virion is composed entirely of viral protein and RNA, with a buoyant density of 1.29 g/cm3 in potassium tartrate/glycerol gradients. Although the capsid icosahedral shape, lack of outer lipid envelope and size of the genome resembles other faecally transmitted viruses including HAV and norovirus, hepatitis E has some distinguishing physicochemical and genetic properties which have led to the virus being assigned its own genus (Hepevirus) and family (Hepeviridae).

It is considered one of the most common, yet underdiagnosed etiologies of acute viral hepatitis with the earliest epidemic going back to 1955 in New Delhi. While it stands as the only member of the genus Hepevirus in the family Hepeviridaea, a plethora of genotypes were identified to date with four of particular clinical significance; genotypes 1 and 2 being confined to humans, while genotypes 3 and 4 can infect humans and animals largely via the faecal-oral route.

The WHO reported an estimated 20 million new hepatitis E virus (HEV) infections every year globally, leading to 3.3 million symptomatic cases of acute hepatitis and in another report to over 44000 deaths in 2015; mainly associated with contaminated water, with the highest prevalence in East and South Asia. In industrialized regions, although the incidence of clinical HEV in humans is low, the seroprevalence of antibodies against HEV is relatively high, indicating a high proportion of subclinical disease and/or underdiagnoses. Although some exposure may arise from travel to endemic regions, HEV can be acquired domestically. Detection of HEV in pork liver sold in retail locations has been reported in Japan, the USA, and the Netherlands. Foodborne HEV outbreaks have been reported, usually from the consumption of raw or undercooked meat. Locally acquired HEV infections have increased in the UK since 2010. A case-control study investigated the association between consumption of pork products, *e.g.*, pies, ham, and sausages, and HEV infection; cooking practices for processed pork may be inadequate, *e.g.*, <70°C for 20 minutes to prevent HEV transmission. The following outbreaks demonstrate how HEV is transmitted through a variety of contaminated foods.

Hepatitis E virus is transmitted mainly via faecal-oral route, with faecally contaminated water providing the most common route of transmission. Vertical transmission of HEV has also been reported.

From the intestinal tract, the virus reaches the liver through an unknown route and mechanism. HEV appears to replicate primarily in liver and gall-bladder cells, but has been noted in the small intestine, lymph nodes, colon and salivary glands. The incubation period following exposure can range from 3 to 8 weeks, with a mean of 6 weeks. The disease usually is mild, asymptomatic and self-resolves in 2 weeks; it is usually seen in age groups 15–40 and can be asymptomatic in children. The clinical signs and symptoms in patients with typical HEV infection are similar to those seen with other forms of acute viral hepatitis. Jaundice is usually accompanied by malaise, anorexia, nausea, vomiting, abdominal pain, fever, and hepatomegaly. Other less common features include diarrhoea, arthralgia, pruritus, and urticarial rash. Some patients have asymptomatic infection. Acute liver failure can occur, resulting in an overall case fatality rate of 0.5 to 3 per cent. Acute liver failure occurs more frequently during pregnancy, resulting in an inordinately high mortality rate of 15 to 25 per cent, primarily in women in the third trimester. Infection with HEV can lead to hepatic decompensation in patients with pre-existing liver disease and those who are malnourished. Chronic hepatitis has been reported in organ transplant recipients and in patients with active HIV infections. Extended faecal shedding when present occurs for approximately 2 weeks after jaundice develops. In immunocompromised patients hepatitis E can be persistent and also associated with increased mortality and morbidity in people with progressed liver disease. Laboratory findings include elevated serum concentrations of bilirubin, alanine aminotransferase (ALT), and aspartate aminotransferase (AST). Resolution of the abnormal biochemical tests generally occurs within one to six weeks after the onset of the illness.

Examples of Foodborne Outbreaks

Hepatitis E has been the cause of sporadic cases and epidemic forms with the primary vehicle being ingestion of faecally contaminated drinking water.

- New Delhi in 1955 following the contamination of the city's drinking water when 29,000 cases of icteric hepatitis occurred.
- Indian subcontinent (1975, 1978, 1980) and USSR (1983): several major outbreaks.
- In 2007 in Corsica, 7 cases of Hepatitis E were attributed to raw figatellu (dish made with pig liver).
- An HEV outbreak was reported from Japan in 2003 associated with consumption of undercooked deer meat 6–7 weeks before symptom onset. Family members who did not eat the meat, or ate very little, did not become infected.
- Four vacationers reported acute HEV infection upon returning from a world cruise to the UK in 2008. Investigation revealed 195 of 789 persons were seropositive for HEV and 33 had immunoglobulin M levels, indicating recent acute infection (eleven were symptomatic). Analysis showed significant association between acute HEV infection and being male, drinking alcohol, and consuming shellfish while on board, demonstrating this was likely a common source foodborne outbreak. The type of shellfish could not be specified, but could have arisen from European shipments contaminated by polluted waters and brought aboard at Southampton.

iii) Rotavirus

Rotaviruses (RVs) belong to genus Rotavirus in the family Reoviridae of order Reovirales under the class Ribohelica. Rotaviruses are 60–80 nm, nonenveloped, linear segmented double-stranded RNA viruses with icosahedral capsid symmetry. There are five species of RV, designated rotavirus A (simian rotavirus) through rotavirus E (porcine rotavirus). Most human infections are caused by rotavirus A, B, and C, but all rotaviral species can infect a range of vertebrates.

RV was first discovered by Bishop *et al.*, using electron micrographs of intestinal and stool biopsies of pediatric gastroenteritis patients that showed a typical wheel shape and hence the Latin name 'rota' meaning wheel. Rotavirus is responsible for profuse diarrhoea and painful gastroenteritis in young children. It has been reported that nearly every child in the world that is under 5 years of age is at risk of infection by group A rotavirus. The virus causes an estimated 450 thousand deaths, over 2 million hospitalizations and 25 million outpatient cases annually, with over 90 per cent incidence in developing countries of Asia and Africa.

Nearly every child in the world under 5 years of age is at risk of infection by group A rotavirus. The burden of disease for this virus is astoundingly high. An estimated 20 per cent to 30 per cent of all cases of acute gastroenteritis at hospitals are attributed to rotavirus infection. One in every 7 premature children deaths in Europe is attributed to rotavirus infection, accounting for a total of 231 deaths, over 87 thousand hospital visits and nearly a million outpatient visits. The mortality rates are highest in South Asian countries such as Bangladesh, India, Pakistan, Nepal, Sri Lanka and African countries such as Nigeria and Congo. A report in 2013 finds half of all rotavirus deaths in the world to take place in India, Nigeria, Pakistan and the Democratic Republic of Congo. Although sporadic and endemic cases of rotavirus infection occur throughout the year, infection rates are highest in the winder when the air the dry and the atmosphere is cool. Developing countries where people are plagued by poor sanitation and waning health facilities, rotavirus induced diarrhoea takes on the most horrific form. Adults are rarely infected due to the immunizing effect of early encounters. Children under 14 months of age are at a higher risk than older children and immune compromised children are at a higher risk than immune competent children. The viral particles are extremely resistant to harsh environmental conditions and dangerously contagious.

The faecal-oral route is the primary mode of transmission. Infections spread through contamination of food items, drinking water, utensils and exposure to contaminated hands and surfaces. In developed countries, the majority of infections are of nosocomial origins. The infectious dose is estimated to be between a hundred and a thousand virus particles.

The incubation period for human RV is reported to be around 1–3 days. After an incubation period of 1 to 3 days, the illness manifests itself with severe impact. Symptoms include abdominal pain, fever, profuse diarrhoea, bulging of the midsection, dehydration, fatigue, dizziness and vomiting. Diarrhoea is caused by disruption of absorptive and secretory functions of the intestinal cells. Death and desquamation of the villous cells and the proliferation of secretory crypt cells induces abdominal discomfort and looseness of the bowels. The illness is exacerbated by the resulting decrease in signaling molecules and digestive enzymes, as well as paracellular leakage from the ransacked tight junctions between the enterocytes.

Although common between other gastrointestinal pathogens, the symptoms tend to be abruptly severe in rotavirus infections. This is why rotavirus induced gastroenteritis patient's end up in the hospital much more frequently than similar symptoms of other pathogenic origins. The diarrhoea characteristic of the disease has been attributed to several different mechanisms. The virus causes destruction of enterocytes and disruption of the absorptive intestinal epithelia. Other effects that may confer diarrheal symptoms include villus ischemia, enterotoxin activity of the viroid, secretion of calcium and chloride ions with intracellular fluid by NSP4, and the activation of distal nerve impulses that indirectly increase enterogastric secretions.

Examples of Rotavirus Foodborne Outbreak

- 1982, outbreak in China due to contamination of water by raw sewage.
- April 2000, 19 confirmed cases and 108 self-reported cases in District of Columbia, United States, possibly from line cooks.
- December 2000, outbreak attributed to contaminated water supplies in Tirane, Albania with 2722 children seen at Tirane Hospital.
- A HRV outbreak occurred on a Washington DC university campus in 2000, with 108 cases associated with eating tuna or chicken salad sandwiches from a campus dining hall. The investigation reported that 40/44 case patients and 27/40 controls consumed at least one deli sandwich from the dining hall during the outbreak period. It was also noted that six employees reported illness. Group A HRV was recovered from 9/27 stool specimens. The outbreak began as foodborne, but spread by person-to-person contact during the second week.
- Italy has witnessed a major viral gastroenteritis through drinking water, which was attributed to contamination of both RoVs and NoVs. The origin of contamination could not be found, but extra chlorination of the water solved the problem.
- In the Netherlands, poor food hygiene was identified as one of the major risk factors for RoV infection
- 9907 patients in 2005 in Malatya City, Turkey, due to possible contamination of a large water depository from a water well, which supplied drinking water to two major districts of the city.

- ✰ 4375 laboratory confirmed outbreaks in England from January-August in England in 2018

iv) Norovirus

Human norovirus, previously known as Norwalk virus, was first identified in stool specimens collected during an outbreak of gastroenteritis in Norwalk, OH, in 1972 and was the first viral agent shown to cause gastroenteritis. It is sometimes called the "stomach flu."

Human noroviruses are the leading cause of epidemic gastroenteritis in all age groups and have been associated with high-profile outbreaks in hospitals, nursing homes, cruise ships, and the military. Anyone can get Norovirus, but it is more common in adults and older children. It is estimated that each year, noroviruses are responsible for 64,000 diarrheal episodes requiring hospitalization, 900,000 clinic visits among children in industrialized nations, and <"200,000 deaths of children <5 years old in the developing world.

Centers for Disease Control and Prevention reports that 50 per cent of foodborne outbreaks are attributed to NoV.[15] NoV causes an estimated one in 15 US residents to become ill annually as well as 56,000–71,000 hospitalizations and 570–800 deaths, predominantly among young children and the elderly. Studies in the UK and the Netherlands showed that NoV was responsible for 3.6 per cent [13] and 15.6 per cent [12] of foodborne illnesses, respectively. NoV-associated deaths are rare, with an incidence of 0.19–0.40 deaths per 10,000 population.

Noroviruses are found in the stool and vomitus of infected people. People get Norovirus infections in several ways such as eating food or drinking liquids contaminated with the Norovirus or touching surfaces or objects contaminated with norovirus and then touching their mouth before washing their hands. Noroviruses are easily transmitted person to person by having direct contact with a person who is ill with norovirus.

The incubation period is relatively brief in most infected individuals who develop symptoms. Examination of the onset of secondary cases in the index outbreak in 1968 led to an estimated incubation period of <"48 h. In a Swedish outbreak involving children and staff at day care centers, the mean incubation period after foodborne transmission was 34 h, while that for secondary person-to-person transmission to household members was <"52 h. A recent systematic review of the literature concluded that the median incubation period for genotype I and II infections is 1.2 days (95 per cent CI, 1.1 to 2.2 days). In contrast to these findings, Rockx and colleagues reported that the median duration of illness was 4 days in community-acquired cases. Longer estimates such as these may, however, be the result of misclassification of secondary transmissions as primary transmissions.

The symptoms of Norovirus are nausea, vomiting, and/or diarrhoea accompanied by abdominal cramps. Some people complain of headache, fever, chills, and muscle aches. Symptoms are usually brief and last only one or two days. However, during that brief period, people can feel very ill and vomit or have diarrhoea, often violently and without warning, many times a day. Symptoms usually begin 24-48 hours after exposure to the virus, but can appear as early as 12 hours after exposure.

In an analysis of 4 outbreaks over 3 years in an inpatient psychiatric unit in Taiwan that affected 172 patients and 7 health care workers, diarrhoea occurred in 87.5 per cent and vomiting occurred in 25.5 per cent of patients, while 4.4 per cent complained of abdominal pain and 2.2 per cent had fever. The mean duration of symptoms was 2.1 ± 1.5 days, with a range of 1.2 to 2.8 days; 86.4 per cent of patients had symptom resolution within 1 to 3 days. The illness may be more severe and prolonged in individuals with medical comorbidities. Norovirus infection-associated illness may also be more prolonged and severe in immunocompromised individuals and may be associated with remarkably persistent viral excretion in some of these individuals.

Examples of Foodborne Norovirus Outbreaks

Food presents two distinct models for norovirus transmission. The first is direct norovirus contamination of food at the point of production, and the second is contamination of food during preparation. In a review of several hundred reported norovirus outbreaks, foodborne transmission (362/666; 54 per cent) and foodservice settings (294/830; 35 per cent) were most commonly implicated. Fruits, vegetables, and shellfish pose a significant risk for disease transmission because they are consumed raw and may be subject to norovirus contamination from water sources. Norovirus has been recovered from a variety of food products ranging from oysters to romaine lettuce, raspberries, and other produce. Additionally, the use of contaminated water for reconstituting food products has been implicated in widespread outbreaks (*e.g.*, custard in Bristol, United Kingdom). Rebedding of shellfish and depuration with purified seawater at increased temperatures have been used to reduce the burden of norovirus from shellfish in contaminated waters. Various food surface decontamination strategies have been explored to enhance the produce safety and decrease viral transmission, including treatment with sodium bicarbonate and chlorine.

- ✰ Lettuce has been implicated in many NoV outbreaks, *e.g.*, in Michigan in 2005 after a national submarine sandwich franchise restaurant catered four luncheons with a total of 88 ill.

- ✰ In Denmark, eleven outbreaks with 260 cases were all linked to lettuce imported from France in 2010.
- ✰ In Europe in 2012, 14 per cent of reported foodborne outbreaks were attributed to viruses, predominantly NoV; mixed foods and buffet meals were the main foodborne vehicles, followed by shellfish, vegetables, juices, and bakery products.
- ✰ Of particular note was a NoV outbreak in Germany associated with frozen Chinese strawberries mainly distributed through one large catering company, with 10,950 people, primarily school/kindergarten children, infected and 38 hospitalized. The acid resistance of NoV and HAV might explain their association with outbreaks where acid fruits were implicated, *e.g.*, raspberries.
- ✰ NoV outbreaks have also been caused by contaminated water from sewage in wells and in recreational water settings, *e.g.*, beaches, pools, and lakes.

Foodborne outbreaks are often linked to hygiene breaches during preparation and are often tied to ill food handlers and common source producers. Unfortunately, it is not just workers with clinical symptoms who transmit NoV, since asymptomatic food handlers were found to be a major contributor to Danish outbreaks, where contamination occurred with equal frequency during production, preparation, and serving, and at self-serve buffets. Some of the workers had been in contact with ill household members before handling the food and remained asymptomatic; others developed symptoms shortly after or were post-symptomatic. However, food may be contaminated by NoV at any point during production, shipping, handling, or preparation. NoV outbreaks can result from faecal contamination of foods at their source, *e.g.*, oysters harvested from contaminated water, environmental sources with leafy greens, and raspberries spray-irrigated with contaminated water. Ill family members of food handlers may represent a reservoir for disease. Norovirus outbreaks in restaurants and canteens have been documented and have been linked to catering companies, with the extent of outbreaks varying from several isolated cases to large multistate events.

Over 90 per cent of diarrheal disease outbreaks on cruise ships are caused by NoV, which is especially challenging to control on ships because of close living quarters, shared dining areas, and rapid turnover of passengers between cruises. Before embarkation, NoV can be brought on board in contaminated food or water or by passengers who are already infected. Repeated outbreaks on consecutive cruises may result from infected crew or environmental contamination not sufficiently decontaminated since NoV can persist on surfaces and is resistant to many common disinfectants. Being aware of the risks, crews try to prevent infected persons from boarding using screening procedures, with limited effectiveness. Many cruise ships do not permit buffet self-service until 24–48 or more hours have elapsed after boarding in an attempt to prevent passenger-to-food transmission.

How can Norovirus Infections be Prevented?

- ✰ Wash hands with soap and warm water after toilet visits and before preparing or eating food.
- ✰ Wash raw fruits and vegetables and steam oysters before eating them.
- ✰ Thoroughly clean and disinfect contaminated surfaces immediately after an episode of illness by using a bleach-based household cleaner.
- ✰ Immediately remove and wash clothing or linens that may be contaminated with the virus after an episode of illness (use hot water and soap).
- ✰ Flush or discard any vomitus and/or stool in the toilet and keep the surrounding area clean.

Food handlers with symptoms of Norovirus should not prepare or touch food during their illness and for three days after they recover.

3. Natural Toxins

Occurrence of natural toxins in food and feed represents undoubtedly an important food safety issue which is of growing concern of both scientists and regulators. Considering the current terminology, several approaches are applied for classification of natural toxins. The narrow definition specifies natural toxins as food constituents endogenously produced by food organisms and capable to give rise to adverse effects when the food organism or product thereof is consumed. The broader definition of natural toxins includes also toxic compounds originating from a contamination of the food.

Natural toxins are of 2 types:

i) Endogenous toxins of plant origin

ii) Natural contaminants

- ✰ Edible plant + toxin
- ✰ Toxic substances by animal
- ✰ microbial toxins (bacterial toxin, mycotoxin)

Plant Toxins

Plants contain a large number of chemical compounds which are distinct from the intermediates and products of primary metabolism. These secondary metabolites that are produced from primary metabolites and their precursor may contribute to various specific features of plants such as odours, tastes and/or colours; some of them are viewed as the key components of active and potent defence mechanisms against bacteria, fungi, insects and/or herbivores. However, the real function of these constituents for plant health and proliferation has not been fully recognized yet. Worth to notice, most of plant species in the world are inedible for humans, many because of the presence of toxic substances; nevertheless, over the millennia the domestification process has gradually reduced their levels in cultural species.

In food plants we eat today there is fairly less compounds hazardous for consumers as compared to their wild relatives. On the other hand, the decrease/elimination of constituents that have evolved as a way to fight off predators unavoidably results in lowered resistance of modern cultivated plants to diseases. In spite of enormous progress in breeding/growing agricultural practices and advances in food processing technologies aiming at production of high quality and safe foods, number of antinutrients (compounds with negative effects on availability of nutrients) and toxic compounds of natural origin still can occur in our diet.

Common classes of plant toxins include alkaloids such as pyrrolizidine alkaloids and glycoalkaloids, cyanogenic glycosides, lectins, saponins, and antinutrients. Some of these toxins are being discussed in the following sections.

i) Glycoalkaloids

One of the widely discussed groups of plant toxins is represented by glycoalkaloids that are contained in all Solanaceous plants including potatoes, an important staple food. Glycoalkaloids consist of a steroidal alkaloid coupled to one or more monosaccharides. All members of the botanical family solanacea produce glycoalkaloid toxins. In potatoes, the major glycoalkaloids are α-solanine and α-chaconine. They are formed in the parenchyma cells of the periderm and the cortex of the tubers. These chemicals are toxic to insects and animals and serve to protect the plant from predators. They are normally present in all potato tubers in small amounts and are concentrated in the skin of the potato and also in the areas of high metabolic activity such as the eye regions. Its level is increased in greened potatoes or blighted potatoes, and can reach very high levels in the sprouts.

'Solanine' content considerably varies among individual varieties and is largely affected by environmental factors such as light, irradiation, mechanical injury, and storage. Potatoes that have been exposed to light in the field or during storage may become green, due to an accumulation of chlorophyll. This greening may affect only the surface (peel) or it may extend into the flesh of the potato. Exposure to light is only one of the stress factors affecting potatoes. Other pre- or post-harvest stress factors are mechanical damage, improper storage conditions, either as a tuber or after partial food processing, and sprouting. As a result of any of the above stress factors, there can be a rapid increase in the concentration of TGA, notably, alpha-solanine and alpha-chaconine, which gives the potatoes a bitter taste.

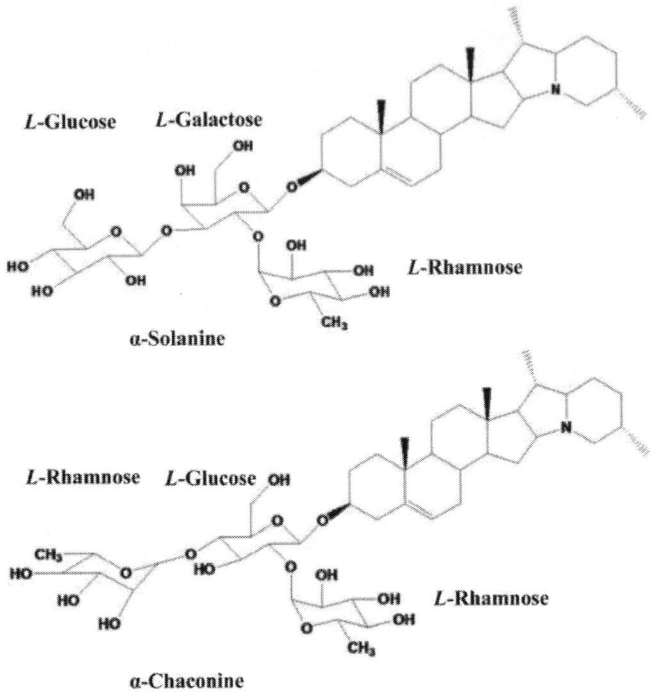

Figure 10.8: Structure of Alpha-solanine and Alpha-chaconine.

Processing of potatoes has been reported to reduce the content of GAs in the final processed product. In general, and according to the literature, the peeling of potatoes reduced the GA content by 25–75 per cent, boiling in water and blanching of peeled potatoes by 5–65 per cent and frying in oil of peeled potatoes by 20–90 per cent. Microwave and oven baking of unpeeled potatoes may cause a reduction in the GA content by 3–45 per cent and by 20–50 per cent, respectively. No information has been found about the chemical nature of the GA degradation products. For the exposure assessment, processing factors for the major food processing steps, comprising peeling and heat processing (boiling, frying, baking), were applied to the occurrence data as follows: processing factors between 0.25 and 0.75 were attributed to the peeling of potatoes, between 0.1 and 0.8 for frying and deep frying, and between 0.35 and 0.95 for all other cooking methods.

Cooked potatoes that contain high level of solanine have a bitter taste and cause a burning sensation in the throat. The most efficient way of lowering dietary exposure is peeling of tubers since their outer layers are rich in these toxins. Depending on tuber size and variety, up to 60 per cent of total content can be removed by manual peeling.

Tomatoes are another important food crop of Solanaceous family contributing to the glycoalkaloids intake. The TGA concentration in tomato depends, as for potato, on the cultivar and the fruit ripening stage. Eltayeb and Roddick (1984) studied the changes in the α-tomatine content as a result of the fruit growth (*e.g.* given by the relative diameter of the berry) and/or the stage of ripening (*e.g.* defined by the development of the fruit colour: from dark green to full red or yellow colour depending on the cultivar), in commercially grown tomato cultivars and mutants with different ripening characteristics. For three normalripening cultivars developing red, orange and yellowripe fruits, respectively, it was found that the highest concentrations occurred in very young green fruits of around 4 days of age (3,100–4,400 mg αtomatine/kg fw). In subsequent stages of fruit development, the concentration dropped to around 700 mg αtomatine/kg fw at day 10. In the same period, the fruit diameter increased from 7 mm at day 4–20 mm at day 20. By 30 days, when the tomato was approximately 30 mm in diameter, 20 g in weight and still green, αtomatine concentrations for the three cultivars had decreased to 3–6 mg/kg fw (Eltayeb and Roddick, 1984). Hence, this crop seems in this respect to be much safer for humans than are potatoes, since tomatine (contains mixture of dehydro tomatine and α-tomatine) is largely degraded as the tomato fruit ripens. The development of transgenic tomatoes has also stimulated interest whether their glycoalkaloids content differs significantly from that occurring in standard varieties during different stages of fruit maturity.

Various toxic effects have been shown for main potato glycoalkaloids, α-chaconine and α-solanine, and their hydrolysis products. Historical reports on acute poisoning of humans by ′solanine′ (its levels are largely elevated in sprouted and/or greened potatoes) are available.

The amount of glycoalkaloids usually found in edible plants that are fresh and undamaged do not normally cause toxicity. However, glycoalkaloids can produce toxic effects at higher doses. The greatest concern for glycoalkaloid toxicity is its acute toxicity. There have been many reported cases of human poisonings (sometimes fatal) due to the ingestion of greened, damaged or sprouted potatoes as a consequence of high levels of glycoalkaloids.

Onset of symptoms has ranged from minutes to 2 days after ingestion of toxic potatoes, with longer incubation periods generally associated with the more severe cases. In humans, acute toxic effects following ingestion of potato GAs include gastrointestinal symptoms of varying severity such as vomiting, diarrhoea and abdominal pain, which may occur from a total potato GAs (potato TGA) intake of 1 mg/kg bw or more. Further symptoms, including drowsiness, apathy, confusion, weakness, vision disturbances, rapid and weak pulse and low blood pressure, may be the consequence of dehydration following vomiting and diarrhoea.

In severe cases, paralysis, respiratory insufficiency, cardiac failure, coma and death have been reported. Doses in the range of 3–6 mg potato TGAs/kg bw are considered to be potentially lethal for humans. Results from limited volunteer studies suggest possible differences in the human population with respect to the individual susceptibility towards adverse effects associated with the intake of potato GAs.

Regarding the mode of action, adverse effects of GAs may be due to their ability to complex with membrane 3βhydroxy sterols, thereby causing disruption and loss of integrity of cell membranes. After oral exposure, these effects may affect the mucosa of the gastrointestinal tract and cause the symptoms observed in intoxicated humans, such as nausea, vomiting and diarrhoea.

GAs inhibit acetylcholinesterase (AChE) and serum butyrylcholinesterase (BuChE) by a reversible, competitive mode of action. The relative potency of inhibition of αsolanine and αchaconine appears to be similar. The aglycones exert weak or no inhibitory effects. The excess of acetylcholine at the neuronal and neuromuscular junctions upon inhibition of the enzymes might also contribute to the symptoms described for intoxications with GAs. Glycoalkaloids also have saponin–like properties and can disrupt membrane function in the gastrointestinal tract leading to haemorrhagic damage. This damage can be severe enough to cause death with the extent of necrosis far outweighing the inhibitory effects on acetylcholinesterase activity.

At high doses, αtomatine may form a nonabsorbable complex with cholesterol and other sterols in the enteral lumen which may impair the absorption of cholesterol. As a consequence, blood cholesterol levels were lowered in rodents.

Because of the lack of chronic toxicity data, an adequate no-observed adverse effect level (NOAEL) for potato glycoalkaloids has not been assessed, and a tolerable daily intake (TDI) for humans has not been determined.

JECFA considered that, despite the long history of consumption of plants containing glycoalkaloids, the available epidemiological and experimental data from human and laboratory animal studies did not permit the determination of a safe level of intake. Based on the available human data, an intake of total glycoalkaloids of 3-6 mg per kg body weight (bw) is considered by JECFA a potentially lethal dose for humans, and total glycoalkaloids of more than 1 to 3 mg per kg bw is considered a toxic dose for humans. Children may be more sensitive than adults. Other factors may be present in suspect potatoes and modulate the toxicity of the steroid glycoalkaloids.

The JECFA Committee recognised that the development of empirical data to support a safe level of intake would require considerable effort. Nevertheless, it felt that the large body of experience with the consumption of potatoes, frequently on a daily basis, indicated that normal glycoalkaloid levels (20-100 mg/kg) found in properly grown and handled tubers were not of concern.

No adequate no-observed-adverse-effect level (NOAEL) is available for determination of Tolerable Daily Intake (TDI) at present. Recommendation for maximum levels of 200 mg α-chaconine and α-solanine (sum) per kg potatoes is based on estimation of 300 g daily intake, nevertheless, doubts exists whether this value is sufficient for consumers protection, since the safety margin in this particular case is rather narrow.

ii) Cyanogenic Glycosides

Glycosides of α-hydroxynitriles (cyanogenic glycosides) releasing hydrogen cyanide via enzymic breakdown catalysed by endogenic β-glycosidases are contained in several economically important plants. Despite great deal of structural diversity in cyanogenic glycosides, almost all of them are believed to be derived from only six different amino acids L valine, Lisoleucine, Lleucine, L phenylalanine, or Ltyrosine, and cyclopentenylglycine (a nonprotein amino acid). Cyanogenic glycosides play pivotal roles in organization of chemical defence system in plants and in plantinsect interactions.

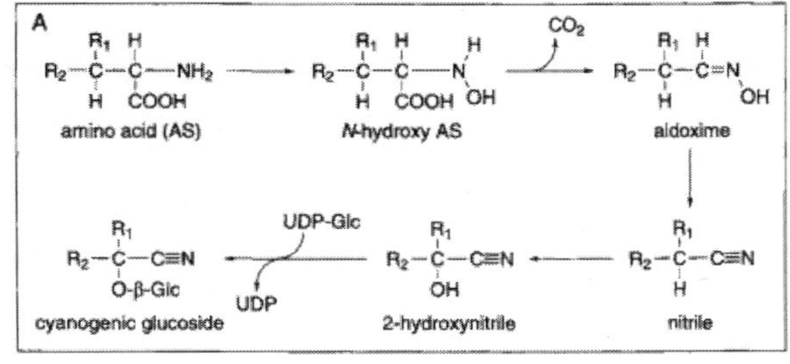

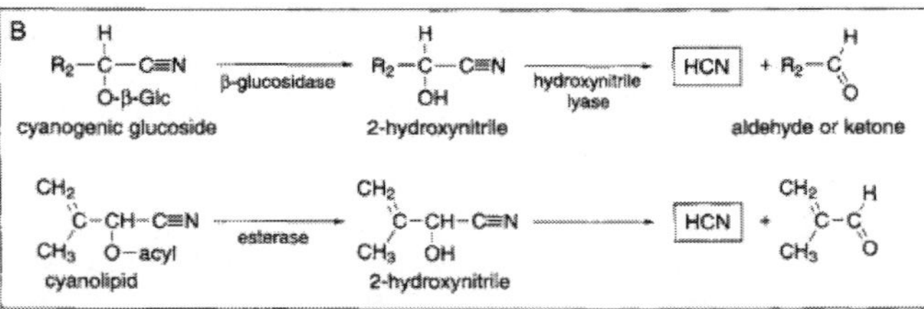

Figure 10.9: Cynogenic Glycosides.

They are widely spread in higher plants. They occur in at least 2000 plant species, of which a number of species are used as food. They are amino-acid-derived constituents of plants produced as secondary metabolites. There are approximately 25 cyanogenic glycosides known. Different kinds of cyanogenic glycosides may be found in different cyanogenic food plants, *e.g.*, taxiphyllin in bamboo shoots, linamarin in cassava. On chewing or getting digested, cyanogenic glycosides present in these foods leads to hydrolysis of the substances, causing cyanide released. Cyanogenic glycosides are common in certain families such as the Fabaceae, Rosaceae, Leguminosae, Linaceae, and Compositae.

Important staple foods for some parts of the world (such as cassava and sorghum) contain cyanogenic glycosides. Other edible plants containing cyanogenic glycosides include bamboo shoot, flaxseeds, and seeds of stone fruits such as apricot and peach, seeds of peas and beans such as chickpeas, lima beans, and shell of soya beans. Other food products that may contain cyanogenic glycosides include some food ingredients with flavouring properties such as ground almonds powder or paste, marzipan, cashews, stone fruit preserves (cherry, plum, apricot, peach), and stone fruit juices, and alcoholic drinks made from stone fruits. These foods therefore represent potential sources of hydrogen cyanide.

Scientific Name	Common Name	Taxonomic Group	Cyanogenic Compound Location	Commercially Produced for Consumption?
Sorghum bicolor	Sorghum	Monocot	young shoots	Y
Manihot esculenta	Cassava	Dicot	entire plant	Y
Davallia trichomanoides	Squirrel's Foot Fern	Pteridophyte	fronds, fiddleheads	N

Figure 10.10: Cyanogenic Glycosides and Plants.

High levels of cyanogenic glycosides have been reported to be present in many nutritious dense plants, thus limiting the use of some of the plants as food. For instance, cassava (*Manihot esculenta*) is one of such plant with high amount of cyanogenic glycosides. It is a crop of economic importance in Africa and a staple food in most African communities. Cassava is also a valued crop in South America and Southeast Asia. The high carbohydrate content of the crop makes it a good source of calorie for feeding billions of people in the world. According to Food and Agriculture organization, cassava is the third most important source of calories in the tropics, after rice and corn. Cassava leaves have higher protein content, contain vitamin C and vitamin A and provide some dietary fiber. Much of the protein in the leaves is made up of linamarase, the enzyme that detoxifies the cyanogenic glycosides in cassava. However, each parts of cassava plants (leaves, stem, root) contains high levels of cyanogenic glycosides; linamarin, lotaustralin, and amygdalin, with linamarin been the most predominant cyanogen.

Cassava: The cyanide level of cassava varies from about 75 to 350 ppm, but can be up to 1000 ppm or more depending on the variety, plant age, soil condition, fertilizer application, weather, and other factors. Studies have shown that the levels of cyanogenic glycosides in cassava roots are generally lower than that in the leaves and stems. Cassava roots have been reported to contain cyanide content of 10–500 mg/kg of dry matter and the leaves were reported to contain 53–1300 cyanide equivalents/kg of dry matter.

Chopping, grinding, soaking in running water as well as fermentation and cooking, are traditional processing practices used by casssava African and South American consumers for detoxification. Cassava root can be boiled and eaten as whole root, or processed into flour for the production of "thick gruel" commonly consumed in Africa. It can also be converted into chips for the production of flour, tapioca, and other foods. Cassava leaves are used in some African countries to produce soup or sauce, following processes such as scrubbing, fermentation, pounding, and boiling. Apart from human consumption, cassava roots and peels are also consumed by animal. The major cyanogenic glycoside in cassava is linamarin, while small amount of lotaustralin (methyl linamarin) and amygdalin are also present, as well as an enzyme linamarase. Linamarin is rapidly hydrolyzed by linamarase to glucose, acetone cyanohydrin, and hydrogen cyanide as shown below in the Figure 10.11.

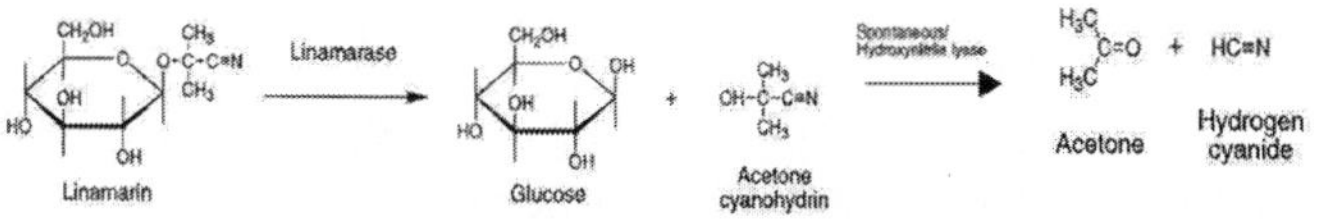

Figure 10.11: Formation of HCN.

Under neutral conditions, acetone cyanohydrin decomposes to acetone and hydrogen cyanide. Although processing methods can reduce linamarin and cyanide in food, improperly processed cassava products would contain some amount of residual linamarin and hydrogen cyanide. This would results in the potential toxicity of the cassava products. Indeed, cases of cyanide toxicity from the consumption of inadequately processed cassava products have been reported.

Sorghum (*Sorghum bicolor*): Dhurrin is the major cyanogenic glucoside in sorghum, representing 30 per cent of the dry weight of shoot tips of seedlings, in addition to amygdalin. In young sorghum leaves, dhurrin is localized in vacuoles of plants and the enzymes responsible for its hydrolysis to hydrogen cyanide is located in cytoplasm. In intact leaf, tissues are free from cyanide due to compartmental separation of the enzyme and the substrate. Concentration of dhurin decreases as plant ages, immature sorghum leaves contain high concentration of dhurrin. The cyanide content of sorghum increases rapidly during early growth stage, after which it declines with plant age. The reports have confirmed the cyanogenic glucoside content in the tip of young seedlings of *Sorghum bicolor* that reaches 6 per cent of the dry weight. Sorghum grains have been reported to contain 122.31 mg/g of amygdalin.

Bamboo shoot: Fresh immature bamboo shoots are consumed as vegetable in some Asian countries. Bamboo shoot contains appreciable quantities of vitamin C, carbohydrates, and protein. Besides its nutritive value, bamboo shoots also contain lethal concentration of cyanogenic glycosides. The cyanogenic glycoside present in bamboo shoot is taxiphyllin, which is decomposed quickly in boiling water. Cyanide content of bamboo shoot ranged from 1000 to 8000mg/kg hydrogen cyanide. Although cyanide content of bamboo shoot is much higher than that of cassava root, the cyanide content in bamboo shoots decreases substantially following harvesting and processing.

Fruits and fruits kernels: The most common cyanogenic glycoside in fruits and fruit kernel is amygdalin. Amygdalin contents of the fruit seeds vary significantly from seed to seed.

Apple is the most cherished fruit among the Rosaceae family. Apples can be consumed raw or processed into alcoholic (cider) or nonalcoholic beverage (apple juice) or apple sauce. Although apples are rich sources of vitamins and other nutrients, apple seeds contain high levels of cyanogenic glycosides. Amygdalin content of apple seeds ranged from 1 to 4mg/g while that of apple juice was reported to be between 0.001 and 0.08 mg/ml.

Apricot fruits are cultivated in the Middle Asia, Africa, America, and Europe. The fruits are consumed raw or used in dry form for confectioneries while the kernels are usually processed before consumption. Apricot kernels are of two varieties; bitter and sweet. Bitter apricot kernels contain high amount of the cyanogenic glycosides, amygdalin, which can cause cyanide toxicity problems at high dose and thus, unsafe for consumption. Sweet apricot kernels as well as apricot flesh are safe for human consumption because of their low level of cyanogens. The concentration of hydrogen cyanide in apricot kernels varies widely (49–4000 mg/kg), depending on whether skin on or off varieties is surveyed. Cyanide level of raw or improperly processed apricot kernels can cause serious acute problems that could lead to death. Studies have shown that apricot kernel contain cyanide (CN) content of 1450mg/kg, approximately 0.5 mg CN/kernel. This value is similar to the toxic dose of cyanide (0.5 mg/kg body weight) reported by previous studies.

Processing methods, such as peeling, drying, grinding, soaking, boiling or cooking, soaking and fermentation have been reported by several studies to cause significant reduction in the cyanogenic glycosides of processed foods. These processes have been applied to food crops such as roots, tubers, cereals, and leaves, to cause significant reduction in the cyanogen contents of the crops. Foodprocessing methods generally disintegrate cyanogens contents of plants, and this leads to the production of cyanide. Since cyanide is volatile, further processing techniques, such as roasting and drying, will volatilize the remaining cyanide to low level.

Toxicity

Cyanogenic glycoside is not toxic on its own. However, when cell structures of plant are disrupted, cyanogenic glycoside will be brought together with the corresponding βglucosidase enzyme. Hydrolysis of cyanogenic glycosides usually occurs when cyanogenic plants are chewed by herbivores or when the plants are disintegrated during processes, such as grinding, pounding or in the presence of water for example during soaking or fermentation. In the presence of water, the non-toxic cyanogenic glycosides are hydrolysed by the enzyme producing cyanohydrins which quickly decompose to the toxic hydrogen cyanide. Hydrolysis is accomplished by the βglucosidase, producing sugars and a cyanohydrin that spontaneously decomposes to HCN and a ketone or aldehyde. In this way, cyanogenic plants are protected against predators. Cyanogenic glycosides, cyanohydrins and hydrogen cyanide are collectively known as cyanogens.

The toxicity of cyanogenic glycosides and their derivatives is dependent on the release of hydrogen cyanide. Toxicity may result in acute cyanide poisoning and has also been implicated in the etiology of several chronic diseases. Cyanide toxicity can occur in animal including humans at doses between 0.5 and 3.5 mg HCN per kilogram body weight. In humans, the clinical signs of acute cyanide intoxication can include rapid respiration, drop in blood pressure, rapid pulse, dizziness, headache, stomach pains, vomiting, diarrhoea, mental confusion, stupor, cyanosis with twitching and convulsions followed by terminal coma. Children are particularly at risk because of their smaller body size.

Death due to cyanide poisoning can occur when the cyanide level exceeds the limit an individual is able to detoxify. The likelihood of cyanide intoxication from consumption of cyanide-containing food is dependent on body weight. For example, it is possible that a child or person of smaller body weight would not be able to detoxify the cyanide resultant from a meal of inadequately prepared cassava or bamboo shoots. The acute lethal dose of hydrogen cyanide for humans is reported to be 0.5-3.5 mg/kg bw. Approximately 50-60 mg of free cyanide constitutes a lethal dose for an adult man.

It is uncommon for dietary cyanide intake to cause chronic diseases. Adverse effects may, however, accompany individuals who have underlying dietary deficiency such as inadequate protein and/or iodine intake. For example, individuals already suffering from goitre and cretinism due to iodine deficiency may sometimes experience exacerbation of the condition following continuous dietary cyanide exposure.

Neurological diseases such as konzo and tropical ataxic neuropathy (TAN) have been reported in individuals with inadequate dietary intake of protein and/or iodine where cassava is consumed as a staple food. Konzo is an upper motor neuron disease characterised by irreversible but non-progressive symmetric spastic paraparesis with an abrupt onset, whereas tropical ataxic neuropathy (TAN) is a severe neurological syndrome with clinical symptoms of optical atrophy, angular stomatitis, sensory gait ataxia, and neurosensory deafness.

Safety of cyanogenic glycosides. The safety of cyanogenic glycosides has been evaluated by JECFA in 1992. However, safe level of intake could not be estimated due to lack of quantitative data.

iii) Lectins or Haemagglutinins

Although they were first discovered more than 100 years ago in plants, lectins are now known to be present throughout nature. It is generally believed that the earliest description of lectin was given by Peter Herman Stillmark in his doctoral thesis presented in 1888 to the University of Dorpat. Stillmark isolated ricin, an extremely toxic hemagglutinin, from seeds of the castor plant (Komath *et al.*, 2006). Stillmark observed that crude extracts of castor beans (Ricinus communis) contained a toxic substance named ricin that agglutinated human and some animal red blood cells.

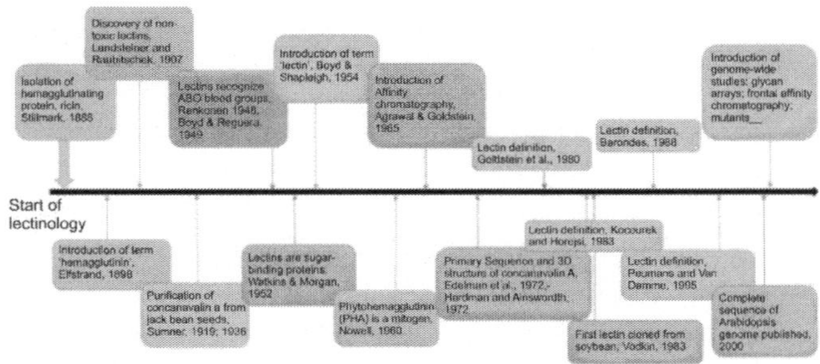

Figure 10.12: Lectins.

"Lectin" has been derived from the Latin word "legere", which means "to select", by William Boyd (Boyd and Shapleigh, 1954). This term was generalized to embrace all sugar-specific agglutinins of nonimmune origin, irrespective of source and blood type specificity. Lectins have the ability to bind carbohydrates and the name "hemagglutinins" is used when the sugar specificity is unknown.

Lectins are the carbohydrate-binding proteins and are known for their toxic relevance. Lectins are proteins/glycoproteins of non-immune origin, which have at least one non-catalytic domain that exhibits reversible binding to specific monosaccharides or oligosaccharides. They can bind to the carbohydrate moieties on the surface of erythrocytes and agglutinate the erythrocytes, without altering the properties of the carbohydrates. Lectins with specific carbohydrate specificity have been purified from various plant tissues and other organisms. They can be classified on the basis of their carbohydrate specificity. They can also be categorized according to the overall structures into merolectins, holoectins, chimerolectins and superlectins, or be grouped into different families (legume lectins, type II ribosome-inactivating. The amount of lectin varies in different organisms.

Due to their high stability, lectins are able to survive digestion by the gastrointestinal tract of consumers and may cause severe gastric upset. Number of outbreaks of food poisoning were reported, typically due to the consumption of uncooked or partially cooked kidney beans (Phaseolus vulgaris). Among many adverse effects described, interferences with digestion and absorption of nutrients have been widely discussed. Pyrrolizidine alkaloids Pyrrolizidine alkaloids are a large group of natural toxins involving almost 350 chemicals, many of them have been shown carcinogenic in animals and are therefore classified as potential human carcinogens. Intoxication of people can occur in several ways, often due to contamination of food supply by seeds containing these alkaloids. Another exposure route is occurrence of plants containing pyrrolizidines in traditional remedies or certain dietary supplements.

Lectins are found in nature. A large number of lectins or hemagglutinins have been purified from different organisms. They are present in the most commonly edible plant foods such as leguminous species and many others hence they are ingested daily in appreciable amounts by both humans and farm animals.

The lectin contents in some parts of plants are higher, *e.g.*, 390 and 75 mg of the purified lectin was recovered from 100 g Remusatia vivipara tubers and Astragalus mongholicus roots, respectively. Lectins are also found in seeds. The lectin content in non-legume plants is low, *e.g.*, 3.3 mg lectin from 100 g *Hibiscus mutabilis* seeds. Lectins are found in abundance in legume seeds. *Phaseolus vulgaris* is an herbaceous annual plant grown worldwide for its edible beans, popular in both dry and green bean forms. The commercial production of beans is well distributed worldwide. There are different varieties, including anasazi bean, black beans, cranberry bean, borlotti beans, pink beans, pinto beans, kidney beans, shell beans, white beans, yellow beans and French beans, *etc.* Lectins or hemagglutinins have been purified from different varieties of *P. vulgaris*. The lectin contents are low in some varieties and high in other varieties.

Yields of lectins from fresh mushrooms are low, *e.g.*, 2.6 mg from 100 g of fresh fruiting bodies of Pleurocybella porrigens. In fact, the water content in fresh mushrooms is very high. Dried fruiting bodies of the mushrooms Russula lepida, Pholiota adiposa, and Inocybe umbrinella yielded 39, 70, and 15 mg lectin per 100 g fruiting bodies, respectively (Zhang *et al.*, 2010). Therefore, production from fresh mushroom is also unpractical.

Lectins are found in different animals. However, the yields are usually extremely low. Mass purification of animal lectins necessitates bulk quantities of raw materials which make it not feasible.

The toxicity of lectin has been identified by consumption of food with high lectin content which can lead to diarrhoea, nausea, bloating, vomiting and even death. Hemagglutinin activity of lectin extracts from different legumes differ in theory potency. Foods with high concentration of lectins may be harmful if consumed in excess in uncooked or improperly processed form. Adverse effects may include nutritional deficiencies and immune reactions. Lectin can cause damage to microvilli, shedding of cells, reduction in the absorption capacity of the intestine, increased turnover of epithelial cells, interference with the immune system, hypersensitivity reactions, interference with microbial ecology of gut and selective overgrowth. Lectins interfere with cell mitosis, cause local haemorrhages, damage kidney, liver and heart and agglutinate red blood cells.

Crude extract of lectin agglutinates the red blood cells of human beings and animals if injected directly into the blood stream. Most effect of lectins is due to gastrointestinal distress through interaction of the lectins with the gut epithelial cells. An *in vitro* study has suggested that mechanism of lectin damage may occur by interfering with the repair of already damaged epithelial cells. Lectin damage to gut wall may allow other non-lectin proteins to cross undigested into general circulation and cause allergic reactions including anaphylaxis.

There has been information that lectins may be inactivated, destroyed or drastically reduced by some processing operation, such as boiling, soaking, malting (germinating) or fermenting. Soaking legumes overnight, draining the water, rinsing and draining again, does not seem to remove many of the lectins. The large concentration of lectins in plant seeds decreases with growth and suggests a role in plant germination and perhaps in the seed's survival itself. Heat processing can reduce the toxicity of lectins, but low temperature or insufficient cooking may not completely eliminate their toxicity, as some plant lectins are resistant to heat. Other conventional processing methods such as cooking, blanching, soaking, sun drying *etc.* have been applied in processing of foods. Cooking with moist heat can reduce the toxicity of lectins. Therefore, the use of lectin-containing plants as food in the human diets is not a cause for concern after adequate cooking. Special attention, however, should be paid when the lectin-containing food is prepared at high altitudes where the boiling point is reduced, when low heat cooking methods are employed, or in situations where heat transfer is uneven. To destroy the toxins, beans should be soaked and boiled thoroughly in fresh water. Beans should not be cooked at a low temperature, for example in a crock pot, since it may not destroy the toxin.

History provides various evidences for their toxicity profile and initially, it was thought that lectins are associated solely with toxic content. However, recent research shows a remarkable development in lectin science and their use in studying glycoprotein biosynthesis, investigating carbohydrates on cells/cell-organelles, mapping neuronal pathways, anti-cancer therapies and in mitogenic stimulation of lymphocytes is well documented. Application of lectins is possible depending on their properties. The antimicrobial and anti-insect activities of lectins can be made use of in the control of pathogens. The production of anti-tumor and antiviral drugs based on lectins may also have a significant utility in therapeutic industry.

Even though lectins have intrigued current researchers for the investigation of their therapeutic values, not enough studies have been conducted so far and knowledge about lectins is limited to certain plants or animal sources and further research is important to identify lectins and their importance in as many sources as possible. Lectins could be the next generation medicines if efficient research is contributed in their understanding.

iv) Pyrrolizidine Alkaloids

Alkaloids are a diverse group of amino acid-derived and nitrogen-bearing molecules that display a wide range of roles in nature, where they occur in plants, microorganisms or animals. In plants, alkaloids can be found in the form of salts of organic acids, mainly malate, acetate and citrate, or combined with other molecules, such as tannins. Most alkaloids display basic properties and present a lipophilic character, being soluble in apolar organic solvents and alcohol.

Within the many known families of alkaloids, pyrrolizidine alkaloids (PA) have been receiving increasing attention due to their occurrence in several species relevant for human and animal nutrition, as well as for their toxicological and pharmacological properties. The increasing awareness of PA contamination in all sorts of foodstuff worldwide justifies the interest and concern around this topic. Although in most cases their levels are insufficient to cause acute poisoning, they are frequently consumed in quantities that exceed the maximum daily intake suggested by authorities, which can be a contributory factor to chronic diseases.

PA are a group of alkaloids derived from ornithine that are distributed in plants of certain taxa, being also found in insects that uptake them for defense against predators. They rarely occur in the free form as a pyrrolizidine base, being instead found as esters (mono-, di- or macrocyclic diesters) formed by a necine base (amino alcohols) and one or more necic acids (mono- or dicarboxylic aliphatic acids), which are responsible for their structural diversity. They are usually found in the form of tertiary bases or pyrrolizidine alkaloids N-oxides (PANO).

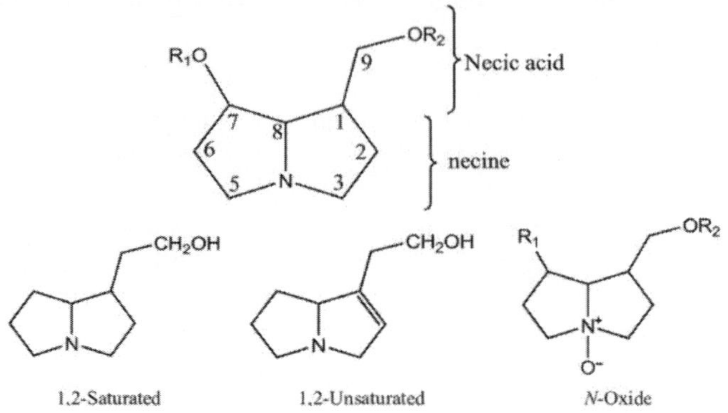

Figure 10.13: Structure of Pyrrolizidine Alkaloids.

PA-containing plants have worldwide distribution, and they probably are the most common poisonous plants affecting livestock, wildlife, and humans. More than 350 PAs have been identified in over 6,000 plants in the Boraginaceae, Compositae, and Leguminosae families. Pyrrolizidine alkaloids occur in an estimated 5 per cent of flowering plants. At least 13 families of flowering plants have been reported to produce pyrrolizidine alklaloids. Of these, 6 families; *viz.* Apocynaceae, Boraginaceae, Asteraceae (Compositae), Leguminosae (Fabaceae), Ranunculaceae and Scrophulariaceae, produce 1,2-dihydropyrrolizidine ester alkaloids that are hepatotoxic, genotoxic, teratogenic, carcinogenic and pneumotoxic. In all, it has been estimated that pyrrolizidine alkaloid producing species represent 3 per cent of all flowering plants. They are widely distributed throughout the world and are found growing in many agricultural production environments. The genera that are primarily responsible for significant poisoning of humans and foraging livestock are Senecio and Eupatorium (Asteraceae), Crotalaria (Leguminosae), Heliotropism and Trichodesma (Boraginaceae). Other toxic pyrrolizidine-producing genera hazardous to humans, where exposure can occur via food contamination *e.g.*, honey and via consumption of traditional herbal medicines, include Echium, Borago, Anchusa, Symphytum, Alkanna, Cynoglossum, (Boraginaceae) as well as Tussilago, Erechthites, Lithospermum, Brachyglot-tis, Cineraria and Petasites (Asteraceae).

About half of the identified PAs are toxic and several have been shown to be carcinogenic in rodents. In many locations, PA-containing plants are introduced species that are considered invasive, noxious weeds. Both native and introduced PA-containing plants often infest open ranges and fields, replacing nutritious plants. Many are not palatable and livestock avoid eating them if other forages are available. However, as they invade fields or crops, plant parts or seeds can contaminate prepared feeds and grains which are then readily eaten by many animals. Human poisonings most often are a result of food contamination or when PA-containing plants are used for medicinal purposes.

History of toxicity: Many episodes of human and animal intoxication, resulting from consumption of certain plant genera belonging to the families Asteraceae, Leguminosae and Boraginaceae, were shown, in the middle of last century, to be caused by PA. Gilruth was the first to confirm earlier suspicions that Senecio jacobaea (Asteraceae) caused chronic liver disease in livestock. Demonstration that other Senecio spp., and Crotalaria spp. (Leguminosae) also produce a similar disease of livestock, soon followed. Toxicity problems were not confined to livestock. It was shown, in 1920, that widespread, chronic liver disease of humans in South Africa was caused by the consumption of bread made from cereals contaminated by seeds of Senecio spp.

In the 1930s it was recognized that endemic liver disease in areas of the former USSR was also caused by consumption of pyrrolizidine alkaloid-contaminated bread. In this case the source of pyrrolizidine alkaloids was seeds and dust from Heliotro-pium lasiocarpum (Boraginaceae), a common contaminant of grain in the affected regions. In more recent times there have been large-scale incidents of acute poisoning from pyrrolizidine alkaloid contamination of grain involving many thousands of people in central and south Asia. In one incident in Afghanistan, approximately 8000 people were acutely poisoned by wheat contaminated by *Heliotropium popovii* subsp. Gillianum; 5000 were seriously affected and many died. Four thousand people were similarly poisoned by *H. lasiocarpum* contamination of wheat in Tadjikistan in 1992.

Seeds and dust from plants producing pyrrolizidine alkaloids growing as weeds in the crop can contaminate cereal grains in many countries. While grain cleaning methods, routinely used in industrialised countries, reduce pyrrolizidine alkaloid contamination to below the level causing acute poisoning by removing the foreign seeds, they do not remove the dust component and there are concerns that diseases such as cirrhosis, cancer and pulmonary arterial hypertension could, in some instances, result from chronic exposure to low levels of pyrrolizidine alkaloids in products such as cereal grains even in the industrialized world.

Medicinal plants containing pyrrolizidine alkaloids have also been found to cause significant liver damage, especially in children. Recognition of this mode of human pyrrolizidine alkaloid intoxication followed investigations into herbal teas and their connection with liver disease in Jamaica and other parts of the West Indies in the 1950s. The use of herbal teas made from pyrrolizidine alkaloid-producing plants to treat minor illnesses is also recognised as a cause of liver disease in parts of Africa and other tropical and subtropical countries. While pyrrolizidine alkaloid poisoning is more common in developing countries, where traditional herbal medicines are widely used, industrialised countries such as the USA and UK have also reported pyrrolizidine alkaloid intoxications from consumption of herbal medicines. Germany, Switzerland and Austria, where it was claimed in the 1980s by some herbal medicine practitioners that traditional medicinal plants had therapeutic benefits without the undesirable side effects reported for other medicines, have also reported fatal cases of pyrrolizidine alkaloid intoxication from consumption of traditional herbal products that contain PA.

As well as the known contamination of grain, there is potential for pyrrolizidine alkaloids to occur in several other foods. For example, honey derived from the flowers of pyrrolizidine alkaloid plants, has been shown to greatly exceed the levels of pyrrolizidine alkaloids deemed tolerable by some health authorities. The highest concentration of pyrrolizidine alkaloids reported in honey originating from Senecio jacobaea is 3900 micrograms/kg. Echium plantagineum honey is said to contain up to 2300 micrograms/kg after correction for extraction efficiency and E. vulgare honey has been reported to contain 500 – 2800 micrograms/kg of pyrrolizidine alkaloids.

Milk has also been demonstrated as a source of pyrrolizidine alkaloids. No human cases of pyrrolizidine alkaloid poisoning via milk have been unequivocally established but Huxtable (1989) refers to several instances where human milk, from women exposed to pyrrolizidine alkaloids, has caused venoocclusive disease in neonates and infants.

Many millions of meat-producing livestock are exposed to plants containing pyrrolizidine alkaloids. Pyrrolizidine alkaloid toxicosis is considered to be the most common poisoning disease of livestock worldwide. It is however not yet clear whether hazardous residues of PA remain in meat entering the human food chain. There are reports of PA in offal from animals fed plants containing pyrrolizidine alkaloids but no evidence has yet been reported for pyrrolizidine alkaloids in meat. Eggs too are a possible source of pyrrolizidine alkaloid exposure. There are a number of reports in the medical literature of liver damage typical of that caused by pyrrolizidine alkaloids, where the source of exposure could not be identified. There is therefore an on-going need for an appropriate control of all sources of dietary for the exposure to PA.

The lowest daily estimated intake causing sub-acute poisoning (veno-occlusive disease) was 14.1μg of pyrrolizidine alkaloids (mainly echimidine) per kg body weight per day, consumed over a period of 120 days (total estimated minimum intake – 85 mg or 1.7 mg/kg) by a 49-year-old woman. Rasenack, Müller *et al.* (2003) estimated that a total mean daily intake of 20–30 µg of pyrrolizidine alkaloids during pregnancy caused liver failure and death of a newborn child. The mother was apparently not affected by this level of exposure. In a poisoning incident in Afghanistan involving 8000 cases and many deaths it was estimated that over a 2-years period the individual uptake of PA (heliotrine from Heliotropium seeds) amounted to 1.46 g.

In acute poisoning, extensive haemorrhagic necrosis of the liver is seen. Clinically, acute poisoning is characterized by hepatomegaly (enlarged liver) and ascites (excess fluid in peritoneal cavity). Death is due to acute liver failure as a result of extensive centrilobular necrosis and massive liver dysfunctions. In a clinical study of acute toxicity from drinking bush teas in the West Indies, it was reported that around 20 per cent of patients died rapidly from acute toxicity and about 50 per cent recovered completely. The remaining 30 per cent of individuals progressed, either directly or after a period of apparent recovery, to sub-acute disease characterized by persistent hepatomegaly and recurrent ascites. In the sub-acute stage endothelial proliferation and medial hypertrophy cause occlusion of the smaller branches of the hepatic vein.

The seriousness of pyrrolizidine alkaloid intoxication is not only related to the level of exposure but also to age and gender. Males are more susceptible than females and foetuses and children are the most sensitive. Other liver damaging agents, nutritional factors, bacterial and viral infections and some medical drugs can also exacerbate the effects of 1,2-dehydropyrrolizidines. Kwashiorkor, frequently observed in children in central Africa, is, for example, exacerbated by liver damage caused by exposure to pyrrolizidine alkaloids and the consequent reduced ability of the liver to synthesise proteins.

As a result of increasing reports of pyrrolizidine alkaloid poisoning and growing awareness of the extent of potential pyrrolizidine alkaloid exposure through agricultural products and herbal medicines, the IPCS held a meeting of experts under the auspices of the United Nations Food and Agriculture Organisation, the World Health Organisation and the International Labour Organisation. The resulting authoritative report on the human health implications of pyrrolizidine alkaloids was published in 1988. In 1992, after several years of investigation, the German Federal Department of Health (1992) promulgated regulations severely restricting manufacture and sale of all herbal medicines containing pyrrolizidine alkaloids with a 1,2-unsaturated necine moiety that confers carcinogenicity and hepatotoxicity. Exempt are pyrrolizidine alkaloid-containing pharmaceuticals with no more than 1 micrograms of 1,2-dehydropyrrolizidine alkaloids per daily oral dose or less than 100 micrograms in the case of a topically applied product. Another regulation implemented was the labeling of these products with the following statement: "Not to be used in pregnancy and during the lactation period.", due to the susceptibility of fetuses and children to diseases instigated by PA. Switzerland and Austria have banned all pyrrolizidine alkaloid-containing herbal medicines and the Netherlands has a regulation limiting pyrrolizidine alkaloids to 1 microgram per kg of food. In the United States of America, the Food and Drug Administration ordered the ban of all PA-containing comfrey preparations from the market. The European Medicines Agency, based on toxicological considerations and the available guidelines for assessment/management of genotoxic carcinogens, also showed concern about the hazards of PA, recommending a maximum daily intake of 0.35 µg PA/day for a person with a body weight of 50 kg and life-long exposure.

v) Flavnoids

Flavonoids are an important class of natural products; particularly, they belong to a class of plant secondary metabolites having a polyphenolic structure, widely found in fruits, vegetables and certain beverages. They consist of a large group of polyphenolic compounds having a benzo-5-pyrone structure and are ubiquitously present in plants. They are synthesized by phenylpropanoid pathway. Flavonoids are hydroxylated phenolic substances and are known to be synthesized by plants in response to microbial infection. Their activities are structure dependent. The chemical nature of flavonoids depends on their structural class, degree of hydroxylation, other substitutions and conjugations, and degree of polymerization.

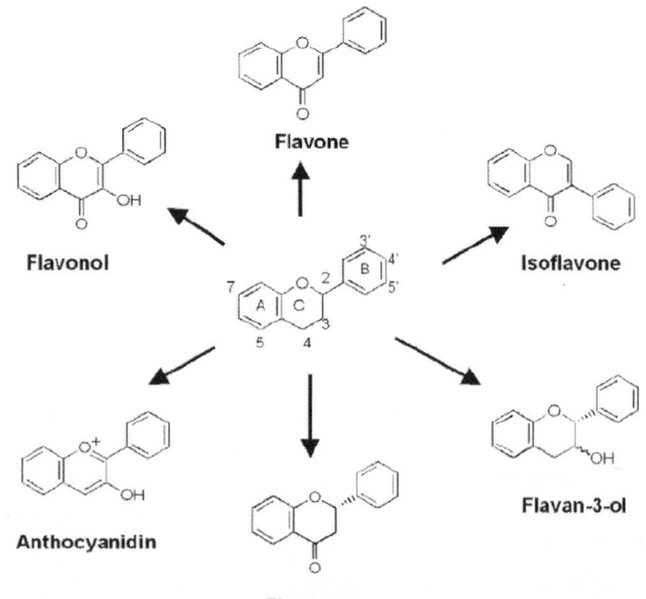

Figure 10.14: Flavnoids.

More than 8000 phenolic compounds as naturally occurring substances from plants have been reported. It is very interesting to note that half of these phenolic compounds are flavonoids presenting as aglycone, glycosides and methylated derivatives. Flavonoid compounds identified in vascular plants vary in type and quantity due to variations in plant growth, conditions, and maturity. Only a small number of plant species have been examined systematically for their flavonoid content and therefore, the identification and quantification of all the types of flavonoids consumed by humans is incomplete.

Flavonoids are used by vegetables for their growth and defence against plaques· They belong to a class of low-molecular-weight phenolic compounds that are widely distributed in the plant kingdom. They constitute one of the most characteristic classes of compounds in higher plants. Many flavonoids are easily recognised as flower pigments in most angiosperm families. However, their occurrence is not restricted to flowers but are found in all parts of plants.

In edible plants, the major flavonoids are such as flavonols, flavones, flavanols, proanthocyanidins (mainly, procyanidin B-1~B-4) and flavanones. In edible plants, the major flavonoids such as flavonols, flavones, proanthocyanidins (mainly, procyanidin B-1~B-4) and flavanones. Flavonols are widely present in whole organs of vegetables, fruits, and grains, and are especially abundant in green leaves. Usually, they are present as glycosides based on aglycons such as kaempferol, quercetin, myricetin, isorhamnetin, and rhamnetin. Among them, quercetin 3-O-glycosides like rutin, isoquercitrin and quercitrin, are major flavonoids in edible plants. In particular, they are found in relatively large amount in cranberry (*Vaccinium macrocarpon*), apple (*Malus sylvestris*), kale (*Brassica oleacea*), lettuce (*Lactuca sativa*), red pepper (*Capsicum annuum*) and leaves of sweet potato (*Ipomoea batatas*).

Compared with flavonols, flavone-containing plants are limited. Major flavones in edible plants are glycosides of apigenin, luteolin, and diosmetin. They are found in umbelliferous vegetables such as parsley (*Petroselinum crispum*) and celery (*Apium graveolens*), and are also present in sorghum (*Sorghum bicolor*). In particular, celery contains apiin and apigenin 7-Oglycoside in relatively large amounts. In addition, glycosides of apigenin and luteolin are constituents of green pepper, paprika, red pepper (*Capsicum annuum*), lettuce, herbs such as oregano (*Origanum vulgare*), thyme (Thymus spp.), rosemary (*Rosmarinus officinalis*), and even citrus fruits. Moreover, citrus fruits contain characteristic polymethoxyflavones with aglycons such as nobiletin and tangeretin. Polymethoxyflavones are particularly rich in shikuwasa (*Citrus depressa*) that is a native citrus in Okinawa, Japan, and it has attracted attention as a functional ingredient. On the other hand, C-glucosylflavones such as vitexin, isovitexin, orientin and isoorientin are contained in citrus fruits and peels, and in cereals.

Flavanols (catechins) are found in fruits and spices, and in high concentrations in tea (*Camellia sinensis*) leaves. Hydroxylation levels of the catechin B-rings in edible plants are usually of either the catechol (catechins) or pyrogallol type (gallocatechins), which exist in either the (-)-epi or (+)-form. Flavanols are widely distributed in apple and grape (Vitis genus) as a component of proanthocyanidins and galloyl esters, or in the free form. Of edible plants in Japan, green tea is the main source of flavanols, composed mainly of four types (-)-epigallocatechin gallate, (-)-epicatechin gallate, (-)-epigallocatechin and (-)-epicatechin. In contrast, the free type, (+)-catechin, (-)-epicatechin and (-)-epigallocatechin, are included in fruits like grape, apple, peach (*Prunus persica*), pear (*Pyrus pyrifolia*), mango (*Mangifera indica*), nectarine (*Prunus persica* var. mucipersica), plum (*Prunus salicina*) and raspberry (*Rubus idaeus*). In addition, (+)-catechin is found in strawberry.

Similarly, proanthocyanidins have been found in pulps and seeds of tree fruits such as persimmon (*Diospyros kaki*), apples, pears, grapes, and cacao (*Theobroma cacao*). Detailed investigations showed that tannins in barley (*Hordeum sativum*) caused the turbidity of beer, and tannins in sorghum caused nutrient reduction in important foods in developing countries. In contrast, procyanidins increased their functionality by polymerization, and demonstrated bioactivity never found in the flavanol monomers. However, their elucidation in other foods is still insufficient. Cacao and tea contain large amounts of tannins, which play important roles in the taste, and colour of their beverages. However, tannin contents are generally low in other agricultural foods. This is believed to be the result of the selection and breeding of crops with low tannin content, because tannins impair the flavour of food by their bitterness and astringency.

Flavanones are present in almost all citrus fruits such as Satsuma mandarin (Citrus unshiu), grapefruit (*C. paradise*) and lemon (*C. limon*), but they are absent from most vegetables. Flavanone glycosides such as naringenin, naringin, and naringenin 7-Oneohesperidoside in grapefruit, hesperetin, hesperidin and hesperetin 7-O-rutinoside in mandarin citrus, and eriodictyol, eriocitrin, and eriodictyol 7-O-rutinoside have been found. On the other hand, a small amount of naringin is present in almonds (*Prunus dulcis*), and also in tomatos. In addition, grape, azuki bean (*Vigna angularis*), and peanut (*Arachis hypogaea*) contain a flavanonol, taxifolin (dihydroquercetin).

Other main flavonoids found in edible plants include isoflavones and anthocyanins. Isoflavones, the major flavonoids of soybean (*Glycine max*) in particular, are found characteristically in legume plants. In Asia, soybean, which has been taken as a valuable source of nutrients, is the primary source of isoflavones. Daidzin (daidzein glycoside), genistin (genistein glycoside) and glycitin (glycitein glycoside), and also malonylated isoflavones are known as the soy isoflavones. However, isoflavones are present in trace amounts in bean (*Phaseolus vulgaris*), broad bean, peanut, and also chickpea (*Cicer arietinum*). Moreover, isoflavones are also contained in many soybean-processed foods, such as soybean flour, tofu, soymilk, soymilk skin, bean paste, soy sauce, natto (fermented soybeans), and deep-fried tofu.

Cyanidin-based anthocyanins are found in many edible plants, and glycosides of peonidin, petunidin, and malvidin in grape and blueberry (*Vaccinium corymbosum*). Cyanidin 3-O-glucoside, the most common anthocyanin, is found in black bean (*Glycine max*) and purple corn (*Zea mays*). Anthocyanins are also contained in the skin of many fruits such as strawberry, grape, apple, peach, cherry (*Prunus avium*) and mulberry (*Morus nigra*). Eggplant (*Solanum melongena*) and crimson glory vine (*Vitis coignetiae*) contain almost exclusively one major anthocyanin, nasunin and malvin, respectively. On the other hand, 15 types of anthocyanins are present in fruits of blueberry and bilberry (*Vaccinium myrtillus*). Also grapes with advanced breeding (varieties, Pione, Fujiminori, and Kyoho), red cabbage (*Brassica oleracea*), red turnip (*Raphanus sativus*) and purple sweet potato (for example, I. batatas variety Ayamurasaki) contain many types of anthocyanins (more than 20). In addition, grape, red cabbage, red turnip, purple sweet potato and colourful potatoes (*Solanum tuberosum*) are rich in acylated anthocyanins. In recent years, the agricultural development of crops containing anthocyanins has been actively carried out. Black rice (*Oryza sativa*), purple sweet potato, and colourful potatoes have been produced and marketed as various processed foods in Japan.

Thus, flavonoids are widely included in agricultural products like vegetables, fruits, grains and herbs. Flavonoids are also contained in many edible plant-used processed foods like chocolate, cocoa, green tea, black tea, and wine. In daily meals, we are taking flavonoids from fruits and vegetables as raw or processed foods mainly, and also from food colours, food additives, and health foods made from extracts of flavonoid rich berries, herbs, green tea, and ginkgo (*Ginkgo biloba*) leaves.

Recent interest in these substances has been stimulated by the potential health benefits arising from the antioxidant activities of these polyphenolic compounds. Functional hydroxyl groups in flavonoids mediate their antioxidant effects by scavenging free radicals and/or by chelating metal ions. The chelation of metals could be crucial in the prevention of radical generation which damage target biomolecules. As a dietary component, flavonoids are thought to have health-promoting properties due to their high antioxidant capacity both *in vivo* and *in vitro* systems. Flavonoids have ability to induce human protective enzyme systems. The number of studies has suggested protective effects of flavonoids against many infectious (bacterial and viral diseases) and degenerative diseases such as cardiovascular diseases, Alzheimer's disease (AD), cancers, and other age-related diseases.

Flavonoids are associated with a broad spectrum of health-promoting effects and are an indispensable component in a variety of nutraceutical, pharmaceutical, medicinal and cosmetic applications. This is because of their antioxidative, anti-inflammatory, anti-mutagenic and anti-carcinogenic properties coupled with their capacity to modulate key cellular enzyme functions. They are also known to be potent inhibitors for several enzymes, such as xanthine oxidase (XO), cyclo-oxygenase (COX), lipoxygenase and phosphoinositide 3-kinase.

Flavonoids play a variety of biological activities in plants, animals and bacteria. In plants, flavonoids have long been known to be synthesised in particular sites and are responsible for the colour and aroma of flowers, and in fruits to attract pollinators and consequently fruit dispersion to help in seed and spore germination, and the growth and development of seedlings. Flavonoids protect plants from different biotic and abiotic stresses and act as unique UV filters, function as signal molecules, allopathic compounds, phytoalexins, detoxifying agents and antimicrobial defensive compounds. Flavonoids have roles against frost hardiness, drought resistance and may play a functional role in plant heat acclimatisation and freezing tolerance.

Adverse reactions from flavonoids have been reported following administration of chronic pharmacological doses that exceed the estimated dietary intake of 2335 to 170 rag/day. Toxic effects that have been documented from doses of 1 to 1.5 g/day of flavonoid drugs such as cianidanol include acute renal failure, hemolytic anemia, thrombocytopenia, hepatitis, fever, and skin reactions. In one study, quercetin is reported to have induced bladder cancer in rats when consumed at the level of 2 per cent in the diet. However, these results were not confirmed in another study where quercetin was administered at doses up to 10 per cent of the diet. Importantly, in a diet containing a wide variety of foods flavonoids are unlikely to be consumed in toxic quantities because foods originating from plants contain many diverse types of flavonoids in varying quantities.

Epidemiological and in vitro evidence of antioxidant and cardioprotective effects support the hypothesis that flavonoids benefit health. The inhibition of LDL oxidation and platelet aggregation by flavonoids suggests that regular consumption of foods containing flavonoids and moderate consumption of red wine may protect against atherosclerosis and thrombotic tendency. The large contribution of flavonoids to the diet from tea, onions, and apples suggests that these foods may have greater nutritional benefits than previously recognized as they appear to constitute a major source of dietary antioxidants. More research is required for further elucidation of the mechanisms of flavonoid absorption, metabolism, and biochemical action and interaction with other nutrients *in vivo*. Furthermore, research is needed to identify the mechanisms by which flavonoids contribute to the amelioration of atherosclerosis and reduce the risk of morbidity and mortality from CHD.

vi) Glucosinolates

The first description of glucosinolates was reported at the beginning of the seventeenth century in mustard seed. They can be considered as natural toxins, but also as antinutritives. Now, these compounds are studied widely as secondary metabolites with complex functions. More than 200 glucosinolates have been identified so far, and most are found in Brassica. They are modified amino acids, carrying an S-glucose functional group and a variety of different side chains. Glucosinolates contain a sulfated isothiocyanate group, thioglucose and an R-group derived from amino acids. Glucosinolates can be divided into three classes based on the R-group as aliphatic, with a methionine, isoleucine, leucine or valine precursor; indole, with a tryptophan precursor; and aromatic, with a phenylalanine or tyrosine precursor.

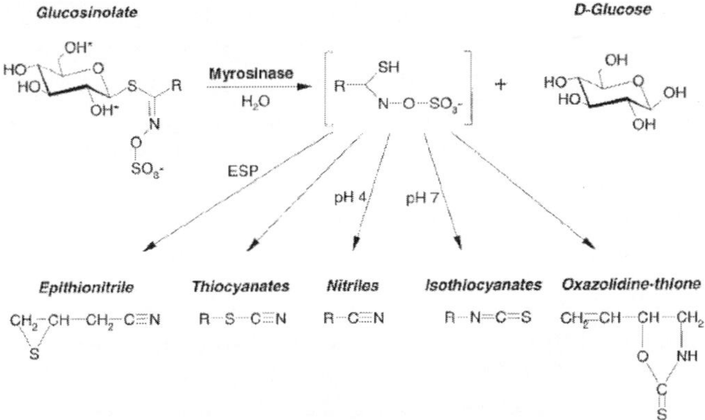

Figure 10.15. Structure of Glucosinolates.

The glucosinolates are chemically stable and biologically inactive whilst they remain sequestered within sub-cellular compartments throughout the plant. However, tissue damage caused by pests, harvesting, food processing or chewing initiates contact with the endogenous enzyme myrosinase, (thioglucoside glycohydrolase; EC 3.2.3.1). This leads to rapid hydrolysis of the glucosidic bond, releasing glucose and an unstable intermediate, which undergoes a spontaneous rearrangement to form a complex variety of breakdown products. Myrosinase is a thioglucosidase stored in a different cell, physically separated from its substrate glucosinolates. Glucosinolates that are not hydrolysed are nontoxic, but when plants tissues are damaged by chewing or other actions, the myrosinase is released to hydrolyse the glucosinolates to glucose, sulfate and several toxic/pungent products, including isothiocyanates. Unstable chemical isothiocyanate products have a demonstrated role in carcinogenesis resistance and inhibition. The isothiocyanates, a group of hot and bitter compounds, commonly called mustard oils are probably the most important and thoroughly investigated of these products.

The species in which glucosinolates occur that are important for animal or human feeding belong to the family Brassicaceae. These include rapeseed, cabbage, cauliflower, brussels sprouts, swede/turnip, calabrese/broccoli and Chinese cabbage, radishes, mustard seed and horse radish. The type and concentration of glucosinolates in food are highly variable depending on several factors, such as genetics, cultivation site, cultivar, growth conditions, developmental stage, plant tissue, post-harvest handling, and food preparation methods. As types and concentration are also the main determinant of their biological activities, estimates of their content in food are essential tool to understand if a certain diet is adequate to deliver qualitatively and quantitatively appropriate glucosinolates and ITCs.

The abundant presence in Brassicaceae vegetables and condiments makes GSLs of interest to human society. To date, the therapeutic benefits of GSLs have drawn more attention to this class of secondary metabolites, alongside with their original food purposes. Although several synthetic approaches have been documented, most natural GSLs reside in plants, with more than 130 different GSLs having been validated. GSL concentration is unequally distributed throughout the plant body. For instance, in Brassica napus, the GSL concentration in the seed is greater than that in leaves. This variation appears to be more relevant in root vegetable crops (Moringacea family) than that in oilseed crops (Brassicaceae family). Moreover, the GSL profile varies depending on the tissue type. Although aliphatic GSLs predominate both in leaves and in seeds, indole GSLs are more abundant in leaves than in seeds. This difference may be related to different functions of different parts of plants. Breeding approaches are often employed to obtain crops with low GSL content for food or feed purposes, while those with high GSL content remain of interest for non-food applications. Therefore, the choice of species should be carefully considered with regard to the downstream purposes of raw materials. We also suggest that growth conditions should be highly regarded in order to adapt the chosen crops to their cultivating environment. The occurrence of GSL varies among different species within the same order. These variations even occur for the same crop depending on the years. It is assumed that the accumulation of GSLs within plants highly depends on environmental factors such as the weather that undergoes slight changes through the years, thus directly impacting the GSL contents of the crops. Therefore, the GSL content of the same crops must be kept updated annually, or more frequently if needed.

Glucosinolates are present at relatively high concentrations in the diet compared to some other phytochemicals. For example, an individual eating 3–4 portions of broccoli per week, which is the level that appears to confer a protective effect against adenomatous polyps, may regularly consume 300-400 mg of glucosinolates. However, the level of exposure varies considerably with the commercial variety of vegetable eaten, agronomic conditions, processing techniques and culinary habits. It is clear that different glucosinolate breakdown products exert a variety of effects on enzyme expression, cell proliferation and cell survival. The epidemiological evidence for protective effects of brassica vegetables reflects a considerable variety of exposures to different vegetables and dietary patterns. In these circumstances it would be unwise to emphasise the importance of any particular variety of vegetable, and dietary advice to the public should emphasise the benefits of consuming a wide variety of brassica vegetables in many different culinary forms. The levels of glucosinolates in brassica vegetables can be manipulated by plant breeders, and their presence influences both the palatability and potentially, the nutritional properties of these crops. Both issues need to be considered when breeding new varieties of vegetable. As with synthetic drugs, plant secondary plant metabolites may, in principle, have adverse side-effects as well as beneficial properties, and glucosinolate breakdown products are no exception. Further research is needed to define the biological activities of the glucosinolate breakdown products in greater detail, so that the balance of benefit, risk and consumer preference can be properly defined.

The levels of glucosinolates ingested depend on variety, agronomic factors, and both storage and processing of the vegetables prior to consumption. Although mechanical damage leads to the rapid hydrolysis and degradation of glucosinolates, cutting and pre-harvest stress has also been shown, under certain circumstances, to increase the concentrations of indole glucosinolates in cabbage. The history of the plant tissue throughout the whole food chain from grower to consumer is therefore a vital influence on its ultimate biological role in human nutrition. Whatever the final level of glucosinolates in the prepared vegetable, the absorption, metabolism and delivery of glucosinolate breakdown products to target tissues depends, to a large extent, upon the residual level of myrosinase activity.

The activation of myrosinase is brought about by the physical disruption of the plant tissue during harvesting, processing, food preparation and consumption. However, chopping induces glucosinolate hydrolysis only at the cut surfaces. Thus, large intact leaves, or florets of broccoli or cauliflower, will undergo only minimal losses of glucosinolates up to the point of cooking. If such vegetables are eaten raw, both intact glucosinolates and active myrosinase are ingested simultaneously, which enables the breakdown of the glucosinolates to occur within the alimentary tract. For example, when rats were fed benzyl glucosinolate in the presence of active myrosinase derived from Brussels sprouts, a substantial proportion of the administered dose appeared as isothiocyanate excretion products in the urine. Some of the ingested glucosinolates were also broken down in the colon, but plant myrosinase appeared to be the dominant factor. Getahun and Chung (1999) observed a similar phenomenon in human subjects, using watercress as a source of PEITC. Volunteers consumed 350 g (475 μmol glucosinolates) of watercress in which the myrosinase had been completely inactivated by cooking, or 150 g (972 μmol glucosinolates) of raw watercress, which retained its myrosinase activity. In the case of cooked watercress, the rate of conversion of glucosinolates to isothiocyanates ranged from 1.2–7.3 per cent, compared to 17.2–77.7 per cent for the raw plant material. Given the considerable loss of glucosinolates that results from cooking in water, it is clear that the method of preparation can make a very large difference, both to the intake of glucosinolates, and to the bioavailability of their breakdown products.

Besides the chemical degradation involving MYR-catalyzed hydrolysis, the thermal degradation of GSLs is often mentioned. Cooking reduces the concentration of glucosinolates in the plant tissue through thermal breakdown and leaching, but also inhibits the activity of myrosinase through denaturation of the enzyme. A mathematical model describing these effects has recently been developed by Dekker and co-workers, and this opens up the possibility of quantifying the losses of glucosinolates, and by extension other phytochemicals, during the storage, processing and culinary preparation of vegetables.

GSL profiles of cooked brassica vegetables are altered at a different level depending on employed culinary techniques, such as cooking, steaming, and microwaving. The reduction of red cabbage (Brassica oleracea) indolic GSL during the cooking process was observed. The content of glucobrassicin and its homologs were drastically declined due to the cooking process performed under 120 °C. On the other hand, aliphatic GSLs appear to be more stable, with only a slight degradation has been observed under the same cooking conditions. The degradation became drastic for all GSL under canning conditions, whereas the process temperature exceeds 120 °C. The total amount of GSL has been reduced by over 70 per cent under these harsh conditions. These observations are drawn from conclusions about the difference in thermal stabilities between aliphatic and indolic GSLs.

A study conducted by Song and Thornalley also reported the thermal degradation of GSL due to the domestic processing of Brassica vegetables, such as

Brussel sprouts, broccoli, cauliflowers, and green cabbage. Moreover, the effects of the cooking method, such as microwave, steam, and stir-fry, on GSL amounts of studied materials were investigated. The results showed that cooking by these cooking methods did not produce a significant loss of GSL, in contrast to boiling, which showed significant losses by leaching of GSL into cooking water at high temperatures. Therefore, boiling Brassica should be avoided in order to preserve intact GSL in raw materials. A recent study on the roasting process of rapeseed seed reported shows that industrial-scale post-harvest treatments, which are often necessary to produce higher quality oil-related products, also impact the GSL profile of plant materials. Up to 29 per cent of the original GSL amount in plant materials have been reduced during the roasting process. The results indicate that the industrial-scale roasting processes reduce the GSL amount of plant materials due to the thermal degradation, with up one-third of GSLs are degraded via thermal degradation. Based on the information outlined above, it is suggested that, with regards to downstream purposes, the selection of plant material should rely on the processing method. Although thermal treatments of plant materials, whereas the GSL content is often reduced, are beneficial for food and feed applications, these should be avoided in order to maintain the desired amount of GSL for non-food purposes.

Calculations of dietary exposure are limited by the shortage of reliable data on levels in plants and also the variation in natural levels between seasons and different strains. Exposure to total glucosinolates by adult New Zealanders was estimated to be 17 mg/day, with approximately one-third of glucosinolate exposure due to consumption of cabbage. UK estimates based on daily intakes of cabbage, cauliflower, Brussels sprouts and swede/turnip were 46 mg/day if all vegetables are considered to have been consumed fresh/raw and 29 mg/day if all vegetables are considered to be consumed cooked. A German study gave similar estimates of 46 and 36 mg/day for winter and summer, respectively.

The main health concern due to the presence of these compounds in foods is the potential for increase in goitre (enlargement of the thyroid gland. Goitre is mainly due to a dietary deficiency of iodine but the differing prevalence of goitre in areas with low levels of iodine indicates that other factors are involved, one of which is the presence of goitrogens in the diet. Isothiocyanates and oxazolidine-2-thiones have been shown to have effects on the thyroid gland in animals and these effects occur even if iodine is sufficient. Isothiocyanates, nitriles and indole products can all be further degraded to produce the thiocyanate ion. This compound is able to cause goitre in humans when dietary iodine is insufficient. However, despite several decades of conjecture, there is no evidence to support a causative role for dietary glucosinolates in human goitre. Glucosinolates may also have beneficial effects. They have been shown to induce the activity of the phase II detoxification enzymes and inhibit phase I (activating) enzymes, exerting an anticancer effect. They have also been shown to provide protection from oxidative stress through elimination of reactive oxygen species. It is thought that this activity may be behind the observed link between vegetable (Brassica) intake and cancer protective effects. On balance, there is greater evidence for a positive impact of glucosinolates on human health than a negative impact. However, strong evidence for a causal relationship with human positive or negative health effects is currently lacking.

Glucosinolates have attracted researcher's attention due to their potential chemopreventive activity. More than 200 different glucosinolates have been identified in plants, and several of these compounds have been studied for the potential anti-cancerogenic effect of their metabolic breakdown products (mainly ITCs). Amino acid-derived glucosinolates (GSLs), which are secondary plant metabolites constituted of a sulfate and thioglucose moiety, play important biological roles in the Brassicaceae family defense system, crops of great relevance to agriculture. The coexistent thioglucosidase myrosinase (MYR) (EC 3.2.1.147) originally segregated within plants, will come in contact with GSL upon tissue disruption. Consequently, the enzymatic hydrolysis of GSL occurs to form glucose, and an unstable aglucone that undergoes degradation to afford a wide range of active components in response to environmental stresses. Along with the aforementioned role in the defense system, GSLs are likely involved in the survival system of the Brassicaceae family. In a study on Arabidopsis thaliana under abiotic stress (*e.g.*, high salt), the overproduction of short-chain aliphatic GSL and underproduction of indolic GSL in leaves occurred, suggesting the adaptation of the plant in response to environmental stresses, and thus demonstrating the biological importance of GSLs in the Brassicaceae survival system, besides their prominent role involved in defense mechanism.

The effects of normal dietary levels of intake of glucosinolates are not known and there is little reliable evidence of harmful dose levels in human beings. Cooking by boiling will reduce exposure through leaching of glucosinolates into water and inactivation of myrosinase. However, steaming, microwaving or stir-frying results in no significant decrease in glucosinolate concentrations.

The safety of glucosinolates in the human diet has not been assessed by any national or international organisation. Safety limits are levels of dietary exposure that are without appreciable risk for a lifetime of exposure. Regulatory limits define the maximum amount of a substance that is permitted in a particular food. No safety limits have been set for glucosinolates.

B. Mycotoxins

Mycotoxins are naturally occurring, toxic secondary metabolites produced by different array of fungi. The name mycotoxin combines the Greek word mykes meaning "fungus" and the Latin word toxicum meaning "poison." The term secondary metabolites are the compounds that are dispensable.

Mycotoxins are compounds produced by moulds under specific environmental conditions. These toxins are harmful to human and animal health. Some of the most important mycotoxins are shown in the Table 10.4.

Mould contamination usually occurs in the field. Mycotoxins can develop at various stages:

- Pre-Harvest
- Harvest
- Storage

Table 10.4: Various Mycotoxins Mould Species

Mycotoxins Mould Species	*Mycotoxins Mould Species*
Aflatoxins	*Aspergillus* spp.
Cyclopiazonic acid	*Aspergillus* and *Penicillium* spp.
Deoxynivalenol (DON)	*Fusarium* spp.
Fumonisin	*Fusarium* spp
Ochratoxin	*Aspergillus* and *Penicillium* spp.
Patulin	*Penicillium* spp.
Sterigmatocystin	*Aspergillus* and *Penicillium* spp.

Mycotoxins occur in many varieties of food and feed. The food crops most often affected are corn, peanuts, cottonseed, sorghum, wheat, barley, coffee, cocoa and tree nuts (pecan, almond, pistachio, hazelnut, walnut and Brazil nut). They have also been found in rice, beer and wine. In Canada, mycotoxins occur mainly in cereal grains and corn, but have been reported in other crops such as alfalfa and oilseed. The following factors enhance mould growth and toxin production:

- High moisture content (20 to 25 per cent)
- High relative humidity (70 to 90 per cent)
- Warm temperatures (22 to 30°C)
- Drought
- Lack of soil fertilization
- High density culture
- Presence of weed, insects or mechanical damage to the orchard or during harvesting, storage and distribution that can weaken the natural defense mechanisms of plants and encourage colonization by mycotoxins producing fungi.

Most mycotoxins are very stable. They can resist high temperatures, up to 180°C and several manufacturing processes such as milling and cooking. It is not surprising to detect small amounts of mycotoxins in processed foods. This is why it is important to prevent the formation of toxins in raw material.

Exposure to these mycotoxins can occur directly or indirectly by contaminated grain and grain products' consumption or eating animals fed with contaminated feed and particularly by the consumption of milk food. It is therefore that the human being has a potential risk to exposure of mycotoxins. High doses of mycotoxins can cause acute illness or death; low doses cause chronic toxicity. The toxicity of mycotoxins may affect the body's nervous system, heart, lungs and digestive tract. Mycotoxins are also associated with acute hepatitis and liver cancer. Some mycotoxins are most dangerous carcinogenic substances produced by the ecosystem, because they are capable of producing mutations in the genetic material of living beings, teratogenic that can cross the placental barrier and damage the foetus, nephrotoxic because in kidney damage, genotoxic as to Cusan DNA damage by changing the transformation of cells and eventually produce malignant tumour, immunosuppressive because it affects and weakens the immune system, hepatotoxic, *i.e.*, cause serious damage to the liver; thus becoming a major public health problem which must be prevented and constantly monitor.

Although the clear-cut toxicity of many mycotoxins for humans has not been demonstrated, the effect of these compounds on experimental animals and their effect in *in vitro* assay systems leaves little doubt about their real and potential toxicity for humans. At least 14 mycotoxins are known to be carcinogens, with the aflatoxins being the most potent. It is generally accepted that about 93 per cent of mutagenic compounds are carcinogens. With mycotoxins, microbial assay systems reveal an 85 per cent level of correlation between carcinogenicity and mutagenesis.

Mycotoxins are difficult, and sometimes impossible to eliminate. The best control is prevention. Mycotoxins can be prevented by reducing the moisture content of food products and controlling storage conditions such as temperature and/or relative humidity. In general, reducing moisture content to the equivalent of less than 0.70 water activity.

Thousands of fungal secondary metabolites have been discovered, but currently 450 classes of mycotoxins are identified and characterized. This section deals with important groups of mycotoxins, such as aflatoxins, ochratoxins A, Zearalenone, fumonisins, and patulin, their toxicity, and health hazards on animals and humans.

i) Aflatoxins

Aflatoxins are the most important mycotoxins, which are produced by certain species of Aspergillus (*A. flavus* and *A. parasiticus*), which develop at high temperatures and humidity levels. The most common commodities contaminated are tree nuts, peanuts, and corn and cottonseed oil. The fungus *Aspergillus flavus* factory mycotoxins called aflatoxins B1, B2, while the fungus *Aspergillus parasiticus* produce mycotoxins known as aflatoxins B1, B2, G1, G2. All are being potent carcinogens. Of these, aflatoxin B1 is considered natural carcinogen, because its toxic mechanism is mediated by the epoxide radical which interacts with fusion proteins, causing toxicity and prevent protein synthesis and immunosuppression, similarly, it is able cause genotoxicity and induce carcinogenic processes. The aflatoxins B1, B2 fluoresce blue and G1, G2 fluoresce green when viewed under a microscope M1 and M2 aflatoxins are metabolic derivatives of aflatoxins B1 and B2, which are eliminated through urine and milk of people who have consumed contaminated feed. Aflatoxins are carcinogenic substances and may be present in a large number of foods. They are colorless, tasteless, and clarified as chemistry, they are solid in food and can cause resistance to normal processes boils, the most difficult to remove once they occur.

More than 18 different aflatoxin types are presently known to occur naturally or as a result of carry over phenomenon in feeds and foods. There are about 13 types of aflatoxins that are naturally produced by toxigenic moulds, some of which can be metabolised by human, animals, or other microorganisms to generate derivatives that retain toxicity, although usually with a lower potency compared with the parent molecules. AFB1, AFB2, AFG1, and AFG2 are of the most concern to economy and public health due to their high incidence and high toxicities, especially AFB1. Meanwhile, aflatoxin M1 (AFM1) is of special concern to the safety of dairy products because it is usually carried over in milk of lactating animals fed on feed contaminated with aflatoxin B1 in addition to its high toxicity and potential carcinogenicity in humans. However, the other aflatoxins should not be overlooked because of their intrinsic toxicity, which may not be negligible, or because they can readily invert to the most potent AFB1. They can also be intermediates for the biosynthesis of more toxic mycotoxins.

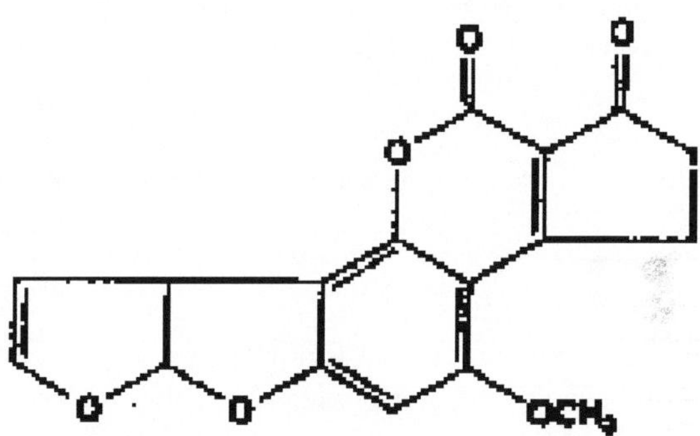

Figure 10.16: Aflatoxin B1.

Table 10.5: Origins of Aflatoxins and the Products most Exposed to Contamination

Aflatoxin	*Source Frequently Contaminated*	*Products*
	Difurocoumarocyclopentenone	
Aflatoxin B1	**Section Flavi:** *A. flavus, A. pseudotamarii, A. togoensis, A. aflatoxiformans, A. austwickii, A. cerealis, A. arachidicola, A. minisclerotigenes, A. mottae, A. luteovirescens (formerly A. bombycis), A. nomius, A. novoparasiticus, A. parasiticus, A. pipericola, A. pseudocaelatus, A. pseudonomius, A. sergii,* and *A. transmontanensis*	Cereals (*e.g.*, sorghum, rice, corn, wheat, barely), oil seeds (*e.g.*, cotton seed, oilseed rape, sunflower seed), nuts (*e.g.*, peanuts, groundnut, pistachio), spices (*e.g.*, turmeric, black and red pepper, ginger, allspices), meats, dairy products, fruit juices, dried fruits, eggs, and feeds and foods derived from these products
	Section Ochraceorosei: *A. ochraceoroseus* and *A. rambellii*	
	Section Nidulantes: *A. astellatus, A. miraensis, A. olivicola,* and*A. venezuelensis*	
Aflatoxin B2	**Section Flavi:** *A. flavus, A. pseudotamarii, A. aflatoxiformans, A. austwickii, A. cerealis, A. arachidicola, A. minisclerotigenes, A. mottae, A. luteovirescens, A. nomius, A. novoparasiticus, A. parasiticus, A. pipericola, A. pseudocaelatus, A. pseudonomius, A. sergii, A. transmontanensis* **Section Ochraceorosei:** *A. ochraceoroseus* and *A. rambellii*	Cereals (*e.g.*, sorghum, rice, corn, wheat, barely), oil seeds (*e.g.*, cotton seed, oilseed rape, sunflower seed), nuts (*e.g.*, peanuts, groundnut, pistachio), spices (*e.g.*, turmeric, black and red pepper, ginger), meats, dairy products, fruit juices, dried fruits, eggs, and feeds and foods derived from these products
Aflatoxin B2a	Hydroxylated metabolite of aflatoxin B1 obtained by water addition to the double bond of the terminal furan under acidic conditions in the liver, the stomach or soil (no evidence for the involvement of specific enzymes) Naturally produced by *A. flavus,* and *A. parasiticus*	NA

Aflatoxin	*Source Frequently Contaminated*	*Products*
Aflatoxin M1	Hydroxylated metabolite of aflatoxin B1 by hepatic microsomal mixed-function oxidase system (MFO), mainly cytochromes, in the liver of mammals Produced *in vitro* from aflatoxin B1 by liver homogenatesNaturally produced by *A. flavus* and *A. parasiticus*	Milk (including human milk) and dairy products Meat products (kidney, liver) Mouldy groundnut and corn
Aflatoxin M2	Hydroxylated metabolite of B2 by hepatic microsomal MFO of mammalsNaturally produced by *A. parasiticus*	Idem as aflatoxin M1
Aflatoxin M2a	Hydration of the terminal furan ring of aflatoxin M1 in dilute acid to yield an hemiketal derivative *In vitro* in liver homogenates	Milk and dairy products
Aflatoxin P1	Demethylated metabolite of aflatoxin B1 by liver microsomal oxidase–catalysed O-demethylase	Mainly excreted in the urine (humans and animals). Dairy products
Aflatoxin Q1	Hydroxylated metabolite of aflatoxin B1 by microsomal enzymes in the liver of higher vertebrates and poultry (main aflatoxin B1 metabolite in monkey)	Assumed to be in edible parts of bovine fed on aflatoxin B1-contaminated feed
Aflatoxin Q2a	Acid hydration of aflatoxin Q1	NA
Aflatoxicol (R_0)	Metabolite of aflatoxin B1 formed by a reversible reduction of the pentanone group in humans, animals and numerous bacteria and molds *In vitro* biotransformation of aflatoxin B1 by a soluble cytoplasm reductase system in fish, rat and human liver preparations Naturally produced by*A. flavus* and*A. parasiticus*	Mainly avian products (major metabolite in avian species fed on B1-contaminated feed). Dairy products Does not accumulate in edible parts of bovine and swine fed on aflatoxin B1-contaminated feed
Aflatoxicol M1	Reduced metabolite of aflatoxin B1, aflatoxin R_0, or aflatoxin M1 catalysed by soluble NADPH-dependent reductases in the liver	Milk and dairy products
Aflatoxicol H1	Reduced metabolite of aflatoxin B1 and aflatoxin Q1 catalysed by soluble NADPH-dependent reductases in the liver	Milk and dairy products
	Difurocoumarolactone	
Aflatoxin G1	*A. flavusa, A. aflatoxiformans, A. austwickii, A. cerealis, A. arachidicola, A. minisclerotigenes, A. mottae, A. luteovirescens, A. nomius, A. novoparasiticus, A. parasiticus, A. pipericola, A. pseudocaelatus, A. pseudonomius, A. sergii, A. transmontanensis,*	Cereals (*e.g.*, sorghum, rice, corn, wheat, barely), oil seeds (*e.g.*, cotton seed, oilseed rape, sunflower seed), nuts (*e.g.*, peanuts, groundnut, pistachio), spices (*e.g.*, turmeric, black and red pepper, ginger), meats, dairy products, fruit juices, dried fruits, eggs, and feeds and foods derived from these products
Aflatoxin G2	*A. flavus1, A. aflatoxiformans, A. austwickii, A. cerealis, A. arachidicola, A. minisclerotigenes, A. mottae, A. luteovirescens, A. nomius, A. novoparasiticus, A. parasiticus, A. pipericola, A. pseudocaelatus, A. pseudonomius, A. sergii, andA. transmontanensis*	Same as aflatoxin G1
Aflatoxin G2a	Hydroxylated metabolite of aflatoxin G1 obtained by catalytic addition of water to the double bond of the terminal furan under acidic conditions in the liver, the stomach or soil (no evidence for the involvement of specific enzymes). Naturally produced by *A. flavus*	NA
Aflatoxin GM1	Hydroxylated metabolite of aflatoxin G1 by MFO in the liver of mammals Produced in vitro by *A. parasiticus* fed aspertoxin as a precursor Naturally produced by *A. flavus*	Milk and dairy products
Aflatoxin GM2	Hydroxylated derivative of aflatoxin G2 by MFO in the liver of mammals Produced in vitro by *A. parasiticus* from dihydro-O-methylsterigmatocystin (DHOMST) Naturally produced by *A. flavus* and *A. parasiticus* and yeast	Milk and dairy products
Aflatoxin GM2a	Metabolite of aflatoxin GM1in the liver of mammals Hydration of the terminal furan ring of aflatoxin M1 in dilute acid to yield an hemiketal *in vitro* in liver homogenates	Milk and dairy products
Parasiticol (aflatoxin B3)	A metabolite of aflatoxin G1 from the biodegradation (hydrolysis and decarboxylation reactions) in *A. flavus, Rhizopus stolonifer, Rhizopus arrhizus,* and *Rhizopus oryzae*	Idem as aflatoxins B1 and G1

Aflatoxin	Source Frequently Contaminated	Products
	Others	
Parasiticol (aflatoxin B3)	A metabolite of aflatoxin G1 from the biodegradation (hydrolysis and decarboxylation reactions) in*A. flavus, Rhizopus stolonifer, Rhizopus arrhizus,* and *Rhizopus oryzae* Naturally produced by*A. parasiticus, A. flavus, A. mottae, A. nomius,* and *A. novoparasiticus*	Idem as aflatoxins B1 and G1
Aspertoxin b	*A. flavus* and *A. parasiticus*	Mainly vegetal products prone to contamination with *A. flavus* and *A. parasiticus*; not considered to be relevant to food products of animal origin

Source: Noreddine Benkerroum (2020). Int. J. Environ. Res. Public Health, 17, 1215.

Aflatoxin-producing moulds, particularly *A. flavus* and *A. parasiticus*, are frequent contaminants of crops where they grow and excrete aflatoxins, which can in turn be found in foods and feeds at high levels making them unfit for consumption. The food and drug administration of the USA (USFDA) considered unavoidable the contamination of agricultural products with aflatoxins that can, at best, be kept at the lowest practical levels to minimize the exposure of humans and animals. However, despite the widespread of aflatoxins throughout the world, their prevalence in foods and feeds is higher in some regions than in others depending on the pedoclimatic conditions, the agricultural practices, the cultivars grown, the mechanical and insect damage of crops, and the awareness of the harmful effects of food-borne toxins on the productivity and safety of produce. In addition to inherent traits that influence the toxigenicity of moulds, such as the species, the strain, the morphotype, and the competitiveness within the microbiome, the level of development of the country; the availability and degree of enforcement of pertaining regulations also account for the extent of food contamination with aflatoxins. Considering these factors, the highest incidence has classically been recorded in SSA (Sub-Saharan Africa) and SEA (South East Asia) countries, owing primarily to the favourable climatic conditions, and then to the low development status and the lack of public awareness of the risk these toxins pose to human and animal health.

Crop residues and by-products from grain mills and/or oil extraction factories are often used as animal feeds. In developing countries, mouldy cereals and nuts of low grades are generally sorted to be fed to animals either directly or as ingredients in manufactured feeds. Therefore, it is reasonable to anticipate that such feeds are more likely to be highly contaminated with aflatoxins than their counterpart crops destined to human consumption. A comprehensive survey on aflatoxin contamination of feeds and feed ingredients in Asia-Oceanian countries, including Malaysia, Philippines, Thailand, Indonesia, and India (SEA) showed that 30.3 per cent of the samples contained AFB1 at an average level of 46.0 µg/kg and a maximum level of 4278.0 µg/kg. Moreover, the levels of aflatoxins in commercial poultry feeds were demonstrated to be significantly higher than the maize used as ingredient in their formulation.

The toxicity of aflatoxins to humans and animals through food and feed consumption and their association with acute and chronic diseases is well established. However, the degree of toxicity and the toxicological effects vary greatly depending mainly on the aflatoxin type and the host. AFB1 is, by far, the most toxic aflatoxin, followed by AFG1, AFB2, and AFG2, while AFM1 has a similar toxicity as AFG1. The other less toxic aflatoxins and those considered to be "non-toxic" or detoxified forms are still of concern to public health due to their inherent, although weak, toxicities with potencies ranging between 0.1 per cent and 50 per cent compared with AFB1. Most importantly, they can invert to their highly toxic precursors in foods or after ingestion. For example, aflatoxicol, which is 25 to 50 per cent as potent as its parent AFB1, is almost entirely converted back in the liver to either the more toxic parent AFB1 or to AFM1.

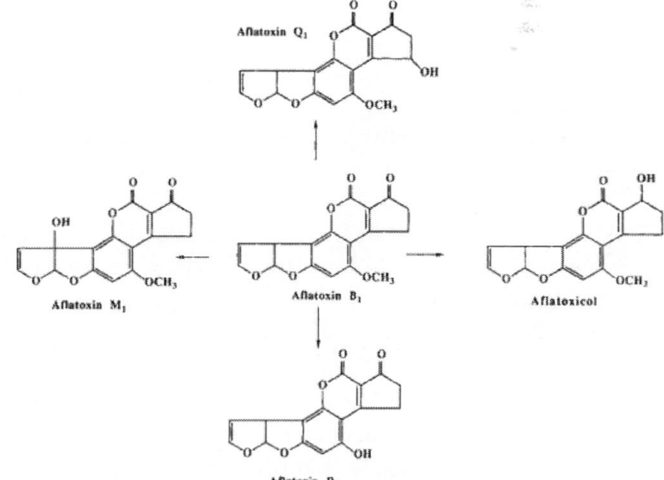

Figure 10.17: Metabolites of Aflatoxin B1.

Although aflatoxin may be considered amongst the most carcinogenic of natural products for some animals, it is still not clear whether it is a human carcinogen. Liver cancer in some parts of the world, such as the African continent, is complex and the initial demonstration of a correlation between exposure to aflatoxin in the diet and the incidence of liver cancer has to be considered with caution. Although liver cancer may be attributable to exposure to aflatoxin in parts of Africa, it is necessary to ask why liver cancer is not also more prevalent in India where dietary exposure to aflatoxin also occurs. In India, cirrhosis of the liver is more common and there is still a lot to learn about the role of aflatoxin in liver cancer and liver damage in different parts of the world.

A diverse range of responses to the toxic effects of a compound may occur because the compound is metabolized in the animal body and the resulting toxicity is influenced by this metabolic activity. This is certainly the case with aflatoxin B1 from which a very wide range of metabolites are formed in the livers of different animal species. Thus, the cow is able to hydroxylate the molecule and secrete the resulting aflatoxin M1 in the milk, hence affording a route for the contamination of milk and milk products in human foods even though these products have not been moulded.

The formation of an epoxide could well be the key to both acute and chronic toxicity and those animals which fail to produce it are relatively resistant to both. Those animals which produce the epoxide but do not effectively metabolize it further, may be at the highest risk to the carcinogenic activity of aflatoxin B1 because the epoxide is known to react with DNA. Those animals which not only produce the epoxide but effectively remove it with a hydrolase enzyme, thus producing a very reactive hydroxyacetal, are most sensitive to the acute toxicity. The hydroxyacetal is known to react with the lysine residues in proteins.

It is now known that aflatoxin B1 epoxide reacts rather specifically with guanine residues of DNA at a number of hot spots, one of which is codon 249 of the *p53* gene. The product of this gene is involved in processes which normally protect against cancer and it is known that the hepatitis B virus binds to the *p53* gene product. Thus, with aflatoxin B1 and hepatitis B interacting with *p53* in different ways it is easy to see that they could act synergistically.

The parent molecule may thus be seen as a very effective delivery system having the right properties for absorption from the gut and transmission to the liver and other organs of the body. It is, however, the manner in which the parent molecule is subsequently metabolized *in vivo* which determines the precise nature of an animal's response. Information available about the metabolic activity in the human liver, suggests that humans are going to be intermediate in sensitivity to the acute toxicity and may show some sensitivity to the chronic toxicity of aflatoxin B1, including carcinogenicity.

The ingestion of aflatoxins at high levels in a single dose or repeatedly for a short period of time induces acute intoxication, hereafter designated aflatoxicosis, in humans and animals with typical symptoms, including jaundice, lethargy, nausea, edema, hemorrhagic necrosis of liver tissues, bile duct hyperplasia, and eventually death (10–60 per cent) subsequent to severe liver damage. Although there is no consensus on the specific dose of aflatoxins that triggers acute toxicity in humans, it is well established that such a dose is highly variable depending on many factors, including the age, gender, health and nutritional status, presence or absence of underlying factors (*e.g.*, chronic viral hepatitis, alcoholism, smoking, cirrhosis, exposure to hepatotoxic microcystins); and it is lowest in youngsters, as substantiated by the highest death rates of this age-group in aflatoxicosis outbreaks.

A rough estimation of the acute dose of AFB1 concerned a case report on a-15-years old Ugandan child weighing 36 kg who has been eating AFB1-contaminated cassava on a daily basis until he died by liver failure. The authors calculated the likely cumulative amount of cassava that caused the child's death to be 3.1 kg contaminated with 1.7 mg/kg, corresponding to a total dose of 146 μg/kg bw that he had eaten in 22 successive days before death. However, these calculations are very approximate, as they were based on the lethal dose of AFB1 to monkeys, and on the assumption that the AFB1 concentration in cassava, determined retrospectively after the death, was constant throughout the whole period of intake preceding the death. Nevertheless, the outcome of these calculations, is in accordance with an estimation of the world health organization (WHO) based on records of aflatoxicosis outbreaks worldwide and *in* vitro tests, which considers that regular consumption of food contaminated with 1 mg AFB1/kg or higher for a short period causes acute intoxication in humans. According to the same report, daily consumption of food contaminated with AFB1 at a dose of 0.02–0.12 mg/kg bw over 1 to 3 weeks causes a life threatening aflatoxicosis. Furthermore, the cumulative lethal dose in humans was suggested to vary from 10 to 20 mg for adults and 3 mg for children, which is also consistent with the estimated total dose of ~5.3 mg ingested in 22 days by the Ugandan teenager. Nonetheless, deliberate ingestion of 5.5 mg chemically pure AFB1 over two days and 35 mg over two weeks in suicide attempts by an adult woman from the USA was reported to cause no serious aflatoxin-related injuries at admission to the hospital for mild symptoms and even 14 years later. Although difficult to explain, this could be due to her overall well-balanced nutritional status, age, and gender, since well-nourished adult females are less susceptible to aflatoxins than males of similar health and nutritional status.

In animals, the lethal dose varies greatly among species, as suggested by the wide variation in their LD50 values ranging between 0.3 and 18.0 mg/kg bw, although values as low as 0.2 mg/kg bw or as high as 60 mg/kg bw were occasionally reported. Animals like ducks, sheep, turkeys, dogs, pigs, and rats are the most susceptible, whereas monkeys, chickens, mice, and ruminants the most resistant. The higher susceptibility of the first group of animals was explained by their ability to metabolize rapidly AFB1 via the phase II metabolism driving towards the formation of aflatoxin-albumin adducts.

In a study on the impact of an orally administered single dose of 18 of 40 AFB1 to mice, 0.66 mg/kg bw induced severe tissue injury 5 days after the ingestion. In poultry, the AFB1 doses that killed all tested birds varied between 0.8 and 4.0 μmg per animal, with turkeys being the most sensitive (0.8 mg) and geese the most resistant (4.0 mg), whereas no death was observed in chickens at the highest doze of 4.0 mg.

Although aflatoxins are most known for chronic diseases, mainly liver cancer, their association with acute intoxications is well established, and acute aflatoxicosis outbreaks making many victims have been reported in endemic countries in SEA and SSA. No aflatoxicosis cases or outbreaks has been reported, to our knowledge, in industrialized countries due to the low exposure which is 100-fold lower than that recorded in developing African and Asian countries (1 ng/kg bw per day vs. 100 ng/kg bw per day).

Although aflatoxins are of a global concern, their negative impact on health, economy, and social life is greater in developing countries located in the tropical and sub-tropical regions. Agricultural products from SSA countries, *e.g.*, Gambia, Uganda, Kenya, and Tanzania, and SEA countries, *e.g.*, China, Thailand, Vietnam, and Indonesia, have classically been associated with the highest incidence of aflatoxins, which paralleled the highest incidence of hepatocellular carcinoma and the occurrence of acute aflatoxicosis outbreak episodes in the region. As matter of fact, these regions have been the primary destination for scientists to carry out epidemiological studies on the relationship between the dietary exposure to aflatoxins and liver cancer, which contributed greatly to the establishment of aflatoxins as an aetiological factor of the disease in humans. Four major types of aflatoxins [aflatoxin B1, aflatoxin B2 (AFB1), aflatoxin G1 (AFG1), and aflatoxin G2 (AFG2)] are the best known and the most studied among more than 18 different types and metabolites presently identified.

Several studies have demonstrated that very young children may be exposed to aflatoxins even before they are weaned because mothers consuming aflatoxin in their food, may secrete aflatoxin M1 in their milk. There is no doubt about the potential danger of aflatoxin in food and every effort should be made to reduce or if possible, eliminate contamination.

In some countries, mainly in Africa and Asia, children become fed almost exclusively with potentially contaminated grain, and after being weaned, this is due to the economic condition. It was shown that the increase in serum concentration of aflatoxin in these children is also related to dwarfism.

The ingestion of aflatoxins is especially relevant in developing countries, where contact with these substances occurs, from the development embryo, into adulthood. One of its effects would be the impairment in the development of children, in addition to the association with nutritional disorders such as Kwashiorkor.

Poor diets, from the nutritional point of view, seem to be related to the increase in the toxic and carcinogenic effects of aflatoxins, a fact that may be related to the low intake of substances with antioxidant potential such as vitamins, a common condition in the poorest populations of countries under development.

The first significant aflatoxicosis outbreak occurred in two Indian regions encompassing more than 200 poor setting ethnic villages with a protein deficient nutritional status who relied mainly on maize as a food source. The climate in these neighbouring Western Indian regions is typically hot desert (BWh) and hot semi-arid (BSh) characterized by low annual rainfall and chronic drought. In 1974, these regions received abundant unseasonal rain (October–November instead of the usual rainfall period of June–September), while the maize standing in the field had attained the full maturity stage to be harvested. Shortly after that, an epidemic struck affecting people in family clusters and the pets sharing the same diet. The possibility of an infectious disease was ruled out, as it was not contagious and the prescription of anti-microbial drugs prove ineffective. Clinical and post-mortem histopathological examinations of dead victims revealed obvious symptoms and liver lesions evoking aflatoxicosis; *i.e.*, periportal hepatic fibrosis, and bile duct proliferation. Thin layer chromatography (TLC) analysis showed the presence of unidentified green and blue spots in extracts of necropsy liver samples and AFB1 in the serum of some patients. Moreover, the suspect maize was highly contaminated with *A. flavus* and contained aflatoxins at concentrations ranging between 6500 and 15,600 μg/kg. Exposure calculations suggested that the populations have been ingesting, through their diet, 2–6 mg of aflatoxins on a daily basis for several weeks from the start of harvest to the depletion of maize stock, which coincided with the end of the outbreak. Together, these data were taken for evidence to ascribe the epidemic to a maize-born aflatoxicosis.

In the SSA region, Kenya has been the most severely afflicted by aflatoxicoses, especially in the East-central region where the prevailing climate is hot semi-arid (BSh), humid subtropical (Cwa), or oceanic tropical highland (Cwb) with frequent alternation of dry and rainy periods. Populations of these regions, mainly of the Akamba/Kamba tribe, grow maize for home consumption as the main staple food and store it by traditional means in containers that they place inside a granary or hung to the ceiling of the kitchen. Two notable aflatoxicosis outbreaks were recorded in the same region of the country. The first one occurred in 1981 after a severe shortage in rainfalls during the year 1980 followed by heavy rainy season that extended from October to May instead the normal period of October to December (short rainy season). Starting from late March to early June 1981, 20 patients, mostly from two family groups of Makueni district, were admitted to the provincial hospital with jaundice and other symptoms suspecting a viral hepatitis. Within 22 days of the early symptom onset (*i.e.*, abdominal discomfort, anorexia, general malaise, and low-grade fever), 12 of the patients developed massive ascites and gastrointestinal haemorrhage before they died from liver failure. Among those, six were from the same family of eight members including two twins who were not affected, as they were not fed the family diet.

The second episode of aflatoxicosis outbreak that occurred in Kenya in 2004 was the most significant worldwide, as it caused 317 cases with 125 deaths. It also covered a larger zone encompassing four districts of more than 40,000 km2 populated by 2.8 million inhabitants, mostly of the Akamba tribe. Makueni and Kitui districts were the most severely affected (47 per cent and 32 per cent of cases, respectively), followed consecutively by Machakos and Thika with 6 per cent and 4 per cent of the total cases. In an almost identical scenario as for the former aflatoxicosis outbreaks in India (1975) and Kenya (1981), this one also occurred after unseasonal heavy rain preceded by a year of severe shortage in rain and foods, which resulted in high aflatoxin contamination of maize and increased susceptibility of nutritionally deficient rural farmers.

In late 1988, during the 9 days of the nine-emperor-gods festival held in Malaysia [197], the consumption of a traditional Chinese dish called Loh See Fun was implicated in a poisoning outbreak resulting in 17 severe cases and 13 deaths. Patients were children of 2.5 to 11 years of age, with one case of a 46-years-old man, and only children died. Other patients (45 in number) who ate the same dish as the affected cases, developed similar but milder and transient symptoms; they were, hence, considered as presumptive cases and discarded from further investigation. The main ingredient of the offending dish consisted of white noodle made with a mixture of rice and corn flour. Boric acid was illicitly added to the dish by the producing factory in Kampar city (Malaysia) to extend its shelf life and enhance its sensory properties. The onset of the poisoning was fairly rapid and the first symptoms evoking a Reye-like syndrome appeared within a mean time of 8.5 h after the ingestion. The patients exhibited different symptoms,

Table 10.6: Significant aflatoxicosis outbreaks in Asia and Africa.

Country	*Region*	*Year (Period)*	*Number of Cases*	*Number of Deaths (per cent of Fatality Rate)*	*Associated Food*	*Level of Contamination (mg/kg)*	*Specific Remark*
India	Western (Rajasthan-Gujarat)	1974 (October–November)	397 106 (26.7) Maize 6.3–15.6	106 (26.7)	Maize	6.3–15.6	Heavy unseasonal rain after drought and faulty storage conditions
Kenya	East-Central (Makueni)	1981 (March–June)	20	12.0	Maize	3.2 and 12.0	Rain shortage in the year preceding the outbreak followed by prolonged high rainy season and faulty storage conditions
	East-Central (Makueni-Kitui-Machakos-Thika)	2004 (January–July) a	317	125 (39.4)	Maize	1.0–46.4	First study relating aflatoxin-albumin adduct to human aflatoxicosis and its use as a biomarker
Malaysia	Perak state	1988	17 b	13 c(76.5)	Chinese noodles d	NS	Possible additive effect of boric acid and aflatoxin
Tanzania	Central	2016 (May–November)	67	20 (30)	Maize	10–51,100	Possible additive effect of fumonisins High titres of aflatoxin-albumin adduct in the serum of patients used as evidence for the aflatoxicosis
	Northeast	2017 (June–July)	8	4 (50)	Maize	NA	Evidence based on symptoms and consumption of maize reported to have been inadequately stored

a: A peak was reached in May to Mid-July; b: 16 children 2.5–11 years of age a 49-years old adult; c: All children, and they died within hours of the intoxication and the percentage of death excludes 45 cases not admitted to the hospital because they developed only mild symptoms; d: Dish called "Loh See Fun" suspected to have been preserved with banned boric acid. Abbreviations: BAP: Boric acid poisoning.

the commonly observed of which were vomiting, seizure, diarrhoea, abdominal pain, anorexia, and coma. Jaundice was generally weak at the beginning and increased in severity with time until the eventual death with liver and kidney failure. Depending on the patient, the survival time varied from 2 to 9 days with a mean of 5 days. The results of clinical, analytical, and histopathological examinations ascribed the intoxication to both boric acid and aflatoxins.

A recent aflatoxicosis outbreak was reported in the central region of Tanzania in 2016. The prevailing climate in the region is hot semi-arid (BSh) and subject to frequent alternations of drought and flood periods caused by ENSO. This phenomenon induces extreme shifts in rainfall and temperature causing both severe drought and rainfall events, usually followed by increased incidence of disease outbreaks, as it creates favourable ecological conditions for microbial pathogens and their vectors to emerge. According to the latter study, a strong El Niño hit Tanzania in 2015–2016 and raised above normal the cases of malaria and cholera in the period of April 2015 to March 2016, which continued through 2017 for cholera. Although not mentioned in the study, this situation applies to the aflatoxicosis outbreak that occurred in the period of 14 May to 14 November 2016. The outbreak affected 68 individuals in family clusters and killed 20 of them. Before death, the patients presented typical symptoms of aflatoxicosis, *i.e.*, jaundice, abdominal pain, vomiting, diarrhoea, and ascites. A house-to-house survey conducted in selected households including case-households with victims and those without (controls), showed that more than 50 per cent of the cases were children below 15 years-old who had eaten home-grown maize contaminated with both aflatoxins and fumonisins at abnormally high levels. Aflatoxin contamination in samples collected from case-households was significantly higher than those of controls (10–51,100 µg/kg versus 2.4–285 µg/kg). Fumonisins were detected in the maize sampled from case-households at concentrations ranging from 945 to 12,630 µg/kg. Of the maize samples contaminated with both mycotoxins, 80 per cent exceeded the regulatory standards of 10 µg/kg and 2000 µg/kg for total aflatoxins and fumonisins, respectively. Moreover, the titres of aflatoxin-albumin adduct in the serum of case-patients usually exceeded 1000 pg/mg and were 3.6 to 8.2 times higher than in the serum of controls (36–32,800 pg/mg vs. 10–4020 pg/mg). The increase in aflatoxin-albumin adduct titre is a strong indication of the causal link between aflatoxins and the outbreak, the severity of which may have been increased by an additive effect of fumonisins.

Considering the well-established adverse health effects of aflatoxins, international and regional organizations have been actively working to provide quantitative estimates of the risk associated with their consumption. SEA and SSA countries were the first target countries for risk assessment studies due to the notoriously high incidence of aflatoxins in the staple crops they produce and for the prevalence of liver cancer among their populations. Therefore, in addition to toxicity data generated from experimental animals, epidemiological studies conducted in these regions were also used to document the earliest quantitative risk assessment on aflatoxins

The risk associated with the consumption of agri-foods of SEA and SSA countries requires to take actions for safety in order to reduce liver cancer incidence. The action can be done through increased control of crop contamination and/or diet change to reduce aflatoxin intake. Different strategies have been used to reduce the levels and incidence of aflatoxins in foods and agricultural products. These strategies can be split into two main actions: (i) Decontamination by degradation/removal of aflatoxins from crops and (ii) Prevention by the implementation of appropriate management systems to reduce crop contamination in the field and during storage.

Various physical, chemical, and microbial treatments have been demonstrated to reduce crop contamination with aflatoxins by degrading, inactivating or removing them from the matrix. However, each of these treatments has drawbacks that limit its wider application, as it may be onerous, require extreme pressure and temperature conditions, yield inconsistent results, release toxic residues, or reduce the sensory, nutritional or functional properties of the crop. Among the chemical treatments, ozonation and ammoniation have been reported to have received high research attention owing to their effective degradation of aflatoxins and potential for use at large scale. Ozonation of various crops reduced to>99 per cent of the initial level of contamination with aflatoxins in a dose- and time-dependent manner without significantly affecting the chemical composition or the safety of the treated product.

Techniques to remove aflatoxins from crops by using natural adsorbents that bind to aflatoxins with high affinity, such as clays and organic adsorbents, including microbial strains or their extracts, have been described. However, despite the well-recognized potential of the latter decontamination means, their actual use remains limited essentially due to inconsistent performances. Difficulties to detoxify aflatoxins is largely related to their chemical nature as a recalcitrant xebobiotic polyheterocyclic toxins that resist biodegradation. Nonetheless, under certain conditions, they can be readily degraded and/or detoxified, usually by disrupting the structure of the terminal furan ring or the coumarin group. Although not toxic by itself, the integrity of the coumarin, in particular its lactone moiety, is necessary for aflatoxins to exert maximum toxicity. Therefore, disruption or removal of the lactone has been associated with drastic reduction or loss of toxicity.

It is well understood now that the global health and economic burden of aflatoxins, with SEA and SSA countries support the highest share of the burden. Therefore, particular attention should be paid to improvement of the situation in these regions where crops prone to aflatoxin contamination such as peanut, maize, sorghum, and sunflower are grown in agroclimatic zones (hot and humid) favourable to aflatoxin production.

After studying about aflatoxins let us learn few more things about ochratoxin.

ii) Ochratoxin A (OTA)

It is a naturally occurring foodborne mycotoxin found in a wide variety of agricultural commodities worldwide, ranging from cereal grains to dried fruits to wine and coffee. It was first isolated from *Aspergillus ochraceus* in South Africa. It is produced by several different fungi including Aspergillus ochraceus, A. carbonarius, A. niger and Penicillium verrucosum. These fungi vary in their optimal growing temperatures and water activity, and contaminate various commodities as summarized in Table 10.7. Contamination generally occurs as a result of poor storage of commodities and suboptimal agricultural practices during the drying of foods. Ingestion is the main source of exposure to OTA, through the foods and beverages listed in Table 10.7.

Table 10.7: Ochratoxin-Producing Fungi, Optimal Growth Conditions, and Commodities Affected

OTA-producing Species	*Optimal Temperature Range (Min-Max) °C*	*Water Activity*	*Commodities Affected*
A. ochraceus	24–31(8–37)	0.95 – 0.99	Smoked and salted dried fish, dried beans, biltong, soya beans, chickpeas, rapeseed, pepper, dried fruit, and sesame seeds, nuts, cereals rice, barley, maize, wheat, flour, and bran, coffee beans
A. carbonarius	32–25 (N/A-40)	0.82	Grapes and grape products, including table grapes, wines, and dried vine fruits
A. niger	35–37(6–47)	0.77	Nuts, apples, pears, peaches, citrus, grapes, figs, strawberries, mangoes, tomatoes, melons, onions, garlic, and yams
P. verrucosum	20(0–30)	0.80	Cereal crops; cheese, meat products

Source: JECFA (Joint FAO/WHO Committee on Food Additives (JECFA), 2001).

OTA is a chemically stable compound; hence, ordinary food processing measures fail to substantially reduce its presence in foods and beverages. Ochratoxin A once been absorbed by the gastrointestinal tract passes into systemic circulation, where there can be detected in tissues and blood; higher concentrations are detected in the most metabolically active organs such as liver, kidney, muscle and fat. Moreover, during distribution, ochratoxin A has higher affinity for adaptation to plasma proteins, presented a half-life of prolonged excretion, *i.e.*, in man of 840 hours (35 days), the free toxin dose <0.2 per cent and excrete their waste by renal and hepatobiliary route. On the other hand, pollutes human foods such as cereals, wheat, barley, corn, rice, grapes, wheat flour, cornmeal, oatmeal, bread, and derivatives, alcoholic beverages, such as red wines, wines pink, white wines, beer with alcohol, non-alcoholic beers, stouts and milling products (coffee, cocoa). Likewise, the values of ochratoxin A in foods are intertwined with processing environments and maintenance. That is, if agricultural practices and environmental conditions (humidity and temperature) during storage and shipping are not adequately performed, it will favour the production of ochratoxin A in these foods. This mycotoxin acts altering the cell oxygen.

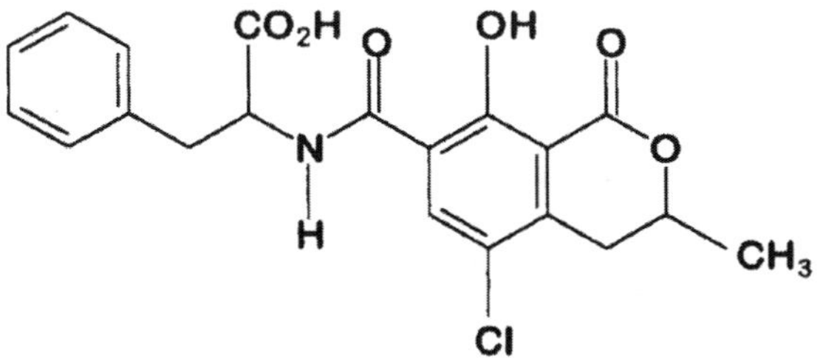

CO_2H O OH O N H O CH_3 Cl

Figure 10.18. Chemical Structure of Ochratoxin A.

OTA has been shown to be toxic and carcinogenic in animals. The kidney is the main target organ for OTA; OTA is a potent renal carcinogen in several animal species. Other adverse effects of OTA include immunotoxicity, inhibition of macromolecular synthesis, increased lipid peroxidation, and inhibition of mitochondrial respiration. OTA has been suspected as a cause of various human nephropathies since the 1970s including Balkan Endemic Nephropathy (BEN) and chronic interstitial nephropathy (CIN). A debilitating human disease known as Balkan endemic nephropathy is associated with diet of those people who have a tradition of storing mould-ripened hams for long periods of time. The International Agency for Research on Cancer (IARC) has classified OTA as a Group 2B possible human carcinogen, based on demonstrated carcinogenicity in animal studies, although OTA-related carcinogenicity has not been conclusively determined in humans. A recent risk assessment on OTA states that OTA was negative in genotoxicity assays with high specificity, and that OTA-DNA adduct levels were low and not typical of genotoxic carcinogens.

Ochratoxin A has long been considered to be associated with the nephropathy of people residing in certain areas of Tunisia, Scandinavia, and the Balkan regions, where exposure may be 'endemic.' The pathological lesions of nephropathy in humans are similar to those observed in endemic porcine nephropathy, which is due to the consumption of feeds contaminated with OA. Since a high proportion of the patients suffering from endemic nephropathy in the Balkan region develop tumours of the renal pelvis and ureter, the possible involvement of nephropathy in tumour development was suggested. Ochratoxin A has been found in human serum in Tunisia and several European countries, including Bulgaria, Poland, Yugoslavia, and Germany. In certain areas of the Balkans and Tunisia, OA levels in the foods and in human serum in endemic regions are higher than in the nonendemic areas. OA also has been found in human milk and kidneys in some endemic regions.

The mode of the toxic effect of OA is not as clearly defined as those of AFB. OA inhibits several enzymes, including phenylalanine hydroxylase, phenylalanine-tRNA synthetase and renal phosphoenolpyruvate carboxykinase, which may play a role in its toxic effects. The synthesis of the latter two enzymes was found to be inhibited by OA. In addition, it is also a strong inhibitor for carboxypeptidase A. Enhancement of lipid peroxidationl is considered as one of the manifestations of cellular damage in OA toxicity. The induction of free radicals in a bacterial model system by OA was found to be regulated by Ca^{2+} ions. Superoxide dismutase and catalase have been shown to have some protective effect for OA-induced nephrotoxicity in rats. OA-induced lipid peroxidation in the Vero cell was also found to be decreased in the presence of these two enzymes, but the effects of aspartame and piroxcam were less pronounced. In the renal epithelial cells, OA interferes with hormonal Ca^{2+} signalling, leading to altered cell proliferation. Thus, OA may impair cellular Ca^{2+} cyclic AMP (cAMP) homeostasis.

Ochratoxin A binds strongly with serum albumin; thus, the toxin has a long half-life in animal (35 days with a single dose of OA) and human bodies. Such an interaction has not only played a significant role in the toxic effect of this mycotoxin because of its heterogenous body distribution, but also been used as a biomarker for human exposure of OA.

Various dose-response studies in animals were the basis for advisory groups' determinations of safe weekly or daily OTA intakes for humans. JECFA first evaluated OTA at its 37th meeting (JECFA, 1991), setting a provisional tolerable weekly intake (PTWI) at 112 ng OTA per kg bodyweight (bw) per week based on a dose-response study of renal function deterioration in pigs, for which the lowest observed adverse effect level (LOAEL) was 8 μg/kg bw/day (Elling, 1979; Krogh, 1974). A combined uncertainty factor (UF) of 500 was applied in the calculation. JECFA re-evaluated OTA at its 44th meeting, taking into account new toxicological data. The PTWI was confirmed, but rounded down to 100 ng/kg bw/week. The most recent assessment of OTA at the 68th meeting in 2008 resulted in retaining the PTWI previously found. JECFA currently estimates OTA exposure from cereals, based on European data, to be about 8-17 ng/kg bw/week: well below the PTWI.

The European Food Safety Authority (EFSA) derived a PTWI for ochatoxin A of 120 ng/kg bw/week, based on the 8 μg/kg bw/day LOAEL used in the JECFA evaluation (European Food Safety Authority, 2006). An uncertainty factor of 450, rather than 500 used in JECFA, was applied to the LOAEL. This composite uncertainty factor was based on an intra-species factor of 10, interspecies factor of 15, and a factor of 3 for use of a LOAEL instead of a no observed adverse effect level (NOAEL). The interspecies factor of 15 was based on the longer OTA half-life in humans and monkeys rather than pigs as determined by Hagelberg *et al.* (1989).

Table 10.8: Summary of Calculated Tolerable Human Intakes of Ochratoxin A (OTA) by International Organization

Organization	*Tolerable Intake Metric**	*Limit*	*Reference*
European Food Safety Authority (EFSA)	PTWI	120 ng/kg bw/week	EFSA, 2006
Health Canada	PTDI	3 ng/kg bw/day	Kuiper-Goodman *et al.*, 2010
Health Canada	NCRI	4 ng/kg bw/day	Kuiper-Goodman *et al.*, 2010
Joint FAO/WHO Expert Committee on Food Additives (JECFA) 2007	PTWI	100 ng/kg bw/week	JECFA, 2007
Nordic Expert Group on Food Safety	TDI	5 ng/kg bw/day	Olson *et al.*, 1991
Scientific Committee of Food (SCF) of the European Union	PTDI	5 ng/kg bw/day	EFSA, 2006

*TDI - tolerable daily intake; PTDI - provisional tolerable daily intake; PTWI - provisional tolerable weekly intake; NCRI - negligible cancer risk intakeSource: Travis R Bui-Klimke and Felicia Wu.2015. Ochratoxin A and human health risk: A review of the evidence. Crit Rev Food Sci Nutr. Nov 10; 55(13): 1860–1869.

A Health Canada risk assessment team (Kuiper-Goodman *et al.*, 2010) chose to re-evaluate EFSA's PTWI for ochratoxin A, positing that the use of LOAEL rather than a NOAEL was not appropriate given the small number of animals per group, and the fact that 4 out of 9 pigs in the lowest dose group showed functional kidney changes. Rather than use a NOAEL or LOAEL, a benchmark dose corresponding to a response of 10 per cent above background (BD^{10}) was derived. Uncertainty factors of 10 for intra-species variability, 25 for interspecies variability, and 2 for use of a sub-chronic rather than chronic study were combined in a composite uncertainty factor of 500. Applying this composite uncertainty factor to the BD_{10} of 1.56 μg/kg bw/day resulted in a TDI of 3.0 ng/kg bw/day after rounding (Kuiper-Goodman *et al.,* 2010), which in practice is considerably stricter than the JECFA or EFSA tolerable limits.

The review of the epidemiological data suggests that, with one exception, there appears to be no statistically significant evidence for human health risks associated with OTA exposure. The one exception concerns an increased risk of nephritic syndrome at very high exposures to OTA, based on case-control studies assessing multiple potential adverse health effects in an Egyptian population. However, the sample size of this studied population was very small, and the urinary OTA levels associated with nephritic syndrome were much higher than urinary OTA levels measured in multiple other world regions, with the exception of Sierra Leone.

iii) Patulin

Patulin is a cyclic γ-lactone mycotoxin found in fruits, particularly moldy apples and apple juice. Several molds produce patulin including, *P. expansum*, *Aspergillus* sp., and *Byssochlamys fulva* and *B. nivea*. Patulin is a small, water-soluble polar compound that is absorbed quickly, metabolized, and excreted in urine.

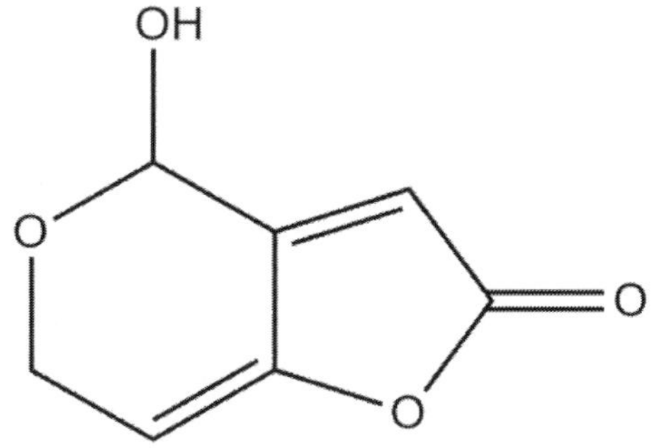

Figure 10.19: Structure of Patulin

The accumulation of PAT in the commodities appears to be independent from the use of fungicides before the storage step, while data support that the temperature of food prior of refrigeration is probably the most important variable in the production and accumulation of PAT.

It is toxic to a wide range of biological systems including microorganisms, plants and animals. The toxicity of patulin has been studied in many experimental models including chicken, quail, cat, cattle, rabbit, mice, and rats. The toxic effects on these animals were found to be oedema and hemorrhage in brain and lungs, capillary damage in the liver, spleen, and kidney, paralysis of motor nerves, and convulsions. Patulin is also immunosuppressive agent which inhibits multiple aspects of macrophage function. Patulin has also been shown to be carcinogenic to mice. The susceptibility of man to this toxin is unknown at this time. However, when the numerous biological systems affected by patulin are considered, it is likely that man would also be susceptible to the toxic effects of this compound in some manner. Thus, patulin must be considered a potential health hazard.

The fact that patulin can be produced by a number of penicillia at refrigeration temperatures is a source of possible concern. Many of these moulds are found on refrigerated food products and the possibility that patulin production could occur during refrigerated storage is very real. Certain foods such as cheeses and cured meats which are low in carbohydrates and high in protein do not appear to support patulin production to any great extent. This combined with the reactivity of patulin with sulfhydryl groups seems to lessen the danger of patulin occurrence in these foods. However, apple products and grain-based products may pose more of a potential problem. Apple products are of particular concern since common apple rotting moulds are frequent producers of patulin and because of the stability of patulin in these products.

It mainly penetrates products of apple as juices, juice or cider, because it is highly soluble solid in aqueous media, and can withstand thermal processing of juices. It produces toxic effects as genotoxicity, immunotoxicity and neurotoxicity in humans and animals in the body causing effects as weakening of the immune system and the nervous system aggression. At the cellular level it has been shown that Patulin causes various toxic effects, including disturbance of the plasma membrane, where there is an impairment of synthesis accompanied protein transport inhibition amino acid linked with Na+, according to the literature via disorder transcription and translation and inhibition of DNA synthesis.

In 2003, Ritieni investigated the presence of PAT in Italian apple products and the positive samples had a concentration ranging between 1.4 and 74.2 μg/L with a mean value of 26.7 μg/L. The PAT concentration in unprocessed fruit stored indoors reached a value of 90 ng/g after 5 days, 400 ng/g after 15 days, and 2.200 ng/g after 33 days. These values decreased to 75 ng/g, 100 ng/g and 695 ng/g, respectively, after washing of fruits before their industrial processing. The washing of the apples with water under high pressure reduced from 10 and up to 100 per cent the initial concentration of PAT depending on the initial level of PAT contamination. For heavily contaminated apples, *i.e.*, 350 μg/L or more, even after washing, the PAT level did not go below 50 μg/L. A refrigerated storage with a controlled atmosphere often does not guarantee the prevention of fungal growth, and

PAT production, as well as additional treatments, including fungicides application.

Exposition to trans-2-hexenal vapour, an aromatic volatile compound produced by many fruits and vegetables, at a concentration of 12.5 µg/L can potentially control blue mould onset, can also minimize the PAT content and maintain the quality of apples of the variety Golden Delicious. Finally, immersion of apples in a sodium hypochlorite 3 per cent solution for five minutes at 25°C completely inhibited fungal growth and the related damages to the fruits. PAT concentration in apple juice increases with the decomposition of the peel of the apple. In fact, the level of PAT found in the juice products prepared from healthy apples are all smallest than 50 ng/g, while the juices produced from apples where more than 30 per cent of peel were decomposed, contained PAT at levels higher than 50 ng/g. PAT can be removed from the juice using granular activated carbon, in particular apple juice can be filtered through activated charcoal (40-60 mesh) and PAT levels are reduced of about 98 per cent with respect to unprocessed apple juice. Another approach used three grams of activated charcoal per liter added to apple juice containing 62 ng/g of PAT, followed by stirring for five minutes. In this case the PAT contamination was reduced of about the 50 per cent. Other carbon-based adsorbents have been studied and can efficiently reduce the PAT levels in apple juice, but these adjuvants also result in a loss of flavour and of the nutritional quality of the final products.

One of the most common approaches to reduce the level of the PAT in the apple juice is the use of pectins that are able to clarify the juice by filtration under vacuum. This this process reduces the level of the PAT of about 39 per cent. Alternatively, a mixture of gelatin/bentonite clay, followed by ultra-filtration, is able to reduce the level of the PAT up to 25 per cent [77]. Different conventional techniques also can reduce PAT levels in clarified apple juice, for example centrifugation is the most effective among the adopted industrial treatments, with a reduction of 21 per cent. The use of refined bentonite clay is the second efficient choice, and a PAT reduction of about 8.5 per cent has been reported.

As patulin occurs mainly in fruits and fruit juices which are often used as baby or infant food; some countries have established regulatory limits. Many research groups have recommended that apple products for human consumption should not contain residual patulin levels in excess of 50 µg l^{-1}.Multiple countries have set regulatory limits for patulin in fruit juice, ranging from 30 to 50 ppb.

iv) Trichothecenes

Trichothecenes belong to a major class of mycotoxins produced by a range of fungi from the order Hypocreales, including those of the genera Fusarium, Myrothecium, Verticimonosporium, Stachybotrys, Trichoderma, Trichothecium, Cephalosporium, and Cylindrocarpon. To date, 148 trichothecenes have been isolated, but only a few have been found to contaminate food and feed. The most frequent contaminants are deoxynivalenol (DON), also known as vomitoxin, nivalenol (NIV), diacetoxyscirpenol (DAS), while T-2 toxin is rarer.

Trichothecenes have been isolated in the past thirty years. The first trichothecene to be isolated was trichothecin from *Trichotheceum roseum*, in 1948 by Freeman and Morrison. Diacetoxyscirpenol (DAS) from *Fusarium equiseti* was preliminarily characterized in 1961 by Brian *et al.*, and was later followed by nivalenol (NIV) and T-2 toxin, both from *F. sporotrichioides*, although they were mis-identified as *F. nivale* and *F. tricinctum*, respectively, in the original articles. However, it was the discovery of 4-deoxynivalenol (DON) from wheat in Eastern North America in 1980, which truly sparked the research into the *Fusarium* species and led to the discovery of trichothecenes from other genera.

In the early stage of trichothecene studies, the first isolate, trichothecin, was discovered as a result of screening of antifungal metabolites in a culture of *Trichothecium roseum*. The following survey demonstrated that *Myrothecium roridum* and *M. verrucaria* produced potent antifungal metabolites, which were designated as verrucarins A through J and roridins A through E. Structural studies on these compounds demonstrated that they are macrocyclic trichothecenes. In the early 1960s another antifungal metabolite was isolated from a strain of Trichoderma, and this compound was called trichodermin. In their chemical work Godtfresden and Vangedal (1964, 1965) related trichodermin to trichothecolone and verrucarol, the alkaline hydrolysates of trichothecin and verrucarin A, respectively. The structures of all these antifungal fungal metabolites contain a modified sesquiterpene skeleton, and they are characterized by the presence of a spiroepoxy group. Another research group in Hungary isolated a new antibiotic called crotocin from Cephalosporium crotocinigenum ScholSchwarz. Crotocin, a diepoxy-containing compound is similar to trichothecin in both structure and biological activity.

Trichothecenes are a large group of sesquiterpenoid fungal metabolites, which share a common core comprised of a rigid tetracyclic ring system consisting of a cyclohexene, A-ring with a double C-C bond occurring between C-9 and C-10; a tetrahydropyranyl B-ring; a cyclopentyl C-ring, and an epoxide at C-12/13. The rigidity of this system results in a distinct stereochemistry for the A- and B-rings. The A-ring adopts a half-chair conformation, and the B-ring is most often found in the chair conformation, although there have been a few odd cases where the B-ring has been shown to adopt a boat conformation. Shank *et al.*, reviewed the types of trichothecenes and showed that they fall into two main classes, simple and macrocyclic.

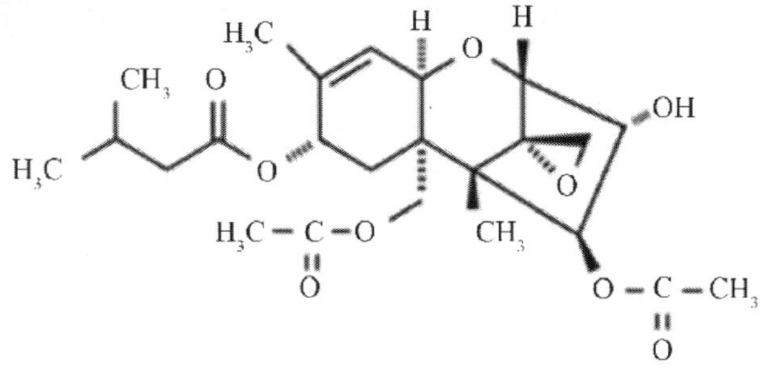

Verrucarin A Chemical Structure

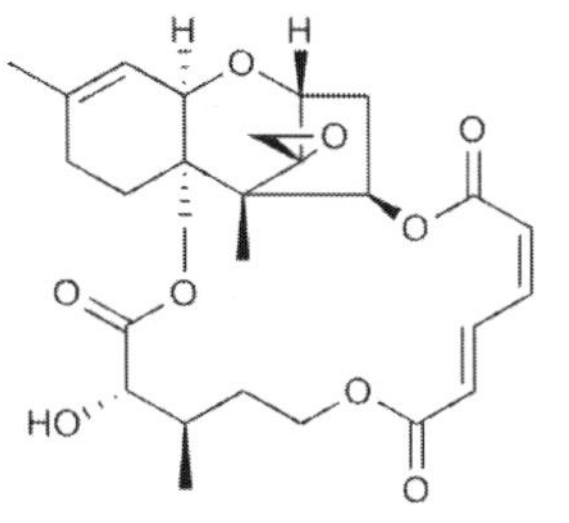

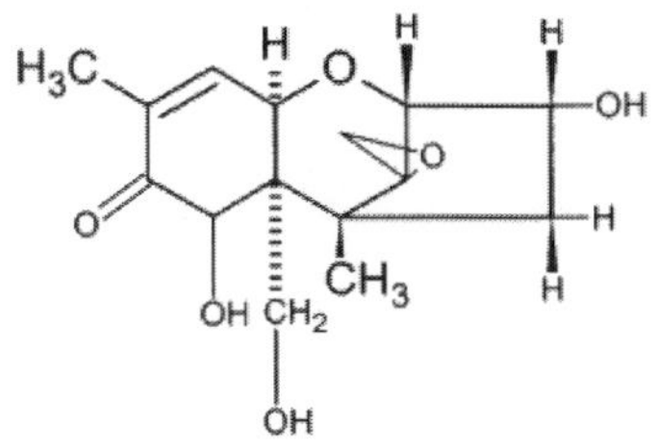

T-2 chemical structure Deoxynivalenol (DON)

Figure 10.20: Chemical Structures of Verrucarin A, T-2 and Deoxynivalenol.

The simple trichothecenes are further divided into three types; A, B and C. Type A trichothecenes are the simplest group, being non-substituted, hydroxylated or esterified. Type B trichothecenes are characterized by a ketone present at C-8. Type A and B trichothecenes, such as T-2 toxin and DON, respectively, are often associated with Fusarium-infected grain. Type C trichothecenes, such as crotocin, are less common than the others, and are distinguished by the presence of a second epoxide ring at C-7/8. A fourth class (Type D), are characterized by the presence of a cyclic diester or triester linkage of C-4 to C-15. These macrocyclic trichothecenes include the satratoxins, verrucarins, roridins, myrotoxins and baccharinoids. While many of the Type D trichothecenes have been isolated from fungi, the baccharinoids were first isolated from the plant Baccharis spp.

TCT have been found to contaminate wheat, barley, corn, rice, rye, oats, and other crops. The effect of TCT has been extensively studied on poultry and farm animals as TCT contaminate feed to a large extent. Human and animals are exposed simultaneously to several trichothecenes because most fusariums are able to produce a number of mycotoxins simultaneously; food commodities can be contaminated by several fungi simultaneously or in quick succession, and a complete diet is made up of various different commodities. Humans may also be exposed to multiple trichothecenes via products from animals that have eaten contaminated feed.

In several cases, trichothecene mycotoxicosis was caused by a single ingestion of bread containing toxic flour or rice. In experimental animals, trichothecenes are 40 times more toxic when inhaled than when given orally. Trichothecenes were found in air samples collected during the drying and milling process on farms, in the ventilation systems of private houses and office buildings, and on the walls of houses with high humidity. There are some reports showing trichothecene involvement in the development of "sick building syndrome". The symptoms of airborne toxicosis disappeared when the buildings and ventilation systems were thoroughly cleaned. There are some reports that indicate that trichothecenes may have been used as chemical warfare agents in South-East Asia (Lao People's Democratic Republic and Cambodia)

The exact mechanisms of trichothecenes' action are not fully understood, but when they are ingested in high doses by animals they cause nausea, vomiting, and diarrhoea.TCT affect cell division in the body, where cells are actively dividing such as the skin, gastrointestinal tract, lymphoid, and erythroid cells. In these cells, TCT inhibit the protein synthesis, resulting in acute necrosis in mucosal lining and skin. More fatal consequences are depressed immune function and reduced bone marrow.

When ingested by pigs and other animals in small doses trichothecenes can cause weight loss and the refusal to eat (vomitoxin). Chronic intake of small amounts of trichothecenes leads to an increased susceptibility to infectious diseases as a result of the suppression of the immune system, as well as other symptoms such as anorexia, anaemia, neuroendocrine changes, and immunologic effects.

The first recognized trichothecene mycotoxicosis was alimentary toxic aleukia in the USSR in 1932; the mortality rate was 60 per cent. In regions where the disease occurred, 5± 40 per cent of grain samples cultured showed the presence of *Fusarium sporotrichoides*, while in those regions where the disease was absent this fungus was found in only 2±8 per cent of samples. The severity of mycotoxicosis was related to the duration of consumption of toxic grain. Such severe trichothecene mycotoxicoses, the consequence of continuous ingestion of toxins, have not been recorded since this outbreak.

In plants, tricothecenes produced by Fusarium spp. cause necrosis, chlorosis, and mortality enabling them to mediate a wide variety of plant diseases, including wilts, stalk rot, root rot and leaf rot in many important crop and ornamental plants. Examples include kudzu, Orobanche, duckweed, and many grain crops

Trichothecenes are associated with worldwide economic and health impacts. While various management strategies have been proposed to reduce the mycotoxin risk, breeding towards FHB-resistance appears to be the most effective means to manage the disease, and reduce trichothecene contamination of cereal-based food products.

v) Deoxynivalenol (DON)

They are produced by fungi of the genus phytotoxic Fusarium. Toxic symptoms attributed to deoxynivalenol are associated with nausea (vomiting), food refusal, gastroenteritis, diarrhea, immunosuppression and hematological disorders. The mycotoxin Deoxynivalenol through the intestinal mucosa through the tight junctions. At the same time, it increases the permeability for Deoxynivalenol intestinal cell by opening the tight junctions, causing this mycotoxin is moved by circulation to other organs leaving toxin and thus produce pathology. In addition, allowing a larger number of bacteria can move through the intestinal epithelium,

Deoxynivalenol (DON) is a toxin formed by fungi of the genus Fusarium. *Fusarium graminearum*, is becoming a big threat to both public health and agro-economy worldwide. It can induce acute and chronic toxic effects on both human and animals through contaminated foods and feeds. Therefore, surveillance and control of DON and its derivatives in foods and feeds are of great importance for producers, regulatory authorities and researchers to protect consumer health and to reduce economic losses.

These fungi occur in the soil and can infect the grown crops. These mycotoxins are common contaminants of cereals such as wheat, corn, barley and silage. The exposure risk to human is directly through foods of plant origin (cereal grains) or indirectly through foods of animal origin (kidney, liver, milk, eggs). It has been detected in buckwheat, popcorn, sorghum, triticale, and other food products including flour, bread, breakfast cereals, noodles, infant foods, pancakes, malt and beer.

The level of Fusaria infections in wheat and the resulting DON content of wheat strongly depends on the weather conditions during the growth and bloom of the cereal and on the agricultural techniques used. DON is a contaminant of wheat products that is not easy to avoid. In 1998, unfavourable weather conditions in Western Europe led to a high level of Fusaria infections in wheat and to correspondingly high DON contents. As a consequence, young children ingested relatively high amounts of DON through wheat products in 1999. In laboratory animals, DON can reduce the rate of weight gain during growth. At higher levels of exposure, it can adversely affect the immune system, fertility and the foetus. There are no data that suggest the presence of such effects in men. One must realise, however, that the possible occurrence of such effects has not been investigated in humans.

DON affects animal and human health causing acute temporary nausea, vomiting, diarrhoea, abdominal pain, headache, dizziness, and fever. The toxic effects induced by DON are well characterized in all animal species, with pigs being the most susceptible (Pestka, 2010). DON can affect the immune system, animals' food consumption, growth, reproduction, neuroendocrine signalling, and also intestinal function (Pestka, 2010; Pinton *et al.*, 2010). In humans, there is historical evidence suggesting that DON causes acute illness and is frequently associated with outbreaks of gastroenteritis. The intestine is the first barrier to food contaminants. Following ingestion of mycotoxin-contaminated food or feed, intestinal epithelial cells can be exposed to high concentrations of toxins. Highly dividing cells, such as intestinal epithelial cells, are especially sensitive to trichothecenes, and the exposure of intestinal epithelial cells to these toxins may alter their capacity to proliferate and to insure a proper barrier function. Exposure to DON can lead to impaired absorption of nutrients and decrease of cell proliferation or differentiation. The entry of luminal antigens and bacteria, normally restricted to the gut lumen, was demonstrated to be a consequence of the disruption of the barrier function in cell culture models or in mice species. It alters the barrier function of intestinal cells directly or indirectly by modulating the expression of tight junction proteins (TJP). It also induces histopathological changes in intestinal tissue as demonstrated ex vivo in cultured jejunal explants and *in vivo* in animal feed– contaminated diet

The legal limits are regulated differently in different countries and are *e.g.*, legally defined EU-wide at 1250 μg/kg for unprocessed cereals, 500 μg/kg for bread or 200 μg/kg for infant food.

Yellow Rice Disease

A complex of disorders recognized in Japan a number of times since the end of the last century has been associated with the presence of several species of penicillia and their toxic metabolites on rice. Although mycotoxin contamination in rice has been reported less often than that in wheat and corn, there have been occasional reports of mycotoxin contamination in stored or imported rice in Japan. The name "yellow rice" (ou-hen-mai; in Japanese; "mai" being Japanese for "rice"), or rice grains affected and coloured by mold, is notorious case in the history of Japanese food hygiene. However, the details of "yellow rice" are not well known.

Yellow rice" is the collective name of rice grains contaminated by Penicillum fungi in Japan. Three kinds of yellow rice, which are caused by different sub-species of toxigenic fungi, are known. Initially, malnutrition resulting from Asian poverty was regarded as the source of illness in people; however, thanks to ongoing research, the close relationship of "Shoshin-kakke (heart-attacking paralysis)", "mycotoxin", and "rice" was elucidated. Subsequently, the illness was shown to derive from "yellow rice", which was due to grains damaged by *Penicillium* spp. After World War II, another two kinds of yellow rice, "Islandia yellow rice" and "Citrinum yellow rice", were found, and the causative toxic compounds were determined through interdisciplinary collaborative studies.

The beginning of the study of mycotoxins in rice in Japan can be regarded as the toxicity study of molded rice **was first** conducted by Junjiro Sakaki in 1891. In those days, people who ate polished rice as a staple diet tended to suffer from Shoshin-kakke (heart-attacking paralysis) in urban area of Asia including Japan. From ethanol extracts of molded rice, Sakaki elucidated the existence of substances that provoked similar symptoms to heart-attacking paralysis in experimental animals. This observation led him to conclude that molded rice contained mycotoxin(s) which could cause paralysis. After that, research on diseased rice grains was started systematically from plant physiological point of view. The name yellow rice refers to grains affected by Penicillium species after harvest). This name can be used not only for imported rice but also for stored domestic rice. In 1934, the Rice Utilization Institute, a predecessor of the National Food Research Institute, was built to conduct research on the utilization and processing of rice. Ichiro Miyake, Hiroshi Naito and Hiroshi Tsunoda were leading researchers on fungi that spoiled rice grains during storage at the Rice Utilization Institute.

Before the discovery of yellow rice in Japan, other mycotoxicoses caused by ergot fungus (*Claviceps purpurea*) and *Fusarium* species were known. Both occurred in the field and not after harvest. Yellow rice was the first example of food mycotoxicosis caused by fungi adherent to crops after harvest. Research on yellow rice was conducted in Japan before the outbreak of aflatoxin mycotoxicosis in 1960 in England. The earlier research related to yellow rice in Japan is an important event in the history of global food hygiene.

Twenty-four kinds of unhealthy rice grains are known, of which 18 are molded rice grains spoiled by fungi). Diseased rice grains include not only grains spoiled in the field but also grains spoiled after harvest. Fungi of Penicillium species are frequent contaminants in stored rice as well as Aspergillus species. After long-term preservation of rice, Versimoss-rice (A. versicolor), Yellow-spot rice (Eurotium amstelodami) (Eurotium is a teleomorph of Aspergillus), and Citrinum yellow rice (P. citrinum) are frequently found. Furthermore, fungi of Penicillium species adherent to stored rice are often found to produce mycotoxins.

This moulded rice is usually discoloured yellow and several of the toxic metabolites implicated are themselves yellow pigments. There was an early awareness that moulds may be responsible for cardiac beriberi and in 1938 it was demonstrated that *Penicillium citreo-viride* (*P. toxicarium*) and its metabolite citreoviridin was responsible. The most toxic of the species of penicillia associated with yellow rice disease is *P. islandicum* which produces two groups of toxins, hepatotoxic chlorinated cyclopeptides such as islanditoxin as well as the much less acutely toxic, but potentially carcinogenic dianthraquinones such as luteoskyrin.

vi) Citrinin

Citrinin (CIT) was first isolated by Hetherington and Raistrick from Penicillium citrinum prior to World War II; subsequently, it was identified in over a dozen species of Penicillium and several species of Aspergillus (*e.g.*, Aspergillus terreus and Aspergillus niveus), including certain strains of Penicillium camemberti (used to produce cheese) and Aspergillus oryzae (used to produce sake, miso, and soy sauce). More recently, citrinin has also been isolated from Monascus ruber and Monascus purpureus, industrial species used to produce red pigments.

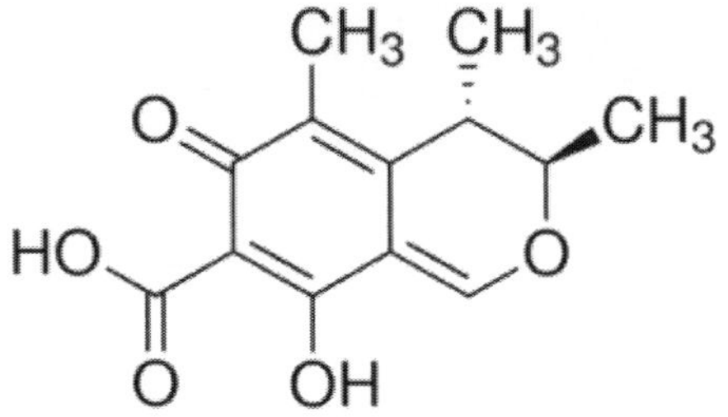

Figure 10.21: Chemical Structure of Citrinin.

The natural occurrence of CIT in rice has been described in different countries, mainly in Asia. Citrinin has been associated with yellow rice disease in Japan. It has also been implicated as a contributor to porcine nephropathy. CIT has been reported to be nephrotoxic, genotoxic, and carcinogenic in all animal species tested, but its acute toxicity varies in different species. The 50 per cent lethal dose for ducks is 57 mg/kg; for chickens it is 95 mg/kg; and for rabbits it is 134 mg/kg. Citrinin can act synergistically with ochratoxin A to depress RNA synthesis in murine kidneys.

Wheat, oats, rye, corn, barley and rice have all been reported to contain citrinin. But compared with other mycotoxins, CTN contamination of food and feed is rather scarce. However, it is reasonable to believe that humans are much more frequently exposed to CTN than generally accepted, because it is produced by the same moulds as OTA, which is a common contaminant of human food all over the world. At present, there are no specific regulations either in Croatia or in the European Union concerning CTN in any kind of commodity.

vii) Alimentary Toxic Aleukia

Alimentary toxic aleukia, is a mycotoxin-induced condition characterized by nausea, vomiting, diarrhea, leukopenia (aleukia), hemorrhaging, skin inflammation, and sometimes death. Alimentary toxic aleukia almost always refers to the human condition associated with presence of T2 Toxin.

Outbreaks of this dreadful disease, which is also known as septic angina and acute myelotoxicosis, occurred during famine conditions in a large area of Russia. A particularly severe outbreak occurred during the period 1942–47 but there had been reports of the disease in Russia since the 19th century.

Studies in Russia itself demonstrated that the disease was associated with the consumption of cereals moulded by *Fusarium sporotrichioides* and *F. poae* but the nature of the toxin remained unknown. A total of 131 isolates of Fusarium poae and F. sporotrichioides from overwintered cereals, which were associated with the alimentary toxic aleukia toxicoses in the Soviet Union, were tested for their ability to produce T-2 toxin [4 beta, 15 diacetoxy-8alpha-(3-methylbutyryloxy)-12,13-epoxytrichothec-9-en 3alpha-ol]. The presence of T-2 toxin was determined by thin-layer chromatography, gas-liquid chromatography, spectroscopic analyses, and the rabbit skin test. A good correlation was demonstrated between T-2 toxin dectetion by thin-layer chromatography and inflammatory skin reactions of rabbits.

Studies of dermonecrosis in cattle in the United States showed it to be caused by a *Fusarium* metabolite called T–2 toxin which is one of the most acutely toxic of a family of compounds called trichothecenes.

The toxic metabolites of fusarial species have been intensively studied in efforts to establish the specific etiology of ATA. Studies with rats, mice, guinea pigs, rabbits, dogs, swine, sheep, poultry, cattle, and horses were unsuccessful in the search for an animal model. The syndrome of ATA has been induced in cats by administration of the sesquiterpene T-2 toxin (3ahydroxy4p, 15-diacetoxy-8a-[3-methylbutyroloxy] 12, 13-epoxytrichothec-9-ene), a naturally occurring trichothecene isolated from Fusarium sporotrichioides.12"3 T-2 toxin in gelatin capsules, administered every 48 hours per os to healthy cats (dosage: 0.08 mg/kg body weight), regularly resulted in a fatal disease (average survival time: 3 weeks) with symptoms of generalized lassitude, weakness, bloody feces, hind-leg ataxia, conjunctivitis, and vomiting. In the terminal stages, anorexia, dyspnea and paralysis were noted, while dehydration and weight loss (average 22 per cent) were more severe.

There is good evidence that T–2 toxin was a major agent in the development of alimentary toxic aleukia in man, the first symptoms of which are associated with damage of the mucosal membranes of mouth, throat and stomach followed by inflammation of the intestinal mucosa. Subsequently petechial hemorrhages of both the skin and mucosa together with enlarged lymph nodes are seen. Pancytopenic abnormalities intensify, and blood coagulation is diminished. The pathologic features of the respiratory system include bronchopneumonia, pulmonary hemorrhages, sepsis, and lung abscesses. Vomiting, bloody feces, lassitude, and incoordination are also observed. All of the clinical features, the pancytopenia, and the gross and microscopic pathologic changes seen in cats receiving T-2 toxin via the alimentary route were strikingly similar to the disease syndrome of alimentary toxic aleukia in man.

Bleeding, vomiting and diarrhoea, which are all associated with damage of mucosal membrane systems, were common but recovery at this stage possible if the patient was given a healthy, uncontaminated, vitamin-rich diet. Continued exposure to the toxin, however, led to damage of the bone marrow and the haematopoietic system followed by anaemia and a decrease in erythrocyte and platelet counts. The occurrence of necrotic tissue and skin haemorrhages were further characteristics of the disease. To summarize, ATA in man is characterized by leukopenia, agranulocytosis, necrotic angina, a haemorrhagic rash, sepsis exhaustion of the bone marrow, bleeding from nose, throat and gums, and fever.

As well as giving rise to this sequence of acute symptoms, the trichothecenes are known to be immunosuppressive and this undoubtedly contributed to victims' sensitivity to relatively trivial infectious agents. Indeed, many people died of bacterial and viral infections before succumbing from the direct effects of the toxin itself. Unlike aflatoxin, the acute toxicity of T–2 toxin is remarkably uniform over a wide range of animal species and it is reasonable to assume that the human LD50 will be in the same range. Although improved harvesting and storage has eliminated alimentary toxic aleukia from Russia yet this disease may still occur in any part of the world ravaged by war and famine.

Three of the most important mycotoxins, aflatoxin, ochratoxin and T–2 toxin, are immunosuppressive but react differently against the immune system. All three inhibit protein biosynthesis, aflatoxin by inhibiting transcription, ochratoxin by inhibiting phenylalanyl tRNA synthetase and T–2 toxin by inhibiting translation through binding with a specific site on the eukaryote ribosome. One consequence of these distinct modes of activity is that mixtures of such mycotoxins are likely to be synergistic in activity and this has been shown experimentally in the case of aflatoxin and T-2 toxin. This observation is significant in the context of the probability that a food which has gone mouldy will probably be infected by several species of mould and may thus be contaminated by several different mycotoxins.

viii) Zearalenone

Zearalenone (6-[10-hydroxy-6-oxo-trans-1-undecenyl]-B-resorcyclic acid lactone), a secondary metabolite from Fusarium graminearum (teleomorph Gibberella zeae) was given the trivial name zearalenone as a combination of G. zeae, resorcylic acid lactone, -ene (for the presence of the C-12 to C-2 double bond), and -one, for the C-62 ketone. Almost simultaneously, a second group isolated, crystallized, and studied the metabolic properties of the same compound and named it F-2. Much of the early literature uses zearalenone and F-2 as synonyms; the family of analogues are known as zearalenones and F-2 toxins, respectively. It is an estrogenic mycotoxin. This mycotoxin causes hyperestrogenism, especially in swine. The first report of moldy corn causing an estrogenic disturbance among swine appeared in 1928. Since then, other reports have come from the United States, Australia, and Ireland. In 1962, Stob *et al.,* demonstrated that an anabolic uterotrophic compound, crystallized and partially characterized by them, was produced by Gibberella zeae, the perfect stage of Fusarium graminearum (F. roseum "Graminearum"), hence this toxin was called zearalenone. Urry *et al.,* determined its chemical structure and named it ZEA. Partial characterization of the compound and environmental conditions suitable for its production were reported by workers at the University of Minnesota, who referred to ZEA as F-2.

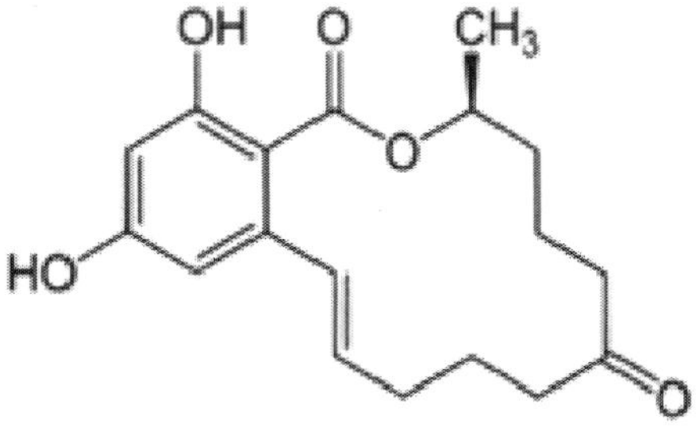

Figure 10. 22: Chemical Structure of Zearalenone.

Fungi-producing ZEA contaminate corn and also colonize, to a lesser extent, barley, oats, wheat, sorghum, millet and rice. In addition, the toxin has been detected in cereals products like flour, malt, soybeans and beer. Fungi of the genus Fusarium infect cereals in the field. Toxin production mainly takes place before harvesting, but may also occurs post-harvest if the crop is not handled and dried properly.

ZEA is produced by a variety of Fusarium fungi, including *F. graminearum, F. culmorum, F. cerealis, F. equiseti, F. crookwellense* and *F. semitectum*, which are common soil fungi, in temperate and warm countries, and are regular contaminants of cereal crops worldwide. Zearalenone, an oestrogenic mycotoxin was first shown to cause vulvovaginitis in pigs fed on mouldy maize. Pigs are especially sensitive to this toxin and, although its acute toxicity is very low. In gilts, the vulva and mammary glands become swollen and, in severe cases, there may be vaginal and rectal prolapse. In older animals there may be infertility, reduced little size and piglets may be born weakened or even deformed. Modern work shows that dietary concentrations of zearalenone as low as 1.0 ppm may lead to hyperestrogenic syndromes in pigs; higher concentrations can lead to disrupted conception, abortion, and other problems. Reproductive problems have also been observed in cattle and sheep. There is concern about the long-term exposure of the human population to such an oestrogen.

ZEA has a less phytotoxicity. It was the least toxic of six metabolites to three varieties of wheat seedlings, where at 100 $\mu g \cdot mL^{-1}$, ZEA did not significantly affect germination or the subsequent root and leaf development. Germinating caryopses of rye, wheat, triticale and field bean were found to differ in their response to treatment with 10 $\mu g \cdot mL^{-1}$ ZEA for 24 h. Mitotic activity increased in rye, but decreased in wheat and field bean. No abnormal metaphase chromosomes were observed. ZEA at 5 $\mu g \cdot mL^{-1}$, had an inhibitory effect on root and shoot elongation and fresh mass accumulation of germinating, excised embryos of Z. mays. Interestingly, 10 and 25 $\mu g \cdot mL^{-1}$ ZEA stimulated these parameters measured. When DON and ZEA were applied simultaneously, to give a final combined toxin concentration of 25 $\mu g \cdot mL^{-1}$, the inhibitory effects on root and shoot elongation and mass were similar to or exceeded the values measured for DON alone. This would suggest an additive and possibly a synergistic effect when ZEA and DON occur in combination. The previous study showed that up to 50 per cent of the mycotoxin becomes bound to starch, hemicellulose, and in particular, lignin fractions, suggesting binding of ZEA metabolites to cell wall components.

Extensive reviews of Canadian and Scandinavian epidemiological data have concluded that the risk to human populations is minimal. The recommended safe human intake of zearalenone is estimated to be 0.05 μg/kg of body weight per day. Zearalenone levels in foodstuffs are not yet regulated anywhere. Nevertheless, because of its genuine biological potency and regular dietary cooccurrence with true mycotoxins such as the fumonisins and trichothecenes, it would not be prudent to ignore the potential of zearalenone to cause adverse health effects.

Zearalenone and the corresponding alcohol zearalenol, are known to have anabolic, or growth promoting activity, and, although its use as a growth promoting agent is banned in some countries, it is permitted in others. This can lead to problems in international trade because zearalenone can be detected in the meat of animals fed on diets containing it.

ix) Sterigmatocystin

Sterigmatocystin (STC) is a mycotoxin produced by fungi of many different Aspergillus species. At least eight derivatives are known. Among the producing organisms are *Aspergillus versicolor, A. nidulans, A. rugulosus*, and others. Other species such as Bipolaris, Chaetomium, Emiricella are also able to produce STC. STC producing fungi were frequently isolated from different foodstuffs, while STC was regularly detected in grains and grain based products, corn, bread, cheese, spices, coffee beans, soybeans, pistachio nuts, beer, animal feed and silage.

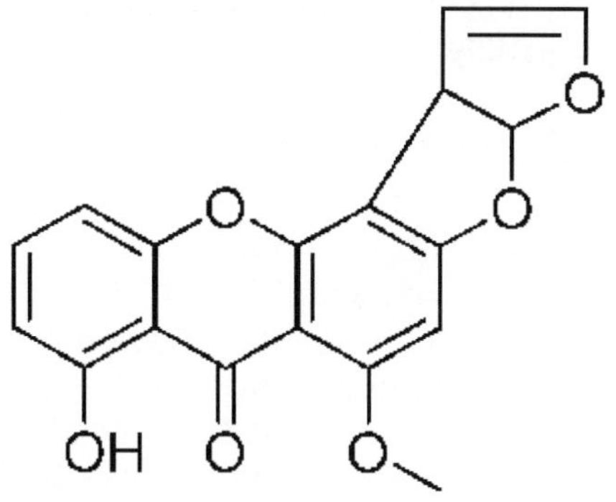

Figure 10.23: Chemical Structure of Sterigmatocystin.

The STE is a known biogenic precursor of aflatoxin B1, sharing with it several structural and biological similarities. The STE has been shown to be hepatotoxic and nephrotoxic in animals and it has been classified as possible human carcinogen (group 2B) by IARC. The STE has been reported to cause a marked decrease in cell proliferation in different mammalian cells. Data available on literature suggest that the cellular mechanisms underlying STE-induced toxicity include the induction of oxidative stress, mitochondrial dysfunction, apoptosis, cell cycle arrest, as well as alteration of immune system function and activation of different signalling pathways. Moreover, STE resulted to be genotoxic, being able to form DNA-adducts and induce DNA damage. Despite its strong cytotoxicity, no risk assessments have been still carried out by authorities due to the lack of toxicity data, so research on STE toxicological impact is still going on.

x) Sambutoxin

The sambutoxin mycotoxin was first reported in 1994 and its structure is shown below. It is associated with dry-rotted potatoes and is produced primarily by strains of *Fusarium sambucinum* and *F. oxysporum*. Of 13 Fusarium species examined, about 90 per cent of strains of the two species noted produced this toxin. It had toxic effects on rats, including body weight loss, feed refusal, hemorrhage in the stomach and intestines, and finally death, when the rats were fed diets supplemented with 0.05 and 0.1 per cent sambutoxin. The toxin was also toxic to cultured human tumor cell lines Molt-4, CCRF-CEM, and KATO III (unpublished data). Most of the toxic isolates of *F. oxysporum* and *F. sambucinum* obtained from rotten potatoes produced sambutoxin. This fact suggests that some Fusarium species frequently produce sambutoxin and that the toxin occurs naturally in agricultural products infected with Fusarium species, including rotten potatoes.

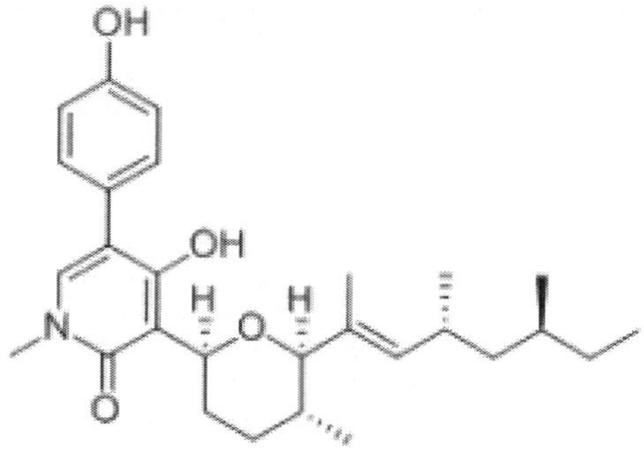

Figure 10. 24: Chemical Structure of Sambutoxin.

From rotten potato samples in Korea, 9 of 21 contained 15.8–78.1 ng/g of sambutoxin with a mean of 49.2 ng/g.46 Using wheat media, levels of 1.1–101 μg/g of sambutoxin were produced. The toxin was found in potatoes from parts of Iran that had a high incidence of esophageal cancer.

xi) Fumonisins

Fumonisins are naturally occurring toxins produced by several species of Fusarium fungi (moulds). Fumonisins were first recognized in 1988. The fumonisins are produced by Fusarium spp. on corn and other grains, and certain diseases of humans and animals are associated with the consumption of grains and grain products that contain high levels of these molds. The species demonstrated to produce fumonisins include *F. sacchari, F. subglutinans, F. thapsinum, F. globosum, F. anthophilum, F. dlamini, F. napiforme, F. nygami, F. moniliforme*, and *F. proliferatum*. The latter species produces large quantities. *F. moniliforme* (formerly *F. verticilliodes*; *Gibberella fujikuroi*) was the first to be associated with the mycotoxin and it is the best studied of the three.

A number of different types of fumonisin are known, but fumonisins B1, B2 and B3 (also named FB1, FB2 and FB3) are the major forms found in food. Fumonisin is frequently detected from maize throughout the world in high concentration. The most major contaminating Fumonisin is Fumonisin B1, followed by B2 and B3. Combined pollution with mycotoxin of trichothecene class and with aflatoxin is also found. Contamination is reported from other cereals and vegetables including milo, wheat, soybeans, rice, and asparagus. However, their contamination level is usually lower than that of maize.

The prevalence of *F. moniliforme* is significantly higher in corn from areas where a high rate of human oesophageal cancer occurred than in low oesophageal cancer rate areas. There are at least 15 fumonisins, with the best known being FB1, FB2, FB3, FB4, FA1, FA2, and FA3. The major ones are FB1–FB3, and the others are considered to be minor and less well characterized. Of the three major toxins, FB1 (also designated macrofusine) is produced in the largest quantities by producing strains. For example, among nine strains of *F. moniliforme*, the range of FB1 produced on autoclaved corn was 960 –2350 µg/g while for FB2 the range was 120–320 µg/g.

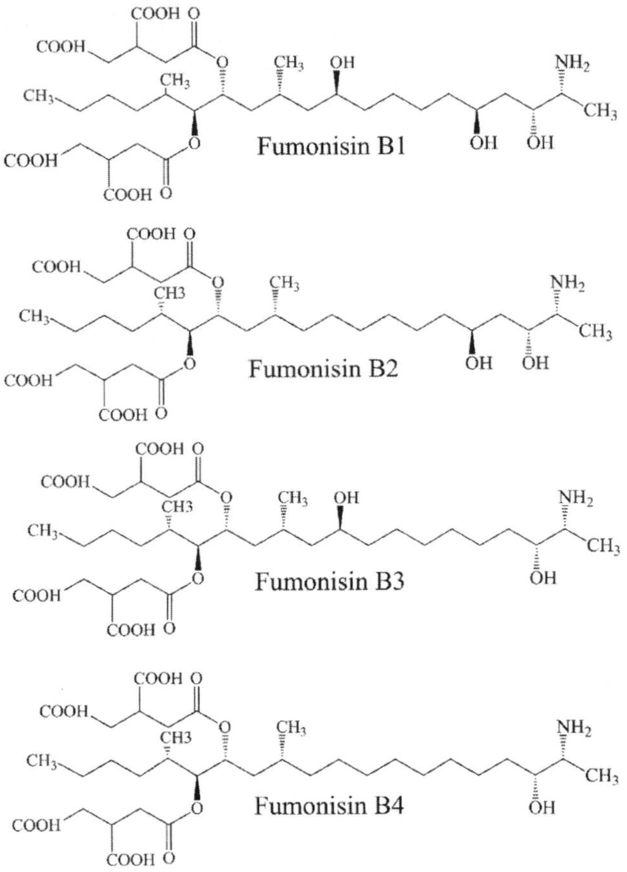

Figure 10.25: Chemical Structures of Funonisins.

The highest rate of human EC in southern Africa occurs in the Transkei where high levels of FB1 and FB2 are found in corn. Concomitant with the occurrence of the fumonisins is the presence of Fusarium spp., especially *F. moniliforme*. The incidence of fumonisin FB1 in a high-risk county in China was about two times higher than in a low-risk area, although the differences were not statistically significant. Trichothecenes (mainly deoxynivalenol) in addition to fumonisins were found in corn from the high risk area. From corn and corn-based products from four states in America, FB1was found in 65 per cent of 34 samples while FB2 was found in only 29 per cent. The upper level for FB1 was 2679 µg/kg, and 797 µg/kg for FB2. Fusarium numbers ranged from 102 to >105.

In 1980, Marasas detected fumonisins as one of the reasons for esophageal cancer regarding many people in South Africa. Fumonisins occur because of these two groups of factors: non-biological factors and biological factors. Fumonisins' structure is similar to sphingolipid, and this toxin in high concentrations inhibits ceramide synthase that is an important enzyme in the de novo pathway of sphingolipid biosynthesis. International Agency for Research on Cancer (IARC) has classified this toxin as a group 2B carcinogen. Consuming any food that is contaminated by this toxin will cause different diseases in animals and humans.

C. Toxigenic Algae

Algae are the green thallophytes, containing a green colouring matter called Chlorophyll, and in many algae the green colouring matter (Chlorophyll) may be masked by other colours, but in all of them, Chlorophyll is always present. They are autotrophic, meaning they manufacture their own food with the help of chlorophyll contained in them. They can affect water quality adversely by lowering the dissolved oxygen in the water. Other factors which are responsible for eutrophication are slow moving water, sunlight, algal spores and nutrients (nitrogen and phosphorus). Eutrophication is caused by excessive algae growth due to the increased availability of one or more limiting growth factors needed for photosynthesis, these include solar radiations, carbon dioxide (CO_2), and dissolved nutrient fertilizers coming from soil surface runoff. Eutrophication may occur naturally after the passage of large time spans over lakes over the stagnant water containing sediments.

A number of algal species produce more potent toxins, if the water contains high concentrations of toxic algae or if their toxin is ingested. It presents a risk to human health; the algal toxins are usually released into the water when the cells of toxic algae rupture or die. Algal blooming and water toxicities is a serious environmental problem.

The toxic algae have been grouped into two; freshwater species which are cyanophyta (blue-green algae) and marine species which are the dinoflagellates (diatoms). Algae have different types of toxicity which are not a phylogenetically conserved feature among the algae. The harmful health effects of the algae depend on the type of algae present and the type of exposure. Two major disease types are recognized; first is paralytic shellfish poisoning and second is called ciguatera seafood poisoning. Red tides, and cyanobacteria are examples of toxic algae that can result multiple serious impacts on fresh water ecosystems, human health and the economy. Some forms of algae produce toxic chemicals the accumulation of this toxic algae is called a harmful eutrophication. These can be found in marine and freshwater environments. The toxins resulting from the spread of algae in the marine environment can affect shellfish and the human body. Under these conditions, algal blooms can be formed on the water surface. The colours vary from dark green to yellow brown or red but cannot be seen immediately.

Direct skin contact with algal toxins may cause skin problems and eye irritation. Inhalation of toxic algae droplets in may have a minor effect on the respiratory tract and may cause hay fever-like symptoms. This usually happens when you come in contact with recreational areas where toxic algae may flourish, such as swimming pools, water skiing and boating areas. Drinking water contaminated with algae and food (such as fish and shellfish) can cause gastroenteritis, which can cause vomiting, diarrhoea, fever and headache. This toxin also affects the nervous system or may affect liver. Preventing sources of nutrients for not reaching the water is the best method against toxic algal blooms.

Planktonic Dinoflagellates

Some algae are injurious to humans as some species produce toxins that can concentrate on crustaceans and fish, making them dangerous or toxic for human consumption. Dinophyceae (the class of dinoflagellates) is the best-known producer of toxins. Planktonic dinoflagellates may occasionally form blooms containing high numbers of organisms when environmental conditions such as temperature, light and nutrients are appropriate. *Gonyaulax catenella* and *G. tamarensis* (now both referred to the genus *Alexandrium*) are the best known of a number of dinoflagellates responsible for paralytic shellfish poisoning.

i) Shellfish Poisoning Syndrome

Paralytic shellfish poisoning is contracted by eating toxic mussels, clams, oysters, scallops, or cockles. These bivalves become toxic after feeding on certain dinoflagellates of which *Gonyaulax catenella* is representative of the United States Pacific Coast biota. Along the North Atlantic Coast of the United States and over to northern Europe, *G. tamarensis* is found, and its poison is more toxic than that of *G. catenella*. *G. acatenella* is found along the coast of British Columbia. Masses or blooms of these toxic dinoflagellates give rise to the red tide condition of seas. In 1996, about 150 manatees were killed during a red tide off the coast of Florida.

Another dinoflagellate, Karenia brevis (= *Gymnodinium breve*), also causes shellfish poisoning. Shellfish poisoning caused by the brevetoxin released by *Gymnodinium breve* is known to kill seafood on the coast. It can cause respiratory distress and food poisoning in humans. It has caused massive fish kills along the east coast of the United States, and has been implicated in the death of bottlenose dolphins and manatees. This can be a serious illness with a high mortality rate. The toxic metabolites of these algae, which include saxitoxin and gonyautoxin block nerve transmission causing symptoms such as tingling and numbness of the fingertips and lips, giddiness and staggering, incoherent speech and respiratory paralysis. Saxitoxin exerts its effect in humans through cardiovascular collapse and respiratory failure. It blocks the propagation of nerve impulses without depolarization, and there is no known antidote. Saxitoxin, so named because it was isolated directly from the Alaskan butter clam, *Saxidomas giganteus*, before it was recognized that it is actually a dinoflagellates metabolite, is a very toxic compound having a lethal dose for the mouse of as little as 0.2 μg. It is heat stable, water soluble, and generally not destroyed by cooking. It can be destroyed by boiling for 3–4 hours at pH of 3.0 Recent studies have even thrown doubt on whether saxitoxin is, in fact, a dinoflagellate metabolite for it has been shown to be produced by a species of bacteria of the genus *Moraxella* isolated from the dinoflagellates.

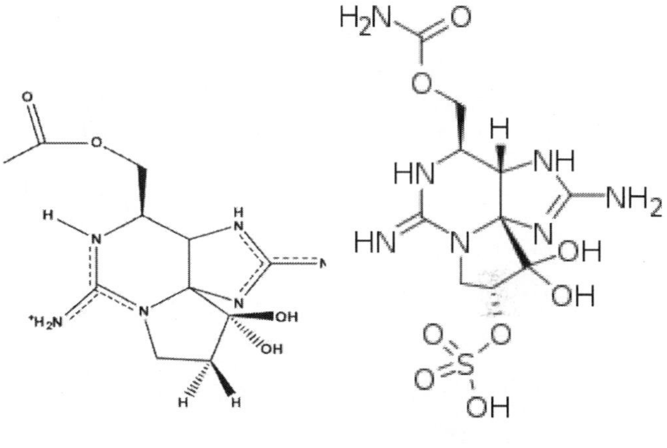

(a) (b)

Figure 10.26: Toxic Metabolites of Algae (a) Saxitoxin (b) Gonyautoxin 2.

Symptoms of PSP develop within 2 hours after ingestion of toxic mollusks, and they are characterized by paresthesia (tingling, numbness, or burning), which begins about the mouth, lips, and tongue and later spreads over the face, scalp, and neck; and to the fingertips and toes. The mortality rate is variously reported to range from 1 to 22 per cent.

Between 1793 and 1958, some 792 cases were recorded, with 173 (22 per cent) deaths. In the 15-year period 1973–1987, 19 outbreaks (with a mean of 8 cases) were reported by state health departments to the CDC. In 1990, there were 19 cases from two outbreaks in the states of Massachusetts and Alaska alone. In the former, six fishermen became ill after eating boiled mussels that contained 4,280 μg/100 g saxitoxin. The raw mussels contained 24,400 μg/100 g. The 13 cases in Alaska resulted in 1 death, and gastric contents from the victim who died contained 370 μg/100 g of PSP toxin, whereas a sample of the butterclam that was consumed contained 2,650 μg/100 g. The maximum safe level of PSP toxin is 80 μg/100 g.

Outbreaks of PSP seem to occur between the months of May and October on the United States West Coast and between August and October on the East Coast. Mollusks may become toxic in the absence of red tides. Detoxification of mollusks can be achieved by their transfer to clean water, and a month or more may be required.

To control paralytic shellfish poisoning, the collection and sale of bivalves in areas affected by algal blooms is banned.

Neurotoxic shellfish poisoning is less severe and less common than paralytic shellfish poisoning and is associated with the brevitoxins, which are complex cyclic polyethers produced by the dinoflagellate *Ptychodiscus brevis* (= *Gymnodinium breve*) one of the species responsible for the phenomenon known as a 'red tide'.

A third quite different form of poisoning associated with eating shellfish which have accumulated dinoflagellates is diarrhoeal shellfish poisoning caused by lipophilic toxins such as dinophysistoxin produced by *Dinophysis fortii*. The major symptoms, which usually occur within an hour or two of consuming toxic shellfish, include diarrhoea, abdominal pain, nausea and vomiting and may persist for several days.

ii) Ciguatera Poisoning

Ciguatera is a food intoxication caused by the consumption of primarily coral fish; these species exist in large numbers in the seas that surround the Colombian territory. The underreported diagnosis of this clinical entity has been widely highlighted due to multiple factors, such as, among others, ignorance by the primary care practitioner consulted for this condition as well as clinical similarity to secondary gastroenteric symptoms and common food poisonings of bacterial, parasitic or viral etiology. Eventually, it was found that people affected by ciguatoxins had trips to coastal areas hours before the onset of symptoms. Thanks to multiple studies over the years, it has been possible to identify the relation between toxigenic dinoflagellates and seagrasses, as well as its incorporation into the food chain, starting by fish primarily inhabiting reef ecosystems and culminating in the intake of these by humans.

This syndrome is contracted from the ingestion of any one of over 300 fish species (barracuda, grouper, sea bass, *etc.*) that feed on herbivorous or reef fishes, which in turn feed on phytoplankton, especially the dinoflagellates. Ciguatera fish poisoning (CFP) is an intoxication due to eating contaminated fish with ciguatoxins. Ciguatoxins are lipid-soluble polyether compounds consisting of 13 to 14 heterocyclic rings whose synthesis is associated with various species of dinoflagellates. The responsible dinoflagellate is *Gambierdiscus toxicus*, which produces ciguatoxin. This toxin is concentrated more in fish organs such as the liver than in muscle tissue. The toxins are concentrated along a food chain starting with the consumption of algae by herbivorous and detritus-feeding reef fish which themselves are consumed by the larger carnivorous fish caught for human consumption.

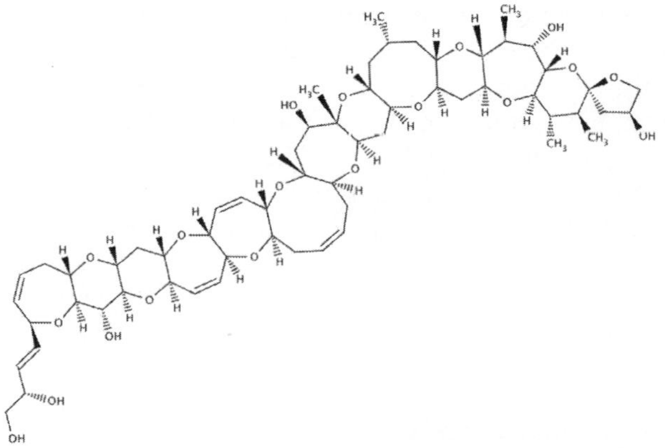

Figure 10.27: Chemical Structure of Ciguatoxin.

The main route of exposure of the toxin is orally; there are other ways of transmission, such as maternal–foetal, for which cases of abortions and preterm birth have been reported. In addition to the oral route is the sexual route, by the presence of the toxin in the seminal fluid, causing pain during ejaculation, vaginal burning and pelvic pain; these symptoms may persist for 2 to 3 weeks.

The intestinal absorption of ciguatoxins is rapid, and it is quickly localized in soft tissues such as skeletal muscle, the heart and the nervous system. Ciguatoxins are transported together with plasma proteins, mainly seroalbumin, and can be found in toxic levels in body fluids such as breast milk and seminal fluid, in addition to their characteristic of liposolubility, making them able to cross the placental barrier and can cause foetal conditions and abortions during the acute phase of the poisoning. After a liver biotransformation, finally the toxin is excreted by bile in the faeces.

The general signs and symptoms of CFP include headache, diarrhoea, vomiting, osteotendinous hyporeflexia, hives, conjunctivitis, pain and decreased visual acuity. These clinical presentations may vary in the acute, chronic and relapsing phases of the illness.

According to the article of Palafox and Buenconsejo-Lum in 2001, the clinical manifestations can be classified into three groups, which are (1) acute (within the first two weeks of exposure of ciguatoxin); (2) chronic (persisting beyond the two weeks of the initial intoxication); and (3) relapsing phases of the illness (occur years after the initial intoxication).

Looking at the first symptoms detected, they frequently appear 4–6 h after toxin consumption, but can occur within minutes with large toxin ingestions or can manifest in 24 h with smaller doses of the toxin.

CFP presents a great variety of symptoms and symptomatological patterns, depending on the region where the contaminated meat has been ingested. It is said that the Caribbean areas have primarily gastrointestinal symptoms of the neurological type, whereas the opposite happens in the Pacific; these differences in symptomatology in different geographic regions are probably related to structural differences between the ciguatoxins of the Pacific, the Caribbean and India. Symptoms usually begin minutes after ingestion; however, cases of late onset, *e.g.*, after more than 36 h, have been described. For unknown reasons it has been observed that people who ingest fish contaminated with ciguatoxins do not present any condition. More than 175 symptoms related to toxic infection have been described, and the following will emphasize the most common ones.

Gastrointestinal tract: It is perhaps the most affected and associated with earlier onset; it is characterized by diarrhoea, nausea, vomiting, metallic taste, abdominal pain and pain with defecation. The gastroenteritis of infectious origin appears easily; however, they have very nonspecific symptoms that do not usually orient the ciguatera disease with certainty, and can take a long time to disappear, usually between 1 and 2 days after ingestion.

Neurological manifestations: Taking into account the pathophysiology of the toxin when producing sodium-channel opening, whereby large intracellular water flow causes oedema at the level of the Schwann cells and the axons generate a decrease in the speed of the nerve conduction, and an increase of the refractory phase of the potential of the action, the patients present paresthesia in their limbs that could start from 30 min or days after the consumption of the contaminated fish, being able to persist for months or years. In addition, cases of odontalgias have been reported, so too ataxia, vertigo, allodynia and headache after consumption of ciguatoxins.

Cardiovascular manifestations: These are the least frequent but the most feared, usually starting 2 and 3 days after consumption. The toxin has a direct effect on the papillary muscles and on the atrium, therefore predisposing the patient to presenting cardiac arrhythmias, extrasystoles and heart failure that can be potentially lethal. Orthostatic hypotension due to stimulation of the parasympathetic system and respiratory arrest due to phrenic nerve block has also been described.

Other manifestations: The most frequent skin conditions of toxic infection are a rash associated with pruritus and are related to acne exacerbations.

CFP is prevalent throughout the world; it is estimated that there are more than 50,000 cases in the world. In the Colombian Caribbean, more than 200 cases of ciguatera have been reported since 1968, without deaths. The symptoms of Colombian CFP patients different from those described in other parts of the world, where 31 per cent presented itching throughout the body and 48 per cent presented a metallic taste in the mouth.

A study carried out by Mancera-Pineda in 2014, where they evaluated the presence of potentially toxic dinoflagellates associated with floating organic material in San Andrés Islas (drif), indicated that drift is an important substrate for dinoflagellates (*Prorocentrum emarginatum* and *Sinophysis microcephala*) and, given its floating nature, it represents perhaps the most important vector for the dispersion of these toxic agents on the island. Regarding the records of ciguatera poisoning on the island of San Andrés, two outbreaks were recorded in 1997, which affected 16 tourists and 9 residents, respectively.

In 2017, Steven and collaborators reported that the highest CFP incidence rates in the Caribbean occur in Antigua and Barbuda (219 per 100,000 (CFP cases per 100,000)), and the 2003–2013 average projected division rate is 100 to 250 div. mo–1 for the Gambierdiscus sp. composite incidence rates; however, this averages zero to 0.02 per 100,000 (CFP cases per 100,000) in Colombia.

Ciguatera poisoning is a disease with great impact on public health, especially in the Colombian coast. The most frequently given advice is not to consume fish weighing more than 2 kg, avoid eating fish such as barracuda and not eating fish parts such as the viscera, brain and gonads, which is where ciguatoxins is mostly accumulated. An invitation is sent to the territorial entities to carry out control measures for the consumption of certain fish; in addition, because this is a medical and environmental alert, the symptoms must be known and an adequate diagnosis of this poisoning must be given, so that CFP does not drastically disrupt our daily life.

iii) Domoic Acid

Amnesic shellfish poisoning (ASP) is caused by consumption of shellfish that have accumulated domoic acid, a neurotoxin produced by some strains of phytoplankton. Domoic acid (DA) was of no special scientific interest until a series of case studies revealed its role as the major marine neurotoxin causing amnesic shellfish poisoning (ASP). This is an uncommon amino acid that antagonizes glutamic acid in the central nervous system. It is produced by a diatom, *Pseudonitzschia pungens*. Diatoms are single-celled algae with walls of silicon.

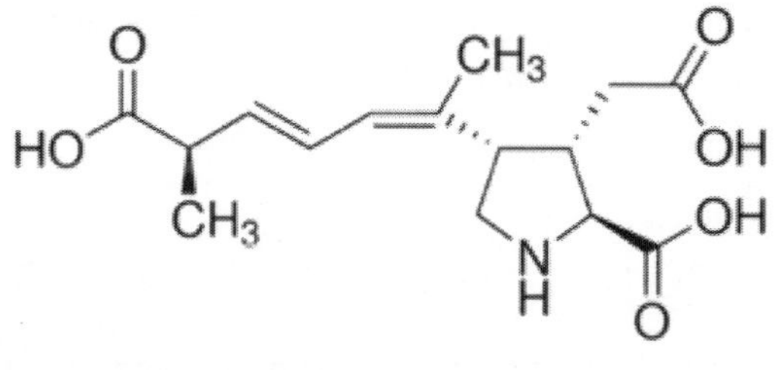

Figure 10.28: Chemical Structure of Domoic Acid.

The toxin is transferred through marine food webs and accumulates in seafood products during harmful algal blooms. Domoic acid causes amnesic shellfish poisoning

(ASP) following the consumption of mussels or scallops harvested from marine waters with a bloom of the diatom noted.

The potential for serious human illness was first discovered in 1987 when 145 people were sickened after consuming blue mussels contaminated with high levels of DA from Prince Edward Island (PEI), Canada. Acute symptoms included vomiting, abdominal cramps, diarrhoea, headache, seizures, coma, and in some cases, death. Among the sentinel cases, several survivors were left with a permanent, severe memory disorder. This remarkable symptom, combined with autopsy results indicating significant cell death in the hippocampus, a cerebral region associated with memory, led to the diagnostic label, Amnesic Shellfish Poisoning (ASP). Initially, the toxin responsible for this illness outbreak was unknown. Rapid follow-up work using rodent models led to the identification of DA as the toxin responsible for ASP and to the extrapolation of regulatory safety levels (20 ppm) for human shellfish consumption

Since this episode, domoic acid-producing diatoms have been found in other parts of the world. An ASP episode affected scallops in northwest Spain in 1996. The largest quantity of domoic acid was found in the hepatopancreas *i.e.*, from 52 per cent to 88 per cent of the total. During frozen storage, some domoic acid transferred to other parts of the scallops. It was found that canning of scallops did not destroy this toxic principal. According to Leira *et al.*, the Canadian regulatory level is 20 μg/g of tissue for fresh bivalve mollusks.

The most persistent DA producing bloom ever recorded, accompanied by the longest lasting DA levels documented in Pacific razor clams (RC, Siliqua patula) occurred during the 2015–2016 RC season, presumably triggered by warm Pacific Ocean anomalies.

The Pacific RC is a targeted vector for dietary exposure. They depurate DA very slowly and can retain the toxin for up to one year in the natural environment, or several years after being processed, canned, or frozen. To date, effective coastal and shellfish monitoring and well-enforced RC harvesting closures when levels reach 20 ppm have been protective, preventing new cases of ASP. Since the regulatory limits, however, were based upon a single, acute exposure to prevent seizures, there may be less dramatic, albeit serious effects associated with repetitive, lower-level exposures.

iv) Pfiesteria piscicida

The emergence of a recently identified toxic dinoflagellate, *Pfiesteria piscicida*, creates new human health risks in the laboratory and natural environment. Toxic *Pfiesteria* heterotrophic dinoflagellates were recognized in the early 1990s to cause kills of millions to billions of finfish and shellfish in coastal waters of the mid-Atlantic and southeastern United States. This potential for human illness was first recognized when marine scientists themselves became ill while studying the effects of this microorganism on fish health (Glasgow *et al.*, 1995, Schmechel and Koltai 2000). One investigator's symptoms were so severe that hospitalization was necessary. Similar illnesses in attenuated form were anecdotally reported in persons with extended contact with North Carolina estuary waters where Pfiesteria-associated toxicity was present (Glasgow *et al.*, 1995).

Ilness is produced by several dinoflagellate species including *Pfiesteria piscicida*, *P. shumwayae*, and perhaps several other unidentified, unnamed dinoflagellates belonging to the potentially toxic *Pfiesteria* complex. The predatory organism, mainly the species *Pfiesteria piscicida*, has also been reported from the Mediterranean Sea, the Gulf of Mexico, and the western Atlantic. It is an animal-like organism that produces potent toxins. The organism exists in several life stages. The organism remains in river and coastal bottoms for years as cysts and, when induced by unknown factors in fish faeces, the cysts bloom into a motile form that swarms to the upper waters and produces very potent toxins. This results in a "feeding frenzy," after which the organism transforms to an amoeba state that feeds on microorganisms and fish remains, followed by reformation of cysts, which settle in the coastal sediment to complete the cycle.

One toxin stuns fish within a few seconds, and the animals die within a few minutes. It is heat stable. Another toxin causes the fish epidermis to slough off. The dinoflagellate reproduces sexually after a fish kill, and it can encyst. In addition to the vicious fish kills, contact with contaminated water and aerosols has been associated with serious adverse health effects in humans. People in close proximity to *Pfiesteria* cultures in at least five different laboratories, or following direct contact with waters at fish kills, have developed ill effects ranging from narcosis, eye irritation, acute burning of skin and skin lesions, stomach cramping, respiratory distress, cognitive impairment, and memory loss lasting up to several months.

Clinical signs of toxicosis include neuropsychological symptoms such as reduced verbal and higher cognitive functions, memory loss, and headache. Skin lesions or a burning sensation of the skin on contact with water may also occur. Asthma-like effects and fatigue as well as sensory symptoms, including tingling or numbness of the lips, hands, and feet, also occur. Fish develop necrotic lesions over their surfaces. Histologic findings of *Pfiesteria* toxicoses have not been reported.

The exact identity of the toxins is unclear; *Pfiesteria* toxins have not been isolated and identified, but evidence indicates that the organism produces metal complexes that form toxic free radicals. Experiments are in progress to determine if the toxic effects in humans are caused by *Pfiesteria* solely or together with associated microorganisms. Currently, *Pfiesteria* should be considered as a cause of human illness from contaminated waters, as well as an occupational and laboratory hazard.

v) Diarrhetic Shellfish Poisoning (DSP)

Another different type of poisoning associated with shellfish which have accumulated dinoflafellates is diarrhoeal shellfish poisoning caused by lipophilic toxins such as dinophysistoxin produced by *Dinophysis fortil.* There are many derivatives of this toxin. Several *Dinophysis* species produce diarrhoetic toxins (okadaic acid and dinophysistoxins) and pectenotoxins.

It is a human intoxication caused by the consumption of shellfish that contain okadaic acid (OA) and its analogs, the dinophysistoxins (DTX1, DTX2), their diol ester precursors (DTX4 and DTX5 groups), and their acyl derivatives (DTX3 group) (okadaates, OAs herein). Okadaates are heat-stable polyether compounds and can be found in various species of shellfish, mainly bivalve molluscs. While OA and DTX2 only differ in the position of one methyl group in the molecule, DTX1 has one additional methyl group, and DTX3 (group) includes a wide range of derivatives of OA, DTX1, and DTX2, esterified with saturated and unsaturated fatty acids, products of metabolic transformations that occur in the shellfish. The acyl derivatives of OA analogs show increased liposolubility compared with the parent (unesterified) compounds, and possess toxic activity following hydrolysis in the human gastrointestinal tract.

These toxins cause gastrointestinal illness, Diarrhetic Shellfish Poisoning (DSP), even at low cell densities (<103 cells·L^{-1}). They are the main threat, in terms of days of harvesting bans, to aquaculture in Northern Japan, Chile, and Europe. Toxicity and toxin profiles are very variable more between strains than species.

Dinophysistoxin-1 (DTX1) was identified in 1982 as the substance responsible for a toxic syndrome (Diarrhetic Shellfish Poisoning, DSP), which affected more than 1600 people in Japan. This toxin is a 35-R-Methyl derivative of okadaic acid, a compound that had previously been isolated from two sponges of the genus Halichondria (*H. okadai* and *H. melanodocia*), and which since then, has been associated with numerous DSP outbreaks occurring all over the world.

The allowable levels in shellfish of this toxin and other toxins or derivatives of the same group have been regulated in many countries. In Europe and other areas, the established regulatory threshold is 160 µg OA-eq/Kg of edible product (quantified together with pectenotoxins). The major symptoms which usually occur within an hour or two of consuming toxic shellfish, include diarrhoea, abdominal pain, nausea and vomiting and may persist for several days.

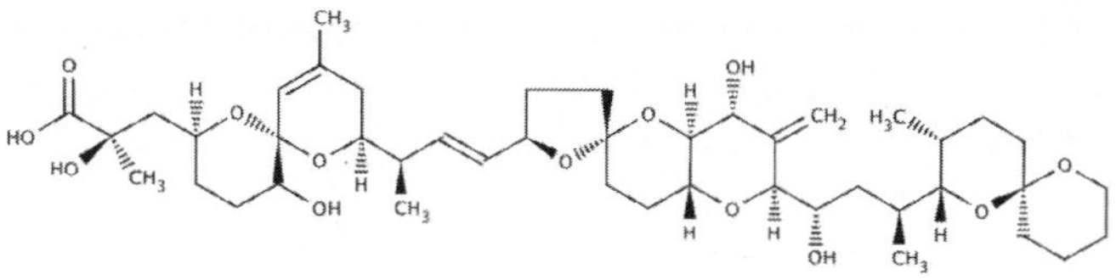

Figure 10.29: Chemical Structure of Dinophysistoxin 1.

A new marine toxin dinophysistoxin-2 (DTX-2), isolated from toxic Irish mussels and biogenetically related to the toxins okadaic acid 1 and dinophysistoxin-1 (DTX-1)2, the principal agents responsible for diarrhetic shellfish poisoning (DSP), has been reported.

Field observations and laboratory experiments have shown that most of the toxins produced by Dinophysis are released into the medium, raising questions about the ecological role of extracellular toxins and their potential uptake by shellfish. Shellfish contamination results from a complex balance between food selection, adsorption, species-specific enzymatic transformations, and allometric processes. Highest risk areas are those combining Dinophysis strains with high cell content of okadaates, aquaculture with predominance of mytilids (good accumulators of toxins), and consumers who frequently include mussels in their diet.

D. Toxigenic Fungi

Now let us try to learn more things about fungi which cause more toxic effects.

The mycotoxins are metabolites of moulds which may contaminate foods, animal feeds, or the raw materials for their manufacture and that happen to be toxic to humans or their domestic animals. The study of mycotoxins and the legislation associated with their control, are based on them being considered as adulterants of foods or animal feeds.

On the other hand, mushrooms and toadstools have provided a traditional source of food in many parts of the world for many thousands of years. Unfortunately, this group of fungi includes a number of species which produce toxic metabolites in their fruiting bodies but, because the toxins are a natural constituent of fruiting bodies deliberately ingested, usually as a result of mistaken identity they are not considered as mycotoxins. This is a somewhat arbitrary distinction based on human behaviour and not on the chemistry, biochemistry of toxicology of the compounds.

There are relatively few species of agarics which can be considered as deadly poisonous but they include the deathcap, *Amanita phalloides,* a quarter of a cap of which can be lethal to a healthy adult and species of *Cortinarius* which are still foolishly mistaken for edible wild fungi. In both these cases the toxins cause irreversible damage to the liver and kidneys and death may follow several weeks after the initial consumption of the poisonous fungi.

It is worth emphasizing a major difference between the toxic metabolites of fungi and the toxins of most of the bacteria associated with food poisoning. The former are relatively low molecular weight compounds, although their chemistry may be very complex, while the latter macromolecules such as polypeptides, proteins or lipopolysaccharides. An exception to this generalization is an unusual bacterial food poisoning associated with a traditional food produced in parts of Indonesia; a form of tempeh is made by inoculating coconut flesh with moulds such as *Rhizopus* and *Mucor.* Occasionally the process becomes contaminated with the bacterium *Burkholderia cocovenans,* previously known as *Pseudomonas cocovenenans,* which produces at least two low molecular weight toxic metabolites, bongkrekic acid and toxoflavin.

Plant-fungal interactions can be complex and there are instances where a toxic plant metabolite is produced in response to fungal attack. Thus, when the sweet potato, *Ipomoea,* is damaged by certain plant pathogens it responds by producing the phytoalexin ipomearanone. This antifungal agent is produced to limit fungal attack but it is also an hepatotoxin to mammals. Further complexity arises when other moulds, such as *Fusarium solani,* degrade ipomeamarone to smaller molecules such as ipomenol which can cause oedema of the lung.

Mushroom Poisoning

Mushrooms have been used as components of human diet and many ancient documents written in oriental countries have already described the medicinal properties of fungal species. Some mushrooms are known because of their nutritional and therapeutical properties and all over the world, some species are known because of their toxicity that causes fatal accidents every year mainly due to misidentification.

Approximately 140.000 species of mushrooms have already been catalogued all over the world, about 2.000 being considered safe for human consumption and about 700 have therapeutic properties. A great variety of species was classified as poisonous and represents risks to health if ingested. Apart from mushrooms that contain psychoactive toxins, ingestion of toxic mushrooms is invariably accidental and caused by misidentification of species. There are some mushrooms that contain exceptionally powerful toxins that represent a real hazard to health even when ingested in small doses. Most toxins were well studied and are described in literature. A wide variety of toxic mushrooms belong to different genus. Mushroom toxins have been divided into seven main categories: amatoxins (cyclopeptides), Orellana's (Cortinarius species), gyromitrin (monomethylhydrazine), muscarine, ibotenic acid, psilocybin,and coprine. Amatoxins that are cytotoxic and cause harm to kidney and liver and orellanine that is nephrotoxic.

Ostreolysin is a cytolytic protein that was isolated from mushrooms of the genus *Pleurotus* that was able to cause cytolytic pore formation when administered by intravenous route to rats. As a consequence, it was observed blood pressure increase, cardiac ischemia, tachycardia, hypoxia and elevated serum potassium. The administration of the mushroom *Phellinuslinteus* to rats bearing experimentally induced prostatic hyperplasia leads to an enlargement of prostate stroma which is involved in transforming growth factor-beta1(TGF-β1) regulation. The prostate is known to be regulated by various growth factors. Among them the TGFs have been reported to play important role in prostate cell growth regulation. The administration of *P. linteus* increased the expression of TGF-β1compared to animals treated with placebo.

A particular syndrome that affected five people in the region of Savoie in France was later identified as intoxication caused by the mushroom *Clitocybe amoenolens.* First symptoms appeared 24 hours after ingestion. Patients presented paresthesia of the toes and fingers followed by paroxysmal burning pain lasting 2-3 hours, notably at night. A sensation of heat, numbness, oedema and local erythema are associated with crises. Symptoms are partially relieved with cold water, acetylsalicylic acid, morphine and clomipramine. Recovery is completely after 1-4 months. The administration of high dose of *C. amoenolens* in rats caused weight loss, locomotor disability and erythema of the toes. Examination of the sciatic nerves showed decreased axon density and neuronal fiber degeneration.

A. muscaria and *A. pantherina* grow in North America, Europe, Africa and Japan, in recent years it has been reported that young people in several countries have intentionally eaten *A. muscaria* to evoke hallucinations. The most common symptoms of intoxication are motor depression, ataxia, changes in mood, perception and feelings, dizziness, euphoria, drowsiness, gastrointestinal disturbances and muscle twitches.

The genus *Cortinarius* comprises between 2,000-3,000 species of mushrooms that were considered as non-toxic until 1950. One hundred-and-thirty-five cases of intoxication caused by *C. orellanus* were described from 1953-1962 in Poland. Poisoning syndrome is characterized by a delayed acute tubulopathy that can progress to chronic renal insufficiency.

The species *Pleucybella porrigens* popularly known as Sugihiratake is a white mushroom widely distributed in the mountain areas of Japan and is commonly used as ingredient to various processed foods, but this was pointed out as hazardous due to the presence of substances analogous to vitamin D that are able to cause cryptogenic encephalopathy in patients with renal failure.

Agaricus bisporus is the most consumed mushroom world-wide but is has been pointed out as potentially carcinogenic due to the substantial amounts of aromatic hydrazines, an established class of direct-acting chemical carcinogens. Life-time administration of *A. bisporus* raw or baked to mice three days a week followed by balanced semi-synthetic diet for the remaining days, induced tumours in a number of tissues.

In some countries, many cases of mushroom poisoning are reported every year. In the year 1998 in France 1,675 cases of intoxications by mushrooms were reported and, in this country, alone it is estimated that 8-10.000 cases are expected to be registered every year. Most of these accidents are due to incorrect identification of species that is often made by empirical and traditional knowledge.

Species of genus *Tricholoma* especially *T. equestre* (*T. flavovirens*), known as yellow tricholoma, has been implicated in 12 human poisonings causing a delayed rhabdomyolysis severe enough to be fatal in 3 cases reported in France. The symptoms were muscular weakness, fatigue and myalgias within 24-72 hours after ingestion. The substance responsible for toxic effects was not identified.

Species of genus *Tricholoma* especially *T. equestre* (*T. flavovirens*), known as yellow tricholoma, has been implicated in 12 human poisonings causing a delayed rhabdomyolysis severe enough to be fatal in 3 cases reported in France. The symptoms were muscular weakness, fatigue and myalgias within 24-72 hours after ingestion. The substance responsible for toxic effects was not identified.

Tricholoma equestre. a wild mushroom is considered in Europe as a delicacy. Toxicity is observed after a consumption of considerable amounts of fresh mushroom which ranges from 100 to 400 g at 3 to 9 consecutive meals.

A neurological syndrome appears after the ingestion of the specie *Hapalopilus rutilans* that is considered edible. Common symptoms consist of visual disturbances, somnolence, hypotonia and hepatic and renal insufficiency. Hepatic cytolysis and renal insufficiency were described in children.

In countries where mushrooms are highly consumed, a number of intoxications are reported every year mainly due to misidentification of species. Hazardous toxins are present in these species and are able to cause different syndromes that can be fatal depending on the amount ingested. Accidental ingestion of mushrooms is difficult to avoid especially in countries where eating wild species is common. Proper identification is important to avoid accidents and the identification of symptoms and signs of intoxication as soon as possible enables the success of treatment.

4. Other Pathogenic Agents

Prion Diseases

The term "prions" refers to abnormal, pathogenic agents that are transmissible and are able to induce abnormal folding of specific normal cellular proteins called prion proteins that are found most abundantly in the brain. The functions of these normal prion proteins are still not completely understood. The abnormal folding of the prion proteins leads to brain damage and the characteristic signs and symptoms of the disease. Prion diseases are usually rapidly progressive and always fatal.

The prion is also defined as the infectious agent that causes transmissible spongiform encephalopathies (TSEs), or prion diseases – transmissible degenerative disorders of the central nervous system. Hence, prions are unique proteins in that they can convert other proteins into damaging ones by causing them to alter their shape. The normal cell prion protein (PrP) exists in the brain cell membrane where it carries out some vital functions and is then degraded by proteases. However, the pathogenic form is distorted and is resistant to proteases, and thus, it accumulates in brain tissue and gives rise to disease.

It has been postulated that the distorted prion molecule, acting as a template, converts normal protein to a distorted form. The normal protein (α-helical form) takes on a protease-resistant β-flat form (PrPSc, PrPres) when it becomes pathogenic. The pathogenic forms tend to aggregate into amyloid fibrils where they cause nerve cell degeneration, which leads to clinical signs of disease.

Prions cause the disease scrapie in sheep, goats, and hamsters; and kuru in humans. The prion agent of each prion disease is named after the disease itself [*e.g.*, Creutzfeldt–Jakob disease (CJD) agent in CJD of human, chronic wasting disease (CWD) agent in CWD of cervids, scrapie agent in scrapie of sheep and goats, and bovine spongiform encephalopathy (BSE) agent in BSE of cattle]. All of these belong to a family of diseases called transmissible spongiform encephalopathies (TSEs).

Although the evidence of a prion etiology of these diseases seems strong, the possibility that a virus is the agent has been raised. These particles were named around 1982 by Stanley Prusiner, who was awarded the 1997 Nobel Prize in physiology for his pioneering work.

i) Creutzfeldt-Jakob Diseases (CJD, vCJD)

The term Creutzfeld-Jakob disease (CJD) was first used in 1922 by Spielmeyer to name the condition described by 2 German physicians. The 6 original subjects (a case report by Creutzfeld and a series of 5 patients by Jakob) were a heterogeneous group of patients with unusual neuropathological findings associated with other disorders, and years later only 2 of the cases were confirmed with modern techniques to actually be CJD. In

1960 the typical clinical picture, electroencephalography (EEG) findings, and classic spongiform changes in the neuropil were recognized as the cardinal features of the disease.

Creutzfeldt-Jakob disease is the most common human form of a group of rare, fatal brain disorders affecting humans and certain other mammals. These disorders are known as prion diseases. They occur when prion protein, which is found throughout the brain and body but whose normal function isn't yet known, begins folding into an abnormal three-dimensional shape. This shape change gradually triggers prion protein in the brain to fold into the same abnormal shape. Through a process scientists don't yet understand, misfolded prion protein destroys brain cells. The damage leads to rapid decline in thinking and reasoning, as well as involuntary muscle movements, confusion, difficulty walking and mood changes.

Since humans are susceptible to the prions that cause CJD, the early concern was whether humans could contract BSE from cattle. Variant CJD (vCJD), first recognized in the United Kingdom in 1996 has been linked to bovine spongiform encephalopathy, (BSE). This new variant of CJD (nvCJD, vCJD) was reported in the United Kingdom in a small group of people, all of whom were much younger than most individuals with CJD. This prompted speculation that vCJD had been contracted from cattle. Normally, CJD appears in persons around age 60 or older, but in the U.K., vCJD was found to afflict individuals in the late teens to the early 40s. It has been concluded that vCJD is the human equivalent of BSE, and the agents for BSE and vCJD appear to be the same based on studies using mice.

Through 2003, around 150 vCJD cases were seen in humans throughout Europe with about one-third in the United Kingdom. For the 5-year period 1991–1995, 94 CJD deaths were recorded in the United States and 9 were below age 55. None conformed to vCJD. Over 85 per cent of CJD patients in the United States die within one year of onset. Between 1991 and 1995, the average annual death rate from CJD in the United States was 1.2/1 million population. BSE is thought to have been contracted by cattle through specified bovine offal (that contained brain, spinal cord, and the like) from infected animals, a practice that was banned in 1989. Using a mouse assay, prions could not be detected in beef muscle and milk from infected cattle. The incubation period for BSE is between 1 and 15 years.

CJD causes a type of dementia that worsens unusually fast. More common causes of dementia, such as Alzheimer's disease, dementia with Lewy bodies and frontotemporal dementia, typically progress more slowly. An especially rare form of CJD can sometimes be transmitted like an infection from one person to another, from one animal to another or from certain animals to people. As a result of this infectious capability, prion diseases are also known as "transmissible spongiform encephalopathies."

CJD progresses rapidly and is fatal. Those affected lose their ability to move or speak and require full-time care to meet their daily needs. An estimated 70 per cent of those diagnosed with CJD die within one year. Those affected by familial CJD tend to develop the disorder at an earlier age and survive somewhat longer than those with the sporadic form, as do those diagnosed with vCJD. Scientists have not yet learned the reason for these differences in survival.

CJD falls into four categories: sporadic, familial, iatrogenic and variant. Sporadic CJD (sCJD), thought to occur worldwide, is the most common form and represents about 85 per cent of all CJD cases; incidence ranges between 0.5 and 1.5 cases per million inhabitants per year. Sporadic CJD has no known cause. Most scientists believe the disease begins when prion protein somewhere in the brain spontaneously misfolds, triggering a "domino effect" that misfolds prion protein throughout the brain. Genetic variation in the prion protein gene at a location called "codon 129" may increase risk of this spontaneous misfolding.

Variation at codon 129 in the prion protein gene may also play a role in making people susceptible to acquired CJD from external sources. Scientists don't yet know why acquired CJD seems to be transmitted through such a limited number of external sources. Researchers have found no evidence that the abnormal protein is commonly transmitted through sexual activity or blood transfusions, although a few cases of vCJD seem to have been spread through blood transfusions. Professionals who regularly encounter blood from a human or animal, such as surgeons, pathologists or butchers, have not been shown to have a higher-than-normal risk through occupational exposure.

Familial CJD is caused by variations in the prion protein gene that increase the likelihood an individual will develop CJD. Researchers have identified more than 50 prion protein mutations in those with inherited CJD. Genetic testing can determine whether family members at risk have inherited a CJD-causing mutation. Experts strongly recommend professional genetic counselling both before and after genetic testing for hereditary CJD.

Age has an influence on sporadic CJD, which tends to develop later in life, usually around age 60. Onset of familial CJD occurs slightly earlier and vCJD has affected people at a much younger age, usually in their late 20s.

Creutzfeld-Jakob disease can be acquired iatrogenically; several outbreaks over the last few decades were associated with intracerebral electrodes, corneal transplantation, dura mater grafts, and growth hormone injections. The first case report, in 1974, was in a patient who received a corneal transplant from an infected cadaver. Dura mater grafts with contaminated material led to more than 60 cases of CJD, with incubation periods lasting between 1 and 14 years. Pooled cadaveric growth hormone was used for injections and led to CJD 5–30 years after the injections. The clinical symptoms are similar to sCJD, as are the MRI and EEG findings. Iatrogenic CJD also has a clinical picture that is similar to sCJD.

Chronic wasting disease is a prion disease similar to mad cow disease that's been found in wild deer, elk and moose in certain U.S. states, Canadian provinces, Korea and Norway. According to the U.S. Centers for Disease Control and Prevention (CDC), there's no evidence to date that chronic wasting disease has been transmitted to humans, including hunters who eat meat from affected animals. There's also no evidence that rates of CJD have increased in states or provinces where chronic wasting disease has been identified. Additional studies are under way to understand what risk, if any, chronic wasting disease poses to humans. The CDC recommends that hunters who plan to eat meat from deer, elk or moose in areas where chronic wasting disease occurs consider having the meat tested by their local state wildlife agency. The CDC also recommends wearing gloves while field dressing these animals to avoid handling the brain or spinal column.

There is no treatment to slow or stop the underlying brain cell destruction caused by CJD and other prion diseases. Various drugs have been tested but have not shown any benefit. Clinical studies of potential CJD treatments are complicated by the rarity of the disease and its rapid progression. Current therapies focus on treating symptoms and supporting individuals and families coping with CJD. Doctors may prescribe painkillers such as opiates to treat pain if it occurs. Muscle stiffness and twitching may be treated with muscle-relaxing medications or antiseizure drugs. In the later stages of the disease, individuals with CJD become completely dependent on others for their daily needs and comfort.

Creutzfeldt-Jakob disease is a fatal neurological condition. That often perplexes the treating physician. Given its rarity, published diagnostic criteria aid in the approach to diagnosis, but these are unfortunately outdated and do not incorporate modern research laboratory methods. The proposed updated diagnostic criteria and treatment algorithm are based on the latest research, and help guide the workup in a disease that can be difficult to diagnose.

ii) Chronic Wasting Disease (CWD)

Chronic wasting disease (CWD) is a prion disease of free-ranging and farmed ungulates (deer, elk, and moose) in North America and South Korea. First described by the late E.S. Williams and colleagues in northern Colorado and southern Wyoming in the 1970s, CWD has increased tremendously both in numerical and geographical distribution, reaching prevalence rates as high as 50 per cent in free-ranging and >90 per cent in captive deer herds in certain areas of USA and Canada.

Chronic wasting disease belongs to a family of diseases known as transmissible spongiform encephalopathies (TSE); a group of neurodegenerative disorders caused by prions. Prions are infectious proteins that appear to replicate by converting a normal cellular protein into copies of the prion. The cellular protein, which is called PrPc, is found on the surface of neurons. Pathogenic isoforms of PrPc are designated PrPres; PrPCWD or PrPTSE are other names for this protein. There may be more than one variant of the CWD prion.

Chronic wasting disease (CWD) is fatal to free-ranging and captive animals in Cervidae (the deer family; referred to as "cervids"). Affected animals include some Odocoileus, Cervus, and Muntiacus species (deer), Cervus canadensis (elk), Alces alces (moose), and Rangifer tarandus (reindeer). Once an animal is infected, CWD typically causes neurological damage that grows more severe until the host animal dies.

Originally identified in 1967, to date (early 2018), the disease has been found in 24 States as well as Canada, South Korea, and Norway. It continues to spread across North America through new and ongoing outbreaks. It has been seen on elk farms in other states. In addition, the proportion of CWD-infected animals is increasing in many areas where the disease is already established. For example, the proportion of adult male and female Odocoileus virginianus (white-tailed deer) in north-central Iowa County, Wisconsin, that tested positive for CWD has more than doubled during a 6-year period (2011–16). Approximately 40–50 per cent of the adult male population and 20-30 per cent of the adult female population was infected by 2016, making north-central Iowa County one of the most heavily affected regions in the United States (Wisconsin Department of Natural Resources, 2017).

CWD is certainly the most contagious prion infection, with significant horizontal transmission of infectious prions by, *e.g.*, urine, faeces, and saliva. Grazing is thought to be important in acquiring these prions from the environment. In deer, prions have been detected in saliva, blood, urine, faeces and antler velvet, and some sources (*e.g.*, saliva, faeces, urine) can contain these agents before the animal develops clinical signs. While the concentration of prions in urine and faeces is very low, the volume of these excretions could contribute significantly to environmental contamination over the course of the infection. Elk, which have smaller amounts of prions in lymphoid tissues than deer, are thought to transmit CWD less efficiently. Nevertheless, these agents have been found in some secretions and excretions of elk, including faeces. Prions have also been detected in the skeletal muscles and fat of deer, and in heart muscle from white-tailed deer, elk and experimentally infected red deer. The occurrence of prions in blood and lymphoid tissues suggests that no tissues from infected cervids should be considered prion free. Vertical transmission has, to date, been documented in the offspring of wild elk, as well as in experimentally infected Reeves' muntjac deer. Whether CWD prions can be shed in milk is unknown.

CWD has a long incubation period and clinical signs may not appear for two to four years after infection. Infectious prions are shed before an animal show symptoms. The minimum incubation period is thought to be approximately 16 months, and the average incubation period is probably 2 to 4 years. The peak incidence occurs between the ages of 2 and 7 years. Much longer incubation periods may also be possible; in herds where CWD is endemic, cases have been reported in animals that were more than 15 years old.

Chronic wasting disease is always fatal. A few apparently asymptomatic animals, or animals with mild clinical signs (*e.g.*, mild ataxia, weight loss), may die suddenly after handling. Such atypical presentations have been reported to be more common in certain populations, such as mule deer with a particular genotype. More typically, cervids develop progressive weight loss, lassitude and behavioral changes over several weeks to months, with many animals becoming severely emaciated before they die. Ataxia, head tremors, teeth grinding, repetitive walking of the enclosure's perimeter, hyperexcitability when handled, or other neurologic signs may be seen; however, neurologic signs and behavioral changes are sometimes subtle, particularly early in the course of the disease or in species such as elk. Some affected animals carry their head low and have a fixed gaze, particularly in the late stages of disease; this can alternate with more normal alertness. Some animals also have difficulty swallowing, leading to excessive salivation. Animals may develop esophageal dilation and regurgitation, or aspiration pneumonia leading to death. Polydipsia/polyuria and syncope may occur late in the disease. Pruritus has not been reported in affected cervids; however, the coat may be rough and dry, with patchy retention of the winter coat in summer. Reproductive losses, including stillbirths and the death of offspring soon after birth, have been reported in experimentally infected muntjac deer. Most animals with clinical signs die within a few months, although a few may live for up to a year or more. Occasionally, the disease may last only a few days, particularly in elk.

Dissemination and persistence of infectivity in the environment combined with the appearance in wild-living and migrating animals make CWD presently uncontrollable, and pose extreme challenges to wild-life disease management. Whereas CWD is extremely transmissible among cervids, its trans-species transmission seems to be restricted, although the possible involvement of rodent and carnivore species in environmental transmission has not been fully evaluated. Whether or not CWD has zoonotic potential as had Bovine spongiform encephalopathy (BSE) has yet to be answered.

iii) Bovine Spongiform Encephalopathy (BSE)

Bovine spongiform encephalopathy (BSE) ("mad cow disease") is a fatal neurological disease of adult cattle that was first recognised in Great Britain (GB) in 1986. Four years later, over 14,000 confirmed cases out of a population of 10 million cattle had been recognized in Great Britain. The epidemic seemed to peak around 1,000 new cases per week in 1993. By February 1998, a total of 172,324 cases were seen in cattle in the United Kingdom. A total 738 of 600 cases were recorded in 8 countries outside the U.K. with 256 (42.7 per cent) in Switzerland. Since 1996, 4.5 million cows have been destroyed. Between 1986 and November 2003, 183,634 cattle were diagnosed with BSE in the U.K. with 4,469 in 22 other countries. BSE has also spread to many European countries, North America, parts of Asia and possibly other areas of the world.

Cases of classical BSE have been reported in indigenous cattle in some European countries, Canada, Israel and Japan. Some of these countries may have eradicated this disease, as it has not been reported in some time. Classical BSE was documented only in imported cattle in some nations, including the U.S., the Falkland Islands and Oman. Other countries, such as Iceland, Australia and New Zealand, seem to have remained completely free of classical BSE. The presence or absence of this disease cannot be determined in countries without adequate surveillance programs. Atypical BSE prions have been reported in Europe, the U.S., Canada, Japan and Brazil, as the result of surveillance programs for BSE. They are also likely to exist in other countries.

Bovine spongiform encephalopathy (BSE) is a fatal neurodegenerative disease, caused by a prion, that mainly affects cattle. BSE mainly occurs in cattle, but the host range of this prion is unusually broad compared to most prions. Rare clinical cases have been reported from goats; exotic ruminants in zoos, including nyala (*Tragelaphus angasi*), kudu (*Tr. strepsiceros*), gemsbok (*Oryx gazella*), eland (*Taurotragus oryx*), Arabian oryx (*O. leucoryx*), scimitar-horned oryx (*O. dammah*), ankole cattle and North American bison (*Bison bison*); various felids including housecats, cheetahs (*Acinonyx jubatus*), pumas (*Felis concolor*), ocelots (*F. pardalis*), tigers (*Panthera tigris*) and Asian golden cats (*Catopuma temminckii*); and captive lemurs, which were apparently infected in contaminated feed. Sheep become ill after experimental inoculation, but no naturally acquired cases have been reported in this species. European red deer (*Cervus elaphus elaphus*) can also develop clinical signs if they are fed a high dose of prions; however, this species does not seem too easy to infect, as only one of 6 orally inoculated red deer became infected. Mink (*Mustela vison*) and cynomolgus macaques (*Macaca fascicularis*) are also susceptible to oral inoculation.

Bovine spongiform encephalopathy (BSE) is a progressive neurological disease of cattle caused by prions, which are infectious agents made up of protein material. The prion proteins affect the brain structure of infected animals, causing the animal to lose motor skills and eventually causing death. Clinical signs include behavioural changes, coordination problems, weight loss and decreased milk production. Once clinical signs appear, the animal's condition deteriorates until it either dies or is euthanized. There is no treatment or vaccine to prevent BSE. There are two general types of BSE - classical and atypical. Neither form is contagious.

BSE is usually transmitted when an animal or human ingests tissues containing the BSE prion. Young animals may be particularly susceptible: some studies suggest that most cattle become infected with BSE during the first six months of life. Sheep are, likewise, most susceptible to experimental (oral) inoculation during the first few months of life, especially during the first few weeks. In cattle, the prions are thought to replicate initially in the Peyer's patches of the ileum, then are transported via the peripheral nerves to the central nervous system (CNS). The highest prion concentrations occur in the CNS (both the brain and spinal cord) and in the ileum. However, very sensitive detection methods have also found this agent in lymphoid tissues associated with the jejunum and colon, various nerve ganglia, peripheral nerves and adrenal glands, and in the optic nerve and retina. The accumulation of BSE in peripheral nerves, nerve ganglia and adrenal gland seems to coincide with or follow prion accumulation in the CNS.

The incubation period for classical BSE is estimated to be 2 to 8 years in cattle, and might be longer than a decade in a few instances. Published incubation periods in sheep fed BSE prions have ranged from approximately 1.5 years to more than 6 years. Other reported incubation periods in animals, after oral inoculation, are 4 years, 9 months in one European red deer, 15 months in mink and several years in experimentally infected macaques.

The clinical signs may include gait abnormalities (particularly hindlimb ataxia) and difficulty negotiating obstacles, low carriage of the head, hyperresponsiveness to stimuli, tremors and behavioral changes such as aggression, nervousness or apprehension, changes in temperament, and even frenzy. A combination of behavioral changes, hyperreactivity to stimuli, and gait abnormalities is highly suggestive of BSE, but some animals exhibit only one category of neurological signs. Behavioral signs are often noted initially, and reluctance to be milked is reported to be a common early sign in dairy cattle. Pacing, a modified gait in which the legs move in lateral pairs, occurred in 25 per cent of the cattle with BSE in one study, and may be suggestive of this disease. Intense pruritus is not usually seen in cattle, but some animals may lick or rub persistently. Nonspecific signs include loss of condition, weight loss, teeth grinding (possibly due to visceral pain or neurological disease) and decreased milk production. Decreased rumination, bradycardia and altered heart rhythms have also been reported. The signs of BSE usually worsen gradually over a few weeks to several months, but rare cases can develop acutely and progress rapidly. Rapid, acute onset neurological disease seems to be particularly common in exotic ruminants in zoos. Once clinical signs appear, BSE is always progressive and fatal. The final stages are characterized by recumbency, coma and death.

Prion agents are highly resistant to various physical and chemical treatments. Normal sterilization procedures are ineffective for the inactivation of prions. Tis resistance to inactivation is at least in part due to the absence of nucleic acid in prions. Prions are mainly composed of protein and lack a genome, which gives them a high level of resistance to a range of physical and chemical treatments. Conventional sterilization procedures, such as exposure to ultraviolet (UV) or γ-ray irradiation as well as autoclaving at 121°C for 20 minutes, fail to inactivate prions. Likewise, treatment with alcohols, such as 70 per cent ethanol, has no effect on the prion agent. Fixation using formaldehyde slightly reduces the infectivity of prions but this is insufficient to eliminate the risk of transmission. Prions can bind tightly to some surfaces, including stainless steel and plastic, without losing infectivity. Prions bound to metal seem to be highly resistant to decontamination.

Control measures, including restrictions on ruminant feed, have now greatly decreased its prevalence, and cases have become uncommon or rare in many areas. Many countries have also passed new regulations to prevent BSE-containing tissues from entering human food supplies. As a result of increased surveillance, BSE prions that differ from the prion causing 'classical' BSE have been identified at very low levels in cattle populations. The leading hypothesis, at present, is that these atypical prions arise spontaneously in cattle. Some experiments suggest that an atypical prion might have given rise to the BSE epizootic when it was amplified in cattle feed.

Testing for prion proteins consists of an immunohistochemistry method (considered to be the gold standard), a conformation-dependent immunoassay (CDI) method (developed in 2003), a screening test developed by the Bio-Rad Corporation (TeSeE), another method developed by the IDEXX Laboratories (HerdchekR).

Some countries test cattle, and sometimes sheep and goats, at slaughter to detect cases of BSE. In most cases, testing in healthy cattle and/or small ruminants intended for human consumption is targeted at animals over a certain age. At one time, Japan tested all cattle, regardless of age. Testing requirements for high-risk animals (*e.g.*, non-ambulatory cattle or those with neurological signs) are usually more stringent. Some countries only (or mainly) test high risk animals. As the prevalence of BSE has decreased, even nations that once tested most or all cattle have scaled back their testing requirements.

5. Epidemiology and Detection for Control of Viral and Parasitic Food Borne Disease

Human enteric viruses and protozoa are parasitic agents that replicate in the intestines of infected hosts and are excreted in the faeces. In general, the viruses are limited to human hosts, while the parasitic agents (in the form of cysts or oocytes) have a variety of human and nonhuman animal hosts. Both are transmitted primarily by the faecal-oral route, and as a result, the major source of contamination for foods and water is through contact with human and animal faecal pollution. This contamination may occur directly, through contaminated meat carcasses or poor personal hygiene practices of infected food handlers, or indirectly, through contact with faecally contaminated water or other cross-contamination routes.

Epidemiology

Human Enteric Viruses

Human enteric viruses are increasingly recognized as important causes of food borne illness. A recent report issued by the Council for Agricultural Science and Technology ranked human enteric viruses as fifth and sixth among identified causes of food borne disease in the United States (Council for Agri. Sci. and Technol., 1994). The most common types of food borne viral disease are infectious hepatitis due to hepatitis A virus and acute viral gastroenteritis associated with the Norwalk agent and other related small, round-structured gastrointestinal viruses.

The human enteroviruses may also be transmitted by foodborne routes and are the most commonly isolated agents in surveys of naturally occurring viral contamination in foods. Foodborne outbreaks due to small round viruses, parvoviruses, and astroviruses are occasionally reported. Rotaviruses, some adenoviruses, and hepatitis E virus are important causes of waterborne disease outbreaks, particularly in developing countries. Foodborne outbreaks associated with human enteric viruses are almost always due to the consumption of faecally contaminated raw or undercooked shellfish and ready-to-eat products contaminated by infected food handlers.

In a current review, Fleet and co-workers at the University of New South Wales (*International Journal of Food Microbiology* 59 2000 127) pointed out that in recent years, Norwalk-like virus (NLV) contamination of oysters and orange juice and hepatitis A virus contamination of oysters have been responsible for large outbreaks of food borne viral disease in Australia. They also drew attention to the fact that gastroenteritis caused by rotavirus is annually responsible for the hospitalisation of about 12,000 children under the age of five years. Outbreaks of this disease are primarily associated with child care centres, pre-schools and other situations where there are large numbers of children. Outbreaks in adults, especially in nursing homes, can also occur. Person to person contact is considered to be the principal mode of spread of the virus in these situations but the initial involvement of faecally contaminated foods or water is often suspected.

Parasitic Protozoa

Since 1981, enteric protozoa have become the leading cause of waterborne disease outbreaks for which an etiologic agent could be determined. Large communitywide waterborne outbreaks of parasitic protozoa are usually associated with surface water supplies that are either unfiltered or subjected to inadequate flocculation and filtration.

The potential for transmission of these agents by foodborne routes is also being increasingly recognized. For instance, from 1988 to 1992, seven food-associated outbreaks of giardiasis, comprising 184 cases, were reported in the United States (Bean *et al.,* 1996). However, since foodborne transmission is only recently documented and more than half of all reported foodborne disease outbreaks have undetermined etiology, the true importance of foodborne transmission of parasitic protozoa is unknown.

The most common human enteric parasitic infections are caused by *Cryptosporidium parvum* and *Giardia lamblia*. *Cyclospora* is also an emerging enteric protozoon that has recently been associated with the consumption of contaminated fruit.

Limitations

Most of the information about viral and parasitic food and waterborne disease comes from outbreak investigations by state and local health departments. However, since many of these diseases are not reportable and surveillance is based on voluntary reporting by state health departments, the magnitude of this disease problem is underestimated. This is exacerbated by a reluctance to use epidemiologic criteria in the classification of food borne viral disease and the failure to report mild outbreaks of gastrointestinal disease. Furthermore, since investigation generally follows an outbreak, important information and samples may have been destroyed, consumed, or lost to inaccurate recall. Since many of these outbreaks are small and confined, epidemiologic investigation may be limited by the resources available to state and local health departments. Difficulties in investigating and reporting are further complicated by the fact that the enteric protozoa cause common opportunistic infections in the immuno-compromised, and the role of foods in these diseases has not been studied. Likewise, the role of the food borne transmission route in sporadic disease and the importance of carrier states and secondary illness after a primary food borne disease outbreak are poorly characterized.

Detection

Clinical Samples

Illness caused by human enteric viruses can be suspected epidemiologically by considering incubation period and illness duration analysis, classic viral gastroenteritis symptoms, and the absence of bacterial or parasitic pathogens in stool samples. Laboratory confirmation of human enteric viral infection has been based on a rise in specific antibody to the virus, or alternatively, the demonstration of virus particles, antigen, or nucleic acid in stools. The detection methods most often applied to clinical samples have included immune electron microscopy, radioimmunoassay, and enzyme immunoassay. The usefulness of these assays has been reduced by low detection limits (>10_4-10_5 particles/ml) and the inability to cultivate Norwalk-like viruses in vitro, which has limited the supply of viral antigen available for developing reagents. In addition, the Norwalk agent is only one of several small round-structured gastrointestinal viruses that cause outbreaks with similar clinical and epidemiologic features.

Before 1981, parasitic disease in humans was diagnosed histologically by identifying the life cycle stages of parasitic agents in the intestinal mucosa. More recently, clinical diagnosis has involved methods to concentrate parasitic agents from stool specimens followed by a variety of fluorescent or immunofluorescent staining techniques and subsequent microscopic examination. Serodiagnostic methods have been developed, but additional evaluations are needed to confirm the diagnostic utility of these methods.

Environmental, Food, and Water Samples

Failure to confirm a viral and parasitic etiology in food borne outbreaks has also been due to the lack of adequate methods to detect the causative agents in environmental samples. Like bacterial pathogens in food and water, viruses and parasites are frequently present in small numbers. However, unlike traditional food microbiologic techniques, which have relied on cultural enrichment and selective plating to increase cell numbers and differentiate pathogens in background microflora, techniques to detect human enteric viruses and parasitic protozoa require live mammalian cells for growth. For this reason, standard methods to detect enteric bacteria in foods cannot be used; instead, detection requires an initial concentration step, often from large volumes of food or water, followed by mammalian cell culture infectivity assay or immunofluorescent staining. Concentration methods are usually cumbersome, and yields are less than optimal. Both cell culture infectivity and immunofluorescent staining are expensive and slow and require highly trained personnel. Furthermore, mammalian cell culture lines are largely unavailable for the epidemiologically important food borne viruses. Alternative detection methods based on immunologic and molecular methods have been reported; recent methodologic developments have focused on overcoming barriers to detection, such as improving recovery efficiencies and detection limits and preventing inhibitions due to food-related compounds.

Human Enteric Viruses in Foods

Concentration

Two general schemes for the concentration of human enteric viruses from foods have been reported: extraction-concentration methods and adsorption-elution-concentration methods (Figure 10.30). The general purpose of concentration is to provide a high recovery of infectious virus in a low-volume aqueous solution free of cytotoxic materials. Both schemes employ conditions favouring the separation of viruses from shellfish tissues (most of the developmental work has used shellfish as model food commodities), primarily through the use of filtration, precipitation, polyelectrolyte flocculation, and solvent extraction. While either method can be used, adsorption-elution-precipitation methods have been favoured in recent years. Virus yields after concentrations are 10 per cent to 90 per cent.

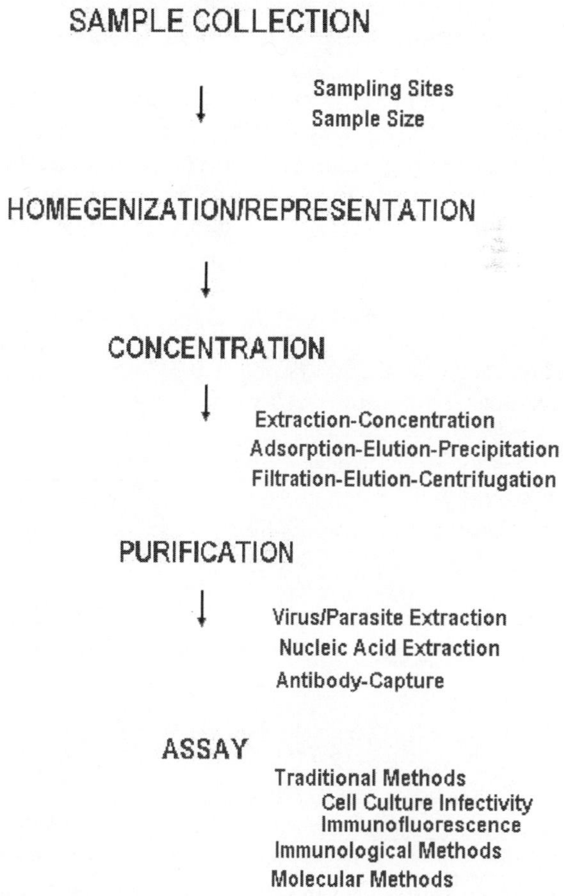

SAMPLE COLLECTION

↓ Sampling Sites
Sample Size

HOMEGENIZATION/REPRESENTATION

↓

CONCENTRATION

↓ Extraction-Concentration
Adsorption-Elution-Precipitation
Filtration-Elution-Centrifugation

PURIFICATION

↓ Virus/Parasite Extraction
Nucleic Acid Extraction
Antibody-Capture

ASSAY

Traditional Methods
Cell Culture Infectivity
Immunofluorescence
Immunological Methods
Molecular Methods

Figure 10.30: General Steps in the Isolation of Human Enteric Viruses and Parasitic Protozoa from Foods.

Detection

Traditional methods to directly detect viruses in foods after concentration have been based on the ability of enteric viruses to infect live mammalian cells in culture. Quantal and enumerative methods using a variety of mammalian cell culture lines, generally from primate kidneys, have been reported. Such approaches have been limited because levels of contaminating virus generally are low (1-200 infectious units per 100 grams of shellfish), residual food components interfere with assays, and the epidemiologically important viruses do not replicate (small round-structured gastrointestinal viruses) or replicate poorly (hepatitis A virus) in mammalian cell culture. Alternative methods such as enzyme-linked immunosorbent assay (ELISA) and DNA/RNA probes have been reported but are limited by high detection limits ($>10_3$ infectious units), unavailability of reagents, and poor sample quality. These difficulties are illustrated by the confirmation of viral contamination in a food in only two reported instances (Xu *et al.,* 1992; Desenclos *et al.,* 1991).

The *in vitro* enzymatic amplification method of polymerase chain reaction (PCR) offers an opportunity to enrich a single specific nucleic acid sequence up to a millionfold and hence provides a sensitive and specific method with a theoretical detection limit of one virus unit. This method is readily adaptable to the detection of RNA viruses by preceding the PCR with a brief reverse transcription (RT) step, hence the designation RT-PCR. The recent cloning of the Norwalk agent and related small round-structured gastrointestinal viruses has provided an opportunity to develop effective molecular detection methods for these previously nondetectable agents (Jiang *et al.,* 1990; Lambden *et al.,* 1993).

The application of PCR methods to the detection of human enteric viruses in foods is an area of active research. However, the development of such methods is complicated by low levels of contamination, high sample volumes, and the presence of food components, which may interfere with enzymatic amplification reactions. To address these issues, three alternative approaches have been used to simultaneously reduce sample volumes and the level of interfering compounds. The most frequently applied approach involves isolating and purifying nucleic acids (RNA) from the food sample before RT-PCR. A second approach combines capture of the virus with specific antibody followed by nucleic acid amplification by using RT-PCR. In the third approach, the intact virus particle is concentrated and purified from the complex food matrix resulting in sample volume reduction and removal of inhibitors, followed by subsequent heat release of viral nucleic acid from the virion capsid and RT-PCR. All three methods have been applied to various shellfish species, and in some cases, to other food commodities and naturally contaminated field shellfish specimens. The methods are summarized in Table 10.9, along with optimized virus detection levels (Dix, 1997; Chung *et al.,* 1996). A combined approach was reported by Chung *et al.* (Chung *et al.,* 1996), who successfully detected human enteroviruses and hepatitis A virus in naturally contaminated oyster samples after viral amplification in mammalian cell cultures. Despite enormous strides in the ability to detect human enteric viruses with PCR, the technique is still limited by the absence of effective concentration methods, the presence of enzymatic inhibitors, and the inability to distinguish between infectious and noninfectious virions.

Table 10.9: Emerging Detection Methods for Human Enteric Viruses in Foods

Pathogen	*Sample*	*Detection Limit*	*Field App.*	*Ref.*
Nucleic Acid	**Extraction/RT-PCR**			
HAV	Clams	2000 particles/g (10 PFU/g)	No	Goswami *et al.,* 1993
Poliovirus	Oysters	38 PFU/20 g (2 PFU/g)	No	Atmar *et al.,* 1993
HAV	Clams/Oysters No	100 PFU/1.5 g (67 PFU/g)	No	Atmar *et al.,* 1995
Norwalk	Clams/Oysters	5-10 PCRU/1.5 g	No	Atmar *et al.,* 1995
		(3-7 PCRU/g)	No	Atmar *et al.,* 1995
Poliovirus	Oysters/Mussels	10 PFU/5 g (2 PFU/g)	No	Lees *et al.,* 1994
SRSV	Oysters/Mussels	Not specified	Yes	Lees *et al.,* 1995
Polivirus, HAV	Clams	100 PFU/50 g (2 PFU/g)	No	Dix, 1997
Norwalk	Clams	1000 PCRU/50 g (20 PCRU/g)	No	Dix, 1997
Norwalk	Various	20-200 PCRU/10 g (2-20 PCRU/g)	No	Gouvea *et al.,* 1994
Antibody-Capture/RT-PCR				
HAV	Clams/Oysters	Not specified	No	Deng *et al.,* 1994
HAV	Oysters	Not specified	Yes	

Pathogen	Sample	Detection Limit	Field App.	Ref.
Virion Concentration				
Poliovirus, HAV	Oysters	10 PFU/50 g (0.02 PFU/g)	Yes	Jaykus *et al.*,1996
Norwalk	Oysters	4500 PCRU/50 g (90 PCRU/g)	No	Jaykus *et al.*, 1996; Chung *et al.*, 1996
Poliovirus, HAV	Clams	1000 PFU/50 g (20 PFU/g)	No	Dix, 1997
Norwalk	Clams	100 PCRU/50 g (2 PCRU/g)	No	Dix, 1997

HAV: Hepatitis A virus; RT-PCR: Reverse transcriptase polymerase chain reaction; PCRU: PCR-amplifiable units; SRSV: Small round-structured gastrointestinal virus.

Enteric Parasitic Agents

Concentration

Like viruses, parasitic protozoa are usually present in low concentrations in contaminated water and hence must be concentrated from large volumes of water before detection. *Giardia* cysts and *Cryptosporidium* oocysts are concentrated from 100 to 1,000 liters of water by filtration through yarn-wound filters. Retained particulates are eluted from the filters and reconcentrated by centrifugation. The pelleted cysts and oocysts are then separated from particulate debris by flotation on Percoll-sucrose gradients, followed by subsequent detection. With this method, a concentrate of less than 5 ml can be provided for final detection (Schaefer, 1996). While recovery efficiencies as high as 100 per cent have been reported (LeChevallier *et al.*, 1995), recovery is generally poor and greatly affected by water quality and particulate matter (Nieminski *et al.*, 1995). Alternative methods using calcium carbonate precipitation can concentrate *Cryptosporidium* oocysts from water with concentration efficiencies as high as 63 per cent (Versey *et al.*, 1993).

Detection

Traditional methods to detect parasitic agents from water sample concentrates have been based on immunofluorescent staining of filtered sample concentrates. The sample concentrate is filtered through cellulose acetate filters and commercial kits that use fluorescein isothiocyante-labeled monoclonal antibodies applied for immunofluorescent staining. The stained filters are examined under an ultraviolet microscope, and cysts and oocysts are classified according to immunofluorescence, size, shape, and internal morphologic characteristics. The results are reported as presumptive and confirmed cysts and oocysts per 100 liters of water. Confirmation is based on the ability to visualize organelles under light microscopy. This method is extremely limited because it is time-consuming and expensive, requires highly skilled personnel, and does not indicate viability of the cysts or oocysts; in addition, cross-reactions of monoclonal antibodies with algal cells and debris interfere with the interpretation of results (Rodgers *et al.*, 1995).

Because of the limitations of immunofluorescence assays, alternative strategies for the detection of protozoal parasites in environmental and water samples are being sought. Most of the approaches use the traditional methods of concentration, in conjunction with alternative detection methods. In many cases, the inclusion/exclusion of fluorogenic dyes is used to enhance morphologic examination; in particular, 4,6-diamidino-2-phenyl indole and propidium iodide have been used to facilitate detection and also assess viability. Two recent methodologic developments include cell sorting/particle counting approaches and molecular approaches. In many cases, a combination of approaches is used. Particle-counting approaches include the Fluorescence-Activated Cell Sorting system, a laser-based particle counter that is able to simultaneously sort particles, sense fluorescence, and determine size (Campbell *et al.*, 1993). This system is being used to detect *Giardia* and *Cryptosporidium* in water samples in England and Australia. Differentially stained cysts and oocysts may be visualized microscopically with cooled charge couple devices (Campbel *et al.*, 1992). Molecular approaches that apply DNA hybridization to the detection of *Giardia* species have been reported (Abbaszadegan *et al.*, 1991). A method that combines fluorescence and in situ hybridization with confocal microscopy has been applied to both detect and speciate *Giardia* cysts (Erlandsen *et al.*, 1994). PCR methods have been developed to detect *Giardia* and *Cryptosporidium* (Mahbubani *et al.*, 1991; Mahbubani *et al.*, 1992; Webster *et al.*, 1993). While PCR methods have the potential to detect one single infectious unit and may be applied to discriminate pathogenic from nonpathogenic species, they remain limited because of enzymatic inhibition, the inability to discriminate between viable and nonviable organisms, and the current absence of quantitative assay.

Several methods in development are combinations of multiple detection approaches. A combined method, designated the electrorotation assay, couples filtration and subsequent elution with affinity immunocapture by using paramagnetic beads. By inducing an electric field on a microscope with a special stage attachment, the organism-bead complexes rotate in a characteristic pattern that enables detection of parasitic protozoa. Several investigators are also working on methods that couple cell culture infectivity with immunostaining, thereby providing detection with simultaneous indication of viability (M. Sobsey, pers. comm.). Clinical ELISA kits have been evaluated for use in environmental water samples with reported detection limits of fewer than 10 cysts or oocyts (de la Cruz and Sivaganesan,1995); however, cross-reaction with algae continues to be a problem. Emerging detection approaches are summarized in Table 10.10. Methods to concentrate and detect parasitic protozoa specifically from foods are under development.

Considerations

Although prototype alternative and rapid methods to detect human enteric viruses and parasitic protozoa in foods and water have been reported, multiple barriers must be overcome before these methods are applicable to routine monitoring. To obtain sample representation and detection sensitivity adequate for the low levels of contamination found with these enteric pathogens in naturally contaminated environmental specimens, large sample volumes of food and water need to be processed. While some of the detection methods begin with large samples, many do not consider sample size, which limits the sensitivity of the assay procedure from the beginning. The approaches then applied to concentrate and purify the pathogens from the samples do not only need to be reasonably efficient but also need to produce a concentrate low in volume and free of inhibitory compounds. To complicate matters further, different food commodities have differing types and levels of inhibitors, many of which have remained recalcitrant to almost all removal processes.

Table 10.10: Emerging Detection Methods for Parasitic Protozoa in Water

Pathogen	*Detection Limit*	*Viability*	*Differentiation*	*Ref.*
Flow Cytometry with Fluorescent Imaging and CCD				
Cryptosporidium	NR	Yes, using differential fluorogenic vital dyes	NR	58, Campbell *et al.,* 1992
DNA Hybridization				
Giardia (16s-like rDNA)	1cyst	NR	No	Abbaszadegan *et al.,* 1991
Giardia (16s-like rDNA)	NR	NR	Yes-*in situ* hybridization with differential fluorescence *G. lamblia; G. muris*	Erlandsen *et al.,* 1994
Polymerase Chain Reaction				
Giardia (Giardin gene)	1 cyst	Partial, using mRNA as target; Depends on inactivation method	NR	Mahbubani *et al.,* 1991
Giardia (Giardin gene)	1 cyst	NR	Yes-primer sequence *G. duodenalis*	Mahbubani *et al.,* 1992
Cryptos-poridium (Target not specified)	20 oocysts	No	Yes-by hybridization *C. parvum*	Webster *et al.,* 1993

CCD: Cooled charge couple device; NR: Not reported.

The relationship between detection by molecular and immunologic approaches and the subsequent viability or infectivity of these enteric pathogens remains a concern. While investigators have addressed these issues using in vitro excystation, inclusion and exclusion of vital dyes, animal infectivity (parasitic protozoa) (Sauch *et al.,* 1991; Schupp and Erlandsen, 1987), and assays combining mammalian cell culture infectivity with alternate detection strategies (viruses) (M. Sobsey, pers. comm.), these methods are not useful for the nonculturable human enteric viruses; they are expensive, time-consuming, and not readily amenable to routine diagnostic work. Establishing quantitative detection methods also needs further research.

The development of detection methods is also limited by the current state of knowledge. For instance, while recent sequencing evidence indicates that the Norwalk-like small round-structured gastrointestinal virus group consists of multiple members having the same physical and genomic characteristics as other viruses in the family Caliciviridae, considerable sequence and antigenic diversity remain among the members of this group. While PCR primers for these genetically diverse agents have been reported (Ando *et al.,* 1995), development of universal detection methods is clearly limited until more complete information about this group of human enteric viruses is available.

Research is needed to develop and refine the prototype protocols into collaboratively tested methods that could be routinely and expeditiously used to evaluate the microbiologic safety of food products. In general, future research needs for the routine application of alternative methods to detect enteric viral and parasitic protozoal contamination in foods requires development of the following: 1) simple, rapid, and cost-effective extraction and concentration procedures; 2) simple and reliable methods for the removal of inhibitors; 3) methods that are not restricted by food product; and 4) quantitative approaches for assessing the relative levels of contamination.

Overall, improved epidemiologic surveillance, predominantly through the creation of population-based centers that will focus on the epidemiology and prevention of food and waterborne infectious diseases, should improve knowledge regarding viral etiology in foodborne disease outbreaks and the clinical and economic importance of viral and parasitic disease agents. Data obtained from these studies may elucidate the role of foods in sporadic disease as well as in secondary spread.

From a food safety standpoint, improved detection methods should eventually provide a regulatory option for monitoring food safety, particularly in the economically important shellfish species and in detecting viral contamination due to human handling, or parasitic protozoal contamination from animal wastes. The development of rapid detection methods will also aid in the evaluation of control strategies for viral and protozoal contamination of foods. For instance, depuration, thermal processing, and irradiation for the control of food borne viral contamination in shellfish need to be further evaluated. In the case of contamination by infected food handlers and animal wastes, the availability of rapid methods will increase our understanding of this transmission route and help determine critical control points for HACCP approaches to control the transmission of these food borne disease agents. This will allow the integration of a true farm-to-table food safety approach for the control of viral and parasitic food and waterborne disease.

The combination of increased surveillance, improved detection methods, and testing requirements should result in a marked improvement in the ability to detect, investigate, and control food and waterborne enteric viral and parasitic protozoal agents. Taken together, these approaches promise to provide increased information necessary to assess risks, control disease, and ultimately improve public health in the next century.

Chapter 11

Food Hazards and Illness: Economic Burden

Introduction

Although food is indispensable to the maintenance of life, it can also be responsible for ill health. Safer food saves lives. With every bite one eats, one is potentially exposed to illness from either microbiological or chemical contamination. Billions of people are at risk and millions fall ill every year; many die as a result of consuming unsafe food. Concerns about food safety have skyrocketed in more affluent societies. However, the real tragedy of foodborne diseases is played out in the developing world. Unsafe water used for the cleaning and processing of food; poor food-production processes and food handling (including inappropriate use of agricultural chemicals); the absence of adequate food storage infrastructure; and inadequate or poorly enforced regulatory standards–these all contribute to a high-risk environment.

While exposed to more hazardous environments, people in developing countries often have difficulty coping with foodborne disease. For many living at or below the poverty line, foodborne illness perpetuates the cycle of poverty. The symptoms of foodborne diseases range from mild and self-limiting (nausea, vomiting and diarrhoea) to debilitating and life-threatening (such as kidney and liver failure, brain and neural disorders, paralysis and potentially cancers), leading to long periods of absenteeism and premature death.

Causative Agents of Foodborne Illnesses

Foods are complex mixtures of chemicals and often contain compounds that are potentially harmful as well as those that are beneficial. Several vitamins are toxic if consumed in excessive amounts and many food plants produce toxic secondary metabolites to discourage their attack by pests.

Potatoes contain the toxic alkaloid solanine. Normally this is more concentrated in aerial parts of the plants and the peel which are not eaten, but high levels are also found in green potatoes and potato sprouts which should be avoided.

Cassava contains cyanogenic glycosides which produce hydrogen cyanide on hydrolysis. Similar compounds are also present in apple seeds, almonds, lima beans, yams and bamboo shoots. The body's detoxification pathway converts cyanide to thiocyanate which can interfere with iodine metabolism giving rise to goitre and cretinism. Traditional methods of preparing cassava eliminate the acute toxicity problem from hydrogen cyanide but the increased incidence of goitre and cretinism in some areas where cassava is a staple may be a reflection of chronic exposure.

Legumes or pulses contain a number of anti-nutritional factors such as phytate, trypsin inhibitors and lectins (haemagglutinins). Many of these are destroyed or removed by normal preparation procedures such as soaking and cooking. Even so, red kidney beans are still responsible for occasional outbreaks of food poisoning when they have been insufficiently cooked to destroy the lectins they contain. Lathyrism is a more sensus condition associated with a toxin in the pulse *Lathyrus sativa* which can be a major food item in North African and Asian communities during times of famine. In favism, an enzyme deficiency predisposes certain individuals to illness caused by glycosides in the broad bean, *Vicia faba.*

A simple insufficiency of food will lead to marasmus (protein-energy deficiency) while over-reliance on staples low in protein, such as cassava, produces the condition known as kwashiorkor. A diet may provide adequate protein and energy but be lacking in specific minerals or vitamins giving rise to characteristic deficiency syndromes such as goitre (iodine deficiency), pellagra (nicotinic acid), beriberi {thiamine) and scurvy (ascorbic acid).

To some extent toxic or antinutritional characteristics can be bred out of cultivars intended for human consumption although the problem cannot be eliminated completely. For affluent consumers in the developed world, particularly toxic foods can be avoided since alternatives are normally available in plenty. This is not always the case in poorer countries where these diet-related conditions are far more common.

In addition to the hazards posed by natural toxins that are an intrinsic feature of their composition, foods may also act as the vehicle which an exogenous harmful agent may be ingested. This may be a pesticide, some other chemical contaminant added by design or accident, a microorganism or its toxin.

The role of plant pathogens in subsequent spoilage of post-harvest crop may be complex, thus *Phytophthora infestans* causes a severe field disease of the potato plant, frequently causing death of the plant, but it may also remain dormant within the tubers and either cause a rot of the tubers during storage, or a new cycle of disease in the next season's crop.

The higher pH values of the tissues of many vegetables makes them more susceptible to bacterial invasion than fruits although there are also a number of important spoilage fungi of stored vegetables. The bacteria involved are usually pectinolytic species of the Gram-negative genera *Erwinia, Pseudomonas* and *Xanthomonas,* although pectinolytic strains of *Clostridium* can also be important in the spoilage of potatoes under some circumstances and the non-sporing Gram-positive organism *Corynebacterium sepedonicum* causes a ring rot of potatoes.

However, the most frequent agents of spoilage are not the plant pathogens themselves but opportunistic microorganisms, which gain access to plant tissue through wounds, cracks, insect damage or even the lesions caused by the plant pathogens. All freshly harvested vegetables have a natural surface flora, including low numbers of pectinolytic bacteria and it is becoming increasingly evident that healthy tissue of the intact plant may also contain very low numbers of viable microorganisms (endophytic). The onset and rate of spoilage will depend on the interactions between the physiological changes occurring in the tissues after harvest and changes in microbial activity. Harvesting itself will produce physiological stress, principally as a result of water loss and wilting and cut surfaces may release nutrients for microbial growth. This stress may also allow growth of the otherwise quiescent endophytic flora.

As described in the case of fruits, the prevention of spoilage during storage and transport of vegetables must involve a range of measures. The control of the relative humidity and the composition of the atmosphere in which vegetables are stored is important but there is a limit to the reduction of relative humidity because at values below 90–95 per cent, loss of water from vegetable tissues will lead to wilting. It is essential to avoid the presence of free water on the surfaces of vegetables and temperature control may be just as important to prevent condensation. The presence of a film of water on the surface will allow access of motile bacteria such as *Erwinia* and pseudomonads to cracks, wounds and natural openings such as stomata.

Vegetables should not normally be a cause of public health concern but the transmission of pathogenic bacteria such as *Salmonella* and *shigella* is possible by direct contamination from faeces of birds and animals, the use of manure or sewage sludge as fertilizer, or the use of contaminated waters for irrigation. Celery, watercress, lettuce, endive, cabbage and beansprouts have all been associated with *Salmonella* infections, including typhoid and paratyphoid fevers and an outbreak of shigellosis has been traced to commercial shredded lettuce.

Table 11.1: Some Microorganisms Involved in the Spoilage of Fresh Vegetables

Microorganism	*Vegetable*	*Symptom*
Bacteria		
Corynebacterium sepedonicum	Potato	Ring rot of tubers
Pseudomonas solanocearum	Potato	Soft rot
Erwinia carotovora var. *atroseptica*	Potato	Soft rot
Streptomyces scabies	Potato	Scab
Xanthomonas campestris	Brassicas	Black rot
Fungi		
Botrytis cinerea	Many	Grey mould
Botrytis allii	Onions	Neck rot
Mycocentrospora acerina	Carrots	Liquorice rot
Trichothecium roseum	Tomato	Pink rot
	Cucurbits	
Fusarium coeruleum	Potato	Dry rot
Aspergillus alliaceus	Onion	Black rot
	Garlic	

Not all pathogens are necessarily transmitted to vegetables by direct or indirect faecal contamination. Organisms such as *Clostridium botulinum* have a natural reservoir in the soil and any products contaminated with soil can be assumed to be contaminated with spores of this organism, possibly in very low numbers. This would not normally present a problem unless processing or storage conditions were sufficiently selective to allow subsequent spore germination, growth and production of toxin. In the past, this has been seen mainly as a problem associated with under processed canned vegetables, but now it must be taken into consideration in the context of sealed, vacuum or modified-atmosphere packs of prepared salads. Those salads containing partly cooked ingredients, where spores may have been activated and potential competitors reduced in numbers, could pose particular problems. In 1987 a case of botulism caused by *Clostridium bolulinum* type A was associated with a pre-packed rice and vegetable salad eaten as part of an airline meal. Similar risks may occur in foil-wrapped or vacuum-packed cooked potatoes or film-wrapped mushrooms and in all these cases adequate refrigeration appears to be the most effective safety factor.

Another group of pathogens naturally associated with the environment includes the psychrotrophic species *Listeria monocytogenes* which is commonly associated with plant material, soil, animals, sewage and a wide range of other environmental sources. Raw celery, tomatoes and lettuce were implicated on epidemiological grounds as a possible cause of listeriosis which occurred in several hospitals in Boston, USA in 1979, although direct microbiological evidence was missing. An outbreak of listeriosis in Canada in 1981 was associated with coleslaw. Strains of *L. monocytogenes* can certainly grow on shredded cabbage and salad vegetables such as lettuce at temperatures as low as 5°C and modified-atmospheres seem to have no effect on this organism.

Food Poisoning: Mechanism

Outbreaks of food poisoning involve a number of people and a common source and are consequently more intensively investigated than the more numerous sporadic cases that occur. Valuable information is derived from these investigations about the most common contributory factors and faults in food hygiene that lead to outbreaks of food-borne illness.

Table 11.2: Factors Contributing to Outbreaks of Food Poisoning

Factor	*England and Wales*	*USA*
Preparation too far in advance	57	29
Storage at ambient temperature	38	63
Inadequate cooling	32	
Contaminated processed food	17	n.i.
Undercooking	15	5
Contaminated canned food	7	n.i.
Inadequate thawing	6	n.i.
Cross contamination	6	15
Food consumed raw	6	n.i.
Improper warm handling	5	27
Infected food handlers	4	26
Use of left overs	4	7
Extra large quantities prepared	3	n.i.

Analysis of the specific failures in food hygiene that have contributed to outbreaks have been conducted on a number of occasions and results of two of these, from the United States and from England and Wales have shown that comparing the two is not entirely straight-forward since, in most outbreaks more than one contributory factor has been identified so that the columns do not add up neatly to 100 per cent. Also, the surveys differ in the categories used and even where they are nominally the same they may still not be equivalent in all respects. Failure to cool foods and hold them at temperatures inimical to microbial growth, or to heat them sufficiently to kill microorganisms, coupled with prolonged storage giving microorganisms time to multiply to dangerous levels.

The alimentary or gastrointestinal tract is not an internal organ of the body but a tube passing through it from the mouth to the anus. Its principal functions are the digestion and absorption of food and the excretion of waste. Unlike most of the body's other external surfaces, it is not lined with a dry protective skin and so, although it possesses some protective features, it offers a more congenial environment for microorganisms and an easier route by which they can penetrate the body.

In the mouth, food is mixed with saliva and broken down mechanically to increase the surface area available for attack by digestive enzymes. Saliva is an alkaline fluid containing starch-degrading (amylase) enzyme and the antimicrobial factors immunoglobulin (IgA), lysozyme, lactoferrin and lactoperoxidase. It provides lubrication to assist chewing and swallowing and performs a cleansing function and rinsing the teeth and mouth to remove debris. On average, an adult secretes and swallows about 1.51 of saliva each day.

The variety of pods consumed and the range of micro-environments in the mouth result in a diverse and continually changing microflora. On the teeth, bacteria are associated with the formation of dental plaque–an organic film in which bacteria are embedded in a matrix derived from salivary glycoproteins and microbial polysaccharides. The microbial composition of plaque varies with its age but filamentous *Fusobacterium* species and streptococci are common components. Plaque offers a protective environment for bacteria and its development is often a prelude to conditions such as dental caries and periodontal disease.

Swallowed food descends via the oesophagus into the stomach; a bulge in the alimentary tract which serves as a balance tank from which food is gradually released into the small intestine for further digestion.

In the stomach, food is blended with gastric juice, an acidic fluid containing hydrochloric acid. Stomach pH can range from 0.8–5.0 (typically 2.0–3.0) and has a marked effect on ingested microorganisms, killing most. Normally only acid-tolerant vegetative cells and spores survive and the microbial count in the stomach is low, although lactobacilli are frequently found in association with the stomach wall. Gastric acidity generally provides very effective protection for subsequent sections of the intestine but is not, as we shall see, an invulnerable defence. Bacteria can evade prolonged exposure to the acid by being sheltered in food particles or as a result of accelerated passage through the stomach as occurs, for instance, when the stomach is full. Alternatively, acidity may be neutralized by the food or absent as a result of illness.

The digestive functions of the stomach are not confined to those of a mechanical churn with antimicrobial features. Proteases, such as pepsin and lipase which can operate at low pH partially digest the stomach contents. The gastric mucosa also secretes a protein responsible for efficient absorption of vitamin B_{12}. Little absorption of nutrients occurs in the stomach, with the notable exception of ethanol, but some material transfer is often necessary to adjust the osmotic pressure of the stomach contents to ensure they are isotonic with body fluids.

In the first section of small intestine, the duodenum, large-scale digestion is initiated by mixing the chyme with digestive juice from the pancreas and bile from the gallbladder which neutralize the chyme's acidity. The pancreatic juice also supplies a battery of digestive enzymes and surfactant bile salts emulsify fats to facilitate their degradation and the absorption of fat soluble vitamins. Further digestive enzymes that break down disaccharides and peptides are secreted by glands in the mucous lining of the duodenum called, with evocations of a Gothic horror, the crypts of Lieberkühn.

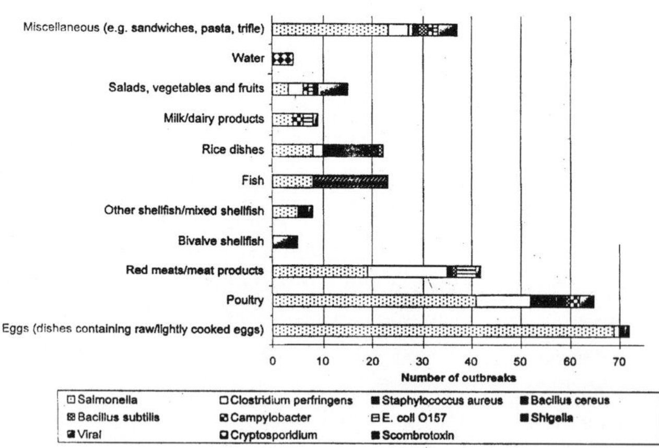

Figure 11.1: Food Vehicles and Causal agents in Food Poisoning.

The duodenum is a relatively sort section of the small intestine, accounting for only about 2 per cent of its overall length. Food is swept along by waves of muscle contraction, known as peristalsis, from the duodenum into the jejunum and thence into the ileum. During this passage, nutrients such as amino acids, sugars, fats, vitamins, minerals and water are absorbed into capillaries in the villi from where they are transported around the body. Absorption is sometimes a result of passive diffusion, but more often involves the movement of nutrients against a concentration gradient; an active process entailing the expenditure of energy.

The microbial population increases down the length of the small intestine; count of 10^2–10^3 ml^{-1} in the duodenum increase to around 10^3–10^4 in the jejunum, 10^5 in the upper ileum and 10^6 in the lower ileum. This corresponds with a decreasing flux of material through the small intestine as water is absorbed along its length. In the higher reaches of the duodenum, the flow rate is such that its flushing effect frequently exceeds the rate at which microorganisms can multiply so that only those with the ability to adhere to the intestinal epithelium can persist for any length of time. As the flow rate decreases further along the small intestine, so the microbial population increases, despite the presence of antimicrobial factors such as lysozyme, secretory immunoglobulin, IgA and bile.

In the healthy individual, the microflora of the small intestine is mainly comprised of lactobacilli and streptococci, although as we shall see, other bacteria have the ability to colonize the epithelium and cause illness as a consequence.

Extensive microbial growth takes place in the colon or large intestine where material can remain for long periods before expulsion as faeces. During this time active absorption of water and salts helps to maintain the body's fluid balance and to dry faecal matter. Bacterial cells account for 25–30 per cent of faeces, amounting to 10^{10}–10^{11} cfu g^{-1}, the remainder is composed of indigestible components or food, epithelial cells shed from the gut, minerals and bile.

Obligate anaerobes such as *Bacteroides* and *Bifidobacterium* make up 99 per cent of the flora of the large intestine and faeces.

The interaction between the gut microflora and its host appears to have both positive and negative aspects and is the subject of much current research and cojecture. Addition of antibiotics to feed has been shown to stimulate the growth of certain animals, suggesting that some gut organisms have a deleterious effect on growth.

A normal gut microflora confers protection against infection. One example of this effect is the inflammatory disease *pseudomembranous colitis* caused by *Clostridium difficile.* Normally this organism is present in the gut in very low numbers, but if the balance of the flora is altered by antibiotic therapy, it can colonize the colon releasing toxins. Similarly, the infective doses of some other enteric pathogens have been shown to be lower in the absence of the normal gut flora.

It appears that protection is not simply a result of the normal flora occupying all available niches, since enterotoxigenic *E. coli* adheres to sites that are normally vacant. Some direct antagonism through the production of organic acids and bacteriocins probably plays a part, but stimulation of the host immune system and its capacity to resist infection also appear to be factors.

Pathogens Related to Foodborne Illnesses

Many foodborne diseases are illnesses that result from ingestion of contaminated food. More than 250 different foodborne hazards have been recognized including infectious bacteria, viruses and parasites, and non-infectious chemicals and toxins. Many of these agents commonly cause diarrhoea and vomiting, but there is no single clinical syndrome for all foodborne diseases. The sources of these agents range from being an inherent constituent of the food to inadvertent (or intentional) addition during food production, processing, or preparation. Moreover, most foodborne agents can be transmitted via routes other than food, including through water or through contact with infected farm animals, pets, and humans. The proportion of disease associated with being foodborne will differ depending on the agent and food handling, with only a few foodborne diseases being transmitted exclusively via foods. For example, *Listeria monocytogenes* is assumed to be almost exclusively foodborne, whereas *Escherichia coli* O157:H7 infections can be acquired by ingesting contaminated food or water or by direct contact with infected animals or persons.

A number of assessments of the relative significance of hazards associated with food have concluded that micro-organisms are of paramount importance. A study conducted in the United States found that, although the attention given to different food hazards by the media, pressure groups and regulatory authorities might differ, as far as the food industry was concerned, microbial hazards were the highest priority. Similarly, it has been estimated that the risk of becoming ill as a result of microbial contamination of food was 100 000 times greater than the risk from pesticide contamination.

Foodborne pathogens take advantage of weak immune systems. Infants and young children, pregnant women, the elderly as well as those immuno-compromised, are particularly at risk of contracting and dying from common food-related diseases. Malnourished infants and children are especially exposed to foodborne hazards and are at higher risk of developing serious forms of foodborne diarrhoeal diseases; these infections in turn exacerbate malnutrition thus leading to a vicious circle of debilitation and mortality. Those who survive may suffer from delayed physical and mental development, depriving them of the opportunity to reach their full potential in society.

Foodborne diseases have been an issue for all societies since the beginning of humanity. The types, severity and impacts of these illnesses have changed through the ages and are still diverse across regions, countries and communities. There were considerable differences in the burden of foodborne disease among subregions delimited on the basis of child and adult mortality. The highest burden per population was observed in Africa (AFR) (AFR D and AFR E subregions), followed by South-East Asia (SEAR) (SEAR B and SEAR D) subregions and the Eastern Mediterranean (EMR) D subregion. Diarrhoeal disease agents were the leading cause of foodborne disease burden in most subregions. NTS was an important burden in all subregions, particularly in Africa. Other main diarrhoeal causes of foodborne disease burden were EPEC, enterotoxigenic *E. coli* (ETEC) and *Vibrio cholerae* in low-income subregions, and *Campylobacter spp.* in high-income subregions.

The burden of aflatoxin was high in the AFR D, Western Pacific (WPR) B and SEAR D subregions. In the SEAR subregions there was a considerable burden of *Salmonella typhi*. The burden of *Opisthorchis* spp. was concentrated in the SEAR B subregion, where the seafood-borne trematodes *Paragonimus* spp. and *Clonorchis sinensis* were also important. In the Americas (AMR) B and D subregions, *Taenia solium* and *Toxoplasma gondii* contributed significantly to the foodborne disease burden. The global burden of foodborne diseases is considerable, with marked regional variations. The burden of foodborne diseases is borne by individuals of all ages, but particularly by children under 5 years of age, and by persons living in low-income subregions of the world (WHO, 2015).

Diarrhoeal Diseases: Main Symptom of Foodborne Illnesses

Diarrhoea is a disease that has different kinds to it, according to ICD-10, such as infective neonatal diarrhoea, unspecified diarrhoea, functional diarrhoea, non-infective neonatal diarrhoea, and irritable bowel syndrome with diarrhoea, *etc.* Along with the diseased person, it also affects the well-being of the ICU health workers due to the increasing workload. Heidegger *et al.*, say diarrhoea episodes have a huge impact on human-related costs and so on the financial burden in ICU

Diarrhoeal diseases are one of the leading causes of morbidity and mortality globally and account for more deaths in early childhood after the neonatal period. Diarrhoeal diseases are associated with an estimated 1.3 million deaths annually, with most occurring in resource-limited countries; note that up to 25 per cent of deaths in young children living in Africa and south-east Asia are attributable to acute gastroenteritis. The youngest children are most vulnerable with the incidence of severe gastroenteritis being highest in the first 2 years of life.

Morbidity due to diarrhoea is further concentrated in marginalized communities within resource-limited countries. Despite improvements in standard of living, advances in sanitation, water treatment and food safety awareness, diarrhoeal disease still accounts for significant economic and societal losses.

It was found that the leading pathogens causing diarrhoea in young children were (in descending order), *Shigella* spp., rotavirus, adenovirus 40/41, ST-positive ETEC, *Cryptosporidium* spp. and *Campylobacter* spp. Foodborne diarrhoeal disease agents caused 230,000 (95 per cent UI 160,000–320,000) deaths, particularly non-typhoidal *Salmonella enterica* (NTS, which causes diarrhoeal and invasive disease). Other major causes of foodborne deaths were *Salmonella Typhi*, *Taenia solium*, hepatitis A virus, and aflatoxin. The global burden of foodborne disease by these 31 hazards was 33 (95 per cent UI 25–46) million DALYs.

In developing world, diarrhoea is caused by a wide range of aetiological agents which include viruses, bacteria and parasites. In low- and middle-income countries (LMICs), bacterial enteritis has been shown to be very common. A study on causes of diarrhoea among children presenting to health facilities in sub-Saharan Africa and Asia prior to rotavirus vaccine introduction found that across all sites, the majority of diarrhoea cases were attributed to *rotavirus*, *Cryptosporidium*, ETEC and *Shigella*. *Rotavirus* and *Escherichia coli*, are the two most common etiological agents of moderate-to-severe diarrhoea in low-income countries and they cause about 40 per cent of hospitalizations for diarrhoea in children under 5. Other pathogens such as *cryptosporidium* and *shigella* species may also be important. About 88 per cent of diarrhoea-associated deaths are attributable to unsafe water, inadequate sanitation, and insufficient hygiene.

Annually in India, rotavirus diarrhoea causes an estimated 122,000-153,000 deaths, 457,000-884,000 hospitalizations, and 2 million outpatient visits in children <5 years of age. India spends Rs 2.0-3.4 billion (US$ 41-72 million) annually in medical costs to treat rotavirus diarrhoea. The use of specific interventions against rotavirus, such as newly available vaccines, would help prevent much of this large disease and economic burden.

In low-income countries like India, children under 3-year-old experience an average of three episodes of diarrhoea every year, contributing to malnutrition as a major cause. Diarrhoea accounts for 13 per cent of all deaths among under-five children and the third-leading cause of childhood mortality in India. Rotavirus infection accounts for 39 per cent of diarrheal admission in India. As per the National Family Health Survey round 4 (2015–2016), in India, the prevalence of diarrhoea was 9.2 per cent among under-five children in the 2 weeks before the survey. A considerable out-of-pocket economic burden is faced by the families for treating diarrhoea in under-five children. Due to a lack of medical insurance, almost all the cost of diarrhoea treatment was borne by the people from their pocket.

Diarrhoea causes death by depleting body fluids resulting in profound dehydration. The actors that are involved in the occurrence of diarrhoea in children are complex, and the relative contribution of each factor varies as a function of interaction between socio-economic, environmental and behavioural variables. The documented risk factors for diarrhoea include younger age, malnutrition, early weaning, seasonal patterns, low maternal education, lack of piped water supply, poor water-storage practices, lack of vigilant hand washing, poor sanitation and not treating water in the home. In LMICs, gastrointestinal infections in children are often associated with poor hygienic conditions; poor sanitation, lack of access to clean water supply and inadequate personal hygiene are responsible for 90 per cent of diarrhoeal disease occurrence. Inadequate drinking water, poor sanitation and hygiene are important risk factors that drive diarrhoea diseases in low-income settings, in addition to socio-economic and age-related risk factors. It has been shown that paediatric diarrhoea incidence is inversely related to socio-economic status, with children in poverty much more vulnerable to acute diarrhoeal episodes, as well as to more severe and longer duration illness. In a study conducted in South Africa, children living in poverty were approximately 10 times more likely to die from diarrhoea than their more privileged counterparts. More than 2.5 billion people lack access to an improved sanitation facility in low-income countries, and inadequate hand hygiene practices have been estimated to affect 80 per cent of the population globally.

Most diarrheal germs are spread from the stool of one person to the mouth of another. These germs are usually spread through contaminated water, food, or objects. Water, food, and objects become contaminated with stool in many ways: people and animals defecate in or near water sources that people drink. Contaminated water is used to irrigate crops. Food preparers do not wash their hands before cooking. People with contaminated hands touch objects, such as doorknobs, tools, or cooking utensils.

Weaning is a particularly hazardous time for the infant. The anti-infective properties of maternal breast milk are lost or diluted and are placed by foods which often have a low nutrient density. At the same time, the immature immune system is exposed to new sources of infection in the environment. Poor hygienic practices in the preparation of weaning foods and the use of contaminated water are often implicated in weaning diarrhoea and it has been estimated that 15–70 per cent of all diarrhoea episodes in young children are food associated.

Diarrhoea is the most common illness in children under 5 years of age. Globally, 1.7 billion cases of childhood diarrheal disease occur annually. Each year diarrhoea kills around 525,000 children under 5 years of age.

According to UNICEF, diarrhoea is a leading killer of children, accounting for approximately 8 per cent of all deaths among children under age 5 worldwide in 2017. This translates to over 1,400 young children dying each day, or about 525,000 children a year, despite the availability of a simple treatment solution. Child mortality or the under-five mortality rate refers to the probability of a child dying between birth and exactly 5 years of age, expressed per 1,000 live births. In 2020, 5.0 million children under 5 years of age died. This translates to 13,800 children under the age of 5 dying every day in 2020. Globally, infectious diseases, including pneumonia, diarrhoea and malaria, remain a leading cause of under-five deaths, along with preterm birth and intrapartum-related complications. In the less developed world, the consequences of food-borne illness are even more serious. Diarrhoeal disease is still a major cause of morbidity and mortality in poor countries, particularly among children. According to UNICEF data, among all the death cases of diarrhoea, in 2019, the most prevalent disease-causing pathogen has been found to be *Campylobacter*. But *Adenovirus* is the major contributor to childhood diarrheal deaths.

Despite the different methods and sources of information, each successive review of the diarrhoea burden over the past three decades has demonstrated declining mortality but relatively stable morbidity. The decline in global diarrhoea mortality is confirmed by detailed information from monitoring in certain countries over time. Increased use of oral rehydration therapy, improved nutrition, increased breastfeeding, better supplemental feeding, female education, measles immunization, and improvements in hygiene and sanitation are believed to have contributed to this decline. Globally, 34·7 per cent of under-5 diarrhoea deaths were caused by rotavirus. Existing rotavirus vaccines prevented about 30000 deaths in 2016. The global under-five mortality rate declined by 61 per cent, from 93 deaths per 1,000 live births in 1990 to 37 in 2020.

Despite this considerable progress, improving child survival remains a matter of urgent concern. In 2020 alone, roughly 13,800 under-five deaths occurred every day, an intolerably high number of largely preventable child deaths; eminently preventable and treatable conditions, such as pneumonia, diarrhoea, malaria, measles, and malnutrition, are still leading killers of children. However, there is variation by country, suggesting that there is no single solution to reduce diarrhoea mortality. Every country must consider their specific context to identify strategies to reduce diarrhoea mortality. There are, however, concerns about potential rebound increases in diarrhoea-associated mortality in the future due to increasing urbanization and climate change.

Accelerating and maintaining these declines rate is essential to meet Sustainable Development Goals for under-5 childhood mortality and ensuring that children everywhere have the opportunity of a full, healthy life.

Burden of Foodborne Illnesses and Hazards

The economic costs associated with foodborne disease can be severe on people, food companies, and country reputation. Foodborne disease globally is still not under control and outbreaks can cause health and economic losses. The causes are unhygienic practices in food production, harvesting, and preparation. Foodborne diseases can be mild with recovery in days, or severe resulting in hospitalization and death in certain patients.

For the food industry, the costs can be huge and it is not unusual for the company producing a product implicated in an outbreak of food poisoning to go bankrupt as a result. Companies not directly involved in an outbreak can also suffer. There is often a general decline in demand for a product prompted by public concern that the same problem could occur with similar products from other manufacturers. There was, for instance, a marked downturn in all yoghurt sales after the hazelnut yoghurt botulism outbreak in England in 1989.

Increased vigilance by companies to ensure that the same process failures responsible for an outbreak do not occur elsewhere, also has its attendant costs. For instance, it was estimated that the costs of checking the integrity of spray-drier cladding by dried-milk manufacturers following a *salmonella* outbreak caused by dried milk were of the order of hundreds of thousands of pounds.

Food retailers can also be affected as a result of a decline in sales, particularly if a suspect product is associated with one particular store.

The global burden of foodborne diseases is considerable, with marked regional variations. The burden of foodborne diseases is borne by individuals of all ages, but particularly by children under 5 years of age, and by persons living in low-income subregions of the world.

Estimates of foodborne disease incidence provide the foundation of other types of analysis needed to effectively manage foodborne diseases. Foodborne illness is prevented by safe production and handling of foods. As a result, food safety controls are usually targeted at particular pathogens on particular foods. Effective food safety policy requires an ability to target resources where they can most effectively reduce the risk of foodborne illness. This requires information on both the relative impact of different foodborne illnesses and information on the costs of control. Food source attribution partitions total foodborne illness to a set of specific food sources.

The concept of burden of disease was developed in the 1990s by the Harvard School of Public Health, the World Bank and the WHO to describe death and loss of health due to diseases, injuries and risk factors for all regions of the world. While burden of disease can be expressed using various indicators, such as incidence, mortality, societal costs and summary measures of population health, this study introduced a new metric, the Disability-Adjusted Life Year (DALY), which combines information on morbidity and mortality caused by diseases. The DALY is now the most widely used public health metric for burden of disease studies, and the key measure in the Global Burden of Disease (GBD) studies.

The first estimates of the burden of foodborne diseases in DALYs were published in 2000, measuring the health burden of a single foodborne pathogen in the Netherlands (Havelaar *et al.,* 2000). Other studies followed in Europe and beyond, using the DALY metric to measure the burden of country-specific single pathogens or a few of foodborne pathogens (Havelaar *et al.,* 2004; Lake *et al.,* 2010; Verhoef *et al.,* 2013). In 2015, the WHO produced the first global DALY estimates of the burden of FBD (WHO, 2015). A few other countries have established burden of FBD studies after that, publishing either routine estimates or ad-hoc reports or articles with estimates for specific pathogens and years. These largely focus on microbiological agents and include mostly high-income countries, such as the Netherlands, Japan, Denmark, Belgium, and the United States of America. Initiatives in low-income countries have taken the first steps to address knowledge gaps. As examples, estimated of the incidence of illness of FBDs from syndromic surveillance data in Rwanda, and the Caribbean Burden of Illness Study estimated the prevalence and incidence of acute gastroenteritis and specific foodborne pathogens in nine countries. These are first steps to national burden of disease studies. Evidence on the burden of individual foodborne chemicals in some countries has also been published. Again, these studies were conducted in high-income countries, where data are more abundant, but the disease burden may be lower when compared to low income and middle income countries (LMIC).

Even if available evidence illustrates that LMICs bear a higher burden than high-income countries, the emphasis is on the data gaps. For example, the WHO's sub-regional DALY estimates for LMICs majorly relied on imputation, where data available from some countries were used to predict data missing from others, as data were scarce in these countries (WHO, 2015; Havelaar *et al.,* 2021; Lake *et al.,* 2015). Given the target of the WHO's global DALY estimates and learning lessons from the few nationwide studies, estimation of the burden of nation-based foodborne diseases would be suitable for informing interventions and policy.

Foodborne diseases (FBD) still cause a substantial public health, economic and social burden worldwide. The COI (Cost of illness) approach traces the flow of resources associated with adverse health outcomes through the quantification of measurable individual and societal level costs. Costs at the individual level are divided into direct and indirect costs. Direct costs represent the value of goods, services, and other resources consumed in providing care due to an illness. These include medical care expenditures associated with the diagnosis, treatment, and management of a disease in an individual. Indirect costs represent productivity losses due to illness or death and intangible costs such as pain and suffering. Costs associated with overhead activities that are shared amongst individuals and expenditures incurred in the process of seeking care are also indirect costs. Operational expenditures for healthcare facilities and personal transportation costs are examples of indirect costs. Costs incurred at the population level are deemed societal costs, which are costs that cannot be completely attributed to an individual's illness but can be incurred when a person or a group of people become ill. Societal costs primarily include expenditures incurred by industry and government. Component costs are the specific costs that make up the above categories, and all of the costs included in a COI estimate comprise the cost inventory for that particular study.

For otherwise healthy, well-nourished people in the developed world, most food poisoning is an unpleasant episode from which recovery a normally complete after a few days. For society as a whole though, it is increasingly being recognized as a largely avoidable economic burden. Costs are incurred in the public sector from the diversion of resources into the treatment of patients and the investigation of the source of infection. To the individual the costs may not always be calculable in strictly financial terms, but could include loss of income, costs of medication and treatment.

Direct comparison of the impact of different foodborne illnesses is difficult because foodborne illnesses vary in severity from a few days of self-limiting diarrhoea to chronic kidney disease and death. Some pathogens are highly infectious but cause mild illness, while other have a low incidence of illness but high mortality rates. Both monetary and nonmonetary metrics have been developed that make it possible to compare the impact of foodborne diseases that have very different physical impacts. Monetary measures can also be used in cost-benefit analyses to assess whether the costs of intervention are justified by expected health benefits. Risk analysis provides a structured means of informing assessment, management, and communication about foodborne disease risk. All of these analyses build on a foundation of epidemiology and incidence estimates.

WHO-FERG's country studies efforts The Foodborne Disease Burden Epidemiology Reference group (FERG) was established by WHO in 2007 to estimate the global and regional burden of FBDs (across the six WHO regions). Other aims of FERG were to strengthen the capacity of countries to assess their burden of FBD, and to increase the number of countries that have undertaken such a study. Activities of FERG to promote national studies involved capacity-building and promotion of the use of information on burden of disease in setting evidence-informed policies. The FERG Country Studies Task Force (CSTF) developed a suite of tools and resources to support national studies. Pilot studies were conducted in Albania, Japan, Thailand and Uganda and provided important practical lessons. In particular, data gaps impeded DALY calculations in several occasions. These gaps included information needed to assign the aetiology for syndromes, such as acute gastrointestinal disease, and data on the incidence of diseases caused by some hazards. The pilot studies also highlighted the need for engagement of stakeholders that can provide access to national data, including public and private data sources. In some countries, private hospitals provide a significant proportion of health care, and may not adhere to the same reporting requirements as public hospitals. Engagement with private hospitals and other facilities may need to be specifically addressed to provide a complete picture of the incidence of FBDs. Data on foodborne hazards may be gathered from primary producers and the food industry, but economic implications, particularly for trade, may constitute a barrier for sharing such data, which therefore requires careful handling.

Recognizing the need to measure the burden and distribution of foodborne diseases, there is a need to encourage evidence-informed policies. Keeping it in view, a number of attempts have been made to quantify these costs and, while the errors must be large, they do at least give an idea of the magnitudes involved. A lack of accurate data on the full extent and cost of foodborne diseases has been a major obstacle to address food safety issues. In order to fill this data gap, WHO together its partners launched the initiative (FERG) to estimate the global burden of foodborne diseases. First-ever estimates, developed by WHO's Foodborne Disease Burden Epidemiology Reference Group (FERG) between 2006 and 2015, were published by WHO in December 2015, presented by age, sex, and region for foodborne diseases caused by a defined list of microbes, parasites, and chemicals (WHO, 2015). These estimates support countries in assessing their own burden of foodborne disease, increase awareness and commitment for the implementation of food safety standards, and, for the first time, highlight the magnitude of this global problem ((Mathers *et al.,* 2020).

This report, resulting from the WHO Initiative to Estimate the Global Burden of Foodborne Diseases and prepared by the WHO (FERG Group), provides the first estimates of global foodborne disease incidence, mortality, and disease burden in terms of Disability Adjusted Life Years (DALYs). For the global estimates, thirty-one foodborne hazards causing 32 diseases are included, being 11 diarrhoeal disease agents (1 virus, 7 bacteria, 3 protozoa), 7 invasive infectious disease agents (1 virus, 5 bacteria, 1 protozoon), 10 helminths and 3 chemicals. The World Health Organization (WHO) estimated that each year, 1 out of 10 people get ill from food contaminated with microbial or chemical agents; 31 foodborne hazards caused 600 million illnesses, 420000 deaths and the loss of 33 million healthy years of life globally in 2010 and this number is likely an underestimation; 30 per cent of foodborne deaths occur among children under 5 years of age (WHO, 2015). While exposed to more hazardous environments, people in developing countries often have difficulty coping with foodborne disease.

In the African region where the FBD burden is the highest, the 31 foodborne hazards included in the WHO estimates of the global burden of FBD have been estimated to cause 1200–1300 DALYs per 100 000 inhabitants in 2010, compared to 35–711 in other regions. Nearly 70 per cent of the burden is due to diarrheal disease agents, particularly to non-typhoidal *Salmonella* (including invasive salmonellosis), and Enteropathogenic and Enterotoxigenic *Escherichia coli*. Other important agents included *Vibrio cholerae* and *Taenia solium*. However, because research and disease surveillance data from Africa are limited, these estimates are subject to uncertainty. The main challenge to estimating burden of FBD in Africa is lack of data, particularly on the incidence of FBDs in the population.

In one study done in Bangladesh, the average cost per episode of diarrhoea for caregivers was US$62, with $29 direct and $34 indirect costs. From the societal perspective, average cost per episode of diarrhoea was $71. In 2018, an estimated $79 million of economic costs were incurred for treating diarrhoea in Bangladesh. Using 10 per cent of income as threshold, over 46 per cent of interviewed households faced catastrophic expenditure from diarrheal disease. The economic costs incurred by caregivers for treating per-episode of diarrhoea was around 4 per cent of the annual national gross domestic product per-capita. Investment in vaccination can help to reduce the prevalence of diarrheal diseases and avert this public health burden.

The finding that the burden per 100,000 population is much higher in low than high income countries, ranging from 1300 disability-adjusted life-years (DALYs) per 100,000 population in low-income sub-Saharan Africa to 35 DALYs per 100,000 population in high-income North American countries, points to the fact that foodborne illnesses are a largely preventable problem. Despite this, foodborne disease is an important problem in developed countries (WHO, 2015; Pires, 2021).

Since the mid-1990s, the USDA, Economic Research Service (ERS) has estimated the economic cost of 15 major foodborne illnesses. U.S. Centers for Disease Control and Prevention (CDC) estimates that one in six people contract a foodborne illness each year in the United States and, of these, 3000 die (Scallan *et al.,* 2015). Disease outcomes include both acute illness and chronic disease that sometimes follow these acute illnesses. These 15 pathogens account for over 95 per cent of the illnesses and deaths from foodborne illnesses acquired in the United States for which the U.S. Centers for Disease Control and Prevention (CDC) can identify a pathogen cause. Monetary measures of these health effects provide a common metric to compare impacts of various pathogens, a way to aggregate impacts across illnesses, and a means of comparing the costs of experiencing those illnesses with the costs of preventing them. Government agencies and private industries rely on cost estimates for foodborne illness as they set priorities, make policies, and develop management initiatives. They also use the ERS estimates to help the public understand the importance of preventing foodborne disease.

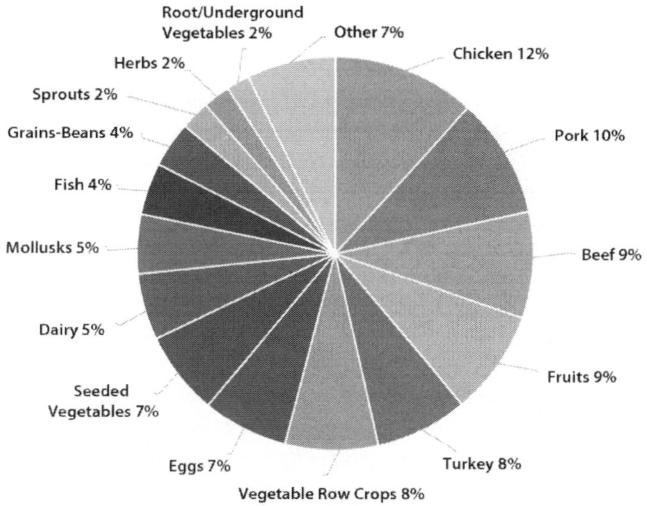

Figure 11.2: Top 15 Foods that Caused Outbreak-Associated Illnesses, 2009–2018. (*Source*: CDC Foodborne Disease Outbreak Surveillance System).

Notes: Vegetable row crops, *e.g.*, leafy vegetables; seeded vegetables, *e.g.*, cucumbers; mollusks, *e.g.*, oysters; root/underground vegetables, *e.g.*, carrots, potatoes.

Unpasteurized dairy products accounted for 80 per cent of outbreak-associated illnesses linked to dairy. Pasteurized dairy products accounted for 10 per cent, and pasteurization status was unknown for 9 per cent.

Other = foods that don't fit in the top 15 categories and other federally regulated items such as alcohol, coffee, other beverages, ice, condiments, and dietary supplements.

Total of percentages does not equal 100 per cent because of rounding.

Five of the pathogens accounted for about 90 per cent of the total costs from these foodborne pathogens ($15.7 billion in 2018 dollars). Three pathogens, *Salmonella*, *Toxoplasma*, and *Listeria*, were responsible for more than 60 per cent of the total economic cost from major foodborne illnesses in 2018 dollars.

ERS last estimated the total cost of the leading U.S. foodborne illnesses at $15.5 billion in 2013 dollars. While overall inflation in the United States has been low in recent years, inflation in the healthcare sector has not been low. From 2013 to 2018, inflation for the entire U.S. economy was 7.8 per cent. During this period, income growth, measured in terms of real per capita Gross Domestic Product (GDP), was 8.8 per cent. At the same time, inpatient hospital costs rose 25 per cent, outpatient hospital costs grew 22.5 per cent, and costs of prescription drugs and medical supplies increased 19.3 per cent. ERS recently released revised cost estimates for these illnesses showing inflation and income growth alone increased the economic cost of these pathogens by $2 billion, to $17.6 billion in 2018 dollars, holding the number of cases constant (Hoffmann and Ahn, 2021).

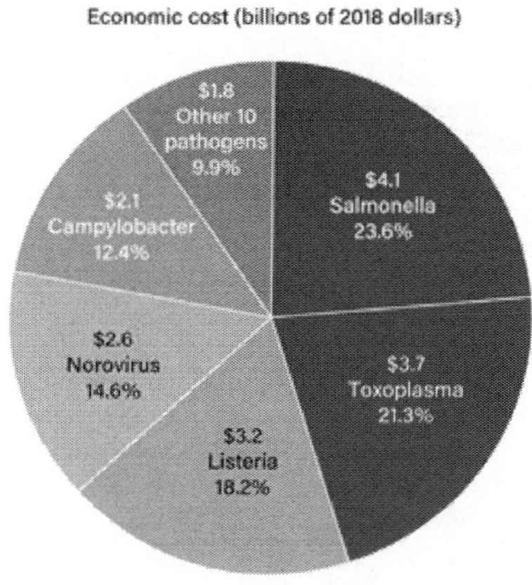

Figure 11.3: Pathogens Associated with Fooborne Illness Expenditure in US.

To determine the economic cost of foodborne illness, ERS looked at the costs of medical care, the value of lost earnings, and a monetary measure of death linked to how much people are willing to pay to reduce risk of dying from foodborne illness. ERS used the Centers for Disease Control and Prevention's (CDC) foodborne illness incidence rates from 2011—the agency's most recent data—to develop the economic cost estimates.

Health Canada estimates that one in eight Canadians are sickened by foodborne illness each year. Estimates made in 2010 show Australians experience an episode of foodborne illness once every 5 years on average. Mangen *et al.* (2015) estimate that foodborne disease imposes a cost of illness of 468 million Euros in the Netherlands each year. For New Zealand, Lake *et al.* (2010) estimate that six major pathogens impose $NZ 86 million in cost of illness annually. Vaillant *et al.* (2005) estimated that 23 pathogens caused about 750,000 cases of foodborne illness in France.

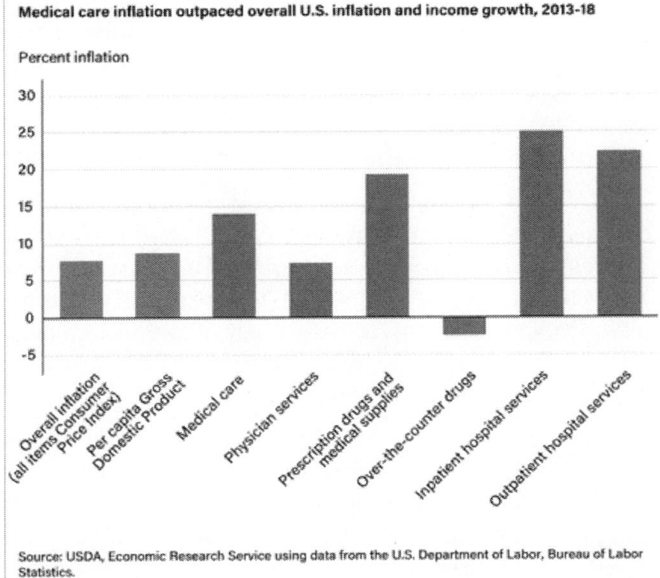

Figure 11.4: Medical Care Inflation Outpaced Overall U.S. Inflation and Income Growth (2013-2018).

In Europe, the annual cost of foodborne illness was estimated to be $171 million USD in Sweden and $2 million USD in Croatia. Many estimates for specific foodborne pathogens, or groups of pathogens, have been published. Although economic estimates for foodborne illness have not been completed in Canada in the past 20 years, it has been recently estimated that 4 million episodes of domestically acquired foodborne illness occur annually in Canada.

Evidence on the burden of individual foodborne chemicals in some countries has also been published. Again, these studies were conducted in high-income countries, where data are more abundant, but the disease burden may be lower when compared to low-income and middle-income countries (LMIC).

In addition to the burden of disease, acute gastroenteritis incurs considerable societal costs. Most cost studies include only healthcare costs and productivity losses due to work absence of persons that were ill themselves or had to take care of a sick child. Especially in the case of community-acquired foodborne outbreaks, however, costs are not limited to the cost-of-illness. Once an outbreak is identified public health authorities conduct investigations to determine the pathogen and the source, and if required, implement control measures to prevent further spread. Not considering these costs would result in an underestimation as they can be quite large. The evaluation of the economic impact of an outbreak is important input for public health policy with respect to prioritization of prevention and control efforts of Salmonellosis. The salmonellosis outbreak was associated with substantial societal cost, accumulating to approximately €1.9 million if not correcting for underestimation and €7.5 million if underestimation is taken into account.

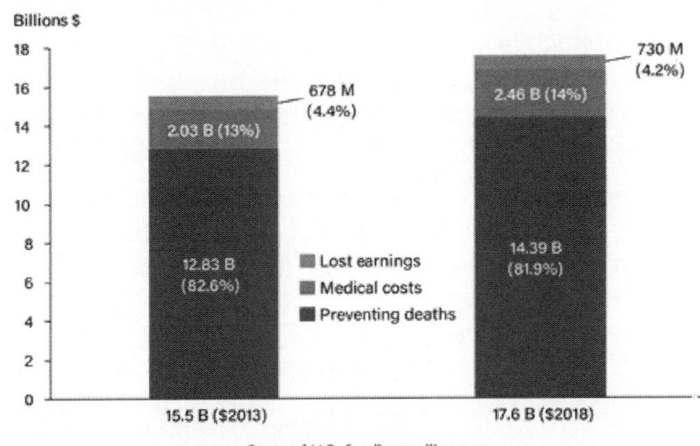

Figure 11.5: Costs of 15 Major Foodborne Illnesses in the U.S. in 2018.

Estimates of the burden of FBDs are useful to prioritize food safety policy and allocate resources to where food safety risks are highest. The experiences of established studies and of their mechanisms of translation of evidence into policy can provide guidance and suggest processes for other national studies. Here we focus on few countries that have been at the forefront of burden of FBD estimation.

The Netherlands

Burden of FBD estimates have been published every year in the Netherlands since 2008. The Dutch Ministry of Health mandates the Dutch National Institute for Public Health and the Environment (RIVM) to provide annual updates of the number of illnesses, disease burdened cost-of-illness caused by an agreed-upon panel of 14 enteric pathogens mainly transmitted by food. The disease burden is expressed in DALYs. The cost-of-illness related to these pathogens is estimated in euros(s), and includes healthcare costs, the costs for the patient, family and caregivers (*e.g.*, travel and external care expenses), and costs in other sectors, for example productivity losses,

Demographic data, as well as data on mortality, live births and stillbirths, are obtained from the Dutch Central Bureau of Statistics. The incidence of infections by pathogen is obtained from various surveillance systems. For instance, data on pathogens like *Listeria monocytogenes*, Shiga-toxin producing *E. coli* (STEC) O157 and hepatitis A virus, which are notifiable, are obtained from case notifications and the laboratory surveillance system, which has national coverage.

From August to December 2012, the Netherlands experienced the largest foodborne outbreak of gastroenteritis due to salmonellosis in humans ever recorded. The cause of the outbreak was smoked salmon processed at a single production site in Greece. In total, 1149 *Salmonella* cases due to *Salmonella Thompson* were identified, laboratory-confirmed and reported to the National Institute for Public Health and the Environment (RIVM).

The most recent estimates, for 2019, show that the 14 pathogens are cumulatively responsible for about 11 000 DALYs and 423 million. The share apportioned to foodborne transmission is estimated at 4200 DALYs and euros 174 million. The largest foodborne burden at population level was caused by *Campylobacter*, followed by *Toxoplasma gondii* and *norovirus*. Regarding other foodborne bacteria, *Salmonella* ranked second after *Campylobacter* spp. Perinatal listeriosis and congenital toxoplasmosis were the diseases with the highest individual burden. The pathogens causing the largest costs were *norovirus, rotavirus, Staphylococcus aureus*, and *Campylobacter*. However, the average cost per case was the largest for perinatal listeriosis (euros 291 000/case). Health-care costs accounted for 21 per cent of the total costs for the 14 pathogens, patient and family costs for 2 per cent, and the costs in other sectors accounted for 77 per cent. About 41 per cent of the foodborne burden was associated with meat, that is, poultry, pork, beef and lamb, which caused 33 per cent of all food-related fatal cases, indicating that the pathogens associated with these foods are responsible for the most severe infections.

Year after year, these national estimates have provided vital insights for policy-making as to guide strategies, such as establishing process hygiene criterion for *Campylobacter* on broiler meat. Yet, they play an even more vital role in resource allocation, such as funding for research and other activities on specific pathogens or conditions that appear to have a higher burden. Burden and cost estimates also enable policy-makers and the scientific community to monitor trends and generate scientific hypotheses. For instance, although the disease burden for *Campylobacter* had continually decreased since 2010, it slightly increased in 2018 and 2019, suggesting a beginning of a reversal of the trend. This calls for more research, such as studies focusing on hygiene measures at primary production, performance of surveillance and diagnostics, risk factor analyses, as well as genomics of circulating strains, to understand the underlying causes.

Denmark

The first Danish burden of FBD study was published in 2014, and has been growing with new hazards and data being added at different points in time (Hald *et al.,* 2016). The study currently includes microbiological and chemical hazards commonly present in foods. The most recent estimates for microbiological agents are from 2017 and cover seven pathogens that are mostly transmitted through foods: *Campylobacter, Salmonella*, STEC, *norovirus, Yersinia enterocolitica, L. monocytogenes*,and *Toxoplasma gondii.*

In 2017, *Campylobacter* caused the highest burden of disease, more than threefold higher than the second highest ranked pathogen, *Salmonella*; *Listeria* and *Yersinia* followed in the ranking. The burden of congenital toxoplasmosis was lower than most of the investigated diseases but was borne by a low number of cases in the population.

The ranking of foodborne pathogens varied substantially when based on reported cases, estimated incidence, and burden of disease estimates. The total estimated incidence was highest for norovirus, but this agent ranked sixth when focusing on foodborne burden measured in DALYs. These differences illustrate the importance of estimating burden of disease in DALYs, particularly when the purpose is to compare across diseases with very diverse severity and duration. *Campylobacter* still led the ranking when excluding DALYs attributable to non-foodborne routes of exposure, but the ranking of some of the other pathogens (particularly norovirus) changed.

The Campylobacter Interest Group, formed in 2017, meets approximately four times a year to discuss how stakeholders can make the best use of the data generated and the knowledge gained from research and surveillance data for a better monitoring of Campylobacter in food and humans (Pires *et al.,* 2021).

Hungary

Salmonellosis is one of the most commonly reported zoonoses in humans in the European Union. Most cases are caused by the *S. Enteritidis* (60.9 per cent) and *S. Typhimurium* (13.8 per cent) subtypes. According to statistics, salmonellosis is one of the most frequent foodborne illnesses in Hungary and is among the nearly 70 notifiable diseases. Based on data from 2014 and 2018, Hungarian rates of confirmed *Salmonella* cases were more than two times higher than the EU average (53.1 and 42.6 vs. 20.7 and 20.1 per 100,000). It should be emphasised that the true occurrence of salmonellosis cases may be several times higher than that registered in the national and international databases. *Salmonella* infections represent significant cost for households, the health care system and companies (because of absent employees). The cost of the disease is often divided into two categories: direct and indirect costs. Both cost categories place a considerable burden on affected households and on society. Direct costs include costs such as cost of treatment and medication at home and in hospitals, while indirect costs include additional costs due to altered consumer behaviour as well as loss due to pain and other psychological suffering. Lack of confidence in certain products due to food contamination can deliver disastrous economic effects at the social level. Based on this data provided by Vajda *et al.* (2021), the total consumer well-being loss could be estimated to be about 7.87 million EUR per year in Hungary.

Australia and New Zealand

In 2000, the OzFoodNet network was established to enhance surveillance of foodborne diseases across Australia. OzFoodNet consists of 7 sites and covers 68 per cent of Australia's population. During 2001, sites reported 15,815 cases of campylobacteriosis, 6,607 cases of salmonellosis, 326 cases of shigellosis, 71 cases of yersiniosis, 61 cases of listeriosis, 47 cases of shiga-toxin producing *E. coli* and 5 cases of haemolytic uraemic syndrome. Sites reported 86 foodborne outbreaks affecting 1,768 people, of whom 4.0 per cent (70/1,768) were hospitalised and one person died. There was a wide range of foods implicated in these outbreaks and the most common agent was *S. Typhimurium*. Sites reported two international outbreaks; one of multi-drug resistant *S. Typhimurium* Definitive Type 104 due to helva imported from Turkey, and one of *S. Stanley* associated with dried peanuts from China. The National Centre for Epidemiology and Population Health conducted a national survey of gastroenteritis. Preliminary data from interviews of 2,417 people suggests that the incidence of foodborne illness is significantly higher than previously thought. OzFoodNet initiated case control studies into risk factors for *Campylobacter*, *Salmonella, Listeria*, and shiga-toxin producing *E. coli*. In Australia and New Zealand, the cost of foodborne illness has been estimated at $1.289 billion and $86 million USD respectively per year.

Japan

In 2011, foodborne disease caused by *Campylobacter* species, *Salmonella* species and EHEC led to an estimated loss of 6099, 3145 and 463 DALYs in Japan, respectively. These estimated burdens are based on the pyramid reconstruction method; are largely due to morbidity rather than mortality; and are much higher than those indicated by routine surveillance data. Routine surveillance data may indicate foodborne disease burdens that are much lower than the true values. Most of the burden posed by foodborne disease in Japan comes from secondary complications.

National burden of foodborne disease studies are essential to establish food safety as a public health priority, identify the most important food safety problems, and inform prevention strategies. Different initiatives are emerging to support international collaboration across countries and public health authorities worldwide.

Chapter 12

Food Spoilage and the Microflora

Introduction

Spoilage is also a subjective quality; what is spoiled for one person may be perfectly acceptable to another. The perception of spoilage is subject to a number of influences, particularly social; foods acceptable in some cultures are unacceptable in others. Some products such as well matured cheeses and game birds that have been hung for extended periods are esteemed by some people and highly objectionable to others. Affluence is another contributory factor–many are not in the position to be able to discard food due to some slight sensory defect. In the 18th and 19th century, navy sailors often preferred to eat in dark corners so that they could not see the weevils and maggots in their food.

According to the Oxford English Dictionary to spoil is to 'deprive of good or effective qualities'. When a food is spoiled, its characteristics are changed so that it is no longer acceptable. Such changes may not always be microbiological in origin; a product may become unacceptable as a result of insect damage, drying out, discolouration, staling or rancidity for instance, but by and large most food spoilage is a result of microbial activity. Microbiological food spoilage can manifest itself in several different ways, some of which often occur in combination. Visible microbial growth may be apparent in the form of surface slime or colonies, degradation of structural components of the food can cause a loss of texture, but the most common manifestation will be chemical products of microbial metabolism, gas, pigments, polysaccharides, off-odours and flavours.

A general feature of microbial spoilage is its relatively sudden onset–it does not appear to develop gradually, day by day a little worse, but more often as an unexpected and unpleasant revelation. This is a reflection of the exponential nature of microbial growth and its consequence that microbial metabolism can also proceed at an exponentially increasing rate. If a microbial produce associated with spoilage, for example an off odour, has a certain detection threshold, the level will be well below this threshold for most of the product's acceptable shelf-life. Once reached, however, it will be rapidly passed so that in a comparatively short time after, levels will be well in excess of the threshold and the product will be profoundly spoiled.

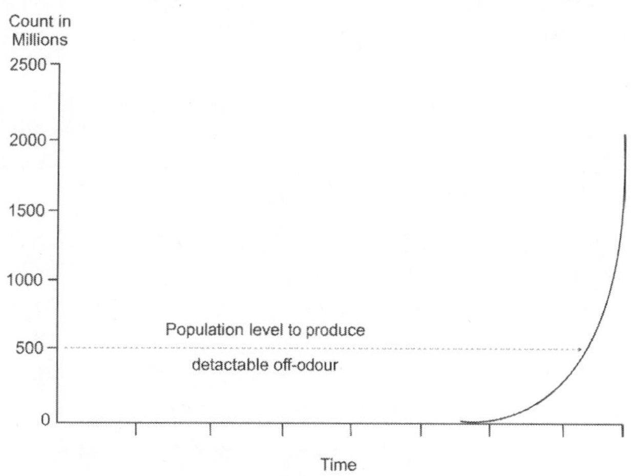

Figure 12.1: Microbial Growth and Spoilage.

Prediction or early detection of spoilage is not always easy since the mechanisms underlying microbiological spoilage can be quite complex. It is generally far easier to identify the chemical responsible for a particular off-odour than to identify the organism(s) responsible. This and the relative speed of chemical analysis have led to the use of chemical indices of spoilage in some areas, but more often the ultimate judgement as to whether a food is spoiled remains subjective. Total microbial counts are generally a poor indicator of spoilage potential. Many of the organisms enumerated may not grow in the food and many of those that do will not be responsible for spoilage.

1. The Microbial Growth in Milk

Historically, milk and dairy products have been significant components in the diets of humans and continue to play an important and increasing role in the diets of the growing population of both rural and urban communities. These are aids for enhancing the economic status of farmers as well as sellers and improve the health of customers.

Milk is the fluid, excluding colostrum, secreted by mammals for the nourishment of their young. Colostrum is a much more concentrated liquid containing up to 25 per cent total solids, mainly protein, secreted immediately after parturition. A number of animals are used to produce milk for human consumption, although the cow is by far the most important in commercial terms.

The principal components of milk are water, fat, protein and lactose. The precise composition varies between species so, for example, human milk has lower protein but higher lactose levels than cow's milk. Generally, the protein content of the milk reflects the growth rate of the young animal–the higher the growth rate, the more protein the milk contains.

There can be considerable compositional difference between breeds of a single species–Jersey and Guernsey milks, for instance, are noted for their higher fat content which is reflected in a richer, creamier taste. Even within a single breed variation in composition can arise depending on factors such as the stage or lactation, the stage of milking, the intervals between milking, the time or day, the number of previous lactations and the general nutritional state and health of the cow.

Generally, milk and dairy products are rich in nutrients, delivering high quality proteins, micronutrients, vitamins and energy-containing fats. Milk, thus, provides an ideal environment for the growth of wide variety of food-borne microorganisms and zoonotic agents.

Table 12.1: Typical Milk Composition [per cent weight $(volume)^{-1}$]

	Fat	*Protein*	*Lactose*	*Total Solids*
Human	3.8	1.0	7.0	12.4
Cow	3.7	3.4	48	12.7
Jersey	5.1	3.8	5.0	14.5
Ayrshire	4.0	3.5	4.8	13.0
Short-horn	3.6	4.9	4.9	12.6
Sheep	7.4	5.5	4.8	19.3
Goat	4.5	2.9	4.1	13.2
Water buffalo	7.4	3.8	4.8	17.2
Horse	1.9	2.5	6.2	11.2

Table 12.2: Composition of Fresh Cow's Milk

	Concentration g/ $litre^{-1}$	
Lipids	37	
	of which	per cent w/w
Triglycerides		95–96
Diglycerides		1.3–1.6
Free fatty acids		0.1–0.5
Total phospholipids		0.8–1.0
Proteins	34	
Casein	26	
αS1	11.1	
αS2	1.7	
β	8.2	
γ	1.2	
k	3.7	
Whey proteins		
α-lactalbumin	0.7	
β-lactoglobulin	3.0	
serum albumin	0.3	
immunoglobulins	0.6	
Non-protein Nitrogen	1.9	
Lactose	48	
Citric Acid	1.75	
Ash	7.0	
Calcium	1.25	
Phosphorus	0.96	

Milk does possess a number of antimicrobial features, present either to protect the udder from infection or to protect the newborn calf. Generally, these are present at too low concentration in cow's milk to have a very marked effect on its keeping quality or safety. In some cases, the antimicrobial activity is antagonized by other milk constituents such as the effect of citrate and bicarbonate on lactoferrin activity. Stimulation of lactoperoxidase activity through the addition of exogenous hydrogen peroxide has been investigated as a means of preserving raw milk in developing countries where ambient temperatures are high and refrigeration is not often available. In one trial in Africa, use of this technique increased the proportion of samples passing the 10 minute resazurin quality test from 26 per cent to 88 per cent.

Food-safety hazard regarding milk generally refers to any biological, chemical or physical agent in in milk, or condition of milk with the potential to cause adverse health consequences for consumers. Such hazards may be introduced into the dairy chain at any time during primary production, milking, formulation and processing, packaging and labelling, transportation, storage, preparation, and serving. Major food safety hazards associated with raw milk production and dairy products may be put into three categories: biological, chemical and physical, as shown in Table 12.3. From raw milk production through processing to the consumer, milk is exposed to several hazards, which ultimately influence the safety and quality of the final product. Some of these hazards may stem from animal husbandry practices, through their feeding, milking and processing.

Table 12.3: Main Categories of Food Safety Hazards Associated with Milk and Dairy Products (Adopted from the Food and Agricultural Organization (FAO)

Biological Hazards	*Chemical Hazards*	*Physical Hazards*
i. Pathogenic bacteria (including toxins produced by bacteria)	i. Naturally occurring toxins	i. Metal fragments
ii. Toxigenic moulds/ fungi	ii. Direct and indirect food additives	ii. Bone fragments
iii. Parasites	iii. Veterinary drug residues	iii. Glass pieces
iv. Viruses	iv. Heavy metals	iv. Insects or their parts
v. Other biological hazards	v. Environmental contaminants	v. Jewellery
	vi. Chemicals from packaging material	vi. Stones/soil/dust
		vii. Hair/fur

Biological Hazards

The microbiological quality of milk, at the point of milking from a healthy animal, is theoretically expected to be safe for human consumption. Generally, pathogenic microorganisms can contaminate raw milk in two ways. First, endogenous contamination occurs when milk is contaminated by a direct transfer of pathogens from the blood (systemic infection) of an infected animal into the milk, or via an infection in the udder. The second means by which fresh milk can be contaminated, known as exogenous contamination, occurs where milk is contaminated during or after collection by animal faeces, the exterior of the udder and teats, the skin, and other environmental sources such as soil, air, feed, water, equipment, animal hides and people.

Bacteria that get on to the outside of the teat may be able to invade the opening and hence the udder interior. Aseptically taken milk from a healthy cow normally contains low numbers of organisms, typically fewer than 102–103 cfu ml^{-1} and milk drawn from some quarters may be sterile. The organisms most commonly isolated are *micrococci*, *streptococci* and the diptheroid *Corynebacterium bovis.* Mastitis, an inflammation of the mammary glands and udder tissues, is caused by a large variety of common bacteria, fungi, mycoplasmas and algae infecting the mammary glands of dairy animals. Mastitis adversely affects animal health, milk quality, consumer safety, and can lead to great economic losses for milk production Counts are frequently higher due to mastitis.

Mastitis among dairy herds is a major constraint, and the disease has been identified as a primary cause of poor-quality and compromises the safety of raw milk in African countries. Generally, milk can easily become contaminated before it is secreted from the udder owing to mastitis. Although microbial contamination of milk during milking is difficult to completely avoid; it is of utmost importance to maintain a very high level of hygiene in dairy farming practices and proper cleaning of teats during milking for good udder health and optimum milk quality, and to ensure safety.

In England and Wales, where it has been estimates to cost the industry around £90 million annually, about 1–2 per cent of cows have a clinical infection at any one time. In the early acute stage of illness, the bacterial count in mastitic milk can exceed 108 cfu ml^{-1} and macroscopic changes are often visible in the milk. Mastitis is also diagnosed by the presence of high numbers of polymorphonuclear leukocytes which can rise to levels of 107 ml^{-1} in infected milk.

Once milk is secreted from the udder, it can be contaminated from several sources including air, faeces, bedding material, soil, feed, water, equipment, animal hides and people. Heavily contaminated teats have been reported to contribute up to 105 cfu ml^{-1} in the milk. This is less of a problem in summer months when cows are allowed to graze in open pasture and is worst when they are housed indoors and under wet conditions.

A critical factor affecting milk safety is milking hygiene. Adequate milking hygiene potentially reduces the contaminating microorganisms and prevents them from inhabiting the immediate environment or skin of the animals, hands of personnel and milking equipment, thereby minimizing their spread during milking. High prevalence of pathogens in raw milk and occurrence of mastitis have previously been recorded in farms that practiced poor milking hygiene. The dairy farm environment can be a reservoir of foodborne pathogens and serve as a major source of microbial contamination of raw milk due to direct contact with the milk. The use of unsterilized collection vessels (containers) and other practices such as milking with unsanitized bare hands and allowing calves to feed without cleaning the teats of udders, expose milk to microbial contamination. In most small-scale milk production farms in developing nations, there are no strict implementations of procedures for cleaning and disinfection of materials used during production processes, from milking to the sales of final products. While most foodborne pathogens such as *E. coli*, *Salmonella spp.* and *Campylobacter* spp. inhabit the ruminant intestinal tract, others including *Listeria spp.* and *Bacillus* spp. are widespread in nature and live in soils and plant environments. Clostridia such as *C. bulyricum* and *C. tyrobutyricum* can get into milk from silage fed to cows and their growth can cause the problem known as late blowing in some cheeses. Thus, these environmental microorganisms can contaminate the milk by direct contact or through milking equipment and personnel on the farm if good hygiene management practices are not followed. Additionally, intentional adulteration of raw milk with contaminated water has been reported, a practice that potentially serves as another major source of pathogenic microorganisms in raw milk

Milk-handling equipment such as teat cups, pipework, milk holders and storage tanks, is the principal source of the microorganisms found in raw milk. As the overall quality of the milk decreases so the proportion of the microflora derived from this source increases. Milk is a nutritious medium and, if equipment is poorly cleaned, milk residues on surfaces that are frequently left wet will act as a focus for microbial growth which can contaminate subsequent batches of milk. Occasional neglect or cleaning and sanitizing procedures is usually less serious since, although it may contribute large numbers of microorganisms to the product, these tend to be fast growing bacteria that are heat sensitive and will be killed by pasteurization. They are also sensitive to sanitizing practices used and will be eliminated once effective cleaning is resumed. If cleaning is persistently neglected though, the hydrophobic, mineral-rich deposit known as milkstone can build up on surfaces, particularly heated ones. This will protect organisms from sanitizers and allow slower growing organisms to develop such as *micrococci* and *enterococci*. Many of these are thermoduric and may not be removed by pasteurization.

Storage and transportation of raw milk immediately after milking through to point of sale or processing are critical for safety and quality. In order to prevent proliferation of pathogens that contaminate freshly collected milk, there should be strict time and temperature controls between the milking and the processing of dairy products. Ideally, raw milk should be immediately cooled to below 4 °C to prevent microbial growth and ensure high-quality, safe milk for processing and consumption. This is in practise not possible for most small-scale producers in developing nations, and therefore pasteurisation and sterilization of the milk are strongly recommended. This is unfortunately often not the case. Modern cooling facilities including mechanical refrigeration or cooling tanks are not available to the many small-scale dairy producers for reasons such as high initial investment and running costs and technical problems, including the lack or unreliable supply of electricity. Additionally, the majority of raw milk producers in these countries are mostly located in remote rural areas with poor road networks making it difficult to transport milk to urban markets and small-scale processing units. Subsequently, raw milk is often transported from the farm to small-scale processing units in urban market centres by bicycles, motorcycles, or by foot. The usual high ambient temperatures, often reaching 35–42 °C in most parts of sub-Sahara Africa and Asia in summer season, highlight the problem by accelerating the growth of spoilage and pathogenic microorganisms during transportation of raw milk.

A number of measures can be taken to minimize milk contamination from the udder exterior and considerable advice on this topic is available to dairy farmers.

Although such procedures certainly improve the microbiological quality of milk, economic constraints such as increasing size of individual dairy herds and decreased manning levels in milking parlours encouraged their neglect. To encourage farmers to apply the available advice on animal husbandry practices, milking procedures, types and design of equipment and cleaning schedules which contribute to good bacteriological quality milk, the Milk Marketing Board (MMB) in England and Wales introduced in 1982 a system of paying farmers based on the total bacterial count (TBC) of their milk. Anticipation of its start, a dramatic fall in the count was noted. Now more than 76 per cent of the milk produced in England and Wales falls into Band A with a mean count of 1.7 × 10 cfu ml^{-1}. The Milk Marketing Board no longer exists as the monopoly purchaser of farm milk in England and Wales, but the bodies that replaced it recognized the value of a payment scheme which includes microbiological quality and have retained similar systems.

In most developed countries milk is chilled almost immediately after it issues from the cow and is held at a low temperature thereafter. It is stored in refrigerated holding tanks before being transported by refrigerated lorry to the dairy where it is kept in chill storage tanks until use. Throughout this time, its temperature remains below 7°C and the only organisms capable of growing will be psychrotrophs. There are many psychrotrophic species, but those most commonly found in raw milk include Gram-negative rods of the genera *Pseudomonas, Acinetobacter, Alcaligenes, Flavobacterium,* psychrotrophic coliforms, predominantly *Aerobacter* spp. and Gram-positive *Bacillus* spp.

One consequence of the current extensive use of refrigeration and the change to a microflora dominated by psychrotrophs is that traditional tests for the microbiological quality of milk based on the reduction of a redox dye such as methylene blue or resazurin have become obsolete. Psychrotrophs tend to reduce these dyes poorly and the tests are not very sensitive to low numbers of bacteria.

The majority of milk produced in South-East Asia is processed into a variety of traditional milk products by small-scale processing units or processors. The final products including spontaneously fermented yoghurt-like milk, traditional *paneer* or cheeses, *khoa*, ghee and butter, are produced with slight variations in processing methods depending on country or local region, which is affected by local tastes, dietary habits or culinary traditions. The production of traditional dairy products is based on recipes handed down from one generation to another, and processors often do not have access to formal training but learn by seeing, hearing and practicing. Small-scale processors of traditional milk products often lack pasteurization, storage and packaging facilities and do not adhere to good manufacturing and hygiene practices (GMP/GHPs), including the implementation of starter culture procedures for milk fermentation. The processing of milk into yoghurts and other fermented products in most parts of the developing nations relies on spontaneous fermentation or back-slopping where a part of a previous batch of a fermented product is used to inoculate the new batch. Consequently, traditional milk products may be susceptible to contamination with human pathogens of public health concerns due to the lack of proper control measures and adherence to good manufacturing practices (GMP) in the traditional fermentation processes.

Dairy Related Situation in Developing Countries

- ✰ Particularly in developing countries, the shelf-life of milk and milk products tends to be shorter as the production of milk and its processing into different milk products takes place under unsanitary conditions and poor production practices.
- ✰ Coliforms are invariably found in raw milk but with good manufacturing practices, their number can be kept very low.
- ✰ Their presence is an indication of unsanitary production practices and/or improper handling of either milk or milk utensils.
- ✰ Even pasteurized milk may have a high bacterial load which may be attributed to defective pasteurization machinery, microbes surviving in pasteurization, and post-pasteurized contamination due to poor processing and handling conditions and/or poor hygienic practices followed by the workers.

Pathogens Occurring in Raw Milk and Dairy Products

Microbes that may be present in milk can include pathogens, spoilage organisms, organisms that may be conditionally beneficial (*e.g.*, lactic acid bacteria), and those that have not been linked to either beneficial or detrimental effects on product quality or human health. Although milk can contain a full range of organisms classified as microbes (*i.e.*, bacteria, viruses, fungi, and protozoans), with few exceptions (*e.g.*, phages that affect fermentations, fungal spoilage organisms, and, to a lesser extent, the protozoan pathogens *Cryptosporidium and Giardia*) dairy microbiology to date has focused predominantly on bacteria. Between 1917 and 2017, our understanding of the microbes present in milk and the tools available for studying those microbes have changed dramatically. Improved microbiological tools have enabled enhanced detection of known microbes in milk and dairy products and have facilitated better identification of pathogens and spoilage organisms that were not known or well recognized in the early 20th century.

A. Raw and Pasteurized Milk

Both raw milk and pasteurized milk contain many types of microorganisms; they are refrigerated, yet they have limited shelf life.

The major pathogens of concern in milk and dairy products have traditionally included *Mycobacterium bovis*, *Brucella abortus* and *Coxiella burnettii*, which are the causative agents of bovine tuberculosis and a form of human tuberculosis, brucellosis and Q fever, respectively. Unfortunately, while these pathogens and their diseases have been reported to be largely eradicated in many developed countries, they still persist or are re-emerging in some of the developing nations.

Mycobacterium bovis, the causative agent for bovine tuberculosis, has been detected in milk and dairy products in different African countries including South Africa (Sichewo *et al.*, 2019), Mozambique, Nigeria, Tunisia and Zambia (Owusu-Kwarteng et al., 2020). Thus, the consumption of unpasteurized raw milk and dairy products continue to be a major risk for exposure to *M. bovis* in Africa. While bovine tuberculosis is known to be widespread in Africa, the limited or lack of sufficient data to expose the true epidemiological picture and burden of the disease in many African countries is a major concern, particularly when the burden of bovine tuberculosis might be considerably underestimated in humans.

Coxiella burnetii, an obligatory intracellular Gram-negative bacterium belonging to the family of Coxiellaceae is the causative organism of Q fever, a zoonosis of almost worldwide distribution except in New Zealand (Melenotte *et al.*, 2016). The most common reservoirs for *C. burnetii* include cattle, sheep and goats, and are considered the main sources of human infection. Thus, consumption of non-pasteurized milk and their products in Africa may be a significant source of human contamination with *C. burnetii*, as this pathogen has been detected in up to 63 per cent of cattle milk samples in Nigeria. While Q fever is usually not considered a tropical disease, *C. burnetii* was found as the etiological agent in 5 per cent of severe pneumonia cases in Tanzania. Additionally, a study of a cohort of severely ill febrile patients in Tanzania revealed 26.2 per cent zoonoses, among which 30 per cent were reportedly Q fever. Furthermore, *C. burnetii* accounts for about 1 to 3 per cent of infective endocarditis in Tunisia and Algeria while Q fever accounts for about 5 per cent of acute febrile illnesses in Burkina Faso (Vanderburg et al., 2014). About 9 per cent of community-acquired pneumonia among patients aged above 15 years in Cameroon tested positive for *C. burnetii*, with *C. burnetii* being the third most frequently isolated agent of pneumonia, after *Streptococcus pneumoniae* and Mycoplasma pneumoniae in Cameroon.

Species of Brucella including *Brucella abortus*, *B. melitensis*, *B. suis* and *B. canis* are all capable of producing brucellosis in humans, with the disease being considered to represent one of the highest public health burdens of any zoonosis globally. Ruminants are the primary hosts for *B. abortus* and *B. melitensis*, and humans become infected by consuming raw milk and dairy products, by direct contact with aborted foetuses, afterbirth and parturition fluids and during slaughter practices. More recently, *Brucella* spp., particularly *B. abortus*, have been reported in milk and dairy products with high prevalence in some African countries. These reports indicate that brucellosis or the causative microorganisms are widespread among dairy supply chains of Africa, and this presents a serious public health threat to local populations, particularly consumers of raw milk and traditional dairy products, as well as dairy farm workers.

Listeria monocytogenes, among other human pathogens, is considered a major microbiological and public threat associated with consumption of raw milk. *Listeria monocytogenes* is ubiquitous in the dairy environment. *Listeria monocytogenes* has a long history of being linked to milk and dairy products, including the fact that this pathogen causes severe disease (*e.g.*, brain infections, abortions) in both cattle and humans. It may contaminate milk during milking and can also be a cause of mastitis in milking animals resulting in direct shedding into raw milk.

L. monocytogenes has traditionally been a major public health issue in temperate regions including Europe and the US, particularly, due to their ability to grow at low temperature environments. However, they have recently been isolated from different animal and milk products across Africa. The first listeriosis outbreak linked to dairy was reported to have occurred in Halle, Germany, from 1949 to 1957; more than 100 stillbirths were suspected to have been caused by raw milk contaminated with *L. monocytogenes*. A comprehensive review on dairy-related listeriosis outbreaks was published in Journal of Dairy Science (2004).

In *broth L. monocytogenes* can grow at temperatures in the range -0.4 to 45°C, with an optimum growth temperature of 37°C. The optimum pH for growth is 7. The maximum growth data combined with data from studies with insufficient data to calculate a maximum growth rate, suggest that the growth of *L. monocytogenes* in raw milk would be very slow at temperatures up to 12-15°C, with less than 1 Log growth in 24 hours. At 20°C, up to 3 Log of growth in 24 hours of *L. monocytogenes* was recorded in raw milk. However, this is based on a small number of data points Growth of *L. monocytogenes* in heat treated milk at temperatures of 10 to 15°C was greater than raw milk with up to 3 Log growth after 24 hours.

Lactic acid bacteria (LAB) are part of the autochthonous microbiota of milk. LAB represent an order of bacteria that share common metabolic and physiological characteristics. Naturally found in milk products and decomposing plants, they produce lactic acid as the main metabolic end product.

Microbial Contamination of Milk

- During refrigerated storage (at dairy farms and processing plants) before pasteurization, only psychrotrophs can grow in refrigerated milk storage such as Pseudomonas, Flavobacterium, Alcaligenes, Listeria monocytogenes, Yersinia enterocolitica, some coliforms, and Bacillus spp.
- Those that spoil milk after heating are the thermoduric microorganisms surviving pasteurization such as Micrococcus, some Enterococcus, Streptococcus, some Lactobacillus, and spores of Bacillus and Clostridium.
- Molds and yeasts are usually eliminated during the pasteurization process therefore they cause milk spoilage after the heat treatment such as Aspergillus, Byssochlamys, Cladosporium, Candida spp.
- To a lesser extent, the protozoan pathogens such as Cryptosporidium and Giardia have been found to contaminate milk.
- The defects that occur in milk due to microorganism are:
    - Off flavours
    - Discoloration
    - Gas production
    - Lactic acid production/Souring
    - Proteolysis
    - Lipolysis with development of rancidity
    - Sweet curdling

Common LAB include the genera *Lactobacillus*, *Lactococcus*, *Enterococcus*, *Streptococcus*, *Pediococcus*. Optimum growth temperatures of LAB range from 30 – 45°C. Growth of lactic acid bacteria was observed in whole cow's milk incubated at 8-10°C and 13-15°C, with approximately 0.7 and 2 Log10 CFU/ml of growth observed after 24 hours, respectively.

The genus *Pseudomonas* represents the majority of psychrotropic bacteria in milk. Raw milk can contain a number of different *Pseudomonas* spp. The growth rates at different temperatures depend on the milk type. The limited data located suggests *Pseudomonas* spp. will grow faster in heat treated milks than in raw milk. One strain, *P. fluorescens*, is considered the main spoilage agent of stored milk (de Oliveira *et al.,* 2015). During late exponential to early stationary phase of cell growth, protease and lipase activity can result in the release of enzymes. Sufficient enzymes to cause spoilage are usually present if the cell population reaches 10^6 – 10^7 CFU/ml (Frank, 2007). Although the bacteria themselves are killed by pasteurisation, their enzymes are often heat resistant and can retain activity after heat treatment. This can lead to spoilage of milk after pasteurisation or render milk unsuitable for the production of certain dairy products. The optimum temperature range for growth of *P. fluorescens* or *P. putida* is 25 to 30°C, while the optimum growth temperature for *P. aeruginosa* is higher at 37°C. The growth rate of *Pseudomonas* spp. increases with temperature over the range of 4 to 20°C. For the data that are available, maximum growth rates of 0.1, 0.17 and 0.18 Log10 CFU/ml hour at 10, 15 and 20°C, respectively, have been observed.

Table 12.4: Bacterial Types Commonly Associated with Milk

Microorganism	*Ill Effects*
Pseudomonas	Spoilage
Brucella	Pathogenic
Enterobacteriaceae	Pathogenic and spoilage
Staphylococci	
Staphylococcus aureus	Pathogenic
Streptococcus	
S. agalactiae	Pathogenic
S. thermophilus	Acid fermentation
S. lactis-diacetyllatic	Flavour production
S. lactis	Acid fermentation
S. cremoris	Acid fermentation
Leuconostoc	Lactic Acid fermentation
Bacillus cereus	Spoilage
Lactobacillus	
L. lactis	Acid production
L. bulgaricus	Acid production
L. acidophilus	Acid production
Propionibacterium	Acid production
Mycobacterium tuberculosis	Pathogenic

B. cereus is one of the most important spoilage microorganisms in the dairy environment and its growth and associated enzyme production may result in a number of defects in dairy products, including milk flavour defects, rancidity, and changes in coagulation time. *B. cereus* can also cause foodborne illness through the ingestion of cells (10^5 – 10^8) or preformed toxins.

The natural reservoirs for *B. cereus* are decaying organic matter, fresh and marine waters, vegetables, fomites and the intestinal flora of different animals, from which soil and food products including milk and dairy products may become contaminated (Kumari and Sarkar, 2016). *B. cereus* is a spore-forming bacterium and spores from some strains are heat-resistant and will survive pasteurisation. Under ideal conditions, the minimal and optimum growth temperatures for *B. cereus* are 4°C and 30 – 40°C, respectively. The optimum pH for growth is 6-7. From the available data for raw and heat-treated milk, there appears to be very limited growth at temperatures below 10°C, with expected growth below 1 Log10 CFU/ml in 24 hours. Above 10°C, the rate of growth increases with temperature, with a 5 Log10 CFU/ml increase observed over 24 hours at 20°C in reconstituted milk or an unknown retail milk.

Campylobacter jejuni is found in the intestinal tract, udder, and faeces of cattle, in poultry and wild birds, and in contaminated water sources. *Campylobacter* spp. may be shed directly into the milk when the animal has clinical or subclinical mastitis due to *Campylobacter* infection, or indirectly through faecal contamination. The prevalence of *C. jejuni* in 25ml samples of bulk tank milk in New Zealand investigated in two studies was 1 out of 296 samples (0.3 per cent) and 0.6 per cent (CI: 0.07, 2.1), respectively (Hill *et al.,* 2012; Marshall *et al.,* 2016). *Campylobacter* spp. do not multiply at temperatures below 30°C (Park, 2002). Although unable to grow below 30°C, they remain metabolically active, and are motile at temperatures as low as 4°C.

Clostridia are thermoduric (survive pasteurisation) spore-forming organisms that can cause craters and cracks in certain cheeses due to gas production. Five hundred spores per litre of milk are sufficient to cause a late gas defect. *Clostridia* are found in the intestinal tract of humans and animals, as well as in the environment and silage. *Cl. perfringens* may also be involved in bovine mastitis so can be part of the flora of raw milk. *Cl. botulinum* is the cause of foodborne botulism, caused by the ingestion of foods (including milk) containing botulinum neurotoxin types (BoNTs) A, B, E and on rare occasions F. The heat stability of BoNT-A and BoNT-B at conventional milk pasteurisation temperatures (63 °C, 30 min) depends on the serotype; BoNT serotype A and E is inactivated but serotype B is heat-stable in milk and not inactivated by pasteurisation.

Most strains of *E. coli* are considered harmless and are part of the normal intestinal microflora of people and animals. However, the species also includes certain strains that can cause severe illness in humans. *E. coli* can be a cause of mastitis in milking animals. The prevalence of *E. coli* O157 in 25ml samples of bulk tank milk in New Zealand investigated in two studies was 1.0 per cent and 0.6 per cent (CI: 0.07, 2.1), respectively. There are limited data available to describe the growth of *E. coli* in raw milk in the temperature range 4 to 20°C. From the data on raw and heat treated milk that are available, there appears to be little growth at temperatures below 10°C, with expected growth below 1 Log10 CFU/ml in 24 hours. Above 10°C, the rate of growth increases with temperature, with 1 to 3 Log10 CFU/ml increase observed over 24 hours at 20°C in raw milk. This is less growth than predicted using the observed maximum growth rate at 20°C in heat-treated milk, which predicts up to 4.8 Log10 CFU/ml increase in cell concentration.

Infected animals can shed large numbers of *Salmonella* cells in their faeces, leading to contamination of hides, fleece, and skin, and the surrounding environment including soil, pasture, streams and lakes. The temperature range for growth of *Salmonella* spp. in broth is 5.2 - 46.2°C, with the optimal temperature being 35 - 43°C. The optimum pH for growth is 7 to 7.5. No growth of *Salmonella* was observed in pasteurised milk or raw milk at 4 to 8°C. Only two studies provided maximum growth rate estimates for *Salmonella* spp. above 8°C. Both of these studies were performed using pasteurised milk. No *Salmonella* was detected in 25ml samples of bulk tank milk in New Zealand investigated in two independent studies (n=297, n=400). There are minimal data available to describe the maximum growth rate of *Salmonella* spp. in milk products over the temperature range of 4 to 20°C, with no maximum growth rate data available for raw cow's milk.

Staphylococcus aureus is a recognised cause of mastitis in dairy herds. *S. aureus* detected in 25ml of bulk tank milk in New Zealand was found to vary, with 21 per cent of samples with no detection, 60 per cent of samples *S. aureus* counts were >1 but 10^4 cfu/ml (Hill *et al.,* 2012). It is recognised that *S. aureus* concentration needs to exceed 10^5 cfu/ml for production of sufficient toxin to cause human illness. *S. aureus* is readily killed at cooking and pasteurisation temperatures. In contrast, *S. aureus* enterotoxins are extremely heat resistant. Heat resistance for enterotoxin B has been reported at D149 = 100 min (aw of 0.99). *S. aureus* enterotoxins were reduced, but still detectable, after pasteurisation for 15 sec at 72°C, 85°C and 92°C (87 per cent, 52 per cent and 45 per cent of samples, respectively). The temperature range for growth of *S. aureus* is 7 - 48°C with optimum growth occurring at 35 - 40°C, the optimum pH for growth is in the range 6-7. The temperature range for toxin production is 10 - 48°C with the optimum temperature range of 40 - 45°C (Paulin *et al.,* 2012). The limited data available to describe the growth of *S. aureus* in milk suggests there will be little growth at temperatures below 10°C, with expected growth below 1 Log10 CFU/ml in 24 hours. Above 10°C, the rate of growth increases with temperature, with 1 to 2.5 Log10 CFU/ml increase observed over 24 hours at 20°C in raw milk or heat-treated milk. Growth rates at temperatures above 12°C were variable between studies and strains.

Starting before 1917, gradual introduction and refinement of pasteurization methods throughout the United States and many other parts of the world have improved the safety and quality of milk and dairy products. In parallel to pasteurization, others strategies for reducing microbial contamination throughout the dairy chain (*e.g.,* improved dairy herd health, raw milk tests, clean-in-place technologies) also played an important role in improving microbial milk quality and safety. Despite tremendous advances in reducing microbial food safety hazards and spoilage issues, the dairy industry still faces important challenges, including but not limited to the need for improved science-based strategies for safety of raw milk cheeses, control of postprocessing contamination, and control of sporeforming pathogens and spoilage organisms.

Chemical Hazards and Contaminants

In addition to microbial hazards, milk and dairy products can also contain chemical hazards and contaminants mainly introduced through the environment, animal feedstuffs, animal husbandry and industry practices. Thus, safety and production are intrinsically linked in the dairy food chain; from production through handling and processing to consumption. Therefore, in order to minimize the food safety risks associated with milk and dairy products, there is the need for a continuous system of preventive measures beginning with safety of animal feed, through good farming practices and on-farm controls, to good manufacturing and hygiene practices, consumers safety awareness, and proper application of food safety management systems throughout the dairy chain.

Milk-borne Diseases

Milk has long been recognized as an agent in the spread of human disease and within a few years it was appreciated that pasteurization was also providing protection against milk-borne disease. Now-a-days it is safety rather than spoilage considerations which determine the minimum legal requirements for pasteurization.

Originally the main health concerns associated with milk were tuberculosis caused by *Mycobacterium bovis* and *Mycobacterium tuberculosis* and brucellosis caused by *Brucella* spp. In some parts of the world milk is still a significant source of these infections but in the UK and some other countries they have now been effectively eliminated from the national dairy herd by a programme of regular testing and culling of infected animals. Such programmes must be constantly maintained to be effective and there have been occasional problems. Initiatives such as the culling of badgers, thought to be a reservoir of *M. bovis* have been the subject of some controversy and in 1993 there was an outbreak of brucellosis in a Welsh dairy herd, although this was the first recorded for three years. Enteric pathogens such as *Salmonella* and *Campylobacter* are still however prevalent m raw milk and pasteurization remains the most effective measure for their control.

Table 12.5: Heat Treatment of Milk

Low Temperature Holding (LTH)	62.8°C for 30 min
High Temperature Short Time (HTST)	71.7°C for 15 s
Ultra High Temperature	135°C for 1 s
'Sterilized'	> 100°C typically 20–40 min

Proposals for the heat treatment of milk were made as early as 1824, forty years before Pasteur's work on the thermal destruction of microorganisms in wine and beer. When milk pasteurization was introduced by the dairy industry around 1890, it was as much to retard souring as to prevent the spread of disease. This had become an important commercial requirement since large quantities of milk were now being transported by rail into large cities rather than being produced locally in cramped and insanitary cowhouses.

A simple test, the phosphatase test, is applied to determine whether milk has been properly pasteurized. Milk contains the enzyme alkaline phosphatase which is inactivated by the time/temperature combinations applied during pasteurization. To determine whether a milk sample has been satisfactorily pasteurized and is free from contaminating raw milk, a chromogenic substrate is added. If active phosphatase is still present then it will hydrolyse the substrate producing a colour which can be compared to standards to determine whether the milk is acceptable or not. The same principle is used in the α-amylase test applied to bulk liquid egg.

The microbiological quality of pasteurized milk is now also governed by EU-based regulations which require pasteurized milk to contain less than 1 coliform ml^{-1}, to have a count at 30° C of less than 3 × 104 and also that, after 5 days storage at 6°C, its count at 21°C should be less than 105 cfu ml^{-1}.

Gram-negative psychotrophs will not survive pasteurization, although some pseudomonads produce extracellular lipases and proteases which are heat resistant. If enough of these bacteria are present (> 105 ml^{-1}), sufficient enzyme can be produced pre-pasteurization to cause rancidity and casein degradation in the processed milk.

Raw milk may also contain a number of organisms known as thermodurics that can survive mild pasteurization treatments. These are generally Gram-positives such as the sporeforming bacteria and members of the genera *Microbacterium, Micrococcus, Enterococcus* and *Lactobacillus*, but 1–10 per cent of strains of the Gram-negative *Alcaligenes tolerans* may also survive.

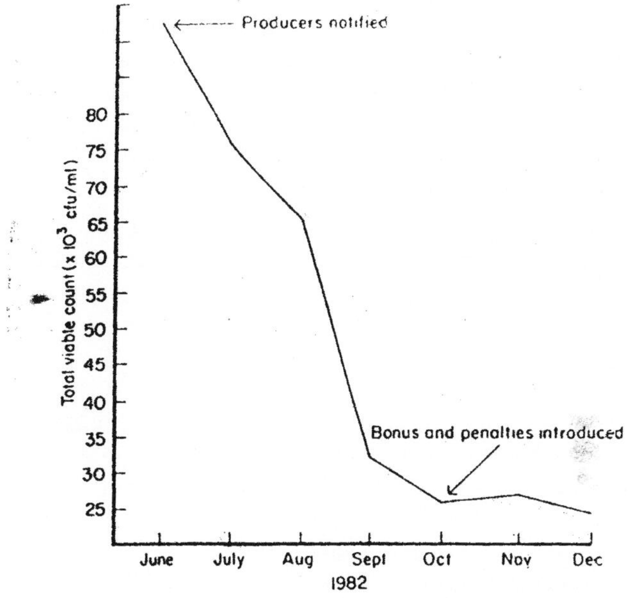

Figure 12.2: Raw Milk Counts and the Bonus Payments Scheme.

In the main, spoilage of pasteurized milk is due to the growth of psychrotrophic Gram-negative rods such as *Pseudomonas, Alcaligenes, Acinetobacter* and *Psychrobacter*, introduced as post-pasteurization contaminants. Product's shelf-life will depend on the number of contaminants introduced, but pasteurized milk produced under conditions of good manufacturing practice should keep for more than 10 days under refrigeration. Spoilage usually manifests itself as off odours and flavours described as fruity or putrid but visual defects such as clotting due to proteolytic activity can also arise. The souring traditionally associated with milk spoilage and due to the growth of lactic acid bacteria is now rare.

In milk which is subject to very low levels of post-pasteurization contamination, spoilage can result from the growth of thermioduric *Bacillus* spp. This may be associated with flavour defects but the most studied example is the bitty cream phenomenon produced by the lecithinase activity of *Bacillus cereus.* This enzyme hydrolyses the phospholipids associated with the milk fat globule membrane to produce small proteinaceous fat particles which float on the surface of hot drinks and adhere to surfaces of crockery and glasses. Bitty cream is associated mostly with milk that has been subject to temperature abuse, although psychrotrophic *Bacillus* species are becoming increasingly associated with the spoilage of refrigerated milk.

Cream

Fat can be concentrated from milk by centrifugal separation to produce a number of different types of cream. These are distinguished primarily by their fat content, which can vary from around 12 per cent for half cream up to 55 per cent in clotted cream, but also by differences in other aspects of their processing such as the heat treatment they receive and whether they are homogenized, whipped or fermented.

After separation the microbial count of the cream fraction is usually higher than that of the skim milk and, despite the fact that some bacteria are removed as slime from the separator, the combined count of skim milk and cream often exceeds that of the original milk. These observations are a result mainly of physical processes for, although mechanical separators operate at temperatures at which growth can occur (25–30°C), there is limited time for microbial growth during efficient processing. During separation it is thought that the fat globules rising through the milk act as a moving sieve to which bacteria adhere and become concentrated in the fat layer. The increase in combined count of the two fractions is attributed to the breaking up of bacterial clumps which increases the number of colony forming units, a phenomenon also noticed during the churning of cream to produce butter.

B. The Microbial Growth in Cheese

Cheese is an ancient traditional fresh or fermented dairy product with a long history of production. Cheese making originated in various West Asian countries about 8,000 years ago (**Sandine and Elliker, 1970**); it is traditionally called "milk pimples" by Mongolian, Kazak, and other nomadic people in northwestern China (**Zheng *et al.*, 2021**). The primary objective of cheese manufacture originally was to extend the shelf life and conserve the nutritious components of milk. Now there are in excess of 1000 cheese varieties worldwide (Sandine and Elliker, 1970), each unique with respect to its flavour and form.

Fermented dairy products with high nutritional value are considered "the pearls in the crown of the dairy industry". With the improvement of the quality of life, the people's dietary needs and demands have been changing in terms of quantity and quality. Furthermore, the lactose content in cheese is low, and its consumption is thus highly suitable for people with lactose intolerance. Currently, nearly 130 countries and regions produce various cheese, and the total global cheese production measures nearly 2,000 × 10^4 tons, with the European Union countries (*e.g.*, Netherlands and Germany) being the largest cheese exporters worldwide. From the development perspective of the global dairy industry, cheese is a very important dairy product. At present, cheese production in developing countries remains in its infancy, with the most individuals from these countries being unfamiliar with cheese. Simultaneously, dairy processing companies face funds and technologies limitations. With the development of dairy industry in various countries, the vigorous development of the cheese industry has become the focus of attention in recent years.

Cheese is made by curdling milk, cream, or partially skimmed buttermilk from cow or goat or a mixture of these products and then separating the whey. In general, it is prepared by adding an appropriate amount of lactic acid bacteria (LAB) starter along with rennet to milk, the fermentation by which transforms the milk proteins (mainly casein), carbohydrates, and fats. Next, the whey is removed and the remaining product is ripened for a certain period. LAB may not be added for making some cheeses.

Cheese is made differently in different regions. For instance, for making cheddar cheese in southwest England, the raw material is sterilized and cooled; then, the fermenting agent, calcium chloride, and rennet are added to ferment the curd. After 30–40 min, the formed clot is cut into 5-mm sized pieces, which are then allowed to stand for 15 min and stirred for another 5–10 min. The clot is then turned over and stacked, broken, salted, molded, and pressed. The pressed cheese is put into a fermentation room for fermenting and ripening after the cloth is changed. In contrast, the milk used to make Parmesan cheese is collected in two separate steps, where the overnight milk and the fresh milk (collected the next morning) are mixed in a copper cheese tank. When the temperature reaches 52°C, the cheese is wrapped with gauze for cutting, molding, pressing, and then, soaking in brine for 3 weeks (**D'Incecco *et al.*, 2020**). Approximately 5 kg of water is evaporated during maturity. On the other hand, soft and semi-hard cheeses, such as feta cheese, are soaked in brine for a short period (**Marino *et al.*, 2017**). For Kazak cheese making, the milk collection process is similar to that of Parmesan cheese, with few differences: In a goatskin bag, old yogurt may or may not be added as a starter, followed by fermentation into yogurt (**Zheng *et al.*, 2021**). Next, this yogurt is boiled with stirring to evaporate water. The remaining curds are placed in a canvas bag, which is then hung outdoors to remove moisture further and solidify into fresh cheese. This fresh cheese is then cut into small pieces or made into a pie shape and finally placed on a bamboo board for 30–90 days of spontaneous ripening (**Zheng *et al.*, 2018**).

Foodborne Illnesses Linked to Cheese Consumption

All foods possess a finite risk of microbiological contamination, the level of which varies with the food.

The highest risk factors include foods of animal origin and foods consumed without prior cooking (either by the manufacturer, the caterer or the consumer). Relatively few foods possess both of these risk factors although cheeses made from unpasteurised milk, fermented meats, smoked or pickled fish, Sushi and some mayonnaise recipes may fall into this commodity category. Although the numbers of food poisoning outbreaks associated with these commodities is not high in comparison with, for example, the number of *Salmonella* infections due to poultry, because of the severity of some reported health hazards, the IFST considers it appropriate to issue a Position Statement on cheese, some varieties of which are examples of this type of commodity.

Spoilage of Cheese

- Cheese is a fermented milk product that is made by coagulating the casein present in milk by using the enzyme rennet.
- The ripening in cheese is achieved due to the proteolytic and lipolytic activities of various microorganisms.
- The low moisture content of hard and semi-hard ripened cheeses makes them susceptible to fungi compared to bacteria.
- Soft and fresh cheeses are spoiled easily due to their higher pH, moisture content, and lower salinity.
- The bacterial cheese spoilage is caused by *Clostridium* spp. (especially *C. pasteurianum*, *C. butyricum*, *C. sporogenes*, and *C. tyrobutyricum*), *Bacillus polymyxa*, *Flavobacterium*, *Pseudomonas* spp., *Alcaligenes*, and *Achromobacter*.
- Yeasts are common cheese spoilage organisms *e.g.*, *Candida* spp., *Debaryomyces hansenii*, *Geotrichum candidum*, and *Pichia* spp.
- The mold spoilage in cheese is caused mainly by *Penicillium* spp. and *Cladosporium* spp.
- Major pathogenic bacteria found in cheese are *Listeria monocytogenes*, *Salmonella* spp., and *Escherichia coli*.

Cheese is generally considered to be a relatively 'safe' food. However, the spread of some diseases by cheese has been demonstrated and, as a result, most cheese is now produced from milk that has been pasteurised. Pasteurisation is one of the major critical control points in the cheesemaking process that prevents pathogenic micro-organisms in the raw milk from contaminating the finished product, which may otherwise contribute a hazard to human health.

Nevertheless, the production of some cheeses still occurs from milk that has not received a 'full' pasteurisation process. Cheese manufactured by 'traditional' processes may be made from raw milk (*i.e.*, milk that has received no heat-treatment at all) whilst some larger-scale operations may use thermised milk (*i.e.*, milk that has been subject to a heat-treatment that is less severe than the full pasteurisation process of 71.7°C for 15 seconds or equivalent). Although thermisation destroys some species of micro-organisms, some pathogenic bacteria, such as *Listeria monocytogenes*, *Salmonella* and *E. coli* O157 may survive and thus, contaminate the final product even though the milk has received a moderate heat process.

Milk produced in Ethiopia is sometimes processed into cottage cheese, also called ayib. Ayib is produced by heating the buttermilk, obtained after churning sour milk, to precipitate the protein fraction. Although ayib has a pH of approximately 3.7, it has a short shelf life due to its high moisture content. Soft cheeses are high moisture content foods that are ripened over a very short time, which does not allow for pathogen die-off. Soft cheeses have intrinsic factors that support microbial growth and they have been implicated in outbreaks of listeriosis in the past. Seven studies that examined the occurrence of Listeria in cottage cheese samples (n = 431) in Ethiopia reported a median prevalence of 4 per cent with prevalence ranging from 0 to 53 per cent for Listeria spp., and a median prevalence of 0 per cent with prevalence ranging from 0 to 5 per cent for *L. monocytogenes* (Keba *et al.*, 2020).

Cheese has a diverse microbial community, which indeed can vary within the cheese from the core to the surface that is greatly influenced by manufacturing including ripening conditions. Understanding the composition of this community (microbiota), and its impact on the quality and safety of cheese products, is of critical importance. In addition to, in the majority of cases, consciously added starter and adjunct bacteria (which are added as a supplement), cheese contains a heterogeneous variety of other, non-starter, microorganisms. These various microorganisms can play vital roles in the development of the organoleptic properties of cheese (**Fox *et al.*, 2000**), nutrient composition, shelf-life, and safety.

Foodborne illnesses related to cheese consumption have occurred in many countries. *S. aureus* infections have been linked to the use of unpasteurized milk or to contamination due to improper handling because the pathogen at more than 5 Log CFU/mL produces heat-resistant enterotoxin (**Delbes *et al.*, 2006; CDC, 2010**). Intake of 20 to 1,000 ng of the enterotoxin can cause typical symptoms of *S. aureus* infection (**Pelisser *et al.*, 2009**). In England, consumption of *S. aureus* enterotoxin contaminated Stilton cheese (internally mold-ripened semisoft blue cheese produced by using *Penicillium roqueforti*) caused 155 illnesses (**Maguire *et al.*, 1991; Johnson, 2001**). In the USA, cheese has also been a food vector for *S. aureus* foodborne outbreaks (**CDC, 2014**). In 1981, improper cheese pasteurization led to 16 cases of *S. aureus* foodborne illness in USA. In 1999, Minas cheese (soft cheese) made from raw bovine milk caused two outbreaks in Brazil affecting 378 individuals (**Simeão do Carmo *et al.*, 2002**). Recently, six *S. aureus* outbreaks occurred in France due to consumption of enterotoxin type E in soft cheese that was processed during weeks 40 and 41 (**Ostyn *et al.*, 2010**).

Cheese-related *L. monocytogenes* outbreaks have had a relatively high fatality rate (15 per cent to 30 per cent), leading to increased public awareness of this infection. Clinically, *L. monocytogenes* causes sepsis and meningitis in immunocompromised individuals (*e.g.*, transplant patients, elderly patients receiving chemotherapy, individuals with diabetes or liver disease), and around 25 per cent of the invasive listeriosis cases occur in pregnant women. Of 13 *L. monocytogenes* serotypes, three (1/2a, 1/2b, and 4b) account for more than 95 per cent of human illnesses due to *L. monocytogenes*. In 2005, 10 cases of *L. monocytogenes*-induced listeriosis occurred in a small northwestern Swiss town with three fatalities due to bacteraemia. Eight of these cases were in older immuno compromised patients, while two cases occurred in pregnant women leading to abortion. This outbreak occurred after these patients had Tomme (soft cheese) contaminated with *L. monocytogenes* serotype 1/2a. **Fretz *et al.* (2010)** reported that Quargel (acid curd cheese) contaminated with *L. monocytogenes* serotype 1/2a caused four deaths out of 14 cases in Austria and Germany in 2009. In the USA, a multistate outbreak of listeriosis among pregnant women occurred between October 2008 and March 2009. These outbreaks were caused by the intake of Mexican-style cheese made from pasteurized milk (**Jackson *et al.*, 2011**). In addition, there was another Mexican-style cheese-related outbreak of listeriosis in North Carolina in the USA between October 2000 and January 2001. Recently, the **CDC (2014)** announced a multistate listeriosis outbreak due to cheese products in California and Maryland. This outbreak affected eight individuals, seven of whom were hospitalized and one died. Even Japan, which has a relatively low cheese consumption compared to the western countries, also had *L. monocytogenes* serotype 1/2b outbreaks due to cheese consumption resulting in 86 cases of *L. monocytogenes* infection (**Makino *et al.*, 2005**).

In cheese, Shiga toxin-producing e. Coli (stec) strains such as e. Coli o157:h7 are important foodborne pathogens. In 2003, e. Coli o157:h7 was linked to an unpasteurized gouda cheese-related outbreak involving 13 cases in Canada. This outbreak resulted in two cases of lethal haemolytic uremic syndrome. Recently, there was a multistate outbreak of e. Coli o157:h7 infections associated with gouda cheese intake in the us. This outbreak affected 38 people in five states (Arizona, California, new Mexico, and Nevada); 15 reported hospitalizations, one showed haemolytic uremic syndrome, but no deaths occurred. Although gouda cheese is a hard cheese (low aw of 0.950 to 0.970, similar to cheddar, colby, edam, and swiss cheeses, e. Coli o157:h7 outbreaks have occurred (choi *et al.*, 2016).

Salmonella is a well-recognized foodborne pathogen. All salmonella strains are gastroenteritis-inducing pathogens, and various salmonella serotypes have been involved in cheese-borne outbreaks. Consumption of Mexican-style cheese in Kane County, Illinois, USA led to an outbreak of salmonella Newport infections during the march 2006 through April 2007 time period, affecting 85 individuals. In the county, S. Newport was a rare, multidrug-resistant serotype. In 2013, raw cashew cheese was linked to a multistate outbreak of *Salmonella* Stanley in California (15 cases), Nevada (one case), and Wyoming (one case) in the US. This outbreak led to three hospitalizations, but no fatalities. Although *Salmonella* Muenster is rare in Europe, the first published outbreak of S. Muenster was reported in Europe in 2008. In France, a nationwide outbreak of gastrointestinal illness due to S *Muenster* occurred during march to April, 2008 after goat cheese consumption in 25 people. Four patients were hospitalized, but no fatalities occurred (van Cauteren *et al.*, 2009).

Some reported cheese-associated outbreaks of foodborne illness (affecting a few or many individuals) are summarised in Table 12.6 but these reports do not include sporadic, individual cases. Taken together, cheese related foodborne illnesses have been generally linked to soft cheese or cheese made from raw or unpasteurized milk, but rarely in hard cheese. Cheeses made with unpasteurised milk appear to have been involved in the majority of reported outbreaks. In the remaining reported outbreaks, it is believed that either milk pasteurisation was not carried out properly (and the heat-treatment was therefore inadequate) or the cheese was made from correctly pasteurised milk subsequently contaminated with pathogenic micro-organisms. Indeed, in the Mexican-style cheese outbreak, the cheese appears to have been made from correctly pasteurised milk to which raw milk had deliberately been added.

It is recognised that, in cheese manufacture, post-pasteurisation contamination can occur whether or not the milk is pasteurised. Nevertheless, the outbreaks shown in Table 12.6 indicate that food-poisoning incidents are more likely to arise with cheese made from unpasteurised milk even if the starting material is of exceptionally high microbiological quality and the manufacturing conditions are the most hygienic possible.

Although these outbreaks do not represent a large proportion of all outbreaks of foodborne illness, on some occasions the consequences for the victims have been particularly severe. Organisms such as *Listeria monocytogenes*, *Salmonella*, *Brucella melitensis* and *Escherichia coli* O157 have all been involved in cheese-associated outbreaks and have caused severe infections which, in a significant proportion of cases, have resulted in death or long term sequelae. *Listeria monocytogenes* can cause meningitis and septicaemia with up to 30 per cent mortality as well as abortion in pregnant women; some *Salmonella* spp. can cause septicaemia and may result in long-term illness such as reactive arthritis; *Brucella melitensis* causes undulant fever, a severe condition that can be long-lasting and incapacitating; Verocytotoxin-producing strains of *Escherichia coli* (including *E. coli* O157) may cause haemorrhagic colitis, haemolytic uraemic syndrome and renal failure which may result in death, particularly in young children.

Hence, manufacturing process and distribution and storage process should be controlled appropriately to prevent foodborne illness by these cheeses. In addition, microbial risk assessment including predictive model for exposure assessment also should be importantly counted for microbiological safety assurance.

Microbial Risk Assessment in Cheese

A microbial risk assessment is considered a scientific and methodical tool for preventing, regulating, and understanding the risk caused by hazardous microorganisms (**EPA, 2012**). The microbial risk assessment consists of four components: hazard identification, exposure assessment, hazard characterization, and risk characterization. This methodology has been used in many countries to prevent the occurrence of foodborne illnesses, to reduce the risk of foodborne illness outbreaks, and to establish microbial criteria. As discussed in previous section, a number of cheese-related foodborne illness outbreaks have occurred. Additionally, cheese consumption has increased despite these foodborne illness outbreaks, which raised the necessity of microbial risk assessments on cheese (**CDC, 2014; USDA/ERS, 2014**).

In New Zealand, the risk profiles prepared by included elements of microbial risk assessment and the information related to risk management for *L. monocytogenes* in soft cheese and low moisture cheese. This work examined both risk assessment outcomes and action, in addition to providing a ranking of food safety issues. Lake *et al.* (2005) suggested that additional risk management measures are unnecessary for low moisture cheeses (<50 per cent of moisture content) such as semi-soft cheese (39 per cent to 50 per cent moisture; stilton, roquefort, gorgonzola, limburger, gouda, edam, *etc.*), hard cheese (less than 39 per cent moisture; cheddar, emmental, gruyere, *etc.*), and very hard cheese (less than 34 per cent moisture content; asiago old, parmesan, romano, grana, *etc.*) because of the low risks associated with *L. monocytogenes* exposure in the general population. A study by lake *et al.* (2005) presented that there is no evidence to link soft cheese consumption with cases of *L. Monocytogenes* foodborne illness in New Zealand, and contamination rates of the pathogen were also very low in domestically produced soft cheese, but they suggested that the contamination with *L. monocytogenes* may occur in soft-ripened cheese after postharvest, especially during handling in ripening rooms, wrapping packing stage, and cutting stage. In addition, surface ripened cheese may be at risk because of increased pH during ripening.

The FDA (2012) completed a *L. monocytogenes* microbial risk assessment on soft-ripened cheese from pasteurized milk and raw milk in the USA and Canada. In this research report, quantitative models were used to predict the risk of *L. monocytogenes* in cheese, and to calculate the probability of invasive listeriosis per serving in soft-ripened cheese (camembert-like cheese) that was produced using raw milk vs pasteurized milk. This report showed that the risk from raw milk cheese was higher than that of pasteurized milk cheese reflecting the higher contamination rate of raw milk.

According to a report by CDC (2013a) soft cheese made with unpasteurized milk was estimated to be 50 to 160 times more likely to have *L. monocytogenes* infection than those made with pasteurized milk. However, only camembert-like cheese was used for a quantitative risk assessment potentially limiting the scope of the model because other cheese may have different properties that affect *L. monocytogenes* growth. The risk of listeriosis from the cheese were 1.4×10^{-10}, 7.2×10^{-9}, 1.8×10^{-8}, and 6.1×10^{-9} for non-susceptible individuals (general), the elderly, pregnant women, and immunocompromised patients in Canada, respectively, and 1.2×10^{-10}, 7.3×10^{-9}, 1.8×10^{-8}, and 5.2×10^{-9} for identical groups in the USA, respectively. In addition, the risk of camembert-like cheese made by raw milk was calculated for non-susceptible (9.5×10^{-9} and 1.8×10^{-8}), elderly (3.8×10^{-7} and 8.2×10^{-7}), pregnant women (9.2×10^{-7} and 1.8×10^{-6}), and immunocompromised (4.2×10^{-7} and 8.1×10^{-7}) individuals in the Canada and the USA, respectively. These results indicate that pregnant women, elderly individuals, and immunocompromised patients had a higher risk for listeriosis from camembert-like cheese consumption. In terms of relative risk, pregnant women are about 10 times more likely than the general population to get *L. monocytogenes* infection while the elderly (65 years and older) are about four times more likely than the general population to get *L. monocytogenes* infection (CDC, 2013b). Therefore, the CDC (2013b) categorized soft cheese made from raw milk such as queso fresco, blue-veined, feta, brie, and camembert cheeses into a higher risk group, but soft cheese made from pasteurized milk such as processed, cream and mozzarella cheese, and hard cheese were categorized into the lower risk group.

These microbial risk assessment studies indicate that raw milk soft cheese has higher risk than low moisture cheese for *L. monocytogenes* infection, especially for the elderly, pregnant women, and immunocompromised patients. Even though milk is pasteurized, soft cheese could be contaminated with *L. monocytogenes* due to postharvest contamination. In addition, the risk of selected hazards such as *C. jejuni*, *E. coli*, *Salmonella* spp., *S. aureus*,

Table 12.6: Reported Outbreaks of Foodborne Disease due to Cheeses Since 1983

Outbreak	Pathogen	No. of Cases	No. of Deaths	Food	Comments	Reference
1983, Netherlands Denmark, Sweden USA	Enterotoxigenic *Escherichia coli*	>3000	NR	Brie cheese	No correlation could be made between high levels of Enterobacteriaceae/*E.coli* and high phosphatase activity (indicating the use of raw milk)	MacDonald *et al.*, 1985; Nooitgedagt and Hartog, 1988
1983-7, Switzerland	*L.monocytogenes*	>122	34	Vacherin Mont d'Or cheese**	Made from thermised milk since 1984	Bille, 1990
1984, Canada	*Salmonella typhimurium*	2700	1	Cheddar cheese**	*Salmonella* survived for up to 8 months in refrigerated storage	D'Aoust *et al.*, 1985; D'Aoust, 1994
1984-5, Scotland,	*Staphylococcus aureus* enterotoxin	>13	0	Sheep milk cheese**	Contamination was associated with clinical mastitis and post-infection carriage by ewes	Bone *et al.*, 1989
1985, Switzerland	*Salmonella typhimurium*	>40	0	Vacherin Mont d'Or cheese**	Hand-borne cross-contamination from a pigsty	Sadik *et al.*, 1986
1985, USA	*L.monocytogenes*	>142	48	Mexican style cheese*	Contamination probably occurred from the addition of raw milk	Linnan *et al.*, 1988
1988-89, England	Unknown cause	155	0	Stilton cheese**	Subsequent decision to pasteurise milk for Stilton manufacture	Maguire *et al.*, 1991
1989, England	*Salmonella dublin*	42	0	Irish soft cheese**	Four cows were asymptomtic excretors. *Salmonella* was detected in the cheese curd but not in the raw milk nor in the factory environment	Maguire *et al.*, 1992
1989, USA	*Salmonella javiana* and *S. oranienberg*	164	0	Mozzarella	The contamination level was in the range 0.36 cells/100 g to 4,3 cells/100 g	Hedberg *et al.*, 1992
1992, England	*Salmonella livingstone*	10	NR	Cheese	–	Djuretic *et al.*, 1997
1992-3, France	Verocytotoxin-forming *E.coli*	NR†	1	Fromage frais**	–	Anon 1994a
1993, France	*Salmonella paratyphi B*	273	1	Goats' milk cheese**	Internal microbiological monitoring failed to detect the contamination for 2 months	Desenclos *et al.*, 1996
1994, Scotland	Verocytotoxin-forming E.coli O157	>20‡	0	Local, farm-produced cheese**	–	Anon 1994b
1995, France	*L. monocytogenes*	20	4	Brie de Meaux cheese**	Disinfection and control measures were reinforced at the plant	Goulet *et al.*, 1995
1995, Malta	*Brucella melitensis*	135	1	Soft cheese**	–	Anon, 1995.
1995, Switzerland and France	*Salmonella dublin*	NR Switz. 25 in France	NR Switz. 5 in France	Cheese from Doubs region of France**	The outbreak was resolved after production control measures (but not pasteurisation) were introduced	Vaillant *et al.*, 1996
1996, England and Scotland	*Salmonella gold-coast*	>84	0	Cheddar cheese+	–	Anon, 1997
1996, Italy	*Clostridium botulinum*	8	1	Mascarpone cheese	"This outbreak of botulism underlines the necessity to apply HACCP principles...."	Aureli *et al.*, 1996
1997, England	*E.coli O157*	2	0	Lancashire-	– type cheese**	Anon, 1997c

*: This cheese was made from pasteurised milk to which raw milk had deliberately been added; **: These products are known to have been produced using unpasteurised milk; ***: Both pasteurised-milk cheese and thermised-milk cheese were contaminated; +: Associated with failure of the pasteurisation process; NR: Not reported; †: All cases of haemolytic uraemic syndrome; ‡: One case of haemolytic uraemic syndrome.

and *L. monocytogenes* of raw milk cheese was analyzed according to different types categorized into 'extra hard', 'swiss', 'cheddar', 'blue', 'feta', and 'camembert'. Extra hard cheeses including parmigiano reggiano, grana padano, sbrinz *etc.* have low moisture content (<36 per cent), and the risk represented a low to negligible. Swiss cheeses including gruyere, emmentaler, appenzeller, tilsiter, tete de de moine, and vacherin fribougeois had low to negligible risk but *L. monocytogenes* represented high risk to susceptible population. Cheddar cheese (moisture: 37 per cent to 42 per cent) had high risk of *E. coli* to general and susceptible population because *E. coli* could survive during cheese making process and maturation, and other pathogens represented low to moderate in case of cheese made from sheep milk. Blue cheese ripened with mold had low risk of *L. monocytogenes* to general population and high risk to susceptible population because *L. monocytogenes* could increase in cheese making process. In cheese making process of feta, *E. coli* might not decrease, which resulted in high risk of *E. coli* and also high risk of *L. monocytogenes* to susceptible population. Camembert (moisture: >55 per cent) cheese showed high risk of *E. coli* to both general and susceptible population and the risks of *L. monocytogenes* were low to general and high to susceptible population, respectively, due to increasing bacterial counts in cheese making process and maturation (Choi *et al.* 2016).

Stec strains such as *E. coli* o157:h7, o26:h11, o111:h8, and o145:h28 are able to survive and grow during cheese processing, especially in soft cheese. Perrin *et al.* (2015) developed a quantitative risk assessment for Stec strains that cause haemolytic uremic syndrome. This research utilized a stochastic model to determine the effect of preharvest and postharvest interventions on the risk of raw milk soft cheese consumption in France. Predicted average percentages of contaminated cheese at baseline were 16.31 per cent and 1.75 per cent for total Stec and five main pathogenic serotypes of Stec (Mps-Stec; o157:h7, o26:h11, o103:h2, o111:h8, and o145:h28), respectively. The levels of contamination of servings were 0.12 cfu/g for Stec and 0.21 cfu/g for Mps-Stec. In addition to this baseline simulation, Perrin *et al.* (2015) simulated preharvest prevention and postharvest prevention, and the risk reduction varied from 1 per cent to 96 per cent with the combination of preventions. Thus, preharvest and postharvest treatments need to be applied to reduce the microbial risk.

The most predictive models for cheese were developed to predict *L. monocytogenes* growth because the pathogen was considered high-risk pathogen when it contaminates in cheese. Thus, many countries applied 'zero-tolerance' policy for *L. monocytogenes* in cheese but non-Stec and *S. aureus* had quantitative microbial criteria rather than 'zero tolerance' policy. According to this fact, it has been suggested that predictive models for non-Stec and *S. aureus* should be developed to improve the accuracy of microbial risk assessment for the pathogens in cheese.

Microbial Safety of Cheese

Traditionally, it has been assumed that pathogenic micro-organisms in raw milk die during cheese manufacture due to the production of high acidity (low pH value) and competition from the 'starter cultures' deliberately added to produce a characteristic flavour and texture. However, it is not possible to generalize about either the composition of cheese or the behaviour of contaminating pathogens.

The compositional characteristics of the soft, mould ripened cheeses (*e.g.*, Brie) are very different from those of the very hard, acid types (*e.g.*, Parmesan). Not only do the pH values of these two products differ widely but also the water activities differ; ripened Brie has a neutral pH and high water activity that permits the growth of many pathogens whilst Parmesan is more acidic with a much lower water activity that inhibits microbial growth. Parameters such as these are paramount in controlling the growth, survival or death of pathogenic micro-organisms, especially if a low storage temperature is not maintained continuously; it therefore cannot be assumed that all pathogenic micro-organisms will die out in all types of cheese. However, even the harsh conditions found in the hard, acid cheeses do not guarantee that all pathogenic bacteria will be completely destroyed and the recent outbreaks of food-poisoning due to *Salmonella* and *E. coli* O157 in semi-hard cheeses bear testimony to this fact.

Pathogenic bacteria vary just as widely as the cheeses they contaminate and their survival characteristics are equally varied. For example, Brie stored under refrigeration will support the growth of *Listeria monocytogenes* whilst Parmesan stored at near-ambient temperature will not; similarly, *Salmonella* may die slowly in Cheddar cheese but *Staphylococcus aureus* may remain viable throughout the shelf life.

The manufacture of cheese is a complex process and the assurance of consumer safety is equally complex. The diverse varieties of cheese are made by a wide range of techniques. Each requires specific procedures that must be (and are) applied for safeguarding the consumer. The manufacture of safe cheese demands the application of systems based on the Hazard Analysis Critical Control Point (HACCP) principles. Such systems must be applied at all points in the process from the rearing of milk-producing animals to cheese consumption by the consumer and indeed, a survey of cheeses sold in the UK has demonstrated a significant improvement in their microbiological quality in recent years.

Control Measures

It is indisputable that some outbreaks of food-borne illness have been clearly linked with the consumption of cheese, the majority of those reported being associated with cheese made from unpasteurised or improperly-pasteurised milk. Whilst pathogens can and do gain access to cheese after curd formation, it is clear that many food-borne pathogens are faecal in origin, it not being possible to milk cows aseptically. Faecal contamination of raw milk, no matter how slight or infrequent, is therefore inevitable. In addition to potential faecal contamination, pathogens may also be excreted into the milk directly from the udder.

Correctly-controlled milk pasteurisation kills such bacteria. Pasteurisation was designed to destroy the vegetative pathogens that may be found in raw milk, and the part that pasteurisation of milk played in dramatically reducing the 19th century diseases such as milk-borne tuberculosis, brucellosis and typhoid fever is widely recognised. Pasteurisation is a Critical Control Point that still provides the simplest means of ensuring the destruction of vegetative pathogens in raw milk. This can clearly be seen from the number of food poisoning outbreaks that have resulted from pasteurisation failure or its deliberate mis-use. Thus, if milk has not been fully and correctly pasteurised, the death of all vegetative pathogens originating from raw milk cannot be guaranteed in all types of cheese on all occasions, no matter how good the control of hygiene during milking and cheese production. Following the major outbreaks of listeriosis due to contaminated cheese many producers of cheese have changed from the use of unpasteurised to pasteurised milk while great improvements have been made in the conditions of hygiene in major cheese factories.

Susceptibility to foodborne pathogens varies greatly from person to person. Young children, the elderly and the immuno-compromised are most at risk whilst pregnant women are particularly susceptible to listeriosis: together these groups form at least 20 per cent of the population. However, it is clear in some of the outbreaks reported that foodborne pathogens can also cause disease in healthy adults. Because of these and other variables, such as the level of contamination in the cheese and the quantity of cheese consumed by different individuals, it is important to miss no opportunity to add another protective control measure to the microbiological safety barrier that the responsible manufacturer will want to build to protect his/her customers and the business. Pasteurisation provides an effective control measure although innovative technologies are being developed that may provide alternatives to heat-treatment for the reduction of pathogens in raw milk.

A 'farm to fork' approach employing HACCP-based systems is widely advocated to control pathogens in food production. Good Manufacturing Practice should be employed at all stages of production and handling, from production of the raw milk at the farm to purchase of the cheese by the consumer. These principles demand that the safety conferred by pasteurisation must be assessed in relation to the quality of the raw milk and the type of cheese being manufactured. The "Code of Best Practice" recently published by the Specialist Cheesemakers' Association is a valuable guide which details numerous good hygienic practices for the production of artisan cheeses, though the IFST considers it is misguided in encouraging the use of unpasteurised milk.

It should not be assumed that all countries have the same level of raw milk hygiene. It is well-recognised that, together with Denmark, UK raw milk has by far the highest microbiological quality in the world. Yet even so, the presence of pathogens in raw milk was demonstrated in a survey of 1673 samples of raw milk from farm bulk tanks in England and Wales in 1992-3 in which 0.36 per cent contained *Salmonella* spp. and 5.08 per cent contained *Listeria monocytogenes*. In addition, the UK Advisory Committee on the Microbiological Safety of Food (18th September 1997) has expressed concern at the presence of pathogens in raw milk for direct consumption in England and Wales, as revealed by surveys carried out by ADAS and the Public Health Laboratory Service (PHLS). In the ADAS survey, *E. coli* (some strains of which are pathogenic and whose presence in foods is considered indicative of faecal contamination) was isolated from 62 per cent of samples, 0.4 per cent having levels of 1000 organisms/ml or more (Anon, 1997a). Moreover, the presence of pathogens may be even more likely in raw milk produced in countries with less strict control of farm hygiene.

Because the pasteurisation of milk forms an important critical control point in the manufacture of safe cheese, it is considered that pathogens in raw milk for cheese manufacture are most effectively eliminated by a full pasteurisation process, unless a cheesemaker can demonstrate that his/her process and product composition together can provide a safe product when manufactured from pathogen-contaminated milk.

Consumer Awareness

The consumer has the right to choose whether to eat cheese made from raw milk or from pasteurised milk. Hence, IFST considers that:

1. All cheeses made from unpasteurised milk should be clearly and adequately labelled as such at the point of sale to the consumer, including delicatessen counters, restaurants and other catering establishments.
2. For those products where a risk assessment indicates a hazard from pathogens in the raw milk, the 'small print' wording, "made from unpasteurised milk" is inadequate. A warning such as that currently required by law for raw, drinking milk sold in England and Wales could be used, *e.g.*, "this cheese is made from milk that has not been pasteurised and may contain organisms harmful to health".

3. The public should be educated as to the potential risks involved, so that the consumer can make an informed choice.
4. Particularly vulnerable people (young children, the elderly, pregnant women and immuno-compromised people) should avoid consuming soft cheeses made with unpasteurised milk

In order to ensure the production of safe cheese, the following measures are therefore important:

1. The raw milk should be collected and maintained in good hygienic conditions,
2. If the raw milk is not to be used immediately, it should be refrigerated to minimise multiplication of bacteria,
3. For those products where a risk assessment indicates a hazard from pathogens in the raw milk, the raw milk should undergo a full pasteurisation or equivalent process,
4. Good conditions of hygiene should be maintained throughout cheese manufacture, ripening, distribution, sale and safe storage until consumption to prevent contamination.

Microbiological tests on finished cheeses are useful for monitoring the successful application of controls but these tests cannot ensure the microbiological safety of the cheese.

In conclusion, although cheese is considered microbiologically safe, cheese-borne outbreaks have still occurred, especially in soft cheese. Predictive models have been used to determine the factors related to bacterial growth such as pH and a_w, and the results from predictive models are used in microbial risk assessment in many countries to uncover cheese that might pose risks to consumers. The microbial risk assessment then showed that microbial risk increased in cheese with high moisture content, especially for raw milk cheese. However, applications of preharvest and postharvest preventions can reduce the microbial risk. Even though microbial risk assessments were conducted for cheese, in some studies limited data were used for predictive models, temperature profile for distribution and consumption patterns as well as dose-response model, especially for susceptible group such as elderly, pregnant women, and immunocompromised individuals. Therefore, more predictive models in cheese-related conditions, dose-response models need to be developed in conjunction with greater collection of consumption data to improve the performance of cheese microbial risk assessment. In addition, the results from microbial risk assessment should be further used for appropriate level of protection and food safety objective, which are for applications on microbial criteria.

II. The Perishability of Fish

Most of the people think about the fish as the swimming teleosts and elasmobranchs. Now let us try to understand more about fish.

Year by year the human population increases in size passing the 7×10^9 mark in 2011, and likely reaching approximately 10^{10} individuals over the course of the coming 30 years (United Nations, Department of Economic and Social Affairs, Population Division, **2015**). This growing population is in need of a steady supply of highquality protein, which to an increasing degree is being supplied by shell and finfish meat. In terms of greenhouse gas emissions, this trend is commendable as the diet of fish eaters on average emits approximately 50 per cent less greenhouse gases than that of meat eaters (Scarborough *et al.,* **2014**).

According to the Food and Agriculture Organization (FAO) (**2020**), world fish production has increased dramatically during the past 60 years, to around 179 million tons in 2018 with a value of $401 billion. Global fish consumption also increased from 9.0 kg per capita in 1961 to 20.5 kg in 2018. Aquaculture production presents 46 per cent of the total production and 62 per cent of the total sale value. Due to the increasing demand for high-quality protein, reduction of the wild fish catch, and advancement in fish farming technologies, global aquaculture production is expected to double by 2050 (Ibrahim *et al.,* **2020**).

China is the major producer of aquaculture fish, contributing more than half of the global production. Carp, tilapia, and salmon are the top three finfish species produced globally. According to the United States National Oceanic and Atmospheric Administration (NOAA) (**2020**), in 2017 the United States ranked 17th in aquaculture production with an estimated 625.7 million pounds valued at $1.47 billion. The leading species for freshwater and saltwater production are catfish (330.4 million pounds) and Atlantic salmon (32.4 million pounds), respectively. The United States has also been the leading importer of fishery commodities, in 2018 importing 6.1 billion pounds of edible fishery products valued at $22.4 billion.

Based on the U.S. Food and Drug Administration (FDA) Code of Federal Regulations Title 21 (21CFR), fish is defined as freshwater and saltwater finfish, crustaceans, mollusks, and other forms of aquatic animal intended for human consumption (21CFR123.3). Fish products represent any human food product containing fish as a characterizing ingredient (21CFR123.3). In 2018, 2,221 fish species were captured, 622 of which belonged to aquaculture production (FAO, **2020**). About 88 per cent of the global fish production are for direct human consumption, 44 per cent of them are in live, fresh, or chilled statuses, forms that are most preferred and highly priced in many markets (FAO, **2020**). The rest of production is processed, with 35 per cent frozen, 11 per cent in prepared and preserved forms, and 10 per cent dried, salted, smoked, or otherwise cured (FAO, **2020**). Because finfish contributes to more than 65 per cent of world food fish production, the fish and fish products discussed here refer to finfish and finfish products. In the United States, 76.5 per cent of domestic landings were consumed fresh and frozen, 1.9 per cent canned, and 1.5 per cent cured (NOAA, **2020**). The top processed fish species in 2018 were Alaska pollock (1.6 billion pounds, $1.98 billion), sockeye salmon (193 million pounds, $940 million), tuna (384 million pounds, $836 million), and cod (292 million pounds, $745 million).

Although broadly similar in composition and structure to meat, fish has a number of distinctive features. Unlike meat, there are no visually obvious deposits of fat. Although the lipid content of fish can be up to 25 per cent, it is largely interspersed between the muscle fibres. A further feature which contributes to the good eating quality of fish is the very low content of connective tissue, approximately 3 per cent of total weight compared with around 15 per cent in meat. This and the lower proportion of body mass contributed by the skeleton, reflect the greater buoyancy in water compared with that in air.

Muscle structure also differs. In land animals it is composed of very long fibres while in fish they form relatively short segments known as myotomes separated by sheets of connective tissue known as myocommata. This gives fish flesh its characteristically flaky texture.

Fish flesh generally contains about 15–20 per cent protein and less than 1 per cent carbohydrate. In non-fatty fish such as the teleosts cod, haddock and whiting, fat levels are only about 0.5 per cent, while in fatty fish such as mackerel and herring, levels can vary between 3 and 25 per cent depending on factors such as the seasons and maturity.

As with meat, the muscle and internal organs of healthy, freshly caught fish are usually sterile but the skin, gills and alimentary tract all carry substantial numbers of bacteria. Reported numbers on the skin have ranged from 10^2–10^7 cfu cm^{-2} and from 10^3–10^9 cfu g^{-1} in the gills and the gut.

Bacteria associated with marine fish should be tolerant of the salt levels found in seawater. Though many do grow best at salt levels 2–3 per cent, the most important organisms are those that are not strictly halophilic but euryhaline, *i.e.,* they can grow over a range of salt concentrations. It is these that will survive and continue to grow as the salt levels associated with the fish decline, for example when the surface is washed melting ice.

After capture at sea, fish are commonly stored in ice or refrigerated sea water until landfall is made. It is important that fresh, clean cooling agent is used as reuse will lead to a rapid build-up of psychrotrophic contaminants and accelerated spoilage of the stored fish. Gutting the fish prior to chilling at sea is not a universal practice, particularly with small fish and where the time between harvest and landing is short. It does however remove a major reservoir of microbial contamination at the price of exposing freshly cut surfaces which will be liable to rapid spoilage. Similarly, any damage to the fish from nets, hooks, *etc.* that breaches the fish's protective skin will provide a focus for spoilage. Subsequent processing operations such as filleting and mincing which increase the surface area to volume of the product also increase the rate of spoilage.

Fish can be further contaminated by handling on board, at the dock and at markets after landing, particularly where they are exposed for sale and are subject to contamination with human pathogens by birds and flies. Generally though, fish have a far better safety record than mammalian meat.

Now let us try to learn about the factors, which contribute to the unique perishability of fish flesh.

In the case of fatty fish, spoilage can be non-microbiological; fish lipids contain a high proportion of polyunsaturated fatty acids which are more reactive chemically than the largely saturated fats that occur in mammalian meat. This makes fish far more susceptible to the development of oxidative rancidity.

In most cases though, spoilage is microbiological in origin. Fish flesh naturally contains very low levels of carbohydrate and these are further depleted during the death struggle of the fish. This has two important consequences for spoilage. Firstly, it limits the degree of post mortem acidification of the tissues so that the ultimate pH of the muscle is 6.2–6.5 compared with around 5.5 in mammalian muscle. Fish which have a lower pH such as halibut tend to have better keeping qualities. Secondly, the absence of carbohydrate means that bacteria present on the fish will immediately resort to using the soluble pool of readily assimilated nitrogenous materials, producing off- odours and flavours far sooner.

The composition of the non-protein nitrogen fraction differs significantly from that in meat. Trimethylamine oxide (TMAO) occurs in appreciable quantities in marine fish as part of the osmoregulatory system. TMAO is used as a terminal electron acceptor by non-fermentative bacteria such as *Shewanella putrefaciens* and this allows them to grow under microaerophilic and anaerobic conditions. The product of this reduction is trimethylamine which is an important component in the characteristic odour of fish. TMAO also contributes to a relatively high redox potential in the flesh since the *E*h of the TMAO/TMA couple is –19 mV.

Elasmobranchs such as dogfish and shark contain high levels of urea. Bacterial urease activity in the flesh can produce ammonia very rapidly giving the product a pungent odour. Not only does this render the flesh itself uneatable but it can also taint the flesh of other fish stored nearby. It is for this reason that in many areas fishermen will discard all but the fins of shark when they catch them.

Shellfish such as lobster have a particularly large pool of nitrogenous extractives and are even more prone to rapid spoilage; a factor which accounts for the common practice of keeping them alive until immediately prior to consumption.

Fish proteins are less stable than mammalian protein. As with meat, extensive proteolysis does not become apparent until the product is already well spoiled, but limited protein degradation may improve bacterial access to the nutrient pool of extractives.

The speed with which a product spoils is also related to the initial microbial load on the product: the higher the count the sooner spoilage occurs. Since fish from cold waters will have a larger proportion of psychrotrophs among their natural microflora, this can shorten the chill shelf-life appreciably.

Spoilage of chilled fish is due principally to the activity of psychrotrophic Gram-negative rods also encountered in meat spoilage: particularly *Shewanella putrefaciens* and *Pseudomonas* spp. The uniquely objectionable smell of decomposing fish is the result of a cocktail of chemicals, many of which also occur in spoiling meat. Sulfurous notes are provided by hydrogen sulfide, methyl mercaptan and dimethyl sulfide and esters contribute the 'fruity' component of the odour. A number of other amines in addition to TMA are produced by bacterial catabolism of amino acids. Skatole, a particularly unpleasant example produced by the degradation of tryptophan, also contributes to the smell of human faeces. The level of volatile bases in fish flesh has provided an index of spoilage, although this and other chemical indices used are often poor substitutes for the trained nose and eyes.

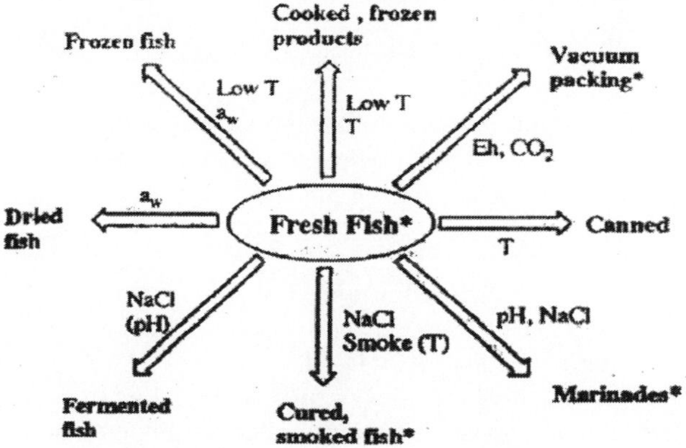

Figure 12.3: Fish and Fish Products. T indicates elevated temperature, *E*h, low redox potential; pH, reduced pH; a_w reduced a_w and * stored at chill temperatures.

The shelf-life extension of vaccum and modified atmosphere-packed cod will vary from less than 3 days to about 2 weeks. *Shewanella putrefaciens* can grow under these conditions producing TMA and hydrogen sulfide to spoil the product. Work in Denmark has also demonstrated that CO_2 tolerant marine vibrios like *Photobac terium phosphoreum* may be responsible for a non-sulfurous spoilage of these products in some instances. This is a large (5 μm diameter), almost yeast-like, bacterium that has been isolated from the intestines of several different fish. Because of its size it produces 10–100-fold more TMA per cell than smaller organisms such as *Shewanella* and therefore can cause spoilage at lower populations, typically around 107 cfu g^{-1} compared with 108 cfu g^{-1} for more conventional bacteria. Whether *Shewanella* or *Photobacterium* is ultimately responsible for spoilage in MAP fish products probably depends on relative numbers present initially and whether any other factors, that might have differential affects on their growth rates, apply.

$$(CH_3)3N \rightarrow O$$

Trimethylamine oxide

↓ TMO reductose

$$(CH_3)_3N$$

Trimethylamine

Figure 12.4: Reduction of Trimethylamine Oxide

Table 12.7 indicates about the nitrogen-containing extractions in fish.

Table 12.7: Nitrogen-Containing Extractions in Fish

	Cod	*Herring*	*Dogfish*	*Lobster*
Total nitrogenous extractives gkg^{-1}	12	12	30	55
Free amino acids (mM l^{-1})	7	30	10	300
TMAO (mM l^{-1})	5	3	10	2
Urea (mM l^{-1})	0	0	33	0
Creatine (mM l^{-1})	3	3	2	0
Betaine (mM l^{-1})	0	0	2	1
Anserine	1	0	0	0

Fish is considered a safe food in general, and the muscles of healthy fish are considered sterile, although debate on this continues (Novoslavskij *et al.*, **2016**). It is well accepted that microorganisms are commonly present on fish surfaces, such as skin and gills, as well as inside of the fish in areas such as the digestive tract and internal organs, for example, the kidney, liver, and spleen. However, fish and fish products, especially raw or undercooked products, have been involved in outbreaks associated with bacterial pathogens, biotoxins, histamine, viruses, and/or parasites (Galaviz-Silva *et al.*, **2009**).

Fish have been identified as reservoirs of bacterial pathogens linked to human diseases including *Mycobacterium* spp., *Streptococcus iniae, Photobacterium damselae, Vibrio alginolyticus, Vibrio vulnificus, Vibrio parahaemolyticus, Vibrio cholerae, Erysipelothrix rhusiopathiae*, pathogenic *Escherichia coli, Aeromonas* spp., *Salmonella* spp., *Staphylococcus aureus, Listeria monocytogenes, Clostridium botulinum, Clostridium perfringens, Campylobacter jejuni, Delftia acidovorans, Edwardsiella tarda, Legionella pneumophila*, and *Plesiomonas shigelloides* (Novotny *et al.*, **2004**). In addition, the presence of antibiotic-resistant genes (ARGs) in these microorganisms has caused concerns about spreading antibiotic resistance (AMR) in the environment and to human beings (Brunton *et al.*, **2019**; Preena *et al.*, **2020**). In addition to human pathogens, bacteria are considered the primary cause of fish spoilage (Rippen and Skonberg, **2012**). Certain spoilage bacteria can produce decarboxylase enzymes and convert free histidine to appreciable quantities of histamine, resulting in scombroid poisoning, another food safety concern.

Table 12.8: Food-borne Illness and Pathogens Associated with Fish

Food-borne Illness and Pathogens
Vibrio cholerae
Vibrio parahaemolyticus
Vibrio vulnificus
Closlridium botulinum Type E
Enteric viruses
Scombroid fish poisoning
Paralytic shellfish poisoning

According to the 2009 to 2018 surveillance report published by the U.S. Centers for Disease Control and Prevention (CDC), fish was the food category most commonly implicated in foodborne outbreaks involving single food categories (CDC, **2018**). From 1998 to 2018, a total of 937 foodborne outbreaks associated with fish were reported, resulting in 5,011 illnesses, 364 hospitalization, and four deaths (CDC, **2018**). The number of total fish-associated outbreaks per year declined from 62 cases in 1998 to 17 cases in 2010. From 2011 to 2018, the number of cases increased, with approximately 37 outbreaks on average each year (Table **12.9**). The numbers and percentages of confirmed fish-associated outbreaks among outbreaks with at least one confirmed etiology showed a similar trend. The percentage of fish outbreaks in total foodborne outbreaks decreased from 11.8 per cent in 1998 to 2.8 per cent in 2010. Percentages increased from 2011 to 2018, with an average of approximately 7.0 per cent of total annual food outbreaks confirmed to be associated with fish. Similar observations have been made by Dewey-Mattiaet *et al.* (**2016, 2018**). Overall, fish and fish products have been continuously implicated in foodborne outbreaks, contributing to some 6 per cent to 8 per cent of total confirmed foodborne-illness outbreaks over recent years. This incidence is higher than the incidence of foodborne-illness outbreaks associated with chicken and beef, which contributed to, on average, 3.6 per cent and 1.9 per cent of total confirmed foodborne outbreaks in 2011 to 2017 (CDC, **2018**). These fish and fish product outbreaks highlight the significant challenges to fish safety and the urgent need for better control strategies.

Table 12.9: The Numbers of Fish-Associated Outbreaks and their Percentages in the Total Foodborne Outbreaks in the United States, 1998 to 2018

Year	*Total Fish-Associated Outbreaks*[a]	*Confirmed Fish-Associated Outbreaks*[b]
	No. (per cent)	*No. (per cent)*
2018	45 (4.3)	39 (8.2)
2017	37 (4.4)	32 (7.9)
2016	26 (3.1)	24 (5.8)
2015	34 (3.7)	30 (6.6)
2014	44 (5.0)	39 (8.2)
2013	50 (6.0)	33 (7.3)
2012	31 (3.7)	24 (5.6)
2011	30 (3.8)	24 (6.4)
2010	17 (2.0)	13 (2.8)
2009	18 (2.7)	16 (4.5)
2008	29 (2.8)	24 (5.0)
2007	41 (3.7)	30 (5.9)
2006	48 (3.8)	42 (6.8)
2005	43 (4.5)	40 (9.8)
2004	52 (3.9)	44 (8.1)
2003	68 (6.2)	50 (11.9)
2002	63 (4.8)	41 (8.4)
2001	73 (5.8)	54(11.6)
2000	65 (4.6)	36 (7.9)
1999	61 (4.6)	30 (8.0)
1998	62 (4.7)	44 (11.8)

Note: Data collected from the National Outbreak Reporting System (NORS).

a: Total fish-associated outbreaks include outbreaks with confirmed, suspected, and undefined etiologies. Percentage (per cent) of the total fish-associated outbreaks = (number of the total fish-associated outbreaks/number of the total food-associated outbreaks) × 100 per cent. b: If at least one etiology was confirmed, the outbreak was considered as a confirmed fish-associated outbreak. Percentage (per cent) of the confirmed fish-associated outbreaks = (number of the confirmed fish-associated outbreaks/number of the total confirmed food-associated outbreaks) × 100 per cent.

Among the 709 fish-associated outbreaks with confirmed etiology, 99.3 per cent were associated with one etiology (Table 12.10). The most common etiologies are scombroid toxin or histamine (377 outbreaks), ciguatoxin (268 outbreaks), *salmonella* spp. (15 outbreaks), and *clostridium botulinum* (14 outbreaks). Scombroid poisoning is a food intoxication caused by the consumption of scombroid marine fish species (*e.g.*, tuna, mahi mahi, and bluefish) that are spoiled by bacteria, which produce decarboxylase enzymes and convert free histidine to large quantities of histamine (Colombo *et al.*, 2018). Ciguatoxin is produced by *gambierdiscus toxicus* and concentrated through the food chain, ultimately causing ciguatera fish poisoning in humans (CDC, 2019). *Salmonella* spp. contributes to a large portion of all hospitalizations and was involved in the largest fish-associated outbreak of foodborne illness, resulting in 425 illnesses and 55 hospitalizations across 28 states in 2012, resulting from the consumption of contaminated raw tuna products (Barrett *et al.*, 2017; CDC, 2012). *Clostridium botulinum* produces a highly toxic neurotoxin and is responsible for two out of four deaths (50 per cent) in fish-associated outbreaks.

Table 12.10: The Numbers and Percentages of Outbreaks, Illnesses, and Hospitalizations Caused by different Etiologies in Confirmed Fish-Associated Outbreaks in the United States, 1998 to 2018

Etiology	*Outbreaks No. (Per cent)*	*Illnesses No. (Per cent)*	*Hospitalizations No. (Per cent)*
Single (only one etiology involved in one outbreak)			
Toxin/chemical			
Scombroid toxin/histamine	377 (53.2)	1,366 (34,2)	46 (14.7)
Ciguatoxin	268 (37.8)	1,019 (25.5)	107 (34.3)
Paralytic shellfish poison	3 (0.4)	6 (0.2)	4 (1.3)
Puffer fish tetrodotoxin	1 (0.1)	5 (0.1)	2 (0.6)
Heavy metals	1 (0.1)	2 (0.1)	0 (0.0)
Others	8 (1.1)	29 (0.7)	6 (1.9)
Bacterial			
Salmonella spp.	15 (2.1)	378 (9.5)	31 (9.9)
Clostridium botulinum	14 (2.0)	38 (1.0)	23 (7.4)
Shigella spp.	3 (0.4)	74 (1.9)	7 (2.2)
Staphylococcus spp.	2 (0.3)	5 (0.1)	0 (0.0)
Bacillus spp.	1 (0.1)	3 (0.1)	0 (0.0)
Campylobacter spp.	1 (0.1)	21 (0.5)	3 (1.0)
Escherichia coli O157:H7	1 (0.1)	7 (0.2)	0 (0.0)
Listeria monocytogenes	1 (0.1)	4 (0.1)	4 (1.3)
Others	1 (0.1)	5 (0.1)	0 (0.0)
Viral			
Norovirus	5 (0.7)	443 (11.1)	0 (0.0)
Rotavirus	1 (0.1)	9 (0.2)	0 (0.0)
Parasitic			
Anisakis	1 (0.1)	14 (0.4)	0 (0.0)
Multiple (two or more etiologies involved in one outbreak)			
	5 (0.7)	561 (14.1)	79 (25.3)
Total	709	3,989	312

Note. Data were collected from the National Outbreak Reporting System (NORS). Outbreaks with at least one confirmed etiology were calculated. Percentage was calculated by dividing the number of outbreaks, illnesses, or hospitalizations caused by a specific etiology by the total number of outbreaks, illnesses, or hospitalizations, respectively.

Table 12.11.: Etiologies Identified from the Top Five Fish Species that were Involved in Confirmed Fish-Associated Outbreaks in the United States, 1998 to 2018

Etiology	*Tuna*[a]	*Mahi mahi*[b]	*Barracuda*	*Jack*[c]	*Salmon*
Single (only one etiology involved in one outbreak)					
Scombroid toxin/ histamine	226 (722)[d]	67 (236)	NA	3 (6)	3 (9)
Ciguatoxin	NA	16 (60)	74 (269)	48 (227)	NA
Salmonella spp.	6 (599)	NA	NA	NA	2 (19)
Clostridium botulinum	2 (5)	NA	NA	NA	5 (14)
Escherichia coli O157:H7	NA	NA	NA	NA	1 (7)
Bacillus cereus	NA	NA	NA	NA	1 (3)
Staphylococcus aureus	1 (3)	NA	NA	NA	NA
Norovirus	2 (15)	NA	NA	NA	1 (23)
Rotavirus	1 (9)	NA	NA	NA	NA
Multiple (two or more etiologies involved in one outbreak)					
	NA	NA	1 (23)[e]	NA	NA
Total	238 (1,353)	83 (296)	75 (292)	51 (233)	13 (75)

Note. Data were collected from the National Outbreak Reporting System (NORS). The top five fish species are determined based on the sum of outbreaks and illnesses caused by them.

a: Tuna includes food vehicles reported as tuna, ahi, albacore, yellowfin, yellowtail, and skipjack; b: Mahi mahi includes food vehicles reported as mahi mahi and roi; c: Jack includes food vehicles reported as Jack, amberjack, almaco jack, Hamachi, ulua, and papio; d: Results are expressed in number of outbreaks (number of illnesses); e: Etiologies involved in this case were ciguatoxin and *Staphylococcus aureus*.

Abbreviation: NA, not available.

Among fish species, tuna, mahi mahi, barracuda, jack, and salmon are responsible for the most outbreaks and illnesses in the united states (Table 12.11). Tuna contributed to about 34 per cent of the total confirmed fish-associated outbreaks (238 out of 709) and illnesses (1,353 out of 3,989). About 97 per cent of tuna-associated outbreaks were caused by scombroid toxin/histamine and *salmonella*. Outbreaks involving mahi mahi were mainly due to scombroid toxin/histamine. Outbreaks associated with barracuda and jack were mainly due to ciguatoxin. Different single etiologies were identified in salmon outbreaks, 69 per cent of which were due to bacteria (nine out of 13), including two outbreaks of *Salmonella* spp., five of *C. botulinum*, one of *E. coli* o157:h7, and one of *Bacillus cereus*.

Based on epidemiological data from the CDC, FDA, and U.S. Department of Agriculture (USDA) as well as survey studies reported in the scientific literature, the prevalence and the types of pathogenic microorganisms in fish and fish products are well documented. A few reviews have summarized the incidence of these pathogens in seafoods, and the top pathogens have been reported to be *Salmonella* spp., *L. monocytogenes*, *Vibrio* spp., *Yersinia* spp., *C. botulinum*, *S. aureus*, and *Aeromonas* spp. (Amagliani *et al.*, **2012**; Fernandes *et al.*, **2018**; Jami *et al.*, **2014**; Novoslavskij *et al.*, **2016**; Vaiyapuri, *et al.*, **2019**). Among them, *Salmonella* has been the leading bacterial etiology for fish-associated outbreaks. *Listeria monocytogenes*, despite a low incidence in fish outbreaks, is the most common bacterial cause for recalls of fish and fish products. In addition, its ubiquity in the processing environment as well as its ability to proliferate at cold temperatures make *L. monocytogenes* a major concern for ready-to-eat (RTE) fish products.

Besides the aforementioned pathogens that are listed as bacterial pathogens of greatest concern in the guidance of Fish and Fishery Products Hazards and Controls (FDA, **2020**), *Aeromonas* spp. have attracted attentions as an emerging human pathogen associated with seafood products. *Aeromonas hydrophila*, which is widely distributed in aquatic environments, can not only cause diseases in both fish and human directly, but also increase the levels of histamine, posing a chemical hazard to human health. Tables **12.12** and **12.13** summarize the latest survey studies about the incidence of *Salmonella* spp., *L. monocytogenes*, and *A. hydrophila* in fish and fish in the world and in the United States. As shown in these two tables, the highest incidence of *Salmonella* spp. identified in the United States was 1.3 per cent. The highest incidence of *Salmonella* spp. identified internationally was 36.6 per cent in Vietnam. The highest incidences of *L. monocytogenes* were 78.7 per cent and 28.1 per cent in the United States (samples were collected from processing plants previously experiencing problems with *L. monocytogenes*) and in Mexico, respectively. The incidence of *A. hydrophila* in the United States was 36.1 per cent based on one study on catfish fillet, and its highest incidence in the world is 47 per cent identified in an Egyptian study.

Table 12.12: Numbers and Percentages of Antibiotic-Resistant *Salmonella* spp., *Listeria monocytogenes* and *Aeromonas hydrophila* Identified in Fish and Fish Products in Scientific Literature

Country	*Fish/fish Products*	*Incidence*[a]	*Numbers of Isolates*	*Antimicrobial Resistance*[b] *(per cent)*	*Reference*
			Salmonella		
Brazil	Retail fish	5.76 (52)	3	Trimethoprim–sulfamethoxazole (100)	Cunha-Neto *et al.* (**2019**)
China	Retail freshwater fish	3.5 (2,328)	NA	NA	Li *et al.* (**2019**)
		18.6 (231)	103[c]	Tetracycline (35.9), ampicillin (28.2), nalidixic acid (26.2), and trimethoprim–sulfamethoxazole (25.2)	Yang *et al.* (**2015**)
	Retail fish	NA	27	Sulfisoxazole (59.2), tetracycline (29.6), nalidixic acid (25.9), streptomycin (22.2), and ampicillin (18.5)	Zhang *et al.* (**2015**)
India	Retail fish	NA	72	Ceftazidime (98.6), streptomycin (98.6), amikacin (91.7), azithromycin (79.2), and trimethoprim (79.2)	Saharan *et al.* (**2019**)
Iran	Wild fish and shrimp	6.9 (274)	19	Nalidixic acid (47.4), tetracycline (36.8), streptomycin (15.8), and trimethoprim (15.8)	Rahimi *et al.* (**2013**)

Country	*Fish/fish Products*	*Incidence[a]*	*Numbers of Isolates*	*Antimicrobial Resistance[b] (per cent)*	*Reference*
	Retail raw/fresh saltwater fish	2.9 (70)	NA	NA	Zarei *et al.* (**2012**)
Italy	Fish products at retail stores and in the processing plants	0.3 (2,086)	NA	NA	Busani *et al.* (**2005**)
Jordan	Imported fish	19 (330)	64	Amoxicillin–clavulanic acid (52.2), cefalothin (52.2), streptomycin (49.3), ampicillin (44.8), and kanamycin (10.4)	Obaidat and Salman (**2017**)
Malaysia	Retail African catfish and African catfish collected from fish farm	20 (30)	6	Penicillin (100), clindamycin (100), tetracycline (100), and rifampicin (100)	Sing *et al.* (**2016**)
	Catfish and tilapia collected from retail stores and ponds	35.9 (64)	32	Clindamycin (100), rifampicin (87.5), tetracycline (56.3), chloramphenicol (40.1), and spectinomycin (31.3)	Budiati *et al.* (**2013**)
Nigeria	Retail catfish and catfish collected from fish farms	11.5 (200)	23	Streptomycin (43.5), sulfamethoxazole (34.8), and trimethoprim (21.7)	Raufu *et al.* (**2014**)
Saudi Arabia	Imported freshwater fish	16.8 (1,678)	140	Tetracycline (90.7), ampicillin (70), and amoxicillin–clavulanic acid (45)	Elhadi *et al.* (**2016**)
Spain	RTE smoked salmon	1.2 (84)	NA	NA	Cabedo *et al.* (**2008**)
Thailand	Retail fish and other seafood	24 (27)	NA	NA	Ananchaipattana *et al.* (**2014**)
	Retail seabass and other seafood	21 (120)	26	Ampicillin (46), sulfisoxazole (46), tetracycline (42), chloramphenicol (31), and trimethoprim–sulfamethoxazole (27)	Woodring *et al.* (**2012**)
Turkey	Wild freshwater common carp	3 (100)	7	Erythromycin (100), streptomycin (57.1), neomycin (42.9), and gentamicin (14.3)	Pamuk *et al.* (**2019**)
	Retail fish	5 (100)	5	Neomycin (80), gentamicin (20), and trimethoprim–sulfamethoxazole (20)	Onmaz *et al.* (**2015**)
Vietnam	Wholesale and retail freshwater fish	36.6 (101)	52	Tetracycline (46.2), ampicillin (34.6), chloramphenicol (30.8), trimethoprim-sulfamethoxazole (25.0), and nalidixic acid (11.5)	Nguyen *et al.* (**2016**)
		Listeria monocytogenes			
Canada	RTE fish products	15 (40)	6	Clindamycin (100), cefoxitin (100), nalidixic acid (100), amikacin (33.3), and streptomycin (33.3)	Kovacevic *et al.* (**2012**)
China	Retail freshwater fish	2.6 (2,328)	NA	NA	Li *et al.* (**2019**)
Estonia	RTE fish products	11.6 (7,652)	NA	NA	Koskar *et al.* (**2019**)
European Union	RTE fish	7.0	NA	NA	EFSA (**2018**)

Country	*Fish/fish Products*	*Incidence*[a]	*Numbers of Isolates*	*Antimicrobial Resistance*[b] *(per cent)*	*Reference*
Greece	RTE fish	6.7 (119)	NA	NA	Soultos *et al.* (**2014**)
India	Retail fish and fish products and fish from the landing centers	2.2 (184)	8	Ampicillin (100), penicillin (100), erythromycin (100), tetracycline (100), and clindamycin (100)	Basha *et al.* (**2019**)
	Retail fish	1.7 (115)	NA	NA	Rodrigues *et al.* (**2015**)
Iran	Retail rainbow trout and rainbow trout collected from fish farm	9.16 (240)	22	Enrofloxacin (72.7), ciprofloxacin (68.2), tetracycline (63.6), erythromycin (40.9), ampicillin (31.8), and chloramphenicol (31.8)	Rezai *et al.* (**2018**)
	Fish, other seafood, and environmental samples	9.5 (662)	278	Ampicillin (38.5), penicillin (38.1), vancomycin (20.9), tetracycline (18.7), and ciprofloxacin (17.6)	Fallah *et al.* (**2013**)
	Retail fish	2.5 (80)	NA	NA	Rahimi *et al.* (**2012**)
		7.72 (220)	NA	NA	Momtaz and Yadollahi (**2013**)
	Retail saltwater fish	1.4 (70)	NA	NA	Zarei *et al.* (**2012**)
Latvia	**RETAIL FRESHWATER FISH** **WILD FRESHWATER FISH**	**26 (106)** **2 (129)**	**NA** **NA**	**NA** **NA**	Terentjeva *et al.* (**2015**)
Mexico	Wild fish and seawater	28.1 (210)	68	Pefloxacin (28.9), cephalothin (26.7), ampicillin (23.7), penicillin (22.5), and ceftazidime (14.4)	Rodas-Suarez *et al.* (**2006**)
Poland	Fish in the processing plants	13.2 (197)	26	Penicillin (15.4), ampicillin (11.5), meropenem (11.5), erythromycin (7.7), and cotrimoxazole (3.8)	Skowron *et al.* (**2019**)
	Retail fresh and smoked fish	18.9 (301)	57	Oxacillin (57.9), ceftriaxone (31.6), and clindamycin (8.8)	Wieczorek and Osek (**2017**)
Spain	RTE smoked salmon	8.6 (803)	69	Ampicillin (1.4) and cephalothin (1.4)	Domenech, *et al.* (**2015**)
Vienna	Retail fish and fish products	4.9 (102)	NA	NA	Nauman *et al.* (**2014**)
Aeromonas hydrophila					
China	Retail fish	3.8 (235)	10	Ampicillin (100), ampicillin-sulbactam (100), cefazolin (100), co-trimoxazole (90), and tetracycline (30)	Wu *et al.* (**2019**)
	Retail grass carp	31.5 (200)	63	Ampicillin (95.24), tetracycline (88.89), streptomycin (79.37), nalidixic acid (73.02), and kanamycin (53.97)	Yang *et al.* (**2018**)

Country	*Fish/fish Products*	*Incidence*[a]	*Numbers of Isolates*	*Antimicrobial Resistance*[b] *(per cent)*	*Reference*
Egypt	Retail frozen fish fillet	6.7 (150)	10	Streptomycin (100), cloxacillin (100), erythromycin (100), sulfamethoxazole (80), and cefotaxime (80)	Hafez *et al.* (**2018**)
	Retail tilapia and mugil	5.4 (294)	16	Cefixime (88), amoxicillin/ clavulanic acid (80), ceftazidime (80), nalidixic acid (76), and tetracycline (72)	Ahmed *et al.* (**2018**)
	Fish collected from fish farms	47 (170)	18	Ampicillin (100), erythromycin (100), nalidixic acid (100), spectinomycin (100), and tetracycline (66.7)	Dahdouh *et al.* (**2016**)
India	Retail fish and fish collected from fish farms	7.5 (173)	13	NA	Rather *et al.* (**2019**)
	Retail fish	33.6 (536)	NA	NA	Vivekanandhan *et al.* (**2005**)
Iran	Wild fish	19.5 (133)	NA	NA	Rahimi *et al.* (**2014**)
Italy	Farmed and wild freshwater fish	NA	53	Cloxacillin (100), spiramycin (100), tilmicosin (100), ampicillin (100), tylosin (100), and sulfadiazine (100)	Borella *et al.* (**2020**)
Malaysia	Retail fish	11.5 (87)	10	Ampicillin (100)	Radu *et al.* (**2003**)
Mexico	Retail frozen tilapia	0.8 (250)	2	Ampicillin (100), penicillin (100), carbenicillin (100), cephalothin (100), streptomycin (100), clindamycin (199), polymycin B (100), and rifampicin (100)	Castro-Escarpulli *et al.* (**2003**)
Poland	Farmed carp	NA	18	Ampicillin (100), penicillin (100), Cephalothin (61), and erythromycin (56)	Guz and Kozinska (**2004**)
Turkey	Retail fish	3.75 (400)	NA	NA	Kahraman, *et al.* (**2017**)

Abbreviations: NA, not available, these studies did not conduct the antibiotic resistance evaluation; RTE, ready-to-eat.

a: Incidence: the percentage of pathogen positive samples (number of total samples tested); b: Name of the antimicrobial (percentage of the antimicrobial-resistant isolates). Percentage of the antimicrobial resistance = (number of antimicrobial-resistant isolates/total number of isolates) × 100 per cent. If more than five antibiotics were tested, the results of the top five antimicrobials are shown.

Table 12.13: Numbers and Percentages of Antibiotic-Resistant *Salmonella* spp., *Listeria monocytogenes* and *Aeromonas hydrophila* in Fish and Fish Products in the United States Based on Scientific Literatures

Pathogen	*Fish*		*Incidence*[a]	*Total Isolates*	*Antimicrobial Resistance*[b] *(per cent)*	*Reference*
	Domestic	*Imported*				
Salmonella						
		Salmon and tilapia	0.6 (133)	3	0 (0/3)	Wang *et al.* (**2011**)
		Fish	NA	17	Tetracycline (5.9) and ampicillin (5.9)	Ponce *et al.* (**2008**)
		Fish	NA	27	Tetracycline (22.2), chloramphenicol (7.4), nalidixic acid (7.4), trimethoprim-sulfamethoxazole (7.4), and sulfamethoxazole (7.4)	Zhao *et al.* (**2006**)
	Fish	Fish	**D: 1.3 (76)[C]I: 12.2 (2,038)**	**NA** **NA**	**NA** **NA**	Heinitz *et al.* (**2000**)
	Smoked fish	Smoked fish	**D: 0I: 3.2 (156)**	NA	NA	Heinitz and Johnson (**1998**)
Listeria monocytogenes						
		Salmon, tilapia, and shrimp	4.1 (171)	20	Nitrofurantoin (75)	Wang *et al.* (**2011**)
	Catfish fillets and processing environmental samples		21.6 (315)	86	Cefotaxime (71) and clindamycin (69)	Chen *et al.* (**2010a**); Chen et al. (**2010c**)
	Catfish fillet		45 (80)	NA	NA	Chen et al. (**2010b**)
	Catfish fillet		37.5 (240)	NA	NA	Chou et al. (**2006**)
	Fish fillet		9.3 (140)	NA	NA	Pao et al. (**2008**)
	RTE smoked salmon[d]		1.4 (72)	NA	NA	Lappi *et al.* (**2004**)
	FISH RTE **SMOKED FISH**		**3.8 (234)** **1.3 (233)**	**NA** **NA**	**NA** **NA**	Thimothe *et al.* (**2004**)
	Smoked fish	Smoked fish	**D: 13.3 (525)** **I: 15.2 (559)**	**NA** **NA**	**NA** **NA**	Heinitz and Johnson (**1998**)
	Frozen raw salmon[d]		36.3 (135)	NA	NA	Eklund *et al.* (**1995**)
	Cold-smoked salmon[d]		78.7 (61)	NA	NA	
Aeromonas hydrophila						
	Catfish fillet		36.1 (238)	80	Ampicillin (100), Bacitracin (100), chlortetracycline (5), oxytetracycline (5), and tetracycline (5)	Wang and Silva (**1999**)

Abbreviations: NA, not available, these studies did not conduct the antibiotic resistance evaluation.; RTE, ready-to-eat.

a: Incidence: percentage of pathogen positive samples (total number of samples tested); b: Name of the antimicrobial (percentage of the antimicrobial-resistance isolates). Percentage of the antimicrobial resistance = (number of antimicrobial-resistant isolates/total number of isolates) × 100 per cent. If more than five antibiotics were tested, the results of the top five antimicrobials are shown; c: D: domestic; I: imported; d: Samples collected from processing plants with a previously documented *L. monocytogenes* contamination.

In recent years, the occurrence of antibiotic-resistant bacteria (ARB) has become a significant challenge for not only aquaculture but the entire food industry (Preena *et al.*, 2020). AMR has become a global public health threat as well as an environment pollution concern. Food has been identified as the most important vehicle for transmitting ARB and ARGs to humans, resulting in serious human health consequences such as treatment failure. Given rising concerns over AMR, the prevalence of antimicrobial-resistant *Salmonella*, *L. monocytogenes*, and *A. hydrophila* identified in fish and fish products is also summarized in Tables 12.12and 12.13. As shown in these two tables, many isolates demonstrated resistance to more than one antibiotic, including critical antibiotics used for treating human and fish diseases. In the following section, incidences of *Salmonella* spp., *L. monocytogenes*, and *A. hydrophila* identified in fish and fish products as well as their antimicrobial-resistant properties are discussed in detail.

Salmonella Spp.

Salmonella spp. are neither natural inhabitants of fish nor a biological contaminant originally reported in fish; however, *S. enterica* subsp. *enterica* serotypes Weltevreden, Typhimurium, Javiana, Enteritidis, Paratyphi B, Braenderup, Bovismorbificans, Thompson, Rubislaw, Barranquilla, Heidelberg, Infantis, Group C1, Bareilly, Nchanga, and Newport have been involved in fish-associated outbreaks (CDC, 2018, 2019). *Salmonella* can be introduced into fish and fish products through contact with contaminated water, inadequate hygiene when handling, and other inappropriate fish breeding, processing, or storage practices (Fernandes *et al.*, 2018).

The prevalence of *Salmonella* contamination in fish and fish products varies among different counties and is also determined by the different fish species, geographic locations, sampling stages (fish farm vs. retail stores), sampling parts (skin vs. intestine), sources (imported vs. domestic), and fish product types (raw vs. RTE). Reported contamination levels vary from 0.3 per cent in Italy to 36.6 per cent in Vietnam. In a recent survey across China, 2,328 retail freshwater fish were collected, and *Salmonella* was present at a level of 3.5 per cent. Carp (*Cyprinus carpio*) had the lowest prevalence of *Salmonella* at 1.8 per cent, and culter alburnus (*Erythroculter ilishaeformis*) had the highest prevalence at 7.7 per cent.

Sing *et al.* (2016) studied African catfish from four farms and four wet markets in Malaysia; the prevalence of *Salmonella* in the surveyed 14 retail fish (28.57 per cent) was higher than that in the 16 fish collected directly from the farms (12.5 per cent). This highlights the fact that *Salmonella* is introduced into fish and fish products during postharvest processing. This study also showed that fish intestines had a higher incidence of *Salmonella* than skin samples. The prevalence was 66.67 per cent for catfish intestine and 33.33 per cent for skin.

In the United States, the overall incidence of *Salmonella* in domestic fish and fish products was lower than the incidence in imported fish and fish products. *Salmonella* was detected in 1.3 per cent of 76 domestic raw fish samples from 1990 to 1998, compared to the 12.2 per cent prevalence in 2,038 samples of imported fish (Heinitz et al. 2000). Another U.S. survey study sampled a total of 156 smoked fish samples from both domestic and foreign origin; the domestic samples tested negative for *Salmonella*, whereas five imported samples tested positive (Heinitz and Johnson, 1998). The prevalence of *Salmonella* was lower in RTE fish products than in raw fish, regardless of origin. For instance, *Salmonella* was not detected in 89 domestic smoked fish samples, whereas 3.9 per cent of 255 imported smoked fish samples tested positive. This underscores the importance of sampling, inspection, and implementation of control strategies not only in domestic but also in foreign fish companies.

The predominant *Salmonella* serotypes isolated from fish and fish products in various reported studies included *Salmonella typhimurium*, *Salmonella* Wandsworth, *Salmonella* Potsdam, *Salmonella* Corvallis, Salmo*nella* Albany, *Salmonella* Hadar, Salmonella Schwarzengrund, *Salmonella enteritidis*, and *Salmonella* Panama. In the United States, the most frequent *Salmonella* serotypes identified in imported fish are as follows: Weltevreden, Senftenberg, Paratyphi-B, Lexington, Enteritidis, Newport, Thompson, Saintpaul, and Bareily. *Salmonella* Tennessee was detected from pollock fillets produced domestically; other serotypes isolated from domestic seafood samples have included Salmonella Newport, Rubislaw and Anatum.

Antibiotic-resistant *Salmonella* is considered as a "Serious Threat Level pathogen". Unfortunately, *Salmonella* strains isolated from fish and fish products have been shown to be resistant to a broad range of antibiotics including ampicillin, tetracycline, sulfisoxazole, streptomycin, trimethoprim–sulfamethoxazole, nalidixic acid, ceftazidime, amikacin, azithromycin, trimethoprim, amoxicillin–clavulanic acid, cefalothin, kanamycin, trimethoprim, clindamycin, rifampicin, chloramphenicol, spectinomycin, erythromycin, neomycin, and gentamicin (Tables 12.12 and 12.13). The most frequently reported antimicrobials to which *Salmonella* has shown resistance are streptomycin, tetracycline, ampicillin, trimethoprim–sulfamethoxazole, nalidixic acid, trimethoprim, and chloramphenicol. Some *Salmonella* isolates have multidrug resistance (MDR), with resistance to at least one antibiotic in three or more drug classes. MDR has been a bigger concern as it can enhance bacterial virulence and result in more severe diseases in patients. Yang *et al.* (2015) reported that 62.5 per cent of *Salmonella* Thompson isolated from retail freshwater fish in China demonstrated resistance to more than 14 antimicrobials. In addition, 62.5 per cent of *Salmonella* Derby and 60 per cent *Salmonella typhimurium* showed resistance to at least six antimicrobials. All the *Salmonella* isolates identified from African catfish in Malaysia were resistant to penicillin, clindamycin, tetracycline, and rifampicin (Sing *et al.,* 2016). About 65.6 per cent of 72 *Salmonella* isolates recovered from fish faecal samples in India showed MDR (Saharan *et al.*, 2020). In the United States, 27 *Salmonella* isolates were identified from imported fish: six showed resistance to antibiotics and five were resistant to at least two antibiotics. *Salmonella* Bareilly isolated from frozen dace fish and *Salmonella* Schwarzengrund isolated from frozen catfish were resistant to more than four antibiotics. *Salmonella* Oslo was resistant to the most antibiotics, including ampicillin, chloramphenicol, streptomycin, sulfisoxazole, tetracycline, and trimethoprim–sulfamethoxazole. One strain out of 17 *Salmonella* Weltevreden isolates from imported fish showed resistance to both ampicillin and tetracycline (Ponce *et al.,* 2008). By contrast, three *Salmonella typhimurium* isolates recovered from imported tilapia did not show antimicrobial resistance (Wang *et al.,* 2011). Though imported fish has repeatedly been identified with MDR bacteria, currently available studies did not check the AMR of *Salmonella* isolates from U.S. domestic fish and fish products, which warrants more attention in future studies.

Listeria monocytogenes

It is a species of Gram-positive bacteria that has a high mortality rate of about 16 per cent. The populations susceptible to *L. monocytogenes* infection include pregnant women, the elderly, and immune-compromised people (Buchanan *et al.*, 2017). *Listeria monocytogenes* is ubiquitous in the environment and tolerant to a broad range of environmental stresses. *Listeria monocytogenes* can grow in salt solutions with up to 14 per cent sodium chloride. It can also survive in freezing temperature and even grow in fish at refrigerator temperatures (Liu *et al.*, 2016). Of particular concern is that *L. monocytogenes* can proliferate in RTE fish products, which lack further inactivation steps before consumption. Because the prevalence of *L. monocytogenes* in aquatic food products worldwide from 1990 to 2013 has been reviewed by Jami *et al.* (2014), this discussion section focuses on the most recent survey studies and the prevalence of antibiotic-resistant *L. monocytogenes* isolates (Tables 12.13).

The occurrence of *L. monocytogenes* in fish and fish products differed among countries and is also influenced by fish species, geographic locations, sampling stages, sampling parts, product types, and sample origins. The incidences of *L. monocytogenes* in fish and fish products range from 1.4 per cent to 78.7 per cent Li *et al.* (2019) reported that the overall incidence of *L. monocytogenes* in freshwater fish across China was 2.6 per cent. No *L. monocytogenes* was detected in Asian swamp eel (*Monopterus albus*), bream (*Parabramis pekinensis*), or culter alburnus (*Erythroculter ilishaeformis*), whereas 4.8 per cent of loach (*Misgurnus anguillicaudatus*) tested positive. The percentage of fish samples that were positive for *L. monocytogenes* in the eastern region was 1.0 per cent and in the central/northwestern region was 3.8 per cent. Retail fish and fish products have been reported to have a higher incidence of *L. monocytogenes* than fish and fish products sampled at any other stage prior to processing.

Basha *et al.* (2019) collected fish and fish products from both retail markets and landing centers in India; *L. monocytogenes* was detected only in retail samples, not in the landing center samples. Similarly, a study in Iran reported that 14.16 per cent of retail rainbow trout were positive for *L. monocytogenes* compared to the 4.16 per cent positive rate of fish samples collected from ponds on fish farms (Rezai *et al.*, 2018). *Listeria monocytogenes* was detected in 26 per cent of retail and 2 per cent of wild-caught freshwater fish in Latvia (Terentjeva *et al.,* 2015).

Fish skin is considered one primary contamination source. In a U.S. study, *L. monocytogenes* was detected on 65.2 per cent of skin samples from raw frozen salmon, whereas absent in flesh samples from underneath the skin. These data suggest that postharvest practices are a major point of *L. monocytogenes* contamination in fish and fish products. During the processing of fish, *L. monocytogenes* present on fish skin or in the environment has numerous chances to be transferred onto fillets through different handling practices and via utensils or media such as cutting knives, gloves of workers, and brine solutions. In a processing plant known to be experiencing problems with *L. monocytogenes*, this pathogen was detected at high levels of 65 per cent to 67 per cent in both raw and finished cold-smoked salmon. Therefore, adequate sanitation and proper control strategies in fish processing plants are vital to reduce the risk of *L. monocytogenes* contamination in fish and fish products.

Additional processing of fish, such as smoking and brining, can control the level of *L. monocytogenes* contamination. The prevalence of *L. monocytogenes* decreased from 3.8 per cent in raw fish to 1.3 per cent in finished smoked fish products. In a survey study conducted in Poland, *L. monocytogenes* was positive in 35.7 per cent of fresh fish and was reduced to 5.3 per cent in smoked fish (Wieczorek and Osek, 2017). However, *L. monocytogenes* remains a big concern in RTE fish products because of the pathogen's ability to grow during cold storage as well as the lack of thermal processing before consumption.

The European Food Safety Authority (EFSA) and European Center for Disease Prevention and Control (ECDC) summarized the data from 18 European countries, and the overall occurrence of *L. monocytogenes* in RTE fish products was 7.0 per cent in 2017 (EFSA, 2018). Kovacevic *et al.* (2012) reported that 15 per cent of RTE fish products were positive for *L. monocytogenes* in Canada. In the United States, the contamination rate of the domestic RTE fish products varied from 1.3 to 1.4 per cent (Lappi *et al.*, 2004) to 78.7 per cent (Eklund *et al.,* 1995). The ultrahigh contamination rate was reported in cold-smoked salmon samples collected from six processing facilities known to have encountered problems with *L. monocytogenes* contamination. A 5-year survey study showed that 13.3 to 15.2 per cent of domestic and imported smoked fish were positive for *L. monocytogenes*, unlike the incidence of *Salmonella*, which is lower in domestic fish and fish products (Heinitz and Johnson, 1998). This indicates a global concern about *L. monocytogenes* and fish safety.

The presence of antibiotic-resistant isolates further complicates *L. monocytogenes* and fish safety. *Listeria monocytogenes* recovered from fish and fish products was frequently resistant to ampicillin, clindamycin, erythromycin, and tetracycline. MDR was also found in 100 per cent of *L. monocytogenes* isolated from fish and fish products in India; 59.1 per cent of strains from raw rainbow trout in Iran; 33 per cent of isolates from RTE fish in Canada; 22.3 per cent of isolates recovered from raw fish, other seafood, and environmental samples in Iran; 6 per cent of strains in raw fish and seawater in Mexico; and 3.5 per cent of isolates identified from fresh and smoked fish in Poland. Although the majority of studies in the United States did not evaluate the AMR of *L. monocytogenes* isolates from fish or fish products regardless of origin, it was reported that 71 per cent and 69 per cent of *L. monocytogenes* isolates recovered from domestic catfish fillets and the processing environment were resistant to cefotaxime and clindamycin, respectively. Among *L. monocytogenes* isolated from imported fish and other seafood, 75 per cent were resistant to nitrofurantoin.

Aeromonas hydrophila

Aeromonas spp. is a common cause of fish disease, with the two most frequent strains causing disease being *A. hydrophila* and *A. salmonicida* (Hayatgheib et al., 2020). *Aeromonas hydrophila* is not only a fish pathogen but also a zoonotic pathogen that can cause human diseases such as gastroenteritis, septicemia, and traumatic and aquatic wound infections. The wide distribution of *A. hydrophila* in the aquatic environment as well as its ability to survive or even grow in foods at cold temperatures further make the pathogen a big concern in food safety. *Aeromonas hydrophila* is frequently isolated from fish and fish products. The incidence varies from 0.8 per cent in retail frozen tilapia collected from Mexico City to 47 per cent in farmed fish from Egypt. A high incidence of over 30 per cent was also reported in catfish fillets collected from processing plants in the United States, retail grass carp in China, and retail fish from India.

The occurrence of *A. hydrophila* in fish varies by season, fish species, sampling parts, sampling stages, and product types. Wang and Silva (1999) reported that the highest incidence of *A. hydrophila* in catfish fillets from the United States was in summer. Vivekanandhan *et al.* (2005) did not find a correlation with sampling time and *A. hydrophila* prevalence in a 3-year survey study of retail fish in India, however. The different results from these two studies might result from the different geographic locations and different weather conditions in these two countries. In a Malaysian survey study, *A. hydrophila* was found in 10 to 16 per cent of tilapia (*Tilapia mossambica*), keli (*Clarias batrachus*), and puyu (*Anabas testudineus*) but not in terubuk (*Tenualosa toli*.

The occurrence of *A. hydrophila* in different parts of fish is also fish-species dependent. For example, *A. hydrophila* was isolated from nine viscera samples (6.3 per cent) but no muscle samples among 144 mugil samples collected from retail stores in Egypt. For tilapia, on the other hand, *A. hydrophila* was detected in 2.7 per cent of muscle and 2.0 per cent of intestine samples. Different processing and handling methods are also factors that determine *A. hydrophila* prevalence in final products. Similar to *Salmonella* and *L. monocytogenes*, *A. hydrophila* has a higher incidence in retail fish than in fish collected from fish farms. An Indian study reported that the prevalence of *A. hydrophila* in swab samples from gills of fish from retail markets and farms was 44 per cent and 28 per cent, respectively (Rather *et al.*, 2019). These surface bacteria can cross-contaminate the fish flesh and the contact surfaces of equipment during processing. It has been reported that fish fillets have a higher incidence of *A. hydrophila* (33.3 per cent) than whole fish (10 per cent to 12.5 per cent). Similarly, when 238 domestic processed catfish fillets were examined for the presence of *A. hydrophila* in the United States, the pathogen was isolated from 36.1 per cent of the samples. This evidence again emphasizes the vital role of processing facilities in ensuring fish product safety.

Resistance to antibiotics in *A. hydrophila* isolates from fish and fish products has been frequently reported internationally (Tables 12.12 and 12.13). Aerom*onas hydrophila* strains are commonly resistant to ampicillin, tetracycline, and streptomycin. A 100 per cent resistance rate to ampicillin in all *A. hydrophila* isolates has been reported repeatedly. In addition, multidrug-resistant strains were also detected at high rates in *A. hydrophila* isolates from fish and fish products. In particular, all *A. hydrophila* isolates recovered from retail fish in Malaysia, retail frozen fish fillets or farmed fish in Egypt, and farmed and wild freshwater fish in Italy were resistant to at least three antibiotics. Multidrug-resistant *A. hydrophila* against ampicillin, bacitracin, and tetracycline (tetracycline, oxytetracycline, or chlortetracycline) has also been detected in U.S. domestic catfish fillets.

To sum up, human pathogens including *Salmonella*, *L. monocytogenes*, and *A. hydrophila* have been detected at varying rates in fish and fish products across the world. In the era of globalization, it is extremely important yet difficult to ensure the safety of fish and other aquaculture products, considering the diverse regulations in different countries. The issue is further complicated by the diversity of fish species, geographic locations, sampling stages, sampling parts, and product types. In general, retail fish and fish products have a higher incidence of human pathogens than fish collected from fish farms, highlighting the vital role of processing plants in fish safety. Fish products can be contaminated not only by the human pathogens originally existing in a processing plant, but by pathogens transferred from contaminated raw materials to finished fish products, processing equipment, and personnel. This results in a cascade of cross-contamination and re-contamination. Once the pathogens are introduced into fish products, they are very challenging to eliminate and can pose huge safety concerns during subsequent distribution and storage.

III. Meat and Poultry

People are now aware that contaminated food items consumed by humans may lead to ill health that may be life-threatening. Foods that are prone to contamination include meat, milk, and fish.

Humankind has consumed animal protein since the dawn of its existence. The archaeological record shows evidence of animal protein consumption as early as 12,500 BC. Consumption of meat is continuously increasing worldwide. The annual per capita consumption increased from 10 kg in the 1960s to 26 kg in 2000 and will reach 37 kg by the year 2030 (Heinz and Hautzinger, 2007). On the other hand, a significant portion of meat and meat products are spoiled every year. Kantor *et al.* (1997) reported that approximately 3.5 billion kg of poultry and meat were wasted at the consumer, retailer and foodservice levels which have a substantial economic and environmental impact. Significant portion of this loss is due to microbial spoilage. Cerveny *et al.* (2009) stated that if 5 per cent of this meat loss is preserved it could satisfy the daily needs of approximately 320,000 people for meat and poultry.

Compared to a multitude of foodstuffs, meat represents one of the most perishable: first, for the presence of chemical and enzymatic activities, and second, because it constitutes a perfect *pabulum* for the growth of a wide variety of microorganisms, especially as a result of its nutrient composition, high water content and moderate pH. Microbial growth, oxidation and enzymatic autolysis are the three basic mechanisms responsible for the spoilage of meat. In addition to lipid oxidation and enzyme reactions, meat spoilage is almost always caused by microbial growth. Meat can act as an ideal substrate for microbial proliferation. Major meat associated pathogenic bacteria include *Clostridium perfringens, Staphylococcus aureus, Salmonella spp*, pathogenic strains *of Escherichia coli, Campylobacter spp, Yersinia enterocolitica, Listeria monocytogenes* and *Aeromonas hydrophila.*

The transformation of animals into meat involves several operations: (a) handling and loading of animals on the farm, (b) transporting animals to slaughterhouses, (c) off-loading and holding of animals and (d) slaughtering of animals. Poor operational techniques and facilities in any of these operations will result in unnecessary suffering and injuries to animals which can lead to loss of meat, reduced meat quality and spoilage of meat. Therefore, prevention of contamination after slaughtering during meat cutting and processing is essential. Storage time can be extended through hygienic slaughtering and clean handling of the carcass.

Different technical operations are involved in slaughtering: (a) stunning, (b) bleeding, (c) skinning, (d) evisceration and (e) carcass splitting. Inadequacy at one stage will result in a rigorous negative impact on the product and/or process in the following stage. In addition to the hygiene and storage temperature, the acidity of the meat and the structure of the muscular tissue also affect the rate of meat spoilage. For example, liver will spoil faster than the firm muscular tissue of beef. After few hours of slaughtering of animals, muscles become firm and rigid, a condition known as rigor mortis. The process of rigor mortis depends on the stress induced on the animals during the slaughtering process. Raw meat quality is reported to be severely affected by the stress conditions during slaughtering process and the slaughtering methods.

The Primary Processing of Meat and Poultry

The tissues of a healthy animal are protected against infection by a combination of physical barriers and the activity of the immune system. Consequently, internal organs and muscles from a freshly slaughtered carcass should be relatively free from microorganisms. Microbial numbers detected in aseptically sampled tissues are usually less than 10 cfu kg^{-1}, although there is evidence that numbers can increase under conditions of stress and they will of course be higher if the animal is suffering from an infection. Since some animal diseases can be transmitted to humans, meat for human consumption should be produced only from healthy animals. Visual inspection before and after slaughter to identify and exclude unfit meat is the general rule, although it will only detect conditions which give some macroscopic pathological sign.

The most heavily colonized areas of the animal that may contaminate meat are the skin (fleece) and gastrointestinal tract. Numbers and types of organisms carried at these sites will reflect both the animal's indigenous microflora and its environment. The animal hide, for example, will carry a mixed microbial population of micrococci, staphylococci, pseudomonads, yeasts and moulds as well as organisms derived from sources such as soil or faeces. Organisms of faecal origin are more likely to be encountered on hides from intensively reared cattle or from those transported or held in crowded conditions.

With reasonable standards of hygienic operations, contamination of meat carcasses from processing equipment, knives and process workers is less important than contamination from the animals themselves. The greatest opportunity for this occurs during dressing, the stages during which the head, feet, hides, excess fat, viscera and offal are separated from the bones and muscular tissues.

Skinning can spread contamination from the hide to the freshly exposed surface of the carcass through direct contact and *via* the skinning knife or handling. Washing the animal prior to slaughter can reduce microbial numbers on the hide but control is most effectively exercised by skilful and hygienic removal of the hide. The viscera contains large numbers of microorganisms, including potential pathogens and great care must be taken to ensure the carcass is not, contaminated with visceral contents either as a result of puncture or leakage from the anus or oesophagus during removal.

After dressing the carcass is cooled to chill temperatures during which cold shock may cause some reduction in numbers. At chill temperatures, microbial growth among the survivors is restricted to those psychrotrophs present and these can be further inhibited by the partial surface drying that takes place. Surface numbers of bacteria at the end of dressing will typically be of the order of 102–104 cfu cm^{-2}. Counts are generally higher in sheep carcasses than beef and higher still in pigs which are processed differently, the skin not being removed from the carcass but scalded and dehaired

Psychrotrophic organisms form only a small percentage of the initial microflora but come to predominate subsequently as the meat is held constantly at chill temperatures. An increase in microbial numbers is seen during cutting and boning, but this is due less to microbial growth, since the operation is usually completed within a few hours at temperatures below 10°C, than to the spreading of contamination to freshly exposed meat surfaces by equipment such as knives, saws and cutting tables.

The primary processing of poultry differs from red meat in a number of respects that have microbiological implications. First among these is the sheer scale of modern poultry operations where processing plants can have production rates up to 12000 birds per hour. This leaves little opportunity for effectively sanitizing equipment and exacerbates problems associated with some of the procedures and equipment used which favour the spread of microorganisms between carcasses.

During transport to the plant contamination can be spread between birds by faeces and feathers and from inadequately cleaned transport cages. Once at the plant, birds are hung by their feet on lines, electrically stunned and killed by cutting the carotid artery. The close proximity of the birds and the flapping wings further contribute to the spread of contamination. This is followed by scalding where the birds are immersed in hot water at about 50°C to facilitate subsequent removal of the feathers. Each bird contributes large numbers of microorganisms to the scald water and these will be spread between birds. This can be reduced to some extent by using a counter-current flow of birds and water so that the birds leaving the scalder are in contact with the cleanest water. Higher scald water temperatures will eliminate most vegetative bacteria but cause an unacceptable loosening of the skin cuticle.

A number of studies have demonstrated how these can pass organisms, for example *Salmonella*, from one carcass to others following it and when the fingers become worn or damaged they are liable to microbial colonization. As the poultry carcass is not skinned, skin-associated organisms will not be removed.

The intestinal tract of poultry will contain high numbers of organisms including pathogens such as *Salmonella* and *Campylobacter*. Poultry evisceration therefore poses similar microbiological hazards to those with other animals but the size and structure of the carcass make it a much more difficult operation to execute hygienically. To allow high processing rates, poultry evisceration is usually automated but this too leads to a high incidence of carcass contamination with gut contents. Since the carcasses are not split like those of sheep and cattle, effective washing of the gut cavity after evisceration is more difficult.

Poultry to be frozen is usually chilled in water and this offers a further opportunity for cross contamination. This is controlled by chlorination of the cooling water, use of a counter-current flow as in scalding and a sufficient flow rate of water to avoid the build up of contamination.

Now let us learn some information about characteristics of some Gram-negative associated with meat.

Non-fermentative Gram-negative rods grown most rapidly under these conditions and come to dominate the spoilage micro flora that develops. Taxonomic description of these organisms has been somewhat unsettled over the years with some being described as *Moraxella* and *Moraxella*-like. Such terms have now been largely abandoned in favour of a concensus that has emerged from numerical taxonomy studies. In this, the principal genera are described as *Pseudomonas, Acinotobacter* and *Psychrobacter* with *Pseudomonas* species such as *P. fragi,*

P. lundensis and *P. fluorescens* generally predominating. Other organisms are usually only a minor component of the spoilage microflora, but include psychrotrophic Enterobacteriaceae such as, *Serratia liquefaciens* and *Enterobacter agglomerans*, lactic acid bacteria and the Gram-positive *Brochothrix thermosphacta*.

Figure 12.5 indicates the Gram-negatives associated with meat.

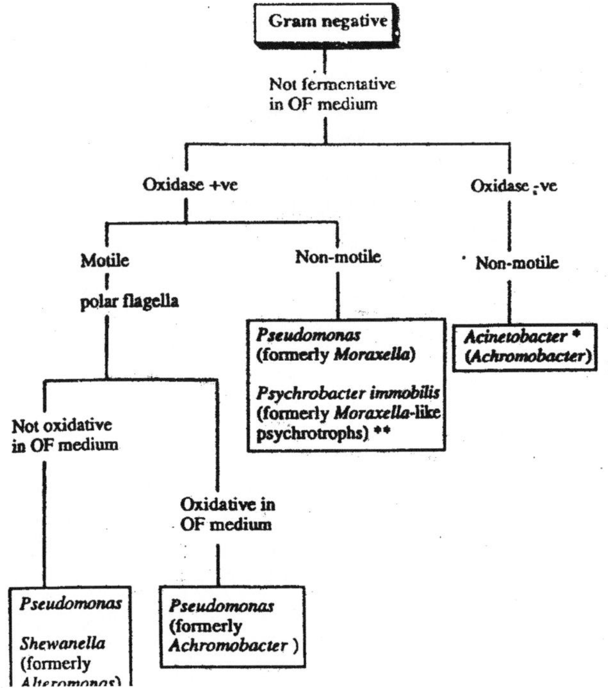

Figure 12.5: Gram-Negatives Associated with Meat.

Bacterial metabolism produces a complex mixture of volatile esters, alcohols, ketones and sulfur-containing compounds which collectively comprise the off odours detected. Such mixtures can be analysed by a combination of gas chromatography and mass spectrometry and the origin of different compounds can be established by pure culture studies. These have confirmed the predominant role of pseudomonads in spoilage of aerobically stored chilled meat. Usually, the different spoilage taints appear in a sequence reflecting the order in which components of the meat are metabolized.

The first indication of spoilage is generally the buttery or cheesy odour associated with production or diacetyl (2,3-butanedione), acetoin (3-hydroxy-2-butanone), 3-methylbutanol and 2-methylpropanol. These compounds are produced from glucose by members of the Enterobacteriaceae, lactic acid bacteria and *Brochothrix thermosphacta*. *Pseudomonads* then begin to increase in importance and the meat develops a sweet or fruity odour. This is due to production of a range of esters by *Pseudomonas* and *Moraxella* species degrading glucose and amino acids and by esterification of acids and alcohols produced during the first phase of spoilage. Ester production is particularly associated with *Pseudomonas fragi* which can produce ethyl esters of acetic, butanoic and hexanoic acids from glucose, but other pseudomonads and *Moraxella* species are also capable of producing esters when grown on minced beef. As the glucose becomes exhausted, the meat develops a putrid odour when *Pseudomonas* species and some *Acinetobacter* and *Moraxella* species turn their entire attention to the amino acid pool, producing volatile sulfur compounds such as methane thiol, dimethyl sulfide and dimethyl disulfide.

In the later stages of spoilage an increase in the meat pH is seen as ammonia and a number of amines are elaborated. Some of these have names highly evocative of decay and corruption such as putrescine and cadaverine but in fact do not contribute to off odour. When microbial numbers reach levels of around 108 cfu cm^{-2}, a further indication of spoilage becomes apparent in the form of a visible surface slime on the meat.

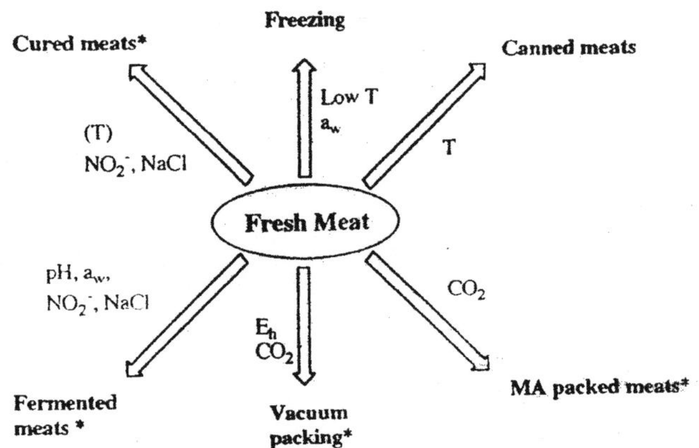

Figure 12.6: Meat and Meat Products. T indicates elevated temperatures; *E*h, low redox potential; pH reduced pH; a_w, reduced a_w and * stored at chill temperatures

Vacuum and modified-atmosphere packing of meat changes the meat microflora and consequently the time-course and character of spoilage. In vacuum packs the accumulation of CO_2 and the absence of oxygen restrict the growth of pseudomonads giving rise to a microflora dominated by Gram-positives, particularly lactic acid bacteria of the genera *Lactobacillus, Carnobacterium* and *Leuconostoc*.

Spoilage of vacuum-packed meat is characterized by the development of sour acid odours which are far less objectionable than the odour associated with aerobically stored meat. The microorganisms reach their maximum population of around 107 cfu cm^{-2} after about a week's storage but the souring develops only slowly thereafter.

Organic acids may contribute to this odour, although the levels produced are generally well below the levels of endogenous lactate already present. Some work has suggested that methane thiol and dimethyl sulfide may contribute to the sour odour.

The extension of shelf-life produced by vaccum packing is not seen with high pH (> 6.0) meat. In this situation *Shewanella putrefaciens*, which cannot grow in normal pH meat and psychrotrophic *Enterobacteriaceae* can grow and these produce high levels of hydrogen sulfide giving the meat an objectionable odour.

In modified-atmospheres containing elevated levels of both CO_2 and O_2 growth of pseudomonads is restricted by the CO_2 while the high levels of O_2 maintain the bright red colour of oxygenated myoglobin in the meat. Here the micro flora depends on the type of meat, its storage temperature and whether it was vacuum packed or aerobically stored, previously. In general terms though, the micro flora and spoilage tend to follow a similar pattern to that of vacuum-packed meat. Heterofermentative lactic acid bacteria can be more numerous due to the stimulatory effect of oxygen on their growth and, under some circumstances, *Brochothrix therfmosphacta*, Enterobacteriaceae and pseudomonads can be more important.

Meat can be processed in a number of different ways which affect its characteristics, shelf-life and microbiology.

Microorganisms Associated with Fresh Meat

In contrast to fruits and vegetables, meats are composed mainly of protein and fats rather than carbohydrates. Water content is 71–76 per cent and therefore, moisture is not an issue except for spoilage microbes on cured meats. Muscles of healthy animals do not contain any bacteria or fungi but as soon as animals are slaughtered, meat is exposed to contaminants and good sanitation practices are essential to produce high quality meats. The number of spoilage organisms on meat just after slaughter is a critical factor in determining shelf life. The surface of beef carcasses may contain anywhere from 10^1 to 10^7 cfu/cm2, most of which are psychrotrophic bacteria.

Chopping and grinding of meats can increase the microbial load as more surface area is exposed and more water and nutrients become available. A large variety of microbes are commonly found on fresh meat, but different microbes become dominant during spoilage of different meats depending on pH, composition and texture of processed meats, temperature and packaging atmosphere. *Pseudomonas* spp. is the predominant spoilage bacteria in aerobically stored raw meat and poultry. Once the initial low levels of glucose are depleted by various microbes, *Pseudomonas* has an advantage because it can catabolize gluconates and amino acids more readily than other microbes. Breakdown of these compounds results in production of malodorous sulfides, ammonia, and amines, including the biogenic amines putrescine and cadaverine. Dark, firm and dry meat with a relatively high pH of 6.0 spoils more rapidly because deamination of amino acids starts earlier. *Shewanella putrefaciens* does not grow on meat at pH<6.0 but can produce sulfides and ammonia even when glucose is still available. These sulfides not only smell bad but also cause colour changes in meat, and therefore Shewanella has a high spoilage potential on fresh, high pH meats stored aerobically even when it is not a dominant microbe. *Brochothrix thermosphacta* is often a significant spoilage organism on fresh meat stored aerobically at refrigeration temperatures. *Enterobacteriaceae*, particularly species of *Serratia*, *Enterobacter*, and *Hafnia*, are major causes of spoilage in vacuum-packed, high pH fresh meats. These organisms are facultative anaerobes that produce organic acids, hydrogen sulfide and greening of meats.

Lactic acid bacteria (LAB) grow on meat and poultry packaged under vacuum and modified atmospheres, producing organic acids from glucose by fermentation. This gives rise to aciduric off-odours which may be accompanied by gas and slime formation and greening of meat. However, LAB are weakly proteolytic and so do not produce large amounts of amines or sulfides, and spoilage of meat by LAB is not as offensive. Psychrophilic, anaerobic *Clostridium* spp. are associated with spoilage of vacuum-packaged meats. "Blown pack" meat spoilage is characterized by excessive gas formation with off odours due to formation of butyric acid, butanol and sulfurous compounds. Yeasts and molds grow relatively slowly on fresh meat and do not compete well with bacteria. Therefore, they are a minor component of spoilage flora.

Microorganisms Associated with Meat During Processing

Meat spoilages indicate (a) colour changes (b) textural changes and (c) development of off-flavour or off-odour or slime as a result of microbial growth. *Salmonella* is the primary microbial challenge for poultry. The primary microbial to the beef industry is *Escherichia coli* O157: H7. *Listeria*, which is an adulterant with zero tolerance, is the major problem for ready to eat meat products.

Some of the factors that influence growth of microorganism in meat include the kind and amount of contamination with microorganism and the spread of these microorganism in the meat; the physical properties of the meat; Chemical properties of the meat; Availability of oxygen; and temperature.

Types of Spoilage of Meats

Spoilage can take place under

1. Aerobic conditions
2. Anaerobic conditions

1. Under Aerobic Conditions, Bacteria may cause the following

Surface slime: It may be caused by species of *Pseudomonas, Acinetobacter, Moraxella, Alcaligenes, streptococcus, Leuconostoc, Bacillus,* and *Micrococcus.* It is an indication of spoilage, often observed before expiry date.

Changes in colour of meat pigment: The red colour of meat, called "bloom", may be changed to shades of green, brown, grey as result of the production of oxidizing compound *e.g.*, hydrogen peroxides, hydrogen sulfide. *Lactobacillus* and *Leuconostoc* are basically responsible.

Changes in fats: The oxidation of unsaturated fats may take place chemically in air and may be catalysed by light and copper. *e.g.*, oxidative rancidity. *Pseudomonas* and *Achromobacter* are responsible for oxidative rancidity or by yeast.

Surface colour discolouration due to pigment forming bacteria: Red spot may be caused by *Serratia marcescens* or other bacteria with red pigment. *Pseudomonas syncyanea* can impart blue colour to the surface; *Micrococcus* or *flavobacterium* imparts yellow colour.

Off odours and off taste: "Taints" or undesirable odours and off taste that appear in a meat. "Souring" that gives the sour odour that may be due to volatile acids. *e.g.*, formic, butyric, propionic acids by the action of microorganisms. Sometimes "Cold storage flavour" or taint observed in meat and it is nothing but stale flavour. In case of musty or earthy flavour, meat contaminated with *Actinomycetes* are the causative agent for the same.

Aerobic growth of moulds may cause:

- **Stickiness**: Surface becomes sticky
- **Whiskers**: When meat is stored in frozen condition, mycelial growth may take place without sporulation.
- **Black spot**: It is caused by *Cladosporium herbarum*
- **White spot**: It is caused by *Sporotricum carnis.*
- **Green patch**: It is caused by *Penicillum* such as *P. expansum, P. oxalicum*
- **Decomposition of fats**: Many molds produce lipases and cause hydrolysis of fats.

2. Spoilage under Anaerobic Conditions

Facultative and anaerobic bacteria are able to grow within the meat under anaerobic conditions and cause spoilage. Following changes occurs in such conditions

Souring: It imparts sour taste to meat due to acids such as formic, propionic, acetic *etc.* Meat's own enzymes are responsible for it.

Putrefaction: True putrefaction is the anaerobic decomposition of protein with the production of foul-smelling compound such as hydrogen sulfide, indole, ammonia, amines due to *Clostridium* species.

Taint: Taints" or undesirable odours and tastes, that appear in a meat.

- *Clostridium* spp. has been associated with the production of large amount of gas in vacuum packaged beef, accompanied by foul off-odours. Gas production (CO_2) by lactic acid bacteria without extensive off odours may be associated with vacuum packaged beef and pork.
- Pasteurized canned product: *Bacillus* sps., *Clostridium* spp., *Enterococci, Streptococci* and heat resistance lactic acid bacteria
- Refrigerated canned product: *Streptococcus* spp.

Treatment with organic acids, hot water steam carcass pasteurization and steam carcass vacuuming, trisodium phosphate, acidified sodium chlorite, chlorine dioxide, lactoferrins, peroxyacetic acid, sodium lactate, sodium acetate and sodium diacetate, ozone and radiation have been used as microbial decontaminants during meat processing operations. Carcass washing with hot water of 80°C for 10 seconds can reduce microbial loads by 2 logs. Current regulatory policies and inspection in the meat industry include the HACCP (Hazard Analysis Critical Control Point) food safety system with an objective to provide safe food for consumption and prevent chemical, physical and biological hazards.

Causes of Meat Spoilage

Preslaughter handling of livestock and post slaughter handling of meat play an important part in deterioration of meat quality. The glycogen content of animal muscles is reduced when the animal is exposed to pre-slaughter stress which changes the pH of the meat, to higher or lower levels, depending on the production level of lactic acid. Lactic acid is produced due to the breakdown of glycogen content of animal muscles via an anaerobic glycolytic pathway. Higher levels of pH (6.4-6.8) result in Dark, Firm and Dry (DFD) meat. Long term stress causes DFD meat which has a shorter shelf life. Severe short term stress results in a Pale, Soft and Exudative (PSE) meat. PSE meat has a pH lower than normal ultimate value of 6.2 which is responsible for the breakdown of proteins, providing a favourable medium for the growth of bacteria. The factors affecting the shelf life of meat and meat products are summarized in Table 12.14 .There are three main mechanisms for meat and meat products spoilage after slaughtering and during processing and storage: (a) microbial spoilage, (b) lipid oxidation and (c) autolytic enzymatic spoilage.

Microbial Spoilage

Meat and meat products provide excellent growth media for a variety of microflora (bacteria, yeasts and molds) some of which are pathogens. The intestinal tract and the skin of the animal are the main sources of these microorganisms. The composition of microflora in meat depends on various factors: (a) preslaughter husbandry practices (free range Vs intensive rearing), (b) age of the animal at the time of slaughtering, (c) handling during slaughtering, evisceration and processing, (d) temperature controls during slaughtering, processing and distribution (e) preservation methods, (f) type of packaging and (g) handling and storage by consumer.

Table 12.14: Factors Affecting Shelf Life of Meat

Type	*Factors*
Intrinsic	✰ Type of animal (bovine, porcine) Breed and fed regime
	✰ Age of animal at time of slaughter
	✰ Initial microflora
	✰ Chemical properties (peroxide value, pH, acidity, redox potential)
	✰ Availability of oxygen
	✰ Processing conditions and control Hygiene (standard of personnel and equipment cleaning)
Extrinsic	✰ Quality- management system
	✰ Temperature control
	✰ Packing system (materials, equipment, gases)
	✰ Storage types

Mold species include *Cladosporium*, *Sporotrichum*, *Geotrichum*, *Penicillium* and *Mucor* while yeasts species include *Candida* spp., *Cryptococcus* spp. and *Rhodotorula* spp. Bacteria species include *Pseudomonas*, *Micrococcus*, *Streptococcus*, *Sarcina*, *Lactobacillus*, *Salmonella*, *Escherichia*, *Clostridium* and *Bacillus*. Hayes *et al.* (2003) found *Enterococcus* spp. to be the most dominant bacteria on 971 of the 981 samples (99 per cent) of all meat (chicken, turkey, pork and beef) in the state of Iowa. About 97 per cent of pork samples contained *Enterococci* with 54 per cent of isolates identified as *Enterococcus faecalis* and 38 per cent as *Enterococcus faecium*, 3.4 per cent as *Enterococcus hirae*, 2.4 per cent as *Enterococcus durans*, 0.8 per cent as *Enterococcus casseliflavus*, 0.4 per cent *Enterococcus gallinarum* and 1 per cent as unidentified. All of beef samples contained *enterococci* with 65 per cent of isolates identified as *Enterococcus faecium*, 17 per cent as *Enterococcus faecalis*, 14 per cent as *Enterococcus hirae*, 2 per cent as *Enterococcus durans* 0.7 per cent, as *Enterococcus casseliflavus*, 0.4 per cent *Enterococcus gallinarum* and 0.9 per cent as unindentified. Cerveny *et al.* (2009) stated that storage conditions affect the type of microbes found in meat and meat products. They reported that *Pseudomonas* spp., *Moraxella* spp., *Psychrobacter* spp., *Acinetobacter* spp. and Gram-negative psychrotrophic members of the family *Enterobacteriaceae* are frequently present on refrigerated meat product. They also indicated that psychrotrophic lactic acid bacteria, *Enterococci*, *Micrococci* and yeasts are predominately found in raw, salted-cured products such as corned beef, uncooked hams and bacon due to their resistance to curing salts.

Garcia-Lopez *et al.* (1998) reported that the growth of *Enterobacteriaceae* and *Pseudomonas* were more prevalent on modified atmosphere packed meat (especially on pork) than on vacuum packed meat, their growth being favoured by storage at 5°C. Sentence (1991) reported that Pseudomonas spp. growth rate was considerably slow at 0°C, but increased at 2°C and affected the shelf life of meat. He also noticed slow *Salmonella* growth below 7°C, which increased above 7°C and affected the shelf life of meat. Borch *et al.* (1996) reported that the growth of lactic acid bacteria on bologna-type sausage was retarded 2 and 4-fold with decreases in temperature from 7-2°C and from 7-0.6°C, respectively. Russell *et al.* (1996) stated that a favourable pH for the growth of spoilage bacteria for meat is in the range of 5.5-7.0. Slime formation, structural components degradation, off odours and appearance change were found in meat as a result of microbial growth within this pH range. Dainty (1996) stated that microbial metabolism produces fatty acids, ketones and alcohols, which exhibit a variety of fruity and sweety odours. Generation of hydrogen sulphide, methylsulphide and dimethylsulphide exhibit putrid and sulphury odours.

Lipid Oxidation

Autoxidation of lipids and the production of free radicals are natural processes which affect fatty acids and lead to oxidative deterioration of meat and off-flavours development. After slaughtering of animals, the fatty acids in tissues undergo oxidation when the blood circulation stops and metabolic processes are blocked. Lipid oxidation is the reaction of oxygen with double bonds of fatty acids. It involves three stage free radical mechanisms: initiation, propagation and termination.

Oxidation of lipids in meat depends on several factors including fatty acid composition, the level of the antioxidant vitamin E (α tocopherol) and prooxidants such as the free iron presence in muscles. Polysaturated fatty acids are more susceptible to lipid oxidation. Hydroperoxides are produced due to the lipid oxidation of highly unsaturated fatty acid fractions of membrane phospholipids, which are susceptible to further oxidation/ decomposition. Their breakage causes secondary reaction products such as pentanal, hexanal, 4- hydroxynonenal and malondialdehyde (MDA) as well as other oxygenated compounds such as aldehydes, acids and ketones. These secondary products can cause loss of colour and nutritive value due to sever effects on lipids, pigments, proteins, carbohydrates and vitamins and are directly related to carcinogenic and mutagenic processes.

In meat, lipid hydrolysis can take place enzymatically or non-enzymatically. The enzymatic hydrolysis of fats is termed lipolysis or fat deterioration and is governed by specific enzymes such as lipases, estarases and phospholipase. Lipases enzymes are present in the skin, blood and tissue of animals. During lipolysis, lipases split the glycerides forming free fatty acids which are responsible for common off-flavour, frequently referred to as rancidity. The main enzymes involved in meat lipid hydrolysis are phospholipase A1 and phospholipase A2.

The non-enzymatic hydrolysis is caused by heme proteins such as hemoglobin, myoglobin and cytochrome which are susceptible to oxidation and produce hydroperoxides. During heme catalysis, a Fe2+ protoporphyrin complex (P–Fe2+), like myoglobin, will be oxidized to P–Fe3+. The formed superoxide radical anion O_2 - reacts with H+ and will yield H_2O_2. Hydrogen peroxide will then oxidize P–Fe3+to the oxene species P–Fe = O. The free iron redox cycle contributed by ascorbic acid is the main initiator of lipid peroxidation in fresh muscle foods and it significantly affects the oxidation of oxymyoglobin.

Autolytic Enzymatic Spoilage

Enzymatic actions are natural process in the muscle cells of the animals after they have been slaughtered and are the leading cause of meat deterioration. The enzymes have the ability to combine chemically with other organic compounds and work as catalysts for chemical reactions that finally end up in meat self-deterioration. In the autolysis process, the complex compounds (carbohydrates, fats and protein) of the tissues are broken down into simpler ones resulting in softening and greenish discoloration of the meat. These autolysis changes include proteolysis and fat hydrolysis which are prerequisite for microbial decomposition. Excessive autolysis is termed "souring". Post mortem breakdown of polypeptides is the result of tissue proteases and is responsible for flavour and is textural changes in meat. Post mortem aging of red meat results in the tenderization process.

Post-mortem autolysis takes place in all animal tissues but at different rates in different organs, quicker in glandular tissue such as the liver and slower in striated muscle. The enzymes calpains, cathepsins and aminopeptidases are found to be responsible for the post mortem autolysis of meat through digestion of the z- line proteins of the myofibril. Among these enzymes, calpains has been described as a preliminary contributor to the proteolytic tenderization process of meat. Cathepsins were, also, found to contribute to tenderization at low pH. Proteolytic enzymes are active at low temperatures (5°C) which lead to deterioration of meat quality due to growth of microbes and biogenic amines production.

Poultry Meat

Poultry meat like meat of other animals is also susceptible to contamination by various sources. Contamination of skin and lining of the body cavity take place during various processing operations. The organisms of great importance in poultry are *Salmonella* spp. and *Campylobacter jejuni.* Several Gram negative psychrotropic bacteria *viz., Pseudomonas, Acenitobacter* and *Flavobacterium* have also been isolated from poultry carasses. It is important to freeze the poultry fast in order to keep it in good condition for several months. Freezing further reduces the number of microorganisms in the poultry meat provided the temperature is maintained quite low (-18 ° C or below).

To summarize, meat spoilage and product shelf-life is an important challenge for all the experts gravitating around this area. The spoilage due to ropy slime-formation has influenced the marketing of vacuum-packed meat products and the use of this technology. The presence of ropy slime-producing bacteria and their associated sensory abnormalities lead to high direct financial losses (waste product) and indirect (such as product selection, disinfection of contaminated surfaces and non-delivery at destination). Although food security is likely to be guaranteed, the macroscopic appearance of the product at the time of packaging is particularly unpleasant, making it unsuitable for further processing or marketing. Food industries and productions must be supported by research, creating a strong link between discoveries and applications. Nowadays, ropy slime-formation on meat products represents a persistent problem, often ignored. It is, therefore, necessary to provide a basis as a starting point to find a beneficial solution.

IV. The Microbiology of Cereals

A cereal is any 'grass' grown for its fruit called grain (botanical term: caryopsis). Cereals include crops, such as wheat, rice, corn, barley, rye, oats and millet, that underpin the world's food supply. The word cereal comes from 'Ceres', the Roman goddess of agriculture. Cereals were first cultivated by ancient Sumerians in the area of the Middle East between the Euphrates and Tigris rivers. Cereal grains provide more food energy content than any other type of crop. Cereal grains in their unprocessed form are called 'whole grains' and are a rich source of carbohydrates, fats, oils, protein, vitamins and minerals.

Cereal grains are the most commonly consumed food group worldwide and they are grown on about 60 per cent of the cultivated land in the world. In order to meet the requirements of a growing world population, worldwide production and yield of cereals has been increased for the last 50 years. Major types of cereal grains include maize, rice, wheat, barley, sorghum, millet, oats, and rye (FAO, 2017).

Table 12.15: Some of the More Important Cereal Crops

Botanical Name	Common Name
Triticum estlivum	Common wheat (bread wheat)
Triticum durum	Durum wheat (pasta wheat)
Hordeum spp.	Barley
Avena sativa	Oats
Secale cereale	Rye
Zea mays	Maize (American corn)
Oryza sativa	Rice
Sorghum vulgare	Sorghum
Panicum miliaceum	Millet
Pennisetum typhoideum	Bulrush millet

Each cereal species is adapted to grow in a particular range of climatic conditions although plant breeding programmes have extended the ranges of several of them. The common wheat is the major cereal of temperate parts of the world and is grown extensively in both northern and southern hemispheres, whereas the durum wheat is grown extensively in the Mediterranean region and in the warmer, drier parts of Asia, North and South America. Rye can be grown in colder parts of the world and is an important crop in central Europe and Russia. Maize, which originated from the New World, is now grown extensively throughout the tropics and subtropical regions of both northern and southern hemispheres.

Although sorghum and some of the millets are not very important in terms of world tonnage, they are especially well adapted growing in warm, dry climates and may be locally the most important millets in such regions as those bordering the southern edge of the Sahara Desert.

Distribution Of Microorganisms within Cereal Grains

The typical structure of a cereal grain constitutes of three edible parts: the bran which consists of the outer coat (pericarp, testa, and aleurone layers), the germ (the embryo) and the starchy endosperm, and an inedible husk that protects the kernel (Figure 12.6). Microbial colonization is generally restricted to the outer layers of cereal grains, that is, the husk, between the husk and pericarp, and within the pericarp tissue. Several studies showed that after debranning, a controlled process in which the outer layers of the grains are removed, cereals are microbiologically purer. However, there are species able to invade the inner part of the grains and penetrate into the endosperm, causing internal infections.

Microbial Contamination within a Cereal Grain

Distribution of the microbial populations on various cereal grains has been studied by several authors. Microscopic observations of barley revealed a high number of microorganisms between the husk and pericarp. It was observed that bacteria clustered as randomly distributed micro-colonies with up to 200 cells. Laca *et al.* (2006) studied distribution of microorganisms within wheat grains and found that most of bacteria and molds were concentrated on the surface of the grain in the pericarp surrounding the endosperm and the germ, therefore, removing some of the outer layers of the grains may be used to substantially reduce the microbial contamination. According to the study, most of the contamination is located in the outer layers, that is, the first and second pearling fractions, which corresponds to a layer thickness of approximately 30 µm. Colonization of the grains by *Alternaria* spp. (black fungi discoloration) is observable on the surface of the kernels as well as beneath the pericarp of wheat, barley and oats, and is believed to be a result of rainfall just before harvest. Andersen and Thrane (2006) reported that wheat and barley surface disinfection with sodium hypochlorite removed only 10–15 per cent of *Alternaria* and *Bipolaris*, which indicates that the grains were contaminated beneath the pericarp. A common result of invasion of the germs by *Aspergillus* species, such as *A. restrictus*, *A. glaucus* and *A. candidus*, is germ-damaged wheat (the fungi grow only in the germ) that can often develop without visible sign of moldiness. Studies based on scanning electron microscopy showed distribution of the fungus mostly over the pericarp, however, contamination of the embryo and endosperm also occurred.

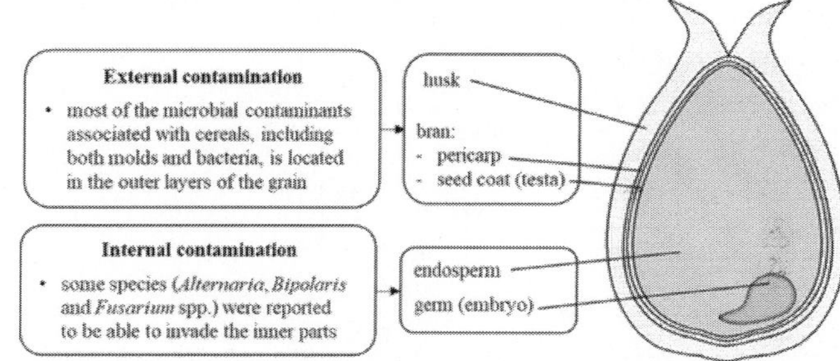

Figure 12.6: Distribution of Microorganisms within Cereal Grains.

Microbial Challenges Associated with Cereal Grains

The Sources of Microbial Contamination of Cereals

Microbial contamination of cereal grains occurs during crop growth, harvesting and postharvest drying and storage and it derives from several sources, including air, dust, water, soil, insects, birds and rodents faeces as well as contaminated equipment and unsanitary handling (Figure 12.7). The type of microbial contamination varies according to the growing region and is heavily influenced by environmental conditions such as drought, rainfall, temperature, and sunlight, as well as unsanitary handling (Figure 12.7, harvesting and processing equipment, and poor storage. High rainfall just before harvest is a factor inducing extensive colonization of the grain ears by *Alternaria* spp., causing black fungi discoloration, that can be observable both on the surface of the kernels and as beneath the pericarp. Doohan *et al.* (2003) investigated the influence of climatic factors on *Fusarium* species pathogenic to cereal grains and found that they differ significantly in their climatic distribution as well as in the optimum climatic conditions required for their persistence.

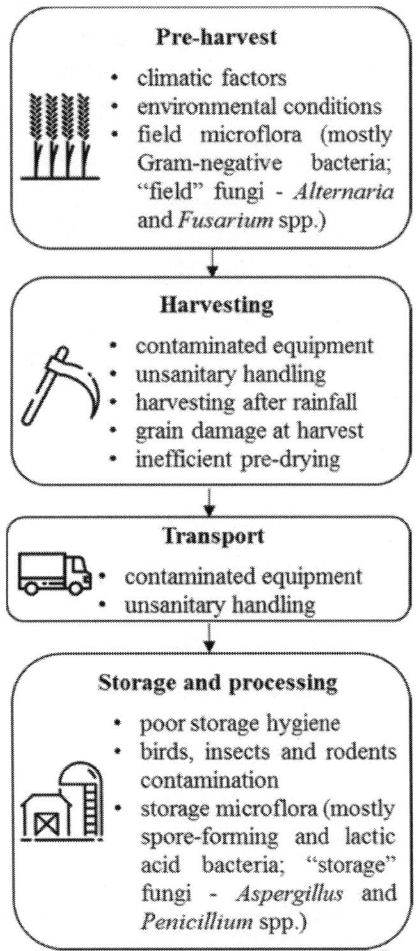

Figure 12.7: Sources and Factors of Microbial ontamination during Cereal Grain Processing.

The Field Microflora

The field microflora consists of microorganisms that occur on or in grains until the time of harvest and depends on the conditions under which the crops were grown. The kernels are numerically dominated by bacteria, with yeasts as the next most abundant component. The number of filamentous fungi increases during the later stage of ripening.

Bacteria

Levels of cereal grains contamination with bacterial pathogens are usually very low and although contamination with species such as *Salmonella*, *Escherichia coli*, and *Bacillus cereus* can occur, bacteria associated with cereals are generally non-pathogenic. The most often they belong to the families *Pseudomonadaceae*, *Micrococcaceae*, *Lactobacillaceae*, and *Bacillaceae* (Laca *et al.,* 2006). Some species of enteric bacteria that are found on cereal grains are plant saprophytes and their presence is not related to fecal contamination (Harris *et al.*, 2013). Gram-negative bacteria numerically dominate the microflora of preharvest barley, with *Erwinia herbicola* (now: *Pantoea agglomerans*) and *Xanthomonas campestris* as predominant bacterial species. Numerous bacteria belonging to *Streptomyces* genus were recently found on barley and spring wheat grains. The authors also reported the presence of antimycin A toxin-producing strains in barley, which is the first report of antimycin A in a food substance (Rasimus-sahari *et al.*, 2016).

Filamentous Fungi and Yeasts

The fungi growing on crops have been traditionally divided into two groups: "field" and "storage" fungi. The main difference between these groups is the time at which they invade the grains and growth conditions, however, the distinction between field and storage fungi is not absolute. It was found that although some field fungi invade the grains on the field, they are still able to grow in storage conditions. Similarly, some fungi commonly classified as storage fungi may invade the grains at earlier stages.

Field fungi, including species such as *Alternaria*, *Cladosporium*, *Fusarium*, and *Helminthosporium*, invade grain in the field at high relative humidities (90 per cent to 100 per cent) when the grain is high in moisture (18 per cent to 30 per cent) (that is, at high a_w) (Bullerman and Bianchini, 2009). Significant increase in a number of infections with *Fusarium* species is often observed when ripening is performed during wet periods. Also, phyllosphere fungi are responsible for preharvest fungal contamination.

As high moisture content is required for field fungi, conditions during the storage are not favourable for their growth, however, some fungi including *Penicillium* and *Fusarium* species and various species of yeasts are able to invade seeds before harvest and continue to grow during storage. Although most *Penicillium* species are xerophiles and they usually are considered to be storage fungi, at certain conditions they can also attack the grains before harvest; it was found that *P. oxalicum* can infect preharvest maize due to insect damage or wounding.

Although field fungi require relatively high-water activities for optimum growth, genera such as *Cladosporium, Alternaria* and *Epicoccum* are able to survive the rapid changes that can occur from the desiccation of a hot sunny day to the cool damp conditions of the night. The genus *Fusarium* includes species which have both pathogenic and saprophytic activities. Thus *F. culmorum* and

F. graminearum can cause both stem rot and head blight of wheat and barley in the field and these field infections may lead to more extensive postharvest spoilage of these commodities if they are stored at too high a water activity.

Similarly, *A. flavus*, usually regarded as a storage fungus, was found in freshly harvested maize. Insect damage to cobs and fungal invasion down thc silks are the main factors increasing the risk of *A. flavus* contamination of corn.

The field fungus most frequently present in both barley and wheat kernels is *Alternaria*. Kosiak *et al.* (2004) reported that the distribution of field fungi contamination varied significantly in wheat, barley and oats samples of reduced quality with dominant *Fusarium* species as compared to acceptable samples for which *Alternaria* was the most numerous. *Fusarium* species are a common contaminant of various cereal grains. While *F. graminearum*, *F. culmorum*, *F. poae*, *F. avenaceum*, and *Microdochium nivale* (formerly known as *F. nivale*) cause diseases of small-grain cereals such as wheat and barley, corn is usually attacked by *F. graminearum*, *F. moniliforme*, *F. proliferatum*, and *F. subglutinans*.

Hill and Lacey (1983) reported that by harvest even 50 per cent to 85 per cent of barley kernels may be colonized by yeasts, with pink yeasts, such as *Sporobolomyces* and *Rhodotorula*, being predominant (Flannigan, 1996). Other species found on barley are *Hansenula*, *Torulopsis*, *Candida*, and *Saccharomyces*, while *Cryptococcus* and *Trichosporon* were isolated from preharvest wheat. Flannigan (1996) suggested that because of the strong resemblance between other components of wheat and barley microfloras, similar yeasts species are likely to be found on both cereals.

The Storage Microflora

Modern methods used for harvesting and proper storage practices ensuring the exclusion of water and lack of possibility for birds, insects and rodents to contaminate the grain, are considered sufficient to prevent the microbial growth. However, these conditions are not always met.

Bacteria

Generally, bacteria are not significantly involved in the spoilage of dry grain due to storage conditions unfavourable for their growth. However, it was found that some bacterial pathogens and spore-forming species are able to survive during storage and may contaminate processed products. Lactic acid bacteria present in the raw grain may be carried over through the processing and spoil doughs prepared from flour and cornmeal. Gram-negative coliforms, pseudomonads and actinomycetes were also found on dry-stored cereals Wachowska *et al.* (2013) reported that the number of bacteria of the genus *Azotobacter* colonizing winter wheat grain was relatively low at harvest, but the counts increased after 6 months of storage.

Filamentous Fungi and Yeasts

Spoilage of grains with filamentous fungi during storage occurs usually due to inefficient drying, what favours microbial growth and may result in increased mycotoxins levels. If drying is delayed and the moisture content of the harvested grain is suitable, growth of the field fungi, for example, *Fusarium* spp., may occur.

Storage fungi seem to be well adapted to the more constant conditions of cereals in storage and generally grow at lower water activities. The most important genera of the storage fungi are *Penicillium* and *Aspergillus*, although species of *Fusarium* may also be involved in spoilage when grain is stored under moist conditions.

Table 12.16: Minimum Water Activity Requirements of some Common Field and Storage Fungi

Species	*Minimum a_w*	*Species*	*Minimum a_w*
Field Fungi			
Fusarium culmorum	0.89	*Fusarium graminearum*	0.89
Alternaria alternala	0.88	*Cladasporium herbarum*	0.85
Storage Fungi			
Penicillium aurantiogriseum	0.82	*Penicillium brevicompactum*	0.80
Aspergillus flavus	0.78	*Aspergillus candidus*	0.75
Eutrotium amstelodami	0.71	*Wallemia sebi*	0.69

Table 12.17: Equilibrium Relative Humidity, Water Activity and Moisture Content (as per cent wet weight) of Cereals at 25°C

Equilibrium Relative Humidity (per cent)	*Water Activity a_w*	*Water Potential (MPa)*	*Barley*	*Water Content (per cent wet weight)*	
				Oats	*Sorghum*
15	0.15	- 261	6.0	5.7	6.4
30	0.30	- 166	8.4	8.0	8.6
45	0.45	- 110	10.0	9.6	10.5
60	0.60	- 70	12.1	11.8	12.0
75	0.75	- 39	14.4	13.8	15.2
90	0.90	- 14.5	19.5	18.5	18.8
100	1.00	0.0	26.8	24.1	21.9

Storage fungi including species of *Eurotium*, *Aspergillus*, *Penicillium*, *Rhizopus*, *Mucor*, and *Wallemia*, invade stored-grains at low relative humidity's (65 per cent to 90 per cent) and lower moisture contents (14 per cent to 16 per cent) of the grains. Storage temperature heavily influences the types and rates of microbial spoilage – *Penicillium* species are dominant at cooler temperatures, while *Aspergillus* and *Eurotium* species are more common at ambient temperatures (20 to 25 °C). Species of *Rhizopus* (common bread mold) and *Neurospora* (red bread mold) can also be found on grains at ambient temperatures, however, they are much more common on baking products.

Among *Penicillium* species, *P. verrucosum* is important as contamination with this fungus may result in the production of carcinogenic mycotoxin - ochratoxin a (OTA), especially in cool climates. It was shown that *P. verrucosum* attacks wheat and barley only after harvest. In tropical conditions other *Penicillium* species are more common; *P. citrinum* is a fungus commonly detected in rice. In maize, different *Penicillium* species occur depending on the stage of processing; some of them invade only preharvest grains (*P. funiculosum*), other are associated only with the stored grain (*P. aurantiogriseum, P. viridicatum*), whereas some species, such as *P. citrinum* and *P. oxalicum*, are present all the time.

Birck *et al.* (2005) compared fungal contamination in wheat grain during 180 days of storage. It was observed that *Fusarium* spp. was the most numerous fungi after harvest and after 30 days of storage, however, the counts decreased gradually until the end of the storage period. After 180 days of storage *Aspergillus*, *Fusarium* and *Penicillium* were found in 96.7 per cent, 46.7 per cent, and 80.0 per cent of wheat samples, respectively. Krnjaja *et al.* (2015) investigated mycobiota of maize; mycological analyses showed the presence of *Aspergillus, Fusarium*, and *Penicillium* on both freshly harvested and stored grains, however, the predominant species varied for each stage of processing. Similar observations were noted for maize; samples analyzed immediately after harvest showed predominance species of *Alternaria*, followed by species of *Cladosporium* and *Penicillium*, whereas after 120 days of storage the maize grains mycoflora consisted of various fungi belonging with predominance to *Fusarium, Aspegillus* and *Penicillum* genera (Dudoiu *et al.*, 2016).

Yeasts found on cereal grains during storage are often amylolytic yeasts. Similarly to lactic acid and spore-forming bacteria, yeasts present on cereals may also be carried through into processed products.

Potential microbiological risks associated with cereal grains remain a major concern of the grain industry as they may hugely affect the quality and properties of the grains. Because of the permanent and ubiquitous presence of microorganisms and fungal spores in the environment, cereals always carry a certain microbial load when harvested. Additionally, climatic conditions, such as temperature and humidity that are not under human control may be crucial for contamination with moulds. Therefore, appropriate post-harvest crop treatment, before and during storage, is as important as pre-harvest strategies in the prevention of microbial spoilage.

In the storage of food and especially cereals, microorganisms are considered unwanted carriers of infections either for the consumer or for the next generation of plants. Some microorganisms can also produce various toxins, *e.g.*, aflatoxin; the production of large quantities of other mycotoxins is also common. Bacterial botulinic neurotoxin may also occur infrequently. The mycotoxins most usually associated with cereal grains are ochratoxins, deoxynivalenol, zearalenone and fumonisins too.

Several methods have been developed to eliminate these undesirable agents. The various food processes having effects on mycotoxins include cleaning, milling, brewing, cooking, baking, frying, roasting, flaking, alkaline cooking, nixtamalization, extrusion and thermal treatment. In addition to chemical agents, they may also be based on the action of various physical phenomena, above all by thermal treatment. Most of the food processes have variable effects on mycotoxins, with those that utilize high temperatures having the greatest effects. In general, the processes reduce mycotoxin concentrations significantly but do not eliminate them completely.

In order to overcome the limitations of conventional technologies, recent works have been focused on developing new techniques, such as microwave treatment, pulsed UV light, cold plasma and organic acids, that can be used for microbial decontamination of cereals. Among methods, the non-thermal plasma (NTP) has been proved to be an effective way to keep the mentioned agents under control. Further studies are needed to ensure that these potential technologies could provide an efficient microbial inactivation and rapid, uniform treatment, whilst at the same time do not affect the grains quality. Prevention of contamination with fungi is the most rational and economical approach to reduce the risks associated with the presence of mycotoxins in cereal food and feed products. However, in current production systems even the best agricultural and manufacturing practices cannot fully prevent mycotoxin contamination, therefore, potential technologies will need to be used for degradation and elimination of these toxic metabolites.

V. The Papilionaceae and Leguminosae

The Papilionaceae and Leguminosae, which form a major source of vegetable proteins and include important crops as peas, beans, soya, groundnuts and lentils. Although many species of peas and beans are familiar to us as fresh vegetables, millions of tons of the mature seeds of soya beans and groundnuts are harvested for longer term storage every year and may be susceptible to mould spoilage if not stored under appropriate conditions. Several of the leguminous seeds, such as groundnuts and soya beans, are also valuable sources of vegetable oils but there are plants from many other diverse families which are now used to provide food quality vegetable oils, rapeseed from the crucifers, sunflower seed from the daisy family, oilpalm and olives to mention just a few. Edible nuts may also come from a botanically wide range of tree species and many of them are rich in oil and give similar microbiological problems as oilseeds.

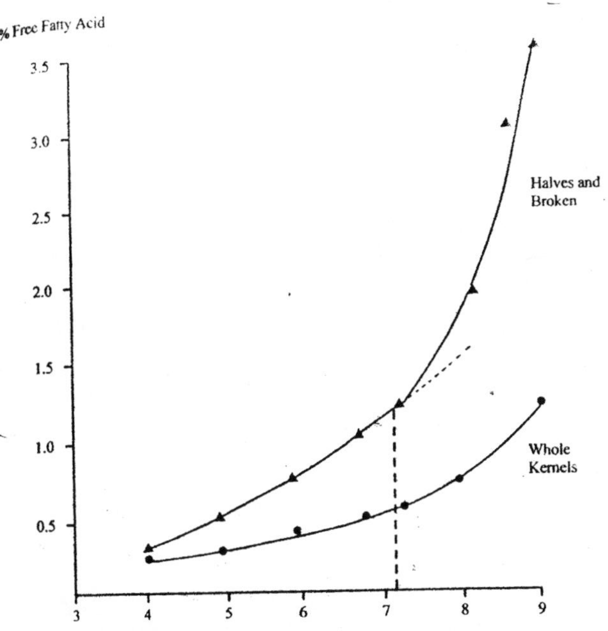

Figure 12.8: The Influence of Moisture Content on Free Fatty Acid Formation in Groundnuts Stored for Four Months

Seeds rich in oil, such as groundnuts, have a much lower water content at a particular water activity than cereals, thus groundnuts with a 7.2 per cent water content have a water activity of about 0.65–0.7 at – 25°C. Apart from the problem of mycotoxin formation in moulded oilseeds, several mould species have strong lipolytic activity leading to the contamination of the extracted oils with free fatty acids which may in turn undergo oxidation to form products contributing to rancidity. The most important lipolytic moulds are species of *Aspergillus*, such as *A. niger* and *A. tamarii, Penicillium* and *Paecilomyces.*

It can be seen that in wholesome nuts there is a steady increase in FFA formation with an increase in moisture content and this is due to the plants own lipolytic enzymes. However, damaged groundnuts show a more rapid rise at low moisture contents presumably due to increased contact of enzymes and substrate as a result of damage, but they also show an especially rapid rise in FFA at moisture contents greater than 7.2 per cent corresponding to the active growth of lipolylic moulds.

It should be noted that, although a commodity may be dry enough to avoid direct microbiological spoilage, it may not be secure against the ravages of pests such as insects and rodents and their activity may lead to secondary invasion and mould spoilage.

VI. Vegetables and Fruits

In the past decade, outbreaks of human illness associated with the consumption of raw vegetables and fruits (or unpasteurized products produced from them) have increased. Changes in the produce industry, social demographics, food consumption patterns, and awareness of fresh fruits and vegetables as potential vehicles of infection may also be contributing to an increase in documented produce-associated outbreaks of human illness. Advances in agronomic, processing, preservation, packaging, shipping, and marketing technologies on a global scale have enabled the fresh fruit and vegetable industry to supply consumers with a wide range of high-quality produce year round. Some of the same technologies and practices have also introduced an increased risk for human illness associated with pathogenic bacteria, mycotoxigenic molds, viruses, and parasites. The use of manure rather than chemical fertilizer, as well as the use of untreated sewage or irrigation water containing pathogens, viruses, or parasites, undoubtedly contributes to this increased risk.

Pathogens such as *Listeria monocytogenes, Clostridium botulinum,* and *Bacillus cereus* are naturally present in some soil, and their presence on fresh produce is not rare. *Salmonella, Escherichia coli* O 157:H7, *Campylobacter jejuni, Vibrio cholerae,* parasites, and viruses are more likely to contaminate fresh produce through vehicles such as raw or improperly composted manure, irrigation water containing untreated sewage, or contaminated wash water.

Sources of Contamination

The presence of pathogenic bacteria, viruses, and parasites on fresh fruits and vegetables has been extensively documented. Contamination of produce can occur in the field or orchard; during harvesting, postharvest handling, processing, shipping, or marketing; or in the home (Table 12.18). Potential pre-harvest sources of contamination include soil, faeces, irrigation water, water used to apply fungicides and insecticides, dust, insects, inadequately composted manure, wild and domestic animals, and human handling (Beuchat, 1996). Post-harvest sources of contamination include faeces, human handling, harvesting equipment, transport containers, wild and domestic animals, insects, dust, rinse water, ice, transport vehicles, and processing equipment (Burnett and Beuchat, 2001). As raw agricultural products, fresh produce should be expected to harbour a wide variety of microorganisms including some pathogenic varieties. After all, with the exception of greenhouse operations, produce is still grown out doors and animals, birds, and insects can all carry human pathogens.

Bacteria such as *Clostridium botulinum*, *Bacillus cereus* and *L. monocytogenes*, all capable of causing illness, are normal inhabitants of many soils, whereas *Salmonella*, *Shigella*, *E. coli* and *Campylobacter* reside in the intestinal tracts of animals, including humans, and are more likely to contaminate raw fruits and vegetables through contact with faeces, sewage, untreated irrigation water or surface water. Numbers of clostridial spores on some types of vegetables appear to increase during the summer. Perhaps the most prevalent disease-causing microorganism in soil is *Listeria monocytogenes* (Beuchat, 1996). A study of soil and domestic animal faeces has shown that *Listeria* is more often present during July to September than other months. *L. monocytogenes* and *L. innocua* were predominant in faeces, whereas *L. ivanovi* and *L. seeligeri* were most common in soil. It has been reported that the predominance of certain serotypes of *L. monocytogenes* may be influenced by the environment and that some strains indigenous to decaying plant vegetation are incapable of causing human illness. It was suggested that *L. monocytogenes* is a saprophyte that lives in a plant-soil environment and could therefore be contracted by humans and animals through many possible routes from many sources. Birds and animals are unlikely to be the only sources responsible for the distribution of *L. monocytogenes* in nature and its presence on fruits and vegetables.

The presence of other pathogenic bacteria, viruses, and parasites in soil likely results largely from application of faeces or untreated sewage, either by chance or design. Whatever the ease, soil on the surface of fruits and vegetables may harbour pathogenic microorganisms that remain viable through subsequent handling to the point of consumption unless effective sanitizing procedures are administered.

Irrigation and surface run-off waters can be sources of pathogenic microorganisms that contaminate fruits and vegetables in the field. Irrigation water containing raw sewage or improperly treated effluents from sewage treatment plants may contain hepatitis A, Norwalk viruses, or enteroviruses (poliomyelitis, echoviruses, and Coxsackie viruses). Rotaviruses are known to retain viability on the surface of vegetables held at 4°C for up to 30 days.

Listeria and other potentially pathogenic bacteria have been reported in sewage. The use of sewage as a fertilizer could contaminate vegetation destined for human consumption. MacGowan *et al.* (1994) examined sewage at 2-month intervals in 1991 to 1992 and found 84 per cent to 100 per cent contained *L. monocytogenes* or *L. innocua*.

Application of sewage sludge or irrigation water to soil is one avenue through which parasites can contaminate fruits and vegetables. It has been observed that *Ascaris* ova sprayed onto tomatoes and lettuce remain viable for up to 1 month, while *Endamoeba histolytica* could not be recovered 1 week after spraying. If sewage irrigation or night soil application is stopped 1 month before harvest, the produce would not likely be vectors for transmission of diseases caused by these parasites.

Table 12.18: Sources of Pathogenic Microorganisms on Fresh Fruits Vegetables*

Preharvest
Faeces
Soil
Irrigation water
Water used to apply fungicides, insecticides
Green or inadequately composted manure
Air (dust)
Wild and domestic animals (including fowl and reptiles)
Insects
Human handling
Postharvest
Faeces
Human handling (workers, consumers)
Harvesting equipment
Transport containers (field to packing shed)
Wild and domestic animals (including fowl and reptiles)
Insects
Air (dust)
Wash and rinse water
Sorting, packing, cutting, and further processing equipment
Ice
Transport vehicles
Improper storage (temperature, physical environment)
Improper packaging (including new packaging technologies)
Cross-contamination (order foods in storage, preparation, and display areas)
Improper display temperature
Improper handling after wholesale or retail purchase

* Adapted from Beuchat.

Faeces have been suspected as sources of pathogens on contaminated fruits, vegetables, or minimally processed produce that have subsequently been associated or confirmed as causes of human disease outbreaks. Among the more recent outbreaks are those linking unpasteurized apple juice to *Escherichia coli* O 157:H7 infections. This pathogen can remain viable in bovine faeces for up to 70 days, depending on inoculum level and temperature (Wang *et al.*, 1996). *Cryptosporidium* infection linked to consumption of unpasteurized apple juice was hypothesized to have been caused by contamination of apples by calf faeces (Millard *et al.*, 1994). Contact of fruits and vegetables by pickers and handlers at the time of harvest also offers a mechanism by which pathogens in faeces can contaminate raw produce.

Wild birds are known to disseminate *Campylobacter*, *Salmonella*, *Vibrio cholerae* and *Listeria* species. Pathogenic bacteria are apparently picked up as a result of birds feeding on garbage, sewage, fish, or lands that are grazed with cattle or have had applications of fresh manure. Control of preharvest contamination of fruits and vegetables with pathogenic bacteria by wild birds would be exceptionally difficult.

Postharvest Sources

Some of the possible preharvest sources of pathogenic microorganisms may also be postharvest sources (Table 12.17). The faecal-oral route of transmission of pathogens broadens to include workers handling fruits and vegetables from the point of removal from the plant through all stages of handling, including preparation at the retail and food service levels and in the home. Changes in eating habits, particularly the increased consumption of meals away from home, must be considered when attempting to provide reasons for increased frequency of outbreaks associated with fresh produce. Proper training of food-service workers in hygienic practices is essential. One cannot assume that newly hired personnel have even rudimentary knowledge of food microbiology. This is particularly critical among teenagers who, partly because they and their parents are eating more- meals away from home, have had minimal or no exposure to proper food-handling practices. Instruction in elementary principles of food hygiene at the high school or middle school levels has diminished greatly in the past two decades.

Traditionally recognized postharvest control points for access of pathogens to whole or cut produce include transport containers and vehicles and sorting, packing, cutting, and further processing equipment. The development of new processing equipment and technologies should include a team of experts in food microbiology as well as engineering. Too often, aspects of sanitizing equipment are not considered or are an afterthought and can increase the risk for contaminated end products. Temperature control is absolutely critical at every stage of postharvest handling if any success is to be achieved in minimizing the growth of pathogens.

Surfaces, including human hands, which come in contact with whole or cut produce, represent potential points of contamination throughout the total system of growing, harvesting, packing, processing, shipping, and preparing produce for consumption.

Potential Food-Borne Illness Outbreaks Associated with Consumption of Fruits and Vegetables

Fresh produce can be a vehicle for the transmission of bacterial, parasitic and viral pathogens capable of causing human illness and a number of reports refer to raw vegetables harbouring potential foodborne pathogens. *L. monocytogenes*, *Salmonella*, and *E. coli* have been isolated from raw vegetables, which can become contaminated while growing or during harvesting, post-harvest handling, or distribution. Table 12.19 shows examples of pathogens that have been isolated from various fruits and vegetables and are involved in outbreaks of foodborne disease reported to date. List of foods and pathogenic microorganism indicated in Table 12.19 is just a representative; many other fresh and minimally processed ready-to-eat fruits and vegetables have been implicated with outbreaks in a number of developed countries. The incidence of foodborne outbreaks caused by contaminated fresh fruit and vegetables has increased in recent years. The pathogens most frequently linked to produce-related outbreaks include bacteria (*Salmonella*, *E. coli*, *L. monocytogenes*, *S. aureus*), viruses (*Norwalklike*, hepatitis A), and parasites (*Cryptosporidium*, *Cyclospora*), with *Salmonella* and *E. coli* 0157:H7 being the leading causes of produce-related outbreaks in the USA. Fresh produce and sprouts have been implicated in a number of documented outbreaks of illness in countries such as Japan and EU. In the U.S. between 1995 and 1998, there were nine outbreaks of foodborne illness caused by *Salmonella* or *E. coli* 0157:H7 due to consumption of fresh vegetable sprouts. These outbreaks involved more than 1234 cases in Missouri, Michigan, California, Washington, Arizona, and Nevada, and in most cases, alfalfa or clover seed were implicated as the initial inoculum source. Also, later in September 2006, an *E. coli* 0157:H7 outbreak affected 26 US States which involved about 200 cases of illness, including some of Hemolytic Uremic Syndrome (HUS) and resulted in three deaths.

Data indicated that fresh spinach grown in three Californian counties was the source of the bacterium. The potential for widespread outbreaks of human infection caused by consumption of raw produce was also dramatically realized during the summer of 1996 in Japan where more than 6000 cases of *E. coli* 0157:H7 infections were reported. The largest outbreak resulted in four deaths and affected more than 4000 school children in and around Sakai City. Raw radish sprouts that had been prepared in central kitchens appear to have transmitted the pathogen, although the mechanism of sprout contamination was not determined. Sprout-related disease outbreaks have also been reported in the United Kingdom, Finland, Denmark, Sweden, and Canada and have involved alfalfa, cress, radish, and mungbean sprouts. Fresh fruit and vegetables may therefore pose a food safety risk because they are consumed raw and are susceptible to be contaminated by faecal material and soil on the farm.

Leafy green vegetables have been identified as the commodity group of highest concern from a microbiological safety perspective. This commodity grouping was considered to include all vegetables of a leafy nature and of which the leaf is the intended for consumption such as lettuce (all varieties), spinach, cabbages, chicory, leafy fresh herbs (*e.g.*, cilantro, basil, parsley) and watercress.

This commodity group does not include green onions which differ in morphology from the above-mentioned vegetables.

Table 12.19: Examples of Pathogens Associated with Fruits and Vegetables Involved in Outbreaks of Foodborne Disease

Agent	Implication/Suspected Food	Reference
Bacillus cereus	Alfalfa sprouts, cress sprouts, cucumbers, mustard sprouts, soyabeans sprouts	Buck *et al.* (2003)
Campylobacter jejuni	Green onions, lettuce, mushroom, potato, parsley, pepper, spinach	CDC (1998), Buck *et al.* (2003)
Clostridium botulinum *Cryptosporidium* *Cyclospora*	Cabbage, mushroom, pepper, Apple cider, Raspberries	Buck *et al.* (2003) CDR (1991) Herwaldt *et al.* (1997)
E. Coli 0157:H7	Alfalfa sprouts, radish sprouts, apple juice, apple cider, cilantro, cress sprouts, lettuce	Buck *et al.* (2003), WHO (1996), CDR (1997)
Hepatitis A virus	Iceberg lettuce, raspberries, strawberries	Rosenblum *et al.* (1990), Niu *et al.* (1992)
Salmonella	Alfalfa sprouts, artichokes, beet leaves, celery, cabbage, cantaloupe, cauliflower, chili, cilantro, egg plant, endive, fennel, green onions, lettuce, mungbean sprouts, mustard cress, orange juice, parsley, pepper, salad greens, spinach, strawberries, tomato, watermelon	Buck *et al.* (2003), Mahon *et al.* (1997)
Shigella	Celery, cantaloupe, lettuce, parsley, scallions, mixed salad	Buck *et al.* (2003), Dunn *et al.* (1995)
Vibrio cholerae *Aeromonas*	Cabbage, coconut milk, lettuce, Alfalfa sprouts, asparagus, broccoli, cauliflower, celery, lettuce, pepper, spinach.	Shuval *et al.* (1989), Buck *et al.* (2003), Buck *et al.* (2003)
Listeria monocytogenes	Beans sprouts, Cabbage, chicory, cucumber, eggplant, lettuce, mushroom, potatoes, radish, salad vegetables, tomato	Buck *et al.* (2003)
Staphylococcus	Alfalfa sprouts, carrot, lettuce, onions sprouts, parsley, radish	Buck *et al.* (2003)

Leafy green vegetables have been associated with multiple outbreaks of foodborne disease with high numbers of illnesses in at least three regions of the world. For example, there have been at least 30 outbreaks linked to these commodities in the USA and 5 in Sweden over the last 10 years. While reports of outbreaks come from developed countries in Europe, North America and the south-west Pacific, it was considered that this may be a reflection of the disease surveillance systems that exist in those regions. A recent outbreak associated with *Salmonella* in fresh basil grown in Israel led to illnesses in at least 4 countries and one reported death. The severity of the disease is often linked to the implicated hazards, and the range of hazards that have been linked with this product group is extensive. For example, enterohaemorrhagic *Escherichia coli* in these products has caused severe illnesses and even deaths.

The second level of priorities identified included berries, green onions, melons, sprouted seeds and tomatoes. Berries are typically acidic and so do not provide a suitable environment for microbial growth. Also, viruses and parasites, the hazards most frequently associated with berries, will not replicate outside their host; however, low numbers of some of these hazards can be adequate to cause illness. As berries are extensively handled, human hands are considered the primary source of contamination and therefore a critical point in terms of hazard control. Other potential areas of control include irrigation water, use of manure as fertilizer, and protection against birds. Outbreaks linked to berries have an economic impact, particularly on producer countries. For instance, an outbreak in North America linked to raspberries from Guatemala resulted in a loss of market for this Central American country.

Green onions have been linked to three outbreaks in the USA, including 4 deaths and 1028 illnesses. While green or spring onions are widely used around the world, outbreaks from other parts of the world were not identified. However, the outbreaks in the USA were linked to imported green onions. Production of green onions was considered to be small compared to some of the other commodities considered here, but the product is widely used, often as a herb, garnish or minor component of a meal. Thus, a small volume of product may lead to exposure of a large number of people. The hepatitis A outbreak in the USA in 2003 led to a decrease in the market price of green onions, and a shutdown of the American market for some Mexican producers. The food safety concerns led to a drop in demand for Mexican green onions and estimated losses of US$ 10.5 million for Mexican growers in a two-week period in November 2003.

The flesh of a melon provides an ideal environment for microbial multiplication. Thus, there is a high risk of amplification of foodborne bacterial pathogens that may be present. In terms of hazard control, the quality of the water used for irrigation, washing and the practice of hydro-cooling is critical. However, the difficulty of preventing soil and dust from getting onto the fruit and possibly contaminating it means that there is still a lack of knowledge as to how to minimize contamination at the farm level. Given that melon flesh is ideal to support microbial growth, refrigeration is critical for pre-cut melon.

Melons are produced in a number of regions in the world and are traded internationally. Some melon types are only available for harvest for a two-week period in any one area, which means that the source of melons on the international market changes regularly. Back-to-back outbreaks in North America linked to melons from Mexico had a large impact on producers in Mexico and resulted in the closure of market access to their biggest customer.

Sprouted seeds include alfalfa sprouts, mung bean sprouts, radish sprouts, sunflower sprouts and clover sprouts, among others. Outbreaks with this commodity have occurred in North America, Europe and Asia. The outbreaks in Japan in 1996, caused by *E. coli* O157:H7, led to 9451 illnesses and 12 deaths—all children—and worldwide attention to potential microbiological hazards in fresh produce. Sprouted seeds are not an important commodity in international trade; however, the seeds themselves, which are often the source of contamination, are traded widely.

This last group of commodities, which were considered to be the lowest priority of the commodities considered, is a mixed group, which includes both fruits and vegetables, namely carrots, cucumbers, almonds, baby corn, sesame seeds, onions and garlic, mango, paw paw, celery and maimai. They are included here as all have been linked to foodborne illness. However, compared to the produce previously addressed they have had a very limited public health impact to date. Information on these produce and the associated hazards and potential control measures is very limited, if it even exists.

Removal of Pathogens

Sanitizer can be used to wash or to assist in lye peeling of fruits and vegetables. No chlorine substitute effective for washing fruits and vegetables is available. Numerous alternatives for sanitizing equipment can be used in a total sanitation program, but none has as broad a spectrum of activity as chlorine.

Chlorine is routinely used as a sanitizer in wash, spray, and flume waters used in the fresh fruit and vegetable industry. Antimicrobial activity depends on the amount of free available chlorine (as hypochlorous acid) in water that comes in contact with microbial cells. The efficacy of chlorine in killing pathogenic microorganisms has been extensively studied. Possible uses in packing houses and during washing, cooling, and transport to control postharvest diseases of whole produce have been reviewed. The effect of chlorine concentration on aerobic microorganisms and faecal coli forms on leafy salad greens was studied by Mazollier (1988). Total counts were markedly reduced with increased concentrations of chlorine up to 50 ppm, but a further increase in concentration up to 200 ppm did not have an additional substantial effect. A standard procedure for washing lettuce leaves in tap water was reported to remove 92.4 per cent of the micro flora (Adams *et al.,* 1989). Including 100 ppm available free chlorine in wash water reduced the count by 97.8 per cent. Adjusting the pH from 9 to 4.5 to 5.0 with inorganic and organic acids resulted in a 1.5- to 4.0-fold increase in microbicidal effect. Increasing the washing time in hypochlorite solution from 5 to 30 minutes did not decrease microbial levels further, whereas extended washing in tap water produced a reduction comparable to hypochlorite. The addition of 100 ppm of a surfactant (Tween 80) to a hypochlorite washing solution enhanced lethality but adversely affected sensory qualities of lettuce.

Populations of *Salmonella* Montevideo on the surface and in the stem core tissue of tomatoes can be substantially reduced by dipping fruits 2 minutes in a solution containing 60 or 110 ppm chlorine, respectively; however, treatment in a solution containing 320 ppm chlorine does not result in complete inactivation.

The efficacy of chlorine, hydrogen peroxide, and ethanol in removing *Salmonella* from injected alfalfa sprouts has been studied. Sprouts were dipped in solutions containing 200, 500, or 2,000 ppm chlorine for 2 minutes. The pathogen was reduced by about 2 log10 CFU/g after treatment with 500 ppm chlorine, compared with the control, and to an undetectable level (<1 CFU/g) after treatment with 2,000 ppm chlorine.

The inaccessibility of chlorine to microbial cells in crevices, creases, pockets, and natural openings in the skin also undoubtedly contributes to the overall lack of effectiveness of chlorine in killing pathogens. The hydrophobic nature of the waxy cuticle on tissue surfaces protects surface contaminants from exposure to chlorine and other produce sanitizers that do not penetrate or dissolve these waxes. Surface-active agents lessen the hydrophobicity of fruit and vegetable skins as well as the surfaces of edible leaves, stems, and flowers, but they may also cause deterioration of sensory qualities. Sanitizers that contain a solvent that would remove the waxy cuticle layer, and with it enmeshed contaminants, without adversely affecting sensory characteristics would hold greater potential than chlorinated water in reducing microbial populations on whole raw produce. Such sanitizers may be limited to use on produce that will be further processed into juice or cut products, or on whole fruits, vegetables, or plant parts destined for immediate consumption, since their application could adversely affect visual appearance. Clearly, chlorine, at concentrations currently permitted for use by the industry to wash fresh fruits and vegetables, cannot be relied upon to eliminate pathogens.

Hence, treatment of produce with chlorinated water reduces populations of pathogenic and other microorganisms on fresh produce but cannot eliminate them. Reduction of risk for human illness associated with raw produce can be better achieved through controlling points of potential contamination in the field; during harvesting; during processing or distribution; or in retail markets, food-service facilities, or the home.

Organic Vegetables – Microbiology

Ready-to-eat fruits and vegetables have been associated with microbiological food poisoning incidents in many countries. This is not in itself surprising but a concern is that with changing production, distribution and marketing procedures, such incidents could become more common and more widespread.

Organic food is a small but growing part of the food industry in many developed countries. Manure and other animal and plant wastes are widely used in both organic and conventional agriculture. This gives rise to the possible contamination of produce with microbial pathogens particularly if the manure and other wastes have not been subjected to accepted procedures to minimize risk of pathogen survival.

A recent paper from the UK, where growth in sales of organic fruits and vegetables through major supermarkets has been particularly strong, has produced information which suggests that despite often stated concerns such produce is usually of satisfactory microbiological quality.

In a major study conducted by the Environmental Surveillance Unit of the Public Health Laboratory Service (*Letters in Applied Microbiology* 33 2001 434), 3200 samples of ready-to- eat vegetables were examined according to a standardized protocol. Samples were collected in England, Wales, Scotland and Northern Ireland. Bacteria tested for were *Escherichia coli*, *Listeria* spp. and *L. monocytogenes*, *Salmonella* spp, *Campylobacter* spp. and *E. coli* 0157.

Based on the PHLS microbiological guidelines for some ready-to-eat foods sampled at the point of sale – for *E. coli* and *Listeria* this is fewer than 20 colony forming units per gram and the other pathogens must not be detectable in 25 grams – three thousand one hundred and forty-six or 98.5 per cent of the 3200 organic vegetables were satisfactory. A further 39 were acceptable and 15 were of unsatisfactory microbiological quality. *Escherichia coli* and *Listeria* spp., not *L. monocytogenes*, were the microorganisms associated with unsatisfactory results, defined as greater than 100 colony forming units per gram.

Minimally Processed Foods and Microorganisms

Changing lifestyle patterns and the increased consumption of healthier foods has increased the demand for freshly prepared, minimally processed salads. An example is minimally processed, cut and packaged lettuce. This is a product that is exposed to a range of conditions during growth, harvest, distribution and processing and these conditions may increase the potential for microbial contamination. Some of the potential contaminants are psychrotrophic bacterial pathogens (such as *Listeria monocytogenes*, *Yersinia enterocolitica* and *Aeromonas hydrophila*).

These pathogens can cause a range of disease manifestations in humans. The diseases caused by *L. monocytogenes* infection vary from a non-specific flu-like illness to sepsis or meningitis, which can be fatal. Those most at risk, particularly for the more severe disease manifestations, include the elderly, foetuses during the third trimester of pregnancy, and the immunocompromised. For *Y. enterocolitica*, illness usually presents as a self-limiting enterocolitis; some long-term complications such as arthritis have been reported. While the role of *Aeromonas* as a cause of foodborne disease is debatable, *Aeromonas* species (*Aer. hydrophila*, *Aer. caviae*, *Aer. veronii* biovars sobria and veronii, *Aer. jandaii*, *Aer. schubertii*, *Aer. trotu*) can cause gastrointestinal and wound infections.

Each of the pathogens noted above are capable of growing at the chill temperatures that the minimally processed food industry relies upon to maintain product quality post-harvest. This coupled with the high moisture content, the absence of a terminal lethality treatment, the potential for temperature abuse of these commodities, and the use of packaging regimes to extend shelf-life, may provide the condition and the time for the survival and growth of psychrotrophic pathogens.

In Australia, (Food Safety and Hygiene, 30 2000 456-60) most lettuce samples (76 per cent) contained between 10^5-10^7 colony forming units (cfu) g^{-1} aerobic bacteria. The aerobic plate counts were not influenced by the time of testing. For example, counts on factory samples, which were tested within 24 hours of processing, ranged from 10^3-10^7 cfu g^{-1}. All samples tested appeared organoleptically acceptable as determined by subjective visual and odour inspection, even those samples with apparent high counts.

Potentially pathogenic psychrotrophic bacterial species were detected in lettuce samples. *L. monocytogenes* was isolated from 3 samples, *A. hydrophila* or *A. caviae* from 66 samples, and *Y. enterocolitica* from 71 samples. Comparable frequencies of isolation have been reported in similar studies undertaken by other research groups in France, England and Germany.

The predominance of *Y. enterocolitica* and *Aeromonas* spp. in packaged lettuce suggested that the product provided an environment that favoured their survival. It was possible to determine the pathogenic potential of the *Y. enterocolitica* isolates by screening for an array of biochemical, serological and genetic traits. The *Y. enterocolitica* isolates lacked many of the phenotypic and genetic markers associated with virulence in primary pathogenic strains. As the roles of the reputed virulence factors of *Aeromonas* spp. in human infection are uncertain, the pathogenic potential of the *Aeromonas* isolates in lettuce remains unclear.

Of the psychrotrophic bacterial pathogens detected in lettuce during the survey, *L. monocytogenes* has presented the greatest concern to both the food industry and regulatory agencies since the early 1980s. Although *L. monocytogenes* was detected at a much lower frequency than *Y. enterocolitica* and *Aeromonas* spp. in the packaged lettuce, its presence required further investigation.

It is difficult to eliminate *L. monocytogenes* from a product like minimally processed lettuce. In order to assess more fully the risks associated with its presence in lettuce, its growth potential in packaged lettuce using bacterial challenge testing was studied. It was found that at 4°C, increases in the number of *L. monocytogenes* were less than 10-fold after storage for 7 days and were less than 100-fold after 14 days storage. The oxygen content of the packages ranged from 0-13 per cent. At 8°C (temperature abuse conditions), *L. monocytogenes* increases were no greater than 100-fold after 7 days and approached 1000-fold increases after 14 days. The oxygen content of temperature abused packaged lettuce ranged from 0-11 per cent.

Implications for Industry

The possibility of contamination and growth of psychrotrophic foodborne pathogens is of concern for the safety of minimally processed, packaged and cut lettuce. The key to minimising health risks associated with the presence of the psychrotrophic pathogens considered are (Szabo, 2000):

1. Strict adherence to good manufacturing practices and sanitation;
2. Effective temperature control during storage and distribution and
3. Limiting the shelf-life expectancy with a 'use-by-date' that does not allow sufficient time for pathogens to grow to elevated levels.

Fresh-Cut Fruits and Vegetables

In developed countries, there has been an ever-growing demand for convenience products and the same trend is followed for fresh-cut fruits and vegetables.

According to International Fresh-cut Produce Association or IFPA, fresh-cut produce is defined as "any fruit or vegetable or combination thereof that has been physically altered from its original form, but remains in a fresh state". There has been a rapid growth of fresh-cut produce industry worldwide in recent years to a multi-billion dollar sector, mainly because of increasing consumer demand for healthy, freshly prepared, convenient fruits and vegetables. Fresh-cut produce may consist of peeled, sliced, shredded, trimmed and/or washed fruits and vegetables (Francis *et al.,* 2012). Conventional preservation methods like freezing, dehydration, curing or salting are never applied to fresh-cut produce because these foods are not heavily processed. Fresh-cut produce on the market in some developed counties include melons, cantaloupe, watermelon, mango, jackfruit, papaya, grapefruit, pineapple, fruit mixes, shredded leafy vegetables and salad mixes, vegetables for cooking like peeled baby carrots, baby corn, broccoli and cauliflower florets, cut celery stalks, shredded cabbage, cut asparagus, cut sweet potatoes and many more (James and Ngarmsak, 2010).

Despite the benefits derived from eating raw fruits and vegetables, quality and safety are still an issue of concern as these foods have long been known to be vehicles for transmitting infectious diseases. The destruction of surface cells during processing (such as peeling, slicing and shredding) of fresh-cut produce exposes the cytoplasm and provides micro-organisms with a richer source of nutrients as compared to intact produce. Besides this, the high water activity and approximately neutral (vegetables) or low acidic (many fruits) tissue pH facilitate rapid microbial growth (Parish *et al.,* 2003). These conditions provide a perfect platform for a number of important human pathogens and spoilage micro-organisms to contaminate fresh-cut produce, which results in a faster deterioration of fresh-cut produce compared to whole fruit or vegetable. A number of physical and physiological changes affecting the viability and quality of the fresh-cut produce are a direct result of the wounding associated with processing. Flaccidity due to loss of water, changes in colour (especially browning at the cut surfaces) and microbial contamination deteriorate the appearance of fresh-cut produce. Wounding of plant tissues may also result in the acceleration of nutrient losses.

Fresh and minimally processed fresh-cut products are naturally contaminated by micro-organisms because of a number of sources, including the farm environment, post-harvest handling and processing. Hence, the microflora linked with fresh-cut produce is diverse. Also, contamination with pathogens can occur at various points during growing, harvesting, processing and handling of the produce. The microflora associated with most vegetables is dominated by gram-negative bacteria, while dominant microflora associated with raw fruits mostly includes yeasts and moulds. The factors that influence the survival and/or growth of the micro-organisms on the fresh produce may include the type of organism, the commodity and environmental conditions in the field and thereafter, including storage conditions.

Packaging is one of the important factors influencing the microbial quality of fresh-cut products. Fresh-cut products are mostly packaged under modified atmospheric conditions and stored under refrigeration which gives rise to a favourable environment and time for proliferation of spoilage micro-organisms and micro-organisms of public health significance. The economic value of fresh-cut products is impaired by micro-organism proliferation because it may lead to decrease in product shelf life, through spoilage, and also pose a risk to public health by causing food-borne illnesses. Gaseous composition of the atmosphere surrounding the fresh-cut produce has a profound effect on the microbial quality and shelf life. Low oxygen-modified atmospheres may inhibit the growth of spoilage micro-organisms and increase the shelf life of packaged produce. Zhang, *et al.* (2013) studied the effect of high carbon dioxide packaging on the microbial spoilage of fresh-cut honeydew melon and suggested that 50 per cent O_2 + 50 per cent CO_2 is a potential gaseous composition in a modified atmosphere for extending the shelf life of fresh-cut honeydew melon.

Microbial contamination of fresh-cut produce may pose threats to human health by causing various diseases like diarrhoea, abdominal cramps, vomiting as well as death. Over the past few decades, there has been a dramatic increase in outbreaks of food-borne illness caused primarily due to *Escherichia coli* O157:H7 and *Salmonella* by the consumption of fresh and minimally processed fruits.

Food-borne pathogens including *E. coli* O157:H7, *Salmonella* and *Listeria monocytogenes* are often present on the surface of fresh produce and may cause public health problems. These organisms have long been recognized as dangerous pathogens causing various food-borne diseases. However, some non-pathogenic bacterial species, unfortunately, will also give positive results on *coliform* detection. Therefore, it is highly encouraged that contamination of raw and freshly processed vegetables with faecal pathogens should not be determined by the *coliform* test. Nonetheless, such tests being simple to perform make them popular in the fresh-cut industry. Microbial counts are reported to vary with type of fruit and vegetable. Gram-negative rods mostly *pseudomonads* have been reported to be the predominant group of micro-organisms on fresh produce. Some antagonistic strains of *pseudomonads* have also been identified and isolated from fresh-cut produce which restrict the growth of pathogenic species of *E. coli*. The strain L-59-66 of *Pseudomonas syringae* has been reported to prevent the growth of *E. coli* on apple wounds. A strain of *Pseudomonas graminis*, isolated from whole apple surface, was investigated for its effectiveness to reduce *E. coli* O157:H7, *Salmonella* and *Listeria innocua* on fresh-cut apples and peaches. Fluorescent *pseudomonads* are the dominant group of micro-organisms isolated from endive, spinach, cauliflower, chicory and mung bean. *Enterobacteriaceae* may also be present on minimally processed products. Another important group of microflora reported to be present on fresh-cut fruits and vegetables constitutes yeast and moulds. For instance, yeast populations of 10^3–10^4 have been reported for processed fruits, and total bacterial populations of 10^6 cfu/g have been reported for cantaloupe. Similarly, a mould population ranging from 10^2 cfu/g on cut lettuce to 10^8 cfu/g in shredded carrots has been reported (Qadri *et al.*, 2015).

Detailed studies are still required to be conducted to explore the knowledge regarding the occurrence of micro-organisms in minimally processed fruits and vegetables. Large number of human pathogens can contaminate fresh-cut produce and past years have witnessed an increase in the number of food-borne outbreaks associated with fresh and fresh-cut produce. For instance, outbreak in 2006 due to spinach-associated *E. coli* O157:H7 caused more than 200 illnesses and even some deaths. Fresh-cut cilantro and parsley have also been reported to be involved in food-borne illnesses. Many cases of salmonellosis have been reported due to consumption of fresh-cut fruits. In certain instances, pathogen levels may decline on the outer surface and rate of decline is influenced by produce type, humidity, temperature, atmosphere and type of packaging. Pathogenic growth on intact surfaces of the produce is not common as they are not able produce the enzymes required to break down the protective outer coverings of the produce. The outer covering acts as a barrier and restricts the availability of nutrients and moisture.

In the fresh-cut produce production chain, washing of produce with sanitizing solutions is considered the only step to achieve a reduction in spoilage micro-organisms and potential pathogens. Proper disinfection or decontamination procedures are considered critical in ensuring the safety of fresh and minimally processed fruits and vegetables. Washing and sanitization steps before cutting and during processing lower the microbial contamination on fruits and vegetables.

Development and application of antimicrobial edible coatings has emerged as an effective means to increase the shelf life of fresh-cut fruits and vegetables. Antimicrobial components from different natural sources have been successfully extracted and incorporated into the coating materials intended to protect the fresh-cut fruits and vegetables. Several categories of such antimicrobials incorporated into edible films and coatings include various organic acids, fatty acid esters, polypeptides, plant essential oils, nitrites and sulphites. Some coating materials having inherent antimicrobial properties are more suitable and cost effective and have already been investigated for coating of various food products including fresh-cut fruits and vegetables.

Fruit Juices and Microorganisms

Unpasteurized fruit juice is defined as the product produced by pressing or squeezing of the fruits. Consumption of fresh juices increased dramatically due to their freshness, high vitamin content, and low caloric consumption. Extracted juices from fruits contain most substances which are found in the original ripe and sound fruit from which the juice is made. The high potassium and low sodium characteristic of most juices help in maintaining a healthy blood pressure. Vitamin C is naturally present in juices which are essential for the body to form collagen, cartilage, muscle, and blood vessels. It also helps in the absorption of iron.

Fruit juices contain a microflora which is normally present on the surface of fruits during harvest and postharvest processing which include transport, storage, and processing. Many microorganisms such as acid tolerant bacteria and fungi (moulds, yeasts) use them as a substrate for their growth. Yeasts form the main flora of fruits before processing because of acidic pH. The major genera include *Candida, Dekkera, Hanseniaspora, Pichia, Saccharomyces,* and *Zygosaccharomyces. Penicillium, Byssochlamys, Aspergillus, Paecilomyces, Mucor, Cladosporium, Fusarium, Botrytis, Talaromyces,* and *Neosartorya* are filamentous fungi most frequently isolated from fresh fruits and juices. Among bacteria, lactic acid bacteria and acetic acid bacteria have been isolated from fruit juices.

The critical factors affecting the spoilage of juices include juice pH, oxidation reduction potential, water activity, availability of nutrients, presence of antimicrobial compounds, and competing microflora. Among these factors, pH and water activity are the most influential factors affecting the spoilage of juices. The spoilage caused by microorganisms in juices includes cloud loss, development of off-flavours, CO_2 production, and changes in colour, texture, and appearance resulting in degradation of product. Commonly reported bacterial genera include *Acetobacter, Alicyclobacillus, Bacillus, Gluconobacter, Lactobacillus, Leuconostoc, Zymomonas,* and *Zymobacter*. Among yeasts *Pichia, Candida, Saccharomyces,* and *Rhodotorula* are commonly encountered genera responsible for spoilage of juices. Certain common moulds such as *Penicillium* sp., *Aspergillus* sp., *Eurotium, Alternaria, Cladosporium, Paecilomyces,* and *Botrytis* have also been reported in spoilage of fruit juices (Aneja *et al*., 2014).

Fruit juices have pH in the acidic range (<4.5). serving as important barrier for microbial growth. However, food borne pathogens such as *E. coli* and *Salmonella* survive in acidic environment of fruit juices due to acid stress response. Therefore, in the last two decades a number of food borne outbreaks associated with unpasteurized fruit juices have been documented in many countries. The source of entry of microorganisms into fresh fruit juices from environment exposure and soil. In developing country like India, a large population of all income and age groups consume freshly squeezed fruit and vegetable juice, but the presence of pathogenic microorganisms in street vended fruit juices has also been reported in various parts of India.

Many reports involving unpasteurised fruit juices in foodborne illness have been published. Blended orange juice from a Mexican supplier was identified as the source of the large outbreak of salmonellosis that occurred in the United States and Canada in June-July, 1999. The illnesses were traced to the relatively rare *Salmonella muenchen*. At least four other *Salmonella* strains were also isolated from product at the plant in Tempe, Arizona where the blended juice was finally packed. The survival of pathogenic bacteria, including *Salmonella* spp., in fruit juices is dependent on the pH, the type of acid in the juice and the temperature of storage.

In an experimental investigation, Mossel and De Bruin studied the survival of a range of *Enterobacteriaceae* including *E. coli* and *S. typhimurium* in tomato, apple, orange and lemon juices as well as mayonnaise and yoghurt. They concluded that with an initial population of 106 organisms/ml, pathogenic *Enterobacteriaceae* do not survive more than one day in lemon juice (pH 2.4–2.6). In the other fruit juices examined, survival occurred for up to 35 days at 5°C and up to 10 days at 24°C. The pH of the products studied is shown below:

pH	*Product*	*Predominant Acid*
2.4–2.6	Lemon juice	Citric
3.0–3.1	Apple juice	Malic
3.1–3.5	Orange juice	Citric and malic
3.8–4.1	Tomato juice	Malic and citric
	Mayonnaise	Acetic
	Yoghurt	Lactic

On the basis of this study, Mossel and De Bruin concluded that fruit juices with a pH greater than 3.0 must be considered as potential sources of pathogenic bacteria. Their findings on survival at refrigeration temperature versus ambient temperature are not in agreement with those of Goverd *et al.,* but because of the long survival periods of potential pathogens at 4°C and 5°C reported in each study, this conflict is of marginal importance.

Mossel and De Bruin also concluded that at a given pH value, in this case pH 3.8–4.1, the order of lethal effectiveness of the organic acids in the study appeared to be: lactic > acetic > malic or citric. This conclusion is debatable given the conditions of the study but, as the authors noted, it confirmed the findings of Shillinglaw and Levine (1943) working under more defined laboratory conditions. Certainly, the results obtained in these two studies with Gram negative bacteria are contrary to comprehensive studies with organic acids on inhibition of yeasts by Pitt (1974) and earlier workers including Fabian and Wadsworth (1939) and Levine and Fellers (1940) who indicated that acetic acid was more inhibitory than other food acids at the same pH level.

Acid Tolerant Microorganisms

The association of *E. coli* 0157: H7 infection with unpasteurised apple juice and mayonnaise has prompted a number of recent studies on the acid tolerance of this organism. Zhao *et al.* (1993) found that *E. coli* 0157: H7 when initially present at a level of 105 organisms/ml survived up to 31 days in apple cider with a pH of 3.7 at 8°C. They concluded that since the expected refrigerated shelf life of apple cider is typically less than 3 weeks, the pathogen is likely to survive during most of the time that the cider would be consumed at numbers close to the size of the initial population introduced into the product. In similar studies at 25°C, survivors were detected at 2 to 3 days but not at 6 days after inoculation.

They also studied the effect on *E. coli* 0157: H7 of 0.1 per cent potassium sorbate, 0.1 per cent sodium benzoate and a combination of 0.1 per cent potassium sorbate and 0.1 per cent sodium benzoate in apple juice. They found that at 8°C, the sodium benzoate and the combination benzoate/sorbate treatment reduced numbers of *E. coli* 0157: H7 but the results obtained do not seem to justify the authors' conclusion that use of 0.1 per cent sodium benzoate is an option processor have, to increase substantially the safety of apple cider.

Miller and Kaspar (1994) conducted further work in this area which confirmed the acid tolerance of two strains of *E. coli* 0157: H7 in apple ciders with pH values of 3.7–4.1. Both strains examined were significantly more acid tolerant than a control strain of *E. coli* although the degree of tolerance differed between strains. Again low storage temperature, 4°C, enhanced survival at low pH levels. The strains investigated were able to survive 14 to 21 days at 4°C when the initial population level was 3.0 x 104 organisms/ml of cider. In studies similar to those of Zhao *et al.* (1993), these authors were unable to confirm that 0.1 per cent sodium benzoate in cider was an effective preservative against the strains of *E. coli* 0157: H7 tested. Both strains investigated survived for 21 days in the presence of the sodium benzoate at 4°C. Miller and Kaspar also concluded that the characteristics which are responsible for acid tolerance in *E. coli* 0157: H7 must be present prior to introduction of the organism into the acid environment. This is in contrast to the situation with *S. typhimurium* where adaptation to the acid environment is reported to occur (Foster and Hall 1990, Leyer and Johnson 1993).

Pasteurization of Fruit Juices

Commercial fruit juices can be prepared either by pasteurization or by adding chemical preservatives. Processed fruit juices rather than pasteurization are more popular due to their fresh tastes. In the industry fruit juice is processed by automatic machine (collection, cleaning, extraction of juices and packaging). There is every possibility of contamination of processed fruit juice at any stage of processing. Contamination may be occurred by spoilage organisms and/or by food borne pathogens.

Food borne disease outbreaks have been documented in different countries. Specially Yeast and molds are the dominant microorganisms in juice because they can thrive the high acidic conditions of the juice. Some examples of them are species of the genera *Cladosporium, Candida, Dekkera, Pichia, Saccharomyces, Aspergillus, Zygosaccharomyces, Penicillium, Byssochlamys, Hanseniaspora, Paecilomyces, Mucor, Fusarium, Botrytis,* and *Neosartorya Talaromyces etc.* Some lactic acid and acetic acid bacteria may be present in fruit juices. Some pathogenic bacteria like *Escherichia coli* 0157, *Salmonella*, and *Cryptosporidium*, faecal *Streptococci* and some spore formers like *Clostridium pasteurianum* and *Bacillus coagulans* may be present in fruit juice if the juice is not processed adequately. Bacterial growth in fruit juice depends on pH, storage temperature, types of packaging material, humidity, water activity, concentration of preservatives, application of UV treatments during production, sugar contents, quality of water used in juice, quality of the machines used in the juice industries, raw materials, quality of fruits *etc.* Mainly improper washing of fruits with poor quality water used in the juice preparation are the major source of contaminating microbes found in the fruit juices. Some fruit juice spoiling organisms are listed in Table 12.20.

High pressure, UV radiation, Gamma radiation, ozonization, electron beam radiation, pasteurization, pulsed electric ofield processing *etc.* and chemical preservatives such as organic acids, hydrogen peroxide, dimethyl dicarbonate, vanillin, monocaprylin, essential oils, vanillic acid, nisin and cinnamon *etc.* are to be applied to preserve fruit juice.

Food borne pathogens if present at sufficient number in fruit juice may cause illness in the consumers. Some microbes only survive in the juice and multiply within the host after consumption and start the disease. Some microbes which can multiply within the juice may yield off odor in juice and make juice unfit for human consumption. *Clostridium perfringens, Staphylococcus aureus, Bacillus cereus, Clostridium botulinum* able to produce toxins in fruit juice and after drinking the juice can cause food poisoning. Fruit juice contaminating pathogens and symptom caused by them are listed in Tables 12.19 and 12.20.

Table 12.20: Disease Symptoms Caused by Contaminating Microbes found in Fruit Juices

Microorganisms	*Examples*	*Disease symptoms*
Bacteria	*Salmonella* spp.	Abdominal pain, diarrhoea, chills, fever, nausea
	Shigella spp.	Abdominal pain, diarrhoea, fever, nausea
	Clostridium botulinum	Nausea, vomiting, fatigue, dizziness, dryness of mouth and throat, muscle paralysis, difficulty swallowing, double or blurred vision.
	*Escherichia coli*0157:H7	Bloody diarrhea, abdominal pain, hemolytic uremic syndrome (HUS), kidney failure.
	Listeria monocytogenes	Gastroenteritis, childbirth in pregnant women, septicaemia, meningitis.

Virus	Hepatitis A	Fever, malaise, anorexia, nausea, abdominal pain, jaundice, dark urine
	Norwalk virus	Vomiting diarrhea, mala ise, fever, nausea, abdominal cramps
Parasites	*Cyclospora* spp.	Watery diarrhoea, nausea, anorexia, abdominal pain
	Cryptosporidium spp.	Profuse watery diarrhoea, abdominal pain, anorexia, vomiting

There have been documented outbreaks of illness due to pathogens which include *Salmonella* and Vero toxin producing. In 1995, unpasteurized fresh orange juice contaminated with *Salmonella* was linked to an outbreak in a Florida Theme Park, USA. More than 60 visitors were affected (Mahale *et al.,* 2008). In Australia, 427 confirmed cases of salmonellosis were reported in 1999 after the consumption of unpasteurized orange juice (Victorian Government Department of Human Services 2005). A total of 48 cases of *Escherichia coli* O157 were reported after drinking unpasteurized apple juice in Washington DC in 1996 (Victorian Government Department of Human Services 2005). *Salmonella* spp. associated food borne disease outbreak occurred before the 20th century in Sarasota County, Florida and United States. Hepatitis A was occurred in Egypt in 2004 due to intake of fruit juice.

The only recent reports of food poisoning caused by fruit juices have been those associated with unpasteurised apple juice in North America. The harvesting and collection of different fruits for processing depends on the type of fruit involved. In the case of the apple juice incidents, it is generally believed that infection of 'drop' apples occurred in the field and this fruit was subsequently collected for processing. This is obviously a risk specifically associated with apples when collected in this manner.

However cross contamination of fruit juice with small numbers of pathogenic bacteria could conceivably have occurred at any point in production. It has been reported (Anon 1996) that the Food and Drug Administration in the USA is considering a requirement that apple juice and possibly other juices must be pasteurised to prevent further food poisoning incidents of the type described above.

In Australia until recently, most fruit juices would have undergone a pasteurisation procedure either at or immediately after the filling operation or at some earlier processing stage, *e.g.,* prior to concentration. The growth of the fresh juice market has seen this change so that in the case of citrus juices in particular, a major share of the market is now occupied by juices stored under refrigeration that have received little or no heat treatment. Fresh juices in Australia are defined in the Australian Food Standards Code as those which:

1. Do not contain preservatives
2. Have not been subjected to a concentration process and
3. Have a durable life of not more than 21 days.

Thus, pasteurisation of juices to be labelled as 'fresh' is permitted but is not a procedure exercised by most processors who are selling their product on the qualities associated with a product which has not been subject to any heating.

Proper handling of food is, however, the preventative measure of choice (Doyle 1991). A correctly constructed HACCP plan for fruit juices and other fruit products should take into account the possibility that under certain conditions these products may act as vehicles for some food poisoning bacteria. In fact, prepared fruit salads, which often contain high pH fruits such as watermelon and rockmelon, may be seen to present a greater potential hazard than fresh juices. This is so because not only may the pH of the product be less inhibitory to pathogens but the opportunity for cross contamination during preparation and storage is far greater (Del Rosario and Beuchat, 1995).

Thermal Resistance of *Salmonellae* in Fruit Juices

Some studies have highlighted that acid tolerant pathogens such as *Salmonella* and enterohaemorrhagic *Escherichia coli* may survive for extended periods in fruit juices. Parish (2000) listed 19 documented disease outbreaks associated with apple or orange juices and this list is already incomplete. According to Parish, normal pasteurisation regimes will eliminate these organisms from juices but that contamination after pasteurisation is still a concern.

It is against this background that it was proposed that:

1. All orange juice, other than freshly squeezed orange juice for immediate consumption on the same premises, either: undergoes a pasteurization process that kills pathogenic bacteria; or be labelled to ensure consumers that the product is unpasteurised and are made aware of the risks associated with its consumption.
2. The juice be heated to not less than 72°C for 15 seconds; or the juice be heated to any other time and temperature that can be demonstrated to produce a 5-log (100,000-fold) reduction in the most resistant microorganism of public health significance likely to occur in the product.

Doyle and Mazzotta (2000) have published a review of the thermal resistance of *Salmonellae*, one of the main acid tolerant pathogens of concern with regard to fruit juices. Perhaps not surprisingly, they reported that there are no data yet reported on decimal reduction times (D-values) for *Salmonella* in fruit juices.

This is important as relatively small changes in the time-temperature schedules to which fruit juices are subjected can have a pronounced effect on the acceptability of the juice.

They do note that the thermal process to inactivate *Salmonella*e (in common with other microorganism) depends on strains of the organism used, their growth phase, the food composition or test media, other limiting physical conditions, *e.g.* water activity, and the presence of other microflora.

They also point out that certain serotypes of *Salmonella enterica* are notorious for resistance to heat treatments. The most prominent of these is *Salmonella* Senftenberg 775W. Although this organism is not usually regarded as an important foodborne pathogen, it is often used as a test organism. The assumption is that if a particular heat process is effective against *Salmonella* Senftenberg 775W, it will also be effective against more common *Salmonella*e in foods.

After reviewing all their collected data excepting that for *Salmonella* Senftenberg and data collected under stress conditions, Doyle and Mazzotta concluded that the highest heat resistance for *Salmonella*e is in liquid eggs and egg yolks. They calculated that a process temperature of 71°C would require 1.2 seconds to bring about a 1-log (10 fold) reduction of *Salmonella* cells.

As a result, Food Standards Code in several countries requires:

1. That untreated fruit juice or vegetable juice must carry a statement to the effect that the product has not been pasteurised or treated to achieve the equivalent result and,
2. Untreated citrus juice must carry a statement to the effect that the product has not been pasteurised or treated to achieve the equivalent result.

Untreated fruit juice or vegetable juice means "juice, other than citrus juice, which, at a single facility, has not undergone a process capable of producing a minimum of a 5-log reduction (1 million fold) in the most treatment resistant organism of public health significance that is likely to occur in the juice." Untreated citrus juice means "citrus juice which, in its production, after cleaning and culling of fruit and at a single facility, has not undergone a process capable of producing a minimum of a 5-log reduction in the most treatment resistant organism of public health significance that is likely to occur in the juice."

Five-log Reduction in Pathogen of Concern in Fruit Juices

In its deliberations on its fruit juice standard, FDA sought the assistance of the National Advisory Committee on Microbiological Criteria for Foods (NACMCF). The NACMCF advised that a tolerable level of risk could be achieved in fruit juice by a validated process that provides a cumulative 5-log reduction in the target pathogen(s).

NACMCF worked from the basis that some samples of apple juice (cider in the US) may contain 1 *Escherichia coli*/ml. If these *E. coli* were a pathogenic strain, then a reduction to fewer than 1 cell per 100 ml serving would be the minimum necessary to provide consumer protection. However, if a 100-fold safety margin is applied then a 5-log reduction to 10–4 *E. coli*/ml would be required. This would be equivalent to one viable *E. coli* in 10 litres of juice.

ANZFA (Australia) has also adopted this reasoning and recommended that the 5-log reduction apply to the juice itself in all juices other than citrus.

In the case of citrus juices, FDA accepted that the risk of internalisation of pathogens in citrus fruit was very slight. Accordingly, surface treatments of whole fruit which alone or cumulatively achieve a 5-log reduction of the target pathogen are permitted. Presumably the relevant authority or its agent would have to accredit treatments nominated by industry as achieving a 5-log reduction in the most treatment resistant organism of public health concern. This would entail reviewing results of experimental studies and deciding if a 5-log reduction would be achieved when the process is applied as described.

It is a standard practice to subject the whole fruit to washing and sanitizing procedures prior to juice extraction. However, the efficacy of chemical sanitizers, including the common chlorine based ones, on the surface microflora of fruits and vegetables is limited or unpredictable.

A report (*Journal of Food Protection*, 1999. 62. 756) compares the effectiveness of hot water treatments and chemical sanitizers in reducing the microbial load on fresh citrus fruit prior to extraction. Trials were conducted on oranges contaminated by the natural microflora and fruit inoculated on the surface with *Escherichia coli*. Chemical sanitizers tested included 200 mg/L chlorine, 100 mg/L chlorine dioxide, 80 mg/L peroxy acetic acid and 2 per cent trisodium phosphate.

Sanitizing capacity of the chemical solutions proved limited in all cases. At the stem-scar area, reduction of *E. coli* after eight minutes in any solution was less than 10-fold. On the remainder of the skin surface, the reduction in *E. coli* after the same time was usually 100 to 1000-fold.

The same organism was effectively reduced by hot water immersion treatments. As with the chemical treatments, hot water was less effective in the stem-scar area. An approximate 100,000-fold reduction in *E. coli* was attained by immersing inoculated fruit in water at 80°C for one minute or 70°C for two minutes. Longer heating times increased the kill but compromised the flavour of the juice subsequently extracted.

It was concluded that a rapid hot water immersion technique may be used on fresh citrus fruit to reduce fruit surface and initial juice microbial loads without altering sensory quality.

In a related study on sanitizing of apple surfaces prior to juicing, (*Journal of Food Protection*, 1997: 62, 778), researchers in the United States found that a hot water dip of 82.2°C for 13 seconds or a chlorine spray wash (initial concentration 160 mg/L, pH 6.5) each resulted in 100 to 1000-fold reduction in coliforms from surface inoculated fruit. None of a number of individual treatments studied achieved a 1000-fold reduction. The authors suggest that a combination of treatments may be more effective in reducing bacterial levels in apples. They also recommend strict adherence to good manufacturing practice and sanitation practice to support a HACCP program.

With a process that is relatively straightforward to monitor, *e.g.* thermal pasteurisation or high pressure processing, compliance by industry should be simple to ascertain either by the manufacturer himself or the relevant authority. Comparable data for high pH vegetable juices do not exist although details of pressure treatment of carrot juice inoculated with *Salmonella* spp. and *E. c*oli 0157: H7 are available in the literature. Whether these are the appropriate target organisms in vegetable juices is debatable.

The situation with regard to citrus fruit surface treatments is rather different. Pao and co-workers at the Florida Department of Citrus have studied both hot water and chemical sanitisers to reduce bacterial numbers on the surface of oranges prior to juice extraction (*Journal of Food Protection* 64 2001 310; *Dairy, Food and Environmental Sanitation* 21 2001 287). Their work shows that greater than 5-log reductions can be achieved in juice extracted from fruit immersed in hot water for 1 or 2 minutes at 80°C. This 5-log reduction was in comparison to the *E. coli* level detected in the control (unheated fruit) juice obtained by homogenisation of fruit inoculated on the surface to give a surface count of 5.4 cfu/sq.cm of *E. coli*. Surface treatment of fruit with a range of chemical sanitisers without heat failed to give a 5-log reduction.

It is worth noting that the level of *E. coli* on non-inoculated fruit surfaces was fewer than 1 cfu/sq.cm and the juices obtained by homogenisation of the non-inoculated fruit had fewer than 1 cfu/ml of *E. coli*. This raises questions about the necessity to effect a 5-log reduction on the surface of oranges to be used for juice extraction.

The processes cited in the references above for thermal pasteurisation and high pressure processing could be used (with others) as validated 5-log reduction processes for the juices examined. They could be monitored and inspected or audited in the same way that milk pasteurisation treatments are. The surface treatment of citrus fruits presents a greater problem. The processes used by Pao and his co-workers were applied to small numbers of fruit. To transfer such a hot water process to the industrial scale with appropriate controls will not be a simple matter. Nor will it be inexpensive.

Manufacturers of untreated citrus juices, orange juice in particular, believe that any heat treatment of the juice diminishes the acceptability of their product. They are therefore reluctant to use any heat treatment on their juice and high pressure processing is not necessarily suitable for their current operations. To avoid having to use a warning label on their product under the proposed legislation, they would be required to effect the 5-log reduction on the surface of the fruit prior to juicing. This too may not be a practical proposition at least in the short term.

Mandatory microbiological criteria are not usually regarded as the most effective way of assuring the safety of food products. Their effectiveness depends to a large extent on how frequently foods are sampled to determine if the criteria are being met. In the case of solid foods, uneven distribution of organisms within a food mass can lead to a false sense of security.

However, in certain circumstances, mandatory microbiological criteria are deemed the most effective way to minimise the risk to consumers. One option previously considered was to insert in Standard 1.6.1 of Volume 2 of the *Food Standards Code* the requirement that:

"Untreated fruit juice or vegetable juice must contain no detectable *E. coli*/ml and *Salmonella* spp. should be absent in 25 g when the sample is analysed by the prescribed method."

This may be a way forward with the fruit juice standard although of course the microbiological criteria could apply to all juices.

The concept of specifying a process severity (*e.g.*, 5-log reduction) and a microbiological criterion is not common in food laws of many countries.

High Pressure Processing of Fruit Juices

In the last few years there has been growing research and commercial interest in non-thermal techniques to pasteurize foods. High pressure and pulsed electric fields are two such technologies.

The driving force to find new processing techniques is the potential market for convenient foods that more closely resemble fresh foods than processed foods but which still have an extended storage life. This represents a considerable challenge for food microbiologists and technologists. Most of the steps we are seeking to remove, *e.g.* a thermal process, chemical preservatives, act as safety or stability factors. Refrigeration is still acceptable but cannot be relied upon as the single hurdle for microbial growth.

High pressure alone or in combination with other technologies offers potential for producing foods with the characteristics outlined above. It has already been commercialized to a limited extent, especially in Japan, with the main drawback being the high capital cost of equipment. This is likely to change in the future with new developments in pump and vessel technology. The primary focus for exploitation to date has been with acid products such as jams and fruit juices and refrigerated foods, notably guacamole and some fish products.

High pressure technology utilizes pressures in the range 3400 to 6800 atmospheres. Normally, water is the transmitting medium and the pressure is applied uniformly and instantaneously. Since the pressure is isostatic (uniform throughout the food), the food is preserved evenly throughout. No permanent deformation of a packaged food occurs because water, the main component of most foods, is relatively incompressible.

The lethality of high pressure on microorganisms is primarily attributed to the disruption of the cell membrane and the denaturation of proteins including some enzymes. Microorganisms vary in their sensitivity to high pressure, with Gram negative bacteria being most sensitive and bacterial spores being the most resistant.

The high resistance of spores has to date limited the application of the technology to acid foods or foods which will be refrigerated to control outgrowth.

Apart from the resistance of bacterial spores, there are other disadvantages associated with high pressure processing. For example:

1. Some food enzymes are very resistant to pressure. These include polyphenoloxidase and pectin esterase, both important in food spoilage;
2. As a result, most foods processed by high pressure require low temperature storage to retain their sensory qualities;
3. Residual oxygen in the food as well as enzyme activity can lead to degradation;
4. Many products may require aseptic packaging.

The potential for improved sensory quality outweighs these disadvantages at least some of which can be overcome by combination processes.

Governing Bodies Selecting the Criteria for Quality of Fruit Juices

According to Fruit Juice and Fruit Nectars (FJFN) Regulations 2013 in England, freezing of fruits before juice production is prohibited, and addition of sugars in juice is also not permitted. According to the opinion of the United State Department of Agriculture (USDA) people consuming 100 per cent juice does not cause obesity in most cases but obese people may gain weight for drinking of 100 per cent juice, whereas, the Academy of Nutrition and Dietetics Evidence Analysis Library (ANDEAL) concluded that in case of children there is no interrelation between weight gain and fruit juice consumption. In Ethiopian Public Health Institute (EPHI) Food Microbiology guideline, it has been suggested that people should not be sold or consumed spoiled, poor quality and harmful toxin containing juices. It is prohibited to manufacture, package, handle and store fruit juice in unhygienic condition. According to FDA, fruit juices should be rejected if it contains toxins, high microbial load and high pesticide residue. Microbial limit in fruit juice is listed in Table 12.21 according to the FDA standard. According to Food Safety and Standard Authority of India (FSSAI, 2011), microbiological limits are listed in Table 12.22.

Table 12.21: Microbial Limits in Fruit Juice

Microorganism	*Maximum Limit*
Total bacterial count/ml	25
Total coliform count/100 ml	Absent
Escherichia coli/100 ml	Absent
Yeast and mold/ml	5

Table 12.22: Microbiological Requirements of Fruit Juices as per FSSAI (2011) *

Count (cfu/ml)	*Maximum Count Permitted*
Total count	Not more than 50
Coliform	Absent in 100 ml
Yeasts	Not more than 2

*FSSAI (Licensing and Registration of Food businesses) Regulation, 2011

Permitted ingredients for juice FSSAI77 include sugar (less than 2 per cent moisture), liquid sucrose/invert sugar/fructose syrup/liquid cane sugar, lime/lemon juice, salts/spices/aromatic herbs for tomato juices, vitamins and minerals.

Chapter 13

The Fermented Foods

Definitions

i) A general definition should define fermentation as a biochemical process through which most microorganisms decompose carbohydrates to produce energy under anaerobic conditions.

ii) Fermented foods are defined as "foods or beverages produced through controlled microbial growth, and the conversion of food components through enzymatic action".

iii) According to Pasteur, "Fermentation is life in the absence of oxygen".

In this somewhat obscure and phenomenological term, various types of fermentation are included, such as yeast fermentation, lactic acid fermentation, butyric acid fermentation, propionic acid fermentation and acetic acid fermentation. There are several variables in the fermentation process including the microorganisms, the nutritional ingredients and the environmental conditions, giving rise to thousands of different variations of fermented foods.

Historically, food fermentation was performed as a method of preservation, as the generation of antimicrobial metabolites (*e.g.*, organic acids, ethanol and bacteriocins) reduces the risk of contamination with pathogenic microorganisms. Fermentation is also used to enhance the organoleptic properties (*e.g.*, taste and texture), with some foods, such as olives, being inedible without fermentation that removes bitter phenolic compounds. It is many of these products that make fermentation so useful and interesting both to science and industry.

Introduction

Fermented foods identify cultures and civilizations. History, climate and the particulars of local production of raw materials have urged humanity to exploit various pathways of fermentation to produce a wide variety of traditional edible products which represent adaptations to specific conditions. Nowadays, industrial-scale production has flooded the markets with ferments. According to recent estimates, the current size of the global market of fermented foods is in the vicinity of USD 30 billion, with increasing trends.

The fermentation of foods and beverages, at the level of know-how, has a past of at least 6000 years and is likely to be the result of natural and spontaneous microbial interactions. Today, food fermentation can be achieved through two main production methods. With the first production method, the fermentation process is driven by the native/indigenous microflora of the raw food materials or processing environment, for instance, sauerkraut, kimchi and certain fermented soy products. These types of fermented foods are often named by researchers, or food technologists, as "wild ferments" or "spontaneous ferments". With the second method, to achieve fermentation, it is necessary to add starter cultures, known as "culture-dependent ferments," for example, kefir, yoghurt, kombucha and natto. One method of performing a culture-dependent ferment is "backslopping," in which a small amount of a previously fermented batch is added to the raw food, for example, sourdough bread. Starters used to initiate fermentation can be either natural (*e.g.*, backslopping), or selected commercial starters to standardize the organoleptic characteristics of the final product. Compiling a geographical list, we can place

Europe, North America, Australia and New Zealand in the areas where fermented products are produced mainly using defined starter cultures, while Asia and Africa are extraordinary ambassadors of spontaneous fermentation.

Moreover, traditional food fermentation processes can be broadly classified into (a) lactic acid fermentation, (b) fungal fermentation and (c) alkaline fermentation. Examples of lactic acid fermented products, *i.e.*, products primarily fermented by lactic acid bacteria, include yoghurt, sausages, cheese, sauerkraut (fermented cabbage from eastern and central Europe), kimchi (fermented and spiced Napa cabbage from Korea) and soy sauce (ganjang).

Yeasts are also involved in the fermentation of many of the lactic acid-fermented products, including kefir (a slightly alcoholic dairy beverage from the Caucasus) and kombucha (a fermented sweetened tea from China). Most of the well-known soy-based fermented foods from Asia, such as tempeh and soy sauce, are produced by fungal fermentation, except natto, which is produced by alkaline fermentation.

Industrial fermentation processes use either submerged or solid-state bioreactors that are conducted in batch, semi-batch or continuous mode. Most food fermentation processes, from *sauerkraut* and *kimchi* to *miso* and *tempeh*, use solid-state fermentation processes conducted in batch mode, where microorganisms are cultivated on the surface of a water insoluble substrate. Submerged fermentation processes are used in the production of yoghurt and other dairy-based beverages, alcoholic beverages and food condiments such as vinegar.

There are two types of fermentation (regarding the way the fermentation microorganisms are utilized):

1. **Natural fermentation**: A fermentation in which the fermentation microorganisms are already part of the natural microflora present in the foodstuff, so they do not need to be added. The only thing required is to create the necessary conditions for their development (*e.g.*, the creation of anaerobic conditions, as in the production of pickles or olives), or the suppression of competing microflora (*e.g.*, with the addition of salt or vinegar to the products).
2. **Controlled fermentation with starter cultures**: A fermentation that starts with the addition of a suitable inoculation with a large population of the desired fermentation microorganisms, necessary when the raw material is pasteurized (*e.g.*, pasteurized milk) or when it is difficult for the desired fermentation microorganisms to prevail over competing microorganisms (*e.g.*, in beer brewing, where wild yeasts may prevent the alcoholic fermentation by the strains of *Saccharomyces*). The starter culture contains the natural fermentation agents already present in the microflora of the food, but in much higher concentration than normal, to ensure that they easily prevail over competing–spoiling microorganisms. Therefore, with the addition of a starter culture, we better ensure a smooth fermentation process, the prevention of spoilage and the standardization of the product (steady qualitative and organoleptic characteristics).

Today, fermentation is no longer simply considered a method of food preservation. Through time, man has managed to control the fermentation process, and fermented food now constitutes a separate sector of the food industry. Fermented foods hold a firm place in cuisine from almost every culture in the world. In the West, there has been a surge in popularity of fermented foods in recent years, major reasons including the proposed health benefits of fermented foods and surging interest in gastrointestinal health. There are several mechanisms through which fermented foods may exert beneficial effects in health and disease. In Germany, for example, approximately 25 per cent of consumers prefer fermented food. As a result of fermentation, the taste, smell, appearance, structure, texture, lifespan and safety of fermented food has different qualities compared to the raw materials from which they are produced. For example, the reduction of pH in meat paste can be achieved with the addition of Glucono-delta-lactone (GDL) or through fermentation. In both cases, the products are safe for consumers and have the same shelf-life and coherence stability, but the distinctive, desired taste of the final product can only be achieved through fermentation.

In addition, consumers have been consuming fermented food and beverages for a very long time, being aware that, in the process, they are ingesting billions of living cells of specific microorganisms or/and the metabolic products of these microorganisms without fear for their health and safety. This fact highlights the resilience of consuming fermented food through time and encourages researchers to continue their never-ending efforts to document the microflora of fermented food and find antimicrobial compounds that may be naturally present in such food. They can then research the possible use of such preserving agents as natural bio-preservatives, instead of chemical preservatives in other food products.

The bibliographical references regarding the microbial ecology of fermented foods and beverages in various regions such as Asia, Africa, Europe, South America and North America are numerous and many species of microorganisms have been documented. In the last few years, the use of molecular techniques and tools have helped clarify, at least partially, the confusion present in the naming of newly isolated microorganisms and the generalisation which is caused by the use of conventional classification interpretation of the documented phenotypical characteristics. In simple terms, these molecular techniques, such as nucleic acid probe, species-specific polymerase chain reaction (PCR), Rep-PCR, multiplex PCR, 16S rDNA sequencing, DGGE and TTGE, offer reliable solutions for microbial identification in fermented foods and are often used to analyse the microbial flora of fermented food products

Almost without exception, fermented foods were discovered before mankind had any knowledge of microorganisms other than as witness to the effects of their activity. It was simply an empirical observation that certain ways of storing food effected desirable changes in its characteristics. Originally the most important of these changes must have been an improvement in the shelf-life and safety of a product, although these became less important in the industrialized world with the advent of alternative preservation methods such as canning, chilling and freezing. Modern technologies have in no way diminished the sensory appeal of fermented products however. This is clear from the way people rarely enthuse over grape juice or milk as some are prone to do over the vast array of wines and cheeses.

Table 13.1: Some Fermented Foods

Food	*Ingredients*	*Geographical Distribution*
Busa	Rice, millet, sugar	Turkey
Beer	Barley	Widespread
Cheese	Milk	Widespread
Chicha	Maize and others	S. America
Dawadawa	Locust beans	W. Africa
Dhokla	Bengal gram flour	India
Gari	Cassava	Nigeria
Idli/dosa	Rice and black gram	India
Imrti	Dehusked black gram	India
Injera	Tef	Ethiopia
I-sushi	Fish	Japan
Kefir	Milk	Eastern Europe
Kenkey	Maize, sorghum	Ghana
Kimchi	Vegetables	Korea
Koko	Maize, sorghum	Ghana
Leavened bread	Wheat	Europe, N. America
Lambic beer	Barley	Belgium
Mahewu	Maize	S. Africa
Nam	Meat	Thailand
Ogi	Maize, sorghum, millet	Nigeria
Olives		Mediterranean area
Palm wine	Palm sap	Widespread
Poi	Taro	Hawaii
Puto	Rice	Philippines
Rabadi	Cereals	India
Salami	Meat	Widespread
Salt stock, cucumbers	Cucumbers	Europe, N. America
Sauerkraut	Cabbage	Europe, N. America
Sorghum beers	Sorghum	S. Africa
Sourdough bread	Wheat, rye	Europe, N, America
Soy sauce miso	Soy beans	S.E. Asia
Tempeh	Soy beans	Indonesia
Tibi	Fruit	Mexico
Wadi	Green gram, black gram	India
Yoghurt	Milk	Widespread

We now know that, in food fermentation, conditions of treatment and storage produce an environment in which certain types of organism can flourish and these have a benign effect on the food rather than spoiling it. The overwhelming majority of fermented foods are produced by the activity of lactic acid bacteria and fungi, principally yeasts but also, to a lesser extent, moulds. Both groups of organisms share a common ecological niche, being able to grow under conditions of low pH and reduced a_w, although only lactic acid bacteria and facultative yeasts will prosper under anaerobic conditions. As a consequence, they frequently occur together in fermented foods; in some cases, members of both groups act in concert to produce a product while in others, one group plays the role of spoilage organisms.

Benefits of Fermented Foods

Fermentation can be viewed as a biological method of food preservation. Foods produced in this way have a reduced risk of contamination when enriched in antimicrobial end-products, such as organic acids, ethanol, and bacteriocins. Advantages of fermented foods also include the new and desirable tastes and textures that are completely unlike those present in the starting materials. Other benefits are more specific to the particular food type. Table olives, for example, are inedible without the microbial (or chemical) — induced removal of bitter-tasting phenolic compounds. Another example is the growth of bakers' yeast (*Saccharomyces cerevisiae*), alone or with lactic acid bacteria, that achieves dough leavening during bread manufacture. Beyond these characteristics, it is increasingly understood that some fermented foods also promote human health in ways not directly attributable to the starting food materials. That is, the outcomes of fermentation, and the contributions of microbes, in particular, can provide additional properties beyond basic nutrition. Recent human clinical studies on fermented foods support this possibility. Large cohort investigations have revealed strong associations between consumption of fermented dairy foods and weight maintenance. Likewise, other long-term prospective studies show reductions in risk of cardiovascular disease (CVD), type 2 diabetes (T2D), and overall mortality from frequent yoghurt consumption. These benefits might extend to immediate physiological responses, a possibility recently indicated by the finding that fermented milk consumption improved glucose metabolism and reduced muscle soreness induced by acute resistance exercise. Similarly, evidence is accumulating for anti-diabetic and anti-obesity benefits of kimchi. In inflammatory bowel diseases and other immune-related pathologies such as arthritis and sclerosis, health benefits of fermented foods have also been proposed, although clinical data have not yet been reported. Lastly, although the microbiota-gut-brain axis is a nascent field of research, there is an indication that fermented food consumption can alter mood and brain activity.

Fermented Soy Products

The first known fermented soy products originated in China and Japan, including fermented black soybean and red fermented tofu. There are many fermented soybean products from different parts of Asia, including tempeh, natto, miso, sufu, douche, soy sauce and doenjang. This section will focus on tempeh, soy sauce, natto and miso. Fermented soy products have been proposed to have beneficial effects on health, including purported "anti-carcinogenic", "anti-diabetic", "antioxidant", "anti-inflammatory" and "anti-hyperlipidaemic" effects, although much of the existing evidence is limited to *in vitro* and animal studies.

Table 13.2: Description and microbial content of common fermented foods.

Name	*Description*	*Region of Origin*	*Source of Microorganisms*	*Microorganisms Identified in Final Product **
Kefir	Fermented milk beverage	Caucasus	Starter culture	*Lactobacillus kefiri, Lactobacillus paracasei, Lactobacillus parabuchneri, Lactobacillus casei, Lactobacillus lactis, Lactococcus lactis, Acetobacter lovaniensis, Kluyveromyces Lactis, Saccharomyces cerevisiae*
Kombucha	Fermented tea beverage	China	Starter culture	*Komagataeibacter xylinus, Saccharomyces cerevisiae, Zygosaccharomyces bailii. Brettanomyces bruxellensis, Acetobacter pasteurianus, Acetobacter aceti, Saccharomyces cerevisiae, Zygosaccharomyces bailii, Brettanomyces bruxellensis, Acetobacter xylinum, Zygosaccharomyces* spp., *Acetobacter, Gluconacetobacter*
Sauerkraut	Fermented cabbage	China	Spontaneous	*Lactobacillus sakei, L. plantarum, Candidatus accumulibacter phosphatis, Thermatoga* spp., *Pseudomonas rhizosphaerae, L. hokkaidonensis, L. rhamnosus, Leuconostoc carnosum, Clostridium saccharobutyrilicum, Rahnella aquatillis, Citrobacter freundii, Enterobacter cloacae, Bifidobacterium dentium, Enterococcus faecalis, Lactobacillus casei, Lactobacillus delbrueckii, Staphylococcus epidermidis, Lactobacillus curvatus, Lactobacillus brevis, Weissella confusa, Lactococcus lactis, Enterobacteriaceae, Leuconostoc* spp., *Yarrowia brassicae*
Tempeh	Fermented boiled and dehulled soybeans	Indonesia	Starter culture (*Rhizopus oligoporus*)	*Enterococcus faecium, Rhizopus oryzae, Rhizopus oligoporus, Mucor indicus, Mucor circinèlloides, Geotrichum candidum, Aureobasidium pullulans, Alternaria alternata, Cladosporium oxysporum, Trichosporon beigelii, Clavispora lusitaniae, Candida maltosa, Candida intermedia, Yarrowia lipolytica, Lodderomyces elongisporus, Rhodotorula mucilaginosa, Candida sake, Hansenula fabiani, Candida tropicalis, Candida parapsilosis, Pichia membranefaciens, Rhodotorula rubra, Candida rugosa, Candida curvata, Hansenula anomola*
Natto	Fermented soybean	Japan	Starter culture (*Bacillus subtilis* natto)	Data not available
Miso	Fermented soybean paste	Japan	Starter culture (*Aspergillus oryzae*)	*Bacillus subtilis, Bacillus amyloliquefaciens, Staphylococcus gallinarum, Staphylococcus kloosii, Lactococcus* sp. GM005

The Tempeh Kedele

Tempeh is a traditional mould-fermented food in Indonesia, though it has also attracted interest in the Netherlands and United States. The most popular type of tempeh is produced from soya beans and is also known as tempeh kedele. This produces a soft white cake with a chewy texture and mushroom-like flavour. The microbial composition of tempeh varies according to variations in production.

Whole clean soya beans are soaked overnight in water to hydrate the beans. A bacterial fermentation occurs during this stage decreasing the pH to 4.5–5.3. The hydrated beans are dehulled and the moist cotyledons cooked; a process which pasteurizes the substrate, destroys the trypsin inhibitor and lectins contained in the bean and releases some of the nutrients required for fungal growth. After cooking, the beans are drained and may be pressed lightly to remove excess moisture before spreading into shallow bamboo trays and allowing to cool. Starter culture is added either by mixing some tempeh within the cooked soya beans prior to packing in the trays or by sprinkling a spore inoculum, prepared by extended incubation of a piece of tempeh, on to the beans.

The fermentation is invariably a mixed culture of moulds, yeasts and bacteria but the most important component appears to be *Rhizopus oligosporus*, although other *Rhizopus* and *Mucor* species are often isolated. Over two days incubation at ambient temperature (30–35°C), the mycelium develops throughout the mass of beans knitting it together. During fermentation the pH, rises to around 7, fungal proteases increase the free amino acid content of the product and lipases hydrolyse over a third of the neutral fat present to free fatty acids.

Unlike many fermented foods, tempeh production is not a means of improving the shelf-life of its raw material which is in any case inherently quite stable. Tempeh contains antioxidants which retard the development of rancidity but will keep for only one to two days as sporulation of the mould discolours the product and a rich ammoniacal odour develops as proteolysis proceeds.

Tempeh production does however improve the acceptability of an otherwise rather unappealing food. Fresh tempeh has a pleasant nutty odour and flavour and can be consumed in a variety of ways, usually after frying in oil.

In addition to improving acceptability, fermentation also improves the nutritional quality of soya beans. In part this stems from the reduction or removal of various anti-nutritional factors at different stages in the processing. Destruction of the trypsin inhibitor and lectins during cooking of the beans has already been mentioned and levels of phytic acid, which can interfere with mineral nutrition, are also reduced by about a third in the course of processing. The notorious ability of beans to produce flatulence, is also regarded as an anti-nutritional property and flatulence-inducing oligosaccharides such as stachyose and raffinose are partially leached out of the beans during the soaking stage.

Figure 13.1 indicates the tempeh production.

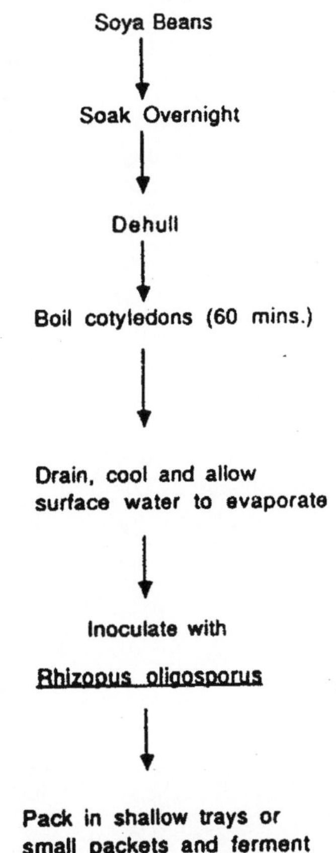

Figure 13.1: Tempeh Production.

Application of soybean and bean tempeh to human microbiota in an *in vitro* gut simulator model increases *Bifidobacterium*, *Lactobacillus*, *Escherichia coli* and *Enterococcus* abundance. In an open-label uncontrolled study of 10 healthy human volunteers, tempeh consumption led to greater stool *Akkermansia muciniphila* abundance and immunoglobulin A concentrations, suggesting that tempeh may influence gut microbiota in humans. However, a larger, controlled study is needed to establish the effects of tempeh on gastrointestinal microbiota composition.

Tempeh has been associated *in vitro* with greater free-radical and superoxide scavenging ability than unfermented soybeans, which may relate to changes in polyphenol content and digestibility in soybeans following fermentation. To date, there are no RCTs of the effects of tempeh consumption in humans. Proposed health effects listed above require investigation in human trials.

Tempeh can be made from a number of different plant materials including other legumes, cereals and agricultural by-products. One variety that has achieved some notoriety is tempeh bongkrek which is made in central Java using the presscake remaining after extraction of coconut oil. Tempeh bongkrek has been associated with occasional serious outbreaks of food poisoning due to the bacterium *Burkholderia cocovenenans* growing in the product and elaborating the toxins bongkrekic acid and toxoflavin. Since 1951, at least 1000 people known to have died as a result of this intoxication and in 1988 the Indonesian Government prohibited the production of tempeh bongkrek.

Despite the extensive proteolytic changes which occur during fermentation, studies have failed to show that the protein in tempeh is more easily digested. With the exception of thiamine which decreases, other vitamins increase to varying degrees during fermentation, Vitamin B_{12}, the anti-pernicious anaemia factor shows the most marked increase and this is associated with the growth of the bacterium *Klebsiella pneumoniae* during fermentation. The usual source of this vitamin in the diet is animal products and it has been suggested that tempeh could be an important source of B_{12} for people subsisting on a largely vegetarian diet.

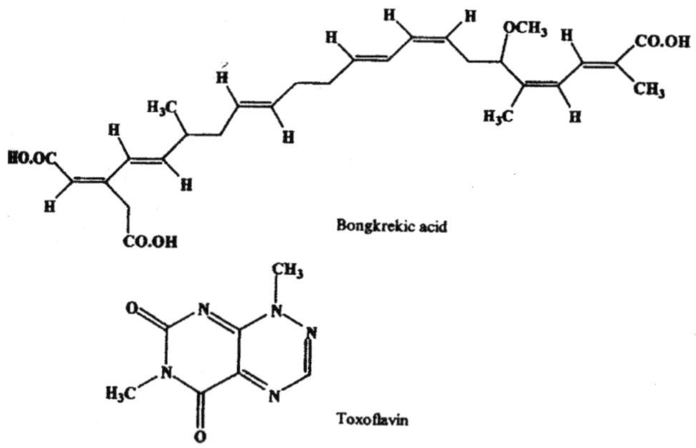

Figure 13.2: Toxoflavin and Bongkrekic Acid.

Two factors are thought to give rise to this problem. Reduction or omission of an initial soaking of the presscake may fail to give a lactic fermentation sufficient to reduce the pH below 6, a level at which the bacterium cannot grow. Also, the fungal inoculum may be too small since it has been shown that *B. cocovenenans* cannot grow if *Rhizopus oligosporus* has more than a tenfold numerical superiority (estimated by plate counts).

Ontjom is a tempeh-like product produced in Indonesia from pean presscake which normally has a fruity/mincemeat character. It can be produced using the tempeh mould but *Neurospora intermedia* is also often used. This mould has strong α-galactosidase activity which can further contribute to the reduction of flatulence-inducing oligosaccharides.

Natto

Natto is a traditional Japanese fermented soybean, of which Itohiki-Natto is the most commonly consumed. Natto is produced through fermentation of cooked yellow soybeans with *Bacillus subtilis* var. natto. This produces a viscous food with a distinct flavour and strong odour. Natto characteristics vary according to soybean steaming time, relative humidity, fermentation time and temperature. The fermentation of Natto produces a number of bioactive factors, including nattokinase, bacillopeptidase F, vitamin K_2 and dipicolinic acid. Furthermore, the quantity of the isoflavone genistein, with purported associations with metabolic and inflammatory disorders and carcinogenesis, is greater in Natto compared to unfermented soy products. A peptide with antibacterial activities against *Streptococcus pneumoniae* and *Bacillus subtilis* has been isolated from Natto, with potential clinical importance in treating *S. pneumoniae* infections, although this has not been investigated in humans.

Nattokinase is an enzyme of the subtilisin family produced by *Bacillus* subtilis var. natto, and can be isolated from Natto. Nattokinase has direct *in vitro* and *in vivo* fibrinolytic activity, in addition to increasing tissue plasminogen activator and reducing platelet aggregation. Anti-thrombotic and anti-hypertensive activities of nattokinase have been demonstrated in small RCTs in humans.

There is limited evidence regarding the effect of Natto on the human GI microbiota. Consumption of Natto-containing miso soup for two weeks led to increased stool *Bacilli* and *Bifidobacteria* and decreased *Clostridia* and *Enterobacteriaceae* in 8 healthy volunteers. Furthermore, stool short-chain fatty acids increased, and ammonia and sulphide declined. However, it is impossible to separate the effect of the miso soup (also a fermented soy product) from possible effects of Natto. In individuals with 3–5 bowel movements per week, consumption of 50 g/day *Bacillus subtilis* K-2 containing Natto for two weeks resulted in greater stool frequency and proportion of stool *Bifidobacteria* compared to consumption of 50 g/day of boiled beans, although only the abstract is available and no sample size is provided.

To date, there is limited evidence from RCTs suggesting Natto might positively influence stool frequency in patients with infrequent bowel motions and influences gastrointestinal microbiota. However, these require confirmation in high quality trials.

Miso

Miso is a traditional Japanese paste of fermented soybean used to make miso soup. Miso is produced by fermenting soybeans with 'Koji', produced from a mould *Aspergillus oryzae*, although *Saccharomyces cerevisiae* and lactic acid bacteria may additionally be used. As with other fermented soy foods, miso production varies greatly in terms of ingredients, temperature and fermentation time, salt concentration and the strain of *A. oryzae* used.

A microbial analysis of miso at different time points following the start of fermentation revealed *Bacillus subtilis*, *Bacillus amyloliquefaciens*, *Staphylococcus gallinarum* and *Staphylococcus kloosii* to be present during fermentation, with only the *Bacillus* species remaining in the final product.

A range of miso samples have also been shown to contain *Lactococcus* sp. GM005, which produces a bacteriocin with strong antibacterial activity that inhibits the growth of a range of bacteria, including *Bacillus subtilis*, *Pediococcus acidilactici* and *Lactobacillus plantarum*.

There is little evidence on the effect of miso intake on gastrointestinal disorders. One cross-sectional study reported an inverse relationship between miso soup intake and subjective gastro-oesophageal reflux disease, functional dyspepsia and reflux scores when adjusted for other dietary factors. This association was hypothesised to relate to histidine, glutamate and aspartate found in miso soup, although no animal or human studies have investigated this mechanism to date.

The high soy intakes in China and Japan have historically been hypothesised to contribute to the relatively low rates colon and prostate cancers in these countries. One proposed mechanism to support this hypothesis is the high concentrations of isoflavones genistein and daidzein found in soybeans. Genistein is structurally similar to oestrogen and may influence breast cancer risk through oestrogen receptor binding, which has been demonstrated *in vitro*. Further *in vitro* studies have demonstrated that genistein may exert effects on cancer risk through promoting cell-cycle arrest inducing apoptosis and reducing cancer cell migration. Genistein and daidzein may be higher in fermented soybean products (miso and natto) compared to unfermented products such as soy milk and tofu.

Numerous Japanese cohort studies have investigated associations between miso intake and cancer risk. These studies are limited in the assessment of dietary intake (food frequency questionnaires, often with limited intake responses) and the presence of a multitude of possible confounding factors. With these methodological limitations in mind, cohort studies have observed inverse associations between frequent miso soup intake and stomach cancer risk in Japanese men. In contrast, cohort and case-control studies have shown positive correlations between frequent miso soup intake and single and multiple stomach cancers in Japanese adults. Furthermore, some cohort studies have shown no association between miso soup intake and risk of various types of cancer.

There are currently no RCTs investigating the effects of miso in functional bowel disorders. There is therefore limited evidence for the effects of miso on gastrointestinal conditions and microbiota. There are several observational studies demonstrating associations between miso intake and the risk of stomach cancer, however the strength and direction of these associations remains unclear.

Soy Sauce Production

Soy sauce also called jiangyou in Chinese, shoyu in Japanese, and soya sauce in British English is the most consumed seasoning in East and Southeast Asian cuisine and is still gaining considerable popularity in as an all-purpose seasoning in more than 100 countries worldwide. It is presumed that "hishio," which is derived from salted fish or grains and originated from China in the 5th century, is the primitive prototype of shoyu and miso (soybean paste). In 1254, the Buddhist monk Kakushin returned to Japan having learned the method of producing Kinzanji miso from China. He later discovered that the liquid pooling at the bottom of the vats in miso fermentation can be used as a tasty condiment, which was later developed into shoyu. Hence, in ancient China, about 2200 years ago, soy sauce was developed as a way to preserve food due to its salt content and also to enhance the flavour of the modest, vegetarian Buddhist diet since salt at that moment was an expensive commodity. Nowadays, soy sauce is used as a food product rather than a food preservative.

It is a dark brown liquid, stable at ambient temperature, which does not require refrigeration during storage due to its low water activity and high salt content. Its distinct flavour is characterized by a strong umami, salty, and caramel-like character which enhances the overall savoury taste and aroma of many kinds of dishes. Soy sauce is used for a wide variety of both cooked (hot and cold) and uncooked foods, such as sushi, sashimi, stir fried noodles, and any type of fish, meat, and/or vegetable stews.

Soy sauce is made using essentially five basic raw ingredients: soybeans/soybean flakes (as the main protein source), wheat/wheat flour (as the main carbohydrate source), salt, water, and *Aspergillus oryzae*/*Aspergillus sojae*, salt-tolerant yeast and lactic acid bacteria. Chinese soy sauce is synthesized from pure soybeans or its mixture with a small amount of wheat, whereas Japanese-style soy sauce uses equal amounts of soybeans and wheat. Many techniques for brewing soy sauce were adopted from sake (Japanese rice wine) production. By the 17th century (end of the Edo era), the basic manufacturing process of soy sauce had been completed and is still mainly utilized at present. Nonetheless, existing fermentation processes are remarkably diverse and complex, making soy sauce a widely varied product with strong local/cultural heritage. Like balsamic vinegar, soy sauce can be found worldwide, from the cheapest hydrolyzed forms to the most expensive premium, aged brands.

Soy sauce and miso are flavouring agents having similar aroma and flavour. Soy sauce is a liquid whereas miso is a paste. Rice wine and soy sauce also share sufficient common features in their production to warrant discussing them together; though they are markedly different in character, both are representatives of products which involve mould activity in a two-stage fermentation process. Soy sauce and miso, each is made by a two-step fermentation process from wheat flour and soy beans with a mixture of molds, yeasts and bacteria. The first step involves fermentation with mold to produce proteolytic and amylolytic enzymes in the Koji which is a culture starter. A good culture starter must give characteristic aroma and flavour to the soy sauce, have high proteolytic and amylolytic activities, and must be easy to culture. In the *koji* stage, aerobic conditions allow moulds to grow on the substrate producing a range of hydrolytic enzymes necessary for utilizing the macromolecular material present. In the case of soy sauce, soaked and cooked soya beans are mixed with roasted cracked wheat in about equal proportions and inoculated with *tane koji* or seed *koji*. This has been previously grown-up on a similar mixture of substrates and contains a mixture of strains of *Aspergillus oryzae*. The moulds are then allowed to grow throughout the mass of material spread as layers about 5 cm deep for 2–3 days at 25–30°C.

First stage is followed by a second fermentation with yeast and bacteria in the presence of 18-20 per cent salt. The microorganisms used in these fermentation steps are not inoculated at the same time, but are applied sequentially. In the second, mash or *moromi* stage, conditions are made anaerobic so no further mould growth can occur. In soy sauce production, this is achieved by mixing the *koji* with an approximately equal volume of brine to give a final salt concentration of 18–20 per cent. Although the moulds can no longer grow, the activity of a whole battery of hydrolytic enzymes continues breaking down proteins, polysaccharides and nucleic acids to produce a liquid rich in soluble nutrients. Yeasts and lactic acid bacteria dominate the microflora producing a number of flavour components and converting roughly half of the soluble sugars to lactic acid and ethanol so that the final soy sauce normally has a pH of 4.5–4.9 and ethanol and lactic acid contents of 2–3 per cent and 1 per cent, respectively. The halophilic lactic acid bacterium *Tetragenococcus halophilus* (formerly *Pediococcus halophilus*) and the yeasts *Zygosaccharomyces rouxii* and *Torulopsis* have been identified as being important in this stage.

The *moromi* stage can be quite protracted, lasting up to a year or more, at the end of which the mash is pressed to remove the solid residues which may then be mixed with brine to undergo a second fermentation and produce a lower grade product.

Rather similar steps are involved in the production of soya bean pastes known as *miso* in Japan and *chiang* in China. These include up to 40 per cent of a grain such as rice or barley, use dry salt rather than brine and employ a shorter fermentation so the product has the consistency of a paste rather than a liquid.

In the brewing of the Japanese rice wine, sake, a *koji* prepared on steamed rice is used. Although the mould used is the same species as in soy sauce production, *Aspergillus oryzae*, the strains used in *sake* production are particularly noted for their ability to produce amylolytic enzymes.

Different Production Methods of Soy Sauce

The production of soy sauce varies greatly in terms of fermentation times, temperatures, and ratios of ingredients soybeans, wheat, salt, water, starter composition, and optionally, other additives. Moreover, the production processes are closely associated with the country of origin. For instance, the Chinese-type uses predominantly more soybeans than wheat (ratios 80:20 and 70:30), whereas the Japanese-type uses equal amounts (ratio 50:50). The presence of wheat is known to influence the final aroma and taste characteristics of the soy sauce. More wheat translates into more carbohydrates, which are degraded to saccharides by the enzymes of the starter microorganisms. These saccharides are the fuel for yeast fermentation, which converts them into various alcohols and other important flavour compounds.

There are essentially two general production processes: traditional fermentation and acid hydrolysis, although in China the latter type is since 2018 no longer recognized as an accepted method to produce soy sauce. Acid hydrolysis, also known as "chemical hydrolysis", uses high concentrations of acid for the initial polymer break down of soybeans and wheat. Traditional fermentation uses starter cultures of microorganisms which secrete enzymes capable of breaking down the proteins, lipids, and starches into peptides, free amino acids, volatiles, and saccharides. Flavour and colour development during the production process are attributed to the enzymatic reactions mentioned above, along with additional non-enzymatic Maillard reactions occurring between amino acids and reducing sugars, as well as Strecker degradation of amino acids, lipid oxidation, and esterification. Maillard reactions, also called browning reactions, occur at higher temperatures and trigger the formation of a wide range of, yet poorly defined, molecules which, in combination, create the characteristic and distinct soy sauce flavour and colour.

In the traditional method, soy sauce is produced by fermenting soybeans, wheat, and salt with koji mold, halophilic lactic acid bacteria, and salt-tolerant yeast. The most typical method of manufacturing soy sauce in Japan is called the "Honjozo" method and has five main processes (Figures 13.3 and 13.4):

1. Treatment of raw materials,
2. Making of koji,
3. Fermentation of moromi mash and aging,
4. Pressing of moromi mash, and
5. Refining.

In the first step, the soybeans are moistened and cooked under steam pressure to denature the proteins, and the wheat is roasted at 160–180 °C and crushed. Approximately equal amounts of soybean and wheat components are then mixed and inoculated with a pure culture of koji mold. This mixture is incubated for two to three days at 30 °C under high humidity to allow the koji mold to grow throughout the mixture and produce various hydrolyzing enzymes, such as proteolytic and amylolytic enzymes, which are responsible for the decomposition of the raw materials. The dry mash formed is called koji (or soy sauce koji). The quality of soy sauce koji heavily relies on the quality of soy sauce. Therefore, the process of making koji, including the processing of raw materials, is the most crucial step in soy sauce brewing.

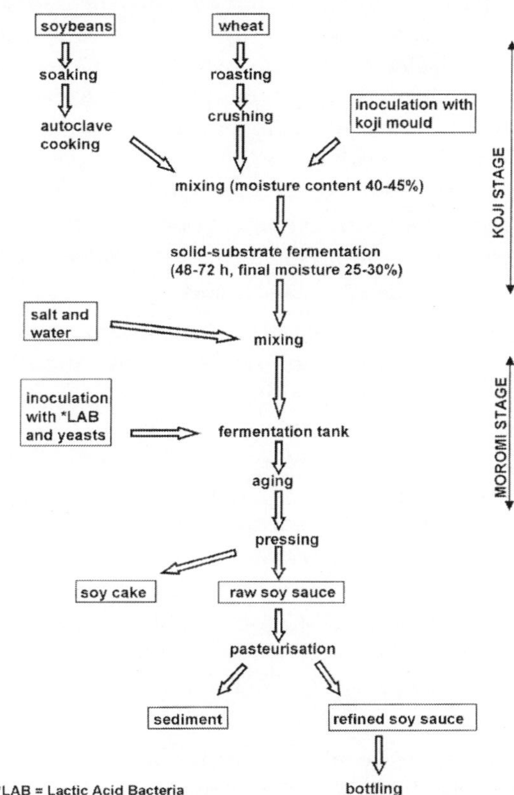

Figure 13.3: Soy Sauce Production.

Soy sauce koji is then added to a concentrated (22–23 per cent) saline solution to produce Moromi or Moromi mash that is traditionally fermented for 6–8 months. In this process, the entire enzyme-rich koji mixture is used, rather than pure enzymes from the raw materials, to decompose the raw materials and yield amino acids and sugars, which are responsible for the distinctive flavour of Japanese soy sauce. In addition, these amino acids and sugars serve as a nutrient source for lactic acid bacteria and yeast that ferment the raw material. The high salt concentration (17–18 per cent) of soy sauce does not allow the growth of non-salt tolerant microorganisms. Soon after brewing, the koji mold dies off and is replaced by salt-tolerant or halophilic microorganisms.

The first microbes that grow in moromi mash are halophilic lactic acid bacteria. Halophilic lactic acid bacteria were first isolated from soy sauce moromi mash in 1907. Initially, it was classified as *Pediococcus soyae*; however, in 1990, it was reclassified as *Tetragenococcus halophilus*. *T. halophilus* is a Gram-positive strain, micrococcus, which often forms a tetrad coccoid, and grows best in an environment containing 5–10 per cent salt. Each strain is highly diverse and exhibits different properties, including amino acid decomposition, sugar-fermentation ability, reducing activity, and organic acid metabolism. *T. halophilus* metabolizes glucose, which is produced by the breakdown of wheat starch by amylolytic enzymes derived from koji mold, into lactic acid and citric acid in soybeans into acetic acid. As a result, the pH of soy sauce moromi is slightly acidic, imparting a slightly sour taste to soy sauce.

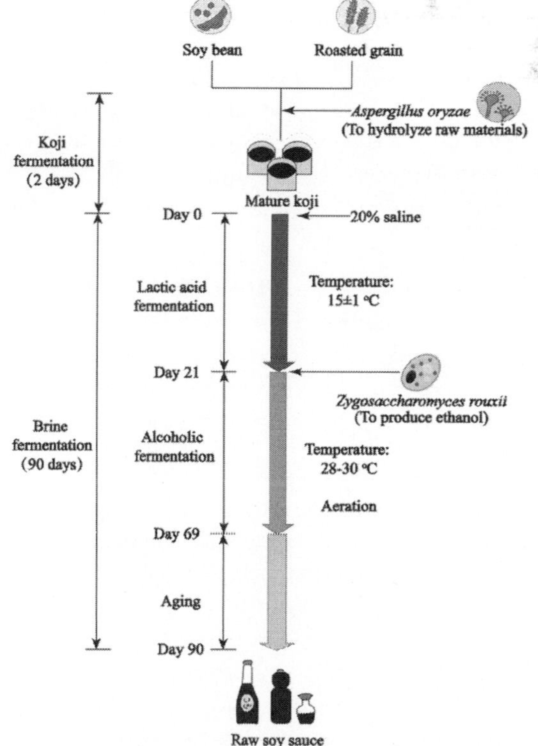

Figure 13.4: Various Stages of Soy Sauce Production.

After lactic acid fermentation by growing halophilic lactic acid bacteria, two types of salt-tolerant yeast grow, namely major fermenting yeast and late maturation yeast. The former is classified as *Zygosaccharomyces rouxii* and mainly affects the flavour of soy sauce. Meanwhile, the latter includes *Candida versatilis* and *C. etc.hellsii* and remains active until the late stage of ripening. *Z. rouxii* produces 2–4 per cent ethanol and a small amount of glycerol from glucose, as well as 4-hydroxy-2 (or 5)-ethyl-5 (or 2)-methyl-3 (2H)- furanone (HEMF), methionol, and many other characteristic aroma constituents of soy sauce. *C. versatilis* produces characteristic aroma compounds, such as 4-ethylguaiacol (4EG), from phenolic compounds derived from wheat lignin. In the presence of a high salt concentration (18 per cent), the pH range in which *Z. rouxii* can grow is narrowed down to 4.0–5.0, and *Z. rouxii* can vigorously grow even when the pH of the moromi mash is lowered by lactic acid fermentation. Therefore, in soy sauce brewing, three kinds of microorganisms work in stages, and multiple parallel fermentation occurs to establish the distinctive flavour of soy sauce. Halophilic lactic acid bacteria and salt-tolerant yeast are mainly present in the environment of koji, brewing containers, and brewing rooms and are mixed into the moromi mash. However, in recent years, microbial strains with excellent fermentation performance, including lactic acid and/or alcohol fermentation, have been isolated and are now widely used in soy sauce brewing. After alcohol fermentation by yeasts, the moromi mash is allowed to mature for a certain period, and various chemical reactions take place among the diverse components of the moromi. For example, amadori compounds are formed by Maillard reactions between sugars, such as glucose, and amino acids produced by the enzymatic decomposition of raw materials. These compounds are responsible for the distinctive colour, mildness, and umami taste of soy sauce. Nearly 300 aromatic compounds have been identified in soy sauce, the majority of which are formed during fermentation and aging.

After fermentation and aging, the moromi mash is wrapped in a square cloth, weighed, and continually pressed. The liquid content produced is raw soy sauce, which undergoes heat treatment for pasteurization and enzyme inactivation. During the heating process, enzymes are denatured, causing particles to coagulate and settle out. Heat also accelerates the chemical reactions involved in soy sauce brewing, thereby contributing to its unique flavour. In addition, only the clear part of the supernatant is extracted to produce the final product. Soy sauce brewing lasts approximately six months.

Koji Mold Species

Koji mold used in soy sauce brewing is a filamentous fungus belonging to the genus Aspergillus and is roughly classified into three species: *Aspergillus oryzae*, *Aspergillus sojae*, and *Aspergillus tamarii*. *A. oryzae* is widely used in the manufacture of sake, amazake (Japanese sweet beverage), mirin (Japanese sweet rice wine), miso, and soy sauce, whereas *A. sojae* is only used in miso and soy sauce production (mostly in soy sauce). Meanwhile, *A. tamarii* is used very infrequently. Therefore, the main koji molds used for soy sauce brewing include *A. oryzae* and *A. sojae* (Ito and Matsuyama, 2021).

A. oryzae was first isolated from rice koji by H. Ahlburg in 1876. Sakaguchi *et al.*, isolated *A. sojae* as distinguishable from *A. oryzae* in koji used for soy sauce production in Japan and found that *A. sojae* produced prominently echinulate conidia and smooth-walled conidiophores. *A. oryzae* and *A. sojae* were selected and bred as strains suitable for soy sauce brewing. Therefore, it is presumed that *A. flavus* and *A. parasiticus*, which are naturally occurring aflatoxin-producing pathogenic fungi, are domesticated. Recently, comparative genomic analysis results have suggested that *A. oryzae* is taxonomically closer to *A. aflatoxiformans* than to *A. flavus*, and that *A. aflatoxiformans* is the ancestor of *A. oryzae*. The conidiophore of selected characteristic koji molds is usually supplied as seed koji (koji starter) to sake, miso, and soy sauce brewers; however, several large manufacturers of soy sauce in Japan produce their own koji mold in-house.

The main role of koji mold in soy sauce brewing is to produce various enzymes that decompose large molecular substances, such as proteins and starches, in the raw materials into smaller substances, such as amino acids and glucose. One main difference between *A. oryzae* and *A. sojae* is their α-amylase and endopolygalacturonase productivity. *A. oryzae* has higher α-amylase productivity, whereas *A. sojae* exhibits higher endopolygalacturonase productivity. Owing to its high amylolytic activity that produces more glucose and subsequently promotes alcoholic fermentation in yeast, *A. oryzae* is widely used in soy sauce, miso, and sake brewing.

The desirable characteristics of koji mold for soy sauce brewing are as follows: (1) high productivity of enzymes, especially proteolytic enzymes, glutaminase, and macerating enzymes; (2) short stalk length and excellent conidiation ability; (3) rapid growth rate and high mycelial content; (4) low consumption of starch during growth; (5) genetically stability; and (6) absence of aflatoxin biosynthesis and little or no production of other mycotoxins. However, selecting a strain exhibiting all these properties is difficult as koji molds have extremely diverse characteristics. Therefore, strains are sometimes bred by mutagenesis to enhance their abilities or modify undesirable characteristics.

The distinctive flavour of soy sauce is attributed to the interplay between taste, aroma, and colour. Soy sauce contains all five basic tastes: sweet, salty, sour, bitter, and umami (savoury), and the harmonious combination and subtle balance of which contributes to its application as all-purpose seasoning. Sweetness comes from sugars, such as glucose, saltiness from salt, and sourness from organic acids, such as lactic and acetic acids. However, sweetness, bitterness, and umami are also derived from amino acids and peptides, which are degradation products of raw material proteins. Sweet amino acids, such as glycine, contribute to sweetness; branched-chain amino acids, such as leucine and peptides, contribute to bitterness; and the 20 amino acids themselves, including acidic amino acids such as glutamic acid and aspartic acid, greatly contribute to the umami taste of soy sauce. Because amino acids and peptides are crucial to the taste of soy sauce, the degradation and solubilization of insoluble raw material proteins have been extensively studied to improve the flavour and yield of soy sauce.

Soy Sauce Production by the Acid hydrolysis Process

In addition to the traditional fermentation method, soy sauce is also made by acid hydrolysis of defatted soya beans with food grade hydrochloric acid as a catalyst in a pressure cooker. The hydrolysate is neutralized with sodium hydroxide or sodium carbonate under careful supervision so that excessive alkali does not come into contact with the neutralized product. The acid-hydrolyzed soy sauce is refined by sedimentation, treatment with active carbon, filtration through a filter press to remove undesirable substances, and high vacuum to remove undesirable volatile compounds. However, the acid hydrolyzed soy sauce is less attractive in aroma and flavour because of lack of aromatic substances such as esters, alcohols and carbonyl compounds which are derived from the fermentation process. In some countries, a combination of fermentation and acid-hydrolysis procedures are used for making less expensive soy sauce. High quality soya sauce is made exclusively by the traditional fermentation process.

Table 13.3: Amylolytic Mould Preparations

koji	Japan	*look pang*	Thailand
ang-kak	Philippines	*ragi-tapai*	Malaysia/ Indonesia
nuruk	Korea	*peh-yueh*	China
bakhar	N. India		

Rice Based Mould Starters

A number of other rice-based mould starters are used in the countries of East and Southeast Asia to fulfil a similar role to *koji.* They are used to produce sweetened rice products which can be consumed fresh or added to other products or can used as a base for the production of rice wine and rice vinegar. The microbiological composition of these generally differs from that of *koji* and comprises primarily *Rhizopus* and *Mucor* species and amylolytic yeasts.

The Product of a Two Stage Fermentation: Vinegar

Vinegar has been part of the human diet since ancient times and has been widely used as a preservative, condiment, aromatizer, and even as a healthy drink. Moreover, it has also been traditionally used in ancient medicine because of its medicinal properties. Vinegar can be made from any carbohydrate source, amylaceous, or sugary substrate through two successive fermentations: alcoholic fermentation, which is carried out by means of yeasts, and acetic fermentation, with acetic bacteria as the protagonist.

Vinegar production ranges from traditional methods employing wood casks and surface culture to submerged fermentation in acetators. Whether naturally produced during fermentation or intentionally added, vinegar retards microbial growth and contributes sensory properties to a number of foods. The wide diversity of products containing vinegar (sauces, ketchup, mayonnaise, *etc.*) and the current fall in wine consumption have favoured an increase in vinegar production. Acetic acid is the predominant flavouring and antimicrobial component in vinegar.

Vinegar is the product of a two-stage fermentation. In the first stage, yeasts convert sugars into ethanol anaerobically, while in the second ethanol is oxidized to acetic (ethanoic) acid aerobically by bacteria of the genera *Acetobacter* and *Gluconobacter.* This second process is a common mechanism of spoilage in alcoholic beverages and the discovery of vinegar was doubtless due to the observation that this product of spoilage could be put to some good use as a flavouring and preservative. The name vinegar is in fact derived from the French *vin aigre* for 'sour wine' and even today the most popular types of vinegar in a region usually reflect the local alcoholic beverage; for example, malt vinegar in the UK, wine vinegar in France and rice vinegar in Japan.

Acetic acid is formed in a four-step reaction involving conversion of starch to sugar by amylases, anaerobic conversion of sugars to ethanol by yeast fermentation, conversion of ethanol to hydrated acetaldehyde, and dehydrogenation to acetic acid by aldehyde dehydrogenase. The last two steps are preformed aerobically with the aid of acetic acid forming bacteria. Acetic acid yield improvements can be achieved using high rates aeration of during continuous production.

Vinegar bacteria, also called acetic acid bacteria, are members of the genus Acetobacter and characterized by their ability to convert ethyl alcohol (C_2H_5OH) into acetic acid (CH_3COOH) by oxidation as shown below:

Anaerobic Aerobic

$$2C_2H_5OH \rightarrow 2CH_3CHO \rightarrow 2CH_3CO_2H + 2H_2O$$

Further processing of vinegar, following substrate conversion to acetic acid may include filtration, clarification, distillation and pasteurization at 165.2° (74°C) before it is bottled. Acetic acid concentration in vinegar may be expressed using the term "grain". For example, 100 grain distilled vinegar is a 10 per cent acetic acid solution.

Vinegar Brewing

In vinegar brewing, the alcoholic substrate, known as vinegar stock, is produced using the same or very similar processes to those used in alcoholic beverage production. Where differences occur, they stem largely from the vinegar brewer's relative disinterest in the flavour of the intermediate and concern to maximize conversion of sugar into ethanol. In the production of malt vinegar for example, hops are not used and the wort is not boiled so the activity of starch-degrading enzymes continues into the fermentation. Here we will concentrate on describing the second stage in the process, acetification.

Acetification

Acetification, the oxidation of ethanol to acetic acid is performed by members of the genera *Acetobacter* and *Gluconobacter*. These are Gram-negative, catalase-positive, oxidase-negative, strictly aerobic bacteria, *Acetobacter* spp. are the better acid producers and are more common in commercial vinegar production, but their ability to oxidize acetic acid to carbon dioxide and water, a property which distinguishes them from *Gluconobacter*, can cause problems in some circumstances when the vinegar brewer will see his key component disappearing into the air as CO_2. Fortunately, over-oxidation, as it is known, is repressed by ethanol and can be controlled by careful monitoring to ensure that ethanol is not completely exhausted during acetification. Most acetifications are run on a semi-continuous basis; when acetification is nearly complete and acetic acid levels are typically around 10–14 per cent w/v, a proportion of the fermenter's contents is reinoved and replaced with an equal volume of fresh alcoholic vinegar stock. Since a substantial amount of finished vinegar is retained in the fermenter, this conserves the culture and means that a relatively high level of acidity is maintained throughout the fermentation, protecting against contamination.

It also protects against over-oxidation as it has been found that *Acetobacter europaeus*, a species commonly found in commercial vinegar fermenters, will not over-oxidize, when the acetic acid concentration is more than 6 per cent.

Many of the acetic acid bacteria associated with commercial acetification are difficult to culture on conventional solidified media, although some success has been enjoyed using a double-layer medium which provides colonies growing on the surface with a constant supply of ethanol and moisture from a lower, semi-solid layer. As a result, vinegar fermentations are usually initiated with seed or mother vinegar, an undefined culture obtained from previous fermentations. Depending on the type of acetification, the culture can be quite heterogeneous and *A. europaeus, A. hansenii, A. acidophilum, A. polyoxogenes* and *A. pasteurianus* have all been isolated from high-acidity fermentations.

Oxidation of ethanol to acetic acid is the relatively simple pathway by which acetic acid bacteria derive their energy. It occurs in two steps mediated by an alcohol dehydrogenase and an aldehyde dehydrogenase. Both enzymes are associated with the cytoplasmic membrane and have pyrroloquinoline quinone (PQQ) as a coenzyme. PQQ acts as a hydrogen acceptor which then reduces a cytochrome *via* an intermediate quinone.

The quick vinegar process derives its name from the faster rates of acetification achieved by increasing the area of active bacterial film and improving oxygen transfer to the acetifying stock. The acetic acid bacteria grow as a surface film on an inert support material packed into a false-bottomed vat. The acetifying stock is sprayed on to the surface of the packing material and trickles down against a counter-current of air which is either pumped through the bed or drawn up by the heat of reaction within it. The packing material normally consists of some lignocellulosic material such as birch twigs, vine twigs, rattan, wood wool, or sugarcane bagasse, although other materials such as coke have also been used. The vinegar stock is collected in a sump at the bottom of the vat and recirculated until the desired level of acidity is reached. The faster rate of reaction achieved means that the wash heats up during passage through the bed and, depending on the size of the fermenter, some cooling may be required.

The process is operated semicontinuously to maintain a high level of acidity throughout and most of the biomass is retained within the packed bed. A well operated quick vinegar process fitted with temperature control and forced aeration can usually acetify a vinegar stock with a GK of 10 and an initial ethanol content of 3 per cent in 4–5 days.

The fastest rates of acetification are achieved using submerged acetification in which acetic acid bacteria grow suspended in a medium which is oxygenated by sparging with air. The most commercially successful technique to have been developed is the Frings Acetator which uses a patented self-priming aerator to achieve very efficient oxygen transfer.

Submerged culture is very efficient and rapid, a semicontinuous run normally takes 24–48 h. It does however require far more careful control than simpler processes. The acetic acid bacteria are very susceptible to interruptions to the air supply, indicating that, in order to survive suspended in a medium with a pH of 2.5 and 10–14 per cent acidity, the bacteria need a constant supply of energy from respiration. A stoppage of only one minute in a stock with a GK of 11.35 is enough to completely arrest acetification which will not resume when aeration is resumed.

Another possible cause of fermentation failure in submerged acetification is phage infection. The presence of bacteriophage particles has been demonstrated in disturbed vinegar fermentations both in submerged acetifiers and the quick vinegar process. The performance of quick vinegar generators appears to be less affected as their acetification rate may slow but rarely stops. This is probably due to the greater heterogeneity of the culture present which allows organisms of different phage susceptibility to take over in the event of phage attack.

Where legal definitions of vinegar exist, it is specified as a fermentation product. 'Artificial vinegars' made by diluting and colouring acetic acid are thus excluded and, in the UK, have to be known rather laboriously as 'non-brewed condiment'. Although vinegar can be made up to 14 per cent acidity, it is usually diluted down to an appropriate strength for bottling. The minimum acetic acid content is usually prescribed to be something between 4 and 6 per cent w/v, but higher strength vinegars are available for pickling.

Though most often thought of in terms of its use as a condiment, vinegar is an important food ingredient. It is used as a preservative and flavouring agent in a large and expanding range of products such as mayonnaise, ketchups, sauces and pickles. In the United States only about 30 per cent of the vinegar produced is sold as table vinegar, the rest being used in food processing.

Production Methods of Vinegar

Vinegar production methods can range from traditional methods employing wood casks (Orleans Process) and surface culture (Generator Process) to submerged fermentation. The quick process and submerged culture process were developed and are used for commercial vinegar production today.

Vinegar is an important ingredient in many food products. The need for large amounts of the vinegar demands industrial fermentation systems that are capable of producing volumes that are reliably controlled. Many technical devices have been developed to improve the industrial production of vinegar. Generally, these improvements increase the speed of the transformation of ethanol into acetic acid in the presence of acetic acid bacteria.

Orleans Process

The slow method of acetifying wine which has been used in France since 1670 is called the French or Orleans process. In this process, alcohol solutions less than 5 per cent in wine cannot be acetified easily. Below this strength, phosphates and nitrogenous substances must be added to the mash and the products have to be sold under the name of "spirit vinegar". The Orleans process was the only method to make pure wine vinegar, and was reported to be the best process to produce fine quality table vinegar. In this process, wood barrels (Figure 13.5) are used and filled with alcohol fermenting liquid to approximately ¾ full. First, holes are drilled at the ends of the barrel a few inches above of the liquid surface. The holes are left open and covered with a fine screen that prevent insects and bugs from entering the cask which will disturb the production. Oxygen is necessary for the production of vinegar due to the fact that the bacteria that turns the solution into vinegar requires oxygen.

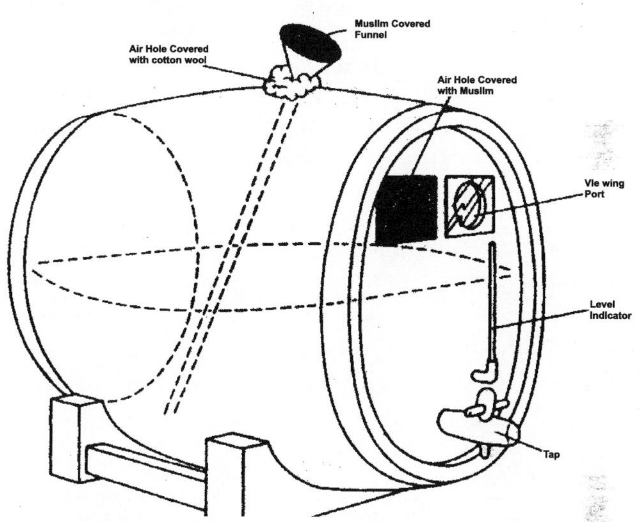

Figure 13.5: The Orleans Process of Vinegar Manufacture.

Secondly approximately 20-25 per cent of fresh vinegar is added into the barrel. The function of adding the fresh vinegar is acidifying the liquid to the point of optimum growth for the vinegar bacteria. Vinegar bacteria settle into the liquid from the air and form a gelatinous slime layer on top of the liquid. The liquid is fermented for about 1 to 3 months at 70°F to 85°F. After this time, 1/4 to 1/3 of the vinegar may then be drawn off for bottling purposes and an equivalent amount of alcoholic liquid added. Alcohol sources must constantly be added to the vinegar or the acetic acid might begin to oxidize.

Generator Fermentation

Early in the nineteenth century, a vinegar-making system called the trickle method [now called generator fermentation or quick process (Schnellessig)] was developed by German chemist Schutzenbach in 1832. According to this process, the bacteria were grown and formed a thick slime coating around a non-compacting material like beech wood shavings, charcoal or coke. The non-compacting material was packed into large 16 upright wood tanks (Figure 13.6) from of 2000 cubic feet capacity above a perforated wood grating floor.

The wood shavings are generally made of air-dried beech wood sliced to form a coil about 2 inches long and 1¼ inches in diameter.

Re-circulated fermenting liquid or mash trickles over the packing material toward the bottom while air moves from the bottom through inlets toward the top. The rate of acetification is dependent upon oxygen concentration. A limited air supply means limited acetic acid production and lower generator temperatures while an overabundant air supply creates over production and higher generator temperatures. The generators must be closely monitored to present over oxidation or unacceptable temperatures. The process takes about 3 to 7 days. Two thirds of the final vinegar product is withdrawn from the tank and fresh mash added. Replacement mash is slowly poured into the tank until the working level for acetification of the solution and a beginning temperature of 70°F (21.1°C) are reached. The optimum temperature for generator operation is 85 to 90°F (30 to 32.2°C). Each gallon of 190 proof alcohol oxidized to acetic acid 18 releases about 30000 to 35000 Btu (32000000 to 37000000 Joules). The optimum temperature for Acetobacter is about 86°F (30°C). A temperature control system is necessary to prevent overheating and consequent inactivation of the bacteria.

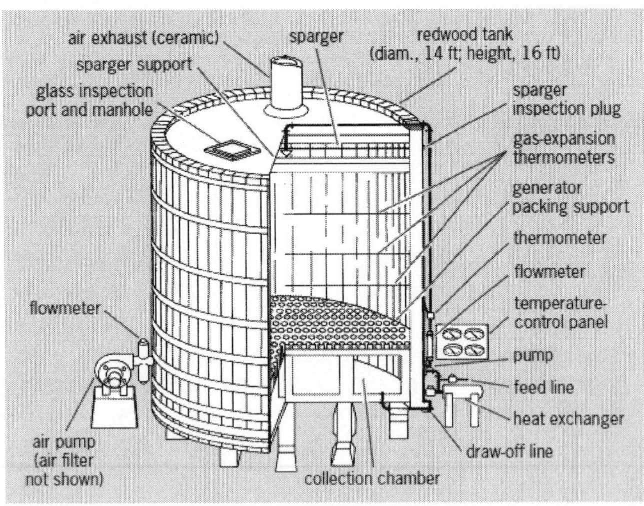

Figure 13.6: The Tricking, Quick Process.

Only a small proportion of the world's vinegar is produced by surface culture today, although it is claimed to produce the finest quality vinegar. More elaborate surface culture techniques based on series of trays have been described but these have received only very limited application.

Submerged Fermentation

Now a days, the most common method used for the fermentation is submerged culture technique having stability to enhance general fermentation attributes (*i.e.*, aeration, stirring, heating *etc.*). This method is the newer, faster, and more efficient methods of creating vinegar. It is mainly used in industrial businesses where the needed equipment is present.

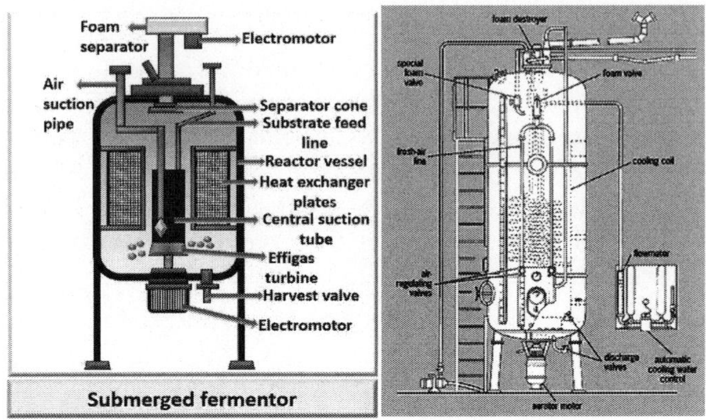

Figure 13.7: The Submerged Fermentation Method.

In this method, a high-speed motor breaks down air that is brought down from a stainless-steel tank into tiny bubbles and is forced into the solution of alcoholic liquid and the bacteria for even faster oxidization. The final steps are filtration and pasteurization of the vinegar to stop any more bacteria growth and enzyme actions. This process usually takes one to two days to process which is why this method is mainly used by big industries.

Dairy Based Fermented Foods

Kefir, yoghurt, cheese, and butter all use bacteria to acidify milk. But, like recipes in the kitchen, the presence of different ingredients–such as the strain of bacteria, other food sources that the microorganisms can use, the temperature, and other factors, all work to produce different products.

Kefir

Kefir is one of those microbe-containing foods that have been part of human history for centuries. Marco Polo spoke of kefir in his epic journeys through the Far East in the 13th century A.D. and in the early 19th century, kefir was in vogue at sanitariums in North America for the treatment of tuberculosis.

The drink originated when Eastern nomadic shepherds accidentally discovered that fresh milk carried in leather pouches would sometimes ferment into a delicious and bubbly beverage. The word kefir likely comes from the Turkish word "Keif" meaning "good feeling", which aptly describes the sense of well-being felt by many having quaffed a glassful of the brew.

The heart of kefir is the group of lactic acid bacteria known as Lactobacillus. In particular, *Lactobacillus acidophilus* is a mainstay of active kefir and other foods including the ever-popular fermented dairy product called yoghurt, as well as sauerkraut. Lactobacillus ferments dairy products and helps preserve other foods because it makes its environment acidic in the absence of alcohol production, via the production of lactic acid. This makes the food taste sour and helps make the environment nasty for the growth of other microbes that would otherwise ruin or contaminate the food.

In the Caucasus Mountains region of the Europe-Asia border between the Black and Caspian Seas, kefir is a popular drink. Perhaps coincidentally, the natives are renowned for their longevity. So, kefir may help put pep in one's step.

Traditional kefir, which originates from the Caucasus Mountains, is a fermented milk drink with a creamy texture, sour taste and subtle effervescence. Kefir varies from other fermented products because of the specific property of its starter: the kefir grains. It is produced by adding a starter culture termed "kefir grains" to milk. Kefir grains range in size from 1 to 4 cm in length and look like small cauliflower florets in shape (irregular and lobed-shaped) and colour (from white to light yellow). This gelatinous and slimy structure is comprised of a natural matrix of exopolysaccharides (EPS) kefiran and proteins in which lactic acid bacteria (LAB), yeasts, and acetic acid bacteria (AAB) co-exist in symbiotic connection. Kefir grains consist of symbiotic lactose-fermenting yeasts (*e.g.*, *Kluyveromyces marxianus*) and non-lactose fermenting yeasts (*e.g.*, *Saccharomyces cerevisiae*, *Saccharomyces unisporus*), as well as lactic and acetic acid producing bacteria, housed within a polysaccharide and protein matrix called kefiran. Lactic acids, flavour-generating components (*e.g.*, acetaldehyde), ethanol and carbon dioxide are all by-products of fermentation and contribute to the organoleptic properties of kefir.

The most pre-dominantly found bacterial species in kefir grains are *Lactobacillus kefiranofaciens*, *Lacticaseibacillus paracasei* (basonym *Lactobacillus paracasei*), *Lactiplantibacillus plantarum*, *Lactobacillus acidophilus*, and *Lactobacillus delbrueckii* subsp. *bulgaricus*. On the other hand, *Saccharomyces cerevisiae*, *S. unisporus*, *Candida kefyr*, and *Kluyveromyces marxianus* ssp. *marxianus* are the predominant yeast species present in kefir. The microbiota of kefir grains may differ depending on the geographical origin of the kefir grains, which are strictly connected to the climate conditions. In fact, the microflora composition in kefir may also differ depending on the substrate used in the fermentation process and culture maintenance method (fermentation time, temperature, degree of agitation, and ratio of kefir grains to substrate). It is recognized that this microbial diversity is responsible for the physicochemical features and biological activities of each kefir, although some major Lactobacillus species always exist because of their probiotic strain-specific properties.

Table 13.4: Microbial Composition of Kefir Samples of different Geographical Locations

Geographic Location of Studied Sample	*Method of Analysis*	*Identified Bacteria or Yeasts*
France, Ireland and the United Kingdom	Metagenomics (16S rRNA and ITS sequencing)	*Lb. kefiranofaciens*, *Leuconostoc* spp., *Lb. helveticus*, *A. pasteurianus*, *Saccharomyces* spp. and *Kazachstania* spp.
Belgium	Metagenomics (16S and 26S rRNA sequencing)	*Lb. kefiranofaciens* or *Lc. lactis*, *Lb. kefiri*, *Acetobacter* spp. and *Enterobacter* spp., *Kl. marxianus*, *Kz. exigua* and *Nauvomozyma* spp.
Malaysia	Metagenomics (16S rRNA sequencing)	*Lb. kefiranofaciens* and *Lb. kefiri*
Italy	Metagenomics (16S rRNA and 26S rRNA sequencing)	*Lb. kefiranofaciens* as dominant and *Lb. kefiri*, *Enterococcus* spp., *Lc. lactis and Acetobacter* spp. as subdominant bacteria, *Dekkera anomalus*, *Kz. exigua*, *S. cerevisiae*
Brazil	PCR-DGGE	*Lb. kefiranofaciens* and *L. kefiri*
	Metagenomics (16S rRNA sequencing)	*Lactobacillus* spp., such as *Lb. kefiranofaciens* subsp. *kefirgranum*, and subsp. *kefiranofaciens*, *Lb. kefiri*, *Lb. parakefiri*, *Lb. parabuchneri*, *Lb. amilovorus*, *Lb. crispatus* and *Lb. buchneri*
Turkey	Metagenomics (16S rRNA sequencing)	*Lb. kefiranofaciens*, *Lb. buchneri*, *Lb. helveticus*
Turkey	PCR-DGGE	*Lb. kefiranofaciens*, *Lb. kefiri*, *Lb. buchneri*, *Lb. sunkii*, *Lb. otakiensis*
	Culture dependent	*Lc. lactis*, *L. mesenteroides*, *Lb. kefiri*
Russia	Culture dependent	*Lb. kefiranofaciens*, *Lb. kefiri*, *Lb. parakefiri*, *Lc. lactis* and *Leuconostoc* spp.

Kefir grain have been reported to be good source of protein, vitamins, minerals, fat, ash with a content of protein (34.3 per cent), fat (4.4 per cent), ash (12.1 per cent), mucopolysaccharide (45.7 per cent); vitamins B and K, calcium, phosphorous and magnesium are also present in kefir grains.

For the production of kefir, different types of milk is used. Researchers used all the types of milk for finding out the different parameters from different kefir samples. This is useful for knowing which milk is better for production of kefir. Basically, milks used are from cow, goat, sheep, buffalo and camel. These all types of milks having their own different nutritional qualities. Raw milk has been traditionally used. According to Tratnik, cow milk has less digestibility than goat milk. Goat milk has better sensory properties than cow milk. He said that, in his paper about the taste and aroma of goat milk. Consumer dislike the taste and aroma of goat milk but they like cow milk. He analysed both cow milk and goat milk at the time of his work. He found different reading of different parameters (Tratnik *et al.,* 2006). In another study, it was reported that, cow milk has lower mineral level than goat milk; they are magnesium, chlorine, potassium. From these, tastes of goat milk get rather salty. It also stated that, the acidity level of cow milk is lower than goat milk.

Types of Kefir

Generally, kefir may well be identified depending on the type of substrate used for fermentation, which are dairy and non-dairy kefir. The majority of the reported kefir studies has been emphasized on the advantages of kefir consumption that used milk substrates for fermentation compared with their non-dairy counterpart. Despite its status as a natural probiotic, the intake of dairy kefir beverage not suitable for lactose intolerant, vegan and dairy-product allergic users. Thus, an alternative method of reaping the health benefits of kefir is through its alteration to non-dairy substrates. The dairy and non-dairy kefir grains are quite similar to each other's in relations of their structure, related microorganisms and their metabolic products during the fermentation procedure. However, the constitution and prevalence of microbial diversity of kefir grains and the concentration of end bioproducts may differ depending on the carbon and energy sources (substrate used) available for grain fermentation.

As an outcome of the diverse microbial constituents that can become proved within kefir grains, varying kefir products with different microbiological, physicochemical, nutritional, and sensorial characterizations of these kefir drinks may be obtained. Nevertheless, both dairy and non-dairy kefir are obtained by inoculating the starter culture, kefir grains, in the substrates at variable ratios (from 1 to 20 per cent w/v) for 18–24 h at 20–25 °C. The fermentation procedure begins when bacteria and yeasts of the kefir grains find the proper culture requirements, which resulted in the increment of 5–7 per cent grains' biomass and the formation of various metabolites. At the end of this procedure, kefir grains may split into new, smaller grains and liberate viable cells into the substrate, wherein the grains are isolated from kefir by sieving and re-use for the next inoculation.

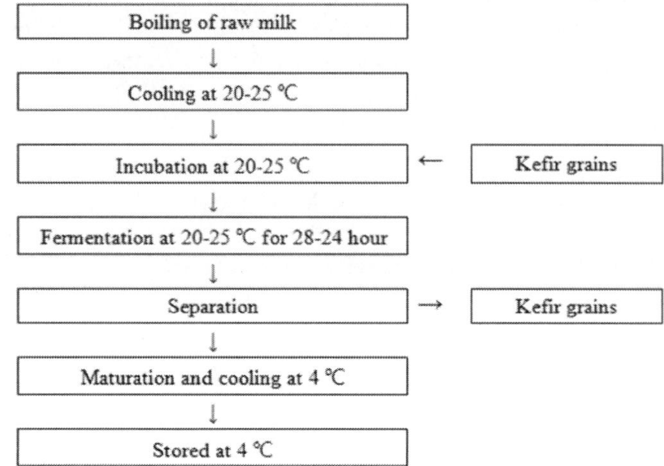

Figure 13.8: Manufacturing of Kefir.

Dairy Kefir

Since the 6th millennium BC, milk has featured in the human diet. In order to increase its shelf life, surplus milk was fermented. Early humans discovered that preserved sour milk maintain their nutrients and is relatively more stable. Kefir is a homemade dairy beverage generated through fermentation of lactose in milk by bacteria and yeasts naturally existent in kefir grains. Traditionally, the fermentation of kefir was performed for 24 h at room temperature in goatskins, clay pots, or wooden buckets. The ruminants' (cows, goats, sheep, camels, or buffalo) milk was applied as the fermentation substrate, in which at the end of this procedure kefir was isolated from the grains by drawing the beverage off.

Kefir can be produced either via the traditional method or a commercial process. The small-scale production calls for direct addition of kefir grains to milk that has been pasteurized and cooled to 20–25 °C. When producing at a large scale, kefir can be manufactured by using commercial starter cultures that are directly inoculated into the milk or by using the "Russian method". This involves a backslopping procedure, a serial fermentation processes that begins with kefir produced from grains that are then used as natural starter cultures for milk fermentation. Although kefir can be produced from different sources of animal milk, kefir made from cow's milk remains the most popular in Eastern Europe. Nevertheless, similar to other fermented milk products, the condition of raw milk is crucial for kefir production. By and large, the raw milk for manufacturing kefir should be: rich in constituent, low bacterial and somatic cell counts, and lack of pathogens or hindering substances, like antibiotics and disinfectant remains.

This yeast-lactic fermented milk produces a viscous and slightly effervescent beverage having a low alcohol content of nearly 0.08–2.0 per cent. Kefir is a non-curdled product and is described by a unique yeast-like flavour and a sparkling sense felt in the mouth. It owns this unique flavour to the symbiotic metabolic activities of several bacterial and yeast species, which include both proteolytic and lipolytic degradation of milk constituents. However, as a result of the complex community of microbiota in kefir grains, type of milk used, and type of manufacturing methods, the fermented kefir product has a highly diverse flavour, composition and metabolic products. The typical composition of kefir contains 80–90 per cent moisture, 0.2 per cent lipid, 3.0 per cent protein, 6.0 per cent sugar, and 0.7 per cent ash, and approximately 1.0 per cent lactic acid and alcohol. This fermentation process also produces metabolic products such as lactic and acetic acids, carbon dioxide, ethanol, acetaldehyde, acetoin, and other volatile compounds, minerals, essential amino acids, vitamins, folic acid, bacteriocins, bioactive peptides and some nutraceutical components.

Non-Dairy Kefir

Currently, kefir or kefir-like beverages are consumed throughout the globe, with information of kefir making in Ireland, Spain, Turkey, Malaysia, Indonesia, Tibet, and North and South America. Water kefir, sugary kefir or tibico (tibico's tepache) became very popular during the 20th century due to the health advantages associated with its intake. Non-dairy kefir is a beverage made from the fermentation of kefir grains with a sugary solution, wherein the brown sugar solution is the main alternative substrate used for kefir fermentation.

Other non-dairy kefir prepared from fruit juices (apple, pineapple, grape, quince, kiwi, pear, pomegranate, melon, strawberry, tomato, coconut), vegetables (ginger, onion, soybean, fennel, carrot), and molasses (sugarcane, honey) are also suitable alternative substrates for the non-milk adaptation of kefir production. These adaptations came about to allow non-dairy consuming and vegan individuals to reap the benefits of drinking kefir. Regular daily intake of vegetables and fruits is strongly advocated for numerous positive health effects and disease prevention. Thus, the production of fruit or vegetable juice-based fermented kefir beverage with may be perceived by consumers as healthy and provides an extra method to boost fruit and vegetable intake. The non-dairy kefir fermentation is carried out by the kefir grains consisting of a consortium of yeasts mainly *Kluyveromyces*, *Candida* and *Saccharomyces*, and lactic acid bacteria (LAB), including the genera *Lactobacillus*, *Lactococcus*, *Leuconostoc,* and *Streptococcus*, embedded in a natural matrix of exopolysaccharides (EPS) kefiran. Various species are found to have symbiotic associations and live or proliferate by sharing their bioproducts as energy supplies or growth-inducing factors, that may vary depending on non-dairy substrates used during kefir fermentation.

Indeed, sugary water or fruit juices contain water, sugar, and a mixture of nutrients; proteins, amino acids, vitamins and minerals that are suitable to prepare fermented beverages like kefir as they provide an ample medium for microbial expansion that could promote a fast rise of kefir grain biomass. Non-dairy kefir beverages are traditionally produced by directly adding kefir grains to the pasteurized and cooled substrate and incubated for around 24 h at 25–30 °C. At the completion of fermentation, the grains are isolated from kefir by sieving, followed by washing, drying at room temperature and storage in a cooling tank for the next round of fermentation procedure. The chemical composition and sensory feature of non-dairy kefir beverages vary corresponding to the substrate used, including sugars (sucrose, glucose, and fructose), organic acids (lactic, acetic, citric, tartaric, butyric, malic, and propionic acids), alcohols (ethanol, hexanol, and glycerol), and esters (ethyl propionate, ethyl hexanoate, octanoate, and decanoate). These metabolites afford distinctive flavour qualities for these products, such as revitalizing taste (due to presence of ethanol), fruity fragrance (due to presence of esters) and body and texture (due to presence of glycerol) that resulted in favourable consumer response for all analyzed products.

Kefir Consumption

In Russia, USA, Japan and Central and Northern Europe, *kefir* has been used in the control of many diseases due to its nutritional and therapeutic aspects. Recently, a new formulation of *kefir* with the addition of enzymes such as lipase or α-amylase to prevent and control obesity was patented in Japan. In the former Soviet Union countries, the consumption of *kefir* has been informally recommended for healthy individuals to reduce the risk of chronic diseases and also for patients with gastrointestinal and metabolic disorders, hypertension, IHD, weight control and allergies.

The industrial production of *kefir* is large in Germany, Austria, France, Luxembourg, Norway, Switzerland, Czech Republic, Slovakia, Poland and Israel. In Brazil, the consumption of *kefir* occurs domestically with spontaneous fermentation of *kefir* grains in milk, without the control of the time or temperature of fermentation. The consumption and industrial production of *kefir* on a larger scale are lacking so far.

Health Benefits Associated with Kefir Consumption

Kefir has a wide spectrum of important health benefits, including physiological, prophylactic and therapeutic properties. These effects are a result of a wide variety of bioactive compounds produced during the fermentation process and the highly diverse microbiota, which act either independently or synergistically to influence these health benefits.

Antimicrobial Properties

Kefir has an antibacterial effect against many pathogenic organisms due to the inherent formation of organic acids, hydrogen peroxide, acetaldehyde, carbon dioxide, and bacteriocins. For example, 3.5 kDa bacteriocin was identified from *Lactobacillus plantarum* ST8KF in kefir. Besides this, hydrogen peroxide is another metabolite produced by some bacteria as an antimicrobial compound. The previous showed that all 21 isolates of lactic acid bacteria from Turkish kefir produced hydrogen peroxide (0.04-0.19 μg/ml). In a later paper, they reported that 11 out of 21 strains of kefir *Lactococci* produced hydrogen peroxide. All Lactococci strains were effective in inhibiting the growth of *Streptococcus aureus*, but were less effective against *Escherichia coli* NRLL B-704 and *Pseudomonas aeruginosa.* These substances also exhibit some effects similar to those of nutraceuticals, preventing gastrointestinal disorders and vaginal infections.

Furthermore, Santos *et al.* (2003) stated that the bacteriocin named lacticin 3147, which was produced by *Lactococcus lactis* strain DPC3147 isolated from kefir grains, had antimicrobial activity against *Escherichia Coli, Listeria monocytogenes, Salmonella typhimurium, S. enteritidis, S. flexneri*, and *Yersinia enterocolitica*.

Moreover, many scientists stated that kefir and its exopolysaccharide, kefiran, had antimicrobial activity. Both were reported to exhibit significant antibiotic activity against Gram-positive and Gram-negative bacteria as well as yeast, Candida albicans. Kefiran protected against *Bacillus cereus* B 10502 damage to Caco-2 cells when introduced at a concentration range from 300 to 1000 mg/l. Hence, kefir may be a good antimicrobial agent in food technology for food safety. More research related to this subject has still to be performed, in order to put the antimicrobial activity of kefir into practice for food technology.

Kefir has been shown to lose its intrinsic inhibitory effect after lyophilisation and re-constitution in distilled water or milk. In an *in vitro* study, antimicrobial activity of kefir grains and kefir suspension was tested against several species of bacteria and fungi and observed higher inhibitory action against *Streptococcus faecalis* and *Fusarium graminearum*. The concentration of kefir from 7 to 10 per cent (w/w) was able to completely inhibit the sporulation of *Aspergillus flavus*, and consequently the production of aflatoxin B1, which demonstrates the antifungal properties of kefir against filamentous fungi. The organic acids produced during fermentation of kefir can change the molecule aflatoxin B1, by converting it into less toxic forms such as aflatoxicol aflatoxin B and B2. In this context, kefir appears as a safe alternative for food preservation, providing protection against poisoning from aflatoxin B1.

Additionally, kefir was able to increase the population of LAB and reduce the levels of *Enterobacteriaceae* and *Clostridium* in the intestinal mucosa of mice. Oral administration of milk kefir or soya milk kefir in mice over a period of 28 d was able to significantly increase *Lactobacillus* and *Bifidobacterium* while reducing *Clostridium perfringens* in animal faeces. Thus, the antimicrobial activity of kefir is differentiated when the fermented milk is used in the reconstituted or liquid form, and also when kefir grains are used. However, it is noted that both kefir grains and fermented milk constitute interesting alternatives that can be used for the prevention of some infections, especially those involving the gastrointestinal tract. It is necessary to perform *in vivo* studies to investigate the antimicrobial properties of kefir, especially with regard to the reduced infection rates and severity of symptoms with kefir consumption in animal and human studies, since the results of *in vitro* studies conducted until now show promising results.

Antioxidative Effect

Harmful biological effects of reactive oxygen species *in vivo* are controlled by a broad spectrum of antioxidant defence mechanisms, including dietary compounds and enzymes with antioxidant activity. In a toxicity test with carbon tetrachloride (CCl_4) in rodents, *kefir* exerted a higher antioxidant effect than vitamin E. Additionally, a study evaluated the effect of *kefir* supplementation in rodents induced to oxidative stress by the use of Pb. After 6 weeks of treatment, the consumption of *kefir* increased glutathione peroxidase and reduced malondialdehyde to levels comparable with those of the non-induced group. The results support the hypothesis that *kefir* is a potential tool in the control of oxidative stress. The study evaluated the antioxidant activity of *kefir* prepared from goat and cow milk was evaluated and it was reported that the kefir had great ability to bind the 1,1-diphenyl-2-picrylhydrazyl (DPPH) radical and superoxide radicals, besides the inhibition of linoleic acid peroxidation. In this situation, the antioxidative activity of *kefir* can reduce DNA damage, which explains its anticarcinogenic potential.

It is known that the increased concentration of free radicals has a strong relationship with an increased risk of chronic diseases. Therefore, *kefir* consumption should be encouraged since it is a natural source of antioxidant compounds and also stimulates the activity of enzymes of the antioxidant system.

Hypocholesterolaemic Effect

The consumption of probiotic dairy products has been proposed as a strategy to reduce levels of circulating cholesterol. Guo *et al.* (2011) in a meta-analysis with thirteen trials, including 485 participants with high, borderline high and normal cholesterol levels, observed that the consumption of probiotic dairy products was able to lower serum cholesterol (mean net change of 6·40 mg/dl; 0·17 mmol/l), LDL-cholesterol (mean net change of 4·90 mg/dl; 0·13 mmol/l) and TAG (mean net change of 3·95 mg/dl; 0·04 mmol/l) concentrations. Some mechanisms are proposed to justify these findings:

1. The LAB inhibit the absorption of exogenous cholesterol in the intestine due to binding and incorporation of cholesterol by the bacterial cells. The high count of LAB present in kefir may directly or indirectly reduce cholesterol in the medium by up to 33 per cent. Vujièiæ *et al.* (1992) verified that after 24 h of fermentation, kefir cultures were able to absorb from 28 to 65 per cent of the cholesterol present in the culture medium.
2. Probiotic bacteria increase the production of SCFA. Among the different SCFA produced, propionate reduces the production of cholesterol by inhibiting hydroxymethylglutaryl CoA (HMG-CoA) reductase activity. Additionally, plasma cholesterol is redistributed to the liver, where the synthesis and secretion of bile acids are increased, since the activity of the 7α-hydrolase enzyme is stimulated. Moreover, propionate inhibits the intestinal expression of genes involved in the biosynthesis of cholesterol.
3. Another possible pathway involves the deconjugation of bile acids, which may be increased in the large intestine, caused by the bile salt hydrolase (BSH) enzyme. The BSH enzyme catalyses the hydrolysis of glycine and/or taurine conjugated to the bile salts in residual amino acids and free bile salts, increasing excretion. With the increasing excretion of bile salts, fewer of them are carried back to the liver by the enterohepatic circulation, which increases the demand for cholesterol for de novo synthesis of bile salts in the liver. Thus, the liver increases the hepatic uptake of LDL from the circulation which leads to the reduction of serum LDL-cholesterol concentrations.

Some animal studies have demonstrated the hypocholesterolaemic effect of kefir. Hamsters fed a hypercholesterolaemic diet supplemented with freeze-dried kefir (milk or soya milk) showed a significant reduction in TAG concentration and in the atherogenic index. In this study, the effects were partially related to increased faecal excretion of neutral sterols and bile acids. Also, *Lactobacillus plantarum* MA2 isolated from kefir grains originated from Tibet was effective in reducing plasma and liver cholesterol and TAG concentrations. This micro-organism was also able to increase faecal excretion of cholesterol and TAG in mice fed a high-fat diet.

Animals that consumed hyperlipidaemic diets associated with kefiran showed a reduction in serum total cholesterol, LDL-cholesterol and TAG concentrations, as well as a reduction in liver cholesterol and TAG concentrations compared with controls. Uchida *et al.* (2010) evaluated the anti-atherogenic effect of kefiran in rabbits fed a high-cholesterol diet and observed lower atherosclerotic lesion in the abdominal aorta and lower concentrations of hepatic cholesterol and lipid peroxidation in the animals fed kefiran in comparison with the control group.

The consumption of kefir (0·5 litres/d) by hypercholesterolaemic adult men for 4 weeks did not affect the circulating concentrations of total cholesterol, HDL-cholesterol, LDL cholesterol or TAG; however, it increased the SCFA concentrations in their faeces (St-Onge *et al.*, 2002). *Ostadrahimi et al.* (2015) conducted a double-blind randomised placebo-controlled clinical trial with diabetic patients, who consumed 600 ml of kefir daily for 8 weeks. Kefir consumption was not able to influence serum TAG, total cholesterol, LDL-cholesterol and HDL-cholesterol levels compared with the control, showing that kefir was unable to reduce plasma lipids in diabetic patients. The benefits of kefir in lowering cholesterol have shown conflicting results. Such inconsistent results may be due to different experimental protocols used, origin of the grains, and fermentation conditions of kefir, and consequently the variety of kefir composition. This scenario may be related to lack of standardisation in nutritional and microbiological composition of kefir used in scientific research.

Improvement in Lactose Intolerance

Lactose maldigestion is the inability to completely digest lactose, the major carbohydrate in virtually all mammalian milks. Lactose maldigestion affects 75 per cent of adults in the world and occurs most often as the result of a genetically programmed decrease in intestinal lactose activity after the age of 3 to 5 years. This enzyme, naturally present in kefir grains, reduces lactose content of the kefir during fermentation, which in turn makes the final product suitable for individuals with lactose intolerance. Moreover, fermented products such as kefir are characterised by a delayed gastric emptying, which helps in lactose digestion.

Hertzler and Clancy (2003) found that the consumption of kefir, which is similar to yoghurt, was able to improve lactose digestion and tolerance in healthy adult subjects clinically diagnosed with lactose intolerance. In this study, yoghurt and kefir were similarly able to reduce the severity of flatulence related to milk by 54 per cent to 71 per cent.

According to Alm (1982), after the fermentation period, kefir has a reduction of 30 per cent in the content of lactose, compared with the unfermented milk, providing better comfort for individuals with lactose intolerance. In addition, enzymes released from the lysed micro-organisms may aid in lactose digestion in the gut in a similar manner to most probiotic preparations containing LAB.

It has also been shown that fermented milk products have a shorter transit time than milk, which may further improve lactose digestion. Furthermore, Rattray and Connell (2011) found that kefir with a diverse microbial population invariably has some degree of β-galactosidase activity, that converts lactose into glucose and galactose, which can then be easily digested. In addition, Steven *et al.* (2003) reported that although it seems plausible that kefir might improve lactose digestion in a manner similar to yoghurt, there is a lack of research to support such a claim. Kefir contains different starter culture microorganisms from yoghurt and the bile acid sensitivity, galactosidase activity, or lactose transport of these organisms may be different. This was the first result found, which demonstrated that plain kefir improved lactose digestion just as well as plain yoghurt.

It is important to note that there are only a few studies on kefir concerning lactose intolerance, and more work is needed to better understand the effects of kefir consumption and its possible effectiveness in reducing the unpleasant symptoms of lactose intolerance in humans. The amount and regularity of consumption of kefir to perform these desirable effects should also be studied.

Control of Plasma Glucose

The regular consumption of probiotics has the ability to improve blood sugar levels. This effect has been attributed mainly to the probiotic ability to positively modulate the composition of the intestinal microbiota and hence reduce intestinal permeability, oxidative stress and inflammation. Similar effects can be observed with regular consumption of kefir.

Different milk sources that were used for kefir fermentation also affect the anti-diabetic activity of kefir. The results obtained by Nurliyani *et al.* (2015) on the anti-diabetic potential of kefir combination from goat milk and soy milk in rats induced with streptozotocin-nicotinamide, demonstrated that diabetic rats fed with the kefir combination had lower plasma glucose than rats fed with goat milk or soy milk kefir alone.

Hadisaputro *et al.* (2012) evaluated the effect of kefir consumption for 30 d in controlling glycaemia in Wistar rats induced to diabetes mellitus by administration of streptozotocin. Kefir supplementation was able to reduce plasma glucose compared with the control group. In a clinical trial, diabetic adults that consumed 600 ml/d kefir, for 8 weeks, showed a significant decrease in fasting glucose levels and glycosylated Hb compared with baseline. Additionally, these same parameters were found to be significantly reduced in individuals who consumed kefir compared with control subjects who consumed a conventional fermented milk (Ostadrahimi *et al.*, 2015).

The regular intake of probiotics can reduce the amount of Gram-negative bacteria in the intestinal lumen, and therefore lower the amount of lipopolysaccharide (LPS). In addition, probiotics can improve intestinal barrier function leading to the reduction of intestinal permeability. The absorption of lower amounts of LPS may therefore diminish the low-grade chronic inflammatory process characteristic of diabetes. Moreover, lower LPS may restore the function of insulin receptors leading to a better control of blood glucose.

Administration of kefir daily intake with metformin among the newly diagnosed type-2 diabetic adult male patients in the Gaza Government, has shown significant differences in some blood biochemical parameters (El-Bashiti *et al.*, 2019). The results demonstrated a decrease in fasting blood sugar (FBS), glycohemoglobin (HbA1c), and phosphorus with an increase in 9 of 26 calcium among diabetic adult males upon kefir intervention. Furthermore, a study on the total incremental plasma glucose area under the curve (iAUC) for strawberry kefir, orange kefir, and low-fat plain kefir showed kefir as a low- to moderate-glycemic index (GI) food, which indicated its suitability for individuals with diabetes (Kong *et al.*, 2012). We can thus conclude that kefir might be used in the prevention of diabetes; however, more studies are needed to demonstrate such effects.

Anti-hypertensive Effect

Some evidence indicates that probiotic bacteria or their fermented products play an important role in controlling blood pressure. The anti-hypertensive effects have been observed in experimental and clinical studies, although the data are limited and controversial. Kefir is able to inhibit the activity of angiotensin-converting enzyme (ACE) through the action of bioactive peptides generated from casein during the milk fermentation process. According to Maeda *et al.* (2004), the antihypertensive activity observed in their study was due to the ability of kefiran to inhibit ACE activity.

The ACE-inhibitory peptides inhibit the production of the vasoconstrictor angiotensin I, and consequently the production of aldosterone, a hormone that stimulates the increase of serum Na concentration, causing an increase in blood pressure. Additionally, ACE-inhibitory peptides also inhibit the breakdown of bradykinin, a hormone that has vasodilating action, contributing to the decrease in blood pressure. Experimental and especially clinical studies that have evaluated the antihypertensive effect of milk kefir are rare in the literature to date. Furthermore, the milk kefir peptides that exhibit the ability to inhibit ACE action have not yet been identified.

Anti-inflammatory Properties

Worldwide, the complications of neuroinflammatory diseases and inflammation in chronic disorders constitute to the primary cause of morbidity and mortality. Over the past years, growing evidence from both *in vitro* and *in vivo* studies displayed conclusive anti-inflammatory and immunomodulatory potentials, where kefir treatment proved to elevate the anti-inflammatory mediators while downregulating the pro-inflammatory cytokines. The immunomodulatory properties of kefir may result from direct action of the microbiota or may be indirect, through different bioactive compounds produced during the fermentation process. The bioactive peptides, produced during milk fermentation by the microbiota present in kefir, are able to activate macrophages, increase phagocytosis, suppress the Th2 immune response, increase the production of NO and cytokines, and stimulate the secretion of IgG and IgA by B lymphocytes in the intestinal lumen.

Vinderola *et al.* (2005) studied the immunomodulatory effect of the exopolysaccharide produced by *Lactobacillus kefiranofaciens* in the intestinal mucosa level in mice by examining the cytokines and immunoglobulins profiles. The oral administration of exopolysaccharide in mice resulted in the gut mucosal response through elevating the production of IgA at both the small and large intestines, and inducing systemic immunity through the production of cytokines in the intestinal fluid and blood serum. These bioactive compounds were able to promote the cell-mediated immune response against infections and intracellular pathogens.

The anti-inflammatory potential of kefir was evaluated in an animal model of asthma, sensitised with ovalbumin. The administration of kefir (50 mg/kg) was found to significantly inhibit the total number of inflammatory cells and eosinophils in the bronchoalveolar fluids. In addition, kefir administration decreased IL-4, IL-13 and IgE to a normal level. Thus, kefir has therapeutic potential for the prevention of allergic bronchial asthma.

Rodrigues *et al.* (2005) evaluated the anti-inflammatory action of kefir in rats using a protocol of oedema and granuloma induction. In this study, water kefir, milk kefir and kefiran extract inhibited the inflammatory process by 41, 44 and 34 per cent, respectively. The treatments also significantly reduced oedema in the animals. The results demonstrate the presence of anti-inflammatory compounds in the symbiotic cultures of kefir. The immunomodulatory effect of kefir can be attributed to the ability of this probiotic to decrease or restore intestinal permeability. Thus, the contact between the host and the antigens present in the intestinal lumen is decreased, which in turn can reduce the inflammatory response. In this situation, kefir may be able to reduce intestinal permeability against food-borne antigens.

Liu *et al.* (2006) observed that animals treated with ovalbumin and consumed milk and soya milk kefir, during 28 d, exhibit lower concentrations of IgE and IgG than the control animals. These results suggest the potential of kefir in the prevention of food allergy and in the improvement of mucosal resistance against pathogen infection. The effect of kefir is not restricted to the modulation of the immune system in the gastrointestinal tract but goes far beyond it. Such effect is a consequence of the micro-organisms and bioactive compounds present in kefir, which positively modulate the composition of the intestinal microbiota and consequently the immune system of the host.

Antioxidative Activity of Kefir

Antioxidants are free-radical scavengers. They prevent damage produced by unstable molecules or free radicals which the body generates under stress and other environmental pressures. Kefir has strong antioxidant potentials, and it has been proven in both *in vitro* and *in vivo* models. In a toxicity test with carbon tetrachloride (CCl4) in rodents, kefir exerted a higher antioxidant effect than vitamin E. Additionally, Ozcan *et al.* (2003) evaluated the effect of kefir supplementation in rodents induced to oxidative stress by the use of Pb. After 6 weeks of treatment, the consumption of kefir increased glutathione peroxidase and reduced malondialdehyde to levels comparable with those of the non-induced group. The results support the hypothesis that kefir is a potential tool in the control of oxidative stress.

Liu *et al.* (2005) examined the antioxidant and antimutagenic properties of kefir fermented in milk and soymilk. The study showed that milk-kefir and soymilk-kefir resulted in a significantly enhanced antimutagenic activity compared to the milk and soymilk controls. The milk-kefir and soymilk-kefir also showed a greater DPPH scavenging activity, decreased glutathione peroxidase activity, and inhibition effect upon linoleic acid peroxidation. The study also demonstrated that fermentation of kefir grains in milk and soymilk did not change the superoxide dismutase activity and ferrous ion chelating ability of the original materials.

Liu *et al.* (2006) evaluated the antioxidant activity of kefir prepared from goat and cow milk. The authors reported the great ability of kefir to bind the 1,1-diphenyl-

2-picrylhydrazyl (DPPH) radical and superoxide radicals, besides the inhibition of linoleic acid peroxidation. In this situation, the antioxidative activity of kefir can reduce DNA damage, which explains its anticarcinogenic potential. It is known that the increased concentration of free radicals has a strong relationship with an increased risk of chronic diseases. Therefore, kefir consumption should be encouraged since it is a natural source of antioxidant compounds and also stimulates the activity of enzymes of the antioxidant system.

A study by Yilmaz-Ersan *et al.* (2018) reflected that kefir samples fermented using kefir grains exhibited better antioxidant potency, which was measured by the 2,2-diphenyl-1-picrylhydrazyl (DPPH) and 2,20-azino-di(3-ethylbenzthiazolin-sulfonate) ABTS assays, than kefir samples fermented by started cultures. The study implied that ewe milk kefir revealed an antioxidant effect in the ABTS assay.

The isolated exopolysaccharide from Tibetan kefir grains during milk fermentation showed high antioxidant activities in *in vitro* and concentration-dependent protection of protein from oxidative injury (Chen et al., 2015). On the other hand, Sabokbar *et al.* (2015) discovered that the addition of kefir to apple juice increased the total phenolic content and antioxidant activities analyzed using diphenyl-2- picrylhydrazyl radical scavenging, reducing power, metal chelating effect, inhibition of linoleic acid autoxidation and inhibition of ascorbate autoxidation assays.

Bensmira and Jiang (2015) reported that kefir fermented in peanut milk demonstrated enhanced antioxidant effects versus peanut milk alone, suggesting the kefir grain's fermentation impact on peanut milk's efficacy. Likewise, kefir fermented with a mixture of cow and soy milk showed improved antioxidant activity compared to kefir fermented with cow's milk alone, and the study found an increased level of phenolic compounds after the fermentation.

Anticarcinogenic Effect

The regular consumption of kefir is able to positively modulate the composition of the intestinal microbiota and immune system of its host. Therefore, it is believed that this fermented milk may play an important role in the modulation of carcinogenesis. Cancer is the second leading cause of death globally, and the burden continues to grow in low- and middle-income countries to have access to timely quality diagnosis and treatment. It is known that genetic factors play a large part in cancer risk. However, it is reported that as much as 50 per cent of cancers may be preventable through various lifestyle modifications, including practicing a healthy eating lifestyle. Therefore, the probiotics dietary aspects of kefir are vital as a potential coadjutant treatment or prevention in cancer. The anti-carcinogenic role of kefir and the fractions of kefir can be related to the prevention of cancer and retardation of tumour growth by apoptosis, immune response, modulation of intestinal microbiota, decreased tumour growth and DNA damage, anti-oxidative process, and inhibition of proliferation, and activation of pro-carcinogens (Sharifi *et al.*, 2017).

The anti-carcinogenic effect of kefir and kefir fractions was studied for different cancer types, such as hematological cancers (leukemias and lymphoma), breast cancer, gastrointestinal system cancers (gastric and colorectal), and sarcoma (connective tissue tumour). In 2002, Liu *et al.* (2002) performed an *in vivo* oral treatment of milk kefir and soymilk kefir in mice inoculated with sarcoma. They found that both types of kefir have resulted in significant suppression of tumour growth through stimulation of apoptotic cell lysis in tumours and a significant rise in IgA levels in mice, proposing that both kefir have anti-cancer attributes and have enhanced the mucosal resistance to gastrointestinal infection after 30 days of consumption.

Another study has shown the immunoregulatory ability of kefir or kefir cell-free fraction (KF) on the immune response in mammary glands to delay tumour growth on a murine hormone-dependent breast cancer model. The results demonstrated the relationship between immune and endocrine systems as shown by the activation of immune cells (increased the number of IgA(+) cells, and decreased the Bcl-2(+) cells) and production of cytokines (increased IL-10, and decreased IL-6).

Study in Syria demonstrated that several kefir namely alkaline kefir (AK), exopolysaccharides (EPS) and alkaline exopolysaccharides (AEPS) which were produced by a novel method at different concentrations exhibited anti-cancer properties against human sarcoma cells *in vitro* as reflected by the initiation of apoptosis. The novel method in this study was described as the new approach in the preparation of kefir being fermented in goat colostrum which involved extensive processes. The study showed that AK was superior for stimulation of apoptosis in sarcoma cells as compared to kefir and other kefir products.

The anti-carcinogenic outcome of kefir and kefir fractions had been demonstrated on gastrointestinal system cancers. Gao *et al.* (2014), studied the anti-proliferative activity of cell-free fraction of Tibetan kefir on human gastric cancer cell SGC7901 *in vitro*. They found that SGC7901 cells treated with kefir were impeded in the G1/S phase, and both early and late apoptotic cells could be distinguished. It was also reported that the induction of apoptosis was affected via the up-regulation of bax and downregulation of bcl-2. In addition, Khoury *et al.* (2014) showed the ability of kefir to inhibit proliferation and induce apoptosis in HT-29 and Caco-2 colorectal cancer cells. Another study on the effect of *Lentilactobacillus kefiri*

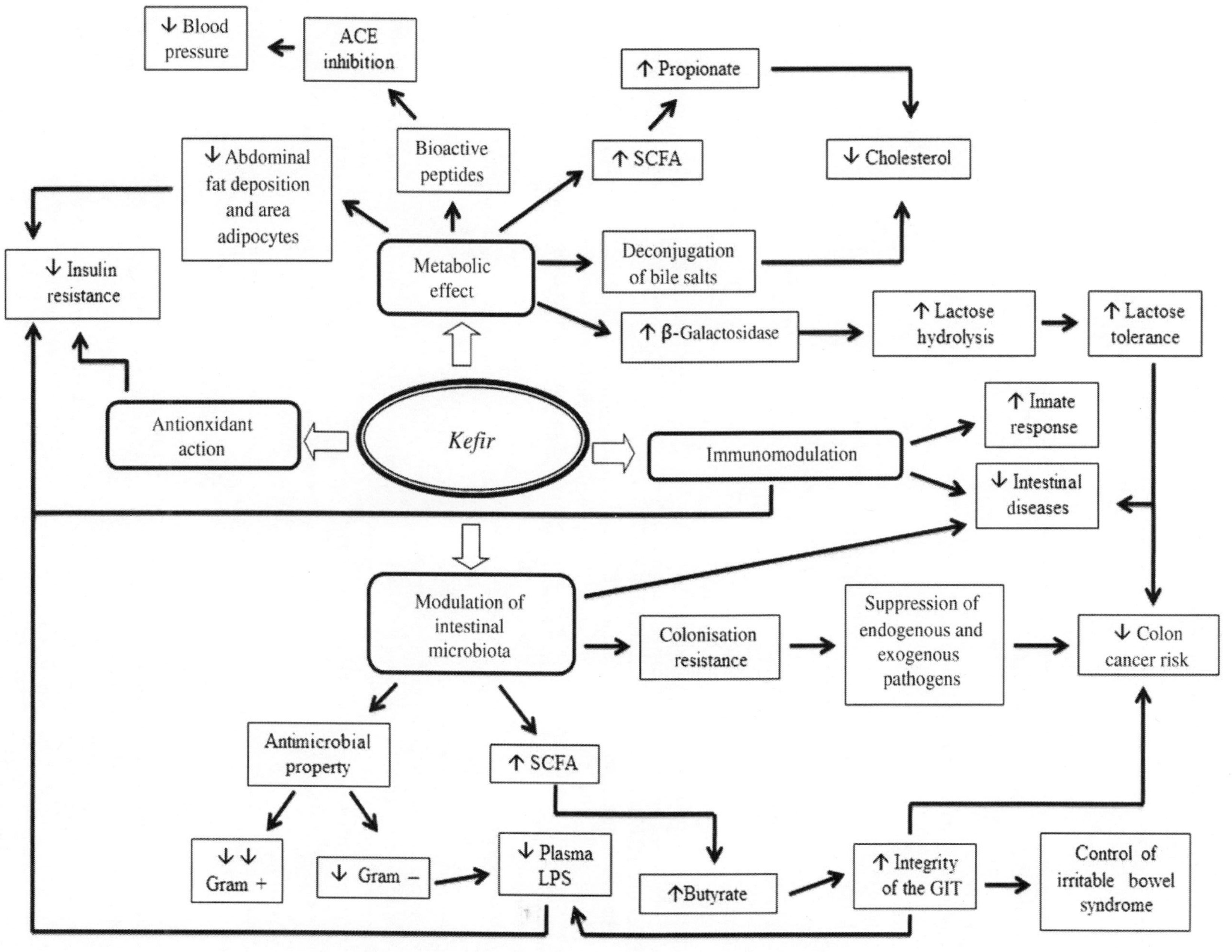

Figure 13.9: Schematic Diagram of the Beneficial Physiological Effects of Kefir on Human Health.

ACE: Angiotensin-converting enzyme; LPS: Lipopolysaccharide; GIT: Gastrointestinal tract.

on the apoptosis induction on gastric cancer cells (AGS) showed its association with the decreased polarization of mitochondrial membrane potential (MMP) and decreased Bcl2 expression, which suggested its therapeutic potential for the treatment of gastric cancers.

In a recent study, Riaz Rajoka *et al.* (2019) characterized the isolated EPS produced by *Lentilactobacillus kefiri* MSR101 (MSR101 EPS) and studied their potential to block colon cancer (HT-29) cells growth. They found that MSR101 EPS illustrated anticancer activity against the HT-29 cancerous cell line and up-regulated the expression of Cyto-c, BAX, BAD, caspase3, caspase 8, and caspase 9. Moreover, the aqueous extract of kefir exhibited potent anti-oxidative activity in UVC-irradiated human melanoma HMV-1 cells and resulted in a remarkable decline in the intracellular reactive oxygen species (ROS) with enhanced DNA repair factors (thymine dimer repair-enhancing activity). The use of kefir supernatant as an adjuvant for doxorubicin (DOX) chemotherapy has also been studied owing to its chemo-sensitizing effects on multidrug-resistant (MDR) human colorectal cancer cells (HT-29). The results revealed that kefir and DOX enhanced the intracellular accumulation of ROS production in HT-29 MDR-developed cells and downregulation of ABC transporters. These outcomes were facilitated by the inactivation of ERK1/2 and NF-κB, and the activation of JNK. Such results indicate that kefir is potentially useful as a non-toxic adjuvant chemotherapy drug in multidrug-resistant developed cells. Altogether, these studies revealed the capability of kefir and kefir fraction as anti-cancer coadjutant in therapy.

Healing Action of Kefir

Recent studies have explored the beneficial effects of probiotics far beyond the intestine. Some of these novel benefits include healthier skin, improvement of eczema, atopic dermatitis and burns, healing of scars, and rejuvenation. Rodrigues *et al.* (2005) tested the scar ability of a 70 per cent kefir and kefiran gel in skin wounds infected with *Staphylococcus aureus* from Wistar rats. The treatment with kefir and kefiran for 7 d showed a protective effect on connective tissue, greatly improving tissue healing compared with treatment with 5 mg/kg neomycin–clostebol emulsion.

The healing properties of kefir were tested in an animal model with experimental burn and contamination with *Pseudomonas aeruginosa*. Kefir grains and gels prepared with kefir culture after 24, 48 and 96 h of incubation were evaluated. After 2 weeks of treatment, the wound area and the percentage of inflammation were reduced in animals treated with kefir grains and gel compared with those treated with silver sulfadiazine cream, used for the topical treatment of burns of second and third degrees. In addition, the percentage of epithelialisation and healing in animals treated with kefir was also improved. It was concluded that treatment with kefir gel was effective in improving outcomes from a severe burn compared with conventional treatment. The ability of kefir to heal wounds can result from its antimicrobial and anti-inflammatory activities, which may act synergistically contributing to the healing (Huseini *et al.*, 2012). Thus, the beneficial health effects provided by kefir go far beyond the gastrointestinal tract, contributing to wound healing.

Dahi (Curd)

Dahi or Indian curd is a set-type fermented dairy product originated in the Indian Subcontinent. The use of dahi has been prevalent since Vedic times, and it is mentioned in ancient scriptures like Vedas, Upanishads and various hymns. It is consumed by large sections of the population throughout the country, either as a part of the daily diet or as a refreshing beverage. Being one of the oldest, most popular and widely consumed fermented milk of India, dahi, resembles Western yoghurt in most aspects. Dahi is considered a yoghurt like-product. However, dahi and yoghurt are different regarding flavour. Dahi flavour is a result of diacetyl metabolite activity of mixed mesophilic lactococci strains. In contrast, acetaldehyde flavour is the main characteristic of yoghurt.

It is largely made at home using traditional kitchen recipes or in the confectionary (Halwais) shop., involving milk of buffaloes, cows and goats. Generally, a mixture of cow and buffalo milk is used. In the household, milk is boiled and cooled to body temperature, inoculated with 0.5-1 per cent dahi starter, usually the left over from the previous day's stock, and incubated undisturbed at ambient temperature for five to six hours or overnight until it acquires a thick consistency. A woollen cloth or a straw or jute bag is used to wrap the vessel in winter season, and it is kept at ambient temperature overnight until the curd is formed. After firm setting of curd, it is transferred to a cooler place or stored under refrigeration.

In the confectionary shops, method employed for preparation of dahi is more or less same except that the milk is concentrated in an open pan before inoculation and usually dahi is set in earthenware.

Dahi may contain additional cane sugar. It should have the same minimum percentage of fat and solids not-fat (SNF) as the milk from which it is prepared. Where dahi or curd, other than skimmed milk dahi, is sold or offered for sale without any indication of the class of milk, the standards prescribed for dahi prepared from buffalo milk shall apply in India.

Standard Operating Procedures (Sop's) For Dahi Manufacturing at The Industrial Level

1. **Raw Milk Reception**: The raw milk received in tankers/food grade barrels is weighed either in weighbridge or in weighing bowl batch wise and

subject to sampling and testing for its parameter as per specified and defined procedures.

2. **Filtration**: The accepted milk is weighed and unloaded in the Dump Tank and Pumped through a chiller after properly filtering, such milk is stored in the silos through the previously cleaned, sterilized/steamed pipe line, and silos act.
3. **Chilling**: Filtered milk is chilled through a chiller ensuring the temperature not more than 5 deg. Chilled milk is stored in the silos through the previously cleaned, sterilized/steamed pipe line, silos *etc.*
4. **Standardization of Milk**: Milk Pasteurization and Standardization of milk is planned as per product requirement and fat and SNF is maintained.
5. **Homogenization and Pasteurization**: Homogenization of milk is done at 150/50 bar in two stages and then Pasteurization at 95°C by passing through automatic holding tube for 6 mins and cooling to 4°C. Milk is then transferred to storage tank.
6. **Inoculation and culture addition**: Now transfer the milk to inoculation tank through PHE for heating milk at 42°C. Take appropriate quantity of milk so that filling is over in one hour. Culture addition is done at recommended doses.
7. **Filling and Sealing**: Milk with culture is filled in required packing sizes and stacked in trolleys.
8. **Incubation**: Trolleys then moved to incubation room maintained a temperature of 42±2°C. Incubation is done for 4 to 6 hours till pH of 4.7±0.5 is achieved.
9. **Blast Cooling**: As soon as pH is achieved, trolleys are transfer to blast cold room at zero degree with direct blowers so that the product temperature of 10°C is achieved within 3.5 to 4 hrs mins.
10. **Cold Storage**: After 3.5 - 4 hrs transfer the trolleys in cold storage under 5°C till the time of dispatch.
11. **Dispatch**: Product is dispatched in refrigerated trucks and vans maintaining a temperature of below 5°C.

The FSSAI defines Dahi or curd as a semisolid product, obtained from pasteurized or boiled milk by souring (natural or otherwise), using a harmless lactic acid or other bacterial cultures. Indigenous microflora isolated from dahi are dominated by lactic acid bacteria. Harun-ur-Rashid *et al.* (2007) reported that *Streptococcus bovis* was the predominant strain. In a review, Kumar *et al.* (2013) reported that some LAB members found in dahi were *Streptococcus*, *Lactobacillus*, *Pediococcus* and *Weissella*. The presence of the genus *Lactococcus* was also reported as the predominant strain.

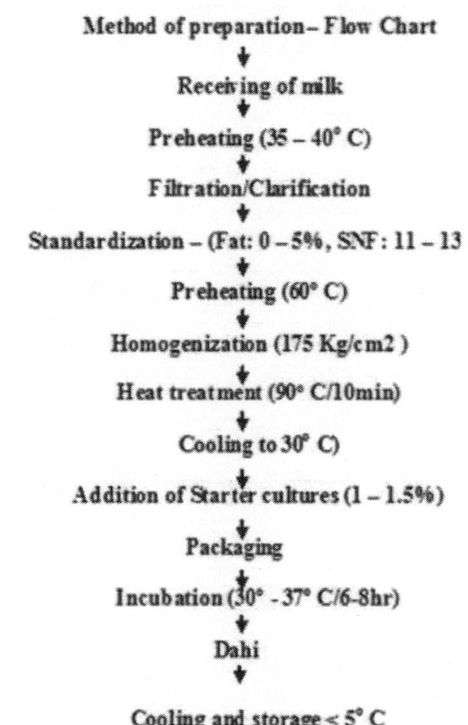

Figure 13.10: Method of Preparation of Curd (*Dahi*).

The Bureau of Indian Standards (BIS) specifications for fermented milk products are based on the type of culture used in their preparation. Mild dahi is made from mesophilic *lactococci*. *Leuconostocs* may be adjunct organisms for added buttery odour and flavour. Sour dahi contains additional cultures belonging to the thermophillic group, which are generally employed in the manufacture of yoghurt. These thermophillic organisms grow rapidly at 37-45°C, producing dahi in less than 4 hours. Standardization of the dahi preparation process was initiated by the Bureau of Indian Standards (BIS). One of the standards implemented is the use of a starter containing *Lactococcus lactis* subsp. Lactis and *L. lactis* subsp. *diacetylactis* or these cultures mixed with *Leuconostoc species*.

In general, dahi may be classified into two types: i) For direct consumption and ii) For churning into desi butter (*makkhan*). The consumers have different taste preferences for traditional products varying from region to region. The technological developments have led to the commercialization of this product. Dahi may be classified on the following basis. Dahi is an intermediate product for manufacture of traditional butter and ghee and is used as a base for preparation of related products such as lassi (Buttermilk), chhash, shrikhand, curd rice *etc.*

Dahi For the Production of Mishthi Dahi, Chakka, Shrikhand, Butter and Butter Milk

Misthi Dahi

The eastern parts of India, especially in West Bengal, Assam, Bihar and Orissa, the sweetened variety of dahi known as Mishti dahi, Lal dahi or Payodhi is quite popular. The product is prepared by the Halwais on a small scale. It is a delicacy of choice during religious festivities and is considered an auspicious item to serve while starting journey or any important work. The product is commonly sold in earthen pots of varying sizes, and served chilled.

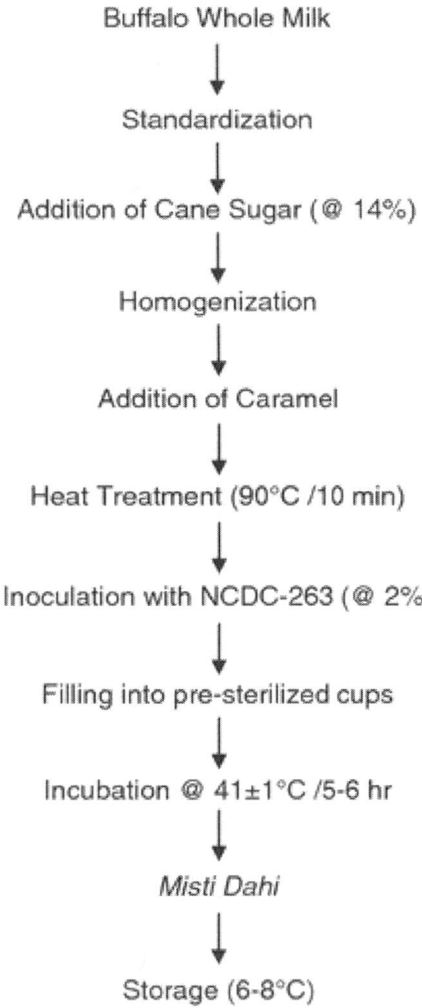

Figure 13.11: Flow Chart for Preparation of Misthi Dahi.

Traditional Method of Preparation

Traditionally, mishti dahi is prepared from cow or mixed milk. The fresh good quality milk is boiled with a required amount of sugar and partially concentrated by simmering over a low fire. This heating is continued for quite some time during which milk develops a distinctive light cream to light brown caramel colour and flavour. The content is then cooled to ambient temperature and cultured with dahi (lactic culture). It is then filled into earthen pots of consumer size or bulk size vessels and incubated overnight. Normally the curd is set within 12-14 hours. After firm setting of curd, it is transferred to a cooler place or stored under refrigeration.

Buttermilk

In India, buttermilk is known with a local name as Lassi, also known as stirred Dahi. Lassi is slight acidic sweet taste, viscous liquid, creamy white colour and can be flavoured with condiments, spices, sugar and salt. Lassi available in two flavours, contains sugar or salt and with condiments; spices depend upon the preference of the consumer. It is manufactured using pasteurized and boiled milk cultured with microorganisms which produce flavour. Originally Lassi is a product of rural part of India and it is a by-product of Dahi obtained while the separation of makkhan (butter) by churning. It was manufactured at cottage level earlier. Salted as well as sweet Lassi are commercially available in market packaged aseptically in UHT boxes/milk cartons/sachets.

Mattha/chhach (Buttermilk without fat)

It is similar to buttermilk, which is drink as refreshing beverage all over the country. Buttermilk, as the name suggests, is the milk that remains after cream is churned for the production of butter. Hence, it is a by-product of Dahi obtained during production of butter by churning. It is a famous product for its slightly sour taste and aroma developed in fermentation by mixed types of microflora used in lactic acid conversion.

Moru

For Moru, a dairy based fermented product prepared in Kerala, dahi is churned to obtain thick buttermilk and diluted with water to decrease acidity along with the addition of salt for flavour. In Moru, when crushed curry leaves, green chilli and ginger is added, it becomes 'shambharaam' which is popular in the southern region of India.

Cultured Buttermilk

This product was originally the fermented by-product of butter manufacture, but today, it is more common to produce cultured buttermilks from skim or whole milk. The culture most frequently used is *S. lactis*, perhaps also spp. *cremoris*. Milk is usually heated to 95°C and cooled to 20–25°C before the addition of the starter culture. Starter is added at 1–2 per cent and the fermentation is allowed to proceed for 16–20 h, to an acidity of 0.9 per cent lactic acid. This product is frequently used as an ingredient in the baking industry, in addition to being packaged for sale in the retail trade.

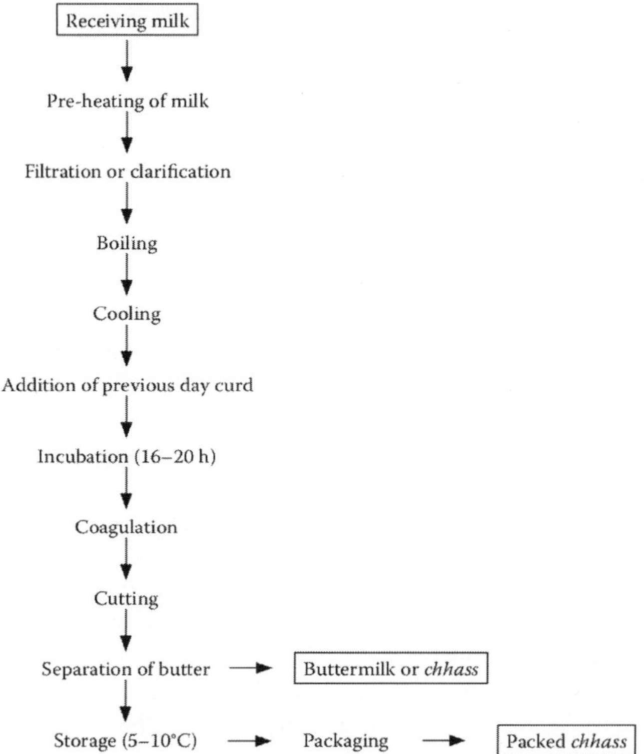

Figure 13.12: Flow Chart for Preparation of Buttermilk or Chhass.

Butter and Sour Cream

Butter, buttermilk, and sour cream are produced generally by inoculating pasteurized cream or milk with a lactic starter culture and holding until the desired amount of acidity is attained. In the case of butter, where cream is inoculated, the acidified cream is then churned to yield butter, which is washed, salted, and packaged.

The commercial product is usually prepared by inoculating skim milk with a lactic or buttermilk starter culture and holding until souring occurs. The resulting curd is broken up into fine particles by agitation, and this product is termed cultured buttermilk. Cultured sour cream is produced generally by fermenting pasteurized and homogenized light cream with a lactic starter. These products owe their tart flavour to lactic acid and their buttery aroma and taste to diacetyl.

Shrikhand

It is an indigenous, sweetish- sour with semi-soft consistency made from whole milk produced from lactic acid fermented curd. The coagulated portion partial stained through a muslin cloth for the removal of whey content and produce solid mass known as 'chakka' used as base ingredient for Shrikhand production. A blend of chakka, sugar, fruits and natural flavouring agents like saffron, cardamom, dry fruits and other spices prepared for Shrikhand manufacturing.

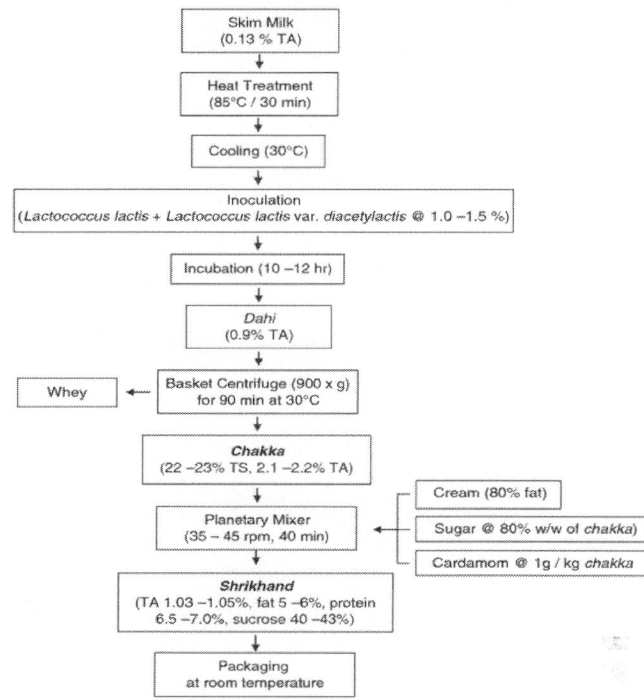

Figure 13.13: Flow Chart for Preparation of *Shrikhand.*

According to the definition of FSSAI 2011 (Food Safety and Standards Authority of India), Shrikhand is obtained from whole milk chakka or skim milk chakka with added fat. Shrikhand may contain natural ingredients like fruits, nuts, sugar, condiments and other spices. It should be free from any type of artificial colour and flavouring agents. The colonies count of yeast, mold and coliform should not be above than 50 CFU/g and 10 CFU/g.

Acidophilus Milk

Acidophilus milk is traditional milk fermented with LA, which has been thought to have therapeutic benefits in the gastrointestinal tract. Skim or whole milk may be used. The milk is heated to high temperature, for example, 95°C for 1 h, to reduce the microbial load and favour the slow growing LA culture. Milk is inoculated at a level of 2–5 per cent and incubated at 37°C until coagulated. Some acidophilus milk has acidity as high as 1 per cent lactic acid, but for therapeutic purposes 0.6–0.7 per cent is more common.

Another variation has been the introduction of a sweet acidophilus milk, one in which the LA culture has been added but there has been no incubation. It is thought that the culture will reach the gastro intestinal (GI) tract where its therapeutic effects will be realized, but the milk has no fermented qualities, thus delivering the benefits without the high acidity and flavour, considered undesirable by some people.

Yoghurt

The word "yoghurt" is derived from Turkish "jugurt", used to describe any fermented food with an acidic taste. Historically, yoghurt was made by fermenting milk with indigenous microorganisms. Yoghurt and yoghurt-like products are known throughout the world. However, they have traditional names other than yoghurt. Greek yoghurt is an extremely popular product in the western markets. It is also called strained yoghurt because whey and other liquids are removed from the traditional yoghurt to produce it. As a result, the Greek yoghurt contains more proteins, less sugars and almost the same amount of fat per 100 gr.

The basic culture is homofermentive and contains *Streptococcus thermophillus* and *Lactobacillus bulgaricus*. Various combinations of starter cultures are selected during manufacturing of yoghurt to achieve desirable characteristics of product and also to provide the consumers with a wide choice of therapeutic benefits. Depending on its activity, manufacturer usually adds 2-4 per cent yoghurt starter culture.

In many products under that commercial name, the microbiome is enriched with the addition of probiotic bacteria of the genera *Bifidobacterium*, *Lactobacillus* and *Lactococcus*. The major product of the fermentation is lactic acid and, in smaller concentrations, other compounds such as diacetyl, acetaldehyde, volatile fatty acids and free amino acids. These compounds are responsible for its exceptional sensorial characteristics.

Yoghurt is only one of several acidified fermented dairy foods with unique organoleptic qualities. This product varies considerably worldwide in composition, flavour, and texture as influenced by the nature (biochemistry) of the fermenting organisms, the type and source of milk base, and the manufacturing process.

Yoghurt is a perishable fermented dairy product which, unheated after fermentation, has a shelf life of up to 3 to 4 wk under refrigeration. Yoghurt is stored at 2- 4°C throughout the distribution chain for avoiding risk of spoilage from yeasts and also for preventing further activity by starter culture.

Shelf life is dependent not only upon storage conditions but also upon type of culture employed in manufacture, its biochemistry during manufacture and storage, and upon standards of sanitation during manufacture and packaging. Shelf life and quality (flavour, body, and texture) are influenced significantly by quality and performance of formula ingredients.

Different Types of Yoghurts

Yoghurt is typically classified into the following groups:

Set Yoghurt: This type of yoghurt is incubated and cooled in the final package and is characterized by a firm jelly like texture.

Stirred Yoghurt: This type of yoghurt is incubated in a tank and the final coagulum is "broken" by stirring before cooling and packing. The texture of stirred yoghurt will be less firm than a set yoghurt somewhat like a very thick cream. A little reformation of coagulum will occur after packaging.

Drinking Yoghurt: It also has the coagulum "broken" prior to cooling. In drinking yoghurt, the agitation used to "break" the coagulum is severe. Very little reformation of coagulum may occur.

Frozen Yoghurt: Frozen yoghurt is inoculated and incubated in the same manner as stirred yoghurt. However, cooling is achieved by pumping through a Whipper/ chiller/freezer in a fashion similar to ice-cream. The texture of the finished product is mainly influenced by the whipper/freezer and the size and distribution of the ice crystals produced.

Concentrated Yoghurt: This type of yoghurt is inoculated and fermented in the same manner as stirred yoghurt. Following the "breaking" of the coagulum the yoghurt is concentrated by boiling off some of the water, this is often done under vacuum to reduce the temperature required. Heating of low pH yoghurt can often lead to protein being totally denatured and producing rough and gritty textures. This is often called strained yoghurt due to the fact that the liquid that is released from the coagulum upon heating used to be "strained" off in a manner similar to making soft cheese.

Flavoured Yoghurt: The flavours are usually added at or just prior to filling into pots. Common additives are fruit or berries, usually as a puree or as whole fruit in syrup. These additives often have as much as 50 per cent sugar in them, however with the trend towards healthy eating gaining momentum, many manufacturers offer a low sugar and low fat version of their products. Low or no sugar yoghurts are often sweetened with saccharin or more commonly aspartame.

Consumption of yoghurt has increased in most countries, and its popularity has spread throughout the world. This widespread popularity and consumption have been influenced immeasurably by introduction of a variety of fruits and flavour combinations. Now a days, there has been increasing trends to fortify the dairy product with fruits (natural fruit juice, pulp, dry fruits). Aesthetic value of new product can be increased by using fruit juice as a functional pigment in fermented milks with array of colours and flavour properties. Also, the intermix of both sundae and Swiss styles as commercial offerings to attract consumer awareness and preference through an aggressive marketing program has done much to increase per capita consumption, especially in those countries where this fermented food was not traditionally part of one's daily diet.

Table 13.5: Classification of Yoghurt

Sl.No.	Basis of Classification	Different Groups of Yoghurt
1.	Chemical standards (Fat)	Full, skimmed/medium or skimmed/low fat
2.	Physical nature of product	Set, stirred or fluid/drinking; the latter is considered Stirred yoghurt of low viscosity
3.	Flavours	Plain/natural, fruit or flavoured;
4.	Post fermentation processing	Enzyme hydrolysis, vitamin fortification, vegetable oils, heat treatment

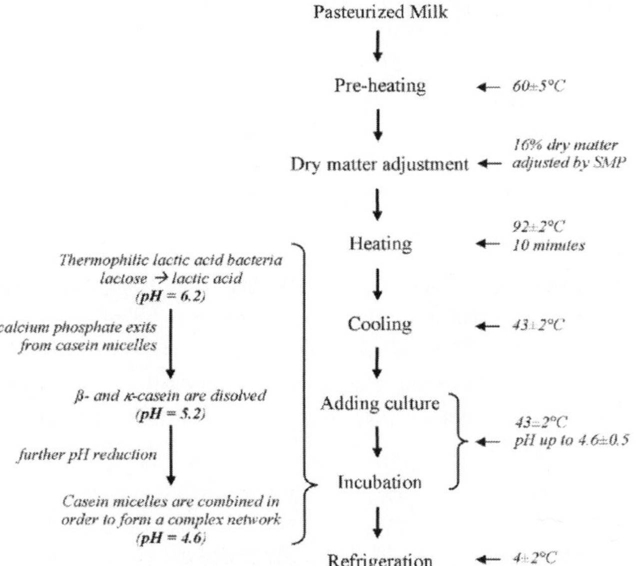

Figure 13.14: Flow Chart for Preparation of Yoghurt.

Yoghurt like curd is a functional food. The functional food includes probiotics, prebiotics and synbiotics. Probiotics can be defined as "live microbial feed supplements that beneficially affect the host animal by improving its intestinal microbial balance. Prebiotics as "non-digestible food ingredient that beneficially affects the host by selectively stimulating the growth and/or activity of one or a limited number of bacteria in the colon". Synbiotic is a combination of probiotics and prebiotics that "beneficially affects the host by improving the survival and the implantation of live microbial dietary supplements in the gastro-intestinal tract by selectively stimulating the growth and/or by activating the metabolism of one or a limited number of health promoting bacteria". Yoghurt or Dahi added with probiotic bacteria provide health benefits such as immune enhancement, blood pressure reduction, antiatherogenic effect, antidiabetic effect, anticarcinogenic effect, antioxidative effect and curing of gastrointestinal disorders.

Greek Cheese

"Tsalafouti" is a Greek cheese made mostly of ewes' and, in some cases, it may also originate from goats' milk. Its technology is very simple and serves as an example of the fermentation capacity of natural microflora. The milk is heated (40–50 °C) until inflation and, after the addition of a small amount of salt and rennet, it is left in a cool and shady place for 5–6 days to acidify and to transform into curd by the naturally occurring microbiome of the environment. After 2–3 days, a thin layer of Penicillium vegetates all over the surface of the curd. Extra salt is added and mixed thoroughly in the curd with "baking-like "movements." The "baking-like movements" are repeated and the *Penicillium* layer is mixed with the curd. This step is repeated two or three more times and, by then, the cheese is ready. The microbiome consists of an impressive variety of lactic acid bacteria (17 species of lactobacilli were isolated) as well as *Penicillium* spp. The cheese is white and creamy, and its flora consists mainly of lactic acid bacteria, namely lactobacilli.

Cheese

Cheese is an ancient traditional fresh or fermented dairy product with a long history of production. Cheese making originated in various West Asian countries about 8,000 years ago; it is traditionally called "milk pimples" by Mongolian, Kazak, and other nomadic people in northwestern China (Zheng *et al.*, 2021). Manufacturing of cheese is the best way of conserving milk solids as milk is highly perishable. Cheese can be used as a snack food, in baked goods or as pre-packaged convenience food.

Cheese is made by curdling milk, cream, or partially skimmed buttermilk from cow or goat or a mixture of these products and then separating the whey. In general, it is prepared by adding an appropriate amount of lactic acid bacteria (LAB) starter along with rennet to milk, the fermentation by which transforms the milk proteins (mainly casein), carbohydrates, and fats. Next, the whey is removed and the remaining product is ripened for a certain period. LAB may not be added for making some cheeses.

Fermented dairy products with high nutritional value are considered "the pearls in the crown of the dairy industry". With the improvement of the quality of life, the people's dietary needs and demands have been changing in terms of quantity and quality. Furthermore, the lactose content in cheese is low, and its consumption is thus highly suitable for people with lactose intolerance. Currently, nearly 130 countries and regions produce various cheese, and the total global cheese production measures nearly 2,000 × 10^4 tons, with the European Union countries (*e.g.*, Netherlands and Germany) being the largest cheese exporters worldwide.

The European Union (EU) is the world's leading cheese producer, followed by the United States (USA).

Combined, the two areas account for around 70 per cent of the global production. In total, EU countries produced 8959 million metric tons of cheese in 2010, and in 2020 the EC production was 10,350 million metric tons. Global cheese production is expected to increase progressively until 2027, with developed countries increasing milk production by 9 per cent. Of this increased milk production, 37 per cent is expected to be used to make cheese. Cheese consumption has grown in all global regions; USA and, in particular, the EU, are the main cheese consuming areas. In 2010, around 17 kg of cheese per person/per year was consumed globally, and this amount increased to 18.44 kg cheese per person in 2020. The EU has recently expanded cheese exports to Canada and the Russian Federation, and it is expected that China and Egypt will double cheese imports by 2027. In addition, other regions, such as the Middle East and North Africa, will become key destinations, accounting for 19 per cent of global cheese imports by 2027, because cheese has been progressively introduced into the diets of their consumers. The increase in global cheese consumption is also due to changes in food habits, particularly in East Asia, where the use of cheese as an ingredient in snacks and processed meals has increased. Thus, cheese preservation methods are highly important for the dairy industry in order to increase cheese consumption.

Cheeses can be made from different types of milk and processing technologies, and ripened for different periods, resulting in numerous varieties with a wide diversity in terms of texture, flavour, and shape. Cow's milk cheeses are produced in the greatest quantity throughout the year, whereas the production of small ruminant milk cheeses is lower and seasonal. Evidently, this seasonality affects the regularity of cheese manufacture during the year, in addition to the milk composition. In this respect, because the manufacturing time is not long, the shelf-life of these cheeses during which they retain their optimal sensory characteristics is short. Therefore, an adequate preservation method is extremely important for hard and semi-hard cheeses, to increase the availability for a longer period without changing their sensory characteristics.

The Codex Alimentarius classifies cheese varieties according to their composition and consistency parameters and ripening times, taking into account the percentage of moisture without fat. The extra-hard specification refers to cheeses with a moisture content less than 51 per cent. Cheeses with a moisture content between 49–56 per cent and 54–69 per cent are called hard and semi-hard cheeses, respectively. Soft cheeses have a moisture content higher than 67 per cent. The Spanish regulation indicates that semi-hard (semi-cured) cheeses must have a minimum ripening time of 20 days (cheese weight less than 1.5 kg) or 35 days (cheese weight greater than 1.5 kg), depending on the cheese weight when the cheeses are marketed. By comparison, hard (cured) cheeses must be ripened for at least 45 or 105 days, depending on whether the cheese weighs less than 1.5 kg

Hard and semi-hard cheeses are versatile nutrient-dense dairy products. Although these are highly valued, from a health perspective, significant controversy exists among consumers and in the scientific community. Cheese contains saturated fatty acids, cholesterol, and salt, which have been associated with cardiovascular disease (CVD) risk; however, cheeses also contain a range of nutrients that are potentially beneficial to health. Recent studies indicate that not all saturated fatty acids raise the content of cholesterol in plasma in the same manner, and that some saturated and trans fatty acids in cheese may play a beneficial role in human health. Moreover, other healthy components present in cheeses are conjugated linoleic acids (CLA) and phospholipids from the fat globular membrane. In this regard, fermented dairy products have been proposed as functional foods with a cholesterol-lowering effect and, therefore, with a protective effect against CVD compared to non-fermented dairy products. The so-called French paradox, in which low mortality from coronary heart disease has been observed despite the high cheese intake by consumers, is an important consideration in the nutritional assessment of cheese consumption. A recent study reported that increased dairy consumption may contribute to a lifestyle associated with a reduction in CVD risk. In addition, bioactive components are generated during cheese fermentation, such as gamma aminobutyric acid, which favours the survival of probiotic microorganisms/

Cheese also contains digestible proteins of high biological value. During cheese ripening and food digestion, caseins are hydrolyzed and peptides with antioxidant capacity are generated. The addition of an adjunct culture and a long ripening time increases the formation of peptides and enhances the antioxidant capacity. Some of these peptides are also a prominent source for nutraceutical functional foods. Many of the fat-soluble vitamins are held in cheese fat. Although some water-soluble vitamins may be lost during whey drainage, folate, niacin, B_{12}, and riboflavin remain in sufficient quantities in the cheese matrix to have a significant effect on human nutrition. In addition, propionic acid bacteria synthesize considerable levels of vitamin B_{12} in hard cheeses.

Ripened cheeses are an important source of minerals, particularly calcium (Ca) and phosphorus (P). The calcium in cheese is highly bioavailable due to the formation of complexes with peptides and its high content promotes fat excretion and reduces blood pressure. Similarly, cheese is a good source of this mineral for lactose intolerant individuals. The acid phosphatase enzymes aid in the generation of phosphorylated peptides, which also have beneficial health effects. In this regard, IDF indicates that the elimination of dairy products from diets may result in a loss of calcium and other essential nutrients for part of the population.

Sodium (Na) is a nutrient that should be reduced in the diet, but it is estimated that cheese adds only about 5–8 per cent of the total Na intake. A recent study reported that the intake of Na from cheese may prevent induced vessel alteration by reducing oxidative stress, rather than the intake of Na from non-dairy foods. Therefore, cheese intake may be an effective dietary strategy to reduce the risk of CVD in healthy older adults without salt-sensitive blood pressure. Therefore, cheese is a highly valued product by consumers and its consumption has increased in recent years because, among other reasons, a significant proportion of consumers perceive it as a healthy food.

Table 13.6: Classification of Cheese Based on Hardness/ Moisture Content

Cheese Type	*Moisture Content (per cent)*	*Examples*
Soft cheese	50–80	Cottage, Quark, Baker's, Mozzarella, Camembert, Feta
Semi-soft cheese	39–50	Blue, Limburger, Provolone, Tilsiter
Hard cheese	Max. 39	Cheddar, Colby, Edam, Swiss, Gouda
Very hard cheese	Max. 34	Parmesan, Romano, Sardo, Grana

Source: Olson (2008) and Fox *et al.* (2016).

More than 400 different varieties of cheese are available worldwide. However, the cheeses are classified (Table 13.6) into specific categories based on several criteria like firmness (moisture content), type of culture, ripened or otherwise, *etc.* Globally, there is a lack of an authoritative cheese-making process, which can be used for cheese production followed in any country abroad. This is due to differences in regions, production methods, and available raw materials. The earliest method of producing cheese in the world was to carry the milk to the animal's internal and the milk was fermented into cheese by constant oscillation during the migration. Cheese is made differently in different regions. For instance, for making cheddar cheese in southwest England, the raw material is sterilized and cooled; then, the fermenting agent, calcium chloride, and rennet are added to ferment the curd. After 30–40 min, the formed clot is cut into 5-mm sized pieces, which are then allowed to stand for 15 min and stirred for another 5–10 min. The clot is then turned over and stacked, broken, salted, molded, and pressed. The pressed cheese is put into a fermentation room for fermenting and ripening after the cloth is changed.

In contrast, the milk used to make Parmesan cheese is collected in two separate steps, where the overnight milk and the fresh milk (collected the next morning) are mixed in a copper cheese tank. When the temperature reaches 52°C, the cheese is wrapped with gauze for cutting, molding, pressing, and then, soaking in brine for 3 weeks. Approximately 5 kg of water is evaporated during maturity. On the other hand, soft and semi-hard cheeses, such as feta cheese, are soaked in brine for a short period. For Kazak cheese making, the milk collection process is similar to that of Parmesan cheese, with few differences: In a goatskin bag, old yogurt may or may not be added as a starter, followed by fermentation into yogurt. Next, this yogurt is boiled with stirring to evaporate water. The remaining curds are placed in a canvas bag, which is then hung outdoors to remove moisture further and solidify into fresh cheese. This fresh cheese is then cut into small pieces or made into a pie shape and finally placed on a bamboo board for 30–90 days of spontaneous ripening.

Mozzarella cheese is very popular soft, unripened variety of Pasta Filata cheese family and the most often used in variety of the food service sector. It has an excellent stretching and melting properties with stringy, elastic textural characteristics. This cheese's clean, mild flavour, shredability and appealing melt and stretch make it ideal for pizza. Mozzarella cheese should exhibit good shredding, melting and stretching properties and be free of off-flavours or textural defects for pizza production. Low moisture part-skim Mozzarella (LMPSM) finds greater application than traditional Mozzarella (up to 60 per cent moisture) on pizza owing to its superior slicing qualities. LMPSM is associated with a moisture content ranging from 45 to 52 per cent and a fat on dry matter (FDM) content ranging from 30 to 45 per cent. Such lower moisture content of LMPSM confers superior shredability and the moderate fat content provides desired fat leakage during its baking applications. Melting, stretching, free-oil formation, elasticity and browning are usually the properties considered to be of significance when used as a pizza topping.

USA leads the global consumption of Pasta filata cheeses at 54.5 per cent of volume, followed by Brazil (6.6 per cent), Italy (5.2 per cent), Germany (5.2 per cent) and Canada (3.7 per cent). Mozzarella is America's favourite cheese accounting for nearly a third of the consumption, since it is one of the main ingredients of pizza. In Argentina, Soft/fresh types dominate the cheese categories comprising 50 per cent of the market. Since 2010, cheese imports in China have tripled which involves fresh and curd type cheese including mozzarella; such cheese type accounted for 58 per cent of cheese imports in 2014. The major players in India for cheese are Amul, Britannia and Dabon International. Other major brands include Vijaya, Verka, Parag Milk Foods Pvt. Ltd. (Gowardhan brand), Nandini and Vadilal. Mozzarella cheese has a share of about 20 per cent of total cheese consumed in India.

Lactic acid bacteria (LAB) help in developing the physical properties of cheese, especially body and texture. The use of starter cultures possessing varying proteolytic activity can alter the stretch, melt and colour of Mozzarella cheese. LAB produces several metabolic products such as organic acids, fatty acids, hydrogen peroxide and bacteriocins that have antimicrobial activity resulting in inhibition of pathogenic and even spoilage bacteria. A high concentration of LAB in the raw materials and in the whey starter ensured safe and good quality Mozzarella cheese. A high LAB concentration (~ 10^7) ensured high quality of Mozzarella cheese with regard to sensory quality, shelf life and safety.

Mozzarella cheese prepared using direct acidification (without whey culture) technique exhibit rapid sensorial quality loss during refrigerated storage as compared to the one made using whey culture. Buffalo milk Mozzarella produced using natural whey starter cultures exhibited higher microbial diversity with lower psychrotrophic bacteria. Cow Mozzarella prepared using commercial starters had high count of *S. thermophilus* and consequently had greater levels of galactose and phenylalanine. Cow milk Mozzarella made using direct acidification method had lower microbial count and higher levels of urea and sugars.

The flow chart for preparing Mozzarella cheese has been depicted in Figure 13.15. Key stages of structure development are as follows:

- ***Curd formation and whey drainage***: Aggregation of *para*-casein micelles with concomitant release of whey into the surrounding spaces.
- ***Cheddaring and salting***: Fusion of *para*-casein micelles into a more continuous protein phase, forming curd granules.
- ***Plasticization***: Linearization on stretching of protein into layers or fibers. Fat globules partially coalesce and align between the protein fibers.
- ***Aging***: Progressive swelling of protein phase and isolation of fat globules. Expressible moisture decreases as the protein hydrates. The increased water binding capacity of the curd improves flow and stretch properties. The pizza maker should specifically use the Mozzarella cheese that has been aged for 1–1.5 weeks to permit improvement in the melting and stretching properties of cheese suitable for its end use on pizza.

Cottage Cheese/*Paneer*

India is the largest milk producer in the world with a production of 112 MT, which increased by 3.3 per cent in the last fiscal. About 55 per cent (61.60 MT) of the total production is the buffalo milk. Traditional dairy products have played an important role in social, economic and nutritional wellbeing of society. The importance of milk and milk products has been recognized since Vedic times and it is considered to be complete food. About half the milk produced is consumed in the liquid form and the remaining is used to prepare products such as *ghee*, curd, *butter*, *khoa*, *paneer*, cheese, *chhana*, ice cream and milk powders.

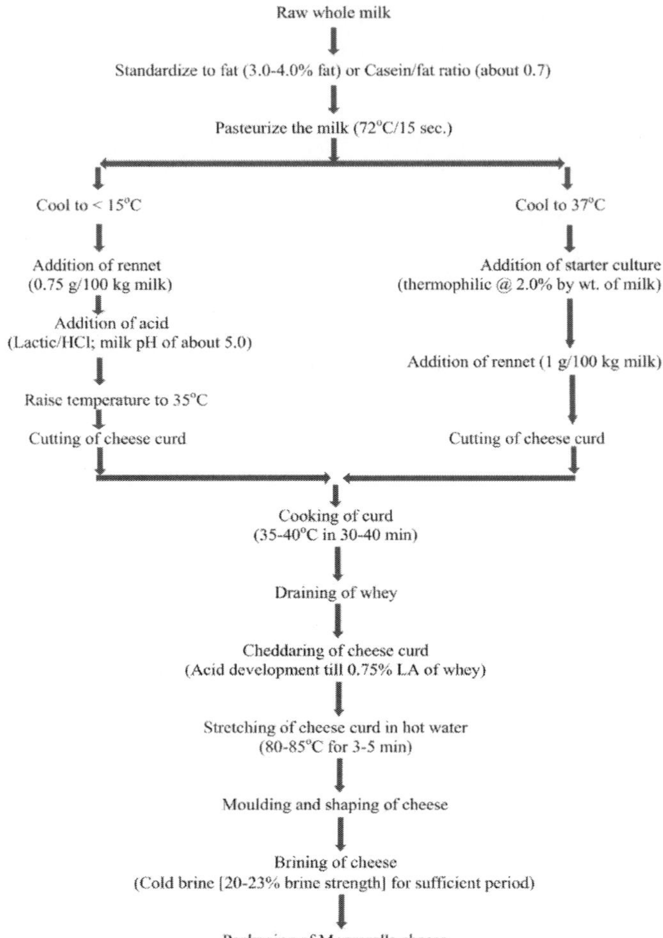

Figure 13.15: Flow Chart of Mozzareella Cheesing Preparation.

Paneer is an important indigenous product which is obtained by heat treating the milk followed by acid coagulation using suitable acid *viz.* citric acid, lactic acid, tartaric acid, alum, sour whey. The whey formed is removed to some extent through filtration and pressing. *Paneer* represents one of the soft varieties of cheese family and is used in culinary dishes/snacks. About 5 per cent of milk produced in India is converted into *paneer*.

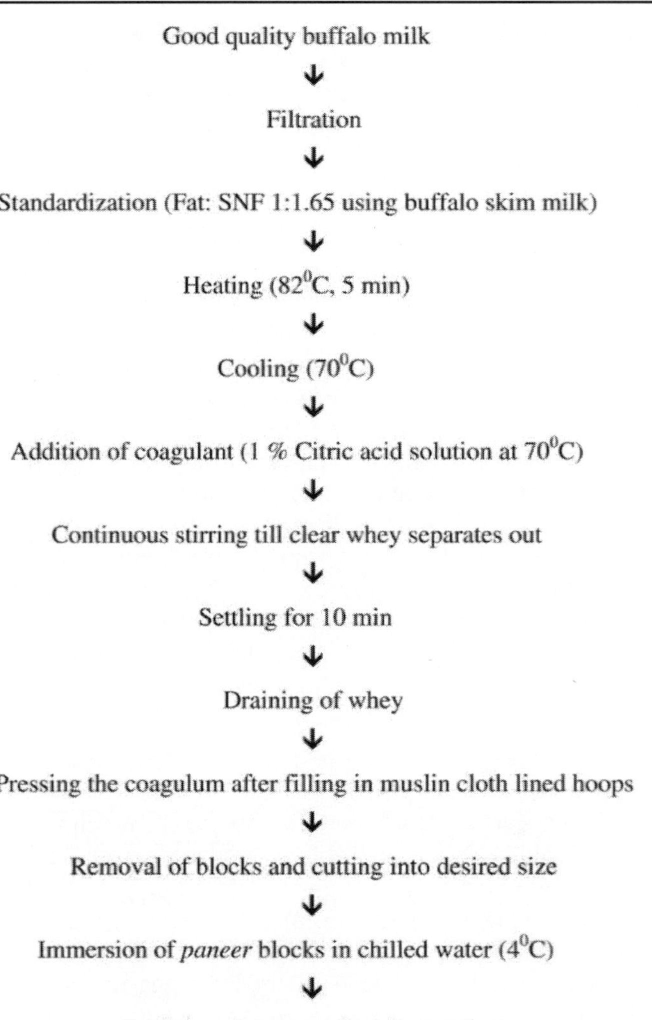

Figure 13.16: Flow Chart for Paneer Preparation.

Paneer has a fairly high level of fat (22–25 per cent) and protein (16–18 per cent) and a low level of lactose (2.0–2.7 per cent). *Paneer* is characterized by a mild acidic flavour with slightly sweet taste, and a soft, cohesive and compact texture. It is an excellent substitute for meat in Indian cuisine.

According to Prevention of Food Adulteration Rules (PFA 2010), *Paneer* means the product obtained from cow or buffalo milk or a combination thereof by precipitation with sour milk, lactic acid or citric acid. It shall not contain more than 70 per cent moisture and the milk fat content shall not be less than 50 per cent of the dry matter. Milk solids may also be used in preparation of the product. Bureau of Indian Standards also specifies a minimum of 50 per cent fat on dry matter basis but a maximum of 60 per cent moisture in *paneer*. In order to achieve these requirements, buffalo milk having 5–6 per cent fat is deemed to be most suitable. *Paneer*. must have a characteristic blend of the flavour of heated milk and acid, *i.e.* pleasant, mildly acidic and sweet (nutty). Its body and texture must be sufficiently firm to hold its shape during cutting/slicing, yet it must be tender enough not to resist crushing during mastication, *i.e.* the texture must be compact and smooth; Its colour and appearance must be uniform, pleasing white, with a greenish tinge in the case of buffalo milk *paneer* and light yellow in the case of cow milk *paneer*.

In the last few decades, consistent efforts have been made for the manufacture of different types of *paneer* like low fat *paneer*, recombined and reconstituted milk *paneer*, dietary fiber enriched low fat paneer, soy *paneer*, filled *paneer*, vegetable impregnated *paneer* and UF *paneer*.

The microbiological quality of *paneer* depends upon the post manufacture conditions, particularly, handling, packaging and storage of the product. Spoilage of *paneer* during storage is mainly due to the growth of spoilage organisms on the surface. Increase in total plate, yeast and mould and coliform counts in stored *paneer* were studied by several workers. In a study carried out on microbiological analysis of 8–24 h old market samples and laboratory made *paneer*. Out of the 54 samples, only 15 had less than 5,000 microorganisms per g and were rated as very good. However, none of the samples was of poor quality as none contained more than 200,000 microbes per g.

Paneer being a perishable commodity is highly susceptible to physicochemical and microbiological changes. Therefore, its packaging must provide protection against these damages while maintaining its quality, sales appeal, freshness and consumer convenience. Various packaging materials utilized for packaging of *paneer* include polythene sachets, coextruded films, laminates, parchment paper *etc.* Most of the *paneer* produced in organized sector is packaged in polyethylene bags because of its better barrier properties in respect of loss of moisture. These bags prove to be a superior packaging material for *paneer* compared to vegetable parchment paper. Packaging of chemical preservatives treated *paneer* with and without vacuum extended its shelf life up to 35 and 50 days, respectively at 8 °C. Vacuum packaging of cow milk *paneer* is reported to have enhanced its shelf life from 1 week to more than 30 days at 6 °C. *Paneer* packaged in high barrier film (EVA/EVA/PVDC/EVA) under vacuum and heat treated at 90 °C for 1 min had a shelf life of 90 days under refrigeration. Low quality milk, faulty method of production, unhygienic conditions, lack of refrigeration facility and poor storage conditions are mainly responsible for defects in *paneer*.

Kombucha

Kombucha is a fermented tea beverage reported to have originated in Northeast China in around 220 BC and consumed extensively during the Qin Dynasty. Similar fermented tea beverages subsequently became popular in Russia and Eastern Europe. In modern societies, a range of kombucha beverages are available commercially, although the microbial and metabolite composition of these products along with methods of production are rarely reported.

Traditional kombucha is produced through aerobic fermentation of black tea (green tea may also be used) and white sugar by a combination of bacteria and yeast, known as the symbiotic culture of bacteria and yeast (SCOBY). The SCOBY is also called the tea mushroom because of its similar shape and appearance to the fruiting caps of macroscopic mushrooms. It is a cellulosic biofilm formed by the polymerization of monosaccharides. The SCOBY microflora includes AAB, yeast, LAB and bifidobacteria. The tea mushroom is used as a starter culture for the production of the kombucha tea drink. It is a drink resulting from the fermentation of a sweetened tea (5 to 15 per cent sucrose) infusion by SCOBY, and produced during 6 to 10 days of fermentation under aerobic conditions, at a temperature range of 20 to 30°C. It is described as "slightly sweet, slightly sour refreshing drink".

The fermentation is two steps fermentation in which, the yeast converts sucrose to ethanol (in addition to organic acids and carbon dioxide) which acetic acid bacteria convert to acetaldehyde and acetic acid which reduces the pH of medium. Besides acetic acid, the fermented liquid contains gluconic, glucuronic and lactic acid, among them all glucuronic acid is the main therapeutic agent in Kombucha. Kombucha metabolism produces glucose, fructose, small amounts of ethanol, carbon-dioxide, vitamins C, B1 B2, B3, B6, B12, folic acid, different organic acids, mainly acetic, gluconic, L-lactic, glucuronic, enzymes and some antibiotically active compounds, and many others. Beverage also contains most of tea ingredients like tea catechines and caffeine.

The microbial and metabolite composition of kombucha varies according to the exact composition of the SCOBY, the type and concentration of tea and sugar, oxygen concentrations, fermentation time, temperature and storage duration. The low pH of kombucha, owing mainly to the production of high concentration of acetic acid, has been shown to prevent the growth of pathogenic bacteria such as *Helicobacter pylori*, *Escherichia coli*, *Salmonella typhimurium* and *Campylobacter jejuni*. Even at neutral pH and after thermal denaturation, kombucha was able to inhibit the growth of pathogens *in vitro*, suggesting that compounds other than acetic acid exert antimicrobial effects.

The bacterial and fungal species constituting the SCOBY typically include acetic acid bacteria (*Acetobacter*, *Gluconobacter*), lactic acid bacteria (*Lactobacillus*, *Lactococcus*) and yeasts (*Saccharomyces*, *Zygosaccharomyces*). Studies utilising high-throughput sequencing analysis have demonstrated that following fermentation, *Candida* and *Zygosaccharomyces* genera are the predominant yeasts in kombucha, while *Komagataeibacter*, *Lyngbya*, *Gluconobacter*, *Lactobacilli* and *Bifidobacteria* are the most abundant bacterial genera.

Kombucha has been shown to exert effects in animal studies on blood glycaemia, oxidative stress, diabetes-induced weight loss, chemically-induced nephrotoxicity, hypercholesterolaemia and indomethacin-induced gastric ulceration. Compounds hypothesised to play a role in these beneficial effects include D-saccharic acid-1,4-lactone (DSL). This is produced by *Gluconobacter* during fermentation, and in rats, inhibits oxidative stress and diabetes-induced renal damage and acetaminophen-induced hepatic injury. Nonetheless, there is no human data on DSL to confirm this proposed mechanism of action. Polyphenol and flavonoid content of tea increases with fermentation. Furthermore, *in vitro* superoxide radical scavenging ability, reducing ability and total phenolic compound concentration increases during kombucha fermentation.

Although kombucha is a rich source of acetic acid and lactic acid bacteria and yeasts, there are no published studies exploring the effect of kombucha consumption on the gastrointestinal microbiota composition or function in either animals or humans. Interestingly, kombucha has been shown to have antimicrobial effects *in vitro*. It is currently unknown whether the proposed physiological effects of kombucha are mediated by the gastrointestinal microbiota or other direct immunological pathways.

Despite evidence of physiological effects of kombucha consumption in animals, the effects in humans remain largely unknown. A recent systematic review did not identify any RCTs of kombucha on gastrointestinal disorders, including any of the functional bowel disorders.

To conclude, there are no studies of the effects of kombucha on gastrointestinal health and microbiota in humans. According to the current state of knowledge, SCOBY and kombucha tea drinks do not have confirmed health-promoting properties for humans. However, there is a large database of results demonstrating the potential wholesome properties that have been obtained through *in vitro* and *in vivo* animal studies.

Fermented Alcoholic Beverages

Fermentation is a well-known natural process used by humanity for thousands of years with the fundamental purpose of making alcoholic beverages, as well as bread and by-products. Upon a strictly biochemical point of view, fermentation is a process of central metabolism in which an organism converts a carbohydrate, such as starch or sugar, into an alcohol or an acid. For example, yeast performs fermentation to obtain energy by converting sugar into alcohol. Fermentation processes were spontaneously carried out before the biochemical process was fully understood.

In the 1850s and 1860s, the French chemist and microbiologist Louis Pasteur became the first scientist to study fermentation, when he demonstrated that this process was performed by living cells. Fermentation processes to produce wines, beers and ciders are traditionally carried out with *Saccharomyces cerevisiae* strains, the most common and commercially available yeast. They are well known for their fermentative behaviour and technological characteristics which allow obtaining products of uniform and standard quality. Upon a biochemical point of view, fermentation is carried out by yeasts (and some bacteria) when pyruvate generated from glucose metabolism is broken into ethanol and carbon dioxide (Figure 13.17).

The schematic chemical equation for the production of ethanol from glucose is as follows:

$C_6H_{12}O_6$(glucose)'$2C_2H_5OH$ (ethanol)+CO_2 (carbon dioxide)

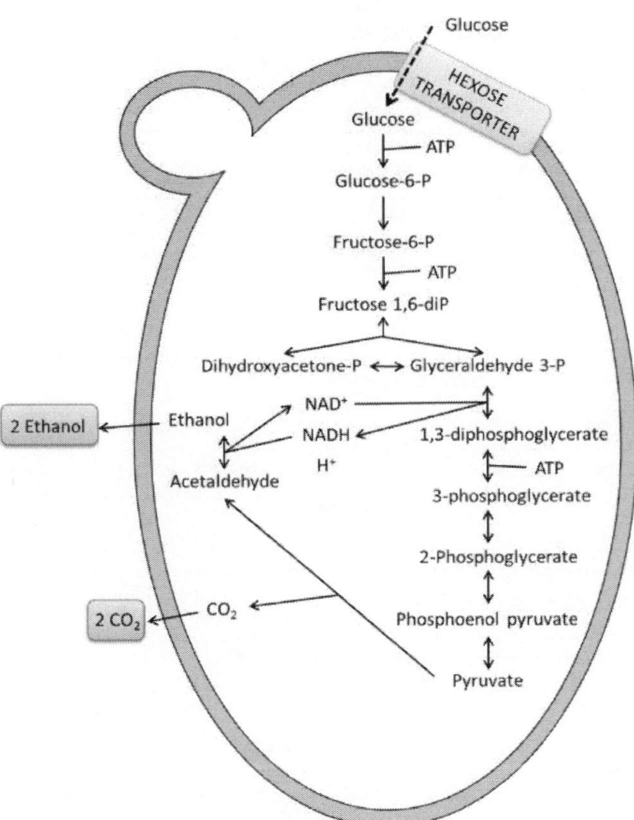

Figure 13.17: Central Metabolism of Fermentation in Yeast.

Fermented beverages include a wide variety of very popular and very famous products. Depending on the substrate and the microorganisms, involved fermentations can take different paths, resulting in different end products. One way to classify them is in two major categories: Alcoholic (more than 0.5 per cent v/v alcohol) and non-alcoholic beverages (less than 0.5 per cent v/v alcohol). A special category of alcoholic beverages are the ones with low alcoholic content (no more than 1.2 per cent v/v alcohol). However, another perhaps more productive way for classification is on the basis of the sources (substrates) from which they originate.

There are five main categories of fermented beverages depending on the raw material: (i) From grains (barley, corn, rye, millet *etc.*), resulting in products such as Irish whiskey, beer, bourbon whiskey, boza, kvass and vodka; (ii) from fruit juice (apples, bananas, cherries, grapes, ginger, *etc.*), producing cider, wines, cherry wine, ginger beer and ginger ale, among other products; (iii) from vegetables (potatoes, sugarcanes, sauerkrauts, *etc.*), producing vodka (in Poland), rum, sauerkraut juice, *etc.*; (iv) from milk, producing many beverages such as ayran, koumis, kefir and buttermilk; and (y) other raw materials, such as tea leaves, honey, sugar and palm sap, producing various beverages (kombucha, mead, *etc.*). Some of these ferments, *e.g.*, wines, are usually consumed for recreational purposes, although, if consumed in reasonable quantities and frequency, they may have certain beneficiary effects to health. Other ferments, however, such asayran, kefir or boza, have exceptional nutritional qualities. Different fermented vegetables have been described for example, soybean tempeh and other soybean paste products, sauerkraut, fermented olives, fermented cucumber and kimchi. In addition, salting procedures were explained along with the role of lactic acid bacteria in fermented vegetables.

Role of Yeast in Alcoholic Fermentations

The production of alcoholic beverages from fermentable carbon sources by yeast is the oldest and most economically important of all ergonomics. Yeast plays a vital role in the production of all alcoholic beverages and the selection of suitable yeast strains is essential not only to maximise alcohol yield, but also to maintain beverage sensory quality. The yeast species that dominates in the production of alcoholic beverages worldwide is *Saccharomyces cerevisiae*, and the particular strains of this species employed in fermentation exert a profound influence on the flavour and aroma characteristics of different beverages. For large-scale beverage fermentations, as in brewing, winemaking and distilled spirit production, pure cultures of selected strains of *S. cerevisiae* are usually used. These strains are either sourced in house or supplied from yeast producing companies. In smaller-scale (artisanal) processes, spontaneous fermentations may be allowed to occur that rely on indigenous microbiological flora (wild yeasts and bacteria) present in the raw material and in the production facility. For example, this would be typical in small distilleries in Mexico (for Tequila and Mezcal production) and in Brazil (for Cachaça production). In some types of alcoholic beverage fermentations, non-*S. cerevisiae* yeasts may be employed either as starter cultures, or occur naturally. For example, in winemaking the *S. cerevisiae* yeast strain used to commence fermentation may be overrun by the indigenous yeast flora associated with the grapes.

Beer Making

In the basic recipe for brewing beer, barley is turned into malt, the malt is cooked in hot water, hops are added as a flavouring and preserving agent, and yeast is introduced to induce fermentation. Modern industrial processes have, in some cases, replaced traditional brewing practices, resulting in substantial variations in composition.

Without malting, the barley grain cannot be fermented; its starches have to be broken down into their component sugars, maltose and dextrin, and it is principally maltose that is later converted to ethanol and carbon dioxide through fermentation. Controlled germination is induced by steeping the barley in water. Traditionally, the grain is spread on a floor during the malting period, which may last for a week or ten days. Germination is then arrested by drying, or by roasting if a dark beer is to be produced. When other grains such as rice and corn are used, the carbohydrates are made soluble by precooking.

The malt and other grains are mixed with hot water to form a mash, which is heated for up to 6 h. A clarified liquid, known as the wort, is extracted and fed into a brewing kettle, where it is heated to boiling point and boiled for 1–2.5 h. Hops are then added. After brewing, the spent hops are removed, and the wort is cooled before being transferred to the fermentation vessel, where the yeast is added.

Fermentation is basically of two types) bottom fermentation (producing lager-type beers) and ii) top fermentation (producing ales and stouts or dark beers), distinguished by the location of the yeast in the fermentation vessel. Brewing yeasts are members of the *Saccharomyces* genus, but two different species are utilised for ale and lager brewing. Regarding the former, these are strains of *S. cerevisiae* that traditionally conduct "top fermentations" where yeasts congregate on the surface of the fermenting wort. Some non-flocculent ale yeasts may be employed if centrifuges are used in yeast harvesting. There are numerous different strains of ale yeasts which have complex genetic (aneuploid or polyploid) characteristics.

Table 13.7: Summary of different Yeast Species Encountered in Alcoholic Beverage Fermentations

Beverage	*Yeast Involved*	*Comments*
Beer	Lager beer: *Saccharomyces pastorianus* Ale: *Saccharomyces cerevisiae* Lamber beer: *Brettanomycis bruxellensis* and other yeasts.	Lager yeasts are likely a natural hybrid (*S.cerevisiae* and *S. eubayanus*). Relatively few strains employed in lager fermentations. Lager strains utilise maltotriose more efficiently than ale strains, and they ferment at cooler temperatures. Ale yeasts are polyploid strains. Numerous strains are employed in ale brewing. Ale yeasts ferment at warmer temperature as compared with lager yeasts. Lamble beer traditionally produced via spontaneous fermentation, but some lambic and Belgian-style beers employ pure starter cultures of *Brettanomyces spp.*
Wine	S. cerevisiae, *Saccharomyces bayanus* (pure cultures) and naturally-occurring yeasts	Traditional winemaking is characterised by spontaneous fermentations of grape must with naturally occurring microflora (the main yeast genera associated with grapes are *Kloeckera* and *Hanseniaspora*, with lesser representations of *Candida*, *Metchnikowia*, *Cryptococcus*, *Pichia* and *Kluyveromyces* and very low populations *Saccharomyces cerevisiae*). Modern, large-scale wineries use specially selected starter cultures of *S. cerevisiae* strains available in dried form (*e.g.*, active dry yeast, ADY) from specialist yeast supply companies. Occasionally, secondary commercial non-*Saccharomyces* starter cultures (*e.g.*, *Candida stellata*) may be employed to impart specific flavour and aroma to wine.
Whisky	*Saccharomyces cerevisiae*	Scotch whisky producers currently use selected distilling strains of *Saccharomyces cerevisiae* in three main formats, cream yeast, pressed (cake) and dried yeast. Malt whisky distilleries traditionally use pressed yeast, but larger grain distillers have now adopted cream yeast. Dried yeast are not as prevalent as pressed and cream formats in whisky fermentations.
Rum	*Saccharomyces cerevisiae* and *Schizosaccharomyces pombe*	*S. cerevisiae* strains in rum fermentations are developed as starter cultures and provide faster fermentation with more higher alcohols and fatty acids, but less esters resulting in lighter style rums.*Schiz. pombe* in rum fermentations provides slower fermentations leading to less higher alcohols and fatty acids, but more esters resulting in heavy, strong aroma rums. Growth of *Schiz. pombe* is favoured by low pH, higher sugar concentration.
Tequila, Mezcal, Bacanora	Natural yeasts in artisanal Agave fermentation	Various yeasts have been isolated from such processes: *S.cerevisiae*, *Kluyveromyces marxianus*, *Pichia spp.*, *Brettanomyces* spp., *Rhodotorula spp. etc.*
Brandies, Gin, Vodka *etc.*	*S. cerevisiae*	For brandies, cognac, *etc.* the base wine is produced by pure starter cultures of *S. cerevisiae*. For gin, vodka *etc.*, selected distilling strains of *S. cerevisiae* will be used.
Cheese whey-derived beverages	*Kluyveromyces marxianus*	Lactose-fermenting yeast to produce ethanol destined for gin, vodka and cream liqueurs, *etc.*

Lager yeast strains are hybrids of the species *S. cerevisiae* and *S. eubayanus* and are now designated as *Saccharomyces pastorianus*. These yeast cultures conduct "bottom fermentations" and usually have a tendency for flocculation. This facilitates cropping of yeast at the end of fermentation on the bottom of a fermenter usually in the cones of cylindro-conical vessels. Bottom fermentation takes place at low temperatures. Primary fermentation (one or two weeks) is carried out at approximately 5–10°C; secondary fermentation ('maturing', 'ageing', 'ripening', 'conditioning', 'lagering') takes place at an even lower temperature ()*2°C) for four weeks or more. When the beer reaches condition, it is usually filtered.

Top fermentation takes place at higher temperatures; primary fermentation, at 15–20°C, lasts for about a week, and maturation at the brewery lasts for only a few days. Traditional English ale, one of the most typical top-fermented beers, is not filtered, so that secondary fermentation continues while the beer is in the barrel.

Beer may or may not be pasteurized; pasteurization stabilizes the beer, but also affects the taste. Bottled beer is more likely to be pasteurized to enable longer storage, and is also more likely to contain preservatives. Draught beer is more commonly sold unpasteurized, but this varies in different parts of the world.

The ethanol content of beer is often measured by weight. Beer may vary in ethanol content from 2.25 per cent by volume to over 10 per cent. German beers usually contain about 5 per cent alcohol, as do US beers; English ales usually contain 2.5–5.5 per cent by volume (mild ale, 2.5–3.5 per cent; bitter ale, 3–5.5 per cent). Recent developments in different countries include stronger beers, also called 'malt liquors', containing up to 10 per cent ethanol, and weaker beers, so-called 'low-alcohol' beers. In western Australia, for instance, a beer with only 0.9 per cent ethanol has taken 15 per cent of the beer market. In many countries, taxation and sales regulations are related to ethanol content; stronger beers are more heavily taxed and sold only in special outlets.

Wine Fermentation

Wines are normal alcoholic fermentations of sound grapes followed by aging. A large number of other fruits such as peaches, pears, and so forth may be fermented for wines, but in these instances the wine is named by the fruit, such as peach wine, pear wine, and the like. Because fruits already contain fermentable sugars, the use of exogenous sources of amylases is not necessary, as it is when grains are used for beers or whiskeys. Wine making begins with the selection of suitable grapes, which are crushed and then treated with a sulfite such as potassium metabisulfite to retard the growth of acetic acid bacteria, wild yeasts, and molds. The pressed juice, called must, is inoculated with a suitable wine strain of S. "ellipsoideus."

During fermentation of grapes, their sugar content, or part of it, is converted into ethanol and carbon dioxide by yeast. Traditionally, the natural yeasts on the grape skins are used, but in modern industrial wine-making yeast may be added. The fermentation is allowed to continue for 3–5 days at temperatures between 70°F and 90°F (21°C and 32°C), and good yeast strains may produce up to 14–18 per cent ethanol. Following fermentation, the wine is racked–that is, drawn off from the lees or sediment, which contains potassium bitartrate (cream of tartar). Winemaking involves the extraction of grape juice ('must') followed by yeast (natural yeasts or commercial starter cultures of *S. cerevisiae*) fermentation. Bacterial malolactic fermentation may also take place (where malic acid is converted to lactic acid), followed by aging, clarification and packaging. Wine yeast starter cultures are predominantly strains of *S. cerevisiae*, but some commercial non-Saccharomyces starter cultures (for example, *Candida stellata*) may be employed to impart specific flavour and aroma characteristics to wine

White wine is made by fermenting only the grape juice obtained by pressing grapes. To make red wine, the grapes are crushed, and fermentation takes place before the juice is separated from the residue. Champagne, a sparkling wine made by a secondary fermentation of wine, is produced by adding sugar, citric acid, and a champagne yeast starter to bottles of a previously prepared, selected table wine. The bottles are corked, clamped, and stored horizontally at suitable temperatures for about 6 months. They are then removed, agitated, and aged for an additional period of up to 4 years. The final sedimentation of yeast cells and tartrates is accelerated by reducing the temperature of the wine to around 25°C and holding for 1–2 weeks. Clarification of the champagne is brought about by working the sediment down the bottle onto the cork over a period of 2–6 weeks by frequent rotation of the bottle. Finally, the sediment is frozen and disgorged upon removal of the cork. Rosé wine is made by fermentation beginning when the grapes are crushed; however, the juice is run into a separate vat after a short time. Red wine gets its colour from the skin of the grapes, and its chemical composition is altered due to the presence of tannins in the skin. The sweetness of the wine depends on the sweetness of the grapes and on the length of fermentation; in the production of sweet wines, fermentation must be stopped before all of the sugar is converted into ethanol. Fermentation can be stopped by adding sulphites or by filtering off the yeast; it can also be stopped by adding alcohol (generally brandy) to raise the level up to or exceeding 15 per cent, as for fortified wines, such as sherry and port.

In wine fermentation, strains with specific characteristics are needed, for instance, highly producers of ethanol to reach values of 11–13 per cent *v*/*v*, typically found in this beverage. On the other hand, beers and ciders contain less amounts of ethanol with a balanced and distinctive sensory profile characteristic of each one. In recent years, new consuming trends and requirements for new and innovative products have emerged. This situation led to rethink about the existing fermented beverages and to meet the demands of consumers. Yeasts are largely responsible for the complexity and sensory quality of fermented beverages. Based on this, current studies are mainly focused on the search of new type of yeasts with technological application.

Non-*Saccharomyces* yeasts have always been considered contaminants in the manufacture of wine and beer. Therefore, procedures for eliminating them are routinely utilized such as must pasteurization, addition of sulfite and sanitization of equipment and processing halls. In recent years, the negative perception about non-*Saccharomyces* yeasts has been changing due to the fact that several studies have shown that during spontaneous fermentations of wine, these yeasts play an important role in the definition of the sensory quality of the final product. Based on this evidence, the fermentative behaviour of some non-*Saccharomyces* yeasts is being studied in deep with the purpose of finding the most adequate conditions and the most suitable strain to be utilized in the production of fermented beverages

Fermentation usually takes about two weeks at 24°C for red wine and four to six weeks at 15.5°C for white wines. This is followed by a second fermentation, the malolactic fermentation, during which bacteria work on the malic acid to convert it to lactic acid, which is less sharp to the taste. The second fermentation may take place in barrels, during which the wine extracts different substances from the wood. In the case of sparkling wines, the second fermentation takes place in bottles and may continue for several months; the carbon dioxide content determines the extent to which the wine will be sparkling. Maturation continues during storage, which may last for several years. Red wines of high quality are often stored for a long time; white wines are generally consumed within a few years, with the exception of wines such as sweet French Sauternes and Italian Tokay wines.

The ethanol content of wine usually varies from 8 to 15 per cent by volume. Recently, light wines with an ethanol content of 4.5 per cent were introduced onto markets such as the USA, primarily as 'low-calorie' wines. In so-called nonalcoholic wines, the ethanol content has been reduced to 0.5 per cent.

Although grapes are by far the most common raw material for wine, it can also be made from other fruits and from berries, particularly in countries where grapes are not grown. Cider, which is fermented from apples or pears, is a common alcoholic beverage in many countries, although its consumption has declined in industrialized areas.

Table 13.8: Some Fermentation Media for Alcoholic Beverages

Media	*Fermentable Sugars*	*Beverage*
Barley malt wort	Glucose, maltose, maltotriose	Ale and lager beer. Scotch malt whisky
Cereal wort based on barley malt and exogenous enzymes plus un-malted starch from wheat, rye, maize, sorghum, *etc.*	Glucose, maltose, maltotriose	Some beers, Scotch grain whisky, Bourbon and Tennessee whiskey, Canadian rye whisky, Irish whiskey, grain neutral spirit (for gin, vodka, *etc.*)
Rice hydrolysate (from Koji enzymes)	Glucose	Sake, Sochu, Arrack, Awamori
Potato hydrolysate (from amylolytic enzymes)	Glucose	Aquavit, vodka
Agave	Fructose	Tequila, mezcal, pulque
Sugar cane molasses	Sucrose	Rum
Sugar cane juice	Sucrose	Cachaça (Brazil), Rhum Agricole
Grape must, fruit juices, honey	Glucose, fructose	Wine, cognac, armagnac, brandy, grappa, kirsch, slivovich, cider, perry, mead
Cheese whey	Lactose	Gin, vodka, cream liquers

Cider

The word cider, in most parts of the world, refers to the fermented, alcoholic beverage made from pressed apple juice. In North America the term for this alcoholic beverage is hard cider. Hard cider is believed to have roots from the winemaking technology in the Basque country of northern Spain. The cider-making technology was spread throughout the northern coast of Europe by Celts as they travelled from Spain to England and Ireland. Colonists who brought apples and cider-making technology with them introduced hard cider to the New World. All cider goes through at least three basic stages. In autumn, the ripe fruit is crushed; then the juice is pressed out. The juice may be fermented for weeks or months, but usually by late winter the finished cider is bottled.

Cider is another alcoholic beverage derived from the apple fruit industry, very popular in different countries in the world, mainly Europe, North America, and Australia. There are up to 12 steps in cider making: the harvest, "sweating", washing, grinding, pressing, blending, testing, fermentation, racking off, filtering or fining, bottling, storage. Each step has its importance in the overall "image" of the final product.

In making apple cider, the fruits are selected, washed, and ground into a pulp. The pulp "cheeses" are pressed to release the juice. The juice is strained and placed in a storage tank, where sedimentation of particulate matter occurs, usually for 12–36 hours or several days if the temperature is kept at 40°F (4.4°C) or below. The clarified juice is cider. If pasteurization is desired, this is accomplished by heating at 170°F (76.7°C) for 10 minutes. The chemical preservative most often used is sodium sorbate at a level of 0.10 per cent. Preservation may also be affected by chilling or freezing. The finished product contains small amounts of ethanol in addition to acetaldehyde.

Although traditional ciders are produced from spontaneous fermentation of juice carried out by autochthonous yeasts, selected *S. cerevisiae* strains are also commonly used to carry out alcoholic fermentation. This ensures a consistent quality of the finished products. Some other non-*Saccharomyces* yeast species are involved in spontaneous fermentation of apple juice for cider production. However, these yeasts contribute at a lesser extent than *Saccharomyces* and can be producers of off-flavours. Research articles on this type of product are scarce compared to wine, especially in phenomena associated with microbial activities. The microbiome of wine fermentation and its dynamics, the organoleptic improvement of healthy and pleasant products and the development of starters are now extensively studied. Although the two beverages seem close in terms of microbiome and process (with both alcoholic and malolactic fermentations), the inherent properties of the raw materials and different production and environmental parameters make it worthwhile research on the specificities of apple fermentation.

In addition to these three worldwide-famous fermented beverages, there are many others made from fruit in various countries in Africa, Asia, and Latin America. Although its consumption is local or regional, in some countries drinks made using fruits such as bananas or grapes as raw materials are very popular. The most widespread alcoholic fruit drink in Eastern Africa is banana beer, which in addition to gastronomic interest is especially culturally relevant. Banana beer is a mixed beverage made from bananas and a cereal flour (often sorghum flour). Dates in North Africa, pineapples and cashew fruits in Latin America and jack fruits in Asia are other of the most relevant products.

Distilled Spirits

Distilled spirits are alcoholic products that result from the distillation of yeast fermentations of grain, grain products, molasses, or fruit or fruit products. Whiskeys, gin, vodka, rum, cordials, and liqueurs are examples of distilled spirits. Although the process for producing most products of these types is quite similar to that for beers, the content of alcohol in the final products is considerably higher than for beers.

Spirits are made by distilling a liquid, or wash, that contains ethanol produced by fermentation of a base ingredient that contains sugar or starch. The purpose of distillation is to increase the ethanol content and to eliminate fractions of the base liquid that are not wanted in the final product. The process of distilling involves heating the liquid so that ethanol and other volatile substances evaporate, passing the vapours through a cooling system, and collecting the liquid which then contains less water and other unwanted substances. The liquid may be redistilled, or rectified, several times to increase its purity: this process leads ultimately to a colourless neutral spirit, which can then be flavoured. In many cases, some of the original flavouring components of the base liquid are retained. The traditional distilling process is carried out in a 'pot-still', and is still used in the production of cognac, malt whisky, most Irish whiskeys and some of the flavouring used in liqueurs. For many other spirits, distillation is performed in a 'continuous still', consisting of two or three columns; the heat of the product at the end of the process is used to warm the cold liquid that enters the system. The pot-still allows greater retention of the original flavouring elements in the base liquid, while continuous distillation is more efficient. After distillation, water is often added to achieve the desired strength of the beverage.

Spirits are made from many raw materials; these may contain sugar, such as sugar cane, molasses, grapes and other fruits, or starch which is converted into sugar by malting, such as grains and potatoes. Wood is also used as a raw material. Brandies are made from a fruit mash or from wine; cognac and other varieties are made from grapes, but numerous other types are made from other fruits, such as apples, pears, pineapples, plums and strawberries. Malted grains are the base of all types of whisky. Scotch malt whisky is made almost entirely from barley, but in other whiskies grains such as maize, oats, rye and wheat are used. Rye and bourbon are examples of whiskeys. In the former, rye and rye malt, or rye and barley malt, are used in different ratios, but at least 51 per cent rye is required by law. Bourbon is made from corn, barley malt, or wheat malt, and usually another grain in different proportions, but at least 51 per cent corn is required by law. Gins and vodkas are produced from grain or potatoes; both are distilled to a fairly high ethanol content, sometimes filtered through charcoal to attain purity, and then, in the case of gin, flavoured. Sugar cane and molasses are used to produce rum; and the Mexican spirit, tequila, is distilled from a mash made from cactus. Brandy is a product prepared by distilling grape or other fruit wines. Palm wine or Nigerian palm wine is an alcoholic beverage consumed throughout the tropics and is produced by a natural fermentation of palm sap. The sap is sweet and dirty brown in color, and it contains 10–12 per cent sugar, mainly sucrose. The fermentation process results in the sap becoming milky white in appearance due to the presence of large numbers of fermenting bacteria and yeasts.

Fermentations for the production of whisky and other distilled spirits derived from cereals are conducted by specific strains of *Saccharomyces cerevisiae* which convert mash sugars into ethanol, carbon dioxide and numerous secondary fermentation metabolites that collectively act as flavour congeners in the final spirit. Therefore, the choice of yeast strain that can contribute significantly to the organoleptic qualities of spirits is critical! The extracted sugars following cereal mashing are predominantly maltose and maltotriose, in contrast to glucose, fructose and sucrose in wine must that are liberated from grape crushing. In malt and grain Scotch whisky production, where no exogenous enzymes are allowed according to the Scotch Whisky Regulations of 2009, small branched maltodextrin molecules remain in the wort following mashing and some whisky yeast strains are available that may further utilise these oligosaccharides. However, unlike brewing, the wort is not boiled and therefore will contain amylolytic activity from malt. A widely used Scotch whisky yeast strain, named "M type"—thought to be a hybrid of *S. cerevisiae* and *S. diastaticus* possesses limited amylolytic activity.

Flavouring is an important part of spirits. The large variety of substances used in flavouring can be grouped into three main categories: herbs and spices, seeds and plants, and fruits. The flavours can be obtained from the raw ingredients by pressure, by extraction or by distillation from an alcoholic solution. One or more ingredients may be soaked in alcohol, which is sometimes warmed to hasten the process. Another method is to mix the ingredient with alcohol and then distil the mixture; the condensed vapour then contains a high proportion of both ethanol and the flavouring components of the added ingredient, provided these components are volatile. Many flavouring ingredients may be used in a single beverage. The recipes are often closely guarded secrets, but it is well known that ingredients such as angelica, aniseed, blackcurrant, camomile, caraway, cinnamon, cloves, coffee, juniper, lavender, lemon, orange and rose petals are used, as are many other fruits and berries. Aperitifs and appetizers are among the richly flavoured spirits, as are liqueurs.

The ethanol content of spirits varies. Generally speaking, whisky, vodka, schnapps, rum, liqueurs and brandies contain around 40 per cent ethanol by volume, but may be stronger. Aperitifs usually have an ethanol content of around 20 per cent. There are three different ways of indicating the ethanol content. American proof is equal to twice the percentage of ethanol by volume; thus, spirits that contain 50 per cent ethanol are 100 proof. The British proof system is slightly more complicated, 100 proof containing 57 per cent ethanol by volume and pure (100 per cent ethanol) alcohol being 175 proof.

Proof, in liquor distilling, a measure of the absolute alcohol content of a distilled liquor, which is a mixture of alcohol and water. The measurement is made by determining the specific gravity of the liquor; that is, the weight per unit volume of the liquid compared to that of water. The measurement of the alcohol content is expressed in terms that vary from country to country: specific gravity, percentage by volume of alcohol, percentage by weight of alcohol, percentage by volume of proof spirit, or by gradations on an arbitrary scale. The measurement is done at an index temperature, as specific gravity varies with temperature.

To summarize, alcoholic beverages are produced following the fermentation of sugars by yeasts, mainly (but not exclusively) strains of the species, *Saccharomyces cerevisiae*. The sugary starting materials may emanate from cereal starches (which require enzymatic pre-hydrolysis) in the case of beers and whiskies, sucrose-rich plants (molasses or sugar juice from sugarcane) in the case of rums, or from fruits (which do not require pre-hydrolysis) in the case of wines and brandies. In the presence of sugars, together with other essential nutrients such as amino acids, minerals and vitamins, *S. cerevisiae* will conduct fermentative metabolism to ethanol and carbon dioxide (as the primary fermentation metabolites) as the cells strive to make energy and regenerate the coenzyme NAD+ under anaerobic conditions. Yeasts will also produce numerous secondary metabolites which act as important beverage flavour congeners, including higher alcohols, esters, carbonyls and sulphur compounds. These are very important in dictating the final flavour and aroma characteristics of beverages such as beer and wine, but also in distilled beverages such as whisky, rum and brandy. Therefore, yeasts are of vital importance in providing the alcohol content and the sensory profiles of such beverages.

Marmite and Vegemite

Not only have yeasts helped us to make and preserve other foods, they have become foods themselves. From England comes marmite and in Australia, vegemite. Both of these products are nutritious pastes, often had for breakfast on toast or bread. With a taste that is best described as "distinctive" for palettes raised on peanut butter, these two products are made from spent brewer's yeast that was previously discarded. It was a German chemist, by the name of Liebig, who found that this yeast could be made into a concentrated vitamin rich food product.

Sourdough Bread

Among fermented foods, sourdough bread is one of the most important baked goods derived from cereal fermentation. Bread, being one of the basic staple foods in many countries and designated the 'staff of life' is believed to be the prior option of sourdough addition from earlier days. The sourdough bread trend was not a major highlight until recently when it re-emerged, and people have become enthused about home-baked sourdough breads.

In 2020, there was a boom in the sourdough bread trend as millions were confined to their homes during the Covid-19 pandemic. The sourdough bread revival has gone into overdrive and was on third Google recipe search globally (https://trends.google.com/trends/yis/2020/GLOBAL/, accessed 21 March 2021). The underlining reasons that have been spurring sourdough bread boom include various health benefits. The estimated annual pro capita intake of bread in European countries is reported to range from 46 (Sweden, Great Britain, Finland and Austria) to 100 kg (Greece, Portugal, Spain and Italy), and about 30 to 50 per cent of bread is manufactured using sourdough.

It is admirable when two simple ingredients, flour and water, and minimal preparation and process are all that are needed to create such an elegant and "alive" final product, sourdough. This unique simplicity, as the main characteristic of fermentation, is "from art to science." However, behind the simplicity lies a complex microbial ecosystem, mainly represented by the species of LAB and yeasts a sym-combiota that, through fermentation, confers flavours, aromas, shelf stability and even nutritional features to the resulting products, dough and bread preparation a famous traveller of the world, with versions of sourdough in Northern Italy, Milan, and San Francisco, as well as selroti and "dhokla" in India; idli in India and Sri Lanka; dosa in India and Sri Lanka; kisra in Sudan; "jalebi" in India, Nepal, and Pakistan; and trahanain in Turkey, Cyprus and Greece.

The use of sourdough as a means of leavening is one of the ancient methods of grain fermentation. Grinding of cereals, pseudo-cereals or legumes followed by addition of water brings about the formation of dough, which subsequently turns into sourdough after a period of time. Making the sourdough starter takes on average seven days and involves replenishing the microbes with fresh flour and water daily. The presence of a symbiotic colony of lactic acid bacteria (LAB) and yeasts inhabiting our diverse ecosystem induce lactic acid fermentation of sourdough which eventually becomes a stable culture after many hours. Once the starter is ready, a small portion is added to the sourdough base ingredients to initiate the sourdough fermentation process; this method is commonly referred to as "backslopping". Unlike standard bread, which is produced through a rapid yeast-only fermentation process, the symbiotic sourdough fermentation of both bacteria and yeast is thought to improve bread quality, including texture, flavour, nutritional content and shelf-life, and replace additives.

Each sourdough harbours different LAB and yeast communities, whose diversity depends on the type of four and water, processing environments and various process parameters such as temperature, time of fermentation, number of refreshments and so on. *Saccharomyces cerevisiae* is the yeast species most frequently retrieved, followed by *Kazachstania humilis* and *Wickerhamomyces anomalus*, while among LAB, *Lactobacillus sanfranciscensis* represents the dominant species. During fermentation, microbial and enzymatic-led conversions of cereal carbohydrates, proteins, lipids and phenolic compounds occur. The microorganisms' and enzymes' activity are interlinked; for example, lactic acid bacteria result in a pH reduction, modulating the activity of cereal enzymes and solubility of substrates (*e.g.*, gluten), and in turn the enzymes can provide substrates to allow the growth of microorganisms.

Different species and strains may have different metabolic patterns, conferring specific properties to each sourdough bread. For example, some strains synthesize essential amino acids or vitamins, *i.e.*, thiamine, vitamin E and folates, others produce prebiotic exopolysaccharides (EPS) and bioactive compounds such as polyphenols, organic acids, peptides and amino acid derivatives, *i.e.*, γ-amino butyric acid (GABA). Moreover, microbial strains may produce enzymes, *i.e.*, proteases, phytases or lipases, and degrade anti-nutritional factors such as raffinose and phytic acid which chelates iron, calcium, magnesium and zinc, thus depleting their bioavailability.

Benefits of Sourdough

Improved Properties of Bread

(a) Extend shelf life
(b) Improve flavour
(c) Enhance aroma
(d) Better texture

Promote Health Benefits

(a) Provide gastrointestinal benefits
    Diminish content of FODMAPs
    - Reduce gluten contents
    - Pre-digest phytic acid-Natural prebiotics and probiotics
(b) Control glycaemic index
    - Better blood glucose control
    - Increase insulin sensitivity
    - Reduce serum cholesterol
    - Reduce risk of developing type 2 diabetes
    - Decreased risk of cardiovascular disease-Improve weight control
(c) Boost amount of vitamins and minerals
    - Regulate metabolism
    - Enhance mood and energy
    - Antioxidant and antihypertensive

Type I sourdough can be categorised as type Ia, containing pure culture sourdough starters with different origin; type Ib, being refreshed every day and fermented spontaneously; and type Ic, which its origin is tropical countries with higher fermentation temperature. At the same time, some researchers also considered another type of sourdough as type 0, referring to pre-doughs or sponge doughs, with addition of baker's yeast (*Saccharomyces cerevisiae*).

To ensure the manufacture of products being optimised and consistent, manufacturers are always keen to identify and develop sourdough types II and III. Despite being relatively unstable, costly and time consuming, the type I sourdough is more commonly utilised for microbiome study due to its natural diversity. Scientists have started to discover that the variability in number and type of microbiota in dough depends not just on the native microbial flora of the baker's environment and hands, but also other factors like choice of flour, when and how often the starter is fed, dough hydration level and type of cereal used, leavening temperature, fermentation time and sourdough maintenance temperature. Different metabolic pathways in sourdough add flavourful metabolic by-products to the mixture, which result in subtler but more complex flavour.

Each microbial community has the ability to produce unique flavour profile as some yield yoghurt-like flavour from lactic acid while others produce sharper, more vinegary note with acetic acid. Lactic acid, together with vinegary acetic acid, contributes to sourdough's tangy characteristic. Hence, with the aid of sourdough, dissatisfying sensory qualities especially the flavour and mouthfeel of some gluten-free bakery products can be enhanced, thus boosting their palatability and market values. As microbial processes show significant potential in improving organoleptic characteristics and design of nutritional quality and health effects of foods and ingredients, sourdough uses have spread widely to other food products *i.e.*, crackers, waffles, pancakes, tortillas, muffin and noodles beyond breads.

The different factors affecting sourdough fermentation and their relationships to the benefits of sourdough intake due to the presence of beneficial microorganisms are also included. Figure 13.19 shows a mindmap guide to sourdough. Sourdough is the result of fermentation from two basic ingredients, mainly flour and water. Microbiome in sourdough is not only affected by various components in different types of flours but also dough hydration level, backslopping time, fermentation time and temperature. Understanding the factors affecting a microbiome enables a greater accrued understanding of microbial ecology, which allows a better selection of starter cultures and fermentation conditions. Microbial succession is also mentioned, which involves the inheritance from ingredients, adaption of acidic conditions and domination by acid-tolerant sourdough bacteria. Three types of carbohydrate fermentation patterns of bacteria are included in sourdough which are obligately homofermentative, obligately heterofermentative and facultatively heterofermentative. With the presence of beneficial microorganisms, sourdough stands out to bring various benefits to properties of bread and health which are also listed above.

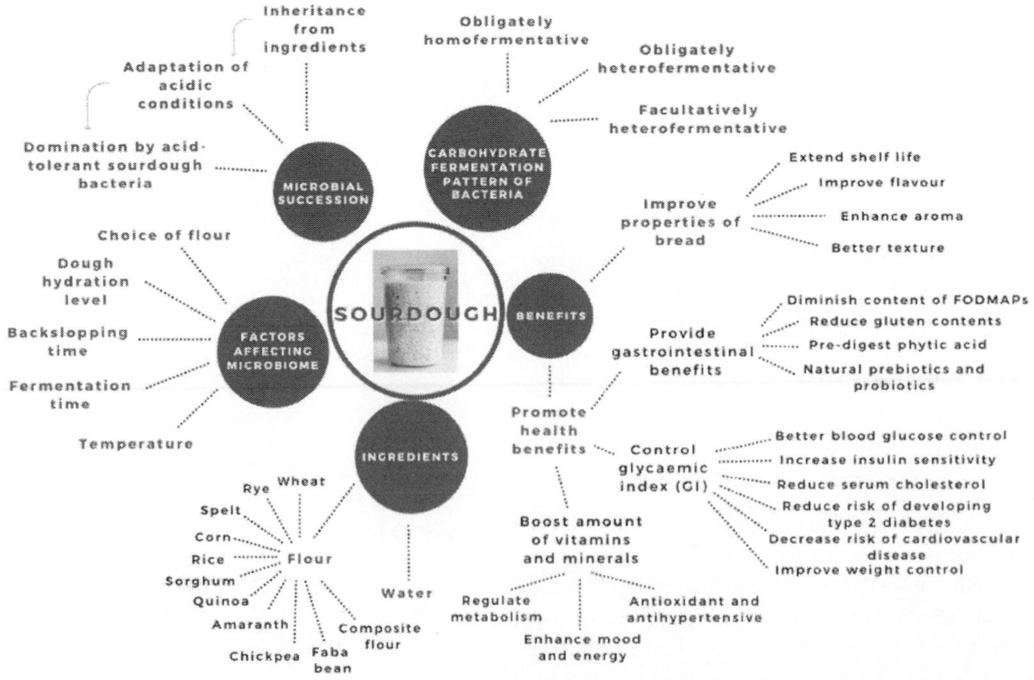

Figure 13.19: A Mind Map guide to Sourdough. (*Source*: Lau *et al.*, 2012. Microorganisms 9, 1355).

Other studies have revealed that, at the genus level, Prevalent strains show a biodiversity in different geographical areas, including *L. plantarum, L. sanfrancisensis, L. pontis, L. brevis, L. curvatus, Lb. sakei, L. panis, L. alimentarius, L. paralimentarius, L. pentosus, L. helveticus, L. casei, L. amylovorus, L. buchneri, L. farciminis, L.fructivorans, L. homohiochii, L. hilgardii, L. johnsonii, L. lactis* subsp. lactis, *L. crispatus, L. delbrueckii, L. fermentum, L. reuteri, L. acidophilus, L. vaginalis, L. gallinarum, L. graminis, Lc. lactis, Leuc. Mesenteroides* subsp. mesenteroides, *Leuc. Mesenteroides* subsp. dextranicum, *Lc. holzapfelii, P. pentosaceus, W. cibaria, W. confuse* and *W. viridescens*, with the yeasts *S. cerevisiae, S. exiguus, C. humilis, C. milleri, Issatchenkia orientalis, Kazachstania barnettii* and *Saccharomyces bayanus/ Kazachstania* sp.

Microbial Communities in Different Sourdough Starters

Enterococcus, Lactococcus, Lactobacillus, Leuconostoc, Pediococcus, Streptococcus and *Weissella* are associated with sourdough production. In the above micro-ecological niche, the genus *Lactobacillus* is dominant. A native strain of *S. cerevisiae* is the principal yeast of most bread fermentations. However, several non-*Saccharomyces* yeasts occur in the fungal consortia in sourdough fermentation, such as *Candida, Rhodotorula, Pichia, Trichosporon, Sporobolomyces* and *Yarrowia*.

Sourdough starter, which is also known as natural yeast, has conducive habitat for microbes that can support growths of more than 50 species of lactic acid bacteria (LAB) and more than 20 species of yeasts. As technology advances, many new species of LAB are discovered. For instance, in March 2020, there are already 262 species being reported under the genus Lactobacillus. Most recently, in study of Zheng *et al.* (2020), the genus Lactobacillus was proposed to be reclassified into 25 genera after reviewing the taxonomy of *Lactobacillaceae* and *Leuconostocaceae*. Consequently, some names of LAB were updated. They included *Fructilactobacillus sanfranciscensis* (formerly known as *Lactobacillus sanfranciscensis*), *Lactiplantibacillus plantarum* subsp. *plantarum* (formerly known as *Lactobacillus plantarum* subsp. *plantarum*) and *Levilactobacillus brevis* (formerly known as *Lactobacillus brevis*).

The different species require different conditions for their optimal growth, hence the proportion of each species found in each sourdough starter also varies. Towards the final stage of fermentation, sourdough starters are asserted to be mature when the cell densities and abundances for all LAB and yeasts reach a plateau. The species that are more competitive and adaptive in mature sourdough are usually those with larger population. The divergence of microbial profile from one sourdough starter to another can also be observed when the types of ingredients (mainly flour and water), number of sourdough propagation steps, fermentation time, fermentation temperature, dough hydration, leavening temperature and baking environments are different.

Cereal and Legumes Based Fermented Foods

Rabadi, wadi, dhokla, dosa, idli etc. are some of the lactic acid fermented foods prepared from cereals and legumes. Several studies have shown improved nutritional value in grains as a result of lactic fermentation, principally through increasing the content of essential amino acids. It has also been reported that fermentation of plant products reduces levels of antinutritional factors which they may contain such as cyanogenic glycosides and phytic acid, although this effect is also due to other aspects of the process such as soaking or crushing rather than microbial action alone. Some have reported an improvement in the bioavailability of minerals and digestibility of starch and protein in the above mentioned fermented foods.

Vegetable Fermented Products

Since ancient times, the fermentation of vegetables has also been practiced by mankind, as proved by a long history of traditional products spread all over the world. Vegetables are mainly fermented by LAB both spontaneously and by means of inoculum and backslopping. Among these lacto-fermented vegetables are fermented cabbage (kimchi and sauerkrauts), fermented leaf (gundruk) and pickles (cucumber, chillies, capers and others). Many of the positive features related to fermented vegetables are derived from the effects of acids and fermentation, which, as a consequence of fermentation, change their form to become more bioavailable, thus increasing their effect and elimination of anti-nutritional compounds.

Fermentation of vegetables has as primary effect of increasing the shelf-life of food. Fermented vegetable products have common characteristics of high acidity and low pH that usually make them safe and microbiologically stable all along their shelf life valuable nutritional and organoleptic attributes, make them foodstuffs from the past and toward the future. However, survival of certain acid-resistant pathogenic bacteria can occur when they are not properly processed or handled. Different additives such as acidifiers (*e.g.*, acetic, lactic, malic, citric, and other acids), preservatives (*e.g.*, sorbates, benzoates, sulfites, in some cases), and others can be added to pickled products to enhance their safety and keeping quality. Nevertheless, the most extended practice is the heat treatment, pasteurization. Heat treatment deactivates vegetative bacteria forms of possible pathogens, preventing their growth in this environment.

It allows to ameliorate the intake of nutrients like fiber, vitamins, and minerals. This effect is particularly useful since it permits the introduction of these micronutrients in periods when vegetables are unavailable. In a recent review, Bousquet *et al.,* tried to find a relation between decease due to COVID-19 and diet of populations, focusing on the consumption of sauerkrauts. From their data analysis emerged how in the areas where the consumption of sauerkrauts is higher the number of deaths is slightly lower. Data anyway do not seem to be correlated and many other factors and bias contribute to the obtained results, thus further studies are needed to confirm any link (Varzakas *et al.,* 2017).

In this section, we focus on the two main products derived from cabbage fermentation, representing a widely consumed staple in western and eastern areas of the world: sauerkrauts and kimchi.

Sauerkraut

Salted and fermented vegetables have a long history in human nutrition since ancient times, with sauerkraut being one of the most popular vegetable fermented products. Sauerkraut, a traditional dish in Central and Eastern Europe, the United States, and Asia, results from the lactic acid fermentation of chopped and salted white cabbage (*Brassica oleracea* var. *capitata*).

The word sauerkraut originated from the German term for "sour cabbage." The first mention of sauerkraut dates from 2000 years ago in China, where cabbage was fermented in rice wine, and was introduced 1000 years later in Europe by Genghis Khan after invading China. Europeans substituted wine with salt in sauerkraut.

The health benefits of sauerkraut were well-known in early civilizations. Hippocrates, a Greek physician, recommended sauerkraut against overweight and Romans consumed it to prevent intestinal infections. Moreover, Captain James Cook, an English navigator and explorer, replenished the food supplies of his ships with sauerkraut on his lengthy voyages, as he noticed that it did not need refrigeration and prevented scurvy in sailors.

Sauerkrauts are the product of cabbage fermentation (Brassica oleracea var. capitata) with 2.3-3.0 per cent salt. Cabbage contains enough lactic acid bacteria in order to ferment and produce sauerkraut with salt alone. In order to obtain product of the highest quality all those bacteria strains must ferment in a certain sequence. This happens naturally as long as sauerkraut is fermented around 65°F (18°C).

Sauerkraut manufacturing can be carried out following spontaneous fermentation, generally involving *Leuconostoc* spp., *Lactobacillus* spp., and *Pediococcus* spp. or fermentations guided by selected and specific bacteria. The low pH of the final product results in a preserved cabbage. Sauerkraut (homemade and shop-bought) has been shown, through culture-dependent techniques, to contain *Bifidobacterium dentium, Enterococcus faecalis, Lactobacillus casei, Lactobacillus delbrueckii, Staphylococcus epidermidis, Lactobacillus sakei, Lactobacillus curvatus, Lactobacillus plantarum, Lactobacillus brevis, Weissella confusa, Lactococcus lactis* and *Enterobacteriaceae.* Adding a starter culture of *Lactobacillus casei* 11MZ-5-1 produced a sauerkraut containing predominantly *Lactobacillus* and *Lactococcus,* compared to spontaneous sauerkraut which, along with *Lactococcus* and *Lactobacillus,* also contained significant *Enterobacter* and *Pseudomonas* and was more variable in microbial composition.

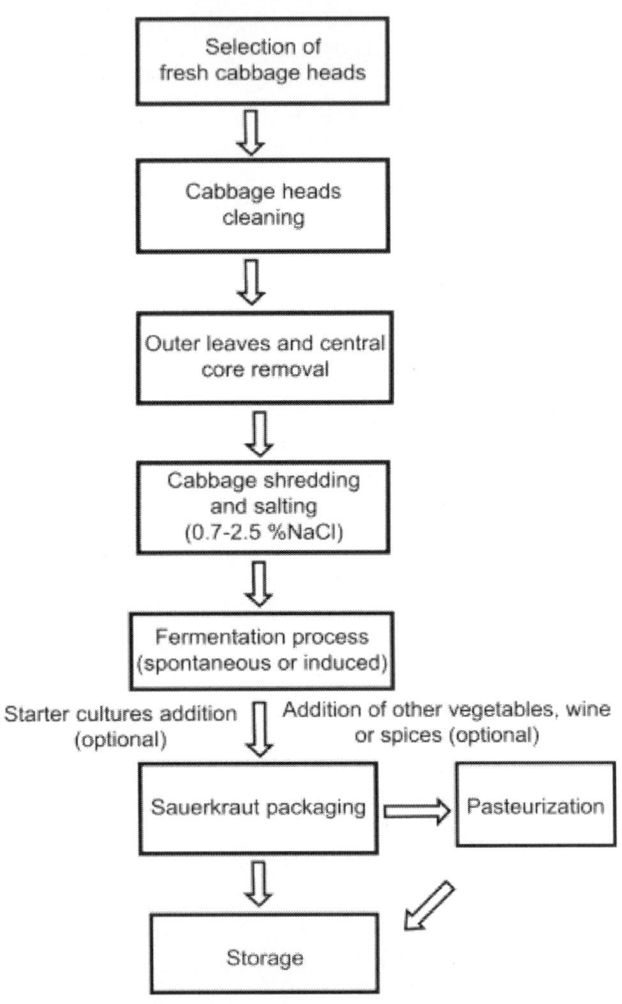

Figure 13.20: Flow Chart for Sauerkraut Production.

Sauerkraut has also been shown to predominantly contain *Leuconostoc* and *Lactobacillus* spp. Certain Lactobacillus species isolated from sauerkraut demonstrate probiotic potential, with tolerance to low pH, adherence to Caco-2 cells and antimicrobial activity against pathogens *in vitro. Lactobacillus paracasei* HD1.7, commonly found in sauerkraut, has been shown to produce a broad-spectrum bacteriocin that may play a role in sauerkraut preservation. Recent studies have also identified the genus *Weissella* as important to early fermentative processes.

Sauerkraut fermentation begins with the initial proliferation of *Leuconostoc mesenteroides*, which rapidly produces carbon dioxide and acid. This quickly lowers the environmental pH, inhibiting the growth of undesirable microorganisms that might cause food spoilage while preserving the colour of the cabbage. The action of *L. mesenteroides* changes the fermentation environment so that it favours the succession of other LAB, such as *Lactobacillus brevis* and *Lactobacillus plantarum*. In traditional sauerkraut production, this process proceeds at 18 °C for roughly one month. The combination of metabolites that these organisms produce leads to favourable sensory qualities, the unique flavours, aromas, and textures associated with fermented foods in the final product. The temperature of fermentation also plays an important role in terms of colour, flavour, and preservability.

The best quality sauerkraut is produced at 65-72°F (18-22°C) temperatures. Temperatures 45.5°F (7.5°C) to 65°F (18°C) favour the growth and metabolism of *L. mesenteroides*. Temperatures higher than 72°F (22°C) favour the growth of *Lactobacillus species*. Generally, lower temperatures produce higher quality sauerkraut, although at 45.5°F (7.5°C) bacteria are growing so slow that the cabbage might need 6 months to complete fermentation. Higher temperatures produce sauerkraut in 7-10 days but of the lesser quality. This creates such a fast fermentation that some types of lactic acid bacteria don't grow at all and less reaction take place inside what results in a less complex flavour.

- Below 45.5°F (7.5°C) fermentation time is up to 6 months.
- At 65°F (18°C) fermentation time is 20 days.
- At 90-96°F (32-36°C) fermentation time is 10 days.

During fermentation, the composition of the product changes and, at the end, aside from macronutrients, it is possible to find a good amount of fiber, vitamin C, organic acids (lactic, acetic, malic and succinic), SCFA (propionic acid), ethanol, and acetaldehyde. Due to the knowledge about bioactive compounds present in fermented foods, in recent years, many efforts were made to improve the general quality of fermented vegetables while creating a product rich in bioactive compounds. For example, the utilization of a nisin-resistant strain of *Leuconostoc mesenteroides* in combination with a nisin-producer strain of *L. lactis* allowed obtaining a product with a suppressed native microflora. Also, *Leu. mesenteroides* in combination with *Pediococus dextrinicus* showed a good potentiality to produce bioactive enriched foods. *Latilactobacillus sakei* showed a predominance in this feature since its utilization in vegetable fermentations allows the obtaining of foods with three times the concentration of bioactive compounds concerning any other studied bacterial strains.

Health Benefits of Sauerkraut Consumption

As fiber is a star in sauerkraut nutrition facts, it is undeniable that one of the best health benefits of sauerkraut is to facilitate the digestion. For long, it has been well-known that the consumption of fiber rich vegetables will boost the digestion effectively. Sauerkraut, a product from cabbage, will supply the body with a high amount of fiber to fight against digestive disorders. The supplement of dietary fiber will have positive effects on the bowel movements, making stools soft and smooth. A study (Nielsen *et al*., 2018) reported that the effect of sauerkraut consumption on irritable bowel syndrome (IBS) affected a patient and reported that consumption of sauerkrauts, both fresh and pasteurized, led to a reduction of symptoms after 6 weeks, with a change in microbial composition of faecal matters of participants. Also, the high presence of dietary fibers seems to be involved in alleviating IBS symptoms.

Since many reports suggest that regular consumption of this product can lead to the intake of a considerable amount of healthy bacteria (>106 log CFU/g), recent studies have focused their attention on the isolation of LAB from sauerkrauts. Strains of *Lactiplantibacillus paraplantarum*, *L. brevis*, and others Lactobacillus strains isolated from sauerkrauts showed adhesion to Caco-2-cells and inhibitory activity towards pathogenic microorganisms.

Cabbages are also rich in phytochemicals with multiple possible bioactivities, but these compounds, mainly glucosinolates, are normally not bioavailable in the fresh product. Hydrolysis of glucosinolates leads to release of isothiocyanates, thiocyanates, epithionitriles, nitriles, and indolic compounds, all recognized for their valuable health boosting activities. Like many other fermented foods, sauerkrauts show antitumoral properties, exerted by activating enzymes that eliminate xenobiotics and increasing apoptosis of cancerous cells. Specifically, indole-3-carbinol (I3C) is deeply investigated since it was shown to exert inflammation-modulating effects, promote cells proliferation, and inhibit tumour invasion in different tissues) (Bousquet *et al.,* 2021).

The presence of vitamins and organic acids gives to sauerkraut a powerful antioxidant feature, but it is also related to reduced inflammation, atherothrombosis, and increased human system efficiency in neutralizing reactive oxygen species. Antioxidant activities are also connected to reduced oxidative damage at the expense of DNA, which can also be due to indolic compounds' ability to scavenge chemicals, avoiding damages to DNA and other structures.

Fermentation enriches sauerkraut with a group of enzymes called Mono Ammino Oxidase Inhibitors (MAOIs), inhibiting Mono Ammino Oxidase (MAOs), which are a family of enzymes involved in arising depressive states, anxiety, obsessive compulsive disorder, and development of Parkinson's disease (Sawicka, 2020).

In this new post- and para-biotic field, sauerkraut has found a niche, since industrial productions require a pasteurization step, resulting in the death of live cells but not hampering the beneficial effects of wild LAB. Sauerkrauts are one of the most studied fermented vegetables, in fact suggestions about the introduction of sauerkraut in the diet are easy to find in the literature, even though intervention studies and dietary supplementation with this fermented food are still lacking and further investigation is surely needed. The literature reveals that sauerkraut possess a vast array of health effects related to glucosinolate compounds and microbial contribution in terms of microbiome, para-, and post-biotics. Despite these incredibly appealing features, there is a possible presence of biogenic ammines (Stiemsma *et al.*, 2020), while paying attention to microbial populations, since some harmful bacteria can survive to harsh conditions that arise during fermentation.

Kimchi

Kimchi is a lactic acid fermented vegetable product that is known to date back at least 2,000 years, as a record of *kimchi* (Zu) appears in the "Book of Odes (Sikyung)" published around 500 BC. *Kimchi's* popularity extends to the present, and it was reported to be one of the world's healthiest foods by the American Journal of Public Health in 2006. Moreover, *kimchi* was registered as a world intangible heritage product in 2013. There are about 200 types of *kimchi* in Korea, which are prepared with fresh vegetables, mainly Chinese cabbage or radish (Kwon *et al.*, 2014).

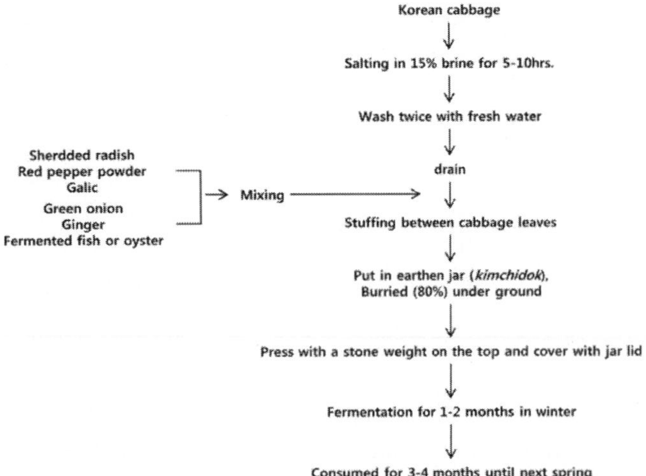

Figure 13.21: Flow Chart for Kimchi Production.

Kimchi is a traditional Korean cuisine made from kimchi cabbage and various seasonings including ginger, garlic, red pepper, spring onion, jeotgal (salted seafood), and salts. The addition of salted and fermented seafood products and other seasonings is optional. There are many types of kimchi in Korea made with various vegetables as the main ingredients. Other vegetables such as radish and cucumber can also be added. In order to eat fresh vegetables that are not available in winter, there is a culture called 'gimjang' in Korea, which manufactures kimchi that can be consumed for a long time (more than a year). In general, kimchi made in early winter is stored at low temperatures and is consumed until the next year.

Prior to fermentation, the kimchi mix contains a variety of different bacterial species within the *Leuconostoc, Lactobacillus, Pseudomonas, Pantoea* and *Weissella* genera. However, once fermentation has started, the bacterial diversity decreases and the bacterial community is rapidly dominated by the genus *Leuconostoc* within only three days of fermentation. The microorganisms essential for kimchi production can be divided into lactic acid bacteria (LAB) and yeast. Kimchi is fermented by lactic acid bacteria (LABs) at low temperatures, ensuring proper ripening and preservation. It is mainly *Leu. mesenteroides* that creates the acidic and anaerobic environment adaptable to the growth of more acid-resistant bacteria like *L. brevis* and *L. plantarum*.

LAB are formed in the early stages of kimchi fermentation, whereas yeast is known to be formed in the middle stages of kimchi fermentation. In many previous studies, LAB have been primarily studied as the key microorganisms responsible for kimchi fermentation. Yeast, known as a symbolic microorganism in many fermented foods due to the production of alcoholic compounds and carbonic acid, can also significantly affect the quality of kimchi. However, the changes in the metabolites by yeast during fermentation and storage of kimchi have not been studied.

Spontaneous fermentation in kimchi leads to the formation of various microorganisms, affecting the sensory qualities of kimchi. The production of organic acids from carbohydrates and the resulting reduction in pH maintain the freshness of the vegetables during storage. It is well known that the composition of metabolites important to kimchi taste and flavour, such as organic acids (lactic acid and acetic acid) and other flavouring compounds (mannitol and amino acids), are directly affected by the kimchi microbial community. LABs produce various compounds in addition to organic acids, including CO_2, ethanol, mannitol, bacteriocins, γ-aminobutyric acid (GABA), ornithine, conjugated linoleic acids, and oligosaccharides; these components contribute to the fermentation characteristics and health functionality of kimchi. Properly fermented kimchi is flavourful, possessing a distinct, savoury combination of sour, spicy, hot, sweet, and fresh tastes. The fermentation is markedly affected by environmental factors, such as temperature, salt concentration, and the presence of certain ingredients.

Kimchi is considered a natural functional food for the high presence of dietary fibers, minerals, vitamins, capsaicin, organic acids, polyphenols, and fermentation by-products (organic acids, bacteriocins, and others). The presence of these compounds is reported in scientific literature to produce positive effects on the health of consumers. Presence of a wild and acid-resistant microbiota is connected to the lowering of pH in the intestinal lumen and in the faeces, which is connected to a better microbiota, with an increased count in LAB and Bifidobacteria and a lower level of harmful and pathogenic microorganisms.

Kimchi is studied, especially in the most recent literature, for its ability to modulate gut microflora, and Park *et al.* (2020) studied the effect of kimchi to exert an anti-obesogenic effect on the microbiome, starting from the assumption that many factors can cause obesity, such as an unbalanced diet; genetic factors; and unhealthy gut microflora resulting in modifying energy intake and accumulation in the adipocytes, increasing obesogenic effect. The intestinal microbiota is more than the sum of its part, it is an organism able to live in symbiosis between the same parts composing it and ourselves. Due to microbial diversity, long-term stability, ease of use, and domestic preparation, kimchi was taken into account to modify gut microbiota, helping pathologically obese subjects, normalizing their lipid levels and modulating their microbiome. Results of these experiments highlighted that supplementation with kimchi in mice fed with an HFD cannot significantly decrease weight gain, with respect to mice fed with just an HFD, indicating that the number of calories introduced is the main factor in weight gain.

The introduction of kimchi anyway showed a reduction in blood glucose, triglycerides, and high- and low-density lipoproteins with respect to mice fed with only HFD. Even if the total weight gain was not significantly decreased by kimchi in HFD mice, other indexes like total body fat gain, liver weight, and adipocytes' dimensions and counts were lowered by the administration of kimchi. Also, gut microflora resulted modulated by the administration of kimchi, Akkermansiaceae, Coriobacteriaceae, and Erysipelotrichaceae, which are normally related to HFD and consequently to obese subjects, were lowered in mice fed with kimchi, while the abundance of Muribaculaceae, negatively correlated with obesity, increased in kimchi-fed mice.

Kimchi, due to the high presence of fibers and nutritional compounds was also studied as a solution to cope with prediabetics patients. Prediabetics are subjects who have blood glucose higher than unaffected subjects, but not high enough to be considered properly diabetics, and are strongly subject to develop this issue later, due to unhealthy lifestyle and diet. Fortunately, a change in dietary and lifestyle habits can slow, and in some case even stop, the progression of prediabetes into diabetes. In an intervention study, An *et al.,* administered 100 grams of kimchi per meal to 21 prediabetic volunteers for 2 weeks, followed by a 4-week washout period. From anthropometric parameters after regular consumption of kimchi, it emerged that insulin sensitivity and resistance and blood pressure were positively affected by introduction of this fermented product. Also, the participants' body mass index (BMI) and weight decreased significatively, together with waist circumference, which is strongly bound to insulin resistance. The consumption of kimchi thus revealed to be a strong ally in fighting the onset of diabetes (Lee *et al.*, 2018).

Being rich in anti-microbial compounds, produced mainly by an active and resistant positive microflora, kimchi is employed from centuries as "medicine" food and can be ascribed in the functional foods group. Functional foods are "foods or dietary components that may provide a health benefit beyond basic nutrition". Some studies focused on the utilization of kimchi to fight infections by *Helicobacter pylori*, which is a well-known contributor to the development of peptic and perforative ulcers and one of the recognized class I carcinogens. The high level of antioxidants, vitamins, and the presence of other phytochemicals, together with the reduction of ingestion of other harmful products prove to be an effective, cheap, and easy way to control the development of *H. pylori* and help eradicate it. During *H. pylori* infection, the body reacts by increasing the expression of Foods pro-inflammatory genes to fight the attack, but the maintenance of an inflammation state for too long in the body can lead to DNA damage through oxidative stress and disruption of cell life cycle. In this way, gastric and intestinal carcinogenesis is facilitated. Therefore, the introduction into the diet of a food possessing antioxidant and anti-inflammatory properties such as kimchi results as being useful. In the optic of cancer's prevention, kimchi has been employed as a tool to fight the development of colon-rectal colitis-associated cancer in patients chronically affected with IBS.

Ingestion of kimchi could be a strong ally, due to ease of use, stability, and of course due to the introduction of a series of nutritional compounds exerting a plethora of positive health-related effects. Kimchi can play an important role also in maintaining under control the degeneration of several chronic diseases like IBS, Crohn's disease, and infections due to external attacks or unhealthy eating habits. However, it has to be considered that, as for other spontaneous fermented foods, kimchi contains a high level of salt involved in the formulation to control negative microflora. In some studies, the kimchi's supplementation was in fact limited to around 100 g of fermented food, matching nutritional suggestions for salt introduction to the diet, while other studies increased to 210 g per day the administration of kimchi, which provides a salt content higher than what the guidelines suggest. All this considered, the introduction of kimchi as a regular meal or side-dish also in Western countries could help populations to control the development of gastro-intestinal issues (Castellone, *et al.*, 2021).

Fermented Meats

Fermented meat products have been known as highly appreciated foods throughout the ages. They are found in many regions worldwide as a part of traditional diets, where they are perceived as attractive gastronomic entities contributing to cultural and geographic distinctiveness. In Europe, fermented meats can be traced back to, at least, classical antiquity where they served to preserve meat after slaughter. The production of dry-fermented sausages, consisting of seasoned and salted minced meat that is stuffed into intestinal casings, has been documented in historical sources as a Roman craft. The technique was, however, largely based on the art of dry ham production developed by the Celts and Gauls, consisting of the salting and drying of hind legs of wild boars and pigs. Over the years, fermented meats came to exist in an overwhelming variety, characterized by large differences in ingredients, shape and caliber, and processing conditions. Several types of meat have been applied, including pork, beef, horse, donkey, deer, poultry, and ostrich.

Well-known examples of dry-fermented sausages include different types of *salami* (*e.g., salame Brianza*), *chorizo* (*e.g., chorizo de Pamplona*), and *saucisson sec* (*e.g., rosette*). In addition, some fermented meats are semidry or even moist, such as German *mettwurst*, which have distinctive texture properties but are generally displaying a decreased stability. Besides Europe and its expansion to the New World, Asia also has a remarkable tradition of meat fermentations, although this has been less documented and explored. Yet, all local fermented meat specialties have origins in more or less remote pasts, with unique stories to tell, but eventually trace back to the same ancestral need for meat preservation. In fact, meat fermentation originally took off as an empirical preservation method of a primary commodity, in a similar manner as the preservation of milk led to the production of cheese and fermented milks of all sorts.

Fermentation modifies raw meat and transforms it into a different product with new organoleptic characteristics. The key point is the acidification, which enhances protein coagulation (enhancing flavour and softness of the tissues) and drying (lowering the pH also lowers the water activity, a_w). Fermented meat products are meat products with longer shelf lives and special flavours, textures and colours that are made of raw meat under specific temperature and humidity conditions by fermentation via microorganisms or enzymes. Fermentation methods include natural fermentation and the artificial addition of a starter to regulate fermentation. The microorganisms involved in the natural fermentation process are primarily derived from raw materials and the environment. Under specific curing and maturing process conditions, typical local microbial communities are formed through competition, thus significantly affecting the product quality and safety. Hence, the fermented meat products are the result of a complex microbiological activity that consists mainly of a lactic fermentation and characteristic biochemical changes for which lactic acid bacteria and coagulase negative cocci are indispensable. The manufacture of fermented meat products is a very important part of the meat industry in many countries, where the use of starter cultures is not a common practice.

Many typical fermented products are still produced using artisanal technologies without selected starters. Artisanal products have higher quality than those of controlled fermentations inoculated with industrial starters and have distinctive qualities, especially due to the specific composition of the microbiota. Manufacturing practices and fermentation techniques lead to obtaining products with specific organoleptic profiles. In recent years, the use of starter cultures with antimicrobial properties has been spread to improve the quality and safety of fermented meat products. Many products already employ starter cultures which are cultivated in SSF. These starter cultures are usually considered as one of the ingredients to prepare meat products.

Meat's indigenous flora is affected by a series of factors, such as the animal species, breed, nutrition, husbandry, slaughter conditions and post slaughter preservation. It consists of lactic acid bacteria (mainly lactobacilli), coagulase negative *staphylococci* (CoNS), *micrococci* and yeasts which can initiate fermentation. Their population size varies significantly from lot to lot, and this makes the control of the process particularly for artisan products seem almost impossible.

The major technological intervention is the addition of starter cultures of microorganisms (at 10^7 CFU/g), which rapidly acidify and ferment in a controllable mode. *Pediococcus acidilactici*, *Pediococcus pentosaceus*, *Lactobacillus pentosus*, *L. casei*, *L. curvatus*, *L. plantarum* and *L. sakei* are the most common bacteria of these starter cultures. The product is anaerobically cased and left to ferment. The pH value in the commercial form is less than 5.0. In general, the microbiome of fermented meats (i) provides a long storage period, (ii) ensures safety from pathogens due to its protective activity, (iii) increases tenderness and palatability, (iv) delivers a high nutritional status due to its enzymatic activity and (v) reduces the cooking time.

Studies have shown that *Staphylococcus carnosus* (*S. carnosus*) and *Staphylococcus xylosus* (*S. xylosus*) can significantly promote the hydrolysis of fat and protein in preserved meat, improve the colour and promote the rapid formation of flavour. The esterases, proteases and catalases produced by microorganisms have synergistic effects with endogenous enzymes, causing biochemical changes in the proteins and fats in raw meat and forming small molecules, such as amino acids, esters, peptides and short-chain volatile fatty acids. The digestive function is greatly improved, and the contents of essential amino acids, vitamins and *bifidobacterium* are increased, enhancing the effects on nutrition and health.

Chen *et al.* (2017) inoculated Harbin dried sausage with a mixed starter, and the free fatty acid (FFA) content in the inoculated group was significantly higher than that in the uninoculated group. The peroxide value and thiobarbituric acid contents in the inoculated group were significantly reduced compared with those in the uninoculated group, and the contents of aldehydes, ketones and hydrocarbons, which are related to lipid oxidation, were significantly reduced compared with those in the uninoculated group. The mixed starter promoted the hydrolysis of fat, inhibited lipid peroxidation and improved the fermentation flavour of Harbin dried sausage. Wang *et al.* (2014) added complex microbial starters to traditional preserved meat, and the contents of free amino acids and the types of volatile flavour substances in the preserved meat increased, which improved its nutritional value, flavour and safety.

Fermented meat products are important not only for their sensory characteristics, nutrients content and cultural heritage, but also for their stability and convenience. They have been negatively linked to potential high content of (curing) salt, sugar, and fat. However, the use of nitrite and nitrate in fermented meat products contribute not only to safety but also to the typical dry cured aroma. The use of nitrite and its effect on aroma improvement is well known; however, the mechanisms involved in cured flavour generation are not well established.

Different mechanisms have been proposed in cured flavour development. The most relevant mechanism is the ability of nitrite to decrease the oxidation in the meat product lessening the rancid odour notes from carbonyl compounds as observed in pork and cooked ham. The application of other chemical antioxidants could not generate the cured aroma produced by nitrite. Furthermore, it has been revealed the role of carbonyl compounds like aldehydes in masking key sulfur compounds, such as 2-methyl-3-(methyldithio)furan. Therefore, the use of nitrite as curing agent was attributed to a balance in odour perception between aldehydes and sulfur compounds.

Microorganisms primarily inhibit the growth and reproduction of harmful bacteria through bacterial competition, acid production and bacteriocin production. Studies have shown that *S. xylosus* can inhibit the growth of *Listeria monocytogenes (L. monocytogenes)* in sausage. When *Latilactobacillus sakei subsp. sakei* (*L. sakei*) was inoculated as a starter culture into Sichuan sausage, and the results showed that the lactic acid bacteria rapidly controlled the growth of the total bacterial colonies during sausage fermentation and inhibited the growth of foodborne pathogenic bacteria, such as *Escherichia coli* (*E. coli*) and enterobacteriaceae. Lactic acid bacteria can decompose carbohydrates in sausages to produce a large number of lactic acid and a small number of acetic acid as well as propionic acid, formic acid, 3-methyl-butyric acid, butyric acid and other organic acids, thus extending the shelf life of sausages. Lactic acid bacteria can reportedly inhibit or kill spoilage microorganisms and pathogenic microorganisms in food.

Microorganisms can improve the quality of fermented meat products to a certain extent. They can not only promote the degradation of protein and fat and accelerate the formation of flavour substances but can also improve the nutritional value of fermented meat products. At the same time, they can inhibit the growth and reproduction of harmful bacteria, reduce the use of salt in fermented meat products, improve the taste, reduce the content of nitrite and BAs, and improve the safety of fermented meat products.

At present, research on fermented meat products largely focuses on the microbial diversity of fermented meat products, the mechanisms by which specific microbes influence the quality of fermented meat products, the characteristic flavour of fermented meat products and their production pathway. A large number of studies have shown that microorganisms can promote the formation flavour of fermented meat products and inhibit the generation of BAs, nitrosamines and other harmful chemicals in fermented meat products, but the specific mechanisms have not been fully reported. In the future, research on these mechanisms will provide effective theoretical support for improving the production process and the modernization and standardization of production of fermented meat products.

Fish Products

In many cultures throughout history, people have developed diverse means to preserve food resources like fish during periods of scarcity following an abundant seasonal supply of fish. The fish had to be preserved via variety of ways due to its perishable nature, including drying, salting, and fermenting.

The usage of fermented fish as a dish in traditional culinary habits dates all the way back to ancient times in many societies throughout the world. The earliest fermented fish sauce ever recorded is *garum* in classical Rome, which was prepared from the viscera and blood of mackerel. Consigned to academic and cultural invisibility in food studies for the most part, this fish sauce has only recently come to be seen as having a striking similarity with fish sauces in Southeast Asia, from its umami taste and flavour and its ability to radically transform everyday dishes to its being cheap and commonly available.

There is a considerable degree of artisanship associated with fermenting fish in all the communities where it is a part of their everyday life and culture. Fish fermentation is a small-scale industry and can provide direct employment to the people involved in it with very simple artisanal equipment. In the developing countries of Southeast Asia, South Asia and Africa, there is a good market for fermented fish products.

Fermented fish is significant traditionally and economically in East and Southeast Asian countries. Because of its unique flavour and taste, its popularity in these countries is far reaching. The fermented fish products are a significant source of protein, forming an integral part of staple diet, and possessing great cultural significance in many Asian countries such as Korea, Japan, China, Thailand, Cambodia, Malaysia, Philippines, and Indonesia and in Asian foodways in general. Apart from East, Southeast and South Asia, countries in other parts of the world like the Middle East, Africa and the Scandinavian countries in Europe also produce and consume fermented fish products, like surstromming (fermented herring) and rakorret (fermented trout) in Scandinavia and fermented anchovy in France and Spain. Other South Asian countries like Sri Lanka, Bangladesh, Bhutan, and Nepal also have their own fermented fish products. In India, fermented fish is popular and various types of fermented fish products have been reported from its Northeastern region. *Matkashidal* is a widely known fermented fish that is economically viable in the Northeast Indian states of Manipur and Tripura. However, these products are typically produced at the family or village level using traditional methods in the region.

Southeast Asia region should be considered a central point of origin of fermented fish products, as the freshwater fishes in the rice producing region (particularly in the rice fields) began to be fermented by the earliest indigenous populations. The primary reasons for the fermentation of fish in this region since ancient times are the seasonal availability of fish and its scarcity during the long dry months. The abundance of water bodies and the fact that the cultivation of rice required a huge and consistent supply of water during its growing period (something that the region's climate ensured) created a conducive environment for fish to flourish during the rainy season. Furthermore, archaeological excavations and studies reveal that Southeast Asia was a hub of salt production (apart from more well-known archaeological sites in China) since prehistoric times. These circumstances together compelled and facilitated the preservation of fish using fermentation methods, given the high degree of humidity in the region, leading to fermented fish and fish products dominating the food cultures in the region. One notable example is North eastern Thailand where rice and fish are staple foods, while salt is the most commonly available preservative. As a result, these resources consequently define the food cultures in this region, in the sense that fish fermented using salt as an additive is a primary side dish that invariably accompanies and renders the staple diet of rice tasteful. Due to the relative lack of economic security and modernized ways of life in many parts of Southeast Asia (especially in the rural areas like in North eastern Thailand), producing and consuming fermented fish were (and still are) a cost-effective and accessible way of including proteins and minerals like calcium and phosphorus in the everyday diet. In addition, fermented fish, with its umami taste, relieved the mundanity of the rice and vegetable diet of the local population.

In the case of India's Northeast, the climate, with its heavy rainfall and high humidity (making this region the wettest in India, and indeed, among the wettest regions in the world), made it impossible to dry the fish, while the sun shone, simply because there was not enough sunlight, unlike in the drier parts of India. The monsoon season from May to September naturally coincided with the fishing season, followed by a long period of scarcity of fresh fish in the region. This necessitated an efficient and relatively faster method of fish preservation through fermenting, and it is thus that this part of India has a central presence in the fermented food map of India.

In many fermented fish products of Asia, *koji* (*Aspergillus* sp.) is added to enhance the fermentation. The Asian climate has a higher humidity and temperature that is optimal for filamentous fungi like *Aspergillus*, *Penicillium*, *Rhizopus*, *Mucor* to thrive, leading to their wide use in fermentation processes in various parts of the continent.

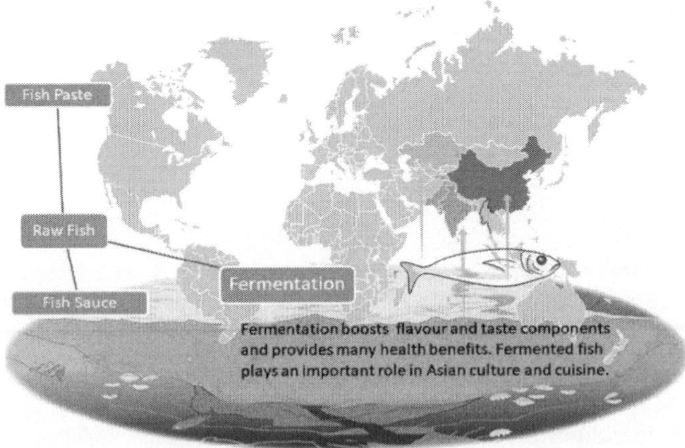

Figure 13.22: Fermentation of Fish in different Regions of the World.

Traditional and non-traditional fish products are classified in different ways according to their respective mechanism of processing, substrates used, salt concentration, and the form of the final product. Such processes and their resultant fermented fish products have been divided into three categories. The first is a purely traditional and natural process based on enzymes present in the fish flesh and entrails with salt as seasoning. In the second category, microbial enzymes and salt are added in the form of 'starter cultures' on the flesh of fish and entrails. The third category consists of non-traditional products manufactured by artificially induced accelerated fermentation.

In Africa, there are three basic techniques, namely fermentation with salting and drying, fermentation and drying without salting, and fermentation with salting but no drying. Moreover, while fermented fish pastes and fish sauces are more popular in Southeast Asia, fermented fish products in Africa are usually presented in the form of the whole fish or large chunks cut from it. Again, Asian and African cultures mostly use the traditional methods of (spontaneous, *i.e.*, without the addition of starter cultures) fermentation unlike those in Europe, North America, Australia, and New Zealand where the fermented food products are normally made with specific starter cultures, *i.e.*, microbial inocula which are used to initiate fermentation. The inoculum used as a starter culture may be a single strain of bacteria or a consortium isolated from natural fermented food products.

In another study, researcher classified traditional fermented fish items in Southeast Asia based on the substrates used or added during the fermentation processes, such as salt and carbohydrate. A classification system was proposed for traditional fermented fish products based on the substrate and source of the enzymes used in the fermentation that comprises three groups: (1) where carbohydrates are not a part of the fermentation and the product is the result of the activity of the bacterial enzyme already present in the fish and salt mixture; (2) where carbohydrates are present in the fish and carbohydrate mixture used in the fermentation process; and (3) where a 'starter culture' is used that brings about the fermentation of the carbohydrates. The most usual process, however, is the first one where the fermentation is brought about by the enzymes present in fish.

Ruddle and Ishige (2010) gave a well-defined generic classification and methods of preparation of fermented fish of East Asia. They classified fermented fish into two groups. In one, freshwater fish was used for fermentation, and the other consisted of marine finfish and shellfish. Normally, traditional fermented fish products are salted products with different salt concentrations. Accordingly, fermented fish can also be classified into highly salted products with 20 per cent salt, low salt products with 6–8 per cent salt, and products without any added salt. Fermented fish products are also categorized according to the form retained by the final product. The main types of fermented fish according to the form are the following:

(a) Fermented fish products retaining their original form which are often salted, like *buro* (Philippines), *makassar* (Indonesia), *pedah-kemburg* (Thailand), *makassar* (Indonesia), *ngari*(India), Colombo cured mackerel (India);

(b) Products that contain fish that have been reduced to a liquid, *e.g.*, *patis* (Philippines), *budu* (Malaysia) *nampla* (Thailand), and *nuoc-mam* (Vietnam); and (c) Products in paste form, *e.g.*, *ngapi* (Myanmar), *bagoong* (Philippines), *prahok* (Cambodia), *trasi* (Indonesia), and *belacan* (Malaysia).

Fermentation Technology and Associated Fish Products in Different Regions

Fish sauces and fish pastes are the most popular fermented fish products in many Asian countries and are regarded as indispensable condiments in these culinary cultures. Fish sauce is produced due to changes in physical and chemical characteristics brought about by microbial activity in high salt concentrations and low oxygen levels. Another form of fermented fish product is one where the fish retains its solid form and does not get reduced to a paste or a liquid form. Based on the country and its traditional practices, fish sauce is made by combining fish and salt in a 2:1 to 6:1 ratio.

Some products like fish sauce, fish pastes and shrimp paste are being produced on an industrial scale and have good market potential. This has led to some nations embarking on improving the quality of such products, like the Cambodian government, which introduced strict industry standards (underlining certain specific characteristics like colour and smell) in 2016 for the production of *prahok*, a type of fermented fish pastes and a national staple food. This was due to the realization that the popularity of *prahok* among both resident and diasporic Cambodians made it necessary to monitor the health and safety issues related to the product.

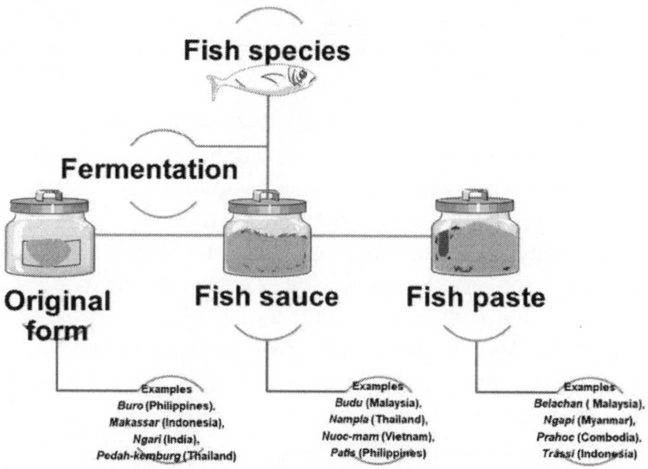

Figure 13.23: Fermented Fish Products.

In Thailand, well-known fermented fish products are *budu, kapi*, and *prala*. *Budu* is an intermediate between fish sauce and fermented fish paste and has a colloidal consistency, and *prala* is fermented fish paste where rice bran is added as a carbohydrate source. *Kapi*, on the other hand, is a fermented shrimp paste using salt in a ratio of 5 (shrimp):1 (salt), that is then dried and compacted. The technology used for the preparation of these products may be similar, but the additives are different depending on the product, with ingredients such as rice and fruits being used in many places. The use of an additive facilitates the growth of lactic acid bacteria, lowering the pH of the solution. Sometimes, similar products may also have diverse names in different localities of Thailand.

The traditional fermented fish products of *Malaysia* are: *pakesan* prepared from freshwater fish mixed with tamarind and salt, *belacan* (shrimp paste), *cincaluk, budu, and kikapikan*. In the Philippines, fermented fish products produced with lactic acid bacteria are consumed in the Central Luzon area. *Bagoong* (fish paste), *balao-balao* (fermented shrimps), *burongisda* (fermented rice and fish), *bagoong alamang* (shrimp paste), and *paite* (fish sauce) are some of the most common and popular fermented fish products in the Philippines. It was reported that many fishes, including herring and anchovy, and shellfish, such as shrimp, are used for the preparation of *bagoong* by mixing salt with well-drained samples. *Bagoong* has a characteristic salty taste, faintly cheese-like odour, and prepared from fish or shrimp. The fish sauce is called *paite* and obtained from thoroughly fermented *bagoong*. *Balao-balao* is a dish made with cooked rice and shrimp (*Penaeus indicus* or *Macrobrachium* sp.), while *burongisda* is made with freshwater fish and rice.

The most popular fermented fish products in Indonesia are *pedah* (fatty, partly dried, salty fish), *trasi* (shrimp paste), *bekasang*, and *bekasam* (freshwater fish that has been fermented), *jambal roti* (marine fish that has been fermented and then dried and salted), *budu, cincaluk, naniura, petis, picungan, pudu, rusip*, and *tukai*. Fish products that have undergone fermentation (*terasi, peda, and kecapikan*) account for less than 2 per cent of all fermented fish products processed.

Myanmar's traditional fermented fish *ngapi* made from salted fish or shrimp is considered the cornerstone of Burmese traditional food. *Nuocmam* is a fish sauce consumed by a huge majority in Vietnam. It is prepared from small fish and shrimps. The fish are layered with salt and kept for fermentation. The supernatant liquor is decanted after it has aged to produce high-quality *nuocmam*. During the fermentation process of fish sauce, the fish tissue is extensively liquefied, and the resulting fluid portion is collected as the sauce.

Fermented fish sauce, like *yu-lu* which is commonly consumed in China as a condiment, is a popular and indispensable seasoning in Chinese cuisine. Anchovies (or other small marine fish) are allowed to ferment by adding salt at the level of 30–40 per cent, salt/fish (*w/w*) in this product. The period of fermentation is 12–18 months.

In Japan, the fermented fish is called *shiokara* (preserves the whole or partial of the fish shape), and *shiokara* paste (yielded by pounding or grating). When, along with fish, additives like salt, vegetables, or steamed rice are used, then it is known as *narezushi*. *Shiokara* and *shiokara* pastes are mostly used as condiments and side dishes.

Korea also possesses a rich dietary tradition of fermented food and fish. Some of the major fermented fish products of Korea are *jeotkal, eo-ganjang*, and *sikhae*. In the context of India, the practice of pit-curing is widespread, particularly along the country's coastlines. The salted fish are buried in pits lined with mats for a period spanning from a few days to two weeks or more in this curing method. The finished product is partially dry and has a distinct flavour and taste that is popular in the eastern region of Tamil Nadu, India.

The 'Colombo cured' method is a prevalent practice in Southern India and Sri Lanka for the preparation of fermented fish. Typically, this method involves mixing mackerel fish and salt in a 3:1 ratio in a concrete tank, along with dried tamarind. Improvements in the quality are obtained if 5 per cent acetic acid is added in place of tamarind.

Fermented fish products, again, are quite prevalent in North East India and West Bengal. It is interesting to note that many such products, like *shidal* and *ngari*, do not use salt. Salt, which has a long and contentious colonial history in India, was introduced in the North eastern region by the British people. In place of salt, sometimes a local alkali called *khar* is used, which helps in alkaline fermentation (*napham, nichow* and *nakham*). The fermented fish pastes prepared in this region are pounded with various herbaceous plants like taro and allowed to age inside bamboo stems. Two types of fermented fish products have been reported in North East India: one that retains its solid form after fermentation and the other in a paste form. They are used as condiments or with other vegetables for their texture and flavour. The local people also associate these products with some health benefits. North East India's fermented fish products are diverse.

Shidal is a popular fermented fish product found in different states of Northeast India. It is prepared from *Puntius* sp. (generally *Puntius sophore*) and *Stipinna phasa* and is known by different names across five states of Northeast India, namely Assam, Mizoram, Arunachal Pradesh, Nagaland and Tripura. *Ngari* is a traditional Manipuri dry, unsalted fermented fish product made solely from sun-dried *Puntius sophore* Ham. *Tungtap* is traditional fermented fish in Meghalaya which is prepared from

Puntius sp. and *Danio* sp. *Hentak, numsing, sukakomacha, sukati, napham, nichow, nakham* are some other fermented fish products which are prepared by different ethnic communities found in Northeast India. *Lonailish* is salted and prepared exclusively from Indian shad (*Tenualosa ilisha*) popular in both Bangladesh and parts of India bordering Bangladesh.

Microbiota of Fermented Fish Products

A broad spectrum of microorganisms has been reported from different fermented fishery products of Asia. The diversity of microbial strains is quite interesting, and their presence is not uniform. In one study, *Pseudomonas aeruginosa* and *Escherichia coli* have been reported in only one product (*Pedah*). Similarly, *Staphylococcus aureus* was discovered in five different products, including *pedah, belacan, ngari, tungtap*, and *shidal*. However, it has been reported that the most prevalent genera are *Lactobacillus, Staphylococcus*, and *Bacillus*, which are found in 12 of the 18 products included in the study. *Micrococcus* and *Staphylococcus* were found in 9 products, making them the second most prevalent bacterial genera in fermented fish. Furthermore, we discovered that *Pseudomonas, Flavobacterium*, and *Brevibacterium* were less dominant strains, with their presence recorded in five, three, and two products, respectively.

Suanyu is a fermented fish product popular among the indigenous people of Southwest China. Suanyu's dominant bacteria were *Lactobacillus, Tetragenococcus*, and *Weissella*, contributing 53.99 per cent t, 35.60 per cent, and 4.10 per cent, respectively. In *budu*, the strains of bacteria isolated were *Tetragenococcus halophilus, Bacillus licheniformis, Staphylococcus piscifermentans*, and *Bacillus subtilis*. In *pedah, Lactobacillus* sp. and *Staphylococcus* sp. were reported. In *ngapi, Bacillus* sp. *Staphylococcus* sp., *Staphylococcus aureus, Micrococcus* sp., and *Pseudomonas* sp. were reported. In *nuoc mam*, a fermented fish from Vietnam, microbial species like *Staphylococcus* sp., *Pediococcus* sp., *Lactobacillus* sp., *Streptococcus* sp., *Halobacterium cutirubrum, Halobacterium salinarum* was isolated.

In the analysis and review of 18 different fermented fish products of South and Southeast Asia, it was found that *Bacillus* sp., *Staphylococcus* sp., *Lactobacillus* sp., *Micrococcus* sp. were the most predominant in fish fermentation. In all these fermented fish products mentioned above, the raw materials used and the manufacturing processes differ, and these factors influence the type and number of microorganisms present in them. In some cases, the addition of certain microorganisms as 'starter cultures' leads to the acceleration in fermentation time resulting in a better product.

The addition of proteolytic enzymes of *Aspergillus oryzae* and *Aspergillus niger* hastened the production of *bagoong* and *patis*, respectively. *Staphylococcus* sp. SK1-1-5 and *Virgibacillus* sp. SK 33 strains, when used as 'starter cultures, ' increase the desired volatile compounds in the fish sauce by inducing a higher production of proteinases. It has been claimed that *Tetragenococcus halophilus* improves the flavour of fish sauce by producing volatile compounds like benzaldehyde, 2 and 3- methylbutanal, and 2-methylpropanol that give the desired flavour to the fish sauce. Fungi like *Aspergillus luchuensis* have been attributed with generating a superior flavour due to their proteolytic and lipolytic activities in low salted fermented fish in a study by Kim. Thus, endogenous enzymes found in a wide spectrum of microbes contained in fermented fish products can impart desirable organoleptic properties and prolong their shelf life. The balanced metabolites present in fermented fish can determine the preferable flavor and texture. Furthermore, the microorganisms also degrade the anti-nutritional components and produce metabolites with less toxic in nature.

Bioactive peptides derived from fermented fish products have been shown to have antioxidant, antagonist, antihypertension, and immunomodulatory, antithrombotic, antimicrobial, and anticancer activity. Fermented fish contained significantly more ascorbic acid, DPPH, and ABTS, which is beneficial, but certain bacteria produce biogenic amines, which may be poisonous if consumed in large quantities. As a result, monitoring the quality of fermented fish before consumption is critical.

Chapter 14

Probiotic Fermented Foods: The New Face of an Old Tradition

Introduction

The distinction between traditional microbe-containing food and probiotics is hazy. For example, the tradition of eating yogurt and kefir (fermented milk with live bacteria) is centuries-old. Yet, these foods have never been subjected to rigorous testing to prove their beneficial health effects, in contrast to probiotics, which must leap through regulatory hoops prior to marketing. So, in that sense, the traditional foods are not probiotics. On the other hand, some yogurts have bacteria added just prior to packaging with the deliberate aim of bettering the health of those who consume it. These are indeed probiotics.

Probiotics have been with us for as long as people have eaten fermented milk, but their association with health benefits dates only from the turn of the last century, when Metchnikoff drew attention to the adverse effects of some gut microflora on the host and suggested that ingestion of fermented milk ameliorated this so called auto intoxication.

During the past few years, there has been a tremendous increase on worldwide basis in the sales of cultured products containing viable probiotic bacteria originating from the human intestine. The physiological and nutritional properties of some strains of probiotic bacteria are now well understood; furthermore, it is possible to verify that some strains are probiotic with documented examples of their ability to maintain and promote the health of the host.

While clinical research has progressed, developments have also occurred in the assessment and improvement of the stability of probiotic bacteria in dairy products and functional foods. With their stability improved and probiotic effect confirmed, we are now approaching the general acceptance of probiotic as documented functional ingredients of foods and beverages.

Meaning of Probiotics

The term probiotic is derived from the Greek word "pro" indicating "promote" and "biotic" representing "life." and has had many definitions in the past. The word "probiotic" ("for life") is opposed to "antibiotics" which means "against life." The history of probiotics began with the consumption of fermented foods by Greek and Romans. In 1908, Russian researcher Elie Metchnikoff, who got a Nobel Prize, firstly proposed the beneficial effects of probiotic microorganisms on human health especially for the treatment of gastrointestinal diseases. Metchnikoff hypothesized that Bulgarians are healthy and long-lived people because of the consumption of fermented milk products which consist of rod-shaped bacteria (*Lactobacillus* spp.). Therefore, these bacteria affect the gut microflora positively and decrease the microbial toxic activity in human intestine.

The term probiotic was first used in 1965 by Lilly and Stillwell to describe substances produced by one protrozoan that stimulated the growth of another. There are a number of disparities in describing probiotics. In 1974, without being aware of the previous definitions, Parker used the term to describe animal feed supplements – organisms and substances that have a beneficial effect on the host animal by contributing to its intestinal microbial balance. However, this general definition was unsatisfactory because the word 'substances' included chemical supplements such as antibiotics. In 1989, Fuller redefined 'probiotics' as a 'live microbial feed supplement which beneficially affects the host animal by improving its intestinal microbial balance'.

Huis in't Veld and Havenaar (1991) have broadened the definition of probiotics as 'a mono – or mixed cultures of live microorganisms which, applied to man or as a fermented product, affects beneficially the host by improving the properties of the indigenous microflora'. This definition implies that probiotic products contain live microorganisms (*e.g.* as freeze dried cells or in a fermented product) and improve the health status of the host by exerting beneficial effects in the mouth or gastrointestinal tract (taken as capsules or food), in the upper respiratory (administered as aerosol) or in the urogenital tract (local application).

ILSI (International Life Sciences Institute) Europe Working Group (1998): "A viable microbial food supplement which beneficially influences the health of the host." Another author in 1999 puts it as "Probiotic food is functional if they have been satisfactorily demonstrated to beneficially affect one or more target functions in the body beyond adequate nutritional effects, in a way that is relevant to either an improved state of health and well-being and/or reduction in the risk of diseases." Naidu *et al.,* in 1999 said "A microbial dietary adjuvant that beneficially affects the host physiology by modulating mucosal and systemic immunity, as well as improving nutritional and microbial balance in the intestinal tract."

Schrezenmeir and de Vrese in 2001 defined probiotics as "A preparation of a product containing viable, defined microorganisms in sufficient numbers, which alter the microflora (by implantation or colonization) in a compartment of the host and by that exert beneficial health effects in this host." FAO/WHO (Food and Agriculture Organization and World Health Organization, 2001) and *Reid et al.* (2003) concentrated exclusively on its health purpose: "Live microorganisms which when administered in adequate amounts confer a health benefit on the host."

These revised definitions stressed the importance of live cells as an essential component of an effective probiotic. Now a day's probiotics are not only the subject of intense medical research but also the source of a multi-billion-dollar global industry, variety of probiotic products are coming in market because of increasing awareness within the consumer about the benefits of probiotic microorganisms.

What is the Need?

The gastrointestinal tract harbours a rich flora of more than 500 different bacterial species, some of which have important health functions. Under natural conditions a protective gut microflora develops and there is no need for bacterial supplements. But today's change in food habits and life style force us to take processed and sterile food. Also, widespread consumption of antibacterial substances affects our access to and colonization by certain type of bacteria. Of course, we cannot relax the high standard of hygiene that are currently practised but we can consider restoring the gut flora by administration of specific bacteria.

Probiotics Currently in Use

The majority of probiotics are bacterial with the species of *Lactobacillus* and *Bifidobacterium* being the most common type of bacteria used. They are listed in Table 14.1. The species currently being used in probiotic preparations are *L. bulgaricus, L. acidophillus, L. casei, L. helveticus, L. lactis, L. salivaris, L. plantarum, Streptococcus thermophilus, Bifidobacterium* sp. *etc.* With two exceptions, they are all intestinal strains. The two exceptions, *L. bulgaricus* and *S. thermophilus* are yoghurt starter organisms and are included on the basis of the claims made for yoghurt. Yoghurt itself can be regarded as a probiotic, although some of the effects which it has may not be dependent on the presence of live bacteria in the gut.

The list although appearing relatively long, is actually a fraction of all known non-pathogenic microorganisms. *L. rahanbisum* strain GG (ATCC 53103) is by far the most thoroughly explored of all lactic acid bacteria.

Table 14.1: Key Genera and Species of Microbes Studied and Used as Probiotics

Genus		*Species*
Lactobacillus	☆	acidophilus
	☆	brevis
	☆	casei
	☆	delbrueckii[a] subsp. bulgaricus
	☆	fermentum
	☆	gasseri
	☆	johnsonii
	☆	paracasei
	☆	plantarum
	☆	reuteri
	☆	rhamnosus
	☆	salivarius
Bifidobacterium	☆	adolescentis
	☆	animalis[b]
	☆	bifidum
	☆	breve
	☆	infantis
	☆	longum
Streptococcus	☆	thermophilus
	☆	salivarius
Enterococcus	☆	faecium
Escherichia	☆	coli
Bacillus	☆	coagulans[c]
	☆	clausii
Saccharomyces	☆	cerevisiae[d]

a: *L. delbrueckii* subsp. *bulgaricus* is typically used as a starter culture for yogurt. b: A subspecies of *B. animalis* is *B. animalis* subsp. *lactis*. Several commercial probiotic strains are members of this subspecies and commonly are referred to as *B. lactis*, although the correct designation is *B. animalis* subsp. *lactis*. c: Some manufacturers market a strain they call "*Lactobacillus sporogenes*." The microbe marketed under this name is likely a *Bacillus coagulans*, not related to the true *Lactobacillus* genus (Sanders, Morelli, and Bush, 2001). d: A prominent probiotic strain marketed as "*Saccharomyces boulardii*" (not a valid species name) is a strain of *Saccharomyces cerevisiae*.

Characteristics and Requirements of a Good Probiotic

Although there are numerous probiotic products on the market for human consumption, there is a lot of scepticism regarding their beneficial effects. This has in part been due to exaggerated and unsubstantiated claims used in the advertising and marketing of these products. Most reports of their positive benefits have been published in commercial brochures rather than in scientific journals. Furthermore, many of the microorganisms included in these products are not viable and have not been selected either for specific beneficial properties (*e.g.* inhibition of enteric pathogens) or for their ability to survive in the GIT (*e.g.* resistance to low pH). If health claims regarding probiotic bacteria are to be substantiated, it is imperative to establish which strains have been obtained. Should freshly isolated strains from humans be used, or strains from culture collections or current products? According to Fuller and Cole (1989), the candidate organisms should have the following characteristics if they are to be used as dietary adjuncts.

Table 14.2: Microorganisms Used as Probiotics

	Bacteria		
(A)	*Lactobacilli*		
1.	*Acidophilus*	2.	*Plantarum*
3.	*Sporogenes*	4.	*Rhamnosum*
5.	*Reuteri*	6.	*Lactis*
7.	*Casei*		
(B)	*Bifidobacterium*		
1.	*Bifidum*	2.	*Infantis*
3.	*Longum*	4.	*Thermophilum*
(C)	*Streptococcus*		
1.	*Lactis*	2.	*Alivarius* subsp. *Thermophillus*
3.	*Intermedius*		
(D)	Enterococcus		
1.	*Faecium*		
(E)	*Propionibacterium*		
(F)	*Leuconostoc*		
(G)	*Pediococcus*		
(H)	*Bacillus*		
	Yeasts and Moulds		
(A)	*A. cerevisiae*		
(B)	*A. niger*		
(C)	*C. pintolopesii*		
(D)	*Saccharomyces boulardii*		

1. The organisms should be normal inhabitants of the site of applications.
2. The organisms should be capable of surviving and growing in the site of application.
3. The organisms should exert beneficial effects on the host.
4. The organisms should maintain their viability and activity throughout manufacture and storage.

The basic requirements for strain selection are given in Table 14.3. It is important that probiotic strains are of human origin because some health promoting effects may be species dependent. Other required properties include acid and bile resistance and the ability to colonize the human intestinal tract, at least temporarily.

For example, resistance to stomach acid and pancreatic secretions such as bile and digestive enzymes would be important for probiotics needing to survive in high numbers through the small intestine. But if the target site for the probiotic is, for example, the mouth, these traits would not be relevant. It is apparent from the broad range of potential probiotic targets that what is required of a probiotic depends on the specific target function. Yet some basic criteria for probiotics can be set: namely, that

- ☆ probiotics must be shown to exert a beneficial effect on the consumer, preferably with a mechanistic explanation of how this occurred;
- ☆ Should be capable of surviving and metabolizing in the gut environment *e.g.* resistant to low pH and organic acids.
- ☆ are non-pathogenic, nontoxic, and free of significant adverse side effects; retain stability during the intended shelf life of the product;
- ☆ contain an adequate number of viable cells in large numbers to confer the health benefit although we do not know the effective dose;
- ☆ Should be stable and capable of remaining viable for long periods under storage and field conditions.
- ☆ are compatible with product format to maintain desired sensory properties; and
- ☆ are labelled in a truthful and informative manner to the consumer.

It is generally accepted that health claims have to be supported by well conducted clinical trials in the targeted population. But the selection of strains from the huge natural reservoir of candidates, the characterization of the purified strain, and the substantiation of a physiological benefit in humans comprise a long path unique to the intended use for the strain. This path is complicated by the lack of rigorous validation of the different laboratory methods conventionally used to select and characterize probiotic candidates.

Adherent strains of probiotic bacteria are favoured because they are more likely to become established in the intestinal tract, and thus have greater probiotic effects. It is also important that probiotic strains to be antagonistic against cariogenic and pathogenic bacteria either by antimicrobial substance production or competition exclusion. Finally, the strains should be safe and tested for human use.

Table 14.3: Features of a Good Probiotic

1.	Should be a strain which is capable of exerting a beneficial effect on the host animal *e.g.* increased growth or resistance to disease.
2.	Should be non-pathogenic and non-toxic.
3.	Should be present as viable cells, preferably in large numbers, although we do not know the effective dose. Should be capable of surviving and metabolizing in the gut environment *e.g.* resistant to low pH and organic acids. Should be stable and capable of remaining viable for long periods under storage and field conditions.

The lactobacilli have a 'generally recognized as safe' status. The safety of lactic acid bacteria has recently been reviewed and it was concluded that the frequency of occurrence of lactic acid bacteria as opportunistic bacteria is extremely rare. However, as new probiotic preparations also include *Bifidobacteria, Enterococci* and even *Sacharomyces* strains, it is important to verify the safety of new probiotic strains.

Selection Criteria for Probiotics

The major selection criteria that have been applied for probiotics include:

1. The ability to survive transit through the gastrointestinal tract
2. Adherence to the intestinal epithelial cell lining
3. Colonization in human intestinal tract
4. Resistance to acid and bile
5. Production of antimicrobial substances towards pathogens
6. Ability to stabilize the intestinal microflora
7. Beneficial effects on human health
8. Non toxigenic and non-pathogenic properties

L. rhamnosus strains GG meet most of these criteria. Another species of *Lactobacillus* called *plantarum* has the unique ability to tolerate low pH, which makes it the dominant species at the last step of natural fermentation. It also has unique adhesiveness to the mucosa and thereby builds an important biological shield to prevent overgrowth by potential microorganisms and resulting microbial translocation.

Recent Developments in the Area of Probiotics and Health

There are only a few well-documented probiotics strains available at present that are applied to food products; their health claims are listed in Table 14.4.

Table 14.4: Health Claims for Probiotics

	Intestinal disorders
1.	Diarrhoea
2.	Antibiotic induced
3.	Travellers
4.	Infantile
5.	Constipation
6.	*Salmonella* and shigella infection
7.	Flatulence
8.	Lactose intolerance
9.	Inflammatory bowel disease
	Other disorders
1.	Vaginitis
2.	Alcohol induced liver disease
3.	Cancer
4.	Hypercholestrolemia
	Other uses
1.	Stabilization of flora
2.	Recolonization of bowel after antibiotic treatment
3.	Treatment of food allergies
4.	Adjuvant for vaccines
5.	Increase weight gain during growth and development

Recently, it has been reported that non-pathogenic yeast, *Saccharomyces boulardii*, is effective in the treatment of antibiotic induced diarrhoea in hospitalized patients. Treatment with this yeast also reduced the incidence of recurring diarrhoea caused by *lostridium difficili* toxin. This condition is also initiated by antibiotic use. *S. boulardii* preparations have also been successfully used for the prophylaxis of traveller's diarrhoea by Austrian tourists.

Lactobacillus strain GG colonizes the human intestinal tract and adheres more tightly to human intestinal and buccal cells than other strains of *Lactobacillus* or *Streptococcus*. When individuals consume GG strain either in a lyophilied form or as a fermented milk product, it is consistently in faeces in concentrations detected $>10^6 g^{-1}$ (Gorbach, 1990). This strain elaborates an antimicrobial substance which has a broad spectrum of activity against a range of bacteria including *Clostridium difficile*.

Beneficial Effect of Probiotic Organisms

The history recording the probiotic properties of live microbial (particularly lactic acid bacteria) food supplements is long. As long as 76 BC, the Roman historian Pliniko advocated the use of fermented milks in the treatment of various pathological forms of gastrointestinal infections. A scientific explanation of the beneficial effects of lactic acid bacteria present in yoghurts was first provided by a Russian biologist Metchnikoff (1907). According to him the consumption of large quantities of yoghurt containing lactobacillus spp. would result in the 'longevity without ageing' – better health and an increased life span.

During the past 80 years, there have been numerous claims advocating nutritional and therapeutic benefits of probiotics in humans (Table 14.5).

Table 14.5: Established and Proposed Health Effects of Probiotics

	Scientifically established effects
1.	Alleviation of the symptoms of lactose intolerance
2.	Reduction of the duration of rotavirus diarrhoea
3.	Prevention of rotavirus diarrhoea
4.	Reduction of the duration of antibiotic associated diarrhoea,
	Future challenges for human studies
1.	Treatment and prevention of food allergy
2.	Reduction in the recurrence of superficial bladder cancer
3.	Reduction in the risk of colon cancer
4.	Treatment of irritable bowel disease and irritable bowel syndrome,
5.	Crohn's disease *Clostridium difficile* diarrhoea, cholesterol control
6.	Inhibition of intestinal pathogens, including *Helicobacter pylori*

Nutritional Benefits

The nutritional benefits of probiotics have been most extensively investigated with regard to the fermentation of food products with lactobacilli. The general parameters that have been are the effect of this fermentation on quantity, availability, digestibility and assimilability of nutrients.

As a result of bacterial proteolysis, yoghurt has higher levels of free amino acids and improved protein digestibility; starch digestibility and mineral bioavailability also enhanced whereas level of antinutrients are reduced in probiotic fermented foods. It has been suggested that a lowering of the gastric pH following the consumption of fermented products improves mineral absorption by the host, however, this remains to be proven. Studies have demonstrated an increase in riboflavin and niacin in yoghurt, B_6 pantothenic in cheddar cheese, B_{12} in cottage cheese and folic acid in a variety of products like yoghurt, *bifidus* milk and *kefir.* Cultural dairy products have higher levels of B-vitamins compared to similar products prepared by direct acidification. Riboflavin and thiamine have also been shown to increase during the preparation of *Lactobacillus* fermented food products including *Idli* and *dosa*. Hence, fermented products contain higher levels of specific vitamins and minerals and protein and carbohydrate in a more digestible form. The changes noted have resulted in better growth of laboratory animals fed on fermented food products compared with the unfermented ones.

Therapeutic Benefits

The strains that have been characterized for their clinical effects and properties have been assessed in recent reviews. A number of health-related effects have been suggested and are partially established, but some can be considered reasonably well established and clinically well documented. Such strain-specific proven health effects are listed in Table 14.5. Other effects reported for specific probiotic strains are included in Table 14.6.

Antimutagenic and Anticarcinogenic Properties

Colorectal cancer is one of the leading causes of mortality in both men and women. It is a complex disorder, characterized by the uncontrolled growth and spread of abnormal cells. There is not yet a cure for cancer largely because the exact causes of most types of cancer are still not known but however, it is a combination of various metabolic and physiologic disturbances in the cell, which are directly or indirectly related to the involvement of genetic makeup. It has been reported that diet makes an important contribution to cancer, *e.g.*, up to 75 per cent of colorectal cancer cases are thought to be associated with diet, implying that risks of cancer are potentially reducible. Evidence from a wide range of sources supports the view that the colonic microflora is involved in the etiology of cancer.

In several countries, a positive correlation between dietary factors, such as consumption of red meat and fat and the incidence of colon cancer has been established. However, Finland is an exception; with a high per capita fat consumption and a relatively low incidence of large-bowel cancer, the Finish people consume a large amount of dairy products, especially yoghurt, and it has been suggested that they harbour high numbers of intestinal lactobacilli that have anticarcinogenic properties. This has led to intense interest in factors such as probiotics that can modulate gut microflora and its metabolism.

It is known that the risk of developing many types of cancer can be reduced by adopting certain lifestyle changes, such as quitting smoking and eating a nutritional balanced diet. The environment delivers risk factors that cause mutations and initiate cancer or enhance growth by genetic and epigenetic mechanisms. Nutrition may supply products that may counteract the causative factors and that can be recommended on the basis of a wholesome and complete diet. Much attention has focused on decreasing cancer risk through diet alterations, particularly consumption of probiotics and increasing intake of dietary fiber (prebiotics).

Probiotics especially lactic acid bacteria have been found to be effective in prevention and suppression of cancer of the colon. At present, several studies have been reported on cancer and intestinal microecology related aspects, but the results have not been conclusive and human studies have been missing. In nutshell, some of the mechanisms by which large-bowel cancer may be prevented are:

1. Metabolic conversion/degradation/absorption of carcinogenic compounds
2. Reduction in levels of faecal bacterial enzymes
3. Stimulation of the immune system

The reports on the benefits of oral administration of probiotic cultured milks and lactic acid bacteria on tumours have been connected with changes related to tumour induction and promotion. The following mechanisms have been indicated in various studies relating lactic acid bacteria intake and cancer.

Table 14.6: Current Probiotic Bacteria and their Reported Effects

Strain	*Reported Effects in Clinical Studies*
L. johnsonii LAI	Adherence to human intestinal cells, balances intestinal microflora, immune enhancement, adjuvant in *H. pylori* treatment
L. acidophilus NCFB 1748	Lowering of faecal enzyme activity, decreasing faecal mutagenicity, prevention of radiotherapy-related diarrhoea, improvement of constipation
L. rhamnosus GG (ATCC 53013)	Treatment and prevention of rotavirus diarrhoea. Prevention of antibiotic-associated diarrhoea, Treatment of food allergy in infants. Treatment of relapsing *C. difficile* diarrhoea
L. aciophilus NFCM	Lowering of faecal enzyme activity, high lactase activity, treatment of lactose intolerance, production of bacteriocins
L. casei Shirota	Prevention of intestinal disturbances, balancing intestinal bacteria, lowering faecal enzyme activities, positive effects on reducing the recurrence of superficial bladder cancer
S. thermophilus; S. bulgaricus	No effect on rotavirus diarrhoea, no immune enhancing effect during rotavirus diarrhoea, no effect on faecal enzymes, strain dependent improvement of lactose intolerance symptoms
L. acidophilus La-5	Balancing intestinal microflora, protection against traveller's diarrhoea, immune enhancement
B. lactis Bb-12	Treatment of viral diarrhoea including rotavirus diarrhoea. Alleviation of symptoms of food allergy. Balancing intestinal microflora
Lactobacillus gassari (ADH)	Faecal enzyme reduction, survival in the intestinal tract
L. reuteri	Shortening of rotavirus diarrhoea. Colonizing the intestinal tract
S. boulardii	Prevention of antibiotic associated diarrhoea. Treatment of *C. difficile colitis*

1. Alteration intestinal microecology (proposed beneficial microflora effects)
2. Altered intestinal metabolic activity (altered conversion of precarcinogens to carcinogens)
3. Normalized intestinal permeability (decrease or delay in the absorption and bioavailability of harmful components and toxins)
4. Enhanced intestinal immunity (enhanced resistance to inflammation, chemicals and other factors)
5. Strengthened intestinal barrier mechanisms (include all of the previous mechanisms 1-4)

Lactic acid bacteria have been reported to inactivate dietary and intestinally generated mutagenic compounds such as azodyes and n-nitrosamines. It has been proposed that peptidoglycan and polysaccharide components of bacterial cell walls have strong binding affinities for mutagens. *Lactobacillus acidophilus, Lactobacillus casei Shirota* strain and *Lactobacillus* GG have been shown to have inhibitory properties on chemically induced tumours in animals. The proposed mechanisms, include: the enteropathogens such as *E. coli* and *C. perfringens* produce enzyme β-glucosidase and urease, these enzymes can convert procarcinogens to proximate carcinogens. Probiotics by lowering the intestinal pH create a bacteriocidal environment and thus, may modulate these bacterial enzymes. Probiotics colonizing at the expense of enteropathogens may bind the ultimate carcinogen by physically removing it via faeces. The cell wall of *B. infants* induces activation of phagocytes to destroy growing tumour cells. Other studies indicate that specific strains of probiotic bacteria may be able to downregulate intestinal microbial enzyme activities. This phenomenon may then decrease carcinogen activating microbial enzymes and has a beneficial effect in the colon, the urinary tract and the bladder. Evidence for the antitumour effect of *L. acidophilus* comes from the studies conducted in laboratory of Goldin and Gorbach (1992). In their experiments, oral supplementation of the diet with viable *L. acidophilus* of human origin, which is bile resistant, caused a significant decline in three different faecal bacterial enzymes *viz.*

β-glucuronidase, azoreductase and nitroreductase. These enzymes catalyse the conversion of procarcinogens to proximal carcinogens in the large bowel.

Ornithine decarboxylase (ODC) is the first and rate limiting enzyme of the polyamine biosynthetic pathway which is responsible for the formation of carcinogenic macromolecules. Some studies have demonstrated that ingestion of lyophilized cultures of *B. longum* was associated with suppression of ODC activity in the colonic mucosa tumors and with the inhibition of colonic mucosal cell proliferation, when compared to control diet. Although the precise mechanism of inhibition of ODC activity by dietary probiotics is not clear, it is likely that these effects may proceed through diverse physiologic and metabolic alterations.

Ras activation represents one of the earliest and most frequently occurring genetic alterations associated with human cancer, especially cancer of colon. Elevated levels of ras p-21 have been correlated with increased cell proliferation, histologic grade, nuclear aplasia and degree of undifferentiation. Reddy *et al.* demonstrated that dietary *B. longum* cultures significantly suppressed the expression of total and mutated ras p-21 in the colonic mucosa and tumors compared with control diet. An additional mechanism of tumor suppression may involve a role for *B. longum* as an immunomodulator and biological response modifier.

The most convincing clinical reports have been published on *Lactobacillus casei* Shirota. There are several mechanistic studies on the effects of the strain reporting decreased urinary mutagen excretion. Other mechanisms have been assessed in mechanistic studies on decreased urinary mutagen excretion, human clinical studies have been conducted using *Lactobacillus casei* Shirota.

Several experimental animal studies clearly demonstrated a protective effect of prebiotics such as oligofructose, probiotics such as some *Lactobacillus* and *Bifidobacterium* strains, or the combination of prebiotics and probiotics on the establishment, growth, and metastasis of transplantable and chemically induced tumours. Human intervention trials to confirm these animal studies are intrinsically difficult because of the natural history of the disease (difficulty in selecting subjects at high risk and requirement of long-term follow-up). A 4-year study of 398 subjects found that *L. casei* Shirota decreased the recurrence of atypical colonic polyp). The European Union (EU)-sponsored "Synbiotics and Cancer Prevention in Humans" project tested a synbiotic (oligofructose plus *L. rhamnosus* GG and *B. animalis* subsp. *lactis Bb12*) in patients at risk for colonic polyps. Among several intermediate end points that were used as biomarkers of colon cancer risk, the study found that the synbiotic decreased uncontrolled growth of intestinal cells.

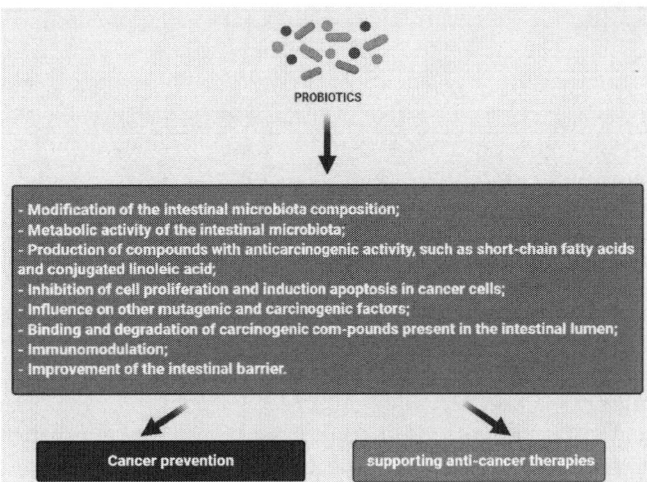

Figure 14.1: Mechanism of Role of Probiotics in Cancer Prevention.

It has been reported that ingestion of probiotics, prebiotics, or combinations of both (synbiotics) plays an important role in the prevention of colorectal cancer (Jain *et al.*, 2010). Goldin and Gorbach (2008) have demonstrated that dietary administration of some specific lactobacilli strains significantly decreased the incidence of 1, 2- dimethylhydrazine-induced experimental colon cancer. The microbial flora and the immune system of the body play an important role in the modulation of carcinogenesis. Both may be influenced by the probiotics. The overall mechanism of probiotic action in the regulation of cancer is not known, however, some of the deduced mechanisms include (Jain *et al.*, 2010):

- Decreasing the exposure of microbial flora to chemical carcinogens
- Detoxifying ingested carcinogens,/inhibition of carcinogens and/or procarcinogens
- Decreasing the population or metabolic activity of bacteria that may generate carcinogenic compounds *i.e.*, inhibition of bacteria that convert pro- carcinogens to carcinogens
- Producing compounds that inhibit the growth of tumour cells
- Stimulating the immune system to defend better against cancer cell proliferation
- Producing metabolic products (*e.g.*, butyrate) which improve programmed cell death (apoptosis).
- Reduction of intestinal pH to reduce microbial activity
- Alteration of colonic motility and transit time.

The preventions or delay in development of intestinal tumours by lactobacilli is ascribed to its binding to mutagenic compounds in the intestine and suppressing the growth of causative bacteria, which convert procarcinogens into carcinogens. The ability of lactobacilli to reduce the risk of cancers has also been based on their ability to modify gut microflora and to decrease B- glucoronidase and other carcinogen levels. There has been evidence that some probiotics produce butyric acid and that this molecule can influence the rate of apoptosis in enterocytes. Probiotics also neutralize the activity of mutagens such as 4- nitroquinoline-N-oxide, 2-nitrofluorene, and benzopyrene (Wollowski *et al.,* 2001). Some probiotics may decrease the faecal concentration of enzymes, mutagens, and secondary bile salts that may be involved in colon carcinogenesis. *L. casei* Shirota strain is reported to possess promising potential for cancer chemoprevention. *L. rhamnosus* GG can protect against the formation of dimethylhydrazine-induced colon cancer in rats.

In one clinical study and another well-designed larger multicentre study the prophylactic effects of oral administration of *Lactobacillus casei* Shirota on the recurrence of superficial bladder cancer have been reported in Japan. Recently, a large Japanese case control study on the habitual intake of lactic acid bacteria and risk reduction of bladder cancer has been conducted in the specific setting of home delivery of the product. This study suggested that the habitual intake of the fermented milk with the strain reduces the risk of bladder cancer in the Japanese population. This result combined with the mechanistic work and human studies warrants further investigations also in other countries.

Reduction in Serum Cholesterol Levels

Cholesterol is a precursor to certain hormones and vitamins and is a component of cell membranes and nerve cells. However, elevated levels of total blood cholesterol (Hypercholesterolemia) or other blood lipids are considered to be a high risk factor for coronary heart disease one of the leading causes of death. However, literature reports of a reduction in serum cholesterol levels following the consumption of skim milk and fermented products have been variable and controversial. It has been suggested that amylo hypocholesterolaemic effect of these products may be due to the presence in fermented milks of organic acids such as uric, orotic and hydroxymethyl-glutaric acids, which inhibit cholesterol synthesis.

Probiotics have been reported to have cholesterol lowering properties in human. The proposed mechanisms include direct assimilation of lipids and conversion of lipids into their metabolites. In addition, the end products of probiotics may affect synthesis of cholesterol. Many studies have been conducted since then to prove this fact, but because of many limitations and confounding factors none of the studies had sufficient statistical power to detect changes in cholesterol of < 15 per cent. More recently, evidences from animal studies have suggested that probiotics not only produce moderate reduction in plasma cholesterol, but also a very marked reduction in fasting and post prandial blood triglycerides.

Rao *et al.* (1981) conducted experiments to study the effect of milk and thermophilus milk on the plasma cholesterol levels and hepatic cholesterogenesis using laboratory rats. White Wistar male rats of 328 g initial average weight were assigned to dietary treatments consisting of: (i) Commercial rat chow + water, (ii) Commercial rat chow + *Thermophilus* milk. Consumption of *Thermophilus* milk by the rats resulted in a significant decrease in the plasma cholesterol levels. The liver cholesterol levels were lower in the group receiving *Thermophilus* milk than the group receiving skim milk.

Two studies carried on 18 and 24 human subjects found no effect from a 'more usual' yogurt intake (450 g/day for 4 weeks and 450 g/day for 4 months) on plasma lipid and lipoprotein levels of the subjects (Halpern *et al.,* 1991; McNamara *et al.,* 1989). Lin *et al.* (1989) did not find any cholesterol-lowering effect from Lactinex$_R$ consumption (10_7 cfu for 16 weeks). In another study, consumption of large quantities of cultured yogurt lowered serum cholesterol levels in human volunteers (Mann, 1977).

The influence of feeding skimmed milk which has been fermented with *L. acidophilus* on serum cholesterol levels was evaluated using rats as animal model (Gruenwald, 1982). All rats had free access to rat chow during the trials. In addition, one group received non-fermented milk and the other received milk fermented with *L. acidophilus.* After 4 weeks, those receiving the fermented milk had lower serum cholesterol levels (65 mg/dl) than did those fed the control milk (79 mg/dl). Weight gain, feed intake, liver lipid content and faecal *Lactobacilli* counts were not different among treatment groups.

In a factorial arrangement of treatments, male rats were given for 6 weeks, (1) a semi-purified diet with maize oil (basal) or (2) with palm oil and cholesterol (high fat, high cholesterol diet HFHC) plus (1) probiotic (Pro; *Bacillus, Lactobacillus, Streptococcus, Saccharomyces* and *Candida* species each at 10_{7-8} cfu/g rice bran) or (2) rice bran (RB, 150 g/kg diet) (Fukushima and Nakano, 1995). Liver weight decreased 35 per cent in rats fed on diets HFHC-Pro. Serum total cholesterol was significantly lowered on diet HFHC-Pro. throughout the experimental period. HDL-cholesterol concentration was higher in the basal-Pro. than in Basal-RB diet. Serum VLDL + LDL + IDL cholesterol concentration in probiotic groups were reduced compared with those of corresponding RB groups.

Zommara *et al.* (1996) fed rats for 6 weeks on a purified diet (reference diet) and on the purified diet supplemented with whey or whey from cultured skim milk with *Bifidobacterium longum, L. acidophilus* or *Streptococcus salivarius* subsp. *thermophilus* resulted in lowest concentration of plasma cholesterol. The non-fermented whey diet was not as effective in lowering plasma cholesterol. The results suggested that whey from cultured milk may exert specific effects on hypercholesterolemia and oxidative stress. Tamai *et al.* (1996) supplemented high cholesterol diets for rats with 0, 1 or 5 per cent *kefir* which was produced using a mixed starter containing 10 types of lactic acid bacteria + *Saccharomyces cerevisiae.* Serum total cholesterol and phospholipid levels were significantly decreased in rats fed on high cholesterol diets containing *kefir.* However, there were non-significant differences in serum-HDL and TG levels among the treatment groups. Total cholesterol and phospholipid levels in the liver showed no significant difference among the treatment groups. Bile acid levels in serum and faeces significantly decreased on diets containing *kefir.*

Fukushima and Nakano (1996) compared the effect of a mixture of probiotic organisms (*Bacillus, Lactobacillus, Streptococcus, Clostridium, Saccharomyces* and *Candida*) on lipid metabolism with that of *L. acidophilus* and *S. faecalis.* Total serum cholesterol concentration of the group fed on the mixture of organisms decreased by 15-33 per cent, compared with the other groups at the end of the 4 week feeding period. This group also had a decreased hepatic cholesterol concentration (36-44 per cent) compared with the 2 single bacteria groups. Faecal cholesterol and bile acid concentration of the mixed organism group increased compared to those of *L. acidophilus* and *S. faecalis* groups.

Akalin *et al.* (1997) studied the effects of yoghurt and acidophilus yoghurt in mice assigned to three dietary groups (n=20/group) for 56 days (i) Commercial rodent chow and water (control) (ii) commercial rodent chow + yoghurt made from milk inoculated with a 3 per cent (v/v) liquid culture of *Streptococcus thermophilus* + *L. delbruckii* subsp. *bulgaricus* and (iii) Commercial rodent chow + yoghurt made from milk inoculated with a 0.01 per cent (w/v) freeze dried culture of *S. thermophilus* + *L. acidophilus.* The live weight gain of (ii) and (iii) were higher than those of (i) mice. The mean values for serum cholesterol concentration LDL concentration were significantly decreased when (iii) was fed on days 28 and 56. HDL and TG were not affected by (ii) or (iii). The highest number of faecal *Lactobacilli* and the lowest number of faecal coliforms were found in (iii) mice.

Thirty two male rats were fed a diet containing 0.500 and 0.125 g cholesterol and sodium cholate/100 g diet, respectively, for 13 days, supplemented with cell wall components from *L. casei* strain Shirota (LC 4 g/100 g diet) or its proteinase-treated preparation (LP 4 g/100 g diet). LP suppressed the increase in serum total cholesterol level at day 7. LC and LP significantly decreased the serum total cholesterol level vs the control group and maintained normal levels of serum HDL at time of sacrifice. LP increased faecal stool excretion, while LC had no effect (Chonan *et al.,* 1997).

Furthermore Gilliland *et al.* (1985) have shown that, in the presence of bile and anaerobic conditions, some strains of lactobacilli and bifidobacteria can assimilate cholesterol from laboratory media. Gilliland *et al.,* have also suggested that deconjugation of bile acids by probiotic bacteria results in less efficient absorption of cholesterol from the intestine.

Schaafsma *et al.* (1998) investigated in adult male volunteers the effect of a new *L. acidophilus* fermented milk product on blood lipids. As compared to the reference product (traditional yogurt), consumption of test product (125 ml, three times daily for three weeks) resulted in significantly lower values for serum total cholesterol, LDL cholesterol and the LDL/HDL ratio by 4.4, 5.4 and 5.3 per cent, respectively. Levels of serum HDL cholesterol, triglycerides and blood glucose remained essentially unchanged.

Kar *et al.* (1998) fed unpasteurized and pasteurized acidophilus yogurt to 2 groups of volunteers for 4 weeks (The unpasteurized yoghurt contained, on average, 7 x 10^8 cells/g). Subjects received 200 g/day of products. Both products caused a significant treatment effect on serum cholesterol concentration. However, there was no significant difference between the effect of unpasteurized and pasteurized yogurt on serum cholesterol.

Some Facts About Role of Probiotics in Lowering Blood Cholesterol Level

- ☆ Assimilation of cholesterol in the small intestine by probiotics and incorporation of cholesterol into the cell membrane of probiotics.
- ☆ The cholesterol-lowering effect of lactic acid bacteria (*Streptococcus, Lactobacillus,* and *Bifidobacterium*) is well established. It has been found that the isolated lactic acid bacteria had an excellent hypocholesterolemic effect. Some strains of lactobacilli have been found to remove cholesterol via various mechanisms and can be used as a dietary adjunct to lower serum cholesterol *in vivo.*
- ☆ *L. acidophilus* deconjugates bile acids into free acids that are rapidly excreted from the intestinal tract. Because free bile salts are excreted from the body, the synthesis of new bile acids from cholesterol lowers its concentration in the body.
- ☆ Isolates of *L. acidophilus* from human intestine are better able to assimilate cholesterol and actively deconjugate bile salts than commercially used cultures of *L. acidophilus.*
- ☆ *Lactobacillus plantarum* PH04 and *L. reuteri* have shown cholesterol-lowering activities.

Hashimoto *et al.* (1998) administered 5.0 per cent of *L. acidophilus* TMC 0330, *L. acidophilus* TMC 0343, *L. casei* TMC 0401 or *L. casei* TMC 0409 to sprague-dawley rats (4 weeks of age) along with 0.5 per cent cholesterol in the diet for 14 days. The control group received 0.5 per cent cholesterol only. Serum cholesterol in experimental rats was suppressed by 13.4, 11.0, 15.2 and 20.4 per cent, respectively as compared to control group. In a dose response study, *L. casei* TMC 0409 suppressed the increase in serum cholesterol when 5.0 per cent was used; however, no hypocholesterolemic effects were observed when 0.02 per cent was used. A study of the stability of *L. casei* TMC 0409 to bile and an acidic environment showed that this bacterium can tolerate exposure to artificial human gastric juice (pH 2, pepsin 0.05 per cent and ox bile 0.5 per cent).

Conclusively, at present, it is not possible to make definite conclusions as to whether cultured products have a hypocholesterolemic effect because many of the studies involved a few subjects and did not establish whether the subjects had elevated or normal serum cholesterol levels at the beginning of the study. In spite of the large number of studies and the great interest in the area, there are currently no consistent studies that relate to the lowering of cholesterol by the use of probiotics or probiotic products. Considerable research is still required to establish whether there is a definite link between the consumption of probiotic bacteria and reduced serum cholesterol levels. Thus, this area remains to be further assessed in the future.

Lactose Intolerance

A large proportion of the world's population is unable to utilize lactose. The enzyme responsible for lactose digestion, although present in the suckling infants, disappears after weaning. Lactose intolerance can be due to either a congenital deficiency of the intestinal mucosal enzyme β-galactosidase (lactase) or a reduction in lactase activity due to intestinal disorders such as gastroenteritis. Lactase deficiency can be primary, secondary, or congenital. The most frequent form is primary lactase deficiency, a consequence of lactase non-persistence characterized by a progressive decline in lactase activity. The prevalence of adult-type lactase deficiency varies among different ethnic groups and geographic locations (following a north–south gradient), ranging from 5 per cent –15 per cent in Northern Central Europe and North American countries to 40 per cent in Mediterranean countries and 65 per cent –90 per cent in African, Asian, and South American countries. In Europe, it is related to the presence of two single nucleotide polymorphisms (SNPs), C/T-13910 and G/A-22018, which mediate lactase downregulation after infancy. Secondary lactase deficiency is caused by pathologies (*e.g.*, celiac disease, Crohn's disease, or infection) and procedures (*e.g.*, surgery) that affect the small intestine and induce a loss of enzyme activity. Congenital lactase deficiency is characterized by the total absence of lactase activity. This form is extremely rare and manifests at birth, soon after the introduction of milk. Lactase levels are minimal or absent in affected infants, which have an otherwise normal intestinal mucosa.

Bacterial fermentation of undigested and unabsorbed lactose leads to an increase in exhaled hydrogen. A deficit in the lactase enzyme leads to lactose malabsorption, since the disaccharide cannot be absorbed and is instead fermented by gut microbiota, leading to the development of clinical signs characteristic of LI. Management of LI typically consists of reducing, or even avoiding, the consumption of dairy products. Such individuals must restrict their intake of milk and dairy products because after consuming unfermented dairy products, lactose intolerant individuals suffer from flatulence, abdominal pain and diarrhoea as a result of the action of the colonic bacteria on undigested lactose.

Fermented foods and dietary supplementation such as probiotics, prebiotics and synbiotics can be used to alter the composition and metabolism of colonic microbiota for the purpose of improving lactose digestion. Previous studies have shown that lactose digestion and the symptoms of lactose intolerance can be improved using probiotics that modify gut pH, express β-gal, and exert positive effects on intestinal activity and overall colonic microbiota. It has been demonstrated that lactic acid bacteria ferment lactose to produce lactate, hydrogen, methane, carbon dioxide and short-chain fatty acids. During fermentation, lactase present in lactic acid bacteria cleaves unabsorbed lactose to glucose and galactose and is then absorbed into the body.

In the small intestine at pH 6–8, lactase activity is optimal, but in the colon, pH decreases to 4 and lactose is left unfermented due to the decrease in bacterial lactase activity. Undigested lactose can thus be considered to be a prebiotic that stimulates the growth of beneficial microflora by fermenting milk containing less lactose. The increase in the number of lactic acid bacteria is due to microbial digestion during fermentation. The degrees of lactose intolerance in different individuals are due to the variability of colonic microflora in fermenting lactose. As a result, dairy foods that contain active cultures can be used to reduce the symptoms of lactose. The consumption of active and live cultures in addition to the use of enzyme β-galactosidase (lactase) that break down lactose and the consumption of probiotics might be beneficial to individuals with lactose-intolerance.

Probiotics exhibit direct effects in the gut in the treatment of lactose digestion. For example, the administration of the probiotic *Lactobacillus* strain *Lactobacillus planatarum* has been shown to improve lactose digestion, decrease diarrhoea and reduce bloating, flatulence and pain during irritable bowel syndrome. In a clinical intervention study of lactose intolerance by

Shaukat *et al.* (2010) and Ibrahim *et al.* (2013), milk-containing *L. acidophilus* was compared to regular milk but was not shown to reduce gastrointestinal symptoms. There is also strong evidence to suggest that some probiotics can improve lactose digestion but do not alleviate the symptoms of lactose intolerance. Intake of yogurt causes fewer symptoms than intake of milk mainly due to the high β-galactosidase activity in yogurt, partial hydrolysis of lactose and slower intestinal transit time that is a result of the digestion of lactose in yogurt This could be related to the particular bacterial strains used in these yogurt products and the viability of the bacterial strains in the final products (Ibrahim and Carr, 2006). A systematic review of one controlled trial study showed the beneficial effects of probiotic supplementation on hydrogen breath test results and on lactose intolerant symptoms. Patients showed reduced exhaled hydrogen and improved abdominal cramping, diarrhoea, vomiting, bloating and/or flatulence (Leis *et al.,* 2020). Similarly, Gingold-Belfer *et al.* (2019) assessed the efficacy of probiotics with a β-galactosidase activity on symptoms of lactose malabsorption and on the lactose hydrogen breath test. Authors found that probiotic formulation with a β-galactosidase activity may serve as an optional treatment in the case of lactose malabsorption.

He *et al.* (2008) and Ibrahim and Gyawali (2013) reported a decrease in lactose content by approximately 20–40 per cent during fermentation by lactic acid bacteria. Their study showed that supplementation of *Bifidobacterium longum* and yogurt with *B. animalis* altered the number of bacteria and increased the β-galactosidase activity in faeces from lactose intolerant individuals. The lactic acid bacteria in fermented milk increases lactase activity in the small intestine and thus, provides beneficial effects. However, it is still not clear whether it is the yogurt that is supplying lactase or if it is bacteria that produce lactase when they enter the gut.

Ibrahim and Gyawali (2013), studied the influence of nonfermented milk containing *L. bulgaricus* on lactose utilization. In this study, nonfermented reduced fat (2 per cent) milk containing *L. bulgaricus* with a final bacterial population of 8 Log CFU/ml was prepared. Lactose maldigestion was then monitored by measuring breath hydrogen levels at hourly intervals for 8 h following consumption of 400 ml of milk samples. This study demonstrated that the lactose maldigestion symptoms significantly improved. The results indicate that *L. bulgaricus* is an effective choice for manufacturing nonfermented milk for lactose maldigesters. This study, among others, also shows clear evidence that consumption of fermented dairy products containing specific probiotic strains in appropriate amounts should be incorporated into the diets of lactose-intolerant subjects. The effectiveness of probiotics also depends on the tolerance of a strain to bile and acid in addition to the strain's lactase level or lactose transport.

Hence, there is a large body of convincing evidence from several studies that lactose-intolerant individuals suffer fewer symptoms if milk in the diet is replaced with fermented dairy products such as yoghurt and acidophilus milk. The reduced levels of lactose in fermented products, due to partial hydrolysis of lactose during fermentation, may contribute to the greater tolerance of fermented dairy products in human studies. Scientists have found that feeding fermented milk to lactose intolerant subjects resulted in a significantly lower level of hydrogen in breath when compared to the hydrogen level for subjects fed unfermented milk. This improved tolerance is due to partial fermentation of lactose by the starter bacteria before ingestion, and release/production of lactase in the intestine by the culture consumed in the fermented products. Furthermore, it has also been suggested that lactase synthesis in the gut mucosa may be induced by live bacteria ingested in cultured products. The bacterial enzyme, β-galactosidase, which can be detected in the duodenum and terminal ileum after consumption of viable yoghurt, is thought to be the major factor that improves digestibility by the hydrolysis of lactose, mainly in the terminal ileum.

In nutshell, the mechanisms of action or probiotics and lactic bacteria include the following:

1. Lower lactose concentration in the fermented product
2. High lactase activity of bacterial preparations used in the production, and
3. Increased active lactase enzyme entering the small intestine with the fermented product or within the viable bacteria able to survive gastric and bile conditions.
4. Other factors suggested to influence lactose digestion include the slower gastric emptying and intestinal transit of semi-solid milk products such as yoghurt. These may influence and modify the intestinal microflora to better tolerate or utilise lactose.

However, most studies are difficult to assess as no indication is given of the specific yoghurt strain, its lactase activity, or the lactose content of the test product. Most data are available in recent human studies, many of them double-blinded and placebo-controlled. However, the lactase activities of the strains in the products are still largely missing as are the strain sources and specifications.

As great strain differences in the ability of even common lactic acid bacteria to utilise lactose have been reported, studies on specific strain and products are most important. Some probiotic strains are not able to utilise lactose as their substrate influencing both the levels of lactose in the final product and the degradation of lactose in the gastrointestinal tract. It is therefore important to conduct studies on final products and related the results to the lactose content in the product and the β-galactosidase activity in the viable or non-viable bacterial cells potentially entering the human gastrointestinal tract. Such an approach should clarify the apparent discrepancies in the published studies. The role of intestinal microflora in lactose utilisation should be clarified comparing the flora in subjects that tolerance no lactose or tolerate small amounts of lactose with the flora of subjects with no lactose malabsorption.

In conclusion, there is good scientific evidence for the alleviation of lactose intolerance symptoms by specific probiotic lactic acid bacteria. Many recent studies have been conducted blinded and placebo-controlled. However, the strain-specific lactase activities may vary over 100-fold. Different products may have varying lactose contents and numbers of bacterial cells which, when released to the duodenum, vary in their lactase activity. Thus, future studies should take these factors into consideration.

Inhibition of Microbial Pathogens

It is now recognized that probiotic bacteria such as lactobacilli and bifidobacteria have antimicrobial properties against a variety of undesirable enteric pathogenic bacteria such as *Salmonella typhimurium, Escherichia coli, Colostridium diffcile* and *Shigeilla* spp. *etc.* Mechanisms responsible for the inhibition of pathogens include:

1. Competition for nutrients and adhesion sites
2. Production of inhibitory metabolites, such as organic acids, hydrogen peroxide, bacterocins and deconjugated bile acids
3. Stimulation of the immune system

There are various studies indicating that consumption/application of probiotic bacteria is effective in the treatment of infantile, traveller's and antibiotic-induced diarrhoea and *Candida* vaginitis. The systemic treatment of AIDS patients with orally administered live bacteria has also been proposed. It has been suggested that hydrogen peroxide produced by lactobacilli acts either alone or in conjunction with a halide and peroxidase of lerkocytic or uterine origin to inactivate the HIV.

Antibiotic-associated Diarrhoea

The incidence of antibiotic associated diarrhoea in children ranges from 20 per cent to 40 per cent of those receiving broad spectrum antibiotics. The infants are more prone to get diarrhoea because of immature intestinal mucosal barrier, immature systemic immunity as well as incompletely evolved intestinal bacterial flora which takes 1 year to reach adult level and composition.

The mechanism by which antibiotic, chemotherapy or radiation associated diarrhoea occurs most likely relates to disturbances of microbial flora in gastrointestinal tract which has profound effect on the protection barrier and results in overgrowth by pathogens, invasion and translocation of toxins and life threatening infections, *e.g. Clostridium difficile, C. difficile* has been reported to colonize 21 per cent of hospitalized patients and is responsible for as many as 20 per cent patients with antibiotic associated diarrhoea. Antibiotics may also disturb the metabolism of carbohydrates, resulting in malabsorption of osmotically active particles.

Antibiotic-associated diarrhoea and gastrointestinal problems are an important area of probiotic application. Reported controlled clinical studies indicate that antibiotic-associated diarrhoea can be prevented by *Lactobacillus* GG strain in a yoghurt form using a dose of two cups of yoghurt daily with about 10_7 cfu/ml in healthy volunteers or as a freeze-dried product using doses varying from 10_9 to 10_{10} cfu/day in infants receiving antibiotics. Supporting studies using *Lactobacillus* GG in the treatment of severe antibiotic-associated form of *Clostridium difficile* colitis indicate some beneficial effects, but more controlled clinical studies in this specific area are still needed.

In a double-blinded study using 20 healthy volunteers treated with 500 mg of ampicillin divided into two groups, half of the volunteers received 4 x 109 cfu live lyophilised *Bifidobacterium lactis* and *L. acidophilus* La5. The volunteers receiving lactic acid bacteria were recolonised faster than those receiving placebo and harboured higher counts than the controls. There are also good studies showing the efficacy of a probiotic yeast *Saccharomyces boulardii* in the treatment of antibiotic-associated diarrhoea and *Clostridium difficile* colitis.

In conclusion, there is scientific evidence from well-controlled studies and different countries for the use of probiotic preparations in the prevention and treatment of antibiotic-associated diarrhoea. However, the studies and the results are strain-and product-specific and cannot be extrapolated to other strains and products.

Pathogen Induced Diarrhoea

Diarrhoea worldwide is a major incapacitating disease in adults and a major cause of death in infants in developing countries. Therefore, the use of probiotics may be an important tool in improving health and nutrition in many developing countries.

Probiotics are considered to be beneficial to the host's health and contain a sufficient amount of non-pathogenic specific live bacteria preparations, such as *Lactobacillus, Yeast, Bifidobacterium, Enterococcus,* and *Bacillus.* Probiotics have been widely reported to treat pathogens-caused diarrhoea by maintaining or improving the balance of gut microbiota and the mechanisms may be associated with the inhibitory effect on the colonization of harmful bacteria by competing for nutrients and producing antibacterial compounds.

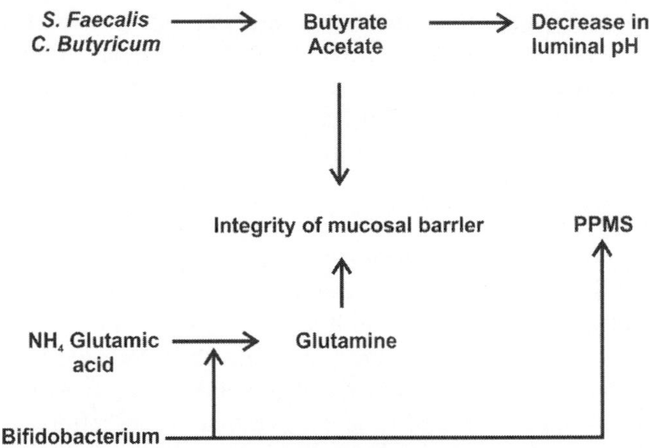

Figure 14.2: Role of Glutamine Produced by *Bifidobacterium* in Maintaining Mucosal Barrier Defenses.

Probiotics play a protective role by replenishing the suppressed bacteria and by inhibiting the growth of more pathogenic flora. Volatile fatty acids produced by some lactic acid bacteria control the colonization of gut by *S. sonnei* and enteropathogenic *E. coli.* Probiotics like *S. faecalis, C. butyricum* and *B. mesentericus* produce butyrate and acetate, which by lowering intestinal pH, create a harmful environment for pathogenic *E. coli, Salmonella* and methicillin resistant *S.aureus.* These metabolites are also required by intestinal epithelial cells as simultant for proliferation.

Probiotics can reduce the severity of *Citrobacter rodentium, Listeria monocytogenes,* EHEC, and *Salmonella typhimurium* infections. For example, *Bifidobacterium breve* and *Bifidobacterium pseudocatenulatum* DSM20439 inhibit the expression of intestinal Shiga toxin EHEC. Glutamine produced by *Bifidobacterium* is also a major supplement for intestinal epithelium, and hence plays a role in maintaining mucosal barrier defenses injured by pathogenic microorganisms and endotoxin.

The administration of fortified milk containing probiotic *Bifidobacterium lactis* HN019 (1.9×10^7CFU) and prebiotic oligosaccharides administered three times a day for a year can reduce the incidence of dysentery in children. Together, the beneficial effects of probiotics on diarrhoea are related to the strain and dosage and the selection and use of the best probiotics for the treatment of diarrhoea need to be determined by more clinical trials.

The anti-diarrheal mechanisms of probiotics have not yet been fully elucidated, the currently considered mechanisms of action of probiotics mainly rely on the following pathways: (1) regulation of the balance of gut microbiota; (2) improvement immunity; (3) manipulation of intestinal defence barrier; (4) metabolites.

Firstly, probiotics have been widely considered to maintain gut microbial hemostasis and a healthy gut microbiota community predicate a low incidence of diarrhoea. In addition, probiotics can produce organic acids in the metabolic process and lower the pH of the gut cavity, thereby inhibiting the growth of pathogens. Secondly, probiotic strains serve as a key activator for the gut innate and adaptive immune systems, thereby signalling antimicrobial and inflammatory responses and enhancing the diarrheal resistance. Thirdly, probiotics can stimulate the production and secretion of mucin, cathelicidins, and defensins from goblet cells and epithelial cells to form an immune barrier, preventing the invasion of pathogenic bacteria.

Probiotics have been found to lower endotoxin levels in serum, secondary to bacterial infection, either by causing decreased number of potentially pathogenic microorganism or by decreasing the incidence of endotoxin transference by a reinforced mucosal barrier and thus, arrest systemic inflammatory response syndrome caused by endotoxin. Probiotics may also inhibit infection by competing with pathogens for the adhesion receptors and nutrients.

Summarily, probiotics not only inhibit the overgrowth of pathogens, but also enhance the anti-pathogenic ability against microbiota associated with diarrhoea.

Rotavirus Diarrhoea

Rotavirus is the leading cause of diarrhoea in infants and young children. Globally, about two million people are admitted each year due to rotavirus diarrhoea (RVD) infection, with 608,000 deaths in infants. The main clinical manifestations of RVD include non-blood watery stools, vomiting, fever, dehydration, nutritional disorders, *etc.*

Rotavirus is mainly transmitted through the faecal-oral route. Even a small amount of virus particles (<100) can induce rotavirus infection. It mainly invades mature epithelial cells and crypt cells of the intestinal villi. Once the virus invades intestinal epithelial cells, it produces NSP4 enterotoxin, further enhancing intracellular calcium ion level through phospholipase C-dependent signalling. Meanwhile, the elevation of cl-secretion leads to secretory diarrhoea.

A few probiotic strains have been reported to be effective in the treatment of rotavirus diarrhoea. Out of the reported probiotics showing potential as gut pathogens antagonists, some species of *Lactobacillus* and *Bifidobacterium* are commonly reported worldwide. Focusing against RV, an *in vivo* evaluation on mouse demonstrated that oral administration of *Bifidobacterium breve* strongly protected against RV-induced diarrhoea, thus observing an anti-RV IgA level increase in faeces, mammary gland and intestine of treated mouse.

In other murine models, pathogen-free rats infected with SA11 RV strain and orally treated with *L. casei*, small intestine lesions, and RV infection level were reduced in all intestine sections, as well as diarrhoea. *In vitro* and *in vivo* studies revealed that some of the mechanisms of probiotics against RV infection are the production of antimicrobial substances (lactic acid, nitric oxide, H_2O_2 and bacteriocins), stimulation of antimicrobial peptides, mucin production by epithelial cells, stimulation of local adaptive (specific IgA response), and innate immune responses.

Probiotics such as *Lactobacillus* and *Bifidobacterium* species, and *Saccharomyces boulardii* have been associated with the prevention of RV infection, to shorter duration and severity of RV diarrhoea, to reduced incidence of reinfections and to the modulation of the immune response and viral shedding.

Shortening of the duration of rotavirus diarrhoea using *Lactobacillus* GG is perhaps the best-documented probiotic effect in human subjects. The best studied strain, *Lactobacillus rhamnosus* strain GG (ATCC 53013) repeatedly reduces the duration of diarrhoea to about half in children with rotavirus diarrhoea in randomised double-blinded and placebo-controlled studies. Studies have been conducted using the strain in a fermented milk and in the freeze-dried form and the results have confirmed the efficacy of both formulations.

The effects of viable and heat-inactivated lactic acid bacteria have been compared in a blinded randomised clinical study. *Lactobacillus* GG administered as a viable preparation during acute rotavirus gastroenteritis resulted in a significant reduction in the duration of diarrhoea and an increase of rotavirus-specific IgA response at convalescence. The heat inactivated *Lactobacillus* GG was clinically as efficient, but the IgA response was not detected. This result suggests that viable and non-viable *Lactobacillus* GG can reduce the duration of rotavirus diarrhoea, but the viability of the strain may be critical in determining the capacity to induce immune stimulation. Also, in a study with different preparations of lactic acid bacteria using the recommended doses (1.25 g dose of freeze-dried preparation twice daily for 5 days) in the treatment of rotavirus diarrhoea it was shown that *Lactobacillus* GG (cell concentration 5 10_9 cfu/g) was clinically effective whilst a preparation containing mixture or *Streptococcus thermophilus* (95 per cent) and *Lactobacillus bulgaricus* (4 per cent) or a *L. rhamnosus* (1 per cent) (2.8 x 10_8 cfu/g) or preparation containing *L. rhamnosus* (2.2 x $10._8$ cfu/g) did not have an effect on the duration of diarrhoea. It has also been reported that *L. reuteri* effectively shortened the duration of watery diarrhoea associated with rotavirus, but multicentre studies are not available for this strain. Further strains studied in this area include *Bifidobacterium lactis* Bb-12 and *Lactobacillus acidophilus* LB1 with both strains studied by different research groups in different geographic areas.

It has been shown that supplementation of infant formula milk with *B. bifidum* and *S. thermophilus* reduced rotavirus shedding and episodes of diarrhoea in children. In a study of HIV associated acute diarrhoea, it was reported that 56 per cent of patients who were treated with *S. boulardii* had their symptoms resolved compared with only 6 per cent of placebo treated patients.

There are also several studies using *Bifidobacterium lactis* Bb-12 and two studies on *Lactobacillus reuteri* as well as studies using the heat-inactivated *Lactobacillus acidophilus* LB1 which reported shortening of the duration of rotavirus diarrhoea in children. As with other documented effects, it is important to remember that these reported studies are specific to the strain used and the strain properties should be carefully safeguarded in industrial use.

When different lactic acid bacteria were compared for their effects on the immune response to rotavirus in children with acute rotavirus gastroenteritis, differences between various strains were observed. Serum antibodies to rotavirus, total number of immunoglobulin-secreting cells (ISC) and specific antibody-secreting cells (ASC) to rotavirus were measured at the acute stage and at convalescence. The treatment with *Lactobacillus* GG was associated with an enhancement of IgA sASC to rotavirus and serum IgA antibody level at convalescence. It was therefore suggested that certain strains of lactic acid bacteria promote systemic and local immune response to rotavirus, which may be of importance for protective immunity against reinfections.

Traveller's Diarrhoea

There are a few studies on the prevention of traveller's diarrhoea and these show positive outcome for *Lactobacillus* GG and a combination of *Lactobacillus acidophilus* LA5 with *Bifidobacterium lactis* Bb-12. There is also a large study on traveller's diarrhoea using the *Saccharomyces boulardii* as an agent with encouraging results, but the study needs to be repeated by other groups. Most of these studies are double-blinded and placebo-controlled and show indications of beneficial effects, and there are some studies with no reported effects, but the information on large good human studies using defined strains on traveller's diarrhoea is still largely lacking. The current data on traveller's diarrhoea does not show scientifically proven effects for any strains used. Thus, further human studies with known bacterial etiology diarrhoea should be conducted to verify the earlier results.

Bowel Transit

Daily consumption of one to three servings of fermented milk containing a probiotic strain, *Bifidobacterium animalis* DN-173 010, decreased the amount of time it took food to travel from the mouth to the anus for people who had longer-than-desired transit time. The effect was more pronounced in elderly subjects and in women. A mixture of eight different strains of lactobacilli, bifidobacteria, and *S. thermophilus* (product name, VSL#3) had no effect on gastrointestinal transit time in irritable bowel syndrome (IBS) subjects. A recent controlled study showed that *L. rhamnosus* Lcr35 improved symptoms of constipation in children.

Celiac Disease

The GI disorder caused by an immunological response after gluten ingestion is known as "celiac disease" and it affects 2 per cent to 5 per cent of genetically susceptible individuals. It has been considered that some micro-environmental conditions such as gut microbiota may influence this response. By performing tests on a humanized mouse model of gluten sensitivity, a new study published by Elsevier found that gut microbiome can play an important role also in the body's response to gluten.

Inflammatory Bowel Disease

Crohn's disease and ulcerative colitis are chronic diseases of the GI tract with no exact known cause that have similar symptoms, such as inflammation and diarrhoea. When patient is suffering from Crohn's disease the inflammation is located in the colon serosa, submucosa and mucosa and in some cases to the entire GI tract, but in ulcerative colitis only the submucosa and mucosa are inflamed. Crohn's disease together with ulcerative colitis are referred to as inflammatory bowel disease (IBD). The Crohn's disease patient also experiences weight loss and abdominal pain; and if ulcerative colitis – bleeding. Hence, Inflammatory bowel disease (IBD), including ulcerative colitis (UC) and Crohn's disease (CD), is a type of chronic bowel inflammation diseases that relapse episodes with unknown aetiology. It has been widely accepted that IBD is the consequence of overly activated response of mucosal immune system to the environmental, dietary, or infectious antigen in a genetically susceptible host. Symptoms of abdominal pain, bloating, and flatulence commonly occur in patients with IBS. These symptoms may result in part from fermentations taking place in the colon that generate gas. Certain gut bacteria process leftover food that reaches the colon without producing gas. Other species even may consume gas, particularly hydrogen. But others produce gas, which is eliminated from the body through flatulence.

The luminal bacterial flora and immunological responses play a major role in initiation and perpetuation of chronic inflammatory bowel disease. Clinical placebo controlled studies showed that probiotics can help ease the associated IBD symptoms. Probiotics by their immunomodulatory and bowel flora manipulating properties show a promising effect in the treatment of chronic inflammatory bowel disease. *S. boulardii* and other LAB or combinations between these were proven to help in ameliorating the symptoms of antibiotic-associated diarrhoea.

In a double-blind, clinical trial of 48 patients with bloating-predominant IBS, the probiotic mixture VSL#3 decreased flatulence scores. Likewise, two other placebo-controlled trials have shown relief of abdominal bloating in patients with IBS treated with VSL#3 or Lactobacillus plantarum 299V. In children, *L. rhamnosus* GG decreased perceived abdominal distension but not abdominal pain. Finally, two large studies in adults showed that either *Bifidobacterium infantis* 35624 or a strain mixture (*L. rhamnosus* GG, *L. rhamnosus* LC705, *Bifidobacterium breve* Bb99, and *Propionibacterium freudenreichii* subsp. shermanii JS) can be effective in alleviating symptoms of IBS. Some probiotics seem useful for controlling the symptoms related to intestinal gas in this group of subjects. But the number of studies is small, and further focused research is needed.

For GI disorders, literature mentions using VSL#3 for diverticulitis and maintenance of remission of chronic relapsing pouchitis after inducing the remission with antibiotics and *B. infantilis* 35624 for IBS. *E. coli* Nissle 1917, LGG and VSL#3 are as effective in inducing or maintaining the remission state for ulcerative colitis or Crohn's disease like the standard therapy with mesalamine – prolonged release anti-inflammatory agent. If they are applied simultaneously their effects will not cumulate. VSL#3 is a highly-concentrated probiotic made of eight bacterial species normally found in the GI microflora – four lactobacilli: *Lactobacillus casei*, *L. plantarum*, *L. acidophilus* and *L. delbrueckii* subsp. bulgaricus; three bifidobacteria: *Bifidobacterium longum*, *B. breve* and *B. infantis* and *Streptococcus salivarius* subsp. thermophilus.

A spectacular breakthrough in GI treatment with probiotics is represented by the use of HuMiX – humanmicrobial crosstalk – which is a modular microfluidics-based model developed for co-culturing human and microbial cells under *in-vivo*-like conditions, specific to the GI humanmicrobe interface. This device allows the re-enactment of transcriptional, metabolic or immunologic conditions after adding *Lactobacillus rhamnosus* GG. A transcriptional response will appear only if another culture is added (*Bacteroides caccae*) along with LGG. HuMiX consists of three chambers (the top one with growth media that circulates among the other chambers, the middle one with a membrane that holds the human intestinal cells, and the lower chamber, being a bacteria growing area) with the ability to recreate the internal aerobic and non-aerobic conditions of the human GI, but also giving the researchers the possibility to monitor oxygen level in real-time, access to every cell compartment and to observe everything through state-of-the-art "microscopes or high-resolution multi-omic analyses".

More clinical trials are needed to evaluate the true place of probiotics within a treatment regimen for chronic inflammatory bowel disease.

Helicobacter Pylori Infection

H. pylori is a microaerophilic and gram-negative bacillus that causes chronic gastritis, peptic ulcer disease and also suspected to be involved in the genesis of malignant gastric ulcer. The adhesion of *H. pylori* to gastric epithelial cells, which is a primary event and resultant secretion of cytokines such as IL-8 and IL-1 play an important role in the occurrence of gastric lesions.

Probiotics have been tested as a strategy for eradication of *H. pylori* infection of the stomach. Some strains of lactic acid bacteria are known to inhibit the growth of *H. pylori* in laboratory experiments. Specific strains of lactic acid bacteria have been reported to inhibit many intestinal pathogens including *Helicobacter pylori.* But results in human studies with different probiotics are mixed. Lactic acid bacteria are often able to survive the acidic gastric conditions and therefore, it has been proposed that they may have a beneficial influence during the eradication of *Helicobacter pylori,* which is involved in the process of gastric ulcer development, possibly by the inhibition of attachment. The inhibitory effect of lactobacilli on the growth of *H. pylori* is dependent on an extracellular secretory product, probably lactic acid which causes inhibition of colonization by *H. pylori.*

It has been shown that the both inhibitory and the specific strains and their supernatants may influence the survival of *Helicobacter pylori,* and studies have been conducted especially with *Lactobacillus johnsonii* strain LA1. It has been shown that there is good *in vitro* inhibition and that the supernatant of the strain has a positive effect when consumed during *Helicobacter* eradication. A recent study shows that the inactivated *L. acidophilus* could be effective in increasing eradication rates of a standard anti-*H. pylori* therapy.

Eradication of *H. pylori* was attempted by feeding yogurt containing probiotic strains selected for their ability to inhibit *H. pylori* in laboratory studies. The strategy was not effective in patients not undergoing simultaneous antibiotic therapy. In contrast, other clinical studies have tested the use of probiotics as a supplement to antibiotic therapy for *H. pylori* eradication. In these studies, the use of probiotics decreased the side effects of antibiotics, improved patient compliance with taking the prescribed therapy, and increased the rate at which *H. pylori* was eradicated.

However, it is clear that more controlled human studies conducted in different populations are needed to assess this effect.

Liver Cirrhosis

Patients with liver cirrhosis have been found to have an altered digestive microflora as urea splitting bacteria and, in these patients, the small intestinal flora is generally increased. It has been. reported that mean blood velocity and flow in portal vein increased and that in splenic vein decreased after administration of lactic acid bacteria, through reduction in plasma endotoxin levels. In experimental alcohol induced liver disease, animal fed lactobacilli had reduced concentration of plasma endotoxin and associated lower liver pathology score. So, it has been suggested that a modification in the composition of the endogenous digestive microflora by oral bacterial therapy with high potency probiotic preparation could be a safe way to regulate the portal pressure.

Hepatic Encephalopathy

It is neurologic disorder associated with liver failure and elevated blood ammonia levels. In patients suffering from liver failure, the detoxification mechanism is impaired and the ammonia levels rise in the circulating blood. Treatment with *L. acidophilus* and/or neomycin results in a lowering of faecal urease and a decrease in ammonia levels of patients with hepatic encephalopathy. These metabolic events are associated with improvement in the clinical status of the patient.

Enhancement of Immune System

Although there is evidence to suggest that LAB reduce the risk of cancer in animal models, there is insufficient clinical evidence available to extrapolate these results to humans. However, there are literature reports of the disappearance of colon carcinomas during prolonged periods of *L. acidophilus* therapy. The factors reported to be involved in stimulation of the immune system by probiotic bacteria are macrophage and lymphocyte activation, enhanced immunoglobulin (A) (Ig A) levels and production of α-interferon.

Probiotics give different results depending on strain or administration method, dosage or the immunological state of the patient. *Lactobacillus acidophilus, Lactobacillus paracasei, Lactobacillus casei, Lactobacillus rhamnosus, Bifidobacterium longum, Lactobacillus johnsonii* EM1, *Lactobacillus gasseri, Bacillus clausii, Escherichia coli* Nissle (EcN) 1917 were firstly used alone but then in combination with other strains (LGG with *L. gasseri, L. acidophilus* and *Bifidobacterium lactis* or the probiotic mix VSL #3) to test their effect on allergic rhinitis. Probiotics from *Lactobacillus* and *Bifidobacterium* gave notable results both because of the fast observed effects but also because of the immunemodulating properties shown in co-culture with mononucleic human peripheral blood cells.

LAB strains act as competitive inhibitors, competing with other bacteria for growth along the digestive tract, thus improving immunity by raising the IgA-producing plasma cells, the number of T lymphocytes and Natural Killer (NK) cells and enhancing phagocytosis. The ESKAPE group (Gram-positive and Gram-negative bacteria like *Enterococcus faecium, Staphylococcus aureus, Klebsiella pneumoniae, Acinetobacter baumanii, Pseudomonas aeruginos* and *Enterobacter* strains) is studied intensively. Probiotic therapy was studied on the ESKAPE pathogens – *lactobacilli, bidifobacteria* and *propionibacteria* were used against *S.aureus* and *P. aeruginosa*, and a study from 2014 proved that *Propionibacterium* acnes killed planktonic *S. aureus* by lowering the bacterial cytoplasms pH: moreover, the propionic acid showed a wide spectrum activity against *E. coli* and *Candida albicans*, inhibiting their growth.

A recent double-blind human trial was conducted to determine a connection between changes in the gut microflora and birch allergy symptoms, and whether probiotic administration has any effect on any of the two. Forty seven allergic children were randomly given either a combination of *L. acidophilus* NCFMTM and *Bifidobacterium lactis* B1-04, or a placebo, the results being monitored during the four months of the experiment. After analysing their blood and nasal secretions (for cytokines and eosinophils) and their faeces (both calprotectin and IgA and for microbiota changes), the study concluded that administrating those probiotics reduced the symptoms associated with birch pollen allergy by lowering the eosinophils levels from nasal secretions and the IgA from patients' faecal samples. The reason the authors used these specific strains was that they were either known for their anti-inflammatory effects or presumed to stimulate Th1 (type 1 helper T cells) response, and their combination might optimize these two effects. The first observed effect was the colonization of the gut microbiota and a high density of these probiotics in the analyzed faecal samples, indicating a positive response to this treatment.

Experiments with human subjects fed a diet supplemented with lyophilized lactobacilli and plain yoghurt showed a progressive increase in blood levels of natural killer cells and a-interferon relative to those of the control group fed skim milk with no lactobacilli. It has recently been suggested that yoghurt consumption can boost the antiallergenic response.

Alleviation of Symptoms of Food Allergy

Probiotics have been suggested to have potential in the management of food allergy. *Bifidobacteria* and lactobacilli have been shown to enhance IgA production in peyer's patches in response to potentially harmful antigens. This enhanced IgA response prevents some potentially allergenic dietary antigens from invading the intestinal mucosa and further alleviates the intestinal inflammation and down regulates the hypersensitivity reaction. Lactobacilli also modify the immuno- modulatory properties of native food protein. Studies with infants with atopic eczema have been given extensively hydrolyzed whey formula or the same formula supplemented with the probiotic *Lactobacillus rhamnosus* strain GG (ATCC 53013). The infants with the probiotic supplementation had a significant improvement of clinical symptoms and alleviation of intestinal inflammation associated with food allergy. In a recent study the first clinical demonstration of specific probiotic strains modifying the markers related to allergic inflammation were reported for *Bifidobacterium lactis* Bb-12 and *Lactobacillus rhamnosus* strain GG (ATCC 53013). The current data indicate that these probiotics may counteract inflammatory responses beyond the intestinal environment. It has been suggested that the combined effects of these probiotics may guide the infant through the weaning period and offer new directions in the search of future foods for allergy treatment and prevention strategies. Further studies should be conducted to determine the interactions between probiotics, gut microflora and the immunological homeostasis of the host. The recent report on probiotics and prevention of atopic disease indicates that there is a great potential in this area of research in the future.

Hypertension

Hypertension is mainly caused by lipid abnormality, hypercholesterolemia and obesity. After Mann and Spoerry first stated that lactobacillus-fermented milk has hypocholesterolemic effects. Scientists have shown that bifidobacteria also reduce the serum cholesterol, as stated both by Sanders (1999) by using *L. casei* YIT9018, *L. helveticus*, and by a human randomized, crossover, and placebo controlled design trial, performed in 2002, by administrating yoghurt containing *L. acidophilus* and *B. longum*. The result consists in a significantly increased HDL (high-density lipoprotein) with beneficial effects in decreasing blood pressure.

Vaginitis

A number of investigations have used lactobacilli with varying success rate for the prevention and treatment of bacterial and yeast vaginitis. In one study, a significantly higher number of uninfected women exposed to gonococci harboured lactobacilli (76 per cent) than did women infected with gonococci (54 per cent). The mechanism of action may involve inhibitory substances which have not been identified or characterized.

Skin Health

By influencing the cytokines, inflammatory mediators, some LAB have been proved to reduce inflammation and hypersensitivity. The healing process of a wound consists in inflammation, cell proliferation (derma fibroblasts and keratinocytes) and extracellular matrix remodeling. A chronic wound has a chronic inflammation and so the cellular proliferation and healing take a very long time.

The bacterial film has a great significance in the dynamic of the chronic inflammation that occurs in non-healing wounds, studies indicating that over 60 per cent of these wounds have a biofilm, and it can eventually lead to the host death because of the infections associated with it. Knowing the interactions between the harmful bacteria and the host, hence the benefit role of the pH and probiotic use is important for the wound management. The pH increases from acidic towards alkaline because of the low oxygen level, healing stage, debridement, worm therapy and bacterial colonization. By manipulating the wound with probiotics it seems that pH drops to normal, the oxygen tension raises, the inflammation is reducing and the pathogen infection is prevented or stopped. Because probiotics also demonstrated their ability to deactivate pathogen bacterial viral factors, it was suggested using them to treat the diabetic feet ulcers.

An Argentinian study compared the effects of applying either *L. plantarum*, or silver sulfazidine cream (plus clorhexidine baths) on infected 2nd degree burns and non-infected 3rd ones, observing the number of bacteria and the healing rate. Although there were no significant differences between the two patient groups, there were no observed cases of sepsis either, suggesting that *L. plantarum* might be used as an alternative because it works similar to the classical treatment for the slow healing burns, without being invasive. The same strain was tested against diabetic and non-diabetic ulcers and, although there were no significant differences between the two study groups, *L. plantarum* topical treatment reduced bacteria, neutrophils, apoptotic/necrotic cells and promoted wound healing.

Diabetes

Recent clinical trials performed in Denmark before 2010 showed a connection between the intestinal microbiota and metabolic diseases such as diabetes or obesity. The microbiota consisted mainly in Firmicutes (Lactobacillus being the most important), Proteobacteria and Bactereodetes (pathogens known for causing tooth decay and gum disease). The researchers discovered that a high amount of Bactereodetes and a low amount of Firmicutes will lead to decreased glucose tolerance which is key problem with diabetes mellitus. By administrating probiotics, the gut microflora is supposed to be balanced once again and therefore the insulin resistance (with *Bifidobacterium* spp.) and hypertensive conditions can be lowered.

Growth during Development

It has been shown that bottle fed infants during the first month of life gained an average 21.9 ozs when fed a standard formula and 26.5 ozs when *L. acidophillus* was added to the formula. The mechanism of this enhanced growth has not been definitively proven but presumably results from infection control and increased bioavailability of nutrients, as they facilitate the breakdown of protein in native milk which releases calcium and magnesium in high concentrations.

Oral Health

Cavities appear because of the acidic environment created by bacteria inside the oral cavity. One of the most harmful bacteria that resides in the oral cavity is Gram-positive anaerobe *Streptococcus mutans*. A group of scientists from the University of Florida discovered a new strain of *Streptococcus*, called *S. A12* which breaks arginine to ammonia and therefore increase the pH level and inhibits the growth of *Streptococcus mutans*. The researchers are now exploring the possibility of implanting this probiotic in a healthy child or adult who might develop cavities, but the frequency of this procedure is yet to be established. Most of the periodontal diseases: gingivitis, periodontitis and pregnancy gingivitis, can be cured or alleviated by using *L. acidophilus*. Gingivitis is treated by using *L. brevis*, *L. casei*, *L. salivarius*, *L. reuteri*, and *Bacillus subtilis*. *L. brevis* has anti-inflammatory proprieties and it also inhibits collagenase (MMP) activity. *B. subtilis* decreases the number of pathogens in periodontal tissues.

Studies should be continued to see how probiotic intake can influence human health, whether the focus is on cancer, chronic wounds management, gastrointestinal problems, obesity, *etc.* by using more strains, by modulating the exposure time, administration technique, target place, study method and the simultaneous interaction between all of the above.

Probiotic Products

Probiotic products for human consumption fall into three categories:

1. Infants
2. Cultured milks
3. Pharmaceutical preparations

There has been a tremendous commercial interest in the incorporation of *bifidobacteria* into baby foods and also in the development of 'bifidus milk' containing growth – promoting bifidogenic factors such as lactulose and fructo-oligosaccharide. Products that have been developed for the baby food market include dried preparations such as 'Lactana-B' and 'Femilact'. Lactana-B developed in Germany in 1964, contains lactulsoe and viable *B. bifidum*. Femilact developed in Czechoslovakia in 1984, contains *B. bifidum*, *L. acidophilus* and *Pediococcus acidilactici* (Table 14.7).

Table 14.7: Commercial Products Containing Probiotic Bacteria

Product	Country of Origin	Probiotic Bacteria
Dairy products		
Acidophilus milk	Many countries	*L. acidophilus*
Bifidus milk	Many countries	*B. bifidum or B. longum*
ACO-Yoghurt	Switzerland	*S. thermophilus*
		L. bulgaricus
		L. acidophilus
Cultura AB-yoghurt	Denmark	*L. acidophilus*
		B. bifidum
Biograde	Germany	*L. acidophilus*
		B. bifidum
		S. thermophilus
Bifighurt	Germany	*B. longum*
		S. thermophilus
Gelilac	Finland	*L. casei*
Yakult	Japan	*L. casei*
Biokys	Czechoslovakia	*B. bifium*
		L. acidophilus
		P. acidilactici
Ofilus	France	*L. bulgaricus*
		L. acidophilus
		S. thermophilus
		B. longum

Many products of dairy origin containing probiotic bacteria are on the market. These include sour cream, ice cream, buttermilk, yoghurt, powdered milk and frozen desserts. The probiotic culture may be added during fermentation with normal yoghurt starters, to the final fermented products, or to the fresh product before shipment. For example, 'Cultura' made in Denmark, is made by fermenting protein – enriched whole milk with *B. bifidum* and *L. acidophilus* and has a shelf life of at least 20 days after production. A drink is obtained by fermenting partially skimmed milk. 'Biogarde', which is produced by more than 45 dairies in Germany, is made by fermenting milk with *S. thermophilus, B. bifidium* and *L. acidophilus.* Biograde is used in the manufacture of other products such as buttermilk, sauces, breakfast cereal, ice cream and beverages. Two dairy products *i.e.,* yoghurt and a fruit flavoured fermented whey drink made with *Lactobacillus* GG (Gefilac) have been reported to have clinically proven health benefits in intestinal disturbances such as infantile diarrhoea caused by rotavirus and traveller's diarrhoea and in the treatment of side effects of antibiotic therapy.

Besides dairy products, an attempt has also been made in our laboratory to develop four probiotic fermented food mixtures for infants and children, which are based upon cereals, legumes, skim milk powder and fresh tomato pulp. These include:

1. BGMT food mixture (Huskless barley flour + Green gram Flour + Skimmed milk powder + Fresh tomato pulp, 2: 1: 1: 1, w/w)
2. PCMT food mixture (Pearl millet flour + Chick pea flour + Skimmed milk powder + Fresh tomato pulp, 2: 1: 1: 1, w/w)
3. RSMT food mixture (Rice flour + Defatted soy flour + Skimmed milk powder + Fresh tomato pulp, 2: 1: 1: 1, w/w)
4. WPMT food mixture (Wheat flour + Pigeon pea flour + Skimmed milk powder + Fresh tomato pulp, 2: 1: 1: 1, w/w)

The cooled and autoclaved slurry of these mixtures was fermented with *L. acidophilus* (10_5 cells/ml) at 37°C for 24 hr. A buttermilk type drink was also prepared from these fresh fermented food mixtures. Feeding of these food mixtures could control diarrhoea among mice and children within 5 days.

In addition to food products containing probiotic bacteria there are various health food adjuncts and pharmaceutical preparations containing probiotics in the market (Table 14.8). These are generally encapsulated freeze-dried bacterial preparations, and are used in the treatment of gastrointestinal disturbances, constipation and certain hepatic diseases.

Table 14.8: Pharmaceutical Preparations

Product	Country of Origin	Probiotic Bacteria
Infloran Berna	Switzerland	*L. acidophilus*
		B. infantis
Euga-lein	Germany	*Bifidobacterium* spp.
Lactopriv	Germany	–do–
Omniflora	Germany	*L. acidophilus*
		B. longum
	E. coli	
Synerlac	France	*L. acidophilus*
		B. bifidum
Lactana-B	Germany	*Bifidobacterium* spp.

Strain Viability and Stability in Retail Product

It is important that the probiotic bacteria maintain their desirable characteristics during product manufacture and storage. Strain survival will depend on:

1. Final product pH
2. Presence of other organisms
3. Storage temperature
4. Presence or absence of microbial inhibitors in the substrate

During the selection of probiotic strains, it is advisable to check their technological requirements. For example, during the manufacture of cultured milk products may be used either as classical liquid cultures grown as bulk starters, or as concentrated, frozen or lyophilized starters. Recent advances in microencapsulation technology may provide protection to acid-sensitive bacteria. By exploiting the latest biotechnological advances in culture production, preservation and storage, it should be possible to maintain high numbers of probiotic bacteria in products marketed in the future.

The composition of a viable cell number of $1 \times 10_6 - 1 \times 10_9$ cells per day is necessary in order for any beneficial effects in human to develop. Of the various probiotic bacteria *L. casei* and *L. plantarum* appear to have longer shelf life than *L. acidophilus, L. reuteri* and *Bifidobacterium* spp. in cultured milk. The reason for the difference in their stability in cultured milk is not entirely clear. Further investigations are necessary to explain the mechanisms that impart various degrees of acidostability to the lactic acid bacteria.

Future Challenges

Relatively few comprehensive studies exist on the development and composition of human intestinal microflora which forms the basis for all intervention studies. However, recent progress in the specific design of probiotic microbes and molecular analysis of human intestinal microflora assessing host microbe relationships have shown that there is a good place for therapeutic management or manipulation of the intestinal microflora and management of microflora-related diseases.

There is a need to further develop and conduct clinical studies with probiotic bacteria since only a few of the claimed effects are hypothesis-based and backed by good clinical studies.

It is important to make sure that the nutritional and human studies are well defined and planned. Each strain and product should be documented and tested independently, since extrapolation of data from closely related strains is not acceptable. Table 14.9 contains suggestions for scientific documentation of health effects in human studies. There is also a need to assess the importance of viability of probiotics for some of the proposed study areas. Non-viable preparations have documented beneficial effects for specific use areas and this form of probiotic product storage and transport conditions for the mostly very temperature and humidity sensitive viable preparations are not available.

Protocols for human nutrition studies need to be developed for probiotics and functional foods. In some cases even post-marketing surveillance studies on intakes and long-term effects are desirable. Such studies have been sued also for the safety assessment of current probiotics.

The design of clinical studies used in pharmaceutical development should serve as a reference point, but also specific protocols and specific criteria relevant to functional foods may be needed. It is necessary to identify specific target groups of individuals who may present higher/lower susceptibilities to potential adverse effects. It is also important to clarify the long-term consequences of the interactions between functional food components and functions in the body and the interactions between components must be carefully monitored. This will enable further beneficial health effects to be scientifically proven for the benefit of the consumers and of probiotic development.

Table 14.9: Requirements for Nutritional and Clinical* Studies of Functional Foods and Probiotic Foods with Health Claims

1.	Each strain documented and tested independently
2.	Exploration of data from closely related strains not acceptable
3.	Well-defined probiotic strains, well-defined study products, well-defined study populations
4.	Double-blind, placebo controlled and randomised human studies
5.	Results confirmed by several independent research groups**
6.	Publication in peer-reviewed journals

*: Clinical studies conducted according to the good clinical practice in human subjects; **: Clinical results should be confirmed by several independent, and at least two research groups.

Safety of Probiotics

Most probiotics have been designed as generally recognized as safe (GRAS) based on their long history of use in food fermentation. However, there have been occasional reports of bacteraemias and endocarditis associated with lactobacilli, generally in severely immunocompromised individuals.

A few workers reviewed the safety data and found no evidence of probiotics being involved in human infections. This is supported by epidemiological data on safety of dairy products. Study with probiotics in patients with AIDS suggested that *L. plantarum* might be administered safely to immunodeficient patients and probably therefore also to immunosuppressed patients. Indeed, probiotics may have a positive effect on immune response.

Negative results do not totally rule out the possibility that current or future probiotics can cause infection. However, to date the probiotics currently being used appear to be safe.

Areas for Future Research

While the rather intuitive hypothesis of the positive health effects of lactic acid bacteria on human health is apparently correct, our understanding of the action mechanisms of health promoting microbes or probiotics is still limited. Areas for future research include the following:

1. Regulation of intestinal motility
2. Modulation of intestinal and systemic immune response
3. Reduction and protection of radiotherapy associated intestinal dysfunction
4. Prevention of intestinal cancers
5. Characterization of the possible antimicrobial properties and their metabolic activities
6. Competitive exclusion
7. Cholesterol lowering effect

Conclusion

Although the concept of probiotics has existed since biblical times, we are still at a very early stage in the development of consistently effective probiotics for use. To derive nutrition and health benefits from probiotic fermented foods, it is imperative that we understand the mechanisms by which these benefits are derived and we ensure that the cultures used maintain these desirable characteristics during product manufacture.

A growing body of experimental and clinical research supports the use of probiotics in various disorders. The ultimate value of LGG and other probiotics as preventive and therapeutic agents in various disorders can only be determined by continued rigorous investigations in well controlled studies. Preliminary data suggest that the continuous consumption of probiotics may well promote overall health in patients who do not have active disease. In an era where medical profession may soon be facing a worldwide epidemic of resistant bacteria, and increasing competition for health care resources, the use of cost effective microbial interference therapy should be a weapon in carrying out these directives.

The current probiotic research has shown that there are possibilities for the increased well-being and therapeutic manipulation of intestinal microflora to promote health. Future studies will provide the information for both food and potentially pharmaceutical uses of probiotic microbes in the future. Co-operation between medical researchers, dairy and food scientists, microbial ecologists and the business community is essential for development and marketing of high quality probiotic food products.

Chapter 15

Food Borne Diseases, Food Hygiene and Sanitation

Food borne diseases may be defined as the disease transmitted or caused by food and represents one of the major problems confronted by a foodservice manager. The term food borne disease encompasses a variety of clinical and etiologic conditions and describes a subset of enteric disease. In food borne disease, the food may act as a vehicle for the transmission of actively growing microorganisms or products of metabolism (toxins), or it may have a passive role as a vehicle for the transmission of nonreplicating bacteria, viruses or protozoa, or stable biologic toxins. In India, the diseases transmitted by food are commonly referred to as food poisoning. A food borne disease generally involves a disturbance of the gastrointestinal tract, with abdominal pain, diarrhoea and sometime vomiting. Gastrointestinal disturbances resulting from the ingestion of food can have a variety of causes, *e.g.*, overeating, allergies, nutritional deficiencies, actual poisoning by chemicals, toxic plants or animals, toxins produced by bacteria, infestation by animal parasites, and infection by microorganisms. Food borne diseases may basically be divided into three main categories (Table 15.1):

1. Biological source, such as illness caused by microorganisms or their toxins.
2. Chemical sources related to constituents that may be directly associated with food borne illnesses such as pesticide, detergent or metal contamination of foods.
3. Physical sources associated with physical objects such as glass or metal in foods.

Table 15.1: Possible Causes of Food-borne Illness

Chemical
Intrinsic, natural toxins, *e.g.*, red kidney bean poisoning, toxic mushrooms Extrinsic contamination
Algae, *e.g.*, paralytic shellfish poisoning
Bacteria (Infection and intoxication)
Fungi (Mycotoxins)
Parasites
Protozoa
Viruses

Food Contamination

Food contamination can be defined as the transference of any objectionable matter into or on the food. There are three types of contamination:

1. Contamination by microorganisms, bacteria, moulds or viruses: This generally occurs in food premises due to ignorance, inadequate space, poor design or because of food handler having short cuts. This type of food contamination is the most serious and may result in food spoilage, food poisoning or even death.
2. Physical contamination by foreign bodies including insects.
3. Chemical contamination.

Food products are mostly contaminated with soil, air and water-borne microorganisms. Harvesting, processing, distribution, and preparation generally lead to contamination of foods. Therefore, food products may transmit certain food poisoning microorganisms that can be classified as either infections or intoxications.

Food Poisoning, Food Borne Intoxications and Infections

Biological sources of food borne illnesses include all those diseases caused by the pathogenic microorganisms. Ordinarily, the term "food poisoning" as applied to diseases caused by microorganisms, is used to include both illnesses caused by the ingestion of toxins elaborated by the organisms and those resulting from infection of host through the intestinal tract. For example, bacterial food borne intoxications are caused by toxins formed in the food prior to consumption while bacterial food borne infections are caused by the activity of large numbers of bacterial cells carried by the food into the gastrointestinal system of the victim. The example of food borne intoxications and infections caused by bacteria has been illustrated in the Figure 15.1.

Food Borne Diseases: Bacterial

Intoxications	*Infections*
1. Staphylococcal intoxication an enterotoxin produced by *Staphylococcus aureus*	1. Salmonellois: enterotoxin and cytotoxin of *Salmonella* spp.
2. Botulism: a neurotoxin produced by *Clostridium botulinum*	2. *Clostridium perfringens* illness: an enterotoxin released during sporulation of *Clostridium perfringens* type A in the intestinal tract.
	3. Enteropathogenic Escherichia coli infection: Several serotypes of *E. coli*, some invasive and some enterotoxigenic
	4. *Bacillus cereus* gastroenteritia: an exoenterotoxin released during lysis of *Bacillus cereus* in the intestinal tract
	5. Others Yersiniosis Shigellosis *Vibrio paranhaemolyticus*

Figure 15.1: Bacterial Foodborne Diseases (*Source*: Frazier and Westhoff, 1988, Food Microbiology).

Food Borne Infections

In addition to food poisoning, other diseases may be spread by food called food-borne infections such as typhoid, paratyphoid fever, dysentery, tuberculosis, giardiasis *etc.* These diseases differ from food poisoning in that:

1. The incubation period is normally longer.
2. A relatively small number of organisms are capable of causing the illness.
3. The food purely acts as a vehicle and multiplication of the organisms within the food is not an important feature of the illness.
4. The infective agent enters the blood stream.
5. Symptoms may or may not include diarrhoea and vomiting.

Causes of Food Borne Diseases

Food borne diseases are generally caused by:

1. Ingestion of toxicants that are found in tissues of certain plants or animals.
2. Toxins formed and excreted by bacteria and fungi while they multiply on or in foods as well toxins formed and excreted by algae and ingested and concentrated by shell fish during their growth in seawater.
3. Poisonous substances that may be intentionally or incidentally added to foods during producing, processing, transporting or storing.

The most common causes of food borne diseases are contamination of food by microorganisms which when ingested lead to the outbreak. The important factors are:

1. Factors which affect the growth of microorganisms.
2. Factors affecting the survival of pathogens.
3. Factors which affect the contamination like food handlers and cross contamination.

High Risk Foods

High Risk Foods are those perishable foods which can support the growth of harmful bacteria and are intended to be eaten without further treatment such as cooking, which would destroy such organisms. They include:

1. All cooked meat and poultry.
2. Cooked meat products including gravy, stock, and roll/sandwich fillings.
3. Milk, cream, artificial cream, custards and dairy products.
4. Cooked eggs and products made with eggs, *e.g. mayonnaise.*
5. Shellfish and other seafood.
6. Cooked rice.

Microbial Sources of Food Borne Illnesses

Microorganisms are often called bugs. As microorganisms are easily dispersed, display physiologic diversity, and tolerate extreme conditions, they are ubiquitous and may contaminate and grow in many food products. Microbial sources of food borne illnesses generally include bacteria, yeast, algae and fungi which we will discuss here in detail.

The behaviour of microbial populations in foods (growth, survival, or death) is determined by the properties of the food (*e.g.*, water activity and pH) and the storage conditions (*e.g.*, temperature, relative humidity, and atmosphere).

Spoilage of any particular food will be by those organisms most suited to the conditions in and around that food. The main groups of concern are:

1. Bacteria
2. Viruses
3. Fungi (yeasts/moulds)
4. Viruses

Bacteria

Bacteria have inhabited the earth for approximately three and a half billion years and have colonized almost every conceivable habitat (Pace, 1996). These are unicellular microorganisms that are approximately 1 mm in diameter with morphology variation from short and elongated rods to spherical or ovoid forms. They are found everywhere in air, soil, water, plants, foods, animals and the human body. If the environmental conditions are favourable, just about any material will support the growth of some bacteria.

Most bacteria are harmless and some are helpful, like those that change milk into cheese or yoghurt. But others cause food spoilage and some known as pathogens are harmful and can cause illness and sometimes death. The diseases that are spread by bacteria that enter the body in food, can multiply at an amazing rate when they are provided with warmth and moisture (especially at room temperature).

Clean food can be contaminated by bacteria from four main sources:

1. The people present in the workplace and their clothing.
2. Other food that is already contaminated.
3. Dirty kitchen or work premises and equipment.
4. Insects and vermin.

Sometimes, harmful bacteria pass directly from the source to high risk food, but usually they rely on other things to transfer them to food. These things are called vehicles. Indirect contamination using an intermediate vehicle is the most common, *e.g.,* the movement of bacteria from the intestine of a food handler to food via their hands, after using the toilet. Where contamination is passed from raw food to high-risk food via for example, a cutting board, this is known as cross contamination. The path that bacteria use to move from the source to the food is known as the route.

Bacterial contamination is the most significant as it results in large amount of spoilt food and large number of food poisoning cases. Certain bacteria release poisons known as toxins. Some toxins which produced by *Clostridium botulinum* are often the cause of death of persons consuming food contaminated by this organism. Some gram-positive bacteria such as *Staphylococcus aureus* and *Bacillus cereus* produce toxins in food during multiplication or sporulation and are occasionally heat resistant so that, although cooking may destroy the bacteria, the toxin is unaffected and can still cause the problems if food is eaten. Endotoxins are generally produced by the gram-negative bacteria such as *Salmonella* in the intestines of persons consuming food contaminated by such organisms.

Yeasts

Yeasts are generally unicellular and differ from bacteria in their larger cell sizes and morphology, and because they produce buds during the process of reproduction by division. These microorganisms grow best in the intermediate acid range, a pH of from 4.0 to 4.5. Some yeasts spoil food, but beneficial uses are in the making of beer, wine and bread. Food that is highly contaminated with yeasts will frequently have a slightly fruity odour.

Moulds

Moulds are multicellular microorganisms with mycelial morphology. These microbes are also characterized by their display of a variety of colours and are generally recognized by their mildewy or fuzzy, cotton like appearance. Usually they look fluffy, being a familiar sight on foods like jam, cheese and bread. Usually they look fluffy, being a familiar sight on foods like jam, cheese and bread. Moulds can develop tiny spores that are found in air and can be spread by air currents. These spores can produce new mould growth if they are transferred to a location that has conditions conducive to germination.

Mould spores are found in the atmosphere, on surfaces, especially damp surfaces and on moldy food. Food must always be covered and moldy food should be isolated. If food is stored at the wrong temperature at high humidity and in excess of the recommended shelf life, there are chances of such foods getting contaminated. Many moulds spoil food and a few can cause disease in plants and man, but beneficial uses are in the ripening of cheeses and production of antibiotics.

Virus

Viruses are organisms much smaller than bacteria. In their pre-infective stage, they are just like a chemical with none of the requirements for life, but once in a living cell they take over and begin to multiply. They can grow only in living tissue, but can be carried in food from one person to another.

Viruses are usually into food premises by food handlers who are carriers or on raw foods such as shell fish which have been grown in sewage polluted water.

Sources of Contaminants in Food Premises

Microorganisms

Among the microorganisms, bacterial contamination is the most common and significant. Various sources by which bacteria may be brought into food premises are as follows:

Man

Food processing organisms are generally present in the nose, mouth, intestine and also on skin. The hands are never free of bacteria and the soiled hands of food handlers are likely to harbour large number of moulds, yeasts and bacteria, some of which may be pathogenic, *e.g. Staphylococcus aureus* (Figure 15.2). Staphylococci are also present in the boils and septic cuts and therefore, it is recommended that food handlers suffering from these conditions should not be allowed to work in food premises. Food handlers should not handle the food directly but they could use tongs and other equipments for this purpose. They must have high standards of personal hygiene.

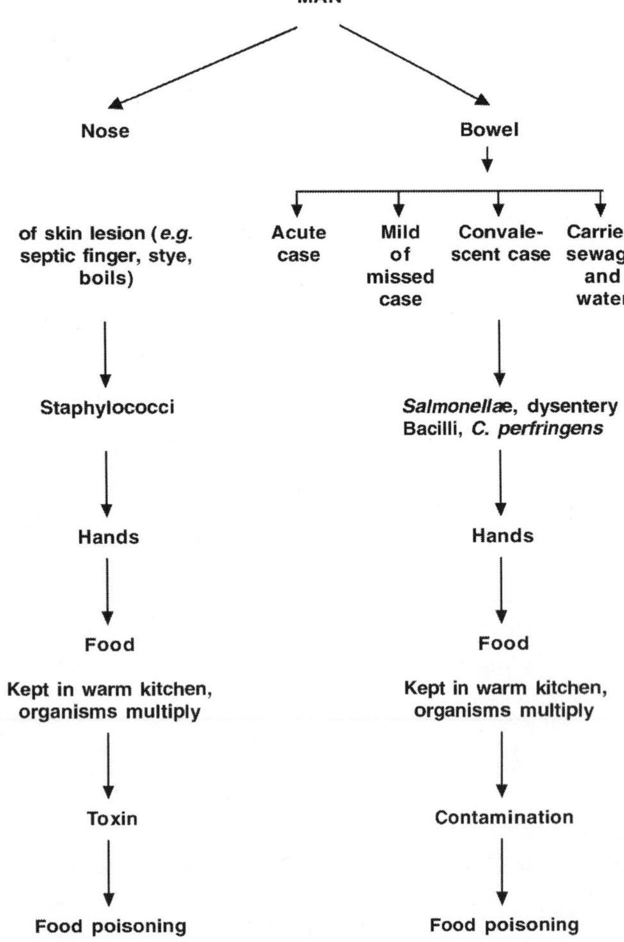

Figure 15.2: Human Reservoirs of Food Poisoning Organisms.

Raw Food

Unfortunately, one of the most viable contamination sources is the food product itself. If raw products are not handled in a sanitary way, they become contaminated and support microbial growth. Raw food such as meat, milk, shell fish, eggs *etc.* are likely to be contaminated with large numbers of bacteria including food poisoning bacteria. Even when they are produced under hygienic conditions, raw foods remain a potential hazard.

Poor hygiene amongst slaughterhouse operatives, dirty equipments, unsatisfactory practices, ill animals *etc.* will undoubtedly contribute to the amount of contaminated raw meat or poultry leaving the slaughterhouse. Microorganisms are introduced in the milk from the animal itself if ill or very dirty, the person carrying out the milking and unhygienic equipment or storage tanks. Shell fish are generally contaminated with food poisoning organisms when grown in sewage polluted waters.

Insects

Insects can contaminate the food by defecation, feeding, walking on or dying in the food. Insects of particular concern with regard to bacterial contamination are flies, cockroaches and ants.

Animals and Birds

Large numbers of *Staphylococcus aureus* are commonly found in the skin and nose of cats and dogs. Persons handling animals should always change their clothing and wash their hands before touching food.

Rodents

Rats and mice generally excrete organisms such as *Salmonella*e. They contaminate the food through defecation, urination or when walk over food or food contact surfaces.

Dust

Dust contains a large number of bacteria and their spores. Open food should always be covered during cleaning.

Soil

The soil contains the greatest variety of microorganisms of any source of contamination. Soil dust is whipped up by air currents, and soil particles are carried by running water to get into or onto foods. Soil harbours a considerable number of pathogens and care must be taken when bringing raw vegetables such as potatoes, carrots and lettuce into food rooms.

Water Supplies and Drainage

Though water is not the ideal medium for the multiplication of bacteria, it can nevertheless transport certain pathogenic organisms such as *Salmonella*e.

Bacterial spores may survive for several months in water. Pollution of waster is usually caused by human sewage or faecal contamination from animals. Safety drains must be installed to carry away waste water. Canned or packaged food which has been in contact with sewage polluted water must be regarded as unfit.

Vehicles and Routes of Bacterial Contamination

The models that relate how food-borne diseases and spoilage microorganisms are transmitted are the 'chain of infection' and 'web of causation'. A chain of infection is a series of related events or factors that must exist or occur and be linked together before an infection will occur. These links can be identified as agents, source, mode of transmission and host. The causative factors that are necessary for the transmission of bacterial food-borne disease include the following:

1. Transmission of the causative agent from the environment in which food is produced, processed or prepared to the food itself.
2. A source and a reservoir of transmission for each agent.
3. Transmission of the agent from the source to a food.
4. Growth support of the microorganisms through the food or host that has been contaminated.

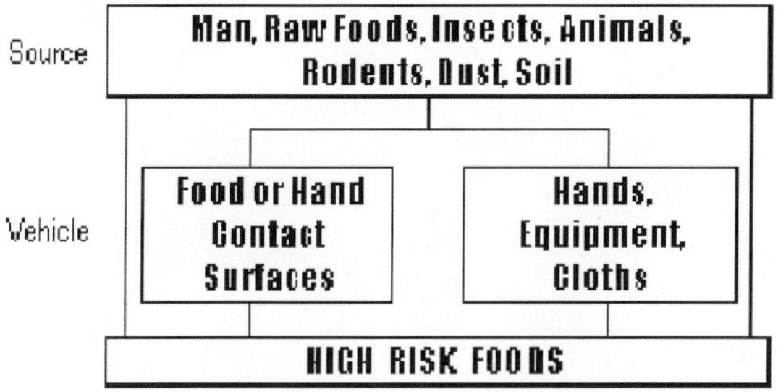

Hands, equipments, clothes and hands or food contact surfaces are the vehicles for food contamination with microorganisms. Knowledge of sources, vehicles and routes is vital to food poisoning prevention, as different controls apply to each. It should be assumed that all the sources *i.e.,* raw foods, animals, insects and the surrounding environment are contaminated. Prevention of contamination depends on either removing the sources or putting barriers between them and food. Thus, human access to food must be restricted, wiping clothes should be clean, hand-contact surfaces such as tap handlers, replaced with knee operated taps and the number of surfaces with which the food comes into contact limited.

Physical Contamination

Foreign bodies such as dust, dirt, stones *etc.* found in food may be brought into food premises with the raw materials or introduced during storage, preparation, service or display. Bolts, nuts, other pieces of metals, staples, cardboard string, polythene, rodents and their hair and droppings, eggs and larvae of insects, sweet papers, cigarette ends, hair, fingernails, buttons, combs and pen tops of personnel handling the food, glass, wood splinters, flaking paint or dust are generally the sources of physical contamination. Care should be taken that they do not contaminate the food.

Chemical Contamination

Unwanted chemicals can enter the food during growth *e.g.,* fertilizers, pesticides, environmental contaminants such as lead or dioxins; during processing *e.g.* oil, cleaning chemicals; during transport as a result of spoilage or leakage and during sale *etc.*

Factors Affecting Microbial Behaviour in Foods

Most studies in food microbiology are concerned with the rapid growth of populations, but in many ecosystems, the survival characteristics of the population also need to be considered. The longevity of bacterial spores and their resistance to harsh conditions are well documented. However, the ability of vegetative cells to resist stressful conditions is increasingly recognized as an important ecologic trait (Archer, 1996). Attention also needs to be given to relatively slow-growing populations in various situations, *e.g.,* when the shelf life of a product is extended by control of rapidly growing spoilage organisms. The behaviour of food borne microorganisms, be it the growth or death of microbial populations, is based on the time of exposure to environmental factors affecting population development; for example, equivalent kills of bacteria in milk are achieved by low temperature–long time pasteurization (60°C/30 min) and high temperature–short time pasteurization (72°C/15 sec). When populations are in the biokinetic range, the rate at which they develop is determined by factors such as temperature, water availability, and pH applied in food preservation procedures. The extent of microbial growth is a function of the time the population is exposed to combinations of intrinsic food properties (*e.g.,* salt concentration and acidity) and extrinsic storage conditions (*e.g.,* temperature, relative humidity, and gaseous atmosphere).

Different factors assume dominance in different foods and preservation strategies. In many foods, the full preservation potential of a single property is restricted because of considerations related to the aesthetic, organoleptic, and nutritional properties of the product. However, several properties or conditions may be combined to provide a desired level of stability (Leistner, 1992). In situations where the preservation strategy is designed to slow the rate of population growth, the effect will always be increased by storage temperature. Temperature control in processing, distribution, and storage (the cold chain) is crucial to ensure the adequate shelf life and safety of many common foods, including meat, fish, poultry, and milk. Newer technologies, including modified atmosphere packaging and sophisticated products such as sous-vide meals, do not obviate the need for strict temperature control. Indeed, the requirement for vigilance increases with increased shelf life and the possibility of growth of psychrotrophic pathogens over an extended period.

Some of the important factors affecting the microbial growth are discussed as under:

Nutrients

High protein foods such as milk, egg, meat and fish are the suitable substrates for bacteria to grow. Foods with high sugar or salt content are unsuitable for the growth of most bacteria. High risk foods include:

Temperature

Temperature values for microbial growth, like pH values, have a minimum and maximum range with an optimum temperature for maximal growth. The rate of growth at extremes of temperature determines the classification of an organism (*e.g.*, psychrotroph, thermotroph). The optimum growth temperature determines its classification as a thermophile, mesophile, or psychrophile.

Depending upon the temperature required for their growth, bacteria can be classified as (Table 15.2):

Table 15.2: Temperature Requirement of Microorganisms

Microorganisms	*Optimum*	*Range*	*Importance*
Psychrophilis (Cold temperature) tolerant m.os)	<20°C	–8° to 25°C	This category includes bacteria which can grow below 20°C and hence cause spoilage in refrigerated and cold stored foods.
Psychrophils (Cold temperature) tolerant m.os)	>20°C	–5°C to 40°C	They include *C. botulinum type E. Aeromones hydrophila etc.*
Mesophiles (Medium temp. loving m.os)	20° to 45°C	10° to 56°C	Include most common pathogens which cause food poisoning.
Thermophiles (High temp. living m.os)	>45°C	35°C to 80°C	Important in canning – some are heat resistant and if not destroyed cause spoilage if cans are stored at high temperature.

The optimum temperature for moulds and yeasts is usually 20° to 30°C although they will grow over a wide range of temperature and may cause problems in refrigerators. Some yeasts can grow at 0°C and below.

Water Activity (a_w)

The amount of moisture in any food available to microorganism is normally considered in terms of water activity (a_w). This is a measure of the availability of water to microorganisms for metabolism (the processes of life). Without water, dehydration (loss of moisture) occurs and the life and growth processes of microorganisms slow down and may stop. The microorganisms might not be destroyed however. The use of salt or sugar in various foods is a way of activating this process. These salts and sugars are crystals that compete with the microorganisms for the available water that they need for survival.

Water molecules are loosely oriented in pure liquid water and can easily rearrange. When other substances (solutes) are added to water, water molecules orient themselves on the surface of the solute and the properties of the solution change dramatically. The microbial cell must compete with solute molecules for free water molecules. Except for *Staphylococcus aureus*, bacteria are rather poor competitors, whereas molds are excellent competitors.

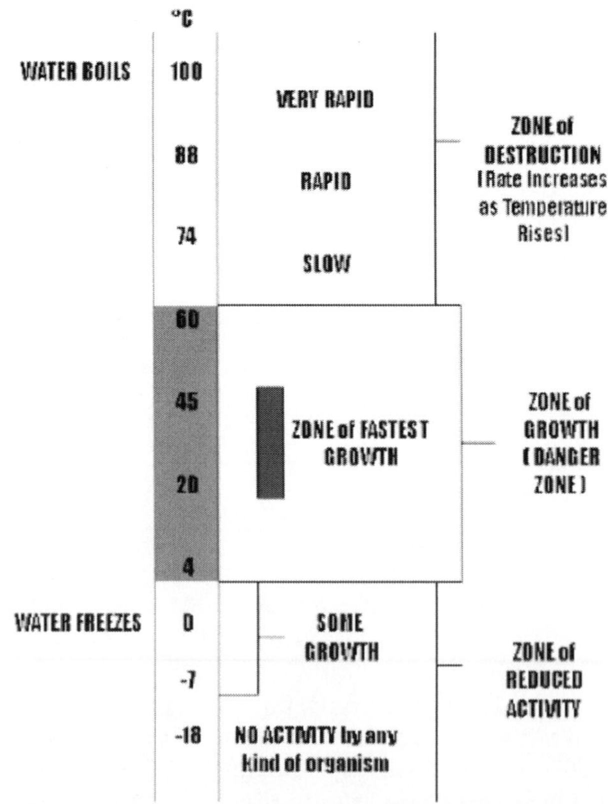

Figure 15.3: Temperature Relationship to Growth and Destruction of Bacteria.

a_w varies very little with temperature over the range of temperatures that support microbial growth. A solution of pure water has an a_w of 1.00. The addition of solute decreases the a_w to less than 1.00 (Table 15.3).

Table 15.3: Water Activity of Various Sodium Chloride Solutions

Water Activity of Various NaCl Solutions		
Per cent NaCl (w/v)	*Molal*	*Water Activity (a_w)*
0.9	0.15	0.995
1.7	0.30	0.99
3.5	0.61	0.98
7.0	1.20	0.96
10.0	1.77	0.94
13.0	2.31	0.92
16.0	2.83	0.90
22.0	3.81	0.86

An a_w value stated for a bacterium is generally the minimum a_w which supports growth. At the minimum a_w, growth is usually minimal, increasing as the a_w increases. At a_w values below the minimum for growth, bacteria do not necessarily die, although some proportion of the population does die. The bacteria may remain dormant, but infectious. Most importantly, a_w is only one factor, and the other factors (*e.g.*, pH, temperature) of the food must be considered. It is the interplay between factors that ultimately determines if a bacterium will grow or not. The a_w of a food may not be a fixed value; it may change over time, or may vary considerably between similar foods from different sources.

Table 15.4: Approximate Minimum (a_w) Values for Growth

Organisms	*Water Activity*
Groups	
Most spoilage bacteria	0.90
Most spoilage yeasts	0.88
Most spoilage moulds	0.80

The a_w of a solution may dramatically affect the ability of heat to kill a bacterium at a given temperature. For example, a population of *Salmonella typhimurium* is reduced tenfold in 0.18 minutes at 60°C if the a_w of the suspending medium is 0.995. If the a_w is lowered to 0.94, 4.3 min are required at 60°C to cause the same tenfold reduction.

Bacteria prefer an a_w around 0.99, many will not grow below 0.95 a_w some bacteria can exist below 0.75. Yeasts and moulds tolerate lower levels of a_w some as low as 0.62.

pH: (Hydrogen ion concentration, relative acidity or alkalinity)

The pH range of a microorganism is defined by a minimum value (at the acidic end of the scale) and a maximum value (at the basic end of the scale). There is a pH optimum for each microorganism at which growth is maximal. Moving away from the pH optimum in either direction slows microbial growth.

The numbers on the pH scale, as shown in the following diagram, indicate the acidity or alkalinity of a fluid. Microorganisms can grow and multiply only within a certain pH range. Most prefer to live in a neutral environment and do not grow in the neutral range. Low pH generally inhibits microbial growth.

Most bacteria will not grow in food with a pH below 4.5 and most of them prefer pH near neutrality for growth in food. Generally, moulds and yeasts are less sensitive to the pH of food than bacteria. Majority of the yeasts prefer acid food (pH 4 to 4.5). Approximate pH Growth Ranges for some Food-borne Illness Causing Micro-organisms have been depicted in Figure 15.4.

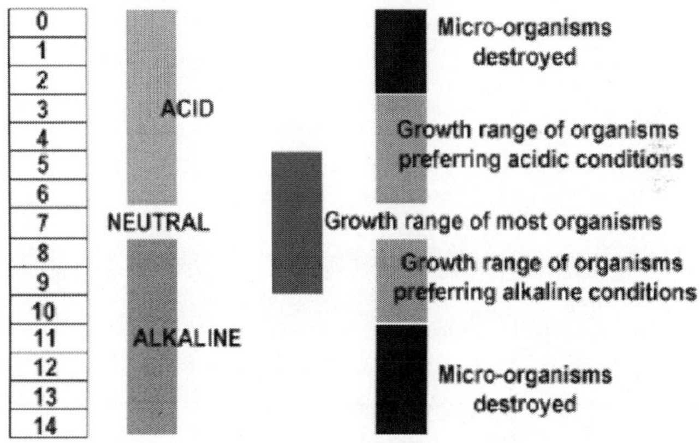

Figure 15.4: pH Growth Ranges for Microorganisms.

Presence of Oxygen

Some microorganisms require oxygen to grow and these are known as obligate aerobes. Other flourish in the absence of oxygen, they are known as obligate anaerobes. There are some bacteria which grow either with or without free oxygen, they are known as facultative. Some food poisoning bacteria such as *Clostridia* are anaerobic whereas others like *Salmonella* and *Staphylococci* are facultative anaerobes.

Free oxygen is normally present in foods with a high liquid content which have been thoroughly boiled. Moulds are aerobic or facultative anaerobes whereas most yeasts grow best in the presence of oxygen, although fermentative types grow slowly anaerobically.

Competition

When there are many different bacteria present, they will compete for the same food. Food is a chemically complex matrix, and predicting whether, or how fast, microorganisms will grow in any given food is difficult. Most foods contain sufficient nutrients to support microbial growth. Several factors encourage, prevent, or limit the growth of microorganisms in foods, the most important are a_w, pH, and temperature.

Most food poisoning bacteria are not as competitive as the normal flora found on food and, unless present in high numbers, will usually die.

Interplay of Factors Affecting Microbial Growth in Foods

Although each of the major factors listed above plays an important role, the interplay between the factors ultimately determines whether a microorganism will grow in a given food. Often, the results of such interplay are unpredictable, as poorly understood synergism or antagonism may occur. Advantage is taken of this interplay with regard to preventing the outgrowth of *C. botulinum*. Food with a pH of 5.0 (within the range for *C. botulinum*) and an a_w of 0.935 (above the minimum for *C. botulinum*) may not support the growth of this bacterium. Certain processed cheese spreads take advantage of this fact and are therefore, shelf stable at room temperature even though each individual factor would permit the outgrowth of *C. botulinum*.

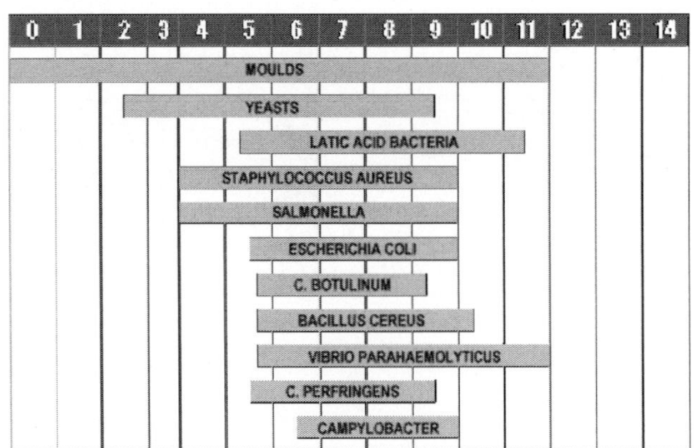

Figure 15.5: Approximate pH Growth Ranges for some Food Borne Illness Causing Microorganisms.

Therefore, predictions about whether or not a particular microorganism will grow in a food can, in general, only be made through experimentation. Also, many microorganisms do not need to multiply in food to cause disease.

Common Food Poisoning Illnesses: Onset, Symptoms and Control

Food poisoning may be one of the most common causes of acute illness; yet cases and outbreaks are generally under-recognised and under-reported.

It is an acute illness, usually of sudden onset, brought about by eating contaminated or poisonous food. Bacteria or their toxins, chemicals, plants, fish, mycotoxins and viruses are the causative agents of food poisoning. The symptoms normally include one or more of the following: abdominal pain, diarrhoea, vomiting and nausea (Figure 15.6).

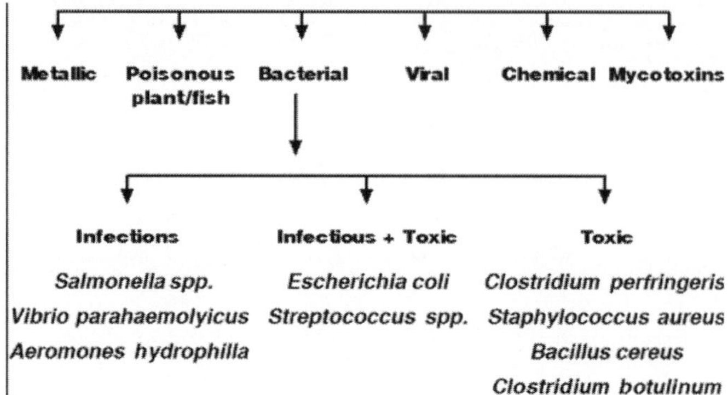

Figure 15.6: Causes of Food Poisoning.

Microorganisms

It is clear from the above figure that bacteria are responsible for most poisoning cases, although poisonous plants, chemicals or metals *etc.* may cause occasional problems. Table 15.5 reveals the various types of bacteria responsible for causing food poisoning, their sources, onset period, symptoms and duration of illness and preventive measures for their control have been discussed.

Mycotoxins are metabolites of some moulds which cause illness and sometimes death, for example aflatoxicosis. Mouldy fruits should not be used for making juice. Now-a-days, outbreaks of food poisoning are increasingly being attributed to microscopic virus particles. Viruses can be destroyed by normal cooking and incidents usually involve raw food, especially shell fish or food contaminated after cooking.

Some of the common food poisoning illnesses caused by bacteria and their toxins are mentioned in Tables 15.5 and 15.6.

Plants

Poisonous plants are rarely the cause of food poisoning in food premises. Plants responsible for causing acute poisoning include deadly nightshade, death cap *etc.* which may be mistaken for edible mushrooms. The gonads, liver and intestines of some fish are highly toxic such as that of puffer fish. Scombrotox fish poisoning is caused by toxins which accumulate in the body of some fish during storage especially above 4°C. Paralytic shell poisoning may result from the consumption of muscles and other bivalves which have fed on poisonous plankton.

Table 15.5: Food Poisoning Bacteria

Bacteria	*Source*	*Onset Period (hours)*	*Symptoms and Duration of illness*	*Control*
Salmonella (infection)	Raw meat, milk, eggs, poultry, carriers, pets, rodents, terrapins, sewage/water	6 to 72 usually 12 to 36	Abdominal pain, diarrhoea, vomiting and fever (1 to 7 days)	Sterilization and strict control of animal feed stuff. Slaughter house hygiene. Safe sewage disposal and chlorination of water. Screening of carrier or suspects. Refrigerated storage of high-risk foods. Thorough thawing and cooking of frozen poultry. Excluding pets, rodents and insects from food. High standards of personal and food hygiene. Avoid raw milk, use pasteurized or sterilized. Separate raw and high-risk foods.
Clostridium perfringens (Toxin in intestine)	Animal and human excreta, soil, dust, insects and raw meat	8 to 22 usually 12 to 18	Abdominal pain and diarrhoea, vomiting is rare (12 to 18 hrs)	Separate raw and cooked foods. High standards of hygiene. Thorough cooking. Rapid cooling and refrigeration within 1.5 hours of reaching 63ºC. Joints should not exceed a size of 2.5 kg.
Staphylo-coccus aureus (Toxin in food)	Human nose, mouth, skin, boils and cuts, raw milk from cows or goats with mastitis	1 to 6	Abdominal pain, vomiting, prostration and sub-normal temperature (6 to 24 hrs)	Avoid handling food, use utensils. Good personal hygiene, especially regarding hand washing. Exclude handlers with respiratory infections involving coughing or sneezing. Cover cuts with waterproof dressings. Exclude persons with boils or septic cuts. Avoid cross-contamination. Avoid the use of raw milk. Rapid cooling and refrigeration of high-risk foods.
Clostridium botulinum (Toxin in food)	Soil, fish meat and vegetables	2 hours to 8 days usually 12 to 36 hrs	Difficulties in swallowing, talking and breathing, double vision and paralysis of the cranial nerves are common and the recovery of survivors may take several months	Strict control over processing low-acid canned foods. Discard blown cans with holes or defective seams. Strict control over smoking and handling of smoked fish. Store smoked fish in a freezer. Avoid cross-contamination. Care in gutting and preparing raw fish. Ensure adequate cooking temperature.
Bacillus cereus				
(*i*) (Toxin in food)	Cereals, especially rice, dust and soil	1 to 5	Vomiting, abdominal pains and some diarrhoea (12 to 24 hrs)	Thorough cooking and rapid cooling. Storing at correct temperatures. Avoid rewarming. Avoid cross-contamination.
(*ii*) (Toxin in intestine)	As above	8 to 16	Abdominal pain, diarrhoea and some vomiting (1 to 2 days)	As above
Vibrio parahaemo-lyticus (infection)	Seafoods	2 to 48 usually 12 to 18	Diarrhoea, vomiting, fever, abdominal pain and dehydration (1 to 7 days)	Avoid cross-contamination. Cook well. Avoid raw shell fish.
Escherichia coli (infection)	Human sewage/ water and raw meat	10 to 72 usually 12 to 24	Abdominal pain, fever, diarrhoea and vomiting (1 to 5 days)	High standards of hygiene. Thorough cooking. Avoid cross-contamination. Storage at correct temperature. Safe sewage disposal and chlorination of water used for drinking or food production.

Table 15.6: The onset, duration and symptoms of various food poisoning illnesses caused by microorganisms, poisonous plants and chemicals.

Approximate Onset Time to Symptoms	*Predominant Symptoms*	*Associated Organism or Toxin*
Upper Gastrointestinal Tract Symptoms (nausea, vomiting) Occur First or Predominate		
Less than 1 h	Nausea, vomiting, unusual taste, burning of mouth.	Metallic salts
1-2 h	Nausea, vomiting, cyanosis, headache, dizziness, dyspnea, trembling, weakness, loss of consciousness.	Nitrites
1-6 h mean 2-4 h	Nausea, vomiting, retching, diarrhoea, abdominal pain, prostration.	*Staphylococcus aureus* and its enterotoxins
8-16 h (2-4 h emesis possible)	Vomiting, abdominal cramps, diarrhea, nausea.	*Bacillus cereus*
6-24 h	Nausea, vomiting, diarrhea, thirst, dilation of pupils, collapse, coma.	Amanita species mushrooms
Sore Throat and Respiratory Symptoms Occur		

Approximate Onset Time to Symptoms	*Predominant Symptoms*	*Associated Organism or Toxin*
12-72 h	Sore throat, fever, nausea, vomiting, rhinorrhea, sometimes a rash.	*Streptococcus pyogenes*
2-5 days	Inflamed throat and nose, spreading grayish exudate, fever, chills, sore throat, malaise, difficulty in swallowing, edema of cervical lymph node.	*Corynebacterium diphtheriae*
Lower Gastrointestinal Tract Symptoms (abdominal cramps, diarrhoea) Occur First or Predominate		
2-36 h, mean 6-12 h	Abdominal cramps, diarrhoea, putrefactive diarrhea associated with *C. perfringens*, sometimes nausea and vomiting.	*Clostridium perfringens, Bacillus cereus, Streptococcus faecalis, S. faecium*
12-74 h, mean 18-36 h	Abdominal cramps, diarrhoea, vomiting, fever, chills, malaise, nausea, headache, possible. Sometimes bloody or mucoid diarrhoea, cutaneous lesions associated with V. vulnificus. Yersinia enterocolitica mimics flu and acute appendicitis.	*Salmonella* species (including *S. arizonae*), *Shigella*, *enteropathogenic Escherichia coli*, other *Enterobacteriacae, Vibrio parahaemolyticus, Yersinia enterocolitica, Aeromonas hydrophila, Plesiomonas shigelloides, Campylobacter jejuni, Vibrio cholerae* (01 and non-01) *V.vulnificus, V. fluvialis*
3-5 days	Diarrhoea, fever, vomiting abdominal pain, respiratory symptoms.	*Enteric viruses*
1-6 weeks	Mucoid diarrhoea (fatty stools) abdominal pain, weight loss.	*Giardia lamblia*
1 to several weeks	Abdominal pain, diarrhoea, constipation, headache, drowsiness, ulcers, variable — often asymptomatic.	*Entamoeba histolytica*
3-6 months	Nervousness, insomnia, hunger pains, anorexia, weight loss, abdominal pain, sometimes gastroenteritis.	*Taenia saginata, T. solium*
Neurological Symptoms (visual disturbances, vertigo, tingling, paralysis) Occur		
Less than 1 h	Gastrointestinal and/or neurologic symptoms (Shellfish Toxins)	Shellfish toxin
1-6 h	Gastroenteritis, nervousness, blurred vision, chest pain, cyanosis, twitching, convulsions.	Organic phosphate
	Excessive salivation, perspiration, gastroenteritis, irregular pulse, pupils constricted, asthmatic breathing.	Muscaria-type mushrooms
	Tingling and numbness, dizziness, pallor, gastro- hemmorrhage, and desquamation of skin, fixed eyes, loss of reflexes, twitching, paralysis.	Tetradon (tetrodotoxin) toxins
	Tingling and numbness, gastroenteritis, dizziness, dry mouth, muscular aches, dilated pupils, blurred vision, paralysis.	Ciguatera toxin
	Nausea, vomiting, tingling, dizziness, weakness, anorexia, weight loss, Chlorinated hydrocarbonsconfusion.	Chlorinated
2 h to 6 days, usually 12-36 h	Vertigo, double or blurred vision, loss of reflex to light, difficulty in swallowing. speaking, and breathing, dry mouth, weakness, respiratory paralysis.	*Clostridium*
More than 72 h	Numbness, weakness of legs, spastic paralysis, impairment of vision, blindness, coma.	Organic mercury
	Gastroenteritis, leg pain, ungainly high-stepping gait, foot and wrist drop.	Triorthocresyl
Allergic Symptoms (facial flushing, itching) Occur		
Less than 1 h	Headache, dizziness, nausea, vomiting, peppery taste, burning of throat, facial swelling and flushing, stomach pain, itching of skin.	Histamine (scombroid)
	Numbness around mouth, tingling sensation, flushing, dizziness, headache, nausea.	Monosodium glutamate
	Flushing, sensation of warmth, itching, abdominal pain, puffing of face and knees.	Nicotinic acid
Generalized Infection Symptoms (fever, chills, malaise, prostration, aches, swollen lymph nodes) Occur		
4-28 days, mean 9 days	Gastroenteritis, fever, edema about eyes, perspiration, muscular pain, chills, prostration, labored breathing.	*Trichinella spiralis*
7-28 days, mean 14 days	Malaise, headache, fever, cough, nausea, vomiting, constipation, abdominal pain, chills, rose spots, bloody stools.	*Salmonella typhi*
10-13 days	Fever, headache, myalgia, rash.	*Toxoplasma gondii*

Approximate Onset Time to Symptoms	*Predominant Symptoms*	*Associated Organism or Toxin*
10-50 days, mean 25-30 days	Fever, malaise, lassitude, anorexia, nausea, abdominal pain, jaundice.	*Etiological agent not yet isolated — probably viral*
Varying periods (depends on specific illness)	Fever, chills, head- or joint ache, prostration, malaise, swollen lymph nodes, and other specific symptoms of disease in question.	*Bacillus anthracis, Brucella melitensis, B. abortus, B. suis, Coxiella burnetii, Francisella tularensis, Listeria monocytogenes, Mycobacterium tuberculosis, Mycobacterium* species, *Pasteurella multocida, Streptobacillus moniliformis, Campylobacter jejuni, Leptospira* species.
Gastrointestinal and/or Neurologic Symptoms - (Shellfish Toxins)		
0.5 to 2 h	Tingling, burning, numbness, drowsiness, incoherent speech, respiratory paralysis	Paralytic Shellfish Poisoning (PSP) (saxitoxins)
2-5 min to 3-4 h	Reversal of hot and cold sensation, tingling; numbness of lips, tongue and throat; muscle aches, dizziness, diarrhea, vomiting	Neurotoxic Shellfish Poisoning (NSP) (brevetoxins)
30 min to 2-3 h	Nausea, vomiting, diarrhoea, abdominal pain, chills, fever	Diarrheic Shellfish Poisoning (DSP) (dinophysis toxin, okadaic acid, pectenotoxin, yessotoxin)
24 h (gastro-intestinal) to 48 h (neurologic)	Vomiting, diarrhea, abdominal pain, confusion, memory loss, disorientation, seizure, coma	Amnesic Shellfish Poisoning (ASP) (domoic acid)

Chemicals

Chemical food poisoning in food premises is rare and is caused by negligence, for example, storing weedkiller, pesticide or cleaning chemicals in unlabelled food containers. Chemical poisoning may also occur because of waste, such as mercury compounds, polluting river water used for drinking or food production. Pesticides sprayed onto fruit and vegetables just prior to harvesting may result in cumulative toxic effects and should be strictly controlled. Several metals are toxic and if ingested in sufficient quantities can give rise to food poisoning. Metals may be absorbed by growing crops or contaminate food during processing. Acid foods should not be cooked or stored in equipments containing metals *i.e.,* antimony, cadmium, copper, lead, tin, iron, zinc, aluminium *etc.*

Chronic Sequelae of Foodborne Disease

In most cases, the clinical conditions usually associated with food borne disease are acute: diarrhoea, vomiting, or other gastrointestinal manifestations such as dysentery. However, other pathophysiologic responses may occur independently or accompany acute-phase responses. A number of chronic sequelae may result from food borne infections, including ankylosing spondylitis, arthropathies, renal disease, cardiac and neurologic disorders, and nutritional and other malabsorptive disorders (incapacitating diarrhoea). The evidence that microorganisms or their products are either directly or indirectly associated with these long-term sequelae ranges from convincing to circumstantial. The reason for this disparity is that, except in rare circumstances, chronic complications are unlikely to be identified or epidemiologically linked to a food borne cause because these data are not systematically collected. Moreover, host symptoms induced by a specific pathogen or product of a pathogen are often wide-ranging and overlapping and therefore difficult to link temporally to a specific incident. These impediments manifest themselves because the problems associated with chronic disease can result from an infection without overt illness. Alternatively, the chronic sequelae may be unrelated to the acute illness and may occur even if the immune system successfully eliminates the primary infection; therefore, activation of the immune system may initiate the chronic condition as a result of an autoimmune response. The variability of the human response from overt illness to chronic carrier status is perhaps the most confounding issue.

Cost of Chronic Sequelae

As the incidence of food borne disease increases, the incidence of chronic sequelae may also rise. Several authors have estimated that chronic sequelae may occur in 2 per cent to 3 per cent of food borne disease cases and suggest that the long-term consequences to human health and the economy may be more detrimental than the acute disease.

Infection: The Microbe/Toxin versus the Host

Several microbial pathogens are highly adapted to parasitization, exhibiting environmentally responsive and adaptive traits that allow attachment, invasion, and replication in the host. Microbial pathogenicity can be viewed solely from the perspective of the microbe; however, this would be not only unidimensional, but also wrong. A major selective force that regulates the phenotype of an infecting microbial pathogen population is the host's immune system, which is also highly adaptive, especially in discriminating self and nonself antigens. When the host-parasite relationship is examined holistically, mechanisms that successful pathogens have apparently evolved to elude the immune system include antigenic heterogeneity or variation; sequestration, either intracellularly or in certain specific host sites; molecular mimicry, through either imitation (cross-reaction) or adsorption of host protein; and direct immune stimulation and/or suppression.

Rheumatoid Disease

Several bacteria, including *Salmonellae*, induce septic arthritis by hematogenous spread to the synovial space, causing inflammation. Viable organisms are recoverable from synovial fluid, and treatment usually involves antibiotic therapy. Prognosis depends on host factors and virulence of the organism; either complete resolution or permanent joint damage can occur.

Yersinia enterocolitica, Y. pseudotuberculosis, Shigella flexneri, Sh. dysenteriae, Salmonella spp., *Campylobacter jejuni,* and *Escherichia coli* initiate aseptic or reactive arthritis, an acute, nonpurulent joint inflammation following infection elsewhere in the body, for example the bowel. *Klebsiella pneumoniae* has been implicated, although it appears now that the bacterium is connected with faecal carriage by ankylosing spondylitis probands. Although a distinct clinical disease, reactive arthritis also occurs in the Reiter syndrome triad with conjunctivitis and uveitis. A subset of patients with symptoms of reactive arthritis and Reiter syndrome get ankylosing spondylitis, a rheumatoid inflammation of synovial joints and entheses within and distal to the spine.

The relative risk of developing these seronegative spondyloarthropathies after a gram-negative enterobacterial infection is high for persons positive for the major histocompatibility class (MHC) antigen B27 and the cross-reacting MHC B7 group. Indeed, persons who are human leukocyte antigen (HLA)-B27 positive have an 18-fold greater risk for reactive arthritis, a 37-fold greater risk for Reiter syndrome, and up to a 126-fold greater risk for ankylosing spondylitis than persons who are HLA-B27 negative and have the same enteric infections. Other genes that may be related or act in concert appear to determine which disease is acquired. These chronic complications are related to a genetically determined host risk factor in combination with an environmental trigger. No cause-and-effect relationship of enteric pathogens in ankylosing spondylitis has been established; however, a low but consistent incidence (0.2 per cent to 2.4 per cent) of reactive arthritis occurs after outbreaks of *S. typhimurium, Sh. flexneri,* and *C. jejuni*. Biotypes and phage types of *Y. enterocolitica* O: 3 and O: 9, endemic to Scandinavia, are either highly arthritogenic or affect a more genetically predisposed population with persistent and debilitating symptoms that may last for years.

The sharing of antigenic determinants by a microbe and its host is a frequent natural occurrence, and bacterial antigens from the pathogens that directly cross-react with MHC B27 have been demonstrated. Additionally, the plasmid-mediated synthesis of bacterial B27 "modifying factor," a protein that binds to and subsequently alters the conformation of B27, has been reported. In both of these models, immune recognition of the foreign antigen leads to an autoimmune anti-B27 response. Alternatively, B27 may act non-immunologically because dissemination of bacterial antigens to infected joints stimulates a local T-cell inflammatory response. Here, B27 may act as a receptor for bacteria or antigens thereof, facilitating invasiveness from mucosal surfaces in the gut. Indeed, transfected B27 on the surface of mouse L cells reportedly can alter bacterial invasion capability.

Despite the strong familial association related to the MHC B27 gene, B27-negative persons are known to become ill, albeit less often, but with apparently equal severity, as shown by an epidemiologic investigation of rheumatoid arthritis following the 1985 *S. typhimurium* gastroenteritis outbreak due to contaminated milk.

Autoimmune Thyroid Disease

Graves disease is an autoimmune disease mediated by autoantibodies to the thyrotropin receptor. The first indication that the disease may be linked to infection was finding antibody titers to *Y. enterocolitica* serotype O: 3 suggestive of molecular mimicry in a majority of patients with Graves disease. Several studies have shown that two low molecular weight envelope proteins of *Yersinia* contain epitopes cross-reactive with the thyrotropin receptor. These proteins are chromosomally encoded, exposed to the surface of the bacterium, and produced by both virulent and avirulent strains of *Yersinia* (*Y. pestis, Y. pseudotuberculosis, Y. enterocolitica* VW+ and WV-). In addition to autoantibody, a suppressor cell dysfunction may be involved in Graves disease. Severe hypothyroidism may also result from chronic intestinal giardiasis due to infection by *Giardia lamblia*; treatment with metronidazole can result in complete elimination of the parasite and recovery of regular intestinal thyroid hormone absorption.

Inflammatory Bowel Disease

Inflammatory bowel disease is the collective term for Crohn disease and ulcerative colitis. While both infections are chronic inflammatory diseases with histologic infiltrates of macrophages and lymphocytes and a prolonged clinical course, the primary clinical and pathologic effects are gastrointestinal. The infections can be difficult to differentiate because the symptoms are often similar. The acute clinical characteristics are diarrhoea, abdominal pain, fever, and weight loss; and the acute pathologic features include a constant flux of neutrophils into inflamed mucosa, eventually penetrating the epithelium into the intestinal lumen. The chronic spontaneously relapsing disorder exhibits many of the symptoms of the acute state; however, this phase has an average symptom duration of 3.2 years before correct diagnosis. Abdominal abscesses are a common and dangerous complication of Crohn disease, while in ulcerative colitis, abdominal perforations may lead to peritonitis. Crohn disease involves the ileum or colon (anaerobes are important), while ulcerative colitis appears restricted to the colon (aerobes are important). Nationality and familial associations suggest a genetic predisposition for the disease.

Although the cause of inflammatory bowel disease and the mechanism(s) for spontaneous exacerbations and remissions are unknown, much research has focused on transmissible agents, including food borne pathogens. An association between bacterial L-forms and inflammatory bowel disease has been sporadically reported, with isolation of *Pseudomonas, Mycobacterium, Enterococcus fecalis,* and *E. coli* from affected tissue but not from appropriate controls. There is considerable debate as to whether L-forms are pathogenic in humans or persist in affected tissue.

Mycobacterium paratuberculosis, the causative agent of Johne disease in ruminants, may be associated with Crohn disease through the production of L-forms of the bacterium. Subclinically infected cows shed *M. paratuberculosis*, and the organism has been identified in pasteurized milk by polymerase chain reaction specific for the *M. paratuberculosis* insertion sequence IS900. The pathogen model suggests that a susceptible human neonate first contracts the organism after ingesting commercial dairy products. This invokes an antigen-poor (lacking a cell wall) L-form that grows slowly and persists in the lamina propria, stimulating a chronic low-grade inflammation. The immune response increases in severity over years without bacterial replication, ultimately producing the pathologic features of Crohn disease. Another model proposes an autoimmune phenomenon mediated by alterations in inflammatory cytokine profiles, possibly as a result of infection.

Recent immunocytochemical techniques demonstrated antigens to *Listeria monocytogenes, E. coli,* and *Streptococcus* spp. in Crohn disease tissues. Macrophages and giant cells immunolabelled for antigen specific to these organisms were found beneath ulcers, around abscesses, along fissures, within the lamina propria, in granulomas, and in germinal centers of mesenteric lymph nodes.

Superantigens and Autoimmunity

In contrast to conventional antigens, superantigens interact with the variable side of the Vβ chain of the T-cell receptor by recognizing elements shared by a subset of T cells. Depending on the type of interaction, recognition can have different consequences, including proliferation and expansion, suppression (clonal deletion), or, alternatively, induction of prolonged unresponsiveness (anergy) or cell death (apoptosis). Superantigens from several food borne bacteria (*e.g., Staphylococcus, Streptococcus, Yersinia,* and *Clostridium*) have been isolated and characterized. Many are thought to be associated with several autoimmune disorders, for example, rheumatic heart disease, rheumatoid arthritis, multiple sclerosis, Graves disease, Sjogren syndrome, autoimmune thyroiditis, psoriasis, Kawasaki disease, Crohn disease, and insulin-dependent diabetes mellitus.

Although it is accepted that superantigens have a role in autoimmune disorders, the acceptance is based on extensive animal model studies but limited human clinical studies. In human diseases where superantigens have been clearly demonstrated as the cause, for example, toxic shock syndrome, initial T-cell proliferation and T-cell receptor-mRNA up-regulation have been observed, but the long-term sequelae in terms of T-cell function are unknown. Recent studies suggest that superantigens may also cause an acute flare of a disease within patients in remission from a pre-existing autoimmune disorder.

Renal Disease

After colitis caused by *E. coli* O157: H7 and other enterohemorrhagic strains of *E. coli*, hemolytic uremic syndrome develops in some patients. The syndrome is characterized by a triad of symptoms: acute renal failure, thrombocytopenia, and microangiopathic hemolytic anemia. Acute renal failure is the leading cause of death in children, and thrombocytopenia is the leading cause of death in adults. Hemolytic uremic syndrome is a worldwide problem that mirrors the distribution of *E. coli* O157: H7 and other Shiga and Shiga-like toxin-producing microorganisms. Outbreaks of hemorrhagic colitis and subsequent cases of hemolytic uremic syndrome have developed as a result of various food vehicles. Besides O157: H7, other Shiga-like toxin-producing *E. coli, Citrobacter, Campylobacter, Shigella, Salmonella,* and *Yersinia* have been linked to the disorder.

The toxin-mediated damage to the kidneys may not be limited to the glomerular endothelial cells as once thought but may include the tubular epithelial cells. Binding studies showed the toxins to be specific for the glycosphingolipid globotriaosylceramide (Gb3), which is present on renal but not umbilical endothelial cells. This may account for the differential sensitivity of renal cells to toxin-induced damage, since Gb3 was present in the glomeruli of infants under 2 years of age but not in the glomeruli of adults. Thus, the presence of Gb3 in the paediatric renal glomerulus may be a risk factor for development of haemolytic uremic syndrome. Characterization of the Shiga toxin receptor has led to a potential preventive treatment.

Neural and Neuromuscular Disorders

Guillain-Barré syndrome is a subacute, acquired, inflammatory demyelinating polyradiculoneuropathy that frequently occurs after acute gastrointestinal infection.

The disease is characterized by alexia, motor paralysis with mild sensory disturbances, and an acellular increase in the total protein content in the cerebrospinal fluid. The disease occurs worldwide and is the most common cause of neuromuscular paralysis. Cases have three dominant characteristics: the predilection to nerve roots, mononuclear infiltration of peripheral nerves, and eventual demyelination (primary axonal degeneration. Severe cases tend to occur in the summer and have been linked to previous infection with *C. jejuni*, although other enteric pathogens may trigger the syndrome.

Some controversy exists regarding whether Guillain-Barré syndrome is an autoimmune disease. Although adequate data exist to classify the syndrome as an autoimmune disease (four major Rose-Witebsky criteria are almost completely met), the immunologic mechanisms at work in Guillain-Barré syndrome triggered by *C. jejuni* are likely to be complex. Studies of the relationship between Guillain-Barré syndrome and *C. jejuni* support the hypothesis of molecular mimicry, since peripheral nerves may share epitopes with surface antigens of *C. jejuni*. This has been supported by studies in which anti-GM1 IgG antibodies recognized surface epitopes on intact *C. jejuni*, and the reaction was strain-specific for certain Penner serotypes. There are inconclusive data with regard to Guillain-Barré syndrome and HLA, although some studies have shown a predilection for the HLA-B35 haplotype. Cytokines may be responsible for inducing the inflammatory process and probably play a role in the response leading to nerve demyelination. Complement has a role in the process leading to nerve damage, possibly through the production of activation products, which lead to an increase in the permeability of the blood nerve barrier, which perpetuates the inflammation. Although Guillain-Barré syndrome might be considered an autoimmune response, it also serves as an example of a disease with an infectious origin, a disease that entails the integrated actions of both humoral and cellular immunity.

Ciguatera poisoning is the most common food borne disease related to the consumption of fin fish; this distinctive clinical syndrome is characterized by a plethora of gastrointestinal, neurologic, and sometimes cardiovascular features. Two toxins are involved in toxicosis. Ciguatoxin-1 (cig-1), the principal toxin, is a heat stable, lipid-soluble polyether that is not inactivated by heat, cold, or gastric juices, nor eliminated by drying, salting, smoking, or marinating. Cig-1 induces membrane depolarization in nerve and muscle tissue by opening voltage-dependent sodium channels. A second toxin, maitotoxin, is water-soluble and opens calcium channels. The role of this second toxin in the pathophysiology of the disease is less well understood. The acute symptoms of the toxicosis are varied and include paraesthesia of the extremities, circumoral paraesthesia, reversal of hot and cold sensations, dental pain, myalgias, arthralgias, generalized pruritus, cranial nerve dysfunction, and dysuria. Severe acute symptoms require urgent care with parenteral atropine for bradyarrhythmias. Mannitol is often administered to counter the effect of the toxin on the sodium channels; however, the mechanism of action is unknown, and the therapy is useless after 24 hours. Many of these symptoms may remain chronic and are often misdiagnosed as chronic fatigue syndrome, brain tumours, or multiple sclerosis. The management of the chronic symptoms is frustrating for the patient and clinician. Interventions include amitriptyline, tocaidine, or mexilitine to modulate sodium channels in conjunction with calcium channel blockers such as nifedipine. Antidepressants such as Prozac also appear to be useful. Patients with chronic symptoms frequently report waxing and waning of symptoms. Activities such as sexual intercourse and drinking alcohol significantly exacerbate expression. Some women with chronic symptoms report worsening during menses. Mood levels, weather conditions, and dietary constituents often exacerbate symptoms. Some clinicians advocate a strict diet that avoids all seafood, fish byproducts, nuts, and alcohol, and in some cases, patients are asked to abstain from sex. One distinctive feature in this toxicosis is that one episode of ciguatera poisoning does not confer immunity. In fact, it is likely to sensitize the patient to otherwise subthreshold doses of toxin

Amnesic shellfish poisoning is caused by domic acid, a conjugate of kainic and glutamic acid. In small quantities domic acid has an excitatory effect, but in large amounts it is neurotoxic. The toxicosis is first characterized by gastrointestinal symptoms followed by neurologic dysfunction. Severe cases may be prolonged and chronic; sequelae include confusion with disorientation, paucity of speech, lack of response to deep pain due to blocking of receptors in the spinal cord, autonomic nervous system dysfunction, seizures, abnormal ocular movements, grimacing posture, myoclonus, loss of reflexes, and coma. Other prominent chronic sequelae include loss of visual-spacial recall and mononeuropathies without dementia, mimicking Alzheimer's disease. The toxicosis is particularly serious in the elderly, and any deaths usually occur within this population. Valium, calcium channel blockers, phenobarbital, diazapam, thiobarbiturates, and hypothermia are treatments for patients with severe and chronic cases.

General Immunity, Organ Impairment, and Neurologic Disorders

Toxoplasmosis due to *Toxoplasma gondii* is a chronic latent parasitic infection. In humans the parasite exists in two forms: the tachyzoite, the rapidly multiplying stage that actively invades host cells and represents the principal form of the acute phase of the disease; and the bradyzoite, the form that multiplies very slowly in host cells, resulting in the formation of cysts that persist in tissues. Toxoplasma infection in humans is usually asymptomatic because of effective immunity involving antibodies, T cells, and cytokines. Activated macrophages, CD4 and CD8 lymphocytes, and the cytokines IFN-g, TNFa, and IL-6 play a major role in control of both the acute infection and maintenance or prevention of the chronic stage. Indeed, treatment with IFN-g is used to control passage into the chronic stage, and treatment with anti-IFN-g reactivates chronic infection. The production of nitric oxide may have opposing effects. Nitric oxide production protects against *T. gondii* and at the same time limits the immune response, probably contributing to the establishment of the chronic state.

The incidence of congenital toxoplasmosis is uncertain but may be as high as 9,500 cases a year. The percentage attributable to food is uncertain; however, consumption or contact with contaminated meat is more important as a cause than is contact with cats. Congenital impairments associated with maternal toxoplasmosis infection passed to the foetus include hearing loss, visual impairment (retinal lesions, strabismus), and slight to severe mental retardation. These impairments are still present in 80 per cent of persons who reach the age of 20 years. Chronic toxoplasmic encephalitis may occur when a person's immune system is impaired. Indeed, toxoplasmic encephalitis marked by dementia and seizures has become the most commonly recognized cause of central nervous system opportunistic infections in AIDS patients. Additionally, it appears that certain cancer treatments weaken the immune system, and old infections in the muscles can become reactivated, causing severe complications or death.

Helminth parasites can cause serious disease in infected persons. The impact of helminth infections is due less to the severity of the diseases they cause, than to the vast number of persons infected. For example, more than one billion people are infected with the largest intestinal nematode, *Ascaris lumbricoides*. Although there is usually no overt clinical sign of infection, disease can arise from an overwhelming infection or an inappropriate immune response. Additionally, infected persons frequently harbour more than one parasite for years. Most intestinal helminth parasites have direct life cycles, with no intermediate host or vector, and are transferred by contaminated food. Some species, such as *Trichuris* (whipworm) and *Enterobius* (pinworm), are restricted to the gut, but others, such as *Ascaris,* have tissue-migrating phases. All, however, induce a dramatic expansion of the Th2 lymphocyte subset. It remains unclear whether these Th2-derived responses (induction of interleukin-4 [IL-4] and down-regulation of IFNg), resulting in stimulation of IgG1 and IgE isotypes, eosinophilia, and mastocytosis are responsible for the immune-mediated pathologic response. Immunologic lesions may occur where early infection is associated with a strong T-cell proliferative response that becomes down-regulated in established chronic disease (evidence of a Th1 defect in the chronic disease). In ascariasis, an allergic response generated by the lung migratory phase (chronic immune sensitization) can cause pneumonia and, in animal models, spontaneous development of idiopathic bronchial asthma. A formative influence on the response of the immune system is the antigenic environment during pregnancy. Children born to infected mothers may have significantly higher susceptibility to the same infection later in life.

Viral agents induce autoimmune disorders, and one potential mechanism of induction is molecular mimicry. Hepatitis A virus infection is a well-recognized cause of acute hepatitis with jaundice in adults. In most affected persons, the course is usually relatively short-lived and benign, and symptoms are usually resolved within weeks. Occasionally, relapses occur after initial recovery, or recovery is marked by severe or prolonged cholestasis. However, even in these cases recovery is usual. Chronic sequelae of hepatitis A virus infections are rare and poorly defined; however, several recent studies suggest that hepatitis A virus infection triggers the onset of (idiopathic) autoimmune chronic active hepatitis within a genetically predisposed subgroup. Apparently, the chronic disorder may develop despite normal serologic response to hepatitis A virus infection. The triggering factors and mechanism of action remain ill-defined; however, in the most recent study, the authors concluded that hepatitis A virus infection may be the precipitating event in the pathogenesis of this disorder.

Metabolic activation and detoxification play a crucial role in determining the toxic response of humans to mycotoxin exposure. These highly toxic secondary metabolites are produced by a wide variety of molds including *Aspergillus, Fusarium,* and *Penicillium*. Mycotoxins exhibit properties of acute, subacute, and chronic toxicities with some molecules being carcinogenic, mutagenic, and teratogenic. Because mycotoxins are resistant to food processing and do not degrade at high temperatures, they enter the human food supply.

In many cases, the relationship between mycotoxins as the causative agent of disease in humans is difficult to determine. Acute effects of gastroenteritis may be easily identified; however, chronic effects often result from ingestion of low to moderate levels and can be difficult to recognize. The most threatening effects of ochratoxin A are its nephrotoxicity and carcinogenicity. Ochratoxin A is increasingly involved in an endemic nephropathy, a human chronic interstitial neuropathy that is usually associated with urinary tract tumours. Aflatoxins have been implicated in both acute and chronic liver disease in humans; however, other organs (kidney, spleen, and pancreas) may also be affected. The best studied chronic effects are those induced by the fumonisins, zearalenone, and trichothecene mycotoxins produced by *Fusarium* sp. Fumonisin levels in corn-based foods have been statistically associated with an increased risk of human esophageal cancer. Zearalenone is an estrogenic mycotoxin. Ingestion by animals, especially swine, causes hyperestrogenism with symptoms of enlargement and prolapse of the uterus, atrophy of ovaries and testicles, enlargement of mammary glands, and infertility. This mycotoxin might add to the estrogen load of humans. Human consumption of trichothecene-contaminated foods causes acute symptoms of headaches, chills, severe nausea, vomiting, and visual disturbances, which may last 7 to 10 days. Since trichothecenes modulate immune function, over time mycotoxicosis could reduce immune resistance to infectious diseases, facilitate tumour growth through reduced immune function, and cause autoimmune disease.

Heart and Vascular Diseases

Several food borne pathogens have been either directly or indirectly associated with endocarditis and myocarditis, and any heart damage appears to be permanent. Persons with ankylosing spondylitis linked to enteric pathogens as the trigger show a high incidence of cardiac conduction abnormalities, which may be sequelae to other seronegative arthropathies. A possible connection between food borne gram-negative bacteria and atherosclerosis has been proposed, suggesting that the bacteria gain access to the lymphoid and general circulation with relative frequency. Endotoxin from degrading bacteria in macrophages may act in concert with the inflammatory factors (cytokines) induced by endotoxin from endothelium and smooth muscle cells. Although the process of atherogenesis is complex and involves many factors, the hypothesis is attractive and provides a model system for further study. Oxidative stress responses by *E. coli* and *S. Typhimurium* and the induction of the peroxide stimulon and the superoxide stimulon have also been recently implicated in atherosclerosis, rheumatoid arthritis, and inflammatory bowel disease.

Nutritional and Gastrointestinal Disturbances

Enteric pathogen-induced diarrhoea may lead to a variety of conditions including loss of fluids, anorexia, and malabsorption of nutrients, all forms of malnutrition. The enteric pathogens that cause malabsorption and nutrient loss vary and include *Enterobacteriaceae*, Rotavirus, *Amoeba, Cryptosporidium,* and *Giardia*. Unless treated with antimicrobial drugs, many diarrheal episodes become chronic; however, the stress of even short periods of diarrhoea may result in subtle changes in immunologic status. The extent of the diseases depends mostly on the immune status of the person and may last for several years or for life. Death due to diarrheal illness in the immunosuppressed and in persons with AIDS is nearly 80 per cent. No effective treatment is available, although treatment with several antibiotics in combination shows promising results. However, AIDS patients may also develop further sequelae. *Cryptosporidium* are host-adapted, which may lead to pulmonary or tracheal cryptosporidiosis accompanied by coughing and frequent low-grade fever. In these cases, there is no effective treatment. Similarly, in cyclosporidiosis, AIDS patients' enteric infection is chronic. Long-term prophylaxis with trimoxazole is required, and discontinuation of the treatment causes severe relapse.

Severe cases of diarrhoea lasting months or years and characterized by dysentery, with foul-smelling, mucous bloody stools accompanied by flatulence and abdominal distention, may result from *C. jejuni, Citrobacter, Enterobacter,* or *Klebsiella* enteric infections. These infections always require extensive antibiotic therapy and usually result in failure to thrive. Enteric infections may alter bowel permeability which allows absorption of otherwise excluded food components. Proteins that can modulate the immune system can be absorbed possibly with deleterious consequences such as the induction of autoimmunity and atopy. Several studies of both human and porcine models indicate that significant quantities of unwanted proteins can be absorbed by damaged gut tissue and that maximum expression of diarrhoea corresponds with peak protein uptake.

Helicobacter pylori is the undisputed cause of chronic gastritis. Environmental sources indicate that *H. pylori* can survive in water, chilled foods, milk, and fresh vegetables for several days because of faecal contamination. The species has never been isolated from these sources; however, infectious, viable but nonculturable (nonspiral coccoid) bodies may survive in fresh water for more than a year. *H. pylori* can be found in human faeces and can be transmitted directly from person to person by the faecal-oral or oral-oral route. *H. pylori* can be found in several animal reservoirs; however, the possibility of animal-to-animal or zoonotic transmission is unknown. Ingestion of *H. pylori* leads to acute gastritis, and colonization of the stomach is virtually always accompanied by chronic inflammation that disappears within 6 to 12 months after eradication of the infection. Infections are generally acquired during childhood or adolescence and result in chronic gastritis lasting for decades or life. On the basis of histologic and serologic follow-up studies, this chronic gastritis has been suggested as an important risk factor (odds ratio 9.0; $p = 0.001$) or first stage in a multistep process leading to gastric mucosal atrophy, intestinal metaplasia, and eventually gastric cancer.

Chronic Sequelae and Personality Changes

One area that has received scant interest is the effect of a chronic infection on human personality factors. Personality changes might be predicted: continual pain from arthritis, an irritable bowel, or chronic diarrhoea would be enough to make anyone temperamental, moody, or depressed. Studies using Cattell's 16 Personality Factor questionnaire showed highly significant correlations ($p < 0.01–0.002$) between chronic toxoplasmosis and several personality factors. Men and women showed distinct differences in behavioural states. For men, low superego strength, protension, guilt-proneness, and group dependency were positively associated, whereas in women the related factors were affectothymia, alexia, untroubled adequacy, and self-sufficiency. A correlation of the intensity of the personality factor-shifts with the duration of the infection suggested that the infection per se induced the shift in personality, not vice versa.

An exploratory study using the 16 Personality Factor questionnaire and the Holmes and Rahe Social Readjustment Rating Scale, which measures stressful life events, was made of patients with rheumatoid arthritis. As a group, patients with rheumatoid arthritis exhibited higher stress at disease onset ($p < 0.01$); a large high-stress-at-onset subgroup of rheumatoid arthritis patients had a worse prognosis. Although there were important personality changes in the high-stress-at-onset-rheumatoid arthritis patients, the study concluded that the interaction between the variables that determined personality changes were very complex and could not simply be referred to as the "rheumatoid arthritis personality" complex.

Food borne diseases are for the most part preventable; however, there is an inherent risk associated with the consumption of certain types of uncooked foods. Recognition by the public health community and the public that many food borne illnesses may have serious chronic sequelae would help eliminate many illnesses and reduce health-care cost. Public health authorities could make a substantial impact by reducing poor or unhygienic food production or food-handling practices and by educating the public about how harmful microorganisms enter the food chain and how they can be avoided.

Impact of Changing Consumer Lifestyles on the Emergence/Re-emergence of Food Borne Pathogens

Foodborne illnesses are a real concern to public health professionals, food scientists, microbiologists, and sanitarians. Today there are more than 200 known diseases transmitted through foods and many of the pathogens of greatest concern were not recognized as causes of foodborne illness 20 years ago. Most cases of foodborne illness involve gastrointestinal symptoms (nausea, vomiting, and diarrhoea), and are usually acute, self-limiting, and of short duration, and can range from mild to severe. Deaths from acute foodborne illnesses are relatively rare and typically occur in the very young, the elderly, or in persons with compromised immune systems.

The U.S. Food and Drug Administration (FDA) estimates that 2 per cent to 3 per cent of all acute foodborne illnesses develop secondary long-term complications often referred to as chronic sequelae as discussed in previous section. These sequelae can occur in any part of the body such as the heart, kidney, nervous system, or joints and can be quite debilitating and, in severe cases, can cause death. There are many factors associated with the emergence of "new" foodborne pathogens and outbreaks of foodborne illnesses. Some of these factors include:

Demographics

Food borne illness of microbial origin is the most serious food safety problem. The population aged 65 and older was 35 million in 2000 in U.S.A. and is expected to more than double by 2050. Significant portions of older Americans suffer from chronic health conditions, including heart disease, cancer, diabetes, and this makes them more susceptible to foodborne illness. For example, persons with AIDS or late-stage HIV infections have a 20 times higher possibility of developing salmonellosis than healthy people. These individuals are also at a 200 to 300 times higher risk to develop listeriosis. As people age, their immune system function decreases, so people have a decreased resistance to pathogens as they get older.

Changes in Consumer's Practices

Practices of consumers have varied levels of awareness of specific microbial hazards, risk factors for foodborne illness, the importance of good personal hygiene during the preparation and serving of foods. Consumers have a relatively poor knowledge of safe food preparation practices in their homes.

The Center for Disease Control and Prevention reports that 79 per cent of outbreaks between 1987 and 1992 were bacterial; improper holding temperature and poor personal hygiene of food handlers contributed most to disease incidence. Some microbes have demonstrated resistance to standard methods of preparation and storage of foods. Nonetheless, food safety and public health officials attribute a rise in incidence of food borne illness to changes in demographics and consumer lifestyles that affect the way food is prepared and stored. Food editors report that fewer than 50 per cent of consumers are concerned about food safety. An American Meat Institute (1996) study details lifestyle changes affecting food behaviour, including an increasing number of women in the workforce, limited commitment to food preparation, and a greater number of single heads of households. Consumers appear to be more interested in convenience and saving time than in proper food handling and preparation.

According to a recent report to Congressional committees (1996), experts believe that the risk for food borne illness is increasing. The food supply is changing in ways that can promote food borne illness, and there are no comprehensive data to explain at what point pathogens are introduced into food. Further, because of demographic changes, more people are at a greater risk of contracting a food borne illness.

According to Ollinger-Snyder and Matthews (1994), changes in agricultural practices, a growing population susceptible to infectious diseases, lifestyle changes, the emergence of new food borne pathogens, and the high turnover rate reported for workers in the food-service industry indicate that new approaches are needed to allay consumers' fears and to prevent the spread of food borne disease in the United States. They recommend implementation of Hazard Analysis and Critical Control Points (HACCP) systems and certification of food-service managers.

The food processing industries are developing and implementing HACCP systems; the meat and poultry industries are mandated to do so beginning in January 1998. Hazard analysis has been defined as the identification of sensitive ingredients, critical processing points, and human factors that affect product safety. Critical control points have been described as processing determinants whose loss of control would result in an unacceptable food-safety risk.

Most contend that the HACCP system approach must be implemented at each stage of the farm-to-family continuum. Where are the critical control points and the HACCP system development in the home, food-service or retail establishments, or the car when food is carried from one location to another? The consumer is a complex and critical control point in the process.

Take the case of the barbecued chicken served to 260 guests at an outdoor barbecue in 1983. Guests were served chicken that was parboiled in the morning by one set of cooks and then placed in a large container and refrigerated. The evening cooks assumed the chicken had been adequately cooked, so they basted it in barbecue sauce and warmed it over the fire. Some 71 per cent of the guests got sick from the chicken that was insufficiently cooked and improperly held. What of the infected bakery worker who stirred a vat full of buttercream frosting with a bare hand and arm? Some 5,000 cases of viral gastroenteritis were caused by the infected worker who claimed he had washed his hands. Other outbreaks which appeared in 1990s are depicted in Table 15.7.

Recent data indicate that 80 per cent of reported food borne illness outbreaks occur outside the home. Every restaurant can take steps to ensure the safety of the food it prepares and serves to its customers. Continuous employee training and institution of HACCP-type systems should assist restaurants and other food-service institutions in improving their food-safety records. Programs available through the national restaurant trade organization could assist even the smallest establishments in achieving food-safety goals.

Table 15.7: Food Borne Illness Reports from Restaurants, 1996

Date	*Description*	*Cause*
6/96	*Salmonella*-contaminated food, 38 cases	Employees did not wash hands before handling food
9/95	*Escherichia coli* O157: H7 "beef," 11 cases	Raw food cross-contaminated other
8/95	*Salmonella* Newport "chicken," >850 cases	Raw meat on cuting board with vegetables
1/95	Hepatitis A, contaminated food, 95 cases	Human fecal matter from handling-handwashing
8/94	*Salmonella*, hollandaise sauce, 56 cases	Holding temperature too low for 9 hours
1/93	*Clostridium botulinum*, canned cheese sauce, 7 cases, 1 death	Unrefrigerated storage of opened container

Source: Center for Science in the Public Interest, 1996

Consumer Knowledge and Concern about Food Borne Illnesses

Consumers are receptive to information about microbiologic hazards. Nationwide surveys by the Food Marketing Institute indicate that more people volunteer concerns about microbiologic hazards than about any other potential food safety issue. From 1992 to 1996, volunteered concern about microbiologic safety increased from 36 per cent to 49 per cent. Specifically, concern about contamination by bacteria or other microorganisms was 77 per cent, more than concern about pesticide residues (66 per cent), product tampering (66 per cent), antibiotic residues (42 per cent), or any other food safety risk.

Food-Handling Practices

Although many consumers recognize the potential seriousness of food borne bacteria, they lack information on safe handling and storage of food products. Williamson *et al.* (1992) found that consumers under 35 years of age knew less about food safety terms and concepts than those over 35. Specific safe food handling was not practiced by 15 per cent to 30 per cent of survey respondents. For example, consumers did not cool cooked food rapidly, with 29 per cent indicating they would let roasted chicken cool completely before refrigerating. Only 32 per cent indicated they would use small, shallow containers to refrigerate leftovers. Consumers did not know that failure to refrigerate may jeopardize safety, with 18 per cent not concerned or uncertain about the safety of cooked meat and 14 per cent not concerned about poultry left unrefrigerated for more than 4 hours. The need for sanitation was not recognized, with only 54 per cent indicating they would wash a cutting board with soap and water between cutting raw meat and chopping vegetables.

Food safety experts have identified the most common food-handling mistakes made by consumers at home. These mistakes included serving contaminated raw food, cooking or heating food inadequately, obtaining food from unsafe sources, cooling food inadequately, allowing 12 hours or more between preparation and eating, and having a colonized person handle implicated food or practice poor hygiene (Bryan, 1988). The same factors were identified in mishandling associated with specific pathogens (Bean and Griffin, 1990).

Nature and Source of Food Borne Illness

Consumer perceptions and behaviour related to food borne illness changed little between 1988 and 1993 (Fein *et al.,* 1995). Consumers misperceived the nature of food borne illness and the most likely pathogen source. Consumer belief about the type of food responsible for food borne illness–meat, poultry, seafood, eggs–was consistent with expert opinions; however, consumers believed that food borne illness was generally mild, without fever, and occurred within a day of eating contaminated food. Infections caused by *Salmonella* and *Campylobacter*, the most common food borne illnesses in the United States, are not consistent with the symptoms consumers described, because these organisms have longer latency and cause fever.

Most consumers believed that their food borne illness was caused by food prepared somewhere other than the home. Williamson *et al.* (1992) found that about one-third of consumers thought food safety problems most likely occurred at food manufacturing facilities, and one-third blamed unsafe restaurant practices. Only 16 per cent thought mishandling was most likely to occur in the home. Fein *et al.* (1995) found that 65 per cent of consumers attributed food borne illness to food prepared at a restaurant, 17 per cent to mishandling at the supermarket, and 17 per cent to mishandling at home. In contrast, food safety experts believe sporadic cases and small outbreaks in the home are far more common than recognized outbreaks.

Failure to recognize the home as a likely source of food borne illness is not unexpected because illness traced to a food establishment affects many people and may receive widespread publicity (Ollinger and Mathews, 1994). Illness that occurs at home is rarely reported unless severe (IFT, 1995).

If consumers misperceive the nature and origin of food borne illness, they underestimate the frequency of serious consequences and are less motivated to change. Schafer *et al.* (1993) found that motivation for proper food handling requires viewing the mishandling of food as a direct threat to one's health. The failure to associate mishandling of food in the home with food borne illness interferes with food borne disease education efforts.

Ubiquity of Organisms

Consumers do not seem to be aware of the ubiquity of microorganisms in the environment. During food borne disease outbreaks, press accounts focus on faecal contamination of food. Government standards classify natural microorganisms as contaminants, which suggest that microorganisms are only present as a result of mishandling. In contrast, Hazard Analysis and Critical Control Points (HACCP) programs recognize and attempt to control potential dangers related to pathogenic microorganisms.

When consumers in a national sampling were asked on whom they rely for product safety, the percentage responding "myself as an individual" decreased from 48 per cent in 1989 to 25 per cent in 1996 (Abt Associates, 1996). As self-reliance decreased, consumer reliance on food manufacturers and supermarkets increased. This may be a response to the message that if raw food contains microorganisms, it is contaminated. It suggests some consumers are shifting the responsibility for safe food to manufacturers and retailers.

Consumers may not realize they can introduce pathogens during food handling. In 1990, the Food Marketing Institute asked consumers what steps they took at home to ensure the safety of food (Food Marketing Institute, 1990)). Respondents volunteered refrigeration (58 per cent), proper storage (35 per cent), checking expiration dates (26 per cent), washing and cleaning the food (25 per cent), cooking properly (22 per cent), and wrapping food properly (20 per cent). No one volunteered washing hands or preventing cross-contamination.

Labelling

Products must contain safety labels instructing consumers how to handle food. In 1989, the National Advisory Committee on Microbiological Criteria for Foods recommended a mandatory uniform logo for perishable refrigerated foods, uniform labelling for frozen food, "Use by" dates, and time/temperature indicators wherever possible (Rhodes, 1991).

Although products are currently labelled when they require refrigeration, the label is ineffective because the warning is difficult to find or read. As the proportion of older people increases, print must be larger. Labels should also display symbols to further enhance the effectiveness of the message.

Safe-handling labels on meat products appear to have made a difference. The Food Marketing Institute (1996) found that 60 per cent of survey respondents had seen the labels. Of those aware of the labels, 65 per cent said the labels increased their awareness of safety, and 43 per cent said they changed their behaviour as a result of the information. The most common volunteered change was washing the counter and utensils after contact with meat (approximately 40 per cent), followed by washing hands more frequently (approximately 20 per cent) and cooking to the proper temperature (approximately 20 per cent).

Labels do not consistently contain needed information. When foodborne illness was related to consumption of unpasteurized apple juice, consumers were not able to determine from the label which products were pasteurized. Many major manufacturers do not indicate whether their fresh juice product is pasteurized.

Consumers are not advised about potential risks for special populations. Raw milk sold in California must contain a warning statement, but other states may not have this requirement. Because of inconsistencies in labelling, unpasteurized juice products may be given to infants. Also, products that contain honey do not include a warning about potential risk for infant botulism.

Changing Lifestyles and Food Habits

Many factors have contributed to consumers' lack of familiarity with safe food handling and increased food borne illnesses. Increased participation in the paid labour force has lessened the exposure of young people to food-handling practices in the home; few schools offer or require food preparation classes; and partially prepared foods may have different, less familiar handling requirements (IFT, 1995; Dumagan and Hackett, 1995).

People eat out more frequently today, thereby increasing their exposure to the food service industry, noted for high turnover rates and minimal job training in personal hygiene. In 2002, U.S. consumers spent 46.1 per cent of their food dollar away from home. The sheer volume of meals prepared each day stresses the need for knowledgeable, well-trained foodservice and retail food store employees who understand the principles of safe food preparation. Furthermore, the population is shifting, with an increased percentage of persons at higher risk for food borne illness because of age or health status. Additionally, some food safety recommendations related to temperature and acidity do not eliminate risks from some pathogens. Food preferences have also changed, with many people now eating raw foods of animal origin or lightly cooked foods that can increase the risk of foodborne illnesses.

Complexity of the Food System

As explained earlier, the food system is a complex, concentrated, and dynamic chain of activities that moves food from farm to table. When errors occur, major food incidents can result. Multiple handling of foods (or ingredients) increases the possibility for contamination and subsequent temperature abuses. The key is to develop close working relationships and strong networks between and among the different sectors of the system to assure a safe and wholesome food supply.

Globalization of the Food Supply

The international sourcing of food and food ingredients has enabled consumers to enjoy a consistent supply of a wide variety of products from around the world. The main concern is that the sanitary standards and safety assurance systems in some countries may not be as stringent as those in other countries. Today, with increasing international travel, a microorganism that causes a problem in one part of the world can be easily transported to another country very quickly. Rapid detection, early intervention, and vigilance are important in preventing the spread of foodborne illness from country to country.

Processing Technologies

Consumers do not realize that pathogens can survive minimal processing, as evidenced by a recent *Escherichia coli* outbreak associated with fresh apple juice, which demonstrated that processors also may not recognize potential risks. A fresh apple juice manufacturer in northern California claimed its product was safe because the juice was squeezed in small batches and frozen immediately.

Freezing is not effective against *E. coli* 0157: H7, but other methods are protective. Several methods have been developed to reduce pathogens and increase the safety of foods. Once these methods are verified as effective and safe, the food industry should be free to use them, and consumers should have the opportunity to select safer foods. In some cases, the regulatory approval process appears to hinder rather than facilitate the safer handling of food.

Food irradiation, exposing food to high levels of electromagnetic energy for specific purposes, has been approved for selected uses. A petition before the Food and Drug Administration to permit irradiation of meat and other muscle foods appears to have satisfied safety concerns, but approval has not yet been granted. The requirement to seek approval for each application of irradiation prevents rapid response in cases of food borne outbreaks. Although this regulatory procedure may have been reasonable when irradiation was first introduced, it warrants a fresh look in view of the wealth of data now available on the safety and wholesomeness of food irradiation (Diehl, 1995).

Attitude surveys and marketing experience consistently demonstrate that consumers will purchase irradiated food (Bruhn, 1995). National surveys indicate consumer concern about irradiation was lower than other food-related concerns. When specifically asked what they considered a serious health hazard, 29 per cent identified irradiation, 77 per cent identified bacteria, and 66 per cent identified pesticides (Food Marketing Institute, 1996). The percentage of consumers concerned about irradiation has decreased significantly over time. In the late 1980s, 42 per cent to 43 per cent classified irradiation as a serious concern, decreasing to 29 per cent in 1996. A relative ranking of food processing methods surveyed by the Gallup Organization found that irradiation, food preservatives, and chlorination generated similar concern (Gallup Organization, 1993).

In a nationwide Food Marketing Institute survey, 69 per cent of consumers indicated they were very or somewhat likely to purchase products irradiated to kill bacteria or other microorganisms (Food Marketing Institute, 1996). Surveys completed in several areas of the country indicate 60 per cent to 70 per cent of consumers would prefer irradiated food (Bruhn, 1995). In one study, information about irradiation increased interest in purchasing to 90 per cent, and education plus food samples increased purchase intent to 99 per cent.

Consumers have purchased irradiated food in select locations across the United States since 1992 (Bruhn, 1995). Fruits from the Mainland and Hawaii have sold well in the Midwest and California (L. Wong, pers. comm.). Irradiated chicken gained over 60 per cent of market share when priced 10 per cent lower than non-irradiated chicken and 47 per cent when priced the same.

Irradiated food is not widely available. Special interest groups threaten companies that exchange information about irradiation processing. Consumers, however, appear to prefer irradiated foods when the benefits of these foods are endorsed by health professionals. Food manufacturers and retailers should offer consumers the choice of safety-enhanced, energy-pasteurized irradiated food.

As the food industry strives for fresher products and longer shelf-life products, product developers must be aware of how composition, processing parameters, packaging systems, and storage conditions influence the microorganisms that are present. Food safety must be built into the product while it is being developed or reformulated. There has been a greater awareness of the environmental conditions in processing plants, retail food stores, and foodservice establishments and the need to ensure that biofilms and microbial niches do not develop.

Changes in Foodborne Pathogens

There have been many changes in the microorganisms that cause foodborne illnesses. Scientists have observed more virulent strains of organisms, where a few cells can cause severe illness. An example is *S. enteritidis* and *E. coli* 0157:H7. Adaptive stress responses have also been observed where organisms have adapted to environmental conditions to survive and grow, such as psychrotropic pathogens that grow (slowly) at refrigerated temperatures. Organisms such as *Yersinia enterocolitica*, *L. monocytogenes*, and *Clostridium botulinum* type E are examples of bacteria capable of growing at refrigerator temperatures.

In recent years increased resistance to antibiotics has been observed in *Salmonella typhimurium* DT104. A number of outbreaks in produce and unpasteurized apple cider have been caused through the protozoan parasites *Cyclospora cayetanensis* and *Cryptosporidium parvum*. All of these factors have played and continue to play a role in the emergence of foodborne pathogens and foodborne illnesses. In a discussion of food safety issues, a chief executive officer (CEO) of a small retail food chain made the following comment: "Today, we're facing a new enemy; it is not business as usual." This statement clearly describes the fact that we live in a changing world and must be proactive in assuring food safety.

Diagnostic Techniques

In the last decade, there have been significant improvements in foodborne disease surveillance and responses to outbreaks, improved diagnostic techniques, and better medical interventions when illnesses occur. More rapid microbial tests have been developed, and electronic data bases such as FoodNet, PulseNet, and ElexNet have been developed to provide better surveillance of foodborne illnesses, improved information sharing, and more rapid responses when outbreaks occur.

Need for Communicating with the Public

To respond to consumers' need for information, a multifaceted program is needed. The HACCP strategy, which teaches consumers to critically think through the food safety process to determine how food borne illness could occur, has been effective. The HACCP approach to home food preservation is logical and highlights key control points.

Consumers should be informed that microorganisms are ubiquitous in the environment, found on raw products of animal or plant origin. Pathogens may survive minimal processing and preservation treatments. People may introduce pathogens during any stage of food processing or handling, including just prior to consumption. Food borne illness can range from mild to severe and life-threatening with chronic complications. People have control and can reduce risks.

Communicating food safety information to the public effectively is another challenge. Consumers obtain most of their information on food, nutrition, and science from the media; television is cited most frequently, and newspapers and magazines follow (USDA, 1996; Reicks *et al.*, 1994). Brochures enforce messages and serve as useful references, although they are not as widely seen as media stories.

Developing messages with the press should be a primary activity of a food safety education program. Consumers judge a message by the credibility of the person conveying it, its appeal to their common sense, and the frequency of the message. Media presentations can motivate people to listen and change behaviour. Consultants from the USDA hotline say, "We've seen an explosion in media coverage of food safety, and callers want more detailed explanations of things they read and hear" (Medeiros *et al.*, 1996).

Information on safe food handling must be motivating and memorable. Stories that capture the public's attention are personal. They relate life experiences of people with whom the public can identify. Stories of the consequences of mistakes are memorable. They can be touching, humorous, or grotesque. It is easy to visualize and remember the infected bakery worker who made 5,000 people ill when he mixed a vat of buttercream frosting with his bare hands and arms despite bouts with diarrhoea (USDA, 1990).

Stories can be heartrending, as in the experiences of a family who lost a child to *E. coli O*157: H7 infection. It is difficult to document the effectiveness of vivid accounts of doing things right or wrong. However, when Washington state carried extensive coverage linking the outbreak to undercooked hamburgers, 13 per cent of the men said they ate undercooked hamburger, compared with 38 per cent in Colorado (USDA, 1990).

Hence, consumer concerns about food borne illness can motivate change in regulatory and industry use of technology, product labelling, and consumer education.

New Technologies

The food industry has both a right and a responsibility to use safe and effective technology to enhance the safety of the food supply. Regulatory authorities should expediently evaluate and facilitate new technologies, such as food irradiation, laser light treatment, and high-pressure processing, which enhance food safety. Health professionals, the food industry, and regulators should challenge special interest groups that distort information and strive to limit consumer choice.

Improved Labelling

Processed and packaged food should bear labels that clearly indicate how food should be handled. Labels should include warnings about special risks to select populations. Benefits from special processing that can reduce microbes should also be encouraged.

Consumer Education

Consumers need to appreciate the seriousness of food borne diseases. They must learn to recognize unsafe food-handling practices, the latency period for some microbes, and the symptoms of food borne disease. They need to understand how to protect themselves through kitchen and personal hygiene, including thoroughness and frequency of hand washing, temperature control, and safe food choices such as foods processed by heat or energy pasteurization. Young people should be reached through age-specific school curricula, such as personal hygiene and special "living skills" units that address food safety and diet. Food industry and health educators should work with the media to develop interesting and timely messages to increase consumer knowledge about safe food handling. Messages must be consistent, science-based, frequent, and personalized.

Communicating Food Borne Disease Risk

The food industry, like many others, has a risk communication problem. That problem is manifested in the public's desire to know the truth about outbreaks of food borne diseases; ongoing concern about the safety of foods, additives, and food-processing procedures; and continued apathy regarding aspects of routine food hygiene. If these concerns are addressed in a coherent and trustworthy way, the public will have better and cheaper food. However, sloppy risk communication can itself cause public health damage. Because citizens are ill-equipped to discriminate among information sources, the food industry as a whole bears responsibility for the successes and failures of its individual members.

When consumers (and producers) need information, they cannot wait for more research. Consumers can read between the lines, especially when they perceive their lives or livelihoods at risk. If they misread messages, the communicators may still be held responsible. Moreover, consumers know that silence is also a form of communication.

At the same time, the guardians of the food supply must wage a continuing struggle to improve the handling of food. In the United States, campaigns are under way for cooking beef more thoroughly, separating raw meat from salad ingredients, and improving the sanitation of food handlers (*e.g.*, Operation Clean Hands). To some extent, these campaigns are the incarnations of old messages that have not been communicated effectively. At the same time, the campaigns are responses to changes in the food supply that have increased the risks associated with conventional practices. For example, as the incidence or severity of food borne disease pathogens increases, the effectiveness of customary food-handling practices decreases.

Like many other risks, emerging food borne pathogens are of primary concern to some specialists but one more thing to worry about for ordinary citizens. The thought processes that people rely on for making decisions are the focus of much research which are given as under:

People simplify: Many decisions require people to deal with more details than they can readily handle at any one time. To cope with the overload, people simplify. People want to know if foods are "safe," rather than treating safety as a continuous variable; they demand proof from scientists who can provide only tentative findings; and they divide the participants in risk disputes into good guys and bad guys. Such simplifications help people cope, yet also lead to predictable biases.

Once people's minds are made up, it's hard to change them: People are adept at maintaining faith in their beliefs unless confronted with overwhelming evidence to the contrary. One psychologic process that helps people to maintain their current beliefs is underestimating the need to seek contrary evidence. Another process is exploiting the uncertainty surrounding negative information to interpret it as consistent with existing beliefs.

People remember what they see: People are good at keeping track of events that come to their attention. As a result, if the appropriate facts reach people in a credible way before their minds are made up, their first impression is likely to be the correct one. Unfortunately, it is hard for people to gain first hand knowledge of many risks, leaving them to decipher the incomplete reports they get.

People cannot readily detect omissions in the evidence they receive: It is unusual both to realize that one's observations may be biased and to undo the effects of such biases. Thus, people's risk perceptions can be manipulated in the short run by selective presentations. People will not know and may not sense how much has been left out. What happens in the long run depends on whether the missing information is revealed by other experiences or sources.

People may disagree more about what "risk" is than about how large it is: One obstacle to determining what people know about specific risks is disagreement about the definition of "risk". For some risk experts, the natural unit of risk is an increase in probability of death; for others, it is reduced life expectancy; for still others, it is the probability of death per unit of exposure. If lay people and risk managers use the term "risk" differently, they may agree on the facts of a hazard, but disagree about its riskiness.

How much does the public know and understand? The answer to this question depends on the risks consumers face and the opportunities they have to learn about them.

Communicating Risk

An overarching theme of risk communication is that people understand risks that draw their attention and are presented comprehensibly. Whether the public's attention is aroused spontaneously or as a result of a message, the opportunity must be seized. The right information must be selected and communicated appropriately. The hallmarks of effective communication should be used.

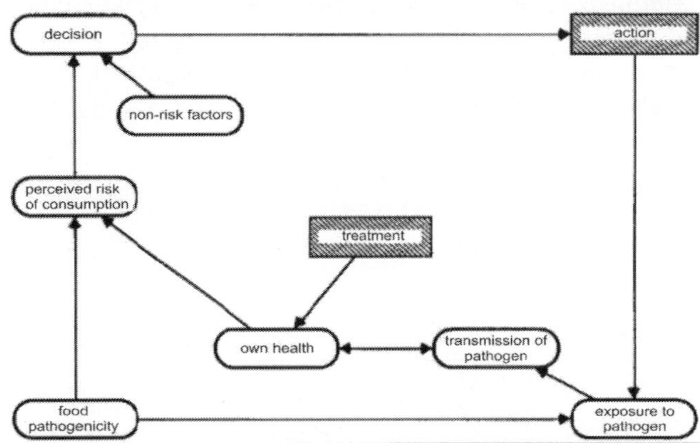

Figure 15.7: Diagram of the General Model

Figure 15.7 shows a simple model for the risks of food borne pathogens. It uses the formalism of the influence diagram. Such a model can be used both to assess risks and to characterize the comprehensiveness of lay understanding. In this model, people incur food-related risks as a result of decisions, which possibly lead to actions or exposures.

These decisions concern such actions as eating a bite of suspicious food, choosing a particular diet, or opting for school (or home) lunch. Those decisions depend, in part, on the perceived risks of those actions as well as other non-risk factors (*i.e.*, other costs and benefits). Exposure may follow, if a pathogen is actually present; it can lead, in turn, to transmission of the pathogen and to changed health states, depending on the resistance to disease that the person's health provides.

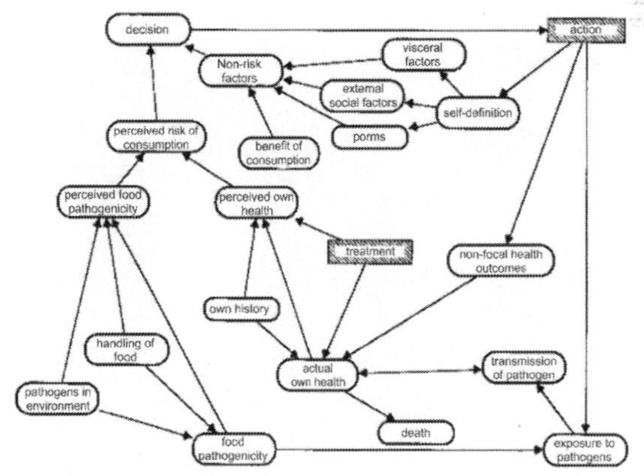

Figure 15.8: Diagram of the Food Borne Model

Figure 15.8 elaborates on this model. It shows that food pathogenicity depends on both the prevalence of pathogens in the environment and the quality of food handling. A person's own health influences risk perceptions through the intermediate variable of perceptions of health, which in turn is influenced by the person's history of food consumption (or avoidance). Actual pathogenicity influences risk perceptions through awareness, a variable that communicators might affect. Nonrisk factors include visceral factors (*e.g.*, hunger), external social factors (*e.g.*, social pressure to eat any food offered by a host), norms (*e.g.*, not eating dog), and the expected benefits of consumption (*e.g.*, taste, texture, and other gustatorial pleasures). Computing risks with this model would require specifying each variable and estimating the contingencies by using statistical sources or expert judgment. For risk communication purposes, even a qualitative model can define the universe of discourse and allow approximate estimates of the most important relationships.

Chapter 16

Food Safety and Quality through Food Hygiene and Sanitation Practices

Food hygiene are the conditions and measures necessary to certify the safety of food from production to consumption. WHO (1984) has defined food hygiene as all conditions and measures that are required during production, processing, storage, distribution and preparation of food to ensure that it is safe, wholesome and fit for human consumption. Lack of requisite food hygiene can lead to foodborne diseases and death of the consumer.

European regulations define food hygiene as: 'the measures and conditions necessary to control hazards and ensure fitness for human consumption of a foodstuff taking into account its intended use'. Food hygiene comprises two components:

- Food safety, which guarantees the harmlessness of food, the absence of adverse effects for the consumer's health when prepared and/or consumed in keeping with its intended use;
- Food suitability, which concerns the intrinsic characteristics of the product, namely taste, smell, texture and presentation, characteristics that can change with the presence of spoilage microbes (bacteria, yeast and mould). Suitability is the assurance that the food is 'acceptable' for human consumption.

Food safety, quality and suitability must be assured at every link of the food chain.

Sanitation: An Applied Science

Sanitation is the creation of a clean and hygienic environment and making it sustainable. Hygiene and sanitation process can't be divided each other in the food sector. Hygiene is the purification of the environment from microorganisms causing diseases, the other hand sanitation is the measure taken for cleaning and hygiene. Sanitation is the purification of foreign substances, microorganisms, drugs, cleaning agents and all visible sources of pollution from the production environment.

Sanitation is an applied science that incorporates the principles of design, development, implementation, maintenance, restoration, and/or improvement of hygienic practices and conditions. Sanitation applications refer to hygienic practices designed to maintain a clean and wholesome environment for food production, processing, preparation, and storage. However, sanitation is more than just cleanliness. Done properly it can improve the aesthetic qualities and hygienic conditions of commercial operations, public facilities, and homes. Also, applied sanitary science can improve waste disposal, which results in less pollution and an improved ecological balance. Therefore, when effectively applied, food sanitation and general sanitary practices have a beneficial effect on our environment.

Sanitation is considered to be an applied science because of its importance to the protection of human health and its relationship with environmental factors that relate to health. Therefore, this applied science relates to control of the biological, chemical, and physical hazards in a food environment. Sanitarians must be familiar with all these hazards and thoroughly understand the basic food microbiology and the organisms that are most likely to affect human health. By identifying, evaluating, and controlling hazards and through the effective application of sanitary practices, a safe and wholesome food supply can be assured.

How to Achieve Safe Food?

Safe food originates from its source, the farm. At this stage Good Agricultural Practices (GAP) are applied to ensure food safety. Subsequent to the farm, Good Manufacturing Practices (GMP) ensure products are consistently produced and controlled according to quality standards at the various stages of processing, storage and transportation till food reaches consumers. General Principles of Food Hygiene apply to both GAP and GMP through the implementation of Good Hygiene Practices (GHP). GHP recommends a HACCP-based approach to reduce risks. Good Hygiene Practices aim to implement the essential principles of food hygiene applicable throughout the food chain (including primary production through to the final consumer), to achieve the goal of ensuring that food is safe and suitable for human consumption.

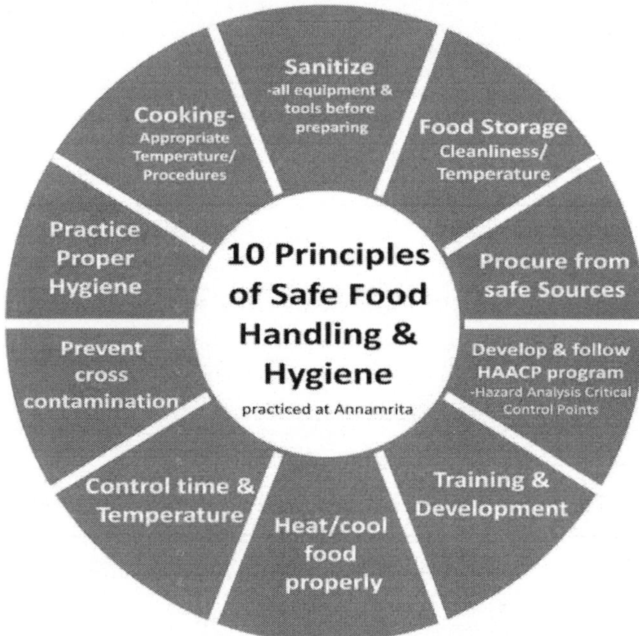

Food Safety is a Basic Human Right

Billions of people in the world are at risk of unsafe food. Many millions become sick while hundreds of thousands die every year because they consume unsafe food. Therefore, safe food saves lives. Safe food enhances individual and population health. Safe food improves economic growth of the region where food safety is practiced and enhanced. Safe and quality food supply depends on both sound science and equitable law enforcement. With technological advances, new regulations must be enacted to protect a continuing supply of food products that are safe and wholesome for the health and wellness of people.

As the standard of living improves, concerns over food safety and potential contaminants will continue to be an important health issue. Consumers demand quality and safety of products they consume because food as energy and nutrient is necessary to sustain life. In general, consumers rely on government to ensure all food products not only are safe but are sold as what they claim to contain. For example, a jar of olive oil labelled as 100 per cent virgin olive oil must contain exactly what the label says except the naturally occurring trace elements that are part of olive oil and which cannot be extracted or eliminated completely without destroying the olive oil.

Challenges and tragedies in food safety include chemical, biological, personal hygiene and environmentally related incidents. Unsafe food poses global health threats. The young, the elderly and the sick are particularly vulnerable. If food supplies are unsecured, population shifts to less healthy diets and consume more "unsafe foods", in which chemical, microbiological and other hazards pose health risks, that in turn costs higher healthcare expenditure and drains national wealth. In light of recurrent food contamination incidents, food safety in the 21st century should expand beyond improving nutritional profile, transparency of ingredients and regulations of unhealthy foods to include regular monitoring, surveillance and enforcement of food products in furtherance of the general public well-being and prevention of foodborne illnesses.

Challenges of Food Safety

They include four major areas:

- ✰ **Microbiological Safety:** Food by nature is biological. It is capable of supporting the growth of microbials that are potential sources of foodborne diseases. Viruses are more responsible for the majority of foodborne illnesses but hospitalizations and deaths associated with foodborne infections are due to bacterial agents. The illnesses range from mild gastroenteritis to neurologic, hepatic, and renal syndromes caused by either toxin from the disease-causing microbe. Foodborne bacterial agents are the leading cause of severe and fatal foodborne illnesses. Over 90 per cent of food-poisoning illnesses are caused by species of *Staphylococcus*, *Salmonella*, *Clostridium*, *Campylobacter*, *Listeria*, *Vibrio*, *Bacillus*, and *E. coli.* For instance, in the US and France, in the last decade of the 20th century, *Salmonella* was the most frequent cause of bacterial foodborne illness accounting for 5700 to 10,200 cases, followed by *Campylobacter* for 2600 to 3500 cases and *Listeria* for 304 cases.
- ✰ **Chemical Safety:** Non-food grade chemical additives, such as colorants and preservatives, and contaminants, such as pesticide residues,

have been found in foods. Some food samples had higher levels of heavy metals such as lead, cadmium, arsenic, mercury, and copper than average food samples, suggesting possible leaching from the utensils and inadequate food hygiene.

- **Personal Hygiene**: The word hygiene is used to describe an application of sanitary principles for the preservation of health. Personal hygiene refers to the cleanliness of a person's body. The health of workers plays an important part in food sanitation. People are potential sources of microorganisms that cause illness in others through the transmission of viruses or through food poisoning. Poor personal hygiene practices of food handler and preparers pose considerable risks to personal and public health. Simple activities such as thorough hand washing and adequate washing facilities can prevent many foodborne illnesses.

  Food handlers can transmit bacteria causing illness. In fact, humans are the major source of food contamination. Their hands, breath, hair, and perspiration contaminate food, as can their unguarded coughs and sneezes, which can transmit microorganisms capable of causing illness. Transfer of human and animal excreta by workers is a potential source of pathogenic microorganisms that can invade the food supply. By necessity, the food industry is focusing more on employee' education and training and emphasizing that supervisors and workers be familiar with the principles of food protection. In multi-unit chain operations, the negative effects of public opinion often spiral outward to uninvolved units.

- **Environmental Sanitation:** The word sanitation is derived from the Latin word sanitas, meaning "health." Applied to the food industry, sanitation is "the creation and maintenance of hygienic and healthful conditions." It is the application of a science to provide wholesome food processed, prepared, merchandised, and sold in a clean environment by healthy workers; to prevent contamination with microorganisms that cause foodborne illness; and to minimize the proliferation of food spoilage microorganisms. Effective sanitation refers to all the procedures that help accomplish these goals.

Inadequate recycling and waste disposal equipment and facilities lead to the accumulation of spoiled and contaminated food. This leads to an increased pest and insect population that can result in risk of food contamination and spoilage. Poor sanitary conditions in the area where foods are processed and prepared contribute to poor food storage and transport as well as selling of unhygienic food.

Why is Safe Food Supply Important?

A safe food supply is important because of significant disease burden as well as economic burden to the society and nation. Worldwide, it has been estimated that more than one billion (1,000,000,000) episodes of food poisoning-related diarrhoea occur annually; these poisonings are responsible for the deaths of about 3 million children a year, mostly in underdeveloped regions.

Foodborne illnesses associated with microbial pathogens or other food contaminants pose serious health threat in developing and developed countries. WHO estimates less than 10 per cent of foodborne illness cases are reported whereas less than 1 per cent of cases are reported in developing nations. In a report, WHO estimated 600 million foodborne illnesses and 420,000 deaths in 2010. The most frequent causes of foodborne illness were diarrheal disease agents, particularly norovirus and *Campylobacter* spp. Other major causes of foodborne deaths were *Salmonella typhi*, *Taenia solium*, hepatitis A virus, and mycotoxins especially aflatoxins.

Children are disproportionately bearing this burden - accounting for an estimated half of foodborne illness cases annually. Children are also among those most at risk of associated death and serious lifelong health complications from foodborne diseases. They are at high risk for foodborne illness for a number of reasons. Children have developing immune systems that are not always well equipped to fight infection; they are often smaller in size than adults, reducing the amount of pathogen needed to make them sick; and children have limited control over their diets and lack the developmental maturity necessary to carefully judge food safety risks.

From economic perspective, access to sufficient amounts of safe and nutritious food is crucial to sustaining life, promoting good health and economic growth. According to one study, the average cost per case of foodborne illness (in US dollars) was $1626 for the enhanced cost-of-illness model and $1068 for the basic model. The resulting aggregated annual cost of illness was $77.7 billion and $51.0 billion for the enhanced and basic models, respectively. The study defines basic cost-of-illness model to include economic estimates for medical costs, productivity losses, and illness-related mortality. The enhanced cost-of-illness model replaces the productivity loss estimates with a more inclusive pain, suffering, and functional disability measure based on monetized quality-adjusted life year estimates.

Table 16.1: Common Foodborne Pathogens and their Medical and Economic Impacts

Foodborne Hazards	*Common Infectious or Toxic Agents*	*Incidence of Foodborne Illness*	*Death Due Foodborne Illness*	*Total DALYs*
Bacteria	*Salmonella, Vibrio, E. coli, Shigella, Listeria, Brucella, Listeria, Campylobacter*	359,747,420	272,554	20,188,792
Virus	Noro virus, Hepatitis A	138,513,782	120,814	3,849,845
Protozoa	Entamoeba, Giardia, Cryptococcus, Toxoplasma	77,462,734	6242	1,311,435
Worms	Cestodes (tapeworms), Nematodes (round worms), Trematodes (flatworms); helminths (parasites)	26,063,664	90,261	11,599,735
Chemicals	Aflatoxins, Cyanogenics, Dioxins, Heavy Metals	217,632	19,712	908,356

The Uncertainty Intervals (UI) are not shown.

Source: Extracted and complied from WHO, 2015.

The Importance of Hygiene and Sanitation in Food Sector

All kinds of food must be purified from any;

- Physical pollution such as glass, metal pieces, stone, sand,
- Chemical pollution such as liquid fuels, detergents and residual of drug,
- Biological pollution sources such as microorganisms and mildew to get good quality, healthy and safe food.

Contaminated food can cause quality problem and the other hand, health related problems can be observed. Hygiene and sanitation should be effectively applied and controlled at every step during manufacturing processes.

If quality control processes are not applied effectively during manufacturing, these problems which are listed below can be observed.

- Decreasing number of sales and consumption
- Exposure to legal proceedings
- Loss of prestige
- Production lag
- Lack of motivation in the company
- Food poisoning or nutritional diseases
- Claim for damage

Food Poisoning Chain, Food Safety and Hygiene

Food poisoning agents especially organisms are present almost everywhere and hence, it is impossible to operate a food business without them being present in either small or large numbers. This is the reason why good hygiene practices are so important to minimise the risks of illness from consuming contaminated food.

In most cases of food poisoning a chain of events takes place, and if we are to reduce the incidence of illness, this chain must be broken.

It is very clear that the food poisoning chain involves:

1. Contamination of high-risk foods with microorganisms
2. Multiplication of these microorganisms
3. Preparation of food too far in advance
4. Foods left at room temperature or food cooked in large pots
5. Improper warm holding *e.g.*, below 60°C
6. Improper cooling *e.g.*, leaving cooked foods at room temperature or storing foods in large containers in refrigerators
7. Inadequate reheating
8. Handling of foods by colonized infected persons
9. Inadequate cleaning of equipment and utensils
10. Cross-contamination from raw to cooked foods
11. Toxic containers
12. Contaminated raw ingredients
13. Consumption of such contaminated foods

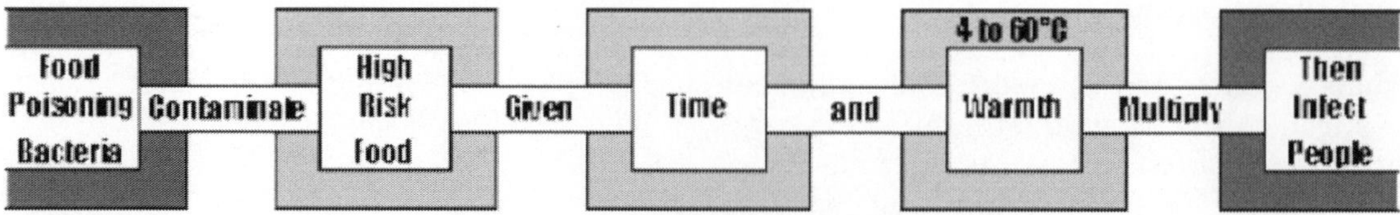

Figure 16.1: Food Poisoning Chain.

Food safety is protection to consumer from risk of food poisoning. Food safety covers all methods for destroying physical, chemical and biological hazards. Food borne disease may ends up with death. The intention of food safety is to prevent food poisoning, (the transmission of disease through food) and to maintain the wholesomeness of the food product through all stages of processing, until it is finally eaten.

Food safety involves more than just cleanliness; it includes all practices involved with:

1. Protecting food from the risk of contamination, including harmful bacteria, poisons and foreign objects.
2. Preventing any bacteria present in the food multiplying to a level that would result in food poisoning, or the early spoilage of the food.
3. Destroying any harmful bacteria in the food by thorough cooking or processing.

Cost effective food safety, hygiene and sanitation may be considered in relation to:

1. Product (food)
2. Personnel
3. Premises and equipments

Food

The highest standards of hygiene and food quality control are essential in all areas of food production, packaging and handling in catering industry. The food poisoning chain involves:

1. Contamination of high-risk foods with food poisoning micro-organisms
2. Multiplication of these micro-organisms
3. Consumption of the foods

This chain must be broken through:

Buy Raw Foods and Ingredients after Examination

Right from the point of receiving food materials, the principles of hygiene and sanitation need to be applied. A strict check on the quality of food received is essential.

There is a need for a structure which is free from physical, chemical and biological pollution sources, provided hygienic production conditions and has a suitable technological infrastructure in order to produce reliable and quality food. Before the manufacturing process get started, all stuff which are used for manufacturing process and manufacturing yard must be cleaned periodically. For this purpose, the check list which is about what the staff has to do before manufacture process starts has to be prepared. The check list must be announced and posted with staff.

According to this program, the hygienic control must carry out periodically and also saved. All "food contact surface areas" including the floor in the environment where foodstuffs are processed, the processing channels, the walls, all kinds of tools, machinery and equipment that come into contact with foodstuffs must be thoroughly cleaned regularly before starting daily work to protect food from contamination. The cleanliness of the ambient air should also be controlled in the workplace where microbiological contamination is important.

The surfaces used for the production and storage of foods with low moisture content should always be dry and hygienic during use. When wet cleaning is required, these surfaces must be sanitized and must be thoroughly dried to use. The entire food contact surface in wet applications must be cleaned and sanitized absolutely before use or at the risk of any contamination. Those with portable properties should be stored in places where they will not cause any contamination risk to the food when not in use.

Protect Foods and Raw Materials from Insects, Rodents and Microbes

Proper storage of raw foods is essential. The presence of rodents, insects and microbes affect both the quantity and quality of food. Frequent disinfestations of the working area, destruction of hideouts, and use of protective mashes are essential.

Taking a series of precautions starting from raw material to reaching the consumer in food production is of great importance in terms of food hygiene. One of these measures is an effective struggle against foodstuffs that damage food and food in various ways. It is called any kind of harmful animal pest that infects the disease agents of our body in various ways. Many pest species cause the disease to spread to humans in a variety of ways, leading to the emergence of important diseases and also to the loss and degradation of economic values of food.

Mouse type rodents, especially seen in food deposits, are at the forefront of pests that need to be struggled against due to damage to raw materials and, if necessary, disease-causing microorganisms. Insects and rodents, flies and cockroaches are associated with living quarters, eating establishments, and food processing facilities, as well as with toilets, garbage, and other filth. These pests transfer filth from contaminated areas to food through their waste products; mouth, feet, and other body parts; while the regurgitation of filth onto clean food during consumption. Rats and mice transmit filth and disease through their feet, fur, and intestinal tract. Like flies and cockroaches, they transfer filth from garbage dumps and sewers to food or food processing and foodservice areas.

Ants, cockroaches, flies, bears, spiders *etc.* creatures carry microorganisms that are transmitted to production and products by means of both other food and equipment, leading to the formation of epidemic diseases, the products being more rapidly deteriorated and economic values being lost. To stop contamination from these pests, eradication is necessary, and food processing, preparation, and serving areas should be protected against their entry.

In general, the following measures must be taken in the fight against pests:

i)All doors and openings to the outside need to be closed to prevent entry of harmful substances. ii) Mechanically or adhesive mouse traps should be placed at appropriate intervals in the production area, traps should be checked regularly. iii)Effective struggle must be carried out to prepare a pest control program. iv)Drugs used against harmful should be stored in separate spaces so that they do not contact with foodstuffs. It should be periodically spraying or supported by a sprayer. If the spraying is to be done by the workplace, the necessary training should be taken in this regard. v)Drug formulation should be recorded. vi)Medicines used in the fight against pests must be approved by the relevant legal authorities. vii)Necessary preventive measures must be taken against the animals and pests that cause contamination. viii) Domestic animals must not enter where the food is prepared or treated. ix)Products that have been exposed to pests must be destroyed.

Cook Food Thoroughly

Many foods are contaminated with disease causing micro-organisms. Thorough cooking will kill the pathogens. Thorough cooking, heat process such as pasteurization, sterilization or canning, bacteria within the food can be destroyed.

Avoid Contact Between Raw Foods and Cooked Foods

Safely cooked foods can be contaminated through even, the slightest contact with raw food. The cross-contamination is more common in non-vegetarian foods. Raw and cooked foods should be separated at all stages of preparation, storage and distribution. The same equipment or cooking surface must not be used to handle raw and high risk foods unless disinfected between uses.

Serve Cooked Foods as Early as Possible

When cooked foods cool to room temperature micro-organisms present in the food may proliferate. The longer the wait, greater the risk. The time between removal of food from the refrigerator and cooking, between cooking and eating or between cooking and returning to the refrigerator must be kept as short as possible.

Store Cooked Food Carefully

If the foods have to be cooked in advance or left overs be kept, they must be stored out of the danger zone *i.e.,* either hot near or above 63°C or below 5°C in refrigerator. Refrigerators to be used for food storage should be maintained at the correct temperatures. Cooling of foods should be done as rapidly as possible to prevent multiplication of bacteria. Dried foods should be stored well so that they do not absorb moisture.

Reheat Cooked Foods Thoroughly

Proper storage slows down microbial growth but does not kill the micro-organisms. Thorough reheating means that all parts of the food must reach at least 70°C. The liquid from thawed, frozen meat and poultry must not come into contact with high-risk foods.

Hand Washing and Drying

The ways in which consumers spread microorganisms to one another and to themselves include more than just coughing and sneezing. Not washing hands before, during, and after handling foods clearly contributes to the spread of food borne infections and intoxication. Hands can spread disease-causing microbes to foods from other foods and from infected persons.

Microorganisms found on the hand surfaces may be transient bacteria or resident bacteria. Transient bacteria are picked up accidentally by food handlers and are transient in that they reside on the hands only temporarily (*e.g., E. coli*). Residual microorganisms permanently reside on the hand surfaces and are the normal or resident microflora of the skin (*e.g., Staphylococcus epidermidis*). The first line of defense against disease is frequent and effective hand washing by food handlers (Taylor, 2000). Approximately 38 per cent of food contamination is attributable to improper hand washing. The most effective method to ensure effective hand washing is through motivation, reinforcement, incentives, and modelling through supervisors and managers practicing appropriate hand washing.

Hand washing is conducted to break the transmission route of the microorganisms from the hands to another source and to reduce resident bacteria. *Pseudomonas aeruginosa, Klebsiella pneumoniae, Serratia marcescens, E. coli,* and *S. aureus* can survive for up to 90 minutes when artificially inoculated on the fingertips. Hand washing for 15 seconds (as opposed to the average of 7 seconds) with soap and water, which act as emulsifying agents to solubilize grease and oils on the hands, will remove transient bacteria. Increased friction through rubbing the hands together or by using a scrub brush with soap can reduce the number of transient and resident bacteria than is done by quick hand washing.

Antimicrobial agents exert a continuous antagonistic action on emerging microbes and enhance the effectiveness of ordinary hand soap at the time of application. The overall efficacy of antimicrobial hand soap depends on continuous use throughout the day. A contact of less than 5 seconds during hand washing has little effect on reducing the microbial load. A cleaning compound will remove more transient bacteria, with subsequent destruction by the sanitizer.

In a comprehensive review of 91 scientific articles published after 1986, Bryan *et al.*(1996) attempted to link hand washing and infections. They reported that hand washing has become an integral component of the tradition and ritual of prevention practice for the spread of infection, but several factors confound the ability to establish the effectiveness of hand washing for reducing infectious disease. Hand-washing practices were shown to significantly reduce infections transmitted by the faecal-oral route and in situations of poor personal hygiene. Hand washing is clearly a critical step in reducing personal contamination of food and cross-contamination between foods. Hand washing is but one practice that could dramatically affect risk, if not incidence, of food borne disease.

According to data provided by the American Society for Microbiology (1996), people do not wash their hands as often as they think they do. In telephone surveys, 94 per cent of respondents claim they always wash up after using the rest room; however, researchers contend that almost one-third of people do not wash their hands after using the bathroom. Of the more than 7,000 people nationwide who participated in the study, 81 per cent said they wash their hands before handling or eating food. However, most say they do not wash up after petting an animal (48 per cent), coughing or sneezing (33 per cent), or handling money (22 per cent).

The key facets to improved hand washing are motivation and training. Although a certain amount of education is essential, a multidisciplinary framework should target institutional or organizational change. Training should be risk based with the consequences of improper hand washing clearly expressed. Overall, some changes in behaviour have occurred, but consumer habits are still frequently less than ideal. A recent study on handwashing habits revealed that only 78 per cent of over 7,500 individuals washed their hands after using public restrooms in airports. This was actually an improvement over a previous study that observed 67 per cent of individuals washed their hands after using public restrooms.

Hands should be washed thoroughly before starting for preparing the food. Even after handling raw foods and before handling cooked foods hands should be washed. The foods should come into contact with hands when unavoidable; otherwise, tongs, plates, trays should be given preference to hands. Disposable clothes for wiping the working surface should be used. Persons with infections should not enter the kitchen and cut/wound on the hands should be appropriately bandaged.

Hand washing and drying efficacy against resident flora ranges from 35 to 60 per cent. All hand washing agents, including water, are effective when the hands are dried with paper towels. Alcohol-based instant hand sanitizers used after hand washing, provide an additional 10- to 100-fold reduction (Anon., 2002). Instant hand sanitizers (protective creams and lotions) should be considered when washing is not possible but they do not have a lasting effect (Taylor, 2000).

Proper Method of Hand Washing Using Mechanized Hand Washers

Because proper hand washing is essential to attain a sanitary operation, mechanized hand washers are being used in food units. A typical unit is located in the processing area. When workers enter the area, they must use the washing unit. This equipment is responsible for increased hand washing frequency by 300 per cent. The user inserts the hands into two cylinders, passing a photo-optic sensor, which activates the cleansing action. High pressure jet sprays within each cylinder spray a mixture of antimicrobial cleansing solution and water on the hands, followed by a potable water rinse. The 10-second (programmable from 10 to 20 seconds), massage like cycle has been clinically proven to be 60 per cent more effective at removing pathogenic bacteria from the hands than the average manual hand washing and reduced water costs. The high-pressure, low volume spray uses approximately 2 L of water per wash cycle, one-third of the amount spent in most manual hand washing methods. Up to a 300 per cent increase in washing frequency is accomplished because this equipment provides an easy-to-use, massaging effect on the hands and is non-irritating. Also, this process can remove contamination from gloves and can accomplish hand or glove washing with approximately 2 L of water or only one-third of the amount used in most manual hand washing methods.

Use of Mounted Hand Sanitizers

A wall-mounted hand sanitizer can be used to reduce microbial contamination of workers. The use of antiseptic products for hand cleansing can reduce bacterial load on the hands and thus decrease the possibility of cross-contamination. These products include soap/detergents, instant hand sanitizers, and antiseptic lotions/creams. Antiseptic soaps or detergents remove surface bacteria and may have a residual effect.

Instant hand sanitizers kill bacteria but do not have a residual effect. The use of protective and antiseptic lotion products after washing may produce a residual antibacterial effect, reduce skin shedding, and protect against the irritating effects of liquids and latex. Alcohol hand rub, gel, or rinse sanitizers that contain at least 60 per cent alcohol have been incorporated as a disinfection step after washing hands with soap and water. The alcohol present will evaporate in approximately 15 seconds. This hand sanitizer is an effective sanitizer that improves personal hygiene and does not contribute to the emergence of microbial resistance. Use of this hand sanitizer before handling food is generally considered to be a safe practice.

Ethanol is more effective at destroying viruses than isopropanol; however, both alcohols are effective for the destruction of bacteria, fungi, and viruses. The most effective regimen for antimicrobial control is the combination of an antibacterial soap hand wash followed by an alcohol gel application (Paulson *et al.,* 1999). Caution is necessary when alcohol is applied because it is flammable at the concentration found in hand sanitizers Other antiseptic hand washing compounds include iodine and iodophors (complexes of iodine with a carrier such as polyvinylpyrrolidone). Although effective, iodine is irritating and may cause allergic reactions (Larson, 1995).

Chlorhexidine gluconate (CHG) is incorporated in both surgical scrubs and hand wash products. It has a broad spectrum of activity, being effective against both gram-positive and gram-negative bacteria with approximately 6 hours of residual activity. Chloroxylenol is less active than CHG, but its activity persists over several hours. Triclosan, another phenolic compound, has a broad spectrum of activity against both gram-positive and gram-negative bacteria.

Use of Gloves for Food Protection

Hand gloves should be used in food premises. Fendler *et al.* (1998) concluded that the scientific evidence is insufficient to support the premise that the use of a physical barrier (gloves) on the hands of food handling personnel prevents the transfer of pathogens to food. These workers then proceeded to conduct a study under simulated food service conditions to try to define the most effective hand hygiene regimens to minimise health risk to the consumer. They used six different washing/gloving protocols and minced beef (inoculated with a non-pathogenic strain of *Escherichia coli*) as the test food. A second experiment involved inoculating the subjects' hands directly with *E. coli* before the test regimen began. The conclusion was that, 'bare hands with a regimen of hourly handwashing and sanitising provided significantly higher hand sanitisation levels than any of the five other regimens including those employing gloves'. When gloves are used, they should be an adjunct to good hand washing practice and in no way a replacement for it.

However, the use of gloves in food service situations is not uncommon in many countries and should be encouraged by some food hygiene training providers. There may be good reasons for wearing gloves in certain food handling situations.

Fendler *et al.* (1998) reported different combinations of gloving and hand washing to protect the hands from *E. coli*-contaminated meat. Bare hands with hourly washing and sanitizing with alcohol had the lowest microbial counts, and low levels were achieved without alcohol alone in spite of the necessity for hand washing due to the physical soil from the meat. Gloves should be put on after the hands are washed and dried. However, gloves may leak and natural rubber latex gloves may cause allergenic reactions from the natural rubber, which may contain traces of proteins that trigger the allergenic response in sensitized people. Furthermore, microorganisms thrive in the warm and moist environment under gloves if a protective, antiseptic lotion is not applied under the gloves (Taylor, 2000). Contamination from unclean hands will be smeared over the gloves. If the hands are not dry, residual moisture forms an incubation environment for bacteria under the gloves. Workers should be reminded that soil on gloves is not as easy to feel as on the bare hands. Non-latex gloves should be considered for handling ready-to-eat foods.

Keep all Kitchen Surface Meticulously Clean

The food preparation area must be clean. Measures should be taken to prevent the entry of insects, animals and birds into food room or coming into contact with food. Food should be stored in rodent-proof containers and ensuring that the lids are tightly replaced after use. Unfit or waste food and refuse should be removed promptly and they should be kept apart from the high-risk food. Wash-hand basins should not be used for washing the food or food equipment. Similarly, food sinks should not be used for hand washing. Clothes that come into contact with dishes and utensils should be changed every day. Separate clothes for cleaning the floors require frequent washing and sanitization. The dust bins should have a proper receptacle. Inspect food areas and processes frequently, act on any defects or unhygienic practices. Smoking should not be allowed in rooms containing open food or whilst handling open food.

Use Pure Water

The potable water is very important for food preparation and for drinking. Unless safe and clean water is used it may serve as a source of contamination.

Water Hygiene

Water is one of the most compelling materials for life. Water hygiene does not engage with the qualities of water used for drinking alone. It also relates to the determination of the qualities of waters to be used in production and domestic works, the prevention of water pollution and the disinfection of waters. The relationship between water and health are related each other. Therefore, in order to be able to decide a clean water with hygienic qualities and evaluate the required qualities thoroughly, the source of the water must be well known in advance.

Biological agents that can be found in the water and are detrimental to human health include pathogenic bacteria, viruses and parasites. The infections caused by the waters are spread around by the patients and the porter. Depending on many factors such as the geographical location of the area, infrastructure facilities, the process of the waste materials, and the socio-economic structure of the community, pathogenic bacteria and other microorganisms reach the waters with faeces and the like. Drinking water is the most important ring of the oral-faecal infection chain. Avoiding bacterial contamination of large quantities of water is possible by disinfecting the water. Scientists and health care providers are working to get clean water, setting water standards, and setting specific criteria for water that is drinkable and usable.

The water properties of chemical, physical and microbiological, should be completely clean, free from any odour, odourless, colourless, harmful substances and no pathogenic substances in it. Taking off this dangerous factor is called water cleaning. Cleaning of the water is divided into 3. These are:

1. Physical cleaning: removal of odour and blur
2. Microbiological cleaning (disinfection of waters)
3. Correction of chemical defects

This phase is the most important phase of cleaning the waters. The disinfection of the processing water is carried out in order to destroy the organisms, which disturb the colour, smell and taste of the water with the pathogenic bacteria that make the disease in the water. Physical and chemical methods are used to kill pathogenic microorganisms in the water.

Physical Methods

With Heat: If water is boiled for 10 minutes at 100 °C, all microorganisms that can cause water epidemics in the body die. Although some sporicidal microorganisms are based on these temperature gradients, they have no hygiene.

With Ultraviolet Light: The ultimate effect of ultraviolet light on microorganisms are very high. Various systems (Nogier and Lacarriere lamp) are available for this purpose. Since the ultraviolet effect lasts for some time after the water has been irradiated, it should not be used immediately to take advantage of this effect.

Chemical Methods

Mostly chemical methods are used for water disinfection. The effect of chemical substances on microorganisms in the water is high, cheap, easy to apply.

Chemicals used for water disinfection include:

Ozone: Ozone is a form of oxygen that is a very active oxidant and a very strong bactericidal gas. It has a superiority over all other disinfectants. The excess is not harmful. Venting is enough to remove the ozone from the water. Ozone can degrade bacteria. Oxidation of organic substances makes it impossible for later bacteria to develop. The ozone's superiority is quite high. They provide a quick disinfection in 10 minutes; the water gives no taste disorder and is harmless. The bactericidal effect is 10 times faster than the chloride.

Iodine: It is applied in hot countries. Bacteria and protozoans are destroyed. The high bactericidal effect of iodine found in nature is exploited. This method is required to be filtered before the water to be applied is clear.

Potassium Permanganate: Especially effective in cholera outbreaks. It can also be applied in fuzzy waters. It does not change the taste of the water. After filtration, clear, colourless and clean water is obtained.

Overall, the food safety can be secured by:

Protecting Food from Contamination by:

1. Inspecting all food items and washing fruit and vegetables before preparation.
2. Separating raw and high risk/ready to eat foods at all stages of preparation, storage, display and distribution. The same equipment, utensils and working surfaces must not be used to handle raw and high risk/ready to eat foods.
3. Only handling food when unavoidable. Gloves, tongs and other utensils, plates and trays should be used in preference to hands (but must be washed or changed frequently).
4. Keeping food covered as much as possible.
5. Preventing insects, animals and birds from entering food rooms.
6. Not using unsuitable, defective, or dirty equipment.
7. Using good personal hygiene practices–always.
8. Not coughing or sneezing over or around food.
9. Not handling the food contact surfaces of crockery, cutlery or utensils.
10. All food handlers wearing suitable protective clothing.
11. Using the correct cleaning procedures.
12. Promptly removing unfit or waste food and refuse from food areas.

Preventing any Bacteria within Food from Multiplying by

1. Keeping high risk foods at temperatures that inhibit the growth of bacteria (ie. out of the danger zone). Food should be kept below 4°C in a refrigerated unit or above 70°C in a suitable warming unit.

2. Ensuring that during preparation, food is in the danger zone for as short a time as possible. High risk foods must not be left sitting out at room temperature.
3. Using suitable preservatives such as salt and sugar.
4. Using various packing methods like gas flushing or vacuum packing.
5. Not allowing dried foods to absorb moisture.

Destroying those Bacteria within Food by

1. Adequately cooking food, ensuring that a minimum internal cooking temperature of 80°C is reached.
2. Heat processing such as pasteurisation, sterilisation or canning.

A combination of a suitable temperature and sufficient time is always required to destroy bacteria. The time and temperature required will depend on the particular organism, (*e.g.* spores of *Clostridium perfringens* are much more heat resistant than *Salmonella* bacteria).

Personnel

The hands, hair, nose, and mouth harbour microorganisms that can be transferred to the food during processing, packaging, preparation, and service by touching, breathing, coughing, or sneezing. Because the human body is warm, microorganisms proliferate rapidly, especially in the absence of hygienic practices. After the chain of infection is broken the spreading of bacteria from one location to another can be prevented. Generally, the mishandling of food by people perpetuates the chain of infection until someone becomes ill or dies before corrective actions are taken to prevent additional outbreaks (Chao, 2003). If every person that handles food could achieve appropriate personal hygiene, food contamination could be minimized. Every employee involved with food manufacturing can play a very important role in preventing food contamination.

Food organizations should establish personal hygiene rules that are clearly defined and uniformly and rigidly enforced. These regulations should be documented, posted, and/or clearly spelled out in booklets. Policy should address personal cleanliness, working attire, acceptable food handling practices, and the use of tobacco and other prohibited practices.

Good grooming habits, neat and sanitary practices and efficient work practices are essential in foodservices. Every employee should be thoroughly checked for medical clearance before being employed. There are chances that food handlers get pathogenic microorganisms through raw food while handling but generally the food handlers being carriers transfer them to foods. The food handlers should be informed about personal habits which help to keep food safe from contamination.

1. Wear and clean white uniform and comfortable non-slippery shoes. He or she must wear suitable, where necessary, protective clothing.
2. Be well groomed and use a hair restraint. Staff should keep hair tied back and wear a suitable head covering, *e.g.*, hat or hairnet, when preparing food.
3. Staff should not wear watches or jewellery when preparing food (except a wedding band).
4. Develop good work habits.
   (a) Wash your hands thoroughly every time you enter in foodservice areas.
   (b) Never touch face, nose, hair while working.
   (c) Use spoon for tasting.
   (d) Wipe hands on paper towel, never on uniform.
   (e) Never touch prepared food with hands. Use proper equipment or sterilized gloves.
   (f) Do not touch food contact surfaces.
   (g) Never eat or drink in the food production area and while on service counters.
   (h) Do not smoke, chew gum, tobacco, betelnut, *paan* in food production and service areas.
   (i) Organise work efficiently.
5. Report all injuries immediately and get first aid.
6. Report colds, sore throat, illnesses, cuts and burns, to the supervisor before entering food production or service area.
7. Any food handler exhibiting food poisoning symptoms *i.e.,* diarrhoea organism should be excluded from any job which would expose food risk contamination.
8. Disposable gloves should be used in both catering and retailing, to prevent the hands coming into direct contact with high-risk foods. The hands should be thoroughly washed before putting on the gloves.

Operating Structure and Design, Premises and Equipments

- ✰ Welfare facilities should be clean, neat, well lighted, and conveniently located away from production areas.
- ✰ Materials, tools and equipment used in accordance with production technology must be resistant to heat, steam, acid, alkali, salt and other substances.

- ❖ Repair, paint, whitewash and periodic maintenance of buildings, installations, materials, tools and equipment should be done without interruption.
- ❖ The workplace should be installed in such a way as to prevent the entry of harmful organisms and environmental pollutants.
- ❖ The area of the production area should be waterproof, non-slip, washable, noncracking, clean and disinfection-compatible, and have sufficient leaning to allow liquid wastes to flow based on the nature of the workplace.
- ❖ The walls should be made of smooth and light-coloured material, not cracked, easily cleanable and disinfected, which does not penetrate water, washable, does not allow harmful living things to settle according to the nature of work done.
- ❖ Windows and similar open spaces must be constructed so as not to allow contamination, thinly porous, easily cleanable, dismountable and permanently cleanable.
- ❖ The doors should have a smooth, waterproof surface, self-sealing and leakproof as the case may be. % Apart from the unavoidable circumstances, cleaning and disinfecting power materials such as untreated wood should not be used.
- ❖ The water used in production must be of a potable and suitable nature. Proper installation must be available for continuous and adequate supply of water, storage, control of pressure and temperature. % The waste system of the workplace must be arranged in such a way that it is not affected by corrosion, its cleaning and maintenance can be easily carried out, and it must be capable of removing the amount of liquid waste.

☆ The social facilities and toilets at the workplace must be separate from the food processing areas. Toilets should not be opened directly to food production areas. Toilets should be designed in such a way that the hygiene rules of the articles are properly removed, and hygiene rules warning signs must be provided in these areas. Restrooms should have self-closing doors. It is also preferred that hand washing stations have foot- or knee operated faucets that supply water at 43°C to 50°C.

☆ In accordance with the nature of the production, it is necessary to install a washbasin with taps suitable for mixing hot and cold water. Liquid soap, dryer or paper towel should be available and measures should be taken to disinfect the hands if necessary. Disposable sanitary towels are best for drying hands. The dining hall must comply with hygiene regulations.

☆ Remotely operated liquid soap dispensers are recommended because bars of soap can increase the transfer of microorganisms.

- ❖ First aid equipment should be available at work.
- ❖ Safety sections with animals must be separate from production and storage facilities.

☆ The workplace should be installed in such a way as to prevent the entry of harmful organisms and pollutants such as dust and fumes.

- ❖ There must be a closed solid waste storage place suitable for washing and disinfection that is suitable for proper collection of solid wastes according to the characteristics of the workplace.
- ❖ There should not be any suitable places for the accumulation of water and harmful living things in the trash and waste piles that will cause pollution around the workplace.
- ❖ The raw material must be stored on a suitable material at a certain height and humidity that does not contact the substrate, additives and other auxiliaries, tools and equipment and the packaging material individually and on conditions to prevent degradation.
- ❖ When manufacturing equipment is placed in the production area, gaps should be left to allow inspection and cleaning.
- ❖ When the machines are installed, all connection points must be closed so that no clearance is left.
- ❖ The ground on which the machines are installed must be accessible for the cleaning operation or must be completely closed.
- ❖ When the equipment passes through sections such as ceilings, floors, walls, *etc.*, adequate clearance must be provided

between the equipment and the wall for cleaning, and the wall must be mounted so that there is no space left in the machine.

- The auxiliary equipment must be detachable and re-mountable to facilitate cleaning.
- The manufacturing area must be covered with hard, smooth, easily cleanable and non-absorbent material.
- For insulation, non-absorbent material should be used. These surfaces must be accessible.
- When products are stored, they should be stacked on pallets with a distance of at least 100 mm from the ground.
- The doors and windows opening out of the production area must be insulated to prevent any contamination from the outside.
- Sufficient ventilation should be provided in the washing rooms.
- Water and water vapor, which come in direct contact with the product, must be drinkable.
- Pipes, valves and fittings that are used for liquid materials should be easily installed and dismantled. % All fixed pipes must be sealed to prevent condensation.
- The drainage systems for the wastewater should be installed in the cleaned areas.
- The ground slope should be such as to prevent water accumulation on the surface.
- The lighting lamps must be protected to prevent falling and breaking.
- Electrical wiring must be secured (ground contact must be avoided).
- The electrical parts of the machines must be closed to prevent dusting.
- Water must be prevented from entering the electroluminescent panels.
- Machine parts, such as fans and motors, must be accessible.
- The motor and other auxiliary equipment must be positioned so that it does not touch the ground.
- Use of wood should be avoided.

Premises

The food premises should be clean and free of waste materials. Choosing a suitable site for new food electricity, gas, water supply, effluent disposal and accessibility for delivery and refuse collection should be available. Clean air and safe water are important for pleasant working environment and hygienic food production. The clean ability of floors, walls, ceilings or any other surfaces built into establishment is the basis for maintaining a structure free from the hazards of infection. The units should be so designed so that work flow is satisfactory and in particular 'dirty flow' lines do not cross 'clean flow' lines in order to minimize the risk of cross contamination. Food preparation areas should be completely separate from wash-up and garbage areas.

Contamination can result from airborne microorganisms in food processing, packaging, storage, and preparation areas. This contamination can result from unclean air surrounding the food plant or from contamination through improper sanitary practices. The most effective methods of reducing air contamination are through sanitary practices, filtering of air entering the food processing and preparation areas, and protection from air by appropriate packaging techniques and materials.

The floors should be durable, smooth but not slippery, non-absorbent and easy to clean and maintain. Walls and ceiling in food preparation, washing areas must be of light colour. The wall surfaces around sinks and equipment must be specially treated to be resistant to heat and moisture. All doors should be self-closing and made rodent proof. The window must be fitted with fly-proof screens. The efficient ventilation must be provided to produce reasonable working and to reduce humidity and temperatures which would assist in the rapid multiplication of bacteria. Suitable and sufficient lighting will produce the environment conducive to clean and safe working. All areas should be well lighted to make dirt, grease and infestation easily detectable. Poor ventilation can lead to headache, giddiness, fainting and even vomiting. Besides producing discomfort, it also affects the efficiency of people at work. Adequate facilities for hand washing and drying should be provided.

The consumption of snacks, beverages, and other foods, as well as smoking, should be confined to a specific area, which should be clean and free of insects and spills.

Kitchen waste comprises peelings, trimmings, plate waste, spillage, empty cans, bottles, packing materials *etc.* Arrangements must be made to collect all waste materials at the end of day and completely destroying them or removing them to garbage disposal centres in sealed, leakproof bags. Improper disposal can pollute water and contaminate equipment and food.

Raw, untreated sewage can contain pathogens that have been eliminated from the human body, as well as other materials of the environment. Examples are microorganisms causing typhoid and paratyphoid fevers, dysentery, and infectious hepatitis. Sewage may contaminate food and equipment through faulty plumbing. If raw sewage drains or flows into potable water lines, wells, rivers, lakes, and ocean bays, the water and living organisms such as seafood are contaminated. To prevent this contamination, privies and septic tanks should be sufficiently separated from wells, streams, and other bodies of water. Raw sewage should not be applied to fields where fruits and vegetables are grown.

In any food establishment, varieties of insects, pests and rodents pose a big threat to the maintenance of hygienic surroundings. Preventive measures should be taken to control pests, rodents and insects in foodservice areas. The clean and pleasant environment in and around the premises is essential for continued patronage of any catering establishment. A cleaning schedule (monthly/weekly/daily) has to be prepared to cover every item of the establishment from kitchen to store-rooms and toilets. The supervisor or any other person assigned the job has to check the cleaning operation to ensure implementation by the cleaning staff of the cleaning schedules.

Transport

Vehicles and/or containers used to transport food must be kept clean and maintained in good repair and condition, to protect food from contamination. Where necessary, they must be designed and constructed to allow adequate cleaning and/or disinfection. Items used to hold food (*e.g.* boxes) in vehicles and/or containers must not be used for transporting anything other than food where this may cause contamination. Where vehicles and/or containers are used for transporting anything other than food, or for transporting different types of food at the same time, you must separate products effectively, where necessary. Where vehicles and/or containers have been used for transporting anything other than food or for transporting different foods, you must clean effectively between loads to avoid the risk of contamination. Food in vehicles and/or containers must be placed and protected in a way that minimises the risk of contamination. Where necessary, vehicles and/or containers used for transporting food must be capable of keeping food at appropriate temperatures and allow those temperatures to be monitored.

Equipment

Hygienic food handling requires appropriate equipment and supplies. Food handling and food processing equipment should be constructed according to regulations of the appropriate regulatory agency.

Contamination of equipment occurs during production, as well as when the equipment is idle. Even with hygienic design features, equipment can collect microorganisms and other debris from the air, as well as from employees and materials. Product contamination of equipment is reduced through improved hygienic design and more effective cleaning.

Regular and effective cleaning and sanitizing of equipment and utensils and also work surfaces reduces the probability of contamination of food during preparation, storage or service. The food contact surfaces of cooking devices and doors *etc.* should be cleaned as often as necessary to keep the equipment free of accumulation of dust, dirt, food particles *etc.* To avoid cross contamination, it is important that the same equipment is not used for handling raw and high-risk products without being disinfected.

In a food operation; all tools, equipment and reusable containers used in production and in contact with food must be easy and well cleanable, disinfectable, smooth and free from contamination, in accordance with the requirements of health. These should always be kept clean and properly disinfected if necessary. Hygienic equipments must be available at the production entrance.

The food contact surfaces of fixed equipment should be designed that all surfaces in contact with food are smooth and non-porous so that particles of food, insect eggs, bacteria are not caught in microscopic cracks and can be cleaned properly. It is necessary to discard/replace any equipment which are chipped, dented or damaged. If possible, equipment should be mobile to facilitate its removal for cleaning. Gas and electricity supply pipes to ovens should be flexible and capable of being disconnected. Handles of knives, brushes and other equipment should all be made from cleansable materials such as stainless steel or polypropylene. Tableware should be washed, rinsed and cleaned after each use. Clothes used for wiping food spills or table ware should be clean, dry and used for no other purpose. The clothes or sponges should be stored in the sanitizing solution between uses.

Food Waste

One must remove food waste and other rubbish from rooms where food is present as quickly as possible, to avoid them building up. One must put food waste and other rubbish in containers that can be closed, unless one can satisfy the local authority that other types of containers or systems of disposing of waste are appropriate. These containers must be of appropriate construction, kept in sound condition, be easy to clean and, where necessary, to disinfect. One must have adequate facilities for storing and disposing of food waste and other rubbish. Stores for waste must be designed and managed in a way that enables them to be kept clean and, where necessary, free of animals and pests. One must get rid of all waste in a hygienic and environmentally friendly way, in accordance with legislation. (There are rules about the way certain types of food waste must be collected and disposed of – contact the local authority for details.). The waste must not be a direct or indirect source of contamination (*e.g.*, touching surfaces that food is prepared on, or attracting pests).

Defrosting

While defrosting any food, it must be done in a way that minimises the risk of harmful bacteria growing, or toxins forming, in the food. While they are being defrosted, one must keep food at a temperature that would not result in a risk to health. Where liquid coming from the defrosting food may present a risk to health (*e.g.* when defrosting raw meat) one must drain it off adequately. Following defrosting, food must be handled in a way that minimises the risk of harmful bacteria growing or toxins forming (*e.g.* keeping it in the fridge).

Wrapping and Packaging

Material used for wrapping and packaging must not be a source of contamination. One must store wrapping materials so they are not exposed to a risk of contamination. One must carry out wrapping and packaging in a way that avoids contamination of products. One must make sure, where appropriate, that the container is clean and not damaged, particularly if using cans and/or glass jars. If re-using any wrapping and/or packaging material for food, it must be easy to clean and, where necessary, to disinfect.

Money Handling in Food Service Operations

An aspect of food service which frequently causes consumer comment is the way a food handler prepares the food, takes money for the purchase, returns change to the customer and then prepares food for the next customer. This is most noticeable in sandwich bar operations but is by no means restricted to these outlets.

Two studies in the United States have endeavoured to determine the likelihood of currency serving as a vehicle for transmitting potential pathogens to hands and food. The first of these involved swabbing and culturing from various coins and paper currency collected at random from doctors and other employees at a New York hospital.

Coagulase-negative staphylococci were recovered from only 3 of 102 coins examined. It was the only organism isolated from 2 coins and on one coin it was mixed with *Bacillus* species, i.e viable bacteria could be cultured from only 3 per cent of coins examined. Eleven of 100 paper notes yielded coagulase negative staphylococci. It was the only organism found on 6 notes while *Bacillus* species were recovered from a further 4 notes. One note was positive for coagulase negative staphylococci and *Bacillus* species in addition to one corynebacterium species.

These results are much lower than and different from those obtained in an earlier study where 70 per cent of coins and 70 per cent of bank notes were contaminated with many types of bacteria.

In the other study, a different experimental approach was taken. The researchers inoculated the surface of coins with the pathogens *Escherichia coli* 0157: H7 and *Salmonella enteritidis* to determine the survival of these organisms on the coins. The inoculated coins were dried under a laminar flow hood for 2.5 hours at 29°C. After this time duplicate samples were aseptically removed and the inoculated bacteria counted. At the completion of drying, coins and controls were held at 25°C and 40 per cent relative humidity for survival studies. *Salmonella enteritidis* cells proved more sensitive to drying than *E. coli* 0157: H7 with a 1000 to 10,000-fold decrease of *Salmonella*e compared to an approximate 100-fold reduction of *E. coli* on the coin surfaces during the 2.5 hour drying time.

During subsequent storage at 25°C, *E. coli* 0157: H7 survived for 7, 9 and 11 days on surfaces of pennies, nickels, and dimes and quarters, respectively. The initial population per coin surface was approximately 10_5 cfus. *Salmonella enteritidis* survived for 1, 2, 4 and 9 days on the surfaces of pennies, nickels, quarters, and dimes, respectively. Dimes and quarters are made up of a pure copper core and an outer layer of 75 per cent copper and 25 per cent zinc alloy. Nickels are an alloy of 75 per cent copper and 25 per cent nickel. Pennies are 95 per cent copper.

It was concluded that the pathogens tested were subject to inactivation on coin surfaces by drying and by the bactericidal effect of copper. They drew attention to the apparent low infectious dose of *E. coli* 0.157: H7 and recommended that food handlers should use an intervention treatment such as washing hands after handling coins and before handling food.

This is not practical in many situations and at least the first study examining natural contamination of coins should give some reassurance to consumers.

HIV and Food Handling

Health authorities around the world all agree that transmission of HIV/AIDS via food and beverages is not a known risk. The statement by the US Centers for Disease Control confirms that there is no known risk of transmission of blood borne and sexually transmitted infections such as HIV/AIDS during the preparation or serving of food or beverages. The statement on "HIV and Its Transmission" includes:

'There is no known risk of HIV transmission to co-workers, clients, or consumers from contact in industries such as food-service establishments. Food-service workers known to be infected with HIV need not be restricted from work unless they have other infections or illnesses (such as diarrhoea or hepatitis A) for which any food-service worker, regardless of HIV infection status, should be restricted.' CDC recommends that all food-service workers follow recommended standards and practices of good personal hygiene and food sanitation.

The Virus and the Mode of Spread

AIDS is caused by one or more of a group of retroviruses which for simplicity have all been called Human Immunodeficiency Virus. Human beings should be regarded as the sole reservoir of the virus; neither foodstuffs, (including food animals), nor animals, nor pets, nor insects are a source. HIV is spread by sexual contact with infected persons, injection of contaminated blood or blood products and by transmission from mother to child. HIV cannot be spread by touching, coughing, sneezing or by insect bites. The virus has an incubation period of 5-8 years and so the progress of the disease varies from person to person; some people can unwittingly carry the virus without showing any symptoms, although it seems probable that all HIV carriers will develop AIDS eventually.

Survival of the Virus Outside the Human Body

The virus dies only slowly at room temperature and it will survive well when frozen. However, it is very sensitive to heat and it is rapidly destroyed at 55 degrees Celsius and above. It is readily inactivated at both low and high pH. The HIV virus is rapidly destroyed by the commonly used concentrations of all biocides, including hypochlorite, ethyl alcohol and isopropyl alcohol. When accidents occur, the precautions used to prevent the spread of viral hepatitis and other blood borne infections will also prevent the spread of HIV. First aiders should cover any exposed cuts or abrasions on their own hands with a waterproof dressing, put on disposable plastic gloves and an apron. Spillages of blood and other body fluids should be flooded with a biocide and mopped up with paper towels. All these disposable items should be placed in plastic bags and safely disposed of, preferably by incineration.

Implications for the Food Industry

Food handlers who carry HIV are not a threat to their workmates or to the products that they handle. They should not be restricted from working with foods or beverages or be restricted from using telephones, machinery, office equipment, toilets, showers, eating facilities or drinking fountains. They may thus work normally, unless of course they develop a secondary infection such as a diarrhoeal disease which will preclude them from handling food. It needs to be emphasised to persons with HIV/AIDS that they must comply with the other health and safety requirements applicable to all food handlers.

Employee Responsibilities

Food processing and foodservice firms should protect their employees and consumers from workers with diseases or other microorganisms of public health concern that can affect the wholesomeness or sanitary quality of food. This precaution is important to maintain a good image and sound operating practices consistent with regulatory organizations. In most communities, local health codes prohibit employees having communicable diseases or those who are carriers of such diseases from handling foods or participating in activities that may result in contamination of food or food contact surfaces.

Responsible employers should exercise caution in selecting employees by screening unhealthy individuals. Although some areas no longer require health cards because of the expense involved, many local health departments require all employees who handle food to be examined by a physician who will issue a health card only to healthy individuals.

Selection of employees should be predicated upon these facts:

1. Absence of communicable diseases should be verified through a county health card or a physician's report.
2. Applicants should not exhibit evidence of a sanitary hazard, such as open sores or presence of excessive skin infections or acne.
3. Applicants who display evidence of respiratory problems should not be hired to handle food or to work in food processing or food preparation areas.
4. Applicants should be clean and neatly groomed and should wear clothing free of unpleasant odour.
5. Applicants should successfully complete a sanitation course and examination such as that provided by the National Restaurant Association.

Although the employer is responsible for the conduct and practices of employees, responsibilities should be assigned to employees at the time employment begins.

Employees should maintain a healthy condition to reduce respiratory or gastrointestinal disorders and other physical ailments.

- ☆ Injuries, including cuts, burns, boils, and skin eruptions, should be reported to the employer.
- ☆ Abnormal conditions, such as respiratory system complications (*e.g.*, head cold, sinus infection, and bronchial and lung disorder), and intestinal disorders, such as diarrhoea, should be reported to the employer.

- Personal cleanliness that should be practiced includes daily bathing, hair washing at least twice a week, daily changing of undergarments, and maintenance of clean fingernails.
- Employees should tell a supervisor that items such as soap or towels in washrooms should be replenished.
- Habits such as scratching the head or other body parts should be stopped.
- The mouth and nose should be covered during coughing or sneezing.
- The hands should be washed after visiting the toilet, using a handkerchief, smoking, handling soiled articles, or handling money.
- Hands should be kept out of food. Food should not be tasted from the hand, nor should it be consumed in food production areas.
- Food should be handled in utensils that are not touched with the mouth.
- Rules related to use of tobacco should be enforced.

Employer Responsibilities

The employer's responsibilities start in the staff recruitment process. Management should require employees to have a pre-employment physical examination to verify that they are in good physical, mental, and emotional health. This is an excellent opportunity to impress the importance of good hygienic habits on a new employee and to emphasize how employees "shed" *Salmonella* and *Shigella*. Furthermore, those with skin infections may be identified before they handle food. All employees who work with food should be checked regularly for signs of illness, infection, and other unhealthy conditions.

The staff should be sure that they have passed the health check and are not infectious. It should be regularly checked by health inspectors and patient personnel should not be allowed to work. The examinations should be strengthened by x-ray, blood analysis, parasite and tuberculin tests.

Several countries have a legal requirement for pre-employment health examinations and require that they be repeated at regular intervals. However, this regulation has been challenged because of the expense of routine medical examinations, the difficulty of administering these plans, and because a clear relationship between food handlers and foodborne disease has not been established.

Personal hygiene is a basic step that should be taken to ensure the production of wholesome food. Requirements for Hygienic Practices Management must establish a protocol to ensure hygienic practices by employees. Supervisors and managers should set an example for employees by their own high levels of hygiene and good health while conveying the importance of these practices to the employees.

Personnel should be supervised during operation and provided compliance with hygiene regulations. Jewellery or accessories should not be allowed inside the premises. Working clothes are so important for hygiene. Employers should provide proper laundry facilities for maintenance of cleanliness through clean dressing rooms, services, and welfare facilities. The employer should ensure that the clothes are clean, ironed and plain. Light coloured dresses should be preferred because they show dirt. The employer should establish periodic training programs that include production and hygiene issues, and develop personal hygiene habits in these training programs. Warning signs, notifications and sanitation programs should be supported in the work environment. Due to good hygiene and sanitation practices, the personnel should be supported and encouraged. One of the most important tasks for the employer to ensure all of this is to have adequate and social facilities designed in accordance with sanitation rules in the organization.

Employers should emphasize hygienic practices of employees as follows:

1. Employees should be provided training in food handling and personal hygiene.
2. A regular inspection of employees and their work habits should be conducted. Violations of practices should be handled as disciplinary violations.
3. Incentives for superior hygiene and sanitary practices should be provided. Food handlers should be responsible for their own health and personal cleanliness.
4. Employers should be responsible for making certain that the public is protected from unsanitary practices that could cause public illness.

Food Safety Management Systems

It is necessary to establish systems related to hygiene, sanitation and preventive maintenance in order to ensure health safety in food industry. Sanitation can be considered as part of quality control. Sanitation programs must be put in place to ensure cleanliness of the installation, microbiological quality of the raw material, health and reliability of the products cause of sanitation has an important role in the formation of quality. Various programs are being implemented to ensure quality, especially in terms of microbiological safety of food. The most common are HACCP, GMP and ISO 9000, 22000 series.

Establishment of Sanitary Practices

Sanitation, good manufacturing practices, and other environmental and operating conditions necessary for the production of safe, wholesome food are known as prerequisite programs. These prerequisite programs provide the foundation for HACCP and are a vital component in a company food safety assurance system. So, the design and development of this entire system in a food facility begins with the establishment of basic sanitary practices. The employer is responsible for establishing and maintaining sanitary practices to protect public health and maintain a positive image. The problem of establishing, implementing, and maintaining sanitary practices within the food industry is certainly a challenge. The sanitarian or person in charge of this important area must assure that the sanitary practices keep low-risk potential hazards from becoming serious hazards that could cause illness or injury.

The sanitarian is both the guardian of public health and the counsellor to company management on quality and safety issues that are influenced by sanitary practices. A large food processing company should have a separate food safety department on the same organizational level as production. It should have a separate food safety department on the same organizational level, as production or research, that is responsible for food safety at all operating plants. A sanitation department or team should exist in a plant on a level with other departments. In a large organization, sanitation should be separated from production and mechanical maintenance, an arrangement that will enable the sanitation department team to exercise company-wide surveillance of sanitary practices and maintain a high level of activity.

Production practices, quality control, and sanitary practices are not always compatible when administered by a single department or individual; but all of these functions are complementary and are best performed when properly coordinated and synchronized. Ideally, an organization should have a fulltime sanitarian with assistants, but this is not always practical. Instead, a trained individual who was originally employed as a quality control technician, a production foreman, a superintendent, or some other individual experienced in production can be charged with the responsibility of the sanitation operation. This situation is common and usually effective. However, unless the sanitarian has an assistant to take care of some of the routine tasks and is given sufficient time for proper attention to sanitary details, the program may not succeed. A one-person safety assurance department with a full schedule of control work will be generally inadequate to assume the tasks of a sanitarian. However, with proper assistance, quality assurance and sanitation supervision can be successfully conducted through a qualified individual that can divide his or her effort between sanitation and quality assurance. It is beneficial for this person to have the advice and service of an outside agency, such as a university, trade association, or private consultant, to avoid becoming submerged in the conflicting interests of different departments. The extra expense can be a worthwhile investment.

A planned sanitation maintenance program is essential to meet legal requirements and protect brand and product reputation, product safety, quality, and freedom from contamination. All phases of food production and plant sanitation should be included in the program to supplement the cleaning and sanitizing procedures for equipment in the facility. A safety assurance program should start with compliance inspection and audit of the entire facility. The inspection and audit should be comprehensive and critical. As each item is considered, the ideal solution should be noted, irrespective of cost. When the audit is completed, all items should be re-evaluated and more practical and/or economic solutions determined. All items that need attention should be prioritized and an action plan for completion should be established. Attention should be clearly focused on critical deficiencies throughout the facility. Aesthetic sanitary practices should not be adopted without clear evidence of their ability to pay dividends in increased sales or because they are necessary to meet competitive sales pressure.

Standard Sanitation Operation Procedures (SSOP)

Standard Sanitation Operation Procedures are written procedures for defining sanitary conditions in food enterprises and as well as for defining sanitary conditions to ensure a healthy and safe product. Sanitation checks can be done in 2 ways:

1. Internal audit: Sanitation should be included in the quality assurance program and explain to audit how could be. When an inspection program is being carried out, areas and storages should be included in this program. In the audit, each section should be handled separately and the status of before and after sanitation must be specified separately on the report.
2. External audit: When the operation is audited by an external auditor or expert, a good review should be made in terms of sanitation.

The areas covered by the audit are:

- Control of the operation environment
- Control of food processing departments
- Warehouse control
- Locations outside the producing areas
- Basement and roof
- Control of personnel and facilities
- Sanitation applications control

Chapter 17

Cleaning and Sanitising Food Premises and Food Equipment

National and international legislation requires the food industry to put safe food on the market. A variety of hazards, including microorganisms and their toxins, previous products and ingredients, colourants, allergens, residues from cleaning agents, sanitisers, disinfectants and lubricants can (cross-) contaminate food during production with food safety or quality consequences. Many stakeholders in the food production chain, from agriculture production to food service, have adopted the Hazard Analysis and Critical Control Points (HACCP) Principles as a tool to identify where hazards (physical, chemical, biological) might occur in the production process. Once the hazards are identified, actions to monitor and control different steps of the process are implemented to significantly mitigate those hazards.

A risk assessment is needed to determine whether an activity that introduces cleaning agents, sanitisers and disinfectants represents a known or reasonably foreseeable chemical hazard. Therefore, the objectives of any preliminary risk assessment are:

1. To identify the need for cleaning, disinfecting or sanitising,
2. To define the best technique and
3. To identify where, why and how sanitisers, disinfectants and cleaning agents are introduced to the food manufacturing process. When necessary, a human risk assessment may be needed to support any risk management decision.

Physical, chemical and microbiological cleanliness are a prerequisite for food safety.

As mentioned above, it is generally considered as part of the Pre-Requisite Programmes (PRPs) or preventative controls. Cleaning can include the use of detergents and can be combined with disinfectants/sanitisers. Various methods of cleaning, with or without disinfection/sanitation are typically practiced.

Terms used in Cleaning and Sanitation

Cleaning: Cleaning is a critical step within the food production industry to maintain and further ensure food safety and quality standards. It is a process of removing unwanted soil (fats, proteins, sugars, scaling, *etc.*) from equipment and/or manufacturing facilities, by application of an effective procedure, either manual or automated. The typical process for cleaning and sanitizing food contact surfaces and equipment is a liquid process and follows this sequence of steps: clean, rinse, sanitize.

- **i) Clean**: Cleaning agents, such as dish soap made from synthetic detergents, are used to remove dirt, germs, objects or impurities from food contact surfaces and equipment. The cleaning agent itself is not required to be organic. Any cleaner or detergent may be used provided that the cleaning agent is disclosed in the handler's organic system plan and also meets the Food and Drug Administration's (FDA) requirements. Unlike sanitizers, cleaners and detergents are designed to be rinsed off, and a subsequent rinse step is sufficient to prevent contamination of organic foods from synthetic cleaner residues.
- **ii) Rinse**: A rinse with potable water removes the cleaning agents. Potable water is simply water deemed safe for drinking and food preparation.

iii) **Sanitize**: Sanitizers are applied to cleaned surfaces to ensure that the surface is free of pathogenic microbes. The sanitizing step is a mandated part of most state and federal food safety protocols for food contact surfaces. Sanitizers merit more scrutiny than cleaners in an organic process because some sanitizers are designed to leave a faint anti-bacterial residue on food contact surfaces. Such residues are usually not allowed in contact with organic food. Therefore, procedures in an organic operation must ensure that organic foods do not contact any prohibited sanitizer residues. Because of the need to ensure a safe food system, the USDA organic regulations allow the use of some synthetic sanitizers for food contact surfaces.

Sanitisation: It is a process of reducing microbiological contamination on an effectively cleaned surface by means of a bactericidal treatment such as heat or chemicals, to a level that is acceptable to local health regulations. For effectiveness, this must be preceded by cleaning.

Disinfection: It is a process that eliminates many or all microorganisms, except bacterial spores, on inanimate objects. Some disinfectant chemicals also have sporicidal activity.

Dry cleaning Process: It is a process of removing unwanted soil, using an effective procedure, from equipment and/or manufacturing facilities, without use of water (including aqueous solutions or steam).

Wet cleaning Process: It is a process of removing unwanted soil, using an effective procedure, from equipment and/or manufacturing facilities, with the use of unrestricted quantities of water. This includes cleaning in place (CIP), cleaning out of place (COP) and other forms of cleaning using unrestricted amounts of water. Wet cleaning can be combined with disinfection/sanitation.

Controlled wet cleaning Process: It is a process of removing unwanted soil, using an effective procedure, from equipment and/or manufacturing facilities with limited amounts of water, applied with strict control or by use of humid clothes. Controlled wet cleaning is followed by immediate active drying of cleaned surfaces. In order to comply with this definition, this type of cleaning is often limited to small surfaces or occasional application on individual pieces of equipment. Controlled wet cleaning can be combined with disinfection/sanitation.

Cleaning in Place (CIP): CIP is a method for cleaning the production lines without dismantling the installation by circulating cleaning solutions according to defined protocols, combining physical, mechanical and chemical energies.

Cleaning out of Place (COP): COP is a method of cleaning whereby equipment or parts of equipment are removed from their normal place of use in a food processing operation specifically for purposes of cleaning.

Cleaning agents, sanitisers and disinfectants are vital to food hygiene and are a global public health protection measure. Limiting microbial antibiotic resistance is also a public health priority. Although many factors contribute to the incidence of antimicrobial resistance, the use of antibiotic compounds in human clinical settings and food-producing animals are primary contributors.

Objective of Cleaning/Disinfection/ Sanitation

For all cleaning/disinfection/sanitation activities, the objectives of the cleaning should be clearly defined, *e.g.*:

- To control hazardous microorganisms;
- To control food chemical contamination;
- To control foreign body contamination;
- To control allergen cross contact;
- To control ingredient/residue/colour/flavour at product changeover;
- To avoid pest infestation;
- To control chemical residues from cleaning/ sanitation regimes;
- To assure mechanical operations of equipment
- To improve process efficiency (*e.g.*, heat transfer efficiency);
- To assure occupational safety;
- To satisfy local regulatory requirements;
- To meet specific customer requirements;
- To meet GFSI requirements.

Cleaning Procedures: Standard Sanitation Operating Procedure (SSOP/SOP)

Cleaning is a complex process. A defined and systematic approach is required to ensure it is conducted correctly. This approach takes the form of Standard Sanitation Operating Procedure (SSOP), usually a legal requirement and a fundamental GFSI requirement. The collection of these cleaning procedures forms a Cleaning Plan or Programme which is specific to a facility.

A typical SSOP includes the following:

- Cleaning frequency/duration/sequence;
- Cleaning agents, sanitisers and disinfectants used (ensuring they are food-grade and fit-for-purpose);
- Cleaning process parameters (equipment used, concentration of chemicals, time, temperature, physical parameters);

- ☆ Safety requirements (assembled/disassembled equipment list, requirements to protect adjacent lines/products);
- ☆ Responsibilities, documentation, visual aids, training/qualification requirements;
- ☆ Necessary monitoring or verification activities. Note that establishing SSOPs require training of cleaning personnel to limit variation between operators.

Additionally, validation procedures are needed to determine the ability of the SSOP to achieve the desired outcome. This consists of microbiological, allergen, cleaning agent/disinfectant residues analyses after each cleaning trail or other tests required on food contact surfaces and other sample points.

How Cleaning Compounds Function

An effective sanitation program includes knowledge of soil deposits and use of the appropriate, versatile cleaning compound for the specific cleaning application. Soil characteristics determine the most appropriate cleaning compound. Generally, an acidic cleaning compound is most effective for removal *of* inorganic deposits, an alkaline cleaner for removing non-petroleum organic soils, and a solvent-type cleaner for removal of petroleum soils.

The major functions of a cleaning compound are to lower the surface tension of water so that soils may be dislodged and loosened and to suspend soil particles for subsequent flushing away. To complete the cleaning process, a sanitizer is applied to destroy residual microorganisms that are exposed through cleaning. One of the oldest and best-known cleaning compounds is plain soap. However, it has limited utility in food processing and foodservice units and is rarely used because it does not clean well and reacts with hard water to form an insoluble curd (such as a bathtub ring). A basic soap contributes to cleaning through the removal of fats, oils, and greases by suspending particles of these water-insoluble materials, although a residual film will exist. The suspension process of water-insoluble materials through interaction with soap is called emulsification. In emulsification, the cleaning compound interacts with water and the soil. Cleaning compound molecule is soluble in water. The hydrophobic portion is soluble in the soil. When the cleaning compound molecules surround the soil, a suspended soil particle results by micelle formation.

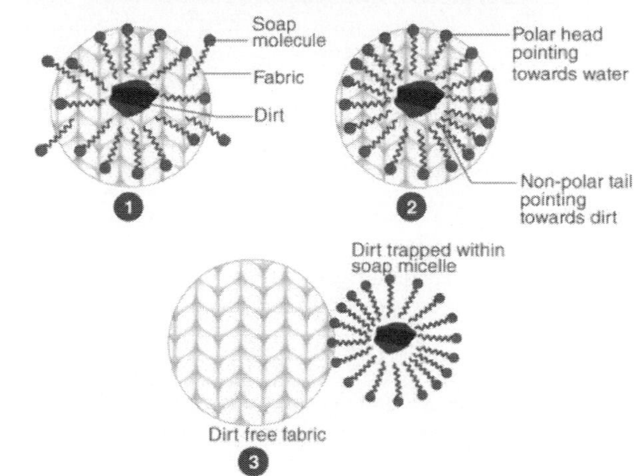

Factors Affecting Cleaning Performance (Anon., 1996)

Time: contact time on the surface being cleaned

Action: physical force exerted onto the surface (velocity or flow)

Concentration: amount of cleaner used

Temperature: amount of energy (as heat) used in the cleaning solution

Water: used to prepare cleaning solution

Individual: worker performing clean-up operation

Nature: composition of the soil

Surface: what material is being cleaned

These factors spell out the acronym TACT WINS and describe important factors involved in cleaning

Classification of Cleaning Compounds

Most cleaning compounds that are used in the food industry are classified as blending products. Ingredients are combined, to produce a single product with specific characteristics that performs a given function for one or more cleaning applications. The following classes of cleaning compounds are most frequently used in connection with foodservice facilities and processing plants:

Table 17.1: Types of Cleaning Compounds for Soil Deposits

Type of Soil	*Required Cleaning Compound*
Inorganic soil	Acid-type cleaner
Organic soil (Non-petroleum)	Alkaline-type cleaner (Petroleum) Solvent-type cleaner

Table 17.2: Characteristics of Various Surfaces of Food Processing Plants

Material	*Characteristics*	*Precautions*
Wood	Previous to moisture, fats, and oils; difficult to maintain; soften by alkali; destroyed by caustics.	Wood should not be used because of its unsanitary conditions. Stainless steel, polyethylene, and rubber materials should be used instead of wood.
Black Metals	Rust may be promoted by acidic acid chlorinated detergents.	Because these metals are prone to rust, they are often tinned or galvanized. Neutral detergents should be used in cleaning these surfaces.
Tin	May be corroded by strong alkaline and acid cleaners.	Tin surfaces should not come in contact with foods.
Concrete	May be etched by acid foods cleaning compounds.	Concrete should be dense, resistant, and nondusting. Acid brick may be used in place of concrete.
Glass	Smooth and impervious; may be etched by strong alkaline cleaning compounds.	Glass should be cleaned with moderately alkaline or neutral cleaning compounds.
Paint	Surface quality depends on the satisfactory method of application *etc.*hed by strong alkaline cleaning compound.	Certain edible paints are satisfactory for food plants.
Rubber	Should be nonporous, nonspongy; and not affected by alkaline detergents; is attacked by organic solvents and strong acids.	Rubber cutting boards can warp, and their surface dulls knife blades.
Stainless Steel	Generally resistant to corrosion; smooth-surfaced and impervious (unless corrosion occurs); resistant to oxidation at high temperatures; easily cleaned; nonmagnetic).	Stainless steel is expensive and may be less plentiful in the future. Certain varieties are attacked by halogens (chlorine, iodine, bromine, and fluorine).

I. Alkaline Cleaning Compounds

Alkaline cleaners are specifically formulated chemical blends consisting of alkaline salts, wetting agents and sequestrant (chelating) agents. They owe their detergency, or cleaning ability, to the displacement of contaminants by surface-active materials and alkaline builders. These constituents remove the contaminants and allow them to be easily rinsed away. Generally, fats, oils, greases, and proteins require alkaline cleaners with a pH of 11 or higher.

The term "alkaline builders" covers a broad group of chemicals: caustic soda (NaOH), caustic potash (KOH), phosphates, silicates and carbonates. These chemicals supply the alkalinity for the cleaner. High-alkalinity products saponify fats and vegetable oils into soluble soaps. These alkaline salts also neutralize acidic contaminants and aid in dispersing oils.

Strongly Alkaline Cleaners

A cleaner that fits in the strong alkali category destroys bacteria and dissolves proteins. Examples of strongly alkaline compounds are caustics (KOH, NaOH) and silicates having high N_2O:SiO_2 ratios. Addition of silicates improves the rinsing properties of sodium hydroxide and reduces its corrosiveness. Caustic soda is not a manual cleaner. These most common type of alkalinity builders, are highly alkaline (pH 12 to 14). They saponify fats and work with surfactants to disperse contaminants. They can remove heavy soil such as from commercial ovens and smokehouses. This type of builder is not safe for use on soft metals like aluminium and zinc.

They have strong dissolving powers, very corrosive, can burn, ulcerate and scar skin. Mixing strong alkaline cleaners with water causes an exothermic reaction; the heat generated may cause the solution to boil or vaporize. Such explosive boiling may spray caustic compound on the bystanders.

Lye (also known as caustic soda or sodium hydroxide) is sometimes found in drain and oven cleaners. These types of cleaners are used to unclog sink drains or in the bathroom. They should not be stored in the kitchen or bathroom cabinets. Lye emits dangerous fumes and can cause skin burns and in some cases blindness if the fumes come in contact with eyes. Products with lye can cause death if swallowed. Wear gloves and safety goggles when using these products. Also, make sure that there is good ventilation in the room when these cleaners are used.

Sodium metasilicate is mixed with water to form a cleaning solution. It makes a highly effective cleaner that can remove dirt and marks from walls in preparation for painting, and can attack stubborn soils on siding when cleaning the exterior of a home. Sodium metasilicate is highly alkaline. Dust from the powder can irritate the nose, eyes, and lungs. Prolonged exposure can cause skin irritation. Wear protective gloves, eye protection, and long-sleeved shirts and pants while using this product. Wear a dust mask while mixing the powder with water if dust irritates nasal passages.

Store the compound in original containers in a dry location. Sodium metasilicate is highly poisonous when ingested. Store it out of reach of children or pets.

Heavy-Duty Alkaline Cleaners

The active ingredients of these cleaners may be sodium metasilicate, sodium hexametaphosphate, sodium pyrophosphate, sodium carbonate, and trisodium phosphate, which are known for its good soil-emulsification activity. The addition of sulfites tends to reduce the corrosion attack on tin and tinned metals. These cleaners are frequently used with high pressure or other mechanized systems.

Sodium carbonate, borax *etc.* function primarily as a buffering agent in many formulations and has a wide range of uses in heavy-duty and manual cleaning applications. These compounds have moderate dissolving powers and are generally slightly corrosive or noncorrosive. They are excellent for removing fats but have no value for mineral deposit control.

Moderate Alkaline Cleaners

Ammonia is a strong, colourless gas. When the gas is dissolved in water it is called liquid ammonia. Ammonia is added to a number of household and commercial cleaners to boost their alkalinity and grease-cutting ability. It dries clear and usually won't streak. Ammonia also is found in glass cleaners and in cleaners used for shiny surfaces, like wax cleaners. Too much of it can ruin or damage some surfaces, so use it with caution.

An effective cleaning agent, ammonia's odour is the main deterrent to its use. Never mix ammonia with chlorine and bleach products because this can release a highly toxic gas. For that matter, never mix any kind of chemicals since this can cause an unstable reaction and perhaps fumes, fire, or volatility (explosions).

- ✰ **Household ammonia**—containing 5 to 10 per cent ammonia gas in water: cleans kitchen range burners, ovens, windows and mirrors.
- ✰ **Sudsy ammonia** has soap or detergent added. Sudsy ammonia cleans garbage pails, kitchen range burners, and sinks.

Mild Alkaline Cleaners

Mild alkaline compounds include sodium bicarbonate, sodium sesquicarbonate, tetrasodium pyrophosphate, phosphate water conditioners (sequesters), and alkyl aryl sulfonates (surfactants).

Mild cleaners frequently exist in solution and are used for hand cleaning of lightly soiled areas. These compounds have good water-softening capabilities but exhibit no value for mineral deposit control.

For the most part, household cleaning chemicals listed in the mild alkali category are ones that contain sodium bicarbonate (baking soda). Mixed with water, this solution is used on areas that require a mild cleaning. It safely cleans glass, wall tile, and porcelain enamels. It also removes coffee and tea stains from china and plastic dishes. Mild alkalis are not corrosive. They are safe to use around children and pets.

Chlorinated Alkaline Cleaners

Hypochlorite is added to these cleaners to peptize the proteins for easier removal. These cleaners are well adapted to cleaning in-place (CIP) of pipes, tanks, and vats and remove effectively fats, oils, grease, and proteins.

Silicates provide medium alkalinity (pH 11 to 12.5) and contribute to detergency. They inhibit attack on soft metals, but they become insoluble at a pH less than 10.

II. Acid Cleaning Compounds

These compounds, especially blends of acids such as phosphoric, nitric, sulfuric, and sulfamic, remove encrusted surface materials and dissolve mineral scale deposits including those formed from using alkaline cleaning compounds or other cleaners. A portion of the minerals found in water may be deposited when heated to 80°C or higher and adhere to metal surfaces and appear as a rusty or whitish scale.

Activity of acid cleaners is expressed through chemical action with minerals found in deposits, making them water soluble and easy to remove. Organic acids, such as citric, tartaric, sulfamic, and gluconic acid, are also excellent water softeners, rinse easily, and are not corrosive or irritating to the skin. Although inorganic acids are excellent for removing and controlling mineral deposits, they can be extremely corrosive and irritating to the skin. Acid cleaning compounds are a specialized type of cleaner and are not recognized as effective, all-purpose cleaning compounds. They are not nearly as effective against soil caused by fats, oils, and proteins, which acts as a binder, as are alkaline cleaning compounds. Alkaline cleaning compounds chemically attack the binder of organic soils, which releases the retaining or tenacious forces. Acid cleaning compounds are not capable of this function.

Strongly Acid Cleaners

They are corrosive to concrete, most metals, and fabrics. Some of these cleaners, on heating, produce corrosive, toxic gases, which can ulcerate lungs. They are used in cleaning operations to remove the encrusted surface matter and mineral scale frequently found on steam-producing equipment, boilers, and some processing equipment.

When the solution temperature is too high, the mineral scale may redeposit and form a tarnish or whitish film on the equipment being cleaned. Strongly acid agents used for cleaning operations in food plants include hydrochloric (muriatic), hydrofluoric, sulfamic, sulfuric, and phosphoric acids. Nitric and sulfuric acids are not used in manual cleaners because of their corrosive properties. Corrosion inhibitors, such as potassium chromate for nitric acid solutions or butylamine for hydrochloric acid detergents, may be added.

Phosphoric acid and hydrofluoric acid both clean and brighten certain metals. However, hydrofluoric acid is corrosive to stainless steel and dangerous to handle because of the tendency toward hydrogen evolution during use.

Phosphoric acid is widely used in many countries as it is relatively low in corrosive properties, compatible with many surfactants, and is used in manual and heavy-duty formulations.

Mildly Acid Cleaners

These compounds are mildly corrosive but may cause allergenic reactions. Examples of mildly acid cleaning compounds are levulinic, hydroxyacetic, acetic, and gluconic acids. Wetting agents and corrosion inhibitors (*i.e.*, 2-naphtoquinoline, acridine, 9-phenylacridine) may be added. The organic acids, which are used as manual cleaning products, are higher in cost than are the other acid cleaning compounds. These mild compounds can also function as water softeners.

III. Abrasive Cleaners

They are used to remove heavy accumulations of soil often found in small areas. The abrasive action is provided by small mineral or metal particles, such as fine steel wool, copper or even nylon.

IV. Cleaners with Active Chlorine

For ages, chlorine compounds have been an essential part of human existence. This got its recognition as an element in 1774 when Scheele found the released gas as chlorine gas while studying the mineral pyrolusite. During experiments, Scheele mixed pyrolusite, primarily manganese dioxide (MnO_2) with hydrochloric acid (HCl), and called it spiritussalis. There he found the release of an offensive suffocating greenish-yellow gas.

He gave it a complex name of dephlogisticated marine acid based of its properties such as reaction with metals, slight dissolvability in water, and affinity towards bleaching flowers and leaves. For several years the nature of his discovery remained misunderstood as few scientists argued that his dephlogisticated marine acid was a compound of a new element and oxygen. In 1807, an English scientist Sir Humphry Davy cleared this confusion by proving that Scheele's substance was a pure element and finally suggested the name chlorine for the element.

Furthermore, around the 1830s, Davy's lab assistant Michael Faraday bestowed the world with his finest work on the electrolytic synthesis of chlorine and its liquefaction procedure. However, the first patent analogous to the preparation of chlorine in an electrolytic cell was acquired by Charles Watt in 1851. And, with the advent of the industrial revolution came the widespread applicability of chlorine around the globe.

Chlorine compounds are broad spectrum compounds that are considered having low toxicity, low cost and easy to use. They do lose potency over time and are not active at temperatures above 110° F or at high pHs (>9). Since these compounds lose activity quickly in the presence of organic debris, sunlight and some metals, they must be applied to thoroughly cleaned surfaces for disinfection.

Active chlorine compounds are those that possess or can produce $OC1^-$ or ROC1 in solution. In addition to its germicidal action, chlorine serves to oxidize organic soils, particularly proteins, down to water-soluble fragments and so reduce foam and spot/film scores. The disulfide linkage in proteins is believed to be particularly susceptible to attack by $OC1^-$. Either chlorinated trisodium phosphate or chlorinated isocyanurates is the chlorine source of choice in machine dishwashing products. The isocyanurates offer better stability and less caking than trisodium phosphate and sodium dichloroisocyanurate dihydrate has additional safety and stability benefits. The structure of this chlorine source is as follows:

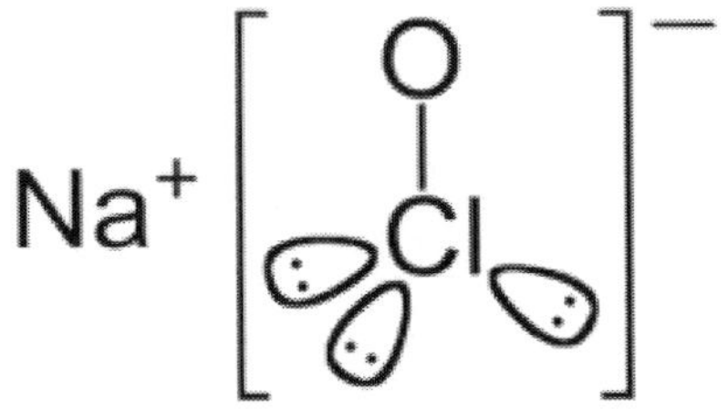

$Na^+ [OCl]^-$

Sodium Hypochlorite Chemical Structure

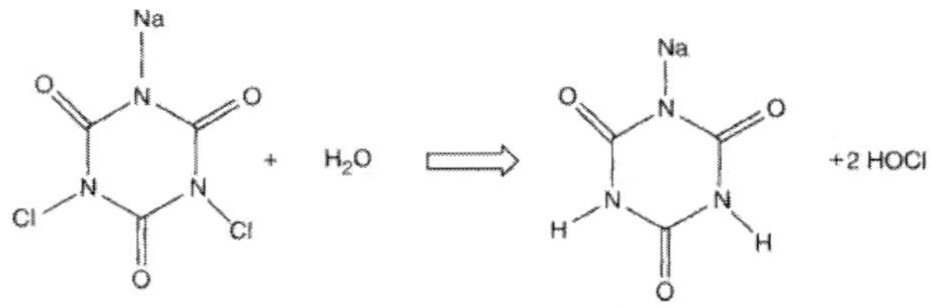

$$(\text{Na dichloroisocyanurate}) + H_2O \Rightarrow (\text{Na isocyanurate}) + 2\,HOCl \quad (4)$$

The species responsible for oxidation is the hypochlorite ion,

$$HOC1 + OH^- \Rightarrow OC1^- + H_2O$$

with the pK_a for the weakly acidic HOC1 lying at around 7.6. The bleaching action for this species is best in the pH range 10–13, where the free chlorine exists predominantly as $OC1^-$. A typical bleach content would be that which provides 5000–25,000 ppm (0.5–2.5 per cent available chlorine), typically 1–2 per cent of the formulation.

One down-side of chlorinated products is a limited storage stability. This instability reduces the chlorine benefit which the detergent can deliver, but also reduces the benefit of the other active ingredients which are destroyed by bleach action. Non-ionic surfactants are particularly susceptible to oxidative degradation when formulated into phosphate and meta-/di-silicate-based powders.

The most commonly used agents are chlorine dioxide, sodium hypochlorite (Chlorine Bleach) and calcium hypochlorite. Cleaners containing active chlorine, such as sodium or potassium hypochlorite, are effective in the removal of carbohydrate and/or proteinaceous soils because they aggressively attack such materials and chemically modify them to render them more susceptible to interaction with the balance of the components.

These products are especially valuable when cleaning a surface in which the soil is derived from a food source comprised of some form of starch or protein. Active chlorine-containing cleaners react with the insoluble proteins and render them soluble and/or readily dispersible through degradation by rapid oxidation of sulfide crosslinks that are present. Because the degradation need not be complete for solubilization to occur, a small amount of hypochlorite will remove a relatively large quantity of protein.

Chlorine compounds are considered broad spectrum, being effective against bacteria, enveloped and non-enveloped viruses, mycobacteria and fungi. At high concentrations, chlorine compounds can be sporicidal.

Because of a form of chemical bonding known as cross-linking, many carbohydrates are such that a large number of the "big" molecules are bonded together. In this instance, they cannot dissolve, which makes cleaning them from a surface very difficult.

The uses of chlorine-based products are:

- Household bleach (5.25 per cent sodium hypochlorite) is the most commonly used chlorine compound and is an excellent, safe, and highly cost-effective disinfectant when used correctly.
- At a dilution of 1:32, bleach is highly effective against bacteria and viruses, including nonenveloped viruses, such as panleukopenia and FCV.
- Solutions must be made fresh daily and stored in opaque containers, because bleach is highly unstable once mixed with water and degrades in the presence of ultraviolet light.
- Surfaces must be thoroughly cleaned with a detergent, rinsed, and dried prior to the application of bleach, because it is ineffective in the presence of detergents and organic material.
- Proper disinfection requires 10 minutes of contact time with a bleach solution.
- Although bleach is not effective when mixed with detergents, it can be safely and effectively mixed with quaternary ammonium compounds, which do provide some cleaning activity. Therefore, this combination can be used for cleaning and disinfection, provided gross organic material is first removed and adequate contact time is allowed. The addition of bleach improves the disinfection properties of the solution, making it effective against nonenveloped viruses, including panleukopenia and FCV.
- Concentrations stronger than a 1:32 dilution can result in respiratory irritation for both animals and people, as well as increased facility corrosion, and are therefore, not recommended for routine use.
- At a dilution of 1:10, bleach will destroy dermatophyte spores.
- The use of calcium hypochlorite is becoming more common and offers the potential advantages of reduced contact time and a neutral pH, which prevents corrosion.

Hypochlorites

Hypochlorite-based products include liquid (sodium hypochlorite), solid or powdered (calcium hypochlorite) formulations. These formulations dissolve in water to create a dilute aqueous chlorine solution in which undissociated hypochlorous acid (HOCl) is active as the antimicrobial compound. Hypochlorite displays a broad spectrum of antimicrobial activity and is effective against several common pathogens at various concentrations. For example, hypochlorite is effective against rotavirus at a concentration of 0.05 per cent (500 ppm), however, higher concentrations of 0.5 per cent (5000 ppm) are required for some highly resistant pathogens in the health-care setting such as *C. auris* and *C. difficile*.

Because hypochlorite is an effective biocide at pH values lower than 8.5, the cleaning reaction rate of this compound is faster at a pH of 8 than at 10. A lower pH accounts for more of the hypochlorite in the form of hypochlorous acid, which diffuses into bacteria and carbohydrate residues faster than the hypochlorite ion, to increase the cleaning reaction rate.

Hypochlorite is rapidly inactivated in the presence of organic material; therefore, regardless of the concentration used, it is important to first clean surfaces thoroughly with soap and water or detergent using mechanical action such as scrubbing or friction. High concentrations of chlorine can lead to corrosion of metal and irritation of skin or mucous membrane, in addition to potential side-effects related to chlorine smell for vulnerable people such as people with asthma.

Commercial sodium hypochlorite products with different levels of concentration may be readily available for use in a variety of settings. Concentration may also vary according to national regulations and manufacturers' formulations. To achieve the desired concentration, it is necessary to prepare sodium hypochlorite by diluting the basic aqueous solution with a given proportion of clean, non-turbid water to produce the final desired concentration.

Chlorine can decay rapidly in solutions depending on the source of chlorine and environmental conditions, for example ambient temperature or UV exposure. Chlorine solutions should be stored in opaque containers, in a well-ventilated, covered area that is not exposed to direct sunlight. Chlorine solutions are most stable at high pH (>9).

Mode of Action of Free Chlorine

The exact mechanism by which free chlorine destroys microorganisms has not been elucidated. Inactivation by chlorine can result from a number of factors: oxidation of sulfhydryl enzymes and amino acids; ring chlorination of amino acids; loss of intracellular contents; decreased uptake of nutrients; inhibition of protein synthesis; decreased oxygen uptake; oxidation of respiratory components; decreased adenosine triphosphate production; breaks in DNA; and depressed DNA synthesis. The actual microbicidal mechanism of chlorine might involve a combination of these factors or the effect of chlorine on critical sites.

Sodium hypochlorite: Sodium hypochlorite (NaOCl) is one of the most widely used chlorine containing disinfectants in the food industry despite the increasing availability of other disinfectants. Sodium hypochlorite fulfills many requirements as the ideal disinfectant and furthermore it has an excellent cleaning action. Biocidal activity is determined by the amount of the available chlorine of the solution.

The effectiveness of sodium hypochlorite in the cleaning and disinfection processes depends on the concentration of available chlorine and the pH of the solution. Hypochlorous acid (HOCl) is a weak acid and dissociates to the hypochlorite ion (-OCl) and proton (H+) depending on the solution pH. It is generally believed that HOCl is the active species in the germicidal action, whereas the concentration of -OCl is a key factor determining the cleaning efficiency. This implies that the optimal pH region of the germicidal activity of sodium hypochlorite differs from that of its cleaning activity. Low concentrations (2 to 500 ppm) are active against vegetative bacteria, fungi and most viruses. Rapid sporicidal action can be obtained around 2500 ppm; however, this concentration is very corrosive so should be limited in its use. High concentrations are also irritating to the mucous membranes, eyes and skin. Chlorine compounds are rapidly inactivated by light and some metals so fresh solutions should always be used. Hypochlorites should never be mixed with acids or ammonia as this will result in the release of toxic chlorine gas.

Sodium dichloroisocyanurate, and chloramine-T.: Alternative compounds that release chlorine and are used in the health-care setting include demand-release chlorine dioxide, sodium dichloroisocyanurate, and chloramine-T. The advantage of these compounds over the hypochlorites is that they retain chlorine longer and so exert a more prolonged bactericidal effect. Sodium dichloroisocyanurate tablets are stable, and for two reasons, the microbicidal activity of solutions prepared from sodium dichloroisocyanurate tablets might be greater than that of sodium hypochlorite solutions containing the same total available chlorine. First, with sodium dichloroisocyanurate, only 50 per cent of the total available chlorine is free (HOCl and OCl^-), whereas the remainder is combined (monochloroisocyanurate or dichloroisocyanurate), and as free available chlorine is used up, the latter is released to restore the equilibrium. Second, solutions of sodium dichloroisocyanurate are acidic, whereas sodium hypochlorite solutions are alkaline, and the more microbicidal type of chlorine (HOCl) is believed to predominate. Chlorine dioxide-based disinfectants are prepared fresh as required by mixing the two components (base solution [citric acid with preservatives and corrosion inhibitors] and the activator solution [sodium chlorite]). In vitro suspension tests showed that solutions containing about 140 ppm chlorine dioxide achieved a reduction factor exceeding 10^6 of *S. aureus* in 1 minute and of *Bacillus atrophaeus* spores in 2.5 minutes in the presence of 3 g/L bovine albumin. The potential for damaging equipment requires consideration because long-term use can damage the outer plastic coat of the insertion tube. In another study, chlorine dioxide solutions at either 600 ppm or 30 ppm killed *Mycobacterium avium-intracellulare* within 60 seconds after contact but contamination by organic material significantly affected the microbicidal properties.

Iodine Compounds

Iodine compounds used in the food industry contain iodine complexed with polyvinylpyrrolidone and other surface-active agents, usually in an acid solution. These are known as iodophors and were first introduced in 1949.

The complexes formed between iodine and carrier molecule are water-soluble and overcome the handling difficulties of iodine whilst retaining the disinfecting power. On dilution, the iodophors release iodine gradually, and it is the free iodine that acts as the disinfecting agent. The optimum pH for microbial activity is pH 5.0.

The surface-active agents provide better wetting and organic soil penetration, thus making iodophors less affected by soil than hypochlorite. The choice of surface-active agent may lead to foam generation in applications such as CIP. Iodines are often formulated with soaps and considered relatively safe.

Iodophors have a broad antimicrobial spectrum that is similar to hypochlorite; they are considered effective for a variety of bacteria, mycobacteria, fungi and viruses although they are less active against bacterial spores. Halogens react not only with living microorganisms but also with their environment, that is, dead bacteria, dissolved proteins, and amino acids. Iodines function by denaturing proteins to interfere with the enzymatic systems of microorganisms. In common with sodium hypochlorite, they are fast-acting but are more expensive. Iodophors are used in soak baths and spray application at up to 10 ppm available iodine. Iodophors can be corrosive; it is therefore necessary to ensure that the correct dilution is used; otherwise, damage to plant and personnel may occur.

V. Synthetic Detergents

Detergents are synthetic cleansing agents made from hydrocarbonds obtained from petroleum fractions *i.e.*, petrochemicals. The major components of synthetic detergents serve essentially the same function as soap-emulsification of fats, oils, and greases-except that there is no reaction to cause a curd formation. **Detergents** contain surfactants to reduce surface tension between food soil and the surface so the detergent can penetrate quickly and lift off the soil from the surface.

The first synthetic detergent was developed by the Germans during World War I and started to develop rapidly since 1949. "Soaps are often replaced by detergents as it has higher washing characteristics in hard water. This is because detergents do not form scum with Mg2+ ions and Ca2+ ions in hard water but soaps do.

Detergents are amphilic molecules meaning that their structures consist of a polar + head, (hydrophilic) and a non-polar+ tail, (hydrophobic). Hydrophilic molecules can dissolve readily in water but not in oils and grease. In contrast, hydrophobic molecules can dissolve readily in oils and grease but not in water. Generally, detergents can be classified into three main types, depending on the charge on the detergent ion:

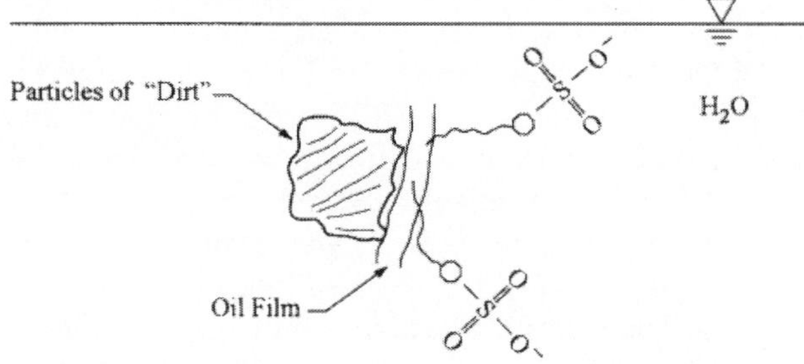

Figure 17.1: Detergent Molecule Forming a Link between Water and Oily Particle with Attached "Dirt."

i) Anionic Detergents

Where the head of the detergent particle contains a negatively-charged ion when in solution. Sulphates, sulphonates, carboxylates (soaps), and phosphates are some of the examples of anionic hydrophiles. They are the most common wetting agents in cleaning compounds because of their compatibility with alkaline cleaning agents and good wetting qualities.

ii) Cationic Detergents

Where the head of the detergent particle contains a positively-charged ion. Cationic hydrophiles are some forms of an amine product. They are normally considered sanitizers rather than wetting agents. They produce positively charged active ions in an aqueous solution. Detergents in this category are poor wetting agents, although they are strong bactericides.

iii) Non-ionic Detergents

Where there is no charged ion on the head of the detergent particle. The hydrophiles can be from hydroxyl group. Wetting agents have no charge associated with them when in aqueous solution. Therefore, they are effective under both acid and alkaline conditions. Their main problem is that they produce foam, which can cause complications in drainage and sewage systems. A cleaning compound does not have to foam to be an effective cleaner. One advantage of non-ionic wetting agents is that they are not affected by water hardness.

VI. Alkaline Soaps

Soap is made by a reaction between natural fats including oils and an alkali, such as lye. As soap dissolves, it reacts with water and the solution becomes alkaline. Some alkalinity is needed for good washing. Acid in soiled clothes lowers the alkalinity of the wash water and reduces the effectiveness of the soap. Minerals in hard water react with soap to form insoluble substances, called soap scum. Scum reduces the amount of soap available for washing, and tends to settle on clothes and give them a grey colour. More soap has to be added to do a good washing job.

Soaps are not popular in industrial cleaning because they are less effective in hard water and are generally inactivated by acid solutions.

VII. Enzyme-Based Cleaners

Over the years, enzymes have been an important factor in the development and improvement of detergent products. In laundering, dishwashing, and in industrial and institutional (I and I) cleaning, they have contributed to shortening washing times, reducing energy and water consumption by lowering washing temperatures, providing environmentally friendlier wash-water effluents, lowering pH levels in wash liquors, and providing fabric care. Enzymes themselves are environmentally attractive since they are derived from renewable sources. They are also highly space-efficient and are thus, of particular advantage in concentrated detergent formulations.

This largest of all industrial enzyme applications began slowly in the early 1930s, based on Röhm's 1913 patent of the use of pancreatic enzymes in pre-soak solutions. Pancreatic enzymes include protease (trypsin and chymotrypsin), carboxypeptidase, α-amylase, lactase, sucrase, maltase, and lipase. Thus, with the exception of cellulases, the foundations were already laid in 1913 for the commercial use of enzymes that continues to be important today. Today the most widely used industrial enzymes are hydrolases, which remove soils based on proteins, lipids, and polysaccharides. Cellulolytic enzymes are another class of hydrolases that provide fabric care through selective reactions not previously possible on fabrics. Research is currently underway into the possibility of using redox enzymes oxidases or peroxidases for bleaching coloured components.

From an enzyme point of view, detergents on the international market contain principal ingredients that operate by almost identical detergency mechanisms. Soil and stains are removed by mechanical action assisted by surfactants, builders, and enzymes. Alkaline proteases, amylases, or lipases in heavy duty detergents hydrolyze and solubilize substrate soil attached to fabrics or hard surfaces (*e.g.*, dishes). Cellulases clean by hydrolysis of glycosidic bonds which removes particulate soils attached to cotton microfibers. Key effects of cellulases are to soften and improve the colour brightness of worn textile surfaces. Surfactants lower the surface tension at interfaces and enhance the repulsive force between the original soil, enzymatically degraded soil, and the fabric. Builders act to chelate, precipitate or ion-exchange calcium and magnesium salts, to provide alkalinity, to prevent soil redeposition, to provide buffering capacity, and to inhibit corrosion.

These cleaners offer potential because they contain no chlorine or phosphates and are less corrosive than chlorine sanitizers. They can lower the pH of effluent. The disadvantages of enzyme-based cleaners are that liquid detergents require injection equipment and a two-part system activation, and they are not as effective on other soils as are chlorine sanitizers.

VIII. Solvent Cleaners

Solvent degreasing (or solvent cleaning) is the physical process of using organic solvents to remove grease, fats, oils, wax or soil from various metal, glass, or plastic items. The types of equipment used in this method are categorized as cold cleaners, open top vapor degreasers, or conveyorized degreasers. Nonaqueous solvents such as petroleum distillates, chlorinated hydrocarbons, ketones, and alcohols are used. Solvent selection is based on the solubility of the substance to be removed and on the toxicity, flammability, flash point, evaporation rate, boiling point, cost, and several other properties of the solvent.

A solvent-type cleaner is frequently required to remove grease type of soil deposit from equipment. This type of soil will not usually be found directly on processing equipment surfaces, but rather in the general area. Heavy-duty solvent cleaners are immiscible with water and frequently form an emulsion when water is added. Heavy-duty solvent cleaners are manufactured for use without water, whereas some solvent cleaners with low solvent content can be combined with water and still exhibit the grease-cutting action expected from a solvent.

Sanitizers

Before knowing about the various types of sanitizers used in food industries, it is important to understand the difference between disinfecting and sanitizing which is explained as under.

Disinfecting versus Sanitizing

Before discussing the chemicals, the differences between sanitizers and disinfectants as used in the food industry must be understood. To disinfect means to destroy or irreversibly inactivate specified infectious fungi and bacteria, but not necessarily the spores, on hard surfaces. To sanitize means to reduce microorganisms of public health importance to levels considered safe, based on established parameters, without adversely affecting either the quality of the product or its safety. While disinfection measures may be employed in food processing and preparation, it is much more common to utilize sanitization methods to reduce microbial presence.

To achieve the required level of sanitization or disinfection, the chemical in question must be applied at a certain concentration for a specified amount of time. These parameters are described on the product label and must be followed to achieve the desired microbial control. In most cases, these products are registered for use as pesticides with the U.S. Environmental Protection Agency (EPA). Once applied, the allowable residues and the monitoring thereof in food processing and preparation are the responsibility of the U.S. Food and Drug Administration. Facilities operating under the jurisdiction of the U.S. Department of Agriculture must additionally use products approved by that agency. Of course, the task of ensuring the chemicals is prepared and applied properly to avoid inappropriate residues rests with the food processor and foodservice operator. The efficacy of a chemical used for sanitizing or disinfection rests upon its ability to reduce the contamination level. The sanitization standard for contamination reduction of food contact surfaces is generally accepted as 99.999 per cent (a 5-log reduction) achieved in 30 seconds (Official Detergent Sanitizer Test). The sanitization standard for non-food contact surfaces is accepted as a reduction of 99.9 per cent (3 logs) within 30 seconds. Disinfection, in contrast, must destroy or irreversibly inactivate all specified organisms within a certain time, usually 10 minutes. Some chemicals may function as both sanitizers and disinfectants.

The process of sanitization depends upon the preparation of the surfaces in question. Most sanitizers must be applied to surfaces that are free of organic matter and cleaner residues. The generally accepted order of events is rinse, clean, rinse and sanitize. The cleaner utilized in the cleaning step needs to be oriented and appropriate for the soil present. For example, alkaline detergents more efficiently remove fat- and protein-based soils, while mineral-based soils require acid cleaners. Thankfully, modern cleaning agents are mixtures of chemical components that can address various cleaning scenarios. Good cleaners must be economical, nontoxic, noncorrosive, noncaking, nondusting, easy to measure or meter, stable during storage, and easily and completely dissolved.

Sanitizing

Sanitization follows cleaning. Sanitization is the application of heat or chemicals to a properly cleaned (and thoroughly rinsed) food-contact surface, yielding a 99.999 per cent reduction of representative pathogenic microorganisms of public health importance. Sanitization is not sterilization. Sterilization is the process of destroying all living microorganisms, not just pathogens.

Sanitizing Methods

Sanitizing is done using heat, radiation, or chemicals. Heat and chemicals are commonly used as a method for sanitizing in a food outlet; radiation rarely is. The item to be sanitized must first be washed properly before it can be properly sanitized. Some chemical sanitizers, such as chlorine and iodine, react with food and soil and so will be less effective on a surface that has not been properly cleaned. Common sanitizing methods are discussed as under:

Heat

There are three methods of using heat to sanitize surfaces *viz.*, steam, hot water, and hot air. Hot water is the most common method used in restaurants.

Heat is an inefficient sanitizer because it takes so much energy. The efficiency of heat depends on the humidity, the temperature required, and the length of time it takes to destroy microbes at that temperature. Heat destroys microorganisms if the temperature is high enough for long enough and if the design of the equipment or plant allows the heat to reach every area. Cleaning staff should use accurate thermometers to measure temperatures during cleaning to make sure that equipment and surfaces are properly sanitized. Steam and hot water are the most common types of heat used for sterilization.

Allowing a food contact surface to be exposed to high heat for a designated period of time will sanitize the surface. An acceptable method of hot water sanitizing is by utilizing the three-compartment sink. If a high-temperature ware washing machine is used to sanitize cleaned dishes, the final sanitizing rinse must be at least 180° F (82° C). For stationary rack, single temperature machines, it must be at least 165° F (74° C). Cleaned items must be exposed to these temperatures for at least 30 seconds. It is important to note that the surface temperature of the object being sanitized must be at 160°F for a long enough time to kill the bacteria. In three-compartment sinks, the water temperature must be at least 171 °F (77 °C).

Steam

Sanitizing with steam is expensive because of high energy costs, and it is usually ineffective. Workers often mistake water vapor for steam, so the temperature usually is not high enough to sterilize the equipment or surface. In case of heavy contamination, a cake is formed on the organic residues on the surface and steam cannot penetrate soil cake; as a result, heat does not reach the microbes; then they stay alive.

Hot Water

Small components such as knives, small parts, eating utensils, and small containers *etc.* can be immersed into water heated to 82°C (180°F) or higher. This is another way to sterilize using heat. Pouring hot water into containers is not a reliable way to sterilize, because of the difficulty of maintaining a water temperature high enough to ensure adequate sterilization. Due to hot water, denaturation of some of the protein molecules in the microbial cell may cause its loss of viability. Hot water is an effective, nonselective sanitizing method for food-contact surfaces; however, spores may survive more than an hour at 100°C. Hot water is frequently used for plate heat exchangers and eating utensils. Various plants utilize 15 minutes of exposure time at 85°C or 20 minutes at 80°C. A shorter time requires a higher temperature.

Radiation

Ultraviolet light units are now commonly used in Europe to disinfect drinking and food processing waters and are being installed in the United States. There are three different sources of ionizing radiation available for the treatment of food products. They are electron beam, (e-beam), gamma rays, and X-rays. Radiation does not penetrate well and its use as an antimicrobial agent should be restricted to microorganisms on sources, in the air, or in clear liquids.

Chemical Sanitization

Sanitizing is also achieved through the use of chemical compounds capable of destroying disease causing bacteria. Chemicals that are approved sanitizers are chlorine, iodine, and quaternary ammonium. Chemical sanitizers have found widespread acceptance in the food service industry. These compounds require labelling with the word "Sanitizer." The labelling should also include what concentration to use, data on minimum effective uses and warnings of possible health hazards.

Chemical sanitizing is performed in two ways; by full immersion or rinsing, swabbing, or spraying. For bleach, objects can be immersed in the three-compartment sink's sanitizer for 7 seconds or it can be wiped down with double the immersion concentration. This is due to the reduced contact time the surface is getting with the sanitizer. When using quaternary ammonium, the concentration should be between 150-400ppm for immersion and in-place cleaning.

Different factors influence the effectiveness of chemical sanitizers. The three factors that must be considered are:

- **Concentration:** Proper concentration is critical. Concentrations below the required amount will result in a failure to sanitize. Concentrations too high are considered toxic and poisonous. It is essential to monitor the sanitizing solution on a regular basis with the proper test strips.
- **Temperature:** The temperature of the solution is crucial. The water must be warm enough to increase the activity of the solution, but not so hot that it increases the evaporation of the sanitizer. Generally, temperatures between 75°F and 120°F allow sanitizers to work properly. At higher temperatures, chlorine compounds may corrode some metal items. Also at higher temperatures, chlorine and iodine compounds may leave the solutions. Generally chemical sanitizers work best in water that is between 55° F (13° C) and 120° F (49° C).
- **Contact time:** In order for the sanitizer to kill harmful microorganisms, the cleaned item must be in contact with the sanitizer (either heat or approved chemical) for the recommended length of time.

Ozone

The ozone molecule (O_3) is an antibacterial agent that is very effective at oxidizing and destroying organic and other compounds on equipment and surfaces. As of June 2001, ozone was approved by the FDA as an additive to kill foodborne pathogens. Because it is a gas, ozone leaves no toxic residues on treated surfaces. However, it could be corrosive to various surfaces at high concentrations, and care must be exercised during its generation because overexposure can result in bodily injury.

Electrolyzed Water

Electrolyzed water is a good sanitization method because it has antimicrobial properties, is not corrosive to skin, mucous membranes or organic material, is safe to handle and has little adverse effect on the environment. Electrolyzed water shows effectiveness against a wide range of microorganisms. It can be produced easily using common salt and an apparatus connected to a power source. Because the size of the machine is quite small, electrolyzed water can be manufactured on site. Although its cost is low, electrolyzed water can be corrosive to certain metal surfaces.

While heat and chemicals are currently the most commonly used sanitization methods, other new methods such as ozone and electrolyzed water show great promise for use in the retail/food service industries.

Sanitizer Testing

Every food industry must have the appropriate testing kit to measure chemical sanitizer concentrations. To accurately test the strength of a sanitizing solution, one must first determine which chemical is being used *i.e.*, chlorine, iodine, or quaternary ammonium. Test kits are not interchangeable so check with the chemical supplier to be certain that you are using the correct kit. The appropriate test kit must then be used throughout the day to measure chemical sanitizer concentrations.

Sanitizing Chemicals

Hypochlorites

Effectiveness, low cost and ease of manufacturing make hypochlorites the most widely used sanitizers. Sodium hypochlorite is the most common compound and is an ideal sanitizer, as it is a strong oxidizer. Hypochlorites cause broad microbial mortality by damaging the outer membrane, likely producing a loss of permeability control and eventual lysis of the cell. In addition, these compounds inhibit cellular enzymes and destroy DNA. Spores, however, are resistant to hypochlorites, as the spore coat is not susceptible to oxidation except at high concentrations coupled with long contact times at elevated temperatures. While hypochlorites are very reactive, their useful properties are negatively impacted by factors such as suspended solids, high temperatures, light, water impurities and improper pH levels. In routine use, surfaces must be as free as possible of organic materials, and the pH must be maintained between 5 to 7 to ensure that the greatest amount of hypochlorous acid is available. As with any sanitizer, measurements must be taken periodically to make certain that the freely available chlorine is at the desired level. For no-rinse applications, the maximum allowable concentration of available chlorine is 200 ppm.

Chlorine Dioxide

Chlorine dioxide plays a significant role in the industrial settings as disinfectants due to its broad antimicrobial property. Despite commonly use as germicide, chlorine dioxide demonstrates a good safety profile, rendering its suitability for use at water treatment and food preparation zones. Protein denaturation including envelope proteins is the major mechanism of chlorine dioxide to inactivate microorganisms even at low concentrations. Adverse reactions are not widely reported due to the typical use at a low concentration. The effectiveness of chlorine dioxide against various microorganisms, in both liquid and gaseous forms, over a wide range of pH and at an extremely low concentration has confirmed chlorine dioxide as a vital and versatile disinfectant.

This inorganic compound is a broad sanitizer effective against bacteria, fungi and viruses. Chlorine dioxide is an oxidizer that reacts with the proteins and fatty acids within the cell membrane, resulting in loss of permeability control and disruption of protein synthesis. While chlorine dioxide is an explosive gas, it is relatively safe in solution. It is produced on-site as it can't be compressed or stored commercially in gaseous form. Most chlorine dioxide generation is accomplished with complex systems. However, recent advances in formulation procedures allow the production of solutions of chlorine dioxide on-site without the use of expensive equipment.

Compared with hypochlorites, chlorine dioxide requires much lower concentrations to achieve microbial mortality. For example, a 5-ppm solution is effective as a sanitizer on food contact surfaces with a contact time of at least 1 minute. Further, disinfection can be achieved with 100 ppm using a contact time of 10 minutes.

Chlorine dioxide reacts more selectively with compounds present in microbial cells as opposed to reacting with organic compounds in general. This ability allows chlorine dioxide to function in more organically loaded solutions, though as organic load increases, efficacy does decrease. Chlorine dioxide functions well over a pH range of about 6 to 10, thus allowing increased mortality of some microbes at higher values. Another advantage is that chlorine dioxide does not form chlorinated organic compounds making it more environmentally friendly.

Iodophors

These compounds are less active than hypochlorites but are effective sanitizers and disinfectants. Iodophors attach to the sulfurs of proteins such as cysteine, causing inactivation and cell wall damage. Carriers with iodophor solutions allow a sustained release effect, resulting in continuous microbial mortality.

Iodophors fare better in situations in which the pH is slightly acidic, as less active forms exist above neutral pH. The common concentration for sanitization is 25 ppm for 1 minute. Unfortunately, iodine compounds easily stain many surfaces, particularly plastics. On the plus side, they are common sanitizers used on glass surfaces, such as in the beer and wine bottling industries.

Peroxyacetic Acid (PAA)

Peroxyacetic acid ($C_2H_4O_3$) is popular by various names as Peracetic Acid, Acetyl Hydroperoxide, Acetic Peroxide, Desoxon, Ethaneperoxic Acid, Peroxyethanoic Acid, Persan and Bioside but basically, it is an aqueous mixture of acetic acid and hydrogen peroxide. PAA is such a chemical that has been tried as sanitizer in food industry. It is emerging as better alternative to the chlorine sanitizers. Properties like low corrosiveness, greater microbial lethality, absence of foam and phosphates, tolerance to hard water and favourable biodegradability adds towards its choice as sanitizer. If proper safety measures are taken, PAA can prove to be a better option as sanitizer.

Table 17.3: Chemical Sanitizers Commonly Used in Food Service Establishments

Chemical Sanitizer	Concentration/ Contact Time	Advantages	Disadvantages
Chlorine	50 mg/L in water between 75 °F (24 °C) and 100 °F (38 °C) for 7 seconds	• effective on a wide range of bacteria • not affected by hard water salts • generally inexpensive	• corrosive, irritating to the skin • effectiveness decreases with increasing pH of solution • deteriorates during storage and when exposed to light • dissipates rapidly • loses activity in presence of organic matter
Iodine	Follow manufacturer's use directions; contact time at least 30 seconds	• forms brown color • not affected by hard water salts • less irritating to skin than chlorine • active against a wide range of non-spore-forming bacteria • activity not lost as rapidly as chlorine in presence of organic matter	• bactericidal effectiveness decreases greatly with increase in pH (most active at pH 3.0 and very low acting at pH 7.0) • less effective against bacterial spores and bacteriophages than chlorine • may discolor equipment and surfaces
Quaternary Ammonium Compounds	Follow manufacturer's use directions; contact time at least 30 seconds	• nontoxic, odorless, colorless, noncorrosive, nonirritating • stable in the presence of heat, relatively stable in the presence of organic matter • active over a wide pH range • active against thermoduric organisms	• slow destruction of coliform and psy-chrophilic organisms • incompatible with anionic detergents and hard-water salts • ineffective against bacteriophages

PAA is an effective sanitizer that is active against many microorganisms and their spores. Mortality is produced by the disruption of chemical bonds within the cell membrane. PAA-based sanitizers are frequently paired with stabilized hydrogen peroxide. These sanitizers function well under cold conditions (~ 4 °C), thus producing acceptable microbial mortality on equipment normally held below ambient temperature. PAA is also effective in removing biofilms and is more active than hypochlorites.

PAA solutions can be attenuated by the organic load and will begin to lose activity as the pH approaches neutral. These solutions are applied at concentrations ranging from about 100 ppm to 200 ppm for peroxyacetic acid and 80 ppm to 600 ppm for hydrogen peroxide.

The primary mode of action is oxidation, PAA disinfects by oxidizing of the outer cell membrane of vegetative bacterial cells, endospores, yeast and mold spores. The mechanism of oxidation is the transfer of electrons, therefore the stronger the oxidizer, the faster electrons are transferred to the microorganism and the faster the microorganism is inactivated or killed. PAA works by penetrating the cell wall and cell membrane and oxidizing the H-S and S-S bonds in the cell's enzyme. Microbes are then unable to function and die. Because the cells are destroyed from the inside-out via oxidation, they can't develop immunity or mutant to counteract it. The PAA is superior to even hydrogen peroxide in its antimicrobial activity.

Peracetic Acid appears to be an effective microbiocide for disinfecting equipment, seeds, plant materials and as a post-harvest treatment of fruits and vegetables. PAA is stabilized equilibrium solution that is approved for numerous uses, including circulation, cleaning and industrial sanitization of equipments such as tanks, pipelines, evaporators, fillers, pasteurizers aseptic machines, *etc.*

PAA-based sanitizers are environmentally friendly as the compounds therein break down into acetic acid, oxygen and water. These sanitizers are also less corrosive to equipment than hypochlorites. As with any highly active oxidizer, concentrated PAA can present a safety hazard. Limitations of PAA include not compatible with chlorine, consumed by sulfites and sulfides, reduced half life above pH 8.5, rapid decomposition in the presence of metal and cannot be stored in mild steel or contaminated containers.

Quaternary Ammonium Compounds (QACS or Quats)

Quaternary ammonium compounds are fairly complex chemicals in which nitrogen is bound to four organic groups. The positively charged cations in the compounds bind with the acidic phospholipids in the microbial cell wall. This action blocks the uptake of nutrients into the microbial cell and prevents the discharge of waste. In general, QACs are effective against a wide range of microbes, although the spore phase is unaffected. At lower concentrations, gram-positive bacteria are more sensitive to QACs than gram-negative bacteria. QACs are formulated in many different variations for specific situations.

QACs may be applied at concentrations varying from about 100 ppm to 400 ppm. As sanitizers, QACs are commonly applied at 200 ppm to food contact surfaces, and the solution is allowed to dry. Once dry, a residue of the QAC compounds remains and provides germicidal activity until degradation occurs. QACs also can function as detergents when present in high concentration because the compounds possess both hydrophilic and lipophilic chemical groups.

QACs are generally very stable, mostly unaffected by pH levels, and remain effective on a food-contact surface for a long time. Their antimicrobial activity is more selective than that of other disinfectants, they are inactivated by organic soil, and they should not be diluted in hard water. QACs are, however, generally very effective against bacterial biofilms. An example of a QAC is benzalkonium chloride, which is often used as a cleaner and sanitizer for various food surfaces, both at home and in industrial applications such as dairy equipment.

QACs are usually odourless, non-staining, noncorrosive and relatively nontoxic to users. They function well over a broad temperature range and a wide pH range, although activity is greater at warmer temperatures and in alkaline situations. While QACs tolerate light organic loads, heavy soil will decrease QAC activity significantly. Some QACs may not function adequately in hard water, but others are formulated with added chelating agents that allow such use.

The quaternaries commonly are used in ordinary environmental sanitation of noncritical surfaces, such as floors, furniture, and walls. EPA-registered quaternary ammonium compounds are appropriate to use for disinfecting medical equipment that contacts intact skin.

Chapter 18

HACCP: A State-of-the-Art Approach to Food Safety

Hazard Analysis Critical Control Points is a state-of-the-art preventive approach to safe food production. This concept is based on the application of prevention and documentation. HACCP is a proactive prevention program based on sound science.

For the food security of more than 1.3 billion people, India has to keep on producing more but safe food. Various techniques have been developed over the time for keeping a check on the quality of food. Of those, HACCP *i.e.*, Hazard Analysis and Critical Control Points is one which ensures the safety of food during the production or the processing of food. The major thrusts of HACCP are to determine how and where food safety hazards may exist and how to prevent their occurrence. The program focuses on safety and not quality and should be considered separate from or a supplement to quality assurance. The objective of HACCP is to ensure that effective sanitation, hygiene and other operational considerations be conducted to produce safe products and to provide proof that safety practices have been followed.

HACCP is based on seven principles and is a risk management system that identifies, evaluates, and controls hazards (biological, chemical and physical) related to food safety throughout the food supply chain. The concept first came to light in 1960s and in India it started early in 2000. HACCP is a protocol for all the stakeholders involved in food industry for the production of safe food. As the HACCP lays emphasis on "Clean Production", therefore, it ensures the use of every single resource whether it may be water, energy or any raw material in an efficient way leading to manufacture of safe food.

The objective of HACCP is to ensure that effective sanitation, hygiene and other operational considerations be conducted to produce safe products and to provide proof that safety practices have been followed.

What is HACCP and its History

The development and the initial use of HACCP can be traced to joint effort of the Pillsbury Company, together with the National Aeronautics and Space Administration (NASA) and the U.S. Army Laboratories at Natick. Their objective was to develop a strategy that would ensure that foods required for the space program were free from any unacceptable health risk. The Pillsbury Company, along with NASA and the US Army Natick laboratories pioneered the development of the HACCP approach for preventing unacceptable levels of food safety hazards. Pillsbury decided that their existing quality control techniques did not provide adequate assurance against contamination during food production. The company found that end-product testing necessary to provide such assurance would be so extensive that little food would be available for space flights. The only way to ensure safety, Pillsbury concluded, would be to develop a preventive system that kept hazards from occurring during production. Since then, Pillsbury's system has been recognised worldwide as the method of choice for control of food safety hazards. It is not a zero-risk system, but it is designed to minimize the risk of food safety hazards. The FDA first required HACCP-based controls for food processing in 1973 for canned foods to protect against *Clostridium botulinum*, the bacterium that causes botulism. HACCP was incorporated to guarantee that food used in the U.S. space program would be 100 per cent free of bacterial pathogens. Clark (1991) described HACCP as a simple but very specific method to identify hazards and for implementing the appropriate control to prevent potential hazards. Because it is designed to prevent and not detect food hazards, HACCP was identified by the U.S. Department of Agriculture (USDA) Food Safety and Inspection Service (FSIS) as a tool to prevent food safety hazards during meat and poultry production.

Table 18.1: Chronology of HACCP Development

Year	*Event*
1958	Foundation of the NASA (National Aeronautics and Space Administration)
1959	Development of the HACCP concept to assure one hundred per cent safety of food to be used in space.
1971	HACCP system was published and documented in the USA
1985	National Academy of Science (NAS) recommended the use of the system. Worldwide the system became used and the FAO/WHO Codex Alimentarius (Food and Agriculture Organisation/World Health Organisation) cited the system in the Codex.
1993	The European regulation 93/43 EG from 14.7.93 provides the use of the system for the production of food.
1997	The Codex Code on General Principles of Food Hygiene was revised and included recommendations for the application of the Codex HACCP Guidelines.
1998	With coming into force on the August the 8th of 1998 the Hygiene Verordnung (German Hygiene Rule) demands the use of the HACCP system in Germany.
2007	Integration of HACCP with ISO 9001 to formulate Food Safety and Standards Act (ISO 22000)

Since then, the HACCP system has become the internationally-recognised and accepted method for food safety assurance. In an assessment of the effectiveness of food regulation in the United States, the National Academy of Sciences (NAS) recommended in 1985 that the HACCP approach be adopted by all regulatory agencies and that it be mandatory for food processors. This recommendation led to the formation of the National Advisory Committee on Microbiological Criteria for Foods (NACMCF). This committee standardized the HACCP principles used by industry and regulatory authorities. The committee's work is the basis of this standardized curriculum. While it was originally developed to ensure microbiological safety of foodstuffs, it has been further broadened to include chemical and physical hazards in foods. The recent growing worldwide concern about food safety by public health authorities, consumers and other concerned parties, to a great extent due to WHO's advocacy in this field, and the continuous reports of food borne outbreaks have been a major impetus in the application of the HACCP system.

The HACCP concept has been embraced and recommended for use by several scientific groups. These include the National Academy of Sciences (NAS) Committee on the Scientific Basis of the Nation's Meat and Poultry Inspection Program and the NAS Subcommittee on Microbiological Criteria of the Committee on Food Protection. These two committees recognized HACCP as a rational and improved approach to food production control

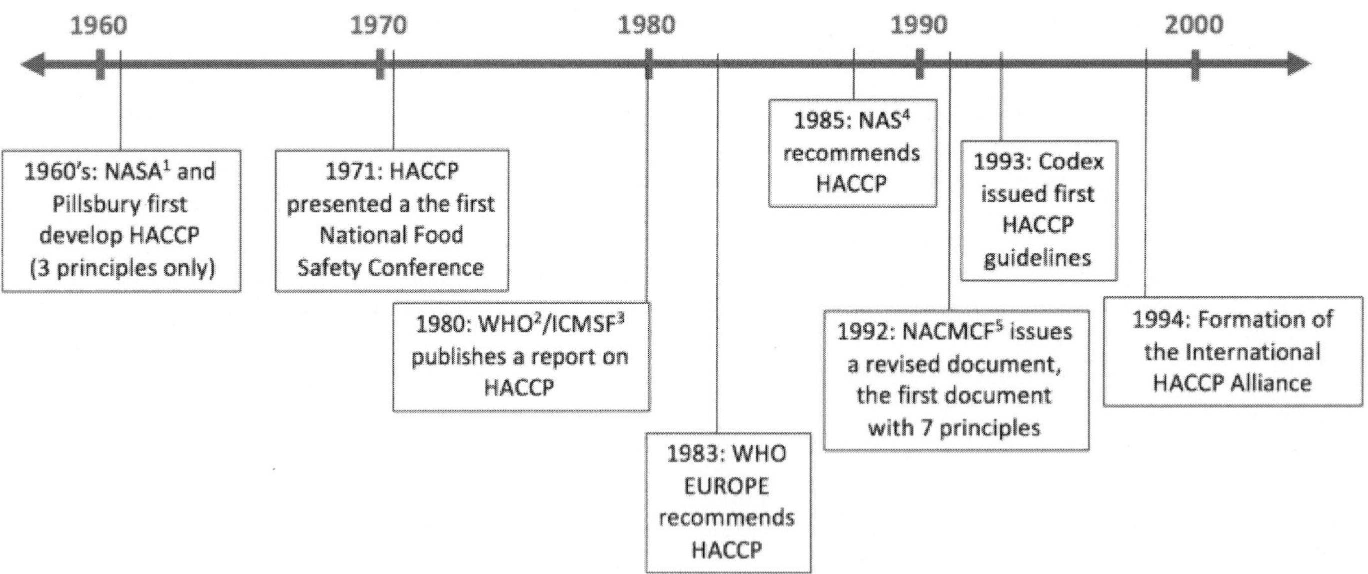

Source: Margaret D Weinroth, Aeriel D Belk, and Keith E Belk History, development, and current status of food safety systems worldwide. Anim Front. 2018 Oct; 8(4): 9–15. Published online 2018 Aug 30. doi: 10.1093/af/vfy016

that can determine those areas where control is most critical to the manufacturing of safe and wholesome food. This technique, which assesses the flow of food through the process, provides a mechanism to monitor these operations frequently and to determine the points that are critical for the control of foodborne disease hazards.

HACCP can be applied by all food businesses, large and small, and also to improving food safety control in developing countries and even to food safety control in the home. The principles of HACCP are applicable to all phases of food production, including basic husbandry practices, food preparation and handling, food processing, food service, distribution systems, and consumer handling and use.

The hazards that HACCP is designed to prevent, fall into two categories: microbiological and chemical. That is to say, HACCP only exists to aid in the control of these contaminants. And, admittedly, neither HACCP nor any other system is perfect. Therefore, future changes in HACCP or development of a new system will be dependent on scientific developments that point the way to a new system or improvements in HACCP. Because HACCP is, in fact, a system and not a scientific development, the basic structure of HACCP may not be improved on. What will change will be the science of contaminant detection and the means of reducing or destroying contaminants.

General Definitions

The HACCP system is a systematic set of activities used to control food production in order to ensure food safety and prevent changes in foodstuffs. The system is based upon the use of control practices in given production steps where there is a greater probability of occurrence of health hazards. The prerequisite programs for HACCP implementation in food industries are GMP and SSOP, which involve several aspects of the food industry, such as physical structure and maintenance, water supply, personal hygiene, pest control, sanitization techniques and equipment, calibration of instruments, and quality control of raw material and ingredients, among others (Barendz, 1998). In general, it is recommended that the system is applied to all steps of the food chain, from the production of raw material to the final product, including aspects related to consumer demands, such as processed products that do not have negative effects on their health (World Health Organization, 1997).

The HACCP system uses predetermined concepts and terms that include, according to Bryan (1992):

Hazard

Unacceptable biological, physical, or chemical contamination that renders food inadequate for consumption. A hazard is the potential to cause harm to the consumer. The HACCP concept has become a valuable program for process control of microbial hazards. This approach is a harbinger of the trend toward more sophistication in food sanitation and inspection. It has been legitimized by governmental regulators and is being adopted by progressive food companies.

- ✰ **Hazard analysis**: Hazard analysis requires a thorough knowledge of food microbiology and knowledge of which microorganisms may be present, and the factors that affect their growth and survival. Food safety and acceptability are most affected by:
  1. Contaminated raw food or adjuncts,
  2. Improper temperature control during processing and storage (time-temperature abuse),
  3. Improper cooling through failure to cool to refrigerated temperature within 2 to 4 hours,
  4. Improper handling after processing, cross-contamination (between products or between raw and processed foods),
  5. Ineffective or improper cleaning of equipment,
  6. Failure to separate raw and cooked products, and
  7. Poor employee hygiene and sanitation practices

Risk

Estimated probability of the occurrence of a hazard.

Control

(a) To manage the conditions of an operation to maintain compliance with established criteria.

(b) The state where correct procedures are being followed and criteria are being met.

- ✰ **Control Measure:** Any action or activity that can be used to prevent, eliminate or reduce a significant hazard.
- ✰ **Control Point:** Any step at which biological, chemical, or physical factors can be controlled.
- ✰ **Corrective Action:** Procedures followed when a deviation occurs.
- ✰ **Critical control point (CCP):** Production step where preventive measures are applied in order to maintain the given product under control, and to eliminate, prevent, or reduce risks to the health of the consumer. There are different types of CCPs according to the level of control of the hazard: CCPe, when hazards are eliminated; CCPp, when hazards are prevented;

and CPPr, when hazards are reduced, minimized, or delayed to significant levels.

- ☆ **Critical limit:** Value or attribute determined for each variable related to a critical point. Noncompliance leads to risks to consumer health. Critical limits are determined by guidelines or legal standards, specialized literature, practical expertise, previous surveys, internal company regulations, and other sources.
- ☆ **Corrective action:** Immediate and specific actions to be put into place when noncompliance with critical limits occurs.
- ☆ **CCP Decision Tree:** A sequence of questions to assist in determining whether a control point is a CCP.
- ☆ **Criterion:** A requirement on which a judgement or decision can be based.
- ☆ **Critical Control Point:** A step at which control can be applied and is essential to prevent or eliminate a food safety hazard or reduce it to an acceptable level.
- ☆ **Critical Limit:** A maximum and/or minimum value to which a biological, chemical or physical parameter must be controlled at a CCP to prevent, eliminate or reduce to an acceptable level the occurrence of a food safety hazard.
- ☆ **Deviation:** Failure to meet a critical limit.
- ☆ **HACCP:** A systematic approach to the identification, evaluation, and control of food safety hazards.
- ☆ **HACCP Plan:** The written document which is based upon the principles of HACCP and which delineates the procedures to be followed.
- ☆ **HACCP System:** The result of the implementation of the HACCP Plan.
- ☆ **HACCP Team:** The group of people who are responsible for developing, implementing and maintaining the HACCP system.
- ☆ **Monitor:** To conduct a planned sequence of observations or measurements to assess whether a CCP is under control and to produce an accurate record for future use in verification.
- ☆ **Prerequisite Programs**: Procedures, including Good Manufacturing Practices, that address operational conditions providing the foundation for the HACCP system.
- ☆ **Severity:** The seriousness of the effect(s) of a hazard.
- ☆ **Step:** A point, procedure, operation or stage in the food system from primary production to final consumption.
- ☆ **Validation:** Use of supplementary tests or review of monitoring records to determine if the HACCP system is functioning according to the plan. That element of verification focused on collecting and evaluating scientific and technical information to determine if the HACCP plan, when properly implemented, will effectively control the hazards.

Q1. Is control of this hazard necessary for food safety?
↓ YES → Go to Q2 | ↓ NO → Not a CCP

Q2. Are there preventive control measures for this hazard?
↓ YES → Go to Q3 | ↓ NO → Modification of the step

Q3. Does this step eliminate, prevent, or reduce the hazard risk to acceptable levels?
↓ YES → CCP | ↓ NO → Go to Q4

Q4. Can contamination occur or reach unacceptable levels?
↓ YES → Go to Q5 | ↓ NO → Not a CCP

Q5. Is there a subsequent step that will eliminate or reduce the hazard risk to acceptable levels?
↓ YES → Not a CCP | ↓ NO → CCP

- ☆ **Verification:** Those activities, other than monitoring, that determine the validity of the HACCP plan and that the system is operating according to the plan.

Principles and Application Guidelines Of HACCP

HACCP is a systematic approach to the identification, evaluation, and control of food safety hazards based on the following seven principles:

Principle 1: **Conduct a hazard analysis**: Potential hazards associated with a food and measures to control those hazards are identified. The hazard could be biological, such as a microbe; chemical, such as a toxin; or physical, such as ground glass or metal fragments.

Principle 2: **Identify the critical control points** (CCPs): These are points in a food's production—from its raw state through processing and shipping to consumption by the consumer—at which the potential hazard can be controlled or eliminated. Examples are cooking, cooling, packaging, and metal detection.

Principle 3: **Establish critical limits**: For a cooked food, for example, this might include setting the minimum cooking temperature and time required to ensure the elimination of any harmful microbes.

Principle 4: **Establish procedures to monitor the critical control points**: Such procedures might include determining how and by whom cooking time and temperature should be monitored.

Principle 5: Establish corrective actions: Establish corrective actions to be taken when monitoring shows that a critical limit has not been met—for example, reprocessing or disposing of food if the minimum cooking temperature is not met.

Principle 6: Establish verification procedures: Establish procedures to verify that the system is working properly—for example, testing time-and-temperature recording devices to verify that a cooking unit is working properly.

Principle 7: Establish record-keeping and documentation procedures: Establish effective recordkeeping to document the HACCP system. This would include records of hazards and their control methods, the monitoring of safety requirements and action taken to correct potential problems.

Each of these principles must be backed by sound scientific knowledge: for example, published microbiological studies on time and temperature factors for controlling foodborne pathogens.

Guidelines for Application of HACCP Principles

The following guidelines will facilitate the development and implementation of effective HACCP plans. These guidelines should be applied as appropriate to each segment of the food industry under consideration.

Prerequisite Programs

The production of safe food products requires that the HACCP system be built upon a solid foundation of prerequisite programs. Each segment of the food industry must provide the conditions necessary to protect food while it is under their control. This has traditionally been accomplished through the application of cGMPs. These conditions and practices are now considered to be prerequisite to the development and implementation of effective HACCP plans.

- Prerequisite programs provide the basic environmental and operating conditions that are necessary for the production of safe, wholesome food. The Codex Alimentarius General Principles of Food Hygiene describe the basic conditions and practices expected for foods intended for international trade.
- In addition to the requirements specified in regulations, industry often adopts policies and procedures that are specific to their operations. Many of these are proprietary.
- While prerequisite programs may impact upon the safety of a food, they also are concerned with ensuring that foods are wholesome and suitable for consumption. HACCP plans are narrower in scope, being limited to ensuring food is safe to consume.
- The existence and effectiveness of prerequisite programs should be assessed during the design and implementation of each HACCP plan.
- All prerequisite programs should be documented and regularly audited.
- Prerequisite programs are established and managed separately from the HACCP plan.
- Certain aspects, however, of a prerequisite program may be incorporated into a HACCP plan. For example, many establishments have preventive maintenance procedures for processing equipment to avoid unexpected equipment failure and loss of production. During the development of a HACCP plan, the HACCP team may decide that the routine maintenance and calibration of an oven should be included in the plan as an activity of verification. This would further ensure that all the food in the oven is cooked to the minimum internal temperature that is necessary for food safety.

Examples of Common Prerequisite Programs

Prerequisite programs provide the basic environmental and operating conditions that are necessary for the production of safe, wholesome food. Common prerequisite programs may include, but are not limited to:

Facilities: The establishment should be located, constructed and maintained according to sanitary design principles. There should be linear product flow and traffic control to minimize cross-contamination from raw to cooked materials.

Supplier Control: Each facility should assure that its suppliers have in place effective GMP and food safety programs. These may be the subject of continuing supplier guarantee and supplier HACCP system verification.

Specifications: There should be written specifications for all ingredients, products, and packaging materials.

Production Equipment: All equipment should be constructed and installed according to sanitary design principles. Preventive maintenance and calibration schedules should be established and documented.

Cleaning and Sanitation: All procedures for cleaning and sanitation of the equipment and the facility should be written and followed. A master sanitation schedule should be in place.

Personal Hygiene: All employees and other persons who enter the manufacturing plant should follow the requirements for personal hygiene.

Training: All employees should receive documented training in personal hygiene, GMP, cleaning and sanitation procedures, personal safety, and their role in the HACCP program.

Chemical Control: Documented procedures must be in place to assure the segregation and proper use of non-food chemicals in the plant. These include cleaning chemicals, fumigants, and pesticides or baits used in or around the plant.

Receiving, Storage and Shipping: All raw materials and products should be stored under sanitary conditions and the proper environmental conditions such as temperature and humidity to assure their safety and wholesomeness.

Traceability and Recall: All raw materials and products should be lot-coded and a recall system in place so that rapid and complete traces and recalls can be done when a product retrieval is necessary.

Pest Control: Effective pest control programs should be in place.

Other examples of prerequisite programs might include quality assurance procedures; standard operating procedures for sanitation, processes, product formulations and recipes; glass control; procedures for receiving, storage and shipping; labelling; and employee food and ingredient handling practices.

Education and Training

- ✰ The success of a HACCP system depends on educating and training management and employees in the importance of their role in producing safe foods.
- ✰ This should also include information about the control of foodborne hazards related to all stages of the food chain.
- ✰ Employees must first understand what HACCP is and then learn the skills necessary to make it function properly.
- ✰ Specific training activities should include working instructions and procedures that outline the tasks of employees monitoring each CCP.
- ✰ Management must provide adequate time for thorough education and training.
- ✰ Personnel must be given the materials and equipment necessary to perform these tasks. Effective training is an important prerequisite to successful implementation of a HACCP plan.

Major Target of HACCP

A major target of HACCP is *Listeria monocytogenes*. HACCP can help prevent the growth of *L. monocytogenes* because it requires steps to confirm the effectiveness of this concept. Samples should be taken from the food facility environment and product lots, to confirm that the control measures are effective. *L. monocytogenes* has been considered to provide the greatest hazard through environmental contamination. Therefore, most samples are tested from environmental sources. Environmental samples should be taken from ceilings, floors, floor drains, water hoses, equipment surfaces, and others on a random basis. Floor drains, which can carry organisms from a large area, should be tested routinely, using a rapid microbial method, such as immunoassay technology.

HACCP Development Plan

Essential steps to the development of a HACCP plan are:

1. **Assembly of an HACCP team**: Assembly of an HACCP team including the person responsible for the plan. Selections should include employees with expertise in sanitation, quality assurance, and plant operations. Also, it is desirable to have expertise in marketing, personnel management, and communications. HACCP should be organized as a part of the firm's quality assurance program.
2. **Description of the food and its distribution:** The name and other descriptors including storage and distribution requirements should be provided. All raw materials and adjuncts should be listed.
3. **Identification of the intended use and consumers of the food:** It is especially important to identify intended consumers if infants and other immunocompromised people are the targeted customers.
4. **Development of a flow diagram**.
5. **Verification of the flow diagram:** The HACCP team should inspect the operation to verify the accuracy and completeness of the flow diagram. Modifications should be made as necessary.
6. **Conduct of a hazard analysis:** a). Identify steps in the process where the hazards of potential significance occur. b). List all identified hazards associated with each step. c). List preventive measures to control hazards.
7. **Identification and documentation** of the CCPs in the process.
8. **Establishment of critical limits** for preventive measures associated with each identified CCP.
9. **Establishment of CCP-monitoring requirements,** including monitoring frequency and person(s) responsible for the specific monitoring activities.
10. **Establishment of corrective action** to be taken when monitoring reveals that a deviation

from an established critical limit exists. The action should include the safe disposition of affected food and the correction of procedures or conditions that caused the out-of-control situation.

11. **Establishment of procedures for verification** that the HACCP system is working correctly. Responsible company personnel should conduct verification of compliance with the HACCP plan on a scheduled basis.
12. **Establishment of effective recordkeeping** procedures that document the HACCP system and update the HACCP plan when a change of products, manufacturing conditions, and evidence of new hazards occurs.

Steps 6 through 12 are known as the seven HACCP principles that have been discussed above.

A food and its raw materials may be classified into categories as follows:

1. **Category 1.** Risk assessment as accomplished through examination of the food for possible hazards.
2. **Category 2**. Assignment of hazard categories through identification of general food hazard characteristics.

Determination of CCPs is also part of the development process. Not all steps in a process should be considered critical, and it is important to separate critical from non-critical points. A practical approach to determining CCPs consists of utilizing a HACCP worksheet with the following headings:

1. Description of the food product and its intended use.
2. Flow diagram with the following components:
   - ✰ raw material handling
   - ✰ in-process preparation, processing, and fabrication steps
   - ✰ finished product packaging and handling steps
   - ✰ storage and distribution
   - ✰ point-of-sale handling

As the flow diagram is made, it is easy to identify CCPs.

A CCP can be a location, practice, procedure, or process, and, if controlled, it can prevent or minimize contamination. Critical control points must be monitored to ensure that the steps are under control. Monitoring may include observation, physical measurements (temperature, pH, a_w), or microbial analysis and most often includes visual and physiochemical measurements because microbial testing is often too time-consuming. Possible exceptions are microbial analysis of the raw materials. Microbial testing may be the only acceptable monitoring procedure when the microbial status of the raw material is a CCP. Microbial methods can be incorporated to determine directly the presence of hazards during processing and in the finished product. They can be used indirectly to monitor effectiveness of control points for cleaning and employee hygiene. Yet, this use of microbiology is a check and does not have to be an ongoing process. Critical limits must be established for each monitoring procedure.

Monitoring must be verified by laboratory analysis to ensure that the process is working. The HACCP concept has been effective because

1. Cooperation existed between the government and industry to develop monitoring procedures for CCPs.
2. Education of processors is required.
3. Use of HACCP is fostered by government agencies.

Marriott *et al.* (1991) suggested that the following must be accomplished for HACCP to function effectively:

1. Food processors and regulators must be educated about HACCP.
2. Technical sophistication must be applied to HACCP through plant personnel.
3. Overuse of HACCP items that are not hazardous should be discontinued.

Developing and Implementing a HACCP Plan

The format of HACCP plans varies. In many cases the plans will be product and process specific. However, some plans may use a unit operations approach. Generic HACCP plans can serve as useful guides in the development of process and product HACCP plans; however, it is essential that the unique conditions within each facility be considered during the development of all components of the HACCP plan.

In the development of a HACCP plan, five preliminary tasks need to be accomplished before the application of the HACCP principles to a specific product and process. The five preliminary tasks are given in Figure 18. 1.

Assemble the HACCP Team

- ✰ The first task in developing a HACCP plan is to assemble a HACCP team consisting of individuals who have specific knowledge and expertise appropriate to the product and process. It is the team's responsibility to develop the HACCP plan.

- Selection criteria should emphasize production and quality assurance knowledge; however, marketing and communication expertise may be appropriate if these employees have an appreciation and understanding of the product and process.

Assemble the HACCP Team
↓
Describe the Food and its Distribution
↓
Describe the Intended Use and Consumers of the Food
↓
Develop a Flow Diagram Which Describes the Process
↓
Verify the Flow Diagram

Figure 18.1: Preliminary Tasks in the Development of the HACCP Plan.

- The team should be multi-disciplinary and include individuals from areas such as engineering, production, sanitation, quality assurance, and food microbiology.
- The team should also include local personnel who are involved in the operation as they are more familiar with the variability and limitations of the operation. In addition, this fosters a sense of ownership among those who must implement the plan.
- The HACCP team may need assistance from outside experts who are knowledgeable in the potential biological, chemical and/or physical hazards associated with the product and the process. However, a plan which is developed totally by outside sources may be erroneous, incomplete, and lacking in support at the local level.
- Due to the technical nature of the information required for hazard analysis, it is recommended that experts who are knowledgeable in the food process should either participate in or verify the completeness of the hazard analysis and the HACCP plan.
- Experts in the team should have the knowledge and experience to correctly: (a) conduct a hazard analysis; (b) identify potential hazards; (c) identify hazards which must be controlled; (d) recommend controls, critical limits, and procedures for monitoring and verification; (e) recommend appropriate corrective actions when a deviation occurs; (f) recommend research related to the HACCP plan if important information is not known; and (g) validate the HACCP plan.

Describe the Food and its Distribution

The HACCP team first describes the food. This consists of a general description of the food, ingredients, and processing methods. The method of distribution should be described along with information on whether the food is to be distributed frozen, refrigerated, or at ambient temperature.

Describe the Intended Use and Consumers of the Food

Describe the normal expected use of the food. The intended consumers may be the general public or a particular segment of the population (*e.g.*, infants, immunocompromised individuals, the elderly, *etc.*).

Develop a Flow Diagram which Describes the Process

- The purpose of a flow diagram is to provide a clear, simple outline of the steps involved in the process.
- The scope of the flow diagram must cover all the steps in the process which are directly under the control of the establishment.
- In addition, the flow diagram can include steps in the food chain which are before and after the processing that occurs in the establishment.
- The flow diagram need not be as complex as engineering drawings.
- A block type flow diagram is sufficiently descriptive. Also, a simple schematic of the facility is often useful in understanding and evaluating product and process flow.

Verify the Flow Diagram

The HACCP team should perform an on-site review of the operation to verify the accuracy and completeness of the flow diagram. Modifications should be made to the flow diagram as necessary and documented.

After these five preliminary tasks have been completed, the seven principles of HACCP are applied.

Implementation and Maintenance of the HACCP Plan

The successful implementation of a HACCP plan is facilitated by commitment from top management. The next step is to establish a plan that describes the individuals responsible for developing, implementing and maintaining the HACCP system. Initially, the HACCP coordinator and team are selected and trained as necessary. The team is then responsible for developing the initial plan and coordinating its implementation. Product teams can be appointed to develop HACCP plans for specific products. An important aspect in developing these teams is to assure that they have appropriate training. The workers who will be responsible for monitoring need to be adequately trained. Upon completion of the HACCP plan, operator procedures, forms and procedures for monitoring and corrective action are developed. Often it is a good idea to develop a timeline for the activities involved in the initial implementation of the HACCP plan. Implementation of the HACCP system involves the continual application of the monitoring, record-keeping, corrective action procedures and other activities as described in the HACCP plan.

Maintaining an effective HACCP system depends largely on regularly scheduled verification activities. The HACCP plan should be updated and revised as needed. An important aspect of maintaining the HACCP system is to assure that all individuals involved are properly trained so they understand their role and can effectively fulfill their responsibilities.

CGMPs: The Building Blocks for HACCP

On April 26, 1969, the FDA published the first Good Manufacturing Practice (GMP) regulations, commonly referred to as the umbrella The Current Good Manufacturing Practices (CGMPs). Emphasis is placed on the prevention of product contamination from direct and indirect sources. These regulations deal primarily with sanitation in manufacturing, processing, packing, or holding food. They are broad and general in nature and can be used to explain tasks that are part of many jobs.

The sanitary operations section establishes basic minimum rules for sanitation in a food establishment. General requirements are provided for the maintenance of physical facilities; cleaning and sanitizing of equipment and utensils; storage and handling of clean equipment and utensils; pest control; and the proper use and storage of cleaning compounds, sanitizers, and pesticides. Minimum demands for sanitary facilities are included through requirements for water, plumbing design, sewage disposal, toilet and hand washing facilities and supplies, and solid waste disposal. There is also a short section on education and training of employees. Specific GMPs supplement the umbrella GMPs and emphasize wholesomeness and safety of several manufactured products. Each regulation covers a specific industry or a closely related class of foods.

Good manufacturing practices in food industry are a quality control system that supervises the processes of production, storage, distribution and consumption of products. This system can be easily applied in many different sectors. The general characteristics that a production facility to be installed according to good production practices should carry are given below in the table.

Good Manufacturing Practices

Building, Ground and Environment Design	*Hygiene and Sanitation*	*Protection*
Water, ice and steam to be used at work	Cleaning and disinfection	Control of domestic animals
Liquid waste lines	Health check	Fighting harmful creatures
Social facilities and toilets	Disease notification	Staff education
Lighting	Personnel statements and behaviors	
Ventilation	Surveillance	
Storage and disposal of solid wastes	Visitors	
Workplace surroundings		
Raw material purchase locations		
Storage		
Lab		
Fuel tanks		

The increased expectation of safe food brought with it many applications. It is possible to increase them as good agricultural practices, good hygiene practices, good production practices. The best production practices, which is the oldest from these 37 applications, are a series of techniques that must be applied seamlessly in raw material, processing, product development, production, packaging, storage and distribution phases to ensure quality in products, which is the basis for the production and distribution of food products. According to Chen and Wang (2003), Good Manufacturing Practices are usually written for each of the following areas:

1. **Personnel:** These practices include direction for disease control, cleanliness, education and training, and supervision.
2. **Buildings and facilities:** The building surrounding grounds, plant construction design, and sanitary operations are included.
3. **Equipment and utensils:** All plant equipment and utensils should be designed of such a material and workmanship that facilitate adequate cleaning and maintenance.
4. **Production and process control:** Sanitation practices for production-related functions *i.e.*, inspection, storage, and cleaning of raw materials ingredients; and procedures for processing operations.
5. **Records and reports:** Records should include filing and maintaining for suppliers, processing and production, and distribution.
6. **Defect action levels:** These levels are defect limits at which the FDA will take action. The levels are set on the basis of no hazards to health.

7. **Miscellaneous:** These include other guidelines such as visitor rules.

However, this approach has limitations because laws that are supposed to be enforced by inspectors are frequently not clearly written, and what constitutes compliance is questionable. Furthermore, it is sometimes difficult to distinguish between requirements critical to safety and those related to aesthetics. In recent years, regulatory agencies have recognized these problems and revised their inspection procedures and forms. Now, many agencies have two major categories to differentiate between food safety items and aesthetic issues. There are critical deficiencies that address items that when left unattended could lead to foodborne illness and general deficiencies related to aesthetic items.

The European Food Authority, HACCP, Good Hygiene Practice (GHP) and GMP systems, which began its activities in 2003 in the EU, have also entered into force legally. This law covers candidate countries and third countries engaged in trade while taking responsibility for EU countries. Good production practices will facilitate the implementation and monitoring of the country's food industry in order to fully demonstrate its performance.

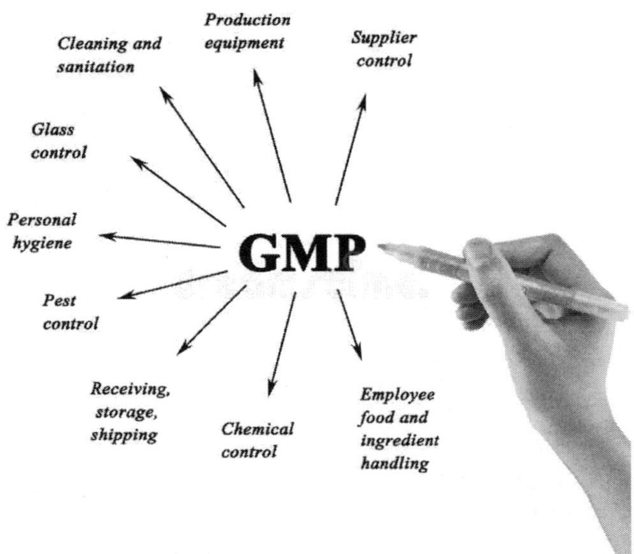

Good manufacturing practices should be selected and adopted before HACCP is implemented. Without the application of CGMP principles, an effective HACCP program cannot be conducted. Furthermore, CGMPs must be applied in the development of sanitation standard operating procedures (SSOPs). A CGMP compliance program should contain documented plans and procedures.

Good manufacturing practices and SSOPs are interrelated and an important part of process control. CGMPs are the minimum sanitary and processing requirements necessary to ensure the production of wholesome food. The areas that should be addressed through CGMPs are personnel hygiene and other practices, buildings and facilities, equipment and utensils, and production and process controls. CGMPs should be broad in nature.

Sanitation Standard Operating Procedures: The Cornerstones of HACCP

Although SSOPs are interrelated with CGMPs, they detail a specific sequence of events necessary to perform a task to ensure sanitary conditions. Standard operating procedures (SOPs) are either SSOPs or manufacturing SOPs. CGMPs should guide the development of SSOPs. SSOPs contain a description of the procedures that an establishment will follow to address the elements of preoperational and operational sanitation relating to the prevention of direct product contamination.

Written SSOPs contain a description of all cleaning procedures necessary to prevent direct contamination or adulteration of products. The frequency with which each procedure in the SSOPs is included along with a designation of the employee(s) responsibility for the implementation and maintenance through actual performance of such activities or that of the person responsible for ensuring that the sanitation procedures are executed.

Interface with GMPS and SSOPS

Sanitation SOPs are a prelude to HACCP. The intent of an HACCP plan is to ensure safety at specific CCPs within specific processes. Sanitation SOPs transcend specific processes. Sanitation SOPs are the cornerstones for an HACCP plan and can serve as a preventive approach to direct product contamination and/or adulteration.

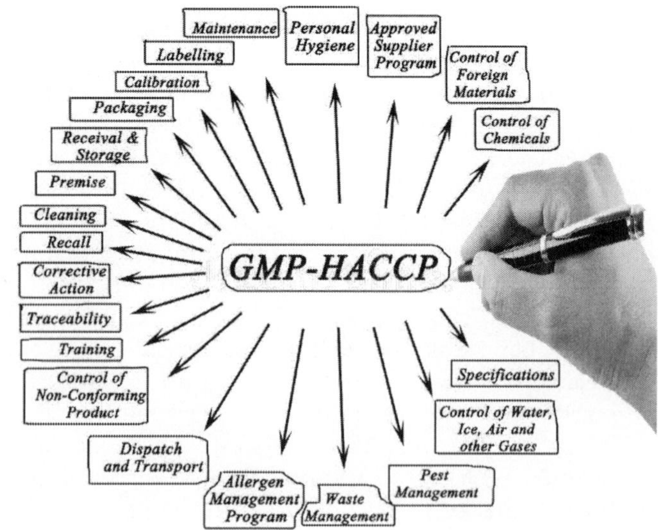

Advantages of HACCP

HACCP offers a number of advantages over the current system. Most importantly, HACCP:

- focuses on identifying and preventing hazards from contaminating food
- is based on sound science
- permits more efficient and effective government oversight, primarily because the recordkeeping allows investigators to see how well a firm is complying with food safety laws over a period rather than how well it is doing on any given day
- places responsibility for ensuring food safety appropriately on the food manufacturer or distributor
- helps food companies compete more effectively in the world market
- reduces barriers to international trade.

Need for HACCP in Future

New challenges to the world food supply have prompted us to consider adopting a HACCP-based food safety system on a wider basis. One of the most important challenges is the increasing number of new food pathogens. For example, between 1973 and 1988, bacteria not previously recognized as important causes of food-borne illness—such as *Escherichia coli* O157:H7 and *Salmonella enteritidis*—became more widespread.

There also is increasing public health concern about chemical contamination of food: for example, the effects of lead in food on the nervous system.

Another important factor is that the size of the food industry and the diversity of products and processes have grown tremendously—in the amount of domestic food manufactured and the number and kinds of foods imported.

The need for HACCP in future is further fuelled by the growing trend in international trade for worldwide equivalence of food products and the Codex Alimentarious Commission's adoption of HACCP as the international standard for food safety

Chapter 19

Food Borne Illness: Implications for the Future

One cannot see microbiologic infections as totally different from other food hazards or microorganisms as totally different from other life forms. We are all hosts and prey, parasites and predators. Even though we differ in size, complexity, and weaponry, we employ many similar strategies–the result of a shared planet, and a substantially shared genetic base. That is also why we have much to learn from animal disease surveillance.

Factors Contributing to Emergence

Outbreaks occur whenever pathogenic agents in sufficient number or quantity encounter a susceptible population without effective interceptive measures. Then, if we did not expect it, we say "it emerged."

Genetic Variability

The large genetic variability of microorganisms is the principal reason why so often some survive after any unfavourable environmental change. Some strains are hypermutable, which reinforces the potential for survival, and have very short generation times, with bacterial minutes comparable to human years. As Dr. Lederberg notes, microorganisms are opponents with whom we cannot race–on their terms.

Environment

Environmental factors also contribute to emergence. Hot, humid climates favour the growth of fungi and the production of mycotoxins. To borrow an example outside food borne pathogens, an unusually wet season produced a sharp increase in the deer mice population and the consequent outbreak of hantavirus in the Four Corners area of the United States.

Behaviour

Human actions and behaviour directly affect food safety. People are vectors for disease, traveling far more often, farther, and more rapidly than ever before, and moving far more swiftly than rats, lice, and mosquitoes. Political boundaries frequently and perversely act as leaky sieves, letting diseases through unimpeded, while blocking measures for disease prevention, control, and treatment.

Urbanization

Urbanization is a major factor in emergence. Crowding increases human contact and opportunities for transmission. Particularly in developing countries, public health services lag far behind the rush from farm to city. Cities, especially in industrialized nations, are economic and governmental centers and harbour institutions of culture and learning. However, cities are also massive projects in the intensive monoculture of humans. With agriculture, it is entirely possible to carry out monoculture effectively and productively–aesthetic considerations aside–if one provides for and tightly controls all essential inputs and conditions, monitors the process closely, and is prepared for prompt and effective intervention if something goes wrong. That hardly describes cities anywhere.

Raw Food Production

The effect of changes in raw food production and harvest practices on opportunities for foodborne outbreaks have been discussed. Centralized processing and wide distribution are the principal characteristics of the "new scenario," as compared to the "old scenario" of local production, home processing, and intrafamily consumption. A classic example is the recent Japanese outbreak of hemorrhagic *Escherichia coli.*

Denial

A behaviour that encourages outbreaks is denying the existence of an epidemic, a practice more common in developing countries concerned about the effect of outbreak publicity on tourist trade and exports. One of our own attitudes is indifference to outbreaks perceived as commonplace and distant.

Economics

War and economic collapse provide unparalleled opportunities for disease outbreaks (*e.g.*, cholera in central Africa). The infrastructure that provides clean water, community medicine, disease surveillance, and food control, even where it exists, is a fragile fabric, easily torn by economic, social, and physical disruption.

Technology

In spite of their benefits, technologies often bring new or enlarged risks. This is not an argument for returning to a state of nature. The invention of sausage doubtless increased the incidence of botulism; indeed, "botulus" is the Latin word for sausage. Without proper processing, any modern packaging that excludes oxygen can have the same effect.

Risk Factors

Factors such as age, illness, and medical treatment increase the risk for foodborne illness. Such increases also result from behaviour that promotes the incidence of other diseases (*e.g.*, AIDS).

Failure to Prevent and Control

The most common human action that adversely affects food safety is the avoidable lack of or failure to use effective prevention and control measures. That failure is why 85 per cent of all outbreaks are traceable, about equally, to mishandling in homes or in food service establishments.

Interacting Factors

In much of the developing world, an interrelated and mutually reinforcing set of problems keeps food borne disease at a high level. Approximately five million children under the age of 5 years living in the tropics die each year of malnutrition and diarrheal disease. Palliation is temporary; only economic and technical development can break through this net.

The contributing factors already mentioned would cause us problems, even acting singly. However, they interact, often synergistically. The combination of bacterial genetic variability and the ease and frequency of mutation of strains present a threat because the process enhances selection for new and more dangerous pathogens. Acid rain and recycling through ruminants may have encouraged the increased environmental durability of acid-tolerant *E. coli* O157: H7. The increasing popularity of marinades in the preparation of foods may have had the same effect.

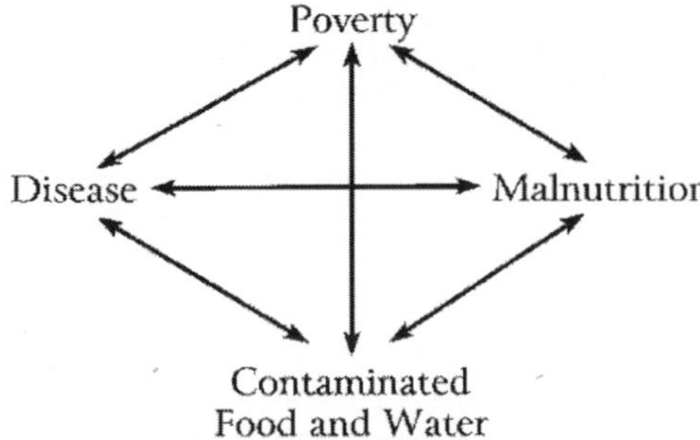

Figure 19.1: Problems Causing Food Borne Disease in Developing Countries.

The globalization of the food trade pulls together several of these contributing factors. One country's contaminated water leads to another country's contaminated raspberries. Refrigeration and controlled atmosphere can preserve pathogens, as well as foods, and spread them all over the world. The *Salmonella* Enteritidis outbreak from contaminated eggs in ice cream was a typically broad problem.

Technologic change combines many of these contributing factors. The intensive monoculture of plants and animals presents concentrated opportunities. For example, an outbreak of southern corn blight was attributed to the narrow genetic base of a popular hybrid corn. Human-guided plant genetics was defeated by the much more rapid genetic adaptability of microorganisms.

Indiscriminate use of antibiotics and the ability of microorganisms to exchange genetic information have led to increasing resistance. Minimal processing of food, warmly received by natural food lovers, is an open invitation to slightly more durable pathogens to take over the food supply. Immunosuppression, due to disease or medication, combined with the failure to take simple sanitary or health precautions makes the emergence of pathogens inevitable.

The Role of Public Health Measures

Achieving optimum public health requires many measures, each necessary but not sufficient in itself. The most basic measure is clean water, as recent incidents with strawberries and other fresh produce washed with contaminated water illustrate.

Effective food control structures (*e.g.*, statutory and regulatory frameworks and inspection and enforcement agencies) are essential. A surveillance system has always been required and should be directed to populations at high risk. Newer methods of typing pathogens, faster response times, and enhanced analytical sensitivity are making those systems far more effective than they have been.

Family health programs and education are equally necessary. Food safety still depends heavily on how each of us chooses and uses (or abuses) our food and each of us is not expert.

The Roles of Food Processing

Food is processed because of the following main reasons: nutrition, safety, preservation, distribution, and aesthetics. Cooking typically increases protein digestibility and destroys antinutrients. Fortification and enrichment add, restore, or increase essential nutrients. Cooking and canning destroy pathogenic organisms; cooking and steeping remove natural toxicants, such as cyanide from cassava. Canning, dehydration, salting, smoking, and preservatives, combined with proper packaging and storage, decrease spoilage. Canning, freezing, refrigeration, and controlled atmosphere storage provide variety and year-round availability. Flavours, colours, pre-sale preparation, and stabilizers that prevent separation or crystallization increase consumer appeal and convenience.

Food Processing Technologies

Food processing technologies reduce our exposure to dietary pathogens in three ways, summarized by Professor E. M. Foster as "the three Ks": keep them out, kill all one can, keep the rest from growing.

One group of technologies removes or destroys microorganisms and naturally occurring toxicants by careful sanitation; heat treatment (*e.g.*, cooking, retorting, pasteurizing, high temperature/short time treatment, ultrahigh temperature treatment, ohmic heating, and other newer techniques); radiation (widely useful and recently applied to all of our current methods); physical separation, (*e.g.*, air separation, gravity tables, visual scanners, and other methods for reducing "natural and unavoidable defects" including gross contamination); dehydration; and new technologies (*e.g.*, hydrostatic pressure, pulsed light).

A second group of technologies keeps contaminants well below dangerous levels. These include raw material quality control; careful sanitation; chemical preservation (*i.e.*, fermentation, bacteriostats, pH control, water activity control, and controlled atmosphere storage); dehydration; and freezing and refrigeration.

The third group of technologies prevents recontamination through sanitation of the general environment and at the point of service, protective packaging, and proper handling to ensure packaging integrity.

Beyond these three groups of technologies are general principles governing the use of all such techniques. All foods require incoming quality control and careful sanitation. For nearly all foods except fresh produce and dried grains, one technology from each group listed above should be used. All technologies used must be applied effectively, or none is effective. These technologies must be used in combination, in a systems approach. Failure to apply properly a preceding technology can cause later technologies to increase microbiologic hazards. If oxygen is later excluded from improperly processed food by packaging (canning, foil, an oil layer), anaerobic pathogens, such as *Botulinum*, are free to multiply.

Future Options

Food safety objectives must be established. There is a clear need for priorities and for employing the "principle of commensurate effort" (*i.e.*, applying effort to risks in the order of their probable impact).

There is also the need for a broad view. The interaction of several factors, not all of which can be seen in advance, makes a team approach essential.

More effort at understanding the evolution of virulence could provide us with insights on the genetic characteristics and the environmental conditions needed to minimize risk. The 1918-19 influenza epidemic can be used as an example; the present search for its genetic makeup suggests the value of prior knowledge. Knowing or estimating the probability of present or future virulence could prepare us for the more effective use of other measures. These include newer, faster, more sensitive methods of detection.

Improved detection would lead to more effective prevention and control measures. The need to establish priorities suggests here, as in cancer research, the value of biomarkers. Biomarkers are genetic or biochemical indicators of impending risk in the potential pathogen, preceding clinical or epidemiologic indications. Such biomarkers might indicate potential virulence, increased environmental durability, or increased resistance to heat, cold, unfavourable pH, preservatives, or antibiotics.

There is a need to apply available knowledge more effectively. Most outbreaks of foodborne disease are due to mishandling food in ways we already know how to avoid. The points that follow are not new, but as our food supply increases in complexity and geographic reach, these weapons must grow apace.

Clean water, public sanitation, disease surveillance, and food control are more important than ever. The value of effective disease surveillance has been demonstrated. But what we now have and do is not enough. A national response team must be created. Data should be shared more broadly (electronic linkage of laboratory and surveillance results) if we are to cope effectively with a global food supply, multistage processing by different firms at different locations, and national and international distribution. The effective application of these resources requires a Hazard Analysis and Critical Control Points (HACCP) approach.

Private health measures must grow in effectiveness and reach. We must make continuing and increasingly effective use of food processing technologies that reduce microbiologic risks. HACCP cannot be generic or static; it must be adapted to each specific product and processing facility and must be updated with continuing feedback on the hazards to be avoided. HACCP, fortified by the public health measures just described and by the results of the research discussed at the conference, will enable the food processing industry to meet the challenges of emerging pathogens.

Even with the steps just described, approximately 85 per cent of all outbreaks occur as a result of food mishandling in food service establishments or homes. There is a need to extend HACCP principles to the food service sector as well. HACCP probably cannot reach into homes. If we are to communicate better, we need first to find out what consumers already know, what they want to know, and what they need to know. In short, if we are to educate effectively, we must have direct evidence of how well the information process is working.

Finally, one can never succeed in achieving desirable consumer risk-management practices until consumers understand the inevitability of some risks and minimize them. Food risks have been categorized into six groups in decreasing order of size: microbiologic, nutritional, natural toxicant, environmental contaminant, pesticide residues, and food additives. The take-home message on food safety for all of us, as consumers, can be summed up in three words: sanitation, variety, moderation. Sanitation deals effectively with microbiologic risks; variety and moderation deal with nutritional risks, minimizing the impact of the four remaining risks to the extent they are even potentially insignificant. Sanitation, variety, and moderation are within our own control as individual consumers. That is doubtless why we have done only a mediocre job of implementing them. Unless all of us can and do pursue these three objectives effectively, we are all at some unnecessarily increased risk.

Identifying and Controlling Emerging Foodborne Pathogens: Research Need

The microbiologic safety of food has been advanced substantially by the introduction and implementation of Hazard Analysis and Critical Control Points (HACCP). HACCP provides a systematic conceptual framework for identifying hazards and focusing efforts on the proper functioning of key food production, processing, and marketing steps. When applied appropriately, HACCP is a cost-effective means of controlling known hazards in foods. Its successful implementation depends on knowledge of such issues as the pathogenic microorganisms' virulence, cultural characteristics, ways in which they contaminate the food, effects of food processing and preparation on their survival, and food consumption patterns. Because it requires substantial knowledge, HACCP cannot be expected to control unknown hazards, such as emerging foodborne pathogens. Therefore, controlling a new foodborne microbial threat requires moving the hazard as quickly as possible from being unknown to being known. The key to this transition is the timely acquisition of needed research data.

Anticipating the Next Emerging Pathogen

Two types of emergence are encountered with pathogenic foodborne microorganisms. A true emergence, where a microorganism that had not been identified as a public health threat begins to cause disease, is relatively rare. More common is re-emergence, where a known microorganism causes disease in a new way, for example, by causing new types of infections, being associated with new foods, or appearing in new geographic locations. For both types, the operational requirement is to control an unanticipated public health threat. The timeliness of response is critical since the public health and economic costs of an emerging pathogen are directly related to the time between its emergence and its control.

The events that lead to emergence are often complex, with the cause often being obscure and only indirectly related to the new agent. Past emergence of foodborne pathogens has been associated with changes in microbial genotypes, demographics, food production and processing methods, marketing and preparation practices, medical diagnostics, globalization of the food industry, changes in consumer education, and general socioeconomic trends. Planning for a microbial threat is a challenge because one does not know what the agent will be, what food it will be associated with, or where or when emergence will occur. While there are several potential ways of anticipating and responding to microbial threats, the contingency planning used by the military to anticipate threats seems well suited for emerging pathogens. Military contingency planning can be viewed as having four major components: intelligence, personnel and facilities, rapid response, and strategic planning.

Intelligence is the gathering of medical, scientific, and other information that allows emergence to be identified. In the United States, this role is filled to a great extent by the Centers for Disease Control and Prevention (CDC). In addition to providing information on known foodborne pathogens, CDC works with local public health agencies and the medical and scientific communities to investigate new disease syndromes and identify unrecognized food borne pathogens. This type of intelligence gathering played a pivotal role in the recent recognition of *Cyclospora* as a cause of food borne gastroenteritis. CDC's new sentinel site program, FoodNet, is expected to greatly enhance the identification of new food borne diseases. However, these surveillance activities are largely limited to the United States, whereas an effective intelligence system for food borne disease must be worldwide. For example, *Cyclospora* was identified as a likely food borne or waterborne pathogen in Asia and South America before an outbreak was reported in the United States. Intelligence related to food borne disease can be acquired from several sources: the World Health Organization's surveillance program, the U.S. military's international network of laboratory and medical investigators, medical and scientific reports, and the Internet. The Internet is increasingly an important source of intelligence related to emerging pathogens; through news groups and bulletin boards such as ProMed, scientists and public health practitioners share their experiences on almost a real-time basis. Such advances in intelligence gathering are critical to reducing the time between emergence and control. However, limiting intelligence to medical considerations is not enough: intelligence gathering must include awareness of changes and advances in food production methods, agricultural practices and conditions, veterinary medicine, environmental and water microbiology, food technologies, consumer trends, and general socioeconomic conditions.

The second component of contingency planning is ensuring sufficient personnel and facilities to characterize a new biologic agent and develop control strategies. The inability to predict the agent or the associated food, coupled with the degree of specialization required of investigators, requires a broad range of capabilities and resources. However, no one organization is likely to maintain the capabilities needed to deal with all contingencies. If we were to follow the military pattern, we would have reserve groups that could be mobilized as needed. However, even this approach requires planning and support to ensure the needed expertise and facilities. For example, the number of researchers and laboratories studying *Clostridium botulinum* has dropped to a point where it would be difficult to rapidly mobilize a research team, despite this pathogen's history of re-emerging in a surprisingly wide range of foods.

Rapid mobilization of resources is the third component. This component is particularly important for free-living infectious agents because one goal is to limit their dissemination to prevent them from establishing secondary reservoirs. It is much easier to fight a small, contained war than a global one. The mobilization of resources to respond to an emergence must be appropriate to the severity of the threat. Overreacting hurts the credibility of the entire system, while underreacting increases both the public health and economic impact. Rapid response efforts have focused at identifying new agents and removing suspect food from the marketplace, two key initial steps. However, research to prevent another occurrence of the emerging pathogen has been much less organized and timely.

The fourth component of contingency planning, strategic planning, is actually the first chronologically. This is the phase where members of war colleges pose "what we would do if" scenarios and plan appropriate responses. This type of contingency planning has generally received attention in relation to emerging pathogens only in connection with the use of biologic warfare agents. This process relies on futurist thinking to consider how changes in society, economics, technology, agriculture, medicine, and international trade may affect the microbiologic safety of the food supply. Such a broad view is needed because more general events or trends in society cause most disease emergence. This type of strategic planning is undertaken with the realization that the probability of any specific "what if" scenario is low, but the probability that one scenario will materialize is extremely high.

Research Needs

Research, an integral part of responding to a new food borne microbial threat, is the key for moving a new or re-emerging biologic agent from being an unknown pathogen to being one for which control measures are available. Two areas of research can be classified on the basis of time constraints. Acute research needs are deficiencies in knowledge that must be addressed to establish control of an emerging pathogen. This research is highly targeted and specific for the microorganism and food of concern; it must be accomplished as quickly as possible. Acute needs generally require applied research, although basic research may have to be conducted if the deficiencies in knowledge are great. The second class encompasses longer-term basic and applied research needs not mandatory to immediate control.

Acute Needs

While the data needed for any single emerging biologic agent are highly specific, acute research needs fall into general categories that are virtually the same for all new pathogens. Common research questions include the following: Are methods available for detecting and categorizing the agent? What food is the vehicle for the pathogen? How do the implicated foods become contaminated? What is the pathogen's reservoir in nature? Is the pathogen's presence in contaminated food the result of an error or breakdown in normal controls? Does the pathogen grow in foods? Does the pathogen survive normal food processing, distribution, and preparation? How infectious/toxigenic is the pathogen? Are there subpopulations of consumers at increased risk for this pathogen? Is the pathogen's ability to cause disease restricted to specific strains with identifiable virulence characteristics? Answering these questions requires specific data that do not differ substantially from pathogen to pathogen (Table 19.1).

Table 19.1: Research Data Needed for Most Emerging Food Borne Pathogens

Research Area	*Knowledge Gaps*
Detection methods	Sampling and enrichment techniques
	Cultivating
	Biochemical/taxonomic char.
	Antibodies for capture and differentiation
	Subtyping
	Virulence-associated char.
	Detecting injured or viable-but-nonculturable cells
Microbial ecology	Contaminated foods
	Reservoirs and routes of transmission
	Life cycles
	Geogr. range and seasonality
	Route of contamination and location of pathogen in food
Pathogenicity	Dis. char. and diagnosis
	Sequelae
	Host range
	Infectious dose
	Subpopulations at risk
	Animal models
Growth characteristics	Free-living vs. obligate parasite
	Growth requirements
	1. Temperature
	2. pH
	3. Water activity
	4. Oxygen
Survival characteristics	Heat resistance
	D-values
	Z-values
	Susceptibility to antimicrobial food additives
	Acid resistance
	Sensitivity to disinfectants or dessication
	Sensitivity to radiation
	1. UV
	2. Ionizing
Control	Effectiveness of food preservation
	Inspection systems to segregate contamination materials

The criteria for classifying needs as acute are reasonably straightforward: Is the research needed to prevent a recurrence of the disease or to modify current HACCP plans? However, these questions have different priorities, which depend on when the information is needed. To deal with emerging pathogens, we should learn from modern business practices, especially the concept of "just-in-time" research. Little consideration has been given to how to assess and set research priorities for emerging foodborne pathogens.

The timeliness of addressing research needs must be an integral part of the planning process, but has been generally overlooked. Past research mobilization efforts for new food borne microbial threats can be best described as haphazard, likely because they reflect the way research is funded. The traditional means of ensuring strong research programs, competitive funding of projects after proposals have undergone extensive peer review, is time-consuming and often not appropriate for the acute phase of responding to an emerging food borne pathogen. Further, the peer-review process tends not to select the often mundane research needed during the acute phase of an emergence. Two alternative approaches may be more effective. The first is to have a series of designated laboratories that have as part of their mission and funding the task of being able to modify their research programs to address acute research needs. Such laboratories would need to have a critical mass of facilities and expertise in various aspects of food safety microbiology. The second approach is to have a group of reserve scientists with unique expertise or access to facilities not available at the designated laboratories or needed to supplement those capabilities. Funds could be earmarked to noncompetitively fund such reserve scientists on an as-needed basis, with the understanding that research needs designated as acute would take precedence over other research needs.

Longer-Term Needs

The three areas of longer-term research associated with emerging pathogens are amenable to more traditional funding. The first area, specific to the new pathogen, consists of research for improvements or alternatives to the detection and control methods initially devised. With initial disease control established, basic and applied research can seek to understand the microorganism and develop more optimal approaches for its prevention, control, or elimination. The second area concerns activities to help reduce the time between the emergence of a pathogen and its initial control (*e.g.*, improved surveillance through the development of new diagnostic methods and further identification and characterization of virulence determinants and modes of pathogenicity to accelerate detection of new agents). Just as important as acquiring research data is rapid data dissemination. The continuing development of computer-based information networks is a component of this second research area.

The third area focuses on identifying research factors that will allow new microbial threats to be anticipated. Of necessity, the current response to emerging pathogens is almost entirely reactive. The public health community detects a new syndrome, and only then is research-mobilized, often during a crisis. While reactive response will always be part of dealing with emerging microbial threats, a more proactive approach is needed if prevention is to be even partially realized. In military terms, war is the last resort and represents the failure of diplomats to predict and prevent a crisis. Microbial threats, like wars, do not spontaneously emerge but are the result of a series of events or conditions. There is a need to reexamine how food is produced, processed, marketed, and prepared to identify conditions that contribute to emergence. For example, organic acids are used extensively throughout the food industry to control spoilage and pathogenic microorganisms. Archer (1996) hypothesized that over time, exposure to pH conditions that stress but do not kill may lead to the emergence of hardier and possibly more virulent food borne pathogens. It is already well established that the induction of acid tolerance can enhance both the survival and virulence of food borne pathogens. Further, one of the basic tenets of microbial genetics is that conditions that kill most, but not all, of a bacterial population foster the development of resistance. This is supported by recent studies that suggest that bacterial stress responses may select for hypermutability. While these findings do not mean that organic acids should not be used as a tool for controlling food borne pathogens, they suggest that proactive research should be conducted to find ways of using these agents that minimize the potential for resistance. Proactive research, including research that might appear unrelated to the emergence of food borne pathogens, can draw on the already substantial body of basic research related to the conditions and requirements for gene transfer among biologic agents. For example, Baur *et al.* (1996) reported on the conditions that led to the competence of *Escherichia coli* for genetic transformation in freshwater environments. Maximal competence occurred when the bacterium was exposed to 2 mM Ca_{2+} as temperatures increased from 10°C to 20°C. With such information, researchers could examine food processing operations to determine the presence and importance of such conditions. For example, fruits and vegetables are often treated with calcium under fluctuating temperatures to enhance the texture during later processing.

A key to being more proactive in addressing the threat of microbial food borne pathogens–consideration of root causes–will likely require food microbiologists to become involved in nontraditional research areas. If new biologic agents arise as the result of changes in technology, society, or global economics, predicting and preventing emergence will ultimately require better understanding of how such factors influence pathogen introduction and dissemination.

Overall, one of the critical lessons of the past 10 years is that we cannot become complacent about infectious diseases. Only a few diseases (*e.g.*, smallpox) have actually been eliminated. The rest, including virtually all food borne diseases, we hold in check, winning battles but not the war. Eventually, our weapons (*e.g.*, antibiotics) become obsolete; pathogens (*e.g.*, *E. coli*) become more dangerous; or we become complacent. Contingency planning must be developed and undertaken with a long-term commitment. Without that commitment and without understanding that planning is successful when problems are avoided or minimized, programs of this type lapse quickly. In the long term, the costs of planning, both in terms of economics and human suffering, are a fraction of those incurred as the result of the emergence of a major microbial threat.

Chapter 20

Food Borne Disease Control: A Transnational Challenge

In the past two to three decades, public health authorities in various countries have been faced with an increasing number of food safety problems. In 1983, a Joint Food and Agriculture Organization/World Health Organization Expert Committee on Food Safety concluded that illness due to contaminated food was perhaps the most widespread health problem in the contemporary world and an important cause of reduced economic productivity. More recent data from industrialized countries indicate that annually up to 10 per cent or more of the population may have a food borne disease. The situation is equally serious in developing countries, where infant diarrhoea causes many illnesses and deaths. In addition to known food borne diseases, public health communities are being challenged by the emergence of new or newly recognized types of food borne illnesses, often with serious and chronic health consequences. Certain populations (*e.g.*, pregnant women, the elderly, infants and children, immunocompromised persons, and the undernourished) are particularly vulnerable. In economic terms, food borne illnesses are very costly for industry, health services, and society as a whole.

Many factors have contributed to the increase in food borne disease. Industrialization, leading to increased wealth and urbanization, has revolutionized the food supply system, resulting in mass production and an explosive increase in the number of food service establishments and food outlets. Mass production, environmental factors, and inadequate knowledge on the part of food handlers have contributed to increased contamination of primary foodstuffs.

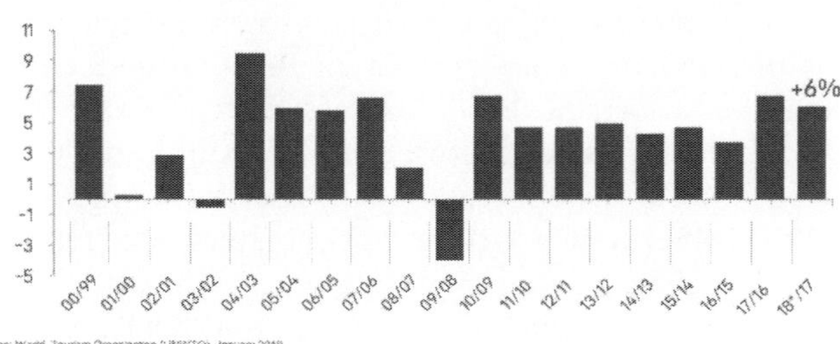

The increase in international trade has increased the risk for cross-border transmission of infectious diseases. The globalization of food (and feed) trade, facilitated by the liberalization of world trade, while offering many benefits and opportunities, also presents new risks. Food, a major trade commodity, is also an important vehicle for transmission of infectious diseases. Because food production, manufacturing, and marketing are now global, infectious agents can be disseminated from the original point of processing and packaging to locations thousands of miles away. This multinational approach to food production and distribution and the progressive opening up of world markets have allowed the international food trade to flourish. The value of food trade, U.S. $266 billion in 1994, was more than 300 per cent greater than it was 20 years ago and continues to grow rapidly.

The globalization of food borne diseases also results from increased travel. International travel is more accessible today. According to The World Tourism Organization, world tourist arrivals were 567 million in

1995, and more than 9 per cent growth in it occurred during 2003-04; 2018 was the second best year sinch 2010 as far as growth (in per cent) for International tourist arrivals was concerned. Over the past 200 years, the average distance travelled and the speed of travel have increased 1,000 times while incubation periods for diseases have not changed. As a result, a person can be exposed to a food borne illness in one country and expose others to the infection in a location thousands of miles from the original source of the infection. Depending on their destination, travellers are estimated to run a 20 per cent to 50 per cent risk of contracting a food borne illness.

As international trade and travel increase, food borne disease outbreaks of the same origin are more likely to occur in different parts of the globe. Food safety in the late 20th century represents a transnational challenge requiring enhanced levels of international cooperation in setting standards and regulations and in strengthening surveillance systems. Effective food safety programs, built on a clear understanding of the epidemiology of food borne disease, must be developed and implemented. The globalization of the world's economy has been accompanied by intense economic competition and increased pressure on governments to downsize. Public sector austerity has reduced disease surveillance in many countries. For example, in Great Britain, the failure to maintain public health infrastructures has, in the words of the British Medical Association, resulted in "Britain returning to the 19th century in terms of public health, with problems such as dirty water, contaminated food, and old infectious diseases re-emerging". Failing a reversal of this trend, public health authorities and health services may be overwhelmed in the near future by outbreaks or epidemics of food borne diseases. The 1991 epidemic of cholera in Peru and the 1996 outbreak of *Escherichia coli* O157 in Japan demonstrate how one single food borne disease epidemic or outbreak may disrupt the functioning of a health-care system.

Epidemiologic surveillance of food borne illness is fundamental to the planning of food safety programs and the development of a strategy for prevention and control. There are different methods of surveillance: death registrations and hospital discharges; disease notification; laboratory-confirmed cases; sentinel surveillance; surveillance of investigated outbreaks; population-based surveillance; and case-control studies of sporadic cases.

Food borne Disease Awareness of Public Health Authorities

Data on the incidence of food borne illnesses collected through notifications, laboratory confirmations, and sentinel or population-based studies can provide a measure of the magnitude of the food borne disease problems, their economic consequences, and over the years, an indication of the trend. Although several weaknesses are associated with the collection of such data particularly those collected through notification and laboratory confirmations (since they represent only the tip of the iceberg) they can nevertheless be useful in raising the awareness of public health authorities about the importance of food safety.

Surveillance data collected in some industrialized countries confirm that food borne diseases constitute one of the most widespread health problems and that they have increased over the last two or three decades of the increase may be attributable to recent improvements in information reporting and collection systems, improved diagnoses, or greater publicity and concern about food safety in general. However, a real increase of food borne disease incidence is not disputed. First, the increase has been steady and cannot be explained by a one-time improvement in the surveillance system. Second, increases have been observed in different countries, including those with no improvement in reporting and surveillance programs. The general increase, as demonstrated by the results of surveillance data, has led many public health authorities to take stringent regulatory and educational measures to improve food safety, with some successful results. For instance, in the United States, active surveillance of food borne listeriosis has led to concerted efforts by industry and government to prevent the disease. As a consequence, the number of cases and deaths has decreased by 44 per cent and 48 per cent, respectively.

Public health authorities must be aware of the magnitude and trend of food borne illness so that necessary resources can be mobilized to improve food safety programs. Lack of reliable epidemiologic data in many parts of the world has impeded the recognition of the public health importance of food safety and consequently the emphasis on food safety programs.

Early Detection of Food Borne Disease Outbreaks

Surveillance of food borne diseases plays an important role in the early detection of food borne disease outbreaks and their control. Early identification of the source of the outbreak is becoming increasingly important as countries move towards industrialization. Increased mass production means outbreaks can change from being small and confined to a family to large, affecting hundreds or even thousands of people.

Table 20.1: Examples of Large Food Borne Disease Outbreaks

Country	*Year*	*Disease*	*No. of Cases*
United Kingdom	1985	Salmonellosis	1,000
United States	1985	Salmonellosis	>168,000
United States	1993	Salmonellosis	224,000

Country	*Year*	*Disease*	*No. of Cases*
China	1988	Hepatitis A	>310,000
Germany	1993	Salmonellosis	1,000
Australia	1991	Norwalk-like agent	>3,050
United States	1992-93	*E. coli* 0157 infection	>500
Japan	1996	*E. coli* 0157 infection	>6,000

Rapid investigation of food borne disease outbreaks is crucial to prevent them from taking on massive proportions. In the 1993 French outbreak of listeriosis due to potted minced pork (affecting 39 persons and causing eight miscarriages and one death), public health authorities traced its source within 1 week and thus prevented the outbreak from spreading by removing the implicated food product from the market and informing the group at risk about its unsafe nature. In an outbreak of botulism in the United Kingdom traced to hazelnut yogurt, the source was identified within 3 days, and the product was withdrawn from the market.

Because of global food distribution and worldwide travel, an international exchange of information on food borne disease incidences and outbreaks and the foods involved is extremely important to identify international clusters originating from a common source. For instance, Salm-Net, a network for the international surveillance of human salmonellosis, has demonstrated the value of such an interactive international collaboration. Individual countries with apparently isolated outbreaks can feed their information into the network and ascertain whether the outbreak is confined to their country or is of wider international importance. The identification and investigation of several international outbreaks have been simplified by the Salm-Net network.

Food: The Transmission of Diseases, and the Identification of Associated Risk Factors

Information collected through investigation of food borne disease outbreaks or case-control studies of sporadic cases provides a better understanding of the role of food in the transmission of communicable diseases and in the identification of risk factors leading to disease. Epidemiologic data from food borne disease surveillance can provide public health authorities with important information about the types of food implicated in outbreaks; populations at risk; practices that lead to contamination, growth, and survival of food borne pathogens; and places where foods are often mishandled. Such data are essential for designing effective intervention programs. Such programs in industrialized countries, for example, have demonstrated the relatively greater prevalence and incidence of food borne diseases of microbial origin over those of chemical origin and the role of food handlers in the transmission of diseases; they have identified campylobacteriosis and salmonellosis (particularly infections caused by *Salmonella Enteritidis*) as the leading food borne diseases. The emergence of other diseases, such as infections due to *E. coli* 0157 and *Listeria monocytogenes*–often with serious sequelae–has been pinpointed as a major public health problem. These surveillance programs have also alerted public health authorities to the foods most often implicated and the major risk factors in food preparation.

Because of the lack of epidemiologic data, the role of food in the transmission of diseases has been poorly acknowledged, particularly in developing countries. Diarrhoeal diseases in infants and children and diseases such as shigellosis and cholera have been perceived as being water-borne for many years. For instance, after the cholera epidemic in Peru (where epidemiologic investigations implicated, among other foods, seafood, and an embargo was placed on trade in foodstuffs), the role that food plays in the transmission of the disease began to be fully recognized.

Increased trade in food, international travel and migration, and economic and technologic development have changed dietary habits. New foods, food preparations, and dietary habits are introduced into different regions, and as a consequence, food borne diseases are emerging or re-emerging. Dietary habits are also changing as a result of nutritional recommendations and campaigns or may be influenced by food policy, production systems, or environmental changes that lead to increased access to certain foods. These changes in dietary habits influence the epidemiology of food borne illnesses and contribute to the emergence of food borne diseases. In the United States, public information campaigns promote an increased consumption of fruits and vegetables. To meet the increased demand, these products have to be imported on a seasonal basis. At certain times of the year, more than 75 per cent of the fresh fruits and vegetables available in grocery stores and restaurants are imported. Epidemiologic data have shown that, partly as a consequence of the increased consumption of fruits and vegetables, the proportion of food borne disease outbreaks has doubled.

Data collected through food borne disease surveillance programs permit the monitoring of changes in the epidemiology of food borne diseases and the identification of new pathogens and new dietary or food preparation habits that may present a health risk. The data can also determine if existing programs need to be readjusted to ensure that the food safety program is adequate and relevant.

A method used in recent years to complement epidemiologic data in identifying risky practices and behaviour is the Hazard Analysis and Critical Control Points system (HACCP). Application of HACCP to food preparation permits the identification of practices that may be potentially hazardous and need to be modified or those that are critical for ensuring the safety of foods and require specific monitoring. However, the first principle of HACCP–to conduct a hazard analysis–calls for epidemiologic data on food borne diseases, as the process involves an appraisal of the possibility of hazards and the severity of their effects; the qualitative and quantitative evaluation of the presence of hazards; the survival and multiplication of microorganisms of concern; the production or persistence of toxins, chemicals, or physical agents in foods; and, conditions leading to the above.

Access to information would be difficult without epidemiologic surveillance of food borne diseases. Similarly, epidemiologic data are also needed to develop sampling plans of food, as demonstrated in the decision tree for *Listeria monocytogenes* sampling plans of foods.

Planning and Evaluating the Effectiveness of Food Safety Programs

The collection of epidemiologic data is important in planning interventions and setting priorities. Countries with scarce resources, facing an abundant number of food borne diseases and food safety problems, need to prioritize food safety issues. Epidemiologic data provide a basis for identifying food borne diseases, groups at risk, or even priority points in the food chain.

Evaluating the effectiveness and impact of an intervention is an important element of any plan. Data collected through disease notification or sentinel studies permit an evaluation of the effectiveness of interventions and their impact on health, and if necessary, the adjustment of a program to improve its efficacy and impact. Data on the rising incidence of food borne illnesses in many countries demonstrate that present prevention strategies, mainly based on regulatory measures, are inadequate and emphasize the need for additional measures (*e.g.*, additional regulatory initiatives and health education about food safety).

Risk Assessment and International Food Standards

The movement of ever-increasing quantities of food across borders has resulted in a transnationalization of disease risk. Therefore, the globalization of food trade and the open access to foreign markets need to be accompanied by effective means of health protection for populations. In the food sector, international regulatory instruments need to be integrated with strengthened surveillance and monitoring.

As a result of the Uruguay Round of Multilateral Trade Negotiations and the increased liberalization of trade facilitated by this agreement, concern about the safety of imported food has grown. However, provisions in the Agreement on the Application of Sanitary and Phytosanitary Measures, which entered into force with the establishment of the World Trade Organization on January 1, 1995, are designed to address these concerns: according to the work of the Codex Alimentarius Commission, its standards, guidelines, and recommendations are recognized as the reference for national food safety requirements. Countries that are members of the World Trade Organization may no longer be able to reject foods that meet Codex standards, guidelines, and recommendations without providing justification.

Moreover, the increased volume of the global food trade underscores the need for sound epidemiologic information and international risk assessment. In this regard, Article 5 of the Sanitary and Phytosanitary Measures agreement explicitly requires World Trade Organization members to conduct scientific and consistent risk assessments. Furthermore, the World Health Organization has recommended that the application of the HACCP system at every stage of the food chain represents an effective approach for governments to meet the terms outlined in the agreement.

Another issue receiving more attention from regulatory agencies and underlined during the Food and Agriculture Organization/World Health Organization Conference on Food Standards, Chemicals in Food, and Food Trade (1991), is the scientific basis of the Codex standards. The Conference recommended that the Codex, in its norm-setting work on health and safety, place greater emphasis on risk assessment. Epidemiologic data on food borne diseases have an important role in risk assessment. One example is assessing the risk of contracting listeriosis associated with different levels of *Listeria monocytogenes* in smoked fish and meat products. However, the need for risk assessment as the basis for setting standards has shown a great gap in knowledge about food borne pathogens and their relation to human illness. To address the national/transnational risks caused by food borne diseases, this gap must be narrowed.

Risk Assessment Approach

Risk assessment is defined as a scientifically based process that has the following steps:

Hazard Identification

The identification of biologic, chemical, and physical agents present in a particular food or group of foods that can cause illness.

Because only some food borne disease outbreaks are adequately investigated and have the etiologic agents identified, many food borne pathogens remain unidentified. Most of the available epidemiologic data are furnished by industrialized countries, while the situation in developing countries is largely unknown. The epidemiologic database must be extended to include information from developing countries. However, investigation and surveillance systems in developing countries need to be strengthened before the database can expand.

Hazard Characterization

The qualitative or quantitative evaluation of the nature of the illness associated with biologic, chemical, and physical agents that may be present in food. For chemical agents, a dose-response assessment should be performed. For biologic or physical agents, a dose-response assessment should be performed if the data are obtainable. For many food borne pathogens, dose-response data are limited or non-existent. Information on which dose-response estimates can be based is difficult to obtain and may be inaccurate for various reasons: host susceptibility to pathogens is highly variable, attack rates from a specific pathogen may vary widely, virulence of a pathogenic species is highly variable, pathogenicity is subject to genetic variation resulting from frequent mutation, antagonism from other bacteria in foods or the digestive system may influence pathogenicity, and foods may modulate the ability of bacteria to infect or otherwise affect the host.

Exposure Assessment

The qualitative or quantitative evaluation of the likely intake of biologic, chemical, and physical agents in food as well as exposures from other sources. An exposure assessment will give an estimate of either the number of pathogenic organisms or the level of toxins consumed in food. Although the levels of chemical agents in food may change only slightly due to processing, the population of bacterial agents is dynamic and may increase or decrease dramatically. Changes in populations of bacteria are affected by complex interactions of these factors: ecology of the bacterial pathogen; processing, packaging, and storing of food; preparation steps, such as cooking, which may inactivate bacterial agents; and cultural factors relating to consumers.

In addition, for some of the emerging food borne pathogens, the sources of exposure are still not fully understood. Information on food borne disease outbreaks provides an opportunity to learn about the types of foods that may harbour the pathogen.

Risk Characterization

The qualitative or quantitative estimation, including uncertainties, of the probability of and severity of known or potential illness in a given population on the basis of hazard identification, hazard characterization, and exposure assessment.

In many cases, data are not available to support a quantitative risk assessment of biologic hazards. We discuss next the types of challenges that make quantitative risk assessment difficult for pathogenic organisms associated with food and the role of epidemiologic surveillance.

Characterizing the risk associated with biologic pathogens depends on information gained in the previous steps. Risk characterization will result in a qualitative or quantitative estimate of the potential for adverse effects from a particular pathogen on a specific population. Whether a quantitative risk assessment approach is possible and appropriate for characterization of risks associated with food borne pathogens is not known. Thus, the qualitative approach to characterizing risk may be the only alternative.

International Travel

International travel and migration are contributing factors in the spread of food borne diseases in some countries. For instance, 80 per cent to 90 per cent of the incidence of salmonellosis in Scandinavian countries is attributed to international travel. Surveillance of travel-related food borne diseases provides a mechanism for appreciating the relative prevalence of food borne diseases in various countries. It also provides a basis for informing physicians and health services about unfamiliar diseases contracted by travellers returning from distant places. In this way, advice on precautionary measures can also be given to travellers. The only food borne disease now covered by the International Health Regulations is cholera, which is reported to the World Health Organization. Since the purpose of these regulations is to help provide maximum security against the international spread of diseases with a minimum of interference with world traffic (*i.e.*, trade and travel), it is timely to consider whether the regulations should cover additional food borne diseases.

Conclusively, the globalization of the risks associated with food borne illness, specifically increased international travel and trade in food, has resulted in greater interdependence in terms of food safety. Therefore, internationally agreed-upon food safety standards and other types of agreements are becoming increasingly important in addressing the complex transnational challenge of food borne disease control. Epidemiologic data provide a common ground for reaching international consensus on food safety issues.

As Morris Potter has said, "If one recognizes that ensuring food safety is inherently uncertain, food borne illnesses become opportunities to learn rather than failures to predict. Food borne disease will occur, and we must be prepared to react quickly to reduce the risk of new food borne hazards".

Bibliography

Abarca, M.L., Accensi, F., Bragulat, M.R., Castell'a G., and Caba ~nes, F. J. 2003. Aspergillus carbonarius as the main source of ochratoxin A contamination in dried vine fruits from the Spanish market. J. Food Protect. 66 (3):504–506.

Abdullah, B. M., Angor, M. M., Al-Ismai, K. M. and Ajo, R. Y. 2011. Reducing fat uptake during deep-frying of minced chicken meat-balls by coating them with some hydrocolloids materials. Italian Journal of Food Science, 23, 331–337.

Abhari, K., Jafarpour, D. and Shekarforoush, S. S. 2018. Effects of in-package pasteurization on prevening spoilage in emulsion vacuum packaged sausages during refrigerated storage. Foods and Raw Materials, 6(1), 40–46.

Abraham, N. M., Liu, L., Jutras, B.L., Yadav, A.K., Narasimhan, S., Gopalakrishnan, V., Ansari, J.M., Jefferson, K.K., Cava, F., Jacobs-Wagner, C. and Fikrig, E. 2017. Pathogen-mediated manipulation of arthropod microbiota to promote infection. Proc. Natl. Acad. Sci. USA, 114:E781–E790. doi:10.1073/pnas.1613422114.

Academy of Nutrition and Dietetics. 2014. Dietary and Metabolic Impact of Fruit Juice Consumption Evidence Analysis Project. Chicago: Academy of Nutrition and Dietetics; Available from: http://andevidencelibrary.com/topic.com?cat=5113.

Adam, K.H., Flint, S.H., Brightwell, G., 2010. Psychrophilic and psychrotrophic clostridia: sporulation and germination processes and their role in the spoilage of chilled, vacuum-packaged beef, lamb and venison. Int. J. Food Sci. Technol. 45, 1539e1544.

Adams, M.R., ýMoss, M.O. and ý McClure, P. 2015 · Food Microbiology 4th Edition. The Royal Society of Chemistry, USA.

Adams, M.R., Hartley, A.D. and Cox, L.J.1989. Factors affecting the efficiency of washing procedures used in the production of prepared salads. Food Microbiology, 6:69-77.

Adiletta, G., Senadeera, W., Liguori, L., Crescitelli, A., Albanese, D. and Russo, P. 2015. The influence of abrasive pretreatment on hot air drying of grape. Food and Nutrition Sciences, 6(3), 355–364.

Ahlawat, S. S. 2012. Development of low-fat emulsion based smoked sausages using different humectants. Journal of Dairying Foods and Home Sciences, 31, 306–310.

Ahmed, H. A., Mohamed, M. E. M., Rezk, M. M., Gharieb, R. M. A. and Abdel-Maksoud, S. A. 2018. *Aeromonas hydrophila* in fish and humans; prevalence, virulotyping and antimicrobial resistance. *Slovenian Veterinary Research*, 55, 113- 124.

Akbar, A. and Anal, A. K. 2014. Zinc oxide nanoparticles loaded active packaging, a challenge study against *Salmonella typhimurium* and *Staphylococcus aureus* in ready-to-eat poultry meat. Food Control, 38(4), 88–95.

Akýr, I.C. 2003. Determination of some probiotic properties on Lactobacilli and Bifidobacteria [Ph.D. thesis], Ankara University.

Alavarado, D.A. and Vera, J.M.R. 2013. Dynamics of the bioaccumulation of ciguatoxins in a food chain. Rev. Cuba. Investig. Biomed. 32, 271–283.

Ali, N. H., Farooqui, A., Khan, A., Khan, A.Y. and Kazmi, S.U. 2010. Microbial contamination of raw meat and its environment in retail shops in Karachi, Pakistan. Journal of Infection in Developing Countries 4: 382-388.

Allwood, M.C. and A.D. Russell. 1967. Mechanism of thermal injury in *Staphylococcus aureus*. I. Relationship between viability and leakage. Appl. Microbiol., 15: 1266–69.

Alm, L. 1982. Effect of fermentation on lactose, glucose, and galactose content in milk and suitability of fermented milk products for lactose intolerant individuals. J Dairy Sci 65, 346–352.

Almonds, B.B., Beetroot, G.E., Belitz, H.D., Grosch, W. and Schieberle, P. 2009. Aroma Compounds, fourth ed. Springer, Berlin (Chapter 5).

Amagliani, G., Brandi, G. and Schiavano, G. F. 2012. Incidence and role of *Salmonella* in seafood safety. Food Research International, 45(2), 780- 788.

An, S.Y., Lee, M.S., Jeon, J.Y., Ha, E.S., Kim, T.H., Yoon, J.Y., Ok, C.O., Lee, H.K.,Hwang, W.S., Choe, S.J. *et al.,* 2013 Beneficial effects of fresh and fermented kimchi in prediabetic individuals. Ann. Nutr. Metab. 63, 111–119.

Andersen, A.A. 1958. New sampler for the collection, sizing and enumeration of viable airborne particles. *J. Bacteriol.*, 76: 471- 84.

Andersen, B. and Thrane, U. 2006. Food-borne fungi in fruit and cereals and their production of mycotoxins. Advances in Experimental Medicine and Biology, 571, 137- 152.

Anderson, K.L. and D.Y.C. Fung. 1983. Anaerobic methods, techniques and principles for food bacteriology: A review. *J. Food Protect.,* 46: 811–22.

Andrews, S., 2010. FastQC: a quality control tool for high throughput sequence data.http://www.bioinformatics.babraham.ac.uk/projects/fastqc/(Nov, 2017).

Andrews, W.H., A.P. Duran, F.D. McClure and D. E. Gentile. 1979. Use of two rapid A-1 methods for the recovery of fecal coliforms and *Escherichia coli* from selected food types. J. Food Sci., 44: 289–91, 293.

Andrews, W.H., C.R. Wilson, P.L. Poelma and A. Romero. 1977. Bacteriological survey of the channel catfish (*Ictalurus punctalus*) at the retail level. J. Food Sci., 42: 359–63.

Angelakis, E. and Raoult, D. Q fever. Vet. Microbiol. 2010, 140, 297–309.

Angelotti, R. and M.I. Foter. 1958. A direct surface agar plate laboratory method for quantitatively detecting bacterial contamination on non-porous surfaces. Food Res., 23: 170–74.

Anon (1994a). Two clusters of haemolytic uraemic syndrome in France. Communicable Diseases Report, 4(7): 29.

Anon. (1994b). *E.coli* O157 phage type 28 infections in Grampian. Communicable Diseases and Environmental Health, Scotland, Weekly Reports 28 (No. 94/46) 1.

Anon. (1995). Brucellosis associated with unpasteurised milk products abroad. Communicable Disease Report 5, (32): 151.

Anon. (1997a) Microbiological Survey: raw cows' milk on retail sale. Food Safety Information Bulletin, December 1997, pp 12-13.

Anon. (1997b) *Salmonella* gold-coast and cheddar cheese: update. Communicable Diseases Report, 7 (11), 93 and 96.

Anon. (1997c) Verocytotoxin producing *Escherichia coli* O157. Communicable Disease Report, Review, 7, (46); 409, 412.

Anon. 1980. Behaviour of pathogens in cheese. Document 122 (revised), International Dairy Federation: Brussels.

Anon. 1998. New Milk Survey. MAFF News Release, 334/98, 10th August 1998, Ministry of Agriculture, Fisheries and Food: London.

Anonymous: Annual Report on Zoonoses in Denmark 2016. NatlFood Institute, Tech Univ Denmark; 2017.

Anonymous: FOCAL - Foodborne Disease Epidemiology, Surveillance and Control in African LMIC. Agriculture, Nutrition and Health Academy; 2019.

Ansorena, D., Gimeno, O., Astiasaran, I. and Bello, J., 2001. Analysis of volatile compounds by GCeMS of a dry fermented sausage: chorizo de Pamplona. Food Res. Int. 34, 67-75.

AOAC. 1996. Salmonella in food, automated conductance methods: AOAC official method 991.38. Official Methods of Analysis of AOAC International, 16th ed., Association of Official Analytical Chemists International, Gaithersburg, MD.

Archer, D.L. 1987. Food borne gram-negative bacteria and atherosclerosis: is there a connection? Journal of Food Protection, 50:783-7.

Archer, D.L. 1996. Preservation microbiology and safety: evidence that stress enhances virulence and triggers adaptive mutations. Trends in Food Science Technol. 7:91-5.

Archer, D.L. 2002. Evidence that ingested nitrate and nitrite are beneficial to health. J. Food Protect. 65: 872–875.

Archer, D.L., Young, F.E. 1988. Contemporary issues: diseases with a food vector. Clin Microbiol Rev.1:377-98.

Ardö, Y. 2006. Flavour formation by amino acid catabolism. Biotechnol. Adv. 24, 238-242.

Askari, G. R., Emam-Djomeh, Z. and Mousavi, S. M. 2013. Heat and mass transfer in apple cubes in a microwave-assisted fluidized bed drier. Food and Bioproducts Processing, 91(3), 207–215.

Association of Official Analytical Chemists. 1984. *Bacteriological analytical manual*, 6th ed., Washington, D.C.: AOAC.

Association of Official Analytical Chemists. 1984. *Official methods of analysis*, 14th ed., Washington, D.C.: AOAC.

Aswal, P., Shukla, A. and Priyadarshi, S. 2012. Yoghurt: preparation, characteristics and recent advancements. Cibtech Journal of Bio-Protocols, 1 (2), 32-44.

Atarés, L., Chiralt, A. and González-Martínez, C. 2008. Effect of solute on osmotic dehydration and rehydration of vacuum impregnated apple cylinders (cv. Granny Smith). Journal of Food Engineering, 89(1), 49–56.

Aureli, P., Franciosa, G. and Poursham, M. 1996. Food borne botulism in Italy. Lancet, 348, 1594.

AUS MEAT, 2005. Handbook of Australian Meat (International Red Meat Manual), seventh ed. Aus. Meat Ltd., Queensland.

Austin, B.L. and B. Thomas. 1972. Dye reduction tests on meat products. *J. Sci. Food Agric.*, 23: 542.

Azarpazhooh, E. and Ramaswamy, H. S. 2012. Modeling and optimization of microwave osmotic dehydration of apple cylinders under continuous-flow spray mode processing conditions. Food and Bioprocess Technology, 5(5), 1486–1501.

Azizi, N.F., Kumar, M.R., Yeap, S.K., Abdullah, J.O., Khalid, M., Omar, A.R., Osman, M.A., Mortadza, S.A.S. and Alitheen, N.B. 2021. Kefir and Its Biological Activities. Foods, 10, 1210.

Baird-Parker, A.C. and E. Davenport. 1965. The effect of recovery medium on the isolation of *Staphylococcus aureus* after heat treatment and after the storage of frozen or dried cells. *J. Appl. Bacteriol.*, 28: 390–402.

Balsam, J., Ossandon, M., Bruck, H.A. *et al.*, 2013. Low-cost technologies for medical diagnostics in low-resource settings. Expert Opin Med Diagn 7:243–55.

Banerjee, R., Kapoor, K.N., Agarwal, R.K. and Ghatak, S. 2001. Verotoxin producing *Escherichia coli* (VETO) in foods of animal origin. J. Food Sci. Technol., 38: 82-84.

Banks, J.G. and Board, R.G. 1983. The classification of pseudomonads and other obligately aerobic Gram-negative bacteria from British pork sausage and ingredients. Syst. Appl. Microbiol. 4, 424-438.

Banwart, G. 1989. Basic Food Microbiology, Chapman and Hall, ITP, USA.

Bär A-K, Phukan N, Pinheiro J, Simoes-Barbosa A. 2015. The interplay of host microbiota and parasitic protozoans at mucosal interfaces: implications for the outcomes of infections and diseases. PLoS Negl Trop Dis 9: e0004176.

Barach, J.T., R.S. Flowers and D.M. Adams. 1975. Repair of heat- injured *Clostridium perfringens* spores during outgrowth. *Appl. Microbiol.*, 30: 873–75.

Baranyi, J. and Roberts, T.A., 1994. A dynamic approach to predicting bacterial growth in food. Int. J. Food Microbiol. 23, 277–294.

Barbe, C.D., R.W. Mandigo and R.L. Henrickson. 1966. Bacterial flora associated with rapid-processed ham. *J. Food Sci.*, 31: 988–93.

Barbosa-Canovas, G.V., Pothakamury, U.R, Palou, E. and Swanson, B.G. 1997. Non-thermal preservation of foods. Marcel Dekker.

Barrett, K. A., Nakao, J. H., Taylor, E. V., Eggers, C. and Gould, L. H. 2017. Fish-associated foodborne disease outbreaks: United States, 1998–2015. Foodborne Pathogens and Disease, 14(9), 537– 543.

Bartz, J.A. and R.K. Showalter. 1981. Infiltration of tomatoes by aqueous bacterial suspensions. Phytopathology 71:515–518.

Batra, N.G., Sharma, A. and Agarwal, N. 2018. Evaluation of microbiological criteria and quality of packed fruit juices. International Food Research Journal 25(2): 458-461.

Batz, M.B., Hoffmann, S.A. and Morris, J.G 2011. Ranking the Risks: The 10 Pathogen-Food Combinations with the Greatest Burden on Public Health. Emerging Pathogens Institute at University of Florida.

Bauer, F.T., J.A. Carpenter and J.O. Reagan. 1981. Prevalence of *Clostridium perfringens* in pork during processing. J. Food Protect., 44: 279–83.

Beck, H.C., Hansen, A.M. and Lauritsen, F.R., 2004. Catabolism of leucine to branchede chain fatty acids in Staphylococcus xylosus. J. Appl. Microbiol. 96, 1185-1193.

Behar, S.M. and Porcelli, S.A. 1995. Mechanisms of autoimmune disease induction: the role of the immune response to microbial pathogens. Arthritis Rheum. 38:458-76.

Belay, E.D. 1999. Transmissible spongiform encephalopathies in humans. Ann. Rev. Microbiol. 53:283–314.

Belitz, H.D., Grosch, W. and Schieberle, P. 2009. Food Chemistry (Chapter 5). Springer, pp. 340-402.

Bell, W.N. and L.A. Shelef. 1978. Availability and microbial stability of retail beef-soy blends. *J. Food Sci.,* 43: 315–18, 333.

Benner Jr. R.A. 2014. Organisms of Concern but not Foodborne or Confirmed Foodborne: Spoilage Microorganisms. Reference Module in Food Science: Encyclopedia of Food Safety. 2: 245-250.

Bensmira, M. and Jiang, B. 2015. Total phenolic compounds and antioxidant activity of a novel peanut based kefir. Food Sci. Biotechnol. 24, 1055–1060.

Berkel, B.M., Boogaard, B.V. and Heijnen, C. 2004. Preservation of fish and meat. Agromisa Foundation, Wageningen, The Netherlands, pp: 8: 78-80.

Berkelman, R.L., Bryan, R.T., Osterholm, M.T., LeDuc, J.W. and Hughes, J.M.1994. Infectious disease surveillance: a crumbling foundation. Science. 264:368-70.

Bernard, D.T. and Scott, V.N. 1995. Risk assessment and food borne micro-organisms: the difficulties of biological diversity. Food Control. 6:329-33.

Berry-Ottaway, P.1992. The Technology of Vitamins in Foods. Blackie Academic and Professional.

Beuchat, L. R., Komitopoulou, E., Beckers, H., Betts, R. P., Bourdichon, F., Fanning, S. and Ter Kuile, B. H. 2013. Low-water activity foods: Increased concern as vehicles of foodborne pathogens. Journal of Food Protection, 76(1), 150–172.

Beuchat, L.R. 1996. Pathogenic microorganisms associated with fresh produce. Journal of Food Protection. 59:204-6.

Beuchat, L.R. 1999. Survival of enterohemorrhagic Escherichia coli O157:H7 in bovine feces applied to lettuce and the effectiveness of chlorinated water as a disinfectant. J. Food Prot., 62: 845-849.

Beuchat, L.R., Nail, B.V., Adler, B.B. and Clavero, M.R.S. 1998. Efficacy of spray application of chlorinated water in killing pathogenic bacteria on raw apples, tomatoes, and lettuce. J. Food Prot., 61: 1305-1311.

Bhadsavle, C.H., T.E. Shehata and E.B. Collins. 1972. Isolation and identification of psychrophilic species of Clostridium from milk. *Appl. Microbiol.,* 24: 699–702.

Bhagya. R., Jonnala, Y., McSweeney, P.L.H., Sheehan J.J. and Cotter, P.D. 2018. Sequencing of the Cheese Microbiome and Its Relevance to Industry. Front. Microbiol., 9: 1020.

Bhavana, J., Sonashree, R., Rashmi, R. Halbavi, Rao, V.V. and Praveena, B. 2019. A Review on the preparation of vinegar from fruit peels. World Journal of Pharmacy and Pharmaceutical Sciences, 8(9), 227-233.

Bille, J.1990. Epidemiology of human listeriosis in Europe with special reference to the Swiss Outbreak. In Foodborne Listeriosis (eds. Miller, A.J., Smith, J.L. and Somkuti,G.A). pp 71-74. Amsterdam: Elsevier.

Birck, N. M. M., Lorini, I. and Scussel, V. M. 2005. Fungus and mycotoxins in wheat grain at post harvest. Paper presented at *International Working Conference on Stored Product Protection*, Sao Paolo, Brazil. Passo Fundo, RS, Brazil: Brazilian Post-harvest Association.

Bishop, M.T., Will, R.G., and Manson, J.C. 2010. Defning sporadic Creutzfeldt-Jakob disease strains and their transmission properties. Proc. Natl. Acad. Sci. U.S.A. 107, 12005–10.

Biswas, A. K., Kondaiah, N., Anjaneyulu, A. S. R. and Mandal, P. K. 2011. Causes, Concerns, Consequences and Control of Microbial Contaminants in Meat-A Review. International Journal of Meat Science, 1(1), 27–35.

Blanchfield, J.R. 1998. Bovine spongiform encephalopathy (BSE)—A review. Int. J. Food Sci. Technol. 33: 81–97.

Blaževic, I., Montaut, S., Burcul, F., Olsen, C.E., Burow, M., Rollin, P. and Agerbirk, N. 2020. Glucosinolate structural diversity, identification, chemical synthesis and metabolism in plants. Phytochemistry, 169, 112100.

Boers, R.H., Dijkmann, K.E., Wijngaards, G., 1994. Shelf-life of vacuum-package wild boar meat in relation to that of vacuum packaged pork: relevance of intrinsic factors. Meat Sci. 37, 91-102.

Bolger, A.M., Lohse, M. and Usadel, B. 2014. Trimmomatic: aflexible trimmer for Illuminasequence data. Bioinformatics 30, 2114–2120.

Bone, F.J., Bogie, D. and Morgan-Jones, S.C.1989. Staphylococcal food poisoning from sheep milk cheese. Epidemiology and Infection 103: 449-458.

Borch, E. and Agerhem, H., 1992. Chemical, microbial and sensory changes during the anaerobic cold storage of beef inoculated with a homofermantative Lactobacillus sp. or a Leuconostoc sp. Int. J. Food Microbiol. 15, 99-08.

Borch, E. and Molin, G., 1989. The aerobic growth and product formation of Lactobacillus, Leuconostoc, Brochothrix, and Carnobacterium in batch cultures. Appl. Microbiol. Biotechnol. 30, 81-88.

Borch, E., Berg, H. and Holst, O. 1991. Heterolactic fermentation by a homofermentative Lactobacillus sp. during limitation in aerobic continuous culture with complete cell recycle. J. Appl. Bacteriol. 71, 26-269.

Borch, E., Kant-Muemans, M.L. and Blixt, Y., 1996. Bacterial spoilage of meat and cured meat products. Int. J. Food Microbiol. 33, 103-120.

Borgdorff, M.W. and Motarjemi Y. 1977. Foodborne disease surveillance: What are the options? World Health Stat Quart. 50(1/2):12-23.

Bórquez, R. M., Canales, E. R. and Redon, J. P. 2010. Osmotic dehydration of raspberries with vacuum pre-treatment followed by microwave-vacuum drying. Journal of Food Engineering, 99(2), 121–127.

Botha, G. E., Oliveira, J. C. and Ahrné, L. 2012. Quality optimisation of combined osmotic dehydration and microwave assisted air drying of pineapple using constant power emission. Food and Bioproducts Processing, 90(C2), 171–179.

Boyd, W.C. and Shapleigh, E. 1954. Specific precipitation activity of plant agglutinins (lectins). Science. 119, 419.

Brennand, Charlotte. 1994. Home Drying of Food. Cooperative Extension Service, Utah State University.

Brewer, M.S. and Vega, J.D. 1995. Detectable odor thresholds of selected lipid oxidation compounds in a meat model system. J. Food Sci. 60, 592-595.

Brightwell, G., Clemens, R., Urlich, S. and Boerema, J. 2007. Possible involvement of psychrotolerant Enterobacteriaceae in blown pack spoilage of vacuum-packed raw meats. Int. J. Food Microbiol. 119, 334-339.

Broda, D.M., DeLacy, K.M., Bell, R.G. and Penney, N. 1996. Association of psychrotrophic Clostridium spp. with deep tissue spoilage of chilled vacuum-packed lamb. Int. J. Food Microbiol. 29, 371-378.

Broda, D.M., Lawson, P.A. and Musgrave, D.R., 1999. Clostridium frigidicarnis sp. nov., a psychrotorelant causing spoilage of vacuum-packed meat. Int. J. Syst. Evol. Microbiol. 49, 1539-1550.

Broda, D.M., Musgrave, D.R. and Bell, R.G. 2003. Molecular differentiation of clostridia associated with 'blown pack' spoilage of vacuum-packed meats using internal transcribed spacer polymorphism analysis. Int. J. Food Microbiol. 84, 71-77.

Broda, D.M., Saul, D.J., Lawson, P.A., Bell, R.G. and Musgrave, D.R., 2000. Clostridium gasigenes sp. nov., a psychrophile causing spoilage of vacuum-packed meat. Int. J. Syst. Evol. Microbiol. 50, 107-118.

Brodsky, M.H. 1984. Bacteriological survey of freshly formed Cheddar cheese. *J. Food Protect.,* 47: 546–48.

Bruckner, D.A. 1992. Amebiasis. Clin Rev Microbiol. 5:356-69.

Bruhn, C.M., Schutz, H.G. and Sommer, R. 1986. Attitude change toward food irradiation among conventional and alternative consumers. Food Technology 40(12):86-91.

Bruijn, J., Rivas, F., Rodriguez, Y., Loyola, C., Flores, A., Melin, P. and Borquez, R. 2015. Effect of vacuum microwave drying on the quality and storage stability of strawberries. Journal of Food Processing and Preservation, 40(5), 1104–1115.

Bryan, F.L., J.C. Ayres and A.A. Kraft. 1968. Salmonellae associated with further-processed turkey products. *Appl. Microbiol.,* 16: 1–9.

Bryan, F.L. 1983. Epidemiology of milk-borne diseases. J. Food. Prot., 46 (7), 637-649.

Buchanan, R. L., Gorris, L. G. M., Hayman, M. M., Jackson, T. C. and Whiting, R. C. 2017. A review of *Listeria monocytogenes*: An update on outbreaks, virulence, dose-response, ecology, and risk assessments. Food Control, 75: 1– 13.

Buck, J.W., Walcott, R. R. and Beuchat, L.R. 2003. Recent trends in microbiological safety of fruits and vegetables. Online. Plant Health Progress doi:10.1094/PHP-2003-0121-01-RV

Buffie, C.G. and Pamer, *e.g.* 2013. Microbiota-mediated colonization resistance against intestinal pathogens. Nat Rev Immunol 13:790–801.

Bullerman, L. B. and Bianchini, A. 2009. Food safety issues and the microbiology of cereals and cereal products. In N. Heredia, I. Wesley, and S. Garcia (Eds.), Microbiologically safe foods (pp. 315– 335). New York, U.S.A.: John Wiley and Sons.

Bunning, V.K. 1994. Immunopathogenic aspects of food borne microbial disease. Food Microbiol. 11:89-95.

Bunning, V.K., Lindsay, J.A., Archer, D.L. 1997. Chronic health effects of food borne microbial disease. World Health Stat Quart. 50: 51-6.

Burdock, G.A. 2005. In: Burdock, G.A. (Ed.), Fenaroli's Handbook of Flavor Ingredients, fifth ed. CRC Press, Boca Raton, p. 937.

Burgess, S.L., Gilchrist, C.A., Lynn, T.C. and Petri, Jr, W.A. 2017. Parasitic Protozoa and Interactions with the Host Intestinal Microbiota. Infect Immun.85(8): e00101-17.

Busani, L., Cigliano, A., Taioli, E., Caligiuri, V., Chiavacci, L., Di Bella, C., ... Epidemiology, I. G. V. 2005. Prevalence of *Salmonella enterica* and *Listeria monocytogenes* contamination in foods of animal origin in Italy. Journal of Food Protection, 68(8), 1729– 1733.

Buttery, R.G., Orts, W.J., Takeoka, G.R. and Nam, Y. 1999. Volatile flavor components of rice cakes. J. Agric. Food Chem. 47, 4353-4356.

Cabedo, L., Barrot, L. P. I. and Canelles, A. T. I. 2008. Prevalence of *Listeria monocytogenes* and *Salmonella* in ready-to-eat food in Catalonia, Spain. Journal of Food Protection, 71(4), 855– 859.

Calkins, C.R., Hodgen, J.M. 2007. A fresh look at meat flavour. Meat Sci. 77, 63-80.

Cantoni, C., Bersani, C., Bregoli, M. and Bernardini, M., 2000. Brochothrix thermosphacta in meat and some meat products. Ind. Aliment. 39, 976-979.

Cao, H., Zhang, M., Mujumdar, A. S., Du, W.-h. and Sun, J.-c. 2006. Optimization of osmotic dehydration of kiwifruit. Drying Technology, 24(1), 89–94.

Carl, K.E. 1975. Oregon's experience with microbiological standards for meat. *J. Milk Food Technol.,* 38: 483–86.

Carrabs, G., Smaldone, G., Carosielli, L., Girasole, M., Iammarino, M. and Chiaravalle, E. 2017. Detection of sulfites in fresh meat preparation commercialised at retail in Lazio Region. Italian Journal of Food Safety, 6: 6482.

Cartea, M.E. and Velasco, P. 2008. Glucosinolates in Brassica foods: Bioavailability in food and significance for human health. Phytochemistry Reviews; 7(2): 213-229.

Casaburi, A., De Filippis, F., Villani, F., Ercolini, D. 2014. Activities of strains of Brochothrix thermosphacta in vitro and in meat. Food Res. 62: 366-374.

Casaburi, A., Piombino, P., Nychas, G., Villani, F. and Ercolini, D. 2015. Bacterial populations and the volatilome associated to meat spoilage. Food Microbiol. 45(Part A), 83-102.

Casewell, M.W., J.E. Cooper and M. Webster. 1981. Enteral feeds contaminated with *Enterobacter cloacae* as a cause of septicaemia. *Brit. Med. J.,* 282: 973.

Castellano, P., Belfiore, C., Fadda, S. and Vignolo, G., 2008. A review of bacteriocinogenic lactic acid bacteria used as bioprotective cultures in fresh meat produced in Argentina. Meat Sci. 79, 483–499.

Castellone, V., Bancalari, E., Rubert, J., Gatti, M., Neviani, E. and Bottari, B. 2021. Eating Fermented: Health Benefits of LAB-Fermented Foods. Foods, 10, 2639.

Castresana, J. 2000. Selection of conserved blocks from multiple alignments for their use in phylogenetic analysis. Mol. Biol. Evol. 17, 540–552.

Cavett, J.J. 1962. The microbiology of vacuum packed sliced bacon. J. appl. Bacteriol., 25, 282 – 289.

CDC (Centers for Disease Control and Prevention). 2013. Cost Analysis – Cost of Illness. 2013, Georgia, United States.

CDC (Centers for Disease Control and Prevention); 1991. Paralytic shellfish poisoning—Massachusetts and Alaska. 1990. Morb. Mort. Wkly. Rep. 40:157–161.

CDC (Centers for Disease Control and Prevention); 1998. Ciguatera fish poisoning—Texas, 1997. Morb. Mort. Wkly. Rep. 47:692–694.

CDC (Centers for Disease Control and Prevention); 2012. Multistate outbreak of Salmonella Bareilly and Salmonella Nchanga infections associated with a raw scraped ground tuna product (final update).

CDC (Centers for Disease Control and Prevention); 2013a. *Listeria (Listeriosis).*

CDC (Centers for Disease Control and Prevention); 2013b.*Listeria and food.*

CDC (Centers for Disease Control and Prevention); 2014. List of Selected Multistate Foodborne Outbreak Investigations.

CDC (Centers for Disease Control and Prevention); 2014b. Multistate outbreak of *Salmonella* Stanley Infections linked to raw Cashew cheese (final update).

CDC (Centers for Disease Control and Prevention); 2014c. List of Selected Multistate Foodborne Outbreak Investigations.

CDC (Centers for Disease Control and Prevention); 2018. National Outbreak Reporting System (NORS).

CDC (Centers for Disease Control and Prevention); 2019a. 2019 AR threats report.

CDC (Centers for Disease Control and Prevention); 2019b. Food poisoning from marine toxins.

CDC (Centers for Disease Control and Prevention); 2019c. Glossary of terms related to antibiotic resistance.

CDC (Centers for Disease Control and Prevention); 2019d. Salmonella *infections linked to frozen raw tuna.*

CDC (Centers for Disease Control and Prevention);1998. Outbreak of Campylobacter enteritis associated with cross-contamination of food, Oklahoma, 1996. Morbidity and mortality weekly report, 47: 129-131.

CDC. (Centers for Disease Control and Prevention); 2003b. Preliminary FoodNet data on the incidence of foodborne illnesses-select sites, United States, 2002. Morb. Mortal. Weekly Rep. 52(15): 340-343.

CDIC (Canadian Dairy Information Centre); Global consumption per capita of dairy products: Total cheese consumption. 2014.

CDR 1997. Hospital outbreak of E. coli O157:H7 associated with a rare phage type, Ontario. Canada communicable disease report., 23:33-39.

Cerveny, J., Meyer, J.D. and Hall, P.A. 2009. Microbiological Spoilage of Meat and Poultry Products In: Compendium of The Microbiological Spoilage, Of Foods and Beverages. Food Microbiology and Food Safety, W.H. Sperber and M.P. Doyle (Eds.). Springer Science and Business Media, NY, pp. 69-868.

Chaijan, M., Panpipat, W. and Nisoa, M. 2016. Chemical deterioration and discoloration of semi-dried tilapia processed by sun drying and microwave drying. Drying Technology, 35(5), 642–649.

Chaillou, S., Chaulot-Talmon, A., Caekebeke, H., Cardinal, M., Christieans, S., Denis, C., *et al.,* 2015. Origin and ecological selection of core and food-specific bacterial communities associated with meat and seafood spoilage. ISME J. 9, 1105-1118.

Chambers, P.G. and Grandin, T. 2001. Guidelines for humane handling, transport and slaughter of livestock. G. Heinz and T. Srisuvan (Eds.). http://www.fao.org/fileadmin/user_upload/animalw elfare/ guidelines per cent 20humane per cent 20handling per cent 20tran sport per cent 20slaughter.pdf.

Chappell, C.L., Matson, C.C. 1992. Giardia antigen detection in patients with chronic gastrointestinal disturbances. J Fam Pract. 35:49-53.

Chen Q., Kong B. H., Han Q., Xia X. F. and Xu L. 2017. The role of bacterial fermentation in lipolysis and lipid oxidation in Harbin dry sausages and its flavour development. *LWT Food Sci. Technol.* 77 389–396.

Chen, B. Y., Pyla, R., Kim, T. J., Silva, J. L. and Jung, Y. S. 2010a. Antibiotic resistance in *Listeria* species isolated from catfish fillets and processing environment. Letters in Applied Microbiology, 50(6), 626– 632.

Chen, Z., Shi, J., Yang, X., Nan, B., Liu, Y. and Wang, Z. 2015. Chemical and physical characteristics and antioxidant activities of the exopolysaccharide produced by Tibetan kefir grains during milk fermentation. Int. Dairy J. 43, 15–21.

Chiodini, R.J. 1989. Crohn's disease and the mycobacterioses: a review and comparison of two disease entities. Clin. Microbiol. Rev. 2: 90-117.

Chiodini, R.J.1989. Crohn's disease and the mycobacteriosis: a review and comparison of two disease entities. Clin Microbiol Rev. 2:90-117.

Chipley, J.R. and E.K. Heaton. 1971. Microbial flora of pecan meat. Appl Microbiol., 22(2): 252–253.

Choi, K.H., Lee, H., Lee, S.L.S. and Yoon, Y. 2016. Cheese Microbial Risk Assessments — A Review. Asian-Australas J Anim Sci., 29(3): 307–314.

Choi, S., Puligundla, P. and Mok, C. 2017. Impact of corona discharge plasma treatment on microbial load and physicochemical and sensory characteristics of semi-dried squid (Todarodes pacificus). Food Science and Biotechnology, 26(4), 1–8.

Chou, C. H., Silva, J. L. and Wang, C. L. 2006. Prevalence and typing of *Listeria monocytogenes* in raw catfish fillets. Journal of Food Protection, 69(4), 815– 819.

Christiansen, L.N. and N.S. King. 1971. The microbial content of some salads and sandwiches at retail outlets. *J. Milk Food Technol.,* 34: 289–93.

Christopher, F.M., S.C. Seideman, Z.L. Carpenter, G.C. Smith and C. Vanderzant. 1979. Microbiology of beef packaged in various gas atmospheres. *J. Food Protect.,* 42: 240–44.

Chwastek, A. 2014. Methods to increase the rate of mass transfer during osmotic dehydration of foods. Acta Scientiarum Polonorum. Technologia Alimentaria, 13(4), 341–350.

Clarke, K.R. 1993. Non-parametric multivariate analysis of changes in community structure. Aust. J. Ecol. 18, 117-143.

Clinical Observations on 3,009 Cases of Ciguatera (Fish Poisoning) in the South Pacific—PubMed. Available online: https://pubmed.ncbi.nlm.nih.gov/574366/

Cliver, D.O. 2004. How now, mad cow? Food Technol. 58(1):100.

CODEX 1991.The Codex Standard for the Labelling of Prepackaged Food (CODEX STAN 106-1985 as amended in 1991).

Collins, M.D., Farrow, J.A.E., Phillips, B.A., Ferusu, S. and Jones, D., 1987. Classification of Lactobacillus divergens, Lactobacillus piscicola, and some catalase-negative, asporogenous, rod-shaped bacteria from poultry in a new genus, Carnobacterium. Int. J.Syst. Bacteriol. 37, 310–316.

Colquhoun, K.O., Timms, S. and Fricker, C.R. 1995. Detection of Escherichia coli in potable water using direct impedance technology. J Appl Bacteriol., 79:635–9.

Corrêa Justus, J. L. G., Oliveira, A. D., Alves, L. F. and Guilherme, E. 2015. Osmotic dehydration of tomato assisted by ultrasound: Evaluation of the liquid media on mass transfer and product quality. International Journal of Food Engineering, 11(4), 505–516.

Cox, N.A., A.I. Mercuri, B.I. Iuven and J.E. Thomson. 1975. *Enterobacteriaceae* at various stages of poultry chilling. *J. Food Sci.,* 40: 44–46.

Creppy, E.E., Baudrimont, I. and Betbeder, A.M. 1995. Prevention of nephrotoxicity of ochratoxin A, a food contaminant. Toxicol Lett. 83:869-77.

Cruess, W.V. 1958. Commercial fruit and vegetable products: Chapter 21 – Vinegar manufacture. 1st ed. New York: McGraw-Hill Book Company, Inc. p 681-707.

Cui, H., Gabriel, A.A. and Nakano, H. 2010. Antimicrobial efficacies of plant extracts and sodium nitrite against *Clostridium botulinum.* Food Control. 21: 1030–1036.

Cui, H., Xue, C., Xue, Y., Wei, S., Li, Z. and Cong, H. 2013. Development of shelf-stable, ready-to-eat (RTE) shrimps (Litopenaeus vannamei) using water activity lowering agent by response surface methodology. Journal of Food Science and Technology, 50(6), 1137–1143.

Curioni, P.M.G. and Bosset, J.O. 2002. Key odorants in various cheese types as determined by gas chromatography-olfactometry. Int. Dairy J. 12, 959-984.

D Gan, S. and Patel, K.R. 2013. Enzyme Immunoassay and Enzyme-Linked Immunosorbent Assay. Journal of Investigative Dermatology, 133 (9): e12.

D'Aoust, J-Y. 1994. Salmonella and international trade. International Journal of Food Microbiol. 24: 11-31.

D'Incecco, P., Limbo, S., Hogenboom, J., Rosi, V., Gobbi, S., and Pellegrino, L. 2020. Impact of extending hard-cheese ripening: a multiparameter characterization of parmigiano reggiano cheese ripened up to 50 months. Foods, 9:268.

Dahdouh, B., Basha, O., Khalil, S. and Tanekhy, M. 2016. Molecular characterization, antimicrobial susceptibility and salt tolerance of *Aeromonas hydrophila* from fresh, brackish and marine fishes., 48(2), 46– 53.

Dainty, R.H. 1996. Chemical/biochemical detection of spoilage. Int. J. Food Microbiol., 33: 19-33.

Dainty, R.H. and Hibbard, C.M. 1983. Precursors of the major end-products of aerobic metabolism of Brochotrix thermosphacta. J. Appl. Bacteriol. 55, 127-133.

Dainty, R.H., Edwards, R.A. and Hibbard, C.M., 1989. Spoilage of vacuum-packed beef by a Clostridium sp. J. Sci. Food Agric. 49, 473-486.

Dainty, R.H., Mackey, B.M., 1992. The relationship between the phenotypic properties of bacteria from chill-stored meat and spoilage processes (Supplement 21). J. Appl. Bacteriol., 103e114 (Symposium).

Dainty, R.H., Shaw, B.G. and Roberts, T.A., 1983. Microbial and chemical changes in chill stored red meats. Soc. Appl. Bacteriol. Symp. 11, 151-178.

Dalgaard P. 1995. "Qualitative and quantitative characterization of spoilage bacteria from packed fish", Int J Food Microbiol., 26(3):319.

Dalgaard, P., Gram, L. and Huss, H.H. 1993. Spoilage and shelf life of cod fillets packed in vacuum or modified atmosphere. Int. J. Food Microbiol. 19, 283-294.

Darling, A.E., Mau, B. and Perna, N.T., 2010. Progressive Mauve: multiple genome alignment with gene gain, loss and rearrangement. PLoS One 5, e11147.

Darvishi, H., Azadbakht, M., Rezaeiasl, A. and Farhang, A. 2013. Drying characteristics of sardine fish dried with microwave heating. Journal of the Saudi Society of Agricultural Sciences, 12(2), 121–127.

Dave, D. and Ghaly, A.E. 2011. Meat Spoilage Mechanisms and Preservation Techniques: A Critical Review. Am. J. Agric. Biol. Sci., 6 (4): 486-510.

de Hoon, M.J.L., Imoto, S., Nolan, J. and Miyano, S. 2004. Open-source clustering software. Bioinformatics 20, 1453-1454.

de Oliveira, G. B., Favarin, L., Luchese, R.H. and McIntosh D. 2015. Psychrotrophic bacteria in milk: How much do we really know? Brazilian Journal of Microbiology, 46(2): 313-321.

Dehghannya, J., Gorbani, R. and Ghanbarzadeh, B. 2015. Effect of ultrasound-assisted osmotic dehydration pretreatment on drying kinetics and effective moisture diffusivity of mirabelle plum. Journal of Food Processing and Preservation, 39(6), 2710–2717.

Dermesonlouoglou, E. K., Giannakourou, M. and Taoukis, P. S. 2016. Kinetic study of the effect of the osmotic dehydration pre-treatment with alternative osmotic solutes to the shelf life of frozen strawberry. Food and Bioproducts Processing, 99, 212–221.

Dermesonlouoglou, E. K., Pourgouri, S. and Taoukis, P. S. 2008. Kinetic study of the effect of the osmotic dehydration pre-treatment to the shelf life of frozen cucumber. Innovative Food Science and Emerging Technologies, 9(4), 542–549.

Derossi, A., Pilli, T. D. and Severini, C. 2010. Reduction in the pH of vegetables by vacuum impregnation: A study on pepper. Journal of Food Engineering, 99(1), 9–15.

Desenclos, J.C. *et al.,* 1996. Large outbreak of Salmonella enterica serotype paratyphi B infection caused by goats' milk cheese, France: a case finding and epidemiological study. British Medical J. 312: 91-94.

Deshpande, S.2002. Fungal toxins. In Handbook of Food Toxicology; CRC Press: Boca Raton, FL, USA, pp. 387–456.

Dethmers, A.E. and Rock, H. 1975. Effect of added sodium nitrite and sodium nitrate on sensory quality and nitrosoamine formation in Thuringer Sausage. J. Food Sci. 40: 491–497.

Dewey-Mattia, D., Roberts, V. A., Vieira, A. and Fullerton, K. E. 2016. Foodborne (1973-2013) and waterborne (1971-2013) disease outbreaks - United States. Morbidity and Mortality Weekly Report, 63(55), 79– 84.

Dhawan, S. S. 1998. Practical Manual on Home scale Processing of fruits and vegetables processing. Haryana Agricultural University, Hisar, India.

Diehl, J.F. 1990. Safety of irradiated foods. Marcel Dekker Inc., NY.

Diehl, J.F. and Josephson, E.S. 1994. Assessment of the wholesomeness of irradiated foods (a review). Acta Alimentaria 23: 195-214.

Dimidi,E., Cox, S.R., Rossi, M. and Kevin Whelan, K. 2019. Fermented Foods: Definitions and Characteristics, Impact on the Gut Microbiota and Effects on Gastrointestinal Health and Disease. Nutrients., 11(8): 1806.

Dinkova-Kostova, A.T. and Kostov, R.V. 2012. Glucosinolates and isothiocyanates in health and disease. Trends Mol. Med., 18, 337–347.

Directive. 2006. Directive 2006/52/EC of the European Parliament and of the Council of 5 July 2006 amending Directive 95/2/EC on food additives other than colours and sweeteners and Directive 95/35/EC on sweeteners for use in foodstuffs. Off. J. Eur. Union. 204: 10–22. 3.

Diwan, J.J. 2007. Glycolysis and Fermentation. http://rpi.edu/dept/bcbp/molbiochem/MBWeb/mb1/part2/glycolysis.htm.

Djuretic, T., Wall, P.G. and Nichols, G. 1997. Large outbreak of *Salmonella enterica* serotype *paratyphi B* infection caused by goats' milk cheese, France; a case finding and epidemiological study Communicable Diseases Report 7, Review No 3, R41-R45.

DoE/DOH. 1995.Cryptosporidium in Water Supplies, Second Report of the Group of Experts, HMSO.

Don Whitley Scientific Limited. 1999. Introduction to principles of impedance. www.dwscientificco.uk

Doohan, F. M., Brennan, J. and Cooke, B. M. 2003. Influence of climatic factors on *Fusarium* species pathogenic to cereals. European Journal of Plant Pathology, 109(7), 755– 768.

Doulgeraki, A.I., Ercolini, D., Villani, F., Nychas, G.J. 2012. Spoilage microbiota associated to the storage of raw meat in different conditions. Int. J. Food Microbiol. 157 (2): 130–141.

Doulgeraki, A.I., Paramithiotis, S., Kagkli, D.M. and Nychas, G.J.E. 2010. Lactic acid bacteria population dynamics during minced beef storage under aerobic or modified atmosphere packaging conditions. Food Microbiol., 27, 1028-1034.

Drake, S.D., I.B. Evans and C.F. Niven, Jr. 1958. Microbial flora of packaged frankfurters and their radiation resistance. *Food Res.,* 23: 291–96.

Draper, H.H. 1980. Advances in Nutritional Research. Vol III (Edited), P.301. Springer Science and Business Media.

Draughon, F.A. 1980. Effect of plant-derived extenders on microbiological stability of foods. *Food Technol.,* 34(10): 69–74.

Drosinos, E.H. and Board, R.G., 1995. Microbial and physicochemical attributes of minced lamb: sources of contamination with pseudomonads. Food Microbiol., 12, 189-197.

Dumont, J.P. and Adda, J. 1978. Occurrence of sesquiterpenes in mountain cheese volatiles. J. Agric. Food Chem. 26, 364-367.

Dupont, J., Menard, D., Herve, C., Chevalier, F., Beliaeff, B. and Minier B. 1996. Rapid estimation of Escherichia coli in live marine bivalve shellfish using automated conductance measurement. J Appl Bacteriol., 80:81–90.

Eden, R. and Eden, G. 1984. Impedance microbiology. Herts, UK: Research Studies Press Ltd.

Edgar, R.C., 2013. UPARSE: highly accurate OTU sequences from microbial amplicon reads. Nat. Methods 10, 996-998.

Edwards, A., Zammit, A. and Farrell, H. 2019. Four recent ciguatera fish poisoning incidents in New South Wales, Australia linked to imported fish. Commun. Dis. Intell., 43.

EFSA BIOHAZ Panel. 2011. Updated technical specifications for harmonised reporting of food-borne outbreaks through the European Union reporting system in accordance with Directive 2003/99/EC.

EFSA Journal, 9(4):2101.

El Sheikha, A. F. 2018. "Revolution in fermented food: from artisan household technology to era of biotechnology," in *Molecular Techniques in Food Biology: Safety, Biotechnology, Authenticity and Traceability.* eds. A. F. El Sheikha, R. Levin, and J. P. Xu (Chichester, UK: John Wiley and Sons Ltd.), 241–260.

Elhadi, N., Aljeldah, M. and Aljindan, R. 2016. Microbiological contamination of imported frozen fish marketed in Eastern Province of Saudi Arabia. International Food Research Journal, 23(6), 2723– 2731.

Elling F. 1979. Ochratoxin A-induced mycotoxic porcine nephropathy: Alterations in enzyme activity in tubular cells. Acta Pathol. Microbiol. Scand., 87:237–243.

Ercolini, D., Casaburi, A., Nasi, A., Ferrocino, I., Di Monaco, R., Ferranti, P., Mauriello, G. and Villani, F. 2010a. Different molecular types of Pseudomonas fragi have the overall behaviour as meat spoilers. Int. J. Food Microbiol., 142, 120-131.

Ercolini, D., Ferrocino, I., La Storia, A., Mauriello, G., Gigli, S., Masi, P. and Villani, F. 2010b. Development of spoilage microbiota in beef stored in nisin activated packaging. Food Microbiol. 27, 137-143.

Ercolini, D., Russo, F., Blaiotta, G., Pepe, O., Mauriello, G. and Villani, F. 2007. Simultaneous detection of Pseudomonas fragi, P. lundensis, and P. putida from meat by use of a multiplex PCR assay targeting the car A gene. Appl. Environ. Microbiol. 73, 2354-2359.

Ercolini, D., Russo, F., Torrieri, E., Masi, P. and Villani, F. 2006. Changes in the spoilage related microbiota of beef during refrigerated storage under different packaging conditions. Appl. Environ. Microbiol. 72, 4663-4671.

Erichsen, I. and Molin, G., 1981. Microbial flora of normal and high pH beef stored at 4°C in different gas environments. J. Food Prot. 44, 866-869.

Etienne, C.F. 2013. Supplement on Caribbean burden of illness study. J Heal Popul Nutr., 31.

European Food Safety Authority (EFSA). 2006. Opinion of the scientific panel on contaminants in the food chain on a request from the commission related to ochratoxin A in food. The EFSA Journal, No. 365:1-56.

Expert Committee on Food Additives (JECFA). World Health Organisation, Geneva.

European Food Safety Authority (EFSA). 2020. Outcome of a public consultation on the risk assessment of ochratoxin A in food. https://doi.org/10.2903/sp.efsa.2020.EN-1845

FAO. 1996. The world food summit. FAO technical background document no. 12.

FAO/WHO. 1991. Report of the Joint FAO/WHO Conference on Food Standards, Chemicals in Food and Food Trade. Rome: FAO.

FAO/WHO. 1995. Report of the Joint FAO/WHO Expert Consultation on Application of Risk Analysis to Food Standards Issues, Geneva.

FAO/WHO. 1997. Report of the Joint FAO/WHO Expert Consultation on Risk Management and Food Safety Issues, Rome, 27-31 Jan. FAO Food Nutr Pap., p. 62.

FAO/WHO.1984. The role of food safety in health and development. Report of Joint FAO/WHO Expert Committee on Food Safety. World Health Organ Tech Rep Ser., 705.

Farmer P. 1996. Social inequalities and emerging infectious diseases.

Farr, S.B. and Kogoma, Y. 1991. Oxidative stress response in *Escherichia coli* and *Salmonella typhimurium*. Microbiol Rev., 55:561-85.

FDA. 1986. Irradiation in the production, processing, and handling of food; final rule. 21 CFR Part 179. Federal Register 51(75): 13376-13399.

FDA. 1990. HHS News Release P90-27. U.S. Dept. of HHS. Rockville, MD. May 1.

Ferriman, A. 1997. Doctors warn of a return of past plagues. The Independent 1997 Mar 7; News section:5.

Fidler, D. 1996. Globalization, international law and emerging infectious diseases. Emerg. Infect. Dis. 2:77-84.

Flegr, J., Zitkova, S., Kodym, P. and Frynta, D. 1996. Induction of changes in human behaviour by the parasitic protozoan *Toxoplasma gondii*. Parasitology.,13:49-54.

Fallah, A. A., Saei-Dehkordi, S. S. and Mahzounieh, M. 2013. Occurrence and antibiotic resistance profiles of *Listeria monocytogenes* isolated from seafood products and market and processing environments in Iran. Food Control, 34(2), 630– 636.

Fang, S., Wang, Z., Hu, X. and Ashimk, D. 2009. Hot-air drying of whole fruit Chinese jujube (Zizyphus jujuba Miller): Physicochemical properties of dried products. International Journal of Food Science and Technology, 44(44), 1415–1421.

Fantasia, L. D., Sperber, W. H. and Deibel, R. H. 1969. Comparison of two procedures for detection of Salmonella in food, feed, and pharmaceutical products. Appl. Microbiol. 17:540-541.

FAO, 1990. Manual on simple methods of meat preservation. FAO animal production and health paper 79. http://www.fao.org/docrep/003/x6932e/X6932E00. htm#TOC.

FAO, 1991. Guidelines for slaughtering, meat cutting and further processing. FAO animal production and health paper 91. http://www.fao.org/docrep/004/t0279e/T0279E00.h tm#TOC.

FAO/WHO, 2002. Guidelines for the evaluation of probiotics in food. http://www.who.int/foodsafety/fs_manageme nt/en/probiotic_guidelines.pdf.

FAO/WHO, 2006; Codex Committee on Food Additives and Contaminants. Proposed Draft Code of Practice for the Prevention and the Reduction of Dioxin and Dioxin-Like PCB Contamination in Foods and Feeds; FAO: Rome, Italy.

FDA (Food and Drug Administration) 2012 b. Bad Bug Book: Handbook of foodborne pathogenic microorganisms and natural toxins.

FDA/CFSAN. 2003b. Foodborne Pathogenic Microorganisms and Natural Toxins Handbook: The "Bad Bug Book." Center for Food Safety and Applied Nutrition, Food and Drug Admin., College Park, Md.

Fehlhaber, K. and Kruger, G. 1998. The study of Salmonella enteritidis growth kinetics using rapid automated bacterial impedance technique. J Appl Microbiol, 84: 945–9.

Felice, C.J., Valentinuzzi, M.E., Vercellone, M.I. and Madrid, R.E. 1992. Impedance bacteriometry: medium and interface contributions during bacterial growth. IEEE Trans Biomed Eng., 39:1310–3.

Fernandes, D. V. G. S., Castro, V. S., Neto, A. D. and Figueiredo, E. E. D. 2018. *Salmonella* spp. in the fish production chain: A review. Ciencia Rural, 48(8), e20180141.

Ferreira, M.A.S.S. and Lund, B.M. 1987. The influence of pH and temperature on initiation of growth of Salmonella spp. Lett Appl Microbiol 5:67-70.

Ferrocino, I., Nasi, A., Ndagijimana, M., Vernocchi, P. and La Storia, A. 2011. Monitoring of microbial metabolites and bacterial diversity in beef stored under different packaging conditions. Appl. Environ. Microbiol. 77, 7372-7381.

Finn, S., Rogers, L., Händler, K., Mcclure, P., Amézquita, A., Hinton, J. C. and Fanning, S. 2015. Exposure of Salmonella enterica Serovar typhimurium to three humectants used in the food industry induces different osmoadaptation systems. Applied and Environmental Microbiology, 81(19): 6800.

Flannigan, B. 1996. The microflora of barley and malt. In F. G. Priest and I. Campbell (Eds.), *Brewing microbiology* (pp. 83– 125).

Fletcher, M.T. and Netzel, G. 2020. Food Safety and Natural Toxins. Toxins (Basel). 12(4): 236.

Flint, S.H. and Brooks, J.D. 2001. Rapid detection of Bacillus stearothermophilus using impedance-splitting. J Microbiol Methods, 44:205–8.

Food and Drug Administration (FDA). June, 2021. Fish and fishery products hazards and controls. Fourth Edition.

Food Safety and Standards (Licensing and Registration of Food businesses) Regulation, 2011.

Forss, D.A. 1972. Odor and flavour compounds from lipids. In: Holman, R.T. (Ed.), Progress in the Chemistry of Fats and Other Lipids. Pergamon Press, London, pp. 181-258.

Fox, P.F., Guinee, T.P., Cogan, T.M. and Mc Sweeney, P.L.H. 2016 In: Fundamentals of cheese science. Principal families of cheese. Chapter 3, Springer, New York, pp 27–69.

Frank, J.F. 2007. Milk and Dairy Products. In: MP Doyle; LR Beuchat (eds). Food Microbiology: Fundamentals and Frontiers. Washington D.C.: ASM Press.

Frazier, W. C and Westhoff, D. C. 2008. Food Microbiology. New Delhi: Tata McGrawHill Publishing Co. Ltd, P. 401-439.

Friedemann, M. 2019. Ciguatera fish poisoning outbreaks from 2012 to 2017 in Germany caused by snappers from India, Indonesia, and Vietnam. J. Consum. Prot. Food Saf., 14, 71–80.

Frisvad, J.C., Hubka, V., Ezekiel, C.N., Hong, S.B., Nováková, A., Chen, A.J., Arzanlou, M, Larsen, T.O., Sklená¡r, F., Mahakarnchanakul, W. *et al.,* 2019. Taxonomy of Aspergillus section Flavi and their production of aflatoxins, ochratoxins and other mycotoxins. Stud. Mycol., 93, 1–63.

Fuller, R.1989. "Probiotics in man and animals," Journal of Applied Bacteriology, 66 (5), 365–378.

Fung, D.Y.C. 1992. Historical development of rapid methods and automation in micro-biology. J Rapid Methods Autom Microbiol., 1:1–14.

Futschik, K., Pfützner, H. and Nussbaum, C. 1995. Selective detection of microorganisms by use of electrode impedance. Proceedings of the IX. International Conference on Electrical Bio-Impedance. Heidelberg, p. 79–82.

Ganesan, B. and Weimer, B.C. 2007. Improving the flavour of cheese. In: Weimer, B.A. (Ed.), Compounds Associated with Cheese Flavor. CRC/ Woodhead Publishing Limited, Cambridge, pp. 26-42.

Garcia, C., Berdague, J.L., Antequera, T. and Lopez-Bote, C. 1991. Volatile components of dry-cured Iberian ham. Food Chem. 41, 23-32.

Gaston, J.H. 1990. How does HLA-B27 confer susceptibility to inflammatory arthritis? Clin Exp Immunol. 82:1-2.

Gedikli S., Tabak Ö., Tomsuk Ö., *et al.,* 2008. Effect of microwaves on some gram negative and gram-positive bacteria. Journal of Applied Biological Science. 2: 67-71.

Genovese, A., Gambuti, A., Piombino, P. and Moio, L., 2007. Sensory properties and aroma compounds of sweet Fiano wine. Food Chem. 103, 1228-1236.

Ghaly, A.E. 2011. Meat Spoilage Mechanisms and Preservation Techniques: A Critical Review. American Journal of Agricultural and Biological Sciences 6 (4): 486-510.

Ghoneum, M. and Gimzewski, J. 2014. Apoptotic effect of a novel kefir product, PFT, on multidrug-resistant myeloid leukemia cells via a hole-piercing mechanism. Int. J. Oncol., 44, 830–837.

Gilch, S., Chitoor, N., Taguchi, Y., Stuart, M., Jewell, J.E. and Schätzl, H. M. 2011.Chronic wasting disease. Top Curr Chem., 305:51-77.

Gill, C.O. 1983. Meat spoilage and evaluation of the potential storage life of fresh meat. J. Food Prot. 46, 444-452.

Gill, C.O. and Molin, G. 1991. Food preservatives. In: Russell, N.J., Gould, G.W. (Eds.), Modified Atmospheres and Vacuum Packaging. Blackie, Glasgow and London, pp. 172-199.

Gill, C.O. and Penney, N. 1985. Modification of in-pack condition to extend the storage life of vacuum packaged lamb. Meat Sci. 14, 43-60.

Gilliland, S.E., Nelson, C.R. and Maxwell, C. 1985. Assimilation of cholesterol by: Lacidophilus. Appl. Environ. Microbiol. 49: 377- 381.

Giraldo, G., Talens, P., Fito, P. and

Chiralt, A. 2003. Influence of sucrose solution concentration on kinetics and yield during osmotic dehydration of mango. Journal of Food Engineering, 58(1), 33–43.

Gkogka, E., Reij, M., Havelaar, A., Zwietering, M. and Gorris, L. 2011. Risk-based estimate of effect of foodborne diseases on public health, Greece. Emerg Infect Dis, 17.

Glasgow, H.B Jr, Burkholder, J.M., Schmechel, D.E., Tester, P.A. and Rublee, P.A. 1995. Insidious effects of a toxic estuarine dinoflagellate on fish survival and human health. Journal of Toxicology and Environmental Health 46: 501–522.

Glassmoyer, K.E. and Russell, S.M. 2001. Evaluation of a selective broth for detection of Staphylococcus aureus using impedance microbiology. J Food Prot., 64:44–50.

Goktepe, I. and Moody, M. W. 2010. Effect of modified atmosphere packaging on the quality of smoked catfish. Journal of Muscle Foods, 9(4), 375–389.

Goldin, B.R. and Gorbach, S.L. 1992. Probiotics for humans. In: Probiotics. The Scientific Basis (Fuller, R. ed.,) pp 355-376. Chapman and Hall.

Goldin, B.R. and Gorbach, S.L. 2008. Clinical indications for probiotics: An Overview. Clinical Infectious Diseases 46(2): S96-S100.

Gomez, R., Bashir, R. and Bhunia, A. 2002. Microscale electronic detection of bacterial metabolism. Sens Actuators B, 86:198–208.

Gonzalez, B. and Diez, V. 2002. The effect of nitrite and starter culture on microbiological quality of "chorizo"—A Spanish dry cured sausage. Meat Sci. 60: 295–298.

Gonzalez-Martnez, C., Escriche, I. and Camacho, M. M. 2001. Use of vacuum impregnation in food salting process. Journal of Food Engineering, 49(2), 141–151.

Goodacre, J.A., Brownlee, C.E.D. and Ross, D.A. 1994. Bacterial superantigens in autoimmune arthritis. Br J Rheumatol., 33: 413-9.

Gorbach, S.L. 1990. Lactic acid bacteria and human health. Ann.Med. 22(1):37-41.

Gould, G. W. 2000. The use of other chemical preservatives: Sulfite and nitrite. In B. M. Lund, T. C. Baird-Parker, and G. W. Gould (Eds.), The microbiological safety and quality of foods (pp. 200–213). Gaithersburg: Aspen.

Goulet V. 1995. Investigation en cas d'épidemie de listériose, Méd Mal Infect 25 Spécial. 184-90.

Goulet, V., Jacquet, C., Vaillant, V., Rebiere, I., Mouret, F., Lorente, C., Maillot, E., Stainer, F. and Rocourt, J. 1995.Listeriosis from consumption of raw-milk cheese. Lancet, 345 (8964), 1581-1582.

Gram, L. and Daglaard, P. 2002. Fish spoilage bacteria-problems and solutions. Curr. Opin. Biotechnol. 13, 262-266.

Gram, L. and Huss H. 1996. Microbiological spoilage of fish and fish products. International Journal of Food Microbiology. pp 121–137.

Gram, L., Ravn, L., Rasch, M., Bruhn, J.B., Christensen, A.B. and Givskov, M. 2002. Food spoilage interactions between food spoilage bacteria. Int. J. Food Microbiol. 78, 79-97.

Grant, I.R., Ball, H.J., Neill, S.D. and Rowe, M.T. 1996. Inactivation of Mycobacterium paratuberculosis in cows' milk at pasteurisation temperatures. Applied and Environmental Microbiology 62, 631-636.

Grau, F.H. 1983. End products of glucose fermentation by Brochothrix thermosphacta. Appl. Environ. Microbiol. 45, 84-90.

Grosch, W. 1982. Food flavours. In: Morton, I.D., MacLeod, A.J. (Eds.), Lipid Degradation Products and Flavour. Elsevier, Amsterdam, pp. 325-398.

Guo, D. J., Xiong, S. B., Song, Z. and Liu, R. 2014. Effect of drying on the quality of low temperature sausage. Science and Technology of Food Industry, 35(18), 843–853.

Guth, H. 1997. Quantitation and sensory studies of character impact odorants of different White wine varieties. J. Agric. Food Chem. 45, 3027-3032.

Güven, A., Güven, A. and Gülmez, M. 2003. The effect of kefir on the activities of GSH-Px, GST, CAT, GSH and LPO levels in carbon tetrachloride-induced mice tissues. J Vet Med B Infect Dis Vet Public Health 50, 412–416.

Guz, L. and Kozinska, A. 2004. Antibiotic susceptibility of *Aeromonas hydrophila* and *A. sobria* isolated from farmed carp (*Cyprinus carpio* L.). Bulletin of the Veterinary Institute in Pulawy, 48(4), 391– 395.

Lancet 345, 1581-1582.

HACCP. 2000. Quality Control/Quality Assurance and International Trade; Good Manufactug. P.G Practices (Gmp). Hygiene Requirements.

Hafez, A. E., Darwish, W. S., Elbayomi, R. M., Hussein, M. A. M. and El Nahal, S. M. 2018. Prevalence, antibiogram and molecular characterization of *Aeromonas hydrophila* isolated from frozen fish market in Egypt. Slovenian Veterinary Research, 55, 445– 454.

Hagelberg, S., Hult, K. and Fuchs, R. 1989. Toxicokinetics of ochratoxin A in several species and its plasma-binding properties. J Appl Toxicol., 9:91–96.

Hald, T., Aspinall, W., Devleesschauwer, B., Cooke, R., Corrigan, T., Havelaar, A.H., Gibb, H.J., Torgerson, P.R., Kirk, M.D., Angulo, F.J. *et al.,* 2016. World Health Organization estimates of the relative contributions of food to the burden of disease due to selected foodborne hazards: a structured expert elicitation. PLoS One, 11:1-35.

Hall, B.G., Acar, H., Nandipati, A. and Barlow, M. 2014. Growth rates made easy. Mol. Biol. Evol. 31, 232–238.

Hamid, R., Masood, A., Wani, I.H. and Rafiq, S. 2013. Lectins: Proteins with Diverse Applications. Journal of Applied Pharmaceutical Science, 3 (4 Suppl 1), S93-S103.

Hammi, I., Delalande, F., Belkhou, R., Marchioni, E., Cianferani, S. and Ennahar, S. 2016. Maltaricin CPN, a new class IIa bacteriocin produced by Carnobacterium mal-taromaticum CPN isolated from mould-ripened cheese. J. Appl. Microbiol. 121,1268–1274.

Harris, L., Shebuski, J., Danyluk, M., Palumbo, M. and Beuchat, L. 2013. Nuts, seeds, and cereals. In M. Doyle and R. Buchanan (Eds.), *Food microbiology* (pp. 203– 221). Wash., U.S.A.: ASM Press.

Hause, L.L., Komorowski, R.A. and Gayon F. 1981. Electrode and electrolyte impedance in the detection of bacteria growth. IEEE Trans Biomed Eng., 28:403–10.

Havelaar, A.H., de Wit, M.A., van Koningsveld, R. and van Kempen, E. 2000. Health burden in the Netherlands due to infection with thermophilic Campylobacter spp. Epidemiol Infect,125:505-522.

Havelaar, A.H., Haagsma, J.A., Mangen, M.J.J., Kemmeren, J.M., Verhoef, L.P.B., Vijgen, S.M.C., Wilson, M., Friesema, I.H.M., Kortbeek, L.M., van Duynhoven, Y.T.H.P. *et al.,* 2012. Disease burden of foodborne pathogens in the Netherlands, 2009. Int J Food Microbiol, 156:231-238.

Hayatgheib, N., Moreau, E., Calvez, S., Lepelletier, D. and Pouliquen, H. 2020. A review of functional feeds and the control of *Aeromonas* infections in freshwater fish. Aquaculture International, 28, 1083–1123.

Hayes, J.R., English, L.L., Carter, P.J., Proescholdt, T., Lee, K.Y. *et al.,* 2003. Prevalence and antimicrobial resistance of Enterococcus species isolated from retail meats. Applied Environmental Microbiology, 69: 7153-7160.

Hayes, P.R. 1998. Food Microbiology and Hygiene, second ed. Elsevier Applied Science, London.

Hazarika, R. A., Singh, D. K., Kapoor, K. N., Agarwal, R. K., Pandey, A. B. and Rajkumar, D. N. 2005. Detection and characterization of vero-toxin producing E. coli (VTEC) isolated from buffalo meat. Journal of Food Safety, 24, 281–290.

Hedberg, C.W. *et al.,* 1992. A multistate outbreak of Salmonella javiana and Salmonella oranienberg infections due to contaminated cheese. Journal of the American Medical Association 268: 3203- 3207.

Hedberg, C.W., MacDonald, K.L. and Osterholm, M.T. 1994. Changing epidemiology of food-borne diseases: a Minnesota perspective. Clinical Infect Dis. 18:671-80.

Heinitz, M. L., Ruble, R. D., Wagner, D. E. and Tatini, S. R. 2000. Incidence of *Salmonella* in fish and seafood. Journal of Food Protection, 63(5), 579– 592.

Henderson, D.M., Denkers, N.D., Hoover, C.E., Garbino, N., Mathiason, C.K. and Hoover, E.A. 2015. Longitudinal detection of prion shedding in saliva and urine by chronic wasting disease-infected deer by real-time quaking-induced conversion. J Virol., 89(18):9338-47.

Hernandez-Macedo, M.L., Contreras-Castillo, C.J., Tsai, S.M., Da Cruz, S.H., Sarantopoulos, C.I., Padula, M. *et al.,* 2012. Gases and volatile compounds associated with micro-organisms in blown pack spoilage of Brazilian vacuum packed beef. Lett. Appl. Microbiol. 55, 467-475.

Hertzler, S.R. and Clancy, S.M. 2003. Kefir improves lactose digestion and tolerance in adults with lactose maldigestion. Journal of the American Dietetic Association. 103, 582 -587.

Hill, B., Smythe, B., Lindsay, D. and Shepherd, J. 2012. Microbiology of raw milk in New Zealand. Int J Food Microbiol., 157(2): 305-308.

Hill, R.A., D.M. Wilson, W.R. Burg and O.L. Shotwell. 1984. Viable fungi in corn dust. *Appl. Environ. Microbiol.,* 47: 84–87.

Hobbie, *i.e.,* R.I. Daley and S. Jasper. 1977. Use of nucleopore filters for counting bacteria by fluorescence microscopy. *Appl. Environ. Microbiol.,* 33: 1225– 28.

Holdstock, J., Aalhus, J.L., Uttaro, B.A., Lopez-Campos, O., Larsen, I.L. and Bruce, H.L. 2014. The impact of ultimate pH on muscle characteristics and sensory attributes of the long is simus thoracis within the dark cutting (Canada B4) beef carcass grade. Meat Sci. 98, 842–849.

Holley, R.A., S.M. Smith and A.G. Kempton. 1977. Rapid measurement of meat quality by resazurin reduction. I. Factors affecting test validity. *Can. Inst. Food Sci. Technol. J.,* 10: 153–57.

Hope, A.F., Tulk, P.A. and Condron, J.R. 1997. Commercial pasteurisation of Mycobacterium paratuberculosis in whole milk. Proceedings of the 5th International Colloquium on Paratuberculosis. International Association for Paratuberculosis Inc.: Rehoboth, MA, USA.

Hospital, X.F., Hierro, E. and Fernàndez, M. 2012. Survival of *Listeria innocua* in dry fermented sausages and changes in the typical microbiota and volatile profile as affected by the concentration of nitrate and nitrite. Int. J. Food Microbiol. 153: 395–401.

Hozbor, M. C., Saiz, A. I., Yeannes, M. I. *et al.,* 2006. "Microbiological changes and its correlation with quality indices during aerobic iced storage of sea salmon (Pseudopercis semifasciata)", LWT-Food Sci. Technol, 39(2): 99-104.

Hui, L., Sun, X., Liao, X. and Gänzle, M. 2020. Control of pathogenic and spoilage bacteria in meat and meat products by high pressure: Challenges and future perspectives. Comprehensive reviews in food science and food safety. 19(20): https://doi.org/10.1111/1541-4337.12617.

Huis in't Veld, J.H.J. 1996. Microbial and biochemical spoilage of foods: an overview. Int. J. Food Microbiol. 33, 1-18.

Hurst, A., A. Hughes, D.L. Collins-Thompson and B.G. Shah. 1974. Relationship between loss of magnesium and loss of salt tolerance after sublethal heating of *Staphylococcus aureus. Can. J. Microbiol.,* 20: 1153–58.

Hurst, A., A. Hughes, I.L. Beare-Rogers and D.L. Collins-Thompson. 1973. Physiological studies on the recovery of salt tolerance by *Staphylococcus aureus* after sublethal heating. *J. Bacteriol.,* 116: 901–7.

Hurst, A., A. Hughes, M. Duckworth and I. Baddiley. 1975. Loss of D-alanine during sublethal heating of *Staphylococcus aureus* S- 6 and magnesium binding during repair. *J. Gen. Microbiol.,* 89: 277–84.

Hurst, A. 1977. Bacterial injury: A review. *Can. J. Microbiol.,* 23: 935– 44.

Huseini H.F., Rahimzadeh, G., Fazeli, M.R., Mehrazma, M. and Salehi, M. 2012. Evaluation of wound healing activities of kefir products. Burns 38 (5), 719–723.

Hussein Hamad, S. 2012. Factors Affecting the Growth of Microorganisms in Food, Chapter 20, In: Progress in Food Preservation (Book Editor(s):Rajeev Bhat, Abd Karim Alias, Gopinadhan Paliyath). Wiley Online Library.

Iandolo, J.J. and Ordal, Z.J. 1966. Repair of thermal injury of ***Staphylococcus aureus.*** J Bacteriol., 91(1): 134–142.

Ibrahim, S.A., Gyawali, R., Awaisheh, S.S., Ayivi, R. D., Silva, R.C., Subedi, K., Aljaloud, S.O., Siddiqui, S.A. and Krastanov, A. 2021. Fermented foods and probiotics: An approach to lactose intolerance. Journal of Dairy Research, 88 (3), 357 - 365.

ICMSF, 1980. International Commission on Microbiological Specification for Foods. Microbial ecology of foods. Volume 1, Factors affecting life and death of microorganisms. Orlando: Academic Pr. p 311.

IFT. 1987. Perspective on food irradiation. Food Technology 41(2):100-101.

International Committee on Microbiological Safety in Foods. HACCP in microbiological safety and quality. In: Microorganisms in Food. Book 4. Oxford: Blackwell Scientific Publications Ltd.; 1988. p. 28-43.

International Consultative Group on Food Irradiation (1991) Facts about Food Irradiation (set of 14 Fact Sheets covering all aspects of food irradiation issued as public information). ICGFI Fact Series 1-14. IAEA, Vienna.

Ismail, S. A., Shehta, A. A. and El-diasty, E.M. 2013. Microbiological quality of some meat products in local markets with special reference to mycotoxins. Global Veterinaria 10: 577-584.

Ito and Matsuyama, A. 2021. Koji Molds for Japanese Soy Sauce Brewing: Characteristics and Key Enzymes Kotaro. J. Fungi, 7, 658.

Iwamoto, M., Ayers, T., Mahon, B.E. and Swerdlow, D.L. 2010. Epidemiology of Seafood-Associated Infections in the United States. Clin Microbiol Rev., 23(2): 399–411.

Jääskeläinen, E., Hultman, J., Parshintsev, J., Riekkola, M.L., Björkroth, J. 2016. Development of spoilage bacterial community and volatile compounds in chilled beef under vacuum or high oxygen atmospheres. Int. J. Food Microbiol. 223, 25-32.

Jääskeläinen, E., Johansson, P., Kostiainen, O., Nieminen, T., Schmidt, G., Somervuo, P., Mohsina, M., Vanninen, P., Auvinen, P. and Björkroth, J. 2013. Significance of heme-based respiration in meat spoilage caused by Leuconostoc gasicomitatum. Appl. Environ. Microbiol. 79, 1078-1085.

Jackson, H. and M. Woodbine. 1963. The effect of sublethal heat treatment on the growth of *Staphylococcus aureus. J. Appl. Bacteriol.,* 26: 152–58.

Jafari, M. and Emam-Djomeh, Z. 2007. Reducing nitrite content in hot dogs by hurdle technology. Food Control, 18(12), 1488–1493.

Jakobsen, L.S., Granby, K., Knudsen, V.K., Nauta, M., Pires, S.M. and Poulsen, M. 2016. Burden of disease of dietary exposure to acrylamide in Denmark. Food Chem Toxicol, 90:151-159.

James, J. B. and Ngarmsak, T. 2010. Processing of fresh-cut tropical fruits and vegetables: A technical guide. 39: 837– 40.

Jana, A.H. and Govind, P. and Tagalpallewar. 2017. Functional properties of Mozzarella cheese for its end use application. J Food Sci Technol., 54(12): 3766–3778.

Jangam, S. V. 2016. Recent critical reviews of drying. Drying Technology, 34(4), 385.

Janssen, D.W. and F.F. Busta. 1973. Influence of milk components on the injury, repair of injury and death of *Salmonella anatum* cells subjected to freezing and thawing. *Appl. Microbiol.,* 26: 725–32.

Jay, J.M. 2000. Modern food microbiology. 6th ed. Gaithersburg (MD): Aspen. p 679.

Jay, J.M., Loessner, M.J. and Golden, D.A. 2005. Modern Food Microbiology. 7 th ed. New York: Springer Science and Business Media, New York, USA.

Jiang, Y., Gao, F., Xu, X., Ye, K. Zhou, G. 2011. Changes in the composition of the bacterial flora on tray-packaged pork during chilled storage analyzed by PCRDGGE and Real-Time PCR. J. Food Sci. 76, 27-33.

Johnson, H.M., Torres, B.A. and Soos, J.M. 1996. Superantigens: structure and relevance to human disease. Proc Soc Exp Biol Med. 212:99-109.

Johnson, M.E. 2001. Cheese products. In: Marth EH, Steele JL, editors. *Applied dairy microbiology.* 2nd ed. Marcel Dekker, Inc; New York, NY, USA. pp. 345–384.

Joint FAO/IAEA/WHO Study Group on High Dose Irradiation (1998). Weekly Epidemiological Record, Jan. 16/98.

Jones, R.J. 2004. Observations on the succession dynamics of lactic acid bacteria populations in chill-stored vacuum-packaged beef. Int. J. Food Microbiol., 90, 273-282.

Joo, S.T. and Kim, G.D. 2011. Control of meat quality. In: Joo, S.T. (Ed.), Meat Quality Traits and Control Technologies. Research Signpost, Kerala, pp. 1-20.

Juffs, H.S. and F.I. Babel. 1975. Rapid enumeration of psychrotrophic bacteria in raw milk by the microscopic colony count. *J. Milk Food Technol.,* 38: 333–36.

Julseth, R.M. and R.H. Deibel. 1974. Microbial profile of selected spices and herbs at import. *J. Milk Food Technol.,* 37: 414–19.

Kadam, S. S. 1997. Processing of Fruits and Vegetables. Chaitanya Pub. India.

Kaferstein, F.K. 1976. The microftora of parsley. *J. Milk Food Technol.,* 39: 837-40

Kahraman, B. B., Dumen, E., Issa, G., Kahraman, T. and Ikiz, S. 2017. Incidence of *Aeromonas hydrophila* and *Plesiomonas shigelloides* in seafoods. Turkish Journal of Fisheries and Aquatic Sciences, 17(6), 1309– 1312.

Kajs, T.M., R. Hagenmaier, C. Vanderzant and K.F. Mattil. 1976. Microbiological evaluation of coconut and coconut products. *J. Food Sci.,* 41: 352–56.

Kalchayanand, N., Ray, B., Field, R.A. and Johnson, M.C. 1989. Spoilage of vacuum packed refrigerated beef by Clostridium. J. Food Prot. 52, 424-426.

Kamboj, A., Gupta, N., Bandral, J.D., Gandotra, G. and Anjum, N. 2020. Food Safety and Hygiene: A Review. Int. J Chem Stud. 8(2): 358-368.

Kampf, N. and Nussinovitch, A. 2000. Hydrocolloid coating of cheeses. Food Hydrocolloids, 14(6), 531–537.

Kanatt, S. R., Chawla, S. P., Chander, R. and Sharma, A. 2006. Development of shelf stable, ready-to-eat (RTE) shrimps (Penaeus indicus) using gamma-radiation as one of the hurdles. LWT – Food Science and Technology, 39(6), 621–626.

Kandler, O. 1983. Carbohydrate metabolism in lactic acid bacteria. Antonie van Leeuwenhoek, 49, 209-224.

Kandler, O. and Weiss, N. 1986. Bergey's manual of systematic bacteriology. In: Sneath, P.H.A., Mair, N.S., Sharpe, M.E., Holt, J.G. (Eds.), Genus Lactobacillus. Williams and Wilkins Co, Baltimore, MD, pp. 1208-1234.

Kantor, L.S., Lipton, K., Manchester, A. and Oliveira, V. 1997. Estimating and addressing America's food losses. Food Rev., 20: 3-11.

Kapasi, K., Inman, R.D. 1992. HLA-B27 expression modulates Gram- negative bacterial invasion into transfected L-cells. J Immunol.148:3554-9.

Karl, D.M. (1980). Cellular nucleotide measurements and applications in microbial ecology. Microbiol. Rev., 44, 739-785.

Kaur, M., Bowman, J.P., Porteus, B., Dann, A.L., Tamplin, M. 2017. Effect of abattoir and cut on variations in microbial communities of vacuum-packaged beef. Meat Sci. 131,34–39.

Kautter, D.A., R.K. Lynt, T. Lilly, Jr. and H.M. Solomon. 1981. Evaluation of the botulism hazard from imitation cheeses. *J. Food Sci.,* 46: 749–50, 764.

Kay, R.A. 1995. The potential role of superantigens in inflammatory bowel disease. Clin Exp Immunol.100:4-6.

Kazanas, N., J.A. Emerson, H.L. Seagram and L.L. Kempe. 1966. Effect of g-irradiation on the microflora of fresh-water fish. I. Microbial load, lag period and rate of growth on yellow perch (*Perca flavescens*) fillets. *Appl. Microbiol.,* 14: 261–66.

Keeton, J.T. and C.C. Melton. 1978. Factors associated with microbial growth in ground beef extended with varying levels of textured soy protein. *J. Food Sci.,* 43: 1125–29.

Kembel, S.W., Wu, M., Eisen, J.A. and Green, J.L. 2012. Incorporating 16S gene copy number information improves estimates of microbial diversity and abundance. PLoS Comp. Biol. 8, e1002743.

Kendall, P. 1985. Food irradiation - promises and controversies. Critical Issues in Food and Nutrition 3(6), Colorado State Univ. Cooperative Extension Service, Fort Collins, CO. 7 pp.

Kenneth, R. and Ishige, N. 2010. On the Origins, Diffusion and Cultural Context of Fermented FishProducts in Southeast Asia. In Globalization, Food and Social Identities in the Asia Pacifc Region.

Kergourlay, G., Taminiau, B., Daube, G. and Champomier Verges, M.C. 2015. Metagenomic insights into the dynamics of microbial communities in food. Int. J. Food Microbiol., 213, 31-39.

Khapre, A. P., Deshpande, H.W. and Katke, S.D. 2020. A review on microbial contamination of Cereal grains. International Journal of Chemical Studies; 8(3): 1829-1832.

Khazaeli, P., Mehrabani, M., Heidari, M.R., Asadikaram, G. and Lari Najafi, M. 2017. Prevalence of aflatoxin contamination in herbs and spices in different regions of Iran. Iran J. Public Health, 46, 1540–1545.

Khoury, N., El-Hayek, S., Tarras, O., El-Sabban, M., El-Sibai, M. and Rizk, S. 2014. Kefir exhibits anti-proliferative and pro-apoptotic effects on colon adenocarcinoma cells with no significant effects on cell migration and invasion. Int. J. Oncol., 45, 2117–2127.

Kim, B., Yun, H., Jung, S., Jung, Y., Jung, H., Choe, W. and Jo, C. 2011. Effect of atmospheric pressure plasma on inactivation of pathogens inoculated onto bacon using two different gas compositions. Food Microbiology, 28(1), 9–13.

Kim, K.M., Lim, J., Lee, J.J. and Hurh, B.S. 2017. Characterization of Aspergillus sojae Isolated from Meju, Korean Traditional Fermented Soybean Brick. J. Microbiol. Biotechnol., 27, 251–261.

Kirk, M.D., Pires, S.M., Black, R.E., Caipo, M., Crump, J. A., Devleesschauwer, B., Do / pfer, D., Fazil, A., Fischer-Walker, C.L., Hald, T. *et al.,* 2015. World Health Organization estimates of the global and regional disease burden of 22 foodborne bacterial, protozoal, and viral diseases, 2010: a data synthesis. PLoS Med, 12:1-21.

Kitchell, A.G. 1962. Micrococci and coagulase negative staphylococci in cured meats and meat products. *J. Appl. Bacteriol.,* 25: 416–31.

Koller, W. 1984. Recovery of test bacteria from surfaces with a simple new swab-rinse technique: A contribution to methods for evaluation of surface disinfectants. *Zent. Bakteriol. Hyg. J. Orig. B.,* 179: 112–124.

Kong, K.L. and Hendrich, S. 2012. Glycemic index, insulinemic index, and satiety index of kefir. J. Am. Coll. Nutr., 31, 280–287.

Kosiak, B., Torp, M., Skjerve, E. and Andersen, B. 2004. *Alternaria* and *Fusarium* in Norwegian grains of reduced quality—A matched pair sample study. International Journal of Food Microbiology, 93(1), 51– 62.

Kotb, M. 1995. Bacterial pyrogenic exotoxins as superantigens. Clin Microbiol Rev. 8:411-26.

Kotb, M. 1995. Infection and autoimmunity: a story of the host, the pathogen and the copathogen. Clin Immunol Immunopathol. 74:10-22.

Kotula, A.W. and B.S. Emswiler-Rose. 1981. Bacteriological quality of hot-boned primal cuts from electrically stimulated beef carcasses. *J. Food Sci.,* 46: 471–74.

Kotula, A.W. 1981. Microbiology of hot–boned and electrostimulated meat. *J. Food Protect.,* 44: 545–49.

Koutsoumanis, K. and Nychas, G.J.E. 2000. Application of a systematic experimental procedure to develop a microbial model for rapid fish shelf life predictions. Int. J. Food Microbiol., 60, 171-184.

Kovacevic, J., Mclntyre, L. F., Henderson, S. B. and Kosatsky, T. 2012. Occurrence and distribution of *Listeria* species in facilities producing ready-to-eat foods in British Columbia, Canada. Journal of Food Protection, 75(2), 216– 224.

Kowalska, H., Lenart, A. and Leszczyk, D. 2008. The effect of blanching and freezing on osmotic dehydration of pumpkin. Journal of Food Engineering, 86(1), 30–38.

Kraft, A.A., J.C. Ayres, G.S. Torrey, R.H. Salzer and G.A.N. DeSilva. 1966. Coryneform bacteria in poultry, eggs and meat. *J. Appl. Bacteriol.,* 29: 161– 66.

Kraft, A.A., V. Reddy, R.I. Hasiak, K.D. Lind and D.E. Galloway. 1982. Microbiological quality of vacuum packaged poultry with or without chlorine treatment. *J. Food Sci.,* 47: 380–85.

Krogh P. 1974. Mycotoxic porcine neprophathy: A possible model for Balkan endemic nephropathy. In: Puhlev A, editor. *Endemic Nephropathy.* Sofia: Publishing House of the Bulgarian Academy of Science. pp. 266–270.

Kuchenmuller, T., Hird, S., Stein, C., Kramarz, P., Nanda, A. and Havelaar, A.H. 2009. Estimating the Global Burden of Foodborne Diseases – a Collaborative Effort. Eurosurveillance, 14 (18): 1-4.

Kuipers, E.J., Uyterlinde, A.M., Pena, A.S., Roosendaal, R., Pals, G. and Nelis, G.F. 1995. Long term sequelae of *Helicobacter pylori* gastritis. Lancet. 345:1525-8.

Kumar, R.S., Kanmani, P., Yuvaraj, N. Paari, K., Pattukumar, V. and Arul, V. 2013. Traditional Indian fermented foods: a rich source of lactic acid bacteria. Int. J. Food Sci. Nutr. 64, 415-428.

Kumari, S. and Sarkar, P.K. 2016. Bacillus cereus hazard and control in industrial dairy processing environment. Food Control., 69: 20-29.

Kunová, S., Èuboò, J., Buèko, O., Kaèániová, M., Tkáèová, J., Hleba, L. and Lopašovský, ¼. 2015. Evaluation of dried salted pork ham and neck quality. Potravinarstvo, 9(1), 509-514.

Kuzdzal-Savoie, S. 1980. Determination of free fatty acids in milk and milk products. Int. Dairy Fed. Annu. Bull. 118, 53-66.

Kwon, D. Y., Jang, D.J., Yang, H.J. and Chung, K.R. History of Korean gochu, gochujang, and kimchi. J. Ethnic Foods, 1(1), 3-7.

Labadie, J.1999. Consequences of packaging on bacterial growth. Meat is an ecological niche. Meat Sci. 52, 299-305.

Laca, A., Mousia, Z., Díaz, M., Webb, C. and Pandiella, S. S. 2006. Distribution of microbial contamination within cereal grains. Journal of Food Engineering, 72(4), 332- 338.

Ladiges, W.C., J.F. Foster and W.M. Ganzi. 1974. Incidence and viability of *Clostridium perfringens* in ground beef. *J. Milk Food Technol.,* 37: 622–23.

Lagesen, K., Hallin, P., Rodland, E.A., Staerfeldt, H.H., Rognes, T. and Ussery, D.W. 2007. RNAmmer: consistent and rapid annotation of ribosomal RNA genes. Nucleic Acids Res., 35(9):3100-8.

Lake, R., Hudson, A., Cressey, P. and Gilbert, S. 2005. Risk profile: *Listeria monocytogenes* in low moisture cheeses. Client report FW0440 of New Zealand Food Safety Authority.

Lake, R.J., Cressey, P.J., Campbell, D.M. and Oakley, E. 2010. Risk ranking for foodborne microbial hazards in New Zealand: burden of disease estimates Risk Anal, 30, 743-752.

Lambert, A.D., Smith, J.P. and Dodds, K.L., 1991. Shelf life extension and microbiological safety of fresh meat d a review. Food Microbiol. 8, 267-297.

Lappi, V. R., Ho, A., Gall, K. and Wiedmann, M. 2004. Prevalence and growth of *Listeria* on naturally contaminated smoked salmon over 28 days of storage at 4 °C. Journal of Food Protection, 67(5), 1022- 1026.

Laroussi, M. 2010. Low temperature plasma-based sterilization: Overview and state-of the-art. Plasma Processes and Polymers, 2(5), 391–400.

Larsen, M. and Poll, L. 1992. Odour thresholds of some important aroma compounds in strawberries. Lebensm Unters Forsch, 195, 120-123 (Zeitschrift for SpringerVerlag).

Lau, S.W., Chong, A.Q., Chin, N.L.,Talib, R.A. and Basha, R.K. 2021. Sourdough Microbiome Comparison and Benefits. Microorganisms, 9, 1355.

Lavieri, N. and Williams, S.K. 2014. Effects of packaging systems and fat concentrations on microbiology, sensory and physical properties of ground beef stored at 4 ± 1 °C for 25 days. Meat Sci. 97, 534–541.

Law, J.W-F., Ab Mutalib, N., Chan, K. and Lee, L-H. 2014. Rapid methods for the detection of foodborne bacterial pathogens: principles, applications, advantages and limitations Front. Microbiol., 5: 770.

Lawrence, D.N. 1980. Ciguatera Fish Poisoning in Miami. J. Am. Med. Assoc., 244, 254.

Lecos, C. 1986. The growing use of irradiation to preserve food. FDA Consumer 20(6):12-15.

Lederberg, J., Shope, R.E. and Oaks, S.C.1992. Emerging infections: microbial threats to health in the United States. Washington (DC): National Academy Press.

Lee, B.H., Shard, R.E., Laleye, L.C. and Holley, R.A. 1985. Effects of temperature and storage duration on the microflora, physicochemical and sensory changes of vacuum- or nitrogen-packed pork. Meat Sci. 13, 99-112.

Lee, C.Y., D.Y.C. Fung and E.L. Kastner. 1982. Computer-assisted identification of bacteria on hot-boned and conventionally processed beef. *J. Food Sci.,* 47: 363–67, 373.

Lee, D. J. 2009. Structure evolution of wastewater sludge during electroosmotic dewatering. Drying Technology, 28(7), 890–900.

Lee, J.S. and D.K. Pfeifer. 1977. Microbiological characteristics of Pacific shrimp (*Pandalus jordani. Appl. Environ. Microbiol.,* 33: 853–59.

Leininger. H.V., L.R. Shelton and K.H. Lewis. 1971. Microbiology of frozen cream-type pies, frozen cooked-peeled shrimp and dry food-grade gelatin. *Food Technol.,* 25: 224–29.

Leira, F.J., Vieites, J.M., Botana, L.M. and Vyeites. M.R. 1998. Domoic acid levels of naturally contaminated scallops as affected by canning. J. Food Sci. 63:1081–1083.

Leisner, J.J., Laursen, B.G., Prevost, H., Drider, D. and Dalgaard, P. 2007. Carnobacterium: positive and negative effects in the environment and in foods. FEMS Microbiol. Rev.31, 592–613.

Leistner L. 1995. Principles and applications of hurdle technology. In: Gould GW, editor. New methods of food preservation. London: Blackie Academic and Professional. p 1-21.

Lekjing, S. 2016. A chitosan-based coating with or without clove oil extends the shelf life of cooked pork sausages in refrigerated storage. Meat Science, 111, 192–197.

Leroy, F., Vasilopoulos, C., Van Hemelryck, S., Falony, G. and De Vuyst, L. 2009. Volatile analysis of spoiled, artisan-type, modified-atmosphere-packaged cooked ham stored under different temperatures. Food Microbiol., 26, 94-102.

Leung, D.Y., Meissner, H.C., Fulton, D.R., Murray, D.L., Kotzin, B.L. and Schlievert, P.M.1993. Toxic shock syndrome toxin secreting *Staphylococcus aureus* in Kawasaki syndrome. Lancet. 342:1385- 8.

Li, X., Xie, X., Zhang, C.-h., Zhen, S. and Jia, W. 2018. Role of mid- and far-infrared for improving dehydration efficiency in beef jerky drying. Drying Technology, 36(3), 283–293.

Li, Y., Pei, X., Yan, J., Liu, D., Zhang, H., Yu, B., ... Yang, D. 2019. Prevalence of foodborne pathogens isolated from retail freshwater fish and shellfish in China. Food Control, 99, 131- 136.

Lillard, H.S. 1971. Occurrence of *Clostridium perfringens* in broiler processing and further processing operations. *J. Food Sci.,* 36: 1008–10.

Lin, C.M. and Wei, C.I. 1997. Transfer of Salmonella montevideo onto the interior surfaces of tomatoes by cutting. J Food Prot 60(7):858-63.

Lin, T. M., Durance, T. D. and Scaman, C. H. 1999. Physical and sensory properties of vacuum microwave dehydrated shrimp. Journal of Aquatic Food Product Technology, 8(4), 41–53.

Lin, X., Luo, C. and Chen, Y. 2016. Effects of vacuum impregnation with sucrose solution on mango tissue. Journal of Food Science, 81(6), E1412–E1418.

Lin. H.-S., D.G. Topel and H.W. Walker. 1979. Influence of prerigor and postrigor muscle on the bacteriological and quality characteristics of pork sausage. *J. Food Sci.*, 44: 1055–57.

Lioe, H.N., Selamat, J. and Yasuda, M. 2010. Soy Sauce and Its Umami Taste: A Link from the Past to Current Situation. J. Food Sci., 75, R71–R76.

Liu, H., Yang, Z., Jin, G., Quan, Y., Zheng, G., Jin, H. and Li, G. 2013. The development of Yanbian yellow dried beef. Food Science and Technology, 38(11), 116–119.

Liu, J.R., Chen, M.J. and Lin, C.W. 2005. Antimutagenic and antioxidant properties of milk-kefir and soymilk-kefir. J. Agric. Food Chem., 53, 2467–2474.

Liu, J.R., Wang, S.Y., Chen, M.J. and Yueh, P.Y. 2006. The antiallergenic properties of milk kefir and soymilk kefir and their beneficial effects on the intestinal microflora. J Sci Food Agric 86, 2527–2533.

Liu, J.R., Wang, S.Y., Lin, Y.Y. and Lin, C.W. 2002. Antitumor activity of milk kefir and soy milk kefir in tumor-bearing mice. Nutr. Cancer, 44, 183–187.

Liu, S., Tang, J., Tadapaneni, R., Yang, R. and Zhu, M. J. 2018. Exponentially increased thermal resistance of Salmonella and Enterococcus faecium at reduced water activity. Applied and Environmental Microbiology, 84(8): e02742–17.

Liu, Y., Sun, Y., Yu, H., Yin, Y., Li, X. and Duan, X. 2016. Hot air drying of purple-fleshed sweet potato with contact ultrasound assistance. Drying Technology, 35, 564–576.

Loch, T.P., Xu, W., Fitzgerald, S.M. and Faisal, M. 2008. Isolation of a Carnobacterium mal-taromaticum-like bacterium from systemically infected lake whitefish (Coregonusclupeaformis). FEMS Microbiol. Lett. 288, 76–84.

Luck, E. and Jager M. 1997. Antimicrobial food additives: characteristics, uses, effects. Springer: Berlin. pp. 260.

Lund, B.M. (1990) The prevention of foodborne listeriosis. *British Food Journal 92* (4), 13-22.

MacDonald, K.L. *et al.,* 1985. A multistate outbreak of gastrointestinal illness caused by enterotoxigenic *Escherichia coli* in imported semisoft cheese. Journal of Infectious Diseases 151, 716-720.

MacGowan AP, Bowker K, McLauchlin J, Bennett PM, Reeves DS.1994. The occurrence and seasonal changes in the isolation of *Listeria* spp. in shop bought food stuffs, human feces, sewage and soil from urban sources. Int J Food Microbiol. 21:325-34.

Maeda, H., Zhu, X., Omura, K., Suzuki, S. and Kitamura, S. 2004. Effects of an exopolysaccharide (kefiran) on lipids, blood pressure, blood glucose, and constipation. Biofactors, 22, 197–200.

Magar, S. T. 2021. Microbial spoilage of meat and meat products. Microbe notes. https://microbenotes.com/preservation-of-meat-and-meat-products/+ April 18, 2021.

Maguire *et al.,* 1992. An outbreak of Salmonella dublin infection in England and Wales associated with a soft, unpasteurized cow's milk cheese. Epidemiology and Infection 109, 389-396.

Mahale, D.P., Khade, R.G. and Vaidya, V.K. 2008. Microbiological analysis of street vended fruit juices from Mumbai city, India. Internet Journal of Food Safety, 10: 31-34.

Maicas, S. 2020. The Role of Yeasts in Fermentation Processes. Microorganisms. 8(8): 1142.

Makino, S.I., Kawamoto, K., Takeshi, K., Okada, Y., Yamasaki, M., Yamamoto, S., Igimi, S. 2005. An outbreak of food-borne listeriosis due to cheese in Japan, during 2001. Int J Food Microbiol., 104:189–196.

Maneffa, A. J., Stenner, R., Matharu, A. S., Clark, J. H., Matubayasi, N. and Shimizu, S. 2017. Water activity in liquid food systems: A molecular scale interpretation. Food Chemistry, 237, 1133–1138.

Mangen, M.J., Friesema, I.H.M., Pijnacker, R., Mughini Gras, L., VanPelt, W. 2018. Disease Burden of Food-Related Pathogens in the Netherlands, 2017.

Marchiore, N. G., Manso, I. J., Kaufmann, K. C., Lemes, G. F., Pizolli, A. P. D. O., Droval, A. A. and Leimann, F. V. 2017. Migration evaluation of silver nanoparticles from antimicrobial edible coating to sausages. LWT – Food Science and Technology, 76, 203–208.

Marco, M.L., Heeney, D., Binda, S., Cifelli, C.J., Cotter, P.D., Foligne, B., Ganzle, M., Kort, R., Pasin, G., Pihlanto, A., Smid, E.J. and Hutkins, R. 2017. Health benefits of fermented foods: microbiota and beyond. Current Opinion in Biotechnology, 44: 94–102.

Marcotte, M. 1988. Consumer acceptance of irradiation to preserve food. FDA Consumer 20(6):12-15.

Marin, S., Sanchis, V., Vines, I., Canella, R. and Magan, N. 1995. Effect of water activity and temperature on growth and fumonisin B1 and B2 production by Fusarium proliferatum and F. moniliforme on maize grain. Lett. Appl. Microbiol. 21:298–301.

Marshall, J.C., Soboleva, T.K., Jamieson, P. and French, N.P. 2016. Estimating Bacterial Pathogen Levels in New Zealand Bulk Tank Milk. J Food Prot., 79(5): 771-780.

Martýn, A., Benito, M.J., Aranda, E., Mayano, S.R., Cordoba, J. and Cordoba, M.G. 2010. Characterization by volatile compounds of microbial deep spoilage in Iberian dry-cured ham. J. Food Sci., 75, 360-366.

Masegosa, R., Delgado-Adámez, J., Contador, R., Sánchez-Íñiguez, F. and Ramírez, R. 2014. Effect of processing by hydrostatic high pressure of two ready to heat vegetable meals and stability after refrigerated storage. Food Science and Technology International, 20(8), 605–615.

Mauricio, J. 2012. Kefir D'Aqua and Its Probiotic Properties. Probiotic Anim., 1–24.

May, K.N. 1962. Bacterial contamination during cutting and packaging chicken in processing plants and retail stores. *Food Technol.,* 16: 89–91.

McFadden, J.J. and Fidler, HM. 1996. Mycobacteria as possible causes of sarcoidosis and Crohn's disease. Society for Applied Bacteriology Symposium Series 25, 47S-52S.

McGuigan, L.E., Prendergast, J.K., Geczy, A.F., Edmonds, J.P., Bashir, H.V. 1986. Significance on non-pathogenic cross-reactive bowel flora in patients with anklosing spondylitis. Ann Rheum Dis. 45:566-71.

McMeekin, T.A. 1982. Developments in food microbiology. In: Davies, R. (Ed.), Microbial Spoilage of Meats. Applied Science Publisher, London, pp. 1-40.

McMullen, L.M. and Stiles, M.E. 1993. Microbial ecology of fresh pork stored under modified atmosphere at–1, 4.4 and 10 degrees C. Int. J. Food Microbiol. 18, 1–14.

Mead, G.C. 1983. Effect of packaging and gaseous environment on the microbiology and shelf life of processed poultry products. In: *Food microbiology: Advances and Prospects*, Ed. T.A. Roberts and F.A. Skinner, New York and London, Academic Press, p. 203–16.

Meilgaard, M.S. and Peppard, T.L. 1986. Food flavours part B. The flavour beverages. In: Morton, I.D., McLeod, A.J. (Eds.), The Flavour of Beer. Elsevier, Amsterdam, pp. 99-170.

Mercuri, A.J., G.J. Banwart, J.A. Kinner and A.R. Sessoms. 1970. Bacteriological examination of commercial precooked Eastern- type turkey rolls. *Appl. Microblol.,* 19: 768–71.

Metris, A., George, S.M. and Baranyi, J. 2006. Use of optical density detection times to assess the effect of acetic acid on single-cell kinetics. Appl. Environ. Microbiol. 72, 6674–6679.

Millar D, Ford J, Sanderson J, Withey S, Tizard M, Doran T, Hermon- Taylor J. 1996. IS900 PCR to detect *Mycobacterium paratuberculosis* in retail supplies of whole pasteurized cow's milk in England and Wales. Appl Environ Microbiol. 62:3446-52.

Miller FW. 1995. Genetics of autoimmune diseases. Exp Clin Immunogenet. :182-90.

Miller, R. and Gladston, G. 1989. Rapid methods for the detection of yeast and lactobacillus by ATP bioluminescence. J. Inst. Brewing, 95, 316-319.

Mills, J., Donnison, A. and Brightwell, G. 2014. Factors affecting microbial spoilage and shelf-life of chilled vacuum-packed lamb transported to distant markets: a review. Meat Sci. 98, 71-80.

Misra, N. N., Tiwari, B. K., Raghavarao, K. S. M. S. and Cullen, P. J. 2011. Nonthermal plasma inactivation of food-borne pathogens. Food Engineering Reviews, 3(3–4), 159–170.

Moio, L., Langlois, D., Etilevant, P.X. and Addeo, F. 1993. Powerful odorants in water buffalo and bovine Mozzarella cheese by use of extract dilution sniffing analysis. Ital. J. Food Sci. 3, 227-37.

Moio, L., Piombino, P. and Addeo, F. 2000. Odour-impact compounds of Gorgonzola cheese. J. Dairy Res. 67, 273-285.

Molimard, P. and Spinnler, H.E. 1996. Review: compounds involved in the flavour of surface mold-ripened cheeses: origins and properties. J. Dairy Sci. 79, 169-184.

Molin, G. and I.-M. Stenstrom. 1984. Effect of temperature on the microbial flora of herring fillets stored in air or carbon dioxide. *Appl. Bacteriol.,* 56: 275– 82.

Molin, G., Ternstrom, A. 1982. Numerical taxonomy of psychrotrophic pseudomonads. J. Gen. Microbiol. 128, 1249-1264.

Molla, L. 2019. Burden of Foodborne Disease: Pilot Study in Albania. FAO/WHO/WTO International Forum on Food Safety and Trade; Geneva, 23-24 April 2019.

Montel, M.C., Masson, F. and Talon, R. 1998. Bacterial role in flavour development. Meat Sci. 49, S111-S123.

Moral, U., Nagar, P., Maan, S. and Kaur, K. 2017. "A Growth of Different Types of Microorganism, Intrinsic and Extrinsic Factors of Microorganism and their Affects in Food: A Review." International Journal of Current Microbiology Applied Science, 6 (1), 290-298.

Moreno, J., Gonzales, M., Zúñiga, P., Petzold, G., Mella, K. and Muñoz, O. 2016. Ohmic heating and pulsed vacuum effect on dehydration processes and polyphenol component retention of osmodehydrated blueberries (cv. Tifblue). Innovative Food Science and Emerging Technologies, 36, 112–119.

Moretti, V.M., Madonia, G., Diaferia, C., Mentasti, T., Paleari, M.A., Panseri, S., Pirone, G. and Gandini, G. 2004. Chemical and microbiological parameters and sensory attributes of a typical Sicilian salami ripened in different conditions. Meat Sci. 66, 845-854.

Morrison, R.M. 1989. An Economics Analysis of Electron Accelerators and Cobalt-60 for Irradiating Food. Technical Bulletin No. 1762. Commodity Economics Division, Economic Research Service, U.S. Department of Agriculture, Rockville, MD. June 1989.

Mossel, D.A.A., Corry, J.E.L., Struijk, C.B. and Baird, R.M. 1995. Essentials of the microbiology of foods: a textbook for advanced studies. Chichester (England): John Wiley and Sons. p 63- 109.

Mottram, D.S. 1998. Flavour formation in meat and meat products: a review. Food Chem. 62, 415-424.

Mundt, J.O., W.F. Graham and *i.e.* McCarty. 1967. Spherical lactic acid– producing bacteria of Southern-grown raw and processed vegetables. *Appl. Microbiol.,* 15: 1303–08.

Muriel, E., Antequera, T., Petron, M.J., Andres, A.I. and Ruiz, J. 2004. Volatile compounds in Iberian dry-cured loin. Meat Sci. 68, 391-400.

Murthy, T.R.K. 1984. Relative numbers of coliforms, *Enterobacteriaceae* (by two methods) and total aerobic bacteria counts as determined from minced goat meat. *J. Food Protect.,* 47: 142–44.

Naidu, A.S., Bidlack, W. R. and Clemens, R. A. 1999. "Probiotic spectra of lactic acid bacteria (LAB)," Critical Reviews in Food Science and Nutrition, 39 (1), 13–126.

Nakamura, M. and K.D. Kelly. 1968. *Clostridium perfringens* in dehydrated soups and sauces. *J. Food Sci.,* 33: 424–26.

Navarro Quiroz, R., Herrera-Usuga, J.C., Osorio-Ospina, L.M., Garcia-Pertuz, K.M., Navarro Quiroz, E. 2020. Review Epidemiology and Toxicology of Ciguatera Poisoning in the Colombian Caribbean. Mar. Drugs, 18 (10), 504.

Neal, G.E. 1995. Genetic implication in the metabolism and toxicity of mycotoxins. Toxicol Lett. 82-83:861-7.

Nejati, F., Junne, S. and Neubauer, P. 2020. A Big World in Small Grain: A Review of Natural Milk Kefir Starters. Microorganisms. 8(2): 192.

New Zealand Food Safety Technical Report No: 2020/04. Feb., 2020. Growth of microorganisms in raw milk: Evaluating the effect of chiller failure. New Zealand Govt.

Newton, K.G. and Gill, C.O. 1980. Control of spoilage in vacuum packaged dark, firm, dry (DFD) meat. Int. J. Food Sci. Tech. 15, 227-234.

Newton, K.G. and W.J. Rigg. 1979. The effect of film permeability on the storage life and microbiology of vacuum-packed meat. *J. Appl. Bacteriol.,* 47: 433–41.

Newton, K.G., J.C.L. Harrison and K.M. Smith. 1977. The e... in various gaseous atmospheres on the microflora of lan... held at –1°C. *J. Appl. Bacteriol.,* 43: 53–59.

Nguyen, D. T., Kanki, M., Nguyen, P. D., Le, H. T., Ngo, P. T., Tran, D. N., le, N.H., Dang, C.V., Kawai, T., Kawahara, R., Yonogi, S., Hirai, Y., Jinnai, M., Kumeda, Y. and Yamamoto, Y. 2016. Prevalence, antibiotic resistance, and extended-spectrum and AmpC beta-lactamase productivity of *Salmonella* isolates from raw meat and seafood samples in Ho Chi Minh City, Vietnam. International Journal of Food Microbiology, 236, 115– 122.

NHFPC – National Health and Family Planning Committee of China. 2015. GB 2760 National food safety standard for uses of food additives. Beijing.

Nichols, G., Greenwood, M. and de Louvois, J. 1996. The microbiological quality of soft cheese. PHLS Microbiology Digest, 13, (2), 68-75.

Nickerson, J.T.R. and S.A. Goldblith. 1964. A study of the microbiological quality of haddock fillets and shucked, soft- shelled clams processed and marketed in the greater Boston area. *J. Milk Food Technol.,* 27: 7–12.

Nielsen, H.-J.S. 1983. Influence of nitrite addition and gas permeability of packaging film on the microflora in a sliced vacuum–packed whole meat product under refrigerated storage. *J. Food Technol.,* 18: 573–85.

Nissen, H., Sorheim, O. and Dainty, R. 1996. Effects of vacuum, modified atmospheres and storage temperature on the microbial flora of packaged beef. Food Microbiol. 13, 183-191.

Nissen, J., Jokelainen, P., Stensvold, C.R., Trevisan, C., Fuchs, J., Burgdorf, K.S., Nielsen, H.V., Pires, S.M. 2017. The disease burden of congenital toxoplasmosis in Denmark, 2014. PLoS One, 12.

NOAA. (2020). Fisheries of the United States, 2018 report. Retrieved from https://www.fisheries.noaa.gov/resource/document/fisheries-united-states-2018-report

Nooitgedagt, A.J. and Hartog, B.J. 1988. A survey of the microbiological quality of Brie and Camembert cheese. Netherlands Milk and Dairy Journal 42, 57-72.

Notermans S, Mead GC. 1996. Incorporation of elements of quantitative risk analysis in the HACCP system. Int J Food Microbiol. 30:157-73.

Notomi, T., Okayama, H., Masubuchi, H., Yonekawa, T., Watanabe, K., Amino, N. and Hase, T. 2000. Loop-mediated isothermal amplification of DNA. Nucleic Acids Res. 28 (12): e63.

Novotny, L., Dvorska, L., Lorencova, A., Beran, V. and Pavlik, I. 2004. Fish: A potential source of bacterial pathogens for human beings. Veterinarni Medicina, 49(9), 343– 358.

Nowacka, M., Tylewicz, U., Laghi, L., Rosa, M. D. and Witrowa-Rajchert, D. 2014. Effect of ultrasound treatment on the water state in kiwifruit during osmotic dehydration. Food Chemistry, 144C (2), 18–25.

Nychas, G., Skandamis, P., Tassou, C. and Koutsoumanis, K. 2008. Meat spoilage during distribution. Meat Sci. 78, 77-89.

Nychas, G.J.E., Dillon, V.M. and Board, R.G. 1988. Glucose the key substrate in the microbiological changes occurring in meat and certain meat products. Biotechnol. Appl. Biochem. 10, 203-231.

Nychas, G.J.E., Drosinos, E.and Board, R.G. 1998. The microbiology of meat and poultry. In: Board, R.G., Davies, A.R. (Eds.), Chemical Changes in Stored Meat. Blackie Academic and Professional, London, pp. 288-326.

Nychas, G.J.E., Marshall, D. and Sofos, J. 2007. Food microbiology fundamentals and frontiers. In: Doyle, M.P., Beuchat, L.R., Montville, T.J. (Eds.), Meat Poultry and Seafood, third ed. ASM Press, pp. 105-140.

Nychas, G.J.E., Skandamis, P.N., Tassou, C.C. and Koutsoumanis, K.P. 2008. Meat spoilage during distribution. Meat Sci. 78, 77-89.

O'Mahony M, Mitchell E, Gilbert RJ, Hutchinson DN, Begg NT, Rodhouse JC, Morris JE. 1990. An outbreak of foodborne botulism associated with contaminated hazelnut yoghurt. Epidemiol Infect.104:389-95.

O'Sullivan, M.G., Thorton, G.O.'Sullivan, G.C. and Collins, J.K. (1992). Probiotic bacteria: myth and reality? Trends in Food Sci. and Technol. 3, 309-314.

Oblinger, J.L., J.E. Kennedy, Jr., C.A. Rothenberg, B.W. Berry and N.J. Stern. 1982. Identification of bacteria isolated from fresh and temperature abused variety meats. *J. Food Protect.,* 45: 650–54.

Ockerman, H.W. and J. Szezawinski. 1983. Effect of electrical stimulation on the microflora of meat. *J. Food Sci.,* 48: 1004– 1005, 1007.

Oliveira, F. I. P., Rodrigues, S. and Fernandes, F. A. N. 2012. Production of low calorie Malay apples by dual stage sugar substitution with Stevia-based sweetener[J]. Food and Bioproducts Processing, 90(4), 713–718.

Olsen, M., Thorup, I., Knudsen, I., Larsen, J.J. and Hald, B.1991. Health evaluation of ochratoxin A in food products. Number 545, Copenhagen (Denmark): Nordic Council of Ministers.

Olson, N.F. 2008. Cheese. In: Rehm HJ, Reed G (eds) Biotechnology, Vol 9, 2nd edn. Wiley-VCH Verlag GmbH, Germany, pp 353–384.

Ostadrahimi, A., Taghizadeh, A., Mobasseri, M., Farrin, N., Payahoo, L., Gheshlaghi, Z.B. and Vahedjabbari, M. 2015. Effect of probiotic fermented milk (kefir) on glycemic control and lipid profile in type 2 diabetic patients: a randomized double-blind placebo-controlled clinical trial. Iran J Public Health 44, 228–237.

Ostyn, A., De Buyser, M.L., Guillier, F., Groult, J., Félix, B., Salah, S., Delmas, G. and Hennekinne, J.A. 2009. First evidence of a food poisoning outbreak due to staphylococcal enterotoxin type E. France, 2009. Euro Surveill., 15 (13):19528.

Owen RJ. 1995. Bacteriology of *Helicobacter pylori*. Bailliere's Clin Microbiol. 9:415-46.

Owusu-Kwarteng, J., Akabanda, F., Agyei, D. and Jespersen, L. 2020. Microbial Safety of Milk Production and Fermented Dairy Products in Africa. Microorganisms, 8, 752.

Paarup, T., Sanchez, J. A., Peláez, C. and Moral, A. 2002. "Sensory, chemical and bacteriological changes in vacuum-packed pressurised squid mantle (Todaropsis eblanae) stored at 4 degrees C". International Journal of Food Microbiology, 74(1-2):1-12.

Pace, P.J. 1975. Bacteriological quality of delicatessen foods: Are standards needed? *J. Milk Food Technol.,* 38: 347–53.

Paelinck, H. 2012. Effect of rosemary extract inclusion level on lipid oxidation, color stability and antioxidant concentrations, in reduced nitrite liver pates. Meat Sci. 90: 925–931.

Pal, A., Labuza, T. P. and Diez-Gonzalez, F. 2008. Shelf life evaluation for ready-to-eat sliced uncured turkey breast and cured ham under probable storage conditions based on *Listeria monocytogenes* and psychrotroph growth. International Journal of Food Microbiology, 126(1–2), 49– 56.

Pal, M., Yodit Ayele, Y., Patel, A. S. and Dulo, F. 2018. Microbiological and hygienic quality of Meat and Meat Products. Beverage and Food World. 45 (5): 21-23; 27.

Palafox, N.A. and Buenconsejo-Lum, L.E. 2001. Ciguatera Fish Poisoning: Review of Clinical Manifestations, J. Toxicol. Toxin Rev., 20 (2), 141–160.

Panades, G., Castro, D., Chiralt, A., Fito, P., Nuñez, M. and Jimenez, R. 2008. Mass transfer mechanisms occurring in osmotic dehydration of guava. Journal of Food Engineering, 87(3), 386–390.

Pao, S., Ettinger, M. R., Khalid, M. F., Reid, A. O. and Nerrie, B. L. 2008. Microbial quality of raw aquacultured fish fillets procured from Internet and local retail markets. Journal of Food Protection, 71(8), 1544– 1549.

Paradis, D.C. and M.E. Stiles. 1978. A study of microbial quality of vacuum packaged, sliced bologna. *J. Food Protect.,* 41: 811–15.

Parish, M.E. 1991. Microbial concern in citrus juice processing. Journal of Food Technology 45(4): 128- 132.

Park, S.E., Kwon, S.J., Cho, K.M., Seo, S.H., Kim, E.J., Unno, T., Bok, S.H., Park, D.H., Son, H.S. 2020. Intervention with kimchi microbial community ameliorates obesity by regulating gut microbiota. J. Microbiol., 58, 859–867.

Parker, R.B. 1974. "Probiotics: the other half of the antibiotic story," Animal Nutrition and Health, 29, 4–8.

Paruthi, J.S. 1999. Quick freezing preservation of foods vol. II. Allied Publishers Ltd. New Delhi.

Patterson, J.T. and P.A. Gibbs. 1977. Incidence and spoilage potential of isolates from vacuum-packaged meat of high pH value. *J. Appl. Bacteriol.,* 43: 25–38.

Pelisser, M.R., Klein, C.S., Ascoli, K.R., Zotti, T.R. and Arisi, A.C. 2009. Ocurrence of *Staphylococcus aureus* and multiplex PCR detection of classic enterotoxin genes in cheese and meat products. Braz J Microbiol., 40:145–148.

Pennacchia, C., Ercolini, D. and Villani, F. 2009. Development of a Real-Time PCR assay for the specific detection of Brochothrix thermosphacta in fresh and spoiled raw meat. Int. J. Food Microbiol. 134, 230-236.

Penner, Karen. 1982. Freezing Fruits and Vegetables. Kansas State University Cooperative Extension Service.

Pereira, N. R., Marsaioli, A. and Ahrné, L. M. 2007. Effect of microwave power, air velocity and temperature on the final drying of osmotically dehydrated bananas. Journal of Food Engineering, 81(1), 79–87.

Pestka, J. J. 2010. Deoxynivalenol: Mechanisms of action, human exposure, and toxicological relevance. Arch. Toxicol. 84, 663–679.

Peter, S., Hoejskov, Minato, Y. and Devleesschauwer, B. 2021. Burden of foodborne diseases: think global, act local. Current Opinion in Food Science, 39:152–159.

Peterson, R.J. and Chang, S.S. 1982. Identification of volatile flavour compounds of fresh, frozen beef stew and comparison of these with those of canned beef stew. J. Food Sci. 47, 1444-1448.

Pham, A.J., Schilling, M.W., Mikel, W.B., Williams, J.B., Martin, J.M. and Coggins, P.C. 2008. Relationships between sensory descriptors, consumer acceptability and volatile flavor compounds of American dry-cured ham. Meat Sci. 80, 728-737.

Pierson, M.D. and Smoot, L.A. 1982. Nitrite, nitrite alternatives, and the control of *Clostridium botulinum* in cured meats. Crit. Rev. Food Sci. Nutr. 17: 141–187.

Pin, C., Garcia de Fernando, G.D. and Ordonez, J.A. 2002. Effect of modified atmosphere composition on the metabolism of glucose by Brochothrix thermosphacta. Appl. Environ. Microbiol. 68, 4441-4447.

Pinton, P., Braicu, C., Nougayrede, J. P., Laffitte, J., Taranu, I. and Oswald, I. P. 2010. Deoxynivalenol impairs porcine intestinal barrier function and decreases the protein expression of claudin-4 through a mitogen-activated protein kinase-dependent mechanism. J. Nutr. 140, 1956–1962.

Piombino, P., Pessina, R., Genovese, A., Lisanti, M.T. and Moio, L. 2008. Sensory profiling, volatiles and odour-active compounds of Canestrato Pugliese PDO cheese made from raw and pasteurized ewes' milk. Ital. J. Food Sci. 2, 225-237.

Pires, S.M., Fischer-Walker, C.L, Lanata, C.F., Devleesschauwer, B., Hall, A.J., Kir, M.D., Duarte, A.S.R., Black, R.E. and Angulo, F.J. 2015. Aetiology-specific estimates of the global and regional incidence and mortality of diarrhoeal diseases commonly transmitted through food. PLoS One, 10:1-17.

Pires, S.M., Mughini-Gras, L., Desta, B.N., Majowicz, S., Hald, T, A. Bevilacqua, M. R. Corbo, D. Campaniello *et al.,* 2011. "Shelf life prolongation of fruit juices through essential oils and homogenization: a review," in Science against Microbial Pathogens: Communicating Current Research and Technological Advances (ed Vilas, A.M.), FORMATEX C/Zurbarán 1, 2° - Oficina 1 06002 Badajoz Spain, pp. 1156–1166.

Pless, P., Futschik, K. and Schopf, E. 1994. Rapid detection of Salmonella by means of a new impedance-splitting method. J Food Prot., 57:369–76.

Post, L.S., D.A. Lee, M. Solberg, D. Furgang, J. Specchio and C. Graham. 1985. Development of botulinal toxin and sensory deterioration during storage of vacuum and modified atmosphere packaged fish fillets. *J. Food Sci.,* 50: 990– 96.

Potter, N.N. and Hotchkiss, J. H. 1996. Food Science. 5th Edition. CBS Publishing and Distributors, New Delhi.

Powers, E.M., C. Ay, H.M. El-Bisi and D.B. Rowley. 1971. Bacteriology of dehydrated space foods. *Appl. Microbiol.,* 22: 441–45.

Powers, E.M., R. Lawyer and Y. Masuoka. 1975. Microbiology of processed spices. *J. Milk Food Technol.,* 38: 683–87.

Powers, E.M., T.G. Latt and T. Brown. 1976. Incidence and levels of *Bacillus cereus* in processed spices. *J. Milk Food Technol.,* 39: 668– 70.

Preena, P. G., Swaminathan, T. R., Kumar, V. J. R. and Singh, I. S. B. 2020. Antimicrobial resistance in aquaculture: A crisis for concern. Biologia, 75, 1497– 1517.

Pulliam, J.D. and D.C. Kelley. 1965. Bacteriological comparisons of hot processed and normally processed hams. *J. Milk Food Technol.,* 28: 285–86.

Qadri, O.S., Yousuf, B., Srivastava, A.K. and Yildiz, F. 2015. Fresh-cut fruits and vegetables: Critical factors influencing microbiology and novel approaches to prevent microbial risks—A review. Cogent Food and Agri. 1: 1.

Qiu, L., Zhang, M., Tang, J., Adhikari, B. and Cao, P. 2019. Innovative technologies for producing and preserving intermediate moisture foods: A review. Food Research International, 116: 90-102.

Que, T. 2013. Effect of Different Drying Methods on the Protein and Product Quality of Hairtail fish Meat Gel. Drying Technology, 31(13–14), 1707–1714.

Raccach, M. and R.C. Baker. 1978. Microbial properties of mechanically de boned fish flesh. *J. Food Sci.,* 43: 1675–77.

Rahimi, E., Raissy, M., Razzaghimanesh, M., Dastgerdi, A. and Shahraki, M. 2014. Occurrence of *Aeromonas hydrophila* in fish, shrimp, lobster and crab in Iran. Kafkas Universitesi Veteriner Fakultesi Dergisi, 20(5), 691– 696.

Rahimi, E., Shakerian, A. and Raissy, M. 2012. Prevalence of *Listeria* species in fresh and frozen fish and shrimp in Iran. Annals of Microbiology, 62(1), 37– 40.

Rahman, M.S. 1999. Post harvest handling of foods of animal origin. In: Handbook of food preservation. CRC Press, Taylor and Francis Group, London, New York.

Rami, J., Sameer, A., Fawzi, B. and Khalid, A. 2005. Electroosmotic dewatering of tomato paste suspension under AC electric field. Drying Technology, 23(7), 1465–1475.

Rampling, A. (1996). Raw milk cheeses and salmonella. British Medical Journal 312, 67-68.

Rashid, Md. H-u., Togo, K., Ueda, M. and Miyamoto, T. 2007. Identification and characterization of lactic acid bacteria isolated from traditional fermented milk 'Dahi' in Bangladesh. World J. Microbiol. Biotechnol. 23 (1), 125-133.

Rather, M. A., Willayat, M. M., Wani, S. A., Hussain, S. A. and Shah, S. A. 2019. Enterotoxin gene profile and molecular epidemiology of *Aeromonas* species from fish and diverse water sources. Journal of Applied Microbiology, 127(3), 921– 931.

Rathore, R.S. 2000. Studies on enterohaemorrhagic *Escherichia coli* with special reference to detection of virulence genes. Ph.D. thesis submitted to Deemed University, IVRI, Izatnagar, India.

Rattanasomboon, N., Bellara, S.R., Harding, C.L., Fryer, P.J., Thomas, C.R., AlRubeai, M. and McFarlane, C.M. 1999. Growth and enumeration of the meat spoilage bacterium Brochothrix thermosphacta. Int. J. Food Microbiol. 51, 145-158.

Rattray, F.P. and Connell, M.J. 2011. Fermented Milks Kefir. In: Fukay, JW, editor. Encyclopedia of Dairy Sciences. 2th ed. Academic Press; San Diego, U.S.A., pp. 518- 524.

Raufu, I. A., Lawan, F. A., Bello, H. S., Musa, A. S., Ameh, J. A. and Ambali, A. G. 2014. Occurrence and antimicrobial susceptibility profiles of *Salmonella* serovars from fish in Maiduguri, sub-Saharah, Nigeria. Egyptian Journal of Aquatic Research, 40, 59– 63.

Ray, B. and R.A. Field. 1983. Bacteriology of restructured lamb roasts made with mechanically deboned meat. *J. Food Protect.,* 46: 26– 28.

Rehberger, T.G., L.A. Wilson and B.A. Glatz. 1984. Microbiological quality of commercial tofu. *J. Food Protect.,* 47: 177–81.

Reij, M.W., Jongenburger, I., Gkogka, E., Gorris, L.G.M., Zwietering, M.H. 2009. Perspective on the risk to infants in the Netherlands associated with Cronobacter spp. occurring in powdered infant formula. Int J Food Microbiol, 136:232-237.

Resconi, V.C., Escudero, A. and Campo, M.M. 2013. The development of aromas in ruminant meat. Molecules, 18, 6748-6781.

Rezai, R., Ahmadi, E. and Salimi, B. 2018. Prevalence and antimicrobial resistance profile of *Listeria* species isolated from farmed and on-sale rainbow trout (*Oncorhynchus mykiss*) in western Iran. Journal of Food Protection, 81(6), 886– 891.

Riadh, M. H., Ahmad, S. A. B., Marhaban, M. H. and Soh, A. C. 2015. Infrared heating in food drying: An overview. Drying Technology, 33(3), 322–335.

Riaz Rajoka, M.S., Mehwish, H.M., Fang, H., Padhiar, A.A., Zeng, X., Khurshid, M., He, Z. and Zhao, L. 2019. Characterization and anti-tumor activity of exopolysaccharide produced by Lactobacillus kefiri isolated from Chinese kefir grains. J. Funct. Foods, 63, 103588.

Rissman, A.I., Mau, B., Biehl, B.S., Darling, A.E., Glasner, J.D. and Perna, N.T., 2009. Reordering contigs of draft genomes using the Mauve aligner. Bioinformatics 25,2071–2073.

Rivas-Cañedo, A., Fernández-García, E. and Nuñez, M. 2009. Volatile compounds in fresh meats subjected to high pressure processing: effect of the packaging material. Meat Sci. 81, 321-328.

Roberts, J.A. 2000. Economic aspects of food-borne outbreaks and their control. Br Med Bull., 56:133–41.

Rødbotten, M., Kubberød, E., Lea, P. and Ueland, O. 2004. A sensory map of the meat universe. Sensory profile of meat from 15 species. Meat Sci. 68, 137-144.

Rodrigues, J., Kalekar, S., Doijad, S., Poharkar, K., D'Costa, D. and Barbuddhe, S. B. 2015. Prevalence and characterization of *Listeria* spp. from seafood. Indian Journal of Fisheries, 62(1), 139– 143.

Rodríguez, M., Medina, L. M. and Jordano, R. 2000. Effect of modified atmosphere packaging on the shelf life of sliced wheat flour bread. Molecular Nutrition and Food Research, 44(4), 247–252.

Rogan, A. and Glaros, G. 1988. Food irradiation: The process and implications for dietitians. JADA 88(7):833-838.

Rosenblum, L.S., Mirkin, I.R., Allen, D.T., Safford, S. and Hadler, S.C. 1990. A multifocal outbreak of hepatitis A traced to commercially distributed lettuce. Am. J. Public Health, 80: 1075-1079.

Rota, V. and Schieberle, P. 2006. Changes in key odorants of sheep meat induced by cooking. ACS Symp. Ser. 920, 73-83.

Roy, H.J. and Kalicki, B. 2009. "Probiotics," Pennington Nutrition, 5, 1–3.

Russell, S.M., Fletcher, D.L., Cox, N.A. 1994. The effect of incubation temperature on recovery of mesophilic bacteria from broiler chicken carcasses subjected to temperature abuse. Poultry Sci., 73: 1144–8.

Russell, S.M., Fletcher, D.L., Cox, N.A. 1996. Spoilage bacteria of fresh broiler chicken carcasses. Poultry Sci., 75: 2041-2047.

Russo, F., Ercolini, D., Mauriello, G. and Villani, F. 2006. Behaviour of Brochotrix thermosphacta in the presence of other meat spoilage microbial group. Food Microbiol. 23, 797-802.

Sadik, C. *et al.,* 1986. An epidemiological investigation following an infection by Salmonella typhimurium due to the ingestion of cheese made from raw milk. In Proceedings of the 2nd World Congress on Foodborne Infections and Intoxications 1, Berlin, pp 280-282.

Sakaki, J. 1891. First report on the toxicological research of molded rice. Tokyo Ishi-kai-shi (in Japanese), 5, 1097- 1104.

Sakala, R.M., Hayashidani, H., Kato, Y., Hirata, T., Makino, Y., Fukushima, A., Yamada, T., Kaneuchi, C. and Ogawa, M. 2002. Change in the composition of the microflora on vacuum-packaged beef during chiller storage. Int. J. Food Microbiol. 74, 87–99.

Saldanha, A.J. 2004. Java Tree view-extensible visualization of microarray data. Bioinformatics 20, 3246-3248.

Samelis, J. 2006. Managing microbial spoilage in meat industry. In: Blackburn, C.W. (Ed.), Food Spoilage Microorganisms. Woodhead Publishing Limited, Cambridge, pp. 213-286.

Sandeep, M., Diwakar, A. and Abhijit, G. 2004. Microbiological analysis of street vended fresh squ

Sanders, M.E. 1999. Probiotics, Scientific Status Summary: A publication of the institute of food technologists expert panel on food safety and nutrition. Food Technology, 53(11).

Sanguinetti, A. M., Caro, A. D., Scanu, A., Fadda, C., Milella, G., Catzeddu, P. and Piga, A. 2016. Extending the shelf life of gluten-free fresh filled pasta by modified atmosphere packaging. LWT – Food Science and Technology, 71, 96–101.

Santos, A., Sanmauro, M., Sanchez, A., Torres, J.M. and Marquina, D. 2003. The antimicrobial properties of different strains of Lactobacillus spp. isolated from kefir. Systematic and Applied Microbiology. 26, 434- 437.

Sawicka, H. 2020. Glucosinolates As Natural Plant Substances-Structural and Application Aspects: A Review. J. Cell Tissue Res., 20, 6919–6928.

Scarborough, P., Appleby, P.N., Mizdrak, A., Briggs, A.D.M., Travis, R.C., Bradbury, K.E. and Key, T.J. 2014. Dietary greenhouse gas emissions of meateaters, fisheaters, vegetarians and vegans in the UK. Clim Change, 125: 179–192.

Scharff, R.L. 2012. Economic Burden from Health Losses Due to Foodborne Illness in the United States. J Food Prot., 75 (1): 123-131.

Schmidt, F. C., Bam, C. and Laurindo, J. B. 2008. Salting operational diagrams for chicken breast cuts: Hydration-dehydration. Journal of Food Engineering, 88(1), 36–44.

Scholtz, V., Jirešová, J., Šerá, B. and Julák, J. 2021. A Review of Microbial Decontamination of Cereals by Non-Thermal Plasma. Foods, 10 (12), 2927.

Schröder, U. 2010. Changes in phosphate and water content during processing of salted Pacific cod (Gadus macrocephalus). Journal of Aquatic Food Product Technology, 19(1), 16–25.

Selby, T. L., Berzins, A., Gerrard, D. E., Corvalan, C. M., Grant, A. L. and Linton, R. H. 2006. Microbial heat resistance of Listeria monocytogenes and the impact on ready to-eat meat quality after post-package pasteurization. Meat Science, 74(3), 425–434.

Sentence, C.B. 1991. Growth of bacteria and spoilage of meat. In: The production of chilled meat for export. http://www.meatupdate.csiro.au/data/Chilled_meat _for_export_02-91.pdf.

Sergio, L., Cantore, V., Spremulli, L., Pinto, L., Baruzzi, F., Venere, D. D. and Boari, F. 2017. Effect of cooking and packaging conditions on quality of semi-dried green asparagus during cold storage. LWT – Food Science and Technology, 89, 712–718.

Severini, C., Rosaria, M., Antonio, C., Antonio, D. and Giuliani, B. 2008. Use of humectants for the stabilization of pesto sauce. International Journal of Food Science and Technology, 43(6), 1041–1046.

Seymour, I.J., Cole, M.B. and Coote, P.J. 1994. A substrate-mediated assay of bacterial proton effux/influx to predict the degree of spoilage of beef mince stored at chill temperatures. J. Appl. Bacteriol. 76, 608-615.

Sharifi, M., Moridnia, A., Mortazavi, D., Salehi, M., Bagheri, M. and Sheikhi, A. 2017. Kefir: A powerful probiotics with anticancer properties. Med. Oncol., 34, 1–7.

Sharma, R., Garg, P., Kumar, P., Bhatia, S.K. and Kulshrestha, S. 2020. Microbial Fermentation and Its Role in Quality Improvement of Fermented Foods. *Fermentation*, *6*(4), 106.

Shaw, B.G. and Latty, J.B. 1982. A numerical taxonomic study of Pseudomonas strains from spoiled meat. J. Appl. Bacteriol. 52, 219-228.

Sichewo, P.R., Etter, E.M.C. and Michel, A.L. 2019. Prevalence of Mycobacterium bovis infection in traditionally managed cattle at the wildlife-livestock interface in South Africa in the absence of control measures. Vet. Res. Commun., 43, 155–164.

Sieper, J. and Braun, J. 1995. Pathogenesis of spondylarthropathies. Arthritis Rheum. 38:1547-54.

Silley P. and Forsythe S. 1996. Impedance microbiology—a rapid change for microbiologists. J Appl Bact., 80:233–43.

Silva, A.R., Paulo, E.N., Sant'Ana, A.S., Chaves, R.D. and Massaguer, P.R. 2011. Involvement of Clostridium gasigenes and C. algidicarnis in 'blown pack' spoilage of Brazilian vacuum-packed beef. Int. J. Food Microbiol. 148, 156-163.

Silva, K. S., Fernandes, M. A. and Mauro, M. A. 2014. Effect of calcium on the osmotic dehydration kinetics and quality of pineapple. Journal of Food Engineering, 134(134), 37–44.

Simal, S., Benedito, J., Sánchez, E. S. and Rosselló, C. 1998. Use of ultrasound to increase mass transport rates during osmotic dehydration. Journal of Food Engineering, 36(3), 323–336.

Simon, J. and Kroneck, P. 2013. Microbial sulfite respiration. Advances in Microbial Physiology, 62: 45–117.

Sims, J.K. 1987. A theoretical discourse on the pharmacology of toxic marine ingestions. Ann. Emerg. Med., 16, 1006–1015.

Sing, C. K., Khan, M. Z. I., Daud, H. H. M. and Aziz, A. R. 2016. Prevalence of *Salmonella* sp in African catfish (*Clarias gariepinus*) obtained from farms and wet markets in Kelantan, Malaysia and their antibiotic resistance. *Sains Malaysiana*, 45(11), 1597– 1602.

Sivamaruthi, B.S., Kesika, P., Prasanth, M.I. and Chaiyasut, C. 2018. A mini review on antidiabetic properties of fermented foods. Nutrients, 10, 1973.

Skandamis, P.N. and Nychas, G.J.E. 2002. Preservation of fresh meat with active and modified atmosphere packaging conditions. Int. J. Food Microbiol. 79, 35-45.

Skinner, M.P., Brewer, T.D., Johnstone, R., Fleming, L.E. and Lewis, R.J. 20111. Ciguatera Fish Poisoning in the Pacific Islands (1998 to 2008). PLoS Negl. Trop. Dis., 5.

Small, A., Jenson, I., Kiermeier, A. and Sumner, J. 2012. Vacuum-packed beef primals with extremely long shelf life have unusual microbiological counts. J. Food Prot. 75, 1524-1527.

Smit, G., Smit, B.A. and Engels, W.J.M. 2005. Flavour formation by lactic acid bacteria and biochemical flavour profiling of cheese products. FEMS Microbiol. Rev. 29, 591-610.

Smith, G., Hole, M. and Hanson, S. W. 1990. Assessment of lipid oxidation in indonesian salted-dried marine catfish (Arius thalassinus). Journal of the Science of Food and Agriculture, 51(2), 193–205.

Smith, J.L. 1994. *Cryptosporidium* and *Giardia* as agents of food borne disease. Journal of Food Protection. 56:451-61.

Smith, J.S. and Pillai, S.D. 2002. Irradiation and food safety. Food Technol. 58(11):48-55.

Snyder, J.D. and Merson, M.H. 1996. The magnitude of the global problem of acute diarrhoeal disease: a review of active surveillance data. Bull World Health Organ. 60(4):605-13.

Soliño, L. and Costa, P.R. 2020. Global impact of ciguatoxins and ciguatera fish poisoning on fish, fisheries and consumers. Environ. Res., 182, 109111.

Somogyi, L.P., Ramaswamy, H.S. and Hai, Y.H. 1996. Processing Fruits: Science and technology, Vol. I. Biology Principles and its Applications. Tectronic Pub. Co. Inc. Lancaster. Basel.

Soncin, S., Chiesa, L.M., Cantoni, C. and Biondi, P.A. 2007. Preliminary study of the volatile fraction in the raw meat of pork, duck and goose. J. Food Compos. Anal. 20, 436-439.

Song, L. and Thornalley, P.J. 2007. Effect of storage, processing and cooking on glucosinolate content of Brassica vegetables. Food and Chemical Toxicology; 45(2): 216-224.

Soni, M.G., Carabin, I.G. and Burdock, G.A. 2005. Safety assessment of esters of *p*-hydroxybenzoic acid (parabens). Food Chem Toxicol., 43:985-1015.

Sospedra, J. Rubert, Soriano, J.M. and Manes, J. 2012. "Incidence of ˜ microorganisms from fresh orange juice processed by squeezing machines," Food Control, 23 (1), 282–285.

Soultos, N., Iossifidou, E., Tzikas, Z., Sergelidis, D., Lazou, T., Drakopoulos, G. and Konstantelis, I. 2014. Prevalence of *Listeria monocytogenes* in ready-to-eat seafood marketed in Thessaloniki (Northern Greece). Veterinary World, 7(11), 1004– 1009.

Speckhahn, A., Srzednicki, G., Desai, D. K. and Devahastin, S. 2010. Drying of beef in superheated steam. Drying Technology, 28(9), 1072–1082.

Srivastava, R.P. and Sanjeev Kumar 1993. Fruit and vegetable preservation. International Book Distribution Co., Lucknow.

Ssemanda, J.N., Reij, M.W., Bagabe, M.C, Muvunyi, C.M., Nyamusore, J., Joosten, H. and Zwietering, M.H. 2018. Estimates of the burden of illnesses related to foodborne pathogens as from the syndromic surveillance data of 2013 in Rwanda. Microb Risk Anal, 9:55-63.

Stabel, JR, Steadham, E and Bolin, CA. 1997. Heat inactivation of Mycobacterium paratuberculosis in raw milk: are current pasteurisation conditions effective? Applied and Environmental Microbiology 63, 4975-4977.

Stamatakis, A. 2014. RAxML version 8: a tool for phylogenetic analysis and post-analysis of large phylogenies. Bioinformatics 30, 1312–1313.

Stanborough, T., Fegan, N., Powell, S.M., Tamplin, M. and Chandry, P.S. 2017. Insight into the genome of Brochothrix the rmosphacta, a problematic meat spoilage bacterium. Appl. Environ. Microbiol. 83(5): e02786-16.

Stanley, G., Shaw, K.J. and Egan, A.F. 1981. Volatile compounds associated with spoilage of vacuum packaged sliced luchen meat by Brochothrix thermosphacta. Appl. Environ. Microbiol. 41, 816-818.

Stanyer, R.Y., Palleroni, N.J. and Doudoroff, M. 1996. The aerobic Pseudomonads: a taxonomy study. J. Gen. Microbiol. 43, 159-271.

Stekelenburg, F. K. 2003. Enhanced inhibition of Listeria monocytogenes in Frankfurter sausage by the addition of potassium lactate and sodium diacetate mixtures. Food Microbiology, 20(1), 133–137.

Stellato, G., La Storia, A., De Filippis, F., Borriello, G., Villani, F. and Ercolini, D. 2016. Overlap of spoilage-associated microbiota between meat and the meat processing environment in small-scale and large-scale retail distributions. Appl. Environ. Microbiol. 82, 4045-4054.

Stiemsma, L.T., Nakamura, R.E., Nguyen, J.G. and Michels, K.B. 2020. Does Consumption of Fermented Foods Modify the Human Gut Microbiota? J. Nutr., 150, 1680–1692.

Stob, M., Baldwin, R.S., Tuite, J., Andrews, F.N. and Gillette, K.G. 1962. Isolation of an anabolic, uterotrophic compound from corn infected with Gibberella zeae. Nature, 196, 13–18.

Stoddard, S.F., Smith, B.J., Hein, R., Roller, B.R.K.and Schmidt, T.M. 2014. rrnDB: improved tools for interpreting rRNA gene abundance in bacteria and archaea and a new foundation for future development. Nucl. Acids Res. 43, D593-D598.

St-Onge, M-P, Farnworth, E.R., Savard, T., Chabot, D. Mafu, A. and Jones, J.H. 2002 Kefir consumption does not alter plasma lipid levels or cholesterol fractional synthesis rates relative to milk in hyperlipidemic men: a randomized controlled trial. BMC Complement Altern Med 2, 1.

Su, Y., Zhang, M. and Mujumdar, A. S. 2015. Recent developments in smart drying technology. Drying Technology, 33(3), 260–276.

Sun, B., Zhao, Y., Yu, J., Ling, J., Shang, H. and Liu, Z. 2017. The combined efficacy of super chilling and high CO_2 modified atmosphere packaging on shelf life and quality of swimming crab (Portunus trituberculatus). Journal of Aquatic Food Product Technology, 26(6), 655–664.

Suppakul, P., Thanathammathorn, T., Samerasut, O. and Khankaew, S. 2016. Shelf life extension of "fios de ovos", an intermediate-moisture egg-based dessert, by active and modified atmosphere packaging. Food Control, 70, 58–63.

Swart, A.N., Mangen, M-J.J., Havelaar, A.H. 2013. Microbiological Criteria as a Decision Tool for Controlling Campylobacter in the Broiler Meat Chain. RIVM Report 330331008/2013.Dutch Ministry of Health, Welfare and Sport. Bilthoven, The Netherlands.

Syamaladevi, R. M., Tang, J., Villa-Rojas, R., Sablani, S., Carter, B. and Campbell, G. 2016. Influence of water activity on thermal resistance of microorganisms in low moisture foods: A review. Comprehensive Reviews in Food Science and Safety, 15: 353–370.

Sydenham, E.W., Shepard, G.S. Thiel, P.G. Marasas, W.F.O. and Stockenstr ¨om, S. 1991. Fumonisin contamination of commercial corn-based human foodstuffs. J. Agric. Food Chem. 39:2014–2018.

Takeoka, G.R., Buttery, R.G., Teranishi, R., Flath, R.A. and Guntert, M. 1991. Identification of additional pineapple volatiles. J. Agric. Food Chem. 39, 1848-1851.

Talaro KP. 2005. Foundations in microbiology. 5th edition. New York, NY: McGraw- Hill; p. 209–10.

Tanaka, N., Traisman, E., Plantong, P., Finn, L., Flom, W., Meskey, L. and Guggisberg, J. 1986. Evaluation of factors involved in antibotulinal properties of pasteurized process cheese spreads. J. Food Prot., 49(7): 526-31.

Tappero, J.W., Schuchat, A., Deaver, K.A., Mascola, L. and Wenger, J.D. 1995. Reduction in the incidence of human listeriosis in the United States. JAMA 273:1118-22.

Taqir, L., Haagsma, J. and Havelaar, 2011. Cost of Illness and Disease Burden in The Netherlands Due to Infections with Shiga Toxin-Producing *Escherichia coli* O157. J Food Prot., 74 (4): 545-552.

Tenhet, V., Finne, G. and Toloday, D. 1981. Penetration of sodium tripolyphosphate into fresh and Prefrozen peeled and deveined shrimp. Journal of Food Science, 46(2), 344–349.

Terentjeva, M., Eizenberga, I., Valcina, O., Novoslavskij, A., Strazdina, V. and Berzins, A. 2015. Prevalence of foodborne pathogens in freshwater fish in Latvia. Journal of Food Protection, 78(11), 2093– 2098.

Thierry, A. and Chicheportiche, R. 1988. Use of ATP bioluminescence measurements for the estimation of biomass during biological humification. J. Appl. Microbiol. Biotechnol., 28, 199-202.

Thimothe, J., Nightingale, K. K., Gall, K., Scott, V. N. and Wiedmann, M. 2004. Tracking of *Listeria monocytogenes* in smoked fish processing plants. Journal of Food Protection, 67(2), 328– 341.

Thippeswamy, L., Venkateshaiah, B. V. and Patil, S. B. 2011. Effect of modified atmospheric packaging on the shelf stability of paneer prepared by adopting hurdle technology. Journal of Food Science and Technology, 48(2), 230–235.

Timbo, B., Koehler, K. M., Wolyniak, C. and Klontz, K. C. 2004. Sulfites—a food and drug administration review of recalls and reported adverse events. Journal of Food Protection, 67(8): 1806– 1811.

Toldra, F. 1998. Proteolysis and lipolysis in flavour development of dry-cured meat products. Meat Sci. 49 (1), 101-110.

Tomasevic, I., Dodevska, M., Simiæ, M., Raicevic, S., Matovic, V. and Djekic, I. 2018. A decade of sulphite control in Serbian meat industry and the effect of HACCP. Food Additives and Contaminants: Part B, 11(1), 49–53.

Torres, M.R., Sanchis, V. and Ramos. A.J. 1998. Occurrence of fumonisins in Spanish beers analyzed by an enzyme-linked immunosorbent assay method. Int. J. Food Microbiol. 39:139–143.

Tortoe, C. 2010. A review of osmodehydration for the food industry. African Journal of Food Science, 4, 303–324.

Tournas, V. H. (2005). Moulds and yeasts in fresh and minimally processed vegetables, and sprouts. International Journal of Food Microbiology, 99, 71–77.

Travis, R., Bui-Klimke and Wu, F. 2015. Ochratoxin A and human health risk: A review of the evidence. Crit Rev Food Sci Nutr., 55(13): 1860–1869.

Tsigarida, E., Boziaris, I.S., Nychas, G.J.E. 2003. Bacterial synergism or antagonism in a gel cassette system. Appl. Environ. Microbiol. 69, 7204-7209.

Uchida, M., Ishii, I., Inoue, C., Akisato, Y., Watanabe, K., Hosoyama, S., Toida, T., Ariyoshi, N. and Kitada, M. 2010. Kefiran reduces atherosclerosis in rabbits fed a high cholesterol diet. J Atheroscler Thomb 17 (9), 980–988.

UFFVA. 1986. Food irradiation for the produce industry. Special Report.

UK Dairy Products (Hygiene) Regulations 1995. SI 1995 No 1086, HMSO.

UK Food Safety (General Food Hygiene) Regulations 1995. SI 1995 No 1763, HMSO.

Ur., A. and Brown, D.F. 1975. Impedence monitoring of bacterial activity. J. Med. Microbiol., 8(1), 19-28.

Urry, W.H., Wehrmeister, H.L., Hodge, E.B. and Hidy, P.H. 1966. The structure of zearalenone. Tetrahedron Lett., 27, 3109–3114.

USDA, 2021. Cost Estimates of Foodborne Illnesses, by Sandra Hoffmann, USDA, Economic Research Service, March, 2021.

USDA/ERS (U.S. Department of Agriculture/Economic Research Service) 2014. Dairy products: Per capita consumption, United State (Annual).

Vaillant, V., Haeghebaert, S., Desenclos, J.C. *et al.,* 1996. Outbreak of *Salmonella dublin* infection in France, November-December 1995. Eurosurvei llance 1, (2) 9-10.

Vajda, A., László Özsvári, L., Szakos, D., Kasza, G., Aguilera, J. M. and Chirife, J. 1994. Combined methods for the preservation of foods in Latin America and the CYTED-D project. Journal of Food Engineering, 22(1-4), 433-444.

Valentinuzzi, M.E., Morucci, J.P. and Felice, C.J. 1996. Monitoring of physiological events by impedance. In: Bourne JR, editor. CRC Reviews in Biomedical Engineering, 24(4–6), Bioelectrical Impedance Techniques in Medicine. New York: Begell House; p. 427–66.

Van Cauteren, D., Silva, N.J., Weill, F.X., King, L., Brisabois, A., Delmas, G., Vaillant, V. and de Valk, H. 2009. Outbreak of *Salmonella* Enterica serotype Muenster infections associated with goat's cheese, France March 2008. Euro Surveill., 14 (31), 19290.

van Lier, E.A., Havelaar, A.H. and Nanda, A. 2007. The burden of infectious diseases in Europe: a pilot study. Eurosurveillance, 12, 3-4.

Van Schothorst M. Sampling plan for Listeria monocytogenes. Food Control, 7:203-8.

Van Spreeken, K.J.A. and Stekelenburg, F.K. 1986. Rapid estimation of the bacteriological quality of fresh fish by impedance measurements. Appl Microbiol Biotechnol, 24, 95–6.

Vanderlinde, P.B., Shay, B. and Murray, J. 1999. Microbiological status of Australian sheep meat. J. Food Prot. 62, 380-385.

Varzakas, T., Zakynthinos, G., Proestos, C. and Radwanska, M. 2017. Fermented Vegetables. In Minimally Processed Refrigerated Fruits and Vegetables; Yildiz, F., Wiliey, R., Eds.; Springer Science: New York, NY, USA, pp. 537–584.

Vasavada, M.N. and Cornforth, D.P. 2005. Evaluation of milk mineral antioxidant activity in beef meatballs and nitrite-cured sausage. J. Food Sci. 70: 250–253.

Vasta, V. and Priolo, A. 2006. Ruminant fat volatiles as affected by diet. A review. Meat Sci. 73, 218-228.

Verhoef, L., Koopmans, M., Van Pelt, W., Duizer, E., Haaqsma, J., Werber, D., Van Asten, L. and Havelaar, A. 2013. The estimated disease burden of norovirus in The Netherlands. Epidemiol Infect., 141 (#), 496-506.

Vermeulen, A., Daelman, J., Van, S. J. and Devlieghere, F. 2012. Screening of different stress factors and development of growth/no growth models for Zygosaccharomyces rouxii in modified Sabouraud medium, mimicking intermediate moisture foods (IMF). Food Microbiology, 32(2): 389–396.

Vetrovský, T. and Baldrian, P. 2013. Analysis of soil fungal communities by amplicon pyrosequencing: current approaches to data analysis and the introduction of the pipeline SEED. Biol. Fertil. Soils 49, 1027e1037.

Vidaver A.K. 1996. Emerging and reemerging infectious diseases. American Society for Microbiology News. 62:583,585.

Vinderola, C.G., Duarte, J., Thangavel, D., Peridigon, G., Farnworth, E. and Matar, C. 2005 Immunomodulating capacity of kefir. J Dairy Res 72 (2), 195–202.

Vogel, G. 1997. Prusiner recognized for once-heretical prion theory. Science 278:214.

Vrese, M.de. and Schrezenmeir, J. 2008. "Probiotics, prebiotics, and synbiotics," Advances in Biochemical Engineering/Biotechnology, 111, 1–66.

Vujièiæ, I.F., Vuliæ, M. and Könyves, T. 1992. Assimilation of cholesterol in milk by kefir cultures. Biotechnol Lett 14, 847 I ¼S4 850.

Wachowska, U., Stasiulewicz-Paluch, A. D., Glowacka, K., Mikolajczyk, W. and Kucharska, K. 2013. Response of epiphytes and endophytes isolated from winter wheat grain to biotechnological and fungicidal treatments. Polish Journal of Environmental Studies, 22(1), 267–273.

Wang, C. and Silva, J. L. 1999. Prevalence and characteristics of *Aeromonas* species isolated from processed channel catfish. Journal of Food Protection, 62(1), 30– 34.

Wang, F., Fu, L., Bao, X. and Wang, Y. 2017. The spoilage microorganisms in seafood with the existed quorum sensing phenomenon. J Food Microbiol. 1 (1): 8-13.

Wang, F., Jiang, L., Yang, Q., Han, F., Chen, S., Pu, S., Vance, A. and Ge, B. 2011. Prevalence and antimicrobial susceptibility of major foodborne pathogens in imported seafood. Journal of Food Protection, 74(9), 1451- 1461.

Wang, G., Tian, Y., Wang, K., Zhang, J. and Fang, J. 2015. Non-thermal plasma activated water inactivation of food-borne pathogen on fresh produce. Journal of Hazardous Materials, 300, 643–651.

Wang, L., Pang, X. F., Wang, G. Z. and Xiao-Bo, H. U. 2012. The drying process optimization of beef jerky. Science and Technology of Food Industry, 33(9), 328–332.

Wang, W., Cao, Z., Sun, Y. and Qin, W. 2012. Effect of antiseptic compound film on the quality of preserved meat. Food Science, 33(2), 276–279.

Wasney, M.A., Holley, R.A. and Jayas, D.S. 2001. Cresol red thallium acetate sucrose inulin (CTSI) agar for the selective recovery of Carnobacterium spp. Int. J. Food Microbiol.64, 167–174.

Water Quality for the Food Industry: Management and Microbiological issues. Guideline No. 27. 2000. Campden and Chorleywood Food Research Association Group.

Wawerla, M., Stolle, A., Schalch, B. and Eisgruber, H. 1999. Impedance microbiology: applications in food hygiene. J Food Prot, 62:1488–96.

Wekell, J.C., Hurst, J. and Lefebvre, K.A. 2004. The origin of the regulatory limits for PSP and ASP toxins in shellfish. J. Shellfish Res., 23, 927–930.

Wheeler, J.L., Harrison, M.A. and Koehler, P.E. 1987. Presence and stability of patulin in pasteurized apple cider. J. Food Sci. 52:479–480.

WHO, 1966. The Technical Basis for Legislation on Irradiated Food. WHO Technical Report Series No. 316.

WHO, 1981. Report of the Working Party on Irradiation of Food. Joint Expert Committee on the Wholesomness of Irradiated Food (JEFF). WHO Technical Report Series No. 659.

WHO, 1994. "Safety and nutritional adequacy of irradiated foods". WHO, Geneva.

WHO, 2010. World Health Organization global estimates and regional comparisons of the burden of foodborne disease in 2010. WHO, Geneva.

WHO, 2013. Initiative to Estimate the Global Burden of Foodborne Diseases. First formal meeting of the Foodborne Disease Burden Epidemiology Reference Group. Geneva, Switzerland.

WHO. 1996. Enterohaemorrhagic Escherichia coli infection. World Health Organisation, Geneva, Switzerland. Weekly epidemiological record, 35: 267-268.

WHO. 2009. "Cardiovascular Disease," Fact sheet 317, WHO, Geneva, Switzerland.

WHO. 2015. WHO estimates of the global burden of foodborne diseases. Foodborne diseases burden epidemiology reference group 2007-2015. WHO, Geneva.

Wiebe, L.A. and Bjeldanes, L.F. 1981. Fusarin C, a mutagen from Fusarium moniliforme grown on corn. J. Food Sci. 46:1424–1426.

Wieczorek, K. and Osek, J. 2017. Prevalence, genetic diversity and antimicrobial resistance of *Listeria monocytogenes* isolated from fresh and smoked fish in Poland. Food Microbiology, 64, 164–171.

Wilson, M. 1995. Travel and the emergence of infectious diseases. Emerg Infect Dis. 1:39-46.

Wojdy³o, A., Figiel, A., Lech, K., Nowicka, P. and Oszmiañski, J. 2014. Effect of convective and vacuum-microwave drying on the bioactive compounds, color, and antioxidant capacity of sour cherries. Food and Bioprocess Technology, 7(3), 829-841.

Wolfe MS. 1992. Giardiasis. Clin Microbiol Rev. 5:93-100.

Wood, R., Foster, L., Damant, A. and Key, P. 2004. Analytical methods for food additives. A volume in Woodhead Publishing Series in Food Science, Technology and Nutrition.

Woodring, J., Srijan, A., Puripunyakom, P., Oransathid, W., Wongstitwilairoong, B. and Mason, C. 2012. Prevalence and antimicrobial susceptibilities of *Vibrio*, *Salmonella*, and *Aeromonas* isolates from various uncooked seafoods in Thailand. Journal of Food Protection, 75(1), 41- 47.

Wu, S., Huang, J., Wu, Q., Zhang, J., Zhang, F., Yang, X., Wu, H., Zeng, H., Chen, M., Ding, Y., Wang, J., Zhang, S. and Xue, L. 2018. *Staphylococcus aureus* isolated from retail meat and meat products in China: Incidence, antibiotic resistance and genetic diversity. Frontiers in Microbiology, 9, 2767.

Wysocki, C.J. and Dalton, P. 1996. Odor and Irritation Thresholds for 1-Butanol in Humans. Monell Chemical Senses Center, Philadelphia.

Xie, X., Li, X., Zhang, C., Jia, W., Li, Y., Sun, H. and Mu, G. 2013. Combined mid-infrared and hot air drying reduces energy-consumption and improves quality of jerky. Transactions of the Chinese Society of Agricultural Engineering, 29(23), 217–226.

Yalow, R.S. and Berson, S.A. 1960. Immunoassay of endogenous plasma insulin in man. J Clin Invest 39:1157–75.

Yang, L. and Bashir, R. 2008. "Electrical/electrochemical impedance for rapid detection of foodborne pathogenic bacteria". Biotechnology Advances, 26, 135 – 150.

Yang, X., Badoni, M., Tran, F. and Gill, C.O. 2015. Microbiological effects of a routine treatment for decontaminating hide-on carcasses at a large beef packing plant. J. Food Prot. 78, 256–263.

Yang, X., Youssef, M.K., Gill, C.O., Badoni, M. and Lopez-Campos, O. 2014. Effects of meat pH on growth of 11 species of psychrotolerant clostridia on vacuum packaged beef and blown pack spoilage of the product. Food Microbiol., 39, 13–18.

Yang, Y., Miao, P., Li, H., Tan, S., Yu, H. and Yu, H. 2018. Antibiotic susceptibility and molecular characterization of *Aeromonas hydrophila* from grass carp. Journal of Food Safety, 38(1), e12393.

Yilmaz-Ersan, L., Ozcan, T., Akpinar-Bayizit, A. and Sahin, S. 2018. Comparison of antioxidant capacity of cow and ewe milk kefirs. J. Dairy Sci., 101, 3788–3798.

Yoshida, H., Yoshikawa, T. and Kawasaki, M. 2013. Evaluation of suitable material properties of sludge for electroosmotic dewatering. Drying Technology, 31(7), 775–784.

Yost, C.K. and Nattress, F.M. 2002. Molecular typing techniques to characterize the development of a lactic acid bacteria community on vacuum-packaged beef. Int. J. Food Microbiol., 72, 97-105.

Youssef, M.K., Gill, C.O., Tran, F., and Yang, X. 2014. Unusual compositions of microflora of vacuum-packaged beef primal cuts of very long storage life. J. Food Prot. 77, 2161–2167.

Yun, H. J., Binna, K., Jung, S., Kruk, Z. A., Danbee, K., Wonho, C. and Cheorun, J. 2010. Inactivation of Listeria monocytogenes inoculated on disposable plastic tray, aluminum foil, and paper cup by atmospheric pressure plasma. Food Control, 21(8), 1182–1186.

Yvon, M. and Rijnen, L. 2001. Cheese flavour formation by amino acid catabolism. Int. Dairy J., 11, 185-201.

Zarein, M., Samadi, S. H. and Ghobadian, B. (2015). Investigation of microwave dryer effect on energy efficiency during drying of apple slices. Journal of the Saudi Society of Agricultural Sciences, 14(1), 41–47.

Zhang, G. 2013. Foodborne pathogenic bacteria detection: an evaluation of current and developing methods. Meducator 1, 15.

Zhang, J., Yang, X., Kuang, D., Shi, X., Xiao, W., Zhang, J., Gu, Z, Xu, X. and Meng, J. 2015. Prevalence of antimicrobial resistance of non-typhoidal Salmonella serovars in retail aquaculture products. International Journal of Food Microbiology, 210, 47– 52.

Zhang, P., Baranyi, J. and Tamplin, M. 2015. Interstrain interactions between bacteria isolated from vacuum-packaged refrigerated beef. Appl. Environ. Microbiol. 81, 2753–2761.

Zhang, P., Badoni, M., Gänzle, M. and Yang, X. 2018. Growth of Carnobacterium spp. isolated from chilled vacuum-packaged meat under relevant acidic conditions. International Journal of Food Microbiology, 286: 120-127.

Zhao, J., Hu, R., Xiao, H., Yang, Y., Liu, F., Gan, Z. and Ni, Y. 2014. Osmotic dehydration pre-treatment for improving the quality attributes of frozen mango: Effects of different osmotic solutes and concentrations on the samples. International Journal of Food Science and Technology, 49(4): 960–968.

Zhao, X., Lin, C. W., Wang, J., and Oh, D. H. 2014. Advances in rapid detection methods for foodborne pathogens. J. Microbiol. Biotechn. 24, 297–312.

Zheng X. C., Ge Z., Lin K., Zhang D., Chen Y., Xiao J., et al., 2021. Dynamic changes in bacterial microbiota succession and flavour development during milk fermentation of Kazak artisanal cheese. Int. Dairy J. 113:104878.

Zheng, X. J., Liu, F., Shi, X., Wang, B., Li, K., Li, B. and Zhuge, B. 2018. Dynamic correlations between microbiota succession and flavor development involved in the ripening of Kazak artisanal cheese. Food Res. Int., 105, 733–742.

Zheng, X., Shi, X. and Wang, B. 2021. A Review on the General Cheese Processing Technology, Flavor Biochemical Pathways and the Influence of Yeasts in Cheese. Front. Microbiol., 12, 703284.

Zhu, J., Zhao, A., Feng, L. and Gao, H. 2016. Quorum sensing signals affect spoilage of refrigerated large yellow croaker (Pseudosciaena crocea) by Shewanella baltica. Int J Food Microbiol., 217:146-55.

Zwirzitz, B., Wetzels, S. U., Dixon, E. D., Stessl, B., Zaiser, A., Rabanser, I., Selberherr, E. 2020. The sources and transmission routes of microbial populations throughout a meat processing facility. NPJ Biofilms and Microbiomes, 6(1), 26.

Index